Statistics for Managers Using Microsoft® Excel®

NINTH EDITION

David M. Levine

Department of Information Systems and Statistics

Zicklin School of Business, Baruch College, City University of New York

David F. Stephan

Two Bridges Instructional Technology

Kathryn A. Szabat

Department of Business Systems and Analytics

School of Business, La Salle University

 Pearson

Library of Congress Cataloging-in-Publication Data

Names: Levine, David M., author. | Stephan, David, author. | Szabat, Kathryn A., author.
Title: Statistics for Managers using Microsoft Excel / David M. Levine, Department of
  Information Systems and Statistics, Zicklin School of Business, Baruch
  College, City University of New York, David F. Stephan, Two Bridges
  Instructional Technology, Kathryn A. Szabat, Department of Business Systems
  and Analytics, School of Business, La Salle University.
Description: Ninth edition. | [Hoboken] : Pearson, [2019] | Includes index.
  | Summary: Statistics for Managers Using Microsoft Excel features a
  more concise style for understanding statistical concepts.-- Provided by publisher.
Identifiers: LCCN 2019043379 | ISBN 9780135969854 (paperback)
Subjects: LCSH: Microsoft Excel (Computer file) | Management--Statistical
  methods. | Commercial statistics. | Electronic spreadsheets. |
  Management--Statistical methods--Computer programs. | Commercial
  statistics--Computer programs.
Classification: LCC HD30.215 .S73 2020 | DDC 519.50285/554--dc23
LC record available at https://lccn.loc.gov/2019043379

7 2022

Instructor's Review Copy
ISBN 10:   0-13-662494-4
ISBN 13: 978-0-13-662494-3
Rental
ISBN 10:   0-13-596985-9
ISBN 13: 978-0-13-596985-4

To our spouses and children,
Marilyn, Mary, Sharyn, and Mark

and to our parents, in loving memory,
Lee, Reuben, Ruth, Francis J., Mary, and William

About the Authors

Kathryn Szabat, David Levine, and David Stephan

David M. Levine, David F. Stephan, and Kathryn A. Szabat are all experienced business school educators committed to innovation and improving instruction in business statistics and related subjects.

David Levine, Professor Emeritus of Statistics and CIS at Baruch College, CUNY, is a nationally recognized innovator in statistics education for more than three decades. Levine has coauthored 14 books, including several business statistics textbooks; textbooks and professional titles that explain and explore quality management and the Six Sigma approach; and, with David Stephan, a trade paperback that explains statistical concepts to a general audience. Levine has presented or chaired numerous sessions about business education at leading conferences conducted by the Decision Sciences Institute (DSI) and the American Statistical Association, and he and his coauthors have been active participants in the annual DSI Data, Analytics, and Statistics Instruction (DASI) mini-conference. During his many years teaching at Baruch College, Levine was recognized for his contributions to teaching and curriculum development with the College's highest distinguished teaching honor. He earned B.B.A. and M.B.A. degrees from CCNY. and a Ph.D. in industrial engineering and operations research from New York University.

Advances in computing have always shaped **David Stephan's** professional life. As an undergraduate, he helped professors use statistics software that was considered advanced even though it could compute *only* several things discussed in Chapter 3, thereby gaining an early appreciation for the benefits of using software to solve problems (and perhaps positively influencing his grades). An early advocate of using computers to support instruction, he developed a prototype of a mainframe-based system that anticipated features found today in Pearson's MathXL and served as special assistant for computing to the Dean and Provost at Baruch College. In his many years teaching at Baruch, Stephan implemented the first computer-based *classroom*, helped redevelop the CIS curriculum, and, as part of a FIPSE project team, designed and implemented a multimedia learning environment. He was also nominated for teaching honors. Stephan has presented at SEDSI and DSI DASI mini-conferences, sometimes with his coauthors. Stephan earned a B.A. from Franklin & Marshall College and an M.S. from Baruch College, CUNY, and completed the instructional technology graduate program at Teachers College, Columbia University.

As Associate Professor of Business Systems and Analytics at La Salle University, **Kathryn Szabat** has transformed several business school majors into one interdisciplinary major that better supports careers in new and emerging disciplines of data analysis including analytics. Szabat strives to inspire, stimulate, challenge, and motivate students through innovation and curricular enhancements and shares her coauthors' commitment to teaching excellence and the continual improvement of statistics presentations. Beyond the classroom she has provided statistical advice to numerous business, nonbusiness, and academic communities, with particular interest in the areas of education, medicine, and nonprofit capacity building. Her research activities have led to journal publications, chapters in scholarly books, and conference presentations. Szabat is a member of the American Statistical Association (ASA), DSI, Institute for Operation Research and Management Sciences (INFORMS), and DSI DASI. She received a B.S. from SUNY-Albany, an M.S. in statistics from the Wharton School of the University of Pennsylvania, and a Ph.D. in statistics, with a cognate in operations research, from the Wharton School of the University of Pennsylvania.

For all three coauthors, continuous improvement is a natural outcome of their curiosity about the world. Their varied backgrounds and many years of teaching experience have come together to shape this book in ways discussed in the Preface.

Brief Contents

Contents

2 Organizing and Visualizing Variables 38

3 Numerical Descriptive Measures 108

4 Basic Probability 152

5 Discrete Probability Distributions 176

6 The Normal Distribution and Other Continuous Distributions 198

7 Sampling Distributions 224

8 Confidence Interval Estimation 244

9 Fundamentals of Hypothesis Testing: One-Sample Tests 275

10 Two-Sample Tests 311

11 Analysis of Variance 352

12 Chi-Square and Nonparametric Tests 389

13 Simple Linear Regression 430

17 Business Analytics 600

18 Getting Ready to Analyze Data in the Future 618

19 Statistical Applications in Quality Management (*online*) 19-1

Preface

As business statistics evolves and becomes an increasingly important part of one's business education, which topics get taught and how those topics are presented becomes all the more important. As authors, we think about these issues as we seek ways to continuously improve the quality of business statistics education. We actively participate in conferences and meetings sponsored by the Decision Sciences Institute, American Statistical Association (ASA), and INFORMS, the Institute for Operations Research and the Management Sciences. We use the ASA's Guidelines for Assessment and Instruction (GAISE) reports and combine them with our experiences teaching business statistics to a diverse student body at several universities.

When writing a book for introductory business statistics students, four learning principles guide us.

Help students see the relevance of statistics to their own careers by using examples from the functional areas that may become their areas of specialization. Students need to learn statistics in the context of the functional areas of business. We discuss every statistical method using an example from a functional area, such as accounting, finance, management, or marketing, and explain the application of methods to specific business activities.

Emphasize interpretation and analysis of statistical results over calculation. We emphasize the interpretation of results, the evaluation of the assumptions, and the discussion of what should be done if the assumptions are violated. We believe that these activities are more important to students' futures and will serve them better than emphasizing tedious manual calculations.

Give students ample practice in understanding how to apply statistics to business. We believe that both classroom examples and homework exercises should involve actual or realistic data, using small and large sets of data, to the extent possible.

Integrate data analysis software with statistical learning. We integrate Microsoft Excel into every statistics method that the book discusses in full. This integration illustrates how software can assist the business decision-making process. In this edition, we also integrate using Tableau into selected topics, where such integration makes best sense. (Integrating data analysis software also supports our second principle about emphasizing interpretation over calculation.)

When thinking about introductory business statistics students using data analysis software, three additional principles guide us.

Using software should model business best practices. We emphasize reusable templates and model solutions over building unaudited solutions from scratch that may contain errors. Using preconstructed and previously validated solutions not only models best practice but reflects regulatory requirements that businesses face today.

Provide detailed sets of instructions that accommodate various levels of software use and familiarity. Instruction sets should accommodate casual software users and as well as users keen to use software to a deeper level. For most topics, we present *PHStat* and *Workbook* instructions, two different sets that create identical statistical results.

Software instruction sets should be complete and contain known starting points. Vague instructions that present statements such as "Use command *X*" that presume students can figure out how to "get to" command *X* are distracting to learning. We provide instruction sets that have a known starting point, typically in the form of "open to a specific worksheet in a specific workbook."

What's New in This Edition?

This ninth edition of *Statistics for Managers Using Microsoft Excel* features many passages rewritten in a more concise style that emphasize definitions as the foundation for understanding statistical concepts. In addition to changes that readers of past editions have come to expect, such as new examples and Using Statistics case scenarios and an extensive number of new end-of-section or end-of-chapter problems, the edition debuts:

- **Tabular Summaries** that state hypothesis test and regression example results along with the conclusions that those results support now appear in Chapters 10 through 13.
- **Tableau Guides** that explain how to use the data visualization software Tableau Public as a complement to Microsoft Excel for visualizing data and regression analysis.
- **A New Business Analytics Chapter (Chapter 17)** that provides a complete introduction to the field of business analytics. The chapter defines terms and categories that introductory business statistics students may encounter in other courses or outside the classroom. This chapter benefits from the insights the authors have gained from teaching and lecturing about business analytics as well as research the authors have done for a forthcoming companion book on business analytics.

Continuing Features That Readers Have Come to Expect

This edition of *Statistics for Managers Using Microsoft Excel* continues to incorporate a number of distinctive features that has led to its wide adoption over the previous editions. Table 1 summaries these carry-over features:

TABLE 1
Distinctive Features Continued in the Ninth Edition

Feature	Details
Using Statistics case scenarios	Each chapter begins with a Using Statistics case scenario that presents a business problem or goal that illustrates the application of business statistics to provide actionable information. For many chapters, scenarios also provide the scaffolding for learning a series of related statistical methods. End-of-chapter "Revisited" sections reinforce the statistical learning of a chapter by discussing how the methods and techniques can be applied to the goal or problem that the case scenario considers.
	In this edition, seven chapters have new or revised case scenarios.
Emphasis on interpretation of the data analysis results	*Statistics for Managers Using Microsoft Excel* was among the first introductory business statistics textbooks to focus on the interpretation of Microsoft Excel statistical results. This tradition continues, now supplemented by Tableau (Public) results for selected methods in which Tableau can enhance or complement Excel results.
Software integration and flexibility	Software instructions feature chapter examples and were personally written by the authors, who collectively have more than one hundred years of experience teaching the application of business software.
	With modularized Wor*kbook,* PHS*tat,* and where applicable, Ana*lysis Toolbook* instructions, both instructors and students can switch among these instruction sets as they use this book with no loss of statistical learning.
Unique Excel workbooks	*Statistics for Managers Using Microsoft Excel* comes with Excel Guide workbooks that illustrate model solutions and provide template solutions to selected methods and Visual Explorations, macro-enhanced workbooks that demonstrate selected basic concepts. This book is fully integrated with PHStat, the Pearson statistical add-in for Excel that places the focus on statistical learning that the authors designed and developed.
	See Appendix H for a complete description of PHStat.

TABLE 1 Distinctive Features Continued in the Ninth Edition (*continued*)

Feature	Details
In-chapter and end-of-chapter reinforcements	Exhibits summarize key processes throughout the book. A key terms list provides an index to the definitions of the important vocabulary of a chapter. "Learning the Basics" questions test the basic concepts of a chapter. "Applying the Concepts" problems test the learner's ability to apply statistical methods to business problems. And, for the more mathematically minded, "Key Equations" list the boxed number equations that appear in a chapter.
End-of-chapter cases	End-of-chapter cases include a case that continues through most chapters and several cases that reoccur throughout the book. "Digital Cases" require students to examine business documents and other information sources to sift through various claims and discover the data most relevant to a business case problem. Many of these cases also illustrate common misuses of statistical information. *The Instructor's Solutions Manual provides instructional tips for using cases as well as solutions to the Digital Cases.*
Answers to even-numbered problems	An appendix provides additional self-study opportunities by provides answers to the "Self-Test" problems and most of the even-numbered problems in this book
Opportunities for additional learning	In-margin student tips and LearnMore references reinforce important points and direct students to additional learning resources. In-chapter *Consider This* essays reinforce important concepts, examine side issues, or answer questions that arise while studying business statistics, such as "What is so 'normal' about the normal distribution?"
Highly tailorable content	With an extensive library of separate online topics, sections, and even two full chapters, instructors can combine these materials and the opportunities for additional learning to meet their curricular needs.

Chapter-by-Chapter Changes Made for This Edition

Because the authors believe in continuous quality improvement, *every* chapter of *Statistics for Managers Using Microsoft Excel* contains changes to enhance, update, or just freshen this book. Table 2 provides a chapter-by-chapter summary of these changes.

TABLE 2
Chapter-by-Chapter
Change Matrix

Chapter	Using Statistics Changed	Problems Changed	Selected Chapter Changes
First Things First		n.a.	Updates opening section. Retitles, revises, and expands old Section FTF.4 as Section FTF.4 and new Section FTF.5 "Starting Point for Using Microsoft Excel." Expands the First Things First Excel Guide. Adds a First Things First Tableau Guide.
1		26%	Old Sections 1.3 and 1.4 revised and expanded as new Section 1.4 "Data Preparation." Adds a Chapter 1 Tableau Guide.
2	•	57%	Uses new samples of 479 retirement funds and 100 restaurant meal costs for in-chapter examples. Includes new examples for organizing and visualizing categorical variables. Uses updated scatter plot and time-series plot examples. Adds new Section 2.8 "Filtering and Querying Data." Adds coverage of bubble charts, treemaps, and (Tableau) colored scatter plots. Revises and expands the Chapter 2 Excel Guide. Adds a Chapter 2 Tableau Guide.

TABLE 2
Chapter-by-Chapter
Change Matrix (*continued*)

Chapter	Using Statistics Changed	Problems Changed	Selected Chapter Changes
3	•	52%	Uses new samples of 479 retirement funds and 100 restaurant meal costs for in-chapter examples. Includes updated Dogs of the Dow NBA team values data sets. Adds a Chapter 3 Tableau Guide.
4	•	41%	Uses the updated Using Statistics scenario for in-chapter examples.
5		32%	Adds a new exhibit that summarizes the binomial distribution.
6		31%	Uses new samples of 479 retirement funds for the normal probability plot example.
7	•	43%	Enhances selected figures for additional clarity.
8		36%	Presents a more realistic chapter opening illustration. Revises Example 8.3 in Section 8.2 "Confidence Interval Estimate for the Mean (σ Unknown)."
9		29%	Adds a new exhibit that summarizes fundamental hypothesis testing concepts. Revises Section 9.2 "t Test of Hypothesis for the Mean (σ Unknown)" example such that the normality assumption is not violated. Revises examples in Section 9.3 "One-Tail Tests" and Section 9.4 "Z Test of Hypothesis for the Proportion."
10		37%	Uses the new tabular test summaries for the two-sample t test, paired t test, and the Z test for the difference between two proportions. Includes new Section 10.1 passage "Evaluating the Normality Assumption." Uses new (market basket) data for the paired "t test example. Enhances selected figures for additional clarity. Contains general writing improvements throughout chapter.
11	•	17%	Presents an updated chapter opening illustration. Uses revised data for the Using Statistics scenario for in-chapter examples. Uses the new tabular test summaries for the one-way ANOVA results. Presents discussion of the Levene test for the homogeneity of variance before the Tukey-Kramer multiple comparisons procedure. Revises Example 11.1 in Section 11.1 "One-Way ANOVA."
12		37%	Uses the new tabular test summaries for the chi-square tests, Wilcoxon rank sum test, and Kruskal-Wallis rank test. Category changes in text of independence example. Uses revised data for Section 12.4 and 12.5 examples for the Wilcoxon and Kruskal-Wallis tests. Contains general writing improvements throughout chapter.

TABLE 2
Chapter-by-Chapter
Change Matrix (*continued*)

Chapter	Using Statistics Changed	Problems Changed	Selected Chapter Changes
13		46%	Reorganizes presentation of basic regression concepts with new "Preliminary Analysis" passage and a revised Section 13.1 retitled as "Simple Linear Regression Models." Revises the exhibit in Section 13.9 that summarizes avoiding potential regression pitfalls. Presents an updated chapter opening illustration. Enhances selected figures for additional clarity. Adds a Chapter 13 Tableau Guide.
14		29%	Retitles Section 14.2 as "Evaluating Multiple Regression Models." Uses tables to summarize the net effects in multiple regression and residual analysis. Uses the new tabular test summaries for the overall F test, the t test for the slope, and logistic regression. Adds the new "What Is Not Normal? (Using a Categorical Dependent Variable)" *Consider This* feature. Adds a new section on cross-validation. Enhances selected figures for additional clarity. Contains general writing improvements throughout chapter.
15		36%	Uses the new tabular test summaries for the quadratic regression results. Revises Section 15.2 "Using Transformations in Regression Models." Replaces Example 5.2 in Section 15.2 with a new sales analysis business example. Reorganizes Section 15.4 "Model Building." Updates the Section 15.4 exhibit concerning steps for successful model building. Contains general writing improvements throughout chapter.
16	•	67%	Combines old Sections 16.1 and 16.2 into a revised Section16.1 "Time-Series Component Factors. Section 16.1 presents a new illustration of time-series components. Uses updated movie attendance time-series data for the Section 16.2 example. Uses new annual time-series revenue data for Alphabet Inc. for Sections 16.3 and 16.5 examples. Uses updated annual time-series revenue data for the Coca-Cola Company in Section 16.4. Uses new quarterly time-series revenue data for Amazon.com, Inc. for the Section 16.6 example. Uses updated data for moving averages and exponential smoothing. Uses updated data for the online Section 16.7 "Index Numbers."
17	•	100%	Completely new "Business Analytics" that expands and updates old Sections 17.3 through 17.5. Includes the new "What's My Major If I Want to Be a Data Miner?" *Consider This* feature.

TABLE 2
Chapter-by-Chapter
Change Matrix (*continued*)

Chapter	Using Statistics Changed	Problems Changed	Selected Chapter Changes
18		55%	Updates old Chapter 17 Sections 17.1 and 17.2 to form the new version of the "Getting Ready to Analyze Data in the Future" chapter.
Appendices		n.a.	Adds new Tableau sections to Appendices B, D, and G. Adds new Appendix B section about using non-numerical labels in time-series plots. Includes updated data files listing in Appendix C.

Serious About Writing Improvements

Ever read a textbook preface that claims writing improvements but offers no evidence? Among the writing improvements in this edition of *Statistics for Managers Using Microsoft Excel,* the authors have used tabular summaries to guide readers to reaching conclusions and making decisions based on statistical information. The authors believe that this writing improvement, which appears in Chapters 9 through 15, adds clarity to the purpose of the statistical method being discussed and better illustrates the role of statistics in business decision-making processes.

For example, consider the following sample passage from Example 10.1 in Chapter 10 that illustrates the use of the new tabular summaries.

Previously, part of the Example 10.1 solution was presented as:
You do not reject the null hypothesis because $t_{STAT} = -1.6341 > -1.7341$. The p-value (as computed in Figure 10.5) is 0.0598. This p-value indicates that the probability that $t_{STAT} < -1.6341$ is equal to 0.0598. In other words, if the population means are equal, the probability that the sample mean delivery time for the local pizza restaurant is at least 2.18 minutes faster than the national chain is 0.0598. Because the p-value is greater than a $= 0.05$, there is insufficient evidence to reject the null hypothesis. Based on these results, there is insufficient evidence for the local pizza restaurant to make the advertising claim that it has a faster delivery time.

In this edition, we present the equivalent solution (on page 316):
Table 10.4 summarizes the results of the pooled-variance t test for the pizza delivery data using the calculation above (*not shown in this sample*) and Figure 10.5 results. Based on the conclusions, the local branch of the national chain and a local pizza restaurant have similar delivery times. Therefore, as part of the last step of the DCOVA framework, you and your friends exclude delivery time as a decision criteria when choosing from which store to order pizza.

TABLE 10.4 Pooled-variance t test summary for the delivery times for the two pizza restaurants

Result	Conclusions
The $t_{STAT} = -1.6341$ is greater than -1.7341. The t test p-value $= 0.0598$ is greater than the level of significance, $\alpha = 0.05$.	1. Do not reject the null hypothesis H_0. 2. Conclude that insufficient evidence exists that the mean delivery time is lower for the local restaurant than for the branch of the national chain. 3. There is a probability of 0.0598 that $t_{STAT} < -1.6341$.

A Note of Thanks

Creating a new edition of a textbook and its ancillaries requires a team effort with our publisher and the authors are fortunate to work with Suzanna Bainbridge, Karen Montgomery, Rachel Reeve, Alicia Wilson, Bob Carroll, Jean Choe, and Aimee Thorne of Pearson Education. Readers who note the quality of the illustrations and software screen shots that appear throughout the book should know that the authors are indebted to Joe Vetere of Pearson Education for his expertise.

We also thank Sharon Cahill of SPi Global US, Inc., and Julie Kidd for their efforts to maintain the highest standard of book production quality (and for being gracious in handling the mistakes we make). We thank Gail Illich of McLennan Community College for preparing the instructor resources and the solutions manual for this book and for helping refine problem questions throughout it. And we thank our colleagues in the Data, Analytics, and Statistics Instruction Specific Interest Group (DASI SIG) of the Decision Sciences Institute for their ongoing encouragement and exchange of ideas that assist us to continuously improve our content.

We strive to publish books without any errors and appreciate the efforts of Jennifer Blue, James Lapp, Sheryl Nelson, and Scott Bennett to copy edit, accuracy check, and proofread our book and to get us closer to our goal.

We thank the RAND Corporation and the American Society for Testing and Materials for their kind permission to publish various tables in Appendix E, and to the American Statistical Association for its permission to publish diagrams from the *American Statistician*. Finally, we would like to thank our families for their patience, understanding, love, and assistance in making this book a reality.

Contact Us!

Please email us at **authors@davidlevinestatistics.com** with your questions about the contents of this book. Please include the hashtag #SMUME9 in the subject line of your email. We always welcome suggestions you may have for a future edition of this book. And while we have strived to make this book as error-free as possible, we also appreciate those who share with us any issues or errors that they discover.

If you need assistance using software, please contact your academic support person or Pearson Technical Support at **support.pearson.com/getsupport/.** They have the resources to resolve and walk you through a solution to many technical issues in a way we do not.

As you use this book, be sure to make use of the "Resources for Success" that Pearson Education supplies for this book (described on the following pages). We also invite you to visit **smume9.davidlevinestatistics.com**, where we may post additional information, corrections, and updated or new content to support this book.

David M. Levine
David F. Stephan
Kathryn A. Szabat

Get the *most* out of
MyLab Statistics

Pearson MyLab

MyLab Statistics Online Course for *Statistics for Managers Using Microsoft® Excel®*, 9th Edition by Levine, Stephan, Szabat
(access code required)

MyLab Statistics is the teaching and learning platform that empowers instructors to reach every student. By combining trusted author content with digital tools and a flexible platform, MyLab Statistics personalizes the learning experience and improves results for each student.

MyLab makes learning and using a variety of statistical programs as seamless and intuitive as possible. Download the data files that this book uses (see Appendix C) in Microsoft Excel. Download supplemental files that support in-book cases or extend learning.

▶ Download the Excel Data Workbooks that contain the data used in chapter examples or named in problems and end-of-chapter cases.

▶ Download the Excel Guide Workbooks that contain the model templates and solutions for statistical methods discussed in the textbook.

▶ Download the JMP Data Tables and Projects that contain the data used in chapter examples or named in problems and end-of-chapter cases.

▶ Download the Minitab Worksheets and Projects that contain the data used in chapter examples or named in problems and end-of-chapter cases.

▶ Download the PHStat readme pdf that explains the technical requirements and getting started instructions for using this Microsoft Excel add-in. To download PHStat, visit the PHStat download page. (Download requires an access code as explained on that page.)

▶ Download the Visual Explorations Workbooks that interactively demonstrate various key statistical concepts.

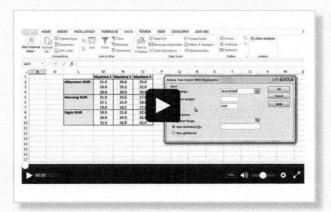

Instructional Videos

Access instructional support videos including Pearson's Business Insight and StatTalk videos, available with assessment questions. Reference technology study cards and instructional videos for Microsoft Excel.

Diverse Question Libraries

Build homework assignments, quizzes, and tests to support your course learning outcomes. From Getting Ready (GR) questions to the Conceptual Question Library (CQL), we have your assessment needs covered from the mechanics to the critical understanding of Statistics. The exercise libraries include technology-led instruction, including new Microsoft Excel-based exercises, and learning aids to reinforce your students' success.

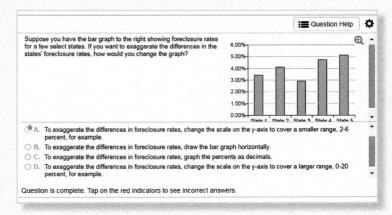

pearson.com/mylab/statistics

Resources for
Success

Pearson
MyLab

Instructor Resources

Instructor's Solutions Manual, presents solutions for end-of-section and end-of-chapter problems and answers to case questions, and provides teaching tips for each chapter. The Instructor's Solutions Manual is available for download at **www.Pearson.com** or in MyLab Statistics.

Lecture PowerPoint Presentations, by Patrick Schur, Miami University (Ohio), are available for each chapter. These presentations provide instructors with individual lecture notes to accompany the text. The slides include many of the figures and tables from the textbook. Instructors can use these lecture notes as is or customize them in Microsoft PowerPoint. The PowerPoint presentations are available for download at **www.Pearson.com** or in MyLab Statistics.

Test Bank, contains true/false, multiple-choice, fill-in, and problem-solving questions based on the definitions, concepts, and ideas developed in each chapter of the text. The Test Bank is available for download at **www.Pearson.com** or in MyLab Statistics.

TestGen® (**www.pearsoned.com/testgen**) enables instructors to build, edit, print, and administer tests using a computerized bank of questions developed to cover all the objectives of the text. TestGen is algorithmically based, allowing instructors to create multiple but equivalent versions of the same question or test with the click of a button. Instructors can also modify test bank questions or add new questions. The software and test bank are available for download from Pearson Education's online catalog.

Student Resources

Student's Solutions Manual, provides detailed solutions to virtually all the even-numbered exercises and worked-out solutions to the self-test problems. (ISBN-10: 0-13-597002-4; ISBN-13: 978-0-13-597002-7)

Online resources complement and extend the study of business statistics and support the content of this book. These resources include data files for in-chapter examples and problems, **templates and model solutions**, and **optional topics and chapters**. (See Appendix C for a complete description of the online resources.)

PHStat helps create Excel worksheet solutions to statistical problems. PHStat uses Excel building blocks to create worksheet solutions. These worksheet solutions illustrate Excel techniques and students can examine them to gain new Excel skills. Additionally, many solutions are what-if templates in which the effects of changing data on the results can be explored. Such templates are fully reusable on any computer on which Excel has been installed. PHStat requires an access code and separate download for use. PHStat access codes can be bundled with this textbook using ISBN-10: 0-13-399058-3; ISBN-13: 978-0-13-399058-4.

pearson.com/mylab/statistics

First Things First

▼ USING **STATISTICS**
"The Price of Admission"

I t's the year 1900, and you are a promoter of theatrical productions, in the business of selling seats for individual performances. Using your knowledge and experience, you establish a selling price for the performances, a price you hope represents a good trade-off between maximizing revenues and not driving away demand for your seats. You print up tickets and flyers, place advertisements in local media, and see what happens. After the event, you review your results and consider the effects of your choices on those results.

Tickets sold very quickly? Next time perhaps you can charge more. The event failed to sell out? Perhaps next time you could charge less or take out more advertisements to drive demand. If you lived over 100 years ago, that's about all you could do.

Jump ahead about 85 years. You're using computer systems that enable you to sell more categories of tickets, such as tickets for premium-priced seat locations. As customers buy tickets over the phone, you can monitor sales through schedule summary reports and, perhaps, add or subtract performance dates using the information in those reports.

Jump ahead to today. Your fully online ticketing system updates seat inventory and enables you to use **dynamic pricing** that automatically alters seat prices based on factors such as increased demand. You also now have the flexibility to set new pricing tables or add special categories that are associated with times of peak demand. Through your sales system you have gained insights about your customers, such as where they live, what other tickets they buy, and their appropriate demographic traits. Because you know more about your customers, you can make your advertising and publicity more efficient by aiming your messages at the types of people more likely to buy your tickets. By using social media networks and other online media, you can also learn almost immediately who is noticing and responding to your advertising messages. You might even run experiments online presenting your advertising in two different ways and seeing which way sells better.

Your current self has capabilities that allow you to be a more effective promoter than any older version of yourself. But just how much better? Turn the page.

Now Appearing on Broadway ... *and* Everywhere Else

In early 2014, Disney Theatrical Productions surprised Broadway when reports revealed that Disney's *17*-year-old *The Lion King* had been the top-grossing Broadway show in 2013. How could such a long-running show earn so much while being so old? Broadway producers "knew" that grosses for a show decline over time and by year 12 (2009) weekly grosses for *The Lion King* had dropped about 25%. But four years *later,* grosses were up 67%, and weekly grosses typically exceeded even the grosses of the opening weeks of the show, adjusted for inflation!

While heavier advertising and some ticket pricing changes helped, the major reason for this change was something else: combining business acumen with an informed application of *business statistics and analytics* to better solve the problem of selling tickets. As a producer of the newest musical at the time said, "We make educated predictions on price. Disney, on the other hand, has turned this into a science" (Healey).

Disney had followed the plan of action that this book presents. It had collected its daily and weekly results and summarized them, using techniques this book introduces in the next three chapters. Disney then analyzed those results by performing experiments and tests on the data collected, using techniques that later chapters introduce. In turn, insights from the results of those analyses led to developing a new interactive seating map that allowed customers to buy tickets for specific seats and permitted Disney to adjust the pricing of each seat for each performance. The whole system was constantly reviewed and refined, using the semiautomated methods that Chapter 17 introduces. The end result was a ticket-selling method that outperformed previous methods used.

Five years after *The Lion King* surprised Broadway, the show continues to average about the same grosses it did in 2014 and still has many weeks in which it grosses $2 million or more, once a rarely achieved Broadway benchmark. However, the show is no longer the top grossing show because another show, about a partially obscure Secretary of the Treasury has grosses that dwarf *The Lion King*'s respectable sales. The producers of *Hamilton* have fully applied the techniques that Disney first used and created a show whose weekly grosses average more than *$3* million and, even after running for four years, can manage record-setting weekly grosses of more than $4 million! As a *Wall Street Journal* article has noted, "It is boom time on Broadway" (Passy).

student TIP

From other business courses, you may recognize that Disney's system uses dynamic pricing.

FTF.1 Think Differently About Statistics

The "Using Statistics" scenario suggests, and the Disney example illustrates, that modern-day information technology has allowed businesses to apply statistics in ways that could not be done years ago. This scenario and example reflect how this book teaches you about statistics. In these first two pages, you may notice

- the lack of calculation details and "math."
- the emphasis on enhancing business methods and management decision making.
- that none of this seems like the content of a middle school or high school statistics class you may have taken.

You may have had some prior knowledge or instruction in *mathematical statistics*. This book discusses *business statistics*. While the boundary between the two can be blurry, business statistics emphasizes business problem solving and shows a preference for using software to perform calculations.

One similarity that you might notice between these first two pages and any prior instruction is *data*. **Data** are the facts about the world that one seeks to study and explore. Some data are unsummarized, such as the facts about a single ticket-selling transaction, whereas other facts, such as weekly ticket grosses, are **summarized**, derived from a set of unsummarized data. While you may think of data as being numbers, such as the cost of a ticket or the percentage that weekly grosses have increased in a year, do not overlook that data can be non-numerical as well, such as ticket-buyer's name, seat location, or method of payment.

Statistics: A Way of Thinking

Statistics are the methods that allow you to work with data effectively. Business statistics focuses on interpreting the results of applying those methods. You interpret those results to help you enhance business processes and make better decisions. Specifically, business statistics provides you with a formal basis to summarize and visualize business data, reach conclusions about that data, make reliable predictions about business activities, and improve business processes.

You must apply this way of thinking correctly. Any "bad" things you may have heard about statistics, including the famous quote "there are lies, damned lies, and statistics" made famous by Mark Twain, speak to the errors that people make when either misusing statistical methods or mistaking statistics as a substitution for, and not an enhancement of, a decision-making process. (Disney Theatrical Productions' success was based on *combining* statistics with business acumen, not *replacing* that acumen.)

DCOVA Framework   To minimize errors, you use a framework that organizes the set of tasks that you follow to apply statistics properly. Five tasks comprise the **DCOVA framework**:

- **D**efine the data that you want to study to solve a problem or meet an objective.
- **C**ollect the data from appropriate sources.
- **O**rganize the data collected, by developing tables.
- **V**isualize the data collected, by developing charts.
- **A**nalyze the data collected, reach conclusions, and present the results.

You must always do the **D**efine and **C**ollect tasks before doing the other three. The order of the other three varies, and sometimes all three are done concurrently. In this book, you will learn more about the **D**efine and **C**ollect tasks in Chapter 1 and then be introduced to the **O**rganize and **V**isualize tasks in Chapter 2. Beginning with Chapter 3, you will learn methods that help complete the **A**nalyze task. Throughout this book, you will see specific examples that apply the DCOVA framework to specific business problems and examples.

Analytical Skills More Important Than Arithmetic Skills   The business preference for using software to automate statistical calculations maximizes the importance of having analytical skills while it minimizes the need for arithmetic skills. With software, you perform calculations faster and more accurately than if you did those calculations by hand, minimizing the need for advanced arithmetic skills. However, with software you can *also* generate inappropriate or meaningless results if you have not fully understood a business problem or goal under study or if you use that software without a proper understanding of statistics.

Therefore, using software to create results that help solve business problems or meet business goals is *always* intertwined with using a framework. And using software does not mean memorizing long lists of software commands or how-to operations, but knowing how to review, modify, and possibly create software solutions. If you can analyze what you need to do and have a general sense of what you need, you can always find instructions or illustrative sample solutions to guide you. (This book provides detailed instructions *as well as* sample solutions for every statistical activity discussed in end-of-chapter software guides and through the use of various downloadable files and sample solutions.)

If you were introduced to using software in an application development setting or an introductory information systems class, do not mistake building applications from scratch as being a necessary skill. A "smart" smartphone user knows how to use apps such as Facebook, Instagram, YouTube, Google Maps, and Gmail effectively to communicate or discover and use information and has no idea how to construct a social media network, create a mapping system, or write an email program. Your approach to using the software in this book should be the same as that smart user. Use your analytical skills to focus on being an effective user and to understand *conceptually* what a statistical method or the software that implements that method does.

Statistics: An Important Part of Your Business Education

Until you read these pages, you may have seen a course in business statistics solely as a required course with little relevance to your overall business education. In just two pages, you have learned that statistics is a way of thinking that can help enhance your effectiveness in business—that is, applying statistics correctly is a fundamental, global skill in your business education.

In the current data-driven environment of business, you need the general analytical skills that allow you to work with data and interpret analytical results regardless of the discipline in which you work. No longer is statistics only for accounting, economics, finance, or other disciplines that directly work with numerical data. As the Disney example illustrates, the decisions you make will be increasingly based on data and not on your gut or intuition supported by past experience. Having a well-balanced mix of statistics, modeling, and basic technical skills as well as managerial skills, such as business acumen and problem-solving and communication skills, will best prepare you for the workplace today ... *and* tomorrow (Advani).

FTF.2 Business Analytics: The Changing Face of Statistics

Of the recent changes that have made statistics an important part of your business education, the emergence of the set of methods collectively known as business analytics may be the most significant change of all. **Business analytics** combine traditional statistical methods with methods from management science and information systems to form an interdisciplinary tool that supports fact-based decision making. Business analytics include

- statistical methods to analyze and explore data that can uncover previously unknown or unforeseen relationships.
- information systems methods to collect and process data sets of all sizes, including very large data sets that would otherwise be hard to use efficiently.
- management science methods to develop optimization models that support all levels of management, from strategic planning to daily operations.

In the Disney Theatrical Productions example, statistical methods helped determine pricing factors, information systems methods made the interactive seating map and pricing analysis possible, and management science methods helped adjust pricing rules to match Disney's goal of sustaining ticket sales into the future. Other businesses use analytics to send custom mailings to their customers, and businesses such as the travel review site tripadvisor.com use analytics to help optimally price advertising as well as generate information that makes a persuasive case for using that advertising.

Generally, studies have shown that businesses that actively use business analytics and combine that use with data-guided management see increases in productivity, innovation, and competition (Advani). Chapter 17 introduces you to the statistical methods typically used in business analytics and shows how these methods are related to statistical methods that the book discusses in earlier chapters.

"Big Data"

Big data is a collection of data that cannot be easily browsed or analyzed using traditional methods. Big data implies data that are being collected in huge volumes, at very fast rates or velocities (typically in near real time), and in a variety of forms that can differ from the structured forms such as records stored in files or rows of data stored in worksheets that businesses use every day. These attributes of volume, velocity, and variety distinguish big data from a "big" (large) set of data that contains numerous records or rows of similar data (Laney). When combined with business analytics and the basic statistical methods discussed in this book, big data presents opportunities to gain new management insights and extract value from the data resources of a business (IBM).

Unstructured Data Big data may also include **unstructured data**, data that have an irregular pattern and contain values that are not comprehensible without additional automated or manual interpretation. Unstructured data take many forms, such as unstructured text, pictures, videos, and audio tracks, with unstructured text, such as social media comments, getting the most immediate attention today for its possible use in customer, branding, or marketing analyses. Unstructured data can be adapted for use with a number of methods, such as regression, which this book illustrates with conventional, structured files and worksheets. Unstructured data may require performing data collection and preparation tasks beyond those tasks that Chapter 1 discusses. While describing all such tasks is beyond the scope of this book, Section 17.1 includes an example of the additional interpretation that is necessary when working with unstructured text.

FTF.3 Starting Point for Learning Statistics

Statistics has its own vocabulary and learning the precise meanings, or **operational definitions**, of basic terms provides the basis for understanding the statistical methods that this book discusses. For example, *in statistics*, a **variable** defines a characteristic, or property, of an item or individual that can vary among the occurrences of those items or individuals. For example, for the item "book," variables would include the title and number of chapters, as these facts can vary from book to book. For a given book, these variables have a specific value. For *this* book, the value of the title variable would be "Statistics for Managers Using Microsoft Excel," and "19" would be the value for the number of chapters variable. Note that a statistical variable is not an algebraic variable, which serves as a stand-in to represent one value in an algebraic statement and could never take a non-numerical value such as the title of this book.

Using the definition of variable, data, in its statistical sense, can be defined as the set of values associated with one or more variables. In statistics, each value for a specific variable is a single fact, not a list of facts. For example, what would be the value of the variable author for this book? Without this rule, you might say that the single list "Levine, Szabat, Stephan" is the value. However, applying this rule, one would say that the variable has three separate values: "Levine", "Stephan", and "Szabat". This distinction of using only *single-value data* has the practical benefit of simplifying the task of entering data for software analysis.

Using the definitions of data and variable, the definition of statistics can be restated as the methods that analyze the data of the variables of interest. The methods that primarily help summarize and present data comprise **descriptive statistics**. Methods that use data collected from a small group to reach conclusions about a larger group comprise **inferential statistics**. Chapters 2 and 3 introduce descriptive methods, many of which are applied to support the inferential methods that the rest of the book presents.

Statistic

The previous section uses *statistics* in the sense of a collective noun, a noun that is the name for a collection of things (methods in this case). The word statistics also serves as the plural form of the noun statistic, as in "one uses methods of descriptive statistics (collective noun) to generate descriptive statistics (plural of the singular noun)." In this sense, a **statistic** refers to a value that summarizes the data of a particular variable. For the Disney Theatrical Productions example, the statement "for 2013, weekly grosses were up 67% from 2009" cites a *statistic* that summarizes the variable weekly grosses using the 2013 data—all 52 values.

When someone warns you of a possible unfortunate outcome by saying, "Don't be a statistic!" you can always reply, "I can't be." *You* always represent one value, and a *statistic* always summarizes multiple values. For the statistic "87% of our employees suffer a workplace accident," you, as an employee, will either have suffered or have not suffered a workplace accident. The "have" or "have not" value contributes to the statistic but cannot be the statistic. A statistic can facilitate preliminary decision making. For example, would you immediately accept a position at a company if you learned that 87% of their employees suffered a workplace accident? (Sounds like this might be a dangerous place to work and that further investigation is necessary.)

Can Statistics (*pl.*, statistic) Lie?

The famous quote "lies, damned lies, and statistics" actually refers to the plural form of *statistic* and does not refer to statistics, the field of study. Can any statistic "lie"? No, faulty or invalid statistics can only be produced through willful misuse of statistics or when DCOVA framework tasks are done incorrectly. For example, many statistical methods are valid only if the data being analyzed have certain properties. To the extent possible, you test the assertion that the data have those properties, which in statistics are called *assumptions*. When an assumption is *violated*, shown to be invalid for the data being analyzed, the methods that require that assumption should not be used.

For the inferential methods that this book discusses in later chapters, you must always look for logical causality. **Logical causality** means that you can plausibly claim something directly causes something else. For example, you wear black shoes today and note that the weather is sunny. The next day, you again wear black shoes and notice that the weather continues to be sunny. The third day, you change to brown shoes and note that the weather is rainy. The fourth day, you wear black shoes again and the weather is again sunny. These four days seem to suggest a strong pattern between your shoe color choice and the type of weather you experience. You begin to think if you wear brown shoes on the fifth day, the weather will be rainy. Then you realize that your shoes cannot plausibly influence weather patterns, that your shoe color choice cannot *logically cause* the weather. What you are seeing is mere coincidence. (On the fifth day, you do wear brown shoes and it happens to rain, but that is just another coincidence.)

You can easily spot the lack of logical causality when trying to correlate shoe color choice with the weather, but in other situations the lack of logical causality may not be so easily seen. Therefore, relying on such correlations by themselves is a fundamental misuse of statistics. When you look for patterns in the data being analyzed, you must *always* be thinking of logical causes. Otherwise, you are misrepresenting your results. Such misrepresentations sometimes cause people to wrongly conclude that all statistics are "lies." Statistics (*pl.*, statistic) are not lies or "damned lies." They play a significant role in *statistics*, the way of thinking that can enhance your decision making and increase your effectiveness in business.

FTF.4 Starting Point for Using Software

learnMORE

About the online supplemental files in Appendix C.

The starting point for using any software is knowledge of basic user interface skills, operations, and vocabulary that Table FTF.1 summarizes and which the supplemental **Basic Computing Skills** online document reviews.

TABLE FTF.1
Basic computing skills

Skill or Operation	Specifics
Identify and use standard window objects	Title bar, minimize/resize/close buttons, scroll bars, mouse pointer, menu bars or ribbons, dialog box, window subdivisions such as areas, panes, or child windows
Identify and use common dialog box items	Command button, list box, drop-down list, edit box, option button, check box, tabs (tabbed panels)
Mouse (or touch) operations	Click, called select in some list or menu contexts and check or clear in some check box contexts; double-click; right-click; drag and drag-and-drop

With such knowledge, learning how to apply any program to a business problem becomes possible. In using Microsoft Excel, this book assumes no more than this starting point. While having prior experience is always useful, readers unfamiliar with Excel can learn about commonly used operations in this book's appendices. Such readers will find instructions for using reusable tools and automated methods, the Excel add-ins Analysis ToolPak and PHStat, throughout the book. Those instructions assume only this starting-point knowledge.

Software-Related Conventions   Table FTF.2 summarizes the typographical conventions that this book uses for its software instructions. These conventions provide a concise and clear way of expressing specific basic-user activities and operations.

TABLE FTF.2 Software typographical conventions in this book

Convention	Example
Special key names appear capitalized and in boldface	Press **Enter**. Press **Command** or **Ctrl**.
Key combinations appear in boldface, with key names linked using this symbol: **+**	Enter the formula and press **Ctrl+Enter**. Press **Ctrl+C**.
Menu or Ribbon selections appear in boldface and sequences of consecutive selections are shown using this symbol: ➔	Select **File➔New**. Select **PHStat➔Descriptive Statistics ➔ Boxplot**.
Target of mouse operations appear in boldface	Click **OK**. Select **Attendance** and then click the **Y button**.
Entries and the location of where entries are made appear in boldface	Enter **450** in cell **B5**. Add **Temperature** to the **Model Effects** list.
Variable or column names sometimes appear capitalized for emphasis	This file contains the Fund Type, Assets, and Expense Ratio for the growth funds.
Placeholders that express a general case appear in italics and may also appear in boldface as part of a function definition	**AVERAGE (*cell range of variable*)** Replace *cell range of variable* with the cell range that contains the Asset variable.
Names of data files mentioned in sections or problems appear in a special font but appear in boldface in end-of-chapter Guide instructions	Retirement Funds Open the **Retirement Funds workbook**.
When current versions of Excel differ in their user interface, alternate instructions appear in a second color immediately following the primary instructions	In the Select Data Source display, click **Edit** that appears under **Horizontal (Category) Axis Labels**. In Excel for Mac, in the Select Data Source display, click the icon inside the **Horizontal (Categories) axis labels**.

Using Software Properly

Learning to use software *properly* can be hard as software has limited ways to provide feedback for user actions that are invalid operations. In addition, no software will ever know if its users are following the proper procedures for use. Exhibit FTF.1 presents a list of guiding principles that, ideally, would govern readers' usage of any software, not just the software used with this book. These principles will minimize the chance of making errors and lessen the frustration that often occurs when these principles are overlooked.

EXHIBIT **FTF.1**

Principles of Using Software Properly

Ensure that software is properly updated. Users who manage their own computers often overlook the importance of ensuring that all installed software is up to date.

Understand the basic operational tasks. Take the time to master the tasks of starting the application, loading and entering data, and how to select or choose commands in a general way.

Understand the statistical concepts that an application uses. Not understanding those concepts can lead to making wrong choices in the application and can make interpreting results difficult.

Know how to review software use for errors. Review and verify that the proper data preparation procedures (see Chapter 1) have been applied to the data before analysis.

(continued)

EXHIBIT FTF.1 *(continued)*

Verify that the correct procedures, commands, and software options have been selected. For information entered for labeling purposes, verify that no typographical errors exist.

Seek reuse of preexisting solutions to solve new problems. Build solutions from scratch only as necessary, particularly if using Excel in which errors can be most easily made. Some solutions, and almost all Excel solutions that this book presents, exist as models or templates that can *and should* be reused because such reuse models best practice.

Understand how to organize and present information from the results that the software produces. Think about the best ways to arrange and label the data. Consider ways to enhance or reorganize results that will facilitate communication with others.

Use self-identifying names, especially for the files that you create and save. Naming files Document 1, Document 2, and so on will impede the later retrieval and use of those files.

FTF.5 Starting Point for Using Microsoft Excel

Microsoft Excel is the data analysis component of Microsoft Office that evolved from earlier electronic spreadsheets first used for financial accounting applications. As such, Excel can be an inadequate tool for *complex* data analysis, even as the program is a convenient way to visually examine data details as well as learn foundational applications of business statistics.

Excel uses **worksheets** to display the contents of a data set and as a means to enter or edit data. Worksheets are containers that present tabular arrangements of data, in which the intersections of rows and columns form **cells**, boxes into which individual entries are made. In the simplest use of Excel, cell entries are single data values that can be either text or numbers. When entering data for a data set to be analyzed, data for each variable are placed in their own column, a vertical group of cells. By convention, the initial row of such columns stores the variable name associated with the column data. Columns so prepared are called either variables, variable columns, or just columns in the context of using Microsoft Excel.

Worksheet cells can also contain **formulas**, instructions to process data or to compute cell values. Formulas can include **functions** that simplify certain arithmetic tasks or provide access to advanced processing or statistical features. Formulas play an important role in designing **templates**, *reusable solutions* that have been previously tested and verified. A hallmark feature of this book is the inclusion of **Excel Guide workbooks**, most of which exist as templates, and all of which simplify the operational details of using Excel for data analysis while demonstrating the application of formulas and functions for such analysis. Appendix C includes a list of the Excel Guide workbooks that this book uses.

Excel stores worksheet data and results as one file called a **workbook**. Workbooks can contain multiple worksheets and chart sheets, sheets that display visualizations of data. Because workbooks contain collections, best practice places data and the results of analyzing that data on separate worksheets, a practice that the Excel Guide workbooks and the Excel data workbooks for this book reflect. Therefore, instructions for using Excel that appear in end-of-chapter Excel Guides often begin with "Open to the *specific* **worksheet** in the *specific* **workbook**" as in "Open to the **DATA worksheet** of the **Retirement Funds workbook**."

An "Excel file" is *always* a workbook file, even if it only contains one worksheet. Excel workbook files have either the file extension **.xlsx**, **.xlsm**, or **.xlam**. The file extension tells users whether the workbook contains data only (.xlsx) or data plus macro or add-in instructions (the other two), which may require an additional step to open, as Appendix D explains.

studentTIP

The Pearson Excel add-in PHStat automates the use of the Excel Guide workbooks, further simplifying the operational details of using Excel for data analysis. Appendix H explains more about PHStat and how it can be used with this book.

Appendix C contains a list of Excel data workbooks, many of which are single-worksheet workbooks.

Tableau Differences   Readers supplementing Excel with Tableau Public need to understand that while Tableau also uses workbooks to store one or more worksheets, Tableau defines these things differently. A Tableau workbook contains worksheets that present tabular and visual summaries, each of which is associated with a *data source*. A **data source** is a pointer to data that can be a complex collection of data or be as simple as an Excel worksheet (the data sources that this book uses for examples). A Tableau workbook can contain more than one data source, and each data source can be used to create multiple visual or tabular summaries.

Tableau workbooks can also include dashboards, a concept that Chapter 17 discusses.

Data sources can be viewed and column formulas can be used to define new columns, but individual values *cannot be edited* because Tableau Public workbooks do not store the data themselves. Therefore, if using a Tableau workbook, the associated data source must be available if data need to be edited or used for additional analyses.

More About the Excel Guide Workbooks

The Excel Guide workbooks contain reusable templates and model solutions for specific statistical analyses. All workbooks are organized similarly, with the data to be analyzed placed on its own page. When working with templates, readers only enter or paste in their data and never enter or edit formulas, thereby greatly reducing the chance that the worksheet will produce erroneous results. When working with a model worksheet solution, readers enter or paste in their data and edit or copy certain formulas, an additional step that the Pearson Excel add-in PHStat can automate.

Not starting from scratch minimizes the chance of errors, and using templates or, when necessary, model solutions reflects best business practice. Allowing individuals to create new solutions from scratch in business can create risk and led to internal control violations. For example, in the aftermath of the 2012 "London Whale" trading debacle, which caused an estimated loss of at least $6.2 billion, JP Morgan Chase discovered a worksheet that could greatly miscalculate the volatility of a trading portfolio because a formula divided a value by the sum of some numbers and not the average of those numbers (Ewok, Hurtado).

Reusability   Templates and model solutions are reusable because they are capable of recalculation. In worksheet **recalculation**, results displayed by formulas automatically change as the data to which the formulas refer change. Reusability also minimizes internal control risk by allowing worksheet formulas and macro to be first tested and audited for correctness before being used for information and decision-making purposes.

Excel Skills That Readers Need

To use Excel effectively with this book, readers will need to know how to make and edit cell entries, how to open to a particular worksheet in a workbook, how to print a worksheet, and how to open and save files. Readers without these skills should review the introduction to these skills that starts in the Excel Guide for this chapter and continues in Appendix B. Readers not using PHStat will also need to modify model worksheet solutions, especially for inferential topics such as ANOVA and regression that later chapters of this book discuss.

Excel Guide Instructions   Chapter Excel Guides provide separate instructions for working with the Excel Guide workbooks directly or for using PHStat for specific statistical analyses that chapter sections discuss. Instructions for working with the workbooks directly have the subhead *Workbook* and instructions for using PHStat have the subhead *PHStat*. Topics that do not include the *Workbook* subhead require solutions that would be too difficult to modify for all but advanced users of Excel to perform. Note that for advanced inferential topics, *Workbook* instructions may contain the advisory that the instructions are for intermediate or advanced Excel users. Such instructions are not designed for readers who are casual or novice Excel users.

For selected topics, Excel Guide instructions include the third subhead *Analysis ToolPak* that provides instructions for using the Excel Analysis Toolpak add-in. Because they explore chapter examples or solve end-of-section or end-of-chapter problems, readers can use a mix of *Workbook*, *PHStat*, or *Analysis ToolPak* instructions because the instructions for a topic have been designed to create identical results.

studentTIP
Readers of past editions of this book who chose not to use PHStat often found the ToolPak instructions, where they exist, more convenient to use than the *Workbook* instructions for a topic.

▼ REFERENCES

Advani, D. "Preparing Students for the Jobs of the Future." *University Business* (2011). **bit.ly/1gNLTJm**.

Davenport, T., J. Harris, and R. Morison. *Analytics at Work*. Boston: Harvard Business School Press, 2010.

Healy, P. "Ticker Pricing Puts 'Lion King' atop Broadway's Circle of Life." *New York Times, New York edition*, March 17, 2014, p. A1. **nyti.ms.1zDkzki**.

Hurtado, P. "The London Whale." *Bloomberg QuickTake*, February 23, 2016. **bloom.bg/2qnsXKM**.

Ewok, J. "The Importance of Excel." *The Baseline Scenario*, February 9, 2013, **bit.ly/1LPeQUy**.

IBM Corporation. "What Is Big Data?" **www.ibm.com/big-data/us/en/**.

Laney, D. *3D Data Management: Controlling Data Volume, Velocity, and Variety*. Stamford, CT: META Group. February 6, 2001.

Levine, D., and D. Stephan. "Teaching Introductory Business Statistics Using the DCOVA Framework." *Decision Sciences Journal of Innovative Education* 9 (Sept. 2011): 393–398.

Passy, C. "Broadway Ticket Sales Sizzle This Summer." *The Wall Street Journal*, August, 20, 2018. **on.wsj.com/2X3ZWRP**.

▼ KEY TERMS

big data   4
cells   8
data   2
business analytics   4
DCOVA framework   3
descriptive statistics   5
formula   8

function   8
inferential statistics   5
logical causality   6
operational definition   5
recalculation   9
reusability   9
statistic   5

statistics   3
summarized data   2
template   8
unstructured data   4
variable   5
workbook   8
worksheets   8

▼EXCEL GUIDE

EG.1 GETTING STARTED with EXCEL

Opening Excel displays a window that contains the Office Ribbon tabs above a worksheet area. When a workbook is opened, Excel displays the name of the workbook centered in the title bar. The top of the worksheet area contains a formula bar that enables one to edit the contents of the currently selected cell (cell A1 in the illustration). At the bottom of the worksheet grid, Excel displays a sheet tab that identifies the current worksheet name (DATA). In workbooks with more than one sheet, clicking a sheet tab makes the worksheet named by the tab the current worksheet.

The illustration below shows the Retirement Funds workbook, one of the Excel data workbooks for this book.

	A	B	C	D	E	F	G	H	I	J	K	L	M	N	O	P	Q
1	Fund Number	Market Cap	Type	Risk	Assets	Turnover Ratio	SD	Sharpe Ratio	Beta	1YrReturn	3YrReturn	5YrReturn	10YrReturn	Expense Ratio	Star Rating		Bins
2	RF001	Large	Growth	High	814.77	228.00	14.13	0.86	1.18	3.22	15.32	14.97	11.93	1.23	Five		
3	RF002	Large	Growth	High	6246.72	24.00	14.11	1.08	1.11	2.56	14.99	15.34	10.50	1.39	Five		
4	RF003	Large	Growth	Average	3874.26	27.00	12.87	1.17	1.05	9.77	14.90	15.66	7.96	0.92	Three		
5	RF004	Large	Growth	Low	3464.59	74.00	11.29	1.31	0.93	3.59	14.86	14.51	10.74	1.21	Five		
6	RF005	Large	Growth	Low	2482.03	95.00	11.88	1.21	0.99	3.75	14.60	14.99	9.51	0.87	Five		
7	RF006	Large	Growth	High	3556.66	34.00	15.24	0.93	1.13	-0.33	14.30	12.43	10.52	0.96	Three		

EG.2 ENTERING DATA

In Excel, enter data into worksheet columns, starting with the leftmost, first column, using the cells in row 1 to enter variable names. Avoid skipping rows or columns as such skipping can disrupt or alter the way certain Excel procedures work. Complete a cell entry by pressing **Tab** or **Enter**, or, if using the formula bar to make a cell entry, by clicking the **check mark icon** in the formula bar. To enter or edit data in a specific cell, either use the cursor keys to move the cell pointer to the cell or select the cell directly.

When using numbers as row 1 variable headings, precede the number with an apostrophe. Pay attention to special instructions in this book that note specific orderings of variable columns that are necessary for some Excel operations. When in doubt, use the DATA worksheets of the Excel Guide Workbooks as the guide for entering and arranging variable data.

EG.3 OPEN or SAVE a WORKBOOK

Use **File→Open** or **File→Save As**.

Open and **Save As** use similar means to open or save the workbook by name while specifying the physical device or network location and folder for that workbook. Save As dialog boxes enable one to save a file in alternate formats for programs that cannot open Excel workbooks (**.xlsx** files) directly. Alternate formats include a simple text file with values delimited with tab characters, **Text (Tab delimited) (*.txt)** that saves the contents of the current worksheet as a simple text file, and **CSV (Comma delimited) (*.csv)** that saves worksheet cell values as text values that are delimited with commas. Excels for Mac list these choices as **Tab Delimited Text (.txt)** and **Windows Comma Separated (.csv)**.

The illustration below shows the part of the Save As dialog box that contains the **Save as type** drop-down list. (Open dialog boxes have a similar drop-down list.) In all Windows Excel versions, you can also select a file format in the Open dialog box. Selecting **All Files (*.*)** from the drop-down list can list files that had been previously saved in unexpected formats.

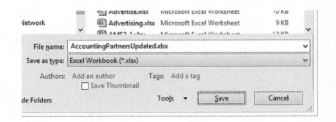

To open a new workbook, select **File→New** (**New Workbook** in Excel for Mac). Excel displays a new workbook with one or more blank worksheets.

EG.4   WORKING with a WORKBOOK

Use **Insert** (or **Insert Sheet**), **Delete**, or **Move or Copy**.

Alter the contents of a workbook by adding a worksheet or by deleting, copying, or rearranging the worksheets and chart sheets that the workbook contains. To perform one of these operations, right-click a sheet tab and select the appropriate choice from the shortcut menu that appears.

To add a worksheet, select **Insert**. In Microsoft Windows Excel, you also click **Worksheet** and then click **OK** in the Insert dialog box. To delete a worksheet or chart sheet, right-click the sheet tab of the worksheet to be deleted and select **Delete**.

To copy or rearrange the position of a worksheet or chart sheet, right-click the sheet tab of the sheet and select **Move or Copy**. In the Move or Copy dialog box, first select the workbook and the position in the workbook for the sheet. If copying a sheet, also check **Create a copy**. Then click **OK**.

EG.5   PRINT a WORKSHEET

Use **File→Print**.

Excel prints worksheets and chart sheets, not workbooks. When you select **Print**, Excel displays a preview of the currently opened sheet in a dialog box or pane that enables worksheet selections to be made. Adjust the print formatting of the worksheet(s) to be printed by clicking **Page Setup**. Typically, in the Page Setup dialog box, one might click the **Sheet** tab and then check or clear the **Gridlines** and **Row and column headings** checkboxes to include or delete these headings from a printout.

EG.6   REVIEWING WORKSHEETS

When reviewing worksheets remember that what is displayed in cells may be the result of either the recalculation of formulas or cell formatting. A cell that displays 4 might contain the value 4, might contain a formula calculation that results in the value 4, or might contain a value such as 3.987 that has been formatted to display as the nearest whole number.

To display and review all formulas, press **Ctrl+`** (grave accent). Excel displays the *formula view* of the worksheet, revealing all formulas. (Pressing **Ctrl+`** a second time restores the worksheet to its normal display.) Note that the Excel Guide Workbooks contain one or more FORMULAS worksheets that provide a second way of viewing all formulas.

The template and model solutions for this book format specific cells by changing text attributes or the background color of cells, with cells containing numeric results typically formatted to display four decimal places. Appendix B discusses these as well as other common worksheet and chart sheet formatting operations.

EG.7   IF YOU USE the *WORKBOOK* INSTRUCTIONS

Excel Guide *Workbook* instructions use the word *display* as in the "Format Axis display" to refer to a user interaction that may be presented by Excel in a **task pane** or a **two-panel dialog box** ("Format Axis task pane" or the "Format Axis dialog box"). Task panes open to the side of the worksheet and can remain onscreen indefinitely. Initially, some parts of a pane may be hidden and an icon or label must be clicked to reveal that hidden part to complete a *Workbook* instruction. Two-panel dialog boxes that open over the worksheet must be *closed* to continue to use Excel. The left panel of such dialog boxes are always visible and clicking entries in the left panel makes visible one of the right panels. (Click the system close button at the top right of a task pane or dialog box to close the display and remove it from the screen.)

Current Excel versions can vary in their menu sequences. Excel Guide instructions show these variations as parenthetical phrases. For example, the menu sequence, "select **Design** (or **Chart Design**)→**Add Chart Element**" tells you to first select **Design** *or* **Chart Design** to begin the sequence and then to continue by selecting **Add Chart Element**. (Microsoft Windows Excel uses **Design** and Excel for Mac uses **Chart Design**.)

For the current Excel versions that this book supports (see the FAQs in Appendix G), the *Workbook* Instructions are generally identical. Occasionally, individual instructions may differ significantly for one (or more) versions. In such cases, the instructions that apply for multiple versions appear first, in normal text, and the instructions for the unique version immediately follows in this text color.

▾TABLEAU GUIDE

TG.1 GETTING STARTED with TABLEAU

The Tableau Guides for this book feature Tableau Public, version 2019, also known as the Tableau Desktop Public Edition. Tableau Public uses a drag-and-drop interface that will be most familiar to users of the Excel PivotTable feature, which Chapter 2 discusses.

Tableau Public uses workbooks to store tabular and visual worksheet summaries, dashboards, and stories. Opening Tableau Public displays the main Tableau window in which the contents of a workbook can be viewed and edited. The main window shown below displays the Mobile Electronics Store worksheet of the Arlingtons National Sales Dashboard Tableau workbook that Chapter 17 uses as an example.

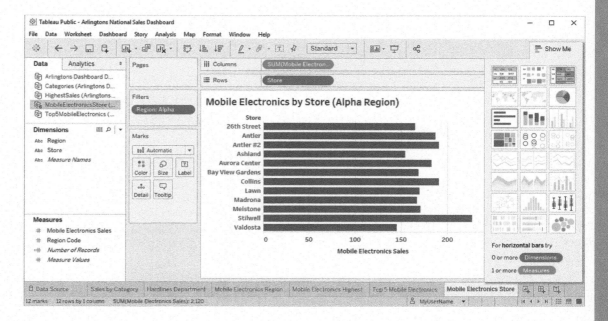

The main Tableau window contains a menu bar and toolbar, a tabbed left area that presents data and formatting details, several special areas that Section TG.4 explains, the worksheet display area, and the Show Me gallery that displays tabular and visual summaries appropriate for the data in the worksheet area. Worksheet tabs appear under these areas, along with a tab for the current data source to the left of the tabs and icon shortcuts for a new worksheet, new dashboard, and new story, respectively, to the right of the tabs. The bottom of the window displays status information at the left, the current signed-in user ("MyUserName"), a drop-down button to sign out of Tableau Public, and four media controls that permit browsing through the worksheet tabs of the current workbook.

The Data tab shown above displays the workbook data sources (five) and lists the "Dimensions" and "Measures," the variable columns of the data source for the worksheet, summary measures, and calculated values for the MobileElectronicsStore data source associated with the displayed worksheet. (Section TG1.1 explains more about the significance of dimensions and measures.)

For the worksheet shown above, the variable Store was dragged-and-dropped in the Rows *shelf* and the variable Mobile Electronics Sales was dragged-and-dropped in the Column *shelf*. The worksheet also contains a *filter* for the Region column that selects only those rows in the source worksheet in which the value in the Region column is Alpha. (To create the bar chart, **horizontal bars icon** was selected from the Show Me gallery.)

TG.2   ENTERING DATA

As Section FTF.5 explains, one defines data sources, which are pointers to Excel worksheets and other data files. Tableau does not support the direct entry of data values.

TG.3   OPEN or SAVE a WORKBOOK

Use **File → Open** or **File → Open from Tableau Public**.
Use **File → Save to Tableau Public As**.

Use Open to import simple data sources such as Microsoft Excel workbooks or text files or to open a *Tableau* workbook that has been previously downloaded and saved. Most Tableau Guide instructions begin by opening an Excel workbook to avoid the limitations of Tableau Public that Appendix Section G.5 discusses.

To open a new workbook, select **File → New**. For a new workbook, the Data tab will display the hyperlink **Connect to Data**. Clicking the hyperlink displays the Connect panel from which data sources can be retrieved. When opening an Excel workbook that contains more than one data worksheet, each Excel worksheet being used must be defined as a separate data source. This Connect panel also appears when opening an Excel workbook directly using the file open command. The illustration below shows the data source linked to the Data worksheet of the Excel Retirement Funds workbook.

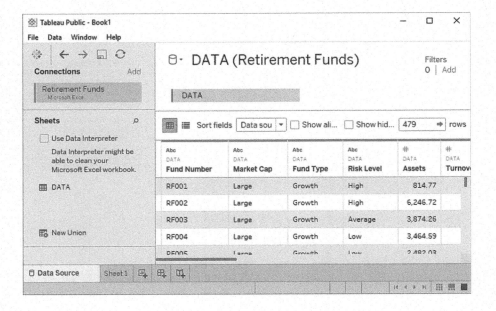

Using the Open from Tableau Public or Save to Tableau Public As command requires a valid Tableau online account. (Accounts are complimentary but require registration.) Using these commands means signing into a Tableau Public account and retrieving (or storing) a Tableau workbook from that account or from an account that has been shared. Save to Tableau Public As stores the Tableau workbook in the account and opens a web browser to display the workbook and to permit its downloading to a local computing device. In paid-subscription editions of Tableau Desktop, these open and save commands appear in the Server menu and not in the File menu.

Although not used in the Tableau Guides, Tableau Public permits *join* and *union* operations to combine columns from two data sources. **Join** operations combine two tables, typically by matching values in a variable column that both original tables share. **Union** operations add rows. Union operations require that tables share columns that hold values for the same variables. Joins and unions can solve problems that arise from seeking to perform an analysis of data on a set of variables stored in two different places, such as two different Excel data worksheets in the same Excel workbook.

TG.4   WORKING with DATA

Tableau Desktop uses the term **data field** to refer to what this book calls a variable or, in some contexts, a column. What Excel calls a formula function, Tableau calls an aggregation. What Excel calls a formula, Tableau calls an aggregate calculation. Tableau also invents its own vocabulary for several user interface elements in the worksheet window that readers may know under more common names. Knowing this vocabulary can be helpful when consulting the Tableau help system or other references.

Tableau calls the Pages, Filters, Columns, and Rows areas, all shown in the Section TG.1 illustration, *shelves*. The Marks area seems to be a shelf, but for reasons that would be only self-evident to a regular user of Tableau, the Marks area is called a *card*. **Shelves** are places into which things can be placed or dropped, such as the *pills* that have been placed in the Filters, Row, and Columns shelves. A **pill** represents some data and is so named because it reminds some of a medicinal capsule. In its simplest form, a pill represents a data field in the data source. However, pills can represent a filtering operation, such as the Filters shelf pill in the illustration below, or a calculated result, similar to a worksheet or data table formula. Pills can be either blue or green, reflecting the type of numerical data, discrete or continuous (see Section TG1.1), or red, reflecting an error condition.

In the Section TG.1 illustration, the Store Dimension has been dragged to the Rows shelf, creating the Store pill and the Mobile Electronics Sales Measure has been dropped on the Columns shelf, creating the SUM(Mobile Electronics Sales) pill (slightly truncated, see figure detail below). Dropping a measure name on a shelf creates an aggregation (formula function). The SUM aggregation pill sums mobile electronic sales values by rows (each store). In the special case where each store is represented by only one row, there is no actual summing of values.

TG.5   PRINT a WORKBOOK

Tableau Public does not contain a print function that would print worksheets. (Commercial versions of Tableau do.) To print a tabular or visual summary, use a screen capture utility to capture the display for later printing. For online worksheets displayed in a web browser, use the print function of the browser.

1

Defining and Collecting Data

OBJECTIVES

- Understand issues that arise when defining variables
- How to define variables
- Understand the different measurement scales
- How to collect data
- Identify the different ways to collect a sample
- Understand the issues involved in data preparation
- Understand the types of survey errors

▼USING STATISTICS
Defining Moments

#1 You're the sales manager in charge of the best-selling beverage in its category. For years, your chief competitor has made sales gains, claiming a better tasting product. Worse, a new sibling product from your company, known for its good taste, has quickly gained significant market share at the expense of your product. Worried that your product may soon lose its number one status, you seek to improve sales by improving the product's taste. You experiment and develop a new beverage formulation. Using methods taught in this book, you conduct surveys and discover that people overwhelmingly like the newer formulation. You decide to use that new formulation going forward, having statistically shown that people prefer it. *What could go wrong?*

#2 You're a senior airline manager who has noticed that your frequent fliers always choose another airline when flying from the United States to Europe. You suspect fliers make that choice because of the other airline's perceived higher quality. You survey those fliers, using techniques taught in this book, and confirm your suspicions. You then design a new survey to collect detailed information about the quality of all components of a flight, from the seats to the meals served to the flight attendants' service. Based on the results of that survey, you approve a costly plan that will enable your airline to match the perceived quality of your competitor. *What could go wrong?*

In both cases, much did go wrong. Both cases serve as cautionary tales that if you choose the wrong variables to study, you may not end up with results that support making better decisions. Defining and collecting data, which at first glance can seem to be the simplest tasks in the DCOVA framework, can often be more challenging than people anticipate.

As the initial chapter notes, statistics is a way of thinking that can help fact-based decision making. But statistics, even properly applied using the DCOVA framework, can never be a substitute for sound management judgment. If you misidentify the business problem or lack proper insight into a problem, statistics cannot help you make a good decision. Case #1 retells the story of one of the most famous marketing blunders ever, the change in the formulation of Coca-Cola in the 1980s. In that case, Coke brand managers were so focused on the taste of Pepsi and the newly successful sibling Diet Coke that they decided only to define a variable and collect data about which drink tasters preferred in a blind taste test. When New Coke was preferred, even over Pepsi, managers rushed the new formulation into production. In doing so, those managers failed to reflect on whether the statistical results about a test that asked people to compare one-ounce samples of several beverages would demonstrate anything about beverage sales. After all, people were asked which beverage tasted better, not whether they would buy that better-tasting beverage in the future. New Coke was an immediate failure, and Coke managers reversed their decision a mere 77 days after introducing their new formulation (Polaris).

Coke managers also overlooked other issues, such as people's emotional connection and brand loyalty to Coca-Cola, issues better discussed in a marketing book than this book.

Case #2 represents a composite story of managerial actions at several airlines. In some cases, managers overlooked the need to state operational definitions for quality factors about which fliers were surveyed. In at least one case, statistics was applied correctly, and an airline spent great sums on upgrades and was able to significantly improve quality. Unfortunately, their frequent fliers still chose the competitor's flights. In this case, no statistical survey about quality could reveal the managerial oversight that given the same level of quality between two airlines, frequent fliers will almost always choose the cheaper airline. While quality was a significant variable of interest, it was not the most significant.

The lessons of these cases apply throughout this book. Due to the necessities of instruction, the examples and problems in all but the last chapter include preidentified business problems and defined variables. Identifying the business problem or objective to be considered is always a prelude to applying the DCOVA framework.

1.1 Defining Variables

Identifying a proper business problem or objective enables one to begin to identify and define the variables for analysis. For each variable identified, assign an **operational definition** that specifies the type of variable and the *scale*, the type of measurement, that the variable uses.

EXAMPLE 1.1

Defining Data at GT&M

You have been hired by Good Tunes & More (GT&M), a local electronics retailer, to assist in establishing a fair and reasonable price for Whitney Wireless, a privately held chain that GT&M seeks to acquire. You need data that would help to analyze and verify the contents of the wireless company's basic financial statements. A GT&M manager suggests that one variable you should use is monthly sales. What do you do?

SOLUTION Having first confirmed with the GT&M financial team that monthly sales is a relevant variable of interest, you develop an operational definition for this variable. Does this variable refer to sales per month for the entire chain or for individual stores? Does the variable refer to net or gross sales? Do the monthly sales data represent number of units sold or currency amounts? If the data are currency amounts, are they expressed in U.S. dollars? After getting answers to these and similar questions, you draft an operational definition for ratification by others working on this project.

Classifying Variables by Type

The type of data that a variable contains determines the statistical methods that are appropriate for a variable. Broadly, all variables are either **numerical**, variables whose data represent a counted or measured quantity, or **categorical**, variables whose data represent categories. Gender

student TIP

Some prefer the terms **quantitative** and **qualitative** over the terms numerical and categorical when describing variables. These two pairs of terms are interchangeable.

with its categories male and female is a categorical variable, as is the variable preferred-New-Coke with its categories yes and no. In Example 1.1, the monthly sales variable is numerical because the data for this variable represent a quantity.

For some statistical methods, numerical variables must be further specified as either being *discrete* or *continuous*. **Discrete** numerical variables have data that arise from a counting process. Discrete numerical variables include variables that represent a "number of something," such as the monthly number of smartphones sold in an electronics store. **Continuous** numerical variables have data that arise from a measuring process. The variable "the time spent waiting in a checkout line" is a continuous numerical variable because its data represent timing measurements. The data for a continuous variable can take on any value within a continuum or an interval, subject to the precision of the measuring instrument. For example, a waiting time could be 1 minute, 1.1 minutes, 1.11 minutes, or 1.113 minutes, depending on the precision of the electronic timing device used.

For a particular variable, one might use a numerical definition for one problem, but use a categorical definition for another problem. For example, a person's age might seem to always be a numerical age variable, but what if one was interested in comparing the buying habits of children, young adults, middle-aged persons, and retirement-age people? In that case, defining age as a categorical variable would make better sense.

Measurement Scales

Determining the **measurement scale** that the data for a variable represent is part of defining a variable. The measurement scale defines the ordering of values and determines if differences among pairs of values for a variable are equivalent and whether one value can be expressed in terms of another. Table 1.1 presents examples of measurement scales, some of which are used in the rest of this section.

TABLE 1.1
Examples of different scales and types

Data	Scale, Type	Values
Cellular provider	nominal, categorical	AT&T, T-Mobile, Verizon, Other, None
Excel skills	ordinal, categorical	novice, intermediate, expert
Temperature (°F)	interval, numerical	−459.67°F or higher
SAT Math score	interval, numerical	a value between 200 and 800, inclusive
Item cost (in $)	ratio, numerical	$0.00 or higher

Define numerical variables as using either an **interval scale**, which expresses a difference between measurements that do not include a true zero point, or a **ratio scale**, an ordered scale that includes a true zero point. Categorical variables use measurement scales that provide less insight into the values for the variable. For data measured on a **nominal scale**, category values express no order or ranking. For data measured on an **ordinal scale**, an ordering or ranking of category values is implied. Ordinal scales contain some information to compare values but not as much as interval or ratio scales. For example, the ordinal scale poor, fair, good, and excellent allows one to know that "good" is better than poor or fair and not better than excellent. But unlike interval and ratio scales, one would not know that the difference from poor to fair is the same as fair to good (or good to excellent).

PROBLEMS FOR SECTION 1.1

LEARNING THE BASICS

1.1 Four different beverages are sold at a fast-food restaurant: soft drinks, tea, coffee, and bottled water.
a. Explain why the type of beverage sold is an example of a categorical variable.
b. Explain why the type of beverage is a nominal-scaled variable.

1.2 U.S. businesses are listed by size: small, medium, and large. Explain why business size is an example of categorical variable.

1.3 The time it takes to download a video from the Internet is measured.
a. Explain why the download time is a continuous numerical variable.
b. Explain why the download time is a ratio-scaled variable.

APPLYING THE CONCEPTS

 1.4 For each of the following variables, determine whether the variable is categorical or numerical. If the variable is numerical, determine whether the variable is discrete or continuous.
a. Number of cell phones in the household
b. Monthly data usage (in MB)
c. Number of text messages exchanged per month
d. Voice usage per month (in minutes)
e. Whether the cell phone is used for streaming video

1.5 The following information is collected from students upon exiting the campus bookstore during the first week of classes.
a. Amount of time spent shopping in the bookstore
b. Number of textbooks purchased
c. Academic major
d. Gender

Classify each variable as categorical or numerical and determine its measurement scale.

1.6 For each of the following variables, determine whether the variable is categorical or numerical and determine its measurement scale. If the variable is numerical, determine whether the variable is discrete or continuous.
a. Name of Internet service provider
b. Time, in hours, spent surfing the Internet per week
c. Whether the individual uses a mobile phone to stream video
d. Number of online purchases made in a month
e. Where the individual accesses social networks to find sought-after information

1.7 For each of the following variables, determine whether the variable is categorical or numerical and determine its measurement scale. If the variable is numerical, determine whether the variable is discrete or continuous.
a. Amount of money spent on electronics equipment in the past month
b. Favorite Internet source for puchases

c. Most likely time period during which Internet shopping takes place (weekday, weeknight, or weekend)
d. Number of computers owned

1.8 Suppose the following information is collected from Robert Keeler on his application for a home mortgage loan at the Metro County Savings and Loan Association.
a. Monthly payments: $2,227
b. Number of jobs in past 10 years: 1
c. Annual family income: $96,000
d. Marital status: Married

Classify each of the responses by type of data and measurement scale.

1.9 One of the variables most often included in surveys is income. Sometimes the question is phrased, "What is your income (in thousands of dollars)?" In other surveys, the respondent is asked to "Select the circle corresponding to your income level" and is given a number of income ranges to choose from.
a. In the first format, explain why income might be considered either discrete or continuous.
b. Which of these two formats would you prefer to use if you were conducting a survey? Why?

1.10 If two students both score 90 on the same examination, what arguments could be used to show that the underlying variable—test score—is continuous?

1.11 The director of market research at a large department store chain wanted to conduct a survey throughout a metropolitan area to determine the amount of time working women spend shopping on the Internet in a typical month.
a. Indicate the type of data the director might want to collect.
b. Develop a first draft of the questionnaire needed by writing three categorical questions and three numerical questions that you feel would be appropriate for this survey.

1.2 Collecting Data

Collecting data using improper methods can spoil any statistical analysis. For example, Coca-Cola managers in the 1980s (see page 17) faced advertisements from their competitor publicizing the results of a "Pepsi Challenge" in which taste testers consistently favored Pepsi over Coke. No wonder—test recruiters deliberately selected tasters they thought would likely be more favorable to Pepsi and served samples of Pepsi chilled, while serving samples of Coke lukewarm (not a very fair comparison!). These introduced biases made the challenge anything but a proper scientific or statistical test. Proper data collection avoids introducing biases and minimizes errors.

Populations and Samples

Data are collected from either a population or a sample. A **population** contains all the items or individuals of interest that one seeks to study. All of the GT&M sales transactions for a specific year, all of the full-time students enrolled in a college, and all of the registered voters in Ohio are examples of populations. A **sample** contains only a portion of a population of interest. One analyzes a sample to estimate characteristics of an entire population. For example, one might select a sample of 200 sales transactions for a retailer or select a sample of 500 registered voters in Ohio in lieu of analyzing the populations of all the sales transactions or all the registered voters.

One uses a sample when selecting a sample will be less time consuming or less cumbersome than selecting every item in the population or when analyzing a sample is less cumbersome or

learnMORE

Read the SHORT TAKES for Chapter 1 for a further discussion about data sources.

more practical than analyzing the entire population. Section FTF.3 defines *statistic* as a "value that summarizes the data of a specific variable." More precisely, a **statistic** summarizes the value of a specific variable for sample data. Correspondingly, a **parameter** summarizes the value of a population for a specific variable.

Data Sources

Data sources arise from the following activities:

- Capturing data generated by ongoing business activities
- Distributing data compiled by an organization or individual
- Compiling the responses from a survey
- Conducting an observational study and recording the results of the study
- Conducting a designed experiment and recording the outcomes of the experiment

When the person conducting an analysis performs one of these activities, the data source is a **primary data source**. When one of these activities is done by someone other than the person conducting an analysis, the data source is a **secondary data source**.

Capturing data can be done as a byproduct of, or as a result of, an organization's transactional information processing, such as the storing of sales transactions at a retailer, or as result of a service provided by a second party, such as customer information that a social media website business collects on behalf of another business. Therefore, such data capture may be either a primary or a secondary source.

Typically, organizations such as market research firms and trade associations distribute complied data, as do businesses that offer syndicated services, such as The Nielsen Company, known for its TV ratings. Therefore, this source of data is usually a secondary source. (If one supervised the distribution of a survey, compiled its results, and then analyzed those results, the survey would be a primary data source.)

Choosing to conduct an observational study or a designed experiment on a variable of interest affects the statistical methods and the decision-making processes that can be used, as Chapters 10–12 and 17 further explain.

In both observational studies and designed experiments, researchers that collect data are looking for the effect of some change, called a **treatment**, on a variable of interest. In an observational study, the researcher collects data in a natural or neutral setting and has no direct control of the treatment. For example, in an observational study of the possible effects on theme park usage patterns that a new electronic payment method might cause, one would take a sample of guests, identify those who use the new method and those who do not, and then "observe" if those who use the new method have different park usage patterns. As a designed experiment, one would select guests to use the new electronic payment method and then discover if those guests have theme park usage patterns that are different from the guests not selected to use the new payment method.

PROBLEMS FOR SECTION 1.2

APPLYING THE CONCEPTS

1.12 The American Community Survey (**www.census.gov/acs**) provides data every year about communities in the United States. Addresses are randomly selected, and respondents are required to supply answers to a series of questions.
a. Which of the sources of data best describe the American Community Survey?
b. Is the American Community Survey based on a sample or a population?

1.13 Visit the website of the Gallup organization at **www.gallup.com**. Read today's top story. What type of data source is the top story based on?

1.14 Visit the website of the Pew Research organization at **www.pewresearch.org**. Read today's top story. What type of data source is the top story based on?

1.15 Transportation engineers and planners want to address the dynamic properties of travel behavior by describing in detail the driving characteristics of drivers over the course of a month. What type of data collection source do you think the transportation engineers and planners should use?

1.16 Visit the Statistics Portal "Statista" at **www.statista.com**. Click **Infographics** on the main menu. On the Infographics web page, examine one of the "Popular Infographic Topics." What type of data source is the information presented here based on?

1.3 Types of Sampling Methods

When selecting a sample to collect data, begin by defining the **frame**. The frame is a complete or partial listing of the items that make up the population from which the sample will be selected. Inaccurate or biased results can occur if a frame excludes certain groups, or portions of the population. Using different frames to collect data can lead to different, even opposite, conclusions.

Using the frame, select either a nonprobability sample or a probability sample. In a **nonprobability sample**, select the items or individuals without knowing their probabilities of selection. In a **probability sample**, select items based on known probabilities. Whenever possible, use a probability sample because such a sample will allow one to make inferences about the population being analyzed.

Nonprobability samples can have certain advantages, such as convenience, speed, and low cost. Such samples are typically used to obtain informal approximations or as small-scale initial or pilot analyses. However, because the theory of statistical inference depends on probability sampling, nonprobability samples *cannot be used* for statistical inference and this more than offsets those advantages in more formal analyses.

Figure 1.1 shows the subcategories of the two types of sampling. A nonprobability sample can be either a convenience sample or a judgment sample. To collect a **convenience sample**, select items that are easy, inexpensive, or convenient to sample. For example, in a warehouse of stacked items, selecting only the items located on the top of each stack and within easy reach would create a convenience sample. So, too, would be the responses to surveys that the websites of many companies offer visitors. While such surveys can provide large amounts of data quickly and inexpensively, the convenience samples selected from these responses will consist of self-selected website visitors. (Read the *Consider This* essay on page 29 for a related story.)

To collect a **judgment sample**, collect the opinions of preselected experts in the subject matter. Although the experts may be well informed, one cannot generalize their results to the population.

The types of probability samples most commonly used include simple random, systematic, stratified, and cluster samples. These four types of probability samples vary in terms of cost, accuracy, and complexity, and they are the subject of the rest of this section.

FIGURE 1.1
Types of samples

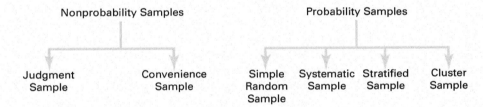

Simple Random Sample

In a **simple random sample**, every item from a frame has the same chance of selection as every other item, and every sample of a fixed size has the same chance of selection as every other sample of that size. Simple random sampling is the most elementary random sampling technique. It forms the basis for the other random sampling techniques. However, simple random sampling has its disadvantages. Its results are often subject to more variation than other sampling methods. In addition, when the frame used is very large, carrying out a simple random sample may be time consuming and expensive.

With simple random sampling, use n to represent the sample size and N to represent the frame size. Number every item in the frame from 1 to N. The chance that any particular member of the frame will be selected during the first selection is $1/N$.

Select samples with replacement or without replacement. **Sampling with replacement** means that selected items are returned to the frame, where it has the same probability of being selected again. For example, imagine a fishbowl containing N business cards, one card for each person. The first selection selects the card for Grace Kim. After her information has been recorded, her business card is placed back in the fishbowl. All cards are thoroughly mixed and a second selection is made. For this second selection, the probability that the card for Grace Kim will be selected remains $1/N$.

Most sampling is *sampling without replacement*. **Sampling without replacement** means that once an item has been selected, the item cannot ever again be selected for the sample.

The chance that any particular item in the frame will be selected—for example, the business card for Grace Kim—on the first selection is $1/N$. The chance that any card not previously chosen will be chosen on the second selection becomes 1 out of $N - 1$.

When creating a simple random sample, avoid the "fishbowl" method of selecting a sample because this method lacks the ability to thoroughly mix items and, therefore, randomly select a sample. Instead, use a more rigorous selection method.

learnMORE

Learn to use a table of random numbers to select a simple random sample in the **Section 1.3 LearnMore** online topic.

One such method is to use a **table of random numbers**, such as Table E.1 in Appendix E, for selecting the sample. A table of random numbers consists of a series of digits listed in a randomly generated sequence. To use a random number table for selecting a sample, assign code numbers to the individual items of the frame. Then generate the random sample by reading the table of random numbers and selecting those individuals from the frame whose assigned code numbers match the digits found in the table. Because every digit or sequence of digits in the table is random, the table can be read either horizontally or vertically. The margins of the table designate row numbers and column numbers, and the digits are grouped into sequences of five in order to make reading the table easier.

Because the number system uses 10 digits (0, 1, 2, ... , 9), the chance that any particular digit will be randomly generated is equal 1 out of 10 and is equal to the probability of generating any other digit. For a generated sequence of 800 digits, one would expect about 80 to be the digit 0, 80 to be the digit 1, and so on.

Systematic Sample

In a **systematic sample**, partition the N items in the frame into n groups of k items, where

$$k = \frac{N}{n}$$

Round k to the nearest integer. To select a systematic sample, choose the first item to be selected at random from the first k items in the frame. Then, select the remaining $n - 1$ items by taking every kth item thereafter from the entire frame.

If the frame consists of a list of prenumbered checks, sales receipts, or invoices, taking a systematic sample is faster and easier than taking a simple random sample. A systematic sample is also a convenient mechanism for collecting data from membership directories, electoral registers, class rosters, and consecutive items coming off an assembly line.

To take a systematic sample of $n = 40$ from the population of $N = 800$ full-time employees, partition the frame of 800 into 40 groups, each of which contains 20 employees. Then select a random number from the first 20 individuals and include every twentieth individual after the first selection in the sample. For example, if the first random number selected is 008, subsequent selections will be 028, 048, 068, 088, 108, ... , 768, and 788.

Simple random sampling and systematic sampling are simpler than other, more sophisticated, probability sampling methods, but they generally require a larger sample size. In addition, systematic sampling is prone to selection bias that can occur when there is a pattern in the frame. To overcome the inefficiency of simple random sampling and the potential selection bias involved with systematic sampling, one can use either stratified sampling methods or cluster sampling methods.

Stratified Sample

learnMORE

Learn how to select a stratified sample in the **Section 1.3 LearnMore** online topic.

In a **stratified sample**, first subdivide the N items in the frame into separate subpopulations, or **strata**. A stratum is defined by some common characteristic, such as gender or year in school. Then select a simple random sample within each of the strata and combine the results from the separate simple random samples. Stratified sampling is more efficient than either simple random sampling or systematic sampling because the representation of items across the entire population is assured. The homogeneity of items within each stratum provides greater precision in the estimates of underlying population parameters. In addition, stratified sampling enables one to reach conclusions about each strata in the frame. However, using a stratified sample requires that one can determine the variable(s) on which to base the stratification and can also be expensive to implement.

Cluster Sample

In a **cluster sample**, divide the N items in the frame into clusters that contain several items. **Clusters** are often naturally occurring groups, such as counties, election districts, city blocks, households, or sales territories. Then take a random sample of one or more clusters and study all items in each selected cluster.

Cluster sampling is often more cost-effective than simple random sampling, particularly if the population is spread over a wide geographic region. However, cluster sampling often requires a larger sample size to produce results as precise as those from simple random sampling or stratified sampling. The Cochran, Groves et al., and Lohr sources discuss systematic sampling, stratified sampling, and cluster sampling procedures.

PROBLEMS FOR SECTION 1.3

LEARNING THE BASICS

1.17 For a population containing $N = 902$ individuals, what code number would you assign for
a. the first person on the list?
b. the fortieth person on the list?
c. the last person on the list?

1.18 For a population of $N = 902$, verify that by starting in row 5, column 1 of the table of random numbers (Table E.1), you need only six rows to select a sample of $N = 60$ *without* replacement.

1.19 Given a population of $N = 93$, starting in row 29, column 1 of the table of random numbers (Table E.1), and reading across the row, select a sample of $N = 15$
a. *without* replacement.
b. *with* replacement.

APPLYING THE CONCEPTS

1.20 For a study that consists of personal interviews with participants (rather than mail or phone surveys), explain why simple random sampling might be less practical than some other sampling methods.

1.21 You want to select a random sample of $n = 1$ from a population of three items (which are called A, B, and C). The rule for selecting the sample is as follows: Flip a coin; if it is heads, pick item A; if it is tails, flip the coin again; this time, if it is heads, choose B; if it is tails, choose C. Explain why this is a probability sample but not a simple random sample.

1.22 A population has four members (called A, B, C, and D). You would like to select a random sample of $n = 2$, which you decide to do in the following way: Flip a coin; if it is heads, the sample will be items A and B; if it is tails, the sample will be items C and D. Although this is a random sample, it is not a simple random sample. Explain why. (Compare the procedure described in Problem 1.21 with the procedure described in this problem.)

1.23 The registrar of a university with a population of $N = 4,000$ full-time students is asked by the president to conduct a survey to measure satisfaction with the quality of life on campus. The following table contains a breakdown of the 4,000 registered full-time students, by gender and class designation:

GENDER	CLASS DESIGNATION				
	Fr.	So.	Jr.	Sr.	Total
Female	700	520	500	480	2,200
Male	560	460	400	380	1,800
Total	1,260	980	900	860	4,000

The registrar intends to take a probability sample of $n = 200$ students and project the results from the sample to the entire population of full-time students.
a. If the frame available from the registrar's files is an alphabetical listing of the names of all $N = 4,000$ registered full-time students, what type of sample could you take? Discuss.
b. What is the advantage of selecting a simple random sample in (a)?
c. What is the advantage of selecting a systematic sample in (a)?
d. If the frame available from the registrar's files is a list of the names of all $N = 4,000$ registered full-time students compiled from eight separate alphabetical lists, based on the gender and class designation breakdowns shown in the class designation table, what type of sample should you take? Discuss.
e. Suppose that each of the $N = 4,000$ registered full-time students lived in one of the 10 campus dormitories. Each dormitory accommodates 400 students. It is college policy to fully integrate students by gender and class designation in each dormitory. If the registrar is able to compile a listing of all students by dormitory, explain how you could take a cluster sample.

✓**SELF TEST** **1.24** Prenumbered sales invoices are kept in a sales journal. The invoices are numbered from 0001 to 5000.
a. Beginning in row 16, column 1, and proceeding horizontally in a table of random numbers (Table E.1), select a simple random sample of 50 invoice numbers.
b. Select a systematic sample of 50 invoice numbers. Use the random numbers in row 20, columns 5–7, as the starting point for your selection.
c. Are the invoices selected in (a) the same as those selected in (b)? Why or why not?

1.25 Suppose that 1,000 customers in an online electronics seller's customer database are categorized by three customer types: 200 potential customers who have visited the site but not purchased, 300 who have purchaesed once, and 200 repeat (loyal) buyers. A sample of 100 customers is needed.
a. What type of sampling should you do? Why?
b. Explain how you would carry out the sampling according to the method stated in (a).
c. Why is the sampling in (a) not simple random sampling?

1.4 Data Cleaning

Even if proper data collection procedures are followed, the collected data may contain incorrect or inconsistent data that could affect statistical results. **Data cleaning** corrects such defects and ensures the data contain suitable *quality* for analysis. Cleaning is the most important data preprocessing task and *must* be done before performing any analysis. Cleaning can take a significant amount of time to do. One survey of big data analysts reported that they spend 60% of their time cleaning data, while only 20% of their time collecting data and a similar percentage for analyzing data (Press).

Data cleaning seeks to correct the following types of irregularities:

- Invalid variable values, including non-numerical data for a numerical variable, invalid categorical values of a categorical variable, and numeric values outside a defined range
- Coding errors, including inconsistent categorical values, inconsistent case for categorical values, and extraneous characters
- Data integration errors, including redundant columns, duplicated rows, differing column lengths, and different units of measure or scale for numerical variables

With the exception of several examples designed for use with this section, data for the problems and examples in this book have already been properly cleaned to allow focus on the statistical concepts and methods that the book discusses.

By its nature, data cleaning cannot be a fully automated process, even in large business systems that contain data cleaning software components. As this chapter's software guides explain, Excel and Tableau contain functionality that lessens the burden of data cleaning (see the Excel and Tableau Guides for this chapter). When performing data cleaning, first preserve a copy of the original data for later reference.

Invalid Variable Values

Invalid variable values can be identified as being incorrect by simple scanning techniques so long as operational definitions for the variables the data represent exist. For any numerical variable, any value that is not a number is clearly an incorrect value. For a categorical variable, a value that does not match any of the predefined categories of the variable is, likewise, clearly an incorrect value. And for numerical variables defined with an explicit range of values, a value outside that range is clearly an error.

Coding Errors

Coding errors can result from poor recording or entry of data values or as the result of computerized operations such as copy-and-paste or data import. While coding errors are literally invalid values, coding errors may be correctable without consulting additional information whereas the invalid variable values *never* are. For example, for a Gender variable with the defined values F and M, the value "Female" is a *coding error* that can be reasonably changed to F. However, the value "New York" for the same variable is an *invalid variable value* that you cannot reasonably change to either F or M.

Unlike invalid variable values, coding errors may be *tolerated* by analysis software. For example, for the same Gender variable, the values M and m might be treated as the "same" value for purposes of an analysis by software that was tolerant of case inconsistencies, an attribute known as being *insensitive* to case.

Perhaps the most frustrating coding errors are extraneous characters in a value. Visual examination may not be able to spot extraneous characters such as nonprinting characters or extra, trailing space characters as one scans data. For example, the value David and the value that is David followed by three space characters may look the same to one casually scanning them but may not be treated the same by software. Likewise, values with nonprinting characters may look correct but may cause software errors or be reported as invalid by analysis software.

Data Integration Errors

Data integration errors arise when data from two different computerized sources, such as two different data repositories are combined into one data set for analysis. Identifying data integration errors may be the most time-consuming data cleaning task. Because spotting these errors requires a type of data interpretation that automated processes of a typical business computer

Perhaps not surprising, supplying business systems with automated data interpretation skills is a goal of many companies that provide data analysis software and services.

systems today cannot supply, spotting these errors using manual methods will be typical for the foreseeable future.

Some data integration errors occur because variable names or definitions for the same item of interest have minor differences across systems. In one system, a customer ID number may be known as Customer ID, whereas in a different system, the same variable is known as Cust Number. A result of combining data from the two systems may result in having both Customer ID and Cust Number variable columns, a redundancy that should be eliminated.

Duplicated rows also occur because of similar inconsistencies across systems. Consider a Customer Name variable with the value that represents the first coauthor of this book, David M. Levine. In one system, this name may have been recorded as David Levine, whereas in another system, the name was recorded as D M Levine. Combining records from both systems may result in two records, where only one should exist. Whether "David Levine" is actually the same person as "D M Levine" requires an interpretation skill that today's software may lack.

Likewise, different units of measurement (or scale) may not be obvious without additional, human interpretation. Consider the variable Air Temperature, recorded in degrees Celsius in one system and degrees Fahrenheit in another. The value 30 would be a plausible value under either measurement system and without further knowledge or context impossible to spot as a Celsius measurement in a column of otherwise Fahrenheit measurements.

Missing Values

Missing values are values that were not collected for a variable. For example, survey data may include answers for which no response was given by the survey taker. Such "no responses" are examples of missing values. Missing values can also result from integrating two data sources that do not have a row-to-row correspondence for each row in both sources. The lack of correspondence creates particular variable columns to be longer, to contain additional rows than the other columns. For these additional rows, *missing* would be the value for the cells in the shorter columns.

Do not confuse missing values with miscoded values. *Unresolved* miscoded values—values that cannot be cleaned by any method—might be changed to *missing* by some researchers or excluded for analysis by others.

Algorithmic Cleaning of Extreme Numerical Values

For numerical variables without a defined range of possible values, one might find **outliers**, values that seem excessively different from most of the other values. Such values may or may not be errors, but all outliers require review. While there is no one standard for defining outliers, most define outliers in terms of descriptive measures such as the standard deviation or the interquartile range that Chapter 3 discusses. Because software can compute such measures, spotting outliers can be automated if a definition of the term that uses such a measure is used. As later chapters note when appropriate, identifying outliers is important as some methods are *sensitive* to outliers and produce very different results when outliers are included in analysis.

1.5 Other Data Preprocessing Tasks

In addition to data cleaning, one might undertake several other data processing tasks before visualizing and analyzing a set of data.

Data Formatting

Data formatting includes rearranging the structure of the data or changing the electronic encoding of the data or both. For example, consider financial data that has been collected for a sample of companies. The collected data may be structured as tables of data, as the contents of standard forms, in a continuous stock ticker stream, or as messages or blog entries that appear on various websites. These data sources have various levels of structure that affect the ease of reformatting them for use.

Because tables of data are highly structured and are similar to the structure of a worksheet, tables would require the least reformatting. In the best case, the rows and columns of a table would become the rows and columns of a worksheet. Unstructured data sources, such as messages and blog entries, often represent the worst case. The data may need to be paraphrased, characterized, or summarized in a way that does not involve a direct transfer. As the use of business analytics grows (see Chapter 17), the use of automated ways to paraphrase or characterize these and other types of unstructured data grows, too.

Independent of the structure, collected data may exist in an electronic form that needs to be changed in order to be analyzed. For example, data presented as a digital picture of Excel worksheets would need to be changed into an actual Excel worksheet before that data could be analyzed. In this example, the electronic encoding of the data changes from a picture format such as jpeg to an Excel workbook format. Sometimes, individual numerical values that have been collected may need to be changed, especially collected values that result from a computational process. Demonstrate this issue in Excel by entering a formula that is equivalent to the expression $1 \times (0.5 - 0.4 - 0.1)$. This should evaluate as 0, but Excel evaluates to a very small negative number. Altering that value to 0 would be part of the data cleaning process.

Stacking and Unstacking Data

When collecting data for a numerical variable, subdividing that data into two or more groups for analysis may be necessary. For example, data about the cost of a restaurant meal in an urban area might be subdivided to consider the cost of meals at restaurants in the center city district separately from the meal costs at metro area restaurants. When using data that represent two or more groups, data can be arranged as either unstacked or stacked.

To use an **unstacked** arrangement, create separate numerical variables for each group. For this example, create a center city meal cost variable and a second variable to hold the meal costs at metro area restaurants. To use a **stacked** arrangement format, pair the single numerical variable meal cost with a second, categorical variable that contains two categories, such as center city and metro area. If collecting data for several numerical variables, each of which will be subdivided in the same way, stacking the data will be the more efficient choice.

When using software to analyze data, a specific procedure may require data to be stacked (or unstacked). When such cases arise using Microsoft Excel for problems or examples that this book discusses, a workbook or project will contain that data in both arrangements. For example, `Restaurants`, that Chapter 2 uses for several examples, contains both the original (stacked) data about restaurants as well as an unstacked worksheet (or data table) that contains the meal cost by location, center city or metro area.

Recoding Variables

After data have been collected, categories defined for a categorical variable may need to be reconsidered or a numerical variable may need to be transformed into a categorical variable by assigning individual numeric values to one of several groups. For either case, define a **recoded variable** that supplements or replaces the original variable in your analysis.

For example, having already defined the variable class standing with the categories freshman, sophomore, junior, and senior, a researcher decides to investigate the differences between lowerclassmen (freshmen or sophomores) and upperclassmen (juniors or seniors). The researcher can define a recoded variable UpperLower and assign the value Upper if a student is a junior or senior and assign the value Lower if the student is a freshman or sophomore.

When recoding variables, make sure that one and only one of the new categories can be assigned to any particular value being recoded and that each value can be recoded successfully by one of your new categories, the properties known as being **mutually exclusive** and **collectively exhaustive**.

When recoding numerical variables, pay particular attention to the operational definitions of the categories created for the recoded variable, especially if the categories are not self-defining ranges. For example, while the recoded categories Under 12, 12–20, 21–34, 35–54, and 55-and-over are self-defining for age, the categories child, youth, young adult, middle aged, and senior each need to be further defined in terms of mutually exclusive and collectively exhaustive numerical ranges.

PROBLEMS FOR SECTIONS 1.4 AND 1.5

APPLYING THE CONCEPTS

1.26 The cell phone brands owned by a sample of 20 respondents were:

Apple, Samsung, Appel, Nokia, Blackberry, HTC, Apple, Samsung, HTC, LG, Blueberry, Samsung, Samsung, APPLE, Motorola, Apple, Samsun, Apple, Samsung

a. Clean these data and identify any irregularities in the data.
b. Are there any missing values in this set of 20 respondents? Identify the missing values.

1.27 The amount of monthly data usage by a sample of 10 cell phone users (in MB) was:

0.4, 2.7MB, 5.6, 4.3, 11.4, 26.8, 1.6, 1,079, 8.3, 4.2

Are there any potential irregularities in the data?

1.28 An amusement park company owns three hotels on an adjoining site. A guest relations manager wants to study the time it takes for shuttle buses to travel from each of the hotels to the amusement park entrance. Data were collected on a particular day for the recorded travel times in minutes.
a. Explain how the data could be organized in an unstacked format.
b. Explain how the data could be organized in a stacked format.

1.29 A hotel management company runs 10 hotels in a resort area. The hotels have a mix of pricing—some hotels have budget-priced rooms, some have moderate-priced rooms, and some have deluxe-priced rooms. Data are collected that indicate the number of rooms that are occupied at each hotel on each day of a month. Explain how the 10 hotels can be recoded into these three price categories.

1.6 Types of Survey Errors

Collected data in the form of compiled responses from a survey must be verifed to ensure that the results can be used in a decision-making process. Verification begins by evaluating the validity of the survey to make sure the survey does not lack objectivity or credibility. To do this, evaluate the purpose of the survey, the reason the survey was conducted, and for whom the survey was conducted.

Having validated the objectivity and credibility of the survey, determine whether the survey was based on a probability sample (see Section 1.3). Surveys that use nonprobability samples are subject to serious biases that render their results useless for decision-making purposes. In the case of the Coca-Cola managers concerned about the "Pepsi Challenge" results (see page 17), the managers failed to reflect on the subjective nature of the challenge as well as the nonprobability sample that this survey used. Had the managers done so, they might not have been so quick to make the reformulation blunder that was reversed just weeks later.

Even after verification, surveys can suffer from any combination of the following types of survey errors: coverage error, nonresponse error, sampling error, or measurement error. Developers of well-designed surveys seek to reduce or minimize these types of errors, often at considerable cost.

Coverage Error

The key to proper sample selection is having an adequate frame. **Coverage error** occurs if certain groups of items are excluded from the frame so that they have no chance of being selected in the sample or if items are included from outside the frame. Coverage error results in a **selection bias**. If the frame is inadequate because certain groups of items in the population were not properly included, any probability sample selected will provide only an estimate of the characteristics of the frame, not the *actual* population.

Nonresponse Error

Not everyone is willing to respond to a survey. **Nonresponse error** arises from failure to collect data on all items in the sample and results in a **nonresponse bias**. Because a researcher cannot always assume that persons who do not respond to surveys are similar to those who do, researchers need to follow up on the nonresponses after a specified period of time. Researchers should make several attempts to convince such individuals to complete the survey and possibly offer an incentive to participate. The follow-up responses are then compared to the initial responses in order to make valid inferences from the survey (see the Cochran, Groves et al., and Lohr sources). The mode of response the researcher uses, such as face-to-face interview, telephone interview, paper questionnaire, or computerized questionnaire, affects the rate of response. Personal interviews and telephone interviews usually produce a higher response rate than do mail surveys—but at a higher cost.

Sampling Error

When conducting a probability sample, chance dictates which individuals or items will or will not be included in the sample. **Sampling error** reflects the variation, or "chance differences," from sample to sample, based on the probability of particular individuals or items being selected in the particular samples.

When there is a news report about the results of surveys or polls in newspapers or on the Internet, there is often a statement regarding a margin of error, such as "the results of this poll are expected to be within ±4 percentage points of the actual value." This **margin of error** is the sampling error. Using larger sample sizes reduces the sampling error. Of course, doing so increases the cost of conducting the survey.

Measurement Error

In the practice of good survey research, design surveys with the intention of gathering meaningful and accurate information. Unfortunately, the survey results are often only a proxy for the ones sought. Unlike height or weight, certain information about behaviors and psychological states is impossible or impractical to obtain directly.

When surveys rely on self-reported information, the mode of data collection, the respondent to the survey, or the survey itself can be possible sources of **measurement error**. Satisficing, social desirability, reading ability, and/or interviewer effects can be dependent on the mode of data collection. The social desirability bias or cognitive/memory limitations of a respondent can affect the results. Vague questions, double-barreled questions that ask about multiple issues but require a single response, or questions that ask the respondent to report something that occurs over time but fail to clearly define the extent of time about which the question asks (the reference period) are some of the survey flaws that can cause errors.

To minimize measurement error, standardize survey administration and respondent understanding of questions, but there are many barriers to this (Bremer, Fowler, Sudman).

Ethical Issues About Surveys

Ethical considerations arise with respect to the four types of survey error. Coverage error can result in selection bias and becomes an ethical issue if particular groups or individuals are purposely excluded from the frame so that the survey results are more favorable to the survey's sponsor. Nonresponse error can lead to nonresponse bias and becomes an ethical issue if the sponsor knowingly designs the survey so that particular groups or individuals are less likely than others to respond. Sampling error becomes an ethical issue if the findings are purposely presented without reference to sample size and margin of error so that the sponsor can promote a viewpoint that might otherwise be inappropriate. Measurement error can become an ethical issue in one of three ways: (1) a survey sponsor chooses leading questions that guide the respondent in a particular direction; (2) an interviewer, through mannerisms and tone, purposely makes a respondent obligated to please the interviewer or otherwise guides the respondent in a particular direction; or (3) a respondent willfully provides false information.

Ethical issues also arise when the results of nonprobability samples are used to form conclusions about the entire population. When using a nonprobability sampling method, explain the sampling procedures and state that the results cannot be generalized beyond the sample.

CONSIDER THIS

New Media Surveys/Old Survey Errors

Software company executives decide to create a "customer experience improvement program" to record how customers use the company's products, with the goal of using the collected data to make product enhancements. Product marketers decide to use social media websites to collect consumer feedback. These people risk making the same type of survey error that led to the quick demise of a very successful magazine nearly 80 years ago.

By 1935, "straw polls" conducted by the magazine *Literary Digest* had successfully predicted five consecutive U.S. presidential elections. For the 1936 election, the magazine promised its largest poll ever and sent about 10 million ballots to people all across the country. After tabulating more than 2.3 million ballots, the *Digest* confidently proclaimed that Alf Landon would be an easy winner over Franklin D. Roosevelt. The actual results: FDR won in a landslide, and Landon received the fewest electoral votes in U.S. history.

Being so wrong ruined the reputation of *Literary Digest*, and it would cease publication less than two years after it made its erroneous claim. A review much later found that the low response rate (less than 25% of the ballots distributed were returned) and nonresponse error (Roosevelt voters were less likely to mail in a ballot than Landon voters) were significant reasons for the failure of the *Literary Digest* poll (Squire).

The *Literary Digest* error proved to be a watershed event in the history of sample surveys. First, the error disproved the assertion that the larger the sample is, the better the results will be—an assertion some people still mistakenly make today. The error paved the way for the modern methods of sampling discussed in this chapter and gave prominence to the more "scientific" methods that George Gallup and Elmo Roper both used to correctly predict the 1936 elections. (Today's Gallup Polls and Roper Reports remember those researchers.)

In more recent times, Microsoft software executives overlooked that experienced users could easily opt out of participating in their improvement program. This created another case of nonresponse error that may have led to the improved product (Microsoft Office) being so poorly received initially by experienced Office users who, by being more likely to opt out of the improvement program, biased the data that Microsoft used to determine Office "improvements."

And while those product marketers may be able to collect a lot of customer feedback data, those data also suffer from nonresponse error. In collecting data from social media websites, the marketers cannot know who chose *not* to leave comments. The marketers also cannot verify if the data collected suffer from a selection bias due to a coverage error.

That you might use media newer than the mailed, dead-tree form that *Literary Digest* used does not mean that you automatically avoid the old survey errors. Just the opposite—the accessibility and reach of new media makes it much easier for unknowing people to commit such errors.

PROBLEMS FOR SECTION 1.6

APPLYING THE CONCEPTS

1.30 A survey indicates that the vast majority of college students own their own smartphones. What information would you want to know before you accepted the results of this survey?

1.31 A simple random sample of $n = 300$ full-time employees is selected from a company list containing the names of all $N = 5,000$ full-time employees in order to evaluate job satisfaction.
a. Give an example of possible coverage error.
b. Give an example of possible nonresponse error.
c. Give an example of possible sampling error.
d. Give an example of possible measurement error.

✓ SELF TEST **1.32** Results of the 2019 annual banking priorities survey of a sample of 220 bankers nationwide by Computer Services, Inc., available at **bit.ly/2POHAlk**, reveal insights on key areas of strategic focus and spending among financial institutions.

The results show that 71% of bankers note new customer acquisition as a major tactic toward reaching their revenue goals. Implementing automated online self-service solutions was highlighted as a strategy to pursue, with 18% of bankers citing self-service account opening as a vital channel for attracting new customers.

Identify *potential* concerns with coverage, nonresponse, sampling, and measurement errors.

1.33 A 2019 PwC survey of 1,000 U.S. executives (available at **pwc.to/2DAHI4q**) indicated that artificial intelligence (AI) is no longer seen as a side project or science experiment. Eighty percent of U.S. CEOs think AI will significantly change the way they will do business in the next five years. At the same time, they are concerned about AI risks that could undermine investments. What risks concern CEOs most? Forty-three percent cite new privacy threats. But CEOs also note growing concerns over how AI could affect cybersecurity, employment, inequality, and the environment. A majority of CEOs are already taking steps to address these concerns by developing and deploying AI systems that are trustworthy.

What additional information would you want to know about the survey before you accepted the results for the study?

1.34 A recent survey (avialable at **bit.ly/2H5mQ6g**) points to the transformation underway in the automotive retail landscape. The 2019 KPMG Global Automotive Executive Study found that automobile executives believe the number of physical automotive retail outlets, as we know them today, will be reduced by 30% to 50%. Eighty-two percent of automobile executives strongly agree that the only viable option for physical retail outlets will be the transformation into service factories, used car hubs, or focusing on an ID-management approach, where the customer is recognized at every single touchpoint.

What additional information would you want to know about the survey before you accepted the results of the study?

▼ USING **STATISTICS**
Defining Moments, Revisited

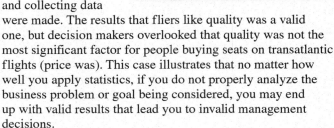

The New Coke and airline quality cases illustrate missteps that can occur during the define and collect tasks of the DCOVA framework. To use statistics effectively, you must properly define a business problem or goal and then collect data that will allow you to make observations and reach conclusions that are relevant to that problem or goal.

In the New Coke case, managers failed to consider that data collected about a taste test would not necessarily provide useful information about the sales issues they faced. The managers also did not realize that the test used improper sampling techniques, deliberately introduced biases, and were subject to coverage and nonresponse errors. Those mistakes invalidated the test, making the conclusion that New Coke tasted better than Pepsi an invalid claim.

In the airline quality case, no mistakes in defining and collecting data were made. The results that fliers like quality was a valid one, but decision makers overlooked that quality was not the most significant factor for people buying seats on transatlantic flights (price was). This case illustrates that no matter how well you apply statistics, if you do not properly analyze the business problem or goal being considered, you may end up with valid results that lead you to invalid management decisions.

▼ SUMMARY

In this chapter, you learned the details about the Define and Collect tasks of the DCOVA framework, which are important first steps to applying statistics properly to decision making. You learned that defining variables means developing an operational definition that includes establishing the type of variable and the measurement scale that the variable uses. You learned important details about data collection as well as some new basic vocabulary terms (sample, population, and parameter) and a more precise definition of statistic. You specifically learned about sampling and the types of sampling methods available to you. Finally, you surveyed data preparation considerations and learned about the type of survey errors you can encounter.

▼ REFERENCES

Biemer, P. B., R. M. Graves, L. E. Lyberg, A. Mathiowetz, and S. Sudman. *Measurement Errors in Surveys*. New York: Wiley Interscience, 2004.

Cochran, W. G. *Sampling Techniques*, 3rd ed. New York: Wiley, 1977.

Fowler, F. J. *Improving Survey Questions: Design and Evaluation*, *Applied Special Research Methods Series*, Vol. 38, Thousand Oaks, CA: Sage Publications, 1995.

Groves R. M., F. J. Fowler, M. P. Couper, J. M. Lepkowski, E. Singer, and R. Tourangeau. *Survey Methodology*, 2nd ed. New York: John Wiley, 2009.

Hellerstein, J. "Quantitative Data Cleaning for Large Databases." **bit.ly/2q7PGIn**.

Lohr, S. L. *Sampling Design and Analysis*, 2nd ed. Boston, MA: Brooks/Cole Cengage Learning, 2010.

Polaris Marketing Research. "Brilliant Marketing Research or What? The New Coke Story," posted September 20, 2011. **bit.ly/1DofHSM** (removed).

Press, G. "Cleaning Big Data: Most Time-Consuming, Least Enjoyable Data Science Task, Survey Says," posted March 23, 2016. **bit.ly/2oNCwzh**.

Rosenbaum, D. "The New Big Data Magic," posted August 20, 2011. **bit.ly/1DUMWzv**.

Osbourne, J. *Best Practices in Data Cleaning.* Thousand Oaks, CA: Sage Publications, 2012.

Squire, P. "Why the 1936 *Literary Digest* Poll Failed." *Public Opinion Quarterly* 52 (1988): 125–133.

Sudman, S., N. M. Bradburn, and N. Schwarz. *Thinking About Answers: The Application of Cognitive Processes to Survey Methodology.* San Francisco, CA: Jossey-Bass, 1993.

▼ KEY TERMS

categorical variable 17
cluster 22
cluster sample 22
collectively exhaustive 26
continuous variable 18
convenience sample 21
coverage error 27
data cleaning 24
discrete variable 18
frame 21
interval scale 18
judgment sample 21
margin of error 28
measurement error 28
measurement scale 18
missing value 25

mutually exclusive 26
nominal scale 18
nonprobability sample 21
nonresponse bias 27
nonresponse error 27
numerical variable 17
operational definition 17
ordinal scale 18
outlier 25
parameter 20
population 19
primary data source 20
probability sample 21
ratio scale 18
recoded variable 26
sample 19

sampling error 28
sampling with replacement 21
sampling without replacement 21
secondary data source 20
selection bias 27
simple random sample 21
stacked 26
statistic 20
strata 22
stratified sample 22
systematic sample 22
table of random numbers 22
treatment 20
unstacked 26

▼ CHECKING YOUR UNDERSTANDING

1.35 What is the difference between a sample and a population?

1.36 What is the difference between a statistic and a parameter?

1.37 What is the difference between a categorical variable and a numerical variable?

1.38 What is the difference between a discrete numerical variable and a continuous numerical variable?

1.39 What is the difference between a nominal scaled variable and an ordinal scaled variable?

1.40 What is the difference between an interval scaled variable and a ratio scaled variable?

1.41 What is the difference between probability sampling and non-probability sampling?

1.42 What is the difference between a missing value and an outlier?

1.43 What is the difference between unstacked and stacked variables?

1.44 What is the difference between coverage error and nonresponse error?

1.45 What is the difference between sampling error and measurement error?

▼ CHAPTER REVIEW PROBLEMS

1.46 Visit the official Microsoft Excel product website, **products .office.com/excel**. Review the features of the program you chose and then state the ways the program could be useful in statistical analysis.

1.47 Results of a 2017 Computer Services, Inc. (CSI) survey of a sample of 163 bank executives reveal insights on banking priorities among financial institutions (**goo.gl/mniYMM**). As financial institutions begin planning for a new year, of utmost importance is boosting profitability and identifying growth areas.

The results show that 55% of bank institutions note customer experience initiatives as an area in which spending is expected to increase. Implementing a customer relationship management (CRM) solution was ranked as the top most important omnichannel strategy to pursue with 41% of institutions citing digital banking enhancements as the greatest anticipated strategy to enhance the customer experience.

a. Describe the population of interest.
b. Describe the sample that was collected.
c. Describe a parameter of interest.
d. Describe the statistic used to estimate the parameter in (c).

1.48 The Gallup organization releases the results of recent polls on its website, **www.gallup.com**. Visit this site and read an article of interest.
a. Describe the population of interest.
b. Describe the sample that was collected.
c. Describe a parameter of interest.
d. Describe the statistic used to estimate the parameter in (c).

1.49 A 2019 PwC survey of 1,000 U.S. executives indicated that artificial intelligence (AI) is no longer seen as a side project. Eighty percent of U.S. CEOs think AI will significantly change the way they will do business in the next five years. At the same time, these CEOs are concerned about AI risks that could undermine investments. What risks concern CEOs most? Forty-three percent cite new privacy threats. But CEOs also note growing concerns over how AI could affect cybersecurity, employment, inequality, and the environment. A majority of CEOs are already taking steps to address these concerns by developing and deploying AI systems that are trustworthy.

Source: "US CEO agenda 2019," PwC, **pwc.to/2UuoVAX**.
a. Describe the population of interest.
b. Describe the sample that was collected.
c. Describe a parameter of interest.
d. Describe the statistic used to estimate the parameter in (c).

1.50 The American Community Survey (**www.census.gov/acs**) provides data every year about communities in the United States. Addresses are randomly selected and respondents are required to supply answers to a series of questions.
a. Describe a variable for which data are collected.
b. Is the variable categorical or numerical?
c. If the variable is numerical, is it discrete or continuous?

1.51 Examine Zarca Interactive's "Sample Employee Satisfaction Survey/Sample Questions for Employee Satisfaction Survey," available at **bit.ly/21qjI6F**.

a. Give an example of a categorical variable included in the survey.
b. Give an example of a numerical variable included in the survey.

1.52 Three professors examined awareness of four widely disseminated retirement rules among employees at the University of Utah. These rules provide simple answers to questions about retirement planning (R. N. Mayer, C. D. Zick, and M. Glaittle, "Public Awareness of Retirement Planning Rules of Thumb," *Journal of Personal Finance*, 2011 10(62), 12–35). At the time of the investigation, there were approximately 10,000 benefited employees, and 3,095 participated in the study. Demographic data collected on these 3,095 employees included gender, age (years), education level (years completed), marital status, household income ($), and employment category.
a. Describe the population of interest.
b. Describe the sample that was collected.
c. Indicate whether each of the demographic variables mentioned is categorical or numerical.

1.53 Social media provides an enormous amount of data about the activities and habits of people using social platforms like Facebook and Twitter. The belief is that mining that data provides a treasure trove for those who seek to quantify and predict future human behavior. A marketer is planning a survey of Internet users in the United States to determine social media usage. The objective of the survey is to gain insight on these three items: key social media platforms used, frequency of social media usage, and demographics of key social media platform users.
a. For each of the three items listed, indicate whether the variables are categorical or numerical. If a variable is numerical, is it discrete or continuous?
b. Develop five categorical questions for the survey.
c. Develop five numerical questions for the survey.

CHAPTER 1

▼ CASES

Managing Ashland MultiComm Services

Ashland MultiComm Services (AMS) provides high-quality telecommunications services in the Greater Ashland area. AMS traces its roots to a small company that redistributed the broadcast television signals from nearby major metropolitan areas but has evolved into a provider of a wide range of broadband services for residential customers.

AMS offers subscription-based services for digital cable television, local and long-distance telephone services, and high-speed Internet access. Recently, AMS has faced competition from other service providers as well as Internet-based, on-demand streaming services that have caused many customers to "cut the cable" and drop their subscription to cable video services.

AMS management believes that a combination of increased promotional expenditures, adjustment in subscription fees, and improved customer service will allow AMS to successfully face these challenges. To help determine the proper mix of strategies to be taken, AMS management has decided to organize a research team to undertake a study.

The managers suggest that the research team examine the company's own historical data for number of subscribers, revenues, and subscription renewal rates for the past few years. They direct the team to examine year-to-date data as well because the managers suspect that some of the changes they have seen have been a relatively recent phenomena.

1. What type of data source would the company's own historical data be? Identify other possible data sources that the research team might use to examine the current marketplace for residential broadband services in a city such as Ashland.

2. What type of data collection techniques might the team employ?

3. In their suggestions and directions, the AMS managers have named a number of possible variables to study but offered no operational definitions for those variables. What types of possible misunderstandings could arise if the team and managers do not first properly define each variable cited?

CardioGood Fitness

CardioGood Fitness is a developer of high-quality cardiovascular exercise equipment. Its products include treadmills, fitness bikes, elliptical machines, and e-glides. CardioGood Fitness looks to increase the sales of its treadmill products and has hired The AdRight Agency, a small advertising firm, to create and implement an advertising program. The AdRight Agency plans to identify particular market segments that are most likely to buy their clients' goods and services and then locate advertising outlets that will reach that market group. This activity includes collecting data on clients' actual sales and on the customers who make the purchases, with the goal of determining whether there is a distinct profile of the typical customer for a particular product or service. If a distinct profile emerges, efforts are made to match that profile to advertising outlets known to reflect the particular profile, thus targeting advertising directly to high-potential customers.

CardioGood Fitness sells three different lines of treadmills. The TM195 is an entry-level treadmill. It is as dependable as other models offered by CardioGood Fitness, but with fewer programs and features. It is suitable for individuals who thrive on minimal programming and the desire for simplicity to initiate their walk or hike. The TM195 sells for $1,500.

The middle-line TM498 adds to the features of the entry-level model: two user programs and up to 15% elevation upgrade. The TM498 is suitable for individuals who are walkers at a transitional stage from walking to running or midlevel runners. The TM498 sells for $1,750.

The top-of-the-line TM798 is structurally larger and heavier and has more features than the other models. Its unique features include a bright blue backlit LCD console, quick speed and incline keys, a wireless heart rate monitor with a telemetric chest strap, remote speed and incline controls, and an anatomical figure that specifies which muscles are minimally and maximally activated. This model features a nonfolding platform base that is designed to handle rigorous, frequent running; the TM798 is therefore appealing to someone who is a power walker or a runner. The selling price is $2,500.

As a first step, the market research team at AdRight is assigned the task of identifying the profile of the typical customer for each treadmill product offered by CardioGood Fitness. The market research team decides to investigate whether there are differences across the product lines with respect to customer characteristics. The team decides to collect data on individuals who purchased a treadmill at a CardioGood Fitness retail store during the prior three months.

The team decides to use both business transactional data and the results of a personal profile survey that every purchaser completes as the team's sources of data. The team identifies the following customer variables to study: product purchased—TM195, TM498, or TM798; gender; age, in years; education, in years; relationship status, single or partnered; annual household income ($); mean number of times the customer plans to use the treadmill each week; mean number of miles the customer expects to walk/run each week; and self-rated fitness on a 1-to-5 scale, where 1 is poor shape and 5 is excellent shape. For this set of variables:

1. Which variables in the survey are categorical?

2. Which variables in the survey are numerical?

3. Which variables are discrete numerical variables?

Clear Mountain State Student Survey

The Student News Service at Clear Mountain State University (CMSU) has decided to gather data about the undergraduate students who attend CMSU. They create and distribute a survey of 14 questions and receive responses from 111 undergraduates (stored in Student Survey).

Download (see Appendix C) and review the survey document **CMUndergradSurvey.pdf**. For each question asked in the survey, determine whether the variable is categorical or numerical. If you determine that the variable is numerical, identify whether it is discrete or continuous.

Learning With the Digital Cases

Identifying and preventing misuses of statistics is an important responsibility for all managers. The Digital Cases allow you to practice the skills necessary for this important task.

Each chapter's Digital Case tests your understanding of how to apply an important statistical concept taught in the chapter. As in many business situations, not all of the information you encounter will be relevant to your task, and you may occasionally discover conflicting information that you have to resolve in order to complete the case.

To assist your learning, each Digital Case begins with a learning objective and a summary of the problem or issue at hand. Each case directs you to the information necessary to reach your own conclusions and to answer the case questions. Many cases, such as the sample case worked out next, extend a chapter's Using Statistics scenario. You can download digital case files, which are PDF format documents that may contain extended features such as interactivity or data file attachments. Open these files with a current version of Adobe Reader, as other PDF programs may not support the extended features. (For more information, see Appendix C.)

To illustrate learning with a Digital Case, open the Digital Case file **WhitneyWireless.pdf** that contains summary information about the Whitney Wireless business. Apparently, from the claim on the title page, this business is celebrating its "best sales year ever."

Review the **Who We Are**, **What We Do**, and **What We Plan to Do** sections on the second page. Do these sections contain any useful information? What *questions* does this passage raise? Did you notice that while many facts are presented, no data that would support the claim of "best sales year ever" are presented? And were those mobile "mobilemobiles" used solely for promotion? Or did they generate any sales? Do you think that a talk-with-your-mouth-full event, however novel, would be a success?

Continue to the third page and the **Our Best Sales Year Ever!** section. How would you support such a claim? With a table of numbers? Remarks attributed to a knowledgeable source? Whitney Wireless has used a chart to present "two years ago" and "latest twelve months" sales data by category. Are there any problems with what the company has done? *Absolutely!*

Take a moment to identify and reflect on those problems. Then turn to pages 4 though 6 that present an annotated version of the first three pages and discusses some of the problems with this document.

In subsequent Digital Cases, you will be asked to provide this type of analysis, using the open-ended case questions as your guide. Not all the cases are as straightforward as this example, and some cases include perfectly appropriate applications of statistical methods. And none have annotated answers!

▾EXCEL GUIDE

EG1.1 DEFINING VARIABLES

Classifying Variables by Type

Microsoft Excel infers the variable type from the data one enters into a column. If Excel discovers a column that contains numbers, it treats the column as a numerical variable. If Excel discovers a column that contains words or alphanumeric entries, it treats the column as a non-numerical (categorical) variable.

This imperfect method works most of the time, especially if one makes sure that the categories for a categorical variables are words or phrases such as "yes" and "no." However, because one cannot explicitly define the variable type, Excel enables one to do nonsensical things such as using a categorical variable with a statistical method designed for numerical variables. If one must use categorical values such as 1, 2, or 3, enter them preceded with an apostrophe, as Excel treats all values that begin with an apostrophe as non-numerical data. (To check whether a cell entry includes a leading apostrophe, select the cell and view its contents in the formula bar.)

EG1.3 TYPES of SAMPLING METHODS

Simple Random Sample

Key Technique Use the **RANDBETWEEN**(*smallest integer, largest integer*) function to generate a random integer that can then be used to select an item from a frame.

Example 1 Create a simple random sample *with* replacement of size 40 from a population of 800 items.

Workbook Enter a formula that uses this function and then copy the formula down a column for as many rows as is necessary. For example, to create a simple random sample with replacement of size 40 from a population of 800 items, open to a new worksheet. Enter **Sample** in cell **A1** and enter the formula **=RANDBETWEEN(1, 800)** in cell **A2**. Then copy the formula down the column to cell **A41**.

Excel contains no functions to select a random sample *without* replacement. Such samples are most easily created using an add-in such as PHStat or the Analysis ToolPak, as described in the following paragraphs.

Analysis ToolPak Use **Sampling** to create a random sample *with replacement*.

For the example, open to the worksheet that contains the population of 800 items in column A and that contains a column heading in cell A1. Select **Data→Data Analysis.** In the Data Analysis dialog box, select **Sampling** from the **Analysis Tools** list and then click **OK**. In the procedure's dialog box (shown at top right):

1. Enter **A1:A801** as the **Input Range** and check **Labels**.
2. Click **Random** and enter **40** as the **Number of Samples**.
3. Click **New Worksheet Ply** and then click **OK**.

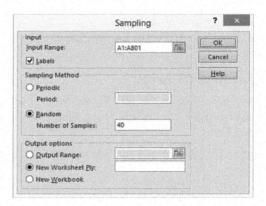

Example 2 Create a simple random sample *without* replacement of size 40 from a population of 800 items.

PHStat Use **Random Sample Generation**.

For the example, select **PHStat → Sampling → Random Sample Generation.** In the procedure's dialog box (shown below):

1. Enter **40** as the **Sample Size**.
2. Click **Generate list of random numbers** and enter **800** as the **Population Size**.
3. Enter a **Title** and click **OK**.

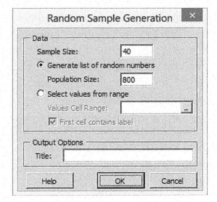

Unlike most other PHStat results worksheets, the worksheet created contains no formulas.

Workbook Use the **COMPUTE worksheet** of the **Random workbook** as a template.

The worksheet already contains 40 copies of the formula **=RANDBETWEEN(1, 800)** in column B. Because the **RANDBETWEEN** function samples *with* replacement as discussed at the start of this section, one may need to add additional copies of the formula in new column B rows until one has 40 unique values.

When the intended sample size is large, spotting duplicate values can be hard. Read the SHORT TAKES for Chapter 1 to learn more about an advanced technique that uses formulas to detect duplicate values.

EG1.4 DATA CLEANING

Key Technique Use a column of formulas to detect invalid variable values in another column.

Example Scan the **DirtyDATA worksheet** in the **Dirty Data workbook** for invalid variable values.

PHStat Use **Data Cleaning**.

For the example, open to the **DirtyData worksheet**. Select **Data Preparation→Numerical Data Scan.** In the procedure's dialog box:

1. Enter a column range as the **Numerical Variable Cell Range**.
2. Click **OK**.

The procedure creates a worksheet that contains a column that identifies every data value as either being numerical or non-numerical and states the minimum and maximum values found in the column. To scan for irregularities in categorical data, use the *Workbook* instructions.

Workbook Use the **ScanData worksheet** of the **Data Cleaning workbook** as a model solution to scan for the following types of irregularities: non-numerical data values for a numerical variable, invalid categorical values of a categorical variable, numerical values outside a defined range, and missing values in individual cells.

The worksheet uses several different Excel functions to detect an irregularity in one column and display a message in another column. For each categorical variable scanned, the worksheet contains a table of valid values that are looked up and compared to cell values to spot inconsistencies. Read the SHORT TAKES for Chapter 1 to learn the specifics of the formulas the worksheet uses to scan data.

EG1.5 OTHER DATA PREPROCESSING

Stacking and Unstacking Variables

PHStat Use **Data Preparation→Stack Data** (or **Unstack Data**).

For **Stack Data**, in the Stack Data dialog box, enter an **Unstacked Data Cell Range** and then click **OK** to create stacked data in a new worksheet. For **Unstack Data**, in the Unstack Data dialog box, enter a **Grouping Variable Cell Range** and a **Stacked Data Cell Range** and then click **OK** to create unstacked data in a new worksheet.

Recoding Variables

Key Technique To recode a categorical variable, first copy the original variable's column of data and then use the find-and-replace function on the copied data. To recode a numerical variable, enter a formula that returns a recoded value in a new column.

Example Using the **DATA worksheet** of the **Recoded workbook**, create the recoded variable UpperLower from the categorical variable Class and create the recoded Variable Dean's List from the numerical variable GPA.

Workbook Use the **RECODED worksheet** of the **Recoded workbook** as a model.

The worksheet already contains UpperLower, a recoded version of Class that uses the operational definitions on page 26, and Dean's List, a recoded version of GPA, in which the value No recodes all GPA values less than 3.3 and Yes recodes all values 3.3 or greater. The **RECODED_ FORMULAS worksheet** in the same workbook shows how formulas in column I use the IF function to recode GPA as the Dean's List variable.

These recoded variables were created by first opening to the **DATA worksheet** in the same workbook and then following these steps:

1. Right-click column **D** (right-click over the shaded "D" at the top of column D) and click **Copy** in the shortcut menu.
2. Right-click column **H** and click the **first choice** in the **Paste Options** gallery.
3. Enter **UpperLower** in cell **H1**.
4. Select column **H**. With column H selected, click **Home→ Find & Select→Replace**.

In the Replace tab of the Find and Replace dialog box:

5. Enter **Senior** as **Find what**, **Upper** as **Replace with**, and then click **Replace All**.
6. Click **OK** to close the dialog box that reports the results of the replacement command.
7. Still in the Find and Replace dialog box, enter **Junior** as **Find what** and then click **Replace All**.
8. Click **OK** to close the dialog box that reports the results of the replacement command.
9. Still in the Find and Replace dialog box, enter **Sophomore** as **Find what**, **Lower** as **Replace with**, and then click **Replace All**.
10. Click **OK** to close the dialog box that reports the results of the replacement command.
11. Still in the Find and Replace dialog box, enter **Freshman** as **Find what** and then click **Replace All**.
12. Click **OK** to close the dialog box that reports the results of the replacement command.

(This creates the recoded variable UpperLower in column H.)

13. Enter **Dean's List** in cell **I1**.
14. Enter the formula =IF(G2 < 3.3, "No", "Yes") in cell **I2**.
15. Copy this formula down the column to the last row that contains student data (row 63).

(This creates the recoded variable Dean's List in column I.)

The RECODED worksheet uses the **IF** function, which Appendix F discusses to recode the numerical variable into two categories. Numerical variables can also be recoded into multiple categories by using the **VLOOKUP** function (see Appendix F).

▾TABLEAU GUIDE

TG1.1 DEFINING VARIABLES

Classifying Variables by Type

Tableau infers three attributes of a column from the data that a column contains. Tableau assigns a *data role* to each column, initially considering any column that contains categorical data to be a Dimension and any column that contains numerical data to be a Measure. The data role of the column, in turn, affects how Tableau processes the column data as well as which choices Tableau presents in the Show Me visualization gallery (see Chapter 2).

Tableau also classifies a column as either having discrete or continuous data. Tableau uses the terms discrete and continuous in the sense that this chapter defines, except that Tableau considers all categorical data as discrete data, too. Tableau colors a column name green if the column contains continuous data and colors a column name blue if the column contains discrete data.

Tableau also determines whether the data type in the column is one of several data types. In the Data tab (shown below), Tableau displays a hashtag icon (#) before the names of columns that contain whole or decimal numbers, displays an Abc icon before the names of columns that contain alphanumeric data, and displays a calendar icon before the names of columns that contain date or date and time data. Tableau precedes a data type icon with an equals sign if the column contains the results of a calculation or represents data copied from another column. Tableau presents a Dimension or Measure name in italics if the name represents a filtered set. In the illustration below, *Number of Records* in the Measures list is a calculation that is also a filtered set.

To change the data role assigned to a column, right-click the column name and select **Convert to Dimension** or **Convert to Measure** from the shortcut menu. To change the assignment of discrete or continuous to a column, right-click the column name and select **Convert to Continuous** or **Convert to Discrete** from the shortcut menu. To change the assigned data type,

right-click the column name and select **Change Data Type** and then select the appropriate choice from the shortcut menu.

In the preceding illustration at left, Tableau has classified Region Code as a continuous (green) measure that has whole or decimal number data. Region Code is a recoding of Region that uses the numerals 1, 2, 3, and 4 to replace categorical values. Therefore, Region Code should be a *Dimension* that contains discrete, alphanumeric data. For the column's shortcut menu, the three right-click selections **Convert to Dimension**, **Convert to Discrete**, and **Change Data Type → String** would be necessary to proper reclassify Region Code. ("String" is the Tableau term for alphanumeric data.)

TG1.4 DATA CLEANING

Use the **Data Interpreter**.

For certain types of structured data, such as Excel worksheets and Google sheets, Tableau can apply its Data Interpreter procedure to examine data for common types of error. For an Excel workbook, the interpreter will not only import corrected values but insert a series of worksheets that document its work into the workbook that is being used.

To use the interpreter, check **Use Data Interpreter** in the left panel of a Data Source display (shown below left). The panel display changes to Cleaned with Data Interpreter (shown below right), when the data cleaning finishes and Tableau has imported the data to be analyzed (in the illustration, the data from the Data worksheet of the Retirement Funds Excel workbook.)

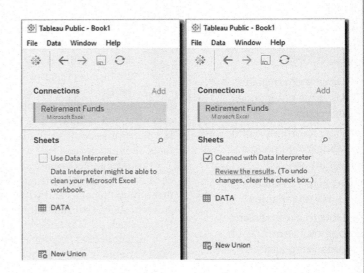

Organizing and Visualizing Variables

CONTENTS

OBJECTIVES

- How to organize and visualize categorical variables
- How to organize and visualize numerical variables
- How to summarize a mix of variables
- How to avoid making common errors when organizing and visualizing variables

▼USING **STATISTICS**
"The Choice Is *Yours"*

The Choice *Is* Yours financial services firm saw its business erode as the firm failed to connect with a rising generation of millennials. Now, unwilling to repeat that past mistake, the firm is preparing a new marketing plan aimed at the emerging Generation Z audience.

To pursue this business objective, a company task force has already selected 479 retirement funds that may prove appropriate for younger investors. You have been asked to define, collect, organize, and visualize data about these funds in ways that could assist prospective clients making decisions about the funds in which they will invest. As a starting point, you think about the facts about each fund that would help customers compare and contrast funds.

You decide to begin by defining the variables for key characteristics of each fund, such as each fund's past performance. You also decide to define variables such as the amount of assets that a fund manages and whether the goal of a fund is to invest in companies whose earnings are expected to substantially increase in future years (a "growth" fund) or invest in companies whose stock price is undervalued, priced low relative to their earnings potential (a "value" fund).

You collect data from appropriate sources and organize the data as a worksheet, placing each variable in its own column. As you think more about your task, you realize that 479 rows of data, one for each fund in the sample, would be hard for prospective clients to review easily.

Is there something else you can do? Can you organize and present these data to prospective clients in a more helpful and comprehensible manner?

Defining the variables of interest and then collecting, preparing, and entering data into worksheets or data tables completes the **Define** and **Collect DCOVA** tasks. The **DCOVA O**rganize task creates summaries of the prepared data that provide initial insights about the variables. These summaries guide further exploration of the data as well as sometimes directly facilitating decision making. For example, in the Choice *Is* Yours scenario, creating a summary of the retirement funds sample that would allow prospective younger investors to quickly identify funds designed for growth and identified as having moderate risk would be useful.

The Organize step uses descriptive statistical methods that produce various types of tabular summaries. Because reviewing tabular *and* visual summaries together can lead to better insights and jumpstart analysis, the DCOVA **V**isualize task is often done concurrent to the **O**rganize task. For that reason, this chapter discusses methods to visualize variables after discussing methods to organize variables. Visual summaries can facilitate the rapid review of larger amounts of data as well as show possible significant patterns to the data. For example, for the retirement funds sample, visualizing the ten-year rate of return and the management expense fees charged by each fund would help identify the funds that would be charging you relatively little in fees for a "good" rate of return as well as the funds whose fees seem excessive given their modest or weak rates of return.

Because the methods used to organize and visualize categorical variables differ from the methods used to organize and visualize numerical variables, this chapter discusses categorical and numerical methods in separate sections. The last section, Section 2.9, discusses the common types of errors that can occur when using descriptive methods to organize or visualize data.

student TIP

Table 2.15 on page 80 lists the methods to organize and visualize variables that this chapter discusses.

2.1 Organizing Categorical Variables

Tallying the set of individual values for the variable by categories and placing the results in tables organize a categorical variable. A *summary table* organizes the data for a single categorical variable and a *contingency table* organizes the data from two or more categorical variables.

The Summary Table

As Appendix C explains, names of data files appear in a special inverse type font.

A **summary table** tallies the set of individual values as frequencies or percentages for each category. A summary table helps reveal differences among the categories by displaying the frequency, amount, or percentage of items in a set of categories in a separate column. Table 2.1 summarizes the results of asking 2,005 employed adults who work outside the home what action they took the last time they were sick. The table shows that 54% of the employees went to work, while only 34% took a sick day.

TABLE 2.1
Actions taken when sick by employees who work outside the home
`Employee Actions`

Action Taken	Percentage
Took a sick day	34%
Went to work	54%
Worked from home	12%

Source: Data extracted from CivicScience, "Working While Sick Is a Slippery Slope to Job Burnout, Unhappiness," **/bit.ly/2P1aNsM.**

EXAMPLE 2.1

Summary Table of
Levels of Risk of
Retirement Funds

The sample of 479 retirement funds in The Choice *Is* Yours scenario, stored in Retirement Funds , includes the variable *risk level* that has the defined categories low, average, and high. Construct a summary table of the retirement funds, categorized by risk.

SOLUTION In Figure 2.1, the percentages for each category are calculated by dividing the number of funds in each category by the total sample size (479). From Figure 2.1, observe that almost half the funds have an average risk, about 30% have low risk, and less than a quarter have high risk.

FIGURE 2.1

Frequency and percentage
summary table of risk
level for 479 retirement
funds

Risk Level	Frequency	Percentage
Low	147	30.69%
Average	224	46.76%
High	108	22.55%
Total	479	100.00%

learnMORE

**The Retirement Funds
Sample online document
defines the variables that
the sample of 479 funds
uses in examples
throughout this book.**

The Contingency Table

A **contingency table** cross-tabulates, or tallies jointly, the data of two or more categorical variables, which enables one to study patterns that may exist between the variables. Tallies can be shown as a frequency, a percentage of the overall total, a percentage of the row total, or a percentage of the column total. Each tally appears in its own **cell**, and there is a cell for each **joint response**, a unique combination of values for the variables being tallied.

In a contingency table, *both* the rows and the columns represent variables. In the simplest case of a contingency table that summarizes two categorical variables, the rows contain the tallies of one variable, and the columns contain the tallies of the other variable. Some use the terms *row variable* and *column variable* to distinguish between the two variables.

For The Choice *Is* Yours scenario, the fund type and risk levels would be one pair of variables that could be summarized for the sample of 479 retirement funds. Because fund type has the defined categories growth and value and the risk level has the categories low, average, and high, this table has six possible joint responses, forming a two row by three column contingency table.

Figure 2.2 contains Excel *PivotTable* and Tableau versions of this contingency table. These summaries show that there are 306 growth and 173 value funds (the row totals) and 147 low risk funds, 224 average risk funds, and 108 high risk funds (the column totals). The tables identify the most frequently encountered joint response in the retirement funds sample as being growth funds with average risk (152).

FIGURE 2.2

Excel (PivotTable) and
Tableau contingency tables
of fund type and risk level
for the sample of 479
retirement funds

Fund Type	Risk Level Low	Average	High	Grand Total
Growth	63	152	91	306
Value	84	72	17	173
Grand Total	147	224	108	479

Fund Type	Risk Level Low	Average	High	Grand ..
Growth	63	152	91	306
Value	84	72	17	173
Grand Total	147	224	108	479

studentTIP

**Remember, each joint
response gets tallied into
only one cell.**

Figure 2.3 on page 41 shows a Tableau contingency table that presents tallies as a percentage of the overall total. From these contingency tables, one concludes that the pattern of risk for growth funds differs from the pattern for value funds.

FIGURE 2.3

Excel and Tableau contingency tables of fund type and risk level for the sample of 479 retirement funds, showing overall total percentages for each joint response

Fund Type	Risk Level			
	Low	Average	High	Grand Total
Growth	13.15%	31.73%	19.00%	63.88%
Value	17.54%	15.03%	3.55%	36.12%
Grand Total	30.69%	46.76%	22.55%	100.00%

Fund Type	Risk Level			
	Low	Average	High	Grand ..
Growth	63	152	91	306
Value	84	72	17	173
Grand Total	147	224	108	479

PROBLEMS FOR SECTION 2.1

LEARNING THE BASICS

2.1 A categorical variable has three categories, with the following frequencies of occurrence:

Category	Frequency
A	13
B	28
C	9

a. Compute the percentage of values in each category.
b. What conclusions can you reach concerning the categories?

2.2 The following data represent the responses to two questions asked in a survey of 40 college students majoring in business: What is your gender? (M = male; F = female), and what is your major? (A = Accounting; C = Computer Information Systems; M = Marketing):

Gender:	M	M	M	F	M	F	F	M	F	M
Major:	A	C	C	M	A	C	A	A	C	C
Gender:	F	M	M	M	M	F	F	M	F	F
Major:	A	A	A	M	C	M	A	A	A	C
Gender:	M	M	M	M	F	M	F	F	M	M
Major:	C	C	A	A	M	M	C	A	A	A
Gender:	F	M	M	M	M	F	M	F	M	M
Major:	C	C	A	A	A	A	C	C	A	C

a. Tally the data into a contingency table where the two rows represent the gender categories and the three columns represent the academic major categories.
b. Construct contingency tables based on percentages of all 40 student responses, based on row percentages and based on column percentages.

APPLYING THE CONCEPTS

2.3 The following table, stored in `Smartphone Sales`, represents the quarterly market share (percentage) of smartphones, by type, for the third quarter of 2017 through the third quarter of 2018, as reported by Counterpoint Research.

Brand	17Q3	17Q4	18Q1	18Q2	18Q3
Apple	33	44	37	40	39
Samsung	23	19	26	25	25
LG	18	14	14	16	17
Motorola	5	5	4	5	8
Others	21	18	19	14	11

Source: Data extracted from **bit.ly/2MvW6fO**.

a. What conclusions can you reach about the market for smartphones in the third quarter of 2017 through the third quarter of 2018?
b. What differences are there in the third quarter of 2017 through the third quarter of 2018?

✓ SELF TEST **2.4** The Consumer Financial Protection Bureau reports on consumer financial product and service complaint submissions by state, category, and company. The following table, stored in `CFPB categories`, represents complaints received from Louisiana consumers by complaint category for a recent year.

Category	Number of Complaints
Bank Account or Service	202
Consumer Loan	132
Credit Card	175
Credit Reporting	581
Debt Collection	486
Mortgage	442
Student Loan	75
Other	72

Source: Data extracted from **bit.ly/2pR7ryO**.

a. Compute the percentage of complaints for each category.
b. What conclusions can you reach about the complaints for the different categories?

(*Problem 2.4 continues on page 42*)

The following table, stored in FinancialComplaints2 , summarizes complaints received from Louisiana consumers by most-complained-about companies for a recent year.

Company	Number of Complaints
Bank of America	42
Capital One	93
Citibank	59
Ditech Financial	31
Equifax	217
Experian	177
JPMorgan	128
Nationstar Mortgage	39
Navient	38
Ocwen	41
Synchrony	43
Trans-Union	168
Wells Fargo	77

c. Compute the percentage of complaints for each company.
d. What conclusions can you reach about the complaints for the different companies?

2.5 A sample of companies were asked three questions about analytics and artificial intelligence usage and their answers stored in Business Analytics , Machine Learning , and Self-learning Robots . The following tables summarize those answers.

Source: Data extracted from A. Ng and N. Jacobstein, "How AI will change everything," *The Wall Street Journal*, March 7, 2017, pp. B1–B2.

Business Analytics Usage	Percentage
Already use	8
Beyond the next five years	12
Don't know	10
No plans to use	9
Within the next five years	61

Machine Language Usage	Percentage
Already use	8
Beyond the next five years	15
Don't know	14
No plans to use	18
Within the next five years	45

Self-Learning Robots Usage	Percentage
Already use	1
Beyond the next five years	17
Don't know	15
No plans to use	46
Within the next five years	21

What conclusions can you reach about the differences in the expected use of business analytics, machine learning, and self-learning robots?

2.6 The following table, stored in Energy Sources , contains the 2018 sources of electricity in the United States as reported by the Energy Information Administration.

Source	Billion kWh
Coal	1,146
Geothermal	17
Hydropower	292
Biomass	63
Natural gas	1,468
Nuclear	807
Other sources	14
Petroleum coke and liquids	26
Solar	67
Wind	275

Source: **www.eia.gov**

What conclusions can you reach about sources of energy in 2018?

2.7 A recent Timetric's survey of insurance professionals explores the use of technology in the industry. The file Technologies contains the responses to the question that asked what technologies these professionals expected to be most used by the insurance industry in the coming year. Those responses are:

Technology	Frequency
Wearable technology	9
Blockchain technology	9
Artificial intelligence	17
IoT: retail insurance	23
IoT: commercial insurance	5
Social media	27

Source: Data extracted from **bit.ly/2qxMFRj**.

a. Compute the percentage of responses for each technology.
b. What conclusions can you reach concerning expected technology usage in the insurance industry in the coming year?

2.8 A survey of 1,520 Americans adults asked "Do you feel overloaded with too much information?" The results indicate that 23% of females feel information overload compared to 17% of males. The results are:

OVERLOADED	GENDER Male	Female	Total
Yes	134	170	304
No	651	565	1,216
Total	785	735	1,520

Source: Data extracted from **bit.ly/2pR5bHZ**.

a. Construct contingency tables based on total percentages, row percentages, and column percentages.
b. What conclusions can you reach from these analyses?

2.9 A study of selected Kickstarter projects showed that overall a majority were successful, achieving their goal and raising, at a minimum, the targeted amounts. In an effort to identify project types that influence success, selected projects were subdivided into project categories (Film & Video, Games, Music, and Technology). The results are as follows:

CATEGORY	OUTCOME		
	Successful	Not Successful	Total
Film & Video	21,759	36,805	58,564
Games	9,329	18,238	27,567
Music	24,285	24,377	48,662
Technology	5,040	20,555	25,595
Total	60,413	99,975	160,388

Source: **Kickstarter.com**, **kickstarter.com/help/stats**.

a. Construct contingency tables based on total percentages, row percentages, and column percentages.

b. Which type of percentage—row, column, or total—do you think is most informative for these data? Explain.
c. What conclusions concerning the pattern of successful Kickstarter projects can you reach?

2.10 An Ipsos poll asked 1,004 adults, "If purchasing a used car made certain upgrades or features more affordable, what would be your preferred luxury upgrade?" The results indicated that 9% of males and 14% of females answered window tinting.

Source: Ipsos, "Safety Technology Tops the List of Most Desired Features Should They Be More Affordable When Purchasing a Used Car—Particularly Collision Avoidance," available at **bit.ly/2ufbS8Z**.

The poll description did not state the sample sizes of males and females. Suppose that both sample sizes were 502 and that 46 of 502 males and 71 of 502 females reported window tinting as their preferred luxury upgrade of choice.

What do these results tell you about luxury upgrade differences between males and females?

2.2 Organizing Numerical Variables

Create ordered arrays and distribution tables to organize numerical variables. Unless the data contain a very large number of values, first arrange the data as an **ordered array**, a list in rank order, from the smallest to the largest value. An ordered array helps see the range of values in the data and is particularly useful when the data contain more than just a handful of values.

When organizing a numerical variable, grouping the data by the value of a categorical variable is sometime necessary for analysis. For example, in collecting meal cost data as part of a study that reviews the travel and entertainment costs that a business incurs in a major city, examining the cost of meals at restaurants located in the center city district separately from the cost at restaurants in the surrounding metropolitan area might reveal cost differences between the two locations. As meal cost data for this study are collected, noting the location can be collected and used later for grouping purposes.

Data for a grouped numerical variable can be stored as stacked or unstacked data, as Section 1.5 discusses. As Section 1.5 notes, requirements for specific software procedures often dictate the choice of using stacked or unstacked data.

Table 2.2A contains the meal cost data collected from a sample of 50 center city restaurants and 50 metro area restaurants that are stored in Restaurants . Table 2.2B presents these two lists of

TABLE 2.2A
Meal cost data for 50 center city and 50 metro area restaurants

Center City Restaurant Meal Costs

81 28 24 38 45 49 36 60 50 41 84 64 78 57 80 69 89 42 55 32 45 71 50 51 50
66 49 91 66 58 80 58 50 44 53 62 40 45 23 66 52 47 70 56 55 52 49 26 79 40

Metro Area Restaurant Meal Costs

54 35 29 24 26 31 42 33 25 47 50 59 35 36 43 40 56 34 41 55 42 43 43 64 46
46 81 33 37 39 54 53 41 39 52 52 42 59 39 69 41 51 36 46 44 75 56 36 33 45

TABLE 2.2B
Ordered array of meal costs for 50 center city and 50 metro area restaurants

Center City Restaurant Meal Costs

23 24 26 28 32 36 38 40 40 41 42 44 45 45 45 47 49 49 49 50 50 50 50 51 52
52 53 55 55 56 57 58 58 60 62 64 66 66 66 69 70 71 78 79 80 80 81 84 89 91

Metro Area Restaurant Meal Costs

24 25 26 29 31 33 33 33 34 35 35 36 36 36 37 39 39 39 40 41 41 41 42 42 42
43 43 43 44 45 46 46 46 47 50 51 52 52 53 54 54 55 56 56 59 59 64 69 75 81

data as two ordered arrays. The ordered arrays in Table 2.2B allow some quick observations about the meal cost data to be made. Using Table 2.2B, one notices that meal costs at center city restaurants range from $23 to $91 and that meal costs at metro area restaurants range from $24 to $81.

When a numerical variable contains a large number of values, using an ordered array to make quick observation or reach conclusions about the data can be difficult. For such a variable, constructing a distribution table would be a better choice. Frequency, relative frequency, percentage, and cumulative distributions are among the types of distribution tables commonly used.

The Frequency Distribution

A **frequency distribution** tallies the values of a numerical variable into a set of numerically ordered **classes**. Each class groups a mutually exclusive range of values, called a **class interval**. Each value can be assigned to only one class, and every value must be contained in one of the class intervals.

To create a useful frequency distribution, one must consider how many classes would be appropriate for the data as well as determine a suitable *width* for each class interval. In general, a frequency distribution should have at least 5 and no more than 15 classes because having too few or too many classes provides little new information. To determine the **class interval width** [see Equation (2.1)], subtract the lowest value from the highest value and divide that result by the number of classes desired for the frequency distribution.

DETERMINING THE CLASS INTERVAL WIDTH

$$\text{Interval width} = \frac{\text{highest value} - \text{lowest value}}{\text{number of classes}} \tag{2.1}$$

For the center city restaurant meal cost data shown in Tables 2.2A and 2.2B, between 5 and 10 classes are acceptable, given the size (50) of that sample. From the center city restaurant meal costs ordered array in Table 2.2B, the difference between the highest value of $91 and the lowest value of $23 is $68. Using Equation (2.1), approximate the class interval width as follows:

$$\frac{68}{10} = 6.8$$

This result suggests an interval width of $6.80. However, the width should always be an amount that simplifies the interpretation of the frequency distribution. For this example, an interval width such as $5 or $10 would be appropriate. However, given the range of values, an interval width of $5 would create 15 classes, too many for the sample size of 50; choosing $10, which creates eight classes, would be the better choice.

Having chosen the class interval width, examine the data to establish **class boundaries** that properly and clearly define each class. To set class boundaries, consider boundary values that are simple to interpret and include all values being summarized. For the meal cost data, with the range of $23 to $91 (center city costs) and $24 to $81 (metro area costs) and a class interval of $10, set the lower class boundary of the first class to $20 for ease of readability. Define the first class as $20 but less than $30, the second class as $30 but less than $40, and so on, ending with the class $90 but less than $100. Table 2.3 uses these class intervals to present frequency distributions for the sample of 50 center city restaurant meal costs and the sample of 50 metro area restaurant meal costs.

Frequency distributions allow one to more easily make observations that support preliminary conclusions about the data. For example, Table 2.3 shows that the cost of center city restaurant meals is concentrated between $40 and $60, while the cost of metro area restaurant meals is concentrated between $30 and $60.

For some charts discussed later in this chapter, class intervals are identified by their **class midpoints**, the values that are halfway between the lower and upper boundaries of each class. For the frequency distributions shown in Table 2.3, the class midpoints are $25, $35, $45, $55,

TABLE 2.3
Frequency distributions for meal costs at 50 center city and 50 metro area restaurants

Meal Cost ($)	Center City Frequency	Metro Area Frequency
20 but less than 30	4	4
30 but less than 40	3	14
40 but less than 50	12	16
50 but less than 60	14	12
60 but less than 70	7	2
70 but less than 80	4	1
80 but less than 90	5	1
90 but less than 100	1	0
Total	50	50

$65, $75, $85, and $95. Note that well-chosen class intervals lead to class midpoints that are simple to read and interpret, as in this example.

If the data collected do not contain a large number of values, different sets of class intervals can create different impressions of the data. Such perceived changes will diminish as more data are collected. Likewise, choosing different lower and upper class boundaries can also affect impressions.

EXAMPLE 2.2

Frequency Distributions of the Three-Year Return Percentages for Growth and Value Funds

As a member of the company task force in The Choice *Is* Yours scenario (see page 38), you are examining the sample of 479 retirement funds stored in Retirement Funds. You want to compare the numerical variable 3YrReturn, the three-year percentage return of a fund, for the two subgroups, growth and value, defined for the categorical variable fund type. You construct separate frequency distributions for the growth funds and the value funds.

SOLUTION The three-year return for the growth funds is concentrated between 2.5 and 15, while the three-year return for the value funds is concentrated between 2.5 and 10.

TABLE 2.4
Frequency distributions of the three-year return percentages for growth and value funds

Three-Year Return Percentage	Growth Frequency	Value Frequency
−5.00 but less than −2.50	1	1
−2.50 but less than 0	0	1
0 but less than 2.50	14	8
2.50 but less than 5.00	27	20
5.00 but less than 7.50	60	69
7.50 but less than 10.00	109	67
10.00 but less than 12.50	68	7
12.50 but less than 15.00	26	0
15.00 but less than 17.50	1	0
Total	306	173

In the solution for Example 2.2, the total frequency is different for each group (306 and 173). When such totals differ among the groups being compared, you cannot compare the distributions directly as was done in Table 2.3 because of the chance that the table will be misinterpreted. For example, the frequencies for the class interval "5.00 but less than 7.50" look similar—60 and 69—but represent two very different parts of a whole: 60 out of 306 and 69 out of 173 or 19.61% and 39.88%, respectively. When the total frequency differs among the groups being compared, you construct either a relative frequency distribution or a percentage distribution.

Classes and Excel Bins

Microsoft Excel creates distribution tables using *bins* rather than classes. A **bin** is a range of values defined by a bin number, the upper boundary of the range. Unlike a class, the lower boundary is not explicitly stated but is deduced by the bin number that defines the preceding bin. Consider the bins defined by the bin numbers 4.99, 9.99, and 14.99. The first bin represents all values up to 4.99, the second bin all values greater than 4.99 (the preceding bin number) through 9.99, and the third bin all values greater than 9.99 (the preceding bin number) through 14.99.

Note that when using bins, the lower boundary of the first bin will always be negative infinity because that bin has no explicit lower boundary.

That makes the first Excel bin always much larger than the rest of the bins and violates the rule about having equal-sized classes. When you translate classes to bins to make use of certain Excel features, you must include an extra bin number as the first bin number. This extra bin number will always be a value slightly less than the lower boundary of your first class.

You translate your classes into a set of bin numbers that you enter into a worksheet column in ascending order. Tables 2.3 through 2.7 use classes stated in the form *"valueA but less than valueB."* For such classes, you create a set of bin numbers that are slightly lower than each *valueB* to approximate each class. For example, you translate the Table 2.4 classes

on page 45 as the set of bin numbers −5.01 (the "extra" first bin number that is slightly lower than −5, the lower boundary value of the first class), −2.51 (slightly less than −2.5 the *valueB* of the first class), −0.01, 2.49, 4.99, 7.49, 9.99, 12.49, 14.99, and 17.49 (slightly less than 17.50, the *valueB* of the eighth class).

For classes stated in the form "all values from *valueA* to *valueB*," you can approximate classes by choosing a bin number slightly more than each *valueB*. For example, you can translate the classes stated as 0.0 through 4.9, 5.0 through 9.9, 10.0 through 14.9, and 15.0 through 19.9, as the bin numbers: −0.01 (the extra first bin number), 4.99 (slightly more than 4.9), 9.99, 14.99, and 19.99 (slightly more than 19.9).

The Relative Frequency Distribution and the Percentage Distribution

Relative frequency and percentage distributions present tallies in ways other than as frequencies. A **relative frequency distribution** presents the relative frequency, or proportion, of the total for each group that each class represents. A **percentage distribution** presents the percentage of the total for each group that each class represents. When comparing two or more groups, knowing the proportion (or percentage) of the total for each group better facilitates comparisons than a table of frequencies for each group would. For example, for comparing meal costs, using Table 2.5 is better than using Table 2.3 on page 45, which displays frequencies.

TABLE 2.5
Relative frequency and percentage distributions of meal costs at center city and metro area restaurants

MEAL COST ($)	CENTER CITY		METRO AREA	
	Relative Frequency	Percentage	Relative Frequency	Percentage
20 but less than 30	0.08	8%	0.08	8%
30 but less than 40	0.06	6%	0.28	28%
40 but less than 50	0.24	24%	0.32	32%
50 but less than 60	0.28	28%	0.24	24%
60 but less than 70	0.14	14%	0.04	4%
70 but less than 80	0.08	8%	0.02	2%
80 but less than 90	0.10	10%	0.02	2%
90 but less than 100	0.02	2%	0.00	0%
Total	1.00	100.0%	1.00	100.0%

student TIP

Relative frequency columns always sum to 1.00. Percentage columns always sum to 100%.

The **proportion**, or **relative frequency**, in each group is equal to the number of *values* in each class divided by the total number of values. The percentage in each group is its proportion multiplied by 100%.

If there are 80 values and the frequency in a certain class is 20, the proportion of values in that class is 0.25 (20/80) and the percentage is 25% (0.25 × 100%).

COMPUTING THE PROPORTION OR RELATIVE FREQUENCY

The proportion, or relative frequency, is the number of *values* in each class divided by the total number of values:

$$\text{Proportion} = \text{relative frequency} = \frac{\text{number of values in each class}}{\text{total number of values}} \qquad \textbf{(2.2)}$$

Construct a relative frequency distribution by first determining the relative frequency in each class. For example, in Table 2.3 on page 45, there are 50 center city restaurants, and the cost per meal at 14 of these restaurants is between $50 and $60. Therefore, as shown in Table 2.5, the proportion (or relative frequency) of meals that cost between $50 and $60 at center city restaurants is

$$\frac{14}{50} = 0.28$$

Construct a percentage distribution by multiplying each proportion (or relative frequency) by 100%. Thus, the proportion of meals at center city restaurants that cost between $50 and $60 is 14 divided by 50, or 0.28, and the percentage is 28%.

Table 2.5 on page 46 presents the relative frequency distribution and percentage distribution of the cost of meals at center city and metro area restaurants. From Table 2.5, one notes that 14% of the center city restaurant meals cost between $60 and $70 as compared to 4% of the metro area restaurant meals, and that 6% of the center city restaurant meals cost between $30 and $40 as compared to 28% of the metro area restaurant meals. These observations support the conclusion that the cost of meals are higher for center city restaurants than for metro area restaurants.

EXAMPLE 2.3

Relative Frequency Distributions and Percentage Distributions of the Three-Year Return Percentages for Growth and Value Funds

As a member of the company task force in The Choice *Is* Yours scenario (see page 38), you want to compare the three-year return percentages for the growth and value retirement funds. You construct relative frequency distributions and percentage distributions for these funds.

SOLUTION From Table 2.6, you conclude that the three-year return percentage is higher for the growth funds than for the value funds. For example, 19.61% of the growth funds have returns between 5.00 and 7.50 as compared to 39.88% of the value funds, while 22.22% of the growth funds have returns between 10.00 and 12.50 as compared to 4.05% of the value funds.

TABLE 2.6
Relative frequency and percentage distributions of the three-year return percentages for growth and value funds

THREE-YEAR RETURN PERCENTAGE	GROWTH Relative Frequency	GROWTH Percentage	VALUE Relative Frequency	VALUE Percentage
−5.00 but less than −2.50	0.0033	0.33%	0.0058	0.58%
−2.50 but less than 0	0.0000	0.00%	0.0058	0.58%
0 but less than 2.50	0.0458	4.58%	0.0462	4.62%
2.50 but less than 5.00	0.0882	8.82%	0.1156	11.56%
5.00 but less than 7.50	0.1961	19.61%	0.3988	39.88%
7.50 but less than 10.00	0.3562	35.62%	0.3873	38.73%
10.00 but less than 12.50	0.2222	22.22%	0.0405	4.05%
12.50 but less than 15.00	0.0850	8.50%	0.0000	0.00%
15.00 but less than 17.50	0.0033	0.33%	0.0000	0.00%
Total	1.0000	100.00%	1.0000	100.00%

The Cumulative Distribution

The **cumulative percentage distribution** provides a way of presenting information about the percentage of values that are less than a specific amount. Use a percentage distribution as the basis to construct a cumulative percentage distribution.

For example, the restaurant meal cost study might seek to determine what percentage of the center city restaurant meals cost less than $40 or what percentage cost less than $50. Starting with the Table 2.5 meal cost percentage distribution for center city restaurants on page 46, combine the percentages of individual class intervals to form the cumulative percentage distribution. Table 2.7 presents this process and displays the cumulative percentages for each class. From this table, one sees that none (0%) of the meals cost less than $20, 8% of meals cost less than $30, 14% of meals cost less than $40 (because 6% of the meals cost between $30 and $40), and so on, until all 100% of the meals cost less than $100.

TABLE 2.7

Developing the cumulative percentage distribution for center city restaurant meal costs

| From Table 2.5: | | Percentage of Meal Costs That Are Less Than |
Class Interval	Percentage	the Class Interval Lower Boundary
20 but less than 30	8%	0% (there are no meals that cost less than 20)
30 but less than 40	6%	8% = 0 + 8
40 but less than 50	24%	14% = 8 + 6
50 but less than 60	28%	38% = 8 + 6 + 24
60 but less than 70	14%	66% = 8 + 6 + 24 + 28
70 but less than 80	8%	80% = 8 + 6 + 24 + 28 + 14
80 but less than 90	10%	88% = 8 + 6 + 24 + 28 + 14 + 8
90 but less than 100	2%	98% = 8 + 6 + 24 + 28 + 14 + 8 + 10
100 but less than 110	0%	100% = 8 + 6 + 24 + 28 + 14 + 8 + 10 + 2

Table 2.8 contains a cumulative percentage distribution for meal costs at the center city restaurants (from Table 2.7) as well as the metro area restaurants (calculations not shown). The cumulative distribution shows that the cost of metro area restaurant meals is lower than the cost of meals in center city restaurants. This distribution shows that 36% of the metro area restaurant meals cost less than $40 as compared to 14% of the meals at center city restaurants; 68% of the metro area restaurant meals cost less than $50, but only 38% of the center city restaurant meals do; and 92% of the metro area restaurant meals cost less than $60 as compared to 66% of such meals at the center city restaurants.

TABLE 2.8

Cumulative percentage distributions of the meal costs for center city and metro area restaurants

Meal Cost ($)	Percentage of Center City Restaurants Meals That Cost Less Than Indicated Amount	Percentage of Metro Area Restaurants Meals That Cost Less Than Indicated Amount
20	0	0
30	8	8
40	14	36
50	38	68
60	66	92
70	80	96
80	88	98
90	98	100
100	100	100

Unlike in other distributions, the rows of a cumulative distribution do not correspond to class intervals. (Recall that class intervals are mutually *exclusive*. The rows of cumulative distributions are not: The next row "down" *includes* all of the rows above it.) To identify a row, use the lower class boundaries from the class intervals of the cumulative percentage distribution as Table 2.8 does and read each row as "less than" an indicated value.

EXAMPLE 2.4

Cumulative Percentage Distributions of the Three-Year Return Percentages for Growth and Value Funds

As a member of the company task force in The Choice *Is* Yours scenario (see page 38), you want to continue comparing the three-year return percentages for the growth and value retirement funds. You construct cumulative percentage distributions for the growth and value funds.

SOLUTION The cumulative distribution in Table 2.9 indicates that returns are higher for the growth funds than for the value funds. The table shows that 33.33% of the growth funds and 57.23% of the value funds have returns below 7.5%. The table also reveals that 68.95% of the growth funds have returns below 10 as compared to 95.95% of the value funds.

TABLE 2.9
Cumulative percentage distributions of the three-year return percentages for growth and value funds

Three-Year Return Percentages	Growth Percentage Less Than Indicated Value	Value Percentage Less Than Indicated Value
−5.0	0.00%	0.00%
−2.5	0.33%	0.58%
0.0	0.33%	1.16%
2.5	4.90%	5.78%
5.0	13.73%	17.34%
7.5	33.33%	57.23%
10.0	68.95%	95.95%
12.5	91.18%	100.00%
15.0	99.67%	100.00%
17.5	100.00%	100.00%

PROBLEMS FOR SECTION 2.2

LEARNING THE BASICS

2.11 Construct an ordered array, given the following data from a sample of $n = 7$ midterm exam scores in accounting:

68 94 63 75 71 88 64

2.12 Construct an ordered array, given the following data from a sample of midterm exam scores in marketing:

88 78 78 73 91 78 85

2.13 Planning and preparing for the unexpected, especially in response to a security incident, is one of the greatest challenges faced by information technology professionals today. An incident is described as any violation of policy, law, or unacceptable act that involves information assets. Incident response (IR) teams should be evaluating themselves on metrics, such as incident detection or dwell time, to determine how quickly they can detect and respond to incidents in the environment. In a recent year, the SANS Institute surveyed organizations about internal response capabilities. The frequency distribution that summarizes the average time organizations took to detect incidents is:

Average Dwell Time	Frequency
Less than 1 day	166
Between 1 and less than 2 days	100
Between 2 and less than 8 days	124
Between 8 and less than 31 days	77
Between 31 and less than 90 days	59
90 days or more	65

Source: **bit.ly/2oZGXGx**.

a. What percentage of organizations took fewer than 2 days, on average, to detect incidents?
b. What percentage of organizations took between 2 and 31 days, on average, to detect incidents?

c. What percentage of organizations took 31 or more days, on average, to detect incidents?

d. What conclusions can you reach about average dwell time of incidents?

2.14 Data were collected on salaries of compliance specialists in corporate accounting firms. The salaries ranged from $61,000 to $261,000.

a. If these salaries were grouped into six class intervals, indicate the class boundaries.

b. What class interval width did you choose?

c. What are the six class midpoints?

APPLYING THE CONCEPTS

2.15 The file Work Hours Needed contains the average number of hours worked necessary to afford the cost of three people attending a National Basketball Association (NBA) game, inclusive of parking and food and beverage costs, at each of the 30 NBA arenas during a recent season.

Source: Data extracted from ValuePenguin, **bit.ly/2mGG2Li**.

a. Organize these costs as an ordered array.

b. Construct a frequency distribution and a percentage distribution for these costs.

c. Around which class grouping, if any, are the costs of attending a basketball game concentrated? Explain.

✓ SELF TEST 2.16 The file Utility contains the following data about the cost of electricity (in $) during July 2018 for a random sample of 50 one-bedroom apartments in a large city.

96	171	202	178	147	102	153	197	127	82
157	185	90	116	172	111	148	213	130	165
141	149	206	175	123	128	144	168	109	167
95	163	150	154	130	143	187	166	139	149
108	119	183	151	114	135	191	137	129	158

a. Construct a frequency distribution and a percentage distribution that have class intervals with the upper class boundaries $99, $119, and so on.

b. Construct a cumulative percentage distribution.

c. Around what amount does the monthly electricity cost seem to be concentrated?

2.17 How much time do commuters living in or near cities spend commuting to work each week? The file Commuting Time contains the average weekly commuting time in 30 U.S. cities.

Source: Data extracted from United States Census Bureau, "American Factfinder," **factfinder.census.gov**.

For the weekly commuting time data,

a. Construct a frequency distribution and a percentage distribution.

b. Construct a cumulative percentage distribution.

c. What conclusions can you reach concerning the weekly commuting time of Americans living in cities?

2.18 How do the average credit scores of people living in different American cities differ? The data in Credit Scores are an ordered array of the average credit scores of 2,570 American cities.

Source: Data extracted from **www.bizjournals.com/sanantonionews/2016/01/11/study-shows-cities-with-highest-and-lowest-credit.htm**.

a. Construct a frequency distribution and a percentage distribution.

b. Construct a cumulative percentage distribution.

c. What conclusions can you reach concerning the average credit scores of people living in different American cities?

2.19 One operation of a mill is to cut pieces of steel into parts that will later be used as the frame for front seats in an automobile. The steel is cut with a diamond saw and requires the resulting parts to be within ± 0.005 inch of the length specified by the automobile company. Data are collected from a sample of 100 steel parts and stored in Steel. The measurement reported is the difference in inches between the actual length of the steel part, as measured by a laser device, and the specified length of the steel part. For example, the first value, -0.002, represents a steel part that is 0.002 inch shorter than the specified length.

a. Construct a frequency distribution and a percentage distribution.

b. Construct a cumulative percentage distribution.

c. Is the steel mill doing a good job meeting the requirements set by the automobile company? Explain.

2.20 Call centers today play an important role in managing day-to-day business communications with customers. Call centers must be monitored with a comprehensive set of metrics so that businesses can better understand the overall performance of those centers. One key metric for measuring overall call center performance is *service level*, the percentage of calls answered by a human agent within a specified number of seconds. The file Service Level contains the following data for time, in seconds, to answer 50 incoming calls to a financial services call center:

16 14 16 19 6 14 15 5 16 18 17 22 6 18 10 15 12 6

19 16 16 15 13 25 9 17 12 10 5 15 23 11 12 14 24 9

10 13 14 26 19 20 13 24 28 15 21 8 16 12

a. Construct a frequency distribution and a percentage distribution.

b. Construct a cumulative percentage distribution.

c. What can you conclude about call center performance if the service level target is set as "80% of calls answered within 20 seconds"?

2.21 The financial services call center in Problem 2.20 also monitors *call duration*, the amount of time spent speaking to customers on the phone. The file Call Duration contains the following data for time, in seconds, spent by agents talking to 50 customers:

243 290 199 240 125 151 158 66 350 1141 251 385 239

139 181 111 136 250 313 154 78 264 123 314 135 99

420 112 239 208 65 133 213 229 154 377 69 170 261

230 273 288 180 296 235 243 167 227 384 331

a. Construct a frequency distribution and a percentage distribution.

b. Construct a cumulative percentage distribution.

c. What can you conclude about call center performance if a call duration target of less than 240 seconds is set?

2.22 The file Bulbs contains the life (in hours) of a sample of forty 6-watt light emitting diode (LED) light bulbs produced by Manufacturer A and a sample of forty 6-watt LED light bulbs produced by Manufacturer B.

a. Construct a frequency distribution and a percentage distribution for each manufacturer, using the class interval widths for each distribution on page 51.

Manufacturer A: 46,500 but less than 47,500; 47,500 but less than 48,500; and so on.

Manufacturer B: 47,500 but less than 48,500; 48,500 but less than 49,500; and so on.

b. Construct cumulative percentage distributions.

c. Which bulbs have a longer life—those from Manufacturer A or Manufacturer B? Explain.

2.23 The file **Drink** contains the following data for the amount of soft drink (in liters) in a sample of fifty 2-liter bottles:

2.109 2.086 2.066 2.075 2.065 2.057 2.052 2.044 2.036 2.038
2.031 2.029 2.025 2.029 2.023 2.020 2.015 2.014 2.013 2.014
2.012 2.012 2.012 2.010 2.005 2.003 1.999 1.996 1.997 1.992
1.994 1.986 1.984 1.981 1.973 1.975 1.971 1.969 1.966 1.967
1.963 1.957 1.951 1.951 1.947 1.941 1.941 1.938 1.908 1.894

a. Construct a cumulative percentage distribution.

b. On the basis of the results of (a), does the amount of soft drink filled in the bottles concentrate around specific values?

2.3 Visualizing Categorical Variables

Visualizing categorical variables involves making choices about data presentation. When visualizing a single categorical variable, think about what is to be highlighted about the data and whether the data are concentrated in only a few categories. To highlight how categories directly compare to each other, use a bar chart. To highlight how categories form parts of a whole, use a pie or doughnut chart. To present data that are concentrated in only a few of your categories, use a Pareto chart.

Thinking about data presentation is also important when visualizing two categorical variables. For two such variables, use a side-by-side chart to highlight direct comparisons and use a doughnut chart to highlight how parts form a whole.

The Bar Chart

A **bar chart** visualizes a categorical variable as a series of bars, each bar separated by space, called a gap. In a bar chart, each bar represents the tallies for a single category, and the length of each bar represents either the frequency or percentage of values for a category gap.

Figure 2.4 includes bar and pie chart visualizations of the Table 2.1 summary table on page 39 that tallies the results of asking 2,005 employed adults who work outside the home what action they took the last time they were sick. By viewing either of these charts, one can conclude that about half of the sick employees went to work and half did not. For this simple example, one could make the same conclusion in the same amount of time reviewing the summary table.

FIGURE 2.4
Table 2.1 summary table visualizations, bar chart (left) and pie chart (right)

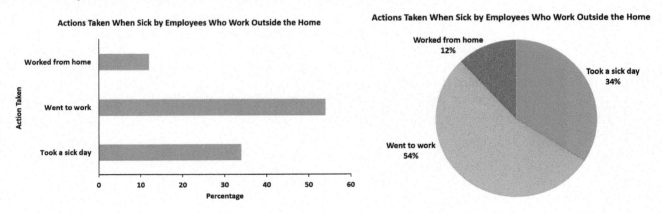

However, for a variable with many categories, or more complex data, visualizations will generally enable discovery of relationships among items to be made faster than from the equivalent tabular summaries.

EXAMPLE 2.5

Bar Chart of Levels of Risk of Retirement Funds

As a member of the company task force in The Choice *Is* Yours scenario (see page 38), you want to examine how the risk level categories in Figure 2.1 on page 39 compare to each other.

SOLUTION You construct the bar chart shown in Figure 2.5. You see that average risk is the largest category, followed by low risk followed by high risk.

FIGURE 2.5
Excel and Tableau bar charts of risk level for the sample of 479 retirement funds

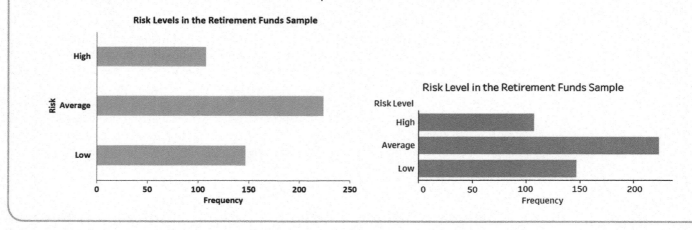

The Pie Chart and the Doughnut Chart

student TIP

Avoid using "3D" charts or any "exploded" chart in which one or more slices has been pulled away from the center. These forms cause visual distortions that can impede understanding the data.

Pie and **doughnut** (or **donut**) charts represent the tallies of each category of a categorical variable as parts of a circle. These parts, or slices, vary by the percentages of the whole for each category. Multiplying category percentages by 360, the number of degrees in a circle, determines the size of each slice, defined as the length of the arc (part of a circle) in degrees. For example, for the Table 2.1 summary table categories, the sizes of the slices would be went to work, 194.4 degrees (54% × 360); took a sick day, 122.4 degrees (34% × 360); and worked from home, 43.2 degrees (12% × 360).

Doughnut charts are pie charts with their centers cut out, creating a hole similar to the holes found in real doughnuts (hence the name). Some believe cutting out centers minimizes a common misperception of pie charts that occurs when people focus on the area of each pie slice and not the length of the arc of each slice. Because most would agree that many pie charts presented together provide an overwhelming visual experience that should be avoided (Edwardtufte), doughnut charts can be useful when more than one chart is presented together. Doughnut charts can also be used to visualize two variables, as this chapter explains later.

EXAMPLE 2.6

Pie Chart and
Doughnut Chart
of the Risk of
Retirement Funds

As a member of the company task force in The Choice *Is* Yours scenario (see page 38), you want to examine how the risk level categories in Figure 2.1 on page 39 form parts of a whole.

SOLUTION   You construct either the Figure 2.6 pie or doughnut chart. You can immediately see that almost half the funds have an average risk and that of the remaining funds, more have low risk than high risk. (A close reading of the labels reveals the actual percentages.)

FIGURE 2.6
Excel pie chart and
doughnut chart of the risk
of retirement funds

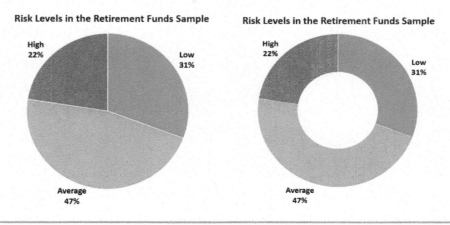

The Pareto Chart

Pareto charts help distinguish the categories that contain the largest tallies from the categories that contain the smallest. Originally developed by the nineteenth-century economist Vilfredo Pareto, these charts help visualize his principle (the **Pareto principle**) that 80% of the consequences result from 20% of the causes. That 20% of the causes are the "vital few" on which one should focus, according to Pareto. While Pareto charts usually do not demonstrate Pareto's 80/20 rule literally, such charts do identify the vital few from the "trivial many" and can be a useful tool today, especially when looking at the frequencies for a large set of categories. In quality management efforts, Pareto charts are very useful tools for prioritizing improvement efforts, such as when data that identify defective or nonconforming items are collected, as in the example that this section uses.

Pareto charts combine two different visualizations: a *vertical* bar chart and a **line graph**, a plot of connected points. The vertical bars represent the tallies for each category, arranged in descending order of the tallies. The line graph represents a cumulative percentage of the tallies from the first category through the last category. The line graph uses a percentage vertical scale, while the bars use either Pareto's original vertical frequency scale or a more recent adaptation that uses a percentage vertical scale line to allow both measurements to share the same scale. In cases with too many categories to display clearly in one chart, categories with the fewest tallies can be combined into a Miscellaneous or Other category and shown as the last (rightmost) bar.

Using Pareto charts can be an effective way to visualize data for studies that seek causes for an observed phenomenon. For example, consider a bank study team that wants to enhance the user experience of automated teller machines (ATMs). After initial investigation, the team identifies incomplete ATM transactions as a significant issue and decides to collect data about the causes of such transactions. Using the bank's own processing systems as a primary data source, causes of incomplete transactions are collected and then organized in the Table 2.10 summary table on page 54.

TABLE 2.10
Summary table of
causes of incomplete
ATM transactions
ATM Transactions

Cause	Frequency	Percentage
ATM malfunctions	32	4.42%
ATM out of cash	28	3.87%
Invalid amount requested	23	3.18%
Lack of funds in account	19	2.62%
Card unreadable	234	32.32%
Warped card jammed	365	50.41%
Wrong keystroke	23	3.18%
Total	724	100.00%

Source: Data extracted from A. Bhalla, "Don't Misuse the Pareto Principle," *Six Sigma Forum Magazine*, May 2009, pp. 15–18.

To separate out the "vital few" causes from the "trivial many" causes, the bank study team creates the Table 2.11 summary table. In this table, causes appear in descending order by frequency, as a Pareto chart requires, and the table includes columns for the percentages and cumulative percentages. The team then uses these columns to construct the Figure 2.7 Pareto chart.

TABLE 2.11
Ordered summary table
of causes of incomplete
ATM transactions

Cause	Frequency	Percentage	Cumulative Percentage
Warped card jammed	365	50.41%	50.41%
Card unreadable	234	32.32%	82.73%
ATM malfunctions	32	4.42%	87.15%
ATM out of cash	28	3.87%	91.02%
Invalid amount requested	23	3.18%	94.20%
Wrong keystroke	23	3.18%	97.38%
Lack of funds in account	19	2.62%	100.00%
Total	724	100.00%	

FIGURE 2.7
Excel Pareto chart
of incomplete ATM
transactions

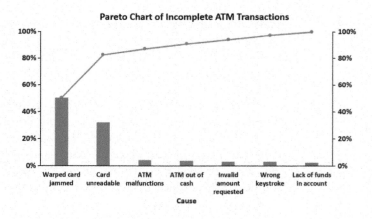

Because the categories in a Pareto chart are ordered by decreasing frequency of occurrence, the team can quickly see which causes contribute most to the problem of incomplete transactions. (Those causes would be the "vital few," and figuring out ways to avoid such causes would be, presumably, a starting point for improving the user experience of ATMs.) By following the cumulative percentage line in Figure 2.7, you see that the first two causes, warped card jammed (50.41%) and card unreadable (32.3%), account for 82.7% of the incomplete transactions. Attempts to reduce incomplete ATM transactions due to warped or unreadable cards should produce the greatest payoff.

EXAMPLE 2.7

Pareto Chart of the Actions Taken When Sick by Employees Who Work Outside the Home

Construct a Pareto Chart based on the Table 2.1 summary table that tallies the results of asking 2,005 employed adults who work outside the home what action they took the last time they were sick.

SOLUTION First, use the Table 2.1 data and create a new table in which the categories are ordered by descending frequency and columns for percentages and cumulative percentages for those ordered categories are included (not shown). From that table, create the Pareto chart. From the Figure 2.8 Pareto chart, one observes that two categories, choosing to go to work and taking a sick day, account for about 90% of employee actions taken.

FIGURE 2.8
Excel and Tableau Pareto charts of actions taken when sick

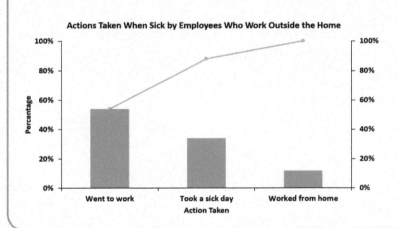
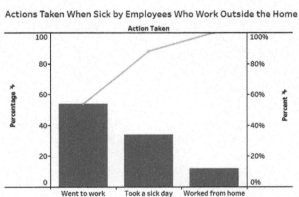

Visualizing Two Categorical Variables

When visualizing two categorical variables, use a side-by-side chart to highlight direct comparisons and use a doughnut chart to highlight how parts form a whole.

The Side-by-Side Chart A **side-by-side chart** visualizes two categorical variables by showing the bars that represent the categories of one variable set grouped by the categories of the second variable. For example, the Figure 2.9 side-by-side chart on page 56 visualizes the data for the levels of risk for growth and value funds shown in Figure 2.2 on page 40. In Figure 2.9, you see that a substantial portion of the growth funds have average risk. However, more of the value funds have low risk than average or high risk.

The Doughnut Chart   When visualizing two variables, the doughnut chart appears as two concentric rings, one inside the other, each ring containing the categories of one variable. In Figure 2.9, the doughnut chart of fund type and risk level highlights that the proportion of funds with average risk (darkest color) is different for growth and value.

FIGURE 2.9
Side-by-side bar chart and doughnut chart of fund type and risk level

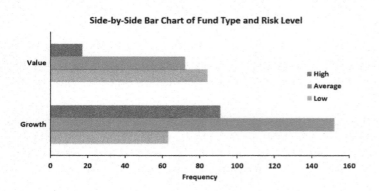

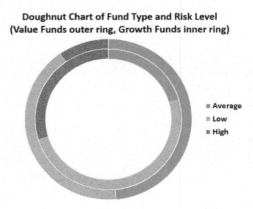

PROBLEMS FOR SECTION 2.3

APPLYING THE CONCEPTS

2.24 A survey of online shoppers revealed that in a recent year they bought more of their purchases online than in stores. The data in Online Shopping reveal how their purchases were made.

a. Construct a bar chart, a pie or doughnut chart, and a Pareto chart.
b. Which graphical method do you think is best for portraying these data?
c. What conclusions can you reach concerning how online shoppers make purchases?

2.25 How do college students spend their day? The 2016 American Time Use Survey for college students found the following results:

Activity	Percentage
Eating and Drinking	4%
Educational Activities	14%
Grooming	3%
Leisure and Sports	17%
Sleeping	37%
Traveling	6%
Working and Related Activities	10%
Other	9%

Source: Data extracted from **bit.ly/2qxIjcH**, accessed February 3, 2017.

a. Construct a bar chart, a pie or doughnut chart, and a Pareto chart.
b. Which graphical method do you think is best for portraying these data?
c. What conclusions can you reach concerning how college students spend their day?

2.26 The following table, stored in Energy Sources, contains the 2018 sources of electricity in the United States as reported by the Energy Information Administration.

Source	MWh (000)
Coal	1,146
Geothermal	17
Hydropower	292
Biomass	63
Natural gas	1,468
Nuclear	807
Other sources	14
Petroleum coke and liquids	26
Solar	67
Wind	275

Source: **www.eia.gov**.

a. Construct a Pareto chart.
b. What percentage of power is derived from coal, nuclear, or natural gas?
c. Construct a pie chart.
d. For this table data, do you prefer using a Pareto chart or a pie chart? Why?

2.27 The Consumer Financial Protection Bureau reports on consumer financial product and service complaint submissions by state, category, and company. The following table, stored in CFPB Categories, represents complaints received from Louisiana consumers by complaint category in a recent year.

Category	Number of Complaints
Bank Account or Service	202
Consumer Loan	132
Credit Card	175
Credit Reporting	581
Debt Collection	486
Mortgage	442
Student Loan	75
Other	72

Source: Data extracted from **bit.ly/2pR7ryO**.

a. Construct a Pareto chart for the categories of complaints.

b. Discuss the "vital few" and "trivial many" reasons for the categories of complaints.

The following table, stored in `CFPB Companies`, represents complaints received from Louisiana consumers by most-complained-about companies in a recent year.

Company	Number of Complaints
Bank of America	42
Capital One	93
Citibank	59
Ditech Financial	31
Equifax	217
Experian	177
JPMorgan	128
Nationstar Mortgage	39
Navient	38
Ocwen	41
Synchrony	43
Trans-Union	168
Wells Fargo	77

c. Construct a bar chart and a pie chart for the complaints by company.

d. What graphical method (bar or pie chart) do you think is best suited for portraying these data?

2.28 The following table, stored in `Residential Electricity Usage`, indicates the percentage of residential electricity consumption in the United States in 2018, organized by type of use.

Type of Use	Percentage
Clothes dryer	60
Clothes washer	10
Computers and related equipment	26
Cooking	16
Dishwashers	7
Freezers	20
Furniture pans and boiler circ. pumps	25
Lighting	91
Refrigerators	87
Space cooling	214
Space heating	207
Televisions and related equipment	62
Water heating	174
Other uses	460

Source: U.S. Energy Information Administration, **bit.ly/2UBvsK0**.

a. Construct a bar chart, a pie chart, and a Pareto chart.

b. Which graphical method do you think is best suited for portraying the table data?

c. What conclusions can you reach concerning residential electricity consumption in the United States?

2.29 A recent survey of insurance professionals explores the use of technology in the industry. The file `Technologies` contains the responses to the question that asked what technologies these professionals expect to be most used by the insurance industry in the coming year.

Technology	Frequency
Wearable technology	9
Blockchain technology	9
Artificial intelligence	17
IoT: retail insurance	23
IoT: commercial insurance	5
Social media	27

Source: Data extracted from **bit.ly/2qxMFRj**.

a. Construct a bar chart and a pie chart.

b. What conclusions can you reach concerning expected technology usage in the insurance industry?

2.30 A survey of 1,520 American adults asked, "Do you feel overloaded with too much information?" The results indicate that 23% of females feel information overload compared to 17% of males. The results are:

	GENDER		
OVERLOADED	Male	Female	Total
Yes	134	170	304
No	651	565	1,216
Total	785	735	1,520

Source: Data extracted from **bit.ly/2pR5bHZ**.

a. Construct a side-by-side bar chart of overloaded with too much information and gender.

b. What conclusions can you reach from this chart?

2.31 A study of selected Kickstarter projects showed that overall a majority were successful, achieving their goal and raising, at a minimum, the targeted amounts. In an effort to identify project types that influence success, selected projects were subdivided into project categories (Film & Video, Games, Music, and Technology). The results are as follows:

	OUTCOME		
CATEGORY	Successful	Not Successful	Total
Film & Video	21,759	36,805	58,564
Games	9,329	18,238	27,567
Music	24,285	24,377	48,662
Technology	5,040	20,555	25,595
Total	60,413	99,975	160,388

Source: **Kickstarter.com, kickstarter.com/help/stats**.

a. Construct a side-by-side bar chart and a doughnut chart of project outcome and category.

b. What conclusions concerning the pattern of successful Kickstarter projects can you reach?

2.32 Ipsos polled 1,004 adults asking them, "If purchasing a used car made certain upgrades or features more affordable, what would be your preferred luxury upgrade?" The summarized results show that 9% of males and 14% of females answered window tinting.

Source: Ipsos, "Safety Technology Tops the List of Most Desired Features Should They Be More Affordable When Purchasing a Used Car—Particularly Collision Avoidance," available at **bit.ly/2ufbS8Z**.

The poll description did not state the sample sizes of males and females. Suppose that both sample sizes were 502 and that 46 of 502 males and 71 of 502 females reported window tinting as their preferred luxury upgrade of choice.

a. Create a side-by-side chart and a doughnut chart of preferred luxury upgrade and gender.

b. What do the chart results tell you about luxury upgrade differences between males and females?

2.4 Visualizing Numerical Variables

A variety of techniques that show the distribution of values visualize the data for a numerical variable. These techniques include the stem-and-leaf display, the histogram, the percentage polygon, and the cumulative percentage polygon (ogive), all discussed in this section, as well as the boxplot, which requires descriptive summary measures that Section 3.3 discusses.

The Stem-and-Leaf Display

A **stem-and-leaf display** visualizes data by presenting the data as one or more row-wise *stems* that represent a range of values. In turn, each stem has one or more *leaves* that branch out to the right of their stem and represent the values found in that stem. For stems with more than one leaf, the leaves are arranged in ascending order.

Stem-and-leaf displays show how the data are distributed and where concentrations of data exist. Leaves typically present the last significant digit of each value, but sometimes values are rounded. For example, the following meal costs (in $) for 15 classmates who had lunch at a fast casual restaurant were collected and stored in Fast Casual :

7.42 6.29 5.83 6.50 8.34 9.51 7.10 6.80 5.90 4.89 6.50 5.52 7.90 8.30 9.60

To construct the stem-and-leaf display, one would use whole dollar amounts as the stems and round the cents to one decimal place as the leaves. For the first value, 7.42, the stem would be 7, and its leaf is 4. For the second value, 6.29, the stem would be 6 and its leaf 3. The completed stem-and-leaf display for these data with the leaves ordered within each stem is:

student TIP

A stem-and-leaf display turned sideways looks like a histogram.

4	9
5	589
6	3558
7	149
8	33
9	56

EXAMPLE 2.8

Stem-and-Leaf Display of the Three-Year Return Percentages for the Value Funds

As a member of the company task force in The Choice *Is* Yours scenario (see page 38), you want to study the past performance of the value funds. One measure of past performance is the numerical variable 3YrReturn, the three-year return percentage. Using the data from the 173 value funds, you want to visualize this variable as a stem-and-leaf display.

SOLUTION Figure 2.10 on page 59 presents an Excel stem-and-leaf display of the three-year return percentage for value funds. You observe

- the lowest three-year return was -2.6.
- the highest three-year return was 11.9.
- the three-year returns were concentrated between 6 and 9.
- very few of the three-year returns were above 11.
- the distribution of the three-year returns appears to have more high values than low values.

▶ *(continued)*

FIGURE 2.10
Excel stem-and-leaf
display of the three-
year return percentages
for value funds

Growth Funds Three-Year Return Percentage

Stem unit 1	
-2	7
-1	0
-0	
0	
1	3799
2	01345566
3	118899
4	1223555566
5	344555566666778999
6	0000111222345555666666678888999
7	0000001112223334445555556688888999
8	000122233444444555556666666666667778888999
9	00133379999
10	00034667
11	9
12	0

The Histogram

A **histogram** visualizes data as a vertical bar chart in which each bar represents a class interval from a frequency or percentage distribution. A histogram displays the numerical variable along the horizontal (X) axis and uses the vertical (Y) axis to represent either the frequency or the percentage of values per class interval. There are never any gaps between adjacent bars in a histogram.

Figure 2.11 visualizes the data of Table 2.3 on page 45, meal costs at center city and metro area restaurants, as a pair of frequency histograms. The histogram for center city restaurants shows that the cost of meals is concentrated between approximately $40 and $60. Ten meals at center city restaurants cost $70 or more. The histogram for metro area restaurants shows that the cost of meals is concentrated between $30 and $60. Very few meals at metro area restaurants cost more than $60.

FIGURE 2.11
Excel (left) and Tableau frequency histograms for meal costs at center city and metro area restaurants

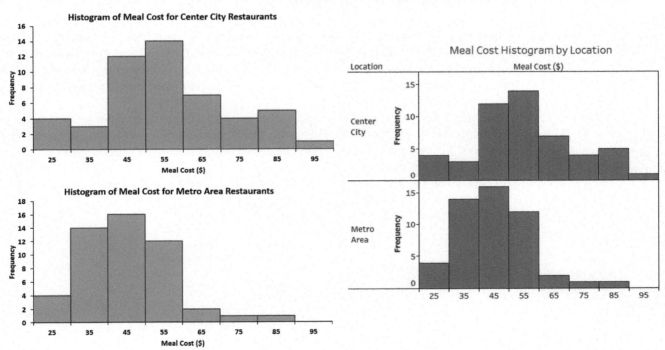

EXAMPLE 2.9

Histograms of the
Three-Year Return
Percentages for the
Growth and Value
Funds

As a member of the company task force in The Choice *Is* Yours scenario (see page 38), you seek
to compare the past performance of the growth funds and the value funds, using the three-year
return percentage variable. Using the data from the sample of 479 funds, you construct histograms for the growth and the value funds to create a visual comparison.

SOLUTION Figure 2.12 displays frequency histograms for the three-year return percentages
for the growth and value funds.

FIGURE 2.12

Excel frequency histograms for the three-year return percentages for the growth and value funds

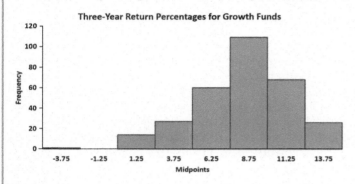
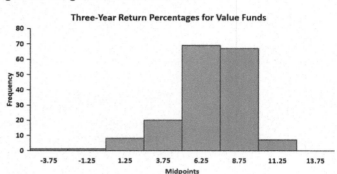

Reviewing the histograms in Figure 2.12 leads you to conclude that the returns were higher
for the growth funds than for value funds. The return for the growth funds is more concentrated
between 5 and 12.5 while the return for the value funds is more concentrated between 5 and 10.

The Percentage Polygon

When using a categorical variable to divide the data of a numerical variable into two or more
groups, you visualize data by constructing a **percentage polygon**. This chart uses the midpoints
of each class interval to represent the data of each class and then plots the midpoints, at their
respective class percentages, as points on a line along the *X* axis. A percentage polygon enables a
direct comparison between groups that is easier to interpret than a pair of histograms. (Compare
Figure 2.11 on page 59 to Figure 2.13.)

Figure 2.13 displays percentage polygons for the cost of meals at center city and metro area
restaurants. The same observations from this chart can be made as were made when examining
the pair of histograms in Figure 2.11. The center city meal cost is concentrated between $40
and $60, while the metro area meal cost is concentrated between $30 and $60. However, the
polygons allow one to more easily identify which class intervals have similar percentages for
the two groups and which do not.

FIGURE 2.13

Excel percentage
polygons of meal costs for
center city and metro area
restaurants

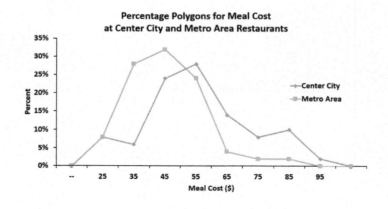

The polygons in Figure 2.13 have points whose values on the X axis represent the midpoint of the class interval. For example, look at the points plotted at $X = 35$ ($35). The point for meal costs at center city restaurants (the lower one) show that 6% of the meals cost between $30 and $40, while the point for the meal costs at metro area restaurants (the higher one) shows that 28% of meals at these restaurants cost between $30 and $40.

When constructing polygons or histograms, the vertical Y axis should include zero to avoid distorting the character of the data. The horizontal X axis does not need to show the zero point for the numerical variable, but a major portion of the axis should be devoted to the entire range of values for the variable.

EXAMPLE 2.10

Percentage Polygons of the Three-Year Return Percentages for the Growth and Value Funds

As a member of the company task force in The Choice *Is* Yours scenario (see page 38), you seek to compare the past performance of the growth funds and the value funds using the three-year return percentage variable. Using the data from the sample of 479 funds, you construct percentage polygons for the growth and value funds to create a visual comparison.

SOLUTION Figure 2.14 displays percentage polygons of the three-year return percentage for the growth and value funds.

FIGURE 2.14
Excel percentage polygons of the three-year return percentages for the growth and value funds

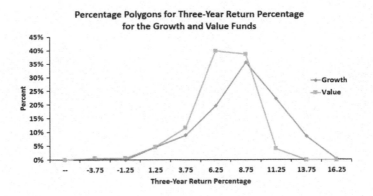

Figure 2.14 shows that the growth funds polygon is to the right of the value funds polygon. This allows you to conclude that the three-year return percentage is higher for growth funds than for value funds. The polygons also show that the return for growth funds is concentrated between 5 and 12.50, and the return for the value funds is concentrated between 5 and 10.

The Cumulative Percentage Polygon (Ogive)

The **cumulative percentage polygon**, or **ogive**, uses the cumulative percentage distribution discussed in Section 2.2 to plot the cumulative percentages along the Y axis. Unlike the percentage polygon, the lower boundaries of the class interval for the numerical variable are plotted, at their respective class percentages as points on a line along the X axis.

Figure 2.15 on page 62 shows cumulative percentage polygons of meal costs for center city and metro area restaurants. In this chart, the lower boundaries of the class intervals (20, 30, 40, etc.) are approximated by the upper boundaries of the previous bins (19.99, 29.99, 39.99, etc.). The curve of the cost of meals at the center city restaurants is located to the right of the curve for the metro area restaurants. This indicates that the center city restaurants have fewer meals that cost less than a particular value. For example, 38% of the meals at center city restaurants cost less than $50, as compared to 68% of the meals at metro area restaurants.

FIGURE 2.15

Excel cumulative percentage polygons of meal costs for center city and metro area restaurants

In Microsoft Excel, you approximate the lower boundary by using the upper boundary of the previous bin.

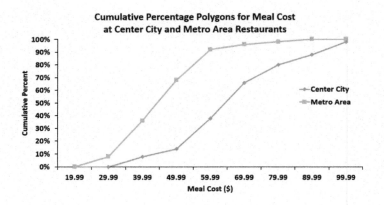

EXAMPLE 2.11

Cumulative Percentage Polygons of the Three-Year Return Percentages for the Growth and Value Funds

As a member of the company task force in The Choice *Is* Yours scenario (see page 38), you seek to compare the past performance of the growth funds and the value funds using the three-year return percentage variable. Using the data from the sample of 479 funds, you construct cumulative percentage polygons for the growth and the value funds.

SOLUTION Figure 2.16 displays cumulative percentage polygons of the three-year return percentages for the growth and value funds.

FIGURE 2.16

Excel cumulative percentage polygons of the three-year return percentages for the growth and value funds

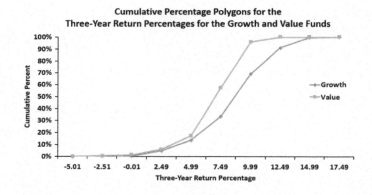

The cumulative percentage polygons in Figure 2.16 show that the curve for the three-year return percentage for the growth funds is located to the right of the curve for the value funds. This allows you to conclude that the growth funds have fewer three-year return percentages that are higher than a particular value. For example, 68.95% of the growth funds had three-year return percentages below 10, as compared to 95.95% of the value funds. You can conclude that, in general, the growth funds outperformed the value funds in their three-year returns.

PROBLEMS FOR SECTION 2.4

LEARNING THE BASICS

2.33 Construct a stem-and-leaf display, given the following data from a sample of midterm exam scores in finance:

54 69 98 93 53 74

2.34 Construct an ordered array, given the following stem-and-leaf display from a sample of $n = 7$ midterm exam scores in information systems:

5	0
6	
7	446
8	19
9	2

APPLYING THE CONCEPTS

2.35 The following is a stem-and-leaf display representing the gallons of gasoline purchased (with leaves in tenths of gallons) for a sample of 25 cars that use a particular service station on the New Jersey Turnpike:

9	147
10	02238
11	125566777
12	223489
13	02

a. Construct an ordered array.
b. Which of these two displays seems to provide more information? Discuss.
c. How many gallons are most likely to be purchased?
d. Is there a concentration of the purchase amounts in the center of the distribution?

✓ SELF TEST **2.36** The file **Work Hours Needed** contains the average number of hours worked necessary for three people to afford to attend an NBA game, inclusive of parking and food and beverage costs, at each of the 30 NBA basketball arenas during a recent season.

Source: from ValuePenguin, **bit.ly/2mGG2Li**.

a. Construct a stem-and-leaf display.
b. Around what value, if any, are the costs of attending a basketball game concentrated? Explain.

2.37 The file **Mobile Speed** contains the overall download and upload speeds in Mbps for nine carriers in the United States.

Source: Data extracted from Tom's Guide, "Fastest Wireless Network 2019: It's Not Even close," **bit.ly/2PcGiQE**.

For download and upload speeds speeds separately:

a. Construct an ordered array.
b. Construct a stem-and-leaf display.
c. Does the ordered array or the stem-and-leaf display provide more information? Discuss.
d. Around what value, if any, are the download and upload speeds concentrated? Explain.

2.38 The file **Utility** contains the following data about the cost of electricity during July of a recent year for a random sample of 50 one-bedroom apartments in a large city:

96	171	202	178	147	102	153	197	127	82
157	185	90	116	172	111	148	213	130	165
141	149	206	175	123	128	144	168	109	167
95	163	150	154	130	143	187	166	139	149
108	119	183	151	114	135	191	137	129	158

a. Construct a histogram and a percentage polygon.
b. Construct a cumulative percentage polygon.
c. Around what amount does the monthly electricity cost seem to be concentrated?

2.39 As player salaries have increased, the cost of attending baseball games has increased dramatically. The following histogram visualizes the total cost (in $) for two tickets, two beers, two hot dogs, and parking for one vehicle at each of the 30 Major League Baseball parks during a recent season stored in **Baseball Game Cost** .

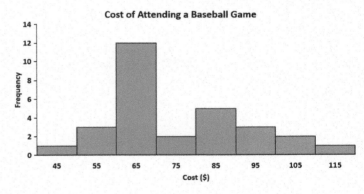

What conclusions can you reach concerning the cost of attending a baseball game at different ballparks?

2.40 The following histogram and cumulative percentage polygon (page 64) visualize the data about the property taxes on a $176K home for the 50 states and the District of Columbia, stored in **Property Taxes** .

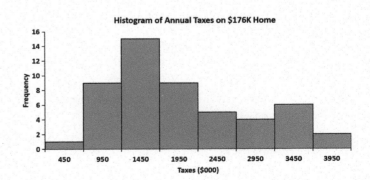

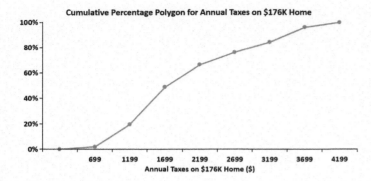

Cumulative Percentage Polygon for Annual Taxes on $176K Home

What conclusions can you reach concerning the property taxes per capita?

2.41 How much time do Americans living in cities spend commuting to work each week? The file **Commuting Time** contains these times for 30 cities.

Source: Data extracted from "American Factfinder," **factfinder.census.gov**.

For the time Americans living in cities spend commuting to work each week:
a. Construct a percentage histogram.
b. Construct a cumulative percentage polygon.
c. What conclusions can you reach concerning the time Americans living in cities spend commuting to work each week?

2.42 How do the average credit scores of people living in various cities differ? The file **Credit Scores** contains an ordered array of the average credit scores of 2,570 American cities.

Source: Data extracted from "Study shows cities with highest and lowest credit scores," accessed at **bit.ly/2uubZfX**.

a. Construct a percentage histogram.
b. Construct a cumulative percentage polygon.
c. What conclusions can you reach concerning the average credit scores of people living in different American cities?

2.43 One operation of a mill is to cut pieces of steel into parts that will later be used as the frame for front seats in an automobile. The steel is cut with a diamond saw and requires the resulting parts to be within ±0.005 inch of the length specified by the automobile company. The measurement reported is the difference in inches between the actual length of the steel part, as measured by a laser device, and the specified length of the steel part. For example, the first value, −0.002, represents a steel part that is 0.002 inch shorter than the specified length. Data are collected from a sample of 100 steel parts and stored in **Steel**.
a. Construct a percentage histogram.
b. Is the steel mill doing a good job meeting the requirements set by the automobile company? Explain.

2.44 Call centers today play an important role in managing day-to-day business communications with customers. Call centers must be monitored with a comprehensive set of metrics so that businesses can better understand the overall performance of those centers. One key metric for measuring overall call center performance is *service level*, the percentage of calls answered by a human agent within a specified number of seconds. The file **Service Level** contains the following data for time, in seconds, to answer 50 incoming calls to a financial services call center:

16 14 16 19 6 14 15 5 16 18 17 22 6 18 10 15 12
 6 19 16 16 15 13 25 9 17 12 10 5 15 23 11 12 14
24 9 10 13 14 26 19 20 13 24 28 15 21 8 16 12

a. Construct a percentage histogram and a percentage polygon.
b. Construct a cumulative percentage polygon.
c. What can you conclude about call center performance if the service level target is set as "80% of calls answered within 20 seconds"?

2.45 The financial services call center in Problem 2.44 also monitors call duration, which is the amount of time spent speaking to customers on the phone. The file **Call Duration** contains the following data for the time, in seconds, spent by agents talking to 50 customers.

243 290 199 240 125 151 158 66 350 1141 251 385 239
139 181 111 136 250 313 154 78 264 123 314 135 99
420 112 239 208 65 133 213 229 154 377 69 170 261
230 273 288 180 296 235 243 167 227 384 331

a. Construct a percentage histogram and a percentage polygon.
b. Construct a cumulative percentage polygon.
c. What can you conclude about call center performance if a call duration target of less than 240 seconds is set?

2.46 The file **Bulbs** contains the life (in hours) of a sample of forty 6-watt LED (light emitting diode) light bulbs produced by Manufacturer A and a sample of forty 6-watt LED light bulbs produced by Manufacturer B.

Use the following class interval widths for each distribution:

Manufacturer A: 46,500 but less than 47,500; 47,500 but less than 48,500; and so on.
Manufacturer B: 47,500 but less than 48,500; 48,500 but less than 49,500; and so on.

a. Construct percentage histograms on separate graphs and plot the percentage polygons on one graph.
b. Plot cumulative percentage polygons on one graph.
c. Which manufacturer has bulbs with a longer life—Manufacturer A or Manufacturer B? Explain.

2.47 The data stored in **Drink** represents the amount of soft drink in a sample of fifty 2-liter bottles.
a. Construct a histogram and a percentage polygon.
b. Construct a cumulative percentage polygon.
c. On the basis of the results in (a) and (b), does the amount of soft drink filled in the bottles concentrate around specific values?

2.5 Visualizing Two Numerical Variables

Visualizing two numerical variables together can reveal possible relationships between the two variables and serve as a basis for applying the methods that Chapters 13–15 discuss. To visualize two numerical variables, use a scatter plot. For the special case in which one of the two variables represents the passage of time, use a time-series plot.

The Scatter Plot

A **scatter plot** explores the possible relationship between two numerical variables by plotting the values of one numerical variable on the horizontal, or X, axis and the values of the second numerical variable on the vertical, or Y, axis. For example, a marketing analyst could study the effectiveness of advertising by comparing advertising expenses and sales revenues of 50 stores by using the X axis to represent advertising expenses and the Y axis to represent sales revenues.

A scatter plot of a pair of variables might show a positive relationship, in which increasing the value of the X variable increases the value of the Y variable; a negative relationship, in which increasing the value of the X variable *decreases* the value of the Y variable; or show no apparent relationship between the variables. Relationships uncovered can be described as either strong or weak, and including a *trend line* (see Chapter 13) in a scatter plot can both highlight the relationship and show whether the relationship is strong (steeper slopes) or weak (shallower slopes).

EXAMPLE 2.12

Scatter Plot for NBA Investment Analysis

Suppose that you are an investment analyst who has been asked to consider the valuations of the 30 NBA professional teams. You seek to know if the value of a team reflects its revenue. You collect revenue (in $millions) and current valuation data (in $billions) for all 30 NBA teams, organize the data as Table 2.12, and store the data in NBA Financial.

TABLE 2.12
Revenue and current valuations for NBA teams

Team Code	Revenue ($millions)	Current Value ($billions)	Team Code	Revenue ($millions)	Current Value ($billions)	Team Code	Revenue ($millions)	Current Value ($billions)
ATL	215	1.3	HOU	323	2.3	OKC	241	1.5
BOS	287	2.8	IND	222	1.4	ORL	223	1.3
BKN	290	2.4	LAC	258	2.2	PHI	268	1.5
CHA	218	1.3	LAL	398	3.7	PHX	235	1.7
CHI	287	2.9	MEM	213	1.2	POR	246	1.6
CLE	302	1.3	MIA	258	1.8	SAC	263	1.6
DAL	287	2.3	MIL	204	1.4	SAS	262	1.7
DEN	222	1.4	MIN	223	1.3	TOR	275	1.7
DET	235	1.3	NOH	214	1.2	UTA	243	1.4
GSW	401	3.5	NYK	443	4.0	WAS	255	1.6

SOLUTION You begin your review by constructing a scatter plot of revenue and current value, to uncover any possible relationships between these variables, placing revenue on the X axis and current value on the Y axis (see Figure 2.17 on page 66). The scatter plot suggests that a strong positive (increasing) relationship exists between revenue and the current value of a team. In other words, teams that generate a smaller amount of revenue have a lower value, while teams that generate higher revenue have a higher value.

The Figure 2.17 Excel scatter plot also includes a trend line. This trend line highlights the strong positive relationship between revenue and current value for NBA teams.

▶(continued)

FIGURE 2.17
Excel (with trend line) and Tableau scatter plots for NBA revenue and current evaluation study

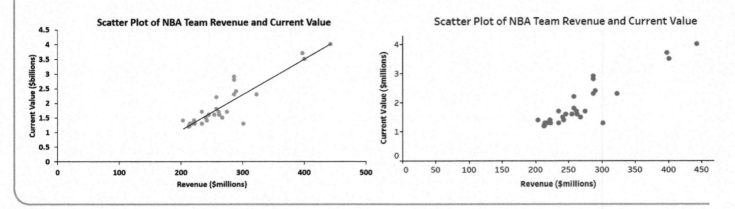

The Time-Series Plot

A **time-series plot** plots the values of a numerical variable on the Y axis and the time period associated with each numerical value on the X axis. A time-series plot can help visualize trends in data that occur over time.

EXAMPLE 2.13

Time-Series Plot for Movie Revenues

As an investment analyst who specializes in the entertainment industry, you are interested in discovering any long-term trends in movie revenues. You collect the annual revenues (in $billions) for movies released from 1995 to 2018, organize the data as Table 2.13, and store the data in Movie Revenues .

To see if there is a trend over time, you construct the time-series plot shown in Figure 2.18 on page 67.

TABLE 2.13
Movie revenues (in $billions) from 1995 to 2018

Year	Revenue ($billions)	Year	Revenue ($billions)	Year	Revenue ($billions)
1995	5.29	2003	9.35	2011	10.19
1996	5.59	2004	9.11	2012	10.83
1997	6.51	2005	8.93	2013	10.90
1998	6.79	2006	9.25	2014	10.36
1999	7.30	2007	9.63	2015	11.13
2000	7.48	2008	9.95	2016	11.38
2001	8.13	2009	10.65	2017	11.07
2002	9.19	2010	10.54	2018	11.89

Source: Data extracted from **www.boxofficemojo.com/yearly/**.

SOLUTION From Figure 2.18, you see that there was a steady increase in the annual movie revenues between 1995 and 2018, followed by an overall upward trend, which includes some downturns, reaching new highs in both 2018. During that time, the revenues increased from under $6 billion in 1995 to almost $12 billion in 2018.

▶(*continued*)

FIGURE 2.18
Excel and Tableau time-series plot of annual movie revenues from 1995 to 2018

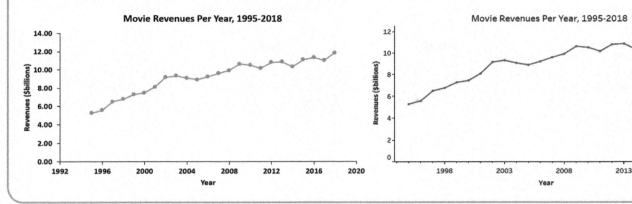

PROBLEMS FOR SECTION 2.5

LEARNING THE BASICS

2.48 The following is a set of data from a sample of $n = 11$ items:

X: 7 5 8 3 6 0 2 4 9 5 8

Y: 1 5 4 9 8 0 6 2 7 5 4

a. Construct a scatter plot.
b. Is there a relationship between X and Y? Explain.

2.49 The following is a series of annual sales (in $millions) over an 11-year period (2008 to 2018):

Year: 2008 2009 2010 2011 2012 2013 2014 2015 2016 2017 2018

Sales: 13.0 17.0 19.0 20.0 20.5 20.5 20.5 20.0 19.0 17.0 13.0

a. Construct a time-series plot.
b. Does there appear to be any change in annual sales over time? Explain.

APPLYING THE CONCEPTS

√SELF TEST **2.50** Movie companies need to predict the gross receipts of individual movies once a movie has debuted. The following results, stored in Potter Movies , are the first weekend gross, the U.S. gross, and the worldwide gross (in $millions) of the eight Harry Potter movies:

Title	First Weekend ($millions)	U.S. Gross ($millions)	Worldwide Gross ($millions)
Sorcerer's Stone	90.295	317.558	976.458
Chamber of Secrets	88.357	261.988	878.988
Prisoner of Azkaban	93.687	249.539	795.539
Goblet of Fire	102.335	290.013	896.013
Order of the Phoenix	77.108	292.005	938.469
Half-Blood Prince	77.836	301.460	934.601
Deathly Hallows Part I	125.017	295.001	955.417
Deathly Hallows Part II	169.189	381.011	1,328.111

Source: Data extracted from **www.the-numbers.com/interactive/comp-Harry-Potter.php**.

a. Construct a scatter plot with first weekend gross on the X axis and U.S. gross on the Y axis.
b. Construct a scatter plot with first weekend gross on the X axis and worldwide gross on the Y axis.
c. What can you say about the relationship between first weekend gross and U.S. gross and first weekend gross and worldwide gross?

2.51 Data were collected on the typical cost of dining at American-cuisine restaurants within a 1-mile walking distance of a hotel located in a large city. The file Bundle contains the typical cost (a per transaction cost in $) as well as a Bundle score, a measure of overall popularity and customer loyalty, for each of 40 selected restaurants.

Source: Data extracted from **www.bundle.com** via the link **on-msn.com/MnlBxo**.

a. Construct a scatter plot with Bundle score on the X axis and typical cost on the Y axis.
b. What conclusions can you reach about the relationship between Bundle score and typical cost?

2.52 The file Mobile Speed contains the overall download and upload speeds in Mbps for eight carriers in the United States.

Source: Data extracted from "Fastest Wireless Network 2019: It's Not Even close." tom's guide Staff, March 22, 2019

a. Do you think that carriers with a higher overall download speed also have a higher overall upload speed?
b. Construct a scatter plot with download speed on the X axis and upload speed on the Y axis.
c. Does the scatter plot confirm or contradict your answer in (a)?

2.53 A Pew Research Center survey found that smartphone ownership is growing rapidly around the world, although not always equally, and social media use is somewhat less widespread, even as Internet use has grown in emerging economies. The file World Smartphone contains the level of smartphone ownership, measured as the percentage of adults polled who report owning a smartphone. The file World Social Media contains the level of social media networking, measured as the percentage of adults who use social media sites, as well as the GDP as purchasing power parity (PPP, current international $) per capita for each of 27 advanced and emerging countries.

Source: Data extracted from Pew Research Center, "Smartphone Ownership is Growing Rapidly Around the World, but Not Always Equally," February 5, 2019, **bit.ly/2IltMOf**.

a. Construct a scatter plot with GDP (PPP) per capita on the X axis and smartphone ownership on the Y axis.
b. What conclusions can you reach about the relationship between GDP and smartphone usage?
c. Construct a scatter plot with GDP (PPP) per capita on the X axis and social media usage on the Y axis.
d. What conclusions can you reach about the relationship between GDP and social media usage?

2.54 How have stocks performed in the past? The following table presents the data stored in Stock Performance and shows the performance of a broad measure of stocks (by percentage) for each decade from the 1830s through 2018:

Decade	Perf (%)	Decade	Perf (%)
1830s	2.8	1920s	13.3
1840s	12.8	1930s	−2.2
1850s	6.6	1940s	9.6
1860s	12.5	1950s	18.2
1870s	7.5	1960s	8.3
1880s	6.0	1970s	6.6
1890s	5.5	1980s	16.6
1900s	10.9	1990s	17.6
1910s	2.2	2000s	1.2
		2010s*	11.0

*Through December 31, 2018.
Source: Data extracted from T. Lauricella, "Investors Hope the '10s Beat the '00s," *The Wall Street Journal*, December 21, 2009, pp. C1, C2 and Moneychimp.com, "CAGR of the Stock Market," **bit.ly/1aiqtiI**.

a. Construct a time-series plot of the stock performance from the 1830s to the 2010s.
b. Does there appear to be any pattern in the data?

2.55 The file New Home Sales contains the number of new homes sold (in thousands) and the median sales price of new single-family houses sold in the United States recorded at the end of each month from January 2000 through December 2017.

Source: Data extracted from **www.census.gov./construction/nrs/pdf/uspricemon/.pdf**.

a. Construct a times series plot of new home sales prices.
b. What pattern, if any, is present in the data?

2.56 The file Movie Attendance contains the yearly movie attendance (in billions) from 2001 through 2018.

Year	Attendance	Year	Attendance
2001	1.44	2010	1.34
2002	1.58	2011	1.28
2003	1.55	2012	1.36
2004	1.47	2013	1.34
2005	1.38	2014	1.27
2006	1.41	2015	1.32
2007	1.40	2016	1.32
2008	1.34	2017	1.23
2009	1.41	2018	1.31

Source: Data extracted from **boxofficemojo.com/yearly**.

a. Construct a time-series plot for the movie attendance (in billions).
b. What pattern, if any, is present in the data?

2.57 The Super Bowl is a big viewing event watched by close to 200 million Americans that is also a big event for advertisers. The file Super Bowl Ad Costs contains the average cost for 30-second ads that ran between the opening kickoff and the final whistle for the 2004 to 2019 Super Bowls.

Source: Data extracted from **www.admeter.usatoday.com/results/2019**.

a. Construct a time-series plot.
b. What pattern, if any, is present in the average cost for 30-second ads that ran between the opening kickoff and the final whistle?

2.6 Organizing a Mix of Variables

Earlier sections of this chapter discuss organizing one or two variables of the same type, either categorical or numeric variables. Organizing a mix of many variables into one tabular summary, called a **multidimensional contingency table**, is also possible. Although any number of variables could be theoretically used in multidimensional contingency tables, using many variables together or using a categorical variable that has many categories will produce results that will be hard to comprehend and interpret. As a practical rule, these tables should be limited to no more than three or four variables, which limits their usefulness when exploring sets of data with many variables or analysis that involves big data.[1]

In typical use, these tables either display statistics about each joint response from multiple categorical variables as frequencies or percentages or display statistics about a numerical variable for each joint response from multiple categorical variables. The first form extends contingency tables (see Section 2.1) to two or more row or column variables. The second form replaces the tallies found in a contingency table with summary information about a numeric variable. Figure 2.19 on page 69 illustrates the first form, adding the variable market cap to the Figure 2.2 PivotTable contingency table of fund type and risk level.

[1] All of the examples in this book follow this rule.

FIGURE 2.19
PivotTables of fund type and risk level (based on Figure 2.2) and fund type, market cap, and risk level for the sample of the 479 retirement funds

Fund Type	Risk Level			
	Low	Average	High	Grand Total
Growth	13.15%	31.73%	19.00%	63.88%
Value	17.54%	15.03%	3.55%	36.12%
Grand Total	30.69%	46.76%	22.55%	100.00%

Fund Type	Risk Level			
	Low	Average	High	Grand Total
Growth	13.2%	31.7%	19.0%	63.88%
Large	9.6%	19.0%	3.5%	32.2%
MidCap	3.3%	9.4%	5.2%	18.0%
Small	0.2%	3.3%	10.2%	13.8%
Value	17.5%	15.0%	3.5%	36.1%
Large	14.6%	7.9%	0.6%	23.2%
MidCap	2.1%	3.5%	0.8%	6.5%
Small	0.8%	3.5%	2.1%	6.5%
Grand Total	30.69%	46.76%	22.55%	100.00%

Entries in this new multidimensional contingency table have been formatted as percentages of the whole with one decimal place to facilitate comparisons. The new table reveals patterns in the sample of retirement funds that a table of just risk level and fund type would not, such as the following:

- The pattern of risk for fund type when market cap is considered can be very different than the summary pattern that Figure 2.2 shows.
- A majority of the large and midcap growth funds have average risk, but most small growth funds have high risk.
- Nearly two-thirds of large market cap value funds have low risk, while a majority of midcap and small value funds have average risk.

Figure 2.20 illustrates the second form of a multidimensional contingency table. To form this table, the numerical variable 10YrReturn has been added to the Figure 2.19 PivotTable of fund type, market cap, and risk level. Note that the numerical variable appears as a statistic that summarizes the variable data, as the mean in these tables. That multidimensional contingency tables can only display a single descriptive statistic for a numerical variable is a limitation of such tables.

studentTIP

Chapter 3 discusses descriptive statistics for numerical variables, including the mean, also known as the average, that the Figure 2.20 table uses.

FIGURE 2.20
PivotTable (in two states) of fund type, market cap, and risk level, displaying the mean ten-year return percentage for the sample of the 479 retirement funds.

Mean 10YrReturn	Risk Level			
Fund Type	Low	Average	High	Grand Total
Growth	8.06	7.78	7.19	7.66
Value	6.45	6.52	5.97	6.43
Grand Total	7.14	7.38	7.00	7.22

Mean 10YrReturn	Risk Level			
Fund Type	Low	Average	High	Grand Total
Growth	8.06	7.78	7.19	7.66
Large	8.04	7.91	7.88	7.94
MidCap	8.10	7.41	6.60	7.30
Small	8.47	8.14	7.25	7.49
Value	6.45	6.52	5.97	6.43
Large	6.30	5.87	4.18	6.10
MidCap	6.99	7.69	6.15	7.27
Small	7.61	6.79	6.43	6.78
Grand Total	7.14	7.38	7.00	7.22

Figure 2.20 shows the same PivotTable in two states, with market cap *collapsed* into fund type (left) and market cap *fully expanded* (right). In the collapsed table, funds with high risk have the lowest mean ten-year return percentages. The expanded table discovers that large growth funds with high risk have one of the *highest* mean ten-year return percentages, something not suggested by the collapsed table. The expanded table also reveals that midcap value funds with average risk have the highest mean ten-year return percentage among all value funds.

Drill-down

In addition to their utility to report summaries of variables, multidimensional contingency tables can **drill down** to reveal the data that the table summarizes. Drilling down reveals a less summarized form of the data. Expanding a collapsed variable, such as Figure 2.20 demonstrates, is an example of drilling down. In Excel, double-clicking a joint response cell in a multidimensional contingency table drills down to the less summarized data. When you double-click a cell, Excel displays the rows of data associated with the joint response in a new worksheet.

Figure 2.21 on page 70 shows the drill-down of the small value funds with low risk cell of the Figure 2.20 PivotTables. This drill-down reveals that the ten-year return percentage for this group of four funds ranges from 4.83% to 9.44% and that the values of some of the other numeric variables also greatly vary.

FIGURE 2.21

Drill-down of the Figure 2.20 PivotTable small value funds with low risk cell (some variable columns not shown)

	Turnover Ratio	SD	Sharpe Ratio	Beta	1YrReturn	3YrReturn	5YrReturn	10YrReturn	Expense Ratio	Star Rating
2	75.00	4.67	0.53	0.20	4.74	2.53	4.82	9.44	1.40	Four
3	23.00	9.61	0.72	0.84	7.74	6.95	8.17	7.30	1.28	Four
4	30.10	10.71	0.70	0.55	5.88	7.60	7.74	4.83	1.70	Two
5	37.00	11.73	0.71	0.79	6.02	8.45	9.29	8.85	0.81	Four

2.7 Visualizing a Mix of Variables

Earlier sections of this chapter discuss visualizing one or two variables of the same type, either categorical or numeric. Visualizing a mix of many variables is also possible and has the following advantages over multidimensional contingency tables:

learnMORE

Chapter 17 discusses business analytics and presents additional visualization techniques that also visualize a mix of variables.

- More data and more variables can be presented in a form that is more manageable to review than a table with many row and column variables.
- The data, not summary descriptive statistics, can be shown for numerical variables.
- Multiple numerical variables can be presented in one summarization.
- Visualizations can reveal patterns that can be hard to see in tables.

These qualities make visualizations of a mix of variables helpful during initial exploratory data analysis and often a necessity in business analytics applications, especially when such techniques are analyzing big data. Because of the relative newness of these visualizations, Excel and Tableau use different ways to visualize a mix of data.

Colored Scatter Plot (Tableau)

A Tableau **colored scatter plot** visualizes two (and sometimes more than two) numerical variables and at least one categorical variable.

For example, Figure 2.22 presents a colored scatter plot of the expense ratio and 3YrReturn numerical variables and the market cap categorical variable for the sample of 479 retirement funds. This visual reveals that for the three-year period, funds with large market capitalizations (blue dots) tend to have the best returns and the lowest cost expense ratios (in other words, plot in the lower right quadrant of the chart). However, a number of large market cap funds plot

FIGURE 2.22

Tableau colored scatter plot of expense ratio, 3YrReturn, and market cap for the sample of 479 retirement funds

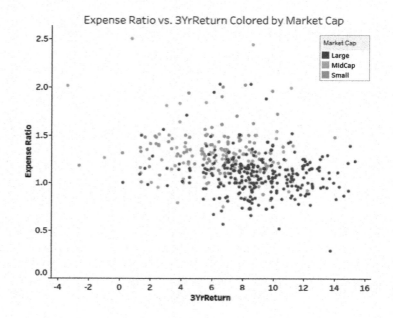

elsewhere on the chart, representing funds with relatively high expense ratios or fair to poor three-year returns. For certain types of analyses, the points representing those funds might be drilled down to determine reasons for their different behavior or to identify such funds as relative laggards in the set of all large market cap funds.

Because they can compare two numerical variables and one categorical variable, colored scatter plots can be considered an "opposite" of multidimensional contingency tables that summarize the two categorical variables and one numerical variable.

Bubble Chart

A **bubble chart** extends color scatter plots by using the size of the points, now called bubbles, to represent an additional variable. In Excel, that additional variable must be numerical, while in Tableau the variable can be either numerical or categorical. Chapter 17 further discusses bubble charts and presents examples of their usage.

PivotChart

Excel **PivotCharts** pull out and visualize specific categories from a PivotTable summary in a way that would otherwise be hard to do in Excel. For example, Figure 2.23 (left) displays a side-by-side PivotChart based on the Figure 2.20 PivotTables of fund type, market cap, and risk level, that display the mean ten-year return percentage for the sample of the 479 retirement funds. Filtering the chart to display the mean ten-year return percentages for only low risk funds, Figure 2.23 (right), highlights that small market cap growth funds have the highest mean ten-year return percentage.

FIGURE 2.23

PivotCharts based on the Figure 2.20 PivotTable of fund type, market cap, and risk level, showing the mean ten-year return percentage

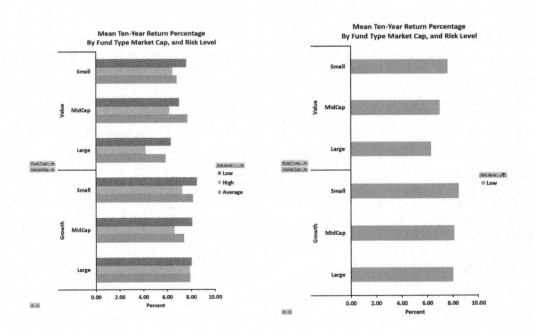

Treemap

Because of the formatting of the Excel treemap, displaying the legend for that treemap is unnecessary in Figure 2.24 but included for comparison purposes.

Treemaps show proportions of the whole of nested categories as colored tiles. In the simplest case of a tree map, the size of tiles corresponds to the tallies in a joint response cell of a contingency table. In more elaborate versions, tiles can be sized to a numerical variable. Figure 2.24 on page 72 presents Excel and Tableau treemaps for fund type and market cap. Note that Excel colors the treemap by the categories of first categorical variable (fund type), while Tableau can color the treemap using the second categorical variable, as the Figure 2.24 treemap illustrates.

FIGURE 2.24
Excel and Tableau treemaps for fund type and market cap

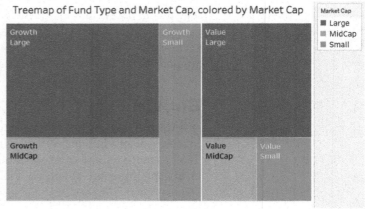

*Section EG2.7 discusses
creating treemaps and
sunburst charts in Excel.*

Excel treemaps display only two categorical variables even if the data contain three or more categorical variables in a hierarchy. (For such cases, the variables corresponding to the highest and lowest levels in the hierarchy are displayed.) Excel includes a **sunburst chart** that shows more than two levels of a hierarchy. While similar to a doughnut chart, a sunburst chart uses the "rings" to represent the different variables, not different categories of a categorical variable, and the segment coloring reflects the categories of the highest level categorical variable.

Sparklines

Sparklines are compact time-series visualizations of numerical variables. This compact form allows you to view all the visualizations together, which can aid in making comparisons among the variables. Sparklines highlight the trends of the plots over the precise graphing of points found in a time-series plot. Although typically used to plot several independent numerical variables, such as several different business indicators, sparklines can also be used to plot time-series data using smaller time units than a time-series plot to reveal patterns that the time-series plot may not.

For example, Figure 2.25 sparklines plot movie revenues for each month for the years 2005 through 2018 (years included in the annual movie revenues Table 2.13 on page 66). The sparklines reveal that monthly movie revenues for September are the very least from year to year, while the summer and end-year months tend to have higher revenues. (These trends reflect the industry practice of opening "big" movies during the summer and end-of-year and not opening such movies during September.)

FIGURE 2.25
Excel and Tableau
sparklines for movie
revenues by month for the
years 2005 through 2018

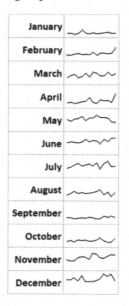

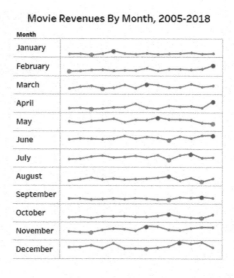

2.8 Filtering and Querying Data

Associated with preparing tabular or visual summaries are two operations that extract subsets of the variables under study. **Data filtering** selects rows of data that match certain criteria, specified values for specific variables. For example, using the filter that selects all rows in which fund type is the value would select 173 rows from the sample of 479 retirement funds that this chapter uses in various examples. In the context of this chapter, **querying** can be a more interactive version of filtering and a method that may not select all of the columns of the matching rows depending how the querying is done.

In Excel, selecting **Data → Filter** displays pull-down menus for each column in row 1 cells. In those menus, check boxes for each unique value in the column appear, and check boxes can be cleared or checked to select specific values or ranges. Excel also contains *slicers* that filter and query data from a PivotTable.

Excel Slicers

A **slicer** is a panel of clickable buttons that appears superimposed over a worksheet and is unique to a variable in the associated PivotTable. Each button in a slicer represents a unique value of the variable as found in the source data for the PivotTable. Slicers can be created for any variable that has been *associated* with a PivotTable, whether or not a variable has been inserted into the PivotTable. Using slicers enables one to work with many variables at once in a way that avoids creating an overly complex multidimensional contingency table that would be hard to comprehend and interpret.

Clicking buttons in the slicer panels queries the data. For example, the Figure 2.26 worksheet contains slicers for the fund type, market cap, star rating, and expense ratio variables and a PivotTable that has been associated with the variables stored in the DATA worksheet of the Retirement Funds workbook. With these four slicers, one can ask questions about the data. For example, "What are the attributes of the fund(s) with the lowest expense ratio?" and "What are the expense ratios associated with large market cap value funds that have a star rating of five?" These questions can be answered by clicking the appropriate buttons of the four slicers.

FIGURE 2.26
PivotTable and slicers for the retirement funds sample data

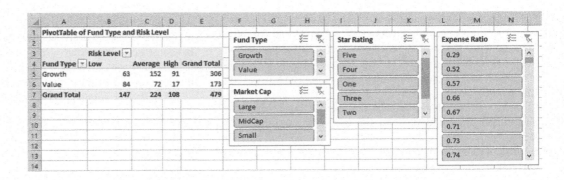

Figure 2.27 on page 74 displays slicers that answer these two questions. Note that Excel dims, or disables, the buttons representing values that the current data filtering excludes in order to highlight answers. For example, the answer to the first question is a growth fund with a large market cap and a five-star rating. (The updated PivotTable display, not shown in Figure 2.27, reveals that there is only one such fund.) For the second question, the answer is that 0.83, 0.94, 1.05, 1.09, 1.18, and 1.19 are the expense ratio percentages associated with large market cap value funds that have a star rating of five. (The updated PivotTable display reveals that there are six funds with those attributes.)

FIGURE 2.27
Slicer displays for answers to questions

PROBLEMS FOR SECTIONS 2.6 THROUGH 2.8

APPLYING THE CONCEPTS

2.58 Using the sample of retirement funds stored in Retirement Funds :
a. Construct a table that tallies fund type, market cap, and star rating.
b. What conclusions can you reach concerning differences among the types of retirement funds (growth and value), based on market cap (small, mid-cap, and large) and star rating (one, two, three, four, and five)?
c. Construct a table that computes the average three-year return percentage for each fund type, market cap, and star rating.
d. Drill down to examine the large cap growth funds with a rating of three. How many funds are there? What conclusions can you reach about these funds?

2.59 Using the sample of retirement funds stored in Retirement Funds :
a. Construct a table that tallies market cap, risk level, and star rating.
b. What conclusions can you reach concerning differences among the funds based on market cap (small, mid-cap, and large), risk level (low, average, and high), and star rating (one, two, three, four, and five)?
c. Construct a table that computes the average three-year return percentage for each market cap, risk level, and star rating.
d. Drill down to examine the large cap funds that are high risk with a rating of three. How many funds are there? What conclusions can you reach about these funds?

2.60 Using the sample of retirement funds stored in Retirement Funds :
a. Construct a table that tallies fund type, risk level, and star rating.
b. What conclusions can you reach concerning differences among the types of retirement funds, based on the risk levels and star ratings?
c. Construct a table that computes the average three-year return percentage for each fund type, risk level, and star rating.
d. Drill down to examine the growth funds with high risk with a rating of three. How many funds are there? What conclusions can you reach about these funds?

2.61 Using the sample of retirement funds stored in Retirement Funds :
a. Construct a table that tallies fund type, market cap, risk level, and star rating.
b. What conclusions can you reach concerning differences among the types of funds based on market cap categories, risk levels, and star ratings?
c. Which do you think is easier to interpret: the table for this problem or the ones for problems 2.58 through 2.60? Explain.
d. Compare the results of this table with those of Figure 2.19 and problems 2.58 through 2.60. What differences can you observe?

2.62 In the sample of 479 retirement funds (Retirement Funds), what are the attributes of the fund with the highest five-year return percentages?

2.63 Using the variables SD and assets in the sample of retirement funds stored in Retirement Funds :
a. Construct a chart that visualizes SD and assets by risk level.
b. Construct a chart that visualizes SD and assets by fund type. Rescale the assets axis, if necessary, to see more detail.
c. How do the patterns that you observe in both charts differ? What data relationships, if any, do those patterns suggest?

2.64 In the sample of 479 retirement funds (Retirement Funds), which funds in the sample have the lowest five-year return percentage?

2.65 Using the sample of retirement funds stored in Retirement Funds :
a. Construct one chart that visualizes 10YrReturn and 1YrReturn by market cap.
b. Construct one chart that visualizes 5YrReturn and 1YrReturn by market cap.
c. How do the patterns of the points of each market cap category change between the two charts?
d. What can you deduce about return percentages in years 6 through 10 included in 10YrReturn but not included in 5YrReturn?

2.66 In the sample of 479 retirement funds (Retirement Funds), what characteristics are associated with the funds that have the lowest five-year return percentages?

2.67 The data in New Home Sales include the median sales price of new single-family houses sold in the United States recorded at the end of each month from January 2000 through December 2017.

Source: Data extracted from "New Residential Sales," **bit.ly/2eEcIBR**.

a. Construct sparklines of new home sales prices by year.
b. What conclusions can you reach concerning the median sales price of new single-family houses sold in the United States from January 2000 through December 2017?
c. Compare the sparklines in (a) to the time-series plot in Problem 2.55 on page 68.

2.68 The file Natural Gas includes the monthly average commercial price for natural gas (dollars per thousand cubic feet) in the United States from January 1, 2008, to December 2018.

Source: Data extracted from Energy Information Administration, "Natural Gas Prices," **bit.ly/2oZIQ5Z**.

a. Construct a sparkline of the monthly average commercial price for natural gas (dollars per thousand cubic feet) by year.
b. What conclusions can you reach concerning the monthly average commercial price for natural gas (dollars per thousand cubic feet)?

2.9 Pitfalls in Organizing and Visualizing Variables

The tabular and visual summaries created when organizing and visualizing variables can jump-start the analysis of the variables. However, care must be taken not to produce results that will be hard to comprehend and interpret or to present the data in ways that undercut the usefulness of the methods this chapter discusses. One can too easily create summaries that obscure the data or create false impressions that would lead to misleading or unproductive analysis. The challenge in organizing and visualizing variables is to avoid such pitfalls.

Obscuring Data

Management specialists have long known that information overload, presenting too many details, can obscure data and hamper decision making (Gross). Both tabular summaries and visualizations can suffer from this problem. For example, consider the Figure 2.28 side-by-side bar chart that shows percentages of the overall total for subgroups formed from combinations of fund type, market cap, risk level, and star rating. While this chart highlights that more large-cap retirement funds have low risk and a three-star rating than any other combination of risk level and star rating, other details about the retirement funds sample are less obvious. The overly complex legend obscures as well and suggests that an equivalent multidimensional contingency table, with 30 joint response cells, would be obscuring, if not overwhelming, for most people.

FIGURE 2.28

Side-by-side bar chart for the retirement funds sample showing percentage of overall total for fund type, market cap, risk level, and star rating

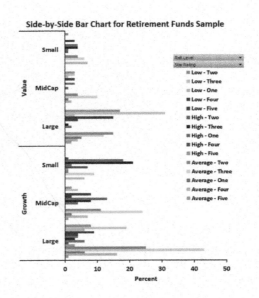

Creating False Impressions

When organizing and visualizing variables, one must be careful not to create false impressions that could affect preliminary conclusions about the data. Selective summarizations and improperly constructed visualizations often create false impressions.

A *selective summarization* is the presentation of only part of the data that have been collected. Frequently, selective summarization occurs when data collected over a long period of time are summarized as percentage changes for a shorter period. For example, Table 2.14 (left) presents the one-year difference in sales of seven auto industry companies for the month of April. The selective summarization tells a different story, particularly for company G, than does Table 2.14 (right) that shows the year-to-year differences for a three-year period that included the historic 2008 economic downturn. The second table reveals that Company D's positive change comes after two very poor years, whereas Company G's gain represents a true gain over the three-year period.

TABLE 2.14

Left: One-year percentage change in year-to-year sales for the month of April

Right: Percentage change for three consecutive years

Company	Change from Prior Year
A	+7.2
B	+24.4
C	+24.9
D	+24.8
E	+12.5
F	+35.1
G	+29.7

Company	Change from Prior Year		
	Year 1	Year 2	Year 3
A	−22.6	−33.2	+7.2
B	−4.5	−41.9	+24.4
C	−18.5	−31.5	+24.9
D	−29.4	−48.1	+24.8
E	−1.9	−25.3	+12.5
F	−1.6	−37.8	+35.1
G	+7.4	−13.6	+29.7

Improperly constructed charts can also create false impressions. Figure 2.29 shows two pie charts that display the market shares of companies in two industries. How many would quickly notice that both pie charts summarize identical data? *(Did you?)*

FIGURE 2.29
Market Shares of companies in "two" industries

If you want to verify that the two pie charts visualize the same data, open the TwoPies worksheet in the Challenging workbook.

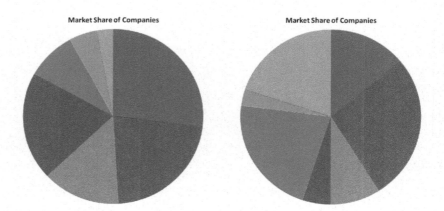

Market Share of Companies   Market Share of Companies

Because of their relative positions and colorings, many people will perceive the dark blue pie slice on the left chart to have a smaller market share than the dark red pie chart on the right chart even though both pie slices represent the company that has 27% market share. In this case, both the ordering of pie slices and the different colorings of the two pie charts contribute to creating the false impression. With other types of charts, improperly scaled axes or a *Y* axis that either does not begin at the origin or is a "broken" axis that is missing intermediate values are other common mistakes that create false impressions.

Chartjunk

Seeking to construct a visualization that can more effectively convey an important point, some people add decorative elements to enhance or replace the simple bar and line shapes of the visualizations discussed in this chapter. While judicious use of such elements may aid in the memorability of a chart (Bateman et al.), most often such elements either obscure the data or, worse, create a false impression of the data. Such elements are called **chartjunk**.

Figure 2.30 presents a visualization that illustrates mistakes that are common ways of creating chartjunk unintentionally. The grapevine with its leaves and bunch of grapes adds to the clutter of decoration without conveying any useful information. The chart inaccurately shows the 1949–1950 measurement (135,326 acres) at a *higher* point on the *Y* axis than larger values such as the 1969–1970 measurement, 150,300 acres. The inconsistent scale of the *X* axis distorts the time variable. (The last two measurements, eight years apart, are drawn about as far apart as the 30-year gap between 1959 and 1989.) All of these errors create a very wrong impression that obscures the important trend of accelerating growth of land planted in the 1990s.

FIGURE 2.30

Two visualizations of the amount of land planted with grapes for the wine industry

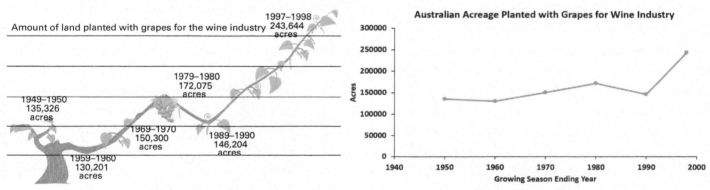

Left illustration adapted from S. Watterson, "Liquid Gold—Australians Are Changing the World of Wine. Even the French Seem Grateful," *Time*, November 22, 1999, p. 68–69.

Exhibit 2.1 summarizes the best practices for creating visual summaries. Microsoft Excel users should be aware that Excel may tempt you to use uncommon chart types and may produce charts that violate some of the best practices that the exhibit lists.

EXHIBIT 2.1

Best Practices for Creating Visual Summaries

- Use the simplest possible visualization.
- Include a title and label all axes.
- Include a scale for each axis if the chart contains axes.
- Begin the scale for a vertical axis at zero and use a constant scale.
- Avoid 3D or "exploded" effects and the use of chartjunk.
- Use consistent colorings in charts meant to be compared.
- Avoid using uncommon chart types, including radar, surface, cone, and pyramid charts.

PROBLEMS FOR SECTION 2.9

APPLYING THE CONCEPTS

2.69 **(Student Project)** Bring to class a chart from a website, newspaper, or magazine published recently that you believe to be a poorly drawn representation of a numerical variable. Be prepared to submit the chart to the instructor with comments about why you believe it is inappropriate. Do you believe that the intent of the chart is to purposely mislead the reader? Also, be prepared to present and comment on this in class.

2.70 **(Student Project)** Bring to class a chart from a website, newspaper, or magazine published this month that you believe to be a poorly drawn representation of a categorical variable. Be prepared to submit the chart to the instructor with comments about why you consider it inappropriate. Do you believe that the intent of the chart is to purposely mislead the reader? Also, be prepared to present and comment on this in class.

2.71 Examine the following visualization, adapted from one that appeared in a post in a digital marketing blog.

a. Describe at least one good feature of this visual display.
b. Describe at least one bad feature of this visual display.
c. Redraw the graph, by using the best practices that Exhibit 2.1 on page 77 presents.

2.72 Examine the following visualization, adapted from one that appeared in the post "Who Are the Comic Book Fans on Facebook?", as reported by **graphicspolicy.com**.
a. Describe at least one good feature of this visual display.
b. Describe at least one bad feature of this visual display.
c. Redraw the graph, by using the Exhibit 2.1 best practices.

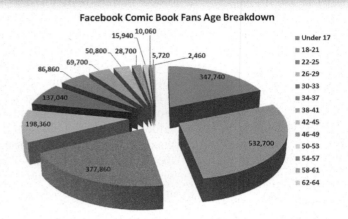

2.73 Examine the following visualization, adapted from a management consulting white paper.

a. Describe at least one good feature of this visual display.
b. Describe at least one bad feature of this visual display.
c. Redraw the graph, by using the Exhibit 2.1 best practices.

2.74 Deanna Oxender Burgess of Florida Gulf Coast University conducted research about corporate annual reports (see D. Rosato, "Worried About Corporate Numbers? How About the Charts?," **nyti.ms/2GedlQu**). Burgess found that even slight distortions in a chart changed readers' perception of the information. Using online or library sources, select a corporation and study its most recent annual report. Find at least one chart in the report that you think needs improvement and develop an improved version of the chart. Explain why you believe the improved chart is better than the one included in the annual report.

2.75 Figure 2.4 on page 51 shows a bar chart and a pie chart for what employed adults did the last time they were sick.

a. Create an exploded pie chart, a cone chart, or a pyramid chart that shows how employed adults who work outside the home responded to being sick.

b. Which graphs either seen in Figure 2.4 or created in (a), do you prefer? Explain.

2.76 Figures 2.5 and 2.6 on pages 52 and 53, show a bar chart and a pie chart for the risk level for the retirement fund data.

a. Create an exploded pie chart, a doughnut chart, a cone chart, and a pyramid chart that show the risk level of retirement funds.

b. Which graphs, either seen in Figures 2.5 and 2.6 or created in (a), do you prefer? Explain.

▼ USING **STATISTICS**
"The Choice *Is* Yours," Revisited

In The Choice *Is* Yours scenario, you were hired to define, collect, organize, and visualize data about a sample of 479 retirement funds in ways that could assist prospective clients to make investment choices. Having verified that each of the 13 variables in the sample were properly defined, you had to consider which tabular and visual summaries were appropriate for each variable and how specific mixes of variables might be used to gain insights about the 479 funds.

From summaries of the risk variable, you learned that nearly half of the funds were of average risk and there are fewer high risk funds than low risk funds. From contingency tables of the fund type and risk level, you observed that high risk funds were proportionally a larger category for growth funds than for value funds. From histograms and percentage polygons of the three-year return percentages, you were able to observe that the three-year returns were higher for the growth funds than for the value funds. Returns for the growth funds were concentrated between 2.5% and 15%, with returns for the value funds concentrated between 2.5% and 10%.

From various multi-dimensional contingency tables, you revealed additional relationships. For example, from a table that summarizes fund type, risk level, and market cap, you discovered that nearly two-thirds of large market cap value funds have low risk, while a majority of small and mid-cap value funds have average risk.

You discovered visual summaries that can combine many variables and present data in a more effective and easier-to-explore way than tables. And you also discovered ways to take subsets of the retirement sample for focused analysis. Finally, you learned to avoid the pitfalls that people experience in organizing and visualizing data. You are better prepared to present fund details to prospective clients.

▼ SUMMARY

Methods to organize and visualize variables vary by the type of variable, as well as the number of variables one seeks to organize and visualize at the same time. Table 2.15 summarizes the methods this chapter discusses.

Using the appropriate methods to organize and visualize data allows one to reach preliminary conclusions about the data. In several different chapter examples, tabular and visual summaries help one to reach conclusions about the primary way people pay for purchases and other transactions, the cost of meals at center city and metro area restaurants, and some of the differences among the 479 funds in a retirement fund sample.

Using the appropriate tabular and visual summaries can provide initial insights about variables and give reason to ask additional questions about the data. Those questions may lead to using interactive techniques to further explore your data or to performing additional analysis at a later time.

Methods to organize and visualize data can be misused, thereby undermining the usefulness of the tabular and visual summaries those methods create. Following the Exhibit 2.1 best practices when making visual summaries can minimize common pitfalls.

For numerical variables, additional ways to summarize data involve computing sample statistics or population parameters. Chapter 3 discusses the most common examples of these *numerical descriptive measures*.

TABLE 2.15
Methods to organize
and visualize variables

Categorical Variables	
Organize	Summary table, contingency table (Section 2.1)
Visualize one variable	Bar chart, pie chart, doughnut chart, Pareto chart (Section 2.3)
Visualize two variables	Side-by-side chart, doughnut chart, sparklines (Sections 2.3 and 2.6)
Numerical Variables	
Organize	Ordered array, frequency distribution, relative frequency distribution, percentage distribution, cumulative percentage distribution (Section 2.2)
Visualize one variable	Stem-and-leaf display, histogram, percentage polygon, cumulative percentage polygon (ogive) (Section 2.4)
Visualize two variables	Scatter plot, time-series plot (Section 2.5)
Mix of Variables	
Organize	Multidimensional tables (Section 2.6)
Visualize	Colored scatter plots, bubble charts, PivotChart (Excel), treemap, sparklines (Section 2.7)
Filter and query	Subsets of tables, slicers (Excel) (Section 2.8)

▼REFERENCES

Batemen, S., R. Mandryk, C. Gutwin, A. Genest, D. McDine, and C. Brooks. "Useful Junk? The Effects of Visual Embellishment on Comprehension and Memorability of Charts," accessed at **bit.ly/1HMDnpc**.

Edwardtufte.com. "Edward Tufte forum: Pie Charts," accessed at **bit.ly/1E3l1Pb**.

Few, S. *Information Dashboard Design: Displaying Data for At-a-Glance Monitoring*, 2nd ed. Burlingame, CA: Analytics Press, 2013.

Gross, B. *The Managing of Organizations: The Administrative Struggle*, Vols. I & II. New York: The Free Press of Glencoe, 1964.

Huff, D. *How to Lie with Statistics*. New York: Norton, 1954.

Tufte, E. R. *The Visual Display of Quantitative Information*, 2nd ed. Cheshire, CT: Graphics Press, 2002.

Tufte, E. R. *Beautiful Evidence*. Cheshire, CT: Graphics Press, 2006.

Wainer, H. *Visual Revelations: Graphical Tales of Fate and Deception from Napoleon Bonaparte to Ross Perot*. Mahwah, NJ: Lawrence Erlbaum Associates, 2000.

▼KEY EQUATIONS

Determining the Class Interval Width

$$\text{Interval width} = \frac{\text{highest value} - \text{lowest value}}{\text{number of classes}} \tag{2.1}$$

Computing the Proportion or Relative Frequency

$$\text{Proportion} = \text{relative frequency} = \frac{\text{number of values in each class}}{\text{total number of values}} \tag{2.2}$$

▼ KEY TERMS

bar chart   51
bins   46
bubble chart   71
cell   40
chartjunk   77
class boundaries   44
class interval   44
class interval width   44
class midpoints   44
classes   44
colored scatter plot   70
contingency table   40
cumulative percentage distribution   48
cumulative percentage polygon (ogive)   61
data filtering   73

doughnut chart   52
drill down   69
frequency distribution   44
histogram   59
joint response   40
line graph   53
multidimensional contingency table   68
ogive (cumulative percentage polygon)   61
ordered array   43
Pareto chart   53
Pareto principle   53
percentage distribution   46
percentage polygon   68
pie chart   52
PivotChart   71

PivotTable   40
proportion   46
querying   73
relative frequency   48
relative frequency distribution   46
scatter plot   65
side-by-side bar chart   55
slicer   73
sparklines   72
stem-and-leaf display   58
summary table   39
sunburst chart   72
time-series plot   86
treemap   71

▼ CHECKING YOUR UNDERSTANDING

2.77 How do histograms and polygons differ in construction and use?

2.78 Why would you construct a summary table?

2.79 What are the advantages and disadvantages of using a bar chart, a pie chart, a doughnut chart, and a Pareto chart?

2.80 Compare and contrast the bar chart for categorical data with the histogram for numerical data.

2.81 What is the difference between a time-series plot and a scatter plot?

2.82 Why is it said that the main feature of a Pareto chart is its ability to separate the "vital few" from the "trivial many"?

2.83 What are the three different ways to break down the percentages in a contingency table?

2.84 How can a multidimensional table differ from a two-variable contingency table?

2.85 What type of insights can you gain from a contingency table that contains three variables that you cannot gain from a contingency table that contains two variables?

2.86 What is the difference between a drill-down and a slicer?

2.87 What is the difference between a time-series plot and sparklines?

▼ CHAPTER REVIEW PROBLEMS

2.88 The following table shown in the file `Textbook Costs` shows the breakdown of the costs of a typical college textbook.

Revenue Category		Percentage %
Publisher		66.06
	Page, printing, ink	4.62
	Editorial production	24.93
	Marketing	11.60
	Freight	0.77
	Misc. overhead	1.74
	Profit	22.4
Bookstore		22.20
	Store personnel	11.40
	Store operations	5.90
	Store income	4.90
Author		11.74

Source: Data extracted from **bit.ly/2ppEetq**.

a. Using the categories of publisher, bookstore, and author, construct a bar chart, a pie chart, and a Pareto chart.

b. Using the subcategories of publisher and the subcategories of bookstore, along with the author category, construct a Pareto chart.

c. Based on the results of (a) and (b), what conclusions can you reach concerning who gets the revenue from the sales of new college textbooks? Do any of these results surprise you? Explain.

2.89 The following table stored in `Movie Types` represents the market share (in number of movies, gross in millions of dollars, and millions of tickets sold) of each type of movie in 2018:

Type	Number	Gross ($millions)	Tickets (millions)
Original screen play	306	3977.5	443.4
Based on fiction book or short story	94	2528.7	281.9
Based on comic or graphic novel	15	2328.5	259.6
Based on real life event	209	841.5	93.8
Spin-off	4	701.4	78.2
Based on TV	5	436.4	48.7
Based on factual book or article	23	364.7	40.7
Remake	10	322.4	35.9
Based on game	2	159.3	17.8
Based on folk tale, legend, or fairytale	7	158.0	17.6

Source: Data Extracted from **www.the-numbers.com/market/2018/summary**.

a. Construct a bar chart, a pie chart, a doughnut chart, and a Pareto chart for the number of movies, gross (in $millions), and number of tickets sold (in millions).
b. What conclusions can you reach about the market shares of the different types of movies in 2018?

2.90 B2B marketers in North America were surveyed about content marketing usage, organization, and success. Content marketers were asked about how content marketing is structured within their organization and how they would describe their organization's commitment to content marketing. The tables in this problem stored in `B2B` summarize the survey results.

B2B Content Marketing Organizational Structure	Percentage
Centralized content marketing group	24%
Each brand has own content marketing group	5%
Both: centralized team and individual teams	13%
Small marketing/content marketing team	55%
Other	3%

Source: Data extracted from **bit.ly/2d98EaN**.

a. Construct a bar chart, a pie or doughnut chart, and a Pareto chart for this table.
b. Which graphical method do you think is best for portraying this table data?

Commitment to Content Marketing	Percentage
Very/Extremely Committed	63%
Somewhat Committed	30%
Not Very/Not at All Important	7%

Source: Data extracted from **bit.ly/2d98EaN**.

c. Construct a bar chart, a pie or doughnut chart, and a Pareto chart for this table.
d. Which graphical method do you think is best for portraying this table data?
e. Based on the two tables, what conclusions can you reach concerning marketer's perspective on content marketing?

2.91 The owner of a restaurant that serves Continental-style entrées has the business objective of learning more about the patterns of patron demand during the Friday-to-Sunday weekend time period. Data were collected from 630 customers on the type of entrée ordered and were organized in the following table (and stored in `Entree`):

Type of Entrée	Number Ordered
Beef	187
Chicken	103
Mixed	30
Duck	25
Fish	122
Pasta	63
Shellfish	74
Veal	26
Total	630

a. Construct a percentage summary table for the types of entrées ordered.
b. Construct a bar chart, a pie chart, doughnut chart, and a Pareto chart for the types of entrées ordered.
c. Do you prefer using a Pareto chart or a pie chart for these data? Why?
d. What conclusions can the restaurant owner reach concerning demand for different types of entrées?

2.92 Suppose that the owner of the restaurant in Problem 2.91 also wants to study the demand for dessert during the same time period. She decides that in addition to studying whether a dessert was ordered, she will also study the gender of the individual and whether a beef entrée was ordered. Data were collected from 630 customers and organized in the following contingency tables:

DESSERT ORDERED	GENDER		
	Male	Female	Total
Yes	96	50	146
No	234	250	484
Total	330	300	630

DESSERT ORDERED	BEEF ENTRÉE		
	Yes	No	Total
Yes	74	68	142
No	113	375	488
Total	187	443	630

a. For each of the two contingency tables, construct contingency tables of row percentages, column percentages, and total percentages.

b. Which type of percentage (row, column, or total) do you think is most informative for each gender? For beef entrée? Explain.

c. What conclusions concerning the pattern of dessert ordering can the restaurant owner reach?

2.93 The following data stored in `Food Consumption` represents the pounds per capita of fresh food and packaged food consumed in the United States, Japan, and Russia in a recent year.

	COUNTRY		
FRESH FOOD	United States	Japan	Russia
Eggs, nuts, and beans	88	94	88
Fruit	124	126	88
Meat and seafood	197	146	125
Vegetables	194	278	335
PACKAGED FOOD			
Bakery goods	108	53	144
Dairy products	298	147	127
Pasta	12	32	16
Processed, frozen, dried, and chilled food, and ready-to-eat meals	183	251	70
Sauces, dressings, and condiments	63	75	49
Snacks and candy	47	19	24
Soup and canned food	77	17	25

a. For each of the three countries, construct a bar chart, a pie or doughnut chart, and a Pareto chart for different types of fresh foods consumed.

b. For each of the three countries, construct a bar chart, a pie or doughnut chart, and a Pareto chart for different types of packaged foods consumed.

c. What conclusions can you reach concerning differences between the United States, Japan, and Russia in the fresh foods and packaged foods consumed?

2.94 The Air Travel Consumer Report, a monthly product of the Department of Transportation's Office of Aviation Enforcement and Proceedings (OAEP), is designed to assist consumers with information on the quality of services provided by airlines. The report includes a summary of consumer complaints by industry group and by complaint category. A breakdown of the 878 January 2019 consumer complaints based on industry group is given in the following table stored in `OAEP Complaints by Group`:

Industry Group	Number of Consumer Complaints
U.S. Airlines	506
Non-U.S. Airlines	336
Travel Agents	27
Miscellaneous	9
Industry Total	878

Source: Data extracted from "The Travel Consumer Report," Office of Aviation Enforcement and Proceedings, January 2019.

a. Construct a Pareto chart for the number of complaints by industry group. What industry group accounts for most of the complaints?

The 878 consumer complaints against airlines are summarized by type in the following table stored in `OAEP Complaints by Category`:

Complaint Category	Complaints
Flight problems	287
Oversales	23
Reservation/ticketing/boarding	99
Fares	53
Refunds	59
Baggage	202
Customer service	97
Disability	48
Advertising	4
Discrimination	7
Other	18
Arrivals	1
Total	878

b. Construct pie and doughnut charts to display the percentage of complaints by type. What complaint category accounts for most of the complaints?

2.95 One of the major measures of the quality of service provided by an organization is the speed with which the organization responds to customer complaints. A large family-held department store selling furniture and flooring, including carpet, had undergone a major expansion in the past several years. In particular, the flooring department had expanded from two installation crews to an installation supervisor, a measurer, and 15 installation crews. A business objective of the company was to reduce the time between when the complaint is received and when it is resolved. During a recent year, the company received 50 complaints concerning carpet installation. The number of days between the receipt of the complaint and the resolution of the complaint for the 50 complaints, stored in `Furniture`, are:

54	5	35	137	31	27	152	2	123	81	74	27
11	19	126	110	110	29	61	35	94	31	26	5
12	4	165	32	29	28	29	26	25	1	14	13
13	10	5	27	4	52	30	22	36	26	20	23
33	68										

a. Construct a frequency distribution and a percentage distribution.

b. Construct a histogram and a percentage polygon.

c. Construct a cumulative percentage distribution and plot a cumulative percentage polygon (ogive).

d. On the basis of the results of (a) through (c), if you had to tell the president of the company how long a customer should expect to wait to have a complaint resolved, what would you say? Explain.

2.96 The file `Domestic Beer` contains the percentage alcohol, number of calories per 12 ounces, and number of carbohydrates (in grams) per 12 ounces for 157 of the best selling domestic beers in the United States.

Source: Data extracted from **www.beer100.com/beercalories.htm**, April 3, 2019.

a. Construct a percentage histogram for percentage alcohol, number of calories per 12 ounces, and number of carbohydrates (in grams) per 12 ounces.

b. Construct three scatter plots: percentage alcohol versus calories, percentage alcohol versus carbohydrates, and calories versus carbohydrates.

c. Discuss what you learned from studying the graphs in (a) and (b).

2.97 The Super Bowl, watched by close to 200 million Americans, is also a big event for advertisers. The file Super Bowl Ad Ratings contains the rating of ads that ran between the opening kickoff and the final whistle.

Source: Data extracted from **www.admeter.usatoday.com/results/2019**.

For ads that ran before halftime and ads that ran at halftime or after separately:

a. Construct an ordered array.

b. Construct a stem-and-leaf display.

c. Plot a percentage histogram.

d. Plot a percentage polygon for both groups on one chart.

e. Plot a cumulative percentage polygon for both groups on one chart.

f. What conclusions can you reach about the difference in ad ratings for ads that ran before halftime compared to ads that ran at halftime or after?

2.98 The file CD Rates contains the yields for one-year certificates of deposit (CDs) and for five-year CDs for 46 financial institutions selling CDs in Lake Worth, Florida, as of April 5, 2019.

Source: Data extracted and compiled from **www.bankrate.com**, April 5, 2019.

a. Construct a stem-and-leaf display for one-year CDs and five-year CDs.

b. What conclusions can you reach about the distributions for the one-year CDs and the five-year CDs?

c. Construct a scatter plot of one-year CDs versus five-year CDs.

d. What is the relationship between the one-year CD rate and the five-year CD rate?

2.99 Download speed of an Internet connection is of great importance to both individuals and businesses in a community. The file City Download Speeds contains the download speed in Mbps for 100 United States cities with the fastest download speeds.

Source: Data extracted from **broadbandnow.com/fastest-cities**.

For download speed:

a. Construct a frequency distribution and a percentage distribution.

b. Construct a histogram and a percentage polygon.

c. Construct a cumulative percentage distribution and plot a cumulative percentage polygon (ogive).

d. Based on (a) through (c), what conclusions can you reach concerning download speed?

e. Construct a scatter plot of download speed and number of providers.

f. What is the relationship between the download speed and number of providers?

2.100 Studies conducted by a manufacturer of Boston and Vermont asphalt shingles have shown product weight to be a major factor in customers' perception of quality. Moreover, the weight represents the amount of raw materials being used and is therefore very important to the company from a cost standpoint. The last stage of the assembly line packages the shingles before the packages are placed on wooden pallets. The variable of interest is the weight in pounds of the pallet, which for most brands holds 16 squares of

shingles. The company expects pallets of its Boston brand-name shingles to weigh at least 3,050 pounds but less than 3,260 pounds. For the company's Vermont brand-name shingles, pallets should weigh at least 3,600 pounds but less than 3,800. Data, collected from a sample of 368 pallets of Boston shingles and 330 pallets of Vermont shingles, are stored in Pallet.

a. For the Boston shingles, construct a frequency distribution and a percentage distribution having eight class intervals, using 3,015, 3,050, 3,085, 3,120, 3,155, 3,190, 3,225, 3,260, and 3,295 as the class boundaries.

b. For the Vermont shingles, construct a frequency distribution and a percentage distribution having seven class intervals, using 3,550, 3,600, 3,650, 3,700, 3,750, 3,800, 3,850, and 3,900 as the class boundaries.

c. Construct percentage histograms for the Boston and Vermont shingles.

d. Comment on the distribution of pallet weights for the Boston and Vermont shingles. Be sure to identify the percentages of pallets that are underweight and overweight.

2.101 College basketball in the United States is big business with television contracts in the billions of dollars and coaches' salaries in the millions of dollars. The file Coaches Pay contains the total pay for coaches for a recent year and whether the school was a member of a major conference (Atlantic Coast, Big 12, Big East, Big Ten, Pac-12, or SEC).

Source: Data extracted from "College Men's Basketball Coaches Compensation, *USA Today*, March 13, 2019, p. C4.

For each of the two groups (member of a major conference or not):

a. Construct frequency and percentage distributions.

b. Construct a histogram and a percentage polygon.

c. Construct a cumulative percentage distribution and plot a cumulative percentage polygon (ogive).

d. Construct a stem-and-leaf display.

e. What conclusions can you reach about the differences in coaches pay between colleges that belonged to a major conference and those that do not.

2.102 The file Protein contains calorie and cholesterol information for popular protein foods (fresh red meats, poultry, and fish).

Source: U.S. Department of Agriculture.

a. Construct frequency and percentage distributions for the number of calories.

b. Construct frequency and percentage distributions for the amount of cholesterol.

c. Construct a percentage histogram for the number of calories.

d. Construct a percentage histogram for the amount of cholesterol.

e. Construct a scatter plot of the number of calories and the amount of cholesterol.

f. What conclusions can you reach from the visualizations?

2.103 The file Natural Gas contains the U.S. monthly average commercial and residential price for natural gas in dollars per thousand cubic feet from January 2008 through December 2018.

Source: Data extracted from "U.S. Natural Gas Prices," **eia.org**, March 30, 2019.

For the commercial price and the residential price:

a. Construct a time-series plot.

b. What pattern, if any, is present in the data?

c. Construct a scatter plot of the commercial price and the residential price.

d. What conclusion can you reach about the relationship between the commercial price and the residential price?

2.104 The data stored in `Drink` represent the amount of soft drink in a sample of 50 consecutively filled 2-liter bottles.
a. Construct a time-series plot for the amount of soft drink on the *Y* axis and the bottle number (going consecutively from 1 to 50) on the *X* axis.
b. What pattern, if any, is present in these data?
c. If you had to make a prediction about the amount of soft drink filled in the next bottle, what would you predict?
d. Based on the results of (a) through (c), explain why it is important to construct a time-series plot and not just a histogram, as was done in Problem 2.47 on page 64.

2.105 The file `Currency` contains the exchange rates of the Canadian dollar, the Japanese yen, and the English pound from 1980 to 2019, where the Canadian dollar, the Japanese yen, and the English pound are expressed in units per U.S. dollar.
a. Construct time-series plots for the yearly closing values of the Canadian dollar, the Japanese yen, and the English pound.
b. Explain any patterns present in the plots.
c. Write a short summary of your findings.
d. Construct separate scatter plots of the value of the Canadian dollar versus the Japanese yen, the Canadian dollar versus the English pound, and the Japanese yen versus the English pound.
e. What conclusions can you reach concerning the value of the Canadian dollar, Japanese yen, and English pound in terms of the U.S. dollar?

2.106 A/B testing allows businesses to test a new design or format for a web page to determine if the new web page is more effective than the current one. Web designers decide to create a new call-to-action button for a web page. Every visitor to the web page was randomly shown either the original call-to-action button (the control) or the new variation. The metric used to measure success was the download rate: the number of people who downloaded the file divided by the number of people who saw that particular call-to-action button. Results of the experiment yielded the following:

Variations	Downloads	Visitors
Original call-to-action button	351	3,642
New call-to-action button	485	3,556

a. Compute the percentage of downloads for the original call-to-action button and the new call-to-action button.
b. Construct a bar chart of the percentage of downloads for the original call-to-action button and the new call-to-action button.
c. What conclusions can you reach concerning the original call-to-action button and the new call-to-action button?

Web designers then created a new page design for a web page. Every visitor to the web page was randomly shown either the original web design (the control) or the new variation. The metric used to measure success was the download rate: the number of people who downloaded the file divided by the number of people who saw that particular web design. Results of the experiment yielded the following:

Variations	Downloads	Visitors
Original web design	305	3,427
New web design	353	3,751

d. Compute the percentage of downloads for the original web design and the new web design.
e. Construct a bar chart of the percentage of downloads for the original web design and the new web design.
f. What conclusions can you reach concerning the original web design and the new web design?
g. Compare your conclusions in (f) with those in (c).

Web designers next tested two factors simultaneously—the call-to-action button and the new page design. Every visitor to the web page was randomly shown one of the following:

> Old call-to-action button with original page design
> New call-to-action button with original page design
> Old call-to-action button with new page design
> New call-to-action button with new page design

Again, the metric used to measure success was the download rate: the number of people who downloaded the file divided by the number of people who saw that particular call-to-action button and web design. Results of the experiment yielded the following:

Call-to-Action Button	Page Design	Downloaded	Declined	Total
Original	Original	83	917	1,000
New	Original	137	863	1,000
Original	New	95	905	1,000
New	New	170	830	1,000
Total		485	3,515	4,000

h. Compute the percentage of downloads for each combination of call-to-action button and web design.
i. What conclusions can you reach concerning the original call to action button and the new call to action button and the original web design and the new web design?
j. Compare your conclusions in (i) with those in (c) and (f).

2.107 (Class Project) Have each student in the class respond to the question, "Which carbonated soft drink do you most prefer?" so that the instructor can tally the results into a summary table.
a. Convert the data to percentages and construct a Pareto chart.
b. Analyze the findings.

2.108 (Class Project) Cross-classify each student in the class by gender (male, female) and current employment status (yes, no) so that the instructor can tally the results.
a. Construct a table with either row or column percentages, depending on which you think is more informative.
b. What would you conclude from this study?
c. What other variables would you want to know regarding employment in order to enhance your findings?

REPORT WRITING EXERCISE

2.109 Referring to the results from Problem 2.100 on page 84 concerning the weights of Boston and Vermont shingles, write a report that evaluates whether the weights of the pallets of the two types of shingles are what the company expects. Be sure to incorporate tables and charts into the report.

▾CASES

Managing Ashland MultiComm Services

Recently, Ashland MultiComm Services has been criticized for its inadequate customer service in responding to questions and problems about its telephone, cable television, and Internet services. Senior management has established a task force charged with the business objective of improving customer service. In response to this charge, the task force collected data about the types of customer service errors, the cost of customer service errors, and the cost of wrong billing errors. The task force compiled the following data:

Types of Customer Service Errors

Type of Errors	Frequency
Incorrect accessory	27
Incorrect address	42
Incorrect contact phone	31
Invalid wiring	9
On-demand programming error	14
Subscription not ordered	8
Suspension error	15
Termination error	22
Website access error	30
Wrong billing	137
Wrong end date	17
Wrong number of connections	19
Wrong price quoted	20
Wrong start date	24
Wrong subscription type	33
Total	448

Cost of Customer Service Errors in the Past Year

Type of Errors	Cost ($thousands)
Incorrect accessory	17.3
Incorrect address	62.4
Incorrect contact phone	21.3
Invalid wiring	40.8
On-demand programming errors	38.8
Subscription not ordered	20.3
Suspension error	46.8
Termination error	50.9
Website access errors	60.7
Wrong billing	121.7
Wrong end date	40.9
Wrong number of connections	28.1
Wrong price quoted	50.3
Wrong start date	40.8
Wrong subscription type	60.1
Total	701.2

Type and Cost of Wrong Billing Errors

Type of Wrong Billing Errors	Cost ($thousands)
Declined or held transactions	7.6
Incorrect account number	104.3
Invalid verification	9.8
Total	121.7

1. Review these data (stored in **AMS2-1**). Identify the variables that are important in describing the customer service problems. For each variable you identify, construct the graphical representation you think is most appropriate and explain your choice. Also, suggest what other information concerning the different types of errors would be useful to examine. Offer possible courses of action for either the task force or management to take that would support the goal of improving customer service.

2. As a follow-up activity, the task force decides to collect data to study the pattern of calls to the help desk (stored in **AMS2-2**). Analyze these data and present your conclusions in a report.

Digital Case

In the Using Statistics scenario, you were asked to gather information to help make wise investment choices. Sources for such information include brokerage firms, investment counselors, and other financial services firms. Apply your knowledge about the proper use of tables and charts in this Digital Case about the claims of foresight and excellence by an Ashland-area financial services firm.

Open **EndRunGuide.pdf**, which contains the EndRun Financial Services "Guide to Investing." Review the guide, paying close attention to the company's investment claims and supporting data and then answer the following.

1. How does the presentation of the general information about EndRun in this guide affect your perception of the business?

2. Is EndRun's claim about having more winners than losers a fair and accurate reflection of the quality of its investment service? If you do not think that the claim is a fair and accurate one, provide an alternate presentation that you think is fair and accurate.

3. Review the discussion about EndRun's "Big Eight Difference" and then open and examine the attached sample of mutual funds. Are there any other relevant data from that file that could have been included in the Big Eight table? How would the new data alter your perception of EndRun's claims?

4. EndRun is proud that all Big Eight funds have gained in value over the past five years. Do you agree that EndRun should be proud of its selections? Why or why not?

CardioGood Fitness

The market research team at AdRight is assigned the task of identifying the profile of the typical customer for each treadmill product offered by CardioGood Fitness. The market research team decides to investigate whether there are differences across the product lines with respect to customer characteristics. The team decides to collect data on individuals who purchased a treadmill at a CardioGood Fitness retail store during the prior three months. The data are stored in the CardioGood Fitness file. The team identifies the following customer variables to study: product purchased, TM195, TM498, or TM798; gender; age, in years; education, in years; relationship status, single or part-nered; annual household income ($); average number of times the customer plans to use the treadmill each week; average number of miles the customer expects to walk/run each week; and self-rated fitness on an 1-to-5 ordinal scale, where 1 is poor shape and 5 is excellent shape.

1. Create a customer profile for each CardioGood Fitness tread-mill product line by developing appropriate tables and charts.

2. Write a report to be presented to the management of Cardio-Good Fitness detailing your findings.

The Choice *Is* Yours Follow-Up

Follow up the Using Statistics Revisited section on page 79 by analyzing the differences in one-year return percentages, five-year return percentages, and ten-year return percentages for the sample of 479 retirement funds stored in Retirement Funds. In your analysis, examine differences between the growth and value funds as well as the differences among the small, mid-cap, and large market cap funds.

Clear Mountain State Student Survey

The student news service at Clear Mountain State University (CMSU) has decided to gather data about the undergraduate students who attend CMSU. They create and distribute a survey of 14 questions (see **CMUndergradSurvey.pdf**) and receive responses from 111 undergraduates, stored in Student Survey. For each question asked in the survey, construct all the appropriate tables and charts and write a report summarizing your conclusions.

▾EXCEL GUIDE

Excel Charts Group Reference

Throughout this book, instructions for creating certain types of visual summaries refer to specific chart icons by name and the number that appears in the following illustration. The illustration displays the current Charts group for Microsoft Windows Excel (left) and very similar current Charts group for Excel for Mac (centered), with the Windows Excel icons numbered.

The numbered icons in the left Charts group represent commands to insert these chart types:

1. Insert Bar or Column Chart
2. Insert Line or Area Chart
3. Insert Pie or Doughnut Chart
4. Insert Hierarchy Chart
5. Insert Statistic Chart
6. Insert Scatter (X,Y) or Bubble Chart

The Excel for Mac tooltip identifies these chart types as Column, Line, Pie, Hierarchy, Statistical, and X Y (Scatter), respectively. The two unnumbered icons, which are also unlabeled by Excel, insert visual summaries that this book does not discuss.

Older Excel versions may display a Charts group similar to the one at right in the illustration above. This older Charts group will contain a seventh icon that inserts a bar chart. (The icon in this older group, which is similar to the #1 icon, inserts only a column chart.)

EG2.1 ORGANIZING CATEGORICAL VARIABLES

The Summary Table

Key Technique Use the PivotTable feature to create a summary table from the set of untallied values for a variable.

Example Create a frequency and percentage summary table similar to Figure 2.1 on page 40.

PHStat Use **One-Way Tables & Charts**.

For the example, open to the **DATA worksheet** of the **Retirement Funds workbook**. Select **PHStat → Descriptive Statistics → One-Way Tables & Charts**. In the procedure's dialog box (shown at right):

1. Click **Raw Categorical Data**.
2. Enter **D1:D480** as the **Raw Data Cell Range** and check **First cell contains label**.
3. Enter a **Title**, check **Percentage Column**, and click **OK**.

PHStat creates a PivotTable summary table on a new worksheet. For problems with tallied data, click **Table of Frequencies** in step 1. Then, in step 2, enter the cell range of the tabular summary as the **Freq. Table Cell Range** (edit box name changes from Raw Data Cell Range).

In the PivotTable, risk categories appear in alphabetical order and not in the order low, average, and high as would normally be expected. To change to the expected order, use steps 9 and 10 of the *Workbook* instructions but change all references to cell A6 to cell A7 and drop the Low label over cell A5, not cell A4.

Workbook (untallied data) Use the **Summary Table workbook** as a model.

For the example, open to the **DATA worksheet** of the **Retirement Funds workbook** and select **Insert→PivotTable**.

In the Create PivotTable dialog box (shown below):

1. Click **Select a table or range** and enter **D1:D480** as the **Table/Range** cell range.
2. Click **New Worksheet** and then click **OK**.

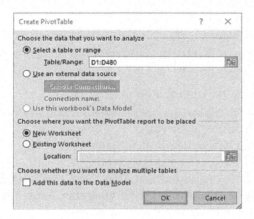

In the PivotTable fields (PivotTable Builder or PivotTable Field List in older Excels) display (shown below):

3. Drag **Risk Level** in the **Choose fields to add to report** box and drop it in the **Rows** (or **Row Labels**) box.
4. **Drag Risk Level in the Choose fields to add to report** box a second time and drop it in the **Σ Values** box. This second label changes to **Count of Risk Level** to indicate that a count, or tally, of the risk categories will be displayed in the PivotTable.

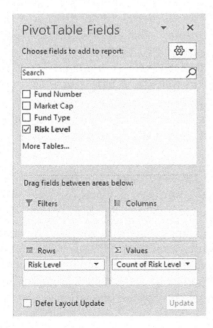

In the PivotTable being created:

5. Enter **Risk Level** in cell **A3** to replace the heading Row Labels.

6. Right-click cell **A3** and then click **PivotTable Options** in the shortcut menu that appears.

In the PivotTable Options dialog box (Windows Excel version shown below):

7. Click the **Layout & Format** tab. Check **For empty cells show** and enter **0** as its value. Leave all other settings unchanged.
   In Excel for Mac, click the **Display** tab. Check **Empty cells as** and enter **0** as its value.
8. Click **OK** to complete the PivotTable.

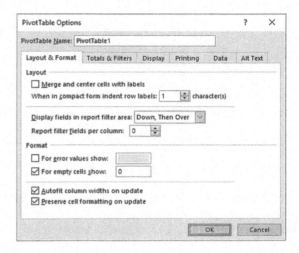

In the PivotTable, the risk level categories appear alphabetically and not the expected order Low, Average, and High. To reorder categories:

9. Click the **Low** label in cell **A6** to highlight cell A6. Move the mouse pointer to the top edge of the cell until the pointer changes to a four-way arrow (hand icon in OS X).
10. Drag the **Low** label and drop the label between cells **A3** and **A4**. The risk categories now appear in the proper order.

To add a percentage frequency column:

11. Enter **Percentage** in cell **C3**. Enter the formula **=B4/B$7** in cell **C4** and copy it down through **row 7**.
12. Select cell range **C4:C7**, right-click, and select **Format Cells** in the shortcut menu.
13. In the **Number** tab of the Format Cells dialog box, select **Percentage** as the **Category** and click **OK**.
14. Adjust the worksheet formatting, if appropriate (see Appendix Section B.4) and enter a title in cell **A1**.

Workbook (tallied data) Use the **SUMMARY_SIMPLE worksheet** of the **Summary Table workbook** as a model for creating a summary table.

The Contingency Table

Key Technique Use the PivotTable feature to create a contingency table from the set of individual values for a variable.
Example Create a contingency table displaying Fund Type and Risk Level similar to Figure 2.2 on page 40.

PHStat (untallied data) Use **Two-Way Tables & Charts**.

For the example, open to the **DATA worksheet** of the **Retirement Funds workbook**. Select **PHStat → Descriptive Statistics → Two-Way Tables & Charts**. In the procedure's dialog box (shown below):

1. Enter **C1:C480** as the **Row Variable Cell Range**.
2. Enter **D1:D480** the **Column Variable Cell Range**.
3. Check **First cell in each range contains label**.
4. Enter a **Title** and click **OK**.

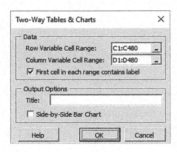

In the PivotTable, risk categories appear in alphabetical order and not in the order low, average, and high as would normally be expected. To change the expected order, use steps 9 and 10 of the following *Workbook* instructions.

Workbook (untallied data) Use the **Contingency Table workbook** as a model.

For the example, open to the **DATA worksheet** of the **Retirement Funds workbook**. Select **Insert → PivotTable**. In the Create PivotTable dialog box:

1. Click **Select a table or range** and enter **A1:N480** as the **Table/Range** cell range.
2. Click **New Worksheet** and then click **OK**.

In the PivotTable Fields (PivotTable Field List in some versions) task pane:

3. Drag **Fund Type** from **Choose fields to add to report** and drop it in the **Rows** (or **Row Labels**) box.
4. Drag **Risk Level** from **Choose fields to add to report** and drop it in the **Columns** (or **Column Labels**) box.
5. Drag **Fund Type** from **Choose fields to add to report** a second time and drop it in the **Σ Values** box. (**Type** changes to **Count of Fund Type**.)

In the PivotTable being created:

6. Select cell **A3** and enter a **space character** to clear the label **Count of Type**.
7. Enter **Fund Type** in cell **A4** to replace the heading Row Labels.
8. Enter **Risk Level** in cell **B3** to replace the heading Column Labels.
9. Click the **Low** label in cell **D4** to highlight cell D4. Move the mouse pointer to the left edge of the cell until

the mouse pointer changes to a four-way arrow (hand icon in Excel for Mac).

10. Drag the **Low** label to the left and drop the label between columns A and B. The Low label appears in B4 and column B now contains the low risk tallies.
11. Right-click over the PivotTable and then click **PivotTable Options** in the shortcut menu that appears.

In the PivotTable Options dialog box:

12. Click the **Layout & Format** tab. Check **For empty cells show** and enter **0** as its value. Leave all other settings unchanged.

    Click the **Display** tab. Check **Empty cells as** and enter 0 as its value. Skip to step 15.

13. Click the **Total & Filters** tab.
14. Check **Show grand totals for columns** and **Show grand totals for rows**.
15. Click **OK** to complete the table.

Workbook (tallied data) Use the **CONTINGENCY_SIMPLE worksheet** of the **Contingency Table workbook** as a model for creating a contingency table.

EG2.2 ORGANIZING NUMERICAL VARIABLES

The Ordered Array

Workbook To create an ordered array, first select the numerical variable to be sorted. Then select **Home → Sort & Filter** (in the Editing group) and in the drop-down menu click **Sort Smallest to Largest**. (You will see **Sort A to Z** as the first drop-down choice if you did not select a cell range of *numerical* data.)

The Frequency Distribution

Key Technique First, establish bins using the workaround that "Classes and Excel Bins" discusses on page 46. Then use the **FREQUENCY** (*untallied data cell range, bins cell range*) *array* function to tally data.

Example Create a frequency, percentage, and cumulative percentage distribution for the restaurant meal cost data that contain the information found in Tables 2.3, 2.5, and 2.8 in Section 2.2.

PHStat (untallied data) Use **Frequency Distribution**. If you plan to construct a histogram or polygon and a frequency distribution, use **Histogram & Polygons** (Section EG2.4).

For the example, open to the **DATA worksheet** of the **Restaurants workbook**. This worksheet contains the meal cost data in stacked format in column B and a set of bin numbers appropriate for those data in column I. Select **PHStat → Descriptive Statistics → Frequency Distribution**. In the procedure's dialog box (shown on the next page):

1. Enter **B1:B101** as the **Variable Cell Range**, enter **I1:I10** as the **Bins Cell Range**, and check **First cell in each range contains label**.

2. Click **Multiple Groups - Stacked** and enter **A1:A101** as the **Grouping Variable Cell Range** (the Location variable.)

3. Enter a **Title** and click **OK**.

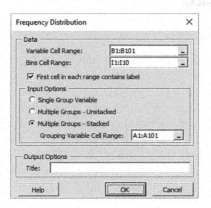

Frequency distributions for the two groups appear on separate worksheets.

To display both distributions on the same worksheet, select the cell range **B3:D12** on one of the worksheets. Right-click that range and click **Copy** in the shortcut menu. Open to the other worksheet. In that other worksheet, right-click cell **E3** and click **Paste Special** in the Paste Special dialog box, click **Values and numbers format** and click **OK**. Adjust the worksheet title and headings. (Appendix Section B.3 further explains the Paste Special command.)

Click **Single Group Variable** in step 2 to construct a distribution from a single group. Click **Multiple Groups - Unstacked** in step 2 if the **Variable Cell Range** contains two or more columns of unstacked data.

Workbook (untallied data) Use the **Center City** and **Metro Area** worksheets of the **Distributions workbook** as models.

For the example, open to the **Unstacked worksheet** of the **Restaurants workbook**. This worksheet contains the unstacked meal cost data in columns A and B and a bin number list in column D. Click the **insert worksheet icon** (the plus sign icon to the right of the sheet tabs, below the bottom of the worksheet) and in the new worksheet:

1. Enter a title in cell **A1**, **Bins** in cell **A3**, and **Frequency** in cell **B3**.

2. Copy the bin number list in the cell range **D2:D10** of the **Unstacked worksheet** and paste this list into cell **A4** of the new worksheet.

3. Select the cell range **B4:B12** that will hold the array formula.

4. Type, but do not press the **Enter** or **Tab** key, the formula **=FREQUENCY(Unstacked!$A$2:$A$51,$A$4:A$12)**. Then, while holding down the **Ctrl** and **Shift** keys, press the **Enter** key to enter the array formula into the cell range **B4:B12**.

5. Adjust the worksheet formatting as necessary.

Note that in step 4, you enter the *absolute* cell range as **Unstacked!$A$2:$A$51** and not as **$A$2:$A$51** because the untallied data are in a second worksheet (Unstacked). (Appendix B discusses both array formulas in Section B.2 and absolute cell references in Section B.3.)

Steps 1 through 5 construct a frequency distribution for the meal costs at center city restaurants. To construct a frequency distribution for the meal costs at metro area restaurants, insert another worksheet and repeat steps 1 through 5, entering **=FREQUENCY(Unstacked!$B$2:$B$51,A$4:A$12)** as the array formula in step 4.

To display both distributions on the same worksheet, select the cell range **B3:B12** on one of the worksheets. Right-click that range and click **Copy** in the shortcut menu. Open to the other worksheet. In that other worksheet, right-click cell **C3** and click **Paste Special** in the shortcut menu. In the Paste Special dialog box, click **Values and numbers format** and click **OK**. Adjust the worksheet title and headings. (Appendix Section B.3 further explains the Paste Special command.)

Analysis ToolPak (untallied data) Use **Histogram**.

For the example, open to the **Unstacked worksheet** of the **Restaurants workbook**. This worksheet contains the meal cost data unstacked in columns A and B and a set of bin numbers appropriate for those data in column D. Then:

1. Select **Data → Data Analysis**. In the Data Analysis dialog box, select **Histogram** from the **Analysis Tools** list and then click **OK**.

In the Histogram dialog box (shown below):

2. Enter **A1:A51** as the **Input Range** and enter **D1:D10** as the **Bin Range**. (If you leave **Bin Range** blank, the procedure creates a set of bins that will not be as well formed as the ones you can specify.)

3. Check **Labels** and click **New Worksheet Ply**.

4. Click **OK** to create the frequency distribution on a new worksheet.

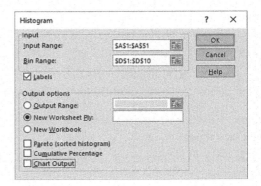

In the new worksheet:

5. Select **row 1**. Right-click this row and click **Insert** in the shortcut menu. Repeat. (This creates two blank rows at the top of the worksheet.)

6. Enter a title in cell **A1**.

The ToolPak creates a frequency distribution that contains an improper bin labeled **More**. Correct this error by using these general instructions:

7. Manually add the frequency count of the **More** row to the frequency count of the preceding row. (For the example, the **More** row contains a zero for the frequency, so the frequency of the preceding row does not change.)

8. Select the worksheet row (for this example, row 14) that contains the More row.

9. Right-click that row and click **Delete** in the shortcut menu.

Steps 1 through 9 construct a frequency distribution for the meal costs at center city restaurants. To construct a frequency distribution for the meal costs at metro area restaurants, repeat these nine steps but in step 2 enter **B1:B51** as the **Input Range**.

The Relative Frequency, Percentage, and Cumulative Distributions

Key Technique Add columns that contain formulas for the relative frequency or percentage and cumulative percentage to a previously constructed frequency distribution.

Example Create a distribution that includes the relative frequency or percentage as well as the cumulative percentage information found in Tables 2.5 (relative frequency and percentage) and 2.8 (cumulative percentage) in Section 2.2 for the restaurant meal cost data.

PHStat (untallied data) Use **Frequency Distribution**.

For the example, use the *PHStat* "The Frequency Distribution" instructions to construct a frequency distribution. PHStat constructs a frequency distribution that also includes columns for the percentages and cumulative percentages. To change the column of percentages to a column of relative frequencies, reformat that column. For example, open to the new worksheet that contains the center city restaurant frequency distribution and:

1. Select the cell range **C4:C12**, right-click, and select **Format Cells** from the shortcut menu.

2. In the **Number** tab of the Format Cells dialog box, select **Number** as the **Category** and click **OK**.

Then repeat these two steps for the new worksheet that contains the metro area restaurant frequency distribution.

Workbook (untallied data) Use the **Distributions workbook** as a model.

For the example, first construct a frequency distribution created using the *Workbook* "The Frequency Distribution" instructions. Open to the worksheet that contains the frequency distribution for the center city restaurants and:

1. Enter **Percentage** in cell **C3** and **Cumulative Pctage** in cell **D3**.

2. Enter **= B4/SUM($B$4:$B$12)** in cell **C4** and copy this formula down through row **12**.

3. Enter **= C4** in cell **D4**.

4. Enter **= C5 + D4** in cell **D5** and copy this formula down through row **12**.

5. Select the cell range **C4:D12**, right-click, and click **Format Cells** in the shortcut menu.

6. In the **Number** tab of the Format Cells dialog box, click **Percentage** in the **Category** list and click **OK**.

Then open to the worksheet that contains the frequency distribution for the metro area restaurants and repeat steps 1 through 6.

If you want column C to display relative frequencies instead of percentages, enter **Rel. Frequencies** in cell **C3**. Select the cell range **C4:C12**, right-click, and click **Format Cells** in the shortcut menu. In the **Number** tab of the Format Cells dialog box, click **Number** in the **Category** list and click **OK**.

Analysis ToolPak Use **Histogram** and then modify the worksheet created.

For the example, first construct the frequency distributions using the *Analysis ToolPak* instructions in "The Frequency Distribution." Then use the *Workbook* instructions to modify those distributions.

EG2.3 VISUALIZING CATEGORICAL VARIABLES

The Bar Chart and the Pie (or Doughnut) Chart

Key Technique Use the Excel bar, pie, or doughnut chart feature with a tabular summary of the variable. If necessary, use the Section EG2.1 "The Summary Table" to first create that summary.

Example Construct a bar or pie (or doughnut) chart from a summary table similar to Figure 2.4 on page 51.

PHStat Use **One-Way Tables & Charts**.

For the example, use the Section EG2.1 "The Summary Table" *PHStat* instructions, but in step 3, check either **Bar Chart** or **Pie Chart** (or both) in addition to entering a **Title**, checking **Percentage Column**, and clicking **OK**.

Workbook Use the **Summary Table workbook** as a model.

For the example, open to the **OneWayTable worksheet** of the **Summary Table workbook**. (The PivotTable in this worksheet was constructed using the Section EG2.1 "The Summary Table" instructions.) To construct a bar chart:

1. Select cell range **A4:B6**. (Begin the selection at cell B6 and not at cell A4.)

2. Select **Insert → Insert Column or Bar Chart** (or **Bar**), labeled #1 in the labeled Charts group that appears at the start of this guide, and select the **Clustered Bar** gallery item, the first selection in the 2-D Bar row.
   In older Excels, select the seventh icon (see page 89).

3. Right-click the **Risk** drop-down button in the chart and click **Hide All Field Buttons** on Chart.
   Step 3 does not apply to Excel for Mac.

4. Select **Design** (or `Chart Design`) → **Add Chart Element** → **Axis Titles** → **Primary Horizontal**. In older Excel, select **Layout** → **Axis Titles** → **Primary Horizontal Axis Title** → **Title Below Axis**.

5. Select the words "Axis Title" and enter **Frequency** as the new axis title.

6. Relocate the chart to a chart sheet and turn off the chart legend and gridlines by using the instructions in Appendix Section B.5.

For other problems, the horizontal axis may not begin at 0. If this occurs, right-click the horizontal axis and click **Format Axis** in the shortcut menu. In the Format Axis display, click **Axis Options**. In the Axis Options, enter **0** as the **Minimum** and then close the display. In Excels with two-panel dialog boxes, in the Axis Options right pane, click the first **Fixed** (for **Minimum**), enter **0** as the value, and then click **Close**.

To construct a pie or doughnut chart, replace steps 2, 4, and 6 with these:

2. Select **Insert** → **Pie** (#3 in the labeled Charts group) and select the **Pie** gallery item (or the **Doughnut** item).

4. Select **Design** (or `Chart Design`) → **Add Chart Element** → **Data Labels** → **More Data Label Options**. In the Format Data Labels display, click **Label Options**. In the Label Options, check **Category Name** and **Percentage** and clear the other Label Contains check boxes. Click **Outside End** under Label Position (pie chart only) and close the display.

6. Relocate the chart to a chart sheet and turn off the chart legend and gridlines by using the instructions in Appendix Section B.5.

To see the Label Options in step 4 in the newest versions of Excel, you may need to first click the chart (fourth) icon at the top of the display. In older versions of Excel, select **Layout** → **Data Labels** → **More Data Label Options** in step 4. To construct a doughnut chart in those Excels, select **Insert** → **Other Charts** and then the **Doughnut** item in step 2.

The Pareto Chart

Key Technique Use the Excel chart feature with a modified summary table.

Example Construct a Pareto chart of the incomplete ATM transactions equivalent to Figure 2.7 on page 54.

PHStat Use **One-Way Tables & Charts**.

For the example, open to the **DATA worksheet** of the **ATM Transactions workbook**. Select **PHStat** → **Descriptive Statistics** → **One-Way Tables & Charts**. In the procedure's dialog box:

1. Click **Table of Frequencies** (because the worksheet contains tallied data).

2. Enter **A1:B8** as the **Freq. Table Cell Range** and check **First cell contains label**.

3. Enter a **Title**, check **Pareto Chart**, and click **OK**.

Workbook Use the **Pareto workbook** as a model.

Note: The following instructions do not use the new Pareto *chart option that Microsoft Windows Excel users can select from the gallery that is displayed when icon #6 is clicked.*

For the example, open to the **ATMTable worksheet** of the **ATM Transactions workbook**. Begin by sorting the modified table by decreasing order of frequency:

1. Select row **11** (the Total row), right-click, and click **Hide** in the shortcut menu. (This prevents the total row from getting sorted.)

2. Select cell **B4** (the first frequency), right-click, and select **Sort** → **Sort Largest to Smallest**.

3. Select rows **10** and **12** (there is no row 11 visible), right-click, and click **Unhide** in the shortcut menu to restore row 11.

Next, add a column for cumulative percentage:

4. Enter **Cumulative Pct.** in cell **D3**. Enter **= C4** in cell **D4**. Enter **= D4 + C5** in cell **D5** and copy this formula down through **row 10**.

5. Adjust the formatting of column D as necessary.

Next, create the Pareto chart:

6. Select the cell range **A3:A10** and while holding down the **Ctrl** key also select the cell range **C3:D10**.
   In Excel for Mac, hold down the **Command** key, not the **Ctrl** key.

7. Select **Insert** → **Insert Column or Bar Chart** (#1 in the labeled Charts group), and select the **Clustered Column** gallery item, the first selection in the 2-D Column row.

8. Select **Format**. In the Current Selection group, select **Series "Cumulative Pct."** from the drop-down list and then click **Format Selection**.
   Select **Series "Cumulative Pct."** from the drop-down list at left and then click **More Formats**.

9. In the Format Data Series display, click **Series Options**. In the Series Options, click **Secondary Axis**, and then close the display. [In a Format Data Series *task pane*, click the third (chart) icon first if the Series Options are not displayed.]

10. With the "Cumulative Pct." series still selected, select **Design** (or `Chart Design`) → **Change Chart Type**. In the Change Chart Type display, click **Combo** in the **All Charts** tab. In the Cumulative Pct. drop-down list, select the **Line with Markers** gallery item. Check **Secondary Axis** for the Cumulative Pct. and click **OK**.
    In Excel for Mac, with the "Cumulative Pct." series still selected, select **Chart Design** → **Change Chart Type** → **Line** → **Line with Markers**.

Next, set the maximum value of the primary and secondary (left and right) *Y* axis scales to 100%. For each *Y* axis:

11. Right-click on the axis and click **Format Axis** in the shortcut menu.
12. In the Format Axis display, click **Axis Options**. In Axis Options, enter **1** as the **Maximum**. Click **Tick Marks**, select **Outside** from the **Major type** dropdown list, and close the display. [In a Format Data Series *task pane*, click the fourth (chart) icon first if the Axis Options are not displayed.]

    In Excel versions with two-panel dialog boxes, in the Axis Options right pane, click **Fixed** for **Maximum**, enter **1** as the value, and then click **Close**.

13. Relocate the chart to a chart sheet, turn off the chart legend and gridlines, and add chart and axis titles by using the instructions in Appendix Section B.5.

If you use a PivotTable as a summary table, replace steps 1 through 6 with these steps:

1. Add a percentage column in column C, using steps 11 through 14 of the *Workbook* "The Summary Table" instructions on page 89.
2. Add a cumulative percentage column in column D. Enter **Cumulative Pctage** in cell **D3**.
3. Enter **= C4** in cell **D4**. Enter **= C5 + D4** in cell **D5**, and copy the formula down through all the rows in the PivotTable.
4. Select the total row, right-click, and click **Hide** in the shortcut menu. (This prevents the total row from getting sorted.)
5. Right-click the cell that contains the first frequency (cell B4 in the example) and select **Sort → Sort Largest to Smallest**.
6. Select the cell range **C3:D10** which contains the percentage and cumulative percentage columns.

This Pareto chart will not include the category labels. To add the category labels:

1. Right-click on the chart and click **Select Data** in the shortcut menu.
2. In the Select Data Source display, click **Edit** that appears under **Horizontal (Category) Axis Labels**. In the Axis Labels display, drag the mouse to select and enter the axis labels cell range (A4:A10) and then click **OK**.

    In Excel for Mac, in the Select Data Source display, click the icon inside the **Horizontal (Category) axis labels** box and drag the mouse to select and enter the axis labels cell range.

3. Click **OK** to close the display.

Do not *type* the axis labels cell range in step 2 for the reasons that Appendix Section B.3 explains.

The Side-by-Side Chart

Key Technique Use an Excel bar chart that is based on a contingency table.

Example Construct a side-by-side chart that displays the Fund Type and Risk Level, similar to Figure 2.9 on page 56.

PHStat Use **Two-Way Tables & Charts**.

For the example, use the Section EG2.1 "The Contingency Table" *PHStat* instructions on page 90 but in step 4, check **Side-by-Side Bar Chart** in addition to entering a **Title** and clicking **OK**.

Workbook Use the **Contingency Table workbook** as a model.

For the example, open to the **TwoWayTable worksheet** of the **Contingency Table workbook** and:

1. Select cell **A3** (or any other cell inside the PivotTable).
2. Select **Insert → Insert Column or Bar Chart** (or **Bar**), labeled #1 in the labeled Charts group, and select the **Clustered Bar** gallery item, the first selection in the 2-D Bar row.

    In older Excels, select the #7 icon, not the #1 icon.

3. Right-click the **Risk** drop-down button in the chart and click **Hide All Field Buttons on Chart**.

    Step 3 does not apply to Excel for Mac.

4. Relocate the chart to a chart sheet, turn off the gridlines, and add chart and axis titles by using the instructions in Appendix Section B.5.

When creating a chart from a contingency table that is not a PivotTable, select the cell range of the contingency table, including row and column headings, but excluding the total row and total column, as step 1.

To switch the row and column variables in a side-by-side chart, right-click the chart and then click **Select Data** in the shortcut menu. In the Select Data Source dialog box, click **Switch Row/Column** and then click **OK**.

EG2.4 VISUALIZING NUMERICAL VARIABLES

The Stem-and-Leaf Display

Key Technique Enter leaves as a string of digits that begin with the ' (apostrophe) character.

Example Construct a stem-and-leaf display of the three-year return percentage for the value retirement funds, similar to Figure 2.10 on page 59.

PHStat Use **Stem-and-Leaf Display**.

For the example, open to the **Unstacked worksheet** of the **Retirement Funds workbook**. Select **PHStat → Descriptive Statistics → Stem-and-Leaf Display**. In the procedure's dialog box (shown top left on next page):

1. Enter **B1:B174** as the **Variable Cell Range** and check **First cell contains label**.
2. Click **Set stem unit as** and enter **10** in its box.
3. Enter a **Title** and click **OK**.

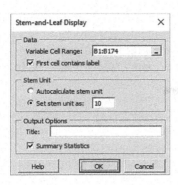

When creating other displays, use the **Set stem unit as** option sparingly and only if Autocalculate stem unit creates a display that has too few or too many stems. (Stem units you specify must be a power of 10.)

Workbook Manually construct the stems and leaves on a new worksheet to create a stem-and-leaf display. Adjust the column width of the column that holds the leaves as necessary.

The Histogram

Key Technique Modify an Excel column chart.

Example Construct histograms for the three-year return percentages for the growth and value retirement funds, similar to Figure 2.12 on page 60.

PHStat Use **Histogram & Polygons**.

For the example, open to the **DATA worksheet** of the **Retirement Funds workbook**. Select **PHStat→Descriptive Statistics→Histogram & Polygons**. In the procedure's dialog box (shown below):

1. Enter **K1:K480** as the **Variable Cell Range**, **Q1:Q11** as the **Bins Cell Range**, **R1:R10** as the **Midpoints Cell Range**, and check **First cell in each range contains label**.
2. Click **Multiple Groups-Stacked** and enter **C1:C480** as the **Grouping Variable Cell Range**. (In the DATA worksheet, the one-year return percentages are stacked. The column C values allow PHStat to unstack the values into growth and value groups.)
3. Enter a **Title**, check **Histogram**, and click **OK**.

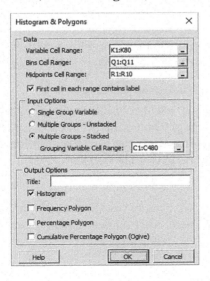

PHStat inserts two new worksheets, each of which contains a frequency distribution and a histogram. To relocate the histograms to their own chart sheets, use the instructions in Appendix Section B.5.

Because you cannot define an explicit lower boundary for the first bin, there can be no midpoint defined for that bin. Therefore, the **Midpoints Cell Range** you enter must have one fewer cell than the **Bins Cell Range**. PHStat uses the first midpoint for the second bin and uses "--" as the label for the first bin.

The example uses the workaround that "Classes and Excel Bins" discusses on page 46. When you use this workaround, the histogram bar labeled—will *always* be a zero bar. Appendix Section B.7 explains how you can delete this unnecessary bar from the histogram, as was done for the Section 2.4 examples.

Workbook Use the **Histogram workbook** as a model.

For the example, first construct frequency distributions for the growth and value funds. Open to the **Unstacked worksheet** of the **Retirement Funds workbook**. This worksheet contains the retirement funds data unstacked in columns A and B and a set of bin numbers and midpoints appropriate for those variables in columns D and E. Click the **insert worksheet icon** to insert a new worksheet.

In the new worksheet:

1. Enter a title in cell **A1**, Bins in cell **A3**, **Frequency** in cell **B3**, and **Midpoints** in cell **C3**.
2. Copy the bin number list in the cell range **D2:D11** of the **Unstacked worksheet** and paste this list into cell **A4** of the new worksheet.
3. Enter '-- in cell **C4**. Copy the midpoints list in the cell range **E2:E13** of the **Unstacked worksheet** and paste this list into cell **C5** of the new worksheet.
4. Select the cell range **B4:B13** that will hold the array formula.
5. Type, but do not press the **Enter** or **Tab** key, the formula **= FREQUENCY(Unstacked!$A$2:$A$307, $A$4: $A$13)**. Then, while holding down the **Ctrl** and **Shift** keys, press the **Enter** key to enter the array formula into the cell range **B4:B13**.
6. Adjust the worksheet formatting as necessary.

Steps 1 through 6 construct a frequency distribution for the growth retirement funds. To construct a frequency distribution for the value retirement funds, insert another worksheet and repeat steps 1 through 6, entering **= FREQUENCY (Unstacked!$B$2:$B$174, $A$4: $A$13)** as the array formula in step 5.

Having constructed the two frequency distributions, continue by constructing the two histograms. Open to the worksheet that contains the frequency distribution for the growth funds and:

1. Select the cell range **B3:B13** (the cell range of the frequencies).
2. Select **Insert→ Insert Column or Bar Chart** (#1 in the labeled Charts group) and select the **Clustered Column** gallery item, the first selection in the 2-D Column row.

3. Right-click the chart and click **Select Data** in the short-cut menu.

In the Select Data Source display:

4. Click **Edit** under the **Horizontal (Categories) Axis Labels** heading. In the Axis Labels display, drag the mouse to select and enter **$C$3:$C$13** as the midpoints cell range and click **OK**.

   In the Select Data Source display, click the icon inside the **Horizontal (Category) axis labels** box and drag the mouse to select and enter the midpoints cell range **$C$3:$C$13**.

5. Click **OK**.

Do not *type* the axis label cell range in step 4 for the reasons that Appendix Section B.3 explains.

In the new chart:

6. Right-click inside a bar and click **Format Data Series** in the shortcut menu.

7. In the Format Data Series display, click **Series Options**. In the Series Options, click **Series Options**, enter **0** as the **Gap Width** and then close the display. (To see the second Series Options, you may have to first click the chart [third] icon near the top of the task pane.)

   In Excel for Mac, there is only one Series Options label, and the Gap Width setting is displayed without having to click Series Options.

   In older Excels, click **Series Options** in the left pane, and in the Series Options right pane, change the **Gap Width** slider to **No Gap** and then click **Close**.

8. Relocate the chart to a chart sheet, turn off the chart legend and gridlines, add axis titles, and modify the chart title by using the Appendix Section B.5 instructions.

This example uses the workaround that "Classes and Excel Bins" discusses on page 46. Appendix Section B.7 explains how the unnecessary histogram bar that this workaround creates, which will *always* be a zero bar, can be deleted, as was done for the Section 2.4 examples.

Analysis ToolPak Use **Histogram**.

For the example, open to the **Unstacked worksheet** of the **Retirement Funds workbook** and:

1. Select **Data → Data Analysis**. In the Data Analysis dialog box, select **Histogram** from the **Analysis Tools** list and then click **OK**.

In the Histogram dialog box:

2. Enter **A1:A307** as the **Input Range** and enter **D1:D11** as the **Bin Range**.

3. Check **Labels**, click **New Worksheet Ply**, and check **Chart Output**.

4. Click **OK** to create the frequency distribution and histogram on a new worksheet.

In the new worksheet:

5. Follow steps 5 through 9 of the *Analysis ToolPak* instructions in "The Frequency Distribution" on pages 91 and 92.

These steps construct a frequency distribution and histogram for the growth funds. To construct a frequency distribution and histogram for the value funds, repeat the nine steps, but in step 2 enter **B1:B174** as the **Input Range**. You will need to correct several formatting errors to the histograms that Excel constructs. For each histogram, first change the gap widths between bars to 0. Follow steps 6 and 7 of the *Workbook* instructions of this section, noting the special instructions that appear after step 8.

Histogram bars are labeled by bin numbers. To change the labeling to midpoints, open to each of the new work-sheets and:

1. Enter **Midpoints** in cell **C3** and '**--** in cell **C4**. Copy the cell range **E2:E10** of the **Unstacked worksheet** and paste this list into cell **C5** of the new worksheet.

2. Right-click the histogram and click **Select Data**.

In the Select Data Source display:

3. Click **Edit** under the **Horizontal (Categories) Axis Labels** heading. In the Axis Labels display, drag the mouse to select and enter the cell range **$C$4:$C$13** and click **OK**.

   In Excel for Mac, in the Select Data Source display, click the icon inside the **Horizontal (Category) axis labels** box and drag the mouse to select and enter the cell range **$C$4:$C$13**.

4. Click **OK**.

5. Relocate the chart to a chart sheet, turn off the chart legend, and modify the chart title by using the instructions in Appendix Section B.5.

Do not type the axis label cell range in step 3 as you would otherwise do for the reasons explained in Appendix Section B.3.

This example uses the workaround that "Classes and Excel Bins" discusses on page 46. Appendix Section B.7 explains how the unnecessary histogram bar that this workaround creates can be deleted, as was done for the Section 2.4 examples.

The Percentage Polygon and the Cumulative Percentage Polygon (Ogive)

Key Technique Modify an Excel line chart that is based on a frequency distribution.

Example Construct percentage polygons and cumulative percentage polygons for the three-year return percentages for the growth and value retirement funds, similar to Figure 2.14 on page 61 and Figure 2.16 on page 62.

PHStat Use **Histogram & Polygons**.

For the example, use the *PHStat* instructions for creating a histogram on page 95 but in step 3 of those instructions, also check **Percentage Polygon** and **Cumulative Percentage Polygon (Ogive)** before clicking **OK**.

Workbook Use the **Polygons workbook** as a model.

For the example, open to the **Unstacked worksheet** of the **Retirement Funds workbook** and follow steps 1 through 6 of the *Workbook* "The Histogram" instructions on page 95 to construct a frequency distribution for the growth funds.

Repeat the steps to construct a frequency distribution for the value funds using the instructions that immediately follow step 6. Open to the worksheet that contains the growth funds frequency distribution and:

1. Select **column C**. Right-click and click **Insert** in the shortcut menu. Right-click and click **Insert** in the shortcut menu a second time. (The worksheet contains new, blank columns C and D and the midpoints column is now column E.)
2. Enter **Percentage** in cell **C3** and **Cumulative Pctage.** in cell **D3**.
3. Enter **= B4/SUM($B$4:$B$13)** in cell **C4** and copy this formula down through **row 13**.
4. Enter **= C4** in cell **D4**.
5. Enter **= C5 + D4** in cell **D5** and copy this formula down through row **13**.
6. Select the cell range **C4:D13** right-click, and click **Format Cells** in the shortcut menu.
7. In the **Number** tab of the Format Cells dialog box, click **Percentage** in the **Category** list and click **OK**.

Open to the worksheet that contains the value funds frequency distribution and repeat steps 1 through 7. To construct the percentage polygons, open to the worksheet that contains the growth funds distribution and:

1. Select cell range **C4:C13**.
2. Select **Insert→Line** (#3 in the labeled Charts group), and select the **Line with Markers** gallery item in the 2-D Line group.
3. Right-click the chart and click **Select Data** in the shortcut menu.

In the Select Data Source display:

4. Click **Edit** under the **Legend Entries (Series)** heading. In the Edit Series dialog box, enter the *formula* **= "Growth Funds"** as the **Series name** and click **OK**.
   In Excel for Mac, enter the formula as the **Name**.
5. Click **Edit** under the **Horizontal (Categories) Axis Labels** heading. In the Axis Labels display, drag the mouse to select and enter the cell range **$E$4:$E$13** and click **OK**.
   In Excel for Mac, in the Select Data Source display, click the icon inside the **Horizontal (Category) axis labels** box and drag the mouse to select and enter the cell range **$E$4:$E$13**.
6. Click **OK**.
7. Relocate the chart to a chart sheet, turn off the chart gridlines, add axis titles, and modify the chart title by using the instructions in Appendix Section B.5.

In the new chart sheet:

8. Right-click the chart and click **Select Data** in the shortcut menu.

In the Select Data Source display:

9. Click **Add** under the **Legend Entries (Series)** heading. In the Edit Series dialog box, enter the

formula **= "Value Funds"** as the **Series name** and press **Tab**.
   In Excel for Mac, click "**+**" icon below the **Legend entries (Series)** list. Enter the formula = "Value Funds" as the **Name**.
10. With the placeholder value in **Series values** highlighted, click the sheet tab for the worksheet that contains the value funds distribution. In that worksheet, drag the mouse to select and enter the cell range **$C$4:$C$13** and click **OK**.
    In Excel for Mac, click the icon in the **Y** values box. Click the sheet tab for the worksheet that contains the value funds distribution and, in that worksheet, drag the mouse to select and enter the cell range **$C$4:$C$13**.
11. Click **Edit** under the **Horizontal (Categories) Axis Labels** heading. In the Axis Labels display, drag the mouse to select and enter the cell range **$E$4:$E$13** and click **OK**.
    In Excel for Mac, in the Select Data Source display, click the icon inside the **Horizontal (Category) axis labels** box and drag the mouse to select and enter the cell range **$E$4:$E$13**. Click **OK**.

Do not *type* the axis label cell range in steps 10 and 11 for the reasons explained in Appendix Section B.3.

To construct the cumulative percentage polygons, open to the worksheet that contains the growth funds distribution and repeat steps 1 through 12, but in step 1, select the cell range **D4:D13;** in step 5, drag the mouse to select and enter the cell range **$A$4:$A$13;** and in step 11, drag the mouse to select and enter the cell range **$D$4:$D$13**.

If the *Y* axis of the cumulative percentage polygon extends past 100%, right-click the axis and click **Format Axis** in the shortcut menu. In the Format Axis display, click **Axis Options**. In the Axis Options, enter **0** as the **Minimum** and then close the display.

In Excels with two-panel dialog boxes, in the Axis Options right pane, click the first **Fixed** (for **Minimum**), enter **0** as the value, and then click **Close**.

EG2.5 VISUALIZING TWO NUMERICAL VARIABLES

The Scatter Plot

Key Technique Use the Excel scatter chart.

Example Construct a scatter plot of revenue and value for NBA teams, similar to Figure 2.17 on page 66.

PHStat Use **Scatter Plot**.

For the example, open to the **DATA worksheet** of the **NBAValues workbook**. Select **PHStat→Descriptive Statistics→Scatter Plot**. In the procedure's dialog box (shown top left on next page):

1. Enter **D1:D31** as the **Y Variable Cell Range**.
2. Enter **C1:C31** as the **X Variable Cell Range**.
3. Check **First cells in each range contains label**.
4. Enter a **Title** and click **OK**.

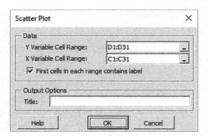

To add a superimposed line like the one shown in Figure 2.17. click the chart and use step 3 of the *Workbook* instructions.

Workbook Use the **Scatter Plot workbook** as a model.

For the example, open to the **DATA worksheet** of the **NBAValues workbook** and:

1. Select the cell range **C1:D31**.
2. Select **Insert→Scatter (X, Y) or Bubble Chart** (#6 in the labeled Charts group) and select the **Scatter** gallery item.
    Excel for Mac labels the #6 icon **X Y (Scatter)**.
3. Select **Design** (or **Chart Design**) → **Add Chart Element→Trendline→Linear**.
4. Relocate the chart to a chart sheet, turn off the chart legend and gridlines, add axis titles, and modify the chart title by using the Appendix Section B.5 instructions.

When constructing Excel scatter charts with other variables, make sure that the X variable column precedes (is to the left of) the Y variable column. (If the worksheet is arranged Y then X, cut and paste so that the Y variable column appears to the right of the X variable column.)

The Time-Series Plot

Key Technique Use the Excel scatter chart.

Example Construct a time-series plot of movie revenue per year from 1995 to 2018, similar to Figure 2.18 on page 67.

Workbook Use the **Time Series workbook** as a model.

For the example, open to the **DATA worksheet** of the **Movie Revenues workbook** and:

1. Select the cell range **A1:B25**.
2. Select **Insert→Scatter (X, Y) or Bubble Chart** (#6 in the labeled Charts group) and select the **Scatter with Straight Lines and Markers** gallery item.
    Excel for Mac labels the #6 icon **X Y (Scatter)**.
3. Relocate the chart to a chart sheet, turn off the chart legend and gridlines, add axis titles, and modify the chart title by using Appendix Section B.5 instructions.

When constructing time-series charts with other variables, make sure that the X variable column precedes (is to the left of) the Y variable column. (If the worksheet is arranged Y then X, cut and paste so that the Y variable column appears to the right of the X variable column.)

Multidimensional Contingency Tables

Key Technique Use the Excel PivotTable feature.

Example Construct a PivotTable showing percentage of overall total for Fund Type, Risk Level, and Market Cap for the retirement funds sample, similar to the one shown at the right in Figure 2.19 on page 69.

Workbook Use the **MCT workbook** as a model.

For the example, open to the **DATA worksheet** of the **Retirement Funds workbook** and:

1. Select **Insert→PivotTable**.

In the Create PivotTable display:

2. Click **Select a table or range** and enter **A1:N480** as the **Table/Range**.
3. Click **New Worksheet** and then click **OK**.

In the PivotTable Fields (PivotTable Builder or PivotTable Field List in older Excels) display (partially shown below):

4. Drag **Fund Type** in the **Choose fields to add to report** box and drop it in the **Rows** (or **Row Labels**) box.
5. Drag **Market Cap** in the **Choose fields to add to report** box and drop it in the **Rows** (or **Row Labels**) box.
6. Drag **Risk Level** in the **Choose fields to add to report** box and drop it in the **Columns** (or **Column Labels**) box.
7. Drag **Fund Type** in the **Choose fields to add to report** box a second time and drop it in the **Σ Values** box. The dropped label changes to **Count of Fund Type**.

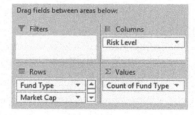

8. Click (not right-click) the dropped label **Count of Fund Type** and then click **Value Field Settings** in the shortcut menu. In the Value Field Settings display, click the **Show Values As** tab and select **% of Grand Total** from the **Show values as** drop-down list, shown at left on the next page.
    Click the "*i*" icon to the right of the dropped label **Count of Fund Type**. In the PivotTable Field dialog box, click the **Show data as** tab and select **% of Grand Total (% of total** in older Excels for Mac) from the drop-down list, shown at right on the next page.
9. Click **OK**.

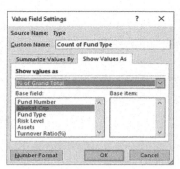

In the PivotTable:

10. Enter a title in cell **A1**.
11. Follow steps 6 through 10 of the *Workbook (untallied data)* "The Contingency Table" instructions on page 90 that relabel the rows and columns and rearrange the order of the risk category columns.

Adding a Numerical Variable

Key Technique Alter the contents of the Σ Values box in the PivotTable Field List pane.

Example Construct the Figure 2.20 PivotTable of Fund Type, Risk Level, and Market Cap on page 69 that shows the mean ten-year return percentage for the retirement funds sample.

Workbook Use the **MCT workbook** as a model.

For the example, first construct the PivotTable showing percentage of overall total for Fund Type, Risk Level, and Market Cap for the retirement funds sample using the 11-step instructions of the "Multidimensional Contingency Table" *Workbook* instructions that starts on page 98. Then continue with these steps:

12. If the PivotTable Fields pane is not visible, right-click cell **A3** and click **Show Field List** in the shortcut menu.
    In older Excels for Mac, if the PivotTable Builder (or Field List) display is not visible, select **PivotTable Analyze → Field List**.

In the display:

13. Drag the blank label (changed from *Count of Fund Type* in a prior step) in the Σ **Values** box and drop it outside the display to delete. In the PivotTable, all of the percentages disappear.
14. Drag **10YrReturn** in the **Choose fields to add to report** box and drop it in the Σ **Values** box. The dropped label changes to *Sum of 10YrReturn*.
15. Click (not right-click) **Sum of 10YrReturn** and then click **Value Field Settings** in the shortcut menu. In the Value Field Settings display, click the **Summarize Values By** tab and select **Average** from the **Summarize value field by** drop-down list.
    Click the "*i*" icon to the right of the label **Sum of 10YrReturn**. In the PivotTable Field dialog box, click the **Summarize by** tab and select **Average** from the list.

16. Click **OK**. The label in the Σ **Values** box changes to *Average of 10YrReturn*.

In the PivotTable:

17. Select cell range **B5:E13**, right-click, and click **Format Cells** in the shortcut menu. In the **Number** tab of the Format Cells dialog box, click Number, set the **Decimal places** to **2**, and click **OK**.

EG2.7 VISUALIZING a MIX of VARIABLES

PivotChart

Key Technique Use the PivotChart feature with a previously constructed PivotTable. (The PivotChart feature is not included in older Excels for Mac.)

Example Construct the PivotChart based on the Figure 2.20 PivotTable of type, risk, and market cap showing mean ten-year return percentage, shown in Figure 2.23 on page 71.

Workbook Use the **MCT workbook** as a model.

For the example, open to the **MCT worksheet** of the **MCT workbook** and:

1. Select cell **A3** (or any other cell inside the PivotTable).
2. Select **Insert → PivotChart**.
3. In the Insert Chart display, click **Bar** in the **All Charts** tab and select the **Clustered Bar** gallery item and click **OK**.
   In Excel for Mac, with the newly inserted chart still selected, select **Insert Column** and select the **Clustered Bar** gallery item in the 2-D Bar group.
4. Relocate the chart to a chart sheet, turn off the gridlines, and add chart and axis titles by using the instructions in Appendix Section B.5.

In the PivotTable, collapse the **Growth** and **Value** categories, hiding the **Market Cap** categories. Note that contents of PivotChart changes to reflect changes made to the PivotTable.

Treemap

Key Technique Use the Excel treemap feature with a specially prepared tabular summary that includes columns that express hierarchical (tree) relationships. (The treemap feature is not included in older Excel versions.)

Example Construct the Figure 2.24 treemap on page 72 that summarizes the sample of 479 retirement funds by Fund Type and Market Cap.

Workbook Use **Treemap**.

For the example, open to the **StackedSummary worksheet** of the **Retirement Funds workbook**. This worksheet contains sorted fund type categories in column A, market cap categories in column B, and frequencies in column C. Select the cell range **A1:C7** and:

1. Select **Insert → Insert Hierarchy Chart** (#4 in the labeled Charts group) and select the **Treemap** gallery item.
2. Click the chart title and enter a new title for the chart.

3. Click one of the tile labels and increase the point size to improve readability. (This change affects all treemap labels.)
4. Select **Design** (or **Chart Design**)→**Quick Layout** and select the **Layout 2** gallery item (which adds the "branch" data labels).
5. Select **Design** (or **Chart Design**)→**Add Chart Element**→**Legend Bottom**.
6. Right-click in the whitespace near the title and select **Move Chart**.
7. In the Move Chart dialog box, click **New Sheet** and click **OK**.

PHStat Use **Treemap**.
As an example, open to the **StackedSummary worksheet** of the **Retirement Funds workbook**. This worksheet contains sorted fund type categories in column A, market cap categories in column B, and frequencies in column C. Select **PHStat→Descriptive Statistics→Treemap**. In the procedure's dialog box:

1. Enter **A1:C7** as the **Hierarchical Cell Range**.
2. Check **First cell in each range contains label**.
3. Enter a **Title** and click **OK**.

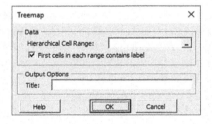

Sunburst Chart

Workbook Use **Sunburst**.
As an example, open to the **StackedHierarchy worksheet** of the **Retirement Funds workbook**. This worksheet contains sorted fund type categories in column A, sorted market cap categories in column B, sorted star rating categories in column C, and frequencies in column D. Select the cell range **A1:D31** and:

1. Select **Insert→Insert Hierarchy Chart** (#4 in the labeled Charts group) and select the **Sunburst** gallery item.
2. Click the chart title and enter a new title for the chart.
3. Click one of the segments and increase the point size to improve readability. (This will change the point size of all labels.)
4. Right-click in the whitespace near the title and select **Move Chart**.
5. In the Move Chart dialog box, click **New Sheet** and click **OK**.

Sparklines

Key Technique Use the sparklines feature.

Example Construct the sparklines for movie revenues per month for the period 2005 to 2018 shown in Figure 2.25 on page 72.

Workbook Use the **Sparklines workbook** as a model.

For the example, open to the **DATA worksheet** of the **Monthly Movie Revenues workbook** and:

1. Select **Insert→Line** (in the **Sparklines group**).
2. In the Create Sparklines dialog box (shown below), enter **B2:M13** as the **Data Range** and **P2:P13** as the **Location Range**.
3. Click **OK**.

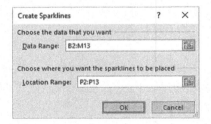

With the sparklines selected:

4. Select **Design→Axis→Same for All Sparklines** under the Vertical Axis Minimum Value Options. Select **Design→Axis→Same for All Sparklines** under the Vertical Axis Maximum Value Options.
   In Excel for Mac, select **Sparkline→Axis**. In the Axes dialog box, click the **Vertical** tab. In that tab, click the two **Same for all sparklines** option buttons and then click **OK**.
5. Select rows 2 through 13. Right-click and click **Row Height** in the shortcut menu. In the Row height dialog box, enter **30** (**0.85** in Excel for Mac) as the **Row Height** and click **OK**.
   In older Excels for Mac, enter **0.85** as the row height.

Optionally, insert one or more rows at the top of the worksheet for a title and copy the month values in column A to column L for easier reference.

PHStat Use **Sparklines**.
For the example, open to the **DATA worksheet** of the **Monthly Movie Revenues workbook**. Select **PHStat→Descriptive Statistics→Sparklines**. In the procedure's dialog box:

1. Enter **A1:A13** as the **Sparkline Labels Cell Range**.
2. Enter **B1:M13** as the **Sparkline Data Cell Range**.
3. Check **First cell in each range contains label**.
4. Select **Separate Chart Sheet**.
5. Enter a **Title** and click **OK**.

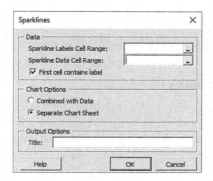

EG2.8 FILTERING and QUERYING DATA

Key Technique Use the Excel data filter feature.

Example Filter the DATA worksheet of the Retirement Funds workbook such that only funds with a four or five star ratings are displayed.

Workbook Use **Filter**.

For the example, open to the **DATA worksheet** of the **Retirement Funds workbook**. Select columns A through O (Fund Number through Star Rating) and:

1. Select **Data→Filter**. Each column displays a pull-down list button.
2. Click the **pull-down button** for column O (Star Rating).

In the pull-down dialog box (shown next column for Windows Excel and Excel for Mac):

3. Clear the **(Select All)** check box to clear all check boxes.
4. Check the **Four** and **Five** check boxes.
5. In Windows Excel, click **OK**.

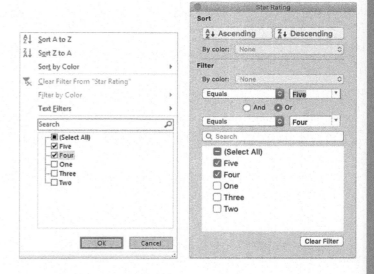

Excel displays the 108 retirement funds that have either a four- or five-star rating. Excel displays the original row number indices of these 108 rows in color and provides visual feedback of the gaps between nonconsecutive row numbers.

To remove this column data filter, click the column O pull-down button and select **Clear Filter from Star Rating**.

CHAPTER 2

▼TABLEAU GUIDE

TG2.1 ORGANIZING CATEGORICAL VARIABLES

The Summary Table

Use **text tables**.

For example, create a frequency and percentage summary table similar to Figure 2.1 on page 40, in a new Tableau workbook, click **Connect to Data** and open the **Retirement Funds For Tableau Excel workbook**. In a new Tableau worksheet:

1. Drag **Risk Level** and drop it in the **Rows** shelf. The shell of a summary table appears in the display area.

2. Drag *Number of Records* and drop it over the "Abc" column in the table. The green pill SUM(Number of Records) appears in the Marks card area.

3. Right-click the green pill and select **Quick Table Calculation→Percent of Total**. The green pill displays a triangle symbol to indicate a table calculation.

4. Drag *Number of Records* and drop it over the **Text icon** in the **Marks** card area to create a second green pill.

5. Select **Analysis→Totals** and check **Show Column Grand Totals**.

Tableau creates the summary table with risk level rows that contain both percentage and frequencies for each category (not separate columns). Tableau lists categories in alphabetical order. To list categories in the order Low, Average, and High, right-click the **Risk Level pill** in the **Rows** shelf and select **Sort** in the shortcut menu. In the Sort [Risk Level] dialog box (shown below):

6. Click **Manual**.
7. In the list box, select **Low** and then click **Up** twice to reorder categories.
8. Click **OK**.

The Risk Level pill displays the three-lines symbol to indicate a sorted list of categories. Enter a worksheet title and, optionally, adjust font and type characteristics using the Appendix Section B.5T instructions.

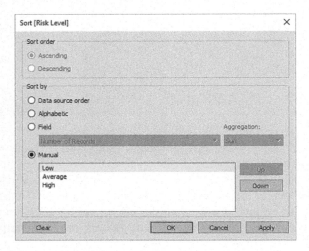

The Contingency Table

Use **text tables**.

For example, to create a contingency table displaying Fund Type and Risk Level similar to Figure 2.2 on page 40, in a new Tableau workbook, click **Connect to Data** and open the **Retirement Funds for Tableau Excel workbook**. In a new Tableau worksheet:

1. Drag **Risk Level** and drop it in the **Columns** shelf. The shell of a summary table appears in the display area.
2. Drag **Fund Type** and drop it in the **Rows** shelf.
3. Drag *Number of Records* and drop it over the "Abc" cells in the table. The green pill SUM(Number of Records) appears in the Marks card area.
4. Select **Analysis ➜ Totals** and check **Show Row Grand Totals**.
5. Select **Analysis ➜ Totals** and check **Show Column Grand Totals**.

Tableau creates the summary table with risk level categories in alphabetical order. To list categories in the order Low, Average, and High, right-click the **Risk Level pill** in the **Columns** shelf and select **Sort** in the shortcut menu. In the Sort [Risk Level] dialog box, follow steps 6 through 8 in the previous section. Enter a worksheet title and, optionally, adjust font and type characteristics using the Appendix Section B.5T instructions.

TG2.2 ORGANIZING NUMERICAL VARIABLES

The Frequency Distribution

Use **View Data** with **histogram**.

For example, to create a restaurant meal cost data frequency distribution similar to Table 2.3 on page 45, first follow the Section TG2.4 "The Histogram" instructions to create histograms for meal costs at center city and metro area restaurants. Right-click the whitespace above the bars in either histogram and select **View Data** in the shortcut menu. In the Summary tab of the View Data window (not shown):

1. Click the **Cost (bin)** column header to sort the bin values in ascending order.
2. Click the **Location** column header to sort the values to form the two frequency distributions (shown below).

Cost (bin)	Location	Count of Cost
20	Center City	4
30	Center City	3
40	Center City	12
50	Center City	14
60	Center City	7
70	Center City	4
80	Center City	5
90	Center City	1
20	Metro Area	4
30	Metro Area	14
40	Metro Area	16
50	Metro Area	12
60	Metro Area	2
70	Metro Area	1
80	Metro Area	1
90	Metro Area	

The Count of Cost column contains the frequency counts. The rows of data in the View Data window are plain text that can be copied, pasted, and rearranged into other programs for better formatting and presentation.

TG2.3 VISUALIZING CATEGORICAL VARIABLES

The Bar Chart or the Pie Chart

Use **horizontal bars** or **pie charts**.

For example, to construct a Tableau bar or pie chart that summarizes Risk Level similar to Figures 2.5 and 2.6 on pages 52 and 53, in a new Tableau workbook, click **Connect to Data** and open the **Retirement Funds for Tableau Excel workbook**. In a new Tableau worksheet:

1. Drag **Risk Level** and drop it in the **Rows** shelf. The shell of a summary table appears in the display area.
2. Drag **Fund Type** and drop it in the **Rows** shelf.
3. Drag *Number of Records* and drop it in the **Columns** shelf.

Tableau displays a bar chart. To change visualization to a pie chart, click the **pie charts icon** in the **Show Me** gallery.

Enter a worksheet title, turn off gridlines, and, optionally, adjust font and type characteristics using the Appendix Section B.5T instructions.

If the Risk Level categories need to be reordered as Low, Average, and High, right-click the **Risk Level pill** in the **Rows** shelf and select **Sort** in the shortcut menu. In the Sort [Risk Level] dialog box, follow steps 6 through 8 on page 102.

The Pareto Chart

Use **horizontal bars** and **lines**.

For example, to construct the Figure 2.7 Pareto chart of the incomplete ATM transactions on page 54, in a new Tableau workbook, click **Connect to Data** and open the **ATM Transactions Excel workbook**. Because this Excel workbook contains two worksheets, drag **DATA** in the **Sheets** list and drop it in the *Drag sheets here* area to manually establish the data source. In a new Tableau worksheet:

1. Drag **Cause** and drop it in the **Columns** shelf.
2. Drag **Frequency** and drop it in the **Rows** shelf.

A vertical bar chart appears and the green pill SUM (Frequency) appears in the Rows shelf.

3. Right-click the **Cause pill** in the **Columns** shelf and select **Sort** in the shortcut menu.

In the Sort[Cause] dialog box (shown below):

4. Select **Field** in the **Sort By** pull-down list.
5. Click **Descending**.
6. Select **Frequency** from the **Field Name** pull-down list.
7. Leave the Aggregation pull-down selected to Sum.
8. Click the **Close icon** for the dialog box (the "X" icon in the upper-right corner of the dialog box).

9. Drag **Frequency** and drop it on the right edge of the chart. (Tableau will reveal a vertical dashed line when the dragged Frequency reaches the right edge.)

The visual appearance of the chart may change.

10. In the Marks area, click the **first SUM(Frequency) entry** and then select **Bar** from the pull-down list below this entry.

11. In the Marks area, click the **second SUM(Frequency) entry** and then select **Line** from the pull-down list below this entry.

The visual appearance of the chart begins to look like a Pareto chart.

12. Right-click the **second SUM(Frequency) pill** in the **Rows** shelf and select **Add Table Calculation** in the shortcut menu.

In the Table Calculation dialog box:

13. Select **Running Total** from the first **Calculation Type** pull-down list.
14. Check **Add secondary calculation**.

Tableau changes (the first) Calculation Type to Primary Calculation Type and widens the dialog box (shown below).

15. Select **Percent of Total** from the **Secondary Calculation Type** pull-down list.
16. Click the **Close icon** for the dialog box.

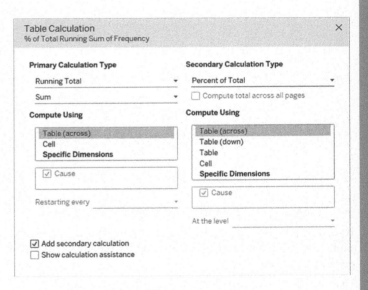

17. Click **Color** in the **Marks** card area.
18. In the Color gallery (shown below), select the **deep orange square icon** and the **second Markers icon** (line with markers) in the **Effects** group.

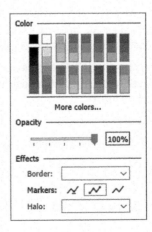

Both the left and right *Y* axes need to be equivalently scaled, with upper end values scaled to 100% or an equivalent amount. Begin with the axis for the line chart.

19. Right-click the **right *Y* axis** and select **Edit Axis**.

In the Edit Axis [% of Total Running Sum of Frequency] dialog box (shown below):

20. Click **Fixed** and enter **1** as the **Fixed end**.
21. Enter **Percent** as the **Title**.
22. Click the **Close icon** for the dialog box (the upper-right "X" icon).

The left *Y* axis scale (for the bars) is not equivalent to the right *Y* axis scale. The upper bound of the left scale needs to equal the sum of all frequencies (724, for this example). To correct this error:

23. Right-click the **left *Y* axis** and select **Edit Axis**.

In the Edit Axis [Frequency] dialog box (not shown):

24. Click **Fixed** and enter **724** as the **Fixed end**.
25. Click the **Close icon** for the dialog box (the upper-right "X" icon).

To express the left *Y* axis scale in terms of percentage, replace Steps 23 through 25 with the following Steps 23 through 26.

23. Right-click the **first SUM(Frequency) pill** in the **Rows** shelf and select **Quick Table Calculation → Percent of Total** in the shortcut menu.

24. Right-click the **left *Y* axis** and select **Edit Axis**.

In the Edit Axis [% of Total Frequency] dialog box:

25. Click **Fixed** and enter **1** (representing 100%) as the **Fixed end**.
26. Click the **Close icon** for the dialog box (the upper-right "X" icon).

Enter a worksheet title, turn off gridlines, and, optionally, adjust font and type characteristics using the Appendix Section B.5T instructions.

Visualizing Two Categorical Variables

Use **horizontal bars**.

For example, to construct a side-by-side chart that displays the fund type and risk level, similar to Figure 2.9 on page 56, in a new Tableau workbook, click **Connect to Data** and open the **Retirement Funds for Excel workbook**. In a new Tableau worksheet:

1. Drag **Fund Type** and drop it in the **Rows** shelf.
2. Drag **Risk Level** and drop it in the **Rows** shelf.
3. Drag *Number of Records* and drop it in the **Columns** shelf.

Tableau creates a side-by-side chart in which all bars have the same color. To color the bars differently for each level of risk, as the Figure 2.9 charts do, drag **Risk Level** a second time and drop it over the **Color icon** in the **Marks** card area. Enter a worksheet title, turn off gridlines, and, optionally, adjust font and type characteristics using the Appendix Section B.5T instructions.

If the Risk Level categories need to be reordered as Low, Average, and High, right-click the **Risk Level pill** in the **Rows** shelf and select **Sort** in the shortcut menu. In the Sort [Risk Level] dialog box, follow steps 6 through 8 on page 102.

TG2.4 VISUALIZING NUMERICAL VARIABLES

The Histogram

Use **histogram**.

For example, to construct the Figure 2.11 histograms for meal costs at center city and metro area restaurants on page 59, in a new Tableau workbook, click **Connect to Data** and open the **Restaurants for Excel workbook**. In a new Tableau worksheet:

1. Drag **Cost** and drop it in the **Columns** shelf.
2. Click the **histogram icon** in the **Show Me** gallery. The green pill CNT(3Cost) appears in the Rows shelf.
3. Right-click **Cost (bin)** in the Dimensions list and select **Edit** in the shortcut menu.
4. In the Edit Bins [Cost] dialog box (shown below), enter **10** in the **Size of bins** box and click **OK**.

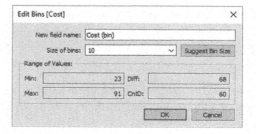

5. Drag **Location** and drop it in the **Rows** shelf. Tableau creates a pair of histograms, the *X* axes of which are mislabeled.

6. Right-click the *X* **axis** and select **Edit Axis** in the shortcut menu.

In the Edit Axis [Cost (bin)] dialog box:

7. In the **General** tab, enter **Meal Cost ($)** as the **Title** and click the **Tick Marks** tab.

8. In the Tick Marks tab (shown below), click **Fixed** in the **Major Tick Marks** group and enter **25** as the **Tick Origin** and click the **Close icon** for the dialog box (the upper-right "X" icon).

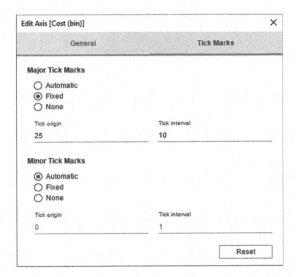

9. Back in the worksheet, right-click the *Y* **axis**, select **Edit Axis** in the shortcut menu.

10. In the Edit Axis [Count of Cost] dialog box, enter **Frequency** as the **Title** and click the **Close icon** for the dialog box (the upper-right "X" icon).

Enter a worksheet title, turn off gridlines, and, optionally, adjust font and type characteristics using the Appendix Section B.5T.

Histogram Classes. Bins that Tableau creates are true classes in the form low value through less than upper value. A bin size of 10 and a first lower class boundary of 20 creates the classes that match the Table 2.7 classes: 20 but less than 30, 30 but less than 40, and so forth. Tableau Desktop Public Edition automatically selects the first lower class boundary—users cannot specify that value as they can in other programs that this book discusses.

TG2.5 VISUALIZING TWO NUMERICAL VARIABLES

The Scatter Plot

Use **scatter plots**.

For example, to construct the Figure 2.17 scatter plot of revenue and value for NBA teams on page 66, in a new Tableau workbook, click **Connect to Data** and open the **NBAValues Excel workbook**. In a new Tableau worksheet:

1. Drag **Revenue** and drop it in the **Columns** shelf.
2. Drag **Current Value** and drop it in the **Rows** shelf.
3. Drag **Team Code** and drop it over the **Detail icon** in the **Marks** card area.
4. In the Marks card area, select **Circle** from the pull-down list.

Enter a worksheet title, turn off gridlines, and, optionally, adjust font and type characteristics using the Appendix Section B.5T instructions.

The Time-Series Plot

Use **lines (discrete)**.

For example, to construct time-series plot of movie revenue per year from 1995 to 2018, similar to Figure 2.18 on page 67, in a new Tableau workbook, click **Connect to Data** and open the **Movie Revenues Excel workbook**. In a new Tableau worksheet:

1. Right-click **Year** in the Dimensions list and select **Change Data Type→Date** in the shortcut menu.
2. Drag **Year** and drop it in the **Columns** shelf.
3. Drag **Revenues** and drop it in the **Rows** shelf.
4. Click the **Color icon** in the **Marks** card area and click the **second Markers icon** (line with markers) in the **Effects** group.
5. Click anywhere in the time-series plot.

Enter a worksheet title, turn off gridlines, and, optionally, adjust font and type characteristics using the Appendix Section B.5T instructions.

TG2.6 ORGANIZING a MIX of VARIABLES

Use **text tables**.

For example, to create a multidimensional contingency table displaying percentage of overall total for fund type, risk level, and market cap for the retirement funds sample similar to Figure 2.19 on page 69, in a new Tableau workbook, click **Connect to Data** and open the **Retirement Funds for Tableau Excel workbook**. In a new Tableau worksheet:

1. Drag **Risk Level** and drop it in the **Columns** shelf. The shell of a summary table appears in the display area.
2. Drag **Fund Type** and drop it in the **Rows** shelf.
3. Drag **Market Cap** and drop it in the **Rows** shelf.
4. Drag *Number of Records* and drop it over the "Abc" cells in the table. The green pill SUM(Number of Records) appears in the Marks card area.
5. Select **Analysis→Totals** and check **Show Row Grand Totals**.
6. Select **Analysis→Totals** and check **Show Column Grand Totals**.
7. Right-click the **SUM(Number of Records) pill** and select **Add Table Calculation** in the shortcut menu.

In the Table Calculation dialog box (not shown):

8. Select **Percent of Total** from the **Calculation Type** pull-down list.
9. Select **Table** in the **Compute Using** list box.
10. Click the **Close icon** for the dialog box (the upper-right "X" icon).

Tableau creates the summary table with risk level categories in alphabetical order. To list categories in the order Low, Average, and High, right-click the **Risk Level pill** in the **Columns** shelf and select **Sort** in the shortcut menu. In the Sort [Risk Level] dialog box:

11. Click **Manual**.
12. In the list box, select **Low** and then click **Up** twice to reorder categories.
13. Click **OK**.

Enter a worksheet title and, optionally, adjust font and type characteristics using the Appendix Section B.5T instructions.

Adding a Numerical Variable. To add a numerical variable to the multidimensional contingency table for Fund Type, Risk Level, and Market Cap to show the mean 10YrReturn, similar to Figure 2.20, continue the 13-step process with steps 14 through 16.

14. Right-click the **SUM(Number of Records) pill** in the Marks area and select **Remove** in the shortcut menu.
15. Right-click **10YrReturn** in the Dimensions list and select **Convert to Measure** in the shortcut menu.
16. Drag **10YrReturn** and drop it over the "Abc" cells in the table. The green pill SUM(10YrReturn) appears in the Marks card area.
17. Right-click the **SUM(10YrReturn) pill** and select **Measure(Sum)→Average** in the shortcut menu.

The mean ten-year return percentage values replace the sums in the multidimensional contingency table. Numerical variables added in this way must be measures and not dimensions, the reason for step 15.

TG2.7 VISUALIZING a MIX of VARIABLES

Colored Scatter Plots

Use **scatter plots**.

For example, to create the Figure 2.22 colored scatter plot of expense ratio, 3YrReturn, and market cap, on page 70, in a new Tableau workbook, click **Connect to Data** and open the **Retirement Funds for Tableau Excel workbook**. In a new Tableau worksheet:

1. Drag **3YrReturn** and drop it in the **Columns** shelf.
2. Drag **Expense Ratio** and drop it in the **Rows** shelf.

3. Drag **Market Cap** and drop it over the **Color icon** in the **Marks** card area.
4. Drag **Fund Number** and drop it over the **Detail icon** in the **Marks** card area.
5. Click the **Fund Number pill** and then select **Circle** from the **Marks** pull-down list. (The Marks card area after step 5 is shown at right.)

Enter a worksheet title, turn off gridlines and the zero line, and, optionally, adjust font and type characteristics using the Appendix Section B.5T instructions.

Treemap

Use **treemaps**.

For example, to construct the Figure 2.24 treemap for Fund Type and Market Cap on page 72, in a new Tableau workbook, click **Connect to Data** and open the **Retirement Funds for Tableau Excel workbook**. In a new Tableau worksheet:

1. Drag **Fund Type** and drop it over the **Text icon** in the **Marks** card area.
2. Drag **Market Cap** and drop it over the **Text icon** in the **Marks** card area.
3. Drag *Number of Records* and drop it over the worksheet display area.
4. Click **Show Me** if the Show Me gallery is not visible.
5. Click the **treemaps icon** in the **Show Me** gallery.

Tableau displays a treemap that is colored by the frequency of each group, which is also represented by the size of each rectangle. To color the treemap by Market Cap:

5. Right-click the (second) **SUM(Number of Records) pill** for Color and select **Remove** in the shortcut menu. Every square becomes the same color.
6. Click the small **Text icon** to the immediate left of the **Market Cap pill** and select **Color** from the popup gallery.
7. Drag **Market Cap** and drop it over the **Text icon** in the **Marks** card area.

Tableau displays a treemap that is labeled by fund type and market cap and colored by market cap. Enter a worksheet title and, optionally, adjust font and type characteristics using the Appendix Section B.5T instructions.

Sparklines

Use **lines (discrete)**.

For example, to construct the Figure 2.25 sparklines for movie revenues per month for the period 2005 to 2018 on page 73, in a new Tableau workbook, click **Connect to Data** and open the **Monthly Movie Revenues Excel workbook**. Because this Excel workbook contains two worksheets, drag **Stacked** in the **Sheets** list and drop it in the *Drag sheets here* area to manually establish the data source. In a new Tableau worksheet:

1. Right-click **Year** in the Dimensions list and select **Change Data Type→Date** in the shortcut menu.
2. Drag **Year** and drop it in the **Columns** shelf.
3. Drag **Month** and drop it in the **Rows** shelf.
4. Drag **Revenue** and drop it in the **Rows** shelf.
5. Click **Color** in the **Marks** card area
6. In the Color gallery, select the **second Markers icon** (line with markers) in the **Effects** group.
7. Select **Format→Lines**.
8. In the Format Lines **Rows** tab, select **None** from the **Grid Lines** pull-down list.
9. In the chart, right-click the **Revenue Y axis** and clear the **Show header** checkmark.
10. Resize the width of the chart and the height of the month line graphs.

To resize the height of the month graphs, place mouse cursor over a horizontal rule that separates any two month labels. When the mouse cursor changes to a two-sided arrow, drag the mouse up to reduce the height of all month rows. Enter a worksheet title and, optionally, adjust font and type characteristics using the Appendix Section B.5T instructions.

Adding the high and low points in the Figure 2.25 sparklines requires defining two calculated fields, an advanced use of Tableau. To add the necessary fields, continue with steps 11 through 19:

11. Select **Analysis→Create Calculated Field**.
12. In the Calculation Editor (shown below), enter **FindMaximum** as the name of the calculation and enter the text **IF [Revenue] = { FIXED [Month] : MAX([Revenue])} THEN [Revenue] END** in the calculation area.
13. Click **OK**.

FindMaximum identifies the maximum value in each monthly time series. Use shortcuts to create a calculated field that identifies the minimum value in each time series.

14. Right-click **FindMaximum** in the Measures list and select **Duplicate** in the shortcut menu.
15. Right-click **FindMaximum (copy)** and select **Edit** in the shortcut menu.
16. In the Calculation Editor, change the name of the measure to **FindMinimum** and change **MAX** to **MIN** in the calculation area.
17. Click **OK**.
18. Drag **FindMaximum** in the Measures list and drop it in the **Rows** shelf. (To see the Measures list, close the Format Lines tab if that tab is still open due to step 7.)
19. Right-click the **FindMaximum pill** and select **Dual Axis** in the shortcut menu.
20. Drag **FindMinimum** in the Measures list and drop it on the on the right edge of the right vertical axis such that the dragged icon changes to two vertical bars.
21. Right-click the **SUM(Revenue) pill** and clear the **Show Header** checkmark in the shortcut menu.

Tableau colors the line graphs to highlight the minimum and maximum values in each graph.

To emphasize the minimum and maximum points, click the **Measure Names** entry in the **Marks** card area and select **Circle** from the pull-down list. Tableau displays a large point if the monthly revenue is either the lowest or highest revenue for the time series.

To change colors of the monthly revenue, maximum, or minimum points, right-click an entry in the legend and select **Edit Colors** in the shortcut menu. In the Edit Colors dialog box (not shown), change individual colors or assign a palette of colors, and then click **OK**. (The treemap that appears in this chapter uses the Tableau Color Blind palette to enhance its accessibility for those that suffer from color vision deficiencies.)

Calculated field formulas are beyond the scope of this book to fully discuss. Calculated field formulas supplement table calculations. Calculated fields help recode, segment, and aggregate data and filter results in addition to performing arithmetic operations.

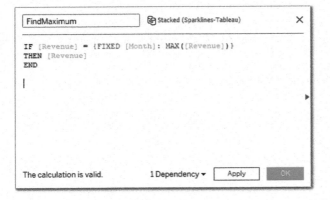

3
Numerical Descriptive Measures

CONTENTS

OBJECTIVES

- Describe the properties of central tendency, variation, and shape in numerical variables
- Construct and interpret a boxplot
- Compute descriptive summary measures for a population
- Compute the covariance and the coefficient of correlation

▼USING **STATISTICS**
More Descriptive Choices

As a member of a Choice *Is* Yours investment service task force, you helped organize and visualize the variables found in a sample of 479 retirement funds. Now, several weeks later, prospective clients are asking for more information on which they can base their investment decisions. In particular, they would like to compare the results of an individual retirement fund to the results of similar funds.

For example, while the earlier work your team did shows how the three-year return percentages are distributed, prospective clients would like to know how the value for a particular mid-cap growth fund compares to the three-year returns of all mid-cap growth funds. They also seek to understand the variation among the returns. Are all the values relatively similar? And does any variable have outlier values that are either extremely small or extremely large?

While doing a complete search of the retirement funds data could lead to answers to the preceding questions, you wonder if there are better ways than extensive searching to uncover those answers. You also wonder if there are other ways of being more *descriptive* about the sample of funds—providing answers to questions not yet raised by prospective clients. If you can help the Choice *Is* Yours investment service provide such answers, prospective clients will be able to better evaluate the retirement funds that your firm features.

he prospective clients in the More Descriptive Choices scenario have begun asking questions about numerical variables such as how the three-year return percentages vary among the individual funds that comprise the sample of 479 retirement funds. Descriptive methods that describe the central tendency, variation, and shape of variables help answer such questions.

Central tendency is the extent to which the values of a numerical variable group around a typical, or central, value. **Variation** measures the amount of dispersion, or scattering, away from a central value that the values of a numerical variable show. The **shape** of a variable is the pattern of the distribution of values from the lowest value to the highest value. This chapter describes the numerical measures that describe these qualities. The chapter also discusses the covariance and the coefficient of correlation, measures that can help show the strength of the association between two numerical variables.

Calculating descriptive measures that this chapter discusses would be one way to help prospective clients of the Choice *Is* Yours service find the answers they seek. More generally, calculating these measures is a typical way to begin *analyzing* numerical data using the DCOVA framework.

3.1 Measures of Central Tendency

Most variables show a distinct tendency to group around a central value. When people talk about an "average value" or the "middle value" or the "most frequent value," they are talking informally about the mean, median, and mode—three measures of central tendency.

The Mean

The **arithmetic mean** (in everyday usage, the **mean**) is the most common measure of central tendency. To calculate a mean, sum the values in a set of data and then divide that sum by the number of values in the set. The mean can suggest a typical or central value and serves as a "balance point" in a set of data, similar to the fulcrum on a seesaw. The mean is the only common measure in which all the values play an equal role.

The symbol $\overline{X}$, read as *X-bar*, represents the mean of a sample. The **sample mean** is the sum of the values in a sample divided by the number of values in the sample. For a sample containing n values, the equation for the sample mean is

$$\overline{X} = \frac{\text{sum of the } n \text{ values}}{n}$$

Using the series $X_1, X_2, \ldots, X_n$ to represent the set of n values, the equation becomes

$$\overline{X} = \frac{X_1 + X_2 + \cdots + X_n}{n}$$

Using summation notation (see Appendix A) to replace the numerator $X_1 + X_2 + \cdots + X_n$ with the term $\sum_{i=1}^{n} X_i$, which means "sum all the X_i values from the first X value, X_1, to the last X value, X_n," forms Equation (3.1), a formal definition of the sample mean.

SAMPLE MEAN

$$\overline{X} = \frac{\sum_{i=1}^{n} X_i}{n} \tag{3.1}$$

where

$$\overline{X} = \text{sample mean}$$

$$n = \text{sample size}$$

$$X_i = i\text{th value of the variable } X$$

$$\sum_{i=1}^{n} X_i = \text{sum of all } X_i \text{ values in the sample}$$

Because all the values play an equal role, a mean is greatly affected by any value that is very different from the others. When a set of data contains extreme values, avoid using the mean as a measure of central tendency.

As an example of using a sample mean, consider that knowing the typical time to get ready in the morning might help people to better manage their weekday schedules. To study this problem using the DCOVA framework, one first defines get-ready time as the elapsed time from when a person wakes up in the morning until the person actually leaves for work, rounded to the nearest minute. For a specific person, one collects the get-ready times for 10 consecutive weekdays and organizes and stores them in `Times`.

Using the collected data, calculating the mean would discover the "typical" get-ready time. For these data:

Day:	1	2	3	4	5	6	7	8	9	10
Time (minutes):	39	29	43	52	39	44	40	31	44	35

the mean time is 39.6 minutes, calculated as:

$$\overline{X} = \frac{\sum_{i=1}^{n} X_i}{n} = \frac{39 + 29 + 43 + 52 + 39 + 44 + 40 + 31 + 44 + 35}{10}$$

$$\overline{X} = \frac{396}{10} = 39.6$$

Even though no individual day in the sample had a value of 39.6 minutes, allotting this amount of time to get ready in the morning would be a reasonable decision to make. The mean is a good measure of central tendency in this case because the data set does not contain any exceptionally small or large values.

To illustrate how the mean can be greatly affected by any value that is very different from the others, imagine that on Day 3, a set of unusual circumstances delayed the person getting ready by an extra hour, so that the time for that day was 103 minutes. This extreme value causes the mean to rise to 45.6 minutes, as follows:

$$\overline{X} = \frac{\sum_{i=1}^{n} X_i}{n} = \frac{39 + 29 + 103 + 52 + 39 + 44 + 40 + 31 + 44 + 35}{10}$$

$$\overline{X} = \frac{456}{10} = 45.6$$

The one extreme value has increased the mean by 6 minutes. The extreme value also moved the position of the mean relative to all the values. The original mean, 39.6 minutes, had a middle, or *central*, position among the data values: five of the times were less than that mean, and five were greater than that mean. In contrast, the mean using the extreme value is greater than 9 of the 10 times, making the new mean a poor measure of central tendency.

EXAMPLE 3.1

Calculating the
Mean Number of
Calories in Cereals

A sample of seven breakfast cereals (stored in `Cereals`) includes nutritional data about the number of calories per serving:

Cereal	Calories
Kellogg's All Bran	80
Kellogg's Corn Flakes	100
Wheaties	100
Nature's Path Organic Multigrain Flakes	110
Kellogg's Rice Krispies	130
Post Shredded Wheat Vanilla Almond	190
Kellogg's Mini Wheats	200

Calculate the mean number of calories in these breakfast cereals.

SOLUTION The mean number of calories is 130, calculated as follows:

$$\overline{X} = \frac{\sum_{i=1}^{n} X_i}{n} = \frac{910}{7} = 130$$

The Median

The **median** is the middle value in an ordered array of data that has been ranked from smallest to largest. Half the values are smaller than or equal to the median, and half the values are larger than or equal to the median. Extreme values do not affect the median, making the median a good alternative to the mean when such values exist in the data.

To calculate the median for a set of data, first rank the values from smallest to largest and then use Equation (3.2) to calculate the rank of the value that is the median.

MEDIAN

$$\text{Median} = \frac{n + 1}{2} th \text{ ranked value} \tag{3.2}$$

Calculate the median by following one of two rules:

- **Rule 1** If the data set contains an *odd* number of values, the median is the measurement associated with the middle-ranked value.
- **Rule 2** If the data set contains an *even* number of values, the median is the measurement associated with the average of the two middle-ranked values.

To further analyze the sample of 10 get-ready times, one can calculate the median. To do so, first rank the daily times:

studentTIP

You must rank the values in order from the smallest to the largest to compute the median.

Ranked values:	29	31	35	39	39	40	43	44	44	52
Ranks:	1	2	3	4	5	6	7	8	9	10

↑
Median = 39.5

Because the result of dividing $n + 1$ by 2 for this sample of 10 is $(10 + 1)/2 = 5.5$, one must use Rule 2 and average the measurements associated with the fifth and sixth ranked values, 39 and 40. Therefore, the median is 39.5. The median of 39.5 means that for half the days, the time to get ready is less than or equal to 39.5 minutes, and for half the days, the time to get ready is greater than or equal to 39.5 minutes. In this case, the median time to get ready of 39.5 minutes is very close to the mean time to get ready of 39.6 minutes.

The previous section noted that substituting 103 minutes for the time of 43 minutes increased the mean by 6 minutes. Doing the same substitution does not affect the value of median, which would remain 39.5. This example illustrates that the median is not affected by extreme values.

EXAMPLE 3.2

Calculating the Median From an Odd-Sized Sample

A sample of seven breakfast cereals (stored in Cereals) includes nutritional data about the number of calories per serving (see Example 3.1 on page 111). Calculate the median number of calories in breakfast cereals.

SOLUTION Because the result of dividing $n + 1$ by 2 for this sample of seven is $(7 + 1)/2 = 4$, using Rule 1, the median is the measurement associated with the fourth-ranked value. The number of calories per serving values are ranked from the smallest to the largest:

Ranked values:	80	100	100	110	130	190	200
Ranks:	1	2	3	4	5	6	7

$$\uparrow$$
$$\text{Median} = 110$$

The median number of calories is 110. Half the breakfast cereals have 110 or less than 110 calories per serving, and half the breakfast cereals have 110 or more than 110 calories per serving.

The Mode

The **mode** is the value that appears most frequently. Like the median and unlike the mean, extreme values do not affect the mode. For a particular variable, there can be several modes or no mode at all. For example, examine the sample of 10 times to get ready in the morning:

$$29 \quad 31 \quad 35 \quad 39 \quad 39 \quad 40 \quad 43 \quad 44 \quad 44 \quad 52$$

There are two modes, 39 minutes and 44 minutes, because each of these values occurs twice.

Sometimes there is no mode because each value in a data set appears the same number of times in the data set. The following sample of 26 net prices for used smartphones collected from an AT&T website during a search of phones has no mode, no value that is "most typical," because every value appears once (Smartphones).

6.00   6.80   7.00   9.00   9.35   10.00   11.67   13.37   14.00   15.00   15.02   16.67   17.00
17.85   19.00   20.00   23.34   25.00   25.17   30.00   33.34   35.57   41.67   43.17   50.00   66.00

EXAMPLE 3.3

Determining the Mode

A systems manager in charge of a company's network keeps track of the number of server failures that occur in a day. Determine the mode for the following data, which represent the number of server failures per day for the past two weeks:

$$1 \quad 3 \quad 0 \quad 3 \quad 26 \quad 2 \quad 7 \quad 4 \quad 0 \quad 2 \quad 3 \quad 3 \quad 6 \quad 3$$

SOLUTION The ordered array for these data is

$$0 \quad 0 \quad 1 \quad 2 \quad 2 \quad 3 \quad 3 \quad 3 \quad 3 \quad 3 \quad 4 \quad 6 \quad 7 \quad 26$$

Because 3 occurs five times, more times than any other value, the mode is 3. Thus, the systems manager can say that the most common occurrence is having three server failures in a day. For this data set, the median is also equal to 3, and the mean is equal to 4.5. The value 26 is an extreme value. For these data, the median and the mode are better measures of central tendency than the mean.

The Geometric Mean

To measure the rate of change of a variable over time, one uses the geometric mean instead of the arithmetic mean. Equation (3.3) defines the geometric mean.

GEOMETRIC MEAN

The **geometric mean** is the nth root of the product of n values:

$$\overline{X}_G = (X_1 \times X_2 \times \cdots \times X_n)^{1/n} \qquad (3.3)$$

The **geometric mean rate of return** measures the mean percentage return of an investment per time period. Equation (3.4) defines the geometric mean rate of return.

GEOMETRIC MEAN RATE OF RETURN

$$\overline{R}_G = [(1 + R_1) \times (1 + R_2) \times \cdots \times (1 + R_n)]^{1/n} - 1 \qquad (3.4)$$

where

$$R_i = \text{rate of return in time period } i$$

To illustrate these measures, consider an investment of \$100,000 that declined to a value of \$50,000 at the end of Year 1 and then rebounded back to its original \$100,000 value at the end of Year 2. The rate of return for this investment per year for the two-year period is 0 because the starting and ending value of the investment is unchanged. However, the arithmetic mean of the yearly rates of return of this investment is

$$\overline{X} = \frac{(-0.50) + (1.00)}{2} = 0.25 \text{ or } 25\%$$

because the rate of return for Year 1 is

$$R_1 = \left(\frac{50,000 - 100,000}{100,000}\right) = -0.50 \text{ or } -50\%$$

and the rate of return for Year 2 is

$$R_2 = \left(\frac{100,000 - 50,000}{50,000}\right) = 1.00 \text{ or } 100\%$$

Using equation (3.4), the geometric mean rate of return per year for the two years is

$$\overline{R}_G = [(1 + R_1) \times (1 + R_2)]^{1/n} - 1$$
$$= [(1 + (-0.50)) \times (1 + (1.0))]^{1/2} - 1$$
$$= [(0.50) \times (2.0)]^{1/2} - 1 = [1.0]^{1/2} - 1$$
$$= 1 - 1 = 0$$

Using the geometric mean rate of return more accurately reflects the (zero) change in the value of the investment per year for the two-year period than does the arithmetic mean.

EXAMPLE 3.4

Computing the Geometric Mean Rate of Return

The percentage change in the Russell 2000 Index of the stock prices of 2,000 small companies was 13.1% in 2017 and −12.2% in 2018. Compute the geometric rate of return.

SOLUTION Using Equation (3.4), the geometric mean rate of return in the Russell 2000 Index for the two years is

$$
\begin{aligned}
\overline{R}_G &= [(1 + R_1) \times (1 + R_2)]^{1/n} - 1 \\
&= [(1 + (0.131)) \times (1 + (-0.122))]^{1/2} - 1 \\
&= [(1.131) \times (0.878)]^{1/2} - 1 = (0.9930)^{1/2} - 1 \\
&= 0.9965 - 1 = -0.0035
\end{aligned}
$$

The geometric mean rate of return in the Russell 2000 Index for the two years is −0.35% per year.

3.2 Measures of Variation and Shape

In addition to central tendency, every variable can be characterized by its variation and shape. Variation measures the **spread**, or **dispersion**, of the values. The shape of a variable represents a pattern of all the values, from the lowest to highest value. As Section 3.3 explains, many variables have a pattern that looks approximately like a bell, with a peak of values somewhere in the middle.

The Range

A simple measure of variation, the **range** is the difference between the largest and smallest value and is the simplest descriptive measure of variation for a numerical variable.

> RANGE
>
> The range is equal to the largest value minus the smallest value.
>
> $$ \text{Range} = X_{\text{largest}} - X_{\text{smallest}} \tag{3.5} $$

Calculating the range would further analyze the sample of 10 get-ready times. To do so, first rank the data from smallest to largest:

$$ 29 \quad 31 \quad 35 \quad 39 \quad 39 \quad 40 \quad 43 \quad 44 \quad 44 \quad 52 $$

Using Equation (3.5), the range is $52 - 29 = 23$ minutes. The range of 23 minutes indicates that the largest difference between any two days in the time to get ready in the morning is 23 minutes.

EXAMPLE 3.5

Calculating the Range of the Number of Calories in Cereals

A sample of seven breakfast cereals (stored in Cereals) includes nutritional data about the number of calories per serving (see Example 3.1 on page 111). Calculate the range of the number of calories for the cereals.

SOLUTION Ranked from smallest to largest, the calories for the seven cereals are

$$ 80 \quad 100 \quad 100 \quad 110 \quad 130 \quad 190 \quad 200 $$

Therefore, using Equation (3.5), the range $= 200 - 80 = 120$. The largest difference in the number of calories between any two cereals is 120.

The range measures the *total spread* in the set of data. However, the range does not take into account *how* the values are distributed between the smallest and largest values. In other words, the range does not indicate whether the values are evenly distributed, clustered near the middle, or clustered near one or both extremes. Thus, using the range as a measure of variation when at least one value is an extreme value is misleading.

The Variance and the Standard Deviation

Being a simple measure of variation, the range does not consider how the values distribute or cluster between the extremes. Two commonly used measures of variation that account for how all the values are distributed are the **variance** and the **standard deviation**. These statistics measure the "average" scatter around the mean—how larger values fluctuate above it and how smaller values fluctuate below it.

A simple calculation of variation around the mean might take the difference between each value and the mean and then sum these differences. However, the sum of these differences would always be zero because the mean is the balance point for *every* numerical variable. A calculation of variation that *differs* from one data set to another *squares* the difference between each value and the mean and then sums those squared differences. This sum of the squared differences, called the **sum of squares (SS)**, forms a basis for calculating the variance and the standard deviation.

For a sample, the **sample variance (S^2)** is the sum of squares divided by the sample size minus 1. The **sample standard deviation (S)** is the square root of the sample variance. Because the sum of squares can never be a negative value, the variance and the standard deviation will always be non-negative values, and in virtually all cases, the variance and standard deviation will be greater than zero. (Both the variance and standard deviation will be zero, meaning no variation, only for the special case in which every value in a sample is the same.)

For a sample containing n values, $X_1, X_2, X_3, \ldots, X_n$, the sample variance ($S^2$) is

$$S^2 = \frac{(X_1 - \overline{X})^2 + (X_2 - \overline{X})^2 + \cdots + (X_n - \overline{X})^2}{n - 1}$$

Equations (3.6) and (3.7) define the sample variance and sample standard deviation using summation notation. The term $\sum_{i=1}^{n} (X_i - \overline{X})^2$ represents the sum of squares.

SAMPLE VARIANCE and SAMPLE STANDARD DEVIATION

$$S^2 = \frac{\sum_{i=1}^{n} (X_i - \overline{X})^2}{n - 1} \tag{3.6}$$

$$S = \sqrt{S^2} = \sqrt{\frac{\sum_{i=1}^{n} (X_i - \overline{X})^2}{n - 1}} \tag{3.7}$$

where

$\overline{X} =$ sample mean

$n =$ sample size

$X_i = i$th value of the variable X

$\sum_{i=1}^{n} (X_i - \overline{X})^2 =$ summation of all the squared differences between the X_i values and $\overline{X}$

Note that in both equations, the sum of squares is divided by the sample size minus 1, $n - 1$. The value is used for reasons related to statistical inference and the properties of sampling distributions that Section 7.2 discusses. For now, observe that the difference between dividing by n and by $n - 1$ becomes smaller as the sample size increases.

Because the sample standard deviation will always be a value expressed in the same units as the original sample data, most use this statistic as the primary measure of variation. (The sample variance is a squared quantity that may have no real-world meaning.) For almost all samples, the majority of the values in a sample will be within an interval of plus and minus 1 standard deviation above and below the mean. Therefore, calculating the sample mean and the sample standard deviation typically helps define where the majority of the values are clustering.

Sample variance can be calculated using this four-step process:

Step 1 Calculate the difference between each value and the mean.
Step 2 Square each difference.
Step 3 Sum the squared differences.
Step 4 Divide this total by $n - 1$.

Table 3.1 illustrates this process for the sample of 10 get-ready times. The middle column performs step 1, the right column performs step 2, the sum of the right column represents step 3, and the division of that sum represents step 4.

TABLE 3.1
Computing the variance of the get-ready times

The mean ($\overline{X}$) equal to 39.6 was calculated previously using the method that page 110 discusses.

Time (X)	Step 1: $(X_i - \overline{X})$	Step 2: $(X_i - \overline{X})^2$
39	−0.60	0.36
29	−10.60	112.36
43	3.40	11.56
52	12.40	153.76
39	−0.60	0.36
44	4.40	19.36
40	0.40	0.16
31	−8.60	73.96
44	4.40	19.36
35	−4.60	21.16

$n = 10$
$\overline{X} = 39.6$

Step 3: **Sum** 412.40
Step 4: **Divide by (n − 1)** 45.82

Sample variance can also be calculated by substituting values for the terms in Equation (3.6):

$$S^2 = \frac{\sum_{i=1}^{n}(X_i - \overline{X})^2}{n - 1} = \frac{(39 - 39.6)^2 + (29 - 39.6)^2 + \cdots + (35 - 39.6)^2}{10 - 1}$$

$$= \frac{412.4}{9} = 45.82$$

The sample standard deviation, S, can be calculated using the square root of the variance or by substituting values for the terms in Equation (3.7) on page 115:

$$S = \sqrt{S^2} = \sqrt{\frac{\sum_{i=1}^{n}(X_i - \overline{X})^2}{n - 1}} = \sqrt{45.82} = 6.77$$

The sample standard deviation indicates that the get-ready times in this sample are clustering within 6.77 minutes around the mean of 39.6 minutes, between $\overline{X} - 1S = 32.83$ and $\overline{X} + 1S = 46.37$ minutes. In fact, 7 out of 10 get-ready times lie within this interval.

While not shown in Table 3.1, the sum of the middle column that represents differences between each value and the mean is zero. For any set of data, this sum will always be zero:

$$\sum_{i=1}^{n}(X_i - \overline{X}) = 0 \text{ for all sets of data}$$

This property is one of the reasons that the mean is used as the most common measure of central tendency.

Example 3.6 illustrates that many applications calculate the sample variance and the sample standard deviation, making hand calculations unnecessary.

EXAMPLE 3.6

Calculating the Variance and Standard Deviation of the Number of Calories in Cereals

A sample of seven breakfast cereals (stored in Cereals) includes nutritional data about the number of calories per serving (see Example 3.1 on page 111). Calculate the variance and standard deviation of the calories in the cereals.

SOLUTION Figure 3.1 contains the Excel results for this example.

FIGURE 3.1
Excel results for the variance and standard deviation of the number of calories in the sample of cereals.

	A	B	C	D
1	Calories		Calculations	
2	80		Variance directly using VAR.S function	
3	100		in formula =VAR.S(A2:A8)	2200
4	100			
5	110		Standard deviation durectly using STDEV.S	
6	130		function in formula =STDEV.S(A2:A8)	46.90
7	190			
8	200			

Alternatively, using Equation (3.6) on page 115:

$$S^2 = \frac{\sum_{i=1}^{n}(X_i - \overline{X})^2}{n - 1} = \frac{(80 - 130)^2 + (100 - 130)^2 + \cdots + (200 - 130)^2}{7 - 1}$$

$$= \frac{13,200}{6} = 2,200$$

Using Equation (3.7) on page 115, the sample standard deviation, S, is

$$S = \sqrt{S^2} = \sqrt{\frac{\sum_{i=1}^{n}(X_i - \overline{X})^2}{n - 1}} = \sqrt{2,200} = 46.9042$$

The standard deviation of 46.9042 indicates that the calories in the cereals are clustering within ± 46.9042 around the mean of 130 (clustering between $\overline{X} - 1S = 83.0958$ and $\overline{X} + 1S = 176.9042$). In fact, 57.1% (four out of seven) of the calories lie within this interval.

The Coefficient of Variation

The coefficient of variation is equal to the standard deviation divided by the mean, multiplied by 100%. Unlike the measures of variation presented previously, the **coefficient of variation (CV)** measures the scatter in the data relative to the mean. The coefficient of variation is a *relative* *measure* of variation that is always expressed as a percentage rather than in terms of the units of the particular data. Equation (3.8) defines the coefficient of variation.

COEFFICIENT OF VARIATION

The coefficient of variation is equal to the standard deviation divided by the mean, multiplied by 100%.

$$CV = \left(\frac{S}{\overline{X}}\right)100\%$$ (3.8)

where

$$S = \text{sample standard deviation}$$
$$\overline{X} = \text{sample mean}$$

For the sample of 10 get-ready times, because $\overline{X} = 39.6$ and $S = 6.77$, the coefficient of variation is

$$CV = \left(\frac{S}{\overline{X}}\right)100\% = \left(\frac{6.77}{39.6}\right)100\% = 17.10\%$$

For the get-ready times, the standard deviation is 17.1% of the size of the mean.

The coefficient of variation is especially useful when comparing two or more sets of data that are measured in different units, as Example 3.7 illustrates.

EXAMPLE 3.7

Comparing Two Coefficients of Variation When the Two Variables Have Different Units of Measurement

Which varies more from cereal to cereal—the number of calories or the amount of sugar (in grams)?

SOLUTION Because calories and the amount of sugar have different units of measurement, you need to compare the relative variability in the two measurements.

For calories, using the mean and variance that Examples 3.1 and 3.6 on pages 111 and 117 calculate, the coefficient of variation is

$$CV_{\text{Calories}} = \left(\frac{46.9042}{130}\right)100\% = 36.08\%$$

For the amount of sugar in grams, the values for the seven cereals are

6   2   4   4   4   11   10

For these data, $\overline{X} = 5.8571$ and $S = 3.3877$. Therefore, the coefficient of variation is

$$CV_{\text{Sugar}} = \left(\frac{3.3877}{5.8571}\right)100\% = 57.84\%$$

You conclude that relative to the mean, the amount of sugar is much more variable than the calories.

Z Scores

The **Z score** of a value is the difference between that value and the mean, divided by the standard deviation. A Z score of 0 indicates that the value is the same as the mean. A positive or negative Z score indicates whether the value is above or below the mean and by how many standard deviations. Z scores help identify **outliers**, the values that seem excessively different from most of the rest of the values (see Section 1.4). Values that are very different from the mean will have either very small (negative) Z scores or very large (positive) Z scores. As a general rule, a Z score that is less than −3.0 or greater than +3.0 indicates an outlier value.

Z SCORE

$$Z = \frac{X - \overline{X}}{S} \tag{3.9}$$

Calculating Z scores would analyze the sample of 10 get-ready times. Because the mean is 39.6 minutes, the standard deviation is 6.77 minutes, and the time to get ready on the first day is 39.0 minutes, the Z score for Day 1 using Equation (3.9) is

$$Z = \frac{X - \overline{X}}{S} = \frac{39.0 - 39.6}{6.77} = -0.09$$

The Z score of −0.09 for the first day indicates that the time to get ready on that day is very close to the mean. Figure 3.2 presents the Z scores for all 10 days.

FIGURE 3.2

Excel worksheet containing the Z scores for 10 get-ready times

	A	B
1	Get-Ready Time	Z Score
2	39	-0.09
3	29	-1.57
4	43	0.50
5	52	1.83
6	39	-0.09
7	44	0.65
8	40	0.06
9	31	-1.27
10	44	0.65
11	35	-0.68

In The Worksheet

Column B contains formulas that use the STANDARDIZE function to calculate the Z scores for the get-ready times. STANDARDIZE requires the mean and standard deviation be known for the data set.

The **Z Scores** and **Variation worksheets** of the Excel Guide **Descriptive workbook** illustrate how STANDARDIZE can use the mean and standard deviation that another worksheet calculates.

The largest Z score is 1.83 for Day 4, on which the time to get ready was 52 minutes. The lowest Z score is −1.57 for Day 2, on which the time to get ready was 29 minutes. Because none of the Z scores are less than −3.0 or greater then +3.0, you conclude that the get-ready times include no apparent outliers.

EXAMPLE 3.8

Calculating the Z Scores of the Number of Calories in Cereals

A sample of seven breakfast cereals (stored in Cereals) includes nutritional data about the number of calories per serving (see Example 3.1 on page 111). Calculate the Z scores of the calories in breakfast cereals.

SOLUTION Figure 3.3 presents the Z scores of the calories for the cereals. The largest Z score is 1.49, for a cereal with 200 calories. The lowest Z score is −1.07, for a cereal with 80 calories. There are no apparent outliers in these data because none of the Z scores are less than −3.0 or greater than +3.0.

FIGURE 3.3

Excel worksheet containing the Z scores for 10 cereals

	A	B
1	Calories	Z Score
2	80	-1.07
3	100	-0.64
4	100	-0.64
5	110	-0.43
6	130	0.00
7	190	1.28
8	200	1.49

Shape: Skewness

Skewness measures the extent to which the data values are not **symmetrical** around the mean. The three possibilities are:

- **Mean < median:** negative, or **left-skewed distribution**
- **Mean = median:** **symmetrical distribution** (zero skewness)
- **Mean > median:** positive, or **right-skewed distribution**

In a *symmetrical* distribution, the values below the mean are distributed in exactly the same way as the values above the mean, and the skewness is zero. In a **skewed** distribution, the data values below and above the mean are imbalanced, and the skewness is a nonzero value (less than zero for a left-skewed distribution, greater than zero for a right-skewed distribution). Figure 3.4 visualizes these possibilities.

FIGURE 3.4
The shapes of three data distributions

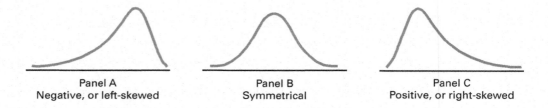

| Panel A | Panel B | Panel C |
| Negative, or left-skewed | Symmetrical | Positive, or right-skewed |

Panel A displays a left-skewed distribution. In a left-skewed distribution, most of the values are in the upper portion of the distribution. Some extremely small values cause the long tail and distortion to the left and cause the mean to be less than the median. Because the skewness statistic for such a distribution will be less than zero, some use the term *negative skew* to describe this distribution.

Panel B displays a symmetrical distribution. In a symmetrical distribution, values are equally distributed in the upper and lower portions of the distribution. This equality causes the portion of the curve below the mean to be the mirror image of the portion of the curve above the mean and makes the mean equal to the median.

Panel C displays a right-skewed distribution. In a right-skewed distribution, most of the values are in the lower portion of the distribution. Some extremely large values cause the long tail and distortion to the right and cause the mean to be greater than the median. Because the skewness statistic for such a distribution will be greater than zero, some use the term *positive skew* to describe this distribution.

Shape: Kurtosis

Kurtosis measures the peakedness of the curve of the distribution—that is, how sharply the curve rises approaching the center of the distribution. Kurtosis compares the shape of the peak to the shape of the peak of a bell-shaped normal distribution (see Chapter 6), which, by definition, has a kurtosis of zero.[1] A distribution that has a sharper-rising center peak than the peak of a normal distribution has *positive* kurtosis, a kurtosis value that is greater than zero, and is called **leptokurtic**. A distribution that has a slower-rising (flatter) center peak than the peak of a normal distribution has *negative* kurtosis, a kurtosis value that is less than zero, and is called **platykurtic**. A leptokurtic distribution has a higher concentration of values near the mean of the distribution compared to a normal distribution, while a platykurtic distribution has a lower concentration compared to a normal distribution.

[1]Several different operational definitions exist for kurtosis. The definition here, used by Excel, is sometimes called *excess kurtosis* to distinguish it from other definitions. Read the SHORT TAKES for Chapter 3 to learn how Excel calculates kurtosis (and skewness).

In affecting the shape of the central peak, the relative concentration of values near the mean also affects the ends, or *tails*, of the curve of a distribution. A leptokurtic distribution has *fatter* tails, many more values in the tails, than a normal distribution has. When an analysis mistakenly assumes that a set of data forms a normal distribution, that analysis will underestimate the occurrence of extreme values if the data actually forms a leptokurtic distribution. Some suggest that such a mistake can explain the unanticipated reverses and collapses that led to the financial crisis of 2007–2008 (Taleb).

EXAMPLE 3.9

Computing Descriptive Statistics for Growth and Value Funds

In the More Descriptive Choices scenario, you are interested in comparing the past performance of the growth and value funds from a sample of 479 funds. One measure of past performance is the three-year return percentage variable. Compute descriptive statistics for the growth and value funds.

SOLUTION Figure 3.5 presents descriptive summary measures for the two types of funds. The results include the mean, median, mode, minimum, maximum, range, variance, standard deviation, coefficient of variation, skewness, kurtosis, count (the sample size), and standard error. The standard error (see Section 7.2) is the standard deviation divided by the square root of the sample size.

In examining the results, you see that there are some differences in the three-year return for the growth and value funds. The growth funds had a mean three-year return of 8.51 and a median return of 8.70. This compares to a mean of 6.84 and a median of 7.07 for the value funds. The medians indicate that half of the growth funds had three-year returns of 8.70 or better, and half the value funds had three-year returns of 6.84 or better. You conclude that the value funds had a lower return than the growth funds.

FIGURE 3.5

Excel descriptive statistics results for the three-year return percentages for the growth and value funds

▲	A	B	C
1	Descriptive Statistics for the 3YrReturn Variable		
2			
3		Growth	Value	Growth
4	Mean	8.51	6.84	=AVERAGE(UNSTACKED!A:A)
5	Median	8.70	7.07	=MEDIAN(UNSTACKED!A:A)
6	Mode	8.71	8.6	=MODE(UNSTACKED!A:A)
7	Minimum	-3.40	-2.65	=MIN(UNSTACKED!A:A)
8	Maximum	15.32	11.98	=MAX(UNSTACKED!A:A)
9	Range	18.72	14.63	=B8 - B7
10	Variance	10.1983	5.4092	=VAR.S(UNSTACKED!A:A)
11	Standard Deviation	3.1935	2.3258	=STDEV.S(UNSTACKED!A:A)
12	Coeff. of Variation	37.53%	34.00%	=B11/B4
13	Skewness	-0.4883	-0.9260	=SKEW(UNSTACKED!A:A)
14	Kurtosis	0.2327	1.6006	=KURT(UNSTACKED!A:A)
15	Count	306	173	=COUNT(UNSTACKED!A:A)
16	Standard Error	0.1826	0.1768	=B11/SQRT(B15)

In the Worksheet

The worksheet uses Excel functions and short arithmetic formulas to calculate descriptive statistics. (Only the formulas for the Growth column are shown at left.)

The **Complete Statistics worksheet** of the Excel Guide **Descriptive workbook** is a similar worksheet that shows calculations for the set of get-ready times.

The growth funds had a higher standard deviation than the value funds (3.1935, as compared to 2.3258). The growth funds and the value funds each showed left or negative skewness. The skewness of the growth funds was −0.4883 and the skewness of the value funds was −0.9260. The kurtosis of the growth funds was slightly positive, indicating a distribution that was more peaked than a normal distribution. The kurtosis of the value funds was positive indicating a distribution that was much more peaked than a normal distribution.

EXAMPLE 3.10

Computing
Descriptive
Statistics Using
Multidimensional
Contingency Tables

Continuing with the More Descriptive Choices scenario, you wish to explore the effects of each combination of Fund Type, Market Cap, and Risk Level on measures of past performance. One measure of past performance is the three-year return percentage. Compute the mean three-year return percentage for each combination of Fund Type, Market Cap, and Risk Level.

SOLUTION A multidimensional contingency table (see Section 2.6) computes the mean three-year return percentage for each combination of the three variables.

FIGURE 3.6
Excel multidimensional contingency tables for the mean three-year return percentages for each combination of fund type, market cap, and risk level

Mean 3YrReturn	Risk Level			
Fund Type	Low	Average	High	Grand Total
⊟Growth	9.87	9.06	6.64	8.51
Large	10.22	10.43	9.79	10.30
MidCap	8.93	6.86	5.78	6.93
Small	9.09	7.43	5.99	6.39
⊟Value	7.76	6.41	4.13	6.84
Large	7.82	6.49	5.02	7.29
MidCap	7.87	7.05	2.22	6.69
Small	6.38	5.60	4.63	5.39
Grand Total	8.66	8.21	6.25	7.91

The three-year return is higher for low-risk funds than average-risk or high-risk funds for both the growth funds and value funds. However, this pattern changes when Market Cap categories are considered. For example, the three-year return percentage for growth funds with average risk is much higher for large cap funds than for midcap or small market cap funds. Also, for value funds with average risk, the three-year return for midcap funds is higher than the return for large funds.

PROBLEMS FOR SECTIONS 3.1 AND 3.2

LEARNING THE BASICS

3.1 The following set of data is from a sample of $n = 5$:

$$7 \ 4 \ 9 \ 8 \ 2$$

a. Calculate the mean, median, and mode.
b. Calculate the range, variance, standard deviation, and coefficient of variation.
c. Calculate the Z scores. Are there any outliers?
d. Describe the shape of the data set.

3.2 The following set of data is from a sample of $n = 6$:

$$7 \ 4 \ 9 \ 7 \ 3 \ 12$$

a. Calculate the mean, median, and mode.
b. Calculate the range, variance, standard deviation, and coefficient of variation.
c. Calculate the Z scores. Are there any outliers?
d. Describe the shape of the data set.

3.3 The following set of data is from a sample of $n = 7$:

$$12 \ 7 \ 4 \ 9 \ 0 \ 7 \ 3$$

a. Calculate the mean, median, and mode.
b. Calculate the range, variance, standard deviation, and coefficient of variation.

c. Calculate the Z scores. Are there any outliers?
d. Describe the shape of the data set.

3.4 The following set of data is from a sample of $n = 5$:

$$7 \ -5 \ -8 \ 7 \ 9$$

a. Calculate the mean, median, and mode.
b. Calculate the range, variance, standard deviation, and coefficient of variation.
c. Calculate the Z scores. Are there any outliers?
d. Describe the shape of the data set.

3.5 Suppose that the rate of return for a particular stock during the past two years was 10% for one of the years and 30% for the other year. Compute the geometric rate of return per year. (A rate of return of 10% is recorded as 0.10, and a rate of return of 30% is recorded as 0.30.)

3.6 Suppose that the rate of return for a particular stock during the past two years was 20% for one of the years and −30% for the other year. Compute the geometric rate of return per year.

APPLYING THE CONCEPTS

3.7 *Wired*, a magazine that delivers a glimpse into the future of business, culture, innovation, and science, reported the following summary for the household incomes of its two types of subscribers, the print reader and the digital reader.

Audience	Median
Wired reader	$99,874
Wired.com user	80,394

Source: Data extracted from "WIRED 2019 Media Kit," **bit.ly/2LqBYiz**.

Interpret the median household income for the *Wired* readers and the Wired.com users.

3.8 The operations manager of a plant that manufactures tires wants to compare the actual inner diameters of two grades of tires, each of which is expected to be 575 millimeters. A sample of five tires of each grade was selected, and the results representing the inner diameters of the tires, ranked from smallest to largest, are as follows:

Grade *X*	Grade *Y*
568 570 575 578 584	573 574 575 577 578

a. For each of the two grades of tires, calculate the mean, median, and standard deviation.
b. Which grade of tire is providing better quality? Explain.
c. What would be the effect on your answers in (a) and (b) if the last value for grade *Y* was 588 instead of 578? Explain.

3.9 According to the U.S. Census Bureau (**census.gov**), in 2018, the median sales price of new houses was $326,200, and the mean sales price was $384,600.
a. Interpret the median sales price.
b. Interpret the mean sales price.
c. Discuss the shape of the distribution of the price of new houses.

✓**SELF TEST** **3.10** The file Mobile Speed contains the overall download and upload speeds in Mbps for eight carriers in the United States.

Carrier	Download Speed	Upload Speed
Verizon	53.3	17.5
T-Mobile	36.3	16.9
AT&T	37.1	12.9
Metro PCS	32.8	13.0
Sprint	32.5	4.0
Boost	29.4	3.7
Straight Talk	31.1	15.6
Cricket	6.5	5.8

Source: Data extracted from Tom's Guide, "Fastest Wireless Network 2019: It's Not Even close," **bit.ly/2PcGiQE**.

For the download speed and the upload speed separately:
a. Calculate the mean and median.
b. Calculate the variance, standard deviation, range, and coefficient of variation.

c. Are the data skewed? If so, how?
d. Based on the results of (a) through (c), what conclusions can you reach concerning the download and upload speeds of various carriers?

3.11 The Super Bowl is a big viewing event watched by close to 200 million Americans, so it is also a big event for advertisers. The file Super Bowl Ad Ratings contains the rating of ads that ran between the opening kickoff and the final whistle.

Source: Data extracted from **www.admeter.usatoday.com/results/2019**.

For ads that ran before halftime and ads that ran at halftime or after separately:
a. Calculate the mean, median, and mode.
b. Calculate the variance, standard deviation, range, coefficient of variation, and *Z* scores.
c. Are the data skewed? If so, how?
d. Based on the results of (a) through (c), what conclusions can you reach about the ads that ran before halftime and ads that ran at halftime or after?

3.12 The file Work Hours Needed contains the average number of hours worked necessary to afford the cost for three people to attend an NBA game, inclusive of parking and food and beverage costs, at each of the 30 NBA arenas during a recent season.
a. Calculate the mean, median, and mode.
b. Calculate the variance, standard deviation, range, and coefficient of variation.
c. Are the data skewed? If so, how?
d. Based on the results of (a) through (c), what conclusions can you reach about the household cost of attending NBA games?

3.13 The file Accounting Partners contains the number of partners in a cohort of rising accounting firms that have been tagged as "firms to watch."

Source: Data extracted from *2019 Top 100 Firms*, Bloomberg BNA, **accountingtoday.com**.

a. Calculate the mean, median, and mode.
b. Calculate the variance, standard deviation, range, coefficient of variation, and *Z* scores. Are there any outliers? Explain.
c. Are the data skewed? If so, how?
d. Based on the results of (a) through (c), what conclusions can you reach concerning the number of partners in rising accounting firms?

3.14 The file Mobile Commerce contains the mobile commerce penetration values, the percentage of the country population that bought something online via a mobile phone in the past month, for 24 of the world's economies.

Source: Data extracted from Statistica.com, **bit.ly/2vsnjv8**.

a. Calculate the mean and median.
b. Calculate the variance, standard deviation, range, coefficient of variation, and *Z* scores. Are there any outliers? Explain.
c. Are the data skewed? If so, how?
d. Based on the results of (a) through (c), what conclusions can you reach concerning mobile commerce population penetration?

3.15 Is there a difference in the variation of the yields of different types of investments? The file CD Rates contains the yields for one-year certificates of deposit (CDs) and five-year CDs for 46 banks listed for West Palm Beach, Florida on April 5, 2019.

Source: Data extracted from **www.Bankrate.com**, April 5, 2019.

a. For one-year and five-year CDs, separately calculate the variance, standard deviation, range, and coefficient of variation.
b. Based on the results of (a), do one-year CDs or five-year CDs have more variation in the yields offered? Explain.

3.16 A financial services call center monitors *call duration*, the amount of time spent speaking to customers on the phone. The file Call Duration contains the data for time, in seconds, agents spent talking to 50 customers:
a. Calculate the mean, median, and mode.
b. Calculate the range, variance, and standard deviation.
c. Based on the results of (a) and (b), what conclusions can you reach concerning the amount of time spent speaking to customers on the phone?

3.17 A bank branch located in a commercial district of a city has the business objective of developing an improved process for serving customers during the noon-to-1:00 P.M. lunch period. The bank collects the waiting time, the number of minutes that elapses from when a customer joins the waiting line to when the customer begins to interact with a teller, for a sample of 15 customers who arrive during the noon hour. The waiting times, stored in Bank Waiting , are:

4.21	5.55	3.02	5.13	4.77	2.34	3.54	3.20
4.50	6.10	0.38	5.12	6.46	6.19	3.79	

a. Calculate the mean and median.
b. Calculate the variance, standard deviation, range, coefficient of variation, and Z scores. Are there any outliers? Explain.
c. Are the data skewed? If so, how?
d. As a customer walks into the branch office during the lunch hour, she asks the branch manager how long she can expect to wait. The branch manager replies, "Almost certainly less than five minutes." On the basis of the results of (a) through (c), evaluate the accuracy of this statement.

3.18 Suppose that another bank branch, located in a residential area, is also concerned about waiting times during the noon hour. That branch collects waiting times, in minutes, for a sample of 15 customers who arrive during the noon hour. These waiting times, stored in Bank Waiting 2 , are:

9.66	5.90	8.02	5.79	8.73	3.82	8.01	8.35
10.49	6.68	5.64	4.08	6.17	9.91	5.47	

a. Calculate the mean and median.
b. Calculate the variance, standard deviation, range, coefficient of variation, and Z scores. Are there any outliers? Explain.
c. Are the data skewed? If so, how?
d. As a customer walks into the branch office during the lunch hour, he asks the branch manager how long he can expect to wait. The branch manager replies, "Almost certainly less than five minutes." On the basis of the results of (a) through (c), evaluate the accuracy of this statement.

3.19 General Electric (GE) is one of the world's largest companies; it develops, manufactures, and markets a wide range of products, including medical diagnostic imaging devices, jet engines, lighting products, and chemicals. In 2017, the stock price rose 0.09%, and in 2018, the stock price declined 56.9%.
a. Compute the geometric mean rate of return per year for the two-year period 2017–2018. (Hint: Denote an increase of 0.09% as $R_2 = 0.09$.)
b. If you purchased $1,000 of GE stock at the start of 2017, what was its value at the end of 2018?
c. Compare the result of (b) to that of Problem 3.20 (b).

 3.20 Facebook's stock price in 2017 increased by 75.8%, and in 2018, it decreased by 25.7%.
a. Compute the geometric mean rate of return per year for the two-year period 2017–2018. (Hint: Denote an increase of 75.8% as $R_1 = 0.7580$.)
b. If you purchased $1,000 of Facebook stock at the start of 2017, what was its value at the end of 2018?
c. Compare the result of (b) to that of Problem 3.19 (b).

3.21 The file Indices contains data that represent the total rate of return percentage for the Dow Jones Industrial Average (DJIA), the Standard & Poor's 500 (S&P 500), and the technology-heavy NASDAQ Composite (NASDAQ) from 2015 through 2018.
a. Compute the geometric mean rate of return per year for the DJIA, S&P 500, and NASDAQ from 2015 through 2018.
b. What conclusions can you reach concerning the geometric mean rates of return per year of the three market indices?
c. Compare the results of (b) to those of Problem 3.22 (b).

3.22 In 2016 through 2018, the value of precious metals fluctuated greatly. The data (stored in Metals) represent the total rate of return (in percentage) for platinum, gold, and silver from 2016 through 2018:
a. Compute the geometric mean rate of return per year for platinum, gold, and silver from 2016 through 2018.
b. What conclusions can you reach concerning the geometric mean rates of return of the three precious metals?
c. Compare the results of (b) to those of Problem 3.21 (b).

3.23 Using the three-year return percentage variable in Retirement Funds :
a. Construct a table that computes the mean for each combination of type, market cap, and risk.
b. Construct a table that computes the standard deviation for each combination of type, market cap, and risk.
c. What conclusions can you reach concerning differences among the types of retirement funds (growth and value), based on market cap (small, midcap, and large) and the risk (low, average, and high)?

3.24 Using the three-year return percentage variable in Retirement Funds :
a. Construct a table that computes the mean for each combination of type, market cap, and rating.
b. Construct a table that computes the standard deviation for each combination of type, market cap, and rating.

c. What conclusions can you reach concerning differences among the types of retirement funds (growth and value), based on market cap (small, mid-cap, and large) and the rating (one, two, three, four, and five)?

3.25 Using the three-year return percentage variable in Retirement Funds :

a. Construct a table that computes the mean for each combination of market cap, risk, and rating.

b. Construct a table that computes the standard deviation for each combination of market cap, risk, and rating.

c. What conclusions can you reach concerning differences based on the market cap (small, mid-cap, and large), risk (low, average, and high), and rating (one, two, three, four, and five)?

3.26 Using the three-year return percentage variable in Retirement Funds :

a. Construct a table that computes the mean for each combination of type, risk, and rating.

b. Construct a table that computes the standard deviation for each combination of type, risk, and rating.

c. What conclusions can you reach concerning differences among the types of retirement funds (growth and value), based on the risk (low, average, and high) and the rating (one, two, three, four, and five)?

3.3 Exploring Numerical Variables

Besides summarizing by calculating the measures of central tendency, variation, and shape, a numerical variable can be explored by examining the distribution of values for the variable. This exploration can include calculating the *quartiles* as well as creating a *boxplot*, a visual summary of the distribution of values.

Quartiles

student TIP

Rank the values in order from smallest to largest *before* calculating the quartiles.

The three **quartiles** split a set of data into four equal parts. The **first quartile** (Q_1) divides the smallest 25% of the values from the other 75% that are larger. The **second quartile** (Q_2), the median, divides the set such that 50% of the values are smaller than or equal to the median, and 50% are larger than or equal to the median. The **third quartile** (Q_3) divides the smallest 75% of the values from the largest 25%.

The boundaries of the four equal parts are the lowest value to Q_1, Q_1 to the median, the median to Q_3, and Q_3 to the highest value. Equations (3.10) and (3.11) define the first and third quartiles.

FIRST QUARTILE, Q_1

$$Q_1 = \frac{n + 1}{4} th \text{ ranked value} \tag{3.10}$$

THIRD QUARTILE, Q_3

$$Q_3 = \frac{3(n + 1)}{4} th \text{ ranked value} \tag{3.11}$$

Exhibit 3.1 on page 126 summarizes the rules for calculating the quartiles. These rules require that the values have been first ranked from smallest to largest.

For example, to calculate the quartiles for the sample of 10 get-ready times, first rank the data from smallest to largest:

Ranked values:	29	31	35	39	39	40	43	44	44	52
Ranks:	1	2	3	4	5	6	7	8	9	10

The first quartile is the $(n + 1)/4 = (10 + 1)/4 = 2.75$ ranked value. Using Rule 3, round up to the third ranked value. The third ranked value for the get-ready times data is 35 minutes. Interpret the first quartile of 35 to mean that on 25% of the days, the time to get ready is less than or equal to 35 minutes, and on 75% of the days, the time to get ready is greater than or equal to 35 minutes.

EXHIBIT 3.1

Rules for Calculating the Quartiles from a Set of Ranked Values

Rule 1   If the ranked value is a whole number, the quartile is equal to the measurement that corresponds to that ranked value.

Example: If the sample size $n = 7$, the first quartile, Q_1, is equal to the measurement associated with the $(7 + 1)/4 =$ second ranked value.

Some sources define different rules for calculating quartiles that may result in slightly different values.

Rule 2   If the ranked value is a fractional half (2.5, 4.5, *etc.*), the quartile is equal to the measurement that corresponds to the average of the measurements corresponding to the two ranked values involved.

Example: If the sample size $n = 9$, the first quartile, Q_1, is equal to the $(9 + 1)/4 = 2.5$ ranked value, halfway between the second ranked value and the third ranked value.

Rule 3   If the ranked value is neither a whole number nor a fractional half, round the result to the nearest integer and select the measurement corresponding to that ranked value.

Example: If the sample size $n = 10$, the first quartile, Q_1, is equal to the $(10 + 1)/4 = 2.75$ ranked value. Round 2.75 to 3 and use the third ranked value.

The third quartile is the $3(n + 1)/4 = 3(10 + 1)/4 = 8.25$ ranked value. Using Rule 3 for quartiles, round this down to the eighth ranked value. The eighth ranked value is 44 minutes. Interpret the third quartile to mean that on 75% of the days, the time to get ready is less than or equal to 44 minutes, and on 25% of the days, the time to get ready is greater than or equal to 44 minutes.

Percentiles   Related to quartiles are **percentiles** that split a variable into 100 equal parts. By this definition, the first quartile is equivalent to the 25th percentile, the second quartile to the 50th percentile, and the third quartile to the 75th percentile.

EXAMPLE 3.11

Calculating the Quartiles for Number of Calories in Cereals

A sample of seven breakfast cereals (stored in `Cereals`) includes nutritional data about the number of calories per serving (see Example 3.1 on page 111). Calculate the first quartile (Q_1) and third quartile (Q_3) of the number of calories for the cereals.

SOLUTION   Ranked from smallest to largest, the number of calories for the seven cereals are as follows:

Ranked values:	80	100	100	110	130	190	200
Ranks:	1	2	3	4	5	6	7

For these data

$$Q_1 = \frac{(n + 1)}{4} \text{ ranked value} = \frac{7 + 1}{4} \text{ ranked value}$$

$$= \text{2nd ranked value}$$

Therefore, using Rule 1, Q_1 is the second ranked value. Because the second ranked value is 100, the first quartile, Q_1, is 100.

▶(*continued*)

To compute the third quartile, Q_3,

$$Q_3 = \frac{3(n+1)}{4} \text{ ranked value} = \frac{3(7+1)}{4} \text{ ranked value}$$

$$= \text{6th ranked value}$$

Therefore, using Rule 1, Q_3 is the sixth ranked value. Because the sixth ranked value is 190, Q_3 is 190.

The first quartile of 100 indicates that 25% of the cereals contain 100 calories or fewer per serving and 75% contain 100 or more calories. The third quartile of 190 indicates that 75% of the cereals contain 190 calories or fewer per serving and 25% contain 190 or more calories.

The Interquartile Range

The **interquartile range** (also called the **midspread**) measures the difference in the center of a distribution between the third and first quartiles.

INTERQUARTILE RANGE

The interquartile range is the difference between the third quartile and the first quartile:

$$\text{Interquartile range} = Q_3 - Q_1 \tag{3.12}$$

The interquartile range measures the spread in the middle 50% of the values and is not influenced by extreme values. The interquartile range can be used to determine whether to classify an extreme value as an outlier. If a value is either more than 1.5 times the interquartile range below the first quartile or more than 1.5 times the interquartile range above the third quartile, that value can be classified as an outlier.

Calculating the interquartile range would further analyze the sample of 10 get-ready times. First order the data as follows:

$$29 \quad 31 \quad 35 \quad 39 \quad 39 \quad 40 \quad 43 \quad 44 \quad 44 \quad 52$$

Using Equation (3.12) and the earlier results on pages 125 and 126, $Q_1 = 35$ and $Q_3 = 44$:

$$\text{Interquartile range} = 44 - 35 = 9 \text{ minutes}$$

Therefore, the interquartile range for the 10 get-ready times is 9 minutes. The interval 35 to 44 is often referred to as the *middle fifty*.

EXAMPLE 3.12
Calculating the Interquartile Range for the Number of Calories in Cereals

A sample of seven breakfast cereals (stored in `Cereals`) includes nutritional data about the number of calories per serving (see Example 3.1 on page 111). Calculate the interquartile range of the number of calories in cereals.

SOLUTION Ranked from smallest to largest, the number of calories for the seven cereals are as follows:

$$80 \quad 100 \quad 100 \quad 110 \quad 130 \quad 190 \quad 200$$

Using Equation (3.12) and the earlier results from Example 3.11 on page 125 $Q_1 = 100$ and $Q_3 = 190$:

$$\text{Interquartile range} = 190 - 100 = 90$$

Therefore, the interquartile range of the number of calories in cereals is 90 calories.

Because the interquartile range does not consider any value smaller than Q_1 or larger than Q_3, it cannot be affected by extreme values. Descriptive statistics such as the median, Q_1, Q_3, and the interquartile range, which are not influenced by extreme values, are called **resistant measures**.

The Five-Number Summary

The **five-number summary** for a variable consists of the smallest value ($X_{smallest}$), the first quartile, the median, the third quartile, and the largest value ($X_{largest}$).

FIVE-NUMBER SUMMARY

$$X_{smallest} \quad Q_1 \quad Median \quad Q_3 \quad X_{largest}$$

The five-number summary provides a way to determine the shape of the distribution for a set of data. Table 3.2 explains how relationships among these five statistics help identify the shape of the distribution.

TABLE 3.2

Relationships among the five-number summary and the types of distribution

	TYPE OF DISTRIBUTION		
COMPARISON	**Left-Skewed**	**Symmetrical**	**Right-Skewed**
The distance from $X_{smallest}$ to the median versus the distance from the median to $X_{largest}$.	The distance from $X_{smallest}$ to the median is greater than the distance from the median to $X_{largest}$.	The two distances are the same.	The distance from $X_{smallest}$ to the median is less than the distance from the median to $X_{largest}$.
The distance from $X_{smallest}$ to Q_1 versus the distance from Q_3 to $X_{largest}$.	The distance from $X_{smallest}$ to Q_1 is greater than the distance from Q_3 to $X_{largest}$.	The two distances are the same.	The distance from $X_{smallest}$ to Q_1 is less than the distance from Q_3 to $X_{largest}$.
The distance from Q_1 to the median versus the distance from the median to Q_3.	The distance from Q_1 to the median is greater than the distance from the median to Q_3.	The two distances are the same.	The distance from Q_1 to the median is less than the distance from the median to Q_3.

Calculating the five-number summary would further analyze the sample of 10 get-ready times. From the ordered array for this sample on page 111, the smallest value is 29 minutes, and the largest value is 52 minutes. Previous calculations in this chapter show that the median $= 39.5$, $Q_1 = 35$, and $Q_3 = 44$. Therefore, the five-number summary is

$$29 \quad 35 \quad 39.5 \quad 44 \quad 52$$

The distance from $X_{smallest}$ to the median ($39.5 - 29 = 10.5$) is slightly less than the distance from the median to $X_{largest}$ ($52 - 39.5 = 12.5$). The distance from $X_{smallest}$ to Q_1 ($35 - 29 = 6$) is slightly less than the distance from Q_3 to $X_{largest}$ ($52 - 44 = 8$). The distance from Q_1 to the median ($39.5 - 35 = 4.5$) is the same as the distance from the median to Q_3($44 - 39.5 = 4.5$). Therefore, the get-ready times are slightly right-skewed.

EXAMPLE 3.13

Calculating the Five-Number Summary of the Number of Calories in Cereals

A sample of seven breakfast cereals (stored in Cereals) includes nutritional data about the number of calories per serving (see Example 3.1 on page 111). Calculate the five-number summary of the number of calories in cereals.

SOLUTION Using earlier results (see pages 112 and 126), the median = 110, Q_1 = 100, and Q_3 = 190. In addition, the smallest value in the data set is 80, and the largest value is 200. Therefore, the five-number summary is

$$80 \quad 100 \quad 110 \quad 190 \quad 200$$

Table 3.2 can be used to evaluate skewness. The distance from X_{smallest} to the median (110 − 80 = 30) is less than the distance (200 − 110 = 90) from the median to X_{largest}. The distance from X_{smallest} to Q_1 (100 − 80 = 20) is greater than the distance from Q_3 to X_{largest} (200 − 190 = 10). The distance from Q_1 to the median (110 − 100 = 10) is less than the distance from the median to Q_3 (190 − 110 = 80).

Two comparisons indicate a right-skewed distribution, whereas the other indicates a left-skewed distribution. Therefore, given the small sample size and the conflicting results, the shape cannot be clearly determined.

The Boxplot

The boxplot is also known as the box-and-whisker chart, a name that Excel uses.

The **boxplot** visualizes the shape of the distribution of the values for a variable. Boxplots get their name from the box that defines the range of the middle 50% of the values and the ends of which correspond to Q_1 and Q_3. Inside the box, an additional line marks a median. Extending in either direction away from the box are whiskers, the ends of which may have dashed lines drawn perpendicular to the whiskers.

In one form of the boxplot, a five-number boxplot, the endpoints of the whiskers represent X_{smallest} and X_{largest}, making the boxplot a visual representation of a five-number summary. In a second form, the endpoints of the whiskers define the smallest and largest values that are within the range of 1.5 times the interquartile range from the box. In this second form, values that are beyond this range in either direction are plotted as points or asterisks and can be considered outliers. Excel versions that contain a boxplot feature construct boxplots that are similar to, but not identical to, this second form. Section EG3.3 *PHStat* and *Workbook* instructions construct five-number boxplots in *any* Excel version, and this book uses such boxplots as illustrations.

Box plots can be drawn either horizontally or vertically. When drawn horizontally, the lowest values appear to the left and Q_1 is to the left of Q_3. When drawn vertically, the lowest values appear toward the bottom and Q_1 is below Q_3. Figure 3.7 contains a horizontal boxplot that visualizes the five-number summary for the sample of 10 times to get ready in the morning.

FIGURE 3.7
Boxplot for the get-ready times

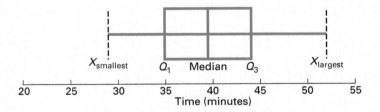

The Figure 3.7 boxplot for the get-ready times shows a slight right-skewness: The distance between the median and the largest value is slightly greater than the distance between the smallest value and the median, and the right tail is slightly longer than the left tail.

EXAMPLE 3.14

Constructing Boxplots of the Three-Year Returns for the Growth and Value Funds

In the More Descriptive Choices scenario, you are interested in comparing the past performance of the growth and value funds from a sample of 479 funds. One measure of past performance is the three-year return percentage (the 3YrReturn variable). Construct the boxplots for this variable for the growth and value funds.

SOLUTION Figure 3.8 contains an Excel five-number summary worksheet and boxplot for the three-year return percentages for the growth and value funds. The five-number summary for the growth funds associated with these boxplots is $X_{\text{smallest}} = -3.4$, $Q_1 = 6.66$, median $= 8.70$, $Q_3 = 10.92$, and $X_{\text{largest}} = 15.32$. The five-number summary for the value funds in this boxplot is $X_{\text{smallest}} = -2.65$, $Q_1 = 5.67$, median $= 7.07$, $Q_3 = 8.5$, and $X_{\text{largest}} = 11.98$.

FIGURE 3.8
Excel five-number summary and boxplot for the three-year return percentage variable

◢	A	B	C
1	Five-Number Summary for 3YrReturn		
2			
3		Growth	Value
4	Minimum	-3.4	-2.65
5	First Quartile	6.66	5.67
6	Median	8.7	7.07
7	Third Quartile	10.92	8.5
8	Maximum	15.32	11.98

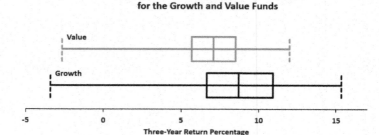

The median return, the quartiles, and the maximum returns are higher for the growth funds than for the value funds. Both the growth and value funds are left-skewed. These results are consistent with the Example 3.9 solution on page 121.

Figure 3.9 demonstrates the relationship between the boxplot and the density curve for four different types of distributions. The area under each density curve is split into quartiles corresponding to the five-number summary for the boxplot.

The distributions in Panels A and D of Figure 3.9 are symmetrical. In these distributions, the mean and median are equal. In addition, the length of the left tail is equal to the length of the right tail, and the median line divides the box in half.

FIGURE 3.9
Five-number summary boxplots and corresponding density curves for four distributions

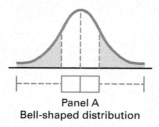

Panel A
Bell-shaped distribution

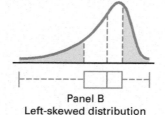

Panel B
Left-skewed distribution

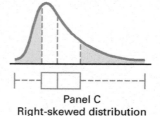

Panel C
Right-skewed distribution

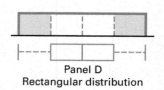

Panel D
Rectangular distribution

student TIP

A long tail on the left side of the boxplot indicates a left-skewed distribution. A long tail on the right side of the boxplot indicates a right-skewed distribution.

The distribution in Panel B of Figure 3.9 is left-skewed. The few small values distort the mean toward the left tail. For this left-skewed distribution, there is a heavy clustering of values at the high end of the scale (i.e., the right side); 75% of all values are found between the left edge of the box (Q_1) and the end of the right tail (X_{largest}). There is a long left tail that contains the smallest 25% of the values, demonstrating the lack of symmetry in this data set.

The distribution in Panel C of Figure 3.9 is right-skewed. The concentration of values is on the low end of the scale, toward the left side of the boxplot. Here, 75% of all values are found between the beginning of the left tail and the right edge of the box (Q_3). There is a long right tail that contains the largest 25% of the values, demonstrating the lack of symmetry in this set of data.

PROBLEMS FOR SECTION 3.3

LEARNING THE BASICS

3.27 The following is a set of data from a sample of $n = 7$:

$$12\ 7\ 4\ 9\ 0\ 7\ 3$$

a. Calculate the first quartile (Q_1), the third quartile (Q_3), and the interquartile range.
b. List the five-number summary.
c. Construct a boxplot and describe its shape.
d. Compare your answer in (c) with that from Problem 3.3 (d) on page 122. Discuss.

3.28 The following is a set of data from a sample of $n = 6$:

$$7\ 4\ 9\ 7\ 3\ 12$$

a. Calculate the first quartile (Q_1), the third quartile (Q_3), and the interquartile range.
b. List the five-number summary.
c. Construct a boxplot and describe its shape.
d. Compare your answer in (c) with that from Problem 3.2 (d) on page 122. Discuss.

3.29 The following is a set of data from a sample of $n = 5$:

$$7\ 4\ 9\ 8\ 2$$

a. Calculate the first quartile (Q_1), the third quartile (Q_3), and the interquartile range.
b. List the five-number summary.
c. Construct a boxplot and describe its shape.
d. Compare your answer in (c) with that from Problem 3.1 (d) on page 122. Discuss.

3.30 The following is a set of data from a sample of $n = 5$:

$$7\ -5\ -8\ 7\ 9$$

a. Calculate the first quartile (Q_1), the third quartile (Q_3), and the interquartile range.
b. List the five-number summary.
c. Construct a boxplot and describe its shape.
d. Compare your answer in (c) with that from Problem 3.4 (d) on page 122. Discuss.

APPLYING THE CONCEPTS

3.31 The file Accounting Partners contains the number of partners in a cohort of rising accounting firms that have been tagged as "firms to watch."

Source: Data extracted from *2019 Top 100 Firms*, Bloomberg BNA, **accounting today.com**.

a. Calculate the first quartile (Q_1), the third quartile (Q_3), and the interquartile range.
b. List the five-number summary.
c. Construct a boxplot and describe its shape.

3.32 The file Mobile Commerce contains the mobile commerce penetration values, the percentage of the country population that bought something online via a mobile phone in the past month, for 24 of the world's economies.

Source: Data extracted from "Mobile commerce share by country 2018," **bit.ly/2vsnjv8**.

a. Calculate the first quartile (Q_1), the third quartile (Q_3), and the interquartile range.

b. List the five-number summary.
c. Construct a boxplot and describe its shape.

3.33 A financial services call center monitors *call duration*, the amount of time spent speaking to customers on the phone. The file Call Duration contains the data for time, in seconds, agents spent talking to 50 customers.
a. Calculate the first quartile (Q_1), the third quartile (Q_3), and the interquartile range.
b. List the five-number summary.
c. Construct a boxplot and describe its shape.

3.34 The Super Bowl is a big viewing event watched by close to 200 million Americans, so it is also a big event for advertisers. The file Super Bowl Ad Ratings contains the rating of ads that ran between the opening kickoff and the final whistle.

Source: Data extracted from **www.admeter.usatoday.com/results/2019**.

For ads that ran before halftime and ads that ran at halftime or after separately:
a. Calculate the first quartile (Q_1), the third quartile (Q_3), and the interquartile range.
b. List the five-number summary.
c. Construct a boxplot and describe its shape.

3.35 The file CD Rates contains the yields for one-year CDs and five-year CDs, for 46 financial institutions selling in Lake Worth, Florida, as of April 5, 2019.

Source: Data extracted from **www.Bankrate.com**, April 5, 2019.

For each type of account:
a. Calculate the first quartile (Q_1), the third quartile (Q_3), and the interquartile range.
b. List the five-number summary.
c. Construct a boxplot and describe its shape.

✓ SELF TEST **3.36** A bank branch located in a commercial district of a city has the business objective of developing an improved process for serving customers during the noon-to-1:00 P.M. lunch period. The bank collects the waiting time, the number of minutes that elapses from when a customer joins the waiting line to when the customer begins to interact with a teller, for a sample of 15 customers who arrive during the noon hour. The waiting times, stored in Bank Waiting, are:

$$4.21\ 5.55\ 3.02\ 5.13\ 4.77\ 2.34\ 3.54\ 3.20$$

$$4.50\ 6.10\ 0.38\ 5.12\ 6.46\ 6.19\ 3.79$$

Another bank branch, located in a residential area, is also concerned about the noon-to-1:00 P.M. lunch hour. That branch collects waiting times, in minutes, for a sample of 15 customers who arrive during the noon hour. These waiting times, stored in Bank Waiting 2, are:

$$9.66\ 5.90\ 8.02\ 5.79\ 8.73\ 3.82\ 8.01\ 8.35$$

$$10.49\ 6.68\ 5.64\ 4.08\ 6.17\ 9.91\ 5.47$$

a. List the five-number summaries of the waiting times at the two bank branches.
b. Construct boxplots and describe the shapes of the distributions for the two bank branches.
c. What similarities and differences are there in the distributions of the waiting times at the two bank branches?

3.4 Numerical Descriptive Measures for a Population

Sections 3.1 and 3.2 discuss the statistics that describe the properties of central tendency and variation for a sample. For data collected from an entire population (see Section 1.2), compute and analyze population *parameters* for these properties, including the population mean, population variance, and population standard deviation.

To help illustrate these parameters, consider the population of stocks for the 10 companies that comprise the "Dogs of the Dow." "Dogs" are the 10 stocks in the Dow Jones Industrial Average (DJIA) that have the highest dividend yields, or dividend-to-price ratios, as of December 31 of the previous year and form the basis for an investment approach developed by Michael O'Higgins. Table 3.3 lists the 2019 "dogs," along with the percentage yield of these stocks during 2018.

TABLE 3.3
Percentage yield for the 2019 "Dogs of the Dow," stored in Dow Dogs

Stock	Percentage Change	Stock	Percentage Change
Chevron	4.12	JP Morgan Chase	3.28
Cisco Systems	3.05	Merck	2.88
Coca-Cola	3.29	Pfizer	3.30
ExxonMobil	4.81	Procter & Gamble	3.12
IBM	5.52	Verizon	4.29

Source: Data extracted from **dogsofthedow.com/2019-dogs-of-the-dow-co.htm**.

The Population Mean

The **population mean**, a measure of central tendency, is the sum of the values in the population divided by the population size, N. The Greek lowercase letter mu, μ, represents this parameter, which Equation (3.13) defines.

POPULATION MEAN

$$\mu = \frac{\sum_{i=1}^{N} X_i}{N} \tag{3.13}$$

where

μ = population mean

X_i = ith value of the variable X

N = number of values in the population

$\sum_{i=1}^{N} X_i$ = summation of all X_i values in the population

To compute the mean one-year percentage change in stock price for the Table 3.3 population of "Dow Dog" stocks, use Equation (3.13):

$$\mu = \frac{\sum\limits_{i=1}^{N} X_i}{N}$$

$$= \frac{4.12 + 3.05 + 3.29 + 4.81 + 5.52 + 3.28 + 2.88 + 3.30 + 3.12 + 4.29}{10}$$

$$= \frac{37.7}{10} = 3.77$$

The mean one-year percentage change in the stock price for the "Dow Dog" stocks is 3.77%.

The Population Variance and Standard Deviation

The population variance and the population standard deviation parameters measure variation in a population. The **population variance** is the sum of the squared differences around the population mean divided by the population size, N, and the **population standard deviation** is the square root of the population variance. In practice, you will most likely use the population standard deviation because, unlike the population variance, the standard deviation will always be a number expressed in the same units as the original population data.

The lowercase Greek letter sigma, σ, represents the population standard deviation, and sigma squared, σ^2, represents the population variance. Equations (3.14) and (3.15) define these parameters. The denominators for the right-side terms in these equations use N and not the $(n - 1)$ term found in Equations (3.6) and (3.7) on page 115, which define the sample variance and standard deviation.

POPULATION VARIANCE

$$\sigma^2 = \frac{\sum\limits_{i=1}^{N}(X_i - \mu)^2}{N} \tag{3.14}$$

where

$$\mu = \text{population mean}$$

$$X_i = i\text{th value of the variable } X$$

$$\sum\limits_{i=1}^{N}(X_i - \mu)^2 = \text{summation of all the squared differences between the } X_i \text{ values and } \mu$$

POPULATION STANDARD DEVIATION

$$\sigma = \sqrt{\frac{\sum\limits_{i=1}^{N}(X_i - \mu)^2}{N}} \tag{3.15}$$

To calculate the population variance for the Table 3.3 data, use Equation (3.14):

$$\sigma^2 = \frac{\sum\limits_{i=1}^{N}(X_i - \mu)^2}{N} = \frac{6.9612}{10} = 0.6961$$

From Equation (3.15), the population sample standard deviation is

$$\sigma = \sqrt{\sigma^2} = \sqrt{\frac{\sum_{i=1}^{N}(X_i - \mu)^2}{N}} = \sqrt{\frac{6.9612}{10}} = 0.8343$$

Therefore, the typical percentage change in stock price differs from the mean of 3.77 by less than one percent (0.8%). This very small amount of variation suggests that the "Dow Dog" stocks produce results that do not differ greatly.

The Empirical Rule

In most data sets, a large portion of the values tend to cluster somewhere near the mean. In right-skewed data sets, this clustering occurs to the left of the mean—that is, at a value less than the mean. In left-skewed data sets, the values tend to cluster to the right of the mean—that is, greater than the mean. In symmetrical data sets, where the median and mean are the same, the values often tend to cluster around the median and mean, often producing a bell-shaped normal distribution (see Chapter 6).

The **empirical rule** states that for population data from a symmetric mound-shaped distribution such as the normal distribution, the following are true:

- Approximately 68% of the values are within ± 1 standard deviation from the mean.
- Approximately 95% of the values are within ± 2 standard deviations from the mean.
- Approximately 99.7% of the values are within ± 3 standard deviations from the mean.

The empirical rule helps examine variability in a population as well as identify outliers. The empirical rule implies that in a normal distribution, only about 1 out of 20 values will be beyond 2 standard deviations from the mean in either direction. As a general rule, consider values not found in the interval $\mu \pm 2\sigma$ as potential outliers. The rule also implies that only about 3 in 1,000 will be beyond 3 standard deviations from the mean. Therefore, values not found in the interval $\mu \pm 3\sigma$ are almost always considered outliers.

EXAMPLE 3.15

Using the Empirical Rule

A population of 2-liter bottles of cola is known to have a mean fill-weight of 2.06 liters and a standard deviation of 0.02 liter. The population is known to be bell-shaped. Describe the distribution of fill-weights. Is it very likely that a bottle will contain less than 2 liters of cola?

SOLUTION

$$\mu \pm \sigma = 2.06 \pm 0.02 = (2.04, 2.08)$$

$$\mu \pm 2\sigma = 2.06 \pm 2(0.02) = (2.02, 2.10)$$

$$\mu \pm 3\sigma = 2.06 \pm 3(0.02) = (2.00, 2.12)$$

Using the empirical rule, you can see that approximately 68% of the bottles will contain between 2.04 and 2.08 liters, approximately 95% will contain between 2.02 and 2.10 liters, and approximately 99.7% will contain between 2.00 and 2.12 liters. Therefore, it is highly unlikely that a bottle will contain less than 2 liters.

Chebyshev's Theorem

For heavily skewed sets of data and data sets that do not appear to be normally distributed, one should use Chebyshev's theorem instead of the empirical rule. **Chebyshev's theorem** (Kendall) states that for any data set, regardless of shape, the percentage of values that are found within distances of k standard deviations from the mean must be at least

$$\left(1 - \frac{1}{k^2}\right) \times 100$$

Use this rule for any value of k greater than 1. For example, consider $k = 2$. Chebyshev's theorem states that at least $[1 - (1/2)^2] \times 100\% = 75\%$ of the values must be found within ± 2 standard deviations of the mean.

Chebyshev's theorem is very general and applies to any distribution. The theorem indicates *at least* what percentage of the values fall within a given distance from the mean. However, if the data set is approximately bell-shaped, the empirical rule will more accurately reflect the greater concentration of data close to the mean. Table 3.4 compares Chebyshev's theorem to the empirical rule.

*Section EG3.4 describes the **VE-Variability Excel** **workbook** that allows one to explore the empirical rule and Chebyshev's theorem.*

TABLE 3.4
How data vary around the mean

Interval	% of Values Found in Intervals Around the Mean	
	Chebyshev's Theorem (any distribution)	**Empirical Rule (normal distribution)**
$(\mu - \sigma, \mu + \sigma)$	At least 0%	Approximately 68%
$(\mu - 2\sigma, \mu + 2\sigma)$	At. least 75%	Approximately 95%
$(\mu - 3\sigma, \mu + 3\sigma)$	At least 88.89%	Approximately 99.7%

Use Chebyshev's theorem and the empirical rule to understand how data are distributed around the mean when you have sample data. With each, use the value calculated for $\overline{X}$ in place of μ and the value calculated for S in place of σ. These results using the sample statistics are *approximations* because population parameters (μ, σ) were not used in the calculations.

EXAMPLE 3.16

Using the Chebyshev Theorem

As in Example 3.15, a population of 2-liter bottles of cola is known to have a mean fill-weight of 2.06 liters and a standard deviation of 0.02 liter. However, the shape of the population is unknown, and you cannot assume that it is bell-shaped. Describe the distribution of fill-weights. Is it very likely that a bottle will contain less than 2 liters of cola?

SOLUTION

$$\mu \pm \sigma = 2.06 \pm 0.02 = (2.04, 2.08)$$

$$\mu \pm 2\sigma = 2.06 \pm 2(0.02) = (2.02, 2.10)$$

$$\mu \pm 3\sigma = 2.06 \pm 3(0.02) = (2.00, 2.12)$$

Because the distribution may be skewed, you cannot use the empirical rule. Using Chebyshev's theorem, you cannot say anything about the percentage of bottles containing between 2.04 and 2.08 liters. You can state that at least 75% of the bottles will contain between 2.02 and 2.10 liters and at least 88.89% will contain between 2.00 and 2.12 liters. Therefore, between 0 and 11.11% of the bottles will contain less than 2 liters.

PROBLEMS FOR SECTION 3.4

LEARNING THE BASICS

3.37 The following is a set of data for a population with $N = 10$:

     7 5 11 8 3 6 2 1 9 8

a. Calculate the population mean.
b. Calculate the population standard deviation.

3.38 The following is a set of data for a population with $N = 10$:

     7 5 6 6 6 4 8 6 9 3

a. Calculate the population mean.
b. Calculate the population standard deviation.

APPLYING THE CONCEPTS

3.39 The file McDonalds Stores contains the number of McDonald's stores per 100,000 population located in each of the 50 U.S. states as of September 13, 2018.

Source: Data extracted from Usatoday.com, "Is your state 'lovin' it'?" **bit.ly/2vFGoay**.

a. Calculate the mean, variance, and standard deviation for this population.
b. What percentage of the 50 states have a number of McDonald's stores within ± 1, ± 2, or ± 3 standard deviations of the mean?
c. Compare your findings with what would be expected on the basis of the empirical rule. Are you surprised at the results in (b)?

3.40 Consider a population of 1,024 mutual funds that primarily invest in large companies. You have determined that μ, the mean one-year total percentage return achieved by all the funds, is 8.20 and that σ, the standard deviation, is 2.75.

a. According to the empirical rule, what percentage of these funds is expected to be within ± 1 standard deviation of the mean?

b. According to the empirical rule, what percentage of these funds is expected to be within ± 2 standard deviations of the mean?

c. According to Chebyshev's theorem, what percentage of these funds is expected to be within ± 1, ± 2, or ± 3 standard deviations of the mean?

d. According to Chebyshev's theorem, at least 93.75% of these funds are expected to have one-year total returns between what two amounts?

3.41 The file **Cigarette Tax** contains the state cigarette tax (in $) for each of the 50 states and the District of Columbia as of January 1, 2019.

a. Calculate the population mean and population standard deviation for the state cigarette tax.

b. Interpret the parameters in (a).

✓**SELF TEST** **3.42** The file **Energy** contains the average residential price for electricity in cents per kilowatt hour in each of the 50 states and the District of Columbia during a recent year.

a. Calculate the mean, variance, and standard deviation for the population.

b. What proportion of these states has an average residential price for electricity within ± 1 standard deviation of the mean, within ± 2 standard deviations of the mean, and within ± 3 standard deviations of the mean?

c. Compare your findings with what would be expected based on the empirical rule. Are you surprised at the results in (b)?

3.43 Thirty companies comprise the DJIA. Just how big are these companies? One common method for measuring the size of a company is to use its market capitalization, the product of multiplying the number of stock shares by the price of a share of stock. On April 24, 2019, the market capitalization of these companies ranged from Traveler's $36.2 billion to Apple's $868.9 billion. The entire population of market capitalization values is stored in **Dow Market Cap**.

Source: Data extracted from "Dow Jones Market capitalization," **bit.ly/2WyjEFD**.

a. Calculate the mean and standard deviation of the market capitalization for this population of 30 companies.

b. Interpret the parameters calculated in (a).

3.5 The Covariance and the Coefficient of Correlation

This section presents two measures of the relationship between two numerical variables: the covariance and the coefficient of correlation.

The Covariance

The **covariance** measures the strength of the linear relationship between two numerical variables (X and Y). Equation (3.16) defines the **sample covariance**, and Example 3.17 illustrates its use.

SAMPLE COVARIANCE

$$\text{cov}(X, Y) = \frac{\displaystyle\sum_{i=1}^{n}(X_i - \bar{X})(Y_i - \bar{Y})}{n - 1} \qquad (3.16)$$

EXAMPLE 3.17

Calculating the Sample Covariance

Section 2.5 uses NBA team revenue and current values stored in **NBA Financial** to construct a scatter plot that showed the relationship between those two variables. Now, measure the association between the team revenue and the current value of a team by calculating the sample covariance.

SOLUTION Figure 3.10 on page 137 contains the data and results worksheets that compute the covariance of revenue and value of 30 NBA teams. From the result in cell B9 of the covariance worksheet, or by using Equation (3.16) directly, you determine that the covariance is 41.3729:

$$\text{cov}(X, Y) = \frac{1199.8133}{30 - 1} = 41.3729$$

▶(*continued*)

FIGURE 3.10
Excel DATA worksheet and covariance results worksheet for the revenue and current value for the 30 NBA teams

	A	B	C	D
1	Revenue	Value	(X-XBar)	(Y-YBar)
2	215	1.3	-52.03	-0.59
3	287	2.8	19.97	0.91
4	290	2.4	22.97	0.51
5	218	1.3	-49.03	-0.59
6	287	2.9	19.97	1.01
7	302	1.3	34.97	-0.59
8	287	2.3	19.97	0.41
9	222	1.4	-45.03	-0.49
10	235	1.3	-32.03	-0.59
11	401	3.5	133.97	1.61
12	323	2.3	55.97	0.41
13	222	1.4	-45.03	-0.49
14	258	2.2	-9.03	0.31
15	398	3.7	130.97	1.81
16	213	1.2	-54.03	-0.69

	A	B	C	D
1	Revenue	Value	(X-XBar)	(Y-YBar)
17	258	1.8	-9.03	-0.09
18	204	1.4	-63.03	-0.49
19	223	1.3	-44.03	-0.59
20	214	1.2	-53.03	-0.69
21	443	4	175.97	2.11
22	241	1.5	-26.03	-0.39
23	223	1.3	-44.03	-0.59
24	268	1.5	0.97	-0.39
25	235	1.7	-32.03	-0.19
26	246	1.6	-21.03	-0.29
27	263	1.6	-4.03	-0.29
28	262	1.7	-5.03	-0.19
29	275	1.7	7.97	-0.19
30	243	1.4	-24.03	-0.49
31	255	1.6	-12.03	-0.29

	A	B	
1	Covariance Analysis of Revenue and Value		
2			
3	Intermediate Calculations		
4	XBar	267.0333	=AVERAGE(DATA!A:A)
5	YBar	1.8867	=AVERAGE(DATA!B:B)
6	Σ(X-XBar)(Y-YBar)	1199.8133	=SUMPRODUCT(DATA!C:C, DATA!D:D)
7	n-1	29	=COUNT(DATA!A:A) - 1
8			
9	Covariance	41.3729	=COVARIANCE.S(DATA!A:A, DATA!B:B)

The covariance has a major flaw as a measure of the linear relationship between two numerical variables. Because the covariance can have any value, the covariance cannot be used to determine the relative strength of the relationship. In Example 3.17, one cannot tell whether the value 41.3729 indicates a strong relationship or a weak relationship between revenue and value. To better determine the relative strength of the relationship, calculate the coefficient of correlation.

The Coefficient of Correlation

The **coefficient of correlation** measures the relative strength of a linear relationship between two numerical variables. The values of the coefficient of correlation range from −1 for a perfect negative correlation to +1 for a perfect positive correlation. *Perfect* in this case means that if the points were plotted on a scatter plot, all the points could be connected with a straight line.

When dealing with population data for two numerical variables, the Greek letter ρ (*rho*) is used as the symbol for the coefficient of correlation. Figure 3.11 illustrates three different types of association between two variables.

In Panel A of Figure 3.11, there is a perfect negative linear relationship between X and Y. Thus, the coefficient of correlation, ρ, equals −1, and when X increases, Y decreases in a perfectly predictable manner. Panel B shows a situation in which there is no relationship between X and Y. In this case, the coefficient of correlation, ρ, equals 0, and as X increases, there is no tendency for Y to increase or decrease. Panel C illustrates a perfect positive relationship where ρ equals +1. In this case, Y increases in a perfectly predictable manner when X increases.

FIGURE 3.11
Types of association between variables

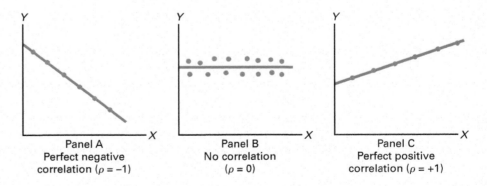

Panel A
Perfect negative
correlation (ρ = −1)

Panel B
No correlation
(ρ = 0)

Panel C
Perfect positive
correlation (ρ = +1)

studentTIP

While causation implies correlation, correlation alone does *not* imply causation.

Correlation alone cannot prove that there is a causation effect—that is, that the change in the value of one variable caused the change in the other variable. A strong correlation can be produced by chance; by the effect of a **lurking variable**, a third variable not considered in the calculation of the correlation; or by a cause-and-effect relationship. One would need to perform additional analysis to determine which of these three situations actually produced the correlation.

Equation (3.17) on page 138 defines the **sample coefficient of correlation (r)**.

SAMPLE COEFFICIENT OF CORRELATION

$$r = \frac{\text{cov}(X, Y)}{S_X S_Y} \qquad (3.17)$$

where

$$\text{cov}(X, Y) = \frac{\sum_{i=1}^{n}(X_i - \overline{X})(Y_i - \overline{Y})}{n - 1}$$

$$S_X = \sqrt{\frac{\sum_{i=1}^{n}(X_i - \overline{X})^2}{n - 1}}$$

$$S_Y = \sqrt{\frac{\sum_{i=1}^{n}(Y_i - \overline{Y})^2}{n - 1}}$$

EXAMPLE 3.18

Calculating the
Sample Coefficient
of Correlation

Example 3.17 on page 136 calculated the covariance of the team revenue and current value for the 30 NBA teams. Now, to measure the relative strength of a linear relationship between the revenue and value, you want to calculate the sample coefficient of correlation.

SOLUTION By using Equation (3.17) directly (shown below) or from cell B14 in the coefficient of correlation worksheet (shown in Figure 3.12), you determine that the sample coefficient of correlation is 0.9104:

$$r = \frac{\text{cov}(X, Y)}{S_X S_Y} = \frac{41.3729}{(58.4445)(0.7776)} = 0.9104$$

FIGURE 3.12

Excel worksheet to
compute the sample
coefficient of correlation
between revenue and
current value

*This worksheet uses
the Figure 3.10 DATA
worksheet (see page 137).*

▲	A	B	
1	**Coefficient of Correlation Analysis**		
2			
3	Intermediate Calculations		
4	XBar	267.0333	=AVERAGE(DATA!A:A)
5	YBar	1.8867	=AVERAGE(DATA!B:B)
6	Σ(X-XBar)²	99056.9667	=DEVSQ(DATA!A:A)
7	Σ(Y-YBar)²	17.5347	=DEVSQ(DATA!B:B)
8	Σ(X-XBar)(Y-YBar)	1199.8133	=SUMPRODUCT(DATA!C:C, DATA!D:D)
9	n -1	29	=COUNT(DATA!A:A) - 1
10	Covariance	41.3729	=COVARIANCE.S(DATA!A:A, DATA!B:B)
11	Sₓ	58.4445	=SQRT(B6/B9)
12	Sᵧ	0.7776	=SQRT(B7/B9)
13			
14	r	0.9104	=CORREL(DATA!A:A, DATA!B:B)

In the Worksheet

The worksheet is the **COMPUTE worksheet** of the Excel Guide **Correlation workbook**. Appendix F explains the DEVSQ and SUMPRODUCT mathematical functions that several formulas use.

The current value and revenue of the NBA teams are very highly correlated. The teams with the lowest revenues have the lowest values. The teams with the highest revenues have the highest values. This relationship is very strong, as indicated by the coefficient of correlation, $r = 0.9104$.

In general, do not assume that just because two variables are correlated, changes in one variable caused changes in the other variable. However, for this example, it makes sense to conclude that changes in revenue would tend to cause changes in the value of a team.

The sample coefficient of correlation is unlikely to be exactly $+1$, 0, or -1. Figure 3.13 on page 139 presents scatter plots along with their respective sample coefficients of correlation, r, for six data sets, each of which contains 100 pairs of X and Y values.

In Panel A, the coefficient of correlation, r, is -0.9. For small values of X, there is a very strong tendency for Y to be large. Likewise, the large values of X tend to be paired with small values of Y. The data do not all fall on a straight line, so the association between X and Y cannot be described as perfect.

FIGURE 3.13
Six scatter plots and their sample coefficients of correlation, *r*

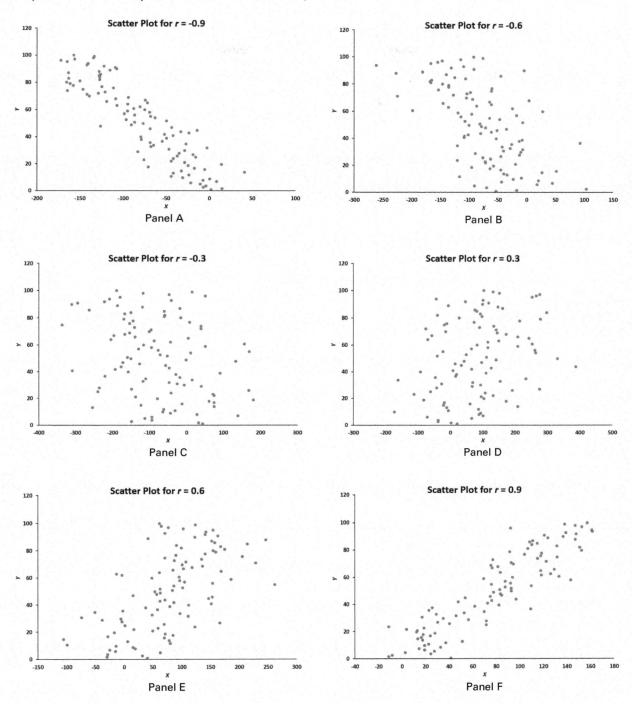

In Panel B, the coefficient of correlation is -0.6, and the small values of X tend to be paired with large values of Y. The linear relationship between X and Y in Panel B is not as strong as that in Panel A. Thus, the coefficient of correlation in Panel B is not as negative as that in Panel A.

In Panel C, the linear relationship between X and Y is very weak, $r = -0.3$, and there is only a slight tendency for the small values of X to be paired with the large values of Y.

Panels D through F depict data sets that have positive coefficients of correlation because small values of X tend to be paired with small values of Y, and large values of X tend to be associated with large values of Y. Panel D shows weak positive correlation, with $r = 0.3$. Panel E shows stronger positive correlation, with $r = 0.6$. Panel F shows very strong positive correlation, with $r = 0.9$.

In summary, the coefficient of correlation indicates the linear relationship, or association, between two numerical variables. When the coefficient of correlation gets closer to $+1$ or -1, the linear relationship between the two variables is stronger. When the coefficient of correlation is near 0, little or no linear relationship exists. The sign of the coefficient of correlation indicates whether the data are positively correlated (the larger values of X are typically paired with the larger values of Y) or negatively correlated (the larger values of X are typically paired with the smaller values of Y). The existence of a strong correlation does not imply a causation effect. It only indicates the tendencies present in the data.

PROBLEMS FOR SECTION 3.5

LEARNING THE BASICS

3.44 The following is a set of data from a sample of $n = 11$ items:

X	7	5	8	3	6	10	12	4	9	15	18
Y	21	15	24	9	18	30	36	12	27	45	54

a. Compute the covariance.
b. Compute the coefficient of correlation.
c. How strong is the relationship between X and Y? Explain.

APPLYING THE CONCEPTS

3.45 A study of 267 college students investigated the impact of smartphones on student connectedness and out-of-class involvement. One finding showed that students reporting a higher perceived usefulness of smartphones in educational settings used their smartphone a higher number of times to send or read email for class purposes than students reporting a lower perceived usefulness of smartphones in educational settings.

Source: Liu X, et al., "The Impact of Smartphone Educational Use on Student Connectedness and Out-of-Class Involvement," *The Electronic Journal of Communication* (2016).

a. Does the study suggest that perceived usefulness of smartphones in educational settings and use of smartphones for class purposes are positively correlated or negatively correlated?
b. Do you think that there might be a cause-and-effect relationship between perceived usefulness of smartphones in educational settings and use of smartphones for class purposes? Explain.

 3.46 The file Cereals lists the calories and sugar, in grams, in one serving of seven breakfast cereals:

Cereal	Calories	Sugar
Kellogg's All Bran	80	6
Kellogg's Corn Flakes	100	2
Wheaties	100	4
Nature's Path Organic Multigrain Flakes	110	4
Kellogg's Rice Krispies	130	4
Post Shredded Wheat Vanilla Almond	190	11
Kellogg's Mini Wheats	200	10

a. Calculate the covariance.
b. Calculate the coefficient of correlation.

c. Which do you think is more valuable in expressing the relationship between calories and sugar—the covariance or the coefficient of correlation? Explain.
d. Based on (a) and (b), what conclusions can you reach about the relationship between calories and sugar?

3.47 Movie companies need to predict the gross receipts of individual movies once a movie has debuted. The data, shown below and stored in Potter Movies , are the first weekend gross, the U.S. gross, and the worldwide gross (in $millions) of the eight Harry Potter movies:

Title	First Weekend	U.S. Gross	Worldwide Gross
Sorcerer's Stone	90.295	317.558	976.458
Chamber of Secrets	88.357	261.988	878.988
Prisoner of Azkaban	93.687	249.539	795.539
Goblet of Fire	102.335	290.013	896.013
Order of the Phoenix	77.108	292.005	938.469
Half-Blood Prince	77.836	301.460	934.601
Deathly Hallows Part 1	125.017	295.001	955.417
Deathly Hallows Part 2	169.189	381.011	1,328.111

Source: Data extracted from **www.the-numbers.com/interactive/comp-Harry-Potter.php**.

a. Calculate the covariance between first weekend gross and U.S. gross, first weekend gross and worldwide gross, and U.S. gross and worldwide gross.
b. Calculate the coefficient of correlation between first weekend gross and U.S. gross, first weekend gross and worldwide gross, and U.S. gross and worldwide gross.
c. Which do you think is more valuable in expressing the relationship between first weekend gross, U.S. gross, and worldwide gross—the covariance or the coefficient of correlation? Explain.
d. Based on (a) and (b), what conclusions can you reach about the relationship between first weekend gross, U.S. gross, and worldwide gross?

3.48 The file Mobile Speed contains the overall download and upload speeds in Mbps for eight carriers in the U.S.

Source: Data extracted from Tom's Guide, "Fastest Wireless Network 2019: It's Not Even Close," **bit.ly/2PcGiQE**.

a. Calculate the covariance between download speed and upload speed.

b. Calculate the coefficient of correlation between download speed and upload speed.

c. Based on (a) and (b), what conclusions can you reach about the relationship between download speed and upload speed?

3.49 A Pew Research Center survey found that smartphone ownership is growing rapidly around the world, although not always equally, and social media use is somewhat less widespread, even as Internet use has grown in emerging economies. The file World Smartphone contains the level of smartphone ownership, measured as the percentage of adults polled who report owning a smartphone. The file World Social Media contains the level of social media networking, measured as the percentage of adults who use social media sites, as well as the GDP at purchasing power parity

(PPP, current international $) per capita for each of 27 advanced and emerging countries.

Source: Data extracted from Pew Research Center, "Smartphone Ownership Is Growing Rapidly Around the World, but Not Always Equally," February 5, 2019, **bit.ly/2IltMOf**.

For the relationship between percentage of Internet users polled who use social networking sites and GDP, and the relationship between the percentage of adults polled who own a smartphone and GDP:

a. Calculate the covariance.

b. Calculate the coefficient of correlation.

c. Based on (a) and (b), what conclusions can you reach about the relationship between the GDP and social media use and the relationship between the percentage of adults polled who own a smartphone and GDP?

3.6 Descriptive Statistics: Pitfalls and Ethical Issues

This chapter describes how a set of numerical data can be characterized by the statistics that measure the properties of central tendency, variation, and shape. In business, reports that are prepared periodically and report summary or status information frequently include descriptive statistics such as the ones this chapter discusses.

The volume of information available from online, broadcast, or print media has produced much skepticism in the minds of many about the objectivity of data. When reading information that contains descriptive statistics, one should keep in mind the quip often attributed to the nineteenth-century British statesman Benjamin Disraeli: "There are three kinds of lies: lies, damned lies, and statistics."

When reviewing statistics for numerical variables, examine both the mean and the median. Are they similar, or are they very different? If only the mean is presented, then one cannot determine whether the data are skewed or symmetrical and whether the median might be a better measure of central tendency than the mean. In addition, one should look to see whether the standard deviation or interquartile range for a very skewed set of data has been included in the statistics provided. Without these, one cannot determine the amount of variation that exists in the data.

Ethical considerations arise when deciding which results to include in a report. One should document both good and bad results. In addition, presentations should report results in a fair, objective, and neutral manner. Unethical behavior occurs when one selectively fails to report pertinent findings that are detrimental to the support of a particular position.

▼ USING STATISTICS
More Descriptive Choices, Revisited

In the More Descriptive Choices scenario, you were hired by the Choice *Is* Yours investment company to assist investors interested in stock mutual funds. A sample of 479 stock mutual funds included 306 growth funds and 173 value funds. By comparing these two categories, you were able to provide investors with valuable insights.

The three-year returns for both the growth funds and the value funds were left-skewed, as the Figure 3.8 boxplot on page 130 reveals. The descriptive statistics (see Figure 3.5 on page 121) allowed you to compare the central tendency, variability, and shape of the returns of the growth funds and the value funds.

The mean indicated that the growth funds returned a mean of 8.51, and the median indicated that half of the growth funds had returns of 8.70 or more. The value funds' central tendencies were

lower than those of the growth funds—they had a mean of 6.84, and half the funds had three-year returns above 7.07.

The growth funds showed more variability than the value funds, with a standard deviation of 3.1935 as compared to 2.3258. The kurtosis of value funds was very positive, indicating a distribution that was much more peaked than a normal distribution.

Although the three-year returns were greater for growth funds than value funds, that relationship may not hold when the one-year, five-year, or ten-year returns for the growth and value fund groups are examined. (Analyze the other return percentage variables in Retirement Funds to see if the relationship holds for these other periods of time.)

▼SUMMARY

This chapter discusses how descriptive statistics such as the mean, median, quartiles, range, and standard deviation describe the characteristics of central tendency, variability, and shape. The chapter also identifies the following basic concepts related to data variation:

- The greater the spread or dispersion of the data, the larger the range, variance, and standard deviation.
- The smaller the spread or dispersion of the data, the smaller the range, variance, and standard deviation.

- If the values are all the same (no variation in the data), the range, variance, and standard deviation will all equal zero.
- Measures of variation (the range, variance, and standard deviation) are never negative.

The chapter also discusses how the coefficient of correlation describes the relationship between two numerical variables and how boxplots visualize the distribution of data. Table 3.5 classifies the methods that this chapter discusses.

TABLE 3.5
Chapter 3 descriptive statistics methods

Type of Analysis	Methods
Central tendency	Mean, median, mode (Section 3.1)
Variation and shape	Quartiles, range, interquartile range, variance, standard deviation, coefficient of variation, Z scores, skewness, kurtosis, boxplot (Sections 3.2 through 3.4)
Describing the relationship between two numerical variables	Covariance, coefficient of correlation (Section 3.5)

▼REFERENCES

Booker, J., and L. Ticknor. "A Brief Overview of Kurtosis." **www .osti.gov/scitech/servlets/purl/677174**.

Kendall, M. G., A. Stuart, and J. K. Ord. *Kendall's Advanced Theory of Statistics, Volume 1: Distribution Theory*, 6th ed. New York: Oxford University Press, 1994.

Taleb, N. *The Black Swan*, 2nd ed. New York: Random House, 2010.

▼KEY EQUATIONS

Sample Mean

$$\overline{X} = \frac{\sum_{i=1}^{n} X_i}{n} \tag{3.1}$$

Median

$$\text{Median} = \frac{n+1}{2} \text{ ranked value} \tag{3.2}$$

Geometric Mean

$$\overline{X}_G = (X_1 \times X_2 \times \cdots \times X_n)^{1/n} \tag{3.3}$$

Geometric Mean Rate of Return

$$\overline{R}_G = [(1 + R_1) \times (1 + R_2) \times \cdots \times (1 + R_n)]^{1/n} - 1 \tag{3.4}$$

Range

$$\text{Range} = X_{\text{largest}} - X_{\text{smallest}} \tag{3.5}$$

Sample Variance

$$S^2 = \frac{\sum_{i=1}^{n} (X_i - \overline{X})^2}{n-1} \tag{3.6}$$

Sample Standard Deviation

$$S = \sqrt{S^2} = \sqrt{\frac{\sum_{i=1}^{n} (X_i - \overline{X})^2}{n-1}} \tag{3.7}$$

Coefficient of Variation

$$CV = \left(\frac{S}{\overline{X}}\right) 100\% \tag{3.8}$$

Z Score

$$Z = \frac{X - \overline{X}}{S} \tag{3.9}$$

First Quartile, Q_1

$$Q_1 = \frac{n + 1}{4} \text{ ranked value} \tag{3.10}$$

Third Quartile, Q_3

$$Q_3 = \frac{3(n + 1)}{4} \text{ ranked value} \tag{3.11}$$

Interquartile Range

$$\text{Interquartile range} = Q_3 - Q_1 \tag{3.12}$$

Population Mean

$$\mu = \frac{\sum_{i=1}^{N} X_i}{N} \tag{3.13}$$

Population Variance

$$\sigma^2 = \frac{\sum_{i=1}^{N}(X_i - \mu)^2}{N} \tag{3.14}$$

Population Standard Deviation

$$\sigma = \sqrt{\frac{\sum_{i=1}^{N}(X_i - \mu)^2}{N}} \tag{3.15}$$

Sample Covariance

$$\text{cov}(X, Y) = \frac{\sum_{i=1}^{n}(X_i - \overline{X})(Y_i - \overline{Y})}{n - 1} \tag{3.16}$$

Sample Coefficient of Correlation

$$r = \frac{\text{cov}(X, Y)}{S_X S_Y} \tag{3.17}$$

▼ KEY TERMS

arithmetic mean (mean)   109
boxplot   129
central tendency   109
Chebyshev's theorem   134
coefficient of correlation   137
coefficient of variation (CV)   117
covariance   136
dispersion (spread)   114
empirical rule   134
five-number summary   128
geometric mean   113
geometric mean rate of return   113
interquartile range (midspread)   127
kurtosis   120
left-skewed   120
leptokurtic   120
lurking variable   137

mean (arithmetic mean)   109
median   111
midspread (interquartile range)   127
mode   112
outliers   119
percentiles   126
platykurtic   120
population mean   132
population standard deviation   133
population variance   133
Q_1: first quartile   125
Q_2: second quartile   125
Q_3: third quartile   125
quartiles   125
range   114
resistant measure   128
right-skewed   120

sample coefficient of correlation (r)   137
sample covariance   136
sample mean   109
sample standard deviation (S)   115
sample variance (S^2)   115
shape   109
skewed   120
skewness   120
spread (dispersion)   114
standard deviation   115
sum of squares (SS)   115
symmetrical   120
variance   115
variation   109
Z score   118

▼ CHECKING YOUR UNDERSTANDING

3.50 What are the properties of a set of numerical data?

3.51 What is meant by the property of central tendency?

3.52 What are the differences among the mean, median, and mode, and what are the advantages and disadvantages of each?

3.53 How do you interpret the first quartile, median, and third quartile?

3.54 What is meant by the property of variation?

3.55 What does the Z score measure?

3.56 What are the differences among the various measures of variation, such as the range, interquartile range, variance, standard deviation, and coefficient of variation, and what are the advantages and disadvantages of each?

3.57 How does the empirical rule help explain the ways in which the values in a set of numerical data cluster and distribute?

3.58 How do the empirical rule and Chebyshev's theorem differ?

3.59 What is meant by the property of shape?

3.60 What is the difference between skewness and kurtosis?

3.61 How do the boxplots for distributions of varying shapes differ?

3.62 How do the covariance and the coefficient of correlation differ?

3.63 What is the difference between the arithmetic mean and the geometric mean?

3.64 What is the difference between the geometric mean and the geometric rate of return?

▼CHAPTER REVIEW PROBLEMS

3.65 The download and upload speeds of a mobile data carrier is of great interest to both individual and business users. The file `City Download Speeds` contains the average download speeds in Mbps for 100 cities in the United States.

Source: Data extracted from BroadBandNow, **bit.ly/2v5X0FM**.

For the average download speed:
a. Calculate the mean, median, first quartile, and third quartile.
b. Calculate the range, interquartile range, variance, standard deviation, and coefficient of variation.
c. Construct a boxplot. Are the data skewed? If so, how?
d. What can you conclude about the download speeds for the various U. S. cities?

3.66 An insurance company has the business objective of reducing the amount of time it takes to approve applications for life insurance. The approval process consists of underwriting, which includes a review of the application, a medical information bureau check, possible requests for additional medical information and medical exams, and a policy compilation stage, in which the policy pages are generated and sent for delivery. The ability to deliver approved policies to customers in a timely manner is critical to the profitability of this service. Using the DCOVA framework you define the variable of interest as the total processing time in days. You collect the data by selecting a random sample of 27 approved policies during a period of one month. You organize the data collected in a worksheet and store them in `Insurance`:
a. Calculate the mean, median, first quartile, and third quartile.
b. Calculate the range, interquartile range, variance, standard deviation, and coefficient of variation.
c. Construct a boxplot. Are the data skewed? If so, how?
d. What would you tell a customer who wishes to purchase this type of insurance policy and asks how long the approval process takes?

3.67 One of the major measures of the quality of service provided by an organization is the speed with which it responds to customer complaints. A large family-held department store selling furniture and flooring, including carpet, had undergone a major expansion in the past several years. In particular, the flooring department had expanded from 2 installation crews to an installation supervisor, a measurer, and 15 installation crews. The business objective of the company was to reduce the time between when a complaint is received and when it is resolved. During a recent year, the company received 50 complaints concerning carpet installation. The data from the 50 complaints, organized in `Furniture`, represent the number of days between the receipt of a complaint and the resolution of the complaint:

54	5	35	137	31	27	152	2	123	81	74	27	11
19	126	110	110	29	61	35	94	31	26	5	12	4
165	32	29	28	29	26	25	1	14	13	13	10	5
27	4	52	30	22	36	26	20	23	33	68		

a. Calculate the mean, median, first quartile, and third quartile.
b. Calculate the range, interquartile range, variance, standard deviation, and coefficient of variation.
c. Construct a boxplot. Are the data skewed? If so, how?
d. On the basis of the results of (a) through (c), if you had to tell the president of the company how long a customer should expect to wait to have a complaint resolved, what would you say? Explain.

3.68 Call centers today play an important role in managing day-to-day business communications with customers. It's important, therefore, to monitor a comprehensive set of metrics, which can help businesses understand the overall performance of a call center. One key metric for measuring overall call center performance is service level which is defined as the percentage of calls answered by a human agent within a specified number of seconds. The file `Service Level` contains the following data for time, in seconds, to answer 50 incoming calls to a financial services call center:

16	14	16	19	6	14	15	5	16	18	17	22	6	18	10
15	12	6	19	16	16	15	13	25	9	17	12	10	5	15
23	11	12	14	24	9	10	13	14	26	19	20	13	24	28
15	21	8	16	12										

a. Calculate the mean, median, range, and standard deviation for the speed of answer, which is the time to answer incoming calls.
b. List the five-number summary.
c. Construct a boxplot and describe its shape.
d. What can you conclude about call center performance if the service level target is set as "75% of calls answered in under 20 seconds?"

3.69 The financial services call center in Problem 3.68 also monitors call duration, which is the amount of time spent speaking to customers on the phone. The file `Call Duration` contains the following data for time, in seconds, spent by agents talking to 50 customers:

243	290	199	240	125	151	158	66	350	1141	251	385	239
139	181	111	136	250	313	154	78	264	123	314	135	99
420	112	239	208	65	133	213	229	154	377	69	170	261
230	273	288	180	296	235	243	167	227	384	331		

a. Calculate the mean, median, range, and standard deviation for the call duration, which is the amount of time spent speaking to customers on the phone. Interpret these measures of central tendency and variability.
b. List the five-number summary.
c. Construct a boxplot and describe its shape.
d. What can you conclude about call center performance if a call duration target of less than 240 seconds is set?

3.70 Data were collected on the typical cost of dining at American-cuisine restaurants within a 1-mile walking distance of a hotel located in a large city. The file `Bundle` contains the typical cost (a per transaction cost in $) as well as a Bundle score, a measure of overall popularity and customer loyalty, for each of 40 selected restaurants.

Source: Data extracted from **www.bundle.com** via the link **on-msn.com/MnlBxo**.

a. For each variable, compute the mean, median, first quartile, and third quartile.
b. For each variable, compute the range, interquartile range, variance, standard deviation, and coefficient of variation.
c. For each variable, construct a boxplot. Are the data skewed? If so, how?
d. Calculate the coefficient of correlation between Bundle score and typical cost.
e. What conclusions can you reach concerning Bundle score and typical cost?

3.71 A quality characteristic of interest for a tea-bag-filling process is the weight of the tea in the individual bags. If the bags are under-filled, two problems arise. First, customers may not be able to brew the tea to be as strong as they wish. Second, the company may be in violation of the truth-in-labeling laws. For this product, the label weight on the package indicates that, on average, there are 5.5 grams of tea in a bag. If the mean amount of tea in a bag exceeds the label weight, the company is giving away product. Getting an exact amount of tea in a bag is problematic because of variation in the temperature and humidity inside the factory, differences in the density of the tea, and the extremely fast filling operation of the machine (approximately 170 bags per minute). The file Teabags contains these weights, in grams, of a sample of 50 tea bags produced in one hour by a single machine:

5.65	5.44	5.42	5.40	5.53	5.34	5.54	5.45	5.52	5.41
5.57	5.40	5.53	5.54	5.55	5.62	5.56	5.46	5.44	5.51
5.47	5.40	5.47	5.61	5.53	5.32	5.67	5.29	5.49	5.55
5.77	5.57	5.42	5.58	5.58	5.50	5.32	5.50	5.53	5.58
5.61	5.45	5.44	5.25	5.56	5.63	5.50	5.57	5.67	5.36

a. Calculate the mean, median, first quartile, and third quartile.
b. Calculate the range, interquartile range, variance, standard deviation, and coefficient of variation.
c. Interpret the measures of central tendency and variation within the context of this problem. Why should the company producing the tea bags be concerned about the central tendency and variation?
d. Construct a boxplot. Are the data skewed? If so, how?
e. Is the company meeting the requirement set forth on the label that, on average, there are 5.5 grams of tea in a bag? If you were in charge of this process, what changes, if any, would you try to make concerning the distribution of weights in the individual bags?

3.72 The manufacturer of Boston and Vermont asphalt shingles provides its customers with a 20-year warranty on most of its products. To determine whether a shingle will last as long as the warranty period, accelerated-life testing is conducted at the manufacturing plant. Accelerated-life testing exposes a shingle to the stresses it would be subject to in a lifetime of normal use via an experiment in a laboratory setting that takes only a few minutes to conduct. In this test, a shingle is repeatedly scraped with a brush for a short period of time, and the shingle granules removed by the brushing are weighed (in grams). Shingles that experience low amounts of granule loss are expected to last longer in normal use than shingles that experience high amounts of granule loss. In this situation, a shingle should experience no more than 0.8 gram of granule loss if it is expected to last the length of the warranty period. The file Granule contains a sample of 170 measurements made on the company's Boston shingles and 140 measurements made on Vermont shingles.
a. List the five-number summaries for the Boston shingles and for the Vermont shingles.
b. Construct side-by-side boxplots for the two brands of shingles and describe the shapes of the distributions.
c. Comment on the ability of each type of shingle to achieve a granule loss of 0.8 gram or less.

3.73 The file Restaurants contains the cost per meal and the ratings of 50 center city and 50 metro area restaurants on their food, décor, and service (and their summated ratings).

Complete the following for the center city and metro area restaurants:
a. Construct the five-number summary of the cost of a meal.
b. Construct a boxplot of the cost of a meal. What is the shape of the distribution?
c. Calculate and interpret the correlation coefficient of the summated rating and the cost of a meal.
d. What conclusions can you reach about the cost of a meal at center city and metro area restaurants?

3.74 The file Protein contains calories, protein, and cholesterol of popular protein foods (fresh red meats, poultry, and fish).
Source: U.S. Department of Agriculture.
a. Calculate the correlation coefficient between calories and protein.
b. Calculate the correlation coefficient between calories and cholesterol.
c. Calculate the correlation coefficient between protein and cholesterol.
d. Based on the results of (a) through (c), what conclusions can you reach concerning calories, protein, and cholesterol of popular protein foods?

3.75 College basketball in the United States is big business with television contracts in the billions of dollars and coaches' salaries in the millions of dollars. The file Coaches Pay contains the total coaches pay for a recent year and whether the school was a member of a major conference (Atlantic Coast, Big 12, Big East, Big Ten, Pac-12, or SEC).
Source: Data extracted from "College Men's Basketball Coaches Compensation, *USA Today*, March 13, 2019, p. C4.

For each of the two groups (major conference member and not a major conference member):
a. Calculate the mean, median, first quartile, and third quartile.
b. Calculate the range, interquartile range, variance, standard deviation, and coefficient of variation.
c. Interpret the measures of central tendency and variation within the context of this problem.
d. Construct a boxplot. Are the data skewed? If so, how?
e. What conclusions can you reach about the differences in coaches pay between colleges that are members of a major conference and those that are not?

3.76 The file Property Taxes contains the property taxes on a $176K home and the median home value for the 50 states and the District of Columbia. For each of these two variables:
a. Calculate the mean, median, first quartile, and third quartile.
b. Calculate the range, interquartile range, variance, standard deviation, and coefficient of variation.
c. Construct a boxplot. Are the data skewed? If so, how?
d. Calculate the coefficient of correlation between the property taxes on a $176K home and the median home value.
e. Based on the results of (a) through (c), what conclusions can you reach concerning property taxes on a $176K home and the median home value for each state and the District of Columbia?

3.77 Have you wondered how Internet connection speed varies around the globe? The file Connection Speed contains the mean connection speed, the mean peak connection speed, the percentage of the time the speed is above 4 Mbps, and the percentage of the time the connection speed is above 10 Mbps for various countries.

Source: Data extracted from **bit.ly/1hHaHVD**.

Answer (a) through (c) for each variable.

a. Calculate the mean, median, first quartile, and third quartile.

b. Calculate the range, interquartile range, variance, standard deviation, and coefficient of variation.

c. Construct a boxplot. Are the data skewed? If so, how?

d. Calculate the coefficient of correlation between mean connection speed, mean peak connection speed, percentage of the time the speed is above 4 Mbps, and the percentage of the time the connection speed is above 10 Mbps.

e. Based on the results of (a) through (c), what conclusions can you reach concerning the connection speed around the globe?

f. Based on (d), what conclusions can your reach about the relationship between mean connection speed, mean peak connection speed, percentage of the time the speed is above 4 Mbps, and the percentage of the time the connection speed is above 10 Mbps?

3.78 311 is Chicago's web and phone portal for government information and nonemergency services. 311 serves as a comprehensive one-stop shop for residents, visitors, and business owners; therefore, it is critical that 311 representatives answer calls and respond to requests in a timely and accurate fashion. The target response time for answering 311 calls is 45 seconds. Agent abandonment rate is one of several call center metrics tracked by 311 officials. This metric tracks the percentage of callers who hang up after the target response time of 45 seconds has elapsed. The file 311 Call Center contains the agent abandonment rate for 22 weeks of call center operation during the 7:00 A.M.–3:00 P.M. shift.

a. Calculate the mean, median, first quartile, and third quartile.

b. Calculate the range, interquartile range, variance, standard deviation, and coefficient of variation.

c. Construct a boxplot. Are the data skewed? If so, how?

d. Calculate the correlation coefficient between day and agent abandonment rate.

e. Based on the results of (a) through (c), what conclusions might you reach concerning 311 call center performance operation?

3.79 How much time do commuters living in or near cities spend commuting to work each week? The file Commuting Time contains the average weekly commuting time in 30 U.S. cities.

Source: Data extracted from **factfinder.census.gov**.

For the weekly commuting time data:

a. Calculate the mean, median, first quartile, and third quartile.

b. Calculate the range, interquartile range, variance, standard deviation, and coefficient of variation.

c. Construct a boxplot. Are the data skewed? If so, how?

d. Based on the results of (a) through (c), what conclusions might you reach concerning the commuting time.

3.80 How do the mean credit scores of people living in various American cities differ? The file Credit Scores is an ordered array of the average credit scores of people living in 2,570 American cities.

Source: Data extracted from "Study shows cities with highest and lowest credit scores," **bit.ly/2uubZfX**.

a. Calculate the mean, median, first quartile, and third quartile.

b. Calculate the range, interquartile range, variance, standard deviation, and coefficient of variation.

c. Construct a boxplot. Are the data skewed? If so, how?

d. Based on the results of (a) through (c), what conclusions might you reach concerning the average credit scores of people living in various American cities?

3.81 You are planning to study for your statistics examination with a group of classmates, one of whom you particularly want to impress. This individual has volunteered to use software to generate the needed summary information, tables, and charts for a data set that contains several numerical and categorical variables assigned by the instructor for study purposes. This person comes over to you with the printout and exclaims, "I've got it all—the means, the medians, the standard deviations, the boxplots, the pie charts—for all our variables. The problem is, some of the output looks weird—like the boxplots for gender and for major and the pie charts for grade point average and for height. Also, I can't understand why Professor Szabat said we can't get the descriptive stats for some of the variables; I got them for everything! See, the mean for height is 68.23, the mean for grade point average is 2.76, the mean for gender is 1.50, the mean for major is 4.33." What is your reply?

REPORT WRITING EXERCISES

3.82 The file Domestic Beer contains the percentage alcohol, number of calories per 12 ounces, and number of carbohydrates (in grams) per 12 ounces for 157 of the best-selling domestic beers in the United States.

Source: Data extracted from **bit.ly/1A4E6AF**, April 3, 2019.

Write a report that includes a complete descriptive evaluation of each of the numerical variables—percentage of alcohol, number of calories per 12 ounces, and number of carbohydrates (in grams) per 12 ounces. Append to your report all appropriate tables, charts, and numerical descriptive measures.

▾CASES

Managing Ashland MultiComm Services

For what variable in the Chapter 2 "Managing Ashland Multi-Comm Services" case (see page 86) are numerical descriptive measures needed?

1. For the variable you identify, compute the appropriate numerical descriptive measures and construct a boxplot.

2. For the variable you identify, construct a graphical display. What conclusions can you reach from this other plot that cannot be made from the boxplot?

3. Summarize your findings in a report that can be included with the task force's study.

Digital Case

Apply your knowledge about the proper use of numerical descriptive measures in this continuing Digital Case.

Open **EndRunGuide.pdf**, the EndRun Financial Services "Guide to Investing." Re-examine EndRun's supporting data for the "More Winners Than Losers" and "The Big Eight Difference" and then answer the following:

1. Can descriptive measures be calculated for any variables? How would such summary statistics support EndRun's claims? How would those summary statistics affect your perception of EndRun's record?

2. Evaluate the methods EndRun used to summarize the results presented on the "Customer Survey Results" page. Is there anything you would do differently to summarize these results?

3. Note that the last question of the survey has fewer responses than the other questions. What factors may have limited the number of responses to that question?

CardioGood Fitness

Return to the CardioGood Fitness case first presented on page 87. Using the data stored in `CardioGood Fitness`:

1. Calculate descriptive statistics to create a customer profile for each CardioGood Fitness treadmill product line.

2. Write a report to be presented to the management of Cardio-Good Fitness, detailing your findings.

More Descriptive Choices Follow-up

Follow up the Using Statistics Revisited section on page 141 by calculating descriptive statistics to analyze the differences in one-year return percentages, five-year return percentages, and ten-year return percentages for the sample of 479 retirement funds stored in `Retirement Funds`. In your analysis, examine differences between the growth and value funds as well as the differences among the small cap, midcap, and large cap market funds.

Clear Mountain State Student Survey

The student news service at Clear Mountain State University (CMSU) has decided to gather data about the undergraduate students who attend CMSU. They create and distribute a survey of 14 questions (see **CMUndergradSurvey.pdf**) and receive responses from 111 undergraduates (stored in `Student Survey`). For each numerical variable included in the survey, calculate all the appropriate descriptive statistics and write a report summarizing your conclusions.

EG3.1 MEASURES of CENTRAL TENDENCY

The Mean, Median, and Mode

Key Technique Use the **AVERAGE(***variable cell range***)**, **MEDIAN(***variable cell range***)**, and **MODE(***variable cell range***)** functions to compute these measures.

Example Compute the mean, median, and mode for the sample of get-ready times that Section 3.1 introduces.

PHStat Use **Descriptive Summary**.

For the example, open to the **DATA worksheet** of the **Times workbook**. Select **PHStat→Descriptive Statistics→Descriptive Summary**. In the procedure's dialog box (shown below):

1. Enter **A1:A11** as the **Raw Data Cell Range** and check **First cell contains label**.
2. Click **Single Group Variable**.
3. Enter a **Title** and click **OK**.

PHStat inserts a new worksheet that contains various measures of central tendency, variation, and shape discussed in Sections 3.1 and 3.2. This worksheet is similar to the CompleteStatistics worksheet of the Descriptive workbook.

Workbook Use the **CentralTendency worksheet** of the **Descriptive workbook** as a model.

For the example, open the **Times workbook**, insert a new worksheet (see Section EG.4), and:

1. Enter a title in cell **A1**.
2. Enter **Get-Ready Times** in cell **B3**, **Mean** in cell **A4**, **Median** in cell **A5**, and **Mode** in cell **A6**.
3. Enter the formula **=AVERAGE(DATA!A:A)** in cell **B4**, the formula **=MEDIAN(DATA!A:A)** in cell **B5**, and the formula **=MODE(DATA!A:A)** in cell **B6**.

For these functions, the *variable cell range* includes the name of the DATA worksheet because the data being summarized appears on the separate DATA worksheet. For another problem, paste the data for the problem into column A of the DATA worksheet, overwriting the existing get-ready times.

Analysis ToolPak Use **Descriptive Statistics.**

For the example, open to the **DATA worksheet** of the **Times workbook** and:

1. Select **Data→Data Analysis**.
2. In the Data Analysis dialog box, select **Descriptive Statistics** from the **Analysis Tools** list and then click **OK**.

In the Descriptive Statistics dialog box (shown below):

1. Enter **A1:A11** as the **Input Range**. Click **Columns** and check **Labels in first row**.
2. Click **New Worksheet Ply** and check **Summary statistics**, **Kth Largest**, and **Kth Smallest**.
3. Click **OK**.

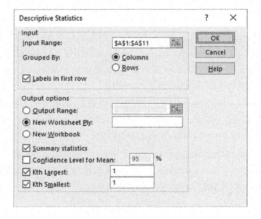

The ToolPak inserts a new worksheet that contains various measures of central tendency, variation, and shape discussed in Sections 3.1 and 3.2.

The Geometric Mean

Key Technique Use the **GEOMEAN((1 + *R*1), (1 + *R*2), …(1 + R*n*)) − 1** function to compute the geometric mean rate of return.

Example Compute the geometric mean rate of return in the Russell 2000 Index for the two years as shown in Example 3.4 on page 114.

Workbook Enter the formula **=GEOMEAN((1 + (0.131)), (1 + (−0.122))) − 1** in any cell.

EG3.2 MEASURES of VARIATION and SHAPE

The Range

Key Technique Use the **MIN**(*variable cell range*) and **MAX** (*variable cell range*) functions to help compute the range.

Example Compute the range for the sample of get-ready times first introduced in Section 3.1.

PHStat Use **Descriptive Summary** (see Section EG3.1).

Workbook Use the **Range worksheet** of the **Descriptive workbook** as a model.

For the example, open the worksheet constructed in the *Workbook* "The Mean, Median, and Mode" instructions. Enter **Minimum** in cell **A7**, **Maximum** in cell **A8**, and **Range** in cell **A9**.

Enter the formula **=MIN(DATA!A:A)** in cell **B7**, the formula **=MAX(DATA!A:A)** in cell **B8**, and the formula **=B8−B7** in cell **B9**.

The Variance, Standard Deviation, Coefficient of Variation, and Z Scores

Key Technique Use the **VAR.S**(*variable cell range*) and **STDEV.S**(*variable cell range*) functions to compute the sample variance and the sample standard deviation, respectively.

Use the AVERAGE and STDEV.S functions for the coefficient of variation. Use the **STANDARDIZE**(*value, mean, standard deviation*) function to compute Z scores.

Example Compute the variance, standard deviation, coefficient of variation, and Z scores for the sample of get-ready times first introduced in Section 3.1.

PHStat Use **Descriptive Summary** (see Section EG3.1).

Workbook Use the **Variation** and **ZScores worksheets** of the **Descriptive workbook** as models.

For the example, the Variation and ZScores worksheets already compute these statistics using the get-ready times in the DATA worksheet. To compute the variance, standard deviation, and coefficient of variation for another problem, paste the data for the problem into column A of the DATA worksheet, overwriting the existing get-ready times.

To compute the Z scores for another problem, copy the updated DATA worksheet. In the new, copied worksheet:

1. Enter *Z Score* in cell **B1**.
2. Enter **=STANDARDIZE(A2, Variation!$B$4, Variation!$B$11)** in cell **B2**.
3. Copy the formula down through row 11.

Analysis ToolPak Use **Descriptive Statistics** (see Section EG3.1). This procedure does not compute Z scores.

Shape: Skewness and Kurtosis

Key Technique Use the **SKEW**(*variable cell range*) and the **KURT**(*variable cell range*) functions to compute these measures.

Example Compute the skewness and kurtosis for the sample of get-ready times first introduced in Section 3.1.

PHStat Use **Descriptive Summary** (see Section EG3.1).

Workbook Use the **Shape worksheet** of the **Descriptive workbook** as a model.

For the example, the Shape worksheet already computes the skewness and kurtosis using the get-ready times in the DATA worksheet. To compute these statistics for another problem, paste the data for the problem into column A of the DATA worksheet, overwriting the existing get-ready times.

Analysis ToolPak Use **Descriptive Statistics** (see Section EG3.1).

EG3.3 EXPLORING NUMERICAL VARIABLES

Quartiles

Key Technique Use the MEDIAN and COUNT, and SMALL, INT, FLOOR, CEILING, and IF functions (see Appendix F) to compute the quartiles. Avoid using any of the Excel quartile functions because they do not use the Section 3.3 rules to calculate quartiles.

Example Compute the quartiles for the sample of get-ready times first introduced in Section 3.1.

PHStat Use **Boxplot** (see page 150).

Workbook Use the **COMPUTE worksheet** of the **Quartiles workbook** as a model.

For the example, the COMPUTE worksheet already computes the quartiles for the get-ready times. To compute the quartiles for another problem, paste the data into column A of the DATA worksheet, overwriting the existing get ready times.

The COMPUTE worksheet uses a number of arithmetic and logical formulas that use the IF function to produce results consistent to the Section 3.3 rules. Open to the **COMPUTE_FORMULAS worksheet** to review these formulas and read the SHORT TAKES for Chapter 3 for a detailed explanation of those formulas.

The **COMPUTE** worksheet avoids using any of the current Excel **QUARTILE** functions because none of them calculate quartiles using the Section 3.3 rules. The COMPARE worksheet compares the **COMPUTE** worksheet results to the quartiles calculated by the Excel QUARTILE.EXC and QUARTILE.INC functions.

The Interquartile Range

Key Technique Use a formula to subtract the first quartile from the third quartile.

Example Compute the interquartile range for the sample of get ready times first introduced in Section 3.1.

Workbook Use the **COMPUTE worksheet** of the **Quartiles workbook** (see previous section) as a model.

For the example, the interquartile range is already computed in cell B19 using the formula $= \textbf{B18} - \textbf{B16}$.

The Five-Number Summary and the Boxplot

Key Technique Plot a series of line segments on the same chart to construct a five-number summary boxplot.

Example Compute the five-number summary and construct the boxplots of the three-year return percentage variable for the growth and value funds used in Example 3.14 on page 130.

PHStat Use **Boxplot**.

For the example, open to the **DATA worksheet** of the **Retirement Funds workbook**. Select **PHStat → Descriptive Statistics → Boxplot**. In the procedure's dialog box (shown below):

1. Enter **K1:K480** as the **Raw Data Cell Range** and check **First cell contains label**.
2. Click **Multiple Groups - Stacked** and enter **C1:C480** as the **Grouping Variable Cell Range**.
3. Enter a **Title**, check **Five-Number Summary**, and click **OK**.

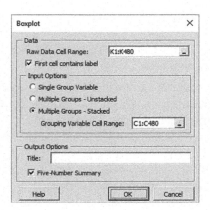

The boxplot appears on its own chart sheet, separate from the worksheet that contains the five-number summary.

Workbook Use the worksheets of the **Boxplot workbook** as templates for five-number summary boxplots.

For the example, use the **PLOT_DATA worksheet** which already shows the five-number summary and boxplot for the value funds. To compute the five-number summary and

construct a boxplot for the growth funds, copy the growth funds from **column A** of the **UNSTACKED worksheet** of the **Retirement Funds workbook** and paste into **column A** of the **DATA worksheet** of the **Boxplot workbook**.

For other problems, use the **PLOT_SUMMARY worksheet** as the template if the five-number summary has already been determined; otherwise, paste your unsummarized data into column A of the DATA worksheet and use the PLOT_DATA worksheet as was done for the example.

The worksheets creatively use charting features to construct a boxplot as the SHORT TAKES for Chapter 3 explains.

EG3.4 NUMERICAL DESCRIPTIVE MEASURES for a POPULATION

The Population Mean, Population Variance, and Population Standard Deviation

Key Technique Use **AVERAGE(*variable cell range*)**, **VAR.P(*variable cell range*)**, and **STDEV.P(*variable cell range*)** to compute these measures.

Example Compute the population mean, population variance, and population standard deviation for the "Dow Dogs" population data of Table 3.3 on page 132.

Workbook Use the **Parameters workbook** as a model.

For the example, the **COMPUTE worksheet** of the **Parameters workbook** already computes the three population parameters for the "Dow Dogs."

The Empirical Rule and Chebyshev's Theorem

Use the **COMPUTE worksheet** of the **VE-Variability workbook** to explore the effects of changing the mean and standard deviation on the ranges associated with ±1 standard deviation, ±2 standard deviations, and ±3 standard deviations from the mean. Change the mean in cell **B4** and the standard deviation in cell **B5** and then note the updated results in rows 9 through 11.

EG3.5 THE COVARIANCE and the COEFFICIENT of CORRELATION

The Covariance

Key Technique Use the **COVARIANCE.S(*variable 1 cell range, variable 2 cell range*)** function to compute this measure.

Example Compute the sample covariance for the NBA team revenue and value shown in Figure 3.10 on page 137.

Workbook Use the **Covariance workbook** as a model.

For the example, the revenue and value have already been placed in columns A and B of the DATA worksheet and the COMPUTE worksheet displays the computed covariance in cell B9. For other problems, paste the data for two variables into columns A and B of the DATA worksheet, overwriting the revenue and value data.

Read the SHORT TAKES for Chapter 3 for an explanation of the formulas found in the DATA and COMPUTE worksheets.

The Coefficient of Correlation

Key Technique Use the **CORREL(***variable 1 cell range, variable 2 cell range***)** function to compute this measure.

Example Compute the coefficient of correlation for the NBA team revenue and value data of Example 3.18 on page 138.

Workbook Use the **Correlation workbook** as a model.

For the example, the revenue and value have already been placed in columns A and B of the DATA worksheet and the COMPUTE worksheet displays the coefficient of correlation in cell B14. For other problems, paste the data for two variables into columns A and B of the DATA worksheet, overwriting the revenue and value data.

The COMPUTE worksheet uses the COVARIANCE.S function to compute the covariance (see the previous section) and also uses the DEVSQ, COUNT, and SUMPRODUCT functions that Appendix F discusses. Open to the **COMPUTE_FORMULAS worksheet** to examine the use of all these functions.

CHAPTER 3

▼TABLEAU GUIDE

TG3.3 EXPLORING NUMERICAL VARIABLES

The Five-Number Summary and the Boxplot

Use **box-and-whisker plots**.

For example, to construct the five-number summary boxplots of the three-year return percentage variable for the growth and value funds, similar to Figure 3.8 on page 130, open to the **Retirement Funds For Tableau Excel workbook**. In a new Tableau worksheet:

1. Right-click **3YrReturn** in the Dimensions list and select **Convert to Measure** in the shortcut menu.
2. Drag **3YrReturn** from the Measures list and drop it in the **Rows** area.
3. Drag **Fund Number** from the Dimensions list and drop it in the **Rows** area.
4. If the ShowMe tab is not visible, click **ShowMe**.
5. In the ShowMe tab, click **box-and-whisker plots**.
6. Drag **Fund Type** from the Dimensions list and drop it in the **Rows** area.
7. Right-click over one of the boxes and select **Edit** from the shortcut menu.

In the Edit Reference Line, Band, or Box dialog box (shown in the next column):

8. Select **Maximum extent of the data** from the **Plot Options** pull-down list.
9. Check **Hide underlying marks (except outliers)**.
10. Click **OK**.

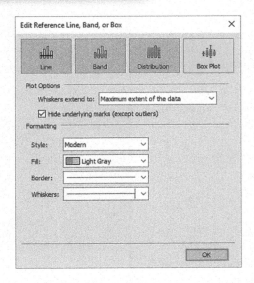

11. Back in the main Tableau window, select **Analysis → Swap Rows and Columns**.

Boxplots change from a vertical to horizontal layout. Moving the mouse pointer over each boxplot, displays the values of the five-number summary of the boxplot. Tableau labels the first quartile, Q_1, the Lower Hinge, and labels the third quartile, Q_3, the Upper Hinge. (The lower and upper hinges are equivalent to these quartiles by the method this book uses to calculate Q_1 and Q_3.)

To construct a boxplot for a numerical variable, the variable must be a measure and not a dimension, the reason for step 1.

4

Basic Probability

CONTENTS

OBJECTIVES

- Understand basic
  probability concepts
- Understand conditional
  probability

▼USING **STATISTICS**
Probable Outcomes at Fredco Warehouse Club

As a Fredco Warehouse Club electronics merchandise manager, you oversee the process that selects, purchases, and markets electronics items that the club sells, and you seek to satisfy customers' needs. Noting new trends in the television marketplace, you have worked with company marketers to conduct an intent-to-purchase study. The study asked the heads of 1,000 households about their intentions to purchase a large TV, a screen size of at least 65 inches, sometime during the next 12 months.

Now 12 months later, you plan a follow-up survey with the same people to see if they did purchase a large TV in the intervening time. For households that made such a purchase, you would like to know if the TV purchased was HDR (high dynamic range) capable, whether they also purchased a streaming media player in the intervening time, and whether they were satisfied with their large TV purchase.

You plan to use survey results to form a new marketing strategy that will enhance sales and better target those households likely to purchase multiple or more expensive products. What questions can you ask in this survey? How can you express the relationships among the various intent-to-purchase responses of individual households?

The principles of probability help bridge the worlds of descriptive statistics and inferential statistics. Probability principles are the foundation for the probability distribution, the concept of mathematical expectation, and the binomial and Poisson distributions. Applying probability to intent-to-purchase survey responses answers purchase behavior questions such as:

- What is the probability that a household is planning to purchase a large TV in the next year?
- What is the probability that a household will actually purchase a large TV?
- What is the probability that a household is planning to purchase a large TV and actually purchases the TV?
- Given that the household is planning to purchase a large TV, what is the probability that the purchase is made?
- Does knowledge of whether a household *plans* to purchase a large TV change the likelihood of predicting whether the household *will* purchase a large TV?
- What is the probability that a household that purchases a large TV will purchase an HDR-capable TV?
- What is the probability that a household that purchases a large TV with HDR will also purchase a streaming media player?
- What is the probability that a household that purchases a large TV will be satisfied with the purchase?

Answers to these question will provide a basis to form a marketing strategy. One can consider whether to target households that have indicated an intent to purchase or to focus on selling TVs that have HDR or both. One can also explore whether households that purchase large TVs with HDR can be easily persuaded to also purchase streaming media players.

4.1 Basic Probability Concepts

In everyday usage, *probability*, according to the Oxford English Dictionary, indicates the extent to which something is likely to occur or exist but can also mean the most likely cause of something. If storm clouds form, the wind shifts, and the barometric pressure drops, the probability of rain coming soon increases (first meaning). If one observes people entering an office building with wet clothes or otherwise drenched, there is a strong probability that it is currently raining outside (second meaning).

In statistics, **probability** is a numerical value that expresses the ratio between the value sought and the set of all possible values that could occur. A six-sided die has faces for 1, 2, 3, 4, 5, and 6. Therefore, for one roll of a *fair* six-sided die, the set of all possible values are the values 1 through 6. If the value sought is "a value greater than 4," then the values 5 or 6 would be sought. One would say the probability of this *event* is 2 outcomes divided by 6 outcomes or 1/3.

Consider tossing a fair coin heads or tails two times. What is the probability of tossing two tails? The set of possible values for tossing a fair coin twice are HH, TT, HT, and TH. Therefore, the probability of tossing two tails is 1/4 because only one value (TT) matches what is being sought, and the set of all possible values has four values.

Events and Sample Spaces

student TIP

Events are represented by letters of the alphabet.

When discussing probability, one uses **outcomes** in place of *values* and calls the set of all possible outcomes the **sample space**. **Events** are subsets of the sample space, the set of all outcomes that produce a specific result. For tossing a fair coin twice, the event "toss at least one head" is the subset of outcomes HH, HT, and TH, and the event "toss two tails" is the subset TT. Both events are examples of a **joint event**, an event that has two or more characteristics. In contrast, a **simple event** has only one characteristic, an outcome that cannot be further subdivided. The event "rolling a value greater than 4" in the first example results in the subset of outcomes 5 and 6 and is an example of a simple event because "5" and "6" represent one characteristic and cannot be further divided.

The **complement** of an event *A*, noted by the symbol *A'*, is the subset of outcomes that are not part of the event. For tossing a fair coin twice, the complement of the event "toss at least one head" is the subset TT, while the complement of the event "toss two tails" is HH, HT, and TH.

A set of events are **mutually exclusive** if they cannot occur at the same. The events "roll a value greater than 4" and "roll a value less than 3" are mutually exclusive when rolling one fair die. However, the events "roll a value greater than 4" and "roll a value greater than 5" are not because both share the outcome of rolling a 6.

A set of events are **collectively exhaustive** if one of the events must occur. For rolling a fair six-sided die, the events "roll a value of 3 or less" and "roll a value of 4 or more" are collectively exhaustive because these two subsets include all possible outcomes in the sample space. However, the set of events "roll a value of 3 or less" and "roll a value greater than 4" is not because this set does not include the outcome of rolling a 4.

Not all sets of collectively exhaustive events are mutually exclusive. For rolling a fair six-sided die, the set of events "roll a value of 3 or less," "roll an even-numbered value," and "roll a value greater than 4" is collectively exhaustive but is not mutually exclusive as, for example, "a value of 3 or less" and "an even-numbered value" could *both* occur if a 2 is rolled.

Certain and *impossible* events represent special cases. A **certain event** is an event that is sure to occur such as "roll a value greater than 0" for rolling one fair die. Because the subset of outcomes for a certain event is the entire set of outcomes in the sample, a certain event has a probability of 1. An **impossible event** is an event that has no chance of occurring, such as "roll a value greater than 6" for rolling one fair die. Because the subset of outcomes for an impossible event is empty—contains no outcomes—an impossible event has a probability of 0.

student TIP

By definition, *an event and its complement* are always both mutually exclusive and collectively exhaustive.

student TIP

A probability cannot be negative or greater than 1.

Types of Probability

The concepts and vocabulary related to events and sample spaces are helpful to understanding how to calculate probabilities. Also affecting such calculations is the type of probability being used: *a priori*, empirical, or subjective.

In ***a priori*** **probability**, the probability of an occurrence is based on having prior knowledge of the outcomes that can occur. Consider a standard deck of cards that has 26 red cards and 26 black cards. The probability of selecting a black card is $26/52 = 0.50$ because there are 26 black cards and 52 total cards. What does this probability mean? If each card is replaced after it is selected, this probability does not mean that one out of the next two cards selected will be black. One cannot say for certain what will happen on the next several selections. However, one can say that in the long run, if this selection process is continually repeated, the proportion of black cards selected will approach 0.50. Example 4.1 shows another example of computing an *a priori* probability.

EXAMPLE 4.1

Finding *A Priori* Probabilities

A standard six-sided die has six faces. Each face of the die contains either one, two, three, four, five, or six dots. If you roll a die, what is the probability that you will get a face with five dots?

SOLUTION Each face is equally likely to occur. Because there are six faces, the probability of getting a face with five dots is 1/6.

The preceding examples use the *a priori* probability approach because the number of ways the event occurs and the total number of possible outcomes are known from the composition of the deck of cards or the faces of the die.

In the **empirical probability** approach, the probabilities are based on observed data, not on prior knowledge of how the outcomes can occur. Surveys are often used to generate empirical probabilities. Examples of this type of probability are the proportion of individuals in the Fredco Warehouse Club scenario who actually purchase a large TV, the proportion of registered voters who prefer a certain political candidate, and the proportion of students who have part-time jobs. For example, if one conducts a survey of students, and 60% state that they have part-time jobs, then there is a 0.60 probability that an individual student has a part-time job.

The third approach to probability, **subjective probability**, differs from the other two approaches because subjective probability differs from person to person. For example, the development team for a new product may assign a probability of 0.60 to the chance of success for the product, while the president of the company may be less optimistic and assign a probability of 0.30. The assignment of subjective probabilities to various outcomes is usually based on a combination of an individual's past experience, personal opinion, and analysis of a particular situation. Subjective probability is especially useful in making decisions in situations in which one cannot use *a priori* probability or empirical probability.

Summarizing Sample Spaces

Sample spaces can be presented in tabular form using contingency tables (see Section 2.1). Table 4.1 in Example 4.2 summarizes a sample space as a contingency table. When used for probability, each cell in a contingency table represents one joint *event*, analogous to the one joint *response* when these tables are used to summarize categorical variables. For example, 200 of the respondents correspond to the joint event "planned to purchase a large TV and subsequently did purchase the large TV."

EXAMPLE 4.2

Events and Sample Spaces

The Fredco Warehouse Club scenario on page 152 involves analyzing the results of an intent-to-purchase study. Table 4.1 presents the results of the sample of 1,000 households surveyed in terms of purchase behavior for large TVs.

TABLE 4.1
Purchase behavior for large TVs

PLANNED TO PURCHASE	ACTUALLY PURCHASED		
	Yes	**No**	**Total**
Yes	200	50	250
No	100	650	750
Total	300	700	1,000

What is the sample space? Give examples of simple events and joint events.

SOLUTION The sample space consists of the 1,000 respondents. Simple events are "planned to purchase," "did not plan to purchase," "purchased," and "did not purchase." The complement of the event "planned to purchase" is "did not plan to purchase." The event "planned to purchase and actually purchased" is a joint event because in this joint event, the respondent must plan to purchase the TV *and* actually purchase it.

Simple Probability

Simple probability is the probability of occurrence of a simple event A, $P(A)$, in which each outcome is *equally likely* to occur. Equation (4.1) defines the probability of occurrence for simple probability.

PROBABILITY OF OCCURRENCE

$$\text{Probability of occurrence} = \frac{X}{T} \tag{4.1}$$

where

X = number of outcomes in which the event occurs
T = total number of possible outcomes

Equation 4.1 represents what some people wrongly think *is* the probability of occurrence for *all* probability problems. (Not all probability problems can be solved by Equation 4.1 as later examples in this chapter illustrate.) In the Fredco Warehouse Club scenario, the collected survey data represent an example of empirical probability. Therefore, one can use Equation (4.1) to determine answers to questions that can be expressed as a simple probability.

For example, one question asked respondents if they planned to purchase a large TV. Using the responses to this question, how can one determine the probability of selecting a household that planned to purchase a large TV? From the Table 4.1 contingency table, determine the value of X as 250, the total of the Planned-to-purchase Yes row and determine the value of T as 1,000, the overall total of respondents located in the lower right corner cell of the table. Using Equation (4.1) and Table 4.1:

$$\text{Probability of occurrence} = \frac{X}{T}$$

$$P(\text{Planned to purchase}) = \frac{\text{Number who planned to purchase}}{\text{Total number of households}}$$

$$= \frac{250}{1,000} = 0.25$$

Thus, there is a 0.25 (or 25%) chance that a household planned to purchase a large TV. Example 4.3 illustrates another application of simple probability.

EXAMPLE 4.3

Computing the Probability That the Large TV Purchased is HDR-capable

In another Fredco Warehouse Club follow-up survey, additional questions were asked of the 300 households that actually purchased large TVs. Table 4.2 indicates the consumers' responses to whether the TV purchased was HDR-capable and whether they also purchased a streaming media player in the past 12 months.

Find the probability that if a household that purchased a large TV is randomly selected, the television purchased was HDR-capable.

TABLE 4.2

Purchase behavior about purchasing an HDR-capable television and a streaming media player

	STREAMING MEDIA PLAYER		
HDR FEATURE	**Yes**	**No**	**Total**
HDR-capable	38	42	80
Not HDR-capable	70	150	220
Total	108	192	300

SOLUTION Using the following definitions:

A = purchased an HDR-capable TV

A' = purchased a television not HDR-capable

B = purchased a streaming media player

B' = did not purchase a streaming media player

$$P(\text{HDR-capable}) = \frac{\text{Number of HDR-capable TVs purchased}}{\text{Total number of TVs}}$$

$$= \frac{80}{300} = 0.267$$

There is a 26.7% chance that a randomly selected large TV purchased is HDR-capable.

Joint Probability

Whereas simple probability refers to the probability of occurrence of simple events, **joint probability** refers to the probability of an occurrence involving two or more events. An example of joint probability is the probability that one will get heads on the first toss of a coin and heads on the second toss of a coin.

In Table 4.1 on page 155, the count of the group of individuals who planned to purchase and actually purchased a large TV corresponds to the cell that represents Planned to purchase Yes and Actually purchased Yes, the upper left numerical cell. Because this group consists of 200 households, the probability of picking a household that planned to purchase *and* actually purchased a large TV is

$$P(\text{Planned to purchase } and \text{ actually purchased}) = \frac{\text{Planned to purchase } and \text{ actually purchased}}{\text{Total number of respondents}}$$

$$= \frac{200}{1,000} = 0.20$$

Example 4.4 also demonstrates how to determine joint probability.

EXAMPLE 4.4

Determining the Joint Probability That a Household Purchased an HDR-capable Large TV and Purchased a Streaming Media Player

In Table 4.2 on page 156, the purchases are cross-classified as being HDR-capable or not and whether the household purchased a streaming media player. Find the probability that a randomly selected household that purchased a large TV also purchased an HDR-capable TV and purchased a streaming media player.

SOLUTION Using Equation (4.1) on page 155 and Table 4.2,

$$P\begin{pmatrix}\text{HDR-capable TV } and \\ \text{streaming media player}\end{pmatrix} = \frac{\begin{array}{c}\text{Number that purchased an HDR-capable TV}\\ and \text{ purchased a streaming media player}\end{array}}{\text{Total number of large TV purchasers}}$$

$$= \frac{38}{300} = 0.127$$

Therefore, there is a 12.7% chance that a randomly selected household that purchased a large TV purchased an HDR-capable TV and purchased a streaming media player.

Marginal Probability

The **marginal probability** of an event consists of a set of joint probabilities. You can determine the marginal probability of a particular event by using the concept of joint probability just discussed. For example, if B consists of two events, B_1 and B_2, then $P(A)$, the probability of event A, consists of the joint probability of event A occurring with event B_1 and the joint probability of event A occurring with event B_2. Use Equation (4.2) to calculate marginal probabilities.

studentTIP

Mutually exclusive events cannot occur simultaneously.

In a collectively exhaustive set of events, one of the events must occur.

MARGINAL PROBABILITY

$$P(A) = P(A \text{ and } B_1) + P(A \text{ and } B_2) + \cdots + P(A \text{ and } B_k) \tag{4.2}$$

where $B_1, B_2, \ldots, B_k$ are k mutually exclusive and collectively exhaustive events

Using Equation (4.2) to calculate the marginal probability of "planned to purchase" a large TV gets the same result as adding the number of outcomes that make up the simple event "planned to purchase":

$$P(\text{Planned to purchase}) = P(\text{Planned to purchase } and \text{ purchased})$$
$$+ P(\text{Planned to purchase } and \text{ did not purchase})$$
$$= \frac{200}{1{,}000} + \frac{50}{1{,}000} = \frac{250}{1{,}000} = 0.25$$

student TIP

The key word when using the addition rule is *or*.

General Addition Rule

The probability of event "*A or B*" considers the occurrence of either event *A* or event *B* or both *A* and *B*. For example, how can one determine the probability that a household planned to purchase *or* actually purchased a large TV?

The event "planned to purchase *or* actually purchased" includes all households that planned to purchase and all households that actually purchased a large TV. Examine each cell of the Table 4.1 contingency table on page 155 to determine whether it is part of this event. From Table 4.1, the cell "planned to purchase *and* did not actually purchase" is part of the event because it includes respondents who planned to purchase. The cell "did not plan to purchase *and* actually purchased" is included because it contains respondents who actually purchased. Finally, the cell "planned to purchase *and* actually purchased" has both characteristics of interest. Therefore, one way to calculate the probability of "planned to purchase *or* actually purchased" is

$$P(\text{Planned to purchase } or \text{ actually purchased}) = P(\text{Planned to purchase } and \text{ did not actually purchase}) + P(\text{Did not plan to purchase } and \text{ actually purchased}) + P(\text{Planned to purchase } and \text{ actually purchased})$$

$$= \frac{50}{1{,}000} + \frac{100}{1{,}000} + \frac{200}{1{,}000}$$

$$= \frac{350}{1{,}000} = 0.35$$

Often, it is easier to determine $P(A \text{ or } B)$, the probability of the event *A or B*, by using the **general addition rule**, defined in Equation (4.3).

GENERAL ADDITION RULE

The probability of *A or B* is equal to the probability of *A* plus the probability of *B* minus the probability of *A and B*.

$$P(A \text{ or } B) = P(A) + P(B) - P(A \text{ and } B) \tag{4.3}$$

Applying Equation (4.3) to the previous example produces the following result:

$$P(\text{Planned to purchase } or \text{ actually purchased}) = P(\text{Planned to purchase}) + P(\text{Actually purchased}) - P(\text{Planned to purchase } and \text{ actually purchased})$$

$$= \frac{250}{1{,}000} + \frac{300}{1{,}000} - \frac{200}{1{,}000}$$

$$= \frac{350}{1{,}000} = 0.35$$

The general addition rule consists of taking the probability of *A* and adding it to the probability of *B* and then subtracting the probability of the joint event *A and B* from this total because the joint event has already been included in computing both the probability of *A* and the probability of *B*. For example, in Table 4.1, if the outcomes of the event "planned to purchase" are added to those of the event "actually purchased," the joint event "planned to purchase *and* actually purchased" has been included in each of these simple events. Therefore, because this joint event has been included twice, you must subtract it to compute the correct result. Example 4.5 illustrates another application of the general addition rule.

EXAMPLE 4.5
Using the General Addition Rule for the Households That Purchased Large TVs

In Example 4.3 on page 156, the purchases were cross-classified in Table 4.2 as TVs that were HDR-capable or not and whether the household purchased a streaming media player. Find the probability that among households that purchased a large TV, they purchased an HDR-capable TV or purchased a streaming media player.

SOLUTION Using Equation (4.3):

$$
\begin{aligned}
P(\text{HDR-capable TV} &= P(\text{HDR-capable TV}) \\
or \text{ purchased a streaming media player}) &\quad + P(\text{purchased a streaming media player}) \\
&\quad - P(\text{HDR-capable TV } and \\
&\quad \text{purchased a streaming media player}) \\
&= \frac{80}{300} + \frac{108}{300} - \frac{38}{300} \\
&= \frac{150}{300} = 0.50
\end{aligned}
$$

Therefore, of households that purchased a large TV, there is a 50% chance that a randomly selected household purchased an HDR-capable TV or purchased a streaming media player.

PROBLEMS FOR SECTION 4.1

LEARNING THE BASICS

4.1 Three coins are tossed.
a. Give an example of a simple event.
b. Give an example of a joint event.
c. What is the complement of a head on the first toss?
d. What does the sample space consist of?

4.2 An urn contains 12 red balls and 8 white balls. One ball is to be selected from the urn.
a. Give an example of a simple event.
b. What is the complement of a red ball?
c. What does the sample space consist of?

4.3 Consider the following contingency table:

	B	*B'*
A	10	20
A'	20	40

What is the probability of event
a. *A*?
b. *A'*?
c. *A and B*?
d. *A or B*?

4.4 Consider the following contingency table:

	B	*B'*
A	10	30
A'	25	35

What is the probability of event
a. *A'*?
b. *A and B*?
c. *A' and B'*?
d. *A' or B'*?

APPLYING THE CONCEPTS

4.5 For each of the following, indicate whether the type of probability involved is an example of *a priori* probability, empirical probability, or subjective probability.
a. The next toss of a fair coin will land on heads.
b. Italy will win soccer's World Cup the next time the competition is held.
c. The sum of the faces of two dice will be seven.
d. The train taking a commuter to work will be more than 10 minutes late.

4.6 For each of the following, state whether the events created are mutually exclusive and whether they are collectively exhaustive.

a. Undergraduate business students were asked whether they were sophomores or juniors.

b. Each respondent was classified by the type of car he or she drives: sedan, SUV, American, European, Asian, or none.

c. People were asked, "Do you currently live in (i) an apartment or (ii) a house?"

d. A product was classified as defective or not defective.

4.7 Which of the following events occur with a probability of zero? For each, state why or why not.

a. A company is listed on the New York Stock Exchange and NASDAQ.

b. A consumer owns a smartphone and a tablet.

c. A cell phone is an Apple and a Samsung.

d. An automobile is a Toyota and was manufactured in the United States.

4.8 Financial experts say retirement accounts are one of the best tools for saving and investing for the future. Are millennials and Gen Xers financially prepared for the future? A survey of American adults conducted by INSIDER and Morning Consult revealed the following:

	HAS A RETIREMENT SAVINGS ACCOUNT	
AGE GROUP	**Yes**	**No**
Millennials	579	628
Gen X	599	532

Source: Data extracted from "Millennials are delusional about the future, but they aren't the only ones," **bit.ly/2KDeDK9**.

a. Give an example of a simple event.

b. Give an example of a joint event.

c. What is the complement of "Has a retirement savings account"?

d. Why is "Millennial and has a retirement savings account" a joint event?

4.9 Referring to the contingency table in Problem 4.8, if an American adult is selected at random, what is the probability that

a. the American adult has a retirement savings account?

b. the American adult was a millennial who has a retirement savings account?

c. the American adult was a millennial *or* has a retirement savings account?

d. Explain the difference in the results in (b) and (c).

4.10 How will marketers change their organic (unpaid) social media postings in the near future? A survey by Social Media Examiner reported that 65% of B2B marketers (marketers who focus primarily on attracting businesses) plan to increase LinkedIn organic social media postings, as compared to 43% of B2C marketers (marketers who primarily target consumers). The survey was based on 1,947 B2B marketers and 3,779 B2C marketers. The following table summarizes the results.

INCREASE LINKEDIN POSTINGS?	BUSINESS FOCUS		
	B2B	**B2C**	**Total**
Yes	1,266	1,625	2,891
No	681	2,154	2,835
Total	1,947	3,779	5,726

Source: Data extracted from "2018 Social Media Marketing Industry Report," **socialmediaexaminer.com**.

a. Give an example of a simple event.

b. Give an example of a joint event.

c. What is the complement of a marketer who plans to increase LinkedIn organic social media postings?

d. Why is a marketer who plans to increase LinkedIn organic social media postings and is a B2C marketer a joint event?

4.11 Referring to the contingency table in Problem 4.10, if a marketer is selected at random, what is the probability that

a. he or she plans to increase LinkedIn organic social media postings?

b. he or she is a B2C marketer?

c. he or she plans to increase LinkedIn organic social media postings *or* is a B2C marketer?

d. Explain the difference in the results in (b) and (c).

✓SELF TEST **4.12** Is there full support for increased use of educational technologies in higher ed? As part of Inside Higher Ed's 2018 Survey of Faculty Attitudes on Technology, academic professionals, namely, full-time faculty members and administrators who oversee their institutions' online learning or instructional technology efforts (digital learning leaders), were asked that question. The following table summarizes their responses.

	ACADEMIC PROFESSIONAL		
FULL SUPPORT	**Faculty Member**	**Digital Learning Leader**	**Total**
Yes	511	175	686
No	1,086	31	1,117
Total	1,597	206	1,803

Source: Data extracted from "The 2018 Inside Higher Ed Survey of Faculty Attitudes on Technology," **bit.ly/301jUit**.

If an academic professional is selected at random, what is the probability that he or she

a. fully supports increased use of educational technologies in higher ed?

b. is a digital learning leader?

c. fully supports increased use of educational technologies in higher ed *or* is a digital learning leader?

d. Explain the difference in the results in (b) and (c).

4.13 Are small and mid-sized business (SMB) owners concerned about limited access to money affecting their businesses? A sample of 150 female SMB owners and 146 male SMB owners revealed the following results:

CONCERNED ABOUT LIMITED ACCESS TO MONEY?	SMB OWNER GENDER		
	Female	**Male**	**Total**
Yes	38	20	58
No	112	126	238
Total	150	146	296

Source: Data extracted from "2019 Small Business Owner Report: The Male vs Female Divide," **bit.ly/2WvByIN**.

If a SMB owner is selected at random, what is the probability that the SMB owner

a. is concerned about limited access to money affecting the business?
b. is a female SMB owner *and* is concerned about limited access to money affecting the business?
c. is a female SMB owner *or* is concerned about limited access to money affecting the business?
d. Explain the difference in the results of (b) and (c).

4.14 Consumers are aware that companies share and sell their personal data in exchange for free services, but is it important to consumers to have a clear understanding of a company's privacy policy before signing up for its service online? According to an Axios-SurveyMonkey poll, 911 of 1,001 older adults (aged 65+) indicate that it is important to have a clear understanding of a company's privacy policy before signing up for its service online and 195 of 260 younger adults (aged 18–24) indicate that it is important to have a clear understanding of a company's privacy policy before signing up for its service online.

Source: Data extracted from "Privacy policies are read by an aging few," **bit.ly/2NyPAWx**.

Construct a contingency table to evaluate the probabilities. What is the probability that a respondent chosen at random

a. indicates it is important to have a clear understanding of a company's privacy policy before signing up for its service online?
b. is an older adult *and* indicates it is important to have a clear understanding of a company's privacy policy before signing up for its service online?

c. is an older adult *or* indicates it is important to have a clear understanding of a company's privacy policy before signing up for its service online?
d. is an older adult *or* a younger adult?

4.15 Each year, ratings are compiled concerning the performance of new cars during the first 90 days of use. Suppose that the cars have been categorized according to whether a car needs a warranty-related repair (yes or no) and the country in which the company manufacturing a car is based (United States or not United States). Based on the data collected, the probability that the new car needs a warranty repair is 0.04, the probability that the car was manufactured by a U.S.-based company is 0.60, and the probability that the new car needs a warranty repair *and* was manufactured by a U.S.-based company is 0.025.

Construct a contingency table to evaluate the probabilities of a warranty-related repair. What is the probability that a new car selected at random

a. needs a warranty repair?
b. needs a warranty repair *and* was manufactured by a U.S.-based company?
c. needs a warranty repair *or* was manufactured by a U.S.-based company?
d. needs a warranty repair *or* was not manufactured by a U.S.-based company?

4.2 Conditional Probability

Each Section 4.1 example involves finding the probability of an event when sampling from the entire sample space. How does one determine the probability of an event if one knows certain information about the events involved?

Calculating Conditional Probabilities

Conditional probability refers to the probability of event A, given information about the occurrence of another event, B.

CONDITIONAL PROBABILITY

The probability of A given B is equal to the probability of A *and* B divided by the probability of B.

$$P(A|B) = \frac{P(A \text{ and } B)}{P(B)} \tag{4.4a}$$

The probability of B given A is equal to the probability of A *and* B divided by the probability of A.

$$P(B|A) = \frac{P(A \text{ and } B)}{P(A)} \tag{4.4b}$$

where

$P(A \text{ and } B)$ = joint probability of A *and* B
$P(A)$ = marginal probability of A
$P(B)$ = marginal probability of B

In the Fredco Warehouse Club scenario, suppose one had been told that a specific household planned to purchase a large TV. What then would be the probability that the household actually purchased the TV?

In this example, the objective is to find P(Actually purchased|Planned to purchase), given the information that a household planned to purchase a large TV. Therefore, the sample space does not consist of all 1,000 households in the survey. It consists of only those households that planned to purchase the large TV. Of 250 such households, 200 actually purchased the large TV. Therefore, based on Table 4.1 on page 155, the probability that a household actually purchased the large TV given that they planned to purchase is

student TIP

The variable that is *given* goes in the denominator of Equation (4.4) denominator. Because planned to purchase was the given, planned to purchase goes in the denominator.

$$P(\text{Actually purchased}|\text{Planned to purchase}) = \frac{\text{Planned to purchase } and \text{ actually purchased}}{\text{Planned to purchase}}$$

$$= \frac{200}{250} = 0.80$$

Defining event A as Planned to purchase and event B as Actually purchased, Equation (4.4b) also calculates this result:

$$P(B|A) = \frac{P(A \text{ and } B)}{P(A)}$$

$$P(\text{Actually purchased}|\text{Planned to purchase}) = \frac{200/1,000}{250/1,000} = \frac{200}{250} = 0.80$$

Example 4.6 further illustrates conditional probability.

EXAMPLE 4.6

Finding the Conditional Probability of Purchasing a Streaming Media Player

Table 4.2 on page 156 is a contingency table for whether a household purchased an HDR-capable TV and whether the household purchased a streaming media player. If a household purchased an HDR-capable TV, what is the probability that it also purchased a streaming media player?

SOLUTION Because you know that the household purchased an HDR-capable TV, the sample space is reduced to 80 households. Of these 80 households, 38 also purchased a streaming media player. Therefore, the probability that a household purchased a streaming media player, given that the household purchased an HDR-capable TV, is

$$P(\text{Purchased streaming media player}|\text{Purchased HDR-capable TV}) = \frac{\text{Number purchasing HDR-capable TV } and \text{ streaming media player}}{\text{Number purchasing HDR-capable TV}}$$

$$= \frac{38}{80} = 0.475$$

Using Equation (4.4b) on page 161 and the following definitions:

A = Purchased an HDR-capable TV

B = Purchased a streaming media player

then

$$P(B|A) = \frac{P(A \text{ and } B)}{P(A)} = \frac{38/300}{80/300} = 0.475$$

Therefore, given that the household purchased an HDR-capable TV, there is a 47.5% chance that the household also purchased a streaming media player. You can compare this conditional probability to the marginal probability of purchasing a streaming media player, which is $108/300 = 0.36$, or 36%. These results tell you that households that purchased HDR-capable TVs are more likely to purchase a streaming media player than are households that purchased large TVs that are not HDR-capable.

Decision Trees

In Table 4.1, households are classified according to whether they planned to purchase and whether they actually purchased large TVs. A **decision tree** is an alternative to the contingency table. Figure 4.1 represents the decision tree for this example.

FIGURE 4.1
Decision tree for planned to purchase and actually purchased

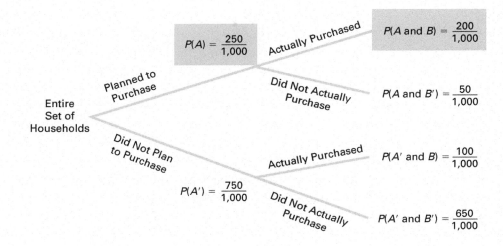

In Figure 4.1, beginning at the left with the entire set of households, there are two "branches" for whether or not the household planned to purchase a large TV. Each of these branches has two subbranches, corresponding to whether the household actually purchased or did not actually purchase the large TV. The probabilities at the end of the initial branches represent the marginal probabilities of A and A'. The probabilities at the end of each of the four subbranches represent the joint probability for each combination of events A and B. You compute the conditional probability by dividing the joint probability by the appropriate marginal probability.

For example, to compute the probability that the household actually purchased, given that the household planned to purchase the large TV, you take P(Planned to purchase *and* actually purchased) and divide by P(Planned to purchase). From Figure 4.1,

$$P(\text{Actually purchased}\mid \text{Planned to purchase}) = \frac{200/1{,}000}{250/1{,}000}$$

$$= \frac{200}{250} = 0.80$$

Example 4.7 illustrates how to construct a decision tree.

EXAMPLE 4.7

Constructing the Decision Tree for the Households That Purchased Large TVs

Using the cross-classified data in Table 4.2 on page 156 construct the decision tree. Use the decision tree to find the probability that a household purchased a streaming media player, given that the household purchased an HDR-capable television.

SOLUTION The decision tree for purchased a streaming media player and an HDR-capable TV is displayed in Figure 4.2.

▶ *(continued)*

FIGURE 4.2

Decision tree for purchased an HDR-capable TV and a streaming media player

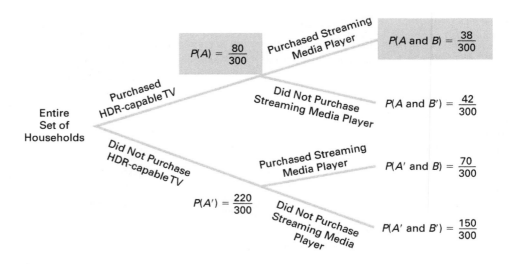

Using Equation (4.4b) on page 161 and the following definitions:

    A = Purchased an HDR-capable TV

    B = Purchased a streaming media player

then

$$P(B|A) = \frac{P(A \text{ and } B)}{P(A)} = \frac{38/300}{80/300} = 0.475$$

Independence

In the example concerning the purchase of large TVs, the conditional probability is $200/250 = 0.80$ that the selected household actually purchased the large TV, given that the household planned to purchase. The simple probability of selecting a household that actually purchased is $300/1{,}000 = 0.30$. This result shows that the prior knowledge that the household planned to purchase affected the probability that the household actually purchased the TV. In other words, the outcome of one event is *dependent* on the outcome of a second event.

When the outcome of one event does *not* affect the probability of occurrence of another event, the events are said to be independent. **Independence** can be determined by using Equation (4.5).

INDEPENDENCE

Two events, A and B, are independent if and only if

$$P(A|B) = P(A) \tag{4.5}$$

    where

        $P(A|B)$ = conditional probability of A given B

        $P(A)$ = marginal probability of A

Example 4.8 on the next page demonstrates the use of Equation (4.5).

EXAMPLE 4.8

Determining
Independence

In the follow-up survey of the 300 households that actually purchased large TVs, the households were asked if they were satisfied with their purchases. Table 4.3 cross-classifies the responses to the satisfaction question with the responses to whether the TV was HDR-capable.

TABLE 4.3
Satisfaction with
purchase of large TVs

HDR FEATURE	SATISFIED WITH PURCHASE?		
	Yes	No	Total
HDR-capable	64	16	80
Not HDR-capable	176	44	220
Total	240	60	300

Determine whether being satisfied with the purchase and the HDR feature of the TV purchased are independent.

SOLUTION For these data:

$$P(\text{Satisfied} \mid \text{HDR-capable}) = \frac{64/300}{80/300} = \frac{64}{80} = 0.80$$

which is equal to

$$P(\text{Satisfied}) = \frac{240}{300} = 0.80$$

Thus, being satisfied with the purchase and the HDR feature of the TV purchased are independent. Knowledge of one event does not affect the probability of the other event.

Multiplication Rules

The **general multiplication rule** is derived using Equation (4.4a) on page 161:

$$P(A \mid B) = \frac{P(A \text{ and } B)}{P(B)}$$

and solving for the joint probability $P(A \text{ and } B)$.

GENERAL MULTIPLICATION RULE

The probability of A and B is equal to the probability of A given B times the probability of B.

$$P(A \text{ and } B) = P(A \mid B)P(B) \tag{4.6}$$

Example 4.9 demonstrates the use of the general multiplication rule.

EXAMPLE 4.9

Using the General
Multiplication Rule

Consider the 80 households that purchased HDR-capable TVs. In Table 4.3 on this page, you see that 64 households are satisfied with their purchase, and 16 households are dissatisfied. Suppose two households are randomly selected from the 80 households. Find the probability that both households are satisfied with their purchase.

SOLUTION Here you can use the multiplication rule in the following way. If

$$A = \text{second household selected is satisfied}$$

$$B = \text{first household selected is satisfied}$$

►*(continued)*

then, using Equation (4.6),

$$P(A \text{ and } B) = P(A|B)P(B)$$

The probability that the first household is satisfied with the purchase is 64/80. However, the probability that the second household is also satisfied with the purchase depends on the result of the first selection. If the first household is not returned to the sample after the satisfaction level is determined (i.e., sampling without replacement), the number of households remaining is 79. If the first household is satisfied, the probability that the second is also satisfied is 63/79 because 63 satisfied households remain in the sample. Therefore,

$$P(A \text{ and } B) = \left(\frac{63}{79}\right)\left(\frac{64}{80}\right) = 0.6380$$

There is a 63.80% chance that both of the households sampled will be satisfied with their purchase.

The **multiplication rule for independent events** is derived by substituting $P(A)$ for $P(A|B)$ in Equation (4.6).

MULTIPLICATION RULE FOR INDEPENDENT EVENTS

If A and B are independent, the probability of A and B is equal to the probability of A times the probability of B.

$$P(A \text{ and } B) = P(A)P(B) \tag{4.7}$$

If this rule holds for two events, A and B, then A and B are independent. Therefore, there are two ways to determine independence:

1. Events A and B are independent if, and only if, $P(A|B) = P(A)$.
2. Events A and B are independent if, and only if, $P(A \text{ and } B) = P(A)P(B)$.

Marginal Probability Using the General Multiplication Rule

Section 4.1 defines the marginal probability using Equation (4.2). One can state the equation for marginal probability by using the general multiplication rule. If

$$P(A) = P(A \text{ and } B_1) + P(A \text{ and } B_2) + \cdots + P(A \text{ and } B_k)$$

then, using the general multiplication rule, Equation (4.8) defines the marginal probability.

MARGINAL PROBABILITY USING THE GENERAL MULTIPLICATION RULE

$$P(A) = P(A|B_1)P(B_1) + P(A|B_2)P(B_2) + \cdots + P(A|B_k)P(B_k) \tag{4.8}$$

where $B_1, B_2, \ldots, B_k$ are k mutually exclusive and collectively exhaustive events.

To illustrate Equation (4.8), refer to Table 4.1 on page 155 and let:

$$P(A) = \text{probability of planned to purchase}$$
$$P(B_1) = \text{probability of actually purchased}$$
$$P(B_2) = \text{probability of did not actually purchase}$$

Then, using Equation (4.8), the probability of planned to purchase is

$$P(A) = P(A \mid B_1)P(B_1) + P(A \mid B_2)P(B_2)$$

$$= \left(\frac{200}{300}\right)\left(\frac{300}{1,000}\right) + \left(\frac{50}{700}\right)\left(\frac{700}{1,000}\right)$$

$$= \frac{200}{1,000} + \frac{50}{1,000} = \frac{250}{1,000} = 0.25$$

PROBLEMS FOR SECTION 4.2

LEARNING THE BASICS

4.16 Consider the following contingency table:

	B	B'
A	10	20
A'	20	40

What is the probability of
a. $A \mid B$?
b. $A \mid B'$?
c. $A' \mid B'$?
d. Are events A and B independent?

4.17 Consider the following contingency table:

	B	B'
A	10	30
A'	25	35

What is the probability of
a. $A \mid B$?
b. $A' \mid B'$?
c. $A \mid B'$?
d. Are events A and B independent?

4.18 If $P(A \text{ and } B) = 0.4$ and $P(B) = 0.8$, find $P(A \mid B)$.

4.19 If $P(A) = 0.7$, $P(B) = 0.6$, and A and B are independent, find $P(A \text{ and } B)$.

4.20 If $P(A) = 0.3$, $P(B) = 0.4$, and $P(A \text{ and } B) = 0.2$, are A and B independent?

APPLYING THE CONCEPTS

4.21 Financial experts say retirement accounts are one of the best tools for saving and investing for the future. Are millennials and Gen Xers financially prepared for the future? A survey of American adults conducted by INSIDER and Morning Consult revealed the following:

AGE GROUP	HAS A RETIREMENT SAVINGS ACCOUNT	
	Yes	No
Millennials	579	628
Gen Xer	599	532

Source: Data extracted from "Millennials are delusional about the future, but they aren't the only ones," **bit.ly/2KDeDK9**.

a. Given that the American adult has a retirement savings account, what is the probability that the American adult is a millennial?
b. Given that the American adult is a millennial, what is the probability that the American adult has a retirement savings account?
c. Explain the difference in the results in (a) and (b).
d. Is having a retirement savings account and age group independent?

4.22 How will marketers change their organic (unpaid) social media postings in the near future? A survey by Social Media Examiner reported that 65% of B2B marketers (marketers who focus primarily on attracting businesses) plan to increase LinkedIn organic social media postings, as compared to 43% of B2C marketers (marketers who primarily target consumers). The survey was based on 1,947 B2B marketers and 3,779 B2C marketers. The following table summarizes the results.

INCREASE LINKEDIN POSTINGS?	BUSINESS FOCUS		
	B2B	B2C	Total
Yes	1,266	1,625	2,891
No	681	2,154	2,835
Total	1,947	3,779	5,726

Source: Data extracted from "2018 Social Media Marketing Industry Report," **socialmediaexaminer.com**.

a. Suppose you know that the marketer is a B2B marketer. What is the probability that he or she plans to increase LinkedIn organic social media postings?

b. Suppose you know that the marketer is a B2C marketer. What is the probability that he or she plans to increase LinkedIn organic social media postings?

c. Are the two events, increase LinkedIn postings and business focus, independent? Explain.

4.23 Are small and mid-sized business (SMB) owners concerned about limited access to money affecting their businesses? A sample of 150 female SMB owners and 146 male SMB owners revealed the following results:

CONCERNED ABOUT LIMITED ACCESS TO MONEY?	SMB OWNER GENDER		
	Female	Male	Total
Yes	38	20	58
No	112	126	238
Total	150	146	296

Source: Data extracted from "2019 Small Business Owner Report: The Male vs Female Divide,"**bit.ly/2WvByIN**.

a. If a SMB owner selected is female, what is the probability that she is concerned about limited access to money affecting the business?

b. If a SMB owner selected is male, what is the probability that he is concerned about limited access to money affecting the business?

c. Is concern about limited access to money independent of SMB owner gender?

✓SELF TEST **4.24** Is there full support for increased use of educational technologies in higher ed? As part of Inside Higher Ed's 2018 Survey of Faculty Attitudes on Technology, academic professionals, namely, full-time faculty members and administrators who oversee their institutions' online learning or instructional technology efforts (digital learning leaders), were asked that question. The following table summarizes their responses:

FULL SUPPORT	ACADEMIC PROFESSIONAL		
	Faculty Member	Digital Learning Leader	Total
Yes	511	175	686
No	1,086	31	1,117
Total	1,597	206	1,803

Source: Data extracted from "The 2018 Inside Higher Ed Survey of Faculty Attitudes on Technology," **bit.ly/301jUit**.

a. Given that an academic professional is a faculty member, what is the probability that the academic professional fully supports increased use of educational technologies in higher ed?

b. Given that an academic professional is a faculty member, what is the probability that the academic professional does not fully support increased use of educational technologies in higher ed?

c. Given that an academic professional is a digital learning leader, what is the probability that the academic professional fully supports increased use of educational technologies in higher ed?

d. Given that an academic professional is a digital learning leader, what is the probability that the academic professional does not fully support increased use of educational technologies in higher ed?

4.25 Consumers are aware that companies share and sell their personal data in exchange for free services, but is it important to consumers to have a clear understanding of a company's privacy policy before signing up for its service online? According to an Axios-SurveyMonkey poll, 911 of 1,001 older adults (aged 65+) and 195 of 260 younger adults (aged 18–24) indicate that it is important to have a clear understanding of a company's privacy policy before signing up for its service online.

Source: Data extracted from "Privacy policies are read by an aging few," **bit.ly/2NyPAW**.

a. Suppose that the respondent chosen is an older adult. What is the probability that the respondent indicates that it is important to have a clear understanding of a company's privacy policy before signing up for its service online?

b. Suppose that the respondent chosen does indicate that it is important to have a clear understanding of a company's privacy policy before signing up for its service online. What is the probability that the respondent is a younger adult?

c. Are importance of a clear understanding and adult age independent? Explain

4.26 Each year, ratings are compiled concerning the performance of new cars during the first 90 days of use. Suppose that the cars have been categorized according to whether a car needs warranty-related repair (yes or no) and the country in which the company manufacturing a car is based (United States or not United States). Based on the data collected, the probability that the new car needs a warranty repair is 0.04, the probability that the car is manufactured by a U.S.-based company is 0.60, and the probability that the new car needs a warranty repair *and* was manufactured by a U.S.-based company is 0.025.

a. Suppose you know that a company based in the United States manufactured a particular car. What is the probability that the car needs a warranty repair?

b. Suppose you know that a company based in the United States did not manufacture a particular car. What is the probability that the car needs a warranty repair?

c. Are the need for a warranty repair and location of the company manufacturing the car independent?

4.27 In 44 of the 68 years from 1950 through 2018 (in 2011 there was virtually no change), the S&P 500 finished higher after the first five days of trading. In 36 out of 44 years, the S&P 500 finished higher for the year. Is a good first week a good omen for the upcoming year? The following table gives the first-week and annual performance over this 68-year period:

FIRST WEEK	S&P 500'S ANNUAL PERFORMANCE	
	Higher	Lower
Higher	36	8
Lower	12	12

a. If a year is selected at random, what is the probability that the S&P 500 finished higher for the year?

b. Given that the S&P 500 finished higher after the first five days of trading, what is the probability that it finished higher for the year?

c. Are the two events "first-week performance" and "annual performance" independent? Explain.

d. Look up the performance after the first five days of 2019 and the 2019 annual performance of the S&P 500 at **finance.yahoo.com**. Comment on the results.

4.28 A standard deck of cards is being used to play a game. There are four suits (hearts, diamonds, clubs, and spades), each having 13 faces (ace, 2, 3, 4, 5, 6, 7, 8, 9, 10, jack, queen, and king), making a total of 52 cards. This complete deck is thoroughly mixed, and you will receive the first 2 cards from the deck, without replacement (the first card is not returned to the deck after it is selected).

a. What is the probability that both cards are queens?

b. What is the probability that the first card is a 10 and the second card is a 5 or 6?

c. If you were sampling with replacement (the first card is returned to the deck after it is selected), what would be the answer in (a)?

d. In the game of blackjack, the face cards (jack, queen, king) count as 10 points, and the ace counts as either 1 or 11 points. All other cards are counted at their face value. Blackjack is achieved if two cards total 21 points. What is the probability of getting blackjack in this problem?

4.29 A box of nine iPhone 7 cell phones contains two red cell phones and seven black cell phones.

a. If two cell phones are randomly selected from the box, without replacement (the first cell phone is not returned to the box after it is selected), what is the probability that both cell phones selected will be red?

b. If two cell phones are randomly selected from the box, without replacement (the first cell phone is not returned to the box after it is selected), what is the probability that there will be one red cell phone and one black cell phone selected?

c. If three cell phones are selected, with replacement (the cell phones are returned to the box after they are selected), what is the probability that all three will be red?

d. If you were sampling with replacement (the first cell phone is returned to the box after it is selected), what would be the answers to (a) and (b)?

4.3 Ethical Issues and Probability

Ethical issues can arise when any statements related to probability are presented to the public, particularly when these statements are part of an advertising campaign for a product or service. Unfortunately, many people are not comfortable with numerical concepts (Paulos) and tend to misinterpret the meaning of the probability. In some instances, the misinterpretation is not intentional, but in other cases, advertisements may unethically try to mislead potential customers.

One example of a potentially unethical application of probability relates to advertisements for state lotteries. When purchasing a lottery ticket, the customer selects a set of numbers (such as 6) from a larger list of numbers (such as 54). Although virtually all participants know that they are unlikely to win the lottery, they also have very little idea of how unlikely it is for them to select all six winning numbers from the list of 54 numbers. They have even less of an idea of the probability of not selecting any winning numbers.

Given this background, one might consider a state lottery commercial that stated, "We won't stop until we have made everyone a millionaire" to be deceptive and possibly unethical. Is it possible that the lottery can make everyone a millionaire? Is it ethical to suggest that the purpose of the lottery is to make everyone a millionaire?

Another example of a potentially unethical application of probability relates to an investment newsletter promising a 90% probability of a 20% annual return on investment. To make the claim in the newsletter an ethical one, the investment service needs to (a) explain the basis on which this probability estimate rests, (b) provide the probability statement in another format, such as 9 chances in 10, and (c) explain what happens to the investment in the 10% of the cases in which a 20% return is not achieved (e.g., is the entire investment lost?).

Probability-related claims should be stated ethically. Readers can consider how one would write an ethical description about the probability of winning a certain prize in a state lottery. Or how one can ethically explain a "90% probability of a 20% annual return" without creating a misleading inference.

4.4 Bayes' Theorem

Developed by Thomas Bayes in the eighteenth century, **Bayes' theorem** builds on conditional probability concepts that Section 4.2 discusses. Bayes' theorem revises previously calculated probabilities using additional information and forms the basis for Bayesian analysis. (Anderson-Cook, Bellhouse, Hooper).

In recent years, Bayesian analysis has gained new prominence for its application to analyzing big data using predictive analytics (see Chapter 17). However, Bayesian analysis does not require big data and can be used in a variety of problems to better determine the *revised probability* of certain events. The **Bayesian Analysis online topic** contains examples that apply Bayes' theorem to a marketing problem and a diagnostic problem and presents a set of additional study problems. The *Consider This* for this section explores an application of Bayes' theorem that many use every day, and references Equation (4.9) that the online topic discusses.

CONSIDER THIS

Divine Providence and Spam

Would you ever guess that the essays *Divine Benevolence: Or, An Attempt to Prove That the Principal End of the Divine Providence and Government Is the Happiness of His Creatures* and *An Essay Towards Solving a Problem in the Doctrine of Chances* were written by the same person? Probably not, and in doing so, you illustrate a modern-day application of Bayesian statistics: spam, or junk mail filters.

In not guessing correctly, you probably looked at the words in the titles of the essays and concluded that they were talking about two different things. An implicit rule you used was that word frequencies vary by subject matter. A statistics essay would very likely contain the word *statistics* as well as words such as *chance*, *problem*, and *solving*. An eighteenth-century essay about theology and religion would be more likely to contain the uppercase forms of *Divine* and *Providence*.

Likewise, there are words you would guess to be very unlikely to appear in either book, such as technical terms from finance, and words that are most likely to appear in both—common words such as *a*, *and*, and *the*. That words would be either likely or unlikely suggests an application of probability theory. Of course, likely and unlikely are fuzzy concepts, and we might occasionally misclassify an essay if we kept things too simple, such as relying solely on the occurrence of the words *Divine* and *Providence*.

For example, a profile of the late Harris Milstead, better known as *Divine*, the star of *Hairspray* and other films, visiting Providence (Rhode Island), would most certainly not be an essay about theology. But if we widened the number of words we examined and found such words as movie or the name John Waters (Divine's director in many films), we probably would quickly realize the essay had something to do with twentieth-century cinema and little to do with theology and religion.

We can use a similar process to try to classify a new email message in your in-box as either spam or a legitimate message (called "ham," in this context). We would first need to add to your email program a "spam filter" that has the ability to track word frequencies associated with spam and ham messages as you identify them on a day-to-day basis. This would allow the filter to constantly update the prior

probabilities necessary to use Bayes' theorem. With these probabilities, the filter can ask, "What is the probability that an email is spam, given the presence of a certain word?"

Applying the terms of Equation (4.9), such a Bayesian spam filter would multiply the probability of finding the word in a spam email, $P(A|B)$, by the probability that the email is spam, $P(B)$, and then divide by the probability of finding the word in an email, the denominator in Equation (4.9). Bayesian spam filters also use shortcuts by focusing on a small set of words that have a high probability of being found in a spam message as well as on a small set of other words that have a low probability of being found in a spam message.

As spammers (people who send junk email) learned of such new filters, they tried to outfox them. Having learned that Bayesian filters might be assigning a high $P(A|B)$, value to words commonly found in spam, such as Viagra, spammers thought they could fool the filter by misspelling the word as Vi@gr@ or V1agra. What they overlooked was that the misspelled variants were even more likely to be found in a spam message than the original word. Thus, the misspelled variants made the job of spotting spam easier for the Bayesian filters.

Other spammers tried to fool the filters by adding "good" words, words that would have a low probability of being found in a spam message, or "rare" words, words not frequently encountered in any message. But these spammers overlooked the fact that the conditional probabilities are constantly updated and that words once considered "good" would be soon discarded from the good list by the filter as their $P(A|B)$ value increased. Likewise, as "rare" words grew more common in spam and yet stayed rare in ham, such words acted like the misspelled variants that others had tried earlier.

Even then, and perhaps after reading about Bayesian statistics, spammers thought that they could "break" Bayesian filters by inserting random words in their messages. Those random words would affect the filter by causing it to see many words whose $P(A|B)$ value would be low. The Bayesian filter would begin to label many spam messages as ham and end up being of no practical use. Spammers again overlooked that conditional probabilities are constantly updated.

Other spammers decided to eliminate all or most of the words in their messages and replace them with graphics so that Bayesian filters would have very few words with which to form conditional probabilities. But this approach failed, too, as Bayesian filters were rewritten to consider things other than words in a message. After all, Bayes' theorem concerns events, and "graphics present with no text" is as valid an event as "some word, *X*, present in a message."

Other future tricks will ultimately fail for the same reason. (By the way, spam filters use non-Bayesian techniques as well, which makes spammers' lives even more difficult.)

Bayesian spam filters are an example of the unexpected way that applications of statistics can show up in your daily life. You will discover more examples as you read the rest of this book. By the way, the author of the two essays mentioned earlier was Thomas Bayes, who is a lot more famous for the second essay than the first essay, a failed attempt to use mathematics and logic to prove the existence of God.

4.5 Counting Rules

In many cases, a large number of outcomes is possible and determining the exact number of outcomes can be difficult. In these situations, rules have been developed for counting the exact number of possible outcomes. The **Section 4.5 online topic** discusses these rules and illustrates their use.

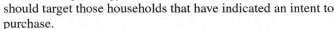

▼ USING **STATISTICS**
Probable Outcomes at Fredco Warehouse Club, Revisited

As a Fredco Warehouse Club electronics merchandise manager, you analyzed an intent-to-purchase-a-large-TV study of the heads of 1,000 households as well as a follow-up study done 12 months later. In that later survey, respondents who answered that they had purchased a large TV were asked additional questions concerning whether the large TV was HDR-capable and whether the respondents had purchased a streaming media player in the past 12 months.

By analyzing the results of these surveys, you were able to uncover many pieces of valuable information that will help you plan a marketing strategy to enhance sales and better target those households likely to purchase multiple or more expensive products. Whereas only 30% of the households actually purchased a large TV, if a household indicated that it planned to purchase a large TV in the next 12 months, there

was an 80% chance that the household actually made the purchase. Thus the marketing strategy should target those households that have indicated an intent to purchase.

You determined that for households that purchased an HDR-capable TV, there was a 47.5% chance that the household also purchased a streaming media player. You then compared this conditional probability to the marginal probability of purchasing a streaming media player, which was 36%. Thus, households that purchased HDR-capable TVs are more likely to purchase a streaming media player than are households that purchased large TVs that were not HDR-capable.

▼ SUMMARY

This chapter develops the basic concepts of probability that serve as a foundation for other concepts that later chapters discuss. Probability is a numeric value from 0 to 1 that represents the chance, likelihood, or possibility that a particular event will occur. In addition to simple probability, the chapter discusses conditional probabilities and independent events. The chapter contingency tables and decision trees summarize and present probability information. The chapter also introduces Bayes' theorem.

▼ REFERENCES

Anderson-Cook, C. M. "Unraveling Bayes' Theorem." *Quality Progress*, March 2014, pp. 52–54.

Bellhouse, D. R. "The Reverend Thomas Bayes, FRS: A Biography to Celebrate the Tercentenary of His Birth." *Statistical Science*, 19 (2004), 3–43.

Hooper, W. "Probing Probabilities." *Quality Progress*, March 2014, pp. 18–22.

Lowd, D., and C. Meek. "Good Word Attacks on Statistical Spam Filters." Presented at the Second Conference on Email and Anti-Spam, 2005.

Paulos, J. A. *Innumeracy*. New York: Hill and Wang, 1988.

Silberman, S. "The Quest for Meaning," *Wired 8.02*, February 2000.

Zeller, T. "The Fight Against V1@gra (and Other Spam)." *The New York Times*, May 21, 2006, pp. B1, B6.

▼ KEY EQUATIONS

Probability of Occurrence

$$\text{Probability of occurrence} = \frac{X}{T} \tag{4.1}$$

Marginal Probability

$$P(A) = P(A \text{ and } B_1) + P(A \text{ and } B_2)$$
$$+ \cdots + P(A \text{ and } B_k) \tag{4.2}$$

General Addition Rule

$$P(A \text{ or } B) = P(A) + P(B) - P(A \text{ and } B) \tag{4.3}$$

Conditional Probability

$$P(A \mid B) = \frac{P(A \text{ and } B)}{P(B)} \tag{4.4a}$$

$$P(B \mid A) = \frac{P(A \text{ and } B)}{P(A)} \tag{4.4b}$$

Independence

$$P(A \mid B) = P(A) \tag{4.5}$$

General Multiplication Rule

$$P(A \text{ and } B) = P(A \mid B)P(B) \tag{4.6}$$

Multiplication Rule for Independent Events

$$P(A \text{ and } B) = P(A)P(B) \tag{4.7}$$

Marginal Probability Using the General Multiplication Rule

$$P(A) = P(A \mid B_1)P(B_1) + P(A \mid B_2)P(B_2)$$
$$+ \cdots + P(A \mid B_k)P(B_k) \tag{4.8}$$

▼ KEY TERMS

▼ CHECKING YOUR UNDERSTANDING

4.30 What are the differences between *a priori* probability, empirical probability, and subjective probability?

4.31 What is the difference between a simple event and a joint event?

4.32 How can you use the general addition rule to find the probability of occurrence of event A or B?

4.33 What is the difference between mutually exclusive events and collectively exhaustive events?

4.34 How does conditional probability relate to the concept of independence?

4.35 How does the multiplication rule differ for events that are and are not independent?

▼ CHAPTER REVIEW PROBLEMS

4.36 A survey by Accenture indicated that 64% of millennials as compared to 28% of baby boomers prefer "hybrid" investment advice—a combination of traditional advisory services and low-cost digital tools—over either a dedicated human advisor or conventional robo-advisory services (computer-generated advice and services without human advisors) alone.

Source: Data extracted from Business Wire, "Majority of Wealthy Investors Prefer a Mix of Human and Robo-Advice, According to Accenture Research," **/bit.ly/2qZY9Ou**.

Suppose that the survey was based on 500 respondents from each of the two generation groups.
a. Construct a contingency table.
b. Give an example of a simple event and a joint event.
c. What is the probability that a randomly selected respondent prefers hybrid investment advice?
d. What is the probability that a randomly selected respondent prefers hybrid investment advice *and* is a baby boomer?
e. Are the events "generation group" and "prefers hybrid investment advice" independent? Explain.

4.37 G2 Crowd provides commentary and insight about employee engagement trends and challenges within organizations in its G2 Crowd Employee Engagement Report. The report represents the results of an online survey conducted in 2019 with employees located across the United States. G2 Crowd was interested in examining differences between HR and non-HR employees. One area of focus was on employees' response to important metrics to consider when evaluating the effectiveness of employee engagement programs. The findings are summarized in the following tables.

Source: Data extracted from "Employee Engagement," *G2 Crowd*, **bit.ly/2WEQkgk**

EMPLOYEE	PRESENTEEISM IS AN IMPORTANT METRIC		
	Yes	No	Total
HR	53	79	132
Non-HR	43	225	268
Total	96	304	400

EMPLOYEE	ABSENTEEISM IS AN IMPORTANT METRIC		
	Yes	No	Total
HR	54	78	132
Non-HR	72	196	268
Total	126	274	400

What is the probability that a randomly chosen employee
a. is an HR employee?
b. is an HR employee *or* indicates that absenteeism is an important metric to consider when evaluating the effectiveness of employee engagement programs?
c. does not indicate that presenteeism is an important metric to consider when evaluating the effectiveness of employee engagement programs *and* is a non-HR employee?
d. does not indicate that presenteeism is an important metric to consider when evaluating the effectiveness of employee engagement programs *or* is a non-HR employee?

e. Suppose the randomly chosen employee does indicate that presenteeism is an important metric to consider when evaluating the effectiveness of employee engagement programs. What is the probability that the employee is a non-HR employee?
f. Are "presenteeism is an important metric" and "employee" independent?
g. Is "absenteeism is an important metric" independent of "employee"?

4.38 To better understand the website builder market, Clutch surveyed individuals who had created a website using a do-it-yourself (DIY) website builder. Respondents, categorized by the type of website they built—business or personal—were asked to indicate the primary purpose for building their website. The following table summarizes the findings:

PRIMARY PURPOSE	TYPE OF WEBSITE		
	Business	Personal	Total
Online Business Presence	52	4	56
Online Sales	32	13	45
Creative Display	28	54	82
Informational Resources	9	24	33
Blog	8	52	60
Total	129	147	276

Source: Data extracted from "How Businesses Use DIY Web Builders: Clutch 2017 Survey," **bit.ly/2qQjXiq**.

If a website builder is selected at random, what is the probability that he or she
a. indicated creative display as the primary purpose for building his/her website?
b. indicated creative display *or* informational resources as the primary purpose for building his/her website?
c. is a business website builder *or* indicated online sales as the primary purpose for building his/her website?
d. is a business website builder *and* indicated online sales as the primary purpose for building his/her website?
e. Given that the website builder selected is a personal website builder, what is the probability that he/she indicated online business presence as the primary purpose for building his/her website?

4.39 The CMO Survey collects and disseminates the opinions of top marketers in order to predict the future of markets, track marketing excellence, and improve the value of marketing in firms and in society. Part of the survey is devoted to the topic of marketing analytics and understanding what factors prevent companies from using more marketing analytics. The following findings are based on responses from 272 senior marketers within B2B firms and 114 senior marketers within B2C firms.

Source: Data extracted from "Results by Firm & Industry Characteristics," *The CMO Survey*, February 2017, p. 148. **bit.ly/2qY3Qvk**.

FIRM	LACK OF PROCESS/TOOLS TO MEASURE SUCCESS		
	Yes	No	Total
B2B	90	182	272
B2C	35	79	114
Total	125	261	386

	LACK OF PEOPLE WHO CAN LINK TO PRACTICE		
FIRM	**Yes**	**No**	**Total**
B2B	75	197	272
B2C	36	78	114
Total	111	275	386

a. What is the probability that a randomly selected senior marketer indicates that lack of process/tools to measure success through analytics is a factor that prevents his/her company from using more marketing analytics?

b. Given that a randomly selected senior marketer is within a B2B firm, what is the probability that the senior marketer indicates that lack of process/tools to measure success through analytics is a factor that prevents his/her company from using more marketing analytics?

c. Given that a randomly selected senior marketer is within a B2C firm, what is the probability that the senior marketer indicates that lack of process/tools to measure success through analytics is a factor that prevents his/her company from using more marketing analytics?

d. What is the probability that a randomly selected senior marketer indicates that lack of people who can link to marketing practice is a factor that prevents his/her company from using more marketing analytics?

e. Given that a randomly selected senior marketer is within a B2B firm, what is the probability that the senior marketer indicates that lack of people who can link to marketing practice is a factor that prevents his/her company from using more marketing analytics?

f. Given that a randomly selected senior marketer is within a B2C firm, what is the probability that the senior marketer indicates that lack of people who can link to marketing practice is a factor that prevents his/her company from using more marketing analytics?

g. Comment on the results in (a) through (f).

CHAPTER 4

▾CASES

Digital Case

Apply your knowledge about contingency tables and the proper application of simple and joint probabilities in this continuing Digital Case from Chapter 3.

Open **EndRunGuide.pdf**, the EndRun Financial Services "Guide to Investing," and read the information about the Guaranteed Investment Package (GIP). Read the claims and examine the supporting data. Then answer the following questions:

How accurate is the claim of the probability of success for EndRun's GIP? In what ways is the claim misleading? How would you calculate and state the probability of having an annual rate of return not less than 15%?

1. Using the table found under the "Show Me the Winning Probabilities" subhead, calculate the proper probabilities for the group of investors. What mistake was made in reporting the 7% probability claim?

2. Are there any probability calculations that would be appropriate for rating an investment service? Why or why not?

CardioGood Fitness

1. For each CardioGood Fitness treadmill product line (see **CardioGood Fitness**), construct two-way contingency tables of gender, education in years, relationship status, and self-rated fitness. (There will be a total of six tables for each treadmill product.)

2. For each table you construct, compute all conditional and marginal probabilities.

3. Write a report detailing your findings to be presented to the management of CardioGood Fitness.

The Choice *Is* Yours Follow-Up

1. Follow up the "Using Statistics: 'The Choice *Is* Yours,' Revisited" on page 79 by constructing contingency tables of market cap and type, market cap and risk, market cap and rating, type and risk, type and rating, and risk and rating for the sample of 479 retirement funds stored in **Retirement Funds** .

2. For each table you construct, compute all conditional and marginal probabilities.

3. Write a report summarizing your conclusions.

Clear Mountain State Student Survey

The Student News Service at Clear Mountain State University (CMSU) has decided to gather data about the undergraduate students who attend CMSU. CMSU creates and distributes a survey of 14 questions (see **CMUndergradSurvey.pdf**) and receives responses from 111 undergraduates (stored in **Student Survey**).

For these data, construct contingency tables of gender and major, gender and graduate school intention, gender and employment status, gender and computer preference, class and graduate school intention, class and employment status, major and graduate school intention, major and employment status, and major and computer preference.

1. For each of these contingency tables, compute all the conditional and marginal probabilities.

2. Write a report summarizing your conclusions.

EXCEL GUIDE

EG4.1 BASIC PROBABILITY CONCEPTS

Simple Probability, Joint Probability, and the General Addition Rule

Key Technique Use Excel arithmetic formulas.

Example Compute simple and joint probabilities for purchase behavior data in Table 4.1 on page 155.

PHStat Use **Simple & Joint Probabilities**.

For the example, select **PHStat → Probability & Prob. Distributions → Simple & Joint Probabilities**. In the new template, similar to the worksheet shown below, fill in the **Sample Space** area with the data.

Workbook Use the **COMPUTE worksheet** of the **Probabilities workbook** as a template.

The worksheet (shown below) already contains the Table 4.1 purchase behavior data. For other problems, change the sample space table entries in the cell ranges **C3:D4** and **A5:D6**.

As you change the event names in cells, B5, B6, C5, and C6, the column A row labels for simple and joint probabilities and the addition rule change as well. These column A labels are *formulas* that use the concatenation operator (&) to form row labels from the event names you enter.

For example, the cell A10 formula **="P("& B5 &")"** combines the two characters **P(** with the **Yes** B5 cell value and the character **)** to form the label **P(Yes)**. To examine all of the COMPUTE worksheet formulas shown below, open to the COMPUTE_FORMULAS worksheet.

	A	B	C	D	E
1	Probabilities				
2					
3	Sample Space		ACTUALLY PURCHASED		
4			Yes	No	Totals
5	PLANNED TO PURCHASE	Yes	200	50	250
6		No	100	650	750
7		Totals	300	700	1000
8					
9	Simple Probabilities				
10	P(Yes)	0.25	=E5/E7		
11	P(No)	0.75	=E6/E7		
12	P(Yes)	0.30	=C7/E7		
13	P(No)	0.70	=D7/E7		
14					
15	Joint Probabilities				
16	P(Yes and Yes)	0.20	=C5/E7		
17	P(Yes and No)	0.05	=D5/E7		
18	P(No and Yes)	0.10	=C6/E7		
19	P(No and No)	0.65	=D6/E7		
20					
21	Addition Rule				
22	P(Yes or Yes)	0.35	=B10 + B12 - B16		
23	P(Yes or No)	0.90	=B10 + B13 - B17		
24	P(No or Yes)	0.95	=B11 + B12 - B18		
25	P(No or No)	0.80	=B11 + B13 - B19		

EG4.4 BAYES' THEOREM

Key Technique Use Excel arithmetic formulas.

Example Apply Bayes' theorem to the television marketing example that the Bayesian Analysis online topic discusses.

Workbook Use the **COMPUTE worksheet** of the **Bayes workbook** as a template.

The worksheet (shown below) already contains the probabilities for the online section example. For other problems, change those probabilities in the cell range **B5:C6**.

	A	B	C	D	E
1	Bayes' Theorem Computations				
2					
3			Probabilities		
4	Event	Prior	Conditional	Joint	Revised
5	S	0.4	0.8	0.32	0.64
6	S'	0.6	0.3	0.18	0.36
7			Total:	0.5	

Joint	Revised
=B5 * C5	=D5/$D$7
=B6 * C6	=D6/$D$7
=D5 + D6	

Open to the **COMPUTE_FORMULAS worksheet** to examine the arithmetic formulas that compute the probabilities, which are also shown as an inset to the worksheet.

5

Discrete Probability Distributions

CONTENTS

OBJECTIVES

- Learn the properties of a
  probability distribution
- Calculate the expected
  value and variance of a
  probability distribution
- Calculate probabilities
  from the binomial and
  Poisson distributions
- Use the binomial and
  Poisson distributions to
  solve business problems

▼USING STATISTICS
Events of Interest at Ricknel Home Centers

L ike most other large businesses, Ricknel Home Centers, LLC, a regional home improvement chain, uses an accounting information system (AIS) to manage its accounting and
financial data. The Ricknel AIS collects, organizes, stores, analyzes, and distributes
financial information to decision makers both inside and outside the firm.

One important function of the Ricknel AIS is to continuously audit accounting information,
looking for errors or incomplete or improbable information. For example, when customers submit orders online, the Ricknel AIS scans orders looking to see which orders have possible mistakes. The system tags those orders and includes them in a daily *exceptions report*. Recent data
collected by the company show that the likelihood is 0.10 that an order form will be tagged.

As a member of the AIS team, you have been asked by Ricknel management to determine
the likelihood of finding a certain number of tagged forms in a sample of a specific size. For
example, what would be the likelihood that none of the order forms are tagged in a sample of
four forms? That one of the order forms is tagged?

How could you determine the solution to this type of probability problem?

Tthis chapter introduces the concept of a probability distribution. Section 1.1 identifies numerical variables as either having *discrete* integer values that represent a count of something, or *continuous* values that arise from a measuring process. This chapter discusses the binomial and Poisson distributions, two probability distributions that represent a discrete numerical variable. In the Ricknel Home Centers scenario, a *probability distribution* could be used as a model that approximates the order process. By using such an approximation, one could make inferences about the actual order process including the likelihood of finding a certain number of tagged forms in a sample.

5.1 The Probability Distribution for a Discrete Variable

A **probability distribution for a discrete variable** is a mutually exclusive list of all the possible numerical outcomes along with the probability of occurrence of each outcome. For example, Table 5.1 gives the distribution of the number of interruptions per day in a large computer network. The list in Table 5.1 is collectively exhaustive because all possible outcomes are included. Thus, the probabilities sum to 1. Figure 5.1 is a graphical representation of Table 5.1.

TABLE 5.1
Probability distribution of the number of interruptions per day

Interruptions per Day	Probability
0	0.35
1	0.25
2	0.20
3	0.10
4	0.05
5	0.05

FIGURE 5.1
Probability distribution of the number of interruptions per day

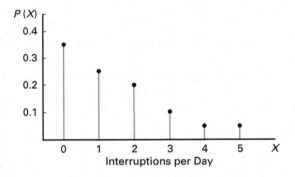

Expected Value of a Discrete Variable

The **expected value** of a discrete variable is the population mean, μ, of its probability distribution. To calculate the expected value, multiply each possible outcome, x_i, by its corresponding probability, $P(X = x_i)$, and then sum these products.

EXPECTED VALUE, μ, OF A DISCRETE VARIABLE

$$\mu = E(X) = \sum_{i=1}^{N} x_i P(X = x_i) \tag{5.1}$$

where

$$x_i = \text{the } i\text{th value of the discrete variable } X$$
$$P(X = x_i) = \text{probability of occurrence of the ith value of } X$$
$$N = \text{number of values of the discrete variable } X$$

For the Table 5.1 probability distribution of the number of interruptions per day in a large computer network, Table 5.2 shows all intermediate steps for calculating the expected value using Equation (5.1).

$$\mu = E(X) = \sum_{i=1}^{N} x_i P(X = x_i)$$

$$= 0 + 0.25 + 0.40 + 0.30 + 0.20 + 0.25$$

$$= 1.40$$

TABLE 5.2
Computing the expected value of the number of interruptions per day

Interruptions per Day (x_i)	$P(X = x_i)$	$x_i P(X = x_i)$
0	0.35	$(0)(0.35) = 0.00$
1	0.25	$(1)(0.25) = 0.25$
2	0.20	$(2)(0.20) = 0.40$
3	0.10	$(3)(0.10) = 0.30$
4	0.05	$(4)(0.05) = 0.20$
5	0.05	$(5)(0.05) = 0.25$
	1.00	$\mu = E(X) = 1.40$

The expected value is 1.40. The expected value of 1.40 interruptions per day represents the *mean* number of interruptions per day even though one cannot have a *fractional* number of interruptions, such as 1.4, on a daily basis.

Variance and Standard Deviation of a Discrete Variable

Compute the variance of a probability distribution by multiplying each possible squared difference $[x_i - E(X)]^2$ by its corresponding probability, $P(X = x_i)$, and then summing the resulting products. Equation (5.2) defines the **variance of a discrete variable**, and Equation (5.3) defines the **standard deviation of a discrete variable**.

VARIANCE OF A DISCRETE VARIABLE

$$\sigma^2 = \sum_{i=1}^{N} [x_i - E(X)]^2 P(X = x_i) \tag{5.2}$$

where

$$x_i = \text{the } i\text{th value of the discrete variable } X$$
$$P(X = x_i) = \text{probability of occurrence of the } i\text{th value of } X$$
$$N = \text{number of values of the discrete variable } X$$

STANDARD DEVIATION OF A DISCRETE VARIABLE

$$\sigma = \sqrt{\sigma^2} = \sqrt{\sum_{i=1}^{N} [x_i - E(X)]^2 P(X = x_i)} \tag{5.3}$$

Table 5.3 shows all intermediate steps for calculating the variance and the standard deviation of the number of interruptions per day using Equations (5.2) and (5.3).

$$\sigma^2 = \sum_{i=1}^{N} [x_i - E(X)]^2 P(X = x_i)$$

$$= 0.686 + 0.040 + 0.072 + 0.256 + 0.338 + 0.648$$

$$= 2.04$$

and

$$\sigma = \sqrt{\sigma^2} = \sqrt{2.04} = 1.4283$$

TABLE 5.3
Calculating the variance and standard deviation of the number of interruptions per day

Interruptions per Day (x_i)	$P(X = x_i)$	$x_i P(X = x_i)$	$[x_i - E(X)]^2$	$[x_i - E(X)]^2 P(X = x_i)$
0	0.35	0.00	$(0 - 1.4)^2 = 1.96$	$(1.96)(0.35) = 0.686$
1	0.25	0.25	$(1 - 1.4)^2 = 0.16$	$(0.16)(0.25) = 0.040$
2	0.20	0.40	$(2 - 1.4)^2 = 0.36$	$(0.36)(0.20) = 0.072$
3	0.10	0.30	$(3 - 1.4)^2 = 2.56$	$(2.56)(0.10) = 0.256$
4	0.05	0.20	$(4 - 1.4)^2 = 6.76$	$(6.76)(0.05) = 0.338$
5	0.05	0.25	$(5 - 1.4)^2 = 12.96$	$(12.96)(0.05) = 0.648$
	1.00	$\mu = E(X) = 1.40$		$\sigma^2 = 2.04$
				$\sigma = \sqrt{\sigma^2} = 1.4283$

Thus, the mean number of interruptions per day is 1.4, the variance is 2.04, and the standard deviation is approximately 1.43 interruptions per day.

PROBLEMS FOR SECTION 5.1

LEARNING THE BASICS

5.1 Given the following probability distributions:

Distribution A		Distribution B	
x_i	$P(X = x_i)$	x_i	$P(X = x_i)$
0	0.50	0	0.05
1	0.20	1	0.10
2	0.15	2	0.15
3	0.10	3	0.20
4	0.05	4	0.50

a. Calculate the expected value for each distribution.
b. Calculate the standard deviation for each distribution.
c. What is the probability that x will be at least 3 in Distribution A and Distribution B?
d. Compare the results of distributions A and B.

APPLYING THE CONCEPTS

 SELF TEST
5.2 The following table contains the probability distribution for the number of traffic accidents per day in a small town:

Number of Accidents Daily (X)	$P(X = x_i)$
0	0.10
1	0.20
2	0.45
3	0.15
4	0.05
5	0.05

a. Calculate the mean number of accidents per day.
b. Calculate the standard deviation.
c. What is the probability that there will be at least two accidents on a given day?

5.3 Recently, a regional automobile dealership sent out fliers to prospective customers indicating that they had already won one of three different prizes: an automobile valued at $25,000, a $100 gas card, or a $5 Walmart shopping card. To claim his or her prize, a prospective customer needed to present the flier at the dealership's showroom. The fine print on the back of the flier listed the probabilities of winning. The chance of winning the car was 1 out of 31,478, the chance of winning the gas card was 1 out of 31,478, and the chance of winning the shopping card was 31,476 out of 31,478.

a. How many fliers do you think the automobile dealership sent out?
b. Using your answer to (a) and the probabilities listed on the flier, what is the expected value of the prize won by a prospective customer receiving a flier?
c. Using your answer to (a) and the probabilities listed on the flier, what is the standard deviation of the value of the prize won by a prospective customer receiving a flier?
d. Do you think this is an effective promotion? Why or why not?

5.4 In the carnival game Under-or-Over-Seven, a pair of fair dice is rolled once, and the resulting sum determines whether the player wins or loses his or her bet. For example, the player can bet $1 that the sum will be under 7—that is, 2, 3, 4, 5, or 6. For this bet, the player wins $1 if the result is under 7 and loses $1 if the outcome equals or is greater than 7. Similarly, the player can bet $1 that the sum will be over 7—that is, 8, 9, 10, 11, or 12. Here, the player wins $1 if the result is over 7 but loses $1 if the result is 7 or under. A third method of play is to bet $1 on the outcome 7. For this bet, the player wins $4 if the result of the roll is 7 and loses $1 otherwise.
a. Construct the probability distribution representing the different outcomes that are possible for a $1 bet on under 7.
b. Construct the probability distribution representing the different outcomes that are possible for a $1 bet on over 7.
c. Construct the probability distribution representing the different outcomes that are possible for a $1 bet on 7.
d. Show that the expected long-run profit (or loss) to the player is the same, no matter which method of play is used.

5.5 The number of arrivals per minute at a bank located in the central business district of a large city was recorded over a period of 200 minutes, with the following results:

Arrivals	Frequency
0	14
1	31
2	47
3	41
4	29
5	21
6	10
7	5
8	2

a. Calculate the expected number of arrivals per minute.
b. Calculate the standard deviation.
c. What is the probability that there will be fewer than two arrivals in a given minute?

5.6 The manager of the commercial mortgage department of a large bank has collected data during the past two years concerning the number of commercial mortgages approved per week. The results from these two years (104 weeks) are as follows:

Number of Commercial Mortgages Approved	Frequency
0	13
1	25
2	32
3	17
4	9
5	6
6	1
7	1

a. Calculate the expected number of mortgages approved per week.
b. Calculate the standard deviation.
c. What is the probability that there will be more than one commercial mortgage approved in a given week?

5.7 You are trying to develop a strategy for investing in two different stocks. The anticipated annual return for a $1,000 investment in each stock under four different economic conditions has the following probability distribution:

		Returns	
Probability	Economic Condition	Stock *X*	Stock *Y*
0.1	Recession	−50	−100
0.3	Slow growth	20	50
0.4	Moderate growth	100	130
0.2	Fast growth	150	200

Calculate the
a. expected return for stock *X* and for stock *Y*.
b. standard deviation for stock *X* and for stock *Y*.
c. Would you invest in stock *X* or stock *Y*? Explain.

5.8 You plan to invest $1,000 in a corporate bond fund or in a common stock fund. The following table presents the annual return (per $1,000) of each of these investments under various economic conditions and the probability that each of those economic conditions will occur.

Probability	Economic Condition	Corporate Bond Fund	Common Stock Fund
0.01	Extreme recession	−300	−999
0.09	Recession	−70	−300
0.15	Stagnation	30	−100
0.35	Slow growth	60	100
0.30	Moderate growth	100	150
0.10	High growth	120	350

Calculate the
a. expected return for the corporate bond fund and for the common stock fund.
b. standard deviation for the corporate bond fund and for the common stock fund.
c. Would you invest in the corporate bond fund or the common stock fund? Explain.
d. If you chose to invest in the common stock fund in (c), what do you think about the possibility of losing $999 of every $1,000 invested if there is an extreme recession?

5.2 Binomial Distribution

In some cases, a mathematical expression or **model** can be used to calculate the probability of a value, or outcome, for a variable of interest. For discrete variables, such mathematical models are also known as **probability distribution functions**. One such function that can be used in many business situations is the **binomial distribution**. Exhibit 5.1 presents the important properties of this distribution.

student TIP

Do not confuse this use of the Greek letter pi, π, to represent the probability of an event of interest with the constant that is the ratio of the circumference to a diameter of a circle— approximately 3.14159.

EXHIBIT 5.1

Properties of the Binomial Distribution

- The sample consists of a fixed number of observations, n.
- Each observation is classified into one of two mutually exclusive and collectively exhaustive categories.
- The probability of an observation being classified as the event of interest, π, is constant from observation to observation. Thus, the probability of an observation being classified as not being the event of interest, $1 - \pi$, is constant over all observations.
- The value of any observation is independent of the value of any other observation.

Use the binomial distribution when the discrete variable is the number of events of interest in a sample of n observations. For example, in the Ricknel Home Improvement scenario suppose the event of interest is a tagged order form and one seeks to determine the number of tagged order forms in a given sample of orders.

What results can occur? If the sample contains four orders, none, one, two, three, or four order forms could be tagged. No other value can occur because the number of tagged order forms cannot be more than the sample size, n, and cannot be less than zero. Therefore, the range of the binomial variable is from 0 to n.

Consider this sample of four orders:

First Order	Second Order	Third Order	Fourth Order
Tagged	Tagged	Not tagged	Tagged

What is the probability of having three tagged order forms in a sample of four orders in this particular sequence? Because the historical probability of a tagged order is 0.10, the probability that each order occurs in the sequence is

First Order	Second Order	Third Order	Fourth Order
$\pi = 0.10$	$\pi = 0.10$	$1 - \pi = 0.90$	$\pi = 0.10$

Each outcome is independent of the others because the order forms were selected from an extremely large or practically infinite population and each order form could only be selected once. Therefore, the probability of having this particular sequence is

$$\pi\pi(1 - \pi)\pi = \pi^3(1 - \pi)^1$$

$$= (0.10)^3(0.90)^1 = (0.10)(0.10)(0.10)(0.90)$$

$$= 0.0009$$

This result indicates only the probability of three tagged order forms (events of interest) from a sample of four order forms in a *specific sequence*. To find the number of ways of selecting x objects from n objects, *irrespective of sequence*, use the **rule of combinations** given in Equation (5.4) on page 182.

COMBINATIONS

The number of combinations of selecting x objects[1] out of n objects is given by

$$_nC_x = \frac{n!}{x!(n-x)!}$$

(5.4)

where

$n! = (n)(n-1) \cdots (1)$ is called n factorial. By definition, $0! = 1$.

With $n = 4$ and $x = 3$, there are four such sequences because

$$_nC_x = \frac{n!}{x!(n-x)!} = \frac{4!}{3!(4-3)!} = \frac{4 \times 3 \times 2 \times 1}{(3 \times 2 \times 1)(1)} = 4$$

The four possible sequences are

    Sequence 1 = (*tagged, tagged, tagged, not tagged*), with probability
$$\pi\pi\pi(1-\pi) = \pi^3(1-\pi)^1 = 0.0009$$
    Sequence 2 = (*tagged, tagged, not tagged, tagged*), with probability
$$\pi\pi(1-\pi)\pi = \pi^3(1-\pi)^1 = 0.0009$$
    Sequence 3 = (*tagged, not tagged, tagged, tagged*), with probability
$$\pi(1-\pi)\pi\pi = \pi^3(1-\pi)^1 = 0.0009$$
    Sequence 4 = (*not tagged, tagged, tagged, tagged*), with probability
$$(1-\pi)\pi\pi\pi = \pi^3(1-\pi)^1 = 0.0009$$

Therefore, the probability of three tagged order forms is equal to 0.0036 (4×0.0009), or more generally, the number of possible sequences $\times$ the probability of a particular sequence.

This straightforward approach can be used for the other possible values of the variable—zero, one, two, and four tagged order forms. However, as the sample size, n, increases, using this straightforward approach becomes increasingly time-consuming. Equation (5.5) provides a general formula for computing any probability from the binomial distribution with the number of events of interest, x, given n and π.

BINOMIAL DISTRIBUTION

$$P(X = x \mid n, \pi) = \frac{n!}{x!(n-x)!} \pi^x (1-\pi)^{n-x}$$

(5.5)

where

$P(X = x \mid n, \pi)$ = probability that $X = x$ events of interest, given n and π

n = number of observations

π = probability of an event of interest

$1 - \pi$ = probability of not having an event of interest

x = number of events of interest in the sample ($X = 0, 1, 2, \ldots, n$)

$\dfrac{n!}{x!(n-x)!}$ = number of combinations of x events of interest out of n observations

Equation (5.5) restates what was intuitively derived previously. The binomial variable X can have any integer value x from 0 through n. In Equation (5.5), the product

$$\pi^x(1-\pi)^{n-x}$$

represents the probability of exactly x events of interest from n observations in a *particular sequence*.

The term

$$\frac{n!}{x!(n-x)!}$$

is the number of *combinations* of the x events of interest from the n observations possible. Hence, given the number of observations, n, and the probability of an event of interest, π, the probability of x events of interest is

$$P(X = x \mid n, \pi) = \text{(number of combinations)} \times \text{(probability of a particular combination)}$$

$$= \frac{n!}{x!(n-x)!}\pi^x(1-\pi)^{n-x}$$

Example 5.1 illustrates the use of Equation (5.5). Examples 5.2 and 5.3 show the computations for other values of X.

EXAMPLE 5.1

Determining $P(X = 3)$, Given $n = 4$ and $\pi = 0.1$

If the likelihood of a tagged order form is 0.1, what is the probability that there are three tagged order forms in the sample of four?

SOLUTION Using Equation (5.5), the probability of three tagged orders from a sample of four is

$$P(X = 3 \mid n = 4, \pi = 0.1) = \frac{4!}{3!(4-3)!}(0.1)^3(1-0.1)^{4-3}$$

$$= \frac{4!}{3!(1)!}(0.1)^3(0.9)^1$$

$$= 4(0.1)(0.1)(0.1)(0.9) = 0.0036$$

EXAMPLE 5.2

Determining $P(X \geq 3)$, Given $n = 4$ and $\pi = 0.1$

studentTIP

Another way of saying "three or more" is "at least three."

If the likelihood of a tagged order form is 0.1, what is the probability that there are three or more tagged order forms in the sample of four?

SOLUTION In Example 5.1, you found that the probability of *exactly* three tagged order forms from a sample of four is 0.0036. To compute the probability of *at least* three tagged order forms, you need to add the probability of three tagged order forms to the probability of four tagged order forms. The probability of four tagged order forms is

$$P(X = 4 \mid n = 4, \pi = 0.1) = \frac{4!}{4!(4-4)!}(0.1)^4(1-0.1)^{4-4}$$

$$= 1(0.1)(0.1)(0.1)(0.1)(1) = 0.0001$$

Thus, the probability of at least three tagged order forms is

$$P(X \geq 3) = P(X = 3) + P(X = 4)$$

$$= 0.0036 + 0.0001 = 0.0037$$

There is a 0.37% chance that there will be at least three tagged order forms in a sample of four.

EXAMPLE 5.3

Determining $P(X < 3)$, Given $n = 4$ and $\pi = 0.1$

If the likelihood of a tagged order form is 0.1, what is the probability that there are less than three tagged order forms in the sample of four?

SOLUTION The probability that there are less than three tagged order forms is

$$P(X < 3) = P(X = 0) + P(X = 1) + P(X = 2)$$

▶(*continued*)

Using Equation (5.5) on page 182, these probabilities are

$$P(X = 0 | n = 4, \pi = 0.1) = \frac{4!}{0!(4-0)!}(0.1)^0(1-0.1)^{4-0} = 0.6561$$

$$P(X = 1 | n = 4, \pi = 0.1) = \frac{4!}{1!(4-1)!}(0.1)^1(1-0.1)^{4-1} = 0.2916$$

$$P(X = 2 | n = 4, \pi = 0.1) = \frac{4!}{2!(4-2)!}(0.1)^2(1-0.1)^{4-2} = 0.0486$$

Therefore, $P(X < 3) = 0.6561 + 0.2916 + 0.0486 = 0.9963$. $P(X < 3)$ could also be calculated from its complement, $P(X \geq 3)$, as follows:

$$P(X < 3) = 1 - P(X \geq 3)$$
$$= 1 - 0.0037 = 0.9963$$

Excel can automate binomial probability calculations, which become tedious as n gets large. Figure 5.2 contains the computed binomial probabilities for $n = 4$ and $\pi = 0.1$.

FIGURE 5.2

Excel worksheet results for computing binomial probabilities with $n = 4$ and $\pi = 0.1$

	A	B	
1	Binomial Probabilities		
2			
3	Data		
4	Sample size	4	
5	Probability of an event of interest	0.1	
6			
7	Parameters		
8	Mean	0.4	=B4 * B5
9	Variance	0.36	=B8 * (1 - B5)
10	Standard deviation	0.6	=SQRT(B9)
11			
12	Binomial Probabilities Table		
13		X	P(X)
14		0	0.6561	=BINOM.DIST(A14, $B$4, $B$5, FALSE)
15		1	0.2916	=BINOM.DIST(A15, $B$4, $B$5, FALSE)
16		2	0.0486	=BINOM.DIST(A16, $B$4, $B$5, FALSE)
17		3	0.0036	=BINOM.DIST(A17, $B$4, $B$5, FALSE)
18		4	0.0001	=BINOM.DIST(A18, $B$4, $B$5, FALSE)

In the Worksheet

The worksheet uses short arithmetic formulas to compute parameters and the BINOM.DIST function to compute probabilities.

The **COMPUTE worksheet** of the Excel Guide **Binomial workbook** is a copy of the Figure 5.2 worksheet.

Histograms for Discrete Variables

Discrete histograms visualize binomial distributions. Figure 5.3 visualizes the binomial probabilities for Example 5.3. Unlike histograms for continuous variables that Section 2.4 discusses, the bars for the values in a discrete histogram are very thin, and there is a large gap between each pair of bars. Ideally, discrete histogram bars would have no width, and some programs suggest that lack of width by graphing vertical lines ("needles") in lieu of solid bars.

FIGURE 5.3

Histogram of the binomial probability with $n = 4$ and $\pi = 0.1$

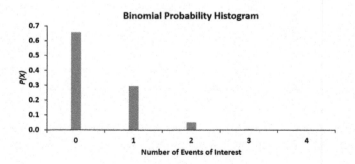

For a binomial probability distribution, the shape depends on the values of π and n. Whenever $\pi = 0.5$, the binomial distribution is symmetrical, regardless of how large or small the value of n. When $\pi \neq 0.5$, both π and n affect the skewness of the distribution.

Figure 5.4 illustrates the effect of π on a binomial distribution. Holding the sample size constant, low values for π, such as 0.2, cause the binomial distribution to be right-skewed (left histogram), while high values, such as 0.8, cause the distribution to be left-skewed (right histogram). Figure 5.5 illustrates that increasing n makes a binomial distribution more symmetrical when π does not equal 0.5. Generally, the closer π is to 0.5 or the larger the number of observations, n, the less skewed the binomial distribution will be.

FIGURE 5.4

Effect of π on the binomial distribution, holding n constant

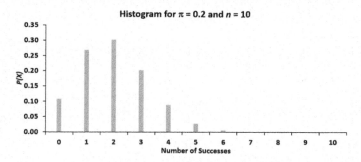
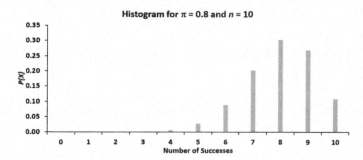

FIGURE 5.5

Effect of n on the binomial distribution, holding π constant

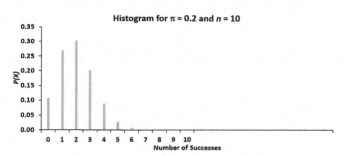
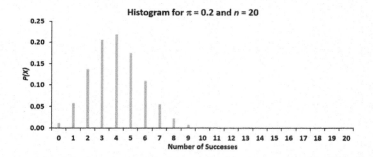

Summary Measures for the Binomial Distribution

The mean, μ, of the binomial distribution is equal to the sample size, n, multiplied by the probability of an event of interest, π. Therefore, Equation (5.6) can be used in lieu of Equation (5.1) to calculate the mean for variables that follow the binomial distribution.

MEAN OF THE BINOMIAL DISTRIBUTION

$$\mu = E(X) = n\pi \tag{5.6}$$

For the tagged orders example, on average and over the long run, one can theoretically expect $\mu = E(X) = n\pi = (4)(0.1) = 0.4$ tagged order forms in a sample of four orders. Equation 5.7 calculates the standard deviation of the binomial distribution.

STANDARD DEVIATION OF THE BINOMIAL DISTRIBUTION

$$\sigma = \sqrt{\sigma^2} = \sqrt{Var(X)} = \sqrt{n\pi(1 - \pi)} \tag{5.7}$$

The standard deviation of the number of tagged order forms is

$$\sigma = \sqrt{4(0.1)(0.9)} = 0.60$$

Using Equation (5.3) on page 178 produces the same result.

Example 5.4 applies the binomial distribution to service at a fast-food restaurant.

EXAMPLE 5.4

Computing Binomial Probabilities for Service at a Fast-Food Restaurant

Accuracy in taking orders at a drive-through window is important for fast-food chains. Periodically, *QSR Magazine* publishes "The Drive-Thru Performance Study: Order Accuracy," which measures the percentage of orders that are filled correctly. In a recent month, the percentage of orders filled correctly at Wendy's was approximately 89.1%.

Suppose that you go to the drive-through window at Wendy's and place an order. Two friends of yours independently place orders at the drive-through window at the same Wendy's. What are the probabilities that all three, that none of the three, and that at least two of the three orders will be filled correctly? What are the mean and standard deviation of the binomial distribution for the number of orders filled correctly?

SOLUTION Because there are three orders and the probability of a correct order is 0.891, $n = 3$, and $\pi = 0.891$, using Equation (5.5) on page 182,

$$P(X = 3 \mid n = 3, \pi = 0.891) = \frac{3!}{3!(3-3)!}(0.891)^3(1-0.891)^{3-3}$$

$$= 1(0.891)(0.891)(0.891)(1) = 0.7073$$

$$P(X = 0 \mid n = 3, \pi = 0.891) = \frac{3!}{0!(3-0)!}(0.891)^0(1-0.891)^{3-0}$$

$$= 1(1)(0.109)(0.109)(0.109) = 0.0013$$

$$P(X = 2 \mid n = 3, \pi = 0.891) = \frac{3!}{2!(3-2)!}(0.891)^2(1-0.891)^{3-2}$$

$$= 3(0.891)(0.891)(0.109) = 0.2596$$

$$P(X \geq 2) = P(X = 2) + P(X = 3)$$

$$= 0.2596 + 0.7073$$

$$= 0.9669$$

Using Equations (5.6) and (5.7),

$$\mu = E(X) = n\pi = 3(0.891) = 2.673$$

$$\sigma = \sqrt{\sigma^2} = \sqrt{Var(X)} = \sqrt{n\pi(1-\pi)}$$

$$= \sqrt{3(0.891)(0.109)}$$

$$= \sqrt{0.2914} = 0.5398$$

The mean number of orders filled correctly in a sample of three orders is 2.673, and the standard deviation is 0.5398. The probability that all three orders are filled correctly is 0.7073, or 70.73%. The probability that none of the orders are filled correctly is 0.0013 (0.13%). The probability that at least two orders are filled correctly is 0.9669 (96.69%).

Figure 5.6 on page 187 shows the Example 5.4 Excel results. The answer to the third question can be found in the last column of the "Binomial Probabilities Table" in the $X = 2$ row.

▶(*continued*)

FIGURE 5.6
Excel results for
computing the
binomial probability
for Example 5.4

	A	B	C	D	E	F
1	Probability of Correct Order at Wendy's					
2						
3	Data					
4	Sample size	3				
5	Probability of an event of interest	0.891				
6						
7	Parameters					
8	Mean	2.673				
9	Variance	0.2914				
10	Standard deviation	0.5398				
11						
12	Binomial Probabilities Table					
13		X	P(X)	P(<=X)	P(<X)	P(>X)	P(>=X)
14		0	0.0013	0.0013	0.0000	0.9987	1.0000
15		1	0.0318	0.0331	0.0013	0.9669	0.9987
16		2	0.2596	0.2927	0.0331	0.7073	0.9669
17		3	0.7073	1.0000	0.2927	0.0000	0.7073

In the Worksheet

The worksheet uses short arithmetic
formulas to compute parameters and the
probabilities in columns D through F.
The worksheet uses the BINOM.DIST
function to compute probabilities in
columns B and C.

The **CUMULATIVE worksheet** of the
Excel Guide **Binomial workbook** is a
copy of the Figure 5.6 worksheet.

PROBLEMS FOR SECTION 5.2

LEARNING THE BASICS

5.9 Determine the following:
a. For $n = 4$ and $\pi = 0.12$, what is $P(X = 0)$?
b. For $n = 10$ and $\pi = 0.40$, what is $P(X = 9)$?
c. For $n = 10$ and $\pi = 0.50$, what is $P(X = 8)$?
d. For $n = 6$ and $\pi = 0.83$, what is $P(X = 5)$?

5.10 Determine the mean and standard deviation of the variable X
in each of the following binomial distributions:
a. $n = 4$ and $\pi = 0.10$
b. $n = 4$ and $\pi = 0.40$
c. $n = 5$ and $\pi = 0.80$
d. $n = 3$ and $\pi = 0.50$

APPLYING THE CONCEPTS

5.11 The increase or decrease in the price of a stock between the
beginning and the end of a trading day is assumed to be an equally
likely random event. What is the probability that a stock will show
an increase in its closing price on five consecutive days?

5.12 In the last quarter of 2018, 47% of U.S. smartphone owners
owned an iPhone.
Source: Data extracted from **pewrsr.ch/2riDGV6**.

Using the binomial distribution, what is the probability that of the
next six Americans who own a smartphone surveyed,
a. four will own an iPhone?
b. all six will own an iPhone?
c. at least four will own an iPhone?
d. What are the mean and standard deviation of the number of American
smartphone owners who own an iPhone in a survey of six?
e. What assumptions do you need to make in (a) through (c)?

5.13 A student is taking a multiple-choice exam in which each
question has four choices. Assume that the student has no knowledge of the correct answers to any of the questions. She has decided
on a strategy in which she will place four balls (marked *A, B, C,* and
D) into a box. She randomly selects one ball for each question and
replaces the ball in the box. The marking on the ball will determine
her answer to the question. There are five multiple-choice questions
on the exam. What is the probability that she will get
a. five questions correct?
b. at least four questions correct?
c. no questions correct?
d. no more than two questions correct?

5.14 A manufacturing company regularly conducts quality control checks at specified periods on the products it manufactures.
Historically, the failure rate for LED light bulbs that the company
manufactures is 3%. Suppose a random sample of 10 LED light
bulbs is selected. What is the probability that
a. none of the LED light bulbs are defective?
b. exactly one of the LED light bulbs is defective?
c. two or fewer of the LED light bulbs are defective?
d. three or more of the LED light bulbs are defective?

5.15 Past records indicate that the probability of online retail orders that
turn out to be fraudulent is 0.08. Suppose that, on a given day, 20 online
retail orders are placed. Assume that the number of online retail orders
that turn out to be fraudulent is distributed as a binomial random variable.
a. What are the mean and standard deviation of the number of online retail orders that turn out to be fraudulent?
b. What is the probability that zero online retail orders will turn
out to be fraudulent?
c. What is the probability that one online retail order will turn out
to be fraudulent?
d. What is the probability that two or more online retail orders will
turn out to be fraudulent?

✓SELF TEST **5.16** In Example 5.4 that begins on page 186, you and
two friends decided to go to Wendy's. Now, suppose that
instead you go to Burger King, which recently filled approximately
90.9% of orders correctly. What is the probability that
a. all three orders will be filled correctly?
b. none of the three will be filled correctly?
c. at least two of the three will be filled correctly?
d. What are the mean and standard deviation of the binomial distribution used in (a) through (c)? Interpret these values.
e. Compare the results of (a) through (d) with those of Wendy's in
Example 5.4 and McDonald's in Problem 5.17.

5.17 In Example 5.4 that begins on page 186, you and two friends
decided to go to Wendy's. Now, suppose that instead you go to
McDonald's, which recently filled approximately 92.9% of the
orders correctly. What is the probability that
a. all three orders will be filled correctly?
b. none of the three will be filled correctly?
c. at least two of the three will be filled correctly?
d. What are the mean and standard deviation of the binomial
distribution used in (a) through (c)? Interpret these values.
e. Compare the results of (a) through (d) with those of Burger King
in Problem 5.16 and Wendy's in Example 5.4.

5.3 Poisson Distribution

Many studies are based on counts of the occurrences of a particular event in an **area of opportunity**, a fixed interval of time or space. In an area of opportunity, there can be more than one occurrence of the event of interest. The **Poisson distribution**, another discrete probability distribution, can be used to compute probabilities for such situations (McGinty). Specifically, the Poisson distribution applies if the following properties hold:

- The counting of the number of times a particular event occurs in a given area of opportunity. The area of opportunity is defined by time, length, surface area, and so forth.
- The probability that an event occurs in a given area of opportunity is the same for all the areas of opportunity.
- The number of events that occur in one area of opportunity is independent of the number of events that occur in any other area of opportunity.
- The probability that two or more events will occur in an area of opportunity approaches zero as the area of opportunity becomes smaller.

Examples of variables that follow the Poisson distribution are the surface defects on a new refrigerator, the number of network failures in a day, the number of people arriving at a bank, and the number of fleas on the body of a dog. Consider a study that seeks to examine the number of customers arriving during the lunch hour at a specific bank branch in the central business district of a large city. If the research problem is stated as the number of customers who arrive each minute, does this study match the four properties needed to use the Poisson distribution?

First, the *event* of interest is the arrival of a customer, and the *given area of opportunity* is defined as a one-minute interval. Will zero customers arrive, one customer arrive, two customers arrive, and so on? Second, a reasonable assumption is that the probability that a customer arrives during a particular one-minute interval is the same as the probability for all the other one-minute intervals. Third, the arrival of one customer in any one-minute interval has no effect on (is independent of) the arrival of any other customer in any other one-minute interval. Finally, the probability that two or more customers will arrive in a given time period approaches zero as the time interval becomes small. For example, the probability is virtually zero that two customers will arrive in a time interval of 0.01 second. Therefore, using the Poisson distribution to determine probabilities involving the number of customers arriving at the bank in a one-minute time interval during the lunch hour is appropriate.

The Poisson distribution has one parameter, called λ (the Greek lowercase letter *lambda*), which is the mean or expected number of events per unit. The variance of a Poisson distribution is also equal to λ, and the standard deviation is equal to $\sqrt{\lambda}$. The number of events, X, of the Poisson variable ranges from 0 to infinity (∞).

Equation (5.8) is the mathematical expression for the Poisson distribution for computing the probability of $X = x$ events, given that λ events are expected.

POISSON DISTRIBUTION

$$P(X = x \mid \lambda) = \frac{e^{-\lambda}\lambda^x}{x!} \tag{5.8}$$

where

$P(X = x \mid \lambda)$ = probability that $X = x$ events in an area of opportunity given λ
λ = expected number of events per unit
e = mathematical constant approximated by 2.71828
x = number of events ($x = 0, 1, 2, \ldots$)

To illustrate an application of the Poisson distribution, suppose that the mean number of customers who arrive per minute at the bank during the noon-to-1 P.M. hour is equal to 3.0. What is the probability that in a given minute, exactly two customers will arrive? And what is the probability that more than two customers will arrive in a given minute?

Using Equation (5.8) and $\lambda = 3$, the probability that in a given minute exactly two customers will arrive is

$$P(X = 2 \mid \lambda = 3) = \frac{e^{-3.0}(3.0)^2}{2!} = \frac{9}{(2.71828)^3(2)} = 0.2240$$

The probability that in any given minute more than two customers will arrive is

$$P(X > 2) = P(X = 3) + P(X = 4) + \cdots$$

Because in a probability distribution all the probabilities must sum to 1, the terms on the right side of the equation $P(X > 2)$ also represent the complement of the probability that X is less than or equal to 2 [i.e., $1 - P(X \le 2)$]. Thus,

$$P(X > 2) = 1 - P(X \le 2) = 1 - [P(X = 0) + P(X = 1) + P(X = 2)]$$

Now, using Equation (5.8),

$$P(X > 2) = 1 - \left[\frac{e^{-3.0}(3.0)^0}{0!} + \frac{e^{-3.0}(3.0)^1}{1!} + \frac{e^{-3.0}(3.0)^2}{2!} \right]$$

$$= 1 - [0.0498 + 0.1494 + 0.2240]$$

$$= 1 - 0.4232 = 0.5768$$

learnMORE

The **Poisson Table online** topic contains a table of Poisson probabilities and explains how to use the table to compute Poisson probabilities.

Thus, there is a 57.68% chance that more than two customers will arrive in the same minute.

Excel can automate Poisson probability calculations, which can be tedious. Figure 5.7 contains the computed Poisson probabilities for the bank customer arrival example.

FIGURE 5.7
Excel worksheet results for computing Poisson probabilities with $\lambda = 3$

	A	B	C	D	E
1	Poisson Probabilities				
2					
3		Data			
4	Mean/Expected number of events of interest:				3
5					
6	Poisson Probabilities Table				
7	X	P(X)			
8	0	0.0498	=POISSON.DIST(A8, $E$4, FALSE)		
9	1	0.1494	=POISSON.DIST(A9, $E$4, FALSE)		
10	2	0.2240	=POISSON.DIST(A10, $E$4, FALSE)		
11	3	0.2240	=POISSON.DIST(A11, $E$4, FALSE)		
12	4	0.1680	=POISSON.DIST(A12, $E$4, FALSE)		
13	5	0.1008	=POISSON.DIST(A13, $E$4, FALSE)		
14	6	0.0504	=POISSON.DIST(A14, $E$4, FALSE)		
15	7	0.0216	=POISSON.DIST(A15, $E$4, FALSE)		
16	8	0.0081	=POISSON.DIST(A16, $E$4, FALSE)		
17	9	0.0027	=POISSON.DIST(A17, $E$4, FALSE)		
18	10	0.0008	=POISSON.DIST(A18, $E$4, FALSE)		
19	11	0.0002	=POISSON.DIST(A19, $E$4, FALSE)		
20	12	0.0001	=POISSON.DIST(A20, $E$4, FALSE)		
21	13	0.0000	=POISSON.DIST(A21, $E$4, FALSE)		
22	14	0.0000	=POISSON.DIST(A22, $E$4, FALSE)		
23	15	0.0000	=POISSON.DIST(A23, $E$4, FALSE)		

In the Worksheet

The worksheet uses the POISSON.DIST function to compute the column B probabilities.

The **COMPUTE worksheet** of the Excel Guide Poisson workbook is a copy of the Figure 5.7 worksheet.

EXAMPLE 5.5

Computing Poisson Probabilities

Assume that the number of new visitors to a website in one minute follows a Poisson distribution with a mean of 2.5. What is the probability that in a given minute, there are no new visitors to the website? That there is at least one new visitor to the website?

SOLUTION Using Equation (5.8) on page 188 with $\lambda = 2.5$ (or Excel or a Poisson table lookup), the probability that there are no new visitors to the website is

$$P(X = 0 \mid \lambda = 2.5) = \frac{e^{-2.5}(2.5)^0}{0!} = \frac{1}{(2.71828)^{2.5}(1)} = 0.0821$$

▶(continued)

The probability that there will be no new visitors to the website in a given minute is 0.0821, or 8.21%. Thus,

$$P(X \geq 1) = 1 - P(X = 0)$$
$$= 1 - 0.0821 = 0.9179$$

The probability that there will be at least one new visitor to the website in a given minute is 0.9179, or 91.79%. Figure 5.8 shows the Example 5.5 Excel results. The answer to the questions can be found in the boldface cells in the "Poisson Probabilities Table."

FIGURE 5.8
Excel worksheet results for computing the Poisson probability for Example 5.5

	A	B	C	D	E
1	Poisson Probabilities for Website Visitors				
2					
3			Data		
4	Mean/Expected number of events of interest:				2.5
5					
6	Poisson Probabilities Table				
7	X	P(X)	P(<=X)	P(<X)	P(>X)
8	0	0.0821	0.0821	0.0000	0.9179

PROBLEMS FOR SECTION 5.3

LEARNING THE BASICS

5.18 Assume a Poisson distribution.
a. If $\lambda = 2.5$, find $P(X = 2)$.
b. If $\lambda = 8.0$, find $P(X = 8)$.
c. If $\lambda = 0.5$, find $P(X = 1)$.
d. If $\lambda = 3.7$, find $P(X = 0)$.

5.19 Assume a Poisson distribution.
a. If $\lambda = 2.0$, find $P(X \geq 2)$.
b. If $\lambda = 8.0$, find $P(X \geq 3)$.
c. If $\lambda = 0.5$, find $P(X \leq 1)$.
d. If $\lambda = 4.0$, find $P(X \geq 1)$.
e. If $\lambda = 5.0$, find $P(X \leq 3)$.

5.20 Assume a Poisson distribution with $\lambda = 5.0$. What is the probability that
a. $X = 1$?
b. $X < 1$?
c. $X > 1$?
d. $X \leq 1$?

APPLYING THE CONCEPTS

5.21 Assume that the number of airline customer service complaints filed with the Department of Transportation's Office of Aviation Enforcement and Proceedings (OAEP) in one day is distributed as a Poisson variable. The mean number of customer service complaints in January 2019 is 3.13 per day.

Source: Data extracted from U.S. Department of Transportation, *Air Travel Consumer Report*, January 2019.

What is the probability that in any given day
a. zero customer service complaints against travel agents will be filed?
b. exactly one customer service complaint against travel agents will be filed?
c. two or more customer service complaints against travel agents will be filed?
d. fewer than three customer service complaints against travel agents will be filed?

5.22 The quality control manager of Marilyn's Cookies is inspecting a batch of chocolate chip cookies that has just been baked. If the production process is in control, the mean number of chocolate chips per cookie is 6.0. What is the probability that in any particular cookie being inspected
a. fewer than five chocolate chips will be found?
b. exactly five chocolate chips will be found?
c. five or more chocolate chips will be found?
d. either four or five chocolate chips will be found?

5.23 Refer to Problem 5.22. How many cookies in a batch of 100 should the manager expect to discard if company policy requires that all chocolate chip cookies sold have at least four chocolate chips?

5.24 The U.S. Department of Transportation maintains statistics for mishandled bags. In January 2019, Delta mishandled 0.32 bags per day. What is the probability that in the next month, Delta will have
a. no mishandled bags?
b. at least one mishandled bag?
c. at least two mishandled bags?

5.25 The U.S. Department of Transportation maintains statistics for mishandled bags. In January 2019, American Airlines mishandled 0.90 bags per day. What is the probability that in the next month, American Airlines will have
a. no mishandled bags?
b. at least one mishandled bag?
c. at least two mishandled bags?

5.26 The Consumer Financial Protection Bureau's Consumer Response team hears directly from consumers about the challenges they face in the marketplace, brings their concerns to the attention of financial institutions, and assists in addressing their complaints. An analysis of complaints registered recently indicates that the mean number of vehicle lease complaints registered by consumers is 3.5 per day.

Sorce: Data extracted from **bit.ly/2nGDsc7**.

Assume that the number of vehicle lease complaints registered by consumers is distributed as a Poisson random variable. What is the probability that in a given day

a. no vehicle lease complaint will be registered by consumers?

b. exactly one vehicle lease complaint will be registered by consumers?

c. more than one vehicle lease complaint will be registered by consumers?

d. fewer than two vehicle lease complaints will be registered by consumers?

5.27 J.D. Power and Associates calculates and publishes various statistics concerning car quality. The dependability score measures problems experienced during the past 12 months by owners of vehicles (2019). For these models of cars, Ford had 1.48 problems per car and Toyota had 1.08 problems per car.

Source: Data extracted from M. Bomey, "Lexus top brand as auto reliability hits an all-time high," *USA Today*, February 14 2009, p. 1B–2B.

Let *X* be equal to the number of problems with a Ford.

a. What assumptions must be made in order for *X* to be distributed as a Poisson random variable? Are these assumptions reasonable?

Making the assumptions as in (a), if you purchased a Ford in the 2019 model year, what is the probability that in the past 12 months, the car had

b. zero problems?

c. two or fewer problems?

d. Give an operational definition for *problem*. Why is the operational definition important in interpreting the initial quality score?

5.28 Refer to Problem 5.27. If you purchased a Toyota in the 2019 model year, what is the probability that in the past 12 months the car had

a. zero problems?

b. two or fewer problems?

c. Compare your answers in (a) and (b) to those for the Ford in Problem 5.27 (b) and (c).

5.29 A toll-free phone number is available from 9 A.M. to 9 P.M. for your customers to register complaints about a product purchased from your company. Past history indicates that a mean of 0.8 calls is received per minute.

a. What properties must be true about the situation described here in order to use the Poisson distribution to calculate probabilities concerning the number of phone calls received in a one-minute period?

Assuming that this situation matches the properties discussed in (a), what is the probability that during a one-minute period

a. zero phone calls will be received?

b. three or more phone calls will be received?

c. What is the maximum number of phone calls that will be received in a one-minute period 99.99% of the time?

5.4 Covariance of a Probability Distribution and Its Application in Finance

Section 5.1 defines the expected value, variance, and standard deviation for the probability distribution of a *single* variable. The **Section 5.4 online topic** discusses covariance between *two* variables and explores how financial analysts apply this method as a tool for modern portfolio management.

5.5 Hypergeometric Distribution

The hypergeometric distribution determines the probability of *x* events of interest when sample data *without* replacement from a *finite* population has been collected. The **Section 5.5 online topic** discusses the hypergeometric distribution and illustrates its use.

▼ USING **STATISTICS**
Events of Interest..., Revisited

In the Ricknel Home Centers scenario at the beginning of this chapter, you were an accountant for the Ricknel Home Centers, LLC. The company's accounting information system automatically reviews order forms from online customers for possible mistakes. Any questionable invoices are tagged and included in a daily exceptions report. Knowing that the probability that an order will be tagged is 0.10, the binomial distribution was able to be used to determine the chance of finding a certain number of tagged forms in a sample of size four. There was a 65.6% chance that none of the forms would be tagged, a 29.2% chance that one would be tagged, and a 5.2% chance that two or more would be tagged.

Other calculations determined that, on average, one would expect 0.4 form to be tagged, and the standard deviation of the number of tagged order forms would be 0.6. Because the binomial distribution can be applied for any known probability and sample size, Ricknel staffers will be able to make inferences about the online ordering process and, more importantly, evaluate any changes or proposed changes to that process.

▼ SUMMARY

This chapter discusses two important discrete probability distributions: the binomial and Poisson distributions. The chapter explains the following selection rules that govern which discrete distribution to select for a particular situation:

- If there is a fixed number of observations, n, each of which is classified as an event of interest or not an event of interest, use the binomial distribution.
- If there is an area of opportunity, use the Poisson distribution.

▼ REFERENCES

Hogg, R. V., J. T. McKean, and A. V. Craig. *Introduction to Mathematical Statistics*, 7th ed. New York: Pearson Education, 2013.

Levine, D. M., P. Ramsey, and R. Smidt. *Applied Statistics for Engineers and Scientists Using Microsoft Excel and Minitab*. Upper Saddle River, NJ: Prentice Hall, 2001.

McGinty, J. "The Science Behind Your Long Wait in Line." *Wall Street Journal*, October 8, 2016, p. A2.

▼ KEY EQUATIONS

Expected Value, μ, of a Discrete Variable

$$\mu = E(X) = \sum_{i=1}^{N} x_i P(X = x_i) \tag{5.1}$$

Variance of a Discrete Variable

$$\sigma^2 = \sum_{i=1}^{N} [x_i - E(X)]^2 P(X = x_i) \tag{5.2}$$

Standard Deviation of a Discrete Variable

$$\sigma = \sqrt{\sigma^2} = \sqrt{\sum_{i=1}^{N} [x_i - E(X)]^2 P(X = x_i)} \tag{5.3}$$

Combinations

$$_nC_x = \frac{n!}{x!(n - x)!} \tag{5.4}$$

Binomial Distribution

$$P(X = x \mid n, \pi) = \frac{n!}{x!(n - x)!} \pi^x (1 - \pi)^{n-x} \tag{5.5}$$

Mean of the Binomial Distribution

$$\mu = E(X) = n\pi \tag{5.6}$$

Standard Deviation of the Binomial Distribution

$$\sigma = \sqrt{\sigma^2} = \sqrt{Var(X)} = \sqrt{n\pi(1 - \pi)} \tag{5.7}$$

Poisson Distribution

$$P(X = x \mid \lambda) = \frac{e^{-\lambda}\lambda^x}{x!} \tag{5.8}$$

▼ KEY TERMS

area of opportunity   188
binomial distribution   181
expected value   177
mathematical model   181

Poisson distribution   188
probability distribution for a discrete
   variable   177
probability distribution function   181

rule of combinations   181
standard deviation of a discrete
   variable   178
variance of a discrete variable   178

▼ CHECKING YOUR UNDERSTANDING

5.30 What is the meaning of the expected value of a variable?

5.31 What are the four properties that must be present in order to use the binomial distribution?

5.32 What are the four properties that must be present in order to use the Poisson distribution?

▼ CHAPTER REVIEW PROBLEMS

5.33 Darwin Head, a 35-year-old sawmill worker, won $1 million and a Chevrolet Malibu Hybrid by scoring 15 goals within 24 seconds at the Vancouver Canucks National Hockey League game (B. Ziemer, "Darwin Evolves into an Instant Millionaire," *Vancouver Sun*, February 28, 2008, p. 1). Head said he would use the money to pay off his mortgage and provide for his children, and he had no plans to quit his job. The contest was part of the Chevrolet Malibu Million Dollar Shootout, sponsored by General Motors Canadian Division. Did GM-Canada risk the $1 million? No! GM-Canada purchased event insurance from a company specializing in promotions at sporting events such as a half-court basketball shot or a hole-in-one giveaway at the local charity golf outing. The event insurance company estimates the probability of a contestant winning the contest and, for a modest charge, insures the event. The promoters pay the insurance premium but take on no added risk as the insurance company will make the large payout in the unlikely event that a contestant wins. To see how it works, suppose that the insurance company estimates that the probability a contestant would win a million-dollar shootout is 0.001 and that the insurance company charges $4,000.

a. Calculate the expected value of the profit made by the insurance company.

b. Many call this kind of situation a win–win opportunity for the insurance company and the promoter. Do you agree? Explain.

5.34 Between 1896—when the Dow Jones index was created—and 2018, the index rose in 66.4% of the years.

Based on this information, and assuming a binomial distribution, what do you think is the probability that the stock market will rise

a. next year?

b. the year after next?

c. in four of the next five years?

d. in none of the next five years?

e. For this situation, what assumption of the binomial distribution might not be valid?

5.35 The Internet has become a crucial marketing channel for business. According to a recent study, 91% of retail brands use two or more social media channels for business.

Source: Data extracted from Brandwatch, "123 Amazing Social Media Statistics and Facts," **bit.ly/2Oh2Crg**.

If a sample of 10 businesses is selected, what is the probability that:

a. 8 use two or more social media channels for business?

b. At least 8 use two or more social media channels for business?

c. At most 6 use two or more social media channels for business?

d. If you selected 10 businesses in a certain geographical area and only three use two or more social media channels for business, what conclusion might you reach about businesses in this geographical area?

5.36 One theory concerning the Dow Jones Industrial Average is that it is likely to increase during U.S. presidential election years. From 1964 through 2016, the Dow Jones Industrial Average increased in 11 of the 14 U.S. presidential election years. Assuming that this indicator is a random event with no predictive value, you would expect that the indicator would be correct 50% of the time.

a. What is the probability of the Dow Jones Industrial Average increasing in 11 or more of the 14 U.S. presidential election years if the probability of an increase in the Dow Jones Industrial Average is 0.50?

b. What is the probability that the Dow Jones Industrial Average will increase in 11 or more of the 14 U.S. presidential election years if the probability of an increase in the Dow Jones Industrial Average in any year is 0.75?

5.37 Medical billing errors and fraud are on the rise. According to Medical Billing Advocates of America, three out of four times, the medical bills that they review contain errors.

Source: Kelly Gooch, "Medical billing errors growing, says Medical Billing Advocates of America," *Becker's Hospital Review*, **bit.ly/2qkA8mR**.

If a sample of 10 medical bills is selected, what is the probability that

a. 0 medical bills will contain errors?

b. exactly 5 medical bills will contain errors?

c. more than 5 medical bills will contain errors?

d. What are the mean and standard deviation of the probability distribution?

5.38 Refer to Problem 5.37. Suppose that a quality improvement initiative has reduced the percentage of medical bills containing errors to 40%. If a sample of 10 medical bills is selected, what is the probability that

a. 0 medical bills will contain errors?

b. exactly 5 medical bills will contain errors?

c. more than 5 medical bills contain errors?

d. What are the mean and standard deviation of the probability distribution?

e. Compare the results of (a) through (c) to those of Problem 5.37 (a) through (c).

5.39 In a recent year, 46% of Google searches were one or two words, while 21% of Google searches were five or six words.

Source: Wordstream Blog, "27 Google Search Statistics You Should Know in 2019 (+ Insights!)," **bit.ly/2GxpUIv**.

If a sample of 10 Google searches is selected, what is the probability that

a. more than 5 are one- or two-word searches?

b. more than 5 are five- or six-word searches?

c. none are one- or two-word searches?

d. What assumptions did you have to make to answer (a) through (c)?

5.40 The Consumer Financial Protection Bureau's Consumer Response Team hears directly from consumers about the challenges they face in the marketplace, brings their concerns to the attention of financial institutions, and assists in addressing their complaints. Of the consumers who registered a bank account and service complaint, 46% cited "account management," complaints related to the marketing or management of an account, as their complaint.

Source: *Consumer Response Annual Report*, **bit.ly/2x4CN5w**.

Consider a sample of 20 consumers who registered bank account and service complaints. Use the binomial model to answer the following questions:

a. What is the expected value, or mean, of the binomial distribution?

b. What is the standard deviation of the binomial distribution?

c. What is the probability that 10 of the 20 consumers cited "account management" as the type of complaint?

d. What is the probability that no more than 5 of the consumers cited "account management" as the type of complaint?

e. What is the probability that 5 or more of the consumers cited "account management" as the type of complaint?

5.41 Refer to Problem 5.40. In the same time period, 24% of the consumers registering a bank account and service compliant cited "deposit and withdrawal" as the type of complaint; these are issues such as transaction holds and unauthorized transactions.

a. What is the expected value, or mean, of the binomial distribution?

b. What is the standard deviation of the binomial distribution?

c. What is the probability that none of the 20 consumers cited "deposit and withdrawal" as the type of complaint?

d. What is the probability that no more than 2 of the consumers cited "deposit and withdrawal" as the type of complaint?

e. What is the probability that 3 or more of the consumers cited "deposit and withdrawal" as the type of complaint?

5.42 One theory concerning the S&P 500 Index is that if it increases during the first five trading days of the year, it is likely to increase during the entire year. From 1950 through 2018, the S&P 500 Index had these early gains in 44 years (in 2011 there was virtually no change). In 36 of these 44 years, the S&P 500 Index increased for the entire year. Assuming that this indicator is a random event with no predictive value, you would expect that the indicator would be correct 50% of the time. What is the probability of the S&P 500 Index increasing in 36 or more years if the true probability of an increase in the S&P 500 Index is

a. 0.50?

b. 0.70?

c. 0.90?

d. Based on the results of (a) through (c), what do you think is the probability that the S&P 500 Index will increase if there is an early gain in the first five trading days of the year? Explain.

5.43 *Spurious correlation* refers to the apparent relationship between variables that either have no true relationship or are related to other variables that have not been measured. One widely publicized stock market indicator in the United States that is an example of spurious correlation is the relationship between the winner of the National Football League Super Bowl and the performance of the Dow Jones Industrial Average in that year. The "indicator" states that when a team that existed before the National Football League merged with the American Football League wins the Super Bowl, the Dow Jones Industrial Average will increase in that year. (Of course, any correlation between these is spurious as one thing has absolutely nothing to do with the other!) Since the first Super Bowl was held in 1967 through 2018, the indicator has been correct 38 out of 52 times. Assuming that this indicator is a random event with no predictive value, you would expect that the indicator would be correct 50% of the time.

a. What is the probability that the indicator would be correct 38 or more times in 50 years?

b. What does this tell you about the usefulness of this indicator?

5.44 The United Auto Courts Reports blog notes that the National Insurance Crime Bureau says that Miami-Dade, Broward, and Palm Beach counties account for a substantial number of questionable insurance claims referred to investigators. Assume that the number of questionable insurance claims referred to investigators by Miami-Dade, Broward, and Palm Beach counties is distributed as a Poisson random variable with a mean of 7 per day.

a. What assumptions need to be made so that the number of questionable insurance claims referred to investigators by Miami-Dade, Broward, and Palm Beach counties is distributed as a Poisson random variable?

Making the assumptions given in (a), what is the probability that

b. 5 questionable insurance claims will be referred to investigators by Miami-Dade, Broward, and Palm Beach counties in a day?

c. 10 or fewer questionable insurance claims will be referred to investigators by Miami-Dade, Broward, and Palm Beach counties in a day?

d. 11 or more questionable insurance claims will be referred to investigators by Miami-Dade, Broward, and Palm Beach counties in a day?

▾CASES

Managing Ashland MultiComm Services

The Ashland MultiComm Services (AMS) marketing department wants to increase subscriptions for its *3-For-All* telephone, cable, and Internet combined service. AMS marketing has been conducting an aggressive direct-marketing campaign that includes postal and electronic mailings and telephone solicitations. Feedback from these efforts indicates that including premium channels in this combined service is a very important factor for both current and prospective subscribers. After several brainstorming sessions, the marketing department has decided to add premium cable channels as a no-cost benefit of subscribing to the *3-For-All* service.

The research director, Mona Fields, is planning to conduct a survey among prospective customers to determine how many premium channels need to be added to the *3-For-All* service in order to generate a subscription to the service. Based on past campaigns and on industry-wide data, she estimates the following:

Number of Free Premium Channels	Probability of Subscriptions
0	0.02
1	0.04
2	0.06
3	0.07
4	0.08
5	0.085

1. If a sample of 50 prospective customers is selected and no free premium channels are included in the *3-For-All* service offer, given past results, what is the probability that
   a. fewer than 3 customers will subscribe to the *3-For-All* service offer?
   b. 0 customers or 1 customer will subscribe to the *3-For-All* service offer?
   c. more than 4 customers will subscribe to the *3-For-All* service offer?
   d. Suppose that in the actual survey of 50 prospective customers, 4 customers subscribe to the *3-For-All* service offer. What does this tell you about the previous estimate of the proportion of customers who would subscribe to the *3-For-All* service offer?

2. Instead of offering no premium free channels as in Problem 1, suppose that two free premium channels are included in the *3-For-All* service offer. Given past results, what is the probability that

a. fewer than 3 customers will subscribe to the *3-For-All* service offer?
b. 0 customers or 1 customer will subscribe to the *3-For-All* service offer?
c. more than 4 customers will subscribe to the *3-For-All* service offer?
d. Compare the results of (a) through (c) to those of Problem 1.
e. Suppose that in the actual survey of 50 prospective customers, 6 customers subscribe to the *3-For-All* service offer. What does this tell you about the previous estimate of the proportion of customers who would subscribe to the *3-For-All* service offer?
f. What do the results in (e) tell you about the effect of offering free premium channels on the likelihood of obtaining subscriptions to the *3-For-All* service?

3. Suppose that additional surveys of 50 prospective customers were conducted in which the number of free premium channels was varied. The results were as follows:

Number of Free Premium Channels	Number of Subscriptions
1	5
3	6
4	6
5	7

How many free premium channels should the research director recommend for inclusion in the *3-For-All* service? Explain.

Digital Case

Apply your knowledge about expected value in this continuing Digital Case from Chapters 3 and 4.

Open **BullsAndBears.pdf**, a marketing brochure from EndRun Financial Services. Read the claims and examine the supporting data. Then answer the following:

1. Are there any "catches" about the claims the brochure makes for the rate of return of Happy Bull and Worried Bear funds?
2. What subjective data influence the rate-of-return analyses of these funds? Could EndRun be accused of making false and misleading statements? Why or why not?
3. The expected-return analysis seems to show that the Worried Bear fund has a greater expected return than the Happy Bull fund. Should a rational investor never invest in the Happy Bull fund? Why or why not?

EG5.1 The PROBABILITY DISTRIBUTION for a DISCRETE VARIABLE

Key Technique Use **SUMPRODUCT(*X cell range, P(X) cell range*)** to compute the expected value. Use **SUMPRODUCT(*squared differences cell range, P(X) cell range*)** to compute the variance.

Example Compute the expected value, variance, and standard deviation for the number of interruptions per day data of Table 5.1 on page 177.

Workbook Use the **Discrete Variable workbook** as a model.

For the example, open to the **DATA worksheet** of the **Discrete Variable workbook**. The worksheet contains the column A and B entries needed to compute the expected value, variance, and standard deviation for the example. Unusual for a DATA worksheet in this book, column C contains formulas. These formulas use the expected value that cell B4 in the COMPUTE worksheet of the same workbook computes (first three rows shown below) and are equivalent to the fourth column calculations in Table 5.3.

	A	B	C
1	X	P(X)	[X-E(X)]^2
2	0	0.35	=(A2 - COMPUTE!$B$4)^2
3	1	0.25	=(A3 - COMPUTE!$B$4)^2
4	2	0.20	=(A4 - COMPUTE!$B$4)^2

For other problems, modify the DATA worksheet. Enter the probability distribution data into columns A and B and, if necessary, extend column C, by first selecting cell C7 and then copying that cell down as many rows as necessary. If the probability distribution has fewer than six outcomes, select the rows that contain the extra, unwanted outcomes, right-click, and then click Delete in the shortcut menu.

Appendix F further explains the SUMPRODUCT function that the COMPUTE worksheet uses to compute the expected value and variance.

EG5.2 BINOMIAL DISTRIBUTION

Key Technique Use the **BINOM.DIST(*number of events of interest, sample size, probability of an event of interest,* FALSE)** function.

Example Compute the binomial probabilities for $n = 4$ and $\pi = 0.1$, and construct a histogram of that probability distribution, similar to Figures 5.2 and 5.3 on page 184.

PHStat Use **Binomial**.

For the example, select **PHStat → Probability & Prob. Distributions → Binomial**. In the procedure's dialog box (shown below):

1. Enter **4** as the **Sample Size**.
2. Enter **0.1** as the **Prob. of an Event of Interest**.
3. Enter **0** as the **Outcomes From** value and enter **4** as the (Outcomes) **To** value.
4. Enter a **Title**, check **Histogram**, and click **OK**.

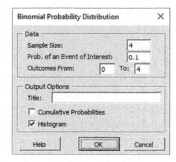

Check **Cumulative Probabilities** before clicking **OK** in step 4 to have the procedure include columns for $P(\leq X)$, $P(<X)$, $P(>X)$, and $P(\geq X)$ in the binomial probabilities table.

Workbook Use the **Binomial workbook** as a template and model.

For the example, open to the **COMPUTE worksheet** of the **Binomial workbook**, shown in Figure 5.2 on page 184. The worksheet already contains the entries needed for the example. For other problems, change the sample size in cell B4 and the probability of an event of interest in cell B5. If necessary, extend the binomial probabilities table by first selecting cell range A18:B18 and then copying that cell range down as many rows as necessary. To construct a histogram of the probability distribution, use the Appendix Section B.6 instructions.

For problems that require cumulative probabilities, use the CUMULATIVE worksheet in the Binomial workbook. The SHORT TAKES for Chapter 5 explains and documents this worksheet.

EG5.3 POISSON DISTRIBUTION

Key Technique Use the **POISSON.DIST(*number of events of interest, the average or expected number of events of interest,* FALSE)** function.

Example Compute the Poisson probabilities for the Figure 5.7 customer arrival problem on page 189.

PHStat Use **Poisson**.

For the example, select **PHStat→Probability & Prob. Distributions→Poisson**. In this procedure's dialog box (shown below):

1. Enter **3** as the **Mean/Expected No. of Events of Interest**.
2. Enter a **Title** and click **OK**.

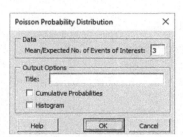

Check **Cumulative Probabilities** before clicking **OK** in step 2 to have the procedure include columns for $P(\leq X)$, $P(<X)$, $P(>X)$, and $P(\geq X)$ in the Poisson probabilities table. Check **Histogram** to construct a histogram of the Poisson probability distribution.

Workbook Use the **Poisson workbook** as a template.

For the example, open to the **COMPUTE worksheet** of the **Poisson workbook**, shown in Figure 5.7 on page 189. The worksheet already contains the entries for the example. For other problems, change the mean or expected number of events of interest in cell E4. To construct a histogram of the probability distribution, use the Appendix Section B.6 instructions.

For problems that require cumulative probabilities, use the CUMULATIVE worksheet in the Binomial workbook. The SHORT TAKES for Chapter 5 explains and documents this worksheet.

6

The Normal Distribution and Other Continuous Distributions

CONTENTS

OBJECTIVES

- Compute probabilities from the normal distribution
- Use the normal distribution to solve business problems
- Use the normal probability plot to determine whether a set of data is approximately normally distributed
- Compute probabilities from the uniform distribution.

▼USING **STATISTICS**
Normal Load Times at MyTVLab

You are the vice president in charge of sales and marketing for MyTVLab, a web-based business that has evolved into a full-fledged, subscription-based streaming video service. To differentiate MyTVLab from the other companies that sell similar services, you decide to create a "Why Choose Us" web page to help educate new and prospective subscribers about all that MyTVLab offers.

As part of that page, you have produced a new video that samples the content MyTVLab streams as well as demonstrates the relative ease of setting up MyTVLab on many types of devices. You want this video to download with the page so that a visitor can jump to different segments immediately or view the video later, when offline.

You know from research (Kishnan and Sitaraman) and past observations, Internet visitors will not tolerate waiting too long for a web page to load. One wait time measure is load time, the time in seconds that passes from first pointing a browser to a web page until the web page is fully loaded and content such as video is ready to be viewed. You have set a goal that the load time for the new sales page should rarely exceed 10 seconds (too long for visitors to wait) and, ideally, should rarely be less than 1 second (a waste of company Internet resources).

To measure this time, you point a web browser at the MyTVLab corporate test center to the new sales web page and record the load time. In your first test, you record a time of 6.67 seconds. You repeat the test and record a time of 7.52 seconds. Though consistent to your goal, you realize that two load times do not constitute strong proof of anything, especially as your assistant has performed his own test and recorded a load time of 8.83 seconds.

Could you use a method based on probability theory to ensure that most load times will be within the range you seek? MyTVLab has recorded past load times of a similar page with a similar video and determined the mean load time of that page is 7 seconds, the standard deviation of those times is 2 seconds, that approximately two-thirds of the load times are between 5 and 9 seconds, and about 95% of the load times are between 3 and 11 seconds.

Could you use these facts to assure yourself that the load time goal you have set for the new sales page is likely to be met?

198

Chapter 5 discusses how to use probability distributions for a *discrete* numerical variable. In the MyTVLab scenario, you are examining the load time, a *continuous* numerical variable. You are no longer considering a table of discrete (specific) values, but a continuous range of values. For example, the phrase "load times are between 5 and 9 seconds" includes *any* value between 5 and 9 and not just the values 5, 6, 7, 8, and 9. If you plotted the phrase on a graph, you would draw a *continuous* line from 5 to 9 and not just plot five discrete points.

When you add information about the shape of the range of values, such as two-thirds of the load times are between 5 and 9 seconds or about 95% of the load times are between 3 and 11 seconds, you can visualize the plot of all values as an area under a curve. If that area under the curve follows the well-known pattern of certain continuous distributions, you can use the continuous probability distribution for that pattern to estimate the likelihood that a load time is within a range of values. In the MyTVLab scenario, the past load times of a similar page describes a pattern that conforms to the pattern associated with the normal distribution, the subject of Section 6.2. That would allow you, as the vice president for sales and marketing, to use the normal distribution with the statistics given to determine if your load time goal is likely to be met.

6.1 Continuous Probability Distributions

Continuous probability distributions vary by the shape of the area under the curve. Figure 6.1 visualizes the normal, uniform, and exponential probability distributions.

FIGURE 6.1
Three continuous probability distributions

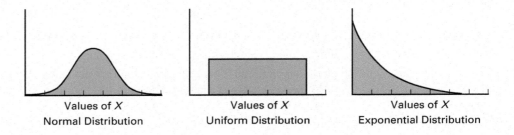

Values of *X*
Normal Distribution

Values of *X*
Uniform Distribution

Values of *X*
Exponential Distribution

Some distributions, including the normal and uniform distributions in Figure 6.1, show a symmetrical shape. Distributions such as the right-skewed exponential distribution do not. In symmetrical distributions the mean equals the median, whereas in a right-skewed distribution the mean is greater than the median. Each of the three distributions also has unique properties.

The **normal distribution** is not only symmetrical, but bell-shaped, a shape that (loosely) suggests the profile of a bell. Being bell-shaped means that most values of the continuous variable will cluster around the mean. Although the values in a normal distribution can range from negative infinity to positive infinity, the shape of the normal distribution makes it very unlikely that extremely large or extremely small values will occur.

Section 6.4 further discusses the uniform distribution.

The online Section 6.5 further discusses the exponential distribution.

The **uniform distribution**, also known as the *rectangular distribution*, contains values that are equally distributed in the range between the smallest value and the largest value. In a uniform distribution, every value is equally likely.

The **exponential distribution** contains values from zero to positive infinity and is right-skewed, making the mean greater than the median. Its shape makes it unlikely that extremely large values will occur.

Besides visualizations such as those in Figure 6.1, a continuous probability distribution can be expressed mathematically as a *probability density function*. A **probability density function** for a specific continuous probability distribution, represented by the symbol $f(X)$, defines the distribution of the values for a continuous variable and can be used as the basis for calculations that determine the likelihood or probability that a value will be within a certain range.

6.2 The Normal Distribution

The most commonly used continuous probability distribution, the normal distribution, plays an important role in statistics and business. Because of its relationship to the Central Limit Theorem (see Section 7.2), the distribution provides the basis for classical statistical inference and can be used to approximate various discrete probability distributions. For business, many continuous variables used in decision making have distributions that closely resemble the normal distribution. The normal distribution can be used to estimate the probability that values occur within a specific range or interval. This probability corresponds to an area under a curve that the normal distribution defines. Because a single point on a curve, representing a specific value, cannot define an area, the area under any single point/specific value will be 0. Therefore, when using the normal distribution to estimate values of a continuous variable, the probability that the variable will be exactly a specified value is always zero.

By the rule the previous paragraph states, the probability that the load time is exactly 7, or any other specific value, is zero.

For the MyTVLab scenario, the load time for the new sales page would be an example of a continuous variable whose distribution approximates the normal distribution. This approximation enables one to estimate probabilities such as the probability that the load time would be between 7 and 10 seconds, the probability that the load time would be between 8 and 9 seconds, or the probability that the load time would be between 7.99 and 8.01 seconds.

Exhibit 6.1 presents four important theoretical properties of the normal distribution. The distributions of many business decision-making continuous variables share the first three properties, sufficient to allow the use of the normal distribution to *estimate* the probability for specific ranges or intervals of values.

EXHIBIT 6.1

Normal Distribution Important Theoretical Properties

Symmetrical distribution. Its mean and median are equal.

Bell-shaped. Values cluster around the mean.

Interquartile range is roughly 1.33 standard deviations. Therefore, the middle 50% of the values are contained within an interval that is approximately two-thirds of a standard deviation below and two-thirds of a standard deviation above the mean.

The distribution has an infinite range ($-\infty < X < \infty$). Six standard deviations approximate this range (see page 205).

Table 6.1 presents the fill amounts, the volume of liquid placed inside a bottle, for a production run of 10,000 one-liter water bottles. Due to minor irregularities in the machinery and the water pressure, the fill amounts will vary slightly from the desired target amount, which is a bit more than 1.0 liters to prevent underfilling of bottles and the subsequent consumer unhappiness that such underfilling would cause.

TABLE 6.1
Fill amounts for 10,000 one-liter water bottles

Fill Amount (liters)	Relative Frequency
< 1.025	48/10,000 = 0.0048
1.025 < 1.030	122/10,000 = 0.0122
1.030 < 1.035	325/10,000 = 0.0325
1.035 < 1.040	695/10,000 = 0.0695
1.040 < 1.045	1,198/10,000 = 0.1198
1.045 < 1.050	1,664/10,000 = 0.1664
1.050 < 1.055	1,896/10,000 = 0.1896
1.055 < 1.060	1,664/10,000 = 0.1664
1.060 < 1.065	1,198/10,000 = 0.1198
1.065 < 1.070	695/10,000 = 0.0695
1.070 < 1.075	325/10,000 = 0.0325
1.075 < 1.080	122/10,000 = 0.0122
1.080 or above	48/10,000 = 0.0048
Total	1.0000

The fill amounts for the 10,000-bottle run cluster in the interval 1.05 to 1.055 liters. The fill amounts distribute symmetrically around that grouping, forming a bell-shaped pattern, which the relative frequency polygon that has been superimposed over the Figure 6.2 histogram highlights. These properties of the fill amount permit the normal distribution to be used to estimate values. Note that the distribution of fill amounts does not have an infinite range as fill amounts can never be less than 0 or more than the entire, fixed volume of a bottle. Therefore, the normal distribution can only be an approximation of the fill amount distribution, a distribution that does not contain an infinite range, an important theoretical property of true normal distributions.

student TIP

Section 2.4 discusses histograms and relative frequency polygons.

FIGURE 6.2
Relative frequency histogram and polygon of the amount filled in 10,000 water bottles

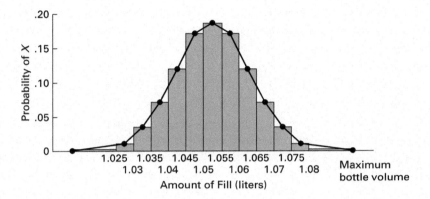

Role of the Mean and the Standard Deviation

Each combination of a mean μ and a standard deviation σ defines a separate normal distribution. Figure 6.3 shows the normal distribution for three such combinations. Distributions A and B have the same mean but have different standard deviations. Distributions A and C have the same standard deviation but have different means. Distributions B and C have different values for both the mean and standard deviation.

FIGURE 6.3
Three normal distributions

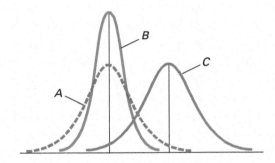

Not surprisingly, both the mean, μ, and the standard deviation, σ, appear in Equation (6.1) that defines the **probability density function for the normal distribution**.

NORMAL PROBABILITY DENSITY FUNCTION

$$f(X) = \frac{1}{\sqrt{2\pi}\sigma} e^{-(1/2)[(X-\mu)/\sigma]^2} \qquad \textbf{(6.1)}$$

where

e = mathematical constant approximated by 2.71828
π = mathematical constant approximated by 3.14159
μ = mean
σ = standard deviation
X = any value of the continuous variable, where $-\infty < X < \infty$

Calculating Normal Probabilities

Examining Equation (6.1) reveals that the only terms that are not numerical constants are the mean, μ, and the standard deviation, σ. This insight allows normal probabilities to be calculated using an alternative method based in part on using the **transformation formula** that Equation (6.2) defines. Using this second method avoids the calculational complexities that the direct use of Equation (6.1) would create.

Z TRANSFORMATION FORMULA

The Z value is equal to the difference between X and the mean, μ, divided by the standard deviation, σ.

$$Z = \frac{X - \mu}{\sigma} \tag{6.2}$$

The transformation formula converts a normally distributed variable, X, to a corresponding **standardized normal variable, Z**. The formula calculates a Z value, called *standardized units*, that expresses the difference of the X value from the mean, μ, in standard deviation units. While a variable, X, has mean, μ, and standard deviation, σ, the standardized variable, Z, always has mean $\mu = 0$ and standard deviation $\sigma = 1$.

With a calculated Z value, use Table E.2, the **cumulative standardized normal distribution**, to determine the probability. For example, recall from the MyTVLab scenario on page 198 that past data indicate that the sales page load time is normally distributed, with a mean $\mu = 7$ seconds and a standard deviation $\sigma = 2$ seconds. Figure 6.4 shows that every measurement X has a corresponding standardized measurement Z, computed from Equation (6.2), the transformation formula.

FIGURE 6.4
Transformation of scales

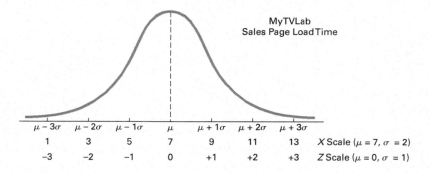

Therefore, a load time of 9 seconds is equivalent to 1 standardized unit (1 standard deviation) above the mean because

$$Z = \frac{9 - 7}{2} = +1$$

A load time of 1 second is equivalent to -3 standardized units (3 standard deviations) below the mean because

$$Z = \frac{1 - 7}{2} = -3$$

In Figure 6.4, the standard deviation is the unit of measurement. In other words, a time of 9 seconds is 2 seconds (1 standard deviation) higher, or *slower*, than the mean time of 7 seconds. Similarly, a time of 1 second is 6 seconds (3 standard deviations) lower, or *faster*, than the mean time.

As another example of applying the transformation formula, suppose that the technical support web page has a load time that is normally distributed, with a mean $\mu = 4$ seconds and a standard deviation $\sigma = 1$ second. Figure 6.5 on the next page shows this distribution.

FIGURE 6.5
A different transformation
of scales

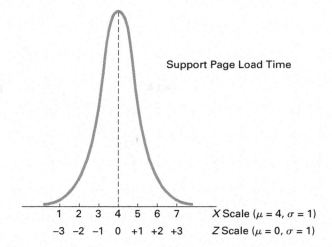

Support Page Load Time

| X Scale ($\mu = 4$, $\sigma = 1$) |
| Z Scale ($\mu = 0$, $\sigma = 1$) |

| 1 | 2 | 3 | 4 | 5 | 6 | 7 |
| −3 | −2 | −1 | 0 | +1 | +2 | +3 |

This transformation shows that a load time of 5 seconds is 1 standard deviation above the mean download time because

$$Z = \frac{5 - 4}{1} = +1$$

A time of 1 second is 3 standard deviations below the mean load time because

$$Z = \frac{1 - 4}{1} = -3$$

Having determined the Z value, use a table of values from the cumulative standardized normal distribution to look up the normal probability. For example, consider that one wanted to find the probability that the load time for the MyTVLab sales page is less than 9 seconds. Recall from page 202 that given a mean $\mu = 7$ seconds and a standard deviation $\sigma = 2$ seconds, transforming $X = 9$ leads to a Z value of $+1.00$.

Use Table E.2 with a calculated Z value to find the cumulative area under the normal curve less than (to the left of) $Z = +1.00$. To read the probability or area under the curve less than $Z = +1.00$, locate the Table E.2 row for the Z value 1.0. Next, locate the entry in the row for the column that contains the 100th place of the Z value. Therefore, in the body of the table, the probability for $Z = 1.00$, that entry is .8413, in the **.00** column.

Table 6.2 highlights this entry and shows how the entry was found. There is an 84.13% chance that the download time will be less than 9 seconds. Figure 6.6 on page 204 visualizes this probability.

studentTIP

When discussing the normal or other continuous distributions, the word *area* has the same meaning as *probability*.

TABLE 6.2
Finding a cumulative area under the normal curve

Source: Extracted from Table E.2.

						Cumulative Probabilities				
Z	**.00**	.01	.02	.03	.04	.05	.06	.07	.08	.09
0.0	.5000	.5040	.5080	.5120	.5160	.5199	.5239	.5279	.5319	.5359
0.1	.5398	.5438	.5478	.5517	.5557	.5596	.5636	.5675	.5714	.5753
0.2	.5793	.5832	.5871	.5910	.5948	.5987	.6026	.6064	.6103	.6141
0.3	.6179	.6217	.6255	.6293	.6331	.6368	.6406	.6443	.6480	.6517
0.4	.6554	.6591	.6628	.6664	.6700	.6736	.6772	.6808	.6844	.6879
0.5	.6915	.6950	.6985	.7019	.7054	.7088	.7123	.7157	.7190	.7224
0.6	.7257	.7291	.7324	.7357	.7389	.7422	.7454	.7486	.7518	.7549
0.7	.7580	.7612	.7642	.7673	.7704	.7734	.7764	.7794	.7823	.7852
0.8	.7881	.7910	.7939	.7967	.7995	.8023	.8051	.8078	.8106	.8133
0.9	.8159	.8186	.8212	.8238	.8264	.8289	.8315	.8340	.8365	.8389
1.0	.8413	.8438	.8461	.8485	.8508	.8531	.8554	.8577	.8599	.8621

FIGURE 6.6
Determining the area less than Z from a cumulative standardized normal distribution

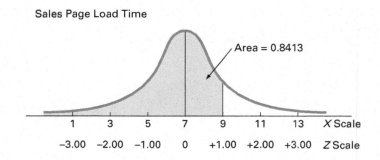

Sales Page Load Time

Area = 0.8413

1	3	5	7	9	11	13	X Scale
−3.00	−2.00	−1.00	0	+1.00	+2.00	+3.00	Z Scale

From Figure 6.5, a *support* page load time of 5 seconds is 1 standardized unit above the mean time of 4 seconds. Thus, the probability that the load time for the support page will be less than 5 seconds is also 0.8413. Figure 6.7 shows that regardless of the value of the mean, μ, and standard deviation, σ, of a normally distributed variable, Equation (6.2) can transform the X value to a Z value.

Using Table E.2 with Equation (6.2) can answer many questions related to the sales page load time, including whether achieving the load time goal is likely, using the normal distribution.

FIGURE 6.7
Demonstrating a transformation of scales for corresponding cumulative portions under two normal curves

student TIP

When calculating probabilities under the normal curve, draw a normal curve and enter the values for the mean and X below the curve as a helpful guide. Shade the desired area to be determined under the curve.

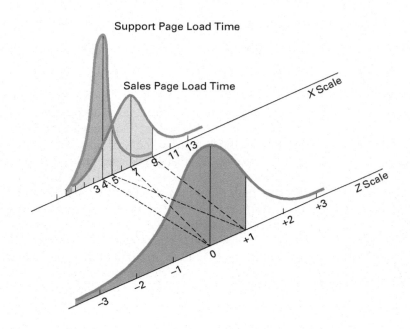

Support Page Load Time

Sales Page Load Time

X Scale

Z Scale

EXAMPLE 6.1

Finding $P(X > 9)$

What is the probability that the load time for the MyTVLab sales page will be more than 9 seconds?

SOLUTION The probability that the load time will be less than 9 seconds is 0.8413 (see Figure 6.6). Thus, the probability that the load time will be more than 9 seconds is the *complement* of less than 9 seconds, $1 - 0.8413 = 0.1587$. Figure 6.8 illustrates this result.

FIGURE 6.8
Finding $P(X > 9)$

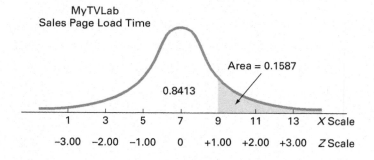

MyTVLab
Sales Page Load Time

Area = 0.1587

0.8413

1	3	5	7	9	11	13	X Scale
−3.00	−2.00	−1.00	0	+1.00	+2.00	+3.00	Z Scale

EXAMPLE 6.2

Finding $P(X < 7$ or $X > 9)$

What is the probability that the load time for the MyTVLab sales page will be less than 7 seconds or more than 9 seconds?

SOLUTION To find this probability, separately calculate the probability of a load time less than 7 seconds and the probability of a load time greater than 9 seconds and then add these two probabilities together. Figure 6.9 illustrates this result.

FIGURE 6.9
Finding
$P(X < 7$ or $X > 9)$

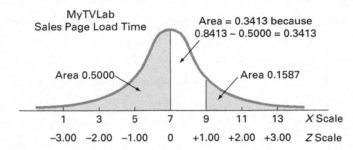

Because the mean is 7 seconds, and because the mean is equal to the median in a normal distribution, 50% of load times are under 7 seconds. From Example 6.1, the probability that the load time is greater than 9 seconds is 0.1587. Therefore, the probability that a load time is less than 7 or more than 9 seconds, $P(X < 7$ or $X > 9)$, is $0.5000 + 0.1587 = 0.6587$.

EXAMPLE 6.3

Finding
$P(5 < X < 9)$

What is the probability that load time for the MyTVLab sales page will be between 5 and 9 seconds—that is, $P(5 < X < 9)$?

SOLUTION In Figure 6.10, the area of interest is shaded dark blue and is located between two values, 5 and 9.

FIGURE 6.10
Finding $P(5 < X < 9)$

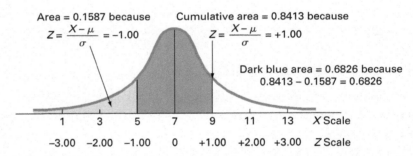

Example 6.1 on page 204, calculates the area under the normal curve less than 9 seconds as 0.8413. To find the area under the normal curve less than 5 seconds,

$$Z = \frac{5 - 7}{2} = -1.00$$

From Table E.2, using $Z = -1.00$, the cumulative probability is 0.1587. Therefore, the probability that the load time will be between 5 and 9 seconds is $0.8413 - 0.1587 = 0.6826$, as Figure 6.10 visualizes.

This result is the justification for the empirical rule presented on page 134. The accuracy of the empirical rule increases the closer the variable follows the normal distribution.

From the Example 6.3 solution and from Figures 6.11 and 6.12 on page 206 that visualize related examples, one can observe that for any normal distribution

- approximately 68.26% of the values fall within ±1 standard deviation of the mean.
- approximately 95.44% of the values fall within ±2 standard deviations of the mean.
- approximately 99.73% of the values fall within ±3 standard deviations of the mean.

From Figure 6.10 in Example 6.3, 68.26% of the values are within ±1 standard deviation of the mean. From Figure 6.11, 95.44% of the values are within ±2 standard deviations of the mean $(0.9772 - 0.0228)$. For the MyTVLab sales page, 95.44% of the download times are between 3 and 11 seconds. From Figure 6.12, 99.73% of the values are within ±3 standard deviations

FIGURE 6.11
Finding $P(3 < X < 11)$

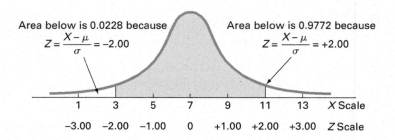

FIGURE 6.12
Finding $P(1 < X < 13)$

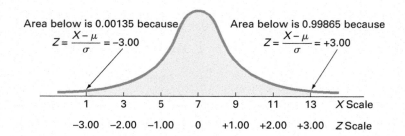

above or below the mean. For the MyTVLab sales page, 99.73% of the load times are between 1 and 13 seconds.

For the MyTVLab sales page, it is unlikely (0.0027, or only 27 in 10,000) that a load time will be so fast or so slow that it will take less than 1 second or more than 13 seconds. In general, use 6σ, 3 standard deviations below the mean to 3 standard deviations above the mean, as a practical approximation of the range for normally distributed data.

VISUAL EXPLORATIONS

Exploring the Normal Distribution

Open the **VE-Normal Distribution add-in workbook** to explore the normal distribution. (For Excel technical requirements, see Appendix D.) When this workbook opens properly, it adds a Normal Distribution menu in the Add-ins tab (Apple menu in Excel for Mac).

To explore the effects of changing the mean and standard deviation on the area under a normal curve, select **Normal Distribution → Probability Density Function**. The add-in displays a normal curve for the MyTVLab example and a floating control panel (top right). Use the control panel spinner buttons to change the values for the mean, standard deviation, and X value and then note the effects of these changes on the probability of $X <$ value and the corresponding shaded area under the curve. To see the normal curve labeled with Z values, click **Z Values**. Click **Reset** to reset the control panel values. Click **Finish** to finish exploring.

To create shaded areas under the curve for problems similar to Examples 6.2 and 6.3, select **Normal Distribution → Areas**. In the Areas dialog box (bottom right), enter values, select an Area Option, and click **OK**.

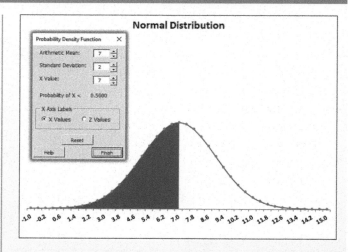

Finding X Values

The solutions to Examples 6.1 through 6.3 require finding the area under the normal curve that corresponds to a specific X value. Other problems require the opposite: Finding the X value that corresponds to a specific area. To do so, first solve Equation (6.2) for X and use that result, Equation (6.3), to find the X value.

FINDING AN X VALUE ASSOCIATED WITH A KNOWN PROBABILITY

The X value is equal to the mean, μ, plus the product of the Z value and the standard deviation, σ.

$$X = \mu + Z\sigma \tag{6.3}$$

To find a *particular* value associated with a known probability, follow these steps:

- Sketch the normal distribution curve and then place the values for the mean and X on the X and Z scales.
- Find the cumulative area less than X.
- Shade the area of interest.
- Using Table E.2, determine the Z value corresponding to the area under the normal curve less than X.
- Using Equation (6.3), solve for X: $X = \mu + Z\sigma$.

Examples 6.4 and 6.5 demonstrate this technique using this five-step procedure to find a particular value associated with a known probability.

EXAMPLE 6.4

Finding the *X* Value for a Cumulative Probability of 0.10

How much time (in seconds) will elapse before the fastest 10% of the MyTVLab sales pages load time occur?

SOLUTION Because 10% of the load times are expected to occur in under X seconds, the area under the normal curve less than this value is 0.1000. Using Table E.2, locate the entry for the area or probability of 0.1000. The closest entry is 0.1003, as Table 6.3 shows.

TABLE 6.3
Finding a *Z* value corresponding to a cumulative area of 0.10 under the normal curve

Source: Extracted from Table E.2.

					Cumulative Probabilities					
Z	**.00**	**.01**	**.02**	**.03**	**.04**	**.05**	**.06**	**.07**	**.08**	**.09**
⋮	⋮	⋮	⋮	⋮	⋮	⋮	⋮	⋮	⋮	⋮
−1.5	.0668	.0655	.0643	.0630	.0618	.0606	.0594	.0582	.0571	.0559
−1.4	.0808	.0793	.0778	.0764	.0749	.0735	.0721	.0708	.0694	.0681
−1.3	.0968	.0951	.0934	.0918	.0901	.0885	.0869	.0853	.0838	.0823
−1.2	.1151	.1131	.1112	.1093	.1075	.1056	.1038	.1020	.1003	.0985

Working from this area to the margins of the table, find the Z value corresponding to the particular Z row (-1.2) and Z column ($.08$) is -1.28, which Figure 6.13 visualizes.

FIGURE 6.13
Finding *Z* to determine *X*

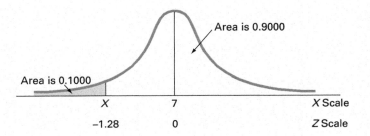

Area is 0.9000

Area is 0.1000

	X	7		X Scale
	−1.28	0		Z Scale

▶*(continued)*

Having determined Z, use Equation (6.3) on page 207 to determine the X value. Substituting $\mu = 7$, $\sigma = 2$, and $Z = -1.28$,

$$X = \mu + Z\sigma$$
$$X = 7 + (-1.28)(2) = 4.44 \text{ seconds}$$

Thus, 10% of the load times are 4.44 seconds or less.

EXAMPLE 6.5

Finding the X Values That Include 95% of the Download Times

What are the lower and upper values of X, symmetrically distributed around the mean, that include 95% of the load times for the MyTVLab sales page?

To answer this question, first find the lower value of X (called X_L) and the upper value of X (called X_U). Because 95% of the values are between X_L and X_U, and because X_L and X_U are equally distant from the mean, 2.5% of the values are below X_L (see Figure 6.14).

FIGURE 6.14
Finding Z to determine X_L

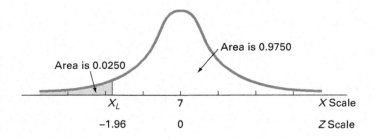

Although X_L is not known, find the corresponding Z value because the area under the normal curve less than this Z is 0.0250. Using Table E.2, locate the entry closest to 0.0250. Table 6.4 shows that Z value for this area is -1.96, as 0.0250 corresponds to the Z row (-1.9) and Z column (.06).

TABLE 6.4
Finding a Z value corresponding to a cumulative area of 0.025 under the normal curve

Source: Extracted from Table E.2.

					Cumulative Probabilities					
Z	.00	.01	.02	.03	.04	.05	.06	.07	.08	.09
⋮	⋮	⋮	⋮	⋮	⋮	⋮	⋮	⋮	⋮	⋮
−2.0	.0228	.0222	.0217	.0212	.0207	.0202	.0197	.0192	.0188	.0183
−1.9	.0287	.0281	.0274	.0268	.0262	.0256	.0250	.0244	.0239	.0233
−1.8	.0359	.0351	.0344	.0336	.0329	.0322	.0314	.0307	.0301	.0294

Having determined Z, use Equation (6.3) on page 207:

$$X = \mu + Z\sigma$$
$$= 7 + (-1.96)(2)$$
$$= 7 - 3.92 = 3.08 \text{ seconds}$$

Use a similar process to find X_U. Because only 2.5% of the load times take longer than X_U seconds, 97.5% of the load times take less than X_U seconds. From the symmetry of the normal distribution, the desired Z value is $+1.96$ (positive because Z lies to the right of the standardized mean of 0), as Figure 6.15 on page 209 shows. Table 6.5 shows that, in Table E.2, 0.975 is the area under the normal curve less than the Z value of $+1.96$.

▶(*continued*)

FIGURE 6.15
Finding Z to determine X_U

Area is 0.9750

Area is 0.0250

7 X_U X Scale

0 +1.96 Z Scale

TABLE 6.5
Finding a Z value corresponding to a cumulative area of 0.975 under the normal curve

Source: Extracted from Table E.2.

					Cumulative Probabilities					
Z	**.00**	**.01**	**.02**	**.03**	**.04**	**.05**	**.06**	**.07**	**.08**	**.09**
⋮	⋮	⋮	⋮	⋮	⋮	⋮	⋮	⋮	⋮	⋮
+1.8	.9641	.9649	.9656	.9664	.9671	.9678	.9686	.9693	.9699	.9706
+1.9	.9713	.9719	.9726	.9732	.9738	.9744	.9750	.9756	.9761	.9767
+2.0	.9772	.9778	.9783	.9788	.9793	.9798	.9803	.9808	.9812	.9817

Using Equation (6.3) on page 207,

$$X = \mu + Z\sigma$$
$$= 7 + (+1.96)(2)$$
$$= 7 + 3.92 = 10.92 \text{ seconds}$$

Therefore, 95% of the load times are between 3.08 and 10.92 seconds.

An Excel worksheet template can automate normal probability calculations. Figure 6.16 shows an example of such a template that an Excel Guide workbook can contain and which PHStat can create. The worksheet template automatically recalculates when new values for the mean, standard deviation, and, if applicable, the X value(s) and percentages are entered.

FIGURE 6.16
Excel worksheet template for computing normal probabilities and finding X values

	A	B	
1	Normal Probabilities		
2			
3	Common Data		
4	Mean	7	
5	Standard Deviation	2	
6			
7	Probability for X <=		
8	X Value	7	
9	Z Value	0	=STANDARDIZE(B8, B4, B5)
10	P(X<=7)	0.5000	=NORM.DIST(B8, B4, B5, TRUE)
11			
12	Probability for X >		
13	X Value	9	
14	Z Value	1	=STANDARDIZE(B13, B4, B5)
15	P(X>9)	0.1587	=1 - NORM.DIST(B13, B4, B5, TRUE)
16			
17	Probability for X<7 or X >9		
18	P(X<7 or X >9)	0.6587	=B10 + B15
19			

	A	B	
20	Probability for a Range		
21	From X Value	5	
22	To X Value	9	
23	Z Value for 5	-1	=STANDARDIZE(B21, B4, B5)
24	Z Value for 9	1	=STANDARDIZE(B22, B4, B5)
25	P(X<=5)	0.1587	=NORM.DIST(B21, B4, B5, TRUE)
26	P(X<=9)	0.8413	=NORM.DIST(B22, B4, B5, TRUE)
27	P(5<=X<=9)	0.6827	=ABS(B26 - B25)
28			
29	Find X and Z Given a Cum. Pctage.		
30	Cumulative Percentage	10.00%	
31	Z Value	-1.28	=NORM.S.INV(B30)
32	X Value	4.44	=NORM.INV(B30, B4, B5)
33			
34	Find X Values Given a Percentage		
35	Percentage	95.00%	
36	Z Value	-1.96	=NORM.S.INV((1 - B35)/2)
37	Lower X Value	3.08	=B4 + (B36 * B5)
38	Upper X Value	10.92	=B4 - (B36 * B5)

In The Worksheet

The worksheet uses arithmetic formulas and the STANDARDIZE, NORM.DIST, NORM.INV, and NORM.S.INV functions to compute values and probabilities.

The **COMPUTE worksheet** of the Excel Guide **Normal workbook** is a copy of the Figure 6.16 worksheet.

CONSIDER THIS

What Is Normal?

Ironically, the statistician who popularized the use of "normal" to describe the distribution that Section 6.2 discussses was someone who saw the distribution as anything but the everyday, anticipated occurrence that the adjective *normal* usually suggests.

Starting with an 1894 paper, Karl Pearson argued that measurements of phenomena do not naturally, or "normally," conform to the classic bell shape. While this principle underlies much of statistics today, Pearson's point of view was radical to contemporaries who saw the world as standardized and normal. Pearson changed minds by showing that some populations are naturally *skewed* (coining that term in passing), and he helped put to rest the notion that the normal distribution underlies all phenomena.

Today, people still make the type of mistake that Pearson refuted. As a student, you are probably familiar with discussions about grade inflation, a real phenomenon at many schools. But have you ever realized that a "proof" of this inflation—that there are "too few" low grades because grades are skewed toward As and Bs—wrongly implies that grades should be "normally" distributed? Because college students represent small *nonrandom* samples, there is good reason to suspect that the distribution of grades would not be "normal."

Misunderstandings about the normal distribution have occurred both in business and in the public sector through the years. These misunderstandings have caused a number of business blunders and have sparked several public policy debates, including the causes of the collapse of large financial institutions in 2008. According to one theory, the investment banking industry's application of the normal distribution to assess risk may have contributed to the global collapse (Hutchinson and Kishnan). Using the normal distribution led these banks to overestimate the probability of having stable market conditions and underestimate the chance of unusually large market losses.

According to this theory, the use of other distributions that have less area in the middle of their curves, and, therefore, more in the "tails" that represent unusual market outcomes, might have led to less serious losses.

As you read this chapter, make sure you understand the assumptions that must hold for the proper use of the "normal" distribution, assumptions that investment bankers did not explicitly verify in 2008. And, most importantly, always remember that the name *normal distribution* does not mean normal in the everyday sense of the word.

PROBLEMS FOR SECTION 6.2

LEARNING THE BASICS

6.1 Given a standardized normal distribution (with a mean of 0 and a standard deviation of 1, as in Table E.2), what is the probability that
a. Z is less than 1.57?
b. Z is greater than 1.84?
c. Z is between 1.57 and 1.84?
d. Z is less than 1.57 or greater than 1.84?

6.2 Given a standardized normal distribution (with a mean of 0 and a standard deviation of 1, as in Table E.2), what is the probability that
a. Z is between -1.57 and 1.84?
b. Z is less than -1.57 or greater than 1.84?
c. What is the value of Z if only 2.5 percent of all possible Z values are larger?
d. Between what two values of Z (symmetrically distributed around the mean) will 68.26 percent of all possible Z values be contained?

6.3 Given a standardized normal distribution (with a mean of 0 and a standard deviation of 1, as in Table E.2), what is the probability that
a. Z is less than 1.08?
b. Z is greater than -0.21?
c. Z is less than -0.21 or greater than the mean?
d. Z is less than -0.21 or greater than 1.08?

6.4 Given a standardized normal distribution (with a mean of 0 and a standard deviation of 1, as in Table E.2), determine the following probabilities:
a. $P(Z > 1.08)$
b. $P(Z < -0.21)$
c. $P(-1.96 < Z < -0.21)$
d. What is the value of Z if only 15.87 percent of all possible Z values are larger?

6.5 Given a normal distribution with $\mu = 100$ and $\sigma = 10$, what is the probability that
a. $X > 75$?
b. $X < 70$?
c. $X < 80$ or $X > 110$?
d. Between what two X values (symmetrically distributed around the mean) are 80 percent of the values?

6.6 Given a normal distribution with $\mu = 50$ and $\sigma = 4$, what is the probability that
a. $X > 43$?
b. $X < 42$?
c. Five percent of the values are less than what X value?
d. Between what two X values (symmetrically distributed around the mean) are 60 percent of the values?

APPLYING THE CONCEPTS

6.7 In 2017, the per capita consumption of bottled water in the United States was reported to be 42.1 gallons.

Source: Data extracted from IBWA, "Bottled Water Market," **bit.ly/2JImynt**.

Assume that the per capita consumption of bottled water in the United States is approximately normally distributed with a mean of 42.1 gallons and a standard deviation of 10 gallons.
a. What is the probability that someone in the United States consumed more than 33 gallons of bottled water in 2017?
b. What is the probability that someone in the United States consumed between 10 and 20 gallons of bottled water in 2017?
c. What is the probability that someone in the United States consumed less than 10 gallons of bottled water in 2017?
d. Ninety-nine percent of the people in the United States consumed less than how many gallons of bottled water?
e. Compare the answers for this problem to the answers for Problem 6.11.

✓ SELF TEST **6.8** Toby's Trucking Company determined that the distance traveled per truck per year is normally distributed, with a mean of 50 thousand miles and a standard deviation of 12 thousand miles.
a. What proportion of trucks can be expected to travel between 34 thousand and 50 thousand miles in a year?
b. What percentage of trucks can be expected to travel either less than 30 thousand or more than 60 thousand miles in a year?
c. How many miles will be traveled by at least 80 percent of the trucks?
d. What would be your answers to (a) through (c) if the standard deviation were 10 thousand miles?

6.9 Millennials spent a mean of $163 per month on dining in 2018.

Source: Data extracted from Shop Tutors, Inc., **bit.ly/2S4xWzg**.

Assume that the amount spent on dining per month is normally distributed and that the standard deviation is $25.
a. What is the probability that a randomly selected millennial spent more than $100 per month?
b. What is the probability that a randomly selected millennial spent between $100 and $200 annually?
c. Between what two values will the middle 95 percent of the amounts spent fall?

6.10 A set of final examination grades in an introductory statistics course is normally distributed, with a mean of 73 and a standard deviation of 8.
a. What is the probability that a student earned below 91 on this exam?
b. What is the probability that a student earned between 65 and 89?
c. The probability is 5 percent that a student taking the test earns higher than what grade?
d. If the professor grades on a curve (gives A's to the top ten percent of the class, regardless of the score), are you better off earning a grade of 81 on this exam or earning a grade of 68 on a different exam that has a mean of 62 and a standard deviation of 3? Explain your answer in statistical terms.

6.11 In 2017, the per capita consumption of bottled water in China was reported to be 25.46 gallons.

Source: Data extracted from IBWA, "Bottled Water Market," **bit.ly/2JImynt**.

Assume that the per capita consumption of bottled water in China is approximately normally distributed with a mean of 25.46 gallons and a standard deviation of 8 gallons.
a. What is the probability that someone in China consumed more than 33 gallons of bottled water in 2017?
b. What is the probability that someone in China consumed between 10 and 20 gallons of bottled water in 2017?
c. What is the probability that someone in China consumed less than 10 gallons of bottled water in 2017?
d. Ninety-nine percent of the people in China consumed less than how many gallons of bottled water?
e. Compare the answers for this problem to the answers for Problem 6.7.

6.12 In 2017, the per capita consumption of soft drinks in the United States was reported to be 37.5 gallons.

Source: Data extracted from Shop Tutors, Inc., **bit.ly/2S4xWzg**.

Assume that the per capita consumption of soft drinks in the United States is approximately normally distributed with a mean of 37.5 gallons and a standard deviation of 8 gallons.
a. What is the probability that someone in the United States consumed more than 50 gallons in 2017?
b. What is the probability that someone in the United States consumed between 25 and 40 gallons in 2017?
c. What is the probability that someone in the United States consumed less than 10 gallons in 2017?
d. Ninety-nine percent of the people in the United States consumed less than how many gallons of soft drinks in 2017?

6.13 Many manufacturing problems involve the matching of machine parts, such as shafts that fit into a valve hole. A particular design requires a shaft with a diameter of 22.000 mm, but shafts with diameters between 21.990 mm and 22.010 mm are acceptable. Suppose that the manufacturing process yields shafts with diameters normally distributed, with a mean of 22.002 mm and a standard deviation of 0.005 mm. For this process, what is
a. the proportion of shafts with a diameter between 21.99 mm and 22.00 mm?
b. the probability that a shaft is acceptable?
c. the diameter that will be exceeded by only two percent of the shafts?
d. What would be your answers in (a) through (c) if the standard deviation of the shaft diameters were 0.004 mm?

6.3 Evaluating Normality

Recall the important theoretical properties of the normal distribution that Exhibit 6.1 lists on page 200. As Section 6.2 notes, many continuous variables used in business closely follow a normal distribution. To determine whether a set of data can be approximated by the normal distribution, either compare the characteristics of the data with the theoretical properties of the normal distribution or construct a normal probability plot.

Comparing Data Characteristics to Theoretical Properties

Many continuous variables have characteristics that approximate theoretical properties. However, other continuous variables are often neither normally distributed nor approximately normally distributed. For such variables, the descriptive characteristics of the data are inconsistent with the properties of a normal distribution. For such a variable, compare the observed characteristics of the variable with what would be expected to occur if the variable follows a normal distribution. To use this method:

- Construct charts and observe their appearance. For small- or moderate-sized data sets, create a stem-and-leaf display or a boxplot. For large data sets, in addition, plot a histogram or polygon.
- Compute descriptive statistics and compare these statistics with the theoretical properties of the normal distribution. Compare the mean and median. Is the interquartile range approximately 1.33 times the standard deviation? Is the range approximately 6 times the standard deviation?
- Evaluate how the values are distributed. Determine whether approximately two-thirds of the values lie between the mean and ± 1 standard deviation. Determine whether approximately four-fifths of the values lie between the mean and ± 1.28 standard deviations. Determine whether approximately 19 out of every 20 values lie between the mean and ± 2 standard deviations.

For example, use these techniques to determine whether the three-year return percentages in the sample of retirement funds that Chapters 2 and 3 discuss follow a normal distribution. Table 6.6 presents the descriptive statistics and the five-number summary for the 3YrReturn variable stored in Retirement Funds that contains those return percentages, and Figure 6.17 on page 213 uses boxplots to visualize the 3YrReturn variable.

TABLE 6.6
Descriptive statistics and five-number summary for the three-year return percentages

Descriptive Statistics		Five-Number Summary	
Mean	7.91	Minimum	−3.40
Median	8.09	First quartile	6.14
Mode	11.93	Median	8.09
Minimum	−3.40	Third quartile	9.86
Maximum	15.32	Maximum	15.32
Range	18.72		
Variance	9.10		
Standard deviation	3.02		
Coeff. of variation	38.15%		
Skewness	−0.33		
Kurtosis	0.42		
Count	479		
Standard error	0.14		

FIGURE 6.17
Excel boxplot for
the three-year return
percentages

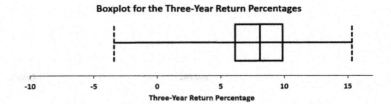

From Table 6.6, Figure 6.17, and from an ordered array of the returns (not shown), one can make these statements about the three-year returns:

- The mean of 7.91 is slightly less than the median of 8.09. (In a normal distribution, the mean and median are equal.)
- The boxplot is slightly left-skewed. (The normal distribution is symmetrical.)
- The interquartile range of 3.72 is approximately 1.23 standard deviations. (In a normal distribution, the interquartile range is 1.33 standard deviations.)
- The range of 18.72 is equal to 6.21 standard deviations. (In a normal distribution, the range is approximately 6 standard deviations.)
- 68.75% of the returns are within ± 1 standard deviation of the mean. (In a normal distribution, 68.26% of the values lie within ± 1 standard deviation of the mean.)
- 79.38% of the returns are within ± 1.28 standard deviations of the mean. (In a normal distribution, 80% of the values lie within ± 1.28 standard deviations of the mean.)
- 94.58% of the returns are within ± 2 standard deviations of the mean. (In a normal distribution, 95.44% of the values lie within ± 2 standard deviations of the mean.)
- The skewness statistic is -0.3288, and the kurtosis statistic is 0.4189. (In a normal distribution, each of these statistics equals zero.)

Based on these statements and the method that page 212 discusses, one can conclude that the three-year returns are approximately normally distributed or, at most, slightly left-skewed. The skewness is slightly negative, and the kurtosis indicates a distribution that is slightly more peaked than a normal distribution.

Constructing the Normal Probability Plot

A **normal probability plot** is a visual display that helps you evaluate whether the data are normally distributed. One common plot is called the **quantile–quantile plot**. To create this plot, you first transform each ordered value to a Z value. For example, for a sample of $n = 19$, the Z value for the smallest value corresponds to a cumulative area of

$$\frac{1}{n + 1} = \frac{1}{19 + 1} = \frac{1}{20} = 0.05$$

The Z value for a cumulative area of 0.05 (from Table E.2) is -1.65. Table 6.7 illustrates the entire set of Z values for a sample of $n = 19$.

TABLE 6.7
Ordered values and
corresponding Z values
for a sample of $n = 19$

Ordered Value	Z Value	Ordered Value	Z Value	Ordered Value	Z Value
1	-1.65	8	-0.25	14	0.52
2	-1.28	9	-0.13	15	0.67
3	-1.04	10	-0.00	16	0.84
4	-0.84	11	0.13	17	1.04
5	-0.67	12	0.25	18	1.28
6	-0.52	13	0.39	19	1.65
7	-0.39				

In a quantile–quantile plot, the Z values are plotted on the X axis, and the corresponding values of the variable are plotted on the Y axis. If the data are normally distributed, the values will plot along an approximately straight line. Figure 6.18 illustrates the typical shape of the quantile–quantile normal probability plot for a left-skewed distribution (Panel A), a normal distribution (Panel B), and a right-skewed distribution (Panel C). If the data are left-skewed, the curve will rise more rapidly at first and then level off. If the data are normally distributed, the points will plot along an approximately straight line. If the data are right-skewed, the data will rise more slowly at first and then rise at a faster rate for higher values of the variable being plotted.

FIGURE 6.18
Normal probability plots for a left-skewed distribution, a normal distribution, and a right-skewed distribution

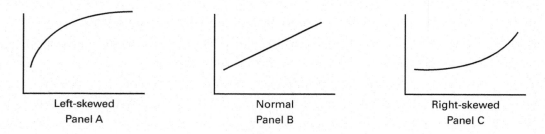

Left-skewed
Panel A

Normal
Panel B

Right-skewed
Panel C

Figure 6.19 shows an Excel normal probability plot for the three-year returns. The Excel plot shows that the bulk of the points approximately follow a straight line except for a few low values.

FIGURE 6.19
Excel normal probability plot for the three-year returns

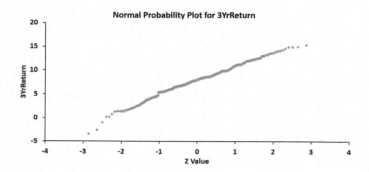

Normal Probability Plot for 3YrReturn

PROBLEMS FOR SECTION 6.3

LEARNING THE BASICS

6.14 Show that for a sample of $n = 39$, the smallest and largest Z values are -1.96 and $+1.96$, and the middle (i.e., 20th) Z value is 0.00.

6.15 For a sample of $n = 6$, list the six Z values.

APPLYING THE CONCEPTS

✓ SELF TEST **6.16** The Super Bowl is a big viewing event watched by close to 200 million Americans, so it is also a big event for advertisers. The file **Super Bowl Ad Ratings** contains the rating of ads that ran between the opening kickoff and the final whistle.
Source: Data extracted from **admeter.usatoday.com/results/2019**.

For ads that ran before halftime and ads that ran at halftime or after separately:

Decide whether the data appear to be approximately normally distributed by
a. comparing data characteristics to theoretical properties.
b. constructing a normal probability plot.

6.17 As player salaries have increased, the cost in household hours worked of attending basketball games has increased dramatically. The file **Work Hours Needed** contains the cost of three tickets purchased, food and beverages, and one parking space at each of the 30 National Basketball Association arenas during a recent season.
Source: Data extracted from ValuePenguin, **bit.ly/2mGG2Li**.

Decide whether the data appear to be approximately normally distributed by
a. comparing data characteristics to theoretical properties.
b. constructing a normal probability plot.

6.18 The file **Property Taxes** contains the property taxes on a $176K home for the 50 states and the District of Columbia. Decide whether the data appear to be approximately normally distributed by
a. comparing data characteristics to theoretical properties.
b. constructing a normal probability plot.

6.19 Thirty companies comprise the DJIA. How big are these companies? One common method for measuring the size of a company is to use its market capitalization, which is computed by multiplying the number of stock shares by the price of a share of stock.

On April 24, 2019, the market capitalization of these companies ranged from Traveler's \$35.07 billion to Apple's \$957.8 billion. The entire population of market capitalization values is stored in Dow Market Cap .

Source: Data extracted from **money.cnn.com**, April 24, 2019.

Decide whether the market capitalization of companies in the DJIA appears to be approximately normally distributed by
a. comparing data characteristics to theoretical properties.
b. constructing a normal probability plot.
c. constructing a histogram.

6.20 One operation of a mill is to cut pieces of steel into parts that will later be used as the frame for front seats in an automotive plant. The steel is cut with a diamond saw, and the resulting parts must be within ± 0.005 inch of the length specified by the automobile company. The data come from a sample of 100 steel parts and are stored in Steel . The measurement reported is the difference, in inches, between the actual length of the steel part, as measured by a laser measurement device, and the specified length of the steel part. Determine whether the data appear to be approximately normally distributed by
a. comparing data characteristics to theoretical properties.
b. constructing a normal probability plot.

6.21 The file CD Rates contains the yields for a one-year certificate of deposit (CD) and a five-year CD for 46 banks listed for West Palm Beach, Florida on April 5, 2019.

Source: Data extracted from **www.Bankrate.com**, April 5, 2019.

For each type of investment, decide whether the data appear to be approximately normally distributed by
a. comparing data characteristics to theoretical properties.
b. constructing a normal probability plot.

6.22 The file Utility contains the electricity costs, in dollars, during July of a recent year for a random sample of 50 one-bedroom apartments in a large city:

96	171	202	178	147	102	153	197	127	82
157	185	90	116	172	111	148	213	130	165
141	149	206	175	123	128	144	168	109	167
95	163	150	154	130	143	187	166	139	149
108	119	183	151	114	135	191	137	129	158

Decide whether the data appear to be approximately normally distributed by
a. comparing data characteristics to theoretical properties.
b. constructing a normal probability plot.

6.4 The Uniform Distribution

In the uniform distribution, the values are evenly distributed in the range between the smallest value, a, and the largest value, b. Selecting random numbers is one of the most common uses of the uniform distribution. When you use simple random sampling (see Section 1.3), you assume that each random digit comes from a uniform distribution that has a minimum value of 0 and a maximum value of 9.

Equation (6.4) defines the probability density function for the uniform distribution.

UNIFORM PROBABILITY DENSITY FUNCTION

$$f(X) = \frac{1}{b - a} \text{ if } a \le X \le b \text{ and 0 elsewhere} \qquad \textbf{(6.4)}$$

where

$$a = \text{minimum value of } X$$
$$b = \text{maximum value of } X$$

Equation (6.5) defines the mean of the uniform distribution, and Equation (6.6) defines the variance and standard deviation of the uniform distribution.

MEAN OF THE UNIFORM DISTRIBUTION

$$\mu = \frac{a + b}{2} \qquad \textbf{(6.5)}$$

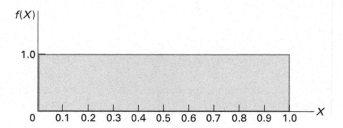

VARIANCE AND STANDARD DEVIATION OF THE UNIFORM DISTRIBUTION

$$\sigma^2 = \frac{(b - a)^2}{12} \tag{6.6a}$$

$$\sigma = \sqrt{\frac{(b - a)^2}{12}} \tag{6.6b}$$

Because of its shape, the uniform distribution is sometimes known as the rectangular distribution, as Section 6.1 notes. Figure 6.20 illustrates the uniform distribution with $a = 0$ and $b = 1$. The total area inside the rectangle is 1.0, equal to the base (1.0) times the height (1.0). Having an area of 1.0 satisfies the requirement that the area under any probability density function equals 1.0.

FIGURE 6.20

Probability density function for a uniform distribution with $a = 0$ and $b = 1$

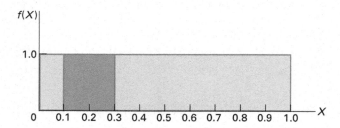

In this uniform distribution, what is the probability of getting a random number between 0.10 and 0.30? The area between 0.10 and 0.30, depicted in Figure 6.21, is equal to the base (which is $0.30 - 0.10 = 0.20$) times the height (1.0). Therefore,

$$P(0.10 < X < 0.30) = (\text{Base})(\text{Height}) = (0.20)(1.0) = 0.20$$

FIGURE 6.21

Finding $P(0.10 < X < 0.30)$ for a uniform distribution with $a = 0$ and $b = 1$

From Equations (6.5) and (6.6), the mean and standard deviation of the uniform distribution for $a = 0$ and $b = 1$ are computed as follows:

$$\mu = \frac{a + b}{2}$$

$$= \frac{0 + 1}{2} = 0.5$$

and

$$\sigma^2 = \frac{(b - a)^2}{12}$$

$$= \frac{(1 - 0)^2}{12}$$

$$= \frac{1}{12} = 0.0833$$

$$\sigma = \sqrt{0.0833} = 0.2887.$$

Thus, the mean is 0.5, and the standard deviation is 0.2887.

Example 6.6 provides another application of the uniform distribution.

EXAMPLE 6.6	In the MyTVLab scenario on page 198, the load time of the new sales page was assumed to be normally distributed with a mean of 7 seconds. Suppose that the load time follows a uniform (instead of a normal) distribution between 4.5 and 9.5 seconds. What is the probability that a load time will take more than 9 seconds?
Computing Uniform Probabilities	

SOLUTION The load time is uniformly distributed from 4.5 to 9.5 seconds. The area between 9 and 9.5 seconds is equal to 0.5 seconds, and the total area in the distribution is $9.5 - 4.5 = 5$ seconds. Therefore, the probability of a load time between 9 and 9.5 seconds is the portion of the area greater than 9, which is equal to $0.5/5.0 = 0.10$. Because 9.5 is the maximum value in this distribution, the probability of a load time above 9 seconds is 0.10. In comparison, if the load time is normally distributed with a mean of 7 seconds and a standard deviation of 2 seconds (see Example 6.1 on page 204), the probability of a load time above 9 seconds is 0.1587.

PROBLEMS FOR SECTION 6.4

LEARNING THE BASICS

6.23 Suppose you select one value from a uniform distribution with $a = 0$ and $b = 10$. What is the probability that the value will be
a. between 5 and 7?
b. between 2 and 3?
c. What is the mean?
d. What is the standard deviation?

APPLYING THE CONCEPTS

✓ SELF TEST **6.24** The time between arrivals of customers at a bank during the noon-to-1 P.M. hour has a uniform distribution between 0 to 120 seconds. What is the probability that the time between the arrival of two customers will be
a. less than 20 seconds?
b. between 10 and 30 seconds?
c. more than 35 seconds?
d. What are the mean and standard deviation of the time between arrivals?

6.25 A study of the time spent shopping in a supermarket for a market basket of 20 specific items showed an approximately uniform distribution between 20 minutes and 40 minutes. What is the probability that the shopping time will be
a. between 25 and 30 minutes?
b. less than 35 minutes?

c. What are the mean and standard deviation of the shopping time?

6.26 How long does it take to download a two-hour HD movie from a streaming media site? According to Fastmetrics's "Download Speed Comparison Table," **bit.ly/1EONXKY**, downloading such a movie using a 100 Mbps Internet connection should take about 4.5 minutes. Assume that the download times are uniformly distributed between 3.5 and 5.5 minutes. If you download a two-hour movie, what is the probability that the download time will be
a. less than 4.5 minutes?
b. more than 4 minutes?
c. between 4.0 and 4.5 minutes?
d. What are the mean and standard deviation of the download times?

6.27 The scheduled commuting time on the Long Island Railroad from Glen Cove to New York City is 65 minutes. Suppose that the actual commuting time is uniformly distributed between 64 and 74 minutes. What is the probability that the commuting time will be
a. less than 70 minutes?
b. between 65 and 70 minutes?
c. more than 65 minutes?
d. What are the mean and standard deviation of the commuting time?

6.5 The Exponential Distribution

The exponential distribution is a continuous distribution that is right-skewed and ranges from 0 to positive infinity (see Figure 6.1 on page 199). The **Section 6.5 online topic** discusses this distribution and illustrates its application.

6.6 The Normal Approximation to the Binomial Distribution

In many circumstances, the normal distribution can be used to approximate the binomial distribution, discussed in Section 5.2. The **Section 6.6 online topic** discusses this technique and illustrates its use.

▼ USING **STATISTICS**
Normal Load Times ... , Revisited

In the Normal Downloading at MyTVLab scenario, you were the sales and marketing vice president for a web-based business. You sought to ensure that the load time for a new sales web page would be within a certain range. By running experiments in the corporate offices, you determined that the amount of time, in seconds, that passes from first pointing a browser to a web page until the web page is fully loaded is a bell-shaped distribution with a mean load time of 7 seconds and standard deviation of 2 seconds. Using the normal distribution, you were able to calculate that approximately 84% of the load times are 9 seconds or less, and 95% of the load times are between 3.08 and 10.92 seconds.

Now that you understand how to calculate probabilities from the normal distribution, you can evaluate load times of similar sales web pages that use other designs. For example, if the standard deviation remained at 2 seconds, lowering the mean to 6 seconds would shift the entire distribution lower by 1 second. Thus, approximately 84% of the load times would be 8 seconds or less, and 95% of the load times would be between 2.08 and 9.92 seconds. Another change that could reduce long load times would be reducing the variation. For example, consider the case where the mean remained at the original 7 seconds but the standard deviation was reduced to 1 second. Again, approximately 84% of the load times would be 8 seconds or less, and 95% of the load times would be between 5.04 and 8.96 seconds.

▼ SUMMARY

This chapter and Chapter 5 discuss probability distributions, mathematical models that can be used to solve business problems. Chapter 5 uses discrete probability distributions for situations where the values come from a counting process such as the number of social media sites to which you belong or the number of tagged order forms in a report generated by an accounting information system. This chapter uses continuous probability distributions for situations where the values come from a measuring process such as your height or the download time of a video.

Continuous probability distributions come in various shapes, but the most common and most important in business is the normal distribution. The normal distribution is symmetrical; thus, its mean and median are equal. It is also bell-shaped, and approximately 68.26% of its values are within ±1 standard deviation of the mean, approximately 95.44% of its values are within ±2 standard deviations of the mean, and approximately 99.73% of its values are within ±3 standard deviations of the mean. Although many variables in business are closely approximated by the normal distribution, not all variables can be approximated by the normal distribution.

Section 6.3 discusses methods for evaluating normality in order to determine whether the normal distribution is a reasonable mathematical model to use in specific situations. Chapter 7 uses the normal distribution that this chapter explains to develop the concept of statistical inference.

▼ REFERENCES

Gunter, B. "Q-Q Plots." *Quality Progress* (February 1994): 81–86.

Hogg, R. V., J. T. McKean, and A. V. Craig. *Introduction to Mathematical Statistics*, 7th ed. New York: Pearson Education, 2013.

Hutchinson, M. "A Finer Formula for Assessing Risk." *The New York Times*, May 11, 2010, p. B2

Kishnan, S. and R. Sitaraman. "Video stream quality impacts viewer behavior: inferring causality using quasi-experimental designs," in *Proceedings of the 2012 ACM conference on Internet measurement conference*: 211–224. New York: ACM.

Levine, D. M., P. Ramsey, and R. Smidt. *Applied Statistics for Engineers and Scientists Using Microsoft Excel and Minitab*. Upper Saddle River, NJ: Prentice Hall, 2001.

Miller, J. "Earliest Known Uses of Some of the Words of Mathematics." **jeff560.tripod.com/mathword.html**.

Pearl, R. "Karl Pearson, 1857–1936." *Journal of the American Statistical Association*, 31 (1936): 653–664.

Pearson, E. S. "Some Incidents in the Early History of Biometry and Statistics, 1890–94." *Biometrika* 52 (1965): 3–18.

Taleb, N. *The Black Swan*, 2nd ed. New York: Random House, 2010.

Walker, H. "The Contributions of Karl Pearson." *Journal of the American Statistical Association* 53 (1958): 11–22.

▼ KEY EQUATIONS

Normal Probability Density Function

$$f(X) = \frac{1}{\sqrt{2\pi}\sigma} e^{-(1/2)[(X-\mu)/\sigma]^2} \qquad (6.1)$$

Z Transformation Formula

$$Z = \frac{X - \mu}{\sigma} \qquad (6.2)$$

Finding an X Value Associated with a Known Probability

$$X = \mu + Z\sigma \qquad (6.3)$$

Uniform Probability Density Function

$$f(X) = \frac{1}{b - a} \qquad (6.4)$$

Mean of the Uniform Distribution

$$\mu = \frac{a + b}{2} \qquad (6.5)$$

Variance and Standard Deviation of the Uniform Distribution

$$\sigma^2 = \frac{(b - a)^2}{12} \qquad (6.6a)$$

$$\sigma = \sqrt{\frac{(b - a)^2}{12}} \qquad (6.6b)$$

▼ KEY TERMS

cumulative standardized normal distribution   202
exponential distribution   199
normal distribution   199

normal probability plot   213
probability density function   199
probability density function for the normal distribution   201

quantile–quantile plot   213
standardized normal variable   202
transformation formula   202
uniform distribution   199

▼ CHECKING YOUR UNDERSTANDING

6.28 How do you find the area between two values under the normal curve?

6.29 How do you find the X value that corresponds to a given percentile of the normal distribution?

6.30 What are some of the distinguishing properties of a normal distribution?

6.31 How does the shape of the normal distribution differ from the shapes of the uniform and exponential distributions?

6.32 How can you use the normal probability plot to evaluate whether a set of data is normally distributed?

▼ CHAPTER REVIEW PROBLEMS

6.33 An industrial sewing machine uses ball bearings that are targeted to have a diameter of 0.75 inch. The lower and upper specification limits under which the ball bearings can operate are 0.74 inch and 0.76 inch, respectively. Past experience has indicated that the actual diameter of the ball bearings is approximately normally distributed, with a mean of 0.753 inch and a standard deviation of 0.004 inch. What is the probability that a ball bearing is
a. between the target and the actual mean?
b. between the lower specification limit and the target?
c. above the upper specification limit?
d. below the lower specification limit?
e. Of all the ball bearings, 93% of the diameters are greater than what value?

6.34 The fill amount in 2-liter soft drink bottles is normally distributed, with a mean of 2.0 liters and a standard deviation of 0.05 liter. If bottles contain less than 95% of the listed net content (1.90 liters, in this case), the manufacturer may be subject to penalty by the state office of consumer affairs. Bottles that have a net content above 2.10 liters may cause excess spillage upon opening. What proportion of the bottles will contain
a. between 1.90 and 2.0 liters?
b. between 1.90 and 2.10 liters?
c. below 1.90 liters or above 2.10 liters?
d. At least how much soft drink is contained in 99% of the bottles?
e. Ninety-nine percent of the bottles contain an amount that is between which two values (symmetrically distributed) around the mean?

6.35 In an effort to reduce the number of bottles that contain less than 1.90 liters, the bottler in Problem 6.34 sets the filling machine so that the mean is 2.02 liters. Under these circumstances, what are your answers in Problem 6.34 (a) through (e)?

6.36 *Webrooming*, researching products online before buying them in store, has become the new norm for some consumers and contrasts with *showrooming*, researching products in a physical store before purchasing online. A recent study by Interactions reported that most shoppers have a specific spending limit in place while shopping online. Findings indicate that men spend an average of $250 online before they decide to visit a store.

Source: Data extracted from **bit.ly/1JEcmqh**.

Assume that the spending limit is normally distributed and that the standard deviation is $20.
a. What is the probability that a male spent less than $210 online before deciding to visit a store?
b. What is the probability that a male spent between $270 and $300 online before deciding to visit a store?
c. Ninety percent of the amounts spent online by a male before deciding to visit a store are less than what value?
d. Eighty percent of the amounts spent online by a male before deciding to visit a store are between what two values symmetrically distributed around the mean?

6.37 The file **Domestic Beer** contains the percentage alcohol, number of calories per 12 ounces, and number of carbohydrates (in grams) per 12 ounces for 157 of the best-selling domestic beers in the United States. Determine whether each of these variables appears to be approximately normally distributed. Support your decision through the use of appropriate statistics and graphs.

Source: Data extracted from **www.Beer100.com**, April 3, 2019.

6.38 The evening manager of a restaurant was very concerned about the length of time some customers were waiting in line to be seated. She also had some concern about the seating times—that is, the length of time between when a customer is seated and the time he or she leaves the restaurant. Over the course of one week, 100 customers (no more than one per party) were randomly selected, and their waiting and seating times (in minutes) were recorded in **Wait** .
a. Think about your favorite restaurant. Do you think waiting times more closely resemble a uniform, an exponential, or a normal distribution?
b. Again, think about your favorite restaurant. Do you think seating times more closely resemble a uniform, an exponential, or a normal distribution?
c. Construct a histogram and a normal probability plot of the waiting times. Do you think these waiting times more closely resemble a uniform, an exponential, or a normal distribution?
d. Construct a histogram and a normal probability plot of the seating times. Do you think these seating times more closely resemble a uniform, an exponential, or a normal distribution?

6.39 The major stock market indexes had weak results in 2018. The mean one-year return for stocks in the S&P 500, a group of 500 very large companies, was −6.24%. The mean one-year return for the NASDAQ, a group of 3,200 small and medium-sized companies, was −3.88%. Historically, the one-year returns are approximately normally distributed, the standard deviation in the S&P 500

is approximately 20%, and the standard deviation in the NASDAQ is approximately 30%.
a. What is the probability that a stock in the S&P 500 gained value in 2018?
b. What is the probability that a stock in the S&P 500 gained 10% or more in 2018?
c. What is the probability that a stock in the S&P 500 lost 20% or more in 2018?
d. What is the probability that a stock in the S&P 500 lost 30% or more in 2018?
e. Repeat (a) through (d) for a stock in the NASDAQ.
f. Write a short summary on your findings. Be sure to include a discussion of the risks associated with a large standard deviation.

6.40 Interns report that when deciding on where to work, career growth, salary and compensation, location and commute, and company culture and values are important factors to them. According to reports by interns to Glassdoor, the mean monthly pay of interns at Intel is $5,940.

Source: Data extracted from **www.glassdoor.com/index.htm**.

Suppose that the intern monthly pay is normally distributed, with a standard deviation of $400. What is the probability that the monthly pay of an intern at Intel is
a. less than $5,900?
b. between $5,700 and $6,100?
c. above $6,500?
d. Ninety-nine percent of the intern monthly pays are higher than what value?
e. Ninety-five percent of the intern monthly pays are between what two values, symmetrically distributed around the mean?

6.41 According to the same Glassdoor source mentioned in Problem 6.40, the mean monthly pay for interns at Facebook is $6,589. Suppose that the intern monthly pay is normally distributed, with a standard deviation of $500. What is the probability that the monthly pay of an intern at Facebook is
a. less than $5,900?
b. between $5,700 and $6,100?
c. above $6,500?
d. Ninety-nine percent of the intern monthly pays are higher than what value?
e. Ninety-five percent of the intern monthly pays are between what two values, symmetrically distributed around the mean?
f. Compare the results for the Intel interns computed in Problem 6.40 to those of the Facebook interns.

6.42 **(Class Project)** One theory about the daily changes in the closing price of a stock is that these changes follow a *random walk*—that is, these daily events are independent of each other and move upward or downward in a random manner—and can be approximated by a normal distribution. To test this theory, use either a newspaper or the Internet to select one company traded on the NYSE, one company traded on the American Stock Exchange, and one company traded on the NASDAQ and then do the following:
1. Record the daily closing stock price of each of these companies for six consecutive weeks (so that you have 30 values per company).
2. Calculate the daily changes in the closing stock price of each of these companies for six consecutive weeks (so that you have 30 values per company).

Note: The random-walk theory pertains to the daily changes in the closing stock price, not the daily closing stock price.

For each of your six data sets, decide whether the data are approximately normally distributed by

a. constructing the stem-and-leaf display, histogram or polygon, and boxplot.

b. comparing data characteristics to theoretical properties.

c. constructing a normal probability plot.

d. Discuss the results of (a) through (c). What can you say about your three stocks with respect to daily closing prices and daily changes in closing prices? Which, if any, of the data sets are approximately normally distributed?

▼CASES

Managing Ashland MultiComm Services

The AMS technical services department has embarked on a quality improvement effort. Its first project relates to maintaining the target upload speed for its Internet service subscribers. Upload speeds are measured on a standard scale in which the target value is 1.0. Data collected over the past year indicate that the upload speed is approximately normally distributed, with a mean of 1.005 and a standard deviation of 0.10. Each day, one upload speed is measured. The upload speed is considered acceptable if the measurement on the standard scale is between 0.95 and 1.05.

1. Assuming that the distribution of upload speed has not changed from what it was in the past year, what is the probability that the upload speed is
   a. less than 1.0?
   b. between 0.95 and 1.0?
   c. between 1.0 and 1.05?
   d. less than 0.95 or greater than 1.05?

2. The objective of the operations team is to reduce the probability that the upload speed is less than 1.0. Should the team focus on process improvement that increases the mean upload speed to 1.05 or on process improvement that reduces the standard deviation of the upload speed to 0.075? Explain.

CardioGood Fitness

Return to the CardioGood Fitness case (stored in CardioGood Fitness) first presented on page 33.

1. For each CardioGood Fitness treadmill product line, determine whether the age, income, usage, and the number of miles the customer expects to walk/run each week can be approximated by the normal distribution.

2. Write a report to be presented to the management of CardioGood Fitness detailing your findings.

More Descriptive Choices Follow-up

Follow up the More Descriptive Choices Revisited Using Statistics scenario on page 147 by constructing normal probability plots for the 1-year return percentages, 5-year return percentages, and 10-year return percentages for the sample of 479 retirement funds stored in Retirement Funds . In your analysis, examine differences between the growth and value funds as well as the differences among the small, mid-cap, and large market cap funds.

Clear Mountain State Student Survey

The Student News Service at Clear Mountain State University (CMSU) has decided to gather data about the undergraduate students who attend CMSU. They create and distribute a survey of 14 questions and receive responses from 111 undergraduates (stored in Student Survey). For each numerical variable in the survey, decide whether the variable is approximately normally distributed by

a. comparing data characteristics to theoretical properties.
b. constructing a normal probability plot.
c. writing a report summarizing your conclusions.

Digital Case

Apply your knowledge about the normal distribution in this Digital Case, which extends the Using Statistics scenario from this chapter.

To satisfy concerns of potential customers, the management of MyTVLab has undertaken a research project to learn how much time it takes users to load a complex video features page. The research team has collected data and has made some claims based on the assertion that the data follow a normal distribution.

Open **MTL_QRTStudy.pdf**, which documents the work of a quality response team at MyTVLab. Read the internal report that documents the work of the team and their conclusions. Then answer the following:

1. Can the collected data be approximated by the normal distribution?

2. Review and evaluate the conclusions made by the MyTVLab research team. Which conclusions are correct? Which ones are incorrect?

3. If MyTVLab could improve the mean time by 5 seconds, how would the probabilities change?

EG6.2   The NORMAL DISTRIBUTION

Key Technique Use the **NORM.DIST**(*X value, mean, standard deviation*, **True**) function to compute normal probabilities and use the **NORM.S.INV**(*percentage*) function and the STANDARDIZE function (see Section EG3.2) to compute the *Z* value.

Example Compute the normal probabilities for Examples 6.1 through 6.3 on pages 204 and 205 and the *X* and *Z* values for Examples 6.4 and 6.5 on pages 207 and 208.

PHStat Use **Normal**.

For the example, select **PHStat→Probability & Prob. Distributions→Normal**. In this procedure's dialog box (shown below):

1. Enter **7** as the **Mean** and **2** as the **Standard Deviation**.
2. Check **Probability for: X <=** and enter **7** in its box.
3. Check **Probability for: X >** and enter **9** in its box.
4. Check **Probability for range** and enter **5** in the first box and **9** in the second box.
5. Check **X for Cumulative Percentage** and enter **10** in its box.
6. Check **X Values for Percentage** and enter **95** in its box.
7. Enter a **Title** and click **OK**.

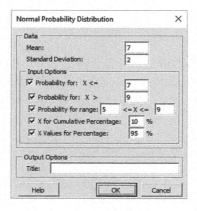

Workbook Use the **COMPUTE worksheet** of the **Normal workbook** as a template.

The worksheet already contains the data for solving the problems in Examples 6.1 through 6.5. For other problems, change the values for the **Mean**, **Standard Deviation**, **X Value**, **From X Value**, **To X Value**, **Cumulative Percentage**, and/or **Percentage**.

Unlike most other Excel Guide COMPUTE worksheets, this worksheet uses formulas in column A to dynamically create labels based on the data values you enter. These formulas make extensive use of the ampersand operator

(&) to construct the actual label. For example, the cell A10 formula **="P(X<="&B8&")"** results in the display of P(X<=7) because the initial contents of cell B8, 7, is combined with "*P(X<=*" and ")". Changing the value in cell B8 to 9, changes the label in cell A10 to P(X<=9).

EG6.3   EVALUATING NORMALITY

Comparing Data Characteristics to Theoretical Properties

Use the Section EG3.1 through EG3.3 instructions to compare data characteristics to theoretical properties.

Constructing the Normal Probability Plot

Key Technique Use a scatter plot with *Z* values computed using the NORM.S.INV function.

Example Construct the Figure 6.19 normal probability plot for three-year return percentages for the sample of 479 retirement funds that is shown on page 214.

PHStat Use **Normal Probability Plot**.

For the example, open to the **DATA worksheet** of the **Retirement Funds workbook**. Select **PHStat→Probability & Prob. Distributions→Normal Probability Plot**. In the procedure's dialog box (shown below):

1. Enter **K1:K480** as the **Variable Cell Range**.
2. Check **First cell contains label**.
3. Enter a **Title** and click **OK**.

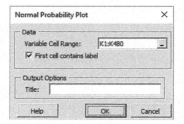

In addition to the chart sheet containing the normal probability plot, the procedure creates a plot data worksheet identical to the PlotData worksheet discussed in the *Worksheet Excel* instructions.

Workbook Use the worksheets of the **NPP workbook** as templates.

The **NormalPlot chart sheet** displays a normal probability plot using the rank, the proportion, the *Z* value, and the variable found in the **PLOT_DATA worksheet**. The PLOT_DATA worksheet already contains the three-year return percentages for the example.

To construct a plot for a different variable, paste the *sorted* values for that variable in **column D** of the **PLOT_DATA worksheet**. Adjust the number of ranks in **column A** and the divisor in the formulas in **column B** to compute cumulative percentages to reflect the quantity $n + 1$ (480 for the example). (Column C formulas use the NORM.S.INV function to compute the Z values for those cumulative percentages.)

If you have fewer than 479 values, delete rows from the bottom up. If you have more than 479 values, select row 480, right-click, click **Insert** in the shortcut menu, and copy down the formulas in columns B and C to the new rows.

To create your own normal probability plot for the 3YrReturn variable, open to the PLOT_DATA worksheet and select the cell range **C1:D480**. Then select **Insert → Scatter (X, Y) or Bubble Chart icon** (#6 in the labeled Chart group shown at the start of the Chapter 2 Excel Guide) and select the **Scatter** gallery item. Excel for Mac labels the same icon as **X Y (Scatter)**. (For more information about creating a scatter plot, see Section EG2.5.)

Relocate the chart to a chart sheet, turn off the chart legend and gridlines, add axis titles, and modify the chart title.

7

Sampling Distributions

OBJECTIVES

- Learn about the concept of the sampling distribution
- Calculate probabilities related to the sample mean and the sample proportion
- Understand the importance of the Central Limit Theorem

▼ USING **STATISTICS**
Sampling Oxford Cereals

As the cereal lines manager for Oxford Cereals Plant 3, you are part of the project team overseeing the installation of three new fill production lines. By automating the bag formation, fill, bag sealing, and weighing operations, three identical lines running at Plant 1 have increased the production of boxes of flaked cereals at that plant by 20%, and similar gains are expected at Plant 3. In the future, these lines will give Oxford Cereals management greater production flexibility by allowing the option to use packaging other than the standard pillow bags long used.

For now, you must verify the calibration of the Plant 3 filling machines. Proper calibration should ensure that filled boxes will contain a mean of 368 grams of cereal, among other attributes. If the calibration is imperfect, the mean weight of the boxes could vary too much from the 368 grams claimed on the preprinted boxes used in the lines. You decide to take samples of the cereal boxes being produced in the initial runs of the new lines. For each sample of cereal boxes you select, you plan to weigh each box in the sample and then calculate a sample mean. You need to determine the probability that such a sample mean could have been randomly selected from a population whose mean is 368 grams. Based on your analysis, you will have to decide whether to maintain, alter, or shut down the cereal-filling process.

Building on the foundation of the normal distribution that Chapter 6 develops, this chapter explores sampling distributions. Sampling distributions enable one to use a statistic, such as the sample mean, to estimate the population parameter, such as the population mean. In the Oxford Cereals scenario, you need to make a decision about a cereal-filling process, based on the weights of a *sample* of filled boxes. Proper application of a sampling distribution can assist you determining whether the calibration of the Plant 3 filling machines is acceptable.

7.1 Sampling Distributions

In many situations, one needs to make inferences that are based on statistics calculated from samples to estimate the values of population parameters. The main focus when using statistical inference is reaching conclusions about a population and *not* reaching conclusions only about the random sample drawn from the population for analysis. For example, a political pollster is interested in the sample results only as a way of estimating the actual proportion of the votes that each candidate will receive from the population of voters. Likewise, as plant operations manager for Oxford Cereals, you seek to use the mean weight calculated from a sample of cereal boxes to estimate the mean weight of a population of boxes.

Hypothetically, to use the sample statistic to estimate the population parameter, one could examine *every* possible sample of a given size that could occur. A **sampling distribution** is the distribution of the results that would occur had one selected all possible samples. In practice, one selects a *single* random sample of a predetermined size from the population. The single result obtained is just one of the results in the sampling distribution.

7.2 Sampling Distribution of the Mean

In Chapter 3, several measures of central tendency, including the mean, median, and mode, were discussed. For several reasons, the mean is the most widely used measure of central tendency, and the sample mean is often used to estimate the population mean. The **sampling distribution of the mean** is the distribution of all possible sample means calculated from all possible samples of a given size.

learnMORE

Learn more about the unbiased property of the sample in the SHORT TAKES for Chapter 7.

The Unbiased Property of the Sample Mean

The sample mean is **unbiased** because the mean of all the possible sample means (of a given sample size, n) is equal to the population mean, μ. A simple example concerning a population of four administrative assistants demonstrates this property. Each assistant is asked to apply the same set of updates to a human resources database. Table 7.1 presents the number of errors made by each of the administrative assistants. This population distribution is shown in Figure 7.1.

TABLE 7.1
Number of errors made by each of four administrative assistants

Administrative Assistant	Number of Errors
Ann	$X_1 = 3$
Bob	$X_2 = 2$
Carla	$X_3 = 1$
Dave	$X_4 = 4$

FIGURE 7.1
Number of errors made by a population of four administrative assistants

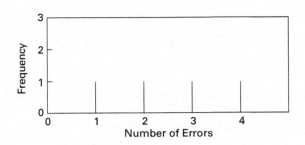

When population data exist, calculate the population mean by using Equation (7.1), and calculate the population standard deviation, σ, by using Equation (7.2).

student TIP

Recall from Section 3.4 that the population mean is the sum of the values in the population divided by the population size, *N*.

POPULATION MEAN

$$\mu = \frac{\sum\limits_{i=1}^{N} X_i}{N} \tag{7.1}$$

POPULATION STANDARD DEVIATION

$$\sigma = \sqrt{\frac{\sum\limits_{i=1}^{N}(X_i - \mu)^2}{N}} \tag{7.2}$$

For the data of Table 7.1,

$$\mu = \frac{3 + 2 + 1 + 4}{4} = 2.5 \text{ errors}$$

and

$$\sigma = \sqrt{\frac{(3 - 2.5)^2 + (2 - 2.5)^2 + (1 - 2.5)^2 + (4 - 2.5)^2}{4}} = 1.12 \text{ errors}$$

If selecting samples of two administrative assistants *with* replacement from this population, 16 samples are possible ($N^n = 4^2 = 16$). Table 7.2 lists the 16 possible sample outcomes. The mean of all 16 of these sample means is 2.5, which is also the mean of the population, μ.

TABLE 7.2

All 16 samples of *n* = 2 administrative assistants from a population of *N* = 4 administrative assistants when sampling with replacement

Sample	Administrative Assistants	Sample Outcomes	Sample Mean
1	Ann, Ann	3, 3	$\bar{X}_1 = 3$
2	Ann, Bob	3, 2	$\bar{X}_2 = 2.5$
3	Ann, Carla	3, 1	$\bar{X}_3 = 2$
4	Ann, Dave	3, 4	$\bar{X}_4 = 3.5$
5	Bob, Ann	2, 3	$\bar{X}_5 = 2.5$
6	Bob, Bob	2, 2	$\bar{X}_6 = 2$
7	Bob, Carla	2, 1	$\bar{X}_7 = 1.5$
8	Bob, Dave	2, 4	$\bar{X}_8 = 3$
9	Carla, Ann	1, 3	$\bar{X}_9 = 2$
10	Carla, Bob	1, 2	$\bar{X}_{10} = 1.5$
11	Carla, Carla	1, 1	$\bar{X}_{11} = 1$
12	Carla, Dave	1, 4	$\bar{X}_{12} = 2.5$
13	Dave, Ann	4, 3	$\bar{X}_{13} = 3.5$
14	Dave, Bob	4, 2	$\bar{X}_{14} = 3$
15	Dave, Carla	4, 1	$\bar{X}_{15} = 2.5$
16	Dave, Dave	4, 4	$\bar{X}_{16} = 4$
			$\mu_{\bar{X}} = 2.5$

Because the mean of the 16 sample means is equal to the population mean, the sample mean is an unbiased estimator of the population mean. Therefore, although one does not know how close the sample mean of any particular sample selected is to the population mean, one can state that the mean of all the possible sample means that could have been selected is equal to the population mean.

Standard Error of the Mean

Figure 7.2 illustrates the variation in the sample means when selecting all 16 possible samples.

FIGURE 7.2
Sampling distribution of the mean, based on all possible samples containing two administrative assistants

Source: Data are from Table 7.2.

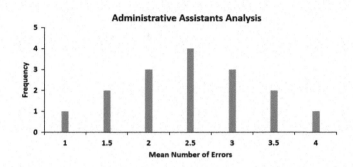

In this small example, although the sample means vary from sample to sample, depending on which two administrative assistants are selected, the sample means do not vary as much as the individual values in the population. That the sample means are less variable than the individual values in the population follows directly from the fact that each sample mean averages together all the values in the sample. A population consists of individual outcomes that can take on a wide range of values, from extremely small to extremely large. However, if a sample contains an extreme value, although this value will have an effect on the sample mean, the effect is reduced because the value is averaged with all the other values in the sample. As the sample size increases, the effect of a single extreme value becomes smaller because it is averaged with more values.

The value of the standard deviation of all possible sample means, called the **standard error of the mean**, expresses how the sample means vary from sample to sample. As the sample size increases, the standard error of the mean decreases by a factor equal to the square root of the sample size. Equation (7.3) defines the standard error of the mean when sampling *with* replacement or sampling *without* replacement from large or infinite populations.

student TIP

Remember, the standard error of the mean measures variation among the means not the individual values.

STANDARD ERROR OF THE MEAN

The standard error of the mean, $\sigma_{\bar{X}}$, is equal to the standard deviation in the population, σ, divided by the square root of the sample size, n.

$$\sigma_{\bar{X}} = \frac{\sigma}{\sqrt{n}} \qquad (7.3)$$

Example 7.1 computes the standard error of the mean when the sample selected without replacement contains less than 5% of the entire population.

EXAMPLE 7.1

Computing the Standard Error of the Mean

Returning to the Oxford Cereals scenario, if you randomly select a sample of 25 boxes without replacement from the thousands of boxes filled during a shift, the sample contains a very small portion of the population. Given that the standard deviation of the cereal-filling process is 15 grams, compute the standard error of the mean.

SOLUTION Using Equation (7.3) with $n = 25$ and $\sigma = 15$ the standard error of the mean is

$$\sigma_{\bar{X}} = \frac{\sigma}{\sqrt{n}} = \frac{15}{\sqrt{25}} = \frac{15}{5} = 3$$

The variation in the sample means for samples of $n = 25$ is much less than the variation in the individual boxes of cereal because $\sigma_{\bar{X}} = 3$, while $\sigma = 15$.

Sampling from Normally Distributed Populations

What distribution does the sample mean, $\overline{X}$, follow? If sampling is done from a population that is normally distributed with mean μ and standard deviation σ, then regardless of the sample size, n, the sampling distribution of the mean is normally distributed, with mean $\mu_{\overline{X}} = \mu$ and standard error of the mean $\sigma_{\overline{X}} = \sigma/\sqrt{n}$.

For the simplest case, in which samples of size $n = 1$ are taken, each possible sample mean is a single value from the population because

$$\overline{X} = \frac{\sum\limits_{i=1}^{n} X_i}{n} = \frac{X_1}{1} = X_1$$

Therefore, if the population is normally distributed, with mean μ and standard deviation σ, the sampling distribution $\overline{X}$ for samples of $n = 1$ must also follow the normal distribution, with mean $\mu_{\overline{X}} = \mu$ and standard error of the mean $\sigma_{\overline{X}} = \sigma/\sqrt{1} = \sigma$. In addition, as the sample size increases, the sampling distribution of the mean still follows a normal distribution, with $\mu_{\overline{X}} = \mu$, but the standard error of the mean decreases so that a larger proportion of sample means are closer to the population mean.

Figure 7.3 illustrates this reduction in variability. Note that 500 samples of size 1, 2, 4, 8, 16, and 32 were randomly selected from a normally distributed population. The Figure 7.3 polygons show that, although the sampling distribution of the mean is approximately[1] normal for each sample size, the sample means are distributed more tightly around the population mean as the sample size increases.

[1] Remember that "only" 500 samples out of an infinite number of samples have been selected, so that the sampling distributions shown are only approximations of the population distribution.

FIGURE 7.3
Sampling distributions of the mean from 500 samples of sizes $n = 1, 2, 4, 8, 16,$ and 32 selected from a normally distributed population

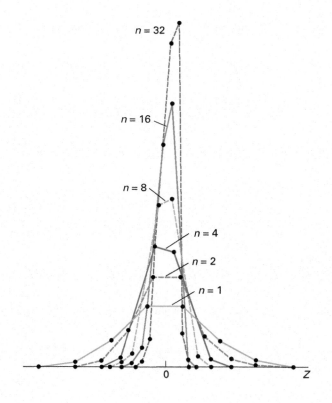

To further examine the concept of the sampling distribution of the mean, return to the Oxford Cereals scenario. The packaging equipment that is filling 368-gram boxes of cereal is set so that the amount of cereal in a box is normally distributed, with a mean of 368 grams. Past experience informs Oxford Cereals that the population standard deviation for this filling process is 15 grams.

If a random sample of 25 boxes is taken from the many thousands that are filled in a day and the sample mean weight calculated, what type of result could one expect? Would the sample mean be 368 grams? 200 grams? 365 grams?

The sample acts as a miniature representation of the population, so if the values in the population are normally distributed, the values in the sample should be approximately normally distributed. If the population mean is 368 grams, the sample mean has a good chance of being close to 368 grams.

How can one determine the probability that the sample of 25 boxes will have a mean below 365 grams? From the application of the normal distribution that Section 6.2 discusses, one can find the area below any value X by converting to standardized Z values:

$$Z = \frac{X - \mu}{\sigma}$$

The Section 6.2 examples show how any single value, X, differs from the population mean. Now, to answer the question, one needs to study how a sample mean, $\overline{X}$, differs from the population mean. Substituting $\overline{X}$ for X, $\mu_{\overline{X}}$ for μ, and $\sigma_{\overline{X}}$ for σ in the equation above produces Equation (7.4).

FINDING Z FOR THE SAMPLING DISTRIBUTION OF THE MEAN

The Z value is equal to the difference between the sample mean, $\overline{X}$, and the population mean, μ, divided by the standard error of the mean, $\sigma_{\overline{X}}$.

$$Z = \frac{\overline{X} - \mu_{\overline{X}}}{\sigma_{\overline{X}}} = \frac{\overline{X} - \mu}{\dfrac{\sigma}{\sqrt{n}}} \tag{7.4}$$

To find the area below 365 grams, from Equation (7.4),

$$Z = \frac{\overline{X} - \mu_{\overline{X}}}{\sigma_{\overline{X}}} = \frac{365 - 368}{\dfrac{15}{\sqrt{25}}} = \frac{-3}{3} = -1.00$$

The area corresponding to $Z = -1.00$ in Table E.2 is 0.1587. Therefore, 15.87% of all the possible samples of 25 boxes have a sample mean below 365 grams.

The preceding statement is not the same as saying that a certain percentage of *individual* boxes will contain less than 365 grams of cereal. That percentage is calculated as follows:

$$Z = \frac{X - \mu}{\sigma} = \frac{365 - 368}{15} = \frac{-3}{15} = -0.20$$

The area corresponding to $Z = -0.20$ in Table E.2 is 0.4207. Therefore, 42.07% of the *individual* boxes are expected to contain less than 365 grams. Comparing these results, you see that many more *individual boxes* than *sample means* are below 365 grams. This result is explained by the fact that each sample consists of 25 different values, some small and some large. The averaging process dilutes the importance of any individual value, particularly when the sample size is large. Therefore, the chance that the sample mean of 25 boxes is very different from the population mean is less than the chance that a *single* box is very different from the population mean.

Examples 7.2 and 7.3 show how these results are affected by using different sample sizes.

EXAMPLE 7.2

The Effect of Sample Size, *n*, on the Computation of $\sigma_{\bar{X}}$

How is the standard error of the mean affected by increasing the sample size from 25 to 100 boxes?

SOLUTION If $n = 100$ boxes, then using Equation (7.3) on page 227,

$$\sigma_{\bar{X}} = \frac{\sigma}{\sqrt{n}} = \frac{15}{\sqrt{100}} = \frac{15}{10} = 1.5$$

The fourfold increase in the sample size from 25 to 100 reduces the standard error of the mean by half—from 3 grams to 1.5 grams. This demonstrates that taking a larger sample results in less variability in the sample means from sample to sample.

EXAMPLE 7.3

The Effect of Sample Size, *n*, on the Clustering of Means in the Sampling Distribution

If you select a sample of 100 boxes, what is the probability that the sample mean is below 365 grams?

SOLUTION Using Equation (7.4) on page 229,

$$Z = \frac{\bar{X} - \mu_{\bar{X}}}{\sigma_{\bar{X}}} = \frac{365 - 368}{\dfrac{15}{\sqrt{100}}} = \frac{-3}{1.5} = -2.00$$

From Table E.2, the area less than $Z = -2.00$ is 0.0228. Therefore, 2.28% of the samples of 100 boxes have means below 365 grams, as compared with 15.87% for samples of 25 boxes.

Sometimes one needs to find the interval that contains a specific proportion of the sample means. To do so, determine a distance below and above the population mean containing a specific area of the normal curve. From Equation (7.4) on page 229,

$$Z = \frac{\bar{X} - \mu}{\dfrac{\sigma}{\sqrt{n}}}$$

Solving for $\bar{X}$ results in Equation (7.5).

FINDING $\bar{X}$ FOR THE SAMPLING DISTRIBUTION OF THE MEAN

$$\bar{X} = \mu + Z\frac{\sigma}{\sqrt{n}} \tag{7.5}$$

Example 7.4 on page 231 illustrates the use of Equation (7.5).

EXAMPLE 7.4

Determining the
Interval That
Includes a Fixed
Proportion of the
Sample Means

For the Oxford Cereals scenario, find an interval symmetrically distributed around the population mean that will include 95% of the sample means, based on samples of 25 boxes.

SOLUTION If 95% of the sample means are in the interval, then 5% are outside the interval. Divide the 5% into two equal parts of 2.5%. The value of Z in Table E.2 corresponding to an area of 0.0250 in the lower tail of the normal curve is -1.96, and the value of Z corresponding to a cumulative area of 0.9750 (0.0250 in the upper tail of the normal curve) is $+1.96$.

The lower value of $\overline{X}$, $\overline{X}_L$, and the upper value of $\overline{X}$, $\overline{X}_U$, are found by using Equation (7.5):

$$\overline{X}_L = 368 + (-1.96)\frac{15}{\sqrt{25}} = 368 - 5.88 = 362.12$$

$$\overline{X}_U = 368 + (1.96)\frac{15}{\sqrt{25}} = 368 + 5.88 = 373.88$$

Therefore, 95% of all sample means, based on samples of 25 boxes, are between 362.12 and 373.88 grams.

Sampling from Non-normally Distributed Populations— The Central Limit Theorem

The sampling distribution of the mean that this section discusses requires a normally distributed population. However, for many analyses, one will either discover that the population is not normally distributed or be able to conclude that the assumption of a normally distributed population is unrealistic. The *Central Limit Theorem* enables one to make inferences about the population mean without having to know the specific shape of the population distribution.

The **Central Limit Theorem** states as the sample size gets *large enough*, the sampling distribution of the mean is approximately normally distributed. This theorem applies regardless of the shape of the distribution of the individual values in the population. This theorem is of crucial importance to using statistical inference to reach conclusions about a population.

What sample size is *large enough*? As a general rule, statisticians have found that for many population distributions, when the sample size is at least 30, the sampling distribution of the mean is approximately normal. However, one can apply the Central Limit Theorem for even smaller sample sizes if the population distribution is approximately bell-shaped. In the case in which the distribution of a variable is extremely skewed or has more than one mode, one needs sample sizes larger than 30 to ensure normality in the sampling distribution of the mean.

Figure 7.4 on page 232 illustrates that the Central Limit Theorem applies to all types of populations, regardless of their shape. In the figure, the effects of increasing sample size are shown for a normally distributed population in the left column; a uniformly distributed population, in which the values are evenly distributed between the smallest and largest values, in the center column; and an exponentially distributed population, in which the values are heavily right-skewed, in the right column. For each population, as the sample size increases, the variation in the sample means decreases, resulting in a narrowing of the width of the graph as the sample size increases from 2 to 30.

Because the sample mean is an unbiased estimator of the population mean, the mean of any sampling distribution in a column will be equal to the mean of the population that the column represents. Because the sampling distribution of the mean is always normally distributed for a normally distributed population, the Column A sampling distribution is always normally distributed.

FIGURE 7.4

Sampling distribution of
the mean for samples of
$n = 2, 5$, and 30, for
three different populations

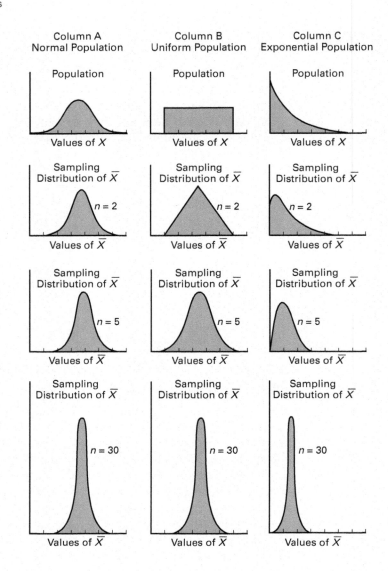

For the other two populations, a *central limiting* effect causes the sample means to become
more similar and the shape of the graphs to become more like a normal distribution. This effect
happens initially more slowly for the heavily skewed exponential distribution than for the uniform distribution, but when the sample size is increased to 30, the sampling distributions of these
two populations converge to the shape of the sampling distribution of the normal population.
Exhibit 7.1 summarizes the conclusions regarding the Central Limit Theorem that Figure 7.4
illustrates.

EXHIBIT **7.1**

Normality and the Sampling Distribution of the Mean

For most distributions, regardless of shape of the population, the sampling distribution of
the mean is approximately normally distributed if samples of at least size 30 are selected.
If the distribution of the population is fairly symmetrical, the sampling distribution of
the mean is approximately normal for samples as small as size 5.
If the population is normally distributed, the sampling distribution of the mean is
normally distributed, regardless of the sample size.

Example 7.5 illustrates a sampling distribution for a skewed population.

EXAMPLE 7.5

Constructing
a Sampling
Distribution for a
Skewed Population

Figure 7.5 shows the distribution of the time it takes to fill orders at a fast-food chain drive-through lane. Note that the probability distribution table is unlike Table 7.1 on page 225, which presents a population in which each value is equally likely to occur.

FIGURE 7.5
Probability distribution and histogram of the service time (in minutes) at a fast-food chain drive-through lane

Service Time (minutes)	Probability
1	0.10
2	0.40
3	0.20
4	0.15
5	0.10
6	0.05

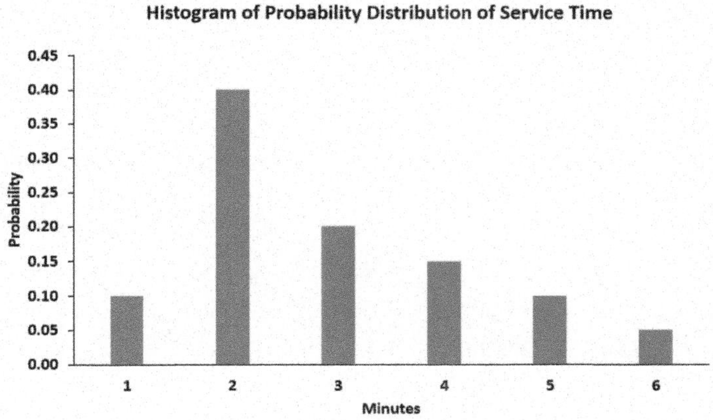

Histogram of Probability Distribution of Service Time

Using Equation (5.1) on page 177, the population mean is 2.9 minutes. Using Equation (5.3) on page 178, the population standard deviation is 1.34. Select 100 samples of $n = 2$, $n = 15$, and $n = 30$. What conclusions can be reached about the sampling distribution of the service time (in minutes) at the fast-food chain drive-through lane?

SOLUTION Table 7.3 represents the mean service time (in minutes) at the fast-food chain drive-through lane for 100 different random samples of $n = 2$. The mean of these 100 sample means is 2.825 minutes, and the standard error of the mean is 0.883.

TABLE 7.3
Mean service times (in minutes) at a fast-food chain drive-through lane for 100 different random samples of $n = 2$

3.5	2.5	3	3.5	4	3	2.5	2	2	2.5
3	3	2.5	2.5	2	2.5	2.5	2	3.5	1.5
2	3	2.5	3	3	2	3.5	3.5	2.5	2
4.5	3.5	4	2	2	4	3.5	2.5	2.5	3.5
3.5	3.5	2	1.5	2.5	2	3.5	3.5	2.5	2.5
2.5	3	3	3.5	2	3.5	2	1.5	5.5	2.5
3.5	3	3	2	1.5	3	2.5	2.5	2.5	2.5
3.5	1.5	6	2	1.5	2.5	3.5	2	3.5	5
2.5	3.5	4.5	3.5	3.5	2	4	2	3	3
4.5	1.5	2.5	2	2.5	2.5	2	2	2	4

Table 7.4 on page 234 represents the mean service time (in minutes) at the fast-food chain drive-through lane for 100 different random samples of $n = 15$. The mean of these 100 sample means is 2.9313 minutes, and the standard error of the mean is 0.3458.

Table 7.5 on page 234 represents the mean service time (in minutes) at the fast-food chain drive-through lane for 100 different random samples of $n = 30$. The mean of these 100 sample means is 2.9527 minutes, and the standard error of the mean is 0.2701.

▶(continued)

TABLE 7.4
Mean service times (in minutes) at a fast-food chain drive-through lane for 100 different random samples of $n = 15$

3.5333	2.8667	3.1333	3.6000	2.5333	2.8000	2.8667	3.1333	3.2667	3.3333
3.0000	3.3333	2.7333	2.6000	2.8667	3.0667	2.1333	2.5333	2.8000	3.1333
2.8000	2.7333	2.6000	3.1333	2.8667	3.4667	2.9333	2.8000	2.2000	3.0000
2.9333	2.6000	2.6000	3.1333	3.1333	3.1333	2.5333	3.0667	3.9333	2.8000
3.0000	2.7333	2.6000	2.4667	3.2000	2.4667	3.2000	2.9333	2.8667	3.4667
2.6667	3.0000	3.1333	3.1333	2.7333	2.7333	3.3333	3.4000	3.2000	3.0000
3.2000	3.0000	2.6000	2.9333	3.0667	2.8667	2.2667	2.5333	2.7333	2.2667
2.8000	2.8000	2.6000	3.1333	2.9333	3.0667	3.6667	2.6667	2.8667	2.6667
3.0000	3.4000	2.7333	3.6000	2.6000	2.7333	3.3333	2.6000	2.8667	2.8000
3.7333	2.9333	3.0667	2.6667	2.8667	2.2667	2.7333	2.8667	3.5333	3.2000

TABLE 7.5
Mean service times (in minutes) at a fast-food chain drive-through lane for 100 different random samples of $n = 30$

3.0000	3.3667	3.0000	3.1333	2.8667	2.8333	3.2667	2.9000	2.7000	3.2000
3.2333	2.7667	3.2333	2.8000	3.4000	3.0333	2.8667	3.0000	3.1333	3.4000
2.3000	3.0000	3.0667	2.9667	3.0333	2.4000	2.8667	2.8000	2.5000	2.7000
2.7000	2.9000	2.8333	3.3000	3.1333	2.8667	2.6667	2.6000	3.2333	2.8667
2.7667	2.9333	2.5667	2.5333	3.0333	3.2333	3.0667	2.9667	2.4000	3.3000
2.8000	3.0667	3.2000	2.9667	2.9667	3.2333	3.3667	2.9000	3.0333	3.1333
3.3333	2.8667	2.8333	3.0667	3.3667	3.0667	3.0667	3.2000	3.1667	3.3667
3.0333	3.1667	2.4667	3.0000	2.6333	2.6667	2.9667	3.1333	2.8000	2.8333
2.9333	2.7000	3.0333	2.7333	2.6667	2.6333	3.1333	3.0667	2.5333	3.3333
3.1000	2.5667	2.9000	3.9333	2.9000	2.7000	2.7333	2.8000	2.6667	2.8333

Figure 7.6 Panels A through C show histograms of the mean service time (in minutes) at the fast-food chain drive-through lane for the three sets of 100 different random samples shown in Tables 7.3 through 7.5. Panel A, the histogram for the mean service time for 100 different random samples of $n = 2$, shows a skewed distribution, but a distribution that is less skewed than the Figure 7.5 population distribution of service times.

FIGURE 7.6
Histograms of the mean service time (in minutes) at the fast-food chain drive-through lane of 100 different random samples of $n = 2$ (Panel A, left), 100 different random samples of $n = 15$ (Panel B, right), and 100 different random samples of $n = 30$ (Panel C, next page).

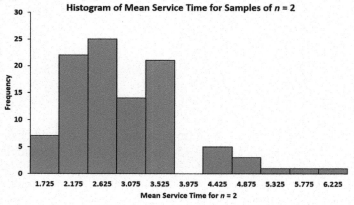

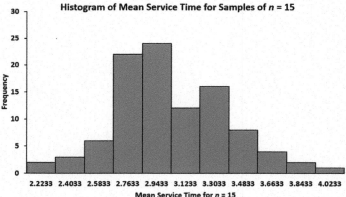

▶(*continued*)

FIGURE 7.6
(*continued*)

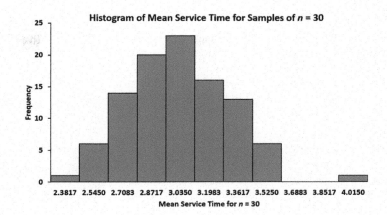

Panel B, the histogram for the mean service time for 100 different random samples of $n = 15$, shows a somewhat symmetrical distribution that contains a concentration of values in the center of the distribution. Panel C, the histogram for the mean service time for 100 different random samples of $n = 30$, shows a distribution that appears to be approximately bell-shaped with a concentration of values in the center of the distribution. The progression of the histograms from a skewed population toward a bell-shaped distribution as the sample size increases is consistent with the Central Limit Theorem.

VISUAL EXPLORATIONS

Exploring Sampling Distributions

Open the **VE-Sampling Distribution add-in workbook** to observe the effects of simulated rolls on the frequency distribution of the sum of two dice. (For Excel technical requirements, see Appendix D.) When this workbook opens properly, it adds a Sampling Distribution menu to the Add-ins tab (Apple menu in Excel for Mac).

To observe the effects of simulated throws on the frequency distribution of the sum of the two dice, select **Sampling Distribution → Two Dice Simulation**. In the Sampling Distribution dialog box, enter the **Number of rolls per tally** and click **Tally**. Click **Finish** when done.

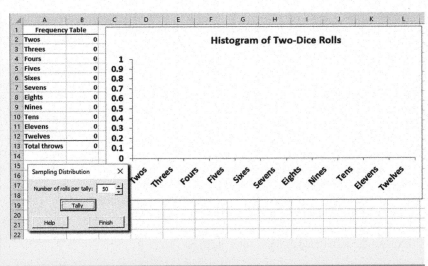

PROBLEMS FOR SECTION 7.2

LEARNING THE BASICS

7.1 Given a normal distribution with $\mu = 100$ and $\sigma = 10$, if you select a sample of $n = 25$, what is the probability that $\overline{X}$ is
a. less than 95?
b. between 95 and 97.5?
c. above 102.2?
d. There is a 65% chance that $\overline{X}$ is above what value?

7.2 Given a normal distribution with $\mu = 50$ and $\sigma = 5$, if you select a sample of $n = 100$, what is the probability that $\overline{X}$ is
a. less than 47?
b. between 47 and 49.5?
c. above 51.1?
d. There is a 35% chance that $\overline{X}$ is above what value?

APPLYING THE CONCEPTS

7.3 For each of the following three populations, indicate what the sampling distribution for samples of 25 would consist of:
a. Customer receipts for a supermarket for a year.
b. Insurance payouts in a particular geographical area in a year.
c. Call center logs of inbound calls tracking handling time for a credit card company during the year.

7.4 The following data represent the number of days absent per year in a population of six employees of a small company:

$$1 \quad 3 \quad 6 \quad 7 \quad 9 \quad 10$$

a. Assuming that you sample without replacement, select all possible samples of $n = 2$ and construct the sampling distribution of the mean. Compute the mean of all the sample means and also compute the population mean. Are they equal? What is this property called?
b. Repeat (a) for all possible samples of $n = 3$.
c. Compare the shape of the sampling distribution of the mean in (a) and (b). Which sampling distribution has less variability? Why?
d. Assuming that you sample with replacement, repeat (a) through (c) and compare the results. Which sampling distributions have the least variability—those in (a) or (b)? Why?

7.5 The amount of water in a two-liter bottle is approximately normally distributed with a mean of 2.05 liters and a standard deviation of 0.025 liter.
a. What is the probability that an individual bottle contains less than 2.03 liters?
b. If a sample of four bottles is selected, what is the probability that the sample mean amount contained is less than 2.03 liters?
c. If a sample of 25 bottles is selected, what is the probability that the sample mean amount contained is less than 2.03 liters?
d. Explain the difference in the results in (a) and (c).
e. Explain the difference in the results in (b) and (c).

7.6 The weight of an energy bar is approximately normally distributed with a mean of 42.05 grams and a standard deviation of 0.025 gram.
a. What is the probability that an individual energy bar contains less than 42.035 grams?
b. If a sample of four energy bars is selected, what is the probability that the sample mean weight is less than 42.035 grams?
c. If a sample of 25 energy bars is selected, what is the probability that the sample mean weight is less than 42.035 grams?
d. Explain the difference in the results in (a) and (c).
e. Explain the difference in the results in (b) and (c).

7.7 The diameter of a brand of tennis balls is approximately normally distributed, with a mean of 2.63 inches and a standard deviation of 0.03 inch. If you select a random sample of nine tennis balls,
a. what is the sampling distribution of the mean?
b. what is the probability that the sample mean is less than 2.61 inches?

c. what is the probability that the sample mean is between 2.62 and 2.64 inches?
d. The probability is 60% that the sample mean will be between what two values symmetrically distributed around the population mean?

7.8 The U.S. Census Bureau announced that the median sales price of new houses sold in January 2018 was $323,000, and the mean sales price was $382,700.

Source: **www.census.gov/newhomesales**, March 8, 2018.

Assume that the standard deviation of the prices is $90,000.
a. If you select samples of $n = 4$, describe the shape of the sampling distribution of $\overline{X}$.
b. If you select samples of $n = 100$, describe the shape of the sampling distribution of $\overline{X}$.
c. If you select a random sample of $n = 100$, what is the probability that the sample mean will be less than $370,000?
d. If you select a random sample of $n = 100$, what is the probability that the sample mean will be between $350,000 and $365,000?

7.9 According to a report by App Annie, a business intelligence company that produces tools and reports for the apps and digital goods industry, smartphone owners are using an average of 30 apps per month.

Source: "Report: Smartphone owners are using 9 apps per day, 30 per month," 2017, **tcrn.ch/2qK4iRr**.

Assume that number of apps used per month by smartphone owners is normally distributed and that the standard deviation is 5. If you select a random sample of 25 smartphone owners,
a. what is the probability that the sample mean is between 29 and 31?
b. what is the probability that the sample mean is between 28 and 32?
c. If you select a random sample of 100 smartphone owners, what is the probability that the sample mean is between 29 and 31?
d. Explain the difference in the results of (a) and (c).

✓SELF TEST **7.10** According to the National Survey of Student Engagement, the average student spends about 15 hours each week preparing for classes; preparation for classes includes homework, reading and any other assignments.

Source: Data extracted from **bit.ly/2qSNwNo**.

Assume the standard deviation of time spent preparing for classes is 4 hours. If you select a random sample of 16 students,
a. what is the probability that the mean time spent preparing for classes is at least 14 hours per week?
b. there is an 85% chance that the sample mean is less than how many hours per week?
c. What assumption must you make in order to solve (a) and (b)?
d. If you select a random sample of 64 students, there is an 85% chance that the sample mean is less than how many hours per week?

7.3 Sampling Distribution of the Proportion

When analyzing a categorical variable, one often wants to know what proportion of the data consists of one specific categorical value, or *characteristic of interest*. In the simplest case, a categorical variable that has only two categories such as yes and no, calculate the sample

studentTIP

Do not confuse this
use of the Greek letter
pi, π, to represent the
population proportion
with the mathematical
constant that represents
the ratio of the
circumference to the
diameter of a circle.

proportion, p, that Equation (7.6) defines, as part of the process to estimate the population proportion, π, the proportion of items in the entire population with the characteristic of interest.

SAMPLE PROPORTION

$$p = \frac{X}{n} = \frac{\text{Number of items having the characteristic of interest}}{\text{Sample size}} \tag{7.6}$$

The sample proportion calculation is a simple fraction. For example, for a yes-no variable in a sample size of five responses, if there are three responses with the characteristic of interest yes, the sample proportion would be 0.6 (three fifths, or 3 divided by 5).

The sample proportion, p, will be between 0 and 1. If all items have the characteristic, p is equal to 1. If half the items have the characteristic, p is equal to 0.5. If none of the items have the characteristic, p is equal to zero.

studentTIP

Remember that the
sample proportion
cannot be negative and
also cannot be greater
than 1.0.

Section 7.2 explains that the sample mean, $\overline{X}$, is an unbiased estimator of the population mean, μ. Similarly, the statistic p is an unbiased estimator of the population proportion, π. By analogy to the sampling distribution of the mean, whose standard error is $\sigma_{\overline{X}} = \dfrac{\sigma}{\sqrt{n}}$, the **standard error of the proportion**, σ_p, is given in Equation (7.7).

STANDARD ERROR OF THE PROPORTION

$$\sigma_p = \sqrt{\frac{\pi(1-\pi)}{n}} \tag{7.7}$$

The **sampling distribution of the proportion** follows the binomial distribution, which Section 5.2 discusses, when sampling with replacement (or without replacement from extremely large populations). However, one can use the normal distribution to approximate the binomial distribution when $n\pi$ and $n(1-\pi)$ are each at least 5. In most cases in which inferences are made about the population proportion, the sample size is substantial enough to meet the conditions for using the normal approximation (Cochran).

Substituting p for $\overline{X}$, π for μ, and $\sqrt{\dfrac{\pi(1-\pi)}{n}}$ for $\dfrac{\sigma}{\sqrt{n}}$ in Equation (7.4) on page 229 results in Equation (7.8).

FINDING Z FOR THE SAMPLING DISTRIBUTION OF THE PROPORTION

$$Z = \frac{p - \pi}{\sqrt{\dfrac{\pi(1-\pi)}{n}}} \tag{7.8}$$

To illustrate the sampling distribution of the proportion, a recent survey of employed adults who work outside the home, 54% reported that they go to work anyway when they are sick (Smith and Padilla). Suppose that one selects a random sample of 200 American workers and seeks to determine the probability that more than 50% of them stated that they go to work anyway when they are sick. Because $n\pi = 200(0.54) = 108 > 5$ and $n(1-\pi) = 200(1-0.54) = 92 > 5$, the sample size is large enough to assume that the sampling distribution of the proportion is approximately normally distributed. Therefore, one can use the survey percentage of 54% as the population proportion and can calculate the probability that more than 50% of American workers say that they go to work even though they are sick using Equation (7.8):

$$Z = \frac{p - \pi}{\sqrt{\dfrac{\pi(1 - \pi)}{n}}}$$

$$= \frac{0.50 - 0.54}{\sqrt{\dfrac{(0.54)(0.46)}{200}}} = \frac{-0.04}{\sqrt{\dfrac{0.2484}{200}}} = \frac{-0.04}{0.0352}$$

$$= -1.14$$

Using Table E.2, the area under the normal curve greater than −1.14 is 0.8729. Therefore, if the population proportion is 0.54, the probability is 87.29% that more than 50% of the 200 American workers in the sample will say that they go to work even though they are sick.

PROBLEMS FOR SECTION 7.3

LEARNING THE BASICS

7.11 In a random sample of 64 people, 48 are classified as "successful."
a. Determine the sample proportion, p, of "successful" people.
b. If the population proportion is 0.70, determine the standard error of the proportion.

7.12 A random sample of 50 households was selected for a phone (landline and cellphone) survey. The key question asked was, "Do you or any member of your household own an Apple product (iPhone, iPod, iPad, or Mac computer)?" Of the 50 respondents, 20 said yes, and 30 said no.
a. Determine the sample proportion, p, of households that own an Apple product.
b. If the population proportion is 0.45, determine the standard error of the proportion.

7.13 The following data represent the yes or no (Y or N) responses from a sample of 40 college students to the question "Do you currently own shares in any stocks?"

N N Y N N Y N Y N Y N N Y N Y Y N N N Y

N Y N N N N Y N N Y Y N N N Y N N Y N N

a. Determine the sample proportion, p, of college students who own shares of stock.
b. If the population proportion is 0.30, determine the standard error of the proportion.

APPLYING THE CONCEPTS

✓ SELF TEST **7.14** A political pollster is conducting an analysis of sample results in order to make predictions on election night. Assuming a two-candidate election, if a specific candidate receives at least 55% of the vote in the sample, that candidate will be forecast as the winner of the election. If you select a random sample of 100 voters, what is the probability that a candidate will be forecast as the winner when
a. the population percentage of her vote is 50.1%?
b. the population percentage of her vote is 60%?
c. the population percentage of her vote is 49% (and she will actually lose the election)?
d. If the sample size is increased to 400, what are your answers to (a) through (c)? Discuss.

7.15 You plan to conduct a marketing experiment in which students are to taste one of two different brands of soft drink. Their task is to correctly identify the brand tasted. You select a random sample of 200 students and assume that the students have no ability to distinguish between the two brands. (Hint: If an individual has no ability to distinguish between the two soft drinks, then the two brands are equally likely to be selected.)
a. What is the probability that the sample will have between 50% and 60% of the identifications correct?
b. The probability is 90% that the sample percentage is contained within what symmetrical limits of the population percentage?
c. What is the probability that the sample percentage of correct identifications is greater than 65%?
d. Which is more likely to occur—more than 60% correct identifications in the sample of 200 or more than 55% correct identifications in a sample of 1,000? Explain.

7.16 What do millennials around the world want in a job? A Deloitte survey of millennials on work-life challenges found that millennials are looking for stability in an uncertain world, with 65% of millennials preferring a permanent, full-time job rather than working freelance or as a consultant on a flexible or short-term basis.

Source: Data extracted from "Freelance flexibility with full-time stability," **bit.ly/2pr6h9r**.

Suppose you select a sample of 100 millennials.
a. What is the probability that in the sample fewer than 70% prefer a permanent, full-time job?
b. What is the probability that in the sample between 60% and 70% prefer a permanent, full-time job?
c. What is the probability that in the sample more than 70% prefer a permanent, full-time job?
d. If a sample of 400 is taken, how does this change your answers to (a) through (c)?

7.17 The goal of corporate sustainability is to manage the environmental, economic, and social effects of a corporation's operations so it is profitable over the long-term while acting in a responsible manner to society. An international study by Unilever reveals that 33% of consumers are choosing to buy from brands they believe are doing social or environmental good.

Source: Data extracted from "Report shows a third of consumers prefer sustainable brands," **bit.ly/2pTyEzO**.

Suppose you select a sample of 100 consumers.
a. What is the probability that in the sample fewer than 30% are choosing to buy from brands they believe are doing social or environmental good?
b. What is the probability that in the sample between 28% and 38% are choosing to buy from brands they believe are doing social or environmental good?
c. What is the probability that in the sample more than 38% are choosing to buy from brands they believe are doing social or environmental good?
d. If a sample of 400 is taken, how does this change your answers to (a) through (c)?

7.18 According to a survey in December 2018 of Women on Boards, women hold 21.7% of director seats on U.S. corporate boards. This study also reports that 39.2% of U.S. companies have three or more female board directors.

Source: Data extracted from **bit.ly/2LU2rEI.**

If you select a random sample of 200 U.S. companies,
a. what is the probability that the sample will have between 30% and 38% companies that have three or more female board directors?
b. the probability is 90% that the sample percentage of companies that have three or more female board directors will be contained within what symmetrical limits of the population percentage?
c. the probability is 95% that the sample percentage of companies that have three or more female board directors will be contained within what symmetrical limits of the population percentage?

7.19 The topic of global warming increasingly appears in the news. It has the potential to impact companies' operations through changes in governmental regulations, new reporting requirements, necessary operational changes, and so on. The Institute of Management Accountants (IMA) conducted a survey of senior finance professionals to gauge members' thoughts on global warming and its impact on their companies. The survey found that 65% of senior finance professionals believe that global warming is having a significant impact on the environment.

Source: Data extracted from "Global Warming: How Has It Affected Your Company?" **bit.ly/2pd341h.**

Suppose that you select a sample of 100 senior finance professionals.
a. What is the probability that the sample percentage indicating global warming is having a significant impact on the environment will be between 64% and 69%?
b. The probability is 90% that the sample percentage will be contained within what symmetrical limits of the population percentage?
c. The probability is 95% that the sample percentage will be contained within what symmetrical limits of the population percentage?
d. Suppose you selected a sample of 400 senior finance professionals. How does this change your answers in (a) through (c)?

7.20 Referring to Problem 7.18, the same survey reported that 20.9% of German companies have women directors and 80% have three or more women on the board.

If you select a random sample of 200 German companies,
a. what is the probability that the sample will have between 70% and 78% companies that have three or more female board directors?
b. the probability is 90% that the sample percentage of companies that have three or more female board directors will be contained within what symmetrical limits of the population percentage?
c. the probability is 95% that the sample percentage of companies that have three or more female board directors will be contained within what symmetrical limits of the population percentage?

7.4 Sampling from Finite Populations

The Central Limit Theorem and the standard errors of the mean and of the proportion are based on samples selected with replacement. However, in nearly all survey research, you sample without replacement from populations that are of a finite size, N. The **Section 7.4 online topic** explains how you use a **finite population correction factor** to compute the standard error of the mean and the standard error of the proportion for such samples.

▼ USING **STATISTICS**
Sampling Oxford Cereals, Revisited

As the plant operations manager for Oxford Cereals, you were responsible for monitoring the amount of cereal placed in each box. To be consistent with package labeling, boxes should contain a mean of 368 grams of cereal. Because weighing each of the thousands of boxes produced each shift would be too time-consuming, costly, and inefficient, you selected a sample of boxes. Based on your analysis of this sample, you had to decide whether to maintain, alter, or shut down the process.

Using the concept of the sampling distribution of the mean, you were able to determine probabilities that such a sample mean could have been randomly selected from a population with a mean of 368 grams. Specifically, if a sample of size $n = 25$ is selected from a population with a mean of 368 and standard deviation of 15, you calculated the probability of selecting a sample with a mean of 365 grams or less to be 15.87%. If a larger sample size is selected, the sample mean should be closer to the population mean. This result was illustrated when you calculated the probability if the sample size were increased to $n = 100$. Using the larger sample size, you determined the probability of selecting a sample with a mean of 365 grams or less to be 2.28%.

▼ SUMMARY

This chapter discusses the sampling distribution of the sample mean and the sampling distribution of the sample proportion. The chapter explains that the sample mean is an unbiased estimator of the population mean, and the sample proportion is an unbiased estimator of the population proportion. The chapter also discusses the Central Limit Theorem, a crucially important theorem in statistical inference, and relates the theorem to the sampling distributions that the chapter identifies.

▼ REFERENCES

Cochran, W. G. *Sampling Techniques*, 3rd ed. New York: Wiley, 1977.

Smith, M. and R. Padilla. "What Working Americans Do When They Are Ill." *USA Today*, February 19, 2019, p. 1A.

▼ KEY EQUATIONS

Population Mean

$$\mu = \frac{\sum_{i=1}^{N} X_i}{N} \tag{7.1}$$

Population Standard Deviation

$$\sigma = \sqrt{\frac{\sum_{i=1}^{N}(X_i - \mu)^2}{N}} \tag{7.2}$$

Standard Error of the Mean

$$\sigma_{\bar{X}} = \frac{\sigma}{\sqrt{n}} \tag{7.3}$$

Finding Z for the Sampling Distribution of the Mean

$$Z = \frac{\bar{X} - \mu_{\bar{X}}}{\sigma_{\bar{X}}} = \frac{\bar{X} - \mu}{\frac{\sigma}{\sqrt{n}}} \tag{7.4}$$

Finding $\bar{X}$ for the Sampling Distribution of the Mean

$$\bar{X} = \mu + Z\frac{\sigma}{\sqrt{n}} \tag{7.5}$$

Sample Proportion

$$p = \frac{X}{n} \tag{7.6}$$

Standard Error of the Proportion

$$\sigma_p = \sqrt{\frac{\pi(1 - \pi)}{n}} \tag{7.7}$$

Finding Z for the Sampling Distribution of the Proportion

$$Z = \frac{p - \pi}{\sqrt{\frac{\pi(1 - \pi)}{n}}} \tag{7.8}$$

▼ KEY TERMS

▼ CHECKING YOUR UNDERSTANDING

7.21 Why is the sample mean an unbiased estimator of the population mean?

7.22 Why does the standard error of the mean decrease as the sample size, *n*, increases?

7.23 Why does the sampling distribution of the mean follow a normal distribution for a large enough sample size, even though the population may not be normally distributed?

7.24 What is the difference between a population distribution and a sampling distribution?

7.25 Under what circumstances does the sampling distribution of the proportion approximately follow the normal distribution?

▼ CHAPTER REVIEW PROBLEMS

7.26 An industrial sewing machine uses ball bearings that are targeted to have a diameter of 0.75 inch. The lower and upper specification limits under which the ball bearing can operate are 0.74 inch (lower) and 0.76 inch (upper). Past experience has indicated that the actual diameter of the ball bearings is approximately normally distributed, with a mean of 0.753 inch and a standard deviation of 0.004 inch. If you select a random sample of 25 ball bearings, what is the probability that the sample mean is
a. between the target and the population mean of 0.753?
b. between the lower specification limit and the target?
c. greater than the upper specification limit?
d. less than the lower specification limit?
e. The probability is 93% that the sample mean diameter will be greater than what value?

7.27 The fill amount of bottles of a soft drink is normally distributed, with a mean of 2.0 liters and a standard deviation of 0.05 liter. If you select a random sample of 25 bottles, what is the probability that the sample mean will be
a. between 1.99 and 2.0 liters?
b. below 1.98 liters?
c. greater than 2.01 liters?
d. The probability is 99% that the sample mean amount of soft drink will be at least how much?
e. The probability is 99% that the sample mean amount of soft drink will be between which two values (symmetrically distributed around the mean)?

7.28 An orange juice producer buys oranges from a large orange grove that has one variety of orange. The amount of juice squeezed from these oranges is approximately normally distributed, with a mean of 4.70 ounces and a standard deviation of 0.40 ounce. Suppose that you select a sample of 25 oranges.
a. What is the probability that the sample mean amount of juice will be at least 4.60 ounces?
b. The probability is 70% that the sample mean amount of juice will be contained between what two values symmetrically distributed around the population mean?
c. The probability is 77% that the sample mean amount of juice will be greater than what value?

7.29 In Problem 7.28, suppose that the mean amount of juice squeezed is 5.0 ounces.
a. What is the probability that the sample mean amount of juice will be at least 4.60 ounces?
b. The probability is 70% that the sample mean amount of juice will be contained between what two values symmetrically distributed around the population mean?
c. The probability is 77% that the sample mean amount of juice will be greater than what value?
d. Compare the results of (a) through (c) with the results of Problem 7.28 (a) through (c).

7.30 The stock market in Qatar reported strong returns in the first 11 months of 2018. The population of stocks earned a mean return of 21.19% in 2018.

Assume that the returns for stocks on the Qatar stock market were distributed as a normal variable, with a mean of 21.19 and a standard

deviation of 20. If you selected a random sample of 16 stocks from this population, what is the probability that the sample would have a mean return
a. less than 0 (i.e., a loss)?
b. between 0 and 10?
c. greater than 10?

7.31 For the year 2018, the stock market in Germany had a mean return of −12.53% in 2018. Assume that the returns for stocks on the German stock market were distributed normally, with a mean of −12.53% and a standard deviation of 10. If you select an individual stock from this population, what is the probability that it would have a return
a. less than 0 (i.e., a loss)?
b. between −10 and −20?
c. greater than −5?

If you selected a random sample of four stocks from this population, what is the probability that the sample would have a mean return
d. less than 0 (a loss)?
e. between −10 and −20?
f. greater than −5?
g. Compare your results in parts (d) through (f) to those in (a) through (c).

7.32 (**Class Project**) The table of random numbers is an example of a uniform distribution because each digit is equally likely to occur. Starting in the row corresponding to the day of the month in which you were born, use a table of random numbers (Table E.1) to take one digit at a time.

Select five different samples each of $n = 2, n = 5,$ and $n = 10$. Compute the sample mean of each sample. Develop a frequency distribution of the sample means for the results of the entire class, based on samples of sizes $n = 2, n = 5,$ and $n = 10$.

What can be said about the shape of the sampling distribution for each of these sample sizes?

7.33 (**Class Project**) Toss a coin 10 times and record the number of heads. If each student performs this experiment five times, a frequency distribution of the number of heads can be developed from the results of the entire class. Does this distribution seem to approximate the normal distribution?

7.34 (**Class Project**) The number of cars waiting in line at a car wash is distributed as follows:

Number of Cars	Probability
0	0.25
1	0.40
2	0.20
3	0.10
4	0.04
5	0.01

You can use a table of random numbers (Table E.1) to select samples from this distribution by assigning numbers as follows:
1. Start in the row corresponding to the day of the month in which you were born.
2. Select a two-digit random number.

3. If you select a random number from 00 to 24, record a waiting line length of 0; if from 25 to 64, record a length of 1; if from 65 to 84, record a length of 2; if from 85 to 94, record a length of 3; if from 95 to 98, record a length of 4; if 99, record a length of 5.

Select samples of $n = 2$, $n = 15$, and $n = 30$. Compute the mean for each sample. For example, if a sample of size 2 results in the random numbers 18 and 46, these would correspond to lengths 0 and 1, respectively, producing a sample mean of 0.5. If each student selects five different samples for each sample size, a frequency distribution of the sample means (for each sample size) can be developed from the results of the entire class. What conclusions can you reach concerning the sampling distribution of the mean as the sample size is increased?

7.35 (Class Project) The file Credit Scores contains the average credit scores of people from 2,750 American cities.

Source: Data extracted from **bit.ly/2oCgnbi**.

a. Select five different samples of $n = 2$, $n = 5$, $n = 15$, and $n = 30$.

b. Compute the sample mean of each sample. Develop a frequency distribution of the sample means for the results of the entire class, based on samples of sizes $n = 2$, $n = 5$, $n = 15$, and $n = 30$.

c. What can be said about the shape of the sampling distribution for each of these sample sizes?

CHAPTER 7

▼ CASES

Managing Ashland MultiComm Services

Continuing the quality improvement effort first described in the Chapter 6 Managing Ashland MultiComm Services case, the target upload speed for AMS Internet service subscribers has been monitored. As before, upload speeds are measured on a standard scale in which the target value is 1.0. Data collected over the past year indicate that the upload speeds are approximately normally distributed, with a mean of 1.005 and a standard deviation of 0.10.

1. Each day, at 25 random times, the upload speed is measured. Assuming that the distribution has not changed from what it was in the past year, what is the probability that the mean upload speed is
   a. less than 1.0?
   b. between 0.95 and 1.0?
   c. between 1.0 and 1.05?
   d. less than 0.95 or greater than 1.05?
   e. Suppose that the mean upload speed of today's sample of 25 is 0.952. What conclusion can you reach about the mean upload speed today based on this result? Explain.

2. Compare the results of AMS Problem 1 (a) through (d) to those of AMS Problem 1 in Chapter 6 on page 221. What conclusions can you reach concerning the differences?

Digital Case

Apply your knowledge about sampling distributions in this Digital Case, which reconsiders the Oxford Cereals Using Statistics scenario.

The advocacy group Consumers Concerned About Cereal Cheaters (CCACC) suspects that cereal companies, including Oxford Cereals, are cheating consumers by packaging cereals at less than labeled weights. Recently, the group investigated the package weights of two popular Oxford brand cereals. Open **CCACC.pdf** to examine the group's claims and supporting data, and then answer the following questions:

1. Are the data collection procedures that the CCACC uses to form its conclusions flawed? What procedures could the group follow to make its analysis more rigorous?

2. Assume that the two samples of five cereal boxes (one sample for each of two cereal varieties) listed on the CCACC website were collected randomly by organization members. For each sample,
   a. calculate the sample mean.
   b. assuming that the standard deviation of the process is 15 grams and the population mean is 368 grams, calculate the percentage of all samples for each process that have a sample mean less than the value you calculated in (a).
   c. assuming that the standard deviation is 15 grams, calculate the percentage of individual boxes of cereal that have a weight less than the value you calculated in (a).

3. What, if any, conclusions can you form by using your calculations about the filling processes for the two different cereals?

4. A representative from Oxford Cereals has asked that the CCACC take down its web page discussing shortages in Oxford Cereals boxes. Is this request reasonable? Why or why not?

5. Can the techniques discussed in this chapter be used to prove cheating in the manner alleged by the CCACC? Why or why not?

▼EXCEL GUIDE

EG7.2 SAMPLING DISTRIBUTION of the MEAN

Key Technique Use an add-in procedure to create a simulated sampling distribution and use the **RAND**() function to create lists of random numbers.

Example Create a simulated sampling distribution that consists of 100 samples of $n = 30$ from a uniformly distributed population.

PHStat Use **Sampling Distributions Simulation**.

For the example, select **PHStat→Sampling→Sampling Distributions Simulation**. In the procedure's dialog box (shown below):

1. Enter **100** as the **Number of Samples**.
2. Enter **30** as the **Sample Size**.
3. Click **Uniform**.
4. Enter a **Title** and click **OK**.

The procedure inserts a new worksheet in which the sample means, overall mean, and standard error of the mean can be found starting in row 34.

Workbook Use the **SDS worksheet** of the **SDS workbook** as a model.

For the example, in a new worksheet, first enter a title in cell A1. Then enter the formula **= RAND**() in cell **A2** and then copy the formula down 30 rows and across 100 columns (through

column **CV**). Then select this cell range (**A2:CV31**) and use **copy and paste values** as discussed in Appendix Section B.4.

Use the formulas that appear in rows 33 through 37 in the **SDS_FORMULAS worksheet** as models if you want to compute sample means, the overall mean, and the standard error of the mean.

Analysis ToolPak Use **Random Number Generation**.

For the example, select **Data→Data Analysis**. In the Data Analysis dialog box, select **Random Number Generation** from the **Analysis Tools** list and then click **OK**.

In the procedure's dialog box (shown below):

1. Enter **100** as the **Number of Variables**.
2. Enter **30** as the **Number of Random Numbers**.
3. Select **Uniform** from the **Distribution** drop-down list.
4. Keep the **Parameters** values as is.
5. Click **New Worksheet Ply** and then click **OK**.

If, for other problems, you select **Discrete** in step 3, you must be open to a worksheet that contains a cell range of X and $P(X)$ values. Enter this cell range as the **Value and Probability Input Range** (not shown when **Uniform** has been selected) in the **Parameters** section of the dialog box.

Use the formulas that appear in rows 33 through 37 in the **SDS_FORMULAS worksheet** of the **SDS workbook** as models if you want to compute sample means, the overall mean, and the standard error of the mean.

8

Confidence Interval Estimation

OBJECTIVES

- Construct and interpret
confidence interval esti-
mates for the mean and
the proportion

- Determine the sample
size necessary to
develop a confidence
interval estimate for the
mean or proportion

▼ USING STATISTICS
Getting Estimates at Ricknel Home Centers

As a member of the AIS team at Ricknel Home Centers, you have already examined the probability of discovering questionable, or *tagged*, invoices. Now you have been assigned the task of auditing the accuracy of the integrated inventory management and point of sale component of the firm's retail management system.

You could review the contents of *every* inventory and sales transaction to check the accuracy of the information system, but such a detailed review would be time-consuming and costly. Could you use statistical inference techniques to reach conclusions about the population of all records from a relatively small sample collected during an audit? At the end of each month, could you select a sample of the sales invoices to estimate population parameters such as

- The mean dollar amount listed on the sales invoices for the month
- The proportion of invoices that contain errors that violate the internal control policy of the company

If you used a sampling technique, how accurate would the results from the sample be? How would you use the results you generate? How could you be certain that the sample size is large enough to give you the information you need?

Section 7.2 explains how the Central Limit Theorem and insight about a population distribution can be used to determine the percentage of sample means that are within certain distances of the population mean. In the Oxford Cereals scenario that Chapter 7 features, Example 7.4 on page 231 uses this knowledge to conclude that 95% of all sample means are between 362.12 and 373.88 grams. That conclusion is an example of *deductive* reasoning, a conclusion based on taking something that is true in general (for the population) and applying it to something specific (the sample means).

Getting the results that Ricknel Home Centers needs requires *inductive* reasoning. Inductive reasoning uses some specifics to make broader generalizations. One cannot guarantee that the broader generalizations are absolutely correct, but with a careful choice of the specifics and a rigorous methodology, one can reach useful conclusions. As a Ricknel AIS team member, you need to use inferential statistics, which uses sample results (the "some specifics") to *estimate* unknown population parameters such as a population mean or a population proportion (the "broader generalizations"). Note that statisticians use the word *estimate* in the same sense of the everyday usage: something about which one is reasonably certain but cannot say is absolutely correct.

One estimates population parameters by using either point estimates or interval estimates. A **point estimate** is the value of a single sample statistic, such as a sample mean. A **confidence interval estimate** is a range of numbers, called an *interval*, constructed around the point estimate. The confidence interval is constructed such that the probability that the interval includes the population parameter is known.

For example, a researcher seeks to estimate the mean GPA of all the students at a public university. The mean GPA for all the students is an unknown population mean, denoted by μ. The researcher selects a sample of students and calculates the sample mean, $\overline{X}$, to be 3.20. As a *point estimate* of the population mean, μ, the researcher seeks to know how accurate is 3.20 as an estimate of the population mean, μ. By taking into account the variability from sample to sample (the sampling distribution of the mean that Section 7.2 discusses), the researcher constructs a *confidence interval estimate* for the population mean to determine this.

A *confidence interval estimate* indicates the confidence of correctly estimating the value of the population parameter, μ. This enables one to say that there is a specified confidence that μ is somewhere in the range of numbers defined by the interval.

Suppose the researcher found that a 95% confidence interval for the mean GPA at that university is $3.15 \leq \mu \leq 3.25$. The researcher can interpret this interval estimate as follows: the researcher is 95% confident that the mean GPA at that university is between 3.15 and 3.25.

8.1 Confidence Interval Estimate for the Mean (σ Known)

How would one estimate the population mean, using the information from a single sample for the Chapter 7 Oxford Cereals scenario? Instead of using $\mu \pm (1.96)(\sigma/\sqrt{n})$ to find the upper and lower limits around μ, as Section 7.2 does, one uses the sample mean, $\overline{X}$, for the unknown μ and uses $\overline{X} \pm (1.96)(\sigma/\sqrt{n})$ as the interval to estimate the unknown μ.

Examining a set of all possible samples of the same sample size helps explain the insight that allows the sample mean to be used in this way. Suppose that a sample of $n = 25$ cereal boxes has a mean of 362.3 grams and a standard deviation of 15 grams. Using the Section 7.2 method, the interval estimate that includes μ is $362.3 \pm (1.96)(15)/(\sqrt{25})$, or 362.3 ± 5.88. Therefore, the estimate of μ is $356.42 \leq \mu \leq 368.18$. This sample results in a correct statement about μ because the population mean, μ, known to be 368 grams, is included within the interval.

Example 7.4 results enable one to conclude that the interval around the population mean from 362.12 through 373.88 grams will hold 95% of all sample means. One can also conclude that 95% of all samples of $n = 25$ will have sample means that can be used to estimate an interval for the population mean.

Figure 8.1 on page 246 shows this interval and visualizes the estimates of μ made using five different samples of $n = 25$, each with their own sample means. Note the third sample of $n = 25$ with a mean of 360 grams does not estimate an interval (shown in red)

FIGURE 8.1

Confidence interval estimates for five different samples of $n = 25$ taken from a population where $\mu = 368$ and $\sigma = 15$

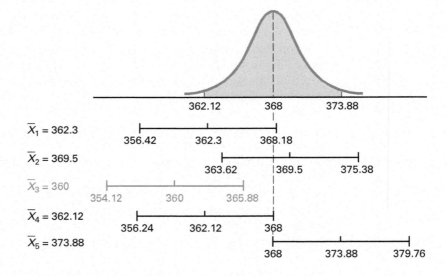

that includes the population mean. Therefore, the estimate of $354.12 \le \mu \le 365.88$ is an incorrect statement.

The fourth and fifth samples of $n = 25$ shown in Figure 8.1 contain the lowest (362.12) and highest (373.88) means for samples of $n = 25$ that can be used to correctly estimate the population mean. Should a subsequent sample of $n = 25$ be taken that has a sample mean less than 362.12 or greater than 373.88, that sample mean will not lead to a correct estimate of the population mean.

In most real-world situations, the population mean, μ, will be unknown, unlike the example that Figure 8.1 illustrates. Therefore, one will not be able to judge whether the estimate of the population mean developed from a single sample mean is a correct statement. However, by the principles that Chapters 6 and 7 discuss, and by using the conclusions of the previous paragraph, one *can* conclude that if one takes all possible samples of a specific sample size n and use their sample means to estimate their 95% confidence intervals for the population mean, 95% of those intervals will include the population mean (and 5% will not). In other words, one has 95% confidence that the population mean is somewhere in an interval estimated by a sample mean.

The intervals shown in Figure 8.1 are properly called 95% confidence intervals. Saying that something is a 95% confidence interval is a shorthand way of saying the following:

> "I am 95% confident that the interval that states that the mean amount of cereal in the population of filled boxes is somewhere between 356.42 and 368.18 grams is an interval that includes the population mean."

Sampling Error

To further understand confidence intervals, consider the order-filling process for an online retailer. Filling orders consists of several steps, including receiving an order, picking the parts of the order, checking the order, packing, and shipping the order. The file Order contains the time, in minutes, to fill orders for a population of $N = 200$ orders on a recent day. Although in practice the population characteristics are rarely known, for this population of orders, the mean, μ, is known to be equal to 69.637 minutes; the standard deviation, σ, is known to be equal to 10.411 minutes; and the population is normally distributed.

To illustrate how the sample mean and sample standard deviation can vary from one sample to another, 20 different samples of $n = 10$ were selected from the population of 200 orders, and the sample mean and sample standard deviation (and other statistics) were calculated for each sample. Figure 8.2 on page 247 shows these results.

FIGURE 8.2

Sample statistics and 95% confidence intervals for 20 samples of $n = 10$ randomly selected from the population of $N = 200$ orders

Sample	n	Mean	Std Dev	Minimum	Median	Maximum	Range	95% Conf. Int.
S01	10	74.15	13.39	56.10	76.85	97.70	41.60	(67.70, 80.60)
S02	10	61.10	10.60	46.80	61.35	79.50	32.70	(54.65, 67.55)
S03	10	74.36	6.50	62.50	74.50	84.00	21.50	(67.91, 80.81)
S04	10	70.40	12.80	47.20	70.95	84.00	36.80	(63.95, 76.85)
S05	10	62.18	10.85	47.10	59.70	84.00	36.90	(55.73, 68.63)
S06	10	67.03	9.68	51.10	69.60	83.30	32.20	(60.58, 73.48)
S07	10	69.03	8.81	56.60	68.85	83.70	27.10	(62.58, 75.48)
S08	10	72.30	11.52	54.20	71.35	87.00	32.80	(65.85, 78.75)
S09	10	68.18	14.10	50.10	69.95	86.20	36.10	(61.73, 74.63)
S10	10	66.67	9.08	57.10	64.65	86.10	29.00	(60.22, 73.12)
S11	10	72.42	9.76	59.60	74.65	86.10	26.50	(65.97, 78.87)
S12	10	76.26	11.69	50.10	80.60	87.00	36.90	(69.81, 82.71)
S13	10	65.74	12.11	47.10	62.15	86.10	39.00	(59.29, 72.19)
S14	10	69.99	10.97	51.00	73.40	84.60	33.60	(63.54, 76.44)
S15	10	75.76	8.60	61.10	75.05	87.80	26.70	(69.31, 82.21)
S16	10	67.94	9.19	56.70	67.70	87.80	31.10	(61.49, 74.39)
S17	10	71.05	10.48	50.10	71.15	86.20	36.10	(64.60, 77.50)
S18	10	71.68	7.96	55.60	72.35	82.60	27.00	(65.23, 78.13)
S19	10	70.97	9.83	54.40	70.05	84.00	30.20	(64.52, 77.42)
S20	10	74.48	8.80	62.00	76.25	85.70	23.70	(68.03, 80.93)

From Figure 8.2, observe:

- The sample statistics differ from sample to sample. The sample means vary from 61.10 to 76.26 minutes, the sample standard deviations vary from 6.50 to 14.10 minutes, the sample medians vary from 59.70 to 80.60 minutes, and the sample ranges vary from 21.50 to 41.60 minutes.
- Some of the sample means are greater than the population mean of 69.637 minutes, and some of the sample means are less than the population mean.
- Some of the sample standard deviations are greater than the population standard deviation of 10.411 minutes, and some of the sample standard deviations are less than the population standard deviation.
- The variation in the sample ranges is much more than the variation in the sample standard deviations.

The variation of sample statistics from sample to sample is called *sampling error*. **Sampling error** is the variation that occurs due to selecting a single sample from the population. The size of the sampling error is primarily based on the amount of variation in the population and on the sample size. Large samples have less sampling error than small samples, but large samples cost more to select.

The last column of Figure 8.2 contains 95% confidence interval estimates of the population mean order-filling time, based on the results of those 20 samples of $n = 10$. Begin by examining the first sample selected. The sample mean is 74.15 minutes, and the interval estimate for the population mean is 67.70 to 80.60 minutes. In a typical study, the value of the population mean is unknown, and therefore one will not know for sure whether this interval estimate is correct. However, for this population of orders, the population mean is known to be 69.637 minutes. The first interval estimate 67.70 to 80.60 minutes includes the population mean of 69.637 minutes. Therefore, the first sample provides a correct estimate of the population mean in the form of an interval estimate. For the other 19 samples, all the other samples *except* samples 2, 5, and 12 have interval estimates that contain the population mean.

For sample 2, the sample mean is 61.10 minutes, and the interval estimate is 54.65 to 67.55 minutes; for sample 5, the sample mean is 62.18, and the interval estimate is between 55.73 and 68.63; for sample 12, the sample mean is 76.26, and the interval estimate is between 69.81 and 82.71 minutes. The population mean of 69.637 minutes is *not* located within any of these intervals, and the estimate of the population mean made using these intervals is incorrect. Although in this example only 17 of the 20 intervals (85%) include the population mean, if one selects all the possible samples of $n = 10$ from a population of $N = 200$, 95% of the intervals would include the population mean.

In some situations, one might want a higher degree of confidence of including the population mean within the interval estimate (such as 99%). In other cases, you might accept less confidence (such as 90%) of correctly estimating the population mean. In general, the **level of confidence** is symbolized by $(1 - \alpha) \times 100\%$, where α is the proportion in the tails of the distribution that is outside the confidence interval. The proportion in the upper tail of the distribution is $\alpha/2$, and the proportion in the lower tail of the distribution is $\alpha/2$. Use Equation (8.1) to construct a $(1 - \alpha) \times 100\%$ confidence interval estimate for the mean with σ known.

CONFIDENCE INTERVAL FOR THE MEAN (σ KNOWN)

$$\overline{X} \pm Z_{\alpha/2}\frac{\sigma}{\sqrt{n}}$$

or

$$\overline{X} - Z_{\alpha/2}\frac{\sigma}{\sqrt{n}} \leq \mu \leq \overline{X} + Z_{\alpha/2}\frac{\sigma}{\sqrt{n}} \tag{8.1}$$

where

$Z_{\alpha/2}$ is the value for an upper-tail probability of $\alpha/2$ from the standardized normal distribution (i.e., a cumulative area of $1 - \alpha/2$)

$Z_{\alpha/2}\dfrac{\sigma}{\sqrt{n}}$ is the sampling error

The value of $Z_{\alpha/2}$ needed for constructing a confidence interval is called the **critical value** for the distribution. 95% confidence corresponds to an α value of 0.05. The critical Z value corresponding to a cumulative area of 0.975 is 1.96 because there is 0.025 in the upper tail of the distribution, and the cumulative area less than $Z = 1.96$ is 0.975.

There is a different critical value for each level of confidence, $1 - \alpha$. A level of confidence of 95% leads to a Z value of 1.96 (see Figure 8.3). 99% confidence corresponds to an α value of 0.01. The Z value is approximately 2.58 because the upper-tail area is 0.005 and the cumulative area less than $Z = 2.58$ is 0.995 (see Figure 8.4).

FIGURE 8.3
Normal curve for determining the Z value needed for 95% confidence

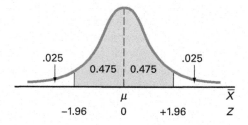

FIGURE 8.4
Normal curve for determining the Z value needed for 99% confidence

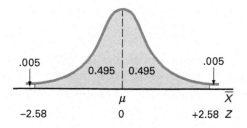

learnMORE

Section 8.4 further discusses the trade-off between the width of the confidence interval and the level of confidence.

On page 249, Example 8.1 illustrates the application of the confidence interval estimate, and Example 8.2 illustrates the effect of using a 99% confidence interval.

Now that various levels of confidence have been considered, why not always make the confidence level as close to 100% as possible? Any increase in the level of confidence widens (and makes less precise) the confidence interval. There is no "free lunch" here. One would have more confidence that the population mean is within a broader range of values; however, the broader range might make the interpretation of the confidence interval less useful.

EXAMPLE 8.1

Estimating the
Mean Cereal Fill
Amount with
95% Confidence

Returning to the Chapter 7 Oxford Cereals scenario, managers must ensure that the mean weight of filled boxes is 368 grams to be consistent with the labeling on those boxes. To determine whether the mean weight is consistent with the expected amount of 368 grams, managers periodically select a random sample of 100 filled boxes from the large number of boxes filled. Past experience states that the standard deviation of the fill amount is 15 grams. One random sample of 100 filled boxes they selected has a sample mean of 369.27 grams. Construct a 95% confidence interval estimate of the mean fill amount.

SOLUTION Using Equation (8.1) on page 248, with $Z_{\alpha/2} = 1.96$ for 95% confidence,

$$\overline{X} \pm Z_{\alpha/2}\frac{\sigma}{\sqrt{n}} = 369.27 \pm (1.96)\frac{15}{\sqrt{100}} = 369.27 \pm 2.94$$
$$366.33 \leq \mu \leq 372.21$$

Thus, with 95% confidence, the population mean is between 366.33 and 372.21 grams. Because the interval includes 368, the value indicating that the cereal-filling process is working properly, there is no evidence to suggest that anything is wrong with the cereal-filling process.

EXAMPLE 8.2

Estimating the
Mean Cereal Fill
Amount with
99% Confidence

Construct a 99% confidence interval estimate for the population mean fill amount.

SOLUTION Using Equation (8.1) on page 248, with $Z_{\alpha/2} = 2.58$ for 99% confidence,

$$\overline{X} \pm Z_{\alpha/2}\frac{\sigma}{\sqrt{n}} = 369.27 \pm (2.58)\frac{15}{\sqrt{100}} = 369.27 \pm 3.87$$
$$365.40 \leq \mu \leq 373.14$$

Once again, because 368 is included within this wider interval, there is no evidence to suggest that anything is wrong with the cereal-filling process.

As Section 7.2 discusses, the sampling distribution of the sample mean, $\overline{X}$, is normally distributed if the population for your characteristic of interest, X, follows a normal distribution. And if the population of X does not follow a normal distribution, the Central Limit Theorem almost always ensures that $\overline{X}$ is approximately normally distributed when n is large. However, when dealing with a small sample size and a population that does not follow a normal distribution, the sampling distribution of $\overline{X}$ is not normally distributed, and therefore the confidence interval that this section discusses is inappropriate. In practice, however, as long as the sample size is large enough and the population is not very skewed, one can use the confidence interval defined in Equation (8.1) to estimate the population mean when σ is known. (To assess the assumption of normality, evaluate the shape of the sample data by constructing a histogram, stem-and-leaf display, boxplot, or normal probability plot.)

Can You Ever Know the Population Standard Deviation?

In most real-world business situations, one will not know the standard deviation of the population because populations are too large to examine.

To use Equation (8.1), one must know the value for σ, the population standard deviation. To know σ implies that one knows all the values in the entire population. (How else would one know the value of this population parameter?) If one knew all the values in the entire population, the population mean could be computed directly. There would be no need to use the *inductive* reasoning of inferential statistics to *estimate* the population mean. In other words, if one knows σ, one really does not have a need to use Equation (8.1) to construct a confidence interval estimate of the mean (σ known).

Given this observation, then why would one study the method? Studying the method is an effective way of understanding the confidence interval concept because the method uses the normal distribution that Chapters 6 and 7 discuss. Understanding the confidence interval concept is very important to understanding concepts that later chapters present.

PROBLEMS FOR SECTION 8.1

LEARNING THE BASICS

8.1 If $\overline{X} = 85$, $\sigma = 8$, and $n = 64$, construct a 95% confidence interval estimate for the population mean, μ.

8.2 If $\overline{X} = 125$, $\sigma = 24$, and $n = 36$, construct a 99% confidence interval estimate for the population mean, μ.

8.3 Why is it not possible in Example 8.1 on page 249 to have 100% confidence? Explain.

8.4 Is it true in Example 8.1 on page 249 that you do not know for sure whether the population mean is between 366.33 and 372.21 grams? Explain.

APPLYING THE CONCEPTS

8.5 A market researcher selects a simple random sample of $n = 100$ Twitter users from a population of more than 100 million Twitter-registered users. After analyzing the sample, she states that she has 95% confidence that the mean time spent on Twitter per day is between 15 and 57 minutes. Explain the meaning of this statement.

8.6 Suppose that you are going to collect a set of data, either from an entire population or from a random sample taken from that population.
a. Which statistical measure would you compute first: the mean or the standard deviation? Explain.
b. What does your answer to (a) tell you about the "practicality" of using the confidence interval estimate formula given in Equation (8.1)?

8.7 Consider the confidence interval estimate discussed in Problem 8.5. Suppose the population mean time spent on Twitter is 46 minutes a day. Is the confidence interval estimate stated in Problem 8.5 correct? Explain.

8.8 You are working as an assistant to the dean of institutional research at your university. The dean wants to survey members of the alumni association who obtained their baccalaureate degrees

five years ago to learn what their starting salaries were in their first full-time job after receiving their degrees. A sample of 100 alumni is to be randomly selected from the list of 2,500 graduates in that class. If the dean's goal is to construct a 95% confidence interval estimate for the population mean starting salary, why is it not possible that you will be able to use Equation (8.1) on page 248 for this purpose? Explain.

8.9 A bottled water distributor wants to estimate the amount of water contained in 1-gallon bottles purchased from a nationally known water bottling company. The water bottling company's specifications state that the standard deviation of the amount of water is equal to 0.02 gallon. A random sample of 50 bottles is selected, and the sample mean amount of water per 1-gallon bottle is 0.993 gallon.
a. Construct a 99% confidence interval estimate for the population mean amount of water included in a 1-gallon bottle.
b. On the basis of these results, do you think that the distributor has a right to complain to the water bottling company about the amount of water that the bottles contain? Why?
c. Must you assume that the population amount of water per bottle is normally distributed here? Explain.
d. Construct a 95% confidence interval estimate. How does this change your answer to (b)?

✓SELF TEST **8.10** The operations manager at a light emitting diode (LED) light bulb factory needs to estimate the mean life of a large shipment of LEDs. The manufacturer's specifications are that the standard deviation is 1,500 hours. A random sample of 64 LEDs indicated a sample mean life of 49,875 hours.
a. Construct a 95% confidence interval estimate for the population mean life of LED light bulbs in this shipment.
b. Do you think that the manufacturer has the right to state that the LED light bulbs have a mean life of 50,000 hours? Explain.
c. Must you assume that the population LED light bulb life is normally distributed? Explain.
d. Suppose that the standard deviation changes to 500 hours. What are your answers in (a) and (b)?

8.2 Confidence Interval Estimate for the Mean (σ Unknown)

Section 8.1 explains that, in most business situations, one does not know σ, the population standard deviation. This section discusses a method of constructing a confidence interval estimate of μ that uses the sample statistic S as an estimate of the population parameter σ.

Student's t Distribution

At the start of the twentieth century, William S. Gosset was working at Guinness in Ireland, trying to help brew better beer less expensively (Kirk). Because he had only small samples to study, he needed to find a way to make inferences about means without having to know σ. Writing under the pen name "Student,"[1] Gosset solved this problem by developing what today is known as the **Student's t distribution**, or the t distribution.

If the variable X is normally distributed, then the following statistic

$$t = \frac{\overline{X} - \mu}{\dfrac{S}{\sqrt{n}}}$$

[1] Guinness considered all research conducted to be proprietary and a trade secret. The firm prohibited its employees from publishing their results. Gosset circumvented this ban by using the pen name "Student" to publish his findings.

has a t distribution with $n - 1$ **degrees of freedom**. This expression has the same form as the Z statistic in Equation (7.4) on page 229, except that S is used to estimate the unknown σ.

The Concept of Degrees of Freedom

Equation 3.6 on page 115 defines the sample variance, S^2, as a fraction, the numerator of which is the sum of squares around the sample mean:

$$\sum_{i=1}^{n}(X_i - \overline{X})^2$$

In order to calculate S^2, one first needs to calculate the sample mean, $\overline{X}$. If one knows $\overline{X}$, then once you know $n - 1$ of the values, the last value is not "free to vary" because the sum of the n values is known from the calculation of $\overline{X}$. This observation is what is meant by saying "having $n - 1$ degrees of freedom." For example, suppose a sample of five values has a mean of 20. How many values does one need to know before one can determine the remainder of the values? From $n = 5$ and $\overline{X} = 20$, one knows

$$\sum_{i=1}^{n}X_i = 100 \text{ because } \frac{\sum_{i=1}^{n}X_i}{n} = \overline{X}$$

If one only knew the first value, e.g., 18, then one would not know what the other four values in the sample would be. However, when one knows four values such as 18, 24, 19, and 16, then one knows that the fifth value must be 23. That value must be 23 in order for the five numbers sum to 100. That value is not "free to vary" from 23. Therefore, this sample has 4 $(n - 1)$ degrees of freedom.

Properties of the t Distribution

The t distribution is very similar in appearance to the standardized normal distribution. Both distributions are symmetrical and bell-shaped, with the mean and the median equal to zero. However, because S is used to estimate the unknown σ, the values of t are more variable than those for Z. Therefore, the t distribution has more area in the tails and less in the center than does the standardized normal distribution (see Figure 8.5).

FIGURE 8.5

Standardized normal distribution and t distribution for 5 degrees of freedom

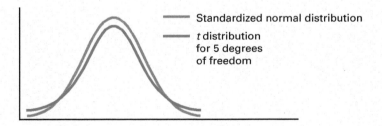

Standardized normal distribution

t distribution for 5 degrees of freedom

As the sample size and degrees of freedom increase, S becomes a better estimate of σ, and the t distribution gradually approaches the standardized normal distribution, until the two are virtually identical. With a sample size of about 120 or more, S estimates σ closely enough so that there is little difference between the t and Z distributions.

As stated earlier, the t distribution assumes that the variable X is normally distributed. In practice, however, when the sample size is large enough and the population is not very skewed, in most cases use the t distribution to estimate the population mean when σ is unknown. When dealing with a small sample size and a skewed population distribution, the confidence interval estimate may not provide a valid estimate of the population mean. To assess the assumption of normality, evaluate the shape of the sample data by constructing a histogram, stem-and-leaf display, boxplot, or normal probability plot. However, the ability of any of these graphs to help evaluate normality is limited when the sample size is small.

Find the critical values of t for the appropriate degrees of freedom from the table of the t distribution (such as Table E.3). The columns of the table present the most commonly used cumulative probabilities and corresponding upper-tail areas. The rows of the table represent the degrees of freedom. The critical t values are found in the cells of the table. For example, for 99 degrees of freedom with 95% confidence, use Table 8.1, a Table E.3 excerpt, to locate the 99 degrees of freedom row. Then find the entry for the column for the cumulative probability of 0.975 (or the upper-tail area of 0.025) because 95% confidence level means that 2.5% of the values (an area of 0.025) are in each tail of the distribution. The entry, the critical value for t for this example, is 1.9842. Figure 8.6 visualizes this value.

TABLE 8.1

Determining the critical value from the t table for an area of 0.025 in each tail with 99 degrees of freedom

		Cumulative Probabilities					
		0.75	0.90	0.95	0.975	0.99	0.995
		Upper-Tail Areas					
Degrees of Freedom		0.25	0.10	0.05	0.025	0.01	0.005
1		1.0000	3.0777	6.3138	12.7062	31.8207	63.6574
2		0.8165	1.8856	2.9200	4.3027	6.9646	9.9248
3		0.7649	1.6377	2.3534	3.1824	4.5407	5.8409
4		0.7407	1.5332	2.1318	2.7764	3.7469	4.6041
5		0.7267	1.4759	2.0150	2.5706	3.3649	4.0322
⋮		⋮	⋮	⋮	⋮	⋮	⋮
96		0.6771	1.2904	1.6609	1.9850	2.3658	2.6280
97		0.6770	1.2903	1.6607	1.9847	2.3654	2.6275
98		0.6770	1.2902	1.6606	1.9845	2.3650	2.6269
99		0.6770	1.2902	1.6604	1.9842	2.3646	2.6264
100		0.6770	1.2901	1.6602	1.9840	2.3642	2.6259

FIGURE 8.6

t distribution with 99 degrees of freedom

Source: Extracted from Table E.3.

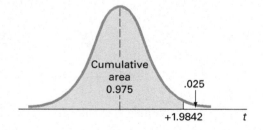

Cumulative area 0.975

.025

+1.9842 t

Because t is a symmetrical distribution with a mean of 0, if the upper-tail value is $+1.9842$, the value for the lower-tail area (lower 0.025) is -1.9842. A t value of -1.9842 means that the probability that t is less than -1.9842 is 0.025, or 2.5%. Note that for a 95% confidence interval, the cumulative probability will always be 0.975, and the upper-tail area will always be 0.025. Similarly, for a 99% confidence interval, the cumulative probability and upper-tail area will always be 0.995 and 0.005, and for a 90% confidence interval, these values will always be 0.95 and 0.05.

The Confidence Interval Statement

Equation (8.2) defines the $(1 - \alpha) \times 100$ confidence interval estimate for the mean with σ unknown.

CONFIDENCE INTERVAL FOR THE MEAN (σ UNKNOWN)

$$\overline{X} \pm t_{\alpha/2}\frac{S}{\sqrt{n}}$$

or

$$\overline{X} - t_{\alpha/2}\frac{S}{\sqrt{n}} \leq \mu \leq \overline{X} + t_{\alpha/2}\frac{S}{\sqrt{n}} \tag{8.2}$$

where

$t_{\alpha/2}$ is the critical value for an upper-tail probability of $\alpha/2$ (i.e., a cumulative area of $1 - \alpha/2$) from the t distribution with $n - 1$ degrees of freedom.

$t_{\alpha/2}\dfrac{S}{\sqrt{n}}$ is the sampling error

To illustrate the application of the confidence interval estimate for the mean when the standard deviation is unknown, recall the Ricknel Home Centers scenario on page 244. In that scenario, the mean dollar amount listed on the sales invoices for the month was one of the population parameters you sought to estimate.

To calculate this estimate, apply the DCOVA framework (see First Things First Chapter) and define the variable of interest as the dollar amount listed on the sales invoices for the month. You then collect data by selecting a sample of 100 sales invoices from the population of sales invoices during the month and organize the data as a worksheet or data table.

Construct various graphs (not shown here) to better visualize the distribution of the dollar amounts. Using the data, calculate the sample mean of the 100 sales invoices as $110.27 and the sample standard deviation as $28.95. For 95% confidence, the critical value from the t distribution (as Table 8.1 shows on page 252) is 1.9842. Using Equation (8.2),

$$\overline{X} \pm t_{\alpha/2}\frac{S}{\sqrt{n}} = 110.27 \pm (1.9842)\frac{28.95}{\sqrt{100}}$$

$$= 110.27 \pm 5.74$$

$$104.53 \leq \mu \leq 116.01$$

Figure 8.7 shows the Excel confidence interval estimate of the mean dollar amount results.

FIGURE 8.7

Excel worksheet results for the confidence interval estimate for the mean sales invoice amount worksheet results for the Ricknel Home Centers example

▲	A	B	
1	Confidence Interval Estimate for the Mean		
2			
3	Data		
4	Sample Standard Deviation	28.95	
5	Sample Mean	110.27	
6	Sample Size	100	
7	Confidence Level	95%	
8			
9	Intermediate Calculations		
10	Standard Error of the Mean	2.895	=B4/SQRT(B6)
11	Degrees of Freedom	99	=B6 - 1
12	t Value	1.9842	=T.INV.2T(1 - B7, B11)
13	Interval Half Width	5.7443	=B12 * B10
14			
15	Confidence Interval		
16	Interval Lower Limit	104.53	=B5 - B13
17	Interval Upper Limit	116.01	=B5 + B13

In the Worksheet

The T.INV.2T function calculates the t critical value in cell B12. That value is multiplied by the standard error of the mean (in cell B10) to calculate the interval half width in cell B13. In turn, that B13 value is subtracted from and added to the sample mean (cell B5) to calculate the lower and upper limits of the confidence interval in cells B16 and B17.

Thus, with 95% confidence, one concludes that the mean amount of all the sales invoices is between $104.53 and $116.01. The 95% confidence level indicates that if all possible samples of 100 were to be selected, 95% of the intervals developed would include the population mean somewhere within the interval. The validity of this confidence interval estimate depends on the assumption of normality for the distribution of the amount of the sales invoices. With a sample of 100, the normality assumption is valid, and the use of the *t* distribution is likely appropriate. Example 8.3 further illustrates how to construct the confidence interval for a mean when the population standard deviation is unknown.

EXAMPLE 8.3

Estimating the Mean Processing Time for Life Insurance Applications

An insurance company has the business objective of reducing the amount of time it takes to approve applications for life insurance. The approval process consists of underwriting, which includes a review of the application, a medical information bureau check, possible requests for additional medical information and medical exams, and a policy compilation stage in which the policy pages are generated and sent for delivery. Using the DCOVA steps first discussed on page 3, you define the variable of interest as the total processing time in days. You collect the data by selecting a random sample of 27 approved policies during a period of one month. You organize the data collected in a worksheet. Table 8.2, stored as **Insurance**, lists the total processing time, in days. To analyze the data, you need to construct a 95% confidence interval estimate for the population mean processing time.

TABLE 8.2

Processing time for life insurance applications

8	11	15	17	19	22	25	27	32	35	38	41	41	45
48	50	51	56	56	60	63	64	69	73	80	84	91	

SOLUTION To visualize the data, you construct a boxplot of the processing time (see Figure 8.8) and a normal probability plot (see Figure 8.9). To analyze the data, you construct the confidence interval estimate (see Figure 8.10 on page 255).

FIGURE 8.8

Excel boxplot for the processing time for life insurance applications

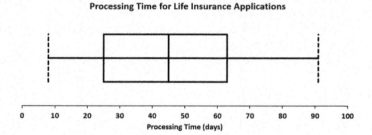

FIGURE 8.9

Excel normal probability plot for the processing time for life insurance applications

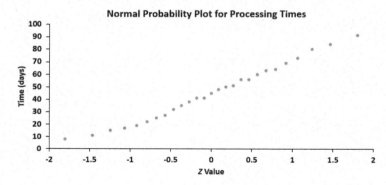

▶(*continued*)

FIGURE 8.10

Excel confidence interval estimate for the mean processing time results for life insurance applications

	A	B
1	Processing Time for Life Insurance Applications	
2		
3	Data	
4	Sample Standard Deviation	23.1472
5	Sample Mean	45.2222
6	Sample Size	27
7	Confidence Level	95%
8		
9	Intermediate Calculations	
10	Standard Error of the Mean	4.4547
11	Degrees of Freedom	26
12	t Value	2.0555
13	Interval Half Width	9.1567
14		
15	Confidence Interval	
16	Interval Lower Limit	36.07
17	Interval Upper Limit	54.38

In the Worksheet

This worksheet solution uses the same template that the Figure 8.7 Ricknel Home Center example on page 253 uses.

Figure 8.10 shows that the sample mean is $\overline{X} = 45.2222$ days and the sample standard deviation is $S = 23.1472$ days. To use Equation (8.2) on page 253 to construct the confidence interval, first determine the critical value from the t table, using the row for 26 degrees of freedom. For 95% confidence, use the column corresponding to an upper-tail area of 0.025 and a cumulative probability of 0.975. From Table E.3, you see that $t_{\alpha/2} = 2.0555$. Thus, using $\overline{X} = 45.2222$, $S = 23.1472$, $n = 27$, and $t_{\alpha/2} = 2.0555$,

$$\overline{X} \pm t_{\alpha/2}\frac{S}{\sqrt{n}} = 45.2222 \pm (2.0555)\frac{23.1472}{\sqrt{27}}$$

$$= 45.2222 \pm 9.1567$$

$$36.07 \leq \mu \leq 54.38$$

Conclude with 95% confidence that the mean processing time for the population of life insurance applications is between 36.07 and 54.38 days. The validity of this confidence interval estimate depends on the assumption that the processing time is normally distributed. From the Figure 8.8 boxplots and the Figure 8.9 normal probability plots, the processing time appears approximately symmetric, so the validity of the confidence interval estimate is not in serious doubt.

The interpretation of the confidence interval when σ is unknown is the same as when σ is known. To illustrate the fact that the confidence interval for the mean varies more when σ is unknown, return to the order-filling times example that the "Sampling Error" section Section 8.1 uses. Suppose that one did *not* know the population standard deviation and instead use the sample standard deviation to construct the confidence interval estimate of the mean. Figure 8.11 on page 256 shows the results for each of 20 samples of $n = 10$ orders.

FIGURE 8.11

Confidence interval estimates of the mean for 20 samples of $n = 10$ randomly selected from the population of $N = 200$ orders with σ unknown

Sample	N	Mean	Std Dev	SE Mean	95% Conf. Int.
S01	10	71.64	7.58	2.40	(66.22, 77.06)
S02	10	67.22	10.95	3.46	(59.39, 75.05)
S03	10	67.97	14.83	4.69	(57.36, 78.58)
S04	10	73.90	10.59	3.35	(66.33, 81.47)
S05	10	67.11	11.12	3.52	(59.15, 75.07)
S06	10	68.12	10.83	3.43	(60.37, 75.87)
S07	10	65.80	10.85	3.43	(58.03, 73.57)
S08	10	77.58	11.04	3.49	(69.68, 85.48)
S09	10	66.69	11.45	3.62	(58.50, 74.88)
S10	10	62.55	8.58	2.71	(56.41, 68.69)
S11	10	71.12	12.82	4.05	(61.95, 80.29)
S12	10	70.55	10.52	3.33	(63.02, 78.08)
S13	10	65.51	8.16	2.58	(59.67, 71.35)
S14	10	64.90	7.55	2.39	(59.50, 70.30)
S15	10	66.22	11.21	3.54	(58.20, 74.24)
S16	10	70.43	10.21	3.23	(63.12, 77.74)
S17	10	72.04	6.25	1.96	(67.57, 76.51)
S18	10	73.91	11.29	3.57	(65.83, 81.99)
S19	10	71.49	9.76	3.09	(64.51, 78.47)
S20	10	70.15	10.84	3.43	(62.39, 77.91)

In Figure 8.11, observe that the standard deviation of the samples varies from 6.25 (sample 17) to 14.83 (sample 3). Thus, the width of the confidence interval developed varies from 8.94 in sample 17 to 21.22 in sample 3. Because one knows that the population mean order time $\mu = 69.637$ minutes, one can determine that the interval for sample 8 (69.68 − 85.48) and the interval for sample 10 (56.41 − 68.69) do not correctly estimate the population mean. All the other intervals correctly estimate the population mean. As section 8.1 observes, in practice, one will select only one sample and will be unable to know for sure whether that sample provides a confidence interval that includes the population mean.

PROBLEMS FOR SECTION 8.2

LEARNING THE BASICS

8.11 If $\overline{X} = 75$, $S = 24$, and $n = 36$, and assuming that the population is normally distributed, construct a 95% confidence interval estimate for the population mean, μ.

8.12 Determine the critical value of t in each of the following circumstances:
a. $1 - \alpha = 0.95, n = 10$
b. $1 - \alpha = 0.99, n = 10$
c. $1 - \alpha = 0.95, n = 32$
d. $1 - \alpha = 0.95, n = 65$
e. $1 - \alpha = 0.90, n = 16$

8.13 Assuming that the population is normally distributed, construct a 95% confidence interval estimate for the population mean for each of the following samples:

Sample A: 1 1 1 1 8 8 8 8

Sample B: 1 2 3 4 5 6 7 8

Explain why these two samples produce different confidence intervals even though they have the same mean and range.

8.14 Assuming that the population is normally distributed, construct a 95% confidence interval for the population mean, based on the following sample of size $n = 7$:

1 2 3 4 5 6 20

Change the value of 20 to 7 and recalculate the confidence interval. Using these results, describe the effect of an outlier (i.e., an extreme value) on the confidence interval.

APPLYING THE CONCEPTS

8.15 A marketing researcher wants to estimate the mean amount spent ($) on Amazon.com by Amazon Prime member shoppers. Suppose a random sample of 100 Amazon Prime member shoppers who recently made a purchase on Amazon.com yielded a mean of $1,500 and a standard deviation of $200.
a. Construct a 95% confidence interval estimate for the mean spending for all Amazon Prime member shoppers.
b. Interpret the interval constructed in (a).

✓SELF TEST **8.16** A survey of nonprofit organizations showed that online fundraising has increased in the past year. Based on a random sample of 133 nonprofits, the mean one-time gift donation resulting from email outreach in the past year was $87. Assume that the sample standard deviation is $9.
a. Construct a 95% confidence interval estimate for the population mean one-time gift donation.
b. Interpret the interval constructed in (a).

8.17 The U.S. Department of Transportation requires tire manufacturers to provide tire performance information on the sidewall of a tire to better inform prospective customers as they make purchasing decisions. One very important measure of tire performance is the tread wear index, which indicates the tire's resistance to tread wear compared with a tire graded with a base of 100. A tire with a grade of 200 should last twice as long, on average, as a tire graded with a base of 100. A consumer organization wants to estimate the actual tread wear index of a brand name of tires that claims "graded 200" on the sidewall of the tire. A random sample of $n = 18$ indicates a sample mean tread wear index of 195.3 and a sample standard deviation of 21.4.

a. Assuming that the population of tread wear indexes is normally distributed, construct a 95% confidence interval estimate for the population mean tread wear index for tires produced by this manufacturer under this brand name.

b. Do you think that the consumer organization should accuse the manufacturer of producing tires that do not meet the performance information provided on the sidewall of the tire? Explain.

c. Explain why an observed tread wear index of 210 for a particular tire is not unusual, even though it is outside the confidence interval developed in (a).

8.18 The file Fast Casual contains the amount that a sample of 15 customers spent for lunch ($) at a fast casual restaurant:

7.42 6.29 5.83 6.50 8.34 9.51 7.10 6.80 5.90
4.89 6.50 5.52 7.90 8.30 9.60

a. Construct a 95% confidence interval estimate for the population mean amount spent for lunch ($) at a fast casual restaurant.

b. Interpret the interval constructed in (a).

c. What assumption must you make about the population distribution in order to construct the confidence interval estimate in (a)?

d. Do you think that the assumption needed in order to construct the confidence interval estimate in (a) is valid? Explain.

8.19 How much time do commuters living in or near cities spend commuting to work each week? The file Commuting Time contains the average weekly commuting time in 30 U.S. cities.

Source: Data extracted from Office of the New York City Comptroller, "NYC Economics Brief," March 2015, p. 3.

a. Construct a 95% confidence interval estimate for the population mean commuting time.

b. Interpret the interval constructed in (a).

c. What assumption must you make about the population distribution in order to construct the confidence interval estimate in (a)?

d. Do you think that the assumption needed in order to construct the confidence interval estimate in (a) is valid? Explain.

8.20 The Super Bowl, watched by close to 200 million Americans, is also a big event for advertisers. The file Super Bowl Ad Ratings contains the rating of ads that ran between the opening kickoff and the final whistle.

Source: Data extracted from **www.admeter.usatoday.com/results/2019**

For ads that ran before halftime and ads that ran at halftime or after separately:

a. Construct a 95% confidence interval estimate for the population mean ad rating.

b. Interpret the interval constructed in (a).

c. What conclusions can you reach about the ad ratings of ads that ran before halftime and ads that ran at halftime or after?

d. What assumption must you make about the population distribution in order to construct the confidence interval estimate in (a)?

e. Do you think that the assumption needed in order to construct the confidence interval estimate in (a) is valid? Explain.

8.21 Is there a difference in the yields of different types of investments? The file CD Rates contains the yields for a one-year certificate of deposit (CD) and a five-year CD sold by 46 financial institutions in Lake Worth, Florida, as of April 5, 2019.

Source: Data extracted from **www.Bankrate.com**, April 5, 2019.

a. Construct a 95% confidence interval estimate for the mean yield of one-year CDs.

b. Construct a 95% confidence interval estimate for the mean yield of five-year CDs.

c. Compare the results of (a) and (b).

8.22 One of the major measures of the quality of service provided by any organization is the speed with which the organization responds to customer complaints. A large family-held department store selling furniture and flooring, including carpet, had undergone a major expansion in the past several years. In particular, the flooring department had expanded from 2 installation crews to an installation supervisor, a measurer, and 15 installation crews. The store had the business objective of improving its response to complaints. The variable of interest was defined as the number of days between when the complaint was made and when it was resolved. Data were collected from 50 complaints that were made in the past year. The data, stored in Furniture, are as follows:

54	5	35	137	31	27	152	2	123	81	74	27	11
19	126	110	110	29	61	35	94	31	26	5	12	4
165	32	29	28	29	26	25	1	14	13	13	10	
5	27	4	52	30	22	36	26	20	23	33	68	

a. Construct a 95% confidence interval estimate for the population mean number of days between the receipt of a complaint and the resolution of the complaint.

b. What assumption must you make about the population distribution in order to construct the confidence interval estimate in (a)?

c. Do you think that the assumption needed in order to construct the confidence interval estimate in (a) is valid? Explain.

d. What effect might your conclusion in (c) have on the validity of the results in (a)?

8.23 A manufacturing company produces electric insulators. You define the variable of interest as the strength of the insulators. If the insulators break when in use, a short circuit is likely. To test the strength of the insulators, you carry out destructive testing to determine how much force is required to break the insulators. You measure force by observing how many pounds are applied to the insulator before it breaks. You collect the force data for 30 insulators selected for the experiment and organize and store these data in Force:

1,870	1,728	1,656	1,610	1,634	1,784	1,552	1,696
1,592	1,662	1,866	1,764	1,734	1,662	1,734	1,774
1,550	1,756	1,762	1,886	1,820	1,744	1,788	1,688
1,810	1,752	1,680	1,810	1,652	1,736		

a. Construct a 95% confidence interval estimate for the population mean breaking force.

b. What assumption must you make about the population distribution in order to construct the confidence interval estimate in (a)?

c. Do you think that the assumption needed in order to construct the confidence interval estimate in (a) is valid? Explain.

8.24 The file Mobile Commerce contains mobile commerce penetration values (the percentage of the country population that bought something online via a mobile phone in the past month) for 24 of the world's economies.

Source: Data extracted from statistica.com, **bit.ly/2IfWSfI**.

a. Construct a 95% confidence interval estimate for the population mean mobile commerce penetration.

b. What assumption do you need to make about the population to construct the interval in (a)?

c. Given the data presented, do you think the assumption needed in (a) is valid? Explain.

8.25 One operation of a mill is to cut pieces of steel into parts that are used in the frame for front seats in automobiles. The steel is cut with a diamond saw, and the resulting parts must be cut to be within ±0.005 inch of the length specified by the automobile company. The measurement reported from a sample of 100 steel parts (stored in Steel) is the difference, in inches, between the actual length of the steel part, as measured by a laser measurement device, and the specified length of the steel part. For example, the first observation,

−0.002, represents a steel part that is 0.002 inch shorter than the specified length.

a. Construct a 95% confidence interval estimate for the population mean difference between the actual length of the steel part and the specified length of the steel part.

b. What assumption must you make about the population distribution in order to construct the confidence interval estimate in (a)?

c. Do you think that the assumption needed in order to construct the confidence interval estimate in (a) is valid? Explain.

d. Compare the conclusions reached in (a) with those of Problem 2.43 on page 64.

8.3 Confidence Interval Estimate for the Proportion

The concept of a confidence interval also applies to categorical data. With categorical data, you want to estimate the proportion of items in a population having a certain characteristic of interest. The unknown population proportion is represented by the Greek letter π. The point estimate for π is the sample proportion, $p = X/n$, where n is the sample size and X is the number of items in the sample having the characteristic of interest. Equation (8.3) defines the confidence interval estimate for the population proportion.

CONFIDENCE INTERVAL ESTIMATE FOR THE PROPORTION

$$p \pm Z_{\alpha/2}\sqrt{\frac{p(1-p)}{n}}$$

or

$$p - Z_{\alpha/2}\sqrt{\frac{p(1-p)}{n}} \leq \pi \leq p + Z_{\alpha/2}\sqrt{\frac{p(1-p)}{n}} \qquad (8.3)$$

where

$$p = \text{sample proportion} = \frac{X}{n} = \frac{\text{Number of items having the characteristic}}{\text{sample size}}$$

π = population proportion

$Z_{\alpha/2}$ = critical value from the standardized normal distribution

n = sample size

Note: To use this equation for the confidence interval, the sample size n must be large enough to ensure that both X and $n - X$ are greater than 5.

Use the confidence interval estimate for the proportion defined in Equation (8.3) to estimate the proportion of sales invoices that contain errors (see the Ricknel Home Centers scenario on page 244). Using the DCOVA steps, first define the variable of interest as whether the invoice contains errors (yes or no). Then, collect the data from a sample of 100 sales invoices and organize and store the results that show 10 invoices contain errors. To analyze the data, compute $p = X/n = 10/100 = 0.10$. Because both $X = 10$ and $n - X = 100 - 10 = 90$ are > 5, using Equation (8.3) and $Z_{\alpha/2} = 1.96$, for 95% confidence,

$$p \pm Z_{\alpha/2}\sqrt{\frac{p(1-p)}{n}}$$

$$= 0.10 \pm (1.96)\sqrt{\frac{(0.10)(0.90)}{100}}$$

$$= 0.10 \pm (1.96)(0.03)$$

$$= 0.10 \pm 0.0588$$

$$0.0412 \le \pi \le 0.1588$$

Therefore, with 95% confidence, the population proportion of all sales invoices containing errors is between 0.0412 and 0.1588. The estimate is that between 4.12% and 15.88% of all the sales invoices contain errors. Figure 8.12 shows an Excel confidence interval estimate for this example.

FIGURE 8.12

Excel confidence interval estimate results for the proportion of sales invoices that contain errors

	A	B
1	Proportion of In-Error Sales Invoices	
2		
3	Data	
4	Sample Size	100
5	Number of Successes	10
6	Confidence Level	95%
7		
8	Intermediate Calculations	
9	Sample Proportion	0.1 =B5/B4
10	Z Value	-1.9600 =NORM.S.INV((1 - B6)/2)
11	Standard Error of the Proportion	0.03 =SQRT(B9 * (1 - B9)/B4)
12	Interval Half Width	0.0588 =ABS(B10 * B11)
13		
14	Confidence Interval	
15	Interval Lower Limit	0.0412 =B9 - B12
16	Interval Upper Limit	0.1588 =B9 + B12

In the Worksheet

The NORM.S.INV function calculates the Z value in cell B10. That value is multiplied by the standard error of the proportion (in cell B11) to calculate the interval half width in cell B12. In turn, that B12 value is subtracted from and added to the sample proportion (cell B9) to calculate the lower and upper limits of the confidence interval in cells B15 and B16.

Example 8.4 illustrates another application of a confidence interval estimate for the proportion.

EXAMPLE 8.4

Estimating the Proportion of Nonconforming Newspapers Printed

The operations manager at a large newspaper wants to estimate the proportion of newspapers printed that have a nonconforming attribute. Using the DCOVA steps, you define the variable of interest as whether the newspaper has excessive rub-off, improper page setup, missing pages, or duplicate pages. You collect the data by selecting a random sample of $n = 200$ newspapers from all the newspapers printed during a single day. You organize the results in a worksheet, which shows that 35 newspapers contain some type of nonconformance. To analyze the data, you need to construct and interpret a 90% confidence interval estimate for the proportion of newspapers printed during the day that have a nonconforming attribute.

SOLUTION Using Equation (8.3),

$$p = \frac{X}{n} = \frac{35}{200} = 0.175, \text{ and with a 90\% level of confidence } Z_{\alpha/2} = 1.645$$

$$p \pm Z_{\alpha/2}\sqrt{\frac{p(1-p)}{n}}$$

$$= 0.175 \pm (1.645)\sqrt{\frac{(0.175)(0.825)}{200}}$$

$$= 0.175 \pm (1.645)(0.0269)$$

$$= 0.175 \pm 0.0442$$

$$0.1308 \le \pi \le 0.2192$$

You conclude with 90% confidence that the population proportion of all newspapers printed that day with nonconformities is between 0.1308 and 0.2192. This means you estimate that between 13.08% and 21.92% of the newspapers printed on that day have some type of nonconformance.

Equation (8.3) contains a Z statistic because you can use the normal distribution to approximate the binomial distribution when the sample size is sufficiently large. In Example 8.4, the confidence interval using Z provides an excellent approximation for the population proportion because both X and $n - X$ are greater than 5. However, if you do not have a sufficiently large sample size, you should use the binomial distribution rather than Equation (8.3) (Cochran, Fisher, and Snedecor). The exact confidence intervals for various sample sizes and proportions of items of interest have been tabulated by Fisher and Yates (Fisher).

PROBLEMS FOR SECTION 8.3

LEARNING THE BASICS

8.26 If $n = 200$ and $X = 50$, construct a 95% confidence interval estimate for the population proportion.

8.27 If $n = 400$ and $X = 25$, construct a 99% confidence interval estimate for the population proportion.

APPLYING THE CONCEPTS

✓SELF TEST **8.28** A cellphone provider has the business objective of wanting to estimate the proportion of subscribers who would upgrade to a new cellphone with improved features if it were made available at a substantially reduced cost. Data are collected from a random sample of 500 subscribers. The results indicate that 135 of the subscribers would upgrade to a new cellphone at a reduced cost.
a. Construct a 99% confidence interval estimate for the population proportion of subscribers that would upgrade to a new cellphone at a reduced cost.
b. How would the manager in charge of promotional programs use the results in (a)?

8.29 An Ipsos poll asked 1,004 adults, "If purchasing a used car made certain upgrades or features more affordable, what would be your preferred luxury upgrade?" The results indicated that 9% of males and 14% of females answered window tinting.

Source: Ipsos, "Safety Technology Tops the List of Most Desired Features Should They Be More Affordable When Purchasing a Used Car—Particularly Collision Avoidance," available at **bit.ly/2ufbS8Z**.

The poll description did not state the sample sizes of males and females. Suppose that both sample sizes were 502 and that 46 of 502 males and 71 of 502 females reported window tinting as their preferred luxury upgrade of choice.

a. Construct a 95% confidence interval estimate of the population proportion of males who said that window tinting would be their favorite upgrade.
b. Construct a 95% confidence interval estimate of the population proportion of females who said that window tinting would be their favorite upgrade.
c. What conclusions do you reach about differences between males and females and their favorite upgrade?

8.30 What do you value most when shopping in a retail store? According to a TimeTrade survey, 26% of consumers value *personalized experience* most.

Source: Data extracted from "The State of Retail, 2017," TimeTrade, **bit.ly/2rFGf7o**.

a. Suppose that the survey had a sample size of $n = 1,000$. Construct a 95% confidence interval estimate for the population proportion of consumers that value *personalized experience* most when shopping in a retail store?

b. Based on (a), can you claim that more than a quarter of all consumers value *personalized experience* most when shopping in a retail store?
c. Repeat parts (a) and (b), assuming that the survey had a sample size of $n = 10,000$.
d. Discuss the effect of sample size on confidence interval estimation.

8.31 A survey of 1,520 American adults asked, "Do you feel overloaded with too much information?" The results indicate that 134 of 304 females feel information overload compared to 651 of 1,216 males.

Source: Data extracted from "Information Overload," **bit.ly/2pR5bHZ**.

a. Construct a 95% confidence interval estimate of the population proportion of males who feel overloaded with too much information.
b. Construct a 95% confidence interval estimate of the population proportion of females who feel overloaded with too much information.
c. What conclusions do you reach about differences between males and females and their information overload?

8.32 A Pew Research Center survey of 4,787 adults found that 4,178 had bought something online. Of these online shoppers, 789 are weekly online shoppers.
a. Construct a 95% confidence interval estimate of the population proportion of adults who had bought something online.
b. Construct a 95% confidence interval estimate of the population proportion of online shoppers who are weekly online shoppers.
c. How would the director of e-commerce sales for a company use the results of (a) and (b)?

8.33 What business, economic, policy, and environmental threats to organization growth are CEOs extremely concerned about? In a survey by PricewaterhouseCoopers (PwC), 57 of 114 U.S. CEOs are extremely concerned about cyberthreats, and 22 are extremely concerned about lack of trust in business.

Source: Data extracted from PWC, "US business leadership in the world in 2017," **pwc.to/2kHRGnE**.

a. Construct a 95% confidence interval estimate for the population proportion of U.S. CEOs who are extremely concerned about cyber threats.
b. Construct a 95% confidence interval estimate for the population proportion of U.S. CEOs who are extremely concerned about lack of trust in business.
c. Interpret the intervals in (a) and (b).

8.4 Determining Sample Size

In each confidence interval developed so far in this chapter, the sample size was reported along with the results, with little discussion of the width of the resulting confidence interval. In the business world, sample sizes are determined prior to data collection to ensure that the confidence interval is narrow enough to be useful in making decisions. Determining the proper sample size is a complicated procedure, subject to the constraints of budget, time, and the amount of acceptable sampling error. In the Ricknel Home Centers scenario, if one wants to estimate the mean dollar amount of the sales invoices, one must first determine two things: how large a sampling error to allow in estimating the population mean and the level of confidence to use (such as 90%, 95%, or 99%) to estimate the population mean.

Sample Size Determination for the Mean

To develop an equation for determining the appropriate sample size needed when constructing a confidence interval estimate for the mean, recall Equation (8.1) on page 248:

$$\overline{X} \pm Z_{\alpha/2}\frac{\sigma}{\sqrt{n}}$$

[2] In this context, some statisticians refer to e as the **margin of error**.

The amount added to or subtracted from $\overline{X}$ is equal to half the width of the interval. This quantity represents the amount of imprecision in the estimate that results from sampling error.[2] The sampling error, e, is defined as

$$e = Z_{\alpha/2}\frac{\sigma}{\sqrt{n}}$$

Solving for n gives the sample size needed to construct the appropriate confidence interval estimate for the mean. "Appropriate" means that the resulting interval will have an acceptable amount of sampling error.

SAMPLE SIZE DETERMINATION FOR THE MEAN

The sample size, n, is equal to the product of the $Z_{\alpha/2}$ value squared and the standard deviation, σ, squared, divided by the square of the sampling error, e.

$$n = \frac{Z_{\alpha/2}^2 \sigma^2}{e^2} \qquad\qquad (8.4)$$

To compute the sample size, three quantities must be known:

[3] One uses Z instead of t because, to determine the critical value of t, one would need to know the sample size, but one does not know that at this point! For most studies, the sample size needed is large enough that the standardized normal distribution is a good approximation of the t distribution.

- The desired confidence level, which determines the value of $Z_{\alpha/2}$, the critical value from the standardized normal distribution[3]
- The acceptable sampling error, e
- The standard deviation, σ

In some business-to-business relationships that require estimation of important parameters, legal contracts specify acceptable levels of sampling error and the confidence level required. For companies in the food and drug sectors, government regulations often specify sampling errors and confidence levels. In general, however, it is usually not easy to specify the three quantities needed to determine the sample size. How can one determine the level of confidence and sampling error? Typically, these questions are answered only by an individual very familiar with the variables under study, such as a subject matter expert. Although 95% is the most common confidence level used, if more confidence is desired, then 99% might be more appropriate; if less confidence is deemed acceptable, then 90% might be used. For the sampling error, one should think not of how much sampling error one would like to have (one really wants no errors) but rather of how much you can tolerate when reaching conclusions from the confidence interval.

In addition to specifying the confidence level and the sampling error, one needs to estimate the standard deviation. Unfortunately, one rarely knows the population standard deviation, σ. In some instances, one can estimate the standard deviation from past data. In other situations, one can make an educated guess by taking into account the range and distribution of the variable. For example, if one assumes a normal distribution, the range is approximately equal to 6σ ($\pm 3\sigma$ around the mean) and σ can be estimated as the range divided by 6. If one cannot estimate σ in this way, one can conduct a small-scale study and estimate the standard deviation from the resulting data.

To explore how to determine the sample size needed for estimating the population mean, consider again the audit at Ricknel Home Centers. In Section 8.2, a sample of 100 sales invoices was selected and a 95% confidence interval estimate for the population mean sales invoice amount. How was this sample size determined? Should a different sample size have been selected?

Suppose that, after consulting with company managers, one determines that a sampling error of no more than $\pm \$5$ is desired, along with 95% confidence. Past data indicate that the standard deviation of the sales amount is approximately \$25. Thus, $e = \$5$, $\sigma = \$25$, and $Z_{\alpha/2} = 1.96$ (for 95% confidence). Using Equation (8.4),

$$n = \frac{Z_{\alpha/2}^2\, \sigma^2}{e^2} = \frac{(1.96)^2(25)^2}{(5)^2}$$

$$= 96.04$$

Because the general rule is to slightly oversatisfy the criteria by rounding the sample size up to the next whole integer, a sample size of 97 is needed. The Section 8.2 example on page 253 uses a sample size $n = 100$, slightly more than what is necessary to satisfy the needs of the company, based on the estimated standard deviation, desired confidence level, and sampling error. Because the calculated sample standard deviation is slightly higher than expected, \$28.95 compared to \$25.00, the confidence interval is slightly wider than desired. Figure 8.13 presents Excel results for determining the sample size.

FIGURE 8.13

Excel results for determining the sample size for estimating the mean sales invoice amount for the Ricknel Home Centers example

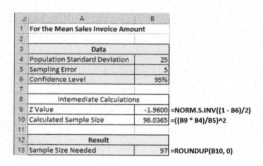

In the Worksheet

The NORM.S.INV function calculates the Z value in cell B9. That value is used in calculated sample size formula in cell B10. The ROUNDUP(B10, 0) function in cell B13 rounds the calculated sample size to the next whole integer (96.0365 to 97).

Example 8.5 illustrates another application of determining the sample size needed to develop a confidence interval estimate for the mean.

EXAMPLE 8.5

Determining the Sample Size for the Mean

Returning to Example 8.3 on page 254, suppose you want to estimate, with 95% confidence, the population mean processing time to within ± 4 days. On the basis of a study conducted the previous year, you believe that the standard deviation is 25 days. Determine the sample size needed.

SOLUTION Using Equation (8.4) on page 261 and $e = 4$, $\sigma = 25$, and $Z_{\alpha/2} = 1.96$ for 95% confidence,

$$n = \frac{Z_{\alpha/2}^2\, \sigma^2}{e^2} = \frac{(1.96)^2(25)^2}{(4)^2}$$

$$= 150.06$$

Therefore, you should select a sample of 151 applications because the general rule for determining sample size is to always round up to the next integer value in order to slightly oversatisfy the criteria desired. An actual sampling error larger than 4 will result if the sample standard deviation calculated in this sample of 151 is greater than 25 and smaller if the sample standard deviation is less than 25.

Sample Size Determination for the Proportion

The previous passage discusses determining the sample size needed for estimating the population mean. A similar method determines the sample size needed to estimate a population proportion, π. Recall that in developing the sample size for a confidence interval for the mean, the sampling error is defined by

$$e = Z_{\alpha/2}\frac{\sigma}{\sqrt{n}}$$

To estimate a proportion, replace σ with $\sqrt{\pi(1-\pi)}$ to calculate sampling error as

$$e = Z_{\alpha/2}\sqrt{\frac{\pi(1-\pi)}{n}}$$

Solving for n produces Equation (8.5), which determines the sample size necessary to develop a confidence interval estimate for a proportion.

SAMPLE SIZE DETERMINATION FOR THE PROPORTION

The sample size n is equal to the product of $Z_{\alpha/2}$ squared, the population proportion, π, and 1 minus the population proportion, π, divided by the square of the sampling error, e.

$$n = \frac{Z_{\alpha/2}^2\pi(1-\pi)}{e^2} \tag{8.5}$$

Therefore, determining the sample size needed to estimate a population proportion requires three quantities:

- The desired confidence level, which determines the value of $Z_{\alpha/2}$, the critical value from the standardized normal distribution
- The acceptable sampling error (or margin of error), e
- The population proportion, π

To solve Equation (8.5), first determine the desired level of confidence in order to be able to find the appropriate $Z_{\alpha/2}$ value from the standardized normal distribution. Then, set the sampling error, e, the amount of error tolerated in estimating the population proportion. However, note that the third required quantity, π, *is* the population parameter that will be estimated later. How does one state a value for something that has yet to be determined?

Past information or relevant experience may provide an educated estimate of π. In the absence of that knowledge, one should provide a value for π that would never *underestimate* the sample size needed. In Equation (8.5), the quantity $\pi(1-\pi)$ appears in the numerator. Therefore, the value of π needs to be the value that will make the quantity $\pi(1-\pi)$ as large as possible. When $\pi = 0.5$, the product $\pi(1-\pi)$ achieves its maximum value, as the following values of π demonstrate:

For $\pi = 0.9$, $\pi(1-\pi) = (0.9)(0.1) = 0.09$
For $\pi = 0.7$, $\pi(1-\pi) = (0.7)(0.3) = 0.21$
For $\pi = 0.5$, $\pi(1-\pi) = (0.5)(0.5) = 0.25$
For $\pi = 0.3$, $\pi(1-\pi) = (0.3)(0.7) = 0.21$
For $\pi = 0.1$, $\pi(1-\pi) = (0.1)(0.9) = 0.09$

Therefore, when no prior knowledge about the population proportion, π, exists, use $\pi = 0.5$ to determine the sample size. Using $\pi = 0.5$ produces the largest possible sample size and results in the narrowest and most precise confidence interval. This increased precision comes at the cost of spending more time and money for an increased sample size. Using $\pi = 0.5$ *overestimates* the sample size needed and will calculate a confidence interval *narrower* than originally intended if the proportion is a value other than 0.5, two outcomes that are acceptable for using the results for decision making.

In the Ricknel Home Centers scenario, suppose that the auditing procedures require you to have 95% confidence in estimating the population proportion of sales invoices with errors to within ± 0.07. The results from past months indicate that the largest proportion has been no more than 0.15. Thus, using Equation (8.5) with $e = 0.07$, $\pi = 0.15$, and $Z_{\alpha/2} = 1.96$ for 95% confidence,

$$n = \frac{Z_{\alpha/2}^2 \pi(1 - \pi)}{e^2} = \frac{(1.96)^2(0.15)(0.85)}{(0.07)^2}$$

$$= 99.96$$

Because the general rule rounds up the sample size to the next whole integer to slightly oversatisfy the criteria, a sample size of 100 is needed. The sample size needed to satisfy the requirements of the company, based on the estimated proportion, desired confidence level, and sampling error, is the same as the sample size that the Section 8.3 example on page 258 uses. The actual confidence interval is narrower than required because the sample proportion is 0.10, whereas 0.15 was used for π in Equation (8.5). Figure 8.14 presents Excel results for determining the sample size.

FIGURE 8.14

Excel for determining the sample size for estimating the proportion of in-error sales invoices for Ricknel Home Centers

	A	B
1	For the Proportion of In-Error Sales Invoices	
2		
3	Data	
4	Estimate of True Proportion	0.15
5	Sampling Error	0.07
6	Confidence Level	95%
7		
8	Intermediate Calculations	
9	Z Value	-1.9600 =NORM.S.INV((1 - B6) / 2)
10	Calculated Sample Size	99.9563 =(B9^2 * B4 * (1 - B4)) / B5^2
11		
12	Result	
13	Sample Size Needed	100 =ROUNDUP(B10, 0)

In the Worksheet

The NORM.S.INV function calculates the Z value in cell B9. That value is used in the calculated sample size formula in cell B10. The ROUNDUP(B10, 0) function in cell B13 rounds the calculated sample size to the next whole integer (99.9563 to 100).

Example 8.6 provides another application of determining the sample size for estimating the population proportion.

EXAMPLE 8.6

Determining the Sample Size for the Population Proportion

You want to have 90% confidence of estimating, to within ± 0.05, the proportion of office workers who respond to email within an hour. Because you have not previously undertaken such a study, no information is available from past data. Determine the sample size needed.

SOLUTION Because no information is available from past data, assume that $\pi = 0.50$. Using Equation (8.5) on page 263 and $e = 0.05$, $\pi = 0.50$, and $Z_{a/2} = 1.645$ for 90% confidence,

$$n = \frac{Z_{\alpha/2}^2 \pi(1 - \pi)}{e^2} = \frac{(1.645)^2(0.50)(0.50)}{(0.05)^2}$$

$$= 270.6$$

Therefore, you need a sample of 271 office workers to estimate the population proportion to within ± 0.05 with 90% confidence.

PROBLEMS FOR SECTION 8.4

LEARNING THE BASICS

8.34 If you want to be 95% confident of estimating the population mean to within a sampling error of ± 5 and the standard deviation is assumed to be 15, what sample size is required?

8.35 If you want to be 99% confident of estimating the population mean to within a sampling error of ± 20 and the standard deviation is assumed to be 100, what sample size is required?

8.36 If you want to be 99% confident of estimating the population proportion to within a sampling error of ± 0.04, what sample size is needed?

8.37 If you want to be 95% confident of estimating the population proportion to within a sampling error of ± 0.02 and there is historical evidence that the population proportion is approximately 0.40, what sample size is needed?

APPLYING THE CONCEPTS

✓ SELF TEST **8.38** A survey is planned to determine the mean annual family medical expenses of employees of a large company. The management of the company wishes to be 95% confident that the sample mean is correct to within $\pm \$50$ of the population mean annual family medical expenses. A previous study indicates that the standard deviation is approximately $400.

a. How large a sample is necessary?

b. If management wants to be correct to within $\pm \$25$, how many employees need to be selected?

8.39 If the manager of a bottled water distributor wants to estimate, with 95% confidence, the mean amount of water in a 1-gallon bottle to within ± 0.004 gallon and also assumes that the standard deviation is 0.02 gallon, what sample size is needed?

8.40 If a light bulb manufacturing company wants to estimate, with 95% confidence, the mean life of light emitting diode (LED) light bulbs to within ± 400 hours and also assumes that the population standard deviation is 1,500 hours, how many LED light bulbs need to be selected?

8.41 If the inspection division of a county weights and measures department wants to estimate the mean amount of soft-drink fill in 2-liter bottles to within ± 0.01 liter with 95% confidence and also assumes that the standard deviation is 0.05 liter, what sample size is needed?

8.42 An advertising media analyst wants to estimate the mean weekly amount of time consumers spend watching television daily. Based on previous studies, the standard deviation is assumed to be 20 minutes. The media analyst wants to estimate, with 99% confidence, the mean weekly amount of time to within ± 5 minutes.

a. What sample size is needed?

b. If 95% confidence is desired, how many consumers need to be selected?

8.43 An advertising media analyst wants to estimate the mean amount of time that consumers spend with digital media daily. From past studies, the standard deviation is estimated as 45 minutes.

a. What sample size is needed if the media analyst wants to be 90% confident of being correct to within ± 5 minutes?

b. If 99% confidence is desired, how many consumers need to be selected?

8.44 Panera Bread is a restaurant chain that features gourmet casual breakfast, lunch, and brunch, a growing niche in the restaurant industry. Suppose that the mean per-person check for breakfast at Panera Bread is approximately $9.50.

a. Assuming a standard deviation of $2.00, what sample size is needed to estimate, with 95% confidence, the mean per-person check for Panera Bread to within $\pm \$0.25$?

b. Assuming a standard deviation of $2.50, what sample size is needed to estimate, with 95% confidence, the mean per-person check for Panera Bread to within $\pm \$0.25$?

c. Assuming a standard deviation of $3.00, what sample size is needed to estimate, with 95% confidence, the mean per-person check for Panera Bread to within $\pm \$0.25$?

d. Discuss the effect of variation on the sample size needed.

8.45 What does brand loyalty mean to consumers? According to a Rare research report, 20% of consumers associate *trust* with brand loyalty.

Source: Data extracted from "Redefining Loyalty," Rare, 2016, **bit.ly/2solA40**.

a. To conduct a follow-up study that would provide 95% confidence that the point estimate is correct to within ± 0.04 of the population proportion, how large a sample size is required?

b. To conduct a follow-up study that would provide 99% confidence that the point estimate is correct to within ± 0.04 of the population proportion, how many consumers need to be sampled?

c. To conduct a follow-up study that would provide 95% confidence that the point estimate is correct to within ± 0.02 of the population proportion, how large a sample size is required?

d. To conduct a follow-up study that would provide 99% confidence that the point estimate is correct to within ± 0.02 of the population proportion, how many consumers need to be sampled?

e. Discuss the effects on sample size requirements of changing the desired confidence level and the acceptable sampling error.

8.46 A Federal Reserve Bank of Atlanta report looks at what strategies and measures financial institutions are pursuing to provide mobile financial services to their customers. When 115 financial institutions were asked about barriers hindering greater consumer adoption of mobile banking, 81 said security concerns is a barrier, 68 said lack of trust in the technology is a barrier, and 16 said difficulty of use was a barrier.

Source: Data extracted from "2016 Mobile Banking and Payments Survey of Financial Institutions in the Sixth District," Federal Reserve Bank of Atlanta **bit.ly/2sfe0co**.

Construct a 95% confidence interval estimate of the population proportion of financial institution who said:

a. security concerns are a barrier hindering greater consumer adoption of mobile banking.

b. lack of trust is a barrier hindering greater consumer adoption of mobile banking.

c. difficulty of use is a barrier hindering greater consumer adoption of mobile banking.

d. You have been asked to update the results of this study. Determine the sample size necessary to estimate, with 95% confidence, the population proportions in (a) through (c) to within ± 0.02.

8.47 In a study of 443 nonprofits nationwide, 130 indicated that the greatest diversity staffing challenge they face is retaining younger staff (those under 30).

Source: Data extracted from "2016 Nonprofit Employment Practices Survey," Nonprofit HR, 2016, **bit.ly/23ZHwhb**.

a. Construct a 95% confidence interval for the population proportion of nonprofits that indicate retaining younger staff is the greatest diversity staffing challenge for their organization.

b. Interpret the interval constructed in (a).

c. If you wanted to conduct a follow-up study to estimate the population proportion of nonprofits that indicate retaining younger staff is the greatest diversity staffing challenge for their organization to within ± 0.01 with 95% confidence, how many nonprofits would you survey?

8.48 Cybersecurity is a critical business issue that demands the attention of business and IT executives. According to a study released by PwC, 38% of surveyed business and IT executives reported phishing scams at their institutions.

Source: Data extracted from "Toward new possibilities in threat management," PwC, 2017 **pwc.to/2kwhPJv**.

a. If you conduct a follow-up study to estimate the population proportion of business and IT executives reporting phishing scams at their institutions, would you use a π of 0.38 or 0.50 in the sample size formula?

b. Using your answer in part (a), find the sample size necessary to estimate, with 95% confidence, the population proportion to within ± 0.03.

8.49 Personal data are the new currency of the digital economy. How do consumers feel about sharing personal data with their communication service providers (CSPs)? A recent IBM report highlights that 40% of 18- to 25-year-old consumers are comfortable sharing personal data with their CSPs.

Source: Data extracted from "The trust factor in the cognitive era," IBM Institute for Business Value, 2017 **ibm.co/2rq48Pd**.

a. To conduct a follow-up study that would provide 99% confidence that the point estimate is correct to within ± 0.03 of the population proportion, how many 18- to 25-year-old consumers need to be sampled?

b. To conduct a follow-up study that would provide 99% confidence that the point estimate is correct to within ± 0.05 of the population proportion, how many 18- to 25-year-old consumers need to be sampled?

c. Compare the results of (a) and (b).

8.5 Confidence Interval Estimation and Ethical Issues

student TIP

When preparing point estimates, always state the interval estimate in a *prominent* place and include a brief explanation of the meaning of the confidence interval. Also, make sure to highlight the sample size and sampling error.

The selection of samples and the inferences that accompany them raise several ethical issues. The major ethical issue concerns whether confidence interval estimates accompany point estimates. Failure to include a confidence interval estimate might mislead the user of the results into thinking that the point estimate is all that is needed to predict the population characteristic with certainty. Confidence interval limits (typically set at 95%), the sample size used, and an interpretation of the meaning of the confidence interval in terms that a person untrained in statistics can understand should always accompany point estimates.

When media outlets publicize the results of a political poll, they often overlook this type of information. Sometimes, the results of a poll include the sampling error, but the sampling error is often presented in fine print or as an afterthought to the story being reported. A fully ethical presentation of poll results would give equal prominence to the confidence levels, sample size, sampling error, and confidence limits of the poll.

8.6 Application of Confidence Interval Estimation in Auditing

Auditing is the collection and evaluation of evidence about information related to an economic entity in order to determine and report on how well the information corresponds to established criteria. Auditing uses probability sampling methods to develop confidence interval estimates. The **Section 8.6 online topic** reviews three common applications of confidence interval estimation in auditing.

8.7 Estimation and Sample Size Determination for Finite Populations

To develop confidence interval estimates for population parameters or determine sample sizes when estimating population parameters, you use the finite population correction factor when samples are selected without replacement from a finite population. The **Section 8.7 online topic** explains how to use the finite population correction factor for these purposes.

8.8 Bootstrapping

The confidence interval estimation procedures discussed in this chapter make assumptions that are often not valid, especially for small samples. Bootstrapping, the selection of an initial sample and repeated sampling from that initial sample, provides an alternative approach that does not rely on those assumptions. The **Section 8.8 online topic** explains this alternative technique.

▼ USING **STATISTICS**
Getting Estimates at Ricknel Home Centers, Revisited

In the Ricknel Home Centers scenario, you were an accountant for a distributor of home improvement supplies in the northeastern United States. You were responsible for the accuracy of the integrated inventory management and sales information system. You used confidence interval estimation techniques to draw conclusions about the population of all records from a relatively small sample collected during an audit.

At the end of the month, you collected a random sample of 100 sales invoices and made the following inferences:

- With 95% confidence, you concluded that the mean amount of all the sales invoices is between $104.53 and $116.01.

- With 95% confidence, you concluded that between 4.12% and 15.88% of all the sales invoices contain errors.

These estimates provide an interval of values that you believe contain the true population parameters. If these intervals are too wide (i.e., the sampling error is too large) for the types of decisions Ricknel Home Centers needs to make, you will need to take a larger sample. You can use the sample size formulas in Section 8.4 to determine the number of sales invoices to sample to ensure that the size of the sampling error is acceptable.

▼ SUMMARY

This chapter discusses confidence intervals for estimating the characteristics of a population, along with how you can determine the necessary sample size. You learned how to apply these methods to numerical and categorical data. Table 8.3 provides a list of topics covered in this chapter.

To determine what equation to use for a particular situation, you need to answer these questions:
- Are you constructing a confidence interval or determining a sample size?
- Do you have a numerical variable or a categorical variable?

TABLE 8.3
Summary of topics in Chapter 8

TYPE OF ANALYSIS	TYPE OF DATA	
	Numerical	**Categorical**
Confidence interval for a population parameter	Confidence interval estimate for the mean (Sections 8.1 and 8.2)	Confidence interval estimate for the proportion (Section 8.3)
Determining sample size	Sample size determination for the mean (Section 8.4)	Sample size determination for the proportion (Section 8.4)

▼ REFERENCES

Cochran, W. G. *Sampling Techniques*, 3rd ed. New York: Wiley, 1977.

Daniel, W. W. *Applied Nonparametric Statistics*, 2nd ed. Boston: PWS Kent, 1990.

Fisher, R. A., and F. Yates. *Statistical Tables for Biological, Agricultural and Medical Research*, 5th ed. Edinburgh: Oliver & Boyd, 1957.

Hahn, G., and W. Meeker. *Statistical Intervals: A Guide for Practitioners*. New York: John Wiley and Sons, Inc., 1991.

Kirk, R. E., Ed. *Statistical Issues: A Reader for the Behavioral Sciences*. Belmont, CA: Wadsworth, 1972.

Larsen, R. L., and M. L. Marx. *An Introduction to Mathematical Statistics and Its Applications*, 5th ed. Upper Saddle River, NJ: Prentice Hall, 2012.

Snedecor, G. W., and W. G. Cochran. *Statistical Methods*, 7th ed. Ames, IA: Iowa State University Press, 1980.

▼ KEY EQUATIONS

Confidence Interval for the Mean (σ Known)

$$\bar{X} \pm Z_{\alpha/2}\frac{\sigma}{\sqrt{n}}$$

or

$$\bar{X} - Z_{\alpha/2}\frac{\sigma}{\sqrt{n}} \leq \mu \leq \bar{X} + Z_{\alpha/2}\frac{\sigma}{\sqrt{n}} \tag{8.1}$$

Confidence Interval for the Mean (σ Unknown)

$$\bar{X} \pm t_{\alpha/2}\frac{S}{\sqrt{n}}$$

or

$$\bar{X} - t_{\alpha/2}\frac{S}{\sqrt{n}} \leq \mu \leq \bar{X} + t_{\alpha/2}\frac{S}{\sqrt{n}} \tag{8.2}$$

Confidence Interval Estimate for the Proportion

$$p \pm Z_{\alpha/2}\sqrt{\frac{p(1-p)}{n}}$$

or

$$p - Z_{\alpha/2}\sqrt{\frac{p(1-p)}{n}} \leq \pi \leq p + Z_{\alpha/2}\sqrt{\frac{p(1-p)}{n}} \tag{8.3}$$

Sample Size Determination for the Mean

$$n = \frac{Z_{\alpha/2}^2 \sigma^2}{e^2} \tag{8.4}$$

Sample Size Determination for the Proportion

$$n = \frac{Z_{\alpha/2}^2 \pi(1-\pi)}{e^2} \tag{8.5}$$

▼ KEY TERMS

confidence interval estimate   245
critical value   248
degrees of freedom   251

level of confidence   247
margin of error   261
point estimate   245

sampling error   247
Student's t distribution   250

▼ CHECKING YOUR UNDERSTANDING

8.50 Why can you never really have 100% confidence of correctly estimating the population characteristic of interest?

8.51 When should you use the t distribution to develop the confidence interval estimate for the mean?

8.52 Why does widening the confidence interval (thereby making it less precise) increase confidence for a given sample size, n?

8.53 Why is the sample size needed to determine the proportion smaller when the population proportion is 0.20 than when the population proportion is 0.50?

▼ CHAPTER REVIEW PROBLEMS

8.54 A GlobalWebIndex study noted the percentage of Internet users that owned various devices. Suppose that a survey of 1,000 Internet users found that 840 own a PC/laptop, 910 own a smartphone, 500 own a tablet, and 100 own a smart watch.

Source: Data extracted from "GWI Device," GlobalWebIndex Quarterly Report, Q1 2017 **bit.ly/2qBks0x**.

a. Construct 95% confidence interval estimates for the population proportion of the devices Internet users own.

b. What conclusions can you reach concerning what devices Internet users own?

8.55 How do smartphone owners use their smartphones when shopping in a grocery store? A sample of 731 smartphone owners in the United States revealed that 358 use their smartphone to access digital coupons, 355 look up recipes, 234 read reviews of products and brands, and 154 locate in-store items.

Source: Data extracted from "U.S. Grocery Shopping Trends, 2016," FMI, **bit.ly/2h9Q4Sl**.

a. For each smartphone user grocery shopping online activity, construct a 95% confidence interval estimate of the population proportion.

b. What conclusions can you reach concerning how smartphone owners use their smartphones when shopping in a grocery store?

8.56 A market researcher for a consumer electronics company wants to study the media viewing behavior of residents of a particular area. A random sample of 40 respondents is selected, and each respondent is instructed to keep a detailed record of time spent engaged viewing content on any screen (TV, smartphone, tablet, etc.) for a specific week. The results of the study are:

Content viewing time per week: $\bar{X} = 51$ hours, $S = 3.5$ hours. 32 respondents have at least one ultra high definition (UHD) screen.

a. Construct a 95% confidence interval estimate for the mean content viewing time per week in this area.

b. Construct a 95% confidence interval estimate for the population proportion of residents who have at least one UHD screen.

Suppose that the market researcher wants to take another survey in a different location. Answer these questions:

c. What sample size is required to be 95% confident of estimating the population mean content viewing time to within ± 2 hours assuming that the population standard deviation is equal to 5 hours?

d. How many respondents need to be selected to be 95% confident of being within ± 0.06 of the population proportion who have UHD on at least one television set if no previous estimate is available?

e. Based on (c) and (d), how many respondents should the market researcher select if a single survey is being conducted?

8.57 An information technology (IT) provider of cloud backup and restore solutions for small to midsize businesses wants to study the consequences of ransomware attacks. A random sample of 50 small to midsized companies in the United States that have experienced a ransomware attack reveals the following:

Time spent dealing with and containing a ransomware incident: $\overline{X} = 42$ hours, $S = 8$ hours.
Thirteen small to midsize companies lost customers as a result of a ransomware incident.

a. Construct a 99% confidence interval estimate for the population mean time spent dealing with and containing a ransomware incident.
b. Construct a 95% confidence interval estimate for the population proportion of small to midsized companies who have lost customers as a result of a ransomware incident.

8.58 The human resource (HR) director of a large corporation wishes to study absenteeism among its mid-level managers at its central office during the year. A random sample of 25 mid-level managers reveals the following:

Absenteeism: $\overline{X} = 6.2$ days, $S = 7.3$ days.
13 mid-level managers cite stress as a cause of absence.

a. Construct a 95% confidence interval estimate for the mean number of absences for mid-level managers during the year.
b. Construct a 95% confidence interval estimate for the population proportion of mid-level managers who cite stress as a cause of absence.

Suppose that the HR director wishes to administer a survey in one of its regional offices. Answer these questions:
c. What sample size is needed to have 95% confidence in estimating the population mean absenteeism to within ± 1.5 days if the population standard deviation is estimated to be 8 days?
d. How many mid-level managers need to be selected to have 90% confidence in estimating the population proportion of mid-level managers who cite stress as a cause of absence to within ± 0.075 if no previous estimate is available?
e. Based on (c) and (d), what sample size is needed if a single survey is being conducted?

8.59 A national association devoted to HR and workplace programs, practices, and training wants to study HR department practices and employee turnover of its member organizations. HR professionals and organization executives focus on turnover not only because it has significant cost implications but also because it affects overall business performance. A survey is designed to estimate the proportion of member organizations that have both talent and development programs in place to drive human-capital management as well as the member organizations' mean annual employee turnover cost (cost to fill a frontline employee position left vacant due to turnover). A random sample of 100 member organizations reveals the following:

Frontline employee turnover cost: $\overline{X} = \$12,500$, $S = \$1,000$.
Thirty member organizations have both talent and development programs in place to drive human-capital management.

a. Construct a 95% confidence interval estimate for the population mean frontline employee turnover cost of member organizations.
b. Construct a 95% confidence interval estimate for the population proportion of member organizations that have both talent and development programs in place to drive human-capital management.
c. What sample size is needed to have 99% confidence of estimating the population mean frontline employee turnover cost to within $\pm \$250$?

d. How many member organizations need to be selected to have 90% confidence of estimating the population proportion of organizations that have both talent and development programs in place to drive human-capital management to within ± 0.045?

8.60 The financial impact of IT systems downtime is a concern of plant operations management today. A survey of manufacturers examined the satisfaction level with the reliability and availability of their manufacturing IT applications. The variables of focus are whether the manufacturer experienced downtime in the past year that affected one or more manufacturing IT applications, the number of downtime incidents that occurred in the past year, and the approximate cost of a typical downtime incident. The results from a sample of 200 manufacturers are as follows:

Sixty-two experienced downtime this year that affected one or more manufacturing applications.
Number of downtime incidents: $\overline{X} = 3.5$, $S = 2.0$.
Cost of downtime incidents: $\overline{X} = \$18,000$, $S = \$3,000$.

a. Construct a 90% confidence interval estimate for the population proportion of manufacturers who experienced downtime in the past year that affected one or more manufacturing IT applications.
b. Construct a 95% confidence interval estimate for the population mean number of downtime incidents experienced by manufacturers in the past year.
c. Construct a 95% confidence interval estimate for the population mean cost of downtime incidents.

8.61 The branch manager of an outlet (Store 1) of a nationwide chain of pet supply stores wants to study characteristics of her customers. In particular, she decides to focus on two variables: the amount of money spent by customers and whether the customers own only one dog, only one cat, or more than one dog and/or cat. The results from a sample of 70 customers are as follows:

- Amount of money spent: $\overline{X} = \$21.34$, $S = \$9.22$.
- Thirty-seven customers own only a dog.
- Twenty-six customers own only a cat.
- Seven customers own more than one dog and/or cat.

a. Construct a 95% confidence interval estimate for the population mean amount spent in the pet supply store.
b. Construct a 90% confidence interval estimate for the population proportion of customers who own only a cat.

The branch manager of another outlet (Store 2) wishes to conduct a similar survey in his store. The manager does not have access to the information generated by the manager of Store 1. Answer the following questions:
c. What sample size is needed to have 95% confidence of estimating the population mean amount spent in this store to within $\pm \$1.50$ if the standard deviation is estimated to be \$10?
d. How many customers need to be selected to have 90% confidence of estimating the population proportion of customers who own only a cat to within ± 0.045?
e. Based on your answers to (c) and (d), how large a sample should the manager take?

8.62 Scarlett and Heather, the owners of an upscale restaurant in Dayton, Ohio, want to study the dining characteristics of their customers. They decide to focus on two variables: the amount of money spent by customers and whether customers order dessert. The results from a sample of 60 customers are as follows:

Amount spent: $\overline{X} = \$38.54$, $S = \$7.26$.
Eighteen customers purchased dessert.

a. Construct a 95% confidence interval estimate for the population mean amount spent per customer in the restaurant.

b. Construct a 90% confidence interval estimate for the population proportion of customers who purchase dessert.

Jeanine, the owner of a competing restaurant, wants to conduct a similar survey in her restaurant. Jeanine does not have access to the information that Scarlett and Heather have obtained from the survey they conducted. Answer the following questions:

c. What sample size is needed to have 95% confidence of estimating the population mean amount spent in her restaurant to within ± $1.50, assuming that the standard deviation is estimated to be $8?

d. How many customers need to be selected to have 90% confidence of estimating the population proportion of customers who purchase dessert to within ± 0.04?

e. Based on your answers to (c) and (d), how large a sample should Jeanine take?

8.63 The manufacturer of Ice Melt claims that its product will melt snow and ice at temperatures as low as 0° Fahrenheit. A representative for a large chain of hardware stores is interested in testing this claim. The chain purchases a large shipment of 5-pound bags for distribution. The representative wants to know, with 95% confidence and within ± 0.05, what proportion of bags of Ice Melt perform the job as claimed by the manufacturer.

a. How many bags does the representative need to test? What assumption should be made concerning the population proportion? (This is called *destructive testing*; i.e., the product being tested is destroyed by the test and is then unavailable to be sold.)

b. Suppose that the representative tests 50 bags, and 42 of them do the job as claimed. Construct a 95% confidence interval estimate for the population proportion that will do the job as claimed.

c. How can the representative use the results of (b) to determine whether to sell the Ice Melt product?

8.64 Claims fraud (illegitimate claims) and buildup (exaggerated loss amounts) continue to be major issues of concern among automobile insurance companies. Fraud is defined as specific material misrepresentation of the facts of a loss; buildup is defined as the inflation of an otherwise legitimate claim. A recent study examined auto injury claims closed with payment under private passenger coverages. Detailed data on injury, medical treatment, claimed losses, and total payments, as well as claim-handling techniques, were collected. In addition, auditors were asked to review the claim files to indicate whether specific elements of fraud or buildup appeared in the claim and, in the case of buildup, to specify the amount of excess payment. The file Insurance Claims contains data for 90 randomly selected auto injury claims. The following variables are included: CLAIM—Claim ID; BUILDUP—1 if buildup indicated, 0 if not; and EXCESSPAYMENT—excess payment amount, in dollars.

a. Construct a 95% confidence interval for the population proportion of all auto injury files that have exaggerated loss amounts.

b. Construct a 95% confidence interval for the population mean dollar excess payment amount.

8.65 A quality characteristic of interest for a tea-bag-filling process is the weight of the tea in the individual bags. In this example, the label weight on the package indicates that the mean amount is 5.5 grams of tea in a bag. If the bags are underfilled, two problems arise. First, customers may not be able to brew the tea to be as strong as they wish. Second, the company may be in violation of the truth-in-labeling laws. On the other hand, if the mean amount of

tea in a bag exceeds the label weight, the company is giving away product. Getting an exact amount of tea in a bag is problematic because of variation in the temperature and humidity inside the factory, differences in the density of the tea, and the extremely fast filling operation of the machine (approximately 170 bags per minute). The following data (stored in Teabags) are the weights, in grams, of a sample of 50 tea bags produced in one hour by a single machine:

5.65	5.44	5.42	5.40	5.53	5.34	5.54	5.45	5.52	5.41
5.57	5.40	5.53	5.54	5.55	5.62	5.56	5.46	5.44	5.51
5.47	5.40	5.47	5.61	5.53	5.32	5.67	5.29	5.49	5.55
5.77	5.57	5.42	5.58	5.58	5.50	5.32	5.50	5.53	5.58
5.61	5.45	5.44	5.25	5.56	5.63	5.50	5.57	5.67	5.36

a. Construct a 99% confidence interval estimate for the population mean weight of the tea bags.

b. Is the company meeting the requirement set forth on the label that the mean amount of tea in a bag is 5.5 grams?

c. Do you think the assumption needed to construct the confidence interval estimate in (a) is valid?

8.66 Call centers today play an important role in managing day-to-day business communications with customers. It's important, therefore, to monitor a comprehensive set of metrics, which can help businesses understand the overall performance of a call center. One key metric for measuring overall call center performance is service level, which is defined as the percentage of calls answered by a human agent within a specified number of seconds. The file Service Level contains the following data for time, in seconds, to answer 50 incoming calls to a financial services call center:

16	14	16	19	6	14	15	5	16	18	17	22	6	18	10	15	12
6	19	16	16	15	13	25	9	17	12	10	5	15	23	11	12	14
24	9	10	13	14	26	19	20	13	24	28	15	21	8	16	12	

a. Construct a 95% confidence interval estimate for the population mean time, in seconds, to answer incoming calls.

b. What assumption do you need to make about the population to construct the interval in (a)?

c. Given the data presented, do you think the assumption needed in (a) is valid? Explain.

8.67 The manufacturer of Boston and Vermont asphalt shingles knows that product weight is a major factor in a customer's perception of quality. The last stage of the assembly line packages the shingles before they are placed on wooden pallets. Once a pallet is full (a pallet for most brands holds 16 squares of shingles), it is weighed, and the measurement is recorded. The file Pallet contains the weight (in pounds) from a sample of 368 pallets of Boston shingles and 330 pallets of Vermont shingles.

a. For the Boston shingles, construct a 95% confidence interval estimate for the mean weight.

b. For the Vermont shingles, construct a 95% confidence interval estimate for the mean weight.

c. Do you think the assumption needed to construct the confidence interval estimates in (a) and (b) is valid?

d. Based on the results of (a) and (b), what conclusions can you reach concerning the mean weight of the Boston and Vermont shingles?

8.68 The manufacturer of Boston and Vermont asphalt shingles provides its customers with a 20-year warranty on most of its products. To determine whether a shingle will last the entire warranty period, accelerated-life testing is conducted at the manufacturing plant. Accelerated-life testing exposes the shingle to the stresses it would be subject to in a lifetime of normal use via a laboratory experiment that takes only a few minutes to conduct. In this test, a shingle is repeatedly scraped with a brush for a short period of time, and the shingle granules removed by the brushing are weighed (in grams). Shingles that experience low amounts of granule loss are expected to last longer in normal use than shingles that experience high amounts of granule loss. In this situation, a shingle should experience no more than 0.8 grams of granule loss if it is expected to last the length of the warranty period. The file Granule contains a sample of 170 measurements made on the company's Boston shingles and 140 measurements made on Vermont shingles.

a. For the Boston shingles, construct a 95% confidence interval estimate for the mean granule loss.

b. For the Vermont shingles, construct a 95% confidence interval estimate for the mean granule loss.

c. Do you think the assumption needed to construct the confidence interval estimates in (a) and (b) is valid?

d. Based on the results of (a) and (b), what conclusions can you reach concerning the mean granule loss of the Boston and Vermont shingles?

REPORT WRITING EXERCISE

8.69 Referring to the results in Problem 8.66 concerning the answer time of calls, write a report that summarizes your conclusions.

CHAPTER 8

▼CASES

Managing Ashland MultiComm Services

Marketing Manager Lauren Adler seeks to increase the number of subscribers to the AMS *3-For-All* cable TV and Internet and smartphone service. Her staff has designed the following 10-question survey to help determine various characteristics of households who subscribe to AMS cable or cellphone services.

1. Does your household subscribe to smartphone service from Ashland?
(1) Yes   (2) No

2. Does your household subscribe to Internet service from Ashland?
(1) Yes   (2) No

3. How often do you watch streaming video on any device?
(1) Every day   (2) Most days
(3) Occasionally or never

4. What type of cable television service do you have?
(1) Basic or none   (2) Enhanced

5. How often do you watch premium content that requires an extra fee?
(1) Almost every day   (2) Several times a week
(3) Rarely   (4) Never

6. Which method did you use to obtain your current AMS subscription?
(1) AMS email/text offer   (2) AMS toll-free number
(3) AMS website   (4) In-store signup
(5) MyTVLab promotion

7. Would you consider subscribing to the *3-For-All* service for a trial period if a discount were offered?
(1) Yes   (2) No
(If no, skip to question 9.)

8. If purchased separately, cable TV and Internet and smartphone service would currently cost $160 per month. How much would you be willing to pay per month for the *3-For-All* service?

9. Does your household use another provider of cellphone services?
(1) Yes   (2) No

10. AMS may distribute vouchers good for one free smartphone for subscribers who agree to a two-year subscription contract to the *3-For-All* service. Would being eligible to receive a voucher cause you to agree to the two-year term?
(1) Yes   (2) No

Of the 500 households selected that subscribe to cable television service from Ashland, 82 households either refused to participate, could not be contacted after repeated attempts, or had telephone numbers that were not in service. The summary results for the 418 households that were contacted are as follows:

Household Has AMS Smartphone Service	Frequency
Yes	83
No	335
Household Has AMS Internet Service	**Frequency**
Yes	262
No	156
Streams Video	**Frequency**
Every day	170
Most days	166
Occasionally or never	82
Type of Cable Service	**Frequency**
Basic or none	164
Enhanced	254
Watches Premium Content	**Frequency**
Almost every day	16
Several times a week	40
Rarely	179
Never	183

Method Used to Obtain Subscription	Frequency
AMS email/text offer	70
AMS toll-free number	64
AMS website	236
In-store signup	36
MyTVLab promotion	12
Would Consider Discounted Trial Offer	**Frequency**
Yes	40
No	378

Trial Monthly Rate ($) Willing to Pay (stored in `AMS8`)

100 79 114 50 91 106 67 110 70 113 90 115 98 75 119
100 90 60 89 105 65 91 86 91 84 92 95 85 80 108
 90 97 79 91 125 99 98 50 77 85

Uses Another Cellphone Provider	Frequency
Yes	369
No	49
Voucher for Two-Year Agreement	**Frequency**
Yes	38
No	380

Analyze the results of the survey of Ashland households that receive AMS cable television service. Write a report that discusses the marketing implications of the survey results for Ashland MultiComm Services.

Digital Case

Apply your knowledge about confidence interval estimation in this Digital Case, which extends the MyTVLab Digital Case from Chapter 6.

Among its other features, the MyTVLab website allows customers to purchase MyTVLab LifeStyles merchandise online. To handle payment processing, the management of MyTVLab has contracted with the following firms:

- **PayAFriend (PAF)**—This is an online payment system with which customers and businesses such as MyTVLab register in order to exchange payments in a secure and convenient manner without the need for a credit card.
- **Continental Banking Company (Conbanco)**—This processing services provider allows MyTVLab customers to pay for merchandise using nationally recognized credit cards issued by a financial institution.

To reduce costs, management is considering eliminating one of these two payment systems. However, Lorraine Hildick of the sales department suspects that customers use the two forms of payment in unequal numbers and that customers display different buying behaviors when using the two forms of payment. Therefore, she would like to first determine the following:

- The proportion of customers using PAF and the proportion of customers using a credit card to pay for their purchases.
- The mean purchase amount when using PAF and the mean purchase amount when using a credit card.

Assist Ms. Hildick by preparing an appropriate analysis. Open **PaymentsSample.pdf**, read Ms. Hildick's comments, and

use her random sample of 50 transactions as the basis for your analysis. Summarize your findings to determine whether Ms. Hildick's conjectures about MyTVLab LifeStyle customer purchasing behaviors are correct. If you want the sampling error to be no more than $3 when estimating the mean purchase amount, is Ms. Hildick's sample large enough to perform a valid analysis?

Sure Value Convenience Stores

You work in the corporate office for a nationwide convenience store franchise that operates nearly 10,000 stores. The per-store daily customer count has been steady, at 900, for some time (i.e., the mean number of customers in a store in one day is 900). To increase the customer count, the franchise is considering cutting coffee prices. The 12-ounce size will now be $0.59 instead of $0.99, and the 16-ounce size will be $0.69 instead of $1.19. Even with this reduction in price, the franchise will have a 40% gross margin on coffee. To test the new initiative, the franchise has reduced coffee prices in a sample of 34 stores, where customer counts have been running almost exactly at the national average of 900. After four weeks, the sample stores stabilize at a mean customer count of 974 and a standard deviation of 96. This increase seems like a substantial amount to you, but it also seems like a pretty small sample. Is there some way to get a feel for what the mean per-store count in all the stores will be if you cut coffee prices nationwide? Do you think reducing coffee prices is a good strategy for increasing the mean number of customers?

CardioGood Fitness

Return to the CardioGood Fitness case first presented on page 33. Using the data stored in `CardioGood Fitness`:

1. Construct 95% confidence interval estimates to create a customer profile for each CardioGood Fitness treadmill product line.

2. Write a report to be presented to the management of Cardio-Good Fitness detailing your findings.

More Descriptive Choices Follow-Up

Follow up the More Descriptive Choices Revisited, Using Statistics scenario on page 147 by constructing 95% confidence intervals estimates of the one-year return percentages, five-year return percentages, and ten-year return percentages for the sample of growth and value funds and for the small, mid-cap, and large market cap funds (stored in `Retirement Funds`). In your analysis, examine differences between the growth and value funds as well as the differences among the small, mid-cap, and large market cap funds.

Clear Mountain State Student Survey

The Student News Service at Clear Mountain State University (CMSU) has decided to gather data about the undergraduate students who attend CMSU. They create and distribute a survey of 14 questions and receive responses from 111 undergraduates (stored in `Student Survey`). For each variable included in the survey, construct a 95% confidence interval estimate for the population characteristic and write a report summarizing your conclusions.

▼EXCEL GUIDE

EG8.1 CONFIDENCE INTERVAL ESTIMATE for the MEAN (σ KNOWN)

Key Technique Use the **NORM.S.INV**(*cumulative percentage*) to compute the Z value for one-half of the $(1 - \alpha)$ value and use the **CONFIDENCE**(*1 – confidence level, population standard deviation, sample size*) function to compute the half-width of a confidence interval.

Example Compute the confidence interval estimate for the mean for the Example 8.1 mean fill amount problem on page 249.

PHStat Use **Estimate for the Mean, sigma known**.

For the example, select **PHStat→ Confidence Intervals→ Estimate for the Mean, sigma known**. In the procedure's dialog box (shown below):

1. Enter **15** as the **Population Standard Deviation**.
2. Enter **95** as the **Confidence Level** percentage.
3. Click **Sample Statistics Known** and enter **100** as the **Sample Size** and **369.27** as the **Sample Mean**.
4. Enter a **Title** and click **OK**.

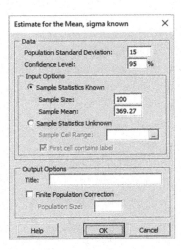

When using unsummarized data, click **Sample Statistics Unknown** and enter the **Sample Cell Range** in step 3.

Workbook Use the **COMPUTE worksheet** of the **CIE sigma known workbook** as a template.

The worksheet already contains the data for the example. For other problems, change the **Population Standard Deviation, Sample Mean, Sample Size**, and **Confidence Level** values in cells B4 through B7.

EG8.2 CONFIDENCE INTERVAL ESTIMATE for the MEAN (σ UNKNOWN)

Key Technique Use the **T.INV.2T**(*1 – confidence level, degrees of freedom*) function to determine the critical value from the t distribution.

Example Compute the Figure 8.7 confidence interval estimate for the mean sales invoice amount on page 253.

PHStat Use **Estimate for the Mean, sigma unknown**.

For the example, select **PHStat→ Confidence Intervals→ Estimate for the Mean, sigma unknown**. In the procedure's dialog box (shown below):

1. Enter **95** as the **Confidence Level** percentage.
2. Click **Sample Statistics Known** and enter **100** as the **Sample Size**, **110.27** as the **Sample Mean**, and **28.95** as the **Sample Std. Deviation**.
3. Enter a **Title** and click **OK**.

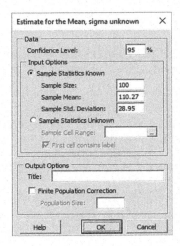

When using unsummarized data, click **Sample Statistics Unknown** and enter the **Sample Cell Range** in step 2.

Workbook Use the **COMPUTE worksheet** of the **CIE sigma unknown workbook** as a template.

The worksheet already contains the data for the example. For other problems, change the **Sample Standard Deviation, Sample Mean, Sample Size**, and **Confidence Level** values in cells B4 through B7.

EG8.3 CONFIDENCE INTERVAL ESTIMATE for the PROPORTION

Key Technique Use the **NORM.S.INV((1–*confidence level*)/2)** function to compute the *Z* value.

Example Compute the Figure 8.12 confidence interval estimate for the proportion of in-error sales invoices on page 259.

PHStat Use **Estimate for the Proportion**.

For the example, select **PHStat→Confidence Intervals→Estimate for the Proportion**. In the procedure's dialog box (shown below):

1. Enter **100** as the **Sample Size**.
2. Enter **10** as the **Number of Successes**.
3. Enter **95** as the **Confidence Level** percentage.
4. Enter a **Title** and click **OK**.

Workbook Use the **COMPUTE worksheet** of the **CIE Proportion workbook** as a template.

The worksheet already contains the data for the example. To compute confidence interval estimates for other problems, change the **Sample Size**, **Number of Successes**, and **Confidence Level** values in cells B4 through B6.

EG8.4 DETERMINING SAMPLE SIZE

Sample Size Determination for the Mean

Key Technique Use the **NORM.S.INV((1–*confidence level*)/2)** function to compute the *Z* value and use the **ROUNDUP(*calculated sample size*, 0)** function to round up the computed sample size to the next higher integer.

Example Determine the sample size for the Figure 8.13 mean sales invoice amount example on page 262.

PHStat Use **Determination for the Mean**.

For the example, select **PHStat→Sample Size→Determination for the Mean**. In the procedure's dialog box (shown at the top right):

1. Enter **25** as the **Population Standard Deviation**.
2. Enter **5** as the **Sampling Error**.
3. Enter **95** as the **Confidence Level** percentage.
4. Enter a **Title** and click **OK**.

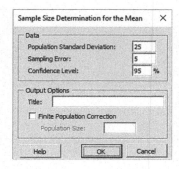

Workbook Use the **COMPUTE worksheet** of the **Sample Size Mean workbook** as a template.

The worksheet already contains the data for the example. For other problems, change the **Population Standard Deviation**, **Sampling Error**, and **Confidence Level** values in cells B4 through B6.

Sample Size Determination for the Proportion

Key Technique Use the **NORM.S.INV** and **ROUNDUP** functions discussed previously to help determine the sample size needed for estimating the proportion.

Example Determine the sample size for the Figure 8.14 proportion of in-error sales invoices example on page 264.

PHStat Use **Determination for the Proportion**.

For the example, select **PHStat→Sample Size→Determination for the Proportion**. In the procedure's dialog box (shown below):

1. Enter **0.15** as the **Estimate of True Proportion**.
2. Enter **0.07** as the **Sampling Error**.
3. Enter **95** as the **Confidence Level** percentage.
4. Enter a **Title** and click **OK**.

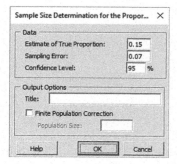

Workbook Use the **COMPUTE worksheet** of the **Sample Size Proportion workbook** as a template.

The worksheet already contains the data for the example. To compute confidence interval estimates for other problems, change the **Estimate of True Proportion**, **Sampling Error**, and **Confidence Level** in cells B4 through B6.

Fundamentals of Hypothesis Testing: One-Sample Tests

9

CONTENTS

OBJECTIVES

- Learn the principles
  of hypothesis testing
- Use hypothesis testing to
  test a mean or proportion
- Evaluate the assumptions
  of each hypothesis-
  testing procedure
  and understand the
  consequences if
  assumptions are seriously
  violated
- Become aware of
  hypothesis testing pitfalls
  and ethical issues
- Avoid the pitfalls involved
  in hypothesis testing

▼ USING **STATISTICS**
Significant Testing at Oxford Cereals

As in Chapter 7, you again find yourself as plant operations manager for Oxford Cereals. Among other duties you are responsible for monitoring the amount in each cereal box filled. Company specifications require a mean weight of 368 grams per box. You must adjust the cereal-filling process when the mean fill-weight in the population of boxes differs from 368 grams. Adjusting the process requires shutting down the cereal production line temporarily, so you do not want to make unnecessary adjustments.

What decision-making method can you use to decide if the cereal-filling process needs to be adjusted? You decide to begin by selecting a random sample of 25 filled boxes and weighing each box. From the weights collected, you compute a sample mean. How could that sample mean be used to help decide whether adjustment is necessary?

Chapter 7 discusses methods to determine whether the value of a sample mean is consistent with a known population mean. In this second Oxford Cereals scenario, you seek to use a sample mean to validate a claim about the population mean, a somewhat different analysis. For such analyses, you use the inferential method called *hypothesis testing*. In hypothesis testing, you state a claim, or *null hypothesis*, unambiguously. You examine a sample statistic to see if it better supports the null hypothesis or a mutually exclusive *alternative hypothesis*.

9.1 Fundamentals of Hypothesis Testing

Hypothesis testing analyzes *differences* between a sample statistic and the results one would expect if a null hypothesis was true. In doing so, hypothesis testing enables one to make inferences about a population parameter that are based on the sample statistic a hypothesis test examines. For the Oxford Cereals scenario, hypothesis testing would permit you to infer from a random sample either

- the mean weight of the cereal boxes in a sample is a value consistent with what you would expect if the mean of the entire population of cereal boxes were 368 grams, *or*
- the population mean is not equal to 368 grams because the sample mean is significantly different from 368 grams.

The **null hypothesis**, represented by the symbol H_0, often states a status quo case. For the Oxford Cereals scenario, a status quo case would be the cereal-filling process is working as intended, and therefore, the population mean fill amount is 368 grams, stated as

$$H_0: \mu = 368$$

The **alternative hypothesis**, represented by the symbol H_1, states a claim that is contrary to the null hypothesis. For the Oxford Cereals scenario, the contrary claim would be stated as

$$H_1: \mu \neq 368$$

student TIP

Hypothesis testing reaches conclusions about parameters, not statistics.

A pair of null and alternative hypotheses are always mutually exclusive—only one of them can be true. To use the hypothesis test methods that this book discusses, a pair of null and alternative hypotheses must also be collectively exhaustive, as the pair for the Oxford Cereals scenario is. Note that the null and alternative hypotheses are always stated in terms of the population parameter because a hypothesis test always examines a sample statistic.

One rejects the null hypothesis in favor of the alternative hypothesis when a hypothesis test provides sufficient evidence from the sample data to show that the null hypothesis is false. The alternative hypothesis is often the focus of underlying research. For example, in new product research, the null hypothesis would be that the new product is as equally effective as existing products, even if the focus of the research was providing evidence that suggests the product is different. In the Oxford Cereals scenario, discovering sufficient evidence that would cause you to reject the null hypothesis would lead to corrective action: stopping production and taking corrective action. In a sense, hypothesis testing for this case is focused on whether these special actions are required.

Finding insufficient evidence causes one not to reject the null hypothesis. This does not mean that hypothesis testing can "prove" that the null hypothesis is true; hypothesis testing can only show that the results have failed to prove that the null hypothesis is false, an important distinction. For the Oxford Cereals example, if one does not reject the null hypothesis, one cannot claim that the hypothesis test "proves" that the population mean fill amount is 368 grams. One can only say that insufficient evidence exists to challenge the assertion that the population mean is 368 grams.

Understanding precisely what hypothesis testing does and avoiding misstatements about hypothesis testing, such as that a test has *proved* a null (or alternative) hypothesis claim to be true, forms the basis for using hypothesis testing correctly. Exhibit 9.1 on page 277 summarizes the concepts that one needs to know to use hypothesis testing knowingly.

> **EXHIBIT 9.1**
>
> **Fundamental Hypothesis Testing Concepts**
>
> The null hypothesis, H_0, states a status quo claim.
>
> The alternative hypothesis, H_1, states a claim that is contrary to the null hypothesis and often represents a research claim or specific inference that an analyst seeks to prove.
>
> A null and alternative pair of hypotheses are always collectively exhaustive.
>
> If one rejects the null hypothesis, one has strong statistical evidence that the alternative hypothesis is correct.
>
> If one does not reject the null hypothesis, one has not proven the null hypothesis. (Rather, one has only failed to prove the alternative hypothesis.)
>
> The null hypothesis always refers to a population parameter such as μ and not a sample statistic such as $\overline{X}$.
>
> The null hypothesis always includes an equals sign when stating a claim about the population parameter, for example, $H_0: \mu = 368$ grams.
>
> The alternative hypothesis never includes an equals sign when stating a claim about the population parameter.

EXAMPLE 9.1

The Null and Alternative Hypotheses

You are the manager of a fast-food restaurant. You want to determine whether the waiting time to place an order has changed in the past month from its previous population mean value of 4.5 minutes. State the null and alternative hypotheses.

SOLUTION The null hypothesis is that the population mean has not changed from its previous value of 4.5 minutes. This is stated as

$$H_0: \mu = 4.5$$

The alternative hypothesis is the opposite of the null hypothesis. Because the null hypothesis is that the population mean is 4.5 minutes, the alternative hypothesis is that the population mean is not 4.5 minutes. This is stated as

$$H_1: \mu \neq 4.5$$

The Critical Value of the Test Statistic

Hypothesis testing uses sample data to determine how likely it is that the null hypothesis is true. In the Oxford Cereals example, the null hypothesis is that the mean amount of cereal per box in the entire filling process is 368 grams, the population parameter specified by the company. To test this hypothesis, one selects a sample of filled boxes, weighs each box, and calculates the sample mean $\overline{X}$.

This sample statistic is an estimate of the corresponding parameter, the population mean, μ. Even if the null hypothesis is true, the sample statistic $\overline{X}$ is likely to differ from the value of the parameter (the population mean, μ) because of variation due to sampling. One does expect the sample statistic to be close to the population parameter if the null hypothesis is true.

If the sample statistic is close to the population parameter, one has insufficient evidence to reject the null hypothesis. For example, if the sample mean is 367.9 grams, one might conclude that the population mean has not changed (that $\mu = 368$) because a sample mean of 367.9 grams is very close to the hypothesized value of 368 grams. However, if there is a large difference between the value of the sample statistic and the hypothesized value of the population parameter, one might conclude that the null hypothesis is false. For example, if the sample mean is 320 grams, one might conclude that the population mean is not 368 grams (i.e., $\mu \neq 368$) because the sample mean is very far from the hypothesized value of 368 grams. Intuitively, one might conclude that getting a sample mean of 320 grams if the population mean is actually

368 grams is very unlikely. Therefore, it is more logical to conclude that the population mean is not equal to 368 grams and to reject the null hypothesis.

However, the decision-making process is not always so clear-cut. Determining what is "very close" and what is "very different" is arbitrary without clear definitions. Hypothesis-testing methodology provides clear definitions for evaluating differences. This methodology quantifies the decision-making process by calculating the probability of getting a certain sample result if the null hypothesis is true. The methodology determines this probability by first calculating the sampling distribution for the sample statistic of interest, such as a sample mean, and then calculating the **test statistic** for the sample. Because the sampling distribution for the test statistic often follows a well-known statistical distribution, such as the standardized normal distribution or *t* distribution, one of these well-known distributions can help determine whether the null hypothesis is true.

student TIP

Every test statistic follows a specific sampling distribution.

Regions of Rejection and Nonrejection

The sampling distribution of the test statistic is divided into two regions, a **region of rejection** (sometimes called the critical region) and a **region of nonrejection** (see Figure 9.1). If the test statistic falls into the region of nonrejection, one does not reject the null hypothesis. For the Oxford Cereals scenario, that outcome would enable one to conclude that there is insufficient evidence that the population mean fill is different from 368 grams. If the test statistic falls into the rejection region, one rejects the null hypothesis. For the Oxford Cereals scenario, that outcome would enable one to conclude that the population mean is not 368 grams.

FIGURE 9.1
Regions of rejection and nonrejection in hypothesis testing

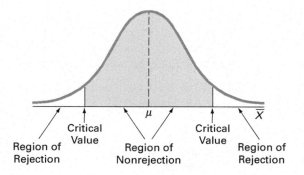

The region of rejection consists of the values of the test statistic that are unlikely to occur if the null hypothesis is true. These values are much more likely to occur if the null hypothesis is false. Therefore, if a value of the test statistic falls into this rejection region, one rejects the null hypothesis because that value is unlikely if the null hypothesis is true.

To make a decision concerning the null hypothesis, one first determines the **critical value** of the test statistic. The critical value divides the nonrejection region from the rejection region. Determining the critical value depends on the size of the rejection region. The size of the rejection region is directly related to the risks involved in using only sample evidence to make decisions about a population parameter.

Risks in Decision Making Using Hypothesis Testing

Using hypothesis testing involves the risk of reaching an incorrect conclusion. One might wrongly reject a true null hypothesis, H_0, or conversely, one might wrongly *not* reject a false null hypothesis, H_0. These types of risk are called Type I and Type II errors.

TYPE I AND TYPE II ERRORS

A **Type I error** occurs if one rejects the null hypothesis, H_0, when it is true and should not be rejected. A Type I error is a "false alarm." The probability of a Type I error occurring is α.

A **Type II error** occurs if one does not reject the null hypothesis, H_0, when it is false and should be rejected. A Type II error represents a "missed opportunity" to take some corrective action. The probability of a Type II error occurring is β.

In the Oxford Cereals scenario, a Type I error would occur if one concluded that the population mean fill is *not* 368 grams when it *is* 368 grams. This error would cause one to needlessly adjust the filling process (the "false alarm") even though the process is working properly. A Type II error would occur if one concluded that the population mean fill *is* 368 grams when it is *not* 368 grams. In this case, one would allow the process to continue without adjustment, even though an adjustment is needed (the "missed opportunity").

Traditionally, one controls the Type I error by determining the risk level, α (the lowercase Greek letter *alpha*), that one is willing to have of rejecting the null hypothesis when it is true. This risk, or probability, of committing a Type I error is known as the **level of significance (α)**. Because one specifies the level of significance before one performs a hypothesis test, one directly controls the risk of committing a Type I error.

Choosing the risk level for making a Type I error depends on the cost of making a Type I error, with risk levels of 0.01, 0.05, or 0.10, being the most common choices. After one specifies the value for α, one can then determine the critical values that divide the rejection and nonrejection regions. One knows the size of the rejection region because α is the probability of rejection when the null hypothesis is true. From this, one can then determine the critical value or values that divide the rejection and nonrejection regions.

The probability of committing a Type II error is called the **β risk**. This probability depends on the difference between the hypothesized and actual values of the population parameter. Unlike the Type I error, one cannot specify this risk. Because large differences are easier to find than small ones, if the difference between the hypothesized and actual values of the population parameter is large, β is small. For example, if the population mean is 330 grams, there is a small chance (β) that one will conclude that the mean has not changed from 368 grams. However, if the difference between the hypothesized and actual values of the parameter is small, β is large. For example, if the population mean is actually 367 grams, there is a large chance (β) that one will conclude that the mean is still 368 grams.

TABLE 9.1
Comparison of Type I and Type II errors

Error Type	Probability of Error	User-specified Risk
Type I	level of significance, α	Yes
Type II	β risk	No

Complements of Type I and Type II Errors The complement of the probability of a Type I error, $(1 - \alpha)$, is called the **confidence coefficient**. The confidence coefficient is the probability that one will not reject the null hypothesis, H_0, when it is true and should not be rejected. For the Oxford Cereals scenario, the confidence coefficient measures the probability of concluding that the population mean fill is 368 grams when it is actually 368 grams.

The complement of the probability of a Type II error, $(1 - \beta)$, is called the **power of a statistical test**. The power of a statistical test is the probability that one will reject the null hypothesis when it is false and should be rejected. For the Oxford Cereals scenario, the power of the test is the probability that one will correctly conclude that the mean fill amount is not 368 grams when it actually is not 368 grams.

Table 9.2 summarizes the outcomes of not rejecting H_0 or rejecting H_0 when using hypothesis testing to support decision making.

TABLE 9.2
Hypothesis testing and decision making

STATISTICAL DECISION	ACTUAL SITUATION	
	H_0 **True**	H_0 **False**
Do not reject H_0	Correct decision Confidence coefficient $= (1 - \alpha)$	Type II error P(Type II error) $= \beta$
Reject H_0	Type I error P(Type I error) $= \alpha$	Correct decision Power $= (1 - \beta)$

One way to reduce the probability of making a Type II error is by increasing the sample size. Large samples generally permit the detection of even very small differences between the hypothesized values and the actual population parameters. For a given level of α, increasing the sample size decreases β, thereby increasing the power of the statistical test to detect that the null hypothesis, H_0, is false.

For any given sample size, one must consider the trade-offs between the two possible types of errors. Because one can directly control the risk of a Type I error, one can reduce this risk by selecting a smaller value for α. For example, if the negative consequences associated with making a Type I error are substantial, one could select $\alpha = 0.01$ instead of 0.05. However, when one decreases α, β increases, so reducing the risk of a Type I error results in an increased risk of a Type II error. However, to reduce β, you could select a larger value for α. If avoiding a Type II error is an important goal, one can increase α, using a value such as 0.05 or 0.10 instead of a value such as 0.01.

For the Oxford Cereals scenario, the risk of a Type I error occurring involves concluding that the mean fill amount has changed from the hypothesized 368 grams when it actually has not changed. The risk of a Type II error occurring involves concluding that the mean fill amount has not changed from the hypothesized 368 grams when it actually has changed. The choice of reasonable values for α and β depends on the costs inherent in each type of error. For example, if it is very costly to change the cereal-filling process, one would want to be very confident that a change is needed before making any changes. In this case, the risk of a Type I error occurring is more important, and one would choose a small α. However, if one wants to be very certain of detecting changes from a mean of 368 grams, the risk of a Type II error occurring is more important, and one would choose a higher level of α.

Now that one has been introduced to hypothesis testing, recall that in the Oxford Cereals scenario, the business problem is to determine if the mean fill-weight in the population of boxes in the cereal-filling process differs from 368 grams. One would first select a random sample of 25 boxes, weigh each box, and calculate the sample mean, $\overline{X}$. Then, one would evaluate the difference between this sample statistic and the hypothesized population parameter by comparing the sample mean weight (in grams) to the expected population mean of 368 grams specified by the company, stating the null and alternative hypotheses as

$$H_0: \mu = 368$$
$$H_1: \mu \neq 368$$

Z Test for the Mean (σ Known)

When the standard deviation, σ, is known (a rare occurrence), one uses the Z **test for the mean** if the population is normally distributed. If the population is not normally distributed, one can still use the Z test if the sample size is large enough for the Central Limit Theorem to take effect (see Section 7.2). Equation (9.1) defines the Z_{STAT} test statistic for determining the difference between the sample mean, $\overline{X}$, and the population mean, μ, when the standard deviation, σ, is known.

Z TEST FOR THE MEAN (σ KNOWN)

$$Z_{STAT} = \frac{\overline{X} - \mu}{\dfrac{\sigma}{\sqrt{n}}} \tag{9.1}$$

In Equation (9.1), the numerator measures the difference between the observed sample mean, $\overline{X}$, and the hypothesized mean, μ. The denominator is the standard error of the mean, so Z_{STAT} represents the difference between $\overline{X}$ and μ in standard error units.

Hypothesis Testing Using the Critical Value Approach

The critical value approach compares the value of the computed Z_{STAT} test statistic from Equation (9.1) to critical values that divide the normal distribution into regions of rejection and nonrejection. The critical values are expressed as standardized Z values that are determined by the level of significance.

For example, if one uses a level of significance of 0.05, the size of the rejection region is 0.05. Because the null hypothesis contains an equal sign and the alternative hypothesis contains a not equal sign, one has a **two-tail test** in which the rejection region is divided equally among the two tails of the distribution, 0.025 for each tail. For this two-tail test, a rejection region of 0.025 in each tail of the normal distribution results in a cumulative area of 0.025 below the lower critical value, a cumulative area of 0.975 (1 − 0.025) below the upper critical value. According to the cumulative standardized normal distribution table (Table E.2), the critical values that divide the rejection and nonrejection regions are −1.96 and +1.96.

Figure 9.2 illustrates that if the mean is actually 368 grams, as H_0 claims, the values of the Z_{STAT} test statistic have a standardized normal distribution centered at $Z = 0$ (which corresponds to an $\overline{X}$ value of 368 grams). Values of Z_{STAT} greater than +1.96 and less than −1.96 indicate that $\overline{X}$ is sufficiently different from the hypothesized $\mu = 368$ that such an $\overline{X}$ value would be unlikely to occur if H_0 were true.

student TIP

Determine the level of significance first. This value enables one to determine the critical value.

FIGURE 9.2
Testing a hypothesis about the mean (σ known) at the 0.05 level of significance

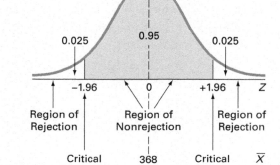

student TIP

In a two-tail test, there is a rejection region in each tail of the distribution.

Therefore, the decision rule is

$$\text{Reject } H_0 \text{ if } Z_{STAT} > +1.96$$
$$\text{or if } Z_{STAT} < -1.96;$$
$$\text{otherwise, do not reject } H_0.$$

student TIP

Remember, the decision rule always concerns H_0. Either you reject H_0 or you do not reject H_0.

Suppose that the sample of 25 cereal boxes indicates a sample mean, $\overline{X}$, of 372.5 grams, and the population standard deviation, σ, is 15 grams. Using Equation (9.1),

$$Z_{STAT} = \frac{\overline{X} - \mu}{\dfrac{\sigma}{\sqrt{n}}} - \frac{372.5 - 368}{\dfrac{15}{\sqrt{25}}} = +1.50$$

Because $Z_{STAT} = +1.50$ is greater than −1.96 and less than +1.96, one does not reject H_0 (see Figure 9.3).

One concludes that there is insufficient evidence that the mean fill amount is not 368 grams. To take into account the possibility of a Type II error, one does *not* say "the mean fill amount is 368 grams."

FIGURE 9.3
Testing a hypothesis about the mean cereal weight (σ known) at the 0.05 level of significance

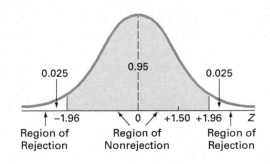

Exhibit 9.2 summarizes the critical value approach to hypothesis testing. Steps 1 and 2 are part of the Define task, step 5 combines the Collect and Organize tasks, and steps 3, 4, and 6 involve the Visualize and Analyze tasks of the DCOVA framework first introduced on page 3. Examples 9.2 and 9.3 apply the critical value approach to hypothesis testing to Oxford Cereals and to a fast-food restaurant.

EXHIBIT 9.2

The Critical Value Approach to Hypothesis Testing

Step 1 State the null hypothesis, H_0, and the alternative hypothesis, H_1.

Step 2 Choose the level of significance, α, and the sample size, n. The level of significance is based on the relative importance of the risks of committing Type I and Type II errors in the problem.

Step 3 Determine the appropriate test statistic and sampling distribution.

Step 4 Determine the critical values that divide the rejection and nonrejection regions.

Step 5 Collect the sample data, organize the results, and determine the value of the test statistic.

Step 6 Make the statistical decision, determine whether the assumptions are valid, and state the managerial conclusion in the context of the theory, claim, or assertion being tested. If the test statistic falls into the nonrejection region, do not reject the null hypothesis. If the test statistic falls into the rejection region, reject the null hypothesis.

EXAMPLE 9.2

Applying the Critical Value Approach to Hypothesis Testing at Oxford Cereals

Apply the critical value approach to hypothesis testing at Oxford Cereals.

SOLUTION

Step 1 State the null and alternative hypotheses. The null hypothesis, H_0, is always stated as a mathematical expression, using population parameters. In testing whether the mean fill is 368 grams, the null hypothesis states that μ equals 368. The alternative hypothesis, H_1, is also stated as a mathematical expression, using population parameters. Therefore, the alternative hypothesis states that μ is not equal to 368 grams.

Step 2 Choose the level of significance and the sample size. Choose the level of significance, α, according to the relative importance of the risks of committing Type I and Type II errors in the problem. The smaller the value of α, the less risk there is of making a Type I error. In this example, making a Type I error means that you conclude that the population mean is not 368 grams when it is 368 grams. You would take corrective action on the filling process even though the process is working properly. In the example, $\alpha = 0.05$, and the sample size, n, is 25.

Step 3 Select the appropriate test statistic. Because σ is known from information about the filling process, you use the Z_{STAT} test statistic because, by the Central Limit Theorem, the sample size is large enough that the sampling distribution is approximately normally distributed.

Step 4 Determine the rejection region. Critical values for the appropriate test statistic are selected so that the rejection region contains a total area of α when H_0 is true and the nonrejection region contains a total area of $1 - \alpha$ when H_0 is true. Because $\alpha = 0.05$ in the cereal example, the critical values of the Z_{STAT} test statistic are -1.96 and $+1.96$. The rejection region is therefore $Z_{STAT} < -1.96$ or $Z_{STAT} > +1.96$. The nonrejection region is $-1.96 \leq Z_{STAT} \leq +1.96$.

▶(continued)

Step 5 Collect the sample data and compute the value of the test statistic. In the cereal example, $\overline{X} = 372.5$, and the value of the test statistic is $Z_{STAT} = +1.50$.

Step 6 State the statistical decision and the managerial conclusion. First, determine whether the test statistic has fallen into the rejection region or the nonrejection region. For the cereal example, $Z_{STAT} = +1.50$ is in the region of nonrejection because $-1.96 \leq Z_{STAT} = +1.50 \leq +1.96$. Because the test statistic falls into the nonrejection region, the statistical decision is to not reject the null hypothesis, H_0. The managerial conclusion is that insufficient evidence exists to prove that the mean fill is different from 368 grams. No corrective action on the filling process is needed.

EXAMPLE 9.3

Testing and Rejecting a Null Hypothesis

You are the manager of a fast-food restaurant. The business problem is to determine whether the population mean waiting time to place an order has changed in the past month from its previous population mean value of 4.5 minutes. From past experience, you can assume that the population is normally distributed, with a population standard deviation of 1.2 minutes. You select a sample of 36 orders during a one-hour period. The sample mean is 5.1 minutes. Use the Exhibit 9.2 six-step approach on page 282 to determine whether there is evidence at the 0.05 level of significance that the population mean waiting time to place an order has changed in the past month from its previous population mean value of 4.5 minutes.

SOLUTION

Step 1 The null hypothesis is that the population mean has not changed from its previous value of 4.5 minutes:

$$H_0: \mu = 4.5$$

Because the null hypothesis is that the population mean is 4.5 minutes, the alternative hypothesis is that the population mean is not 4.5 minutes:

$$H_1: \mu \neq 4.5$$

Step 2 You have selected a sample of $n = 36$. The level of significance is 0.05 ($\alpha = 0.05$).

Step 3 Because σ is assumed to be known, you use the Z_{STAT} test statistic because the sample size is large enough so that the Central Limit Theorem tells you that the sampling distribution is approximately normally distributed.

Step 4 Because $\alpha = 0.05$, the critical values of the Z_{STAT} test statistic are -1.96 and $+1.96$. The rejection region is $Z_{STAT} < -1.96$ or $Z_{STAT} > +1.96$. The nonrejection region is $-1.96 \leq Z_{STAT} \leq +1.96$.

Step 5 You collect the sample data and compute $\overline{X} = 5.1$. Using Equation (9.1) on page 280, you compute the test statistic:

$$Z_{STAT} = \frac{\overline{X} - \mu}{\dfrac{\sigma}{\sqrt{n}}} = \frac{5.1 - 4.5}{\dfrac{1.2}{\sqrt{36}}} = +3.00$$

Step 6 Because $Z_{STAT} = +3.00 > +1.96$, you reject the null hypothesis. You conclude that there is evidence that the population mean waiting time to place an order has changed from its previous value of 4.5 minutes. The mean waiting time for customers is longer now than it was last month. As the manager, you would now want to determine how waiting time could be reduced to improve service.

Hypothesis Testing Using the *p*-Value Approach

The **p-value** is the probability of getting a test statistic equal to or more extreme than the sample result, given that the null hypothesis, H_0, is true. The *p*-value is also known as the *observed level of significance*. Using the *p*-value to determine rejection and nonrejection is another approach to hypothesis testing.

The decision rules for rejecting H_0 in the *p*-value approach are

1. If the *p*-value is greater than or equal to α, do not reject the null hypothesis.
2. If the *p*-value is less than α, reject the null hypothesis.

Many people confuse these rules, mistakenly believing that a high *p*-value is reason for rejection. This mistake can be avoided by remembering

If the *p*-value is low, then H_0 must go.

To understand the *p*-value approach, recall the Oxford Cereals scenario. One tested whether the mean fill was equal to 368 grams. The test statistic resulted in a Z_{STAT} value of $+1.50$, and one did not reject the null hypothesis because $+1.50$ was less than the upper critical value of $+1.96$ and greater than the lower critical value of -1.96.

To use the *p*-value approach for the *two-tail test*, find the probability that the test statistic Z_{STAT} is equal to or *more extreme than* 1.50 standard error units from the center of a standardized normal distribution. In other words, one determines the probability that the Z_{STAT} value is greater than $+1.50$ and the probability that the Z_{STAT} value is less than -1.50. Using Table E.2, the probability of a Z_{STAT} value below -1.50 is 0.0668, and the probability of a value below $+1.50$ is 0.9332. From the latter value, the probability of a value above $+1.50$ is 0.0668 $(1 - 0.9332)$, as Figure 9.4 shows. Therefore, the *p*-value for this two-tail test is 0.1336 $(0.0668 + 0.0668)$. The probability of a test statistic equal to or more extreme than the sample result is 0.1336. Because 0.1336 is greater than $\alpha = 0.05$, one does not reject the null hypothesis.

studentTIP

A small (low) *p*-value indicates a small probability that H_0 is true. A large (high) *p*-value indicates a large probability that H_0 is true.

FIGURE 9.4
Finding a *p*-value for a two-tail test

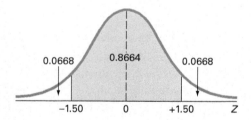

In this example, the observed sample mean is 372.5 grams, 4.5 grams above the hypothesized value, and the *p*-value is 0.1336. If the population mean is 368 grams, there is a 13.36% chance that the sample mean differs from 368 grams by at least 4.5 grams and, therefore, is ≥ 372.5 grams or ≤ 363.5 grams. Therefore, even though 372.5 grams is above the hypothesized value of 368 grams, a result as extreme as or more extreme than 372.5 grams is not highly unlikely when the population mean is 368 grams.

Unless one is using a test statistic that follows the normal distribution, one will only be able to approximate the *p*-value from the tables of the distribution. Programs such as Excel can compute the *p*-value for any hypothesis test with greater precision. Such computed *p*-values enable one to substitute the *p*-value approach for the critical value approach when one conducts hypothesis testing.

Figure 9.5 presents the Z test for the mean results for the cereal-filling example that this section uses.

FIGURE 9.5

Excel Z test for the mean (σ known) results for the cereal-filling example

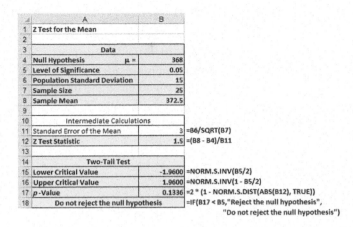

Exhibit 9.3 summarizes the p-value approach to hypothesis testing. Note that the first three steps are identical to the first three steps of the critical value approach.

EXHIBIT 9.3

The p-Value Approach to Hypothesis Testing

Step 1   State the null hypothesis, H_0, and the alternative hypothesis, H_1.

Step 2   Choose the level of significance, α, and the sample size, n. The level of significance is based on the relative importance of the risks of committing Type I and Type II errors in the problem.

Step 3   Determine the appropriate test statistic and the sampling distribution.

Step 4   Collect the sample data and calculate the value of the test statistic.

Step 5   Calculate the p-value.

Step 6   Make the statistical decision and state the managerial conclusion in the context of the theory, claim, or assertion being tested. If the p-value is greater than or equal to α, do not reject the null hypothesis. If the p-value is less than α, reject the null hypothesis.

Example 9.4 applies the p-value approach to the fast-food restaurant example.

EXAMPLE 9.4

Testing and Rejecting a Null Hypothesis Using the p-Value Approach

You are the manager of a fast-food restaurant. The business problem is to determine whether the population mean waiting time to place an order has changed in the past month from its previous value of 4.5 minutes. From past experience, you can assume that the population standard deviation is 1.2 minutes and the population waiting time is normally distributed. You select a sample of 36 orders during a one-hour period. The sample mean is 5.1 minutes. Use the Exhibit 9.3 six-step p-value approach to determine whether there is evidence that the population mean waiting time to place an order has changed in the past month from its previous population mean value of 4.5 minutes.

SOLUTION

Step 1   The null hypothesis is that the population mean has not changed from its previous value of 4.5 minutes:

$$H_0: \mu = 4.5$$

▶*(continued)*

Because the null hypothesis is that the population mean is 4.5 minutes, the alternative hypothesis is that the population mean is not 4.5 minutes:

$$H_1: \mu \neq 4.5$$

Step 2 You have selected a sample of $n = 36$ and you have chosen a 0.05 level of significance (i.e., $\alpha = 0.05$).

Step 3 Select the appropriate test statistic. Because σ is assumed known, you use the normal distribution and the Z_{STAT} test statistic.

Step 4 You collect the sample data and compute $\overline{X} = 5.1$. Using Equation (9.1) on page 280, you compute the test statistic as follows:

$$Z_{STAT} = \frac{\overline{X} - \mu}{\dfrac{\sigma}{\sqrt{n}}} = \frac{5.1 - 4.5}{\dfrac{1.2}{\sqrt{36}}} = +3.00$$

Step 5 To find the probability of getting a Z_{STAT} test statistic that is equal to or more extreme than 3.00 standard error units from the center of a standardized normal distribution, you determine the probability of a Z_{STAT} value greater than $+3.00$ along with the probability of a Z_{STAT} value less than -3.00. From Table E.2, the probability of a Z_{STAT} value below -3.00 is 0.00135. The probability of a value below $+3.00$ is 0.99865. Therefore, the probability of a value above $+3.00$ is $1 - 0.99865 = 0.00135$. Thus, the p-value for this two-tail test is $0.00135 + 0.00135 = 0.0027$.

Step 6 Because the p-value $= 0.0027 < \alpha = 0.05$, you reject the null hypothesis. You conclude that there is evidence that the population mean waiting time to place an order has changed from its previous population mean value of 4.5 minutes. The mean waiting time for customers is longer now than it was last month.

A Connection Between Confidence Interval Estimation and Hypothesis Testing

This chapter and Chapter 8 discuss confidence interval estimation and hypothesis testing, the two major elements of statistical inference. Although confidence interval estimation and hypothesis testing share the same conceptual foundation, they are used for different purposes. In Chapter 8, confidence intervals estimated parameters. In this chapter, hypothesis testing makes decisions about specified values of population parameters. Hypothesis tests are used when trying to determine whether a parameter is less than, more than, or not equal to a specified value. Proper interpretation of a confidence interval, however, can also indicate whether a parameter is less than, more than, or not equal to a specified value. For example, in this section, whether the population mean fill amount was different from 368 grams was tested by using Equation (9.1) on page 280:

$$Z_{STAT} = \frac{\overline{X} - \mu}{\dfrac{\sigma}{\sqrt{n}}}$$

Instead of testing the null hypothesis that $\mu = 368$ grams, one can reach the same conclusion by constructing a confidence interval estimate of μ. If the hypothesized value of $\mu = 368$ is contained within the interval, one does not reject the null hypothesis because 368 would not be considered an unusual value. However, if the hypothesized value does not fall into the interval, one rejects the null hypothesis because $\mu = 368$ grams is then considered an unusual value. Using these results:

$$n = 25, \overline{X} = 372.5 \text{ grams}, \sigma = 15 \text{ grams}$$

and Equation (8.1) on page 248, for a confidence level of 95% ($\alpha = 0.05$),

$$\bar{X} \pm Z_{\alpha/2} \frac{\sigma}{\sqrt{n}}$$

$$372.5 \pm (1.96) \frac{15}{\sqrt{25}}$$

$$372.5 \pm 5.88$$

so that

$$366.62 \leq \mu \leq 378.38$$

Because the interval includes the hypothesized value of 368 grams, one does not reject the null hypothesis. There is insufficient evidence that the mean fill amount for the entire filling process is not 368 grams, the same conclusion reached by using a two-tail hypothesis test.

Can You Ever Know the Population Standard Deviation?

Section 8.1 concludes with the thought that one would be unlikely to use a confidence interval estimation method that required knowing σ, the population standard deviation, because if one knew the population standard deviation, one could directly calculate the population mean—eliminating the need to use a method to estimate that parameter!

Likewise, for most practical applications, one is unlikely to use a hypothesis-testing method that requires knowing σ. If one knew the population standard deviation, one would also know the population mean and therefore have no need to perform a test. This observation raises the question "Why, then, does this section discuss the method?" Discussing the test makes explaining the fundamentals of hypothesis testing simpler. With a known population standard deviation, one can use the normal distribution and calculate p-values using the tables of the normal distribution.

PROBLEMS FOR SECTION 9.1

LEARNING THE BASICS

9.1 If you use a 0.05 level of significance in a two-tail hypothesis test, what decision will you make if $Z_{STAT} = -1.76$?

9.2 If you use a 0.05 level of significance in a two-tail hypothesis test, what decision will you make if $Z_{STAT} = +2.21$?

9.3 If you use a 0.05 level of significance in a two-tail hypothesis test, what is your decision rule for rejecting a null hypothesis that the population mean equals 500 if you use the Z test?

9.4 If you use a 0.01 level of significance in a two-tail hypothesis test, what is your decision rule for rejecting $H_0 : \mu = 12.5$ if you use the Z test?

9.5 What is your decision in Problem 9.4 if $Z_{STAT} = -2.81$?

9.6 What is the p-value if, in a two-tail hypothesis test, $Z_{STAT} = +2.00$?

9.7 In Problem 9.6, what is your statistical decision if you test the null hypothesis at the 0.05 level of significance?

9.8 What is the p-value if, in a two-tail hypothesis test, $Z_{STAT} = -1.38$?

APPLYING THE CONCEPTS

9.9 In the U.S. legal system, a defendant is presumed innocent until proven guilty. Consider a null hypothesis, H_0, that a defendant is innocent, and an alternative hypothesis, H_1, that the defendant is guilty. A jury has two possible decisions: Convict the defendant (i.e., reject the null hypothesis) or do not convict the defendant (i.e., do not reject the null hypothesis). Explain the meaning of the risks of committing either a Type I or Type II error in this example.

9.10 Suppose the defendant in Problem 9.9 is presumed guilty until proven innocent. How do the null and alternative hypotheses differ from those in Problem 9.9? What are the meanings of the risks of committing either a Type I or Type II error here?

9.11 Many consumer groups feel that the U.S. Food and Drug Administration (FDA) drug approval process is too easy and, as a result, too many drugs are approved that are later found to be unsafe. On the other hand, a number of industry lobbyists have pushed for a more lenient approval process so that pharmaceutical companies can get new drugs approved more easily and quickly. Consider a null hypothesis that a new, unapproved drug is unsafe and an alternative hypothesis that a new, unapproved drug is safe.
a. Explain the risks of committing a Type I or Type II error.
b. Which type of error are the consumer groups trying to avoid? Explain.
c. Which type of error are the industry lobbyists trying to avoid? Explain.
d. How would it be possible to lower the chances of both Type I and Type II errors?

9.12 As a result of complaints from both students and faculty about lateness, the registrar at a large university is ready to undertake a study to determine whether the scheduled break between classes should be changed. Until now, the registrar has believed that there should be 20 minutes between scheduled classes. State the null hypothesis, H_0, and the alternative hypothesis, H_1.

9.13 Do business seniors at your school prepare for class more than, less than, or about the same as business seniors at other schools? The National Survey of Student Engagement (NSSE) annual results found that business seniors spent a mean of 13 hours per week preparing for class.
a. State the null and alternative hypotheses to try to prove that the mean number of hours preparing for class by business seniors at your school is different from the 13-hour-per-week benchmark reported by the NSSE.
b. What is a Type I error for your test?
c. What is a Type II error for your test?

✓ SELF TEST **9.14** The quality-control manager at a light emitting diode (LED) factory needs to determine whether the mean life of a large shipment of LEDs is equal to 50,000 hours. The population standard deviation is 1,500 hours. A random sample of 64 LEDs indicates a sample mean life of 49,875 hours.
a. At the 0.05 level of significance, is there evidence that the mean life is different from 50,000 hours?
b. Compute the p-value and interpret its meaning.
c. Construct a 95% confidence interval estimate of the population mean life of the LEDs.
d. Compare the results of (a) and (c). What conclusions do you reach?

9.15 Suppose that in Problem 9.14, the standard deviation is 500 hours.
a. Repeat (a) through (d) of Problem 9.14, assuming a standard deviation of 500 hours.
b. Compare the results of (a) to those of Problem 9.14.

9.16 A bottled water distributor wants to determine whether the mean amount of water contained in 1-gallon bottles purchased from a nationally known water bottling company is actually 1 gallon. You know from the water bottling company specifications that the standard deviation of the amount of water per bottle is 0.02 gallon. You select a random sample of 50 bottles, and the mean amount of water per 1-gallon bottle is 0.995 gallon.
a. Is there evidence that the mean amount is different from 1.0 gallon? (Use $\alpha = 0.01$.)
b. Compute the p-value and interpret its meaning.
c. Construct a 99% confidence interval estimate of the population mean amount of water per bottle.
d. Compare the results of (a) and (c). What conclusions do you reach?

9.17 Suppose that in Problem 9.16, the standard deviation is 0.012 gallon.
a. Repeat (a) through (d) of Problem 9.16, assuming a standard deviation of 0.012 gallon.
b. Compare the results of (a) to those of Problem 9.16.

9.2 t Test of Hypothesis for the Mean (σ Unknown)

In most hypothesis-testing situations concerning the population mean, μ, one will not know the population standard deviation, σ. However, one will always be able to calculate the sample standard deviation, S. If one assumes that the population is normally distributed, then the sampling distribution of the mean will follow a t distribution with $n - 1$ degrees of freedom and the **t test for the mean** can be used. If the population is not normally distributed, one can still use the t test if the population is not too skewed and the sample size is not too small. Equation (9.2) defines the test statistic for determining the difference between the sample mean, $\overline{X}$, and the population mean, μ, when using the sample standard deviation, S.

t TEST FOR THE MEAN (σ UNKNOWN)

$$t_{STAT} = \frac{\overline{X} - \mu}{\frac{S}{\sqrt{n}}} \tag{9.2}$$

where the t_{STAT} test statistic follows a t distribution having $n - 1$ degrees of freedom.

To illustrate the use of the t test for the mean, return to the Chapter 8 Ricknel Home Centers scenario on page 244. Members of the AIS team might have been assigned the business problem to determine if the mean amount per sales invoice has not changed from the $120 of the past five years.

A t test for the mean (σ unknown) would enable team members to determine whether the mean amount per sales invoice is increasing or decreasing. Either the critical value or p-value approach could be used to evaluate this test.

To perform this two-tail hypothesis test using either approach, begin with steps 1 through 3 of the six-step methods that Exhibits 9.2 and 9.3 on pages 282 and 285 summarize.

studentTIP

Remember, the null hypothesis uses an equals sign and the alternative hypothesis *never* uses an equals sign.

Step 1 Define the following hypotheses:

$$H_0: \mu = 120$$
$$H_1: \mu \neq 120$$

The alternative hypothesis contains the statement for which team members seek to find evidence. If the null hypothesis is rejected, then there is statistical evidence that the population mean amount per sales invoice is no longer $120. If the statistical conclusion is "do not reject H_0," then the team will conclude that there is insufficient evidence to prove that the mean amount differs from the long-term mean of $120.

Step 2 Collect the data from a sample of $n = 12$ sales invoices. The team decides to use $\alpha = 0.05$.

Step 3 Because σ is unknown, the team uses the *t* distribution and the t_{STAT} test statistic. Team members assume that the population of sales invoices is approximately normally distributed in order to use the *t* distribution because the sample size is only 12. (See "Checking the Normality Assumption" on page 291.)

Continue with either the critical value approach on page 282 or the *p*-value approach on page 285.

studentTIP

Because this is a two-tail test, the level of significance, $\alpha = 0.05$, is divided into two equal 0.025 parts, in each of the two tails of the distribution.

Using the Critical Value Approach

Having completed steps 1 through 3:

Step 4 For a given sample size, n, the test statistic t_{STAT} follows a *t* distribution with $n - 1$ degrees of freedom. For this problem, use the critical values of the *t* distribution with $12 - 1 = 11$ degrees of freedom. The alternative hypothesis, $H_1: \mu \neq 120$, has two tails. The area in the rejection region of the *t* distribution's left (lower) tail is 0.025, and the area in the rejection region of the *t* distribution's right (upper) tail is also 0.025.

From Table E.3, the critical values of *t*, a portion of which Table 9.3 shows, the critical values are ± 2.2010. The decision rule is

$$\text{Reject } H_0 \text{ if } t_{STAT} < -2.2010$$
$$\text{or if } t_{STAT} > +2.2010;$$
$$\text{otherwise, do not reject } H_0.$$

TABLE 9.3

Determining the critical value from the *t* table for an area of 0.025 in each tail, with 11 degrees of freedom

	Cumulative Probabilities					
	.75	.90	.95	.975	.99	.995
	Upper-Tail Areas					
Degrees of Freedom	.25	.10	.05	.025	.01	.005
1	1.0000	3.0777	6.3138	12.7062	31.8207	63.6574
2	0.8165	1.8856	2.9200	4.3027	6.9646	9.9248
3	0.7649	1.6377	2.3534	3.1824	4.5407	5.8409
4	0.7407	1.5332	2.1318	2.7764	3.7469	4.6041
5	0.7267	1.4759	2.0150	2.5706	3.3649	4.0322
6	0.7176	1.4398	1.9432	2.4469	3.1427	3.7074
7	0.7111	1.4149	1.8946	2.3646	2.9980	3.4995
8	0.7064	1.3968	1.8595	2.3060	2.8965	3.3554
9	0.7027	1.3830	1.8331	2.2622	2.8214	3.2498
10	0.6998	1.3722	1.8125	2.2281	2.7638	3.1693
11	0.6974	1.3634	1.7959	2.2010	2.7181	3.1058

Source: Extracted from Table E.3.

FIGURE 9.6

Testing a hypothesis for the mean (σ unknown) at the 0.05 level of significance with 11 degrees of freedom

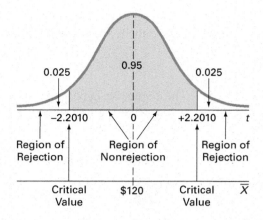

Step 5 The team organizes and stores the data from a random sample of 12 sales invoices in Invoices:

108.98	152.22	111.45	110.59	127.46	107.26
93.32	91.97	111.56	75.71	128.58	135.11

Using Equations (3.1) and (3.7) on pages 109 and 115,

$$\overline{X} = \$112.85 \text{ and } S = \$20.80$$

From Equation (9.2) on page 288,

$$t_{STAT} = \frac{\overline{X} - \mu}{\frac{S}{\sqrt{n}}} = \frac{112.85 - 120}{\frac{20.80}{\sqrt{12}}} = -1.1908$$

The test statistic can also be determined by software results, as the Figure 9.7 Excel results for this test illustrate.

Step 6 Because $-2.2010 < t_{STAT} = -1.1908 < 2.2010$, the team does not reject H_0. The team has insufficient evidence to conclude that the mean amount per sales invoice differs from \$120. The audit suggests that the mean amount per invoice has not changed.

FIGURE 9.7

Excel results for the sales invoices example *t* test

	A	B	
1	t Test for the Hypothesis of the Mean		
2			
3	Data		
4	Null Hypothesis $\mu=$	120	
5	Level of Significance	0.05	
6	Sample Size	12	
7	Sample Mean	112.85	
8	Sample Standard Deviation	20.8	
9			
10	Intermediate Calculations		
11	Standard Error of the Mean	6.0044	=B8/SQRT(B6)
12	Degrees of Freedom	11	=B6 - 1
13	t Test Statistic	-1.1908	=(B7 - B4)/B11
14			
15	Two-Tail Test		
16	Lower Critical Value	-2.2010	=-T.INV.2T(B5, B12)
17	Upper Critical Value	2.2010	=T.INV.2T(B5, B12)
18	p-Value	0.2588	=T.DIST.2T(ABS(B13), B12)
19	Do not reject the null hypothesis		=IF(B18 < B5,"Reject the null hypothesis", "Do not reject the null hypothesis")

Using the *p*-Value Approach

Having completed steps 1 through 3 on page 289:

Step 4 Using the Figure 9.7 software results, the test statistic $t_{STAT} = -1.19$.

Step 5 Using the software results, the *p*-value = 0.2588.

Step 6 Because the *p*-value of 0.2588 is greater than $\alpha = 0.05$, the team does not reject H_0. The data provide insufficient evidence to conclude that the mean amount per sales invoice differs from \$120. The audit suggests that the mean amount per invoice has not changed.

The *p*-value indicates that if the null hypothesis is true, the probability that a sample of 12 invoices could have a sample mean that differs by $7.15 or more from the stated $120 is 0.2588. In other words, if the mean amount per sales invoice is truly $120, then there is a 25.88% chance of observing a sample mean below $112.85 or above $127.15.

For this example, it is incorrect to state that there is a 25.88% chance that the null hypothesis is true. Remember that the *p*-value is a conditional probability, calculated by *assuming* that the null hypothesis is true. In general, it is proper to state the following:

If the null hypothesis is true, there is a (*p*-value) × 100% chance of observing a test statistic at least as contradictory to the null hypothesis as the sample result.

Checking the Normality Assumption

Use the *t* test when the population standard deviation, σ, is not known and is estimated using the sample standard deviation, *S*. To use the *t* test, assume that the data represent a random sample from a population that is normally distributed. In practice, as long as the sample size is not very small and the population is not very skewed, the *t* distribution provides a good approximation of the sampling distribution of the mean when σ is unknown.

There are several ways to evaluate the normality assumption necessary for using the *t* test. One can examine how closely the sample statistics match the normal distribution's theoretical properties. One can also construct a histogram, stem-and-leaf display, boxplot, or normal probability plot to visualize the distribution of the sales invoice amounts. For details on evaluating normality, see Section 6.3.

Figure 9.8 presents descriptive statistics and a boxplot, and Figure 9.9 presents a normal probability plot for the sales invoice data.

FIGURE 9.8

Excel descriptive statistics and boxplot for the sales invoice data

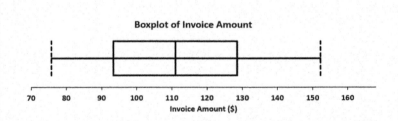

	Invoice Amount
Mean	112.8508
Median	111.02
Mode	#N/A
Minimum	75.71
Maximum	152.22
Range	76.51
Variance	432.5565
Standard Deviation	20.7980
Coeff. of Variation	18.43%
Skewness	0.1336
Kurtosis	0.1727
Count	12
Standard Error	6.0039

FIGURE 9.9

Excel normal probability plot for the sales invoice data

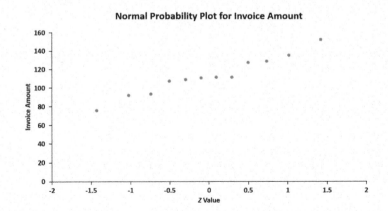

The mean is very close to the median, and the points on the normal probability plot appear to be increasing approximately in a straight line. The boxplot appears to be approximately symmetrical. Thus, one can assume that the population of sales invoices is approximately normally distributed. The normality assumption is valid, and therefore the auditor's results are valid.

The t test is a **robust** test. A robust test does not lose power if the shape of the population departs somewhat from a normal distribution, particularly when the sample size is large enough to enable the test statistic t to follow the t distribution. However, one can reach erroneous conclusions and can lose statistical power if one uses the t test incorrectly. If the sample size, n, is small (less than 30) and one cannot easily make the assumption that the underlying population is at least approximately normally distributed, then using *nonparametric* tests would be more appropriate (Bradley, Daniel).

PROBLEMS FOR SECTION 9.2

LEARNING THE BASICS

9.18 If, in a sample of $n = 16$ selected from a normal population, $\overline{X} = 56$ and $S = 12$, what is the value of t_{STAT} if you are testing the null hypothesis $H_0: \mu = 50$?

9.19 In Problem 9.18, how many degrees of freedom does the t test have?

9.20 In Problems 9.18 and 9.19, what are the critical values of t if the level of significance, α, is 0.05 and the alternative hypothesis, H_1, is $\mu \neq 50$?

9.21 In Problems 9.18, 9.19, and 9.20, what is your statistical decision if the alternative hypothesis, H_1, is $\mu \neq 50$?

9.22 If, in a sample of $n = 16$ selected from a left-skewed population, $\overline{X} = 65$, and $S = 21$, would you use the t test to test the null hypothesis $H_0: \mu = 60$? Discuss.

9.23 If, in a sample of $n = 160$ selected from a left-skewed population, $\overline{X} = 65$, and $S = 21$, would you use the t test to test the null hypothesis $H_0: \mu = 60$? Discuss.

APPLYING THE CONCEPTS

✓SELF TEST **9.24** You are the manager of a restaurant for a fast-food franchise. Last month, the mean waiting time at the drive-through window for branches in your geographic region, as measured from the time a customer places an order until the time the customer receives the order, was 3.7 minutes. You select a random sample of 64 orders. The sample mean waiting time is 3.57 minutes, with a sample standard deviation of 0.8 minute.
a. At the 0.05 level of significance, is there evidence that the population mean waiting time is different from 3.7 minutes?
b. Because the sample size is 64, do you need to be concerned about the shape of the population distribution when conducting the t test in (a)? Explain.

9.25 A manufacturer of chocolate candies uses machines to package candies as they move along a filling line. Although the packages are labeled as 8 ounces, the company wants the packages to contain a mean of 8.15 ounces so that virtually none of the packages contain less than 8 ounces. A sample of 50 packages is selected periodically, and the packaging process is stopped if there is evidence that the mean amount packaged is different from 8.15 ounces. Suppose that in a particular sample of 50 packages, the mean amount dispensed is 8.159 ounces, with a sample standard deviation of 0.051 ounce.
a. Is there evidence that the population mean amount is different from 8.15 ounces? (Use a 0.05 level of significance.)
b. Determine the p-value and interpret its meaning.

9.26 A marketing researcher wants to estimate the mean amount spent per year (\$) on Amazon.com by Amazon Prime member shoppers. Suppose a random sample of 100 Amazon Prime member shoppers who recently made a purchase on Amazon.com yielded a mean amount spent of \$1,500 and a standard deviation of \$200.
a. Is there evidence that the population mean amount spent per year on Amazon.com by Amazon Prime member shoppers is different from \$1,475? (Use a 0.05 level of significance.)
b. Determine the p-value and interpret its meaning.

9.27 The U.S. Department of Transportation requires tire manufacturers to provide performance information on tire sidewalls to help prospective buyers make their purchasing decisions. One very important piece of information is the tread wear index, which indicates the tire's resistance to tread wear. A tire with a grade of 200 should last twice as long, on average, as a tire with a grade of 100.

A consumer organization wants to test the actual tread wear index of a brand name of tires that claims "graded 200" on the sidewall of the tire. A random sample of $n = 18$ indicates a sample mean tread wear index of 198.8 and a sample standard deviation of 21.4.
a. Is there evidence that the population mean tread wear index is different from 200? (Use a 0.05 level of significance.)
b. Determine the p-value and interpret its meaning.

9.28 The file **Fast Casual** contains the amount that a sample of 15 customers spent for lunch (\$) at a fast-food restaurant:

7.42	6.29	5.83	6.50	8.34	9.51	7.10	6.80	5.90
4.89	6.50	5.52	7.90	8.30	9.60			

a. At the 0.05 level of significance, is there evidence that the mean amount spent for lunch is different from \$6.50?
b. Determine the p-value in (a) and interpret its meaning.
c. What assumption must you make about the population distribution in order to conduct the t test in (a) and (b)?
d. Because the sample size is 15, do you need to be concerned about the shape of the population distribution when conducting the t test in (a)? Explain.

9.29 An insurance company has the business objective of reducing the amount of time it takes to approve applications for life insurance. The approval process consists of underwriting, which includes a review of the application, a medical information bureau check, possible requests for additional medical information and medical exams, and a policy compilation stage in which the policy pages are generated and sent for delivery. The ability to deliver approved policies to customers in a timely manner is critical to the profitability of this service. During a period of one month, you collect a random sample of 27 approved policies and store their total processing times, in days, in Insurance .

a. In the past, the mean processing time was 45 days. At the 0.05 level of significance, is there evidence that the mean processing time has changed from 45 days?

b. What assumption about the population distribution is needed in order to conduct the *t* test in (a)?

c. Construct a boxplot or a normal probability plot to evaluate the assumption made in (b).

d. Do you think that the assumption needed in order to conduct the *t* test in (a) is valid? Explain.

9.30 The following data (in Drink) represent the amount of soft drink filled in a sample of 50 consecutive 2-liter bottles. The results, listed horizontally in the order of being filled, were:

2.109	2.086	2.066	2.075	2.065	2.057	2.052	2.044
2.036	2.038	2.031	2.029	2.025	2.029	2.023	2.020
2.015	2.014	2.013	2.014	2.012	2.012	2.012	2.010
2.005	2.003	1.999	1.996	1.997	1.992	1.994	1.986
1.984	1.981	1.973	1.975	1.971	1.969	1.966	1.967
1.963	1.957	1.951	1.951	1.947	1.941	1.941	1.938
1.908	1.894						

a. At the 0.05 level of significance, is there evidence that the mean amount of soft drink filled is different from 2.0 liters?

b. Determine the *p*-value in (a) and interpret its meaning.

c. In (a), you assumed that the distribution of the amount of soft drink filled was normally distributed. Evaluate this assumption by constructing a boxplot or a normal probability plot.

d. Do you think that the assumption needed in order to conduct the *t* test in (a) is valid? Explain.

e. Examine the values of the 50 bottles in their sequential order, as given in the problem. Does there appear to be a pattern to the results? If so, what impact might this pattern have on the validity of the results in (a)?

9.31 One of the major measures of the quality of service provided by any organization is the speed with which it responds to customer complaints. A large family-held department store selling furniture and flooring, including carpet, had undergone a major expansion in the past several years. In particular, the flooring department had expanded from two installation crews to an installation supervisor, a measurer, and 15 installation crews. The store had the business objective of improving its response to complaints. The variable of interest was defined as the number of days between when the complaint was made and when it was resolved. Data were collected from 50 complaints that were made in the past year. These data, stored in Furniture , are:

54	5	35	137	31	27	152	2	123	81	74	27
11	19	126	110	110	29	61	35	94	31	26	5
12	4	165	32	29	28	29	26	25	1	14	13
13	10	5	27	4	52	30	22	36	26	20	23
33	68										

a. The installation supervisor claims that the mean number of days between the receipt of a complaint and the resolution of the complaint is 20 days. At the 0.05 level of significance, is there evidence that the claim is not true (i.e., the mean number of days is different from 20)?

b. What assumption about the population distribution is needed in order to conduct the *t* test in (a)?

c. Construct a boxplot or a normal probability plot to evaluate the assumption made in (b).

d. Do you think that the assumption needed in order to conduct the *t* test in (a) is valid? Explain.

9.32 A manufacturing company produces steel housings for electrical equipment. The main component part of the housing is a steel trough that is made out of a 14-gauge steel coil. It is produced using a 250-ton progressive punch press with a wipe-down operation that puts two 90-degree forms in the flat steel to make the trough. The distance from one side of the form to the other is critical because of weatherproofing in outdoor applications. The company requires that the width of the trough be between 8.31 inches and 8.61 inches. The file Trough contains the widths of the troughs, in inches, for a sample of $n = 49$:

8.312	8.343	8.317	8.383	8.348	8.410	8.351	8.373	8.481	8.422
8.476	8.382	8.484	8.403	8.414	8.419	8.385	8.465	8.498	8.447
8.436	8.413	8.489	8.414	8.481	8.415	8.479	8.429	8.458	8.462
8.460	8.444	8.429	8.460	8.412	8.420	8.410	8.405	8.323	8.420
8.396	8.447	8.405	8.439	8.411	8.427	8.420	8.498	8.409	

a. At the 0.05 level of significance, is there evidence that the mean width of the troughs is different from 8.46 inches?

b. What assumption about the population distribution is needed in order to conduct the *t* test in (a)?

c. Evaluate the assumption made in (b).

d. Do you think that the assumption needed in order to conduct the *t* test in (a) is valid? Explain.

9.33 One operation of a steel mill is to cut pieces of steel into parts that are used in the frame for front seats in an automobile. The steel is cut with a diamond saw and the resulting parts must be cut to be within ± 0.005 inch of the length specified by the automobile company. The file Steel contains a sample of 100 steel parts. The measurement reported is the difference, in inches, between the actual length of the steel part, as measured by a laser measurement device, and the specified length of the steel part. For example, a value of -0.002 represents a steel part that is 0.002 inch shorter than the specified length.

a. At the 0.05 level of significance, is there evidence that the mean difference is different from 0.0 inches?

b. Construct a 95% confidence interval estimate of the population mean. Interpret this interval.

c. Compare the conclusions reached in (a) and (b).

d. Because $n = 100$, do you have to be concerned about the normality assumption needed for the *t* test and *t* interval?

9.34 In Problem 3.71 on page 145, you were introduced to a teabag-filling operation. An important quality characteristic of interest for this process is the weight of the tea in the individual bags. The file Teabags contains an ordered array of the weight, in grams, of a sample of 50 tea bags produced during an 8-hour shift.

a. Is there evidence that the mean amount of tea per bag is different from 5.5 grams? (Use $\alpha = 0.01$.)

b. Construct a 99% confidence interval estimate of the population mean amount of tea per bag. Interpret this interval.

c. Compare the conclusions reached in (a) and (b).

9.35 We Are Social and Hootsuite reported that the typical American spends 2.02 hours (121 minutes) per day accessing the Internet via mobile devices.

Source: *Digital in 2017 Global Overview*, available at **bit.ly/2jXeS3F**

In order to test the validity of this statement, you select a sample of 30 friends and family. The results for the time spent per day accessing the Internet via mobile devices (in minutes) are stored in Mobile Device Study .

a. Is there evidence that the population mean time spent per day accessing the Internet via mobile devices is different from 121 minutes? Use the *p*-value approach and a level of significance of 0.05.

b. What assumption about the population distribution is needed in order to conduct the *t* test in (a)?

c. Make a list of the various ways you could evaluate the assumption noted in (b).

d. Evaluate the assumption noted in (b) and determine whether the test in (a) is valid.

9.3 One-Tail Tests

The examples of hypothesis testing in Sections 9.1 and 9.2 are called two-tail tests because the rejection region is divided into the two tails of the sampling distribution of the mean. In contrast, some hypothesis tests are one-tail tests because they require an alternative hypothesis that focuses on a *particular direction*. Either the critical value or *p*-value approach can be used to evaluate one-tail hypothesis tests, using one of the six-step methods that Exhibits 9.2 and 9.3 summarize, with a minor adjustment that reflects the one tail.

One example of a one-tail hypothesis test would be a test to determine whether the population mean is *less than* a specified value. For example, a quick-service or fast-casual restaurant might undertake a quality improvement effort to improve the speed of drive-through service using as a benchmark the mean service time as reported in a recent drive-through study conducted by *QSR* magazine (Oches). In that study, the mean drive-through service time for McDonald's was 208.16 seconds, which was fifth fastest out of the 15 chains surveyed. Suppose that McDonald's began an effort to improve service by reducing the service time and had deployed an improved drive-through order fulfillment process in a sample of 25 stores. Because McDonald's would want to institute the new process in all of its stores only if the test sample saw a *decreased* drive-through time, the entire rejection region is located in the lower tail of the distribution.

For this problem, McDonald's seeks to determine whether the new drive-through process has a mean that is less than 208.16 seconds. To perform this one-tail hypothesis test, McDonald's begins with the steps 1 through 3 that the critical value and *p*-value approach share (see Exhibits 9.2 and 9.3).

student TIP

The rejection region matches the direction of the alternative hypothesis. If the alternative hypothesis contains a < sign, the rejection region is in the lower tail. If the alternative hypothesis contains a > sign, the rejection region is in the upper tail.

Step 1 McDonald's defines the null and alternative hypotheses:

$$H_0: \mu \geq 208.16$$
$$H_1: \mu < 208.16$$

The alternative hypothesis contains the statement for which McDonald's seeks to find evidence. If the conclusion of the test is "reject H_0," there is statistical evidence that the mean drive-through time is less than the drive-through time in the old process. This would be reason to change the drive-through process for the entire population of stores. If the conclusion of the test is "do not reject H_0," then there is insufficient evidence that the mean drive-through time in the new process is significantly less than the drive-through time in the old process. If this occurs, there would be insufficient reason to institute the new drive-through process in the population of stores.

Step 2 McDonald's collects the data by selecting a sample of $n = 25$ stores. McDonald's decides to use $\alpha = 0.05$.

Step 3 Because σ is unknown, McDonald's uses the *t* distribution and the t_{STAT} test statistic. McDonald's needs to assume that the drive-through time is normally distributed because a sample of only 25 drive-through times is selected.

Continue with either the critical value approach on page 282 or the *p*-value approach on page 285.

Using the Critical Value Approach

Having completed steps 1 through 3:

Step 4 The rejection region is entirely contained in the lower tail of the sampling distribution of the mean because, for this problem, one rejects H_0 only when the sample mean is significantly less than 208.16 seconds. When the entire rejection region is contained in one tail of the sampling distribution of the test statistic, the test is called a **one-tail test**, or **directional test**. If the alternative hypothesis includes the *less than* sign, the critical value of t is negative.

Because the entire rejection region is in the lower tail of the t distribution and contains an area of 0.05 (see Figure 9.10), due to the symmetry of the t distribution, the critical value of the t test statistic with $25 - 1 = 24$ degrees of freedom is -1.7109. (see Table 9.4, a portion of Table E.3). The decision rule is

$$\text{Reject } H_0 \text{ if } t_{STAT} < -1.7109;$$

$$\text{otherwise, do not reject } H_0.$$

TABLE 9.4
Determining the critical value from the t table for an area of 0.05 in the lower tail, with 24 degrees of freedom

	Cumulative Probabilities					
	.75	.90	.95	.975	.99	.995
	Upper-Tail Areas					
Degrees of Freedom	.25	.10	.05	.025	.01	.005
1	1.0000	3.0777	6.3138	12.7062	31.8207	63.6574
2	0.8165	1.8856	2.9200	4.3027	6.9646	9.9248
3	0.7649	1.6377	2.3534	3.1824	4.5407	5.8409
⋮	⋮	⋮	⋮	⋮	⋮	⋮
23	0.6853	1.3195	1.7139	2.0687	2.4999	2.8073
24	0.6848	1.3178	1.7109	2.0639	2.4922	2.7969
25	0.6844	1.3163	1.7081	2.0595	2.4851	2.7874

Source: Extracted from Table E.3.

FIGURE 9.10
One-tail test of hypothesis for a mean (σ unknown) at the 0.05 level of significance

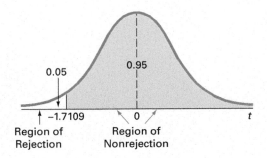

0.05   0.95

−1.7109   0 t

Region of Rejection   Region of Nonrejection

Step 5   For the sample of 25 stores, McDonald's finds that the sample mean service time at the drive-through equals 195.6 seconds and the sample standard deviation equals 22.1 seconds. Using $n = 25$, $\overline{X} = 195.6$, $S = 22.1$, and Equation (9.2) on page 288,

$$t_{STAT} = \frac{\overline{X} - \mu}{\dfrac{S}{\sqrt{n}}} = \frac{195.6 - 208.16}{\dfrac{22.1}{\sqrt{25}}} = -2.8416$$

The test statistic can also be determined by software results, as the Figure 9.11 on page 296 Excel results for this test illustrate.

Step 6   Because $t_{STAT} = -2.8416 < -1.7109$ (see Figure 9.10), McDonald's rejects the null hypothesis. McDonald's concludes that the mean service time at the drive-through is less than 208.16 seconds. There is sufficient evidence to change the drive-through process for the entire population of stores.

Using the *p*-Value Approach

Having completed steps 1 through 3 on page 294:

Step 4 Using the Figure 9.11 software results, the test statistic $t_{STAT} = -2.8416$. Because the alternative hypothesis indicates a rejection region entirely in the lower tail of the sampling distribution, to calculate the *p*-value, McDonald's needs to find the probability that the t_{STAT} test statistic will be less than -2.8416.

Step 5 Using the software results, the *p*-value = 0.0045.

Step 6 The *p*-value of 0.0045 is less than $\alpha = 0.05$ (see Figure 9.12). McDonald's rejects H_0 and concludes that the mean service time at the drive-through is less than 208.16 seconds. There is sufficient evidence to change the drive-through process for the entire population of stores.

FIGURE 9.11
Excel *t* test worksheet results for the drive-through time study

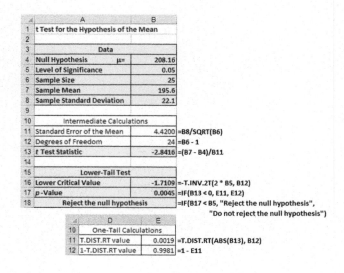

FIGURE 9.12
Determining the *p*-value for a one-tail test

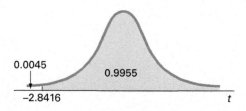

Example 9.5 illustrates a one-tail test in which the rejection region is in the upper tail.

EXAMPLE 9.5

A One-Tail Test for the Mean

A company that manufactures chocolate bars is particularly concerned that the mean weight of a chocolate bar is not greater than 6.03 ounces. A sample of 50 chocolate bars is selected; the sample mean is 6.034 ounces, and the sample standard deviation is 0.02 ounce. Using the $\alpha = 0.01$ level of significance, is there evidence that the population mean weight of the chocolate bars is greater than 6.03 ounces?

SOLUTION Using the Exhibit 9.2 critical value approach on page 282,

Step 1 First, define the null and alternative hypotheses:

$$H_0: \mu \leq 6.03$$
$$H_1: \mu > 6.03$$

Step 2 Collect the data from a sample of $n = 50$. You decide to use $\alpha = 0.01$.

Step 3 Because σ is unknown, you use the *t* distribution and the t_{STAT} test statistic.

▶(*continued*)

Step 4 The rejection region is entirely contained in the upper tail of the sampling distribution of the mean because you want to reject H_0 only when the sample mean is significantly greater than 6.03 ounces. Because the entire rejection region is in the upper tail of the t distribution and contains an area of 0.01, the critical value of the t distribution with $50 - 1 = 49$ degrees of freedom is 2.4049 (see Table E.3).

The decision rule is

$$\text{Reject } H_0 \text{ if } t_{STAT} > 2.4049;$$

$$\text{otherwise, do not reject } H_0.$$

Step 5 From your sample of 50 chocolate bars, you find that the sample mean weight is 6.034 ounces, and the sample standard deviation is 0.02 ounces. Using $n = 50$, $\overline{X} = 6.034$, $S = 0.02$, and Equation (9.2) on page 288,

$$t_{STAT} = \frac{\overline{X} - \mu}{\dfrac{S}{\sqrt{n}}} = \frac{6.034 - 6.03}{\dfrac{0.02}{\sqrt{50}}} = 1.414$$

Step 6 Because $t_{STAT} = 1.414 < 2.4049$ or the p-value (from Excel) is $0.0818 > 0.01$, you do not reject the null hypothesis. There is insufficient evidence to conclude that the population mean weight is greater than 6.03 ounces.

To perform one-tail tests of hypotheses, one must properly formulate H_0 and H_1. Exhibit 9.4 summarizes the key points about the null and alternative hypotheses for one-tail tests.

EXHIBIT 9.4

The Null and Alternative Hypotheses in One-Tail Tests

The null hypothesis, H_0, states a status quo claim.

The alternative hypothesis, H_1, states a claim that is contrary to the null hypothesis and often represents a research claim or specific inference that an analyst seeks to prove.

A null and alternative pair of hypotheses are always collectively exhaustive.

If one rejects the null hypothesis, one has strong statistical evidence that the alternative hypothesis is correct.

If one does not reject the null hypothesis, one has not proven the null hypothesis. (Rather, one has only failed to prove the alternative hypothesis.)

The null hypothesis always refers to a population parameter such as μ and not a sample statistic such as $\overline{X}$.

The null hypothesis always includes an equals sign when stating a claim about the population parameter, for example, H_0: $\mu \geq 208.16$ grams.

The alternative hypothesis never includes an equals sign when stating a claim about the population parameter, for example, H_1: $\mu < 208.16$ grams.

PROBLEMS FOR SECTION 9.3

LEARNING THE BASICS

9.36 In a one-tail hypothesis test where you reject H_0 only in the *upper* tail, what is the p-value if $Z_{STAT} = +2.00$?

9.37 In Problem 9.36, what is your statistical decision if you test the null hypothesis at the 0.01 level of significance?

9.38 In a one-tail hypothesis test where you reject H_0 only in the *lower* tail, what is the p-value if $Z_{STAT} = -1.38$?

9.39 In Problem 9.38, what is your statistical decision if you test the null hypothesis at the 0.05 level of significance?

9.40 In a one-tail hypothesis test where you reject H_0 only in the *lower* tail, what is the *p*-value if $Z_{STAT} = +1.38$?

9.41 In Problem 9.40, what is the statistical decision if you test the null hypothesis at the 0.05 level of significance?

9.42 In a one-tail hypothesis test where you reject H_0 only in the *upper* tail, what is the critical value of the *t*-test statistic with 10 degrees of freedom at the 0.01 level of significance?

9.43 In Problem 9.42, what is your statistical decision if $t_{STAT} = +2.79$?

9.44 In a one-tail hypothesis test where you reject H_0 only in the *lower* tail, what is the critical value of the t_{STAT} test statistic with 20 degrees of freedom at the 0.01 level of significance?

9.45 In Problem 9.44, what is your statistical decision if $t_{STAT} = -3.15$?

APPLYING THE CONCEPTS

9.46 The Washington Metropolitan Area Transit Authority has set a bus fleet reliability goal of 8,000 bus-miles. Bus reliability is measured specifically as the number of bus-miles traveled before a mechanical breakdown that requires the bus to be removed from service or deviate from the schedule. Suppose a sample of 64 buses resulted in a sample mean of 8,210 bus-miles and a sample standard deviation of 625 bus-miles.
a. Is there evidence that the population mean bus-miles is greater than 8,000 bus-miles? (Use a 0.05 level of significance.)
b. Determine the *p*-value and interpret its meaning.

9.47 *CarMD* reports that the mean repair cost of a Toyota in 2018 for a "check engine" light was $482.

Source: **www.carmd.com/wp/vehicle-health-index-introduction/2018-carmd-manufacturer-vehicle-rankings**

Suppose a sample of 100 "check engine" light repairs completed in the last month was selected. The sample mean repair cost was $271 with the sample standard deviation of $100.
a. Is there evidence that the population mean repair cost is less than $482? (Use a 0.05 level of significance.)
b. Determine the *p*-value and interpret its meaning.

✓ SELF TEST **9.48** Patient waiting is a common phenomenon in the doctor's waiting room. One acceptable standard of practice states that waiting time for patients to be seen by the first provider in hospital outpatient and public health clinics should be less than 30 minutes. A study was conducted to assess patient waiting at a primary healthcare clinic. Data were collected on a sample of 860 patients. In this sample, the mean wait time was 24.05 minutes, with a standard deviation of 16.5 minutes.

Source: Data extracted from BA Ahmad, K. Khairatul, and A. Farnazza, "An assessment of patient waiting and consultation time in a primary healthcare clinic," *Malaysian Family Practice*, 2017, 12(1), pp. 14–21.

a. If you test the null hypothesis at the 0.01 level of significance, is there evidence that the population mean wait time is less than 30 minutes?
b. Interpret the meaning of the *p*-value in this problem.

9.49 You are the manager of a restaurant that delivers pizza to college dormitory rooms. You have just changed your delivery process in an effort to reduce the mean time between the order and completion of delivery from the current 25 minutes. A sample of 36 orders using the new delivery process yields a sample mean of 22.4 minutes and a sample standard deviation of 6 minutes.
a. Using the six-step critical value approach, at the 0.05 level of significance, is there evidence that the population mean delivery time has been reduced below the previous population mean value of 25 minutes?
b. At the 0.05 level of significance, use the six-step *p*-value approach.
c. Interpret the meaning of the *p*-value in (b).
d. Compare your conclusions in (a) and (b).

9.50 A survey of nonprofit organizations showed that online fundraising has increased in the past year. Based on a random sample of 133 nonprofits, the mean one-time gift donation resulting from email outreach in the past year was $87. Assume that the sample standard deviation is $9.
a. If you test the null hypothesis at the 0.01 level of significance, is there evidence that the mean one-time gift donation is greater than $85.50?
b. Interpret the meaning of the *p*-value in this problem.

9.51 The population mean waiting time to check out of a supermarket has been 4 minutes. Recently, in an effort to reduce the waiting time, the supermarket has experimented with a system in which infrared cameras use body heat and in-store software to determine how many lanes should be opened. A sample of 100 customers was selected, and their mean waiting time to check out was 3.10 minutes, with a sample standard deviation of 2.5 minutes.
a. At the 0.05 level of significance, using the critical value approach to hypothesis testing, is there evidence that the population mean waiting time to check out is less than 4 minutes?
b. At the 0.05 level of significance, using the *p*-value approach to hypothesis testing, is there evidence that the population mean waiting time to check out is less than 4 minutes?
c. Interpret the meaning of the *p*-value in this problem.
d. Compare your conclusions in (a) and (b).

9.4 Z Test of Hypothesis for the Proportion

In some situations, one seeks to test a hypothesis about the proportion of events of interest in the population, π, rather than test the population mean. To begin, one selects a random sample and calculates the **sample proportion**, $p = X/n$. Then, one compares the value of this statistic to the hypothesized value of the parameter, π, in order to decide whether to reject the null hypothesis.

If the number of events of interest (X) and the number of events that are not of interest ($n - X$) are each at least five, the sampling distribution of a proportion approximately follows

studentTIP

Do not confuse this use of the Greek letter pi, π, to represent the population proportion with the constant that is the ratio of the circumference to the diameter of a circle—approximately 3.14159.

a normal distribution, and one can use the *Z* **test for the proportion**. Equation (9.3) defines this hypothesis test for the difference between the sample proportion, *p*, and the hypothesized population proportion, π. Equation (9.4) provides an alternate definition that uses the number of events of interest, *X*, instead of the sample proportion, *p*.

Z TEST FOR THE PROPORTION

$$Z_{STAT} = \frac{p - \pi}{\sqrt{\dfrac{\pi(1 - \pi)}{n}}} \tag{9.3}$$

where

$$p = \text{sample proportion} = \frac{X}{n} = \frac{\text{number of events of interest in the sample}}{\text{sample size}}$$

$$\pi = \text{hypothesized proportion of events of interest in the population}$$

**Z TEST FOR THE PROPORTION
IN TERMS OF THE NUMBER OF EVENTS OF INTEREST**

$$Z_{STAT} = \frac{X - n\pi}{\sqrt{n\pi(1 - \pi)}} \tag{9.4}$$

The Z_{STAT} test statistic approximately follows a standardized normal distribution when *X* and $(n - X)$ are each at least 5 for this test.

Either the critical value or *p*-value approach can be used to evaluate the *Z* test for the proportion. To illustrate both approaches, consider a 2016 survey conducted by CareerBuilder in which 45% of American workers reported that they work during nonbusiness hours.[1] Suppose a research firm decides to take a new survey to determine whether the proportion has changed since 2016. The firm conducts this new survey and finds that 208 of 400 American workers reported that they work during nonbusiness hours. To determine whether the proportion has changed, the firm defines the null and alternative hypotheses:

[1]as reported by L. Petrecca in "Always 'on': How you can disconnect from work," *USA Today*, January 16, 2017.

H_0: $\pi = 0.45$ (the proportion of American workers who reported that they work during nonbusiness hours has not changed since 2016)

H_1: $\pi \neq 0.45$ (the proportion of American workers who reported that they work during nonbusiness hours has changed since 2016)

Because the firm seeks to determine whether the population proportion of American workers who reported that they work during nonbusiness hours has changed from 0.45 since 2016, the firm uses a two-tail test with either the critical value approach or the *p*-value approach.

Using the Critical Value Approach

The firm selects an $\alpha = 0.05$ level of significance that defines the rejection and nonrejection regions that Figure 9.13 visualizes. The decision rule is

Reject H_0 if $Z_{STAT} < -1.96$ or if $Z_{STAT} > +1.96$;

otherwise, do not reject H_0.

FIGURE 9.13
Two-tail test of hypothesis for the proportion at the 0.05 level of significance

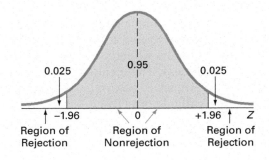

0.025   0.95   0.025

−1.96   0   +1.96   *Z*

Region of   Region of   Region of
Rejection   Nonrejection   Rejection

Having determined the null and alternate hypotheses and the Z test statistic, the research firm first calculates the sample proportion, p. In the new survey, 208 of 400 American workers reported that they work during nonbusiness hours, which makes $p = 0.52$ ($208 \div 400$). Because $X = 208$ and $n - X = 192$, and each value is greater than 5, either Equation (9.3) or (9.4) can be used:

$$Z_{STAT} = \frac{p - \pi}{\sqrt{\dfrac{\pi(1 - \pi)}{n}}} = \frac{0.52 - 0.45}{\sqrt{\dfrac{0.45(1 - 0.45)}{400}}} = \frac{0.0700}{0.0249} = 2.8141$$

$$Z_{STAT} = \frac{X - n\pi}{\sqrt{n\pi(1 - \pi)}} = \frac{208 - (400)(0.45)}{\sqrt{(400)(0.45)(0.55)}} = \frac{28}{9.9499} = 2.8141$$

The Z test statistic can also be determined by software results, as the Figure 9.14 Excel results for this test illustrate.

Because $Z_{STAT} = 2.8141 > 1.96$, the firm rejects H_0. There is evidence that the population proportion of American workers who reported that they work during nonbusiness hours has changed from 0.46 since 2016.

FIGURE 9.14

Excel Z test results for whether the proportion of American workers who reported that they work during nonbusiness hours has changed from 0.45 since 2016

	A	B	
1	Z Test of Hypothesis for the Proportion		
2			
3	Data		
4	Null Hypothesis $\pi=$	0.45	
5	Level of Significance	0.05	
6	Number of Items of Interest	208	
7	Sample Size	400	
8			
9	Intermediate Calculations		
10	Sample Proportion	0.5200	=B6/B7
11	Standard Error	0.0249	=SQRT(B4 * (1 - B4)/B7)
12	Z Test Statistic	2.8141	=(B10 - B4)/B11
13			
14	Two-Tail Test		
15	Lower Critical Value	-1.9600	=NORM.S.INV(B5/2)
16	Upper Critical value	1.9600	=NORM.S.INV(1 - B5/2)
17	p-Value	0.0049	=2 * (1 - NORM.S.DIST(ABS(B12), TRUE))
18	Reject the null hypothesis		=IF(B17 < B5,"Reject the null hypothesis",)
			"Do not reject the null hypothesis")

Using the *p*-Value Approach

Having determined the null and alternate hypotheses, the research firm uses the Figure 9.14 software results to determine the Z value and the p-value $= 0.0049$. Because this value is less than the selected level of significance ($\alpha = 0.05$), the firm rejects the null hypothesis.

Example 9.6 illustrates a one-tail test for a proportion.

EXAMPLE 9.6

Testing a Hypothesis for a Proportion

▶(*continued*)

In addition to the business problem of the speed of service at the drive-through, quick-service restaurant chains want to fill orders correctly. The same study that reported that McDonald's had a drive-through service time of 208.16 seconds also reported that McDonald's filled 92.9% of its drive-through orders correctly (Oches). Suppose that McDonald's implements a new procedure to ensure that orders at the drive-through are filled correctly and seeks to determine whether the new process can increase the percentage of orders filled correctly. Data are collected from a sample of 500 orders using the new process. The results indicate that 476 orders were filled correctly. At the 0.01 level of significance, can you conclude that the new process has increased the proportion of orders filled correctly?

SOLUTION The null and alternative hypotheses are

H_0: $\pi \leq 0.929$ (the population proportion of orders filled correctly using the new process is less than or equal to 0.929)

H_1: $\pi > 0.929$ (the population proportion of orders filled correctly using the new process is greater than 0.929)

Because $X = 476$ and $n - X = 24$, both > 5, using Equation (9.3) on page 299,

$$p = \frac{X}{n} = \frac{476}{500} = 0.952$$

$$Z_{STAT} = \frac{p - \pi}{\sqrt{\frac{\pi(1 - \pi)}{n}}} = \frac{0.952 - 0.929}{\sqrt{\frac{0.929(1 - 0.929)}{500}}} = \frac{0.0230}{0.0115} = 2.00$$

The *p*-value $Z_{STAT} > 2.00$ is 0.0228.

Using the critical value approach, you reject H_0 if $Z_{STAT} > 2.33$. Using the *p*-value approach, you reject H_0 if the *p*-value < 0.01. Because $Z_{STAT} = 2.00 < 2.33$ or the *p*-value $= 0.0228 > 0.01$, you DO NOT reject H_0. There is insufficient evidence that the new process has increased the proportion of correct orders above 0.929 or 92.9%. (McDonald's management would not be able to use this result to support a decision to implement the new process at additional stores.)

PROBLEMS FOR SECTION 9.4

LEARNING THE BASICS

9.52 If, in a random sample of 400 items, 88 are defective, what is the sample proportion of defective items?

9.53 In Problem 9.52, if the null hypothesis is that 20% of the items in the population are defective, what is the value of Z_{STAT}?

9.54 In Problems 9.52 and 9.53, suppose you are testing the null hypothesis H_0: $\pi = 0.20$ against the two-tail alternative hypothesis H_1: $\pi \neq 0.20$ and you choose the level of significance $\alpha = 0.05$. What is your statistical decision?

APPLYING THE CONCEPTS

9.55 According to a recent National Association of Colleges and Employers (NACE) report, 44% of college students who had unpaid internships received full-time job offers post-graduation compared to 72% of college students who had paid internships.

Source: Data extracted from "Here's Why You May Want to Rethink That Unpaid Internship," available at **for.tn/29CAnU9**.

A recent survey of 60 college unpaid interns at a local university found that 30 received full-time job offers post-graduation.
a. Use the six-step *p*-value approach to hypothesis testing and a 0.05 level of significance to determine whether the proportion of college unpaid interns that received full-time job offers post-graduation is different from 0.44.
b. Assume that the study found that 35 of the 60 college unpaid interns had received full-time job offers post-graduation and repeat (a). Are the conclusions the same?

9.56 The worldwide market share for the Chrome web browser was 69.52% in March 2019.

Source: Data extracted from **netmarketshare.com**.

Suppose that you decide to select a sample of 100 students at your university and you find that 60 use the Chrome web browser.
a. Use the six-step *p*-value approach to determine whether there is evidence that the market share for the Chrome web browser at your university is less than the worldwide market share of 69.52%. (Use the 0.05 level of significance.)
b. Suppose that the sample size is $n = 600$, and you find that 60% of the sample of students at your university (360 out of 600) use the Chrome web browser. Use the six-step *p*-value approach to try to determine whether there is evidence that the market share for the Chrome web browser at your university is greater than the worldwide market share of 69.52%. (Use the 0.05 level of significance.)
c. Discuss the effect that sample size has on hypothesis testing.
d. What do you think are your chances of rejecting any null hypothesis concerning a population proportion if a sample size of $n = 20$ is used?

9.57 One of the issues facing organizations is increasing diversity throughout an organization. One of the ways to evaluate an organization's success at increasing diversity is to compare the percentage of employees in the organization in a particular position with a specific background to the percentage in a particular position with that specific background in the general workforce. Recently, a large academic medical center determined that 9 of 17 employees in a particular position were female, whereas 55% of the employees for this position in the general workforce were female. At the 0.05 level of significance, is there evidence that the proportion of females in this position at this medical center is different from what would be expected in the general workforce?

✓ SELF TEST **9.58** What are companies' biggest obstacles to attracting the best talent? Of 703 surveyed U.S. and Canadian talent acquisition professionals, 464 reported that competition for talent is the biggest obstacle at their company.

Source: *U.S. and Canadian Recruiting Trends 2017*, LinkedIn Talent Solutions, **bit.ly/2s2S6Mc**.

At the 0.05 level of significance, is there evidence that the proportion of all talent acquisition professionals who report competition is the biggest obstacle to attracting the best talent at their company is different from 60%?

9.59 A cellphone provider has the business objective of wanting to determine the proportion of subscribers who would upgrade to a new cellphone with improved features if it were made available at a substantially reduced cost. Data are collected from a random sample of 500 subscribers. The results indicate that 135 of the subscribers would upgrade to a new cellphone at a reduced cost.

a. At the 0.05 level of significance, is there evidence that more than 20% of the customers would upgrade to a new cellphone at a reduced cost?

b. How would the manager in charge of promotional programs concerning residential customers use the results in (a)?

9.60 Actuation Consulting conducted a global survey of product teams with the goal of better understanding the dynamics of product team performance and uncovering the practices that make these teams successful. Having a clear definition of "done" is a basic element of successful product management process. One of the survey findings was that 29.4% of organizations indicated that a collective decision by the product team established this important definition of "done."

Source: *The Study of Product Team Performance, 2016*, available at **bit.ly/2rAGhMT**.

Suppose another study is conducted to check the validity of this result, with the goal of proving that the percentage is less than 29.4%.

a. State the null and alternate hypotheses.

b. A sample of 100 organizations is selected, and results show that 27 indicated that a collective decision by the product team established this important definition of "done." Use either the six-step critical value hypothesis testing approach or the six-step p-value approach to determine at the 0.05 level of significance whether there is evidence that the percentage is less than 29.4%.

9.5 Potential Hypothesis-Testing Pitfalls and Ethical Issues

Using hypothesis testing comes with potential pitfalls and raises the ethical issues that this section summarizes.

Important Planning Stage Questions

When using hypothesis testing with data collected from a survey, research study, or a designed experiment, one must answer the questions that Exhibit 9.5 lists.

EXHIBIT 9.5

Questions for the Planning Stage of Hypothesis Testing

1. What is the goal of the survey, study, or experiment? How can you translate the goal into a null hypothesis and an alternative hypothesis?
2. Is the hypothesis test a two-tail test or one-tail test?
3. Can you select a random sample from the underlying population of interest?
4. What types of data will you collect in the sample? Are the variables numerical or categorical?
5. At what level of significance should you conduct the hypothesis test?
6. Is the intended sample size large enough to achieve the desired power of the test for the level of significance chosen?
7. Which statistical test procedure should you use and why?
8. What conclusions and interpretations can you reach from the results of the hypothesis test?

Failing to consider these questions early in the planning process can lead to biased or incomplete results. Proper planning can help ensure that the statistical study will provide objective information needed to make good business decisions.

Statistical Significance Versus Practical Significance

One must make a distinction between the existence of a statistically significant result and its practical significance in a field of application. Sometimes, due to a very large sample size, one may get a result that is statistically significant but has little practical significance.

For example, suppose that prior to a national marketing campaign focusing on a series of expensive television commercials, one believes that the proportion of people who recognize a specific brand is 0.30. At the completion of the campaign, a survey of 20,000 people indicates that 6,168 recognized that brand. A one-tail test trying to prove that the proportion is now greater than 0.30 results in a p-value of 0.0048, and the correct statistical conclusion is that the proportion of consumers recognizing that brand name has now increased.

Was the campaign successful? The result of the hypothesis test indicates a statistically significant increase in brand awareness, but is this increase practically important? The population proportion is now estimated at $6,168/20,000 = 0.3084 = 0.3084$ or 30.84%. This increase is less than 1% above the hypothesized value of 30%. Did the large expenses associated with the marketing campaign produce a result with a meaningful increase in brand awareness? Because of the minimal real-world impact that an increase of less than 1% has on the overall marketing strategy and the huge expenses associated with the marketing campaign, one should conclude that the campaign was *not* successful. On the other hand, if the campaign increased brand awareness from 30% to 50%, one could conclude that the campaign was successful.

Statistical *Insignificance* Versus Importance

Some results may be important in a business sense even when the results are not statistically significance. In *Matrixx Initiatives, Inc. v. Siracusano*, the U.S. Supreme Court ruled that companies cannot rely solely on whether the result of a study is statistically significant when determining what study results they communicate to investors (Bialik).

In some situations, the lack of a sufficiently large sample size may result in a nonsignificant result when in fact an important difference does exist (Seaman 2011). A study that compared male and female entrepreneurship rates globally and within Massachusetts found a significant difference globally but not within Massachusetts, even though the entrepreneurship rates for females and for males in the two geographic areas were similar (8.8% for males in Massachusetts as compared to 8.4% globally; 5% for females in both geographic areas). The difference was due to the fact that the global sample size was 20 times larger than the Massachusetts sample size.

Reporting of Findings

In conducting research, one should document both good and bad results. One should not just report the results of hypothesis tests that show statistical significance but omit those for which there is insufficient evidence in the findings. In instances in which there is insufficient evidence to reject H_0, one must make it clear that this does not prove that the null hypothesis is true. What the result indicates is that with the sample size used, there is not enough information to *disprove* the null hypothesis.

Ethical Issues

One needs to distinguish between poor research methodology and unethical behavior. Ethical considerations arise when the hypothesis-testing process is manipulated. Some of the areas where ethical issues can arise include the use of human subjects in experiments, the data collection method, the type of test (one-tail or two-tail test), the choice of the level of significance, the cleansing and discarding of data, and the failure to report pertinent findings.

9.6 Power of the Test

The power of a hypothesis test is the probability that one correctly rejects a false null hypothesis. The power of the test is affected by the level of significance, the sample size, and whether the test is one-tail or two-tail. The **Section 9.6 online topic** further explains the power of the test and illustrates its use.

▼ USING **STATISTICS**
Significant Testing..., Revisited

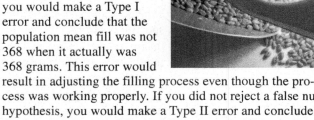

A s the plant operations manager for Oxford Cereals, you were responsible for the cereal-filling process. It was your responsibility to adjust the process when the mean fill-weight in the population of boxes deviated from the company specification of 368 grams. You chose to conduct a hypothesis test.

You determined that the null hypothesis should be that the population mean fill was 368 grams. If the mean weight of the sampled boxes was sufficiently above or below the expected 368-gram mean specified by Oxford Cereals, you would reject the null hypothesis in favor of the alternative hypothesis that the mean fill was different from 368 grams. If this happened, you would stop production and take whatever action was necessary to correct the problem. If the null hypothesis was not rejected, you would continue to believe in the status quo—that the process was working correctly—and therefore take no corrective action.

Before proceeding, you considered the risks involved with hypothesis tests. If you rejected a true null hypothesis, you would make a Type I error and conclude that the population mean fill was not 368 when it actually was 368 grams. This error would result in adjusting the filling process even though the process was working properly. If you did not reject a false null hypothesis, you would make a Type II error and conclude that the population mean fill was 368 grams when it actually was not 368 grams. Here, you would allow the process to continue without adjustment even though the process was not working properly.

After collecting a random sample of 25 cereal boxes, you used either the six-step critical value or *p*-value approaches to hypothesis testing. Because the test statistic fell into the nonrejection region, you did not reject the null hypothesis. You concluded that there was insufficient evidence to prove that the mean fill differed from 368 grams. No corrective action on the filling process was needed.

▼ SUMMARY

Table 9.5 lists the hypothesis tests that this chapter discusses. This chapter presents the foundations of hypothesis testing. The chapter discusses how to perform tests on the population mean and on the population proportion. The chapter develops both the critical value approach and the *p*-value approach to hypothesis testing.

In deciding which test to use, one must ask the following question: Does the test involves a numerical variable or a categorical variable? If the test involves a numerical variable, you use the *t* test for the mean. If the test involves a categorical variable, you use the *Z* test for the proportion.

TABLE 9.5
Summary of topics

TYPE OF ANALYSIS	TYPE OF DATA	
	Numerical	**Categorical**
Hypothesis test concerning a single parameter	*Z* test of hypothesis for the mean (Section 9.1) *t* test of hypothesis for the mean (Section 9.2)	*Z* test of hypothesis for the proportion (Section 9.4)

▼ REFERENCES

Bialik, C. "Making a Stat Less Significant." *The Wall Street Journal*, April 2, 2011, A5.

Bradley, J. V. *Distribution-Free Statistical Tests*. Upper Saddle River, NJ: Prentice Hall, 1968.

Daniel, W. *Applied Nonparametric Statistics*, 2nd ed. Boston: Houghton Mifflin, 1990.

Oches, S. "The 2018 QSR Drive-Thru Study." *QSR*, **bit.ly/2Adr1cz**.

Seaman, J., and E. Allen. "Not Significant, But Important?" *Quality Progress*, August 2011, 57–59.

Seaman, J., and E. Allen. "The Significance of Power." *Quality Progress*, July 2015, 51–53.

▼KEY EQUATIONS

Z Test for the Mean (σ Known)

$$Z_{STAT} = \frac{\overline{X} - \mu}{\dfrac{\sigma}{\sqrt{n}}} \qquad (9.1)$$

t Test for the Mean (σ Unknown)

$$t_{STAT} = \frac{\overline{X} - \mu}{\dfrac{S}{\sqrt{n}}} \qquad (9.2)$$

Z Test for the Proportion

$$Z_{STAT} = \frac{p - \pi}{\sqrt{\dfrac{\pi(1 - \pi)}{n}}} \qquad (9.3)$$

Z Test for the Proportion in Terms of the Number of Events of Interest

$$Z_{STAT} = \frac{X - n\pi}{\sqrt{n\pi(1 - \pi)}} \qquad (9.4)$$

▼KEY TERMS

alternative hypothesis (H_1)   276
β risk   279
confidence coefficient   279
critical value   278
directional test   295
hypothesis testing   276
level of significance (α)   279
null hypothesis (H_0)   276

one-tail test   295
p-value   284
power of a statistical test   279
region of nonrejection   278
region of rejection   278
robust   292
sample proportion   298

t test for the mean   288
test statistic   278
two-tail test   281
Type I error   278
Type II error   278
Z test for the mean   280
Z test for the proportion   299

▼CHECKING YOUR UNDERSTANDING

9.61 What is the difference between a null hypothesis, H_0, and an alternative hypothesis, H_1?

9.62 What is the difference between a Type I error and a Type II error?

9.63 What is meant by the power of a test?

9.64 What is the difference between a one-tail test and a two-tail test?

9.65 What is meant by a p-value?

9.66 How can a confidence interval estimate for the population mean provide conclusions for the corresponding two-tail hypothesis test for the population mean?

9.67 What is the six-step critical value approach to hypothesis testing?

9.68 What is the six-step p-value approach to hypothesis testing?

▼CHAPTER REVIEW PROBLEMS

9.69 In hypothesis testing, the common level of significance is $\alpha = 0.05$. Some might argue for a level of significance greater than 0.05. Suppose that web designers tested the proportion of potential web page visitors with a preference for a new web design over the existing web design. The null hypothesis was that the population proportion of web page visitors preferring the new design was 0.60, and the alternative hypothesis was that it was not equal to 0.60. The p-value for the test was 0.20.
a. State, in statistical terms, the null and alternative hypotheses for this example.
b. Explain the risks associated with Type I and Type II errors in this case.
c. What would be the consequences if you rejected the null hypothesis for a p-value of 0.20?

d. What might be an argument for raising the value of α?
e. What would you do in this situation?
f. What is your answer in (e) if the p-value equals 0.12? What if it equals 0.01?

9.70 Financial institutions utilize prediction models to predict bankruptcy. One such model is the Altman Z-score model, which uses multiple corporate income and balance sheet values to measure the financial health of a company. If the model predicts a low Z-score value, the firm is in financial stress and is predicted to go bankrupt within the next two years. If the model predicts a moderate or high Z-score value, the firm is financially healthy and is predicted to be a nonbankrupt firm. This decision-making procedure can be expressed in the hypothesis-testing framework. The null hypothesis is that a firm is predicted to be a nonbankrupt firm. The alternative hypothesis is that the firm is predicted to be a bankrupt firm.

a. Explain the risks associated with committing a Type I error in this case.
b. Explain the risks associated with committing a Type II error in this case.
c. Which type of error do you think executives want to avoid? Explain.
d. How would changes in the model affect the probabilities of committing Type I and Type II errors?

9.71 IAB conducted a study of 821 U.S. adults to understand the behavioral shift of consumers' TV viewing experience. The study found that 460 of U.S. adults own streaming enabled TVs, including smart TVs and video streaming devices.

Source: *The Changing TV Experience: 2017,* available at **bit.ly/2sz4Mal**.

The authors of the report imply that the survey proves that more than half of all U.S. adults own streaming enabled TVs, including smart TVs and video streaming devices.

a. Use the six-step *p*-value approach to hypothesis testing and a 0.05 level of significance to try to prove that more than half of all U.S. adults own streaming enabled TVs, including smart TVs and video streaming devices.
b. Based on your result in (a), is the claim implied by the authors valid?
c. Suppose the study found that 428 of U.S. adults own streaming enabled TVs, including smart TVs and video streaming devices. Repeat parts (a) and (b).
d. Compare the results of (b) and (c).

9.72 The owner of a specialty coffee shop wants to study coffee purchasing habits of customers at her shop. She selects a random sample of 60 customers during a certain week, with the following results:

- The amount spent was $\overline{X} = \$7.25, S = \$1.75$.
- Thirty-one customers say they "definitely will" recommend the specialty coffee shop to family and friends.

a. At the 0.05 level of significance, is there evidence that the population mean amount spent was different from $6.50?
b. Determine the *p*-value in (a).
c. At the 0.05 level of significance, is there evidence that more than 50% of all the customers say they "definitely will" recommend the specialty coffee shop to family and friends?
d. What is your answer to (a) if the sample mean equals $6.25?
e. What is your answer to (c) if 39 customers say they "definitely will" recommend the specialty coffee shop to family and friends?

9.73 An auditor for a government agency was assigned the task of evaluating reimbursement for office visits to physicians paid by Medicare. The audit was conducted on a sample of 75 reimbursements, with the following results:

- In 17 of the office visits, there was an incorrect amount of reimbursement.
- The amount of reimbursement was $\overline{X} = \$93.70, S = \$34.55$.

a. At the 0.05 level of significance, is there evidence that the population mean reimbursement was less than $100?
b. At the 0.05 level of significance, is there evidence that the proportion of incorrect reimbursements in the population was greater than 0.10?
c. Discuss the underlying assumptions of the test used in (a).
d. What is your answer to (a) if the sample mean equals $90?
e. What is your answer to (b) if 15 office visits had incorrect reimbursements?

9.74 A bank branch located in a commercial district of a city has the business objective of improving the process for serving customers during the noon-to-1:00 p.m. lunch period. The waiting time (defined as the time the customer enters the line until he or she reaches the teller window) of a random sample of 15 customers is collected, and the results are organized and stored in **Bank Waiting**. These data are:

4.21	5.55	3.02	5.13	4.77	2.34	3.54	3.20
4.50	6.10	0.38	5.12	6.46	6.19	3.79	

a. At the 0.05 level of significance, is there evidence that the population mean waiting time is less than 5 minutes?
b. What assumption about the population distribution is needed in order to conduct the *t* test in (a)?
c. Construct a boxplot or a normal probability plot to evaluate the assumption made in (b).
d. Do you think that the assumption needed in order to conduct the *t* test in (a) is valid? Explain.
e. As a customer walks into the branch office during the lunch hour, she asks the branch manager how long she can expect to wait. The branch manager replies, "Almost certainly not longer than 5 minutes." On the basis of the results of (a), evaluate this statement.

9.75 Call centers today play an important role in managing day-to-day business communications with customers. It's important, therefore, to monitor a comprehensive set of metrics, which can help businesses understand the overall performance of a call center. One key metric for measuring overall call center performance is service level, which is defined as the percentage of calls answered by a human agent within a specified number of seconds. The file **Service Level** contains the following data for time, in seconds, to answer 50 incoming calls to a financial services call center:

16	14	16	19	6	14	15	5	16	18	17	22	6	18	10
15	12	6	19	16	16	15	13	25	9	17	12	10	5	15
23	11	12	14	24	9	10	13	14	26	19	20	13	24	28
15	21	8	16	12										

a. At the 0.05 level of significance, is there evidence that the population mean time to answer calls is less than 20 seconds?
b. What assumption about the population distribution is needed in order to conduct the *t* test in (a)?
c. Construct a histogram, boxplot, or normal probability plot to evaluate the assumption made in (b).
d. Do you think that the assumption needed in order to conduct the *t* test in (a) is valid? Explain.

9.76 An important quality characteristic used by the manufacturer of Boston and Vermont asphalt shingles is the amount of moisture the shingles contain when they are packaged. Customers may feel that they have purchased a product lacking in quality if they find moisture and wet shingles inside the packaging. In some cases, excessive moisture can cause the granules attached to the shingles for texture and coloring purposes to fall off the shingles, resulting in appearance problems. To monitor the amount of moisture present, the company conducts moisture tests. A shingle is weighed and then dried. The shingle is then reweighed, and based on the amount of moisture taken out of the product, the pounds of moisture per 100 square feet are calculated. The company would like to show that the mean moisture content is less than 0.35 pound per 100 square feet. The file **Moisture** includes 36 measurements (in pounds per 100 square feet) for Boston shingles and 31 for Vermont shingles.

a. For the Boston shingles, is there evidence at the 0.05 level of significance that the population mean moisture content is less than 0.35 pound per 100 square feet?
b. Interpret the meaning of the *p*-value in (a).
c. For the Vermont shingles, is there evidence at the 0.05 level of significance that the population mean moisture content is less than 0.35 pound per 100 square feet?
d. Interpret the meaning of the *p*-value in (c).
e. What assumption about the population distribution is needed in order to conduct the *t* tests in (a) and (c)?
f. Construct histograms, boxplots, or normal probability plots to evaluate the assumption made in (a) and (c).
g. Do you think that the assumption needed in order to conduct the *t* tests in (a) and (c) is valid? Explain.

9.77 Studies conducted by the manufacturer of Boston and Vermont asphalt shingles have shown product weight to be a major factor in the customer's perception of quality. Moreover, the weight represents the amount of raw materials being used and is therefore very important to the company from a cost standpoint. The last stage of the assembly line packages the shingles before the packages are placed on wooden pallets. Once a pallet is full (a pallet for most brands holds 16 squares of shingles), it is weighed, and the measurement is recorded. The file Pallet contains the weight (in pounds) from a sample of 368 pallets of Boston shingles and 330 pallets of Vermont shingles.
a. For the Boston shingles, is there evidence at the 0.05 level of significance that the population mean weight is different from 3,150 pounds?
b. Interpret the meaning of the *p*-value in (a).
c. For the Vermont shingles, is there evidence at the 0.05 level of significance that the population mean weight is different from 3,700 pounds?
d. Interpret the meaning of the *p*-value in (c).

e. In (a) through (d), do you have to be concerned with the normality assumption? Explain.

9.78 The manufacturer of Boston and Vermont asphalt shingles provides its customers with a 20-year warranty on most of its products. To determine whether a shingle will last through the warranty period, accelerated-life testing is conducted at the manufacturing plant. Accelerated-life testing exposes the shingle to the stresses it would be subject to in a lifetime of normal use in a laboratory setting via an experiment that takes only a few minutes to conduct. In this test, a shingle is repeatedly scraped with a brush for a short period of time, and the shingle granules removed by the brushing are weighed (in grams). Shingles that experience low amounts of granule loss are expected to last longer in normal use than shingles that experience high amounts of granule loss. The file Granule contains a sample of 170 measurements made on the company's Boston shingles and 140 measurements made on Vermont shingles.
a. For the Boston shingles, is there evidence at the 0.05 level of significance that the population mean granule loss is different from 0.30 grams?
b. Interpret the meaning of the *p*-value in (a).
c. For the Vermont shingles, is there evidence at the 0.05 level of significance that the population mean granule loss is different from 0.30 grams?
d. Interpret the meaning of the *p*-value in (c).
e. In (a) through (d), do you have to be concerned with the normality assumption? Explain.

REPORT WRITING EXERCISE

9.79 Referring to the results of Problems 9.76 through 9.78 concerning Boston and Vermont shingles, write a report that evaluates the moisture level, weight, and granule loss of the two types of shingles.

CHAPTER

▼CASES

9

Managing Ashland MultiComm Services

Continuing its monitoring of the upload speed first described in the Chapter 6 Managing Ashland MultiComm Services case on page 221, the technical operations department wants to ensure that the mean target upload speed for all Internet service subscribers is at least 0.97 on a standard scale in which the target value is 1.0. Each day, upload speed was measured 50 times, with the following results (stored in AMS9).

```
0.854 1.023 1.005 1.030 1.219 0.977 1.044 0.778 1.122 1.114
1.091 1.086 1.141 0.931 0.723 0.934 1.060 1.047 0.800 0.889
1.012 0.695 0.869 0.734 1.131 0.993 0.762 0.814 1.108 0.805
1.223 1.024 0.884 0.799 0.870 0.898 0.621 0.818 1.113 1.286
1.052 0.678 1.162 0.808 1.012 0.859 0.951 1.112 1.003 0.972
```

1. Compute the sample statistics and determine whether there is evidence that the population mean upload speed is less than 0.97.
2. Write a memo to management that summarizes your conclusions.

Digital Case

Apply your knowledge about hypothesis testing in this Digital Case, which continues the cereal-fill-packaging dispute first discussed in the Digital Case from Chapter 7.

In response to the negative statements made by the Concerned Consumers About Cereal Cheaters (CCACC) in the Chapter 7 Digital Case, Oxford Cereals recently conducted an experiment concerning cereal packaging. The company claims that the results of the experiment refute the CCACC allegations that Oxford Cereals has been cheating consumers by packaging cereals at less than labeled weights.

Open **OxfordCurrentNews.pdf**, a portfolio of current news releases from Oxford Cereals. Review the relevant press releases and supporting documents. Then answer the following questions:

1. Are the results of the experiment valid? Why or why not? If you were conducting the experiment, is there anything you would change?

2. Do the results support the claim that Oxford Cereals is not cheating its customers?

3. Is the claim of the Oxford Cereals CEO that many cereal boxes contain *more* than 368 grams surprising? Is it true?

4. Could there ever be a circumstance in which the results of the Oxford Cereals experiment *and* the CCACC's results are both correct? Explain.

Sure Value Convenience Stores

You work in the corporate office for a nationwide convenience store franchise that operates nearly 10,000 stores. The per-store daily customer count (defined as the mean number of customers in a store in one day) has been steady, at 900, for some time.

To increase the customer count, the chain is considering cutting prices for coffee beverages. The small size will now be $0.59 instead of $0.99, and the medium size will be $0.69 instead of $1.19. Even with this reduction in price, the chain will have a 40% gross margin on coffee.

To test the new initiative, the chain has reduced coffee prices in a sample of 34 stores, where customer counts have been running almost exactly at the national average of 900. After four weeks, the stores sampled stabilize at a mean customer count of 974 and a standard deviation of 96. This increase seems like a substantial amount to you, but it also seems like a pretty small sample. Is there statistical evidence that reducing coffee prices is a good strategy for increasing the mean customer count? Be prepared to explain your conclusion.

▾EXCEL GUIDE

EG9.1 FUNDAMENTALS of HYPOTHESIS TESTING

Key Technique Use the **NORM.S.INV**(*level of significance/2*) and **NORM.S.INV**(1 – *level of significance/2*) functions to compute the lower and upper critical values.

Use **NORM.S.DIST** (*absolute value of the Z test statistic*, **True**) as part of a formula to compute the *p*-value.

Example Perform the Figure 9.5 two-tail *Z* test for the mean for the cereal-filling example on page 285.

PHStat Use **Z Test for the Mean, sigma known**.

For the example, select **PHStat → One-Sample Tests → Z Test for the Mean, sigma known**. In the procedure's dialog box (shown below):

1. Enter **368** as the **Null Hypothesis**.
2. Enter **0.05** as the **Level of Significance**.
3. Enter **15** as the **Population Standard Deviation**.
4. Click **Sample Statistics Known** and enter **25** as the **Sample Size** and **372.5** as the **Sample Mean**.
5. Click **Two-Tail Test**.
6. Enter a **Title** and click **OK**.

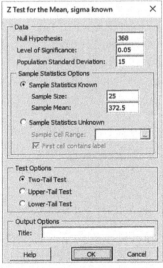

When using unsummarized data, click **Sample Statistics Unknown** in step 4 and enter the cell range of the unsummarized data as the **Sample Cell Range**.

Workbook Use the **COMPUTE worksheet** of the **Z Mean workbook** as a template.

The worksheet already contains the data for the example. For other problems, change the null hypothesis, level of significance, population standard deviation, sample size, and sample mean values in cells B4 through B8 as necessary.

EG9.2 *t* TEST of HYPOTHESIS for the MEAN (σ UNKNOWN)

Key Technique Use the **T.INV.2T**(*level of significance, degrees of freedom*) function to compute the lower and upper critical values.

Use **T.DIST.2T**(*absolute value of the t test statistic, degrees of freedom*) to compute the *p*-value.

Example Perform the Figure 9.7 two-tail *t* test for the mean for the sales invoices example on page 290.

PHStat Use **t Test for the Mean, sigma unknown**.

For the example, select **PHStat → One-Sample Tests → t Test for the Mean, sigma unknown**. In the procedure's dialog box (shown below):

1. Enter **120** as the **Null Hypothesis**.
2. Enter **0.05** as the **Level of Significance**.
3. Click **Sample Statistics Known** and enter **12** as the **Sample Size**, **112.85** as the **Sample Mean**, and **20.8** as the **Sample Standard Deviation**.
4. Click **Two-Tail Test**.
5. Enter a **Title** and click **OK**.

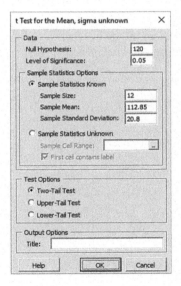

When using unsummarized data, click **Sample Statistics Unknown** in step 3 and enter the cell range of the unsummarized data as the **Sample Cell Range**.

Workbook Use the **COMPUTE worksheet** of the **T mean workbook**, as a template.

The worksheet already contains the data for the example. For other problems, change the values in cells B4 through B8 as necessary.

EG9.3 ONE-TAIL TESTS

Key Technique Use the **NORM.S.INV** with (level of significance) and (*1 – level of significance*) to compute the lower and upper critical values.

Use **NORM.S.DIST(Z** *test statistic*, **True)** and **1 – NORM.S.DIST(Z** *test statistic*, **True)** to compute the lower-tail and upper-tail *p*-values.

Key Technique (t test for the mean) Use the **–T.INV.2T** and **T.INV.2T** functions with (*2 * level of significance, degrees of freedom*) to compute the lower and upper critical values.

Use an IF function that tests the *t* test statistic to determine whether **T.DIST.RT(***absolute value of the t test statistic, degrees of freedom***)** or **1 – T.DIST.RT(***absolute value of the t test statistic, degrees of freedom***)** computes the *p*-value.

Example Perform the Figure 9.11 lower-tail *t* test for the mean for the drive-through time study example on page 296.

PHStat Click either **Lower-Tail Test** or **Upper-Tail Test** in the procedure dialog boxes discussed in Sections EG9.1 and EG9.2 to perform a one-tail test.

For the example, select **PHStat → One-Sample Tests → t Test for the Mean, sigma unknown**. In the procedure's dialog box (shown below):

1. Enter **208.16** as the **Null Hypothesis**.
2. Enter **0.05** as the **Level of Significance**.
3. Click **Sample Statistics Known** and enter **25** as the **Sample Size**, **195.6** as the **Sample Mean**, and **22.1** as the **Sample Standard Deviation**.
4. Click **Lower-Tail Test**.
5. Enter a **Title** and click **OK**.

Workbook Use the **COMPUTE_LOWER worksheet** or the **COMPUTE_UPPER worksheet** of the **Z Mean** and **T mean workbooks** as templates.

For the example, open to the **COMPUTE_LOWER worksheet** of the **T mean workbook**. For other problems that require a *t* test, open to the appropriate worksheet and change the **Null Hypothesis, Level of Significance, Sample Size, Sample Mean**, and **Sample Standard Deviation** in the cell range B4:B8.

For other problems that require a *Z* test, open to the appropriate worksheet and change the **Null Hypothesis, Level of Significance, Population Standard Deviation, Sample Size**, and **Sample Mean** in the cell range B4:B8.

To see the all of the formulas used in the one-tail test worksheets, open to the COMPUTE_ALL_FORMULAS worksheet.

EG9.4 Z TEST of HYPOTHESIS for the PROPORTION

Key Technique Use the **NORM.S.INV(***level of significance/2***)** and **NORM.S.INV(1 – *level of significance/2***)** functions to compute the lower and upper critical values.

Use **NORM.S.DIST(***absolute value of the Z test statistic*, **True)** as part of a formula to compute the *p*-value.

Example Perform the Figure 9.14 two-tail *Z* test for the proportion for whether the proportion of American workers who reported that they work during nonbusiness hours has changed on page 300.

PHStat Use **Z Test for the Proportion**.

For the example, select **PHStat → One-Sample Tests → Z Test for the Proportion**. In the procedure's dialog box (shown below):

1. Enter **0.45** as the **Null Hypothesis**.
2. Enter **0.05** as the **Level of Significance**.
3. Enter **208** as the **Number of Items of Interest**.
4. Enter **400** as the **Sample Size**.
5. Click **Two-Tail Test**.
6. Enter a **Title** and click **OK**.

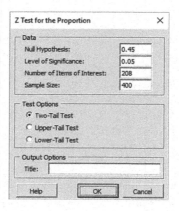

Workbook Use the **COMPUTE worksheet** of the **Z Proportion workbook** as a template.

The worksheet already contains the data for the example. For other problems, change the null hypothesis, level of significance, population standard deviation, sample size, and sample mean values in cells B4 through B7 as necessary.

Use the COMPUTE_LOWER or COMPUTE_UPPER worksheets as templates for performing one-tail tests.

Two-Sample Tests

OBJECTIVES

- Compare the means
or proportions of two
independent populations
- Compare the means of
two related populations
- Compare the variances
of two independent
populations

▼ USING **STATISTICS**
Differing Means for Selling Streaming Media Players at Arlingtons?

To what extent does the location of products in a store affect sales? At Arlingtons, a general merchandiser that competes with discount and wholesale club retailers, management has been considering this question as part of a general review. Seeking to enhance revenues, managers have decided to create a special sales area at the front of the each store. Arlingtons plans to charge product manufacturers a placement fee for placing specific products in this front area, but first needs to demonstrate that the area would boost sales.

While some manufacturers refuse to pay such placement fees, Arlingtons has found a willing partner in Pierrsöhn Technologies. Pierrsöhn wants to introduce VLABGo, their new mobile streaming player, and is willing to pay a placement fee to be featured at the front of each Arlingtons store. However, Pierrsöhn management wants reassurance that the front of the store will be worth the placement fee. As the retail operations chief at Arlingtons, you have been asked to negotiate with Pierrsöhn. You propose a test that will involve 20 Arlingtons locations, all with similar storewide sales volumes and shopper demographics. You explain that you will randomly select 10 stores to sell the VLABGo player among other, similar items in the mobile electronics aisle in those Arlingtons stores. For the other 10 stores, you will place the VLABGo players in a special area at the front of the store.

At the end of the one-month test period, the sales of VLABGo players from the two store samples will be recorded and compared. You wonder how you could determine whether the sales in the in-aisle stores are different from the sales in the stores where the VLABGo players appear in the special front area. You also would like to decide if the variability in sales from store to store is different for the two types of sales location. If you can demonstrate a difference in sales, you will have a stronger case for asking for a special front of the store placement fee from Pierrsöhn. What should you do?

311

C hapter 9 discusses several hypothesis-testing procedures commonly used to test a single sample of data selected from a single population. Hypothesis testing can be extended to **two-sample tests** that compare statistics from samples selected from *two* populations. In the Arlingtons scenario, one such test would be, "Are the mean VLABGo player monthly sales at the special front location (one population) different from the mean VLABGo player monthly sales at the in-aisle location (a second population)?"

10.1 Comparing the Means of Two Independent Populations

Using the correct two-sample test to compare the means of samples selected from each of two independent populations requires first establishing whether the assumption holds that the variances in the two populations are equal. If the assumption holds, you use a *pooled-variance t test*, otherwise you use a *separate-variance t test*. Determining whether the assumption that the two variances are equal can be complicated because when sampling from two independent populations, one almost always does not know the standard deviation of either population, as Sections 8.1 and 9.1 note. However, using the sample variances, one can test whether the two population variances are equal using the method that Section 10.4 discusses.

Pooled-Variance *t* Test for the Difference Between Two Means Assuming Equal Variances

If one assumes that the random samples are independently selected from two populations and that the populations are normally distributed and have equal variances, one can use a **pooled-variance *t* test** to determine whether there is a significant difference between the means. If the populations do not differ greatly from a normal distribution, one can still use the pooled-variance *t* test, especially if the sample sizes are large enough (typically ≥ 30 for each sample).

This *t* test is called *pooled-variance* because the test statistic pools, or combines, the two sample variances S_1^2 and S_2^2 to calculate S_p^2, the best estimate of the variance common to both populations, under the assumption that the two population variances are equal. Equation (10.1) defines the pooled-variance *t* test.[1]

[1] When $n_1 = n_2$, the pooled variance equation simplifies as

$$S_p^2 = \frac{S_1^2 + S_2^2}{2}$$

POOLED-VARIANCE *t* TEST FOR THE DIFFERENCE BETWEEN TWO MEANS

$$t_{STAT} = \frac{(\overline{X}_1 - \overline{X}_2) - (\mu_1 - \mu_2)}{\sqrt{S_p^2\left(\dfrac{1}{n_1} + \dfrac{1}{n_2}\right)}} \tag{10.1}$$

where

S_p^2 = pooled variance

$\overline{X}_1$ = mean of the population 1 sample

μ_1 = mean of population 1

n_1 = size of the population 1 sample

$\overline{X}_2$ = mean of the population 2 sample

μ_2 = mean of population 2

n_2 = size of the population 2 sample

and $S_p^2 = \dfrac{(n_1 - 1)S_1^2 + (n_2 - 1)S_2^2}{(n_1 - 1) + (n_2 - 1)}$

S_1^2 = variance of the population 1 sample

S_2^2 = variance of the population 2 sample

The t_{STAT} test statistic follows a *t* distribution with $n_1 + n_2 - 2$ degrees of freedom.

student TIP

The population defined as population 1 in the null and alternative hypotheses must be defined as population 1 in Equation (10.1). The population defined as population 2 in the null and alternative hypotheses must be defined as population 2 in Equation (10.1).

For the pooled-variance t test, the null and alternative hypotheses of a test for difference in the means of two independent populations can be stated in one of two ways:

$$H_0: \mu_1 = \mu_2 \quad \text{or} \quad \mu_1 - \mu_2 = 0$$
$$H_1: \mu_1 \neq \mu_2 \quad \text{or} \quad \mu_1 - \mu_2 \neq 0$$

For a given level of significance, α, in a two-tail test, one rejects the null hypothesis if the t_{STAT} test statistic is greater than the upper-tail critical value from the t distribution or if the t_{STAT} test statistic is less than the lower-tail critical value from the t distribution (see Figure 10.1).

FIGURE 10.1
Regions of rejection and nonrejection for the pooled-variance t test for the difference between the means (two-tail test)

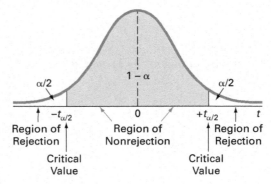

In a one-tail test in which the rejection region is in the lower tail, one rejects the null hypothesis if the t_{STAT} test statistic is less than the lower-tail critical value from the t distribution. In a one-tail test in which the rejection region is in the upper tail, one rejects the null hypothesis if the t_{STAT} test statistic is greater than the upper-tail critical value from the t distribution.

To demonstrate the pooled-variance t test, recall the Using Statistics scenario on page 311. Two populations are of interest: the set of all possible VLABGo player monthly sales at the special front location (population 1) and the set of all possible VLABGo player monthly sales at the in-aisle location (population 2). Using the DCOVA framework, one defines the business objective as determining whether there is a difference in the mean VLABGo player monthly sales at the special front and in-aisle locations. One collects the data from a sample of 10 Arlingtons stores that have been assigned the special front location and another sample of 10 Arlingtons stores that have been assigned the in-aisle location. Table 10.1 presents the organized data, stored in `VLABGo`.

student TIP

A lower or less than comparison indicates a lower-tail test. An upper or more than comparison indicates an upper-tail test. A different or not the same as comparison indicates a two-tail test.

TABLE 10.1
Comparing VLABGo player sales from two different locations

SALES LOCATION									
Special Front					In-Aisle				
224	189	248	285	273	192	236	164	154	189
190	243	215	280	317	220	261	186	219	202

The null and alternative hypotheses are

$$H_0: \mu_1 = \mu_2 \quad \text{or} \quad \mu_1 - \mu_2 = 0$$
$$H_1: \mu_1 \neq \mu_2 \quad \text{or} \quad \mu_1 - \mu_2 \neq 0$$

Assuming that the samples are from normal populations having equal variances, one can use the pooled-variance t test. The t_{STAT} test statistic follows a t distribution with $10 + 10 - 2 = 18$ degrees of freedom. Using an $\alpha = 0.05$ level of significance, you divide the rejection region into two tails of 0.025 each. Table E.3 shows that the critical values for this two-tail test are -2.1009 and $+2.1009$ (see Figure 10.2 on page 314). The decision rule is

$$\text{Reject } H_0 \text{ if } t_{STAT} > +2.1009$$
$$\text{or if } t_{STAT} < -2.1009;$$
$$\text{otherwise, do not reject } H_0.$$

FIGURE 10.2

Two-tail test of hypothesis for the difference between the means at the 0.05 level of significance with 18 degrees of freedom

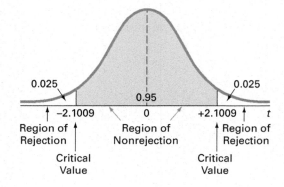

Using Equation (10.1) on page 312 and the Figure 10.3 descriptive statistics,

$$t_{STAT} = \frac{(\bar{X}_1 - \bar{X}_2) - (\mu_1 - \mu_2)}{\sqrt{S_p^2\left(\dfrac{1}{n_1} + \dfrac{1}{n_2}\right)}}$$

where

$$S_p^2 = \frac{(n_1 - 1)S_1^2 + (n_2 - 1)S_2^2}{(n_1 - 1) + (n_2 - 1)} = \frac{9(42.5420)^2 + 9(32.5271)^2}{9 + 9} = 1{,}433.9167$$

the result is

$$t_{STAT} = \frac{(246.4 - 202.3) - 0.0}{\sqrt{1{,}433.9167\left(\dfrac{1}{10} + \dfrac{1}{10}\right)}} = \frac{44.1}{\sqrt{286.7833}} = 2.6041$$

Figure 10.3 shows the Excel results for the two different sales locations data.

FIGURE 10.3

Excel pooled-variance *t* test results with confidence interval estimate for the two different sales locations data

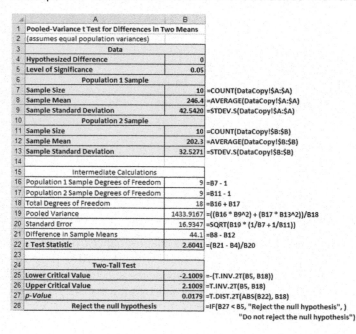

In the Worksheet

The worksheet uses the T.INV.2T functions to compute the upper and lower critical values.

The **COMPUTE worksheet** of the Excel Guide **Polled-Variance T workbook** is a copy of the Figure 10.3 worksheet.

Table 10.2 summarizes the results of the pooled-variance t test for the difference between the two sales locations using the calculations on page 314 and the Figure 10.3 results. Based on the conclusions, the special front location generates significantly higher sales. Therefore, as part of the last step of the DCOVA framework, one can offer a justification for charging a placement fee for the special front location.

TABLE 10.2

Pooled-variance t test summary for the two sales locations

Result	Conclusions
The $t_{STAT} = 2.6041$ is greater than 2.1009.	1. Reject the null hypothesis H_0. 2. Conclude that evidence exists that the mean sales are different for the two sales locations.
The t test p-value $= 0.0179$ is less than the level of significance, $\alpha = 0.05$.	3. The probability of observing a difference in the two sample means this large or larger is 0.0179.
The t_{STAT} is positive.	4. Conclude that the mean sales are higher for the special front location.

Evaluating the Normality Assumption

The pooled-variance t test assumes that the two populations are normally distributed, with equal variances. When the two populations have equal variances, the pooled-variance t test is **robust** (not sensitive) to moderate departures from the assumption of normality, provided that the sample sizes are large. In such situations, one can use the pooled-variance t test without serious effects on its *power*, the probability that one correctly rejects a false null hypothesis. However, for cases in which one cannot assume that both populations are normally distributed, two alternatives exist. One can use a nonparametric procedure, such as the Wilcoxon rank sum test (see Section 12.4), that does not depend on the assumption of normality for the two populations, or use a normalizing transformation on each of the values before using the pooled-variance t test.

To check the assumption of normality in each of the two populations, one can construct a boxplot of the sales for the two display locations (see Figure 10.4). For these two small samples, there appears to be only a slight departure from normality, so the assumption of normality needed for the t test is not seriously violated.

FIGURE 10.4

Excel boxplot for sales at the special front and in-aisle locations

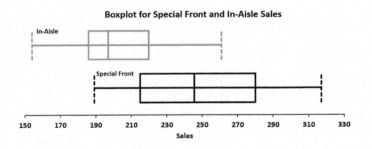

Example 10.1 provides another application of the pooled-variance t test.

EXAMPLE 10.1

Testing for the Difference in the Mean Delivery Times

You and some friends have decided to test the validity of an advertisement by a local pizza restaurant, which says it delivers to the dormitories faster than a local branch of a national chain. Both the local pizza restaurant and the national chain are located across the street from your college campus. You define the variable of interest as the delivery time, in minutes, from the time the pizza is ordered to when it is delivered. You collect the data by ordering 10 pizzas from the local pizza restaurant and 10 pizzas from the national chain at different times. You organize and store the data in **Pizza Delivery**. Table 10.3 shows the delivery times.

TABLE 10.3

Delivery times (in minutes) for a local pizza restaurant and a national pizza chain

Local		Chain	
16.8	18.1	22.0	19.5
11.7	14.1	15.2	17.0
15.6	21.8	18.7	19.5
16.7	13.9	15.6	16.5
17.5	20.8	20.8	24.0

At the 0.05 level of significance, is there evidence that the mean delivery time for the local pizza restaurant is less than the mean delivery time for the national pizza chain?

SOLUTION Because you want to know whether the mean is *lower* for the local pizza restaurant than for the national pizza chain, you have a one-tail test with the following null and alternative hypotheses:

$H_0: \mu_1 \geq \mu_2$ (The mean delivery time for the local pizza restaurant is equal to or greater than the mean delivery time for the national pizza chain.)

$H_1: \mu_1 < \mu_2$ (The mean delivery time for the local pizza restaurant is less than the mean delivery time for the national pizza chain.)

Figure 10.5 displays the results for the pooled-variance t test for these data.

FIGURE 10.5

Excel pooled-variance t test results for the pizza delivery time data

	A	B
1	Pooled-Variance t Test for Differences in Two Means	
2	(assumes equal population variances)	
3	Data	
4	Hypothesized Difference	0
5	Level of Significance	0.05
6	Population 1 Sample	
7	Sample Size	10
8	Sample Mean	16.7
9	Sample Standard Deviation	3.0955
10	Population 2 Sample	
11	Sample Size	10
12	Sample Mean	18.88
13	Sample Standard Deviation	2.8662
14		
15	Intermediate Calculations	
16	Population 1 Sample Degrees of Freedom	9
17	Population 2 Sample Degrees of Freedom	9
18	Total Degrees of Freedom	18
19	Pooled Variance	8.8986
20	Standard Error	1.3341
21	Difference in Sample Means	-2.18
22	t Test Statistic	-1.6341
23		
24	Lower-Tail Test	
25	Lower Critical Value	-1.7341
26	p-Value	0.0598
27	Do not reject the null hypothesis	

In the Worksheet

The worksheet uses a design similar to the design of the COMPUTE worksheet that Figure 10.3 presents.

The **COMPUTE_LOWER worksheet** of the Excel Guide **Pooled-Variance T workbook** is a copy of the Figure 10.5 worksheet.

▶(*continued*)

Using Equation (10.1) on page 312,

$$t_{STAT} = \frac{(\bar{X}_1 - \bar{X}_2) - (\mu_1 - \mu_2)}{\sqrt{S_p^2\left(\dfrac{1}{n_1} + \dfrac{1}{n_2}\right)}}$$

where

$$S_p^2 = \frac{(n_1 - 1)S_1^2 + (n_2 - 1)S_2^2}{(n_1 - 1) + (n_2 - 1)} = \frac{9(3.0955)^2 + 9(2.8662)^2}{9 + 9} = 8.8986$$

the result is

$$t_{STAT} = \frac{(16.7 - 18.88) - 0.0}{\sqrt{8.8986\left(\dfrac{1}{10} + \dfrac{1}{10}\right)}} = \frac{-2.18}{\sqrt{1.7797}} = -1.6341$$

Table 10.4 summarizes the results of the pooled-variance t test for the pizza delivery data using the calculations above and Figure 10.5 results. Based on the conclusions, the local branch of the national chain and a local pizza restaurant have similar delivery times. Therefore, as part of the last step of the DCOVA framework, you and your friends exclude delivery time as a decision criteria when choosing from which store to order pizza.

TABLE 10.4

Pooled-variance t test summary for the delivery times for the two pizza restaurants

Result	Conclusions
The $t_{STAT} = -1.6341$ is greater than -1.7341. The t test p-value $= 0.0598$ is greater than the level of significance, $\alpha = 0.05$.	1. Do not reject the null hypothesis H_0. 2. Conclude that insufficient evidence exists that the mean delivery time is lower for the local restaurant than for the branch of the national chain. 3. There is a probability of 0.0598 that $t_{STAT} < -1.6341$.

Confidence Interval Estimate for the Difference Between Two Means

Instead of, or in addition to, testing for the difference between the means of two independent populations, one can use Equation (10.2) to develop a confidence interval estimate of the difference in the means.

CONFIDENCE INTERVAL ESTIMATE FOR THE DIFFERENCE BETWEEN THE MEANS OF TWO INDEPENDENT POPULATIONS

$$\left(\bar{X}_1 - \bar{X}_2\right) \pm t_{\alpha/2}\sqrt{S_p^2\left(\frac{1}{n_1} + \frac{1}{n_2}\right)} \tag{10.2}$$

or

$$\left(\bar{X}_1 - \bar{X}_2\right) - t_{\alpha/2}\sqrt{S_p^2\left(\frac{1}{n_1} + \frac{1}{n_2}\right)} \leq \mu_1 - \mu_2 \leq \left(\bar{X}_1 - \bar{X}_2\right) + t_{\alpha/2}\sqrt{S_p^2\left(\frac{1}{n_1} + \frac{1}{n_2}\right)}$$

where $t_{\alpha/2}$ is the critical value of the t distribution, with $n_1 + n_2 - 2$ degrees of freedom, for an area of $\alpha/2$ in the upper tail.

Using the Figure 10.3 sample statistics results on page 314 and using 95% confidence,

$$\bar{X}_1 = 246.4 \,, n_1 = 10 \,, \bar{X}_2 = 202.3 \,, n_2 = 10 \,, S_p^2 = 1{,}433.9167, \text{ with } 10 + 10 - 2$$
$$= 18 \text{ degrees of freedom, } t_{0.025} = 2.1009$$

Using Equation (10.2)

$$(246.4 - 202.3) \pm (2.1009)\sqrt{1{,}433.9167\left(\frac{1}{10} + \frac{1}{10}\right)}$$

$$44.10 \pm (2.1009)(16.9347)$$

$$44.10 \pm 35.5784$$

$$8.5216 \le \mu_1 - \mu_2 \le 79.6784$$

Therefore, one can be 95% confident that the difference in mean sales between the special front and in-aisle locations is between 8.5216 and 79.6784 VLABGo players sold. In other words, one estimates, with 95% confidence, that the special front location has mean sales of between 8.5216 and 79.6784 more VLABGo players than the in-aisle location. Because the interval does not include zero, one rejects the null hypothesis of no difference between the means of the two populations.

Separate-Variance t Test for the Difference Between Two Means, Assuming Unequal Variances

For situations in which the two independent populations can be assumed to be normally distributed but cannot be assumed to have equal variances, use the **separate-variance t test** that Satterthwaite developed to use the two separate sample variances (Satterthwaite). Figure 10.6 displays the separate-variance t test results for the two different sales locations data. Observe that the test statistic $t_{STAT} = 2.6041$ and the p-value is $0.019 < 0.05$. The results for the separate-variance t test are nearly the same as those of the pooled-variance t test. The assumption of equality of population variances had no appreciable effect on the results.

FIGURE 10.6

Excel separate-variance t test results for the two different sales locations data

	A	B	
1	Separate-Variances t Test		
2	(assumes unequal population variances)		
3	Data		
4	Hypothesized Difference	0	
5	Level of Significance	0.05	
6	Population 1 Sample		
7	Sample Size	10	=COUNT(DataCopy2!$A:$A)
8	Sample Mean	246.4	=AVERAGE(DataCopy2!$A:$A)
9	Sample Standard Deviation	42.5420	=STDEV.S(DataCopy2!$A:$A)
10	Population 2 Sample		
11	Sample Size	10	=COUNT(DataCopy2!$B:$B)
12	Sample Mean	202.3	=AVERAGE(DataCopy2!$B:$B)
13	Sample Standard Deviation	32.5271	=STDEV.S(DataCopy2!$B:$B)
14			
15	Intermediate Calculations		
16	Numerator of Degrees of Freedom	82244.6803	=(E18 + E19)^2
17	Denominator of Degrees of Freedom	4883.1600	=(E18^2)/(B7 - 1) + (E19^2)/(B11 - 1)
18	Total Degrees of Freedom	16.8425	=B16/B17
19	Degrees of Freedom	16	=INT(B18)
20	Standard Error	16.9347	=SQRT(E18 + E19)
21	Difference in Sample Means	44.1000	=B8 - B12
22	Separate-Variance t Test Statistic	2.6041	=(B21 - B4)/B20
23			
24	Two-Tail Test		
25	Lower Critical Value	-2.1199	= -(T.INV.2T(B5, B19))
26	Upper Critical Value	2.1199	=T.INV.2T(B5, B19)
27	p-Value	0.0192	=T.DIST.2T(ABS(B22), B19)
28	Reject the null hypothesis		=IF(B27 < B5,"Reject the null hypothesis", "Do not reject the null hypothesis")

	D	E
15	Calculations Area	
16	Pop. 1 Sample Variance	1809.8222
17	Pop. 2 Sample Variance	1058.0111
18	Pop. 1 Sample Var./Sample Size	180.9822
19	Pop. 2 Sample Var./Sample Size	105.8011
20	For one-tailed tests:	
21	T.DIST.RT value	0.0096
22	1-T.DIST.RT value	0.9904

In the Worksheet

The worksheet uses a calculations area in cell range D15:E22 to simplify other calculations in the worksheet.

The **COMPUTE worksheet** of the Excel Guide **Separate-Variance T workbook** is a copy of the Figure 10.6 worksheet.

Sometimes, the results from the pooled-variance and separate-variance t tests conflict because the assumption of equal variances is violated. Therefore, one must evaluate the assumptions and use those results as a guide in selecting a test procedure. In Section 10.4, the F test for the ratio of two variances determines whether there is evidence of a difference in the two population variances. The results of that test can help one decide which of the t tests—pooled-variance or separate-variance—is more appropriate.

CONSIDER THIS

Do People Really Do This?

Some question whether decision makers really use confirmatory methods such as hypothesis testing. The following real case study, contributed by a student, reveals a role that such methods still play in business and also answers the question: "Do businesses really monitor their customer service calls for quality assurance purposes as they sometimes claim?"

In her first full-time job at a financial services company, a student was asked to improve a training program for new hires at a call center that handled customer questions about outstanding loans. For feedback and evaluation, she planned to randomly select phone calls received by each new employee and rate the employee on 10 aspects of the call, including whether the employee maintained a pleasant tone with the customer. When she presented her plan to her boss for approval, her boss wanted proof that her new training program would improve customer service. The boss, quoting a famous statistician, said "In God we trust; all others must bring data." Faced with this request, she called her business statistics professor. "Hello, Professor, you'll never believe why I called. I work for a large company, and in the project I am currently working on, I have to put some of the statistics you taught us to work! Can you help?" Together they formulated this test:

- Randomly assign the 60 most recent hires to two training programs. Assign half to the preexisting training program and the other half to the new training program.
- At the end of the first month, compare the mean score for the 30 employees in the new training program against the mean score for the 30 employees in the preexisting training program.

She listened as her professor explained, "What you are trying to show is that the mean score from the new training program is higher than the mean score from the current program. You can make the null hypothesis that the means are equal and see if you can reject it in favor of the alternative that the mean score from the new program is higher."

"Or, as you used to say, 'if the p-value is low, H_0 must go!'—yes, I do remember!" she replied. Her professor chuckled and added, "If you can reject H_0 you will have the evidence to present to your boss." She thanked him for his help and got back to work, with the newfound confidence that she would be able to successfully apply the t test that compares the means of two independent populations.

PROBLEMS FOR SECTION 10.1

LEARNING THE BASICS

10.1 If you have samples of $n_1 = 12$ and $n_2 = 15$, in performing the pooled-variance t test, how many degrees of freedom do you have?

10.2 Assume that you have a sample of $n_1 = 8$ with the sample mean $\overline{X}_1 = 42$, and a sample standard deviation $S_1 = 4$, and you have an independent sample of $n_2 = 15$ from another population with a sample mean of $\overline{X}_2 = 34$ and a sample standard deviation $S_2 = 5$.
a. What is the value of the pooled-variance t_{STAT} test statistic for testing $H_0: \mu_1 = \mu_2$?
b. In finding the critical value, how many degrees of freedom are there?
c. Using the level of significance $\alpha = 0.01$, what is the critical value for a one-tail test of the hypothesis $H_0: \mu_1 \leq \mu_2$ against the alternative, $H_1: \mu_1 > \mu_2$?
d. What is your statistical decision?

10.3 What assumptions about the two populations are necessary in Problem 10.2?

10.4 Referring to Problem 10.2, construct a 95% confidence interval estimate of the population mean difference between μ_1 and μ_2.

10.5 Referring to Problem 10.2, if $n_1 = 5$ and $n_2 = 4$, how many degrees of freedom do you have?

10.6 Referring to Problem 10.2, if $n_1 = 5$ and $n_2 = 4$, at the 0.01 level of significance, is there evidence that $\mu_1 > \mu_2$?

APPLYING THE CONCEPTS

10.7 When people make estimates, they are influenced by anchors to their estimates. A study was conducted in which students were asked to estimate the number of calories in a cheeseburger. One group was asked to do this after thinking about a calorie-laden cheesecake. A second group was asked to do this after thinking about an organic fruit salad. The mean number of calories estimated in a cheeseburger was 780 for the group that thought about the cheesecake and 1,041 for the group that thought about the organic fruit salad.

Source: Data extracted from "Drilling Down, Sizing Up a Cheeseburger's Caloric Heft," *New York Times*, October 4, 2010, p. B2.

Suppose that the study was based on a sample of 20 people who thought about the cheesecake first and 20 people who thought about the organic fruit salad first, and the standard deviation of the number of calories in the cheeseburger was 128 for the people who thought about the cheesecake first and 140 for the people who thought about the organic fruit salad first.

a. State the null and alternative hypotheses if you want to determine whether the mean estimated number of calories in the cheeseburger is lower for the people who thought about the cheesecake first than for the people who thought about the organic fruit salad first.

b. In the context of this study, what is the meaning of the Type I error?

c. In the context of this study, what is the meaning of the Type II error?

d. At the 0.01 level of significance, is there evidence that the mean estimated number of calories in the cheeseburger is lower for the people who thought about the cheesecake first than for the people who thought about the organic fruit salad first?

e. If you were developing a commercial for a cheeseburger, based on the results of (d), what other foods might you show in the commercial?

10.8 A study found that 51 children who watched a commercial for Walker Crisps (potato chips) featuring a long-standing sports celebrity endorser ate a mean of 36 grams of Walker Crisps as compared to a mean of 25 grams of Walker Crisps for 41 children who watched a commercial for an alternative food snack.

Source: Data extracted from E. J. Boyland et al., "Food Choice and Overconsumption: Effect of a Premium Sports Celebrity Endorser," *Journal of Pediatrics*, March 13, 2013, **bit.ly/16NR4Bi**.

Suppose that the sample standard deviation for the children who watched the sports celebrity–endorsed Walker Crisps commercial was 21.4 grams and the sample standard deviation for the children who watched the alternative food snack commercial was 12.8 grams.

a. Assuming that the population variances are equal and $\alpha = 0.05$, is there evidence that the mean amount of Walker Crisps eaten was significantly higher for the children who watched the sports celebrity–endorsed Walker Crisps commercial?

b. Assuming that the population variances are equal, construct a 95% confidence interval estimate of the difference between the mean amount of Walker Crisps eaten by children who watched the sports celebrity–endorsed Walker Crisps commercial and children who watched the alternative food snack commercial.

c. Compare and discuss the results of (a) and (b).

d. Based on the results of (a) and (b), if you wanted to increase the consumption of Walker Crisps, which commercial would you choose?

10.9 Is there a difference in the satisfaction rating of traditional cellphone providers who bill for service at the end of a month often under a contract and prepaid cellphone service providers who bill in advance without a contract? The file Cellphone Providers contains the satisfaction rating for 10 traditional cellphone providers and 13 prepaid cellphone service providers.

Source: Data extracted from "Carrier Ratings: Why It Pays to Think Small," *Consumer Reports*, February 2016, p. 51.

a. Assuming that the population variances from both types of cellphone providers are equal, is there evidence of a difference in the mean ratings between the two types of cellphone providers? (Use $\alpha = 0.05$.)

b. Find the *p*-value in (a) and interpret its meaning.

c. What other assumption is necessary in (a)?

d. Assuming that the population variances from both cellphone providers are equal, construct and interpret a 95% confidence interval estimate of the difference between the population means of the two cellphone providers.

e. What conclusions can you reach about the satisfaction rating of traditional cellphone providers who bill for service at the end of a month often under a contract and prepaid cellphone service providers who bill in advance without a contract?

✓SELF TEST **10.10** *Accounting Today* identified the top accounting firms in 10 geographic regions across the United States. All 10 regions reported growth in a recent year. The Southeast and Gulf Coast regions reported growth of 12.49% and 9.78%, respectively. A characteristic description of the accounting firms in the Southeast and Gulf Coast regions included the number of partners in the firm. The file Accounting Partners 2 contains the number of partners.

Source: Data extracted from Accounting Today, *The 2019 Top Firms and Regional Leaders*, available at **accountingtoday.com/collections/the-2019-top-100-firms-and-regional-leaders**.

a. At the 0.05 level of significance, is there evidence of a difference between Southeast region accounting firms and Gulf Coast accounting firms with respect to the mean number of partners?

b. Determine the *p*-value and interpret its meaning.

c. What assumptions do you have to make about the two populations in order to justify the use of the *t* test?

10.11 The Super Bowl is a big viewing event watched by close to 200 million Americans, so it is also a big event for advertisers. The file Super Bowl Ad Ratings contains the rating of ads that ran between the opening kickoff and the final whistle.

Source: Data extracted from **www.admeter.usatoday.com/results/2019**.

a. Assuming that the population variances from the ads that ran before halftime and the ads that ran at halftime or after are equal, is there evidence of a difference in the mean score between the two types of ads? (Use $\alpha = 0.05$.)

b. Determine the *p*-value in (a) and interpret its meaning.

c. Assuming that the population variances from both types of ads are equal, construct and interpret a 95% confidence interval estimate of the difference between the population mean score of the two types of ads.

10.12 A bank with a branch located in a commercial district of a city has the business objective of developing an improved process for serving customers during the noon-to-1 P.M. lunch period. Management decides to first study the waiting time in the current process. The waiting time is defined as the number of minutes that elapses from when the customer enters the line until he or she reaches the teller window. Data are collected from a random sample of 15 customers and stored in Bank Waiting. These data are:

4.21	5.55	3.02	5.13	4.77	2.34	3.54	3.20
4.50	6.10	0.38	5.12	6.46	6.19	3.79	

Suppose that another branch, located in a residential area, is also concerned with improving the process of serving customers in the noon-to-1 P.M. lunch period. Data are collected from a random sample of 15 customers and stored in Bank Waiting 2 . These data are:

9.66   5.90   8.02   5.79   8.73   3.82   8.01   8.35
10.49   6.68   5.64   4.08   6.17   9.91   5.47

a. Assuming that the population variances from both banks are equal, is there evidence of a difference in the mean waiting time between the two branches? (Use $\alpha = 0.05$.)

b. Determine the p-value in (a) and interpret its meaning.

c. In addition to equal variances, what other assumption is necessary in (a)?

d. Construct and interpret a 95% confidence interval estimate of the difference between the population means in the two branches.

10.13 Repeat Problem 10.12 (a), assuming that the population variances in the two branches are not equal. Compare these results with those of Problem 10.12 (a).

10.14 As a member of the international strategic management team in your company, you are assigned the task of exploring potential international market entry. As part of your initial investigation, you want to know if there is a difference between developed markets and emerging markets with respect to the time required to start a business. You select 15 developed countries and 15 emerging countries. The time required to start a business, defined as the number of days needed to complete the procedures to legally operate a business in these countries, is stored in International Market .

Source: Data extracted from **data.worldbank.org**.

a. Assuming that the population variances for developed countries and emerging countries are equal, is there evidence of a difference in the mean time required to start a business between developed countries and emerging countries? (Use $\alpha = 0.05$.)

b. Determine the p-value in (a) and interpret its meaning.

c. In addition to equal variances, what other assumption is necessary in (a)?

d. Construct a 95% confidence interval estimate of the difference between the population means of developed countries and emerging countries.

10.15 Repeat Problem 10.14 (a), assuming that the population variances from developed and emerging countries are not equal. Compare these results with those of Problem 10.14 (a).

10.16 We Are Social and Hootsuite reported that the typical American spends 2.02 hours (121 minutes) per day accessing the Internet through a mobile device.

Source: *Digital in 2017 Global Overview*, available at **bit.ly/2jXeS3F**.

You wonder if males and females spend differing amounts of time per day accessing the Internet through a mobile device.

You select a sample of 60 friends and family (30 males and 30 females), collect times spent per day accessing the Internet through a mobile device (in minutes), and store the data collected in Mobile Device Study 2 .

a. Assuming that the variances in the population of times spent per day accessing the Internet via a mobile device are equal, is there evidence of a difference between males and females in the mean time spent per day accessing the Internet via a mobile device? (Use a 0.05 level of significance.)

b. In addition to equal variances, what other assumption is necessary in (a)?

10.17 Brand valuations are critical to CEOs, financial and marketing executives, security analysts, institutional investors, and others who depend on well-researched, reliable information needed for assessments and comparisons in decision making. Millward Brown Optimor has developed the BrandZ Top 100 Most Valuable Global Brands for WPP, the world's largest communications services group. Unlike other studies, the BrandZ Top 100 Most Valuable Global Brands fuses consumer measures of brand equity with financial measures to place a financial value on brands. The file BrandZ TechFin contains the brand values for the technology sector and the financial institution sector in the BrandZ Top 100 Most Valuable Global Brands for a recent year.

Source: Data extracted from *BrandZ Top100 Most Valuable Global Brands 2019*, available at **millwardbrown.com/brandz/rankings-and-reports/top-us-brands/2019**.

a. Assuming that the population variances are equal, is there evidence of a difference between the technology sector and the financial institutions sector with respect to mean brand value? (Use $\alpha = 0.05$.)

b. Repeat (a), assuming that the population variances are not equal.

c. Compare the results of (a) and (b).

10.2 Comparing the Means of Two Related Populations

Section 10.1 hypothesis-testing procedures examine differences between the means of two *independent* populations. This section discusses a procedure for examining the mean difference between two populations when samples are collected from populations that are **related**, that is, when results of the first population are *not* independent of the results of the second population.

There are two cases in which related data are used: when one takes repeated measurements from the same set of items or individuals or when one matches items or individuals according to some characteristic. In these situations, one examines the *difference between the two related values* rather than the *individual values* themselves.

When one takes **repeated measurements** on the same items or individuals, one assumes that the same items or individuals will behave alike if treated alike. The objective is to show that any differences between two measurements of the same items or individuals are due to different treatments that have been applied to the items or individuals. For example, to conduct an experiment that compares the prices of items from two retailers, one collects the prices of equivalent items that the retailers sell. For each item, the two prices are the "repeated measurements" of the item.

Using repeated measurements enables one to answer questions such as "Do prices for the same items differ between two retailers?" By collecting the prices of the *same* items from both sellers, one creates two related samples and can use a test that is more powerful than the tests Section 10.1 discusses. Those tests use two *independent* samples that most likely will not contain the same sample of items. That means that differences observed might be due to one sample having products that are inherently costlier than the other.

Matched samples represent another type of related data between populations. In matched samples, items or individuals are paired together according to some characteristic of interest. For example, in test marketing a product in two different advertising campaigns, a sample of test markets can be *matched* on the basis of the test-market population size and/or demographic variables. By accounting for the differences in test-market population size and/or demographic variables, one can better measure the effects of the two different advertising campaigns.

Whether using matched samples or repeated measurements, the objective is to study the difference between two measurements by reducing the effect of the variability that is due to the items or individuals themselves. Table 10.5 shows the differences between the individual values for two related populations. In this table, $X_{11}, X_{12}, \ldots, X_{1n}$ represent the n values from the first sample and $X_{21}, X_{22}, \ldots, X_{2n}$ represent either the corresponding n matched values from a second sample or the corresponding n repeated measurements from the initial sample. The set of values $D_1, D_2, \ldots, D_n$ in the last column represents the corresponding set of n **difference scores**. To test for the mean difference between two related populations, one treats the difference scores, each D_i, as values from a single sample.

student TIP

Which sample you define as sample 1 determines which one-tail test (lower or upper) to perform, if a one-tail test is needed.

TABLE 10.5
Determining the difference between two related samples

Value	Sample 1	Sample 2	Difference Score
1	X_{11}	X_{21}	$D_1 = X_{11} - X_{21}$
2	X_{12}	X_{22}	$D_2 = X_{12} - X_{22}$
$\vdots$	$\vdots$	$\vdots$	$\vdots$
i	X_{1i}	X_{2i}	$D_i = X_{1i} - X_{2i}$
$\vdots$	$\vdots$	$\vdots$	$\vdots$
n	X_{1n}	X_{2n}	$D_n = X_{1n} - X_{2n}$

Paired t Test

If one can assume that the difference scores are randomly and independently selected from a population that is normally distributed, one can use the **paired t test for the mean difference** in related populations to determine whether there is a significant population mean difference. As with the one-sample t test that Section 9.2 discusses and which Equation (9.2) defines (see page 288), the paired t test statistic follows the t distribution with $n - 1$ degrees of freedom. Although the paired t test assumes that the population is normally distributed, one can use this test as long as the sample size is not very small and the population is not highly skewed because this test is robust.

Equation (10.3) on page 323 defines the paired t test for the mean difference.

PAIRED t TEST FOR THE MEAN DIFFERENCE

$$t_{STAT} = \frac{\overline{D} - \mu_D}{\dfrac{S_D}{\sqrt{n}}} \qquad\qquad \textbf{(10.3)}$$

where

$$\mu_D = \text{hypothesized mean difference, } \mu_1 - \mu_2$$

$$\overline{D} = \frac{\displaystyle\sum_{i=1}^{n} D_i}{n}$$

$$D_i = i\text{th difference score}$$

$$S_D = \sqrt{\frac{\displaystyle\sum_{i=1}^{n}(D_i - \overline{D})^2}{n - 1}}$$

The t_{STAT} test statistic follows a t distribution with $n - 1$ degrees of freedom.

The null and alternative hypotheses of a test for difference in the means of two related populations are

$$H_0: \mu_D = 0$$
$$H_1: \mu_D \neq 0$$

For a two-tail test with a given level of significance, α, you reject the null hypothesis if the t_{STAT} test statistic is greater than the upper-tail critical value $t_{\alpha/2}$ from the t distribution or, if the t_{STAT} test statistic is less than the lower-tail critical value $-t_{\alpha/2}$, from the t distribution. The decision rule is

$$\text{Reject } H_0 \text{ if } t_{STAT} > t_{\alpha/2}$$

$$\text{or if } t_{STAT} < -t_{\alpha/2};$$

$$\text{otherwise, do not reject } H_0.$$

To illustrate the use of the paired t test for the mean difference, consider a researcher who seeks to determine if the prices of the same or equivalent grocery items differ between Costco, a warehouse club that sells only to members who pay an annual fee, and Walmart, a large general retailer that sells groceries. In this repeated measures experiment, the researcher uses one market basket (set) of products. For each product, the researcher determines the price of the item at Costco and the price of the same or equivalent item at Walmart. By using the same market basket, the researcher reduces the variability in the prices that would occur if the researcher used two market baskets that contained different sets of items and can focus on the differences between the prices of the equivalent products offered by the two retailers.

Table 10.6, stored in Market Basket , contains market basket prices for the $n = 7$ selected items from Costco and Walmart.

Using the DCOVA framework, the researcher defines the business objective as determining if there is any difference between the mean price at Costco and Walmart. In other words, is there evidence that the mean price is different between the two retailers? The null and alternative hypotheses are

$H_0: \mu_D = 0$ (There is no difference in the mean price between Costco and Walmart.)
$H_1: \mu_D \neq 0$ (There is a difference in the mean price between Costco and Walmart.)

TABLE 10.6

Prices (in dollars) of equivalent items at Costco and Walmart

Equivalent Item	Costco	Walmart
Chicken broth per 128 oz.	5.98	5.88
Vanilla ice cream per 96 oz.	8.59	7.19
Dishwasher detergent per 100 loads	9.00	17.00
Laundry detergent per 100 loads	11.00	12.00
Paper towels per 100 square feet	1.47	2.09
Toilet paper per 10 rolls	12.00	27.00
Tissues per 100 tissues	1.23	1.12

Source: Data extracted and adapted from "The Best Everyday Products," *Consumer Reports*, January 2015, p. 29.

Choosing the level of significance $\alpha = 0.05$ and assuming that the differences are normally distributed, the researcher uses the paired t test for mean difference. For a sample of $n = 7$ items, there are $n - 1 = 6$ degrees of freedom. As Figure 10.7 visualizes, using Table E.3, the decision rule is

$$\text{Reject } H_0 \text{ if } t_{STAT} > 2.4469$$
$$\text{or if } t_{STAT} < -2.4469;$$
$$\text{otherwise, do not reject } H_0.$$

FIGURE 10.7

Two-tail paired t test at the 0.05 level of significance with 6 degrees of freedom

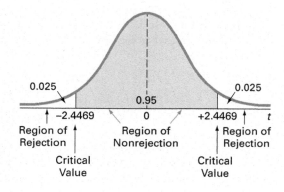

For the $n = 7$ differences (calculated from the Table 10.6 data), the sample mean difference is

$$\overline{D} = \frac{\sum_{i=1}^{n} D_i}{n} = \frac{-23.01}{7} = -3.2871$$

and

$$S_D = \sqrt{\frac{\sum_{i=1}^{n}(D_i - \overline{D})^2}{n - 1}} = 6.0101$$

From Equation (10.3) on page 323,

$$t_{STAT} = \frac{\overline{D} - \mu_D}{\dfrac{S_D}{\sqrt{n}}} = \frac{-3.2871 - 0}{\dfrac{6.0101}{\sqrt{7}}} = -1.4471$$

Table 10.7 summarizes the results of the paired t test for the difference between the two sales locations using the calculation on page 324 and Figure 10.8 results. Based on the conclusions, Costco and Walmart sell equivalent items at similar prices. Therefore, as part of the last step of the DCOVA framework, a researcher can state that a price-conscious shopper would do equally well shopping for the market basket of items at either retailer.

TABLE 10.7

Paired t test summary for the prices of equivalent items at Costco and Walmart

Result	Conclusions
The $t_{STAT} = -1.4471$ is greater than -2.4469 and less than 2.4469.	1. Do not reject the null hypothesis H_0.
The t test p-value $= 0.1980$ is greater than the level of significance, $\alpha = 0.05$.	2. Conclude that no evidence exists that there is a difference in the mean price of equivalent items purchased at Costco and Walmart.

FIGURE 10.8

Excel, paired t test results for the equivalent products price data

▲	A	B	
1	Paired t Test		
2			
3	Data		
4	Hypothesized Mean Difference	0	
5	Level of Significance	0.05	
6			
7	Intermediate Calculations		
8	Sample Size	7	=COUNT(PtCalcs!$A:$A)
9	DBar	-3.2871	=AVERAGE(PtCalcs!$C:$C)
10	degrees of freedom	6	=B8 - 1
11	S_D	6.0101	=SQRT(DEVSQ(PtCalcs!C:C)/B10)
12	Standard Error	2.2716	=B11/SQRT(B8)
13	t Test Statistic	-1.4471	=(B9 - B4)/B12
14			
15	Two-Tailed Test		
16	Lower Critical Value	-2.4469	=-T.INV.2T(B5, B10)
17	Upper Critical Value	2.4469	=T.INV.2T(B5, B10)
18	p-Value	0.1980	=T.DIST.2T(ABS(B13), B10)
19	Do not reject the null hypothesis		=IF(B18 < B5, "Reject the null hypothesis",)
			"Do not reject the null hypothesis")

In the Worksheet

The worksheet uses the DEVSQ function in cell B11 to compute the sum of the squares of the differences to simplify other calculations in the worksheet.

The **COMPUTE worksheet** of the Excel Guide **Paired T workbook** is a copy of the Figure 10.8 worksheet.

To evaluate the validity of the assumption of normality, if the sample size is sufficiently large (which is not the case with the preceding example), one can construct a boxplot, histogram, or normal probability plot. If these plots reveal that the assumption of underlying normality in the population is severely violated, then the t test may be inappropriate, especially if the sample size is small. If one concludes that the t test is inappropriate, one can use either a *nonparametric* procedure that does not make the assumption of underlying normality (see Section 12.5) or make a data transformation (Snedecor) and then check the assumptions again to determine whether the t test can be used.

EXAMPLE 10.2

Paired t Test of Pizza Delivery Times

Recall from Example 10.1 on page 316 a local pizza restaurant situated across the street from your college campus advertises that it delivers to the dormitories faster than the local branch of a national pizza chain. In order to determine whether this advertisement is valid, you and some friends decided to order 10 pizzas from the local pizza restaurant and 10 pizzas from the national chain. In fact, each time you ordered a pizza from the local pizza restaurant, at the same time, your friends ordered a pizza from the national pizza chain. Therefore, you have matched samples (because each pair of pizzas was ordered at the same time). For each of the 10 times that pizzas were ordered, you have one measurement from the local pizza restaurant and one from the national chain. At the 0.05 level of significance, is the mean delivery time for the local pizza restaurant less than the mean delivery time for the national pizza chain?

SOLUTION Use the paired t test to analyze the Table 10.8 data (stored in `Pizza Delivery`). Figure 10.9 on page 326 shows the paired t test results for the pizza delivery data.

▶*(continued)*

TABLE 10.8
Delivery times for local pizza restaurant and national pizza chain

Time	Local	Chain	Difference
1	16.8	22.0	−5.2
2	11.7	15.2	−3.5
3	15.6	18.7	−3.1
4	16.7	15.6	1.1
5	17.5	20.8	−3.3
6	18.1	19.5	−1.4
7	14.1	17.0	−2.9
8	21.8	19.5	2.3
9	13.9	16.5	−2.6
10	20.8	24.0	−3.2
			$\overline{-21.8}$

FIGURE 10.9
Excel paired t test results for the pizza delivery data

⊿	A	B	C	D	E
1	Paired t Test				
2					
3	Data				
4	Hypothesized Mean Diff.	0			
5	Level of significance	0.05			
6					
7	Intermediate Calculations				
8	Sample Size	10			
9	DBar	-2.1800			
10	degrees of freedom	9			
11	S_D	2.2641			
12	Standard Error	0.7160			
13	t Test Statistic	-3.0448			
14					
15	Lower-Tail Test			One-Tail Calculations	
16	Lower Critical Value	-1.8331		T.DIST.RT	0.0070
17	p -Value	0.0070		1 - T.DIST.RT	0.9930
18	Reject the null hypothesis				

In the Worksheet

The worksheet uses a design similar to the design of the COMPUTE worksheet that Figure 10.8 presents.

The **COMPUTE_LOWER worksheet** of the Excel Guide **Paired T workbook** is a copy of the Figure 10.9 worksheet.

The null and alternative hypotheses are:

H_0: $\mu_D \geq 0$ (Mean difference in the delivery time between the local pizza restaurant and the national pizza chain is greater than or equal to 0.)

H_1: $\mu_D < 0$ (Mean difference in the delivery time between the local pizza restaurant and the national pizza chain is less than 0.)

Choosing the level of significance $\alpha = 0.05$ and assuming that the differences are normally distributed, you use the paired t test for mean difference. For a sample of $n = 10$ delivery times, there are $n - 1 = 9$ degrees of freedom. Using Table E.3, the decision rule is

$$\text{Reject } H_0 \text{ if } t_{STAT} < -t_{0.05} = -1.8331;$$
$$\text{otherwise, do not reject } H_0.$$

For $n = 10$ differences (see Table 10.8), the sample mean difference is

$$\overline{D} = \frac{\sum_{i=1}^{n} D_i}{n} = \frac{-21.8}{10} = -2.18$$

and the sample standard deviation of the difference is

$$S_D = \sqrt{\frac{\sum_{i=1}^{n} (D_i - \overline{D})^2}{n - 1}} = 2.2641$$

▶(continued)

From Equation (10.3) on page 323,

$$t_{STAT} = \frac{\overline{D} - \mu_D}{\dfrac{S_D}{\sqrt{n}}} = \frac{-2.18 - 0}{\dfrac{2.2641}{\sqrt{10}}} = -3.0448$$

Table 10.9 summarizes the results of the paired t test for the pizza delivery data using the calculation above and Figure 10.9 results. Based on the conclusions, the local pizza restaurant has a faster (lower) delivery time than the branch of the national chain. Therefore, as part of the last step of the DCOVA framework, you and your friends should order from the local pizza restaurant if delivery time is an important decision-making criterion for choosing a restaurant.

TABLE 10.9
Paired t test summary for the delivery times for the two pizza restaurants

Result	Conclusions
The $t_{STAT} = -3.0448$ is less than -1.8331. The t test p-value $= 0.0070$ is less than the level of significance, $\alpha = 0.05$.	1. Reject the null hypothesis H_0. 2. Conclude that evidence exists that the mean delivery time is lower for the local restaurant than for the branch of the national chain. 3. There is a probability of 0.0070 that $t_{STAT} < -3.0448$

This conclusion differs from the conclusion reached when using the pooled-variance t test for these data (see Example 10.1 on page 316). By pairing the delivery times, you are able to focus on the differences between the two pizza delivery services and not the variability created by ordering pizzas at different times of day. The paired t test is a more powerful statistical procedure that reduces the variability in the delivery time because you are controlling for the time of day the pizza was ordered.

Confidence Interval Estimate for the Mean Difference

When conducting a two-tail test for the mean difference between two related populations, one can use Equation (10.4) to construct a confidence interval estimate for the population mean difference.

CONFIDENCE INTERVAL ESTIMATE FOR THE MEAN DIFFERENCE

$$\overline{D} \pm t_{\alpha/2}\frac{S_D}{\sqrt{n}} \tag{10.4}$$

or

$$\overline{D} - t_{\alpha/2}\frac{S_D}{\sqrt{n}} \le \mu_D \le \overline{D} + t_{\alpha/2}\frac{S_D}{\sqrt{n}}$$

where $t_{\alpha/2}$ is the critical value of the t distribution, with $n - 1$ degrees of freedom, for an area of $\alpha/2$ in the upper tail.

Recall the example comparing equivalent item prices at Costco and Walmart that begins on page 323. Using Equation (10.4), $\overline{D} = -3.2871$, $S_D = 6.0101$, $n = 7$, and $t_{\alpha/2} = 2.4469$ (for 95% confidence and $n - 1 = 7$ degrees of freedom),

$$-3.2871 \pm (2.4469)\frac{6.0101}{\sqrt{7}}$$

$$-3.2871 \pm 5.5583$$

$$-8.8455 \leq \mu_D \leq 2.2713$$

Thus, with 95% confidence, you estimate that the population mean difference in equivalent item prices between Costco and Walmart is between $-\$8.8455$ and $\$2.2713$. Because the interval estimate contains zero, using the 0.05 level of significance and a two-tail test, the researcher can conclude that there is no evidence of a difference in the mean item prices between Costco and Walmart, the same as the the t test results (see Table 10.7 on page 325).

PROBLEMS FOR SECTION 10.2

LEARNING THE BASICS

10.18 An experimental design for a paired t test has 20 pairs of identical twins. How many degrees of freedom are there in this t test?

10.19 Fifteen volunteers are recruited to participate in an experiment. A measurement is made (such as blood pressure) before each volunteer is asked to read a particularly upsetting passage from a book and and then again after each volunteer reads the passage from the book. In the analysis of the data collected from this experiment, how many degrees of freedom are there in the test?

APPLYING THE CONCEPTS

✓SELF TEST **10.20** Nine experts rated two brands of coffee in a taste-testing experiment. A rating on a 7-point scale (1 = extremely unpleasing, 7 = extremely pleasing) is given for each of four characteristics: taste, aroma, richness, and acidity. The following data stored in Coffee contain the ratings accumulated over all four characteristics:

EXPERT	BRAND A	BRAND B
C.C.	24	26
S.E.	27	27
E.G.	19	22
B.L.	24	27
C.M.	22	25
C.N.	26	27
G.N.	27	26
R.M.	25	27
P.V.	22	23

a. At the 0.05 level of significance, is there evidence of a difference in the mean ratings between the two brands?
b. What assumption is necessary in order to perform this test?
c. Determine the p-value in (a) and interpret its meaning.
d. Construct and interpret a 95% confidence interval estimate of the difference in the mean ratings between the two brands.

10.21 How do the ratings of TV and Internet services compare? The file Telecomm contains the ratings of 10 different providers.

Source: Data extracted from *ACSI Telecommunication Report 2017*, available at **bit.ly/2syfcbA**.

a. At the 0.05 level of significance, is there evidence of a difference in the mean service rating between TV and Internet services?
b. What assumption is necessary in order to perform this test?
c. Use a graphical method to evaluate the validity of the assumption in (a).
d. Construct and interpret a 95% confidence interval estimate of the difference in the mean service rating between TV and Internet services.

10.22 Do upload and download speeds vary between cities and Internet service providers? The file Internet Provider Speeds contains the upload and download speeds in Mbps in 40 different cities for four Internet service providers, AT&T, T-Mobile, Sprint, and Verizon, as measured in November 2017. In this problem, you have decided to focus on the difference in the download times between AT&T and Verizon.

Source: Data extracted from "Speedsmart Wireless Speed Index," **bit.ly/2FjoRdt** .

a. At the 0.05 level of significance, is there evidence of a difference in the mean download speed between AT&T and Verizon?
b. What assumption is necessary to perform this test?
c. Use a graphical method to evaluate the assumption made in (a).
d. Construct and interpret a 95% confidence interval estimate of the difference in the mean download speed between AT&T and Verizon.

10.23 Do upload and download speeds vary between cities and Internet service providers? The file Internet Provider Speeds contains the upload and download speeds in Mbps in 40 different cities for four Internet service providers, AT&T, T-Mobile, Sprint, and Verizon, as measured in November 2017. In this problem, you have decided to focus on the difference in the upload times between AT&T and Verizon.

Source: Data extracted from "Speedsmart Wireless Speed Index," **bit.ly/2FjoRdt**.

a. At the 0.05 level of significance, is there evidence of a difference in the mean upload speed between AT&T and Verizon?
b. What assumption is necessary to perform this test?
c. Use a graphical method to evaluate the assumption made in (a).
d. Construct and interpret a 95% confidence interval estimate of the difference in the mean upload speed between AT&T and Verizon.

10.24 Multiple myeloma, or blood plasma cancer, is characterized by increased blood vessel formulation (angiogenesis) in the bone marrow that is a predictive factor in survival. One treatment approach used for multiple myeloma is stem cell transplantation with the patient's own stem cells. The data stored in `Myeloma` and shown below represent the bone marrow microvessel density for patients who had a complete response to the stem cell transplant (as measured by blood and urine tests). The measurements were taken immediately prior to the stem cell transplant and at the time the complete response was determined.

Patient	Before	After
1	158	284
2	189	214
3	202	101
4	353	227
5	416	290
6	426	176
7	441	290

Data extracted from S. V. Rajkumar, R. Fonseca, T. E. Witzig, M. A. Gertz, and P. R. Greipp, "Bone Marrow Angiogenesis in Patients Achieving Complete Response After Stem Cell Transplantation for Multiple Myeloma," *Leukemia* 13 (1999): 469–472.

a. At the 0.05 level of significance, is there evidence that the mean bone marrow microvessel density is higher before the stem cell transplant than after the stem cell transplant?
b. Interpret the meaning of the *p*-value in (a).
c. Construct and interpret a 95% confidence interval estimate of the mean difference in bone marrow microvessel density before and after the stem cell transplant.
d. What assumption is necessary in order to perform the test in (a)?

10.25 To assess the effectiveness of a cola video ad, a random sample of 38 individuals from a target audience was selected to participate in a copy test. Participants viewed two ads, one of which was the ad being tested. Participants then answered a series of questions about how much they liked the ads. An adindex measure was created and stored in `Adindex`; the higher the adindex value, the more likeable the ad. Compute descriptive statistics and perform a paired *t* test. State your findings and conclusions in a report. (Use the 0.05 level of significance.)

10.26 The file `Concrete Strength` contains the compressive strength, in thousands of pounds per square inch (psi), of 40 samples of concrete taken two and seven days after pouring.

Source: Data extracted from O. Carrillo-Gamboa and R. F. Gunst, "Measurement-Error-Model Collinearities," *Technometrics*, 34 (1992): 454–464.

a. At the 0.01 level of significance, is there evidence that the mean strength is lower at two days than at seven days?
b. What assumption is necessary in order to perform this test?
c. Find the *p*-value in (a) and interpret its meaning.

10.3 Comparing the Proportions of Two Independent Populations

Often, one needs to make comparisons and analyze differences between two population proportions. Two different procedures perform a test for the difference between two proportions selected from independent populations. This section presents a procedure whose test statistic, Z_{STAT}, is approximated by a standardized normal distribution. (Section 12.1 discusses an equivalent procedure in which the test statistic, χ^2_{STAT}, is approximated by a chi-square distribution.)

Z Test for the Difference Between Two Proportions

To evaluate differences between two population proportions, one uses a *Z* **test for the difference between two proportions**. Equation (10.5) defines this test. The Z_{STAT} test statistic is based on the difference between two sample proportions $(p_1 - p_2)$, and the test statistic approximately follows a standardized normal distribution for large enough sample sizes.

Z TEST FOR THE DIFFERENCE BETWEEN TWO PROPORTIONS

$$Z_{STAT} = \frac{(p_1 - p_2) - (\pi_1 - \pi_2)}{\sqrt{\bar{p}(1 - \bar{p})\left(\dfrac{1}{n_1} + \dfrac{1}{n_2}\right)}} \tag{10.5}$$

where

$$\bar{p} = \frac{X_1 + X_2}{n_1 + n_2} \quad p_1 = \frac{X_1}{n_1} \quad p_2 = \frac{X_2}{n_2}$$

and

$$
\begin{aligned}
p_1 &= \text{proportion of items of interest in sample 1} \\
X_1 &= \text{number of items of interest in sample 1} \\
n_1 &= \text{sample size of sample 1} \\
\pi_1 &= \text{proportion of items of interest in population 1} \\
p_2 &= \text{proportion of items of interest in sample 2} \\
X_2 &= \text{number of items of interest in sample 2} \\
n_2 &= \text{sample size of sample 2} \\
\pi_2 &= \text{proportion of items of interest in population 2} \\
\bar{p} &= \text{pooled estimate of the population proportion of items of interest}
\end{aligned}
$$

The Z_{STAT} test statistic approximately follows a standardized normal distribution.

studentTIP

Do not confuse this use of the Greek letter pi, π, to represent the population proportion with the mathematical constant that is approximately 3.14159.

The null hypothesis in the Z test for the difference between two proportions states that the two population proportions are equal ($\pi_1 = \pi_2$). Because the pooled estimate for the population proportion is based on the null hypothesis, you combine, or pool, the two sample proportions to compute $\bar{p}$, an overall estimate of the common population proportion. This estimate is equal to the number of items of interest in the two samples ($X_1 + X_2$) divided by the total sample size from the two samples ($n_1 + n_2$).

As the following table summarizes, the Z test for the difference between population proportions can be used to determine whether there is a difference in the proportion of items of interest in the two populations (two-tail test) or whether one population has a higher proportion of items of interest than the other population (one-tail test).

Two-Tail Test	One-Tail Test	One-Tail Test
$H_0: \pi_1 = \pi_2$	$H_0: \pi_1 \geq \pi_2$	$H_0: \pi_1 \leq \pi_2$
$H_1: \pi_1 \neq \pi_2$	$H_1: \pi_1 < \pi_2$	$H_1: \pi_1 > \pi_2$

For a given level of significance, α, one rejects the null hypothesis if the Z_{STAT} test statistic that Equation (10.5) defines is greater than the upper-tail critical value from the standardized normal distribution or if the Z_{STAT} test statistic is less than the lower-tail critical value from the standardized normal distribution.

To illustrate the use of the Z test for the equality of the two proportions, recall the Chapter 6 Using Statistics MyTVLab scenario. MyTVLab web designers face a new task to revise the signup page for the website. The designers ponder whether this page should ask for many personal details or just a few. They decide to design both types of signup pages and devise an experiment. Every visitor going to the signup page will be randomly shown one of the two new designs. Effectiveness will be measured by whether the visitor clicks the signup button that appears on the page displayed.

Using the DCOVA framework, the designers define the business objective as determining if there is evidence of a significant difference in signups generated by the two pages. The results of the experiment showed that of 4,325 visitors to the signup page that asks only a few personal details, 387 clicked the signup button while of 4,639 visitors to the signup page that asks for many personal details, 283 clicked the signup button. At the 0.05 level of significance, is there evidence of a significant difference in signup between a signup page that asks only a few personal details and a signup page that asks for many personal details?

The null and alternative hypotheses are

$$
\begin{aligned}
H_0: \pi_1 = \pi_2 \quad &\text{or} \quad \pi_1 - \pi_2 = 0 \\
H_1: \pi_1 \neq \pi_2 \quad &\text{or} \quad \pi_1 - \pi_2 \neq 0
\end{aligned}
$$

Using the 0.05 level of significance, the critical values are -1.96 and $+1.96$ (see Figure 10.10), and the decision rule is

$$\text{Reject } H_0 \text{ if } Z_{STAT} < -1.96$$
$$\text{or if } Z_{STAT} > +1.96;$$
$$\text{otherwise, do not reject } H_0.$$

FIGURE 10.10
Regions of rejection and nonrejection when testing a hypothesis for the difference between two proportions at the 0.05 level of significance.

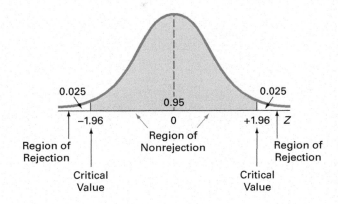

Using Equation (10.5) on page 329,

$$Z_{STAT} = \frac{(p_1 - p_2) - (\pi_1 - \pi_2)}{\sqrt{\bar{p}(1 - \bar{p})\left(\dfrac{1}{n_1} + \dfrac{1}{n_2}\right)}}$$

where

$$p_1 = \frac{X_1}{n_1} = \frac{387}{4{,}325} = 0.0895 \quad p_2 = \frac{X_2}{n_2} = \frac{283}{4{,}639} = 0.0610$$

$$\bar{p} = \frac{X_1 + X_2}{n_1 + n_2} = \frac{387 + 283}{4{,}325 + 4{,}639} = 0.0747$$

the result is

$$Z_{STAT} = \frac{(0.0895 - 0.0610) - (0)}{\sqrt{0.0747(1 - 0.0747)\left(\dfrac{1}{4{,}325} + \dfrac{1}{4{,}639}\right)}} = \frac{0.0285}{\sqrt{0.000308}} = +5.1228$$

Table 10.10 summarizes the results of the Z test for the difference between the two signup proportions using the calculation above and Figure 10.11 results on page 332. Based on the conclusions, the signup page that asks fewer personal details generates a significantly greater proportion of visitors who sign up. Therefore, as part of the last step of the DCOVA framework, the designers would recommend that MyTVLab use the signup page that asks the fewer number of personal questions.

TABLE 10.10
Z test summary for the two signup pages problem

Result	Conclusions
The $Z_{STAT} = +5.1228$ is greater than $+1.96$.	1. Reject the null hypothesis H_0. 2. Conclude that evidence exists that the signup pages are significantly different with respect to signups.
The Z test p-value $= 0.0000$ is less than the level of significance, $\alpha = 0.05$.	3. There is a probability of 0.0000 that $Z_{STAT} > 5.1228$ or < -5.1228.

FIGURE 10.11

Excel Z test worksheet results for the difference between two proportions for the two signup pages problem

	A	B
1	Z Test for Differences in Two Proportions	
2		
3	**Data**	
4	Hypothesized Difference	0
5	Level of Significance	0.05
6	**Group 1**	
7	Number of Successes	387
8	Sample Size	4325
9	**Group 2**	
10	Number of Successes	283
11	Sample Size	4639
12		
13	**Intermediate Calculations**	
14	Group 1 Proportion	0.0895 =B7/B8
15	Group 2 Proportion	0.0610 =B10/B11
16	Difference in Two Proportions	0.0285 =B14 - B15
17	Average Proportion	0.0747 =(B7 + B10)/(B8 + B11)
18	Z Test Statistic	5.1228 =(B16 - B4)/SQRT(B17 * (1 - B17) * (1/B8 + 1/B11))
19		
20	**Two-Tail Test**	
21	Lower Critical Value	-1.9600 =NORM.S.INV(B5/2)
22	Upper Critical Value	1.9600 =NORM.S.INV(1 - B5/2)
23	p -Value	0.0000 =2 * (1 - NORM.S.DIST(ABS(B18), TRUE))
24	**Reject the null hypothesis**	=IF(B23 < B5, "Reject the null hypothesis", "Do not reject the null hypothesis")

In the Worksheet

The worksheet uses the NORM.S.INV function in cells B21 and B22 to compute the lower and upper critical values.

The **COMPUTE worksheet** of the Excel Guide **Z Two Proportions** is a copy of the Figure 10.11 worksheet.

EXAMPLE 10.3

Testing for the Difference Between Two Proportions

Are men less likely than women to say that a major reason they use Facebook is to share with many people at once? A survey reported that 42% of men (193 out of 459 sampled) and 50% of women (250 out of 501 sampled) said that a major reason they use Facebook is to share with many people at once. (Source: "6 new facts about Facebook," **bit.ly/1kENZcA**.)

SOLUTION Because you want to know whether there is evidence that the proportion of men who say that a major reason they use Facebook is to share with many people at once is *less* than the proportion of women who say that a major reason they use Facebook is to share with many people at once, you have a one-tail test. The null and alternative hypotheses are

$H_0: \pi_1 \geq \pi_2$ (The proportion of men who say that a major reason they use Facebook is to share with many people at once is greater than or equal to the proportion of women who say that a major reason they use Facebook is to share with many people at once.)

$H_1: \pi_1 < \pi_2$ (The proportion of men who say that a major reason they use Facebook is to share with many people at once is less than the proportion of women who say that a major reason they use Facebook is to share with many people at once.)

Using the 0.05 level of significance, for the one-tail test in the lower tail, the critical value is -1.645. The decision rule is

$$\text{Reject } H_0 \text{ if } Z_{STAT} < -1.645;$$

$$\text{otherwise, do not reject } H_0.$$

Using Equation (10.5) on page 329,

$$Z_{STAT} = \frac{(p_1 - p_2) - (\pi_1 - \pi_2)}{\sqrt{\bar{p}(1 - \bar{p})\left(\frac{1}{n_1} + \frac{1}{n_2}\right)}}$$

where

$$p_1 = \frac{X_1}{n_1} = \frac{193}{459} = 0.4205 \quad p_2 = \frac{X_2}{n_2} = \frac{250}{501} = 0.4990$$

$$\bar{p} = \frac{X_1 + X_2}{n_1 + n_2} = \frac{193 + 250}{459 + 501} = 0.4615$$

the result is

$$Z_{STAT} = \frac{(0.4205 - 0.4990) - (0)}{\sqrt{0.4615(1 - 0.4615)\left(\frac{1}{459} + \frac{1}{501}\right)}} = \frac{-0.0785}{\sqrt{0.0010437}} = -2.4379$$

▶*(continued)*

Table 10.11 summarizes the results of the Z test for the difference between the gender proportions using the calculation above and Figure 10.12 results. Based on the conclusions, the proportion of men who say that sharing with many people at once is a major reason they use Facebook is less than the proportion of women who say that. Therefore, as part of the last step of the DCOVA framework, you might recommend that Facebook explore other ways in which the two genders differently view the usefulness of Facebook.

TABLE 10.11

Z test summary for the difference in the proportion of men and women who say sharing with many people at once is a major reason they use Facebook

Result	Conclusions
The $Z_{STAT} = -2.4379$ is less than -1.645. The Z test p-value $= 0.0074$ is less than the level of significance, $\alpha = 0.05$.	1. Reject the null hypothesis H_0. 2. Conclude that evidence exists that men are significantly less likely than women to say that sharing with many people at once is a major reason they use Facebook. 3. There is a probability of 0.0074 that $Z_{STAT} < -2.4379$.

FIGURE 10.12

Excel Z test results for the major reason men and women use Facebook

	A	B
1	Z Test for Differences in Two Proportions	
2		
3	Data	
4	Hypothesized Difference	0
5	Level of Significance	0.05
6	Group 1	
7	Number of Items of Interest	193
8	Sample Size	459
9	Group 2	
10	Number of Items of Interest	250
11	Sample Size	501
12		
13	Intermediate Calculations	
14	Group 1 Proportion	0.4205
15	Group 2 Proportion	0.4990
16	Difference in Two Proportions	-0.0785
17	Average Proportion	0.4615
18	Z Test Statistic	-2.4379
19		
20	Lower-Tail Test	
21	Lower Critical Value	-1.6449
22	p-Value	0.0074
23	Reject the null hypothesis	

In the Worksheet

The worksheet uses a design similar to the design of the COMPUTE worksheet that Figure 10.11 presents.

The **COMPUTE_LOWER worksheet** of the Excel Guide **Two Proportions** is a copy of the Figure 10.12 worksheet.

Confidence Interval Estimate for the Difference Between Two Proportions

Instead of, or in addition to, testing for the difference between the proportions of two independent populations, one can construct a confidence interval estimate for the difference between the two proportions using Equation (10.6) on page 334.

CONFIDENCE INTERVAL ESTIMATE FOR THE DIFFERENCE BETWEEN TWO PROPORTIONS

$$(p_1 - p_2) \pm Z_{\alpha/2}\sqrt{\frac{p_1(1 - p_1)}{n_1} + \frac{p_2(1 - p_2)}{n_2}} \qquad (10.6)$$

or

$$(p_1 - p_2) - Z_{\alpha/2}\sqrt{\frac{p_1(1 - p_1)}{n_1} + \frac{p_2(1 - p_2)}{n_2}} \leq (\pi_1 - \pi_2)$$

$$\leq (p_1 - p_2) + Z_{\alpha/2}\sqrt{\frac{p_1(1 - p_1)}{n_1} + \frac{p_2(1 - p_2)}{n_2}}$$

Recall the MyTVLab example that begins on page 330 that seeks to determine if there is a significant population difference between the proportion of visitors to the few personal details page who signed up and the proportion of visitors to the many personal details page who signed up. Using these Figure 10.11 software results (see page 332)

$$p_1 = \frac{X_1}{n_1} = \frac{387}{4,325} = 0.0895 \quad p_2 = \frac{X_2}{n_2} = \frac{283}{4,639} = 0.0610$$

and Equation (10.6) with 95% confidence, results in

$$(0.0895 - 0.0610) \pm (1.96)\sqrt{\frac{0.0895(1 - 0.0895)}{4,325} + \frac{0.0610(1 - 0.0610)}{4,639}}$$

$$0.0285 \pm (1.96)(0.0056)$$

$$0.0285 \pm 0.0109$$

$$0.0175 \leq (\pi_1 - \pi_2) \leq 0.0394$$

The designers have 95% confidence that the difference between the population proportion of visitors to the few personal details page who signed up and the population proportion of visitors to the many personal details page who signed up is between 0.0175 and 0.0394. In percentages, the difference is between 1.75% and 3.94%. Visitors are more likely to sign up if they are presented with a signup page that asks for fewer personal details.

PROBLEMS FOR SECTION 10.3

LEARNING THE BASICS

10.27 Let $n_1 = 100$, $X_1 = 50$, $n_2 = 100$, and $X_2 = 30$.
a. At the 0.05 level of significance, is there evidence of a significant difference between the two population proportions?
b. Construct a 95% confidence interval estimate for the difference between the two population proportions.

10.28 Let $n_1 = 100$, $X_1 = 45$, $n_2 = 50$, and $X_2 = 25$.
a. At the 0.01 level of significance, is there evidence of a significant difference between the two population proportions?
b. Construct a 99% confidence interval estimate for the difference between the two population proportions.

APPLYING THE CONCEPTS

10.29 An online survey asked 1,004 adults, "If purchasing a used car made certain upgrades or features more affordable, what would be your preferred luxury upgrade?" The results indicated that 9% of the males and 14% of the females answered window tinting.

Source: Data extracted from Ipsos, "Safety Technology Tops the List of Most Desired Features Should They Be More Affordable When Purchasing a Used Car—Particularly Collision Avoidance," **bit.ly/1RCcc1L**.

The sample sizes of males and females were not provided. Suppose that both sample sizes were 502 and that 46 of 502 males and 71 of 502 females reported window tinting as their preferred luxury upgrade of choice.

a. Is there evidence of a difference between males and females in the proportion who said they prefer window tinting as a luxury upgrade at the 0.01 level of significance?

b. Find the *p*-value in (a) and interpret its meaning.

c. Construct and interpret a 99% confidence interval estimate for the difference between the proportion of males and females who said they prefer window tinting as a luxury upgrade.

d. What are your answers to (a) through (c) if 60 males said they prefer window tinting as a luxury upgrade?

10.30 Does Cable Video on Demand (VOD D4+) increase ad effectiveness? A recent VOD study compared general TV and VOD D4+ audiences after viewing a brand ad. Data were collected on whether the viewer indicated that the ad made them want to visit the brand website. The results were:

VIEWING AUDIENCE	MADE ME WANT TO VISIT THE BRAND WEBSITE	
	Yes	No
VOD D4+	147	103
General TV	35	166

Source: Data extracted from Canoe Ventures, *Understanding VOD Advertising Effectiveness*, **bit.ly/1JnmMup, removed**.

a. Set up the null and alternative hypotheses to try to determine whether ad impact is stronger following VOD D4+ viewing than following general TV viewing.

b. Conduct the hypothesis test defined in (a), using the 0.05 level of significance.

c. Does the result of your test in (b) make it appropriate to claim that ad impact is stronger following VOD D4+ than following general TV viewing?

10.31 Are you an impulse shopper? A survey of 500 grocery shoppers indicated that 29% of males and 40% of females make an impulse purchase every time they shop. Assume that the survey consisted of 250 males and 250 females.

Source: Data extracted from *Women shoppers are impulsive while men snap up bargains*, available at **bit.ly/2sLYmVx**.

a. At the 0.05 level of significance, is there evidence of a difference in the proportion of males and females who make an impulse purchase every time they shop?

b. Find the *p*-values and interpret its meaning.

10.32 The Society for Human Resource Management (SHRM) collaborated with Globoforce on a series of organizational surveys with the goal of identifying challenges that HR leaders face and what strategies help them conquer those challenges. A recent survey indicates that employee retention/turnover (46%) and employee engagement (36%) were cited as the most important organizational challenges currently faced by HR professionals.

One strategy that may have an impact on employee retention, turnover, and engagement is a successful employee recognition program. Surveying small organizations, those with 500 to 2,499 employees, and large organizations, those with 10,000 or more employees, SHRM and Globoforce showed that 326 (77%) of the 423 small organizations have employee retention programs as compared to 167 (87%) of the 192 large organizations.

Source: Data extracted from *SHRM Survey Finding: Influencing Workplace Culture Through Employee Retention and Other Efforts*, **bit.ly/2rFvE9w**.

a. At the 0.01 level of significance, is there evidence of a significant difference between organizations with 500 to 2,499 employees and organizations with 10,000 or more employees with respect to the proportion that have employee recognition programs?

b. Find the *p*-value in (a) and interpret its meaning.

c. Construct and interpret a 99% confidence interval estimate for the difference between organizations with 500 to 2,499 employees and organizations with 10,000 or more employees with respect to the proportion that have employee recognition programs.

10.33 What social media tools do marketers commonly use? A survey by Social Media Examiner of B2B marketers (marketers that focus primarily on attracting businesses) and B2C marketers (marketers that primarily target consumers) reported that 267 (81%) of B2B marketers and 295 (44%) of B2C marketers commonly use LinkedIn as a social media tool. The study also revealed that 149 (45%) of B2B marketers and 308 (46%) of B2C marketers commonly use YouTube as a social media tool.

Data extracted from *2017 Social Media Marketing Industry Report*, **bit.ly/2rFmLzh**.

Suppose the survey was based on 330 B2B marketers and 670 B2C marketers.

a. At the 0.05 level of significance, is there evidence of a difference between B2B marketers and B2C marketers in the proportion that commonly use LinkedIn as a social media tool?

b. Find the *p*-value in (a) and interpret its value.

c. At the 0.05 level of significance, is there evidence of a difference between B2B marketers and B2C marketers in the proportion that commonly use YouTube as a social media tool?

10.34 Does cobrowsing have positive effects on the customer experience? Cobrowsing refers to the ability to have a contact center agent and customer jointly navigate an online document or mobile application on a real-time basis through the web. A study of businesses indicates that 81 of 129 cobrowsing organizations use skills-based routing to match the caller with the *right* agent, whereas 65 of 176 non-cobrowsing organizations use skills-based routing to match the caller with the *right* agent.

Source: *Cobrowsing Presents a 'Lucrative' Customer Service Opportunity*, available at **bit.ly/1wwALWr**.

a. At the 0.05 level of significance, is there evidence of a difference between cobrowsing organizations and non-cobrowsing organizations in the proportion that use skills-based routing to match the caller with the *right* agent?

b. Find the *p*-value in (a) and interpret its meaning.

10.35 One of the most innovative advances in online fundraising during the past decade has been the rise of crowdfunding websites. While features differ from site to site, crowdfunding websites give people an opportunity to set up an online fundraising web page and to accept money directly from that page through an online payments system. Kickstarter, one such crowdfunding website, reported that 72 of 415 *technology* crowdfunding projects in Canada were successfully funded in the past year and 88 of 300 *film and video* crowdfunding projects were successfully funded in Canada in the past year.

Source: Data extracted from **bit.ly/2rSCPtp**.

a. Is there evidence of a significant difference in the proportion of *technology* crowdfunding projects and *film and video* crowdfunding projects that were successful? (Use $\alpha = 0.05$.)

b. Determine the *p*-value in (a) and interpret its meaning.

c. Construct and interpret a 95% confidence interval estimate for the difference between the proportion of *technology* crowdfunding projects and *film and video* crowdfunding projects that are successful.

10.4 *F* Test for the Ratio of Two Variances

Examining the variances of two populations determines which Section 10.1 *t* test for the differences in two means to use: the pooled-variance *t* test, which assumes equal variances, or the separate-variance *t* test, which does not assume equal variances.

The test for the difference between the variances of two independent populations is based on the ratio of the two sample variances. If one assumes that each population is normally distributed, then the sampling distribution of the ratio S_1^2/S_2^2 is distributed as an *F* distribution. Unlike the normal and *t* distributions, which are symmetric, the *F* distribution is right-skewed. Equation (10.7) defines the **F test for the ratio of two variances** that uses the *F* distribution. In this test, the sample with the *larger* sample variance is defined as the first sample and the sample with the *smaller* sample variance is defined as the second sample. The population from which the first sample was drawn is defined population 1 and the population from which the second sample was drawn is defined as population 2.

F TEST STATISTIC FOR TESTING THE RATIO OF TWO VARIANCES

The F_{STAT} test statistic is equal to the variance of sample 1 (the larger sample variance) divided by the variance of sample 2 (the smaller sample variance).

$$F_{STAT} = \frac{S_1^2}{S_2^2} \qquad (10.7)$$

studentTIP

Because the numerator in Equation (10.7) contains the larger variance, the F_{STAT} statistic is always greater than or equal to 1.0.

where

$S_1^2 = $ variance of sample 1 (the larger sample variance)

$S_2^2 = $ variance of sample 2 (the smaller sample variance)

$n_1 = $ population 1 sample size

$n_2 = $ population 2 sample size

$n_1 - 1 = $ degrees of freedom from sample 1 (the numerator degrees of freedom)

$n_2 - 1 = $ degrees of freedom from sample 2 (the denominator degrees of freedom)

The F_{STAT} test statistic follows an *F* distribution with $n_1 - 1$ and $n_2 - 1$ degrees of freedom.

The critical values of the **F distribution** (see Table E.5) depend on the degrees of freedom in the two samples. The numerator degrees of freedom are the degrees of freedom for the first sample, and the denominator degrees of freedom are the degrees of freedom for the second sample.

For a given level of significance, α, the null and alternate hypotheses of a test for equality of population variances are

$$H_0: \sigma_1^2 = \sigma_2^2$$
$$H_1: \sigma_1^2 \neq \sigma_2^2$$

One rejects the null hypothesis if the F_{STAT} test statistic is greater than the upper-tail critical value, $F_{\alpha/2}$, from the *F* distribution, with $n_1 - 1$ degrees of freedom in the numerator and $n_2 - 1$ degrees of freedom in the denominator. The decision rule is

Reject H_0 if $F_{STAT} > F_{\alpha/2}$;

otherwise, do not reject H_0.

To illustrate how to use the *F* test to determine whether the two variances are equal, recall the Using Statistics scenario on page 311 that concerns the sales of VLABGo players in two different sales locations. To determine whether to use the pooled-variance *t* test or the separate-variance *t* test, one first tests the equality of the two population variances. The null and alternative hypotheses are

$$H_0: \sigma_1^2 = \sigma_2^2$$

$$H_1: \sigma_1^2 \neq \sigma_2^2$$

Because sample 1 is defined as having the larger sample variance, the rejection region in the upper tail of the *F* distribution contains $\alpha/2$. Using the level of significance $\alpha = 0.05$, the rejection region in the upper tail contains 0.025 of the distribution.

Because there are samples of 10 stores for each of the two sales locations, there are $10 - 1 = 9$ degrees of freedom for both the numerator (the sample with the larger variance) and the denominator (the sample with the smaller variance). Use Table E.5 to determine $F_{\alpha/2}$, the upper-tail critical value of the *F* distribution. From Table E.5, a portion of which Table 10.12 shows, the upper-tail critical value, $F_{\alpha/2}$, is 4.03. Therefore, the decision rule is

Reject H_0 if $F_{STAT} > F_{0.025} = 4.03$;

otherwise, do not reject H_0.

TABLE 10.12

Finding the upper-tail critical value of *F* with 9 denominator and numerator degrees of freedom for an upper-tail area of 0.025

	Cumulative Probabilities = 0.975 **Upper-Tail Area = 0.025** **Numerator df_1**						
Denominator df_2	**1**	**2**	**3**	**...**	**7**	**8**	**9**
1	647.80	799.50	864.20	...	948.20	956.70	963.30
2	38.51	39.00	39.17	...	39.36	39.37	39.39
3	17.44	16.04	15.44	...	14.62	14.54	14.47
⋮	⋮	⋮	⋮	⋮	⋮	⋮	⋮
7	8.07	6.54	5.89	...	4.99	4.90	4.82
8	7.57	6.06	5.42	...	4.53	4.43	4.36
9	7.21	5.71	5.08	...	4.20	4.10	4.03

Source: Extracted from Table E.5.

Using the Table 10.1 VLABGo sales data on page 313 and Equation (10.7) on page 336

$$S_1^2 = (42.5420)^2 = 1,809.8222 \quad S_2^2 = (32.5271)^2 = 1,058.0111$$

the result is

$$F_{STAT} = \frac{S_1^2}{S_2^2} = \frac{1,809.8222}{1,058.0111} = 1.7106$$

Because $F_{STAT} = 1.7106 < 4.03$, you do not reject H_0. Figure 10.13 on page 338 shows the results for this test, including the *p*-value, 0.4361. Because $0.4361 > 0.05$, one concludes that there is no evidence of a significant difference in the variability of the sales of the VLABGo players for the two sales locations.

In testing for a difference between two variances using the *F* test, one assumes that each of the two populations is normally distributed. The *F* test is very sensitive to the normality assumption. If boxplots or normal probability plots suggest even a mild departure from normality for either of the two populations, one should not use the *F* test. Instead, use the Levene test (see Section 11.1) or a nonparametric approach (Corder and Daniel).

In testing for the equality of variances as part of assessing the appropriateness of the pooled-variance *t* test procedure, the *F* test is a two-tail test with $\alpha/2$ in the upper tail. However, when one examines the variability in situations other than the pooled-variance *t* test, the *F* test is often a one-tail test. Example 10.4 illustrates a one-tail test.

FIGURE 10.13
Excel *F* test results for
the two different sales
locations data.

	A	B
1	F Test for Differences in Two Variances	
2		
3	Data	
4	Level of Significance	0.05
5	Larger-Variance Sample	
6	Sample Size	10 =COUNT(DataCopy!$A:$A)
7	Sample Variance	1809.822 =VAR.S(DataCopy!$A:$A)
8	Smaller-Variance Sample	
9	Sample Size	10 =COUNT(DataCopy!$B:$B)
10	Sample Variance	1058.011 =VAR.S(DataCopy!$B:$B)
11		
12	Intermediate Calculations	
13	F Test Statistic	1.7106 =B7/B10
14	Population 1 Sample Degrees of Freedom	9 =B6 - 1
15	Population 2 Sample Degrees of Freedom	9 =B9 - 1
16		
17	Two-Tail Test	
18	Upper Critical Value	4.0260 =F.INV.RT(B4/2, B14, B15)
19	p-Value	0.4361 =2 * F.DIST.RT(B13, B14, B15)
20	Do not reject the null hypothesis	=IF(B19 < B4, "Reject the null hypothesis",)
		"Do not reject the null hypothesis")

In the Worksheet

The worksheet uses the F.INV.RT
function to compute the upper critical
value in cell B18 and the F.DIST.RT
funciton to help compute the *p*-value.

The **COMPUTE worksheet** of the Excel
Guide **F Two Variances workbook** is a
copy of the Figure 10.8 worksheet.

EXAMPLE 10.4

**A One-Tail Test
for the Difference
Between Two
Variances**

Waiting time is a critical issue at fast-food chains, which not only want to minimize the mean
service time but also want to minimize the variation in the service time from customer to customer. One fast-food chain carried out a study to measure the variability in the waiting time
(defined as the time in minutes from when an order was completed to when it was delivered to
the customer) at lunch and breakfast at one of the chain's stores. The results were as follows:

$$\text{Lunch: } n_1 = 25 \quad S_1^2 = 4.4$$
$$\text{Breakfast: } n_2 = 21 \quad S_2^2 = 1.9$$

At the 0.05 level of significance, is there evidence that there is more variability in the service time
at lunch than at breakfast? Assume that the population service times are normally distributed.

SOLUTION The null and alternative hypotheses are

$$H_0: \sigma_L^2 \le \sigma_B^2$$
$$H_1: \sigma_L^2 > \sigma_B^2$$

Equation (10.7) on page 336 calculates the F_{STAT} test statistic as

$$F_{STAT} = \frac{S_1^2}{S_2^2}$$

One uses Table E.5 to find the upper critical value of the F distribution. With $n_1 - 1 = 25 - 1 = 24$
degrees of freedom in the numerator, $n_2 - 1 = 21 - 1 = 20$ degrees of freedom in the denominator, and $\alpha = 0.05$, the upper-tail critical value, $F_{0.05}$, is 2.08. The decision rule is

$$\text{Reject } H_0 \text{ if } F_{STAT} > 2.08;$$
$$\text{otherwise, do not reject } H_0.$$

From Equation (10.7)

$$F_{STAT} = \frac{S_1^2}{S_2^2}$$
$$= \frac{4.4}{1.9} = 2.3158$$

Because $F_{STAT} = 2.3158 > 2.08$, you reject H_0. Using a 0.05 level of significance, one concludes
that there is evidence that there is more variability in the service time at lunch than at breakfast.

PROBLEMS FOR SECTION 10.4

LEARNING THE BASICS

10.36 Determine the upper-tail critical values of *F* in each of the following two-tail tests.
a. $\alpha = 0.10, n_1 = 16, n_2 = 21$
b. $\alpha = 0.05, n_1 = 16, n_2 = 21$
c. $\alpha = 0.01, n_1 = 16, n_2 = 21$

10.37 Determine the upper-tail critical value of *F* in each of the following one-tail tests.
a. $\alpha = 0.05, n_1 = 16, n_2 = 21$
b. $\alpha = 0.01, n_1 = 16, n_2 = 21$

10.38 The following information is available for two samples selected from independent normally distributed populations:

$$\text{Population A: } n_1 = 25 \quad S_1^2 = 16$$
$$\text{Population B: } n_2 = 25 \quad S_2^2 = 25$$

a. Which sample variance do you place in the numerator of F_{STAT}?
b. What is the value of F_{STAT}?

10.39 The following information is available for two samples selected from independent normally distributed populations:

$$\text{Population A: } n_1 = 25 \quad S_1^2 = 161.9$$
$$\text{Population B: } n_2 = 25 \quad S_2^2 = 133.7$$

What is the value of F_{STAT} if you are testing the null hypothesis $H_0: \sigma_1^2 = \sigma_2^2$?

10.40 In Problem 10.39, how many degrees of freedom are there in the numerator and denominator of the *F* test?

10.41 In Problems 10.38 and 10.39, what is the upper-tail critical value for *F* if the level of significance, α, is 0.05 and the alternative hypothesis is $H_1: \sigma_1^2 \neq \sigma_2^2$?

10.42 In Problem 10.39, what is your statistical decision?

10.43 The following information is available for two samples selected from independent but very right-skewed populations:

$$\text{Population A: } n_1 = 16 \quad S_1^2 = 47.3$$
$$\text{Population B: } n_2 = 13 \quad S_2^2 = 36.4$$

Should you use the *F* test to test the null hypothesis of equality of variances? Discuss.

10.44 In Problem 10.43, assume that two samples are selected from independent normally distributed populations.
a. At the 0.05 level of significance, is there evidence of a difference between σ_1^2 and σ_2^2?
b. Suppose that you want to perform a one-tail test. At the 0.05 level of significance, what is the upper-tail critical value of *F* to determine whether there is evidence that $\sigma_1^2 > \sigma_2^2$? What is your statistical decision?

APPLYING THE CONCEPTS

10.45 Is there a difference in the variance of the satisfaction rating of traditional cellphone providers who bill for service at the end of a month (often under a contract) and prepaid cellphone service providers who bill in advance without a contract? The file

Cellphone Providers contains the satisfaction rating for 10 traditional cellphone providers and 13 prepaid cellphone service providers.

Source: Data extracted from "Carrier Ratings: Why It Pays to Think Small," *Consumer Reports*, February 2016, p. 51.

a. At the 0.05 level of significance, is there evidence of a difference in the variability of the satisfaction rating between the types of cellphone providers?
b. Determine the *p*-value in (a) and interpret its meaning.
c. What assumption do you need to make in (a) about the two populations in order to justify your use of the *F* test?
d. Based on the results of (a) and (b), which *t* test defined in Section 10.1 should you use to compare the mean satisfaction rating of the two types of cellphone providers?

✓ SELF TEST **10.46** *Accounting Today* identified top accounting firms in 10 geographic regions across the United States. All 10 regions reported growth in a recent year. The Southeast and Gulf Coast regions reported growths of 12.49% and 9.78%, respectively. A characteristic description of the accounting firms in the Southeast and Gulf Coast regions included the number of partners in the firm. The file **Accounting Partners 2** contains the number of partners.

Source: Data extracted from *Accounting Today*, "The 2019 Top Firms and Regional Leaders," available at **bit.ly/2XdQVVW**.

a. At the 0.05 level of significance, is there evidence of a difference in the variability in numbers of partners for Southeast region accounting firms and Gulf Coast accounting firms?
b. Determine the *p*-value in (a) and interpret its meaning.
c. What assumption do you have to make about the two populations in order to justify the use of the *F* test?
d. Based on (a) and (b), which *t* test defined in Section 10.1 should you use to test whether there is a significant difference in the mean number of partners for Southeast region accounting firms and Gulf Coast accounting firms?

10.47 A bank with a branch located in a commercial district of a city has the business objective of improving the process for serving customers during the noon-to-1 P.M. lunch period. To do so, the waiting time (defined as the number of minutes that elapses from when the customer enters the line until he or she reaches the teller window) needs to be shortened to increase customer satisfaction. A random sample of 15 customers is selected and the waiting times are collected and stored in **Bank Waiting**. These data are:

4.21	5.55	3.02	5.13	4.77	2.34	3.54	3.20
4.50	6.10	0.38	5.12	6.46	6.19	3.79	

Suppose that another branch, located in a residential area, is also concerned with the noon-to-1 P.M. lunch period. A random sample of 15 customers is selected and the waiting times are collected and stored in **Bank Waiting 2**. These data are:

9.66	5.90	8.02	5.79	8.73	3.82	8.01	8.35
10.49	6.68	5.64	4.08	6.17	9.91	5.47	

a. Is there evidence of a difference in the variability of the waiting time between the two branches? (Use $\alpha = 0.05$.)
b. Determine the p-value in (a) and interpret its meaning.
c. What assumption about the population distribution of each bank is necessary in (a)? Is the assumption valid for these data?
d. Based on the results of (a), is it appropriate to use the pooled-variance t test to compare the means of the two branches?

10.48 The Super Bowl is a big viewing event watched by close to 200 million Americans, so it is also a big event for advertisers. The file **Super Bowl Ad Ratings** contains the rating of ads that ran between the opening kickoff and the final whistle.

Source: Data extracted from **www.admeter.usatoday.com/results/2019**.

a. Is there evidence of a difference in the variability of the scores between the ads that ran in the first half and the ads that did not? (Use $\alpha = 0.05$.)
b. Determine the p-value in (a) and interpret its meaning.
c. What assumption about the population distribution of the two types of ads is necessary in (a)? Is the assumption valid for these data?
d. Based on the results of (a), which t test defined in Section 10.1 should you use to compare the mean scores of the ads that ran in the first half and the ads that did not?

10.49 We Are Social and Hootsuite reported that the typical American spends 2.02 hours (121 minutes) per day accessing the Internet through a mobile device.

Source: *Digital in 2017 Global Overview*, available at **bit.ly/2jXeS3F**.

You wonder if males and females spend differing amounts of time per day accessing the Internet through a mobile device.

You select a sample of 60 friends and family (30 males and 30 females), collect times spent per day accessing the Internet through a mobile device (in minutes), and store the data collected in **Mobile Device Study 2**.
a. Using a 0.05 level of significance, is there evidence of a difference in the variances of time spent per day accessing the Internet via mobile device between males and females?
b. On the basis of the results in (a), which t test that Section 10.1 defines should you use to compare the means of males and females? Discuss.

10.50 A taxicab company has been receiving an increasing number of complaints concerning the delay time between when a call for a taxicab is received and when the passenger is picked up. The file **Taxi Delays** contains the delay times (in minutes) for two drivers on the same route for a period of 35 days.

Source: Data extracted from M. Sharma, "Above and Beyond," *Six Sigma Forum Magazine*, May 2015, p. 21–25.

a. At the 0.05 level of significance, is there evidence of a difference in the variance of the delay times between the two drivers?
b. What assumption do you need to make in order to do (a)?
c. Evaluate the validity of the assumption in (a).
d. Based on the results of (a), which t-test for the difference between the means from Section 10.1 should you use to determine whether there is evidence of a difference in the mean delay time between the two drivers?

10.5 Effect Size

Section 9.5 discusses the issue of the practical significance of a statistically significant test and explains that when a very large sample is selected, a statistically significant result can be of limited importance. The **Section 10.5 online topic** shows how to measure the effect size of a statistical test.

▼USING **STATISTICS**
Differing Means for Selling..., Revisited

In the Using Statistics scenario, you sought to show that the sales location in a store could affect sales of a product. If you could show such an effect, you would have an argument for charging a placement fee for the better location. You designed an experiment that would sell the new VLABGo mobile streaming media player in one of two sales locations, at a special front of store location or in the mobile electronics aisle. In the experiment, 10 stores used the special front location to sell VLABGo players and 10 stores used the mobile electronics aisle.

Using a t test for the difference between two means, you conclude that the mean sales using the special front location

are higher than the mean sales for the in-aisle location. A confidence interval enabled you to infer with 95% confidence that population mean amount sold at the special front location was between 8.52 and 79.68 more than the in-aisle location. The F test for the difference between two variances enables you to conclude that there was no significant difference in the variability of the sales for the two sales locations. That you now have evidence that sales are higher in the special front location gives you one argument for charging manufacturers a placement fee for that location.

▼ SUMMARY

This chapter discusses statistical test procedures for analyzing possible differences between means, proportions, and variances. The chapter also discusses a test procedure that is frequently used when analyzing differences between the means of two related samples. Selecting the most appropriate hypothesis-testing procedure requires investigating the validity of the assumptions underlying each of the possible hypothesis-testing procedures that might be used. Table 10.13 lists the analyses that compare two populations which this chapter discusses.

TABLE 10.13
Analyses that compare two populations

Numerical Data	Categorical Data
t tests for the difference in the means of two independent populations (Section 10.1)	Z test for the difference between two proportions (Section 10.3)
Paired t test (Section 10.2)	
F test for the difference between two variances (Section 10.4)	

The Figure 10.14 roadmap guides readers to select the most appropriate procedure that compares two populations. The roadmap presents these questions:

1. What type of variables do you have?
2. If you have a numerical variable, do you have independent samples or related samples?
3. If you have independent samples, is your focus on variability or central tendency?
4. If your focus is central tendency and you can assume approximate normality, can you assume that the variances of the two populations are equal?

FIGURE 10.14
Roadmap for selecting a test of hypothesis for two populations

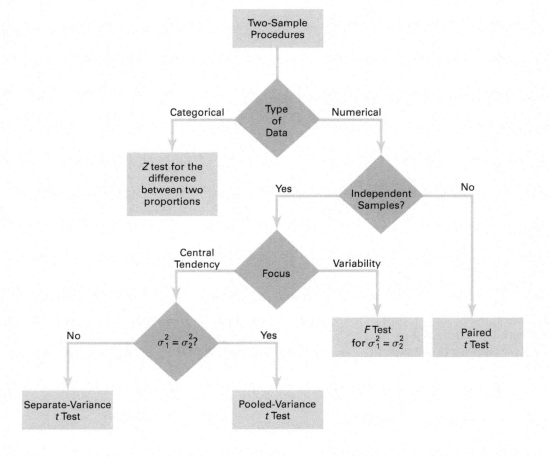

▼ REFERENCES

Corder, G. W., and D. I. Foreman, *Nonparametric Statistics: A Step-by-Step Approach*. New York: Wiley, 2014.

Daniel, W. *Applied Nonparametric Statistics*, 2nd ed. Boston: Houghton Mifflin, 1990.

Satterthwaite, F. E. "An Approximate Distribution of Estimates of Variance Components." *Biometrics Bulletin*, 2(1946): 110–114.

Snedecor, G. W., and W. G. Cochran. *Statistical Methods*, 8th ed. Ames, IA: Iowa State University Press, 1989.

▼ KEY EQUATIONS

Pooled-Variance t Test for the Difference Between Two Means

$$t_{STAT} = \frac{(\bar{X}_1 - \bar{X}_2) - (\mu_1 - \mu_2)}{\sqrt{S_p^2\left(\frac{1}{n_1} + \frac{1}{n_2}\right)}} \qquad (10.1)$$

Confidence Interval Estimate for the Difference Between the Means of Two Independent Populations

$$(\bar{X}_1 - \bar{X}_2) \pm t_{\alpha/2}\sqrt{S_p^2\left(\frac{1}{n_1} + \frac{1}{n_2}\right)} \qquad (10.2)$$

or

$$(\bar{X}_1 - \bar{X}_2) - t_{\alpha/2}\sqrt{S_p^2\left(\frac{1}{n_1} + \frac{1}{n_2}\right)} \leq \mu_1 - \mu_2$$

$$\leq (\bar{X}_1 - \bar{X}_2) + t_{\alpha/2}\sqrt{S_p^2\left(\frac{1}{n_1} + \frac{1}{n_2}\right)}$$

Paired t Test for the Mean Difference

$$t_{STAT} = \frac{\bar{D} - \mu_D}{\frac{S_D}{\sqrt{n}}} \qquad (10.3)$$

Confidence Interval Estimate for the Mean Difference

$$\bar{D} \pm t_{\alpha/2}\frac{S_D}{\sqrt{n}} \qquad (10.4)$$

or

$$\bar{D} - t_{\alpha/2}\frac{S_D}{\sqrt{n}} \leq \mu_D \leq \bar{D} + t_{\alpha/2}\frac{S_D}{\sqrt{n}}$$

Z Test for the Difference Between Two Proportions

$$Z_{STAT} = \frac{(p_1 - p_2) - (\pi_1 - \pi_2)}{\sqrt{\bar{p}(1 - \bar{p})\left(\frac{1}{n_1} + \frac{1}{n_2}\right)}} \qquad (10.5)$$

Confidence Interval Estimate for the Difference Between Two Proportions

$$(p_1 - p_2) \pm Z_{\alpha/2}\sqrt{\frac{p_1(1 - p_1)}{n_1} + \frac{p_2(1 - p_2)}{n_2}} \qquad (10.6)$$

or

$$(p_1 - p_2) - Z_{\alpha/2}\sqrt{\frac{p_1(1 - p_1)}{n_1} + \frac{p_2(1 - p_2)}{n_2}} \leq (\pi_1 - \pi_2)$$

$$\leq (p_1 - p_2) + Z_{\alpha/2}\sqrt{\frac{p_1(1 - p_1)}{n_1} + \frac{p_2(1 - p_2)}{n_2}}$$

F Test Statistic for Testing the Ratio of Two Variances

$$F_{STAT} = \frac{S_1^2}{S_2^2} \qquad (10.7)$$

▼ KEY TERMS

difference scores   322
F distribution   336
F test for the ratio of two variances   336
matched samples   322

paired t test for the mean difference   322
pooled-variance t test   312
related (populations)   321
repeated measurements   321

robust   315
separate-variance t test   318
two-sample tests   312
Z test for the difference between two proportions   329

▼CHECKING YOUR UNDERSTANDING

10.51 What are some of the criteria used in the selection of a particular hypothesis-testing procedure?

10.52 Under what conditions should you use the pooled-variance *t* test to examine possible differences in the means of two independent populations?

10.53 Under what conditions should you use the *F* test to examine possible differences in the variances of two independent populations?

10.54 What is the distinction between two independent populations and two related populations?

10.55 What is the distinction between repeated measurements and matched items?

10.56 When you have two independent populations, explain the similarities and differences between the test of hypothesis for the difference between the means and the confidence interval estimate for the difference between the means.

10.57 Under what conditions should you use the paired *t* test for the mean difference between two related populations?

▼CHAPTER REVIEW PROBLEMS

10.58 The American Society for Quality (ASQ) conducted a salary survey of all its members. ASQ members work in all areas of manufacturing and service-related institutions, with a common theme of an interest in quality. Two job titles are black belt and green belt. (See Section 15.6 for a description of these titles in a Six Sigma quality improvement initiative.) Descriptive statistics concerning salaries for these two job titles are given in the following table:

Job Title	Sample Size	Mean	Standard Deviation
Black belt	96	101.785	33.354
Green belt	29	72.345	19.344

Source: Data extracted from "QP Salary Survey," *Quality Progress*, December 2017, p. 34.

a. Using a 0.05 level of significance, is there a difference in the variability of salaries between black belts and green belts?
b. Based on the result of (a), which *t* test defined in Section 10.1 is appropriate for comparing mean salaries?
c. Using a 0.05 level of significance, is the mean salary of black belts greater than the mean salary of green belts?

10.59 How do private universities and public colleges compare with respect to debt at graduation incurred by students? The file **College Debt** contains the average debt at graduation incurred by students for 100 private universities and 100 public colleges, as reported by Kiplinger.

Source: Data extracted from "Kiplinger's Best College Values," available at **bit.ly/1z39qsT**.

a. At the 0.05 level of significance, is there a difference in the variance of average graduation debt incurred by students for private universities and public colleges?
b. Using the results of (a), which *t* test is appropriate for comparing mean debt at graduation incurred by students at private universities and public colleges.

c. At the 0.05 level of significance, conduct the test selected in (b).
d. Write a short summary of your findings.

10.60 Do males and females differ in the amount of time they spend online and the amount of time they spend playing games while online? A study reported that women spent a mean of 1,254 minutes per week online as compared to 1,344 minutes per week for men. Suppose that the sample sizes were 100 each for women and men and that the standard deviation for women was 60 minutes per week as compared to 70 minutes per week for men.

Source: Data extracted from Ofcom, *Adults' Media Use and Attitudes, Report 2016*, **bit.ly/2emgWRk**.

a. Using a 0.01 level of significance, is there evidence of a difference in the variances of the amount of time spent online between women and men?
b. To test for a difference in the mean online time of women and men, is it most appropriate to use the pooled-variance *t* test or the separate-variance *t* test? Using a 0.01 level of significance, use the most appropriate test to determine if there is a difference in the mean amount of time spent online between women and men.

The report found that women spent a mean of 294 minutes per week playing games while online compared to a mean of 360 minutes per week for men. Suppose that the standard deviation for women was 15 minutes per week compared to 20 minutes per week for men.

c. Using a 0.01 level of significance, is there evidence of a difference in the variances of the amount of time spent playing games while online per week by women and men?
d. Based on the results of (c), use the most appropriate test to determine, at the 0.01 level of significance, whether there is evidence of a difference in the mean amount of time spent playing games online per week by women and men.

10.61 The file **Restaurants** contains the ratings for food, décor, service, and the price per person for a sample of 50 restaurants located in a center city and 50 restaurants located in an outlying area. Completely analyze the differences between center city and outlying area restaurants for the variables food rating, décor rating, service rating, and cost per person, using $\alpha = 0.05$.

Source: Data extracted from *Zagat Survey 2016 New York City Restaurants*.

10.62 A computer information systems professor is interested in studying the amount of time students in a programming course take to write a small Python program. The professor hires you to analyze the following results (in minutes), stored in Python , from a random sample of nine students:

$$10 \quad 13 \quad 9 \quad 15 \quad 12 \quad 13 \quad 11 \quad 13 \quad 12$$

a. At the 0.05 level of significance, is there evidence that the population mean time is greater than 10 minutes? What will you tell the professor?
b. Suppose that the professor, when checking her results, realizes that the fourth student needed 51 minutes rather than the recorded 15 minutes to write the Python program. At the 0.05 level of significance, reanalyze the question posed in (a), using the revised data. What will you tell the professor now?
c. The professor is perplexed by these paradoxical results and requests an explanation from you regarding the justification for the difference in your findings in (a) and (b). Discuss.
d. A few days later, the professor calls to tell you that the dilemma is completely resolved. The original number 15 (the fourth data value) was correct, and therefore your findings in (a) are being used in the article she is writing for a computer journal. Now she wants to hire you to compare the results from that group of Introduction to Computers students against those from a sample of 11 computer majors in order to determine whether there is evidence that computer majors can write a Python program in less time than introductory students. For the computer majors, the sample mean is 8.5 minutes, and the sample standard deviation is 2.0 minutes. At the 0.05 level of significance, completely analyze these data. What will you tell the professor?
e. A few days later, the professor calls again to tell you that a reviewer of her article wants her to include the *p*-value for the "correct" result in (a). In addition, the professor inquires about an unequal-variances problem, which the reviewer wants her to discuss in her article. In your own words, discuss the concept of *p*-value and also describe the unequal-variances problem. Then, determine the *p*-value in (a) and discuss whether the unequal-variances problem had any meaning in the professor's study.

10.63 Do social shoppers differ from other online consumers with respect to spending behavior? A study of browser-based shopping sessions reported that social shoppers, consumers who click away from social networks to retail sites or share an item on a social network, spent a mean of $126.12 on a retail site in a 30-day period compared to other online shoppers who spent a mean of $115.55.

Source: Data extracted from "Social shoppers spend 8% more than other online consumers," **bit.ly/1FyyXP5**.

Suppose that the study consisted of 500 social shoppers and 500 other online shoppers and the standard deviation of the order value was $40 for social shoppers and $10 for other online shoppers. Assume a level of significance of 0.05.

a. Is there evidence of a difference in the variances of the order values between social shoppers and other online shoppers?
b. Is there evidence of a difference in the mean order value between social shoppers and other online shoppers?
c. Construct a 95% confidence interval estimate for the difference in mean order value between social shoppers and other online shoppers.

10.64 The lengths of life (in hours) of a sample of 40 six-watt light emitting diode (LED) light bulbs produced by manufacturer A and a sample of 40 six-watt light emitting diode (LED) light bulbs produced by manufacturer B are stored in Bulbs . Completely analyze the differences between the lengths of life of the light emitting diode (LED) light bulbs produced by the two manufacturers. (Use $\alpha = 0.05$.)

10.65 A hotel manager looks to enhance the initial impressions that hotel guests have when they check in. Contributing to initial impressions is the time it takes to deliver a guest's luggage to the room after check-in. A random sample of 20 deliveries on a particular day were selected in Wing A of the hotel, and a random sample of 20 deliveries were selected in Wing B. The results are stored in Luggage . Analyze the data and determine whether there is a difference between the mean delivery times in the two wings of the hotel. (Use $\alpha = 0.05$.)

10.66 The owner of a restaurant that serves Continental-style entrées has the business objective of learning more about the patterns of patron demand during the Friday-to-Sunday weekend time period. She decided to study the demand for dessert during this time period. In addition to studying whether a dessert was ordered, she will study the gender of the individual and whether a beef entrée was ordered. Data were collected from 630 customers and organized in the following contingency tables:

DESSERT ORDERED	GENDER		
	Male	Female	Total
Yes	96	50	146
No	234	250	484
Total	330	300	630

DESSERT ORDERED	BEEF ENTRÉE		
	Yes	No	Total
Yes	74	68	142
No	123	365	488
Total	197	433	630

a. At the 0.05 level of significance, is there evidence of a difference between males and females in the proportion who order dessert?
b. At the 0.05 level of significance, is there evidence of a difference in the proportion who order dessert based on whether a beef entrée has been ordered?

10.67 The manufacturer of Boston and Vermont asphalt shingles knows that product weight is a major factor in the customer's perception of quality. Moreover, the weight represents the amount of raw materials being used and is therefore very important to the company from a cost standpoint. The last stage of the assembly line packages the shingles before they are placed on wooden pallets. Once a pallet is full (a pallet for most brands holds 16 squares of shingles), it is weighed, and the measurement is recorded. The file Pallet contains the weights (in pounds) from a sample of 368 pallets of Boston shingles and 330 pallets of Vermont shingles. Completely analyze the differences in the weights of the Boston and Vermont shingles, using $\alpha = 0.05$.

10.68 The manufacturer of Boston and Vermont asphalt shingles provides its customers with a 20-year warranty on most of its products. To determine whether a shingle will last as long as the warranty period, the manufacturer conducts accelerated-life testing. Accelerated-life testing exposes the shingle to the stresses it would be subject to in a lifetime of normal use in a laboratory setting via an experiment that takes only a few minutes to conduct. In this test, a shingle is repeatedly scraped with a brush for a short period of time, and the shingle granules removed by the brushing are weighed (in grams). Shingles that experience low amounts of granule loss are expected to last longer in normal use than shingles that experience high amounts of granule loss. In this situation, a shingle should experience no more than 0.8 grams of granule loss if it is expected to last the length of the warranty period. The file Granule contains a sample of 170 measurements made on the company's Boston shingles and 140 measurements made on Vermont shingles. Completely analyze the differences in the granule loss of the Boston and Vermont shingles, using $\alpha = 0.05$.

10.69 Market data indicates that smartphone users are very concerned about the battery life of their smartphones. An experiment is conducted in which the battery life of a newly designed smartphone battery is compared to the battery life of an existing smartphone battery. The following table summarizes the results of the experiment.

Design	Sample Size	Mean (hours)	Standard Deviation (hours)
Existing	30	18.45	0.35
New	30	16.10	0.15

Source: Data extracted from L. Ferryanto, "Are These The Same?," *Quality Progress*, May 2017, 29–36.

Completely analyze these data and indicate which battery design you prefer.

REPORT WRITING EXERCISE

10.70 Referring to the results of Problems 10.67 and 10.68 concerning the weight and granule loss of Boston and Vermont shingles, write a report that summarizes your conclusions.

▾CASES

Managing Ashland MultiComm Services

AMS communicates with customers who subscribe to telecommunications services through a special secured email system that sends messages about service changes, new features, and billing information to in-home digital set-top boxes for later display. To enhance customer service, the operations department established the business objective of reducing the amount of time to fully update each subscriber's set of messages. The department selected two candidate messaging systems and conducted an experiment in which 30 randomly chosen cable subscribers were assigned one of the two systems (15 assigned to each system). Update times were measured, and the results are organized in Table AMS 10.1 and stored in AMS10.

1. Analyze the data in Table AMS 10.1 and write a report to the computer operations department that indicates your findings. Include an appendix in which you discuss the reason you selected a particular statistical test to compare the two independent groups of callers.

2. Suppose that instead of the research design described in the case, there were only 15 subscribers sampled, and the update process for each subscriber email was measured for each of the two messaging systems. Suppose that the results were organized in Table AMS 10.1—making each row in the table a pair of values for an individual subscriber. Using these suppositions, reanalyze the Table AMS 10.1 data and write a report for presentation to the team that indicates your findings.

TABLE AMS 10.1
Update times (in seconds) for two different email interfaces

Email Interface 1	Email Interface 2
4.13	3.71
3.75	3.89
3.93	4.22
3.74	4.57
3.36	4.24
3.85	3.90
3.26	4.09
3.73	4.05
4.06	4.07
3.33	3.80
3.96	4.36
3.57	4.38
3.13	3.49
3.68	3.57
3.63	4.74

Digital Case

Apply your knowledge about hypothesis testing in this Digital Case, which continues the cereal-fill packaging dispute Digital Case from Chapters 7 and 9.

Even after the recent public experiment about cereal box weights, Consumers Concerned About Cereal Cheaters (CCACC) remains convinced that Oxford Cereals has misled the public. The group has created and circulated **MoreCheating.pdf**, a document in which it claims that cereal boxes produced at Plant Number 2 in Springville weigh less than the claimed mean of 368 grams. Review this document and then answer the following questions:

1. Do the CCACC's results prove that there is a statistically significant difference in the mean weights of cereal boxes produced at Plant Numbers 1 and 2?

2. Perform the appropriate analysis to test the CCACC's hypothesis. What conclusions can you reach based on the data?

Sure Value Convenience Stores

You continue to work in the corporate office for a nationwide convenience store franchise that operates nearly 10,000 stores. The per-store daily customer count (the mean number of customers in a store in one day) has been steady, at 900, for some time. To increase the customer count, the chain is considering cutting prices for coffee beverages. The small size will now be either $0.59 or $0.79 instead of $0.99. Even with this reduction in price, the chain will have a 40% gross margin on coffee.

The question to be answered is how much to cut prices to increase the daily customer count without reducing the gross margin on coffee sales too much. The chain decides to carry out an experiment in a sample of 30 stores where customer counts have been running almost exactly at the national average of 900. In 15 of the stores, the price of a small coffee will now be $0.59 instead of $0.99, and in 15 other stores, the price of a small coffee will now be $0.79. After four weeks, the 15 stores that priced the small coffee at $0.59 had a mean daily customer count of 964 and a standard deviation of 88, and the 15 stores that priced the small coffee at $0.79 had a mean daily customer count of 941 and a standard deviation of 76. Analyze these data (using the 0.05 level of significance) and answer the following questions.

1. Does reducing the price of a small coffee to either $0.59 or $0.79 increase the mean per-store daily customer count?

2. If reducing the price of a small coffee to either $0.59 or $0.79 increases the mean per-store daily customer count, is there any difference in the mean per-store daily customer count

between stores in which a small coffee was priced at $0.59 and stores in which a small coffee was priced at $0.79?

3. What price do you recommend for a small coffee?

CardioGood Fitness

Return to the CardioGood Fitness case first presented on page 33. Using the data stored in CardioGood Fitness :

1. Determine whether differences exist between males and females in their age in years, education in years, annual household income ($), mean number of times the customer plans to use the treadmill each week, and mean number of miles the customer expects to walk or run each week.

2. Write a report to be presented to the management of Cardio-Good Fitness detailing your findings.

More Descriptive Choices Follow-Up

Use the "More Descriptive Choices, Revisited" section on page 147 to answer the following question.

Determine whether there is a difference in the 1-year return percentage, 5-year return percentages, and 10-year return percentages of the growth and value funds (stored in Retirement Funds).

Clear Mountain State Student Survey

The Student News Service at Clear Mountain State University (CMSU) has decided to gather data about the undergraduate students who attend CMSU. It creates and distributes a survey of 14 questions and receives responses from 111 undergraduates (stored in Student Survey).

1. At the 0.05 level of significance, is there evidence of a difference between males and females in grade point average, expected starting salary, number of social networking sites registered for, age, spending on textbooks and supplies, text messages sent in a week, and the wealth needed to feel rich?

2. At the 0.05 level of significance, is there evidence of a difference between students who plan to go to graduate school and those who do not plan to go to graduate school in grade point average, expected starting salary, number of social networking sites registered for, age, spending on textbooks and supplies, text messages sent in a week, and the wealth needed to feel rich?

▼ EXCEL GUIDE

EG10.1 COMPARING the MEANS of TWO INDEPENDENT POPULATIONS

Pooled-Variance *t* Test for the Difference Between Two Means

Key Technique Use the **T.INV.2T**(*level of significance, total degrees of freedom*) function to compute the lower and upper critical values.

Use the **T.DIST.2T**(*absolute value of the t test statistic, total degrees of freedom*) to compute the *p*-value.

Example Perform the Figure 10.3 pooled-variance *t* test for the Table 10.1 Arlingtons sales data for the two in-store sales locations.

PHStat Use **Pooled-Variance t Test**.

For the example, open to the **DATA worksheet** of the **VLABGo workbook**. Select **PHStat → Two-Sample Tests (Unsummarized Data) → Pooled-Variance t Test**. In the procedure's dialog box (shown below):

1. Enter **0** as the **Hypothesized Difference**.
2. Enter **0.05** as the **Level of Significance**.
3. Enter **A1:A11** as the **Population 1 Sample Cell Range**.
4. Enter **B1:B11** as the **Population 2 Sample Cell Range**.
5. Check **First cells in both ranges contain label**.
6. Click **Two-Tail Test**.
7. Check **Confidence Interval Estimate** and enter **95** as the **Confidence level**.
8. Enter a **Title** and click **OK**.

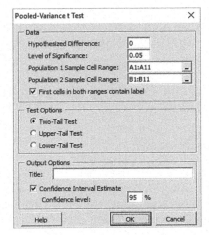

When using summarized data, select **PHStat → Two-Sample Tests (Summarized Data) → Pooled-Variance t Test**. In that procedure's dialog box, enter the hypothesized difference and level of significance, as well as the sample size, sample mean, and sample standard deviation for each sample.

Workbook Use the **COMPUTE worksheet** of the **Pooled-Variance T workbook** as a template.

The worksheet already contains the data and formulas to use the unsummarized data for the example. For other problems, use this worksheet with either unsummarized or summarized data.

For unsummarized data, paste the data in columns A and B in the **DataCopy** worksheet and keep the COMPUTE worksheet formulas that compute the sample size, sample mean, and sample standard deviation in the cell range B7:B13. For summarized data, replace the formulas in the cell range B7:B13 with the sample statistics and ignore the DataCopy worksheet.

Use the **COMPUTE_LOWER** or **COMPUTE_ UPPER** worksheets in the same workbook as templates for performing one-tail pooled-variance *t* tests with either unsummarized or summarized data. For unsummarized data, paste the new data into the DataCopy worksheet. For summarized data, replace COMPUTE worksheet formulas with sample statistics.

Analysis ToolPak Use **t-Test: Two-Sample Assuming Equal Variances**.

For the example, open to the **DATA worksheet** of the **VLABGo workbook** and:

1. Select **Data → Data Analysis**.
2. In the Data Analysis dialog box, select **t-Test: Two-Sample Assuming Equal Variances** from the **Analysis Tools** list and then click **OK**.

In the procedure's dialog box (shown below):

3. Enter **A1:A11** as the **Variable 1 Range**.
4. Enter **B1:B11** as the **Variable 2 Range**.
5. Enter **0** as the **Hypothesized Mean Difference**.
6. Check **Labels** and enter **0.05** as **Alpha**.
7. Click **New Worksheet Ply**.
8. Click **OK**.

Results (shown below) appear in a new worksheet that contains both two-tail and one-tail test critical values and *p*-values. Unlike the results shown in Figure 10.3, only the positive (upper) critical value is listed for the two-tail test.

	A	B	C
1	t-Test: Two-Sample Assuming Equal Variances		
2			
3		Front	In-Aisle
4	Mean	246.4	202.3
5	Variance	1809.8222	1058.0111
6	Observations	10	10
7	Pooled Variance	1433.9167	
8	Hypothesized Mean Difference	0	
9	df	18	
10	t Stat	2.6041	
11	P(T<=t) one-tail	0.0090	
12	t Critical one-tail	1.7341	
13	P(T<=t) two-tail	0.0179	
14	t Critical two-tail	2.1009	

Confidence Interval Estimate for the Difference Between Two Means

PHStat The *PHStat* instructions for the pooled-variance *t* test includes a step to create a confidence interval estimate.

Workbook Use the *Workbook* instructions for the pooled-variance *t* test. The COMPUTE worksheet of the Pooled-Variance T workbook includes confidence interval estimate calculations in columns D and E.

Separate-Variance *t* Test for the Difference Between Two Means, Assuming Unequal Variances

Key Technique Use the **T.INV.2T**(*level of significance, degrees of freedom*) function to compute the lower and upper critical values.

Use the **T.DIST.2T**(*absolute value of the t test statistic, degrees of freedom*) to compute the *p*-value.

Example Perform the Figure 10.6 separate-variance *t* test for the two in-store sales locations data on page 318.

PHStat Use **Separate-Variance t Test**.

For the example, open to the **DATA worksheet** of the **VLABGo workbook**. Select **PHStat → Two-Sample Tests (Unsummarized Data) → Separate-Variance t Test**. In the procedure's dialog box (shown in the right column):

1. Enter **0** as the **Hypothesized Difference**.
2. Enter **0.05** as the **Level of Significance**.
3. Enter **A1:A11** as the **Population 1 Sample Cell Range**.
4. Enter **B1:B11** as the **Population 2 Sample Cell Range**.
5. Check **First cells in both ranges contain label**.
6. Click **Two-Tail Test**.
7. Enter a **Title** and click **OK**.

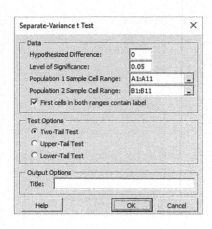

When using summarized data, select **PHStat → Two-Sample Tests (Summarized Data) → Separate-Variance t Test**. In that procedure's dialog box, enter the hypothesized difference and the level of significance, as well as the sample size, sample mean, and sample standard deviation for each group.

Workbook Use the **COMPUTE worksheet** of the **Separate-Variance T workbook** as a template.

The worksheet already contains the data and formulas to use the unsummarized data for the example. For other problems, use this worksheet with either unsummarized or summarized data.

For unsummarized data, paste the data in columns A and B in the **DataCopy worksheet** and keep the COMPUTE worksheet formulas that compute the sample size, sample mean, and sample standard deviation in the cell range B7:B13. For summarized data, replace those formulas in the cell range B7:B13 with the sample statistics and ignore the DataCopy worksheet.

Use the **COMPUTE_LOWER** or **COMPUTE_UPPER worksheets** in the same workbook as templates for performing one-tail pooled-variance *t* tests with either unsummarized or summarized data. For unsummarized data, paste the new data into the DataCopy worksheet. For summarized data, replace the COMPUTE worksheet formulas with sample statistics.

Analysis ToolPak Use **t-Test: Two-Sample Assuming Unequal Variances**.

For the example, open to the **DATA worksheet** of the **VLabGo workbook** and:

1. Select **Data → Data Analysis**.
2. In the Data Analysis dialog box, select **t-Test: Two-Sample Assuming Unequal Variances** from the **Analysis Tools** list and then click **OK**.

In the procedure's dialog box (shown on page 349):

3. Enter **A1:A11** as the **Variable 1 Range**.
4. Enter **B1:B11** as the **Variable 2 Range**.
5. Enter **0** as the **Hypothesized Mean Difference**.
6. Check **Labels** and enter **0.05** as **Alpha**.

7. Click **New Worksheet Ply**.

8. Click **OK**.

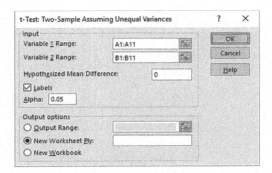

Results (shown below) appear in a new worksheet that contains both two-tail and one-tail test critical values and *p*-values. Unlike the results shown in Figure 10.6, only the positive (upper) critical value is listed for the two-tail test. Because the Analysis ToolPak uses table lookups to approximate the critical values and the *p*-value, the results will differ slightly from the values shown in Figure 10.6.

▲	A	B	C
1	t-Test: Two-Sample Assuming Unequal Variances		
2			
3		Special Front	In-Aisle
4	Mean	246.4	202.3
5	Variance	1809.8222	1058.0111
6	Observations	10	10
7	Hypothesized Mean Difference	0	
8	df	17	
9	t Stat	2.6041	
10	P(T<=t) one-tail	0.0093	
11	t Critical one-tail	1.7396	
12	P(T<=t) two-tail	0.0185	
13	t Critical two-tail	2.1098	

EG10.2 COMPARING the MEANS of TWO RELATED POPULATIONS

Paired *t* Test

Key Technique Use the **T.INV.2T**(*level of significance, degrees of freedom*) function to compute the lower and upper critical values.

Use the **T.DIST.2T**(*absolute value of the t test statistic, degrees of freedom*) to compute the *p*-value.

Example Perform the Figure 10.8 paired *t* test for the equivalent products price data on page 325.

PHStat Use **Paired t Test**.

For the example, open to the **DATA worksheet** of the **Market Basket workbook**. Select **PHStat → Two-Sample Tests (Unsummarized Data) → Paired t Test**. In the procedure's dialog box (shown in the right column):

1. Enter **0** as the **Hypothesized Mean Difference**.
2. Enter **0.05** as the **Level of Significance**.
3. Enter **B1:B8** as the **Population 1 Sample Cell Range**.
4. Enter **C1:C8** as the **Population 2 Sample Cell Range**.
5. Check **First cells in both ranges contain label**.

6. Click **Two-Tail Test**.

7. Enter a **Title** and click **OK**.

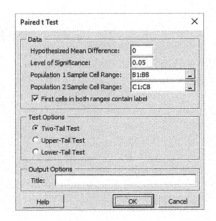

The procedure creates two worksheets, one of which is similar to the PtCalcs worksheet discussed in the following Workbook section. When using summarized data, select **PHStat→Two-Sample Tests (Summarized Data)→Paired t Test t Test**. In that procedure's dialog box, enter the hypothesized mean difference, the level of significance, and the differences cell range.

Workbook Use the **COMPUTE** and **PtCalcs worksheets** of the **Paired T workbook** as a template.

The COMPUTE and supporting PtCalcs worksheets already contain the equivalent products price data for the example. The PtCalcs worksheet also computes the differences that allow the COMPUTE worksheet to compute the S_D in cell B11.

For other problems, paste the unsummarized data into columns A and B of the PtCalcs worksheet. For sample sizes greater than 7, select cell C8 and copy the formula in that cell down through the last data row. For sample sizes less than 7, delete the column C formulas for which there are no column A and B values.

If you know the sample size, $\overline{D}$, and S_D values, you can ignore the PtCalcs worksheet and enter the values in cells B8, B9, and B11 of the COMPUTE worksheet, overwriting the formulas that those cells contain.

Use the similar **COMPUTE_LOWER** and **COMPUTE_UPPER worksheets** in the same workbook as templates for performing one-tail tests. For unsummarized data, paste the new data into the DataCopy worksheet. For summarized data, replace COMPUTE worksheet formulas with sample statistics.

Analysis ToolPak Use **t-Test: Paired Two Sample for Means**.

For the example, open to the **DATA worksheet** of the **MarketBasket workbook** and:

1. Select **Data → Data Analysis**.
2. In the Data Analysis dialog box, select **t-Test: Paired Two Sample for Means** from the **Analysis Tools** list and then click **OK**.

In the procedure's dialog box (shown below):

3. Enter **B1:B8** as the **Variable 1 Range**.
4. Enter **C1:C8** as the **Variable 2 Range**.
5. Enter **0** as the **Hypothesized Mean Difference**.
6. Check **Labels** and enter **0.05** as **Alpha**.
7. Click **New Worksheet Ply**.
8. Click **OK**.

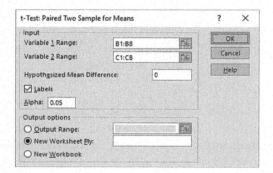

Results (shown below) appear in a new worksheet that contains both two-tail and one-tail test critical values and *p*-values. Unlike in Figure 10.8, only the positive (upper) critical value is listed for the two-tail test.

	A	B	C
1	t-Test: Paired Two Sample for Means		
2			
3		Costco	Walmart
4	Mean	2.8514	2.9986
5	Variance	14.7542	17.0431
6	Observations	7	7
7	Pearson Correlation	0.9935	
8	Hypothesized Mean Difference	0	
9	df	6	
10	t Stat	-0.7235	
11	P(T<=t) one-tail	0.2483	
12	t Critical one-tail	1.9432	
13	P(T<=t) two-tail	0.4966	
14	t Critical two-tail	2.4469	

EG10.3 COMPARING the PROPORTIONS of TWO INDEPENDENT POPULATIONS

Z Test for the Difference Between Two Proportions

Key Technique Use the **NORM.S.INV** (*percentage*) function to compute the critical values.

Use the **NORM.S.DIST** (*absolute value of the Z test statistic, True*) function to compute the *p*-value.

Example Perform the Figure 10.11 *Z* test for the web signup page experiment.

PHStat Use **Z Test for Differences in Two Proportions**.

For the example, select **PHStat→Two-Sample Tests (Summarized Data)→Z Test for Differences in Two Proportions**. In the procedure's dialog box (shown in the right column):

1. Enter **0** as the **Hypothesized Difference**.
2. Enter **0.05** as the **Level of Significance**.

3. For the Population 1 Sample, enter **387** as the **Number of Items of Interest** and **4325** as the **Sample Size**.
4. For the Population 2 Sample, enter **283** as the **Number of Items of Interest** and **4639** as the **Sample Size**.
5. Click **Two-Tail Test**.
6. Enter a **Title** and click **OK**.

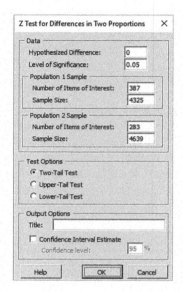

Workbook Use the **COMPUTE worksheet** of the **Z Two Proportions workbook** as a template.

The worksheet already contains data for the website signup survey. For other problems, change the hypothesized difference, the level of significance, and the number of items of interest and sample size for each group in the cell range B4:B11.

Use the similar **COMPUTE_LOWER** and **COMPUTE_UPPER worksheets** in the same workbook as templates for performing one-tail *Z* tests for the difference between two proportions. For unsummarized data, paste the new data into the DataCopy worksheet. For summarized data, replace COMPUTE worksheet formulas with sample statistics.

Confidence Interval Estimate for the Difference Between Two Proportions

PHStat Modify the *PHStat* instructions for the *Z* test for the difference between two proportions. In step 6, also check **Confidence Interval Estimate** and enter a **Confidence Level** in its box, in addition to entering a **Title** and clicking **OK**.

Workbook Use the "Z Test for the Difference Between Two Proportions" *Workbook* instructions in this section. The Z Two Proportions workbook worksheets include a confidence interval estimate for the difference between two means in the cell range D3:E16.

EG10.4 *F* TEST for the RATIO of TWO VARIANCES

Key Technique Use the **F.INV.RT(***level of significance/2, population 1 sample degrees of freedom, population 2 sample degrees of freedom***)** function to compute the upper critical value.

Use the **F.DIST.RT(***F test statistic, population 1 sample degrees of freedom, population 2 sample degrees of freedom***)** function to compute the *p*-values.

Example Perform the Figure 10.13 *F* test for the ratio of two variances for the Table 10.1 Arlingtons sales data for two in-store locations.

PHStat Use **F Test for Differences in Two Variances**.

For the example, open to the **DATA worksheet** of the **VLABGo workbook**. Select **PHStat → Two-Sample Tests (Unsummarized Data) → F Test for Differences in Two Variances**. In the procedure's dialog box (shown below):

1. Enter **0.05** as the **Level of Significance**.
2. Enter **A1:A11** as the **Population 1 Sample Cell Range**.
3. Enter **B1:B11** as the **Population 2 Sample Cell Range**.
4. Check **First cells in both ranges contain label**.
5. Click **Two-Tail Test**.
6. Enter a **Title** and click **OK**.

When using summarized data, select **PHStat → Two-Sample Tests (Summarized Data) → F Test for Differences in Two Variances**. In that procedure's dialog box, enter the level of significance and the sample size and sample variance for each sample.

Workbook Use the **COMPUTE worksheet** of the **F Two Variances workbook** as a template.

The worksheet already contains the data and formulas for using the unsummarized data for the example. For unsummarized data, paste the data in columns A and B in the **DataCopy** worksheet and keep the COMPUTE worksheet formulas that compute the sample size and sample variance for the two samples in cell range B4:B10. For summarized data, replace the COMPUTE worksheet formulas in cell ranges B4:B10 with the sample statistics and ignore the DataCopy worksheet.

Use the similar **COMPUTE_UPPER** worksheet in the same workbook as a template for performing the upper-tail test. For unsummarized data, paste the new data into the Data-Copy worksheet. For summarized data, replace COMPUTE worksheet formulas with sample statistics.

Analysis ToolPak Use **F-Test Two-Sample for Variances**.

For the example, open to the **DATA worksheet** of the **VALBGo workbook** and:

1. Select **Data → Data Analysis**.
2. In the Data Analysis dialog box, select **F-Test Two-Sample for Variances** from the **Analysis Tools** list and then click **OK**.

In the procedure's dialog box (shown below):

3. Enter **A1:A11** as the **Variable 1 Range** and enter **B1:B11** as the **Variable 2 Range**.
4. Check **Labels** and enter **0.05** as **Alpha**.
5. Click **New Worksheet Ply**.
6. Click **OK**.

Results (shown below) appear in a new worksheet and include only the one-tail test *p*-value (0.2181), which must be doubled for the two-tail test shown in Figure 10.13 on page 338.

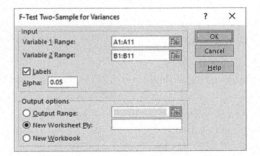

	A	B	C
1	F-Test Two-Sample for Variances		
2			
3		Special Front	In-Aisle
4	Mean	246.4	202.3
5	Variance	1809.8222	1058.0111
6	Observations	10	10
7	df	9.0000	9
8	F	1.7106	
9	P(F<=f) one-tail	0.2181	
10	F Critical one-tail	3.1789	

11

Analysis of Variance

CONTENTS

OBJECTIVES

- Introduce the basic concepts of experimental design

- Learn to use the one-way analysis of variance to test for differences among the means of several groups

- Learn to use the two-way analysis of variance and interpret the interaction effect

- Learn to perform multiple comparisons in a one-way analysis of variance and a two-way analysis of variance

▼USING STATISTICS
The Means to Find Differences at Arlingtons

Boosting sales of mobile electronics items is a key goal in a new strategic plan just issued by the senior management at the general merchandiser Arlingtons. Having helped to conduct an experiment that provided evidence that sales of a mobile streaming media player are higher in a special front location than in the mobile electronics aisle (see Chapter 10), you wonder if moving all mobile electronics items to another in-store location might also affect sales.

The strategic plan also encourages managers to make better use of kiosks, the special end-of-aisle endcap areas that face customers as customers enter aisles and suggests that managers create *expert counters*, positions that customers can visit to ask specially trained staff questions about items for sale in a specific department. Might these two in-store locations be combined with the two locations of the previous experiment (the special front location and the regular in-aisle location) into one larger experiment?

You propose an experiment in which mobile electronics in selected Arlingtons stores will be sold at one of four in-store locations: the current in-aisle location, the special front of the store location, in a special endcap kiosk, or at the expert counter position for mobile electronics. You suggest that Arlingtons select 20 stores that have similar annual sales and divide the stores into four groups of five stores each and assign to each group a different in-store sales location for mobile electronics: current in-aisle, special front, kiosk, or expert counter.

How would you determine if varying the locations had an effect on mobile electronics sales? As you consider this, another manager suggests that customers who use mobile payment methods might be more likely to buy mobile electronics items. If you also wanted to later explore the effects of permitting customers to use mobile payment methods to purchase mobile electronics items, could you design an experiment that examined this second factor while it was examining the effects of in-store location?

Comparing possible differences has been the subject of the statistical methods discussed in the previous two chapters. In the one-sample tests of Chapter 9, the comparison is to a standard, such as a certain mean weight for a cereal box being filled by a production line. In Chapter 10, the comparison is between samples taken from two populations. **Analysis of variance**, known by the acronym **ANOVA**, allows statistical comparison among samples taken from many populations.

In ANOVA, the comparison is typically the result of an experiment. For example, the management of a general merchandiser might be brainstorming ways of improving sales of mobile electronics items. At Arlingtons, the management decided to try selling those items in four different in-store locations and then observe what the sales would be in each of those locations. The basis for an ANOVA experiment is called the **factor**, which in the Arlingtons scenario is in-store location. The statistical use of the word "factor" complements the everyday usage, illustrated by a question such as "How much of a *factor* is in-store location in determining mobile electronics sales?"

The actual different locations (in-aisle, special front, kiosk, and expert counter) are the **levels** of the factor. Levels of a factor are analogous to the categories of a categorical variable, but you call in-store location a *factor* and not a categorical variable because the variable under study is mobile electronics sales. Levels provide the basis of comparison by dividing the variable under study into **groups**. In the Arlingtons scenario, the groups are the stores selling the mobile electronics items in the mobile electronics aisle, the stores selling those items at the special front location, the stores selling those items at the kiosk location, and the stores selling those items at the expert counter.

The **completely randomized design** is the ANOVA method that analyzes a single factor. One executes this design using the statistical method **one-way ANOVA**. One-way ANOVA is a two-part process. In the first part, one determines if there is a significant difference among the group means. If one rejects the null hypothesis that there is no difference among the means, one proceeds with a second method that seeks to identify the groups whose means are significantly different from the other group means.

The **factorial design** is the ANOVA method that analyzes two or more factors. This design uses a two-way ANOVA if there are two factors and a general ANOVA if there are more than two factors. With a factorial design, the ANOVA method includes interaction terms that measure the interrelationship of the factors.

Randomized block design is a method that evaluates differences among the means of more than two independent groups. In this design, members of each group have been placed in blocks either by being matched or subjected to repeated measurement. Blocking removes variability due to individual differences so that the differences among the groups are more evident. This ANOVA method can also include different levels or one or more factors for each group.

While ANOVA literally does analyze variation, the purpose of ANOVA is to reach conclusions about possible differences among the *means* of each group, analogous to the hypothesis tests of the previous chapter. Every ANOVA design uses samples that represent each group and subdivides the total variation observed across all samples (all groups) toward the goal of analyzing possible differences among the means of each group. How this subdivision, called *partitioning*, works is a function of the design being used, but total variation, represented by the quantity **sum of squares total (SST)**, will always be the starting point. As with other statistical methods, ANOVA requires making assumptions about the populations that the groups represent.

student TIP

Understanding ANOVA is important to understanding both hypothesis testing and regression, the subject of Chapters 13 through 17.

11.1 One-Way ANOVA

In one-way ANOVA, to analyze variation toward the goal of determining possible differences among the group means, you partition the total variation into variation that is due to differences among the groups and variation that is due to differences within the groups (see Figure 11.1 on page 354). The **within-group variation (SSW)** measures random variation. The **among-group variation (SSA)** measures differences from group to group. The symbol n represents the number of values in all groups and the symbol c represents the number of groups.

FIGURE 11.1
Partitioning the total variation in a completely randomized design

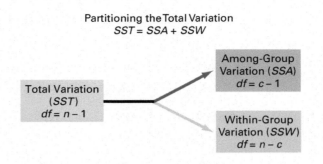

Partitioning the Total Variation
$SST = SSA + SSW$

When using Excel, always organize multiple-sample data as unstacked data, one column per group.

Assuming that the c groups represent populations whose values are randomly and independently selected, follow a normal distribution, and have equal variances, the null hypothesis of no differences in the population means:

$$H_0: \mu_1 = \mu_2 = \cdots = \mu_c$$

is tested against the alternative that not all the c population means are equal:

$$H_1: \text{Not all } \mu_j \text{ are equal (where } j = 1, 2, \ldots, c).$$

student TIP

Another way of stating the alternative hypothesis, H_1, is that at least one population mean is different from the others.

To perform an ANOVA test of equality of population means, one subdivides the total variation in the values into two parts—that which is due to variation among the groups and that which is due to variation within the groups. The **total variation** is represented by the sum of squares total (*SST*). Because the population means of the c groups are assumed to be equal under the null hypothesis, one calculates the total variation among all the values by summing the squared differences between each individual value and the **grand mean**, $\overline{\overline{X}}$. The grand mean is the mean of all the values in all the groups combined. Equation (11.1) defines the total variation.

TOTAL VARIATION IN ONE-WAY ANOVA

$$SST = \sum_{j=1}^{c} \sum_{i=1}^{n_j} (X_{ij} - \overline{\overline{X}})^2 \tag{11.1}$$

where

$$\overline{\overline{X}} = \frac{\displaystyle\sum_{j=1}^{c} \sum_{i=1}^{n_j} X_{ij}}{n} = \text{grand mean}$$

$X_{ij} = i$th value in group j

$n_j = $ number of values in group j

$n = $ total number of values in all groups combined

(that is, $n = n_1 + n_2 + \cdots + n_c$)

$c = $ number of groups

student TIP

Remember that a sum of squares (SS) cannot be negative.

One calculates the among-group variation, usually called the **sum of squares among groups (SSA)**, by summing the squared differences between the sample mean of each group, $\overline{X}_j$, and the grand mean, $\overline{\overline{X}}$, weighted by the sample size, n_j, in each group. Equation (11.2) on page 355 defines the among-group variation.

AMONG-GROUP VARIATION IN ONE-WAY ANOVA

$$SSA = \sum_{j=1}^{c} n_j \left(\overline{X}_j - \overline{\overline{X}} \right)^2 \tag{11.2}$$

where

c = number of groups

n_j = number of values in group j

$\overline{X}_j$ = sample mean of group j

$\overline{\overline{X}}$ = grand mean

The within-group variation, usually called the **sum of squares within groups (SSW)**, measures the difference between each value and the mean of its own group and sums the squares of these differences over all groups. Equation (11.3) defines the within-group variation.

WITHIN-GROUP VARIATION IN ONE-WAY ANOVA

$$SSW = \sum_{j=1}^{c} \sum_{i=1}^{n_j} \left(X_{ij} - \overline{X}_j \right)^2 \tag{11.3}$$

where

X_{ij} = ith value in group j

$\overline{X}_j$ = sample mean of group j

Because one compares c groups, there are $c - 1$ degrees of freedom associated with the sum of squares among groups. Because each of the c groups contributes $n_j - 1$ degrees of freedom, there are $n - c$ degrees of freedom associated with the sum of squares within groups. In addition, there are $n - 1$ degrees of freedom associated with the sum of squares total because you are comparing each value, X_{ij}, to the grand mean, $\overline{\overline{X}}$, based on all n values.

Dividing each of these sums of squares by its respective degrees of freedom computes three variances, which in ANOVA are known as **mean squares**: MSA (mean square among), MSW (mean square within), and MST (mean square total).

student TIP

Remember, *mean square* is just another term for *variance* that is used in the analysis of variance. Also, because the mean square is equal to the sum of squares divided by the degrees of freedom, a mean square can never be negative.

MEAN SQUARES IN ONE-WAY ANOVA

$$MSA = \frac{SSA}{c - 1} \tag{11.4a}$$

$$MSW = \frac{SSW}{n - c} \tag{11.4b}$$

$$MST = \frac{SST}{n - 1} \tag{11.4c}$$

F Test for Differences Among More Than Two Means

To determine if there is a significant difference among the *c* group means, use the *F* test for differences among more than two means. As Section 10.4 discusses, the *F* distribution is right-skewed with a minimum value of 0. If the null hypothesis is true and there are no differences among the *c* group means, *MSA*, *MSW*, and *MST*, will provide estimates of the overall variance in the population. Equation (11.5) defines the one-way ANOVA F_{STAT} test statistic as the ratio of *MSA* to *MSW*.

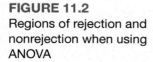

student TIP

The test statistic compares mean squares (the variances) because one-way ANOVA reaches conclusions about possible differences among the *means* of *c* groups by examining variances.

ONE-WAY ANOVA F_{STAT} TEST STATISTIC

$$F_{STAT} = \frac{MSA}{MSW} \tag{11.5}$$

The F_{STAT} test statistic follows an *F* distribution, with $c - 1$ numerator degrees of freedom and $n - c$ denominator degrees of freedom.

The null and alternative hypotheses for this test are

$$H_0: \mu_1 = \mu_2 = \cdots = \mu_c$$
$$H_1: \text{Not all } \mu_j \text{ are equal (where } j = 1, 2, \ldots, c)$$

For a given level of significance, α, one rejects the null hypothesis if the F_{STAT} test statistic is greater than the upper-tail critical value, F_α, from the *F* distribution with $c - 1$ numerator degrees of freedom and $n - c$ denominator degrees of freedom (see Table E.5). As Figure 11.2 visualizes, the decision rule is

Reject H_0 if $F_{STAT} > F_\alpha$;
otherwise, do not reject H_0.

FIGURE 11.2
Regions of rejection and nonrejection when using ANOVA

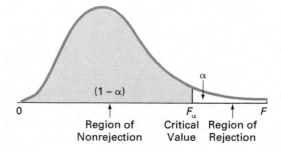

If the null hypothesis is true, the F_{STAT} test statistic is expected to be approximately equal to 1 because both the numerator and denominator mean square terms are estimating the overall variance in the population. If H_0 is false (and there are differences in the group means), the F_{STAT} test statistic is expected to be larger than 1 because the numerator, *MSA*, is estimating the differences among groups in addition to the overall variability in the values, while the denominator, *MSW*, is measuring only the overall variability in the values. Therefore, one rejects the null hypothesis at a selected level of significance, α, only if the computed F_{STAT} test statistic is *greater than* F_α, the upper-tail critical value of the *F* distribution having $c - 1$ and $n - c$ degrees of freedom.

Table 11.1 on page 357 presents a generalized **ANOVA summary table** that is commonly used to summarize the results of a one-way ANOVA. The table includes entries for the sources of variation (among groups, within groups, and total), the degrees of freedom, the sums of squares, the mean squares (the variances), and the computed F_{STAT} test statistic. The table may also include the *p*-value, the probability of having an F_{STAT} value as large as or larger than the one computed, given that the null hypothesis is true. The *p*-value enables one to reach conclusions about the null hypothesis without needing to refer to a table of critical values of the *F* distribution. If the *p*-value is less than the chosen level of significance, α, you reject the null hypothesis.

TABLE 11.1
ANOVA summary table

Source	Degrees of Freedom	Sum of Squares	Mean Square (Variance)	F
Among groups	$c - 1$	SSA	$MSA = \dfrac{SSA}{c - 1}$	$F_{STAT} = \dfrac{MSA}{MSW}$
Within groups	$n - c$	SSW	$MSW = \dfrac{SSW}{n - c}$	
Total	$n - 1$	SST		

To illustrate the one-way ANOVA *F* test, return to the Using Statistics scenario (on page 352). Arlingtons has accepted your proposal and has defined the business objective as determining whether significant differences exist in the mobile electronics sales for the four different in-store locations.

To test the comparative effectiveness of the four in-store locations, Arlingtons conducts a 60-day experiment at 20 same-sized stores that have similar storewide net sales. Arlingtons randomly assigns five stores to use the current mobile electronics aisle (in-aisle), five stores to use the special front location (front), five stores to use the kiosk location (kiosk), and five stores to use the expert counter (expert). At the end of the experiment, researchers organize the mobile electronics sales data by group and store the data in unstacked format in Mobile Electronics . Figure 11.3 presents that unstacked data, along with the sample mean and the sample standard deviation for each group.

FIGURE 11.3
Mobile electronics sales (in thousands of dollars), sample means, and sample standard deviations for four different in-store locations

	In-aisle	Front	Kiosk	Expert
	30.06	32.22	30.78	30.33
	29.96	31.47	30.91	30.29
	30.19	32.13	30.79	30.25
	29.96	31.86	30.95	30.25
	27.74	32.29	31.13	30.55
Sample Mean	29.582	31.994	30.912	30.334
Sample Standard Deviation	1.034	0.335	0.143	0.125

Figure 11.3 shows differences among the sample means for the mobile electronics sales for the four in-store locations. For the original in-aisle location, mean sales were $29.582 thousands, whereas mean sales at the three new locations varied from $30.334 thousands ("expert" location) to $30.912 thousands ("kiosk" location) to $31.994 thousands ("front" location).

Differences in the mobile electronics sales for the four in-store locations can also be presented visually. The Figure 11.4 scatter plot presents the mobile electronics sales at each store in each group that visualizes differences *within* each location as well as among the four locations.

FIGURE 11.4
Excel scatter plot of mobile electronics sales for four in-store locations

For the chart, the locations have been relabeled 1, 2, 3, and 4 in order to use the scatter plot chart type and the Y axis minimum value has been set to 27.

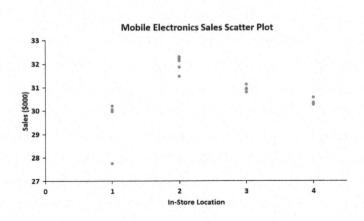

studentTIP

If the sample sizes in each group were larger, you could construct stem-and-leaf displays, boxplots, and normal probability plots as additional ways of visualizing the sales data.

Having observed that the four sample means appear to be different, researchers use the F test for differences among more than two means to determine if these sample means are sufficiently different to conclude that the *population* means are not all equal. The null hypothesis states that there is no difference in the mean sales among the four in-store locations:

$$H_0: \mu_1 = \mu_2 = \mu_3 = \mu_4$$

The alternative hypothesis states that at least one of the in-store location mean sales differs from the other means:

$$H_1: \text{Not all the means are equal.}$$

To determine whether to reject the null hypothesis, researchers use the ANOVA summary table in the software results (see Figure 11.6 on page 360) or construct an ANOVA summary table. To construct the table, researchers first calculate the sample means in each group (see Figure 11.3 on page 357). Then researchers calculate the grand mean:

$$\overline{\overline{X}} = \frac{\sum\limits_{j=1}^{c}\sum\limits_{i=1}^{n_j} X_{ij}}{n} = \frac{614.12}{20} = 30.706$$

Then, using Equations (11.1) through (11.3) on pages 354–355, researchers calculate the sum of squares:

$$SSA = \sum_{j=1}^{c} n_j(\overline{X}_j - \overline{\overline{X}})^2 = (5)(29.582 - 30.706)^2 + (5)(31.994 - 30.706)^2$$

$$+ (5)(30.912 - 30.706)^2 + (5)(30.334 - 30.706)^2$$

$$= 15.5157$$

$$SSW = \sum_{j=1}^{c}\sum_{i=1}^{n_j}(X_{ij} - \overline{X}_j)^2$$

$$= (30.06 - 29.582)^2 + (29.96 - 29.582)^2 + (30.19 - 29.582)^2$$

$$+ (29.96 - 29.582)^2 + (27.74 - 29.582)^2 + \cdots + (30.55 - 30.334)^2$$

$$= 4.8706$$

$$SST = \sum_{j=1}^{c}\sum_{i=1}^{n_j}(X_{ij} - \overline{\overline{X}})^2$$

$$= (30.06 - 30.706)^2 + (29.96 - 30.706)^2 + \cdots + (30.55 - 30.706)^2$$

$$= 20.3863$$

Researchers could have also determined the *SSA, SSW,* and *SST* values from software results. These three values are in the *SS* column in the Excel ANOVA table (see Figure 11.6 on page 360).

Researchers then calculate the mean squares by dividing the sum of squares by the corresponding degrees of freedom using Equation (11.4) on page 355. Because $c = 4$ and $n = 20$,

$$MSA = \frac{SSA}{c - 1} = \frac{15.5157}{4 - 1} = 5.1719$$

$$MSW = \frac{SSW}{n - c} = \frac{4.8706}{20 - 4} = 0.3044$$

Using Equation (11.5) on page 356, the result is

$$F_{STAT} = \frac{MSA}{MSW} = \frac{5.1719}{0.3044} = 16.9898$$

Researchers could have also determined the F_{STAT} value from software results. This value appears in the F Excel column in the summary table.

Because the researchers are seeking to determine whether *MSA* is greater than *MSW*, the researchers only reject H_0 if F_{STAT} is greater than the upper critical value of F. For a selected level of significance, α, you find the upper-tail critical value, F_α, from the F distribution. For the in-store location sales experiment, there are 3 numerator degrees of freedom and 16 denominator degrees of freedom. From Table E.5, a portion of which Table 11.2 shows, F_α, the upper-tail critical value at the 0.05 level of significance, is 3.24.

TABLE 11.2

Finding the critical value of F with 3 and 16 degrees of freedom at the 0.05 level of significance

			Cumulative Probabilities = 0.95 Upper-Tail Area = 0.05 Numerator df_1						
Denominator df_2	**1**	**2**	**3**	**4**	**5**	**6**	**7**	**8**	**9**
:	:	:	:	:	:	:	:	:	:
11	4.84	3.98	3.59	3.36	3.20	3.09	3.01	2.95	2.90
12	4.75	3.89	3.49	3.26	3.11	3.00	2.91	2.85	2.80
13	4.67	3.81	3.41	3.18	3.03	2.92	2.83	2.77	2.71
14	4.60	3.74	3.34	3.11	2.96	2.85	2.76	2.70	2.65
15	4.54	3.68	3.29	3.06	2.90	2.79	2.71	2.64	2.59
16	4.49	3.63	3.24	3.01	2.85	2.74	2.66	2.59	2.54

Source: Extracted from Table E.5.

Because $F_{STAT} = 16.9898$ is greater than $F_\alpha = 3.24$, the Arlingtons researchers reject the null hypothesis (see Figure 11.5). They conclude that there is a significant difference in the mean sales for the four in-store locations.

FIGURE 11.5

Regions of rejection and nonrejection for the one-way ANOVA at the 0.05 level of significance, with 3 and 16 degrees of freedom

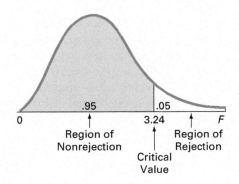

Figure 11.6 on page 360 shows the ANOVA results for the in-store location sales experiment, including the *p*-value. In Figure 11.6, what Table 11.1 on page 357 labels Among Groups is labeled Between Groups in the Excel table.

Table 11.3 on page 360 summarizes the results of the one-way ANOVA for the mobile electronics data. Based on the conclusions, there is a difference in sales among the in-store locations. However, researchers still do not know *which* in-store locations differ. All they know is that there is sufficient evidence to state that the population means are not all the same. In other words, one or more population means are significantly different. To determine which in-store locations differ, one can use a multiple comparisons procedure such as the Tukey-Kramer procedure that this section discusses later.

TABLE 11.3
One-Way ANOVA summary for the mobile electronics data

Result	Conclusions
The $F_{STAT} = 16.9898$ is greater than 3.24. The F test p-value $= 0.0000$ is less than the level of significance, $\alpha = 0.05$.	1. Reject the null hypothesis H_0. 2. Conclude that evidence exists that the mean sales are different at the in-store locations. 3. There is a probability of 0.0000 that $F_{STAT} > 16.9898$.

FIGURE 11.6
Excel worksheet results for the in-store location sales experiment

	A	B	C	D	E	F	G
1	One-Way ANOVA (ANOVA: Single Factor)						
2							
3	SUMMARY						
4	Groups	Count	Sum	Average	Variance		
5	In-aisle	5	147.91	29.582	1.06922		
6	Front	5	159.97	31.994	0.11243		
7	Kiosk	5	154.56	30.912	0.02032		
8	Expert	5	151.67	30.334	0.01568		
9							
10							
11	ANOVA						
12	Source of Variation	SS	df	MS	F	P-value	F crit
13	Between Groups	15.5157	3	5.1719	16.9898	0.0000	3.2389
14	Within Groups	4.8706	16	0.3044			
15							
16	Total	20.3863	19				
17						Level of significance	0.05

One-Way ANOVA *F* Test Assumptions

To use the one-way ANOVA F test, one must make three assumptions:

- **Randomness and independence** of the samples selected
- **Normality** of the c groups from which the samples are selected
- **Homogeneity of variance** (the variances of the c groups are equal)

student TIP

To use the one-way ANOVA *F* test, the variable to be analyzed must either be interval or ratio scaled.

Most critical of all is the first assumption. The validity of any experiment depends on random sampling or a randomization process. To avoid biases in the outcomes, one needs to select random samples from the c groups or use a randomization process to randomly assign the items to the c levels of the factor. Selecting a random sample or randomly assigning the levels ensures that a value from one group is independent of any other value in the experiment. Departures from this assumption can seriously affect inferences made using the ANOVA results. These problems are discussed more thoroughly in Daniel and in Gitlow, Melnyck, and Levine.

As for the second assumption, normality, the one-way ANOVA F test is fairly robust against departures from the normal distribution. As long as the distributions are not extremely different from a normal distribution, the level of significance of the ANOVA F test is usually not greatly affected, particularly for large samples. One can assess the normality of each of the c samples by constructing a normal probability plot or a boxplot.

As for the third assumption, homogeneity of variance, if each group has the same sample size, inferences based on the F distribution are not seriously affected by unequal variances. Whenever possible, groups should have equal sample sizes because with unequal sample sizes, unequal variances can have a serious effect on inferences made using the ANOVA results. (Use the Levene test for homogeneity of variance, which the next section discusses, to test whether the variances of the c groups are equal.)

When only the normality assumption is violated, one can use the Kruskal-Wallis rank test, a nonparametric procedure. When only the homogeneity-of-variance assumption is violated, one can use procedures similar to those used in the separate-variance test that Section 10.1 discusses. When both the normality and homogeneity-of-variance assumptions have been violated, one needs to use an appropriate data transformation that both normalizes the data and reduces the differences in variances or use a more general nonparametric procedure. The Berenson, Corder, and Kutner references contain more information about these alternative tests.

Levene Test for Homogeneity of Variance

Although the one-way ANOVA F test is relatively robust with respect to the assumption of equal group variances, large differences in the group variances can seriously affect the level of significance and the power of the F test. One powerful yet simple procedure for testing the equality of the variances is the modified **Levene test** to test for the homogeneity of variance. The null and alternative hypotheses for this test are

$$H_0: \sigma_1^2 = \sigma_2^2 = \cdots = \sigma_c^2$$
$$H_1: \text{Not all } \sigma_j^2 \text{ are equal } (j = 1, 2, 3, \ldots, c)$$

student TIP

The Levene test performs a one-way ANOVA on the absolute differences from the median in each group, not on the actual values in each group.

To test the null hypothesis of equal variances, one first calculates the absolute value of the difference between each value and the median of the group. Then one performs a one-way ANOVA using these *absolute differences*, typically using a level of significance of $\alpha = 0.05$.

To illustrate the modified Levene test, recall the Arlingtons scenario and the Figure 11.6 data on page 360 for the in-store location sales experiment. Table 11.4 calculates the absolute differences from the median of each location.

TABLE 11.4
Absolute differences from the median sales for four locations

In-Aisle (Median = 29.96)	Front (Median = 32.13)	Kiosk (Median = 30.91)	Expert (Median = 30.29)
$\|30.06 - 29.96\| = 0.10$	$\|32.22 - 32.13\| = 0.09$	$\|30.78 - 30.91\| = 0.13$	$\|30.33 - 30.29\| = 0.04$
$\|29.96 - 29.96\| = 0.00$	$\|31.47 - 32.13\| = 0.66$	$\|30.91 - 30.91\| = 0.00$	$\|30.29 - 30.29\| = 0.00$
$\|30.19 - 29.96\| = 0.23$	$\|32.13 - 32.13\| = 0.00$	$\|30.79 - 30.91\| = 0.12$	$\|30.25 - 30.29\| = 0.04$
$\|29.96 - 29.96\| = 0.00$	$\|31.86 - 32.13\| = 0.27$	$\|30.95 - 30.91\| = 0.04$	$\|30.25 - 30.29\| = 0.04$
$\|27.74 - 29.96\| = 2.22$	$\|32.29 - 32.13\| = 0.16$	$\|31.13 - 30.91\| = 0.22$	$\|30.55 - 30.29\| = 0.26$

Figure 11.7 presents the results of performing a one-way ANOVA using the Table 11.4 absolute differences. From those results, observe that the F_{STAT} test statistic = 0.7849. $F_{STAT} = 0.7849 < 3.2389$ (or because the p-value $= 0.5197 > 0.05$), one does not reject H_0. One concludes that insufficient evidence of a significant difference among the four variances exists. One can claim that because the four in-store locations have an equal amount of variability in sales, the homogeneity-of-variance assumption has not been violated.

FIGURE 11.7
Excel Levene test results for the absolute differences for the in-store location sales experiment

	A	B	C	D	E	F	G
1	ANOVA: Levene Test						
2							
3	SUMMARY						
4	*Groups*	*Count*	*Sum*	*Average*	*Variance*		
5	In-aisle	5	2.55	0.51	0.9227		
6	Front	5	1.18	0.236	0.06593		
7	Kiosk	5	0.51	0.102	0.00732		
8	Expert	5	0.38	0.076	0.01088		
9							
10							
11	ANOVA						
12	*Source of Variation*	*SS*	*df*	*MS*	*F*	*P - value*	*F crit*
13	Between Groups	0.59266	3	0.1976	0.7849	0.5197	3.2389
14	Within Groups	4.02732	16	0.2517			
15							
16	Total	4.61998	19				
17					*Level of significance*		0.05

Multiple Comparisons: The Tukey-Kramer Procedure

The one-way ANOVA F test indicates if there is a difference among the c groups. For example, for the Arlingtons scenario, this test indicates a difference among the in-store locations. When a difference is discovered, the next step is to construct **multiple comparisons** to test the null hypothesis that the differences in the means of all pairs of in-store locations are equal to 0.

For alternative methods to examine multiple comparisons, see Hicks and Turner, Kutner et al., and Montgomery.

Although many methods exist to examine multiple comparisons, one commonly used procedure is the **Tukey-Kramer multiple comparisons procedure for one-way ANOVA** that determines which of the c means are significantly different. This procedure enables one to simultaneously make comparisons between *all* pairs of groups. The procedure consists of the four steps:

1. Calculate the absolute mean differences, $|\overline{X}_j - \overline{X}_{j'}|$ (where j refers to group j, j' refers to group j', and $j \neq j'$), among all pairs of sample means [$c(c - 1)/2$ pairs].
2. Calculate the **critical range** for the Tukey-Kramer procedure that Equation (11.6) defines. If the sample sizes differ, calculate a critical range for each pairwise comparison of sample means.
3. Compare each of the $c(c - 1)/2$ pairs of means against its corresponding critical range. Declare a specific pair significantly different if the absolute difference in the sample means, $|\overline{X}_j - \overline{X}_{j'}|$, is greater than the critical range.
4. Interpret the results.

CRITICAL RANGE FOR THE TUKEY-KRAMER PROCEDURE

$$\text{Critical range} = Q_\alpha \sqrt{\frac{MSW}{2}\left(\frac{1}{n_j} + \frac{1}{n_{j'}}\right)} \qquad \textbf{(11.6)}$$

where

n_j = the sample size in group j

$n_{j'}$ = the sample size in group j'

Q_α = the upper-tail critical value from a **Studentized range distribution** having c degrees of freedom in the numerator and $n - c$ degrees of freedom in the denominator. (Obtain critical values for the Studentized range distribution from Table E.7.)

The mobile electronics sales example compares four in-store locations. Four groups make $4(4 - 1)/2 = 6$ pairwise comparisons. To apply the Tukey-Kramer multiple comparisons procedure, one first calculates the absolute mean differences for all six pairwise comparisons:

student TIP

You have an α level of risk in the entire set of comparisons not just a single comparison.

$|\overline{X}_1 - \overline{X}_2| = |29.582 - 31.994| = 2.412$

$|\overline{X}_1 - \overline{X}_3| = |29.582 - 30.912| = 1.330$

$|\overline{X}_1 - \overline{X}_4| = |29.582 - 30.334| = 0.752$

$|\overline{X}_2 - \overline{X}_3| = |31.994 - 30.912| = 1.082$

$|\overline{X}_2 - \overline{X}_4| = |31.994 - 30.334| = 1.660$

$|\overline{X}_3 - \overline{X}_4| = |30.912 - 30.334| = 0.578$

For this example, because the sample sizes in the four groups are equal, only one critical range needs to be calculated. (Had the sample sizes in some of the groups been different, one would need to calculate several critical ranges.) From the Figure 11.6 ANOVA summary table on page 360, $MSW = 0.3044$ and $n_j = n_{j'} = 5$. From Table 11.5 on page 363, a portion of Table E.7, for $\alpha = 0.05$, $c = 4$, and $n - c = 20 - 4 = 16$, Q_α, the upper-tail critical value of the test statistic, is 4.05.

TABLE 11.5
Finding the Studentized range, Q_α, statistic for $\alpha = 0.05$, with 4 and 16 degrees of freedom

	Cumulative Probabilities = 0.95 **Upper-Tail Area = 0.05** **Numerator df_1**							
Denominator df_2	**2**	**3**	**4**	**5**	**6**	**7**	**8**	**9**
⋮	⋮	⋮	⋮	⋮	⋮	⋮	⋮	⋮
11	3.11	3.82	4.26	4.57	4.82	5.03	5.20	5.35
12	3.08	3.77	4.20	4.51	4.75	4.95	5.12	5.27
13	3.06	3.73	4.15	4.45	4.69	4.88	5.05	5.19
14	3.03	3.70	4.11	4.41	4.64	4.83	4.99	5.13
15	3.01	3.67	4.08	4.37	4.60	4.78	4.94	5.08
16	3.00	3.65	4.05	4.33	4.56	4.74	4.90	5.03

Source: Extracted from Table E.7.

Using Equation (11.6),

$$\text{Critical range} = 4.05\sqrt{\left(\frac{0.3044}{2}\right)\left(\frac{1}{5} + \frac{1}{5}\right)} = 0.9993$$

Figure 11.8 presents the Tukey-Kramer procedure results for the mobile electronics sales in-store location experiment. By using $\alpha = 0.05$, all six of the comparisons can be made with an overall error rate of only 5%.

FIGURE 11.8
Excel Tukey-Kramer procedure results for the in-store location sales experiment

	A	B	C	D	E	F	G	H	I
1	Tukey Kramer Multiple Comparisons								
2									
3		Sample	Sample			Absolute	Std. Error	Critical	
4	Group	Mean	Size		Comparison	Difference	of Difference	Range	Results
5	1: In-aisle	29.582	5		Group 1 to Group 2	2.412	0.2467	0.9993	Means are different
6	2: Front	31.994	5		Group 1 to Group 3	1.33	0.2467	0.9993	Means are different
7	3: Kiosk	30.912	5		Group 1 to Group 4	0.752	0.2467	0.9993	Means are not different
8	4: Expert	30.334	5		Group 2 to Group 3	1.082	0.2467	0.9993	Means are different
9					Group 2 to Group 4	1.66	0.2467	0.9993	Means are different
10	Other Data				Group 3 to Group 4	0.578	0.2467	0.9993	Means are not different
11	Level of significance	0.05							
12	Numerator d.f.	4							
13	Denominator d.f.	16							
14	MSW	0.3044							
15	Q Statistic	4.05							

Because the absolute mean difference for four pairs (1 and 2, 1 and 3, 2 and 3, 2 and 4) is greater than 0.9993, one concludes that there is a significant difference between the mobile electronics sales means of those pairs. Because the absolute mean difference for pair 3 (in-aisle and expert locations) is 0.752, which is less than 0.9993, one concludes that there is no evidence of a difference in the means of those two locations. Also, because the absolute mean difference for pair 6 (kiosk and expert) is 0.578, which is less than 0.9993, one concludes that there is no evidence of a difference in the means of these two locations.

These results enable one to estimate that the population mean sales for mobile electronics items will be higher at the front location than any other location *and* that the population mean sales for mobile electronics items at kiosk locations will be higher when compared to the in-aisle location. Arlingtons management can conclude that selling mobile electronics items at the special front location would increase sales the most, but that selling those items at the kiosk location would also improve sales. (These results also present additional evidence for charging a placement fee for the special front location, the subject of the Chapter 10 Using Statistics scenario.)

Example 11.1 illustrates another example of the one-way ANOVA.

<table>
<tr><td>

EXAMPLE 11.1

ANOVA of the Speed of Drive-Through Service at Fast-Food Chains
</td><td>

For fast-food restaurants, the drive-through window is an important revenue source. The chain that offers the fastest service is likely to attract additional customers. Each year *QSR Magazine*, www.qsrmagazine.com, publishes its results of a survey of drive-through service times (from menu board to departure) at fast-food chains. In a recent year, the mean time was 226.07 seconds for Wendy's, 236.50 seconds for Taco Bell, 193.31 seconds for Burger King, 273.29 seconds for McDonald's, and 218.85 seconds for Chick-fil-A. Suppose the study was based on 20 customers for each fast-food chain. At the 0.05 level of significance, is there evidence of a difference in the mean drive-through service times of the five chains?

Table 11.6 contains the ANOVA table for this problem.
</td></tr>
</table>

TABLE 11.6
ANOVA summary table of drive-through service times at fast-food chains

Source	Degrees of Freedom	Sum of Squares	Mean Squares	F	p-value
Among chains	4	68,028.2704	17,007.0676	130.2226	0.0000
Within chains	95	12,407.00	130.60		

SOLUTION

H_0: $\mu_1 = \mu_2 = \mu_3 = \mu_4 = \mu_5$ where 1 = Wendy's, 2 = Taco Bell, 3 = Burger King,
$\qquad\qquad\qquad\qquad\qquad\qquad$ 4 = McDonald's, 5 = Chick-fil-A

$\quad H_1$: Not all μ_j are equal $\qquad$ where $j = 1, 2, 3, 4, 5$

Decision rule: If the p-value < 0.05, reject H_0. Because the p-value is 0.0000, which is less than $\alpha = 0.05$, reject H_0. You have sufficient evidence to conclude that the mean drive-through times of the five chains are not all equal.

To determine which of the means are significantly different from one another, use the Tukey-Kramer procedure [Equation (11.6) on page 362] to establish the critical range:

Critical value of Q with 5 and 95 degrees of freedom ≈ 3.92

$$\text{Critical range} = Q_\alpha \sqrt{\left(\frac{MSW}{2}\right)\left(\frac{1}{n_j} + \frac{1}{n_{j'}}\right)} = (3.92)\sqrt{\left(\frac{130.6}{2}\right)\left(\frac{1}{20} + \frac{1}{20}\right)}$$
$$= 10.02$$

Any observed difference greater than 10.02 is considered significant. The mean drive-through service times are different between Wendy's (mean of 226.07 seconds) and Taco Bell, Burger King, and McDonald's, and also between Taco Bell (mean of 236.5) and Burger King, McDonald's, and Chick-fil-A. In addition, the mean drive-through service time is different between McDonald's (273.29) and Burger King and Chick-fil-A. Also, Burger King (193.31) is different from Chick-fil-A. Therefore, with 95% confidence, you can conclude that the service time for Burger King is faster than for the other four chains, that Chick-fil-A is faster than Taco Bell and McDonald's, that Wendy's is faster than Taco Bell and McDonald's, and Taco Bell is faster than McDonald's.

PROBLEMS FOR SECTION 11.1

LEARNING THE BASICS

11.1 An experiment has a single factor with five groups and seven values in each group.

a. How many degrees of freedom are there in determining the among-group variation?

b. How many degrees of freedom are there in determining the within-group variation?

c. How many degrees of freedom are there in determining the total variation?

11.2 You are working with the same experiment as in Problem 11.1.

a. If $SSA = 60$ and $SST = 210$, what is SSW?

b. What is MSA?

c. What is MSW?

d. What is the value of F_{STAT}?

11.3 You are working with the same experiment as in Problems 11.1 and 11.2.

a. Construct the ANOVA summary table and fill in all values in the table.

b. At the 0.05 level of significance, what is the upper-tail critical value from the F distribution?

c. State the decision rule for testing the null hypothesis that all five groups have equal population means.

d. What is your statistical decision?

11.4 Consider an experiment with three groups, with seven values in each.

a. How many degrees of freedom are there in determining the among-group variation?

b. How many degrees of freedom are there in determining the within-group variation?

c. How many degrees of freedom are there in determining the total variation?

11.5 Consider an experiment with four groups, with eight values in each. For the ANOVA summary table below, fill in all the missing results:

Source	Degrees of Freedom	Sum of Squares	Mean Square (Variance)	F
Among groups	$c - 1 = ?$	$SSA = ?$	$MSA = 80$	$F_{STAT} = ?$
Within groups	$n - c = ?$	$SSW = 560$	$MSW = ?$	
Total	$n - 1 = ?$	$SST = ?$		

11.6 You are working with the same experiment as in Problem 11.5.

a. At the 0.05 level of significance, state the decision rule for testing the null hypothesis that all four groups have equal population means.

b. What is your statistical decision?

c. At the 0.05 level of significance, what is the upper-tail critical value from the Studentized range distribution?

d. To perform the Tukey-Kramer procedure, what is the critical range?

APPLYING THE CONCEPTS

11.7 One of the steps involved in the processing of corn flakes for cereals involves toasting the flakes. The file **Corn Flakes** contains the following data for corn flakes thickness (mm) for four different toasting times (seconds).

20 sec	40 sec	60 sec	80 sec
1.6	1.7	2.0	1.0
0.6	1.6	1.2	0.7
0.7	0.8	0.7	0.3

Source: Data extracted from C. Borror, "Blocking benefits," *Quality Progress*, November 2015, pp. 60–62.

a. At the 0.05 level of significance, is there evidence of a difference in the mean thickness of the corn flakes for the different toasting times?

b. If appropriate, determine which toasting times differ in mean thickness.

c. At the 0.05 level of significance, is there evidence of a difference in the variation in the mean thickness of the corn flakes?

d. Which toasting times differ in thickness of the corn flakes? Explain.

✓ SELF TEST **11.8** The more costly and time-consuming it is to export and import, the more difficult it is for local companies to be competitive and to reach international markets. As part of an initial investigation exploring foreign market entry, 10 countries were selected from each of four global regions. The cost associated with compliance of the economy's customs regulations to import a shipment in these countries (in US$), is stored in **International Market 2**.

Source: Data extracted from **doingbusiness.org/data**.

a. At the 0.05 level of significance, is there evidence of a difference in the mean cost of importing across the four global regions?

b. If appropriate, determine which global regions differ in mean cost of importing.

c. At the 0.05 level of significance, is there evidence of a difference in the variation in cost of importing among the four global regions?

d. Which global region(s) should you consider for foreign market entry? Explain.

11.9 A hospital conducted a study of the waiting time in its emergency room. The hospital has a main campus and three affiliated locations. Management had a business objective of reducing waiting time for emergency room cases that did not require immediate attention. To study this, a random sample of 15 emergency room cases that did not require immediate attention at each location were selected on a particular day, and the waiting times (measured from check-in to when the patient was called into the clinic area) were collected and stored in **ER Waiting**.

a. At the 0.05 level of significance, is there evidence of a difference in the mean waiting times in the four locations?

b. If appropriate, determine which locations differ in mean waiting time.

c. At the 0.05 level of significance, is there evidence of a difference in the variation in waiting time among the four locations?

11.10 A manufacturer of pens has hired an advertising agency to develop an advertising campaign for the upcoming holiday season. To prepare for this project, the research director decides to initiate a study of the effect of advertising on product perception. An experiment is designed to compare five different advertisements. Advertisement A greatly undersells the pen's characteristics. Advertisement B slightly undersells the pen's characteristics. Advertisement C slightly oversells the pen's characteristics. Advertisement D greatly oversells the pen's characteristics. Advertisement E attempts to correctly state the pen's characteristics. A sample of 30 adult respondents, taken from a larger focus group, is randomly assigned to the five advertisements (so that there are six respondents to each advertisement). After reading the advertisement and developing a sense of "product expectation," all respondents unknowingly receive the same pen to evaluate. The respondents are permitted to test the pen and the plausibility of the advertising copy. The respondents are then asked to rate the pen from 1 to 7 (lowest to highest) on the product characteristic scales of appearance, durability, and writing performance. The *combined* scores of these three ratings for the 30 respondents, stored in **Pen**, are as follows:

A	B	C	D	E
15	16	8	5	12
18	17	7	6	19
17	21	10	13	18
19	16	15	11	12
19	19	14	9	17
20	17	14	10	14

a. At the 0.05 level of significance, is there evidence of a difference in the mean rating of the pens following exposure to five advertisements?

b. If appropriate, determine which advertisements differ in mean ratings.

c. At the 0.05 level of significance, is there evidence of a difference in the variation in ratings among the five advertisements?

d. Which advertisement(s) should you use, and which advertisement(s) should you avoid? Explain.

11.11 *QSR* reports on the largest quick-serve and fast-casual brands in the United States. The file `Quick-Serve Sales` contains the segment (burger, chicken, ethnic, pizza, sandwich, or snack) and U.S. mean sales per unit ($thousands) for 50 quick-service brands.
Source: Data extracted from "The QSR 50," **bit.ly/2xN4q4v**.

a. At the 0.05 level of significance, is there evidence of a difference in the mean U.S. mean sales per unit ($thousands) among the segments?

b. At the 0.05 level of significance, is there a difference in the variation in U.S. average sales per unit ($thousands) among the segments?

c. What effect does your result in (b) have on the validity of the results in (a)?

d. If appropriate, determine which segments differ in mean sales.

11.12 Brand valuations are critical to CEOs, financial and marketing executives, security analysts, institutional investors, and others who depend on well-researched, reliable information needed for assessments and comparisons in decision making. Millward Brown Optimor has developed the BrandZ Top 100 Most Valuable Global Brands for WPP, the world's largest communications services group. Unlike other studies, the BrandZ Top 100 Most Valuable Global Brands fuses consumer measures of brand equity with financial measures to place a financial value on brands. A research assistant compared brand values for three sectors in the BrandZ Top 100 Most Valuable Global Brands for a recent year: the financial institution sector, the technology sector, and the telecom sector. The research assistant's findings were as follows:

Source	Degrees of Freedom	Sums of Squares	Mean Squares	F
Among groups	2	15,671,226,037.72		
Within groups	50	230,181,430,044.96		
Total	52	245,852,656,082.68		

Group	n	Mean
Financial Institution	19	36,004.95
Technology	23	73,266.91
Telecom	11	44,426.27

Source: Data extracted from *BrandZ Top100 Most Valuable Global Brands 2019*, available at **bit.ly/2JHQZZn**.

a. Complete the ANOVA summary table.

b. At the 0.05 level of significance, is there evidence of a difference in mean brand value among the sectors?

c. If the results in (b) indicate that it is appropriate, use the Tukey-Kramer procedure to determine which sectors differ in mean brand value. Discuss your findings.

11.13 A pet food company has a business objective of expanding its product line beyond its current kidney and shrimp-based cat foods. The company developed two new products, one based on chicken liver and the other based on salmon. The company conducted an experiment to compare the two new products with its two existing ones, as well as a generic beef-based product sold at a supermarket chain.

For the experiment, a sample of 50 cats from the population at a local animal shelter was selected. Ten cats were randomly assigned to each of the five products being tested. Each of the cats was then presented with 3 ounces of the selected food in a dish at feeding time. The researchers defined the variable to be measured as the number of ounces of food that the cat consumed within a 10-minute time interval that began when the filled dish was presented. The results for this experiment are summarized in the following table and stored in `Cat Food`.

Kidney	Shrimp	Chicken Liver	Salmon	Beef
2.37	2.26	2.29	1.79	2.09
2.62	2.69	2.23	2.33	1.87
2.31	2.25	2.41	1.96	1.67
2.47	2.45	2.68	2.05	1.64
2.59	2.34	2.25	2.26	2.16
2.62	2.37	2.17	2.24	1.75
2.34	2.22	2.37	1.96	1.18
2.47	2.56	2.26	1.58	1.92
2.45	2.36	2.45	2.18	1.32
2.32	2.59	2.57	1.93	1.94

a. At the 0.05 level of significance, is there evidence of a difference in the mean amount of food eaten among the various products?

b. If appropriate, determine which products appear to differ significantly in the mean amount of food eaten.

c. At the 0.05 level of significance, is there evidence of a difference in the variation in the amount of food eaten among the various products?

d. What should the pet food company conclude? Fully describe the pet food company's options with respect to the products.

11.14 A transportation strategist wanted to compare the traffic congestion levels across four continents: Asia, Europe, North America, and South America. The file `Congestion Level` contains congestion level, defined as the increase (%) in overall travel time when compared to a free flow situation (an uncongested situation) for 10 cities in each continent.

Source: Data extracted from "TomTom Traffic Index," **bit.ly/32mD3fM**

a. At the 0.05 level of significance, is there evidence of a difference in the mean congestion level across continents?
b. If the results in (a) indicate that it is appropriate to do so, use the Tukey-Kramer procedure to determine which continents differ in congestion level.
c. What assumptions are necessary in (a)?
d. At the 0.05 level of significance, is there evidence of a difference in the variation of the congestion level across continents?

11.2 Two-Way ANOVA

Two-way ANOVA evaluates two factors simultaneously. Each factor is evaluated at two or more levels. For example, in the Arlingtons scenario on page 352, the company faces the business problem of simultaneously evaluating four locations and the effectiveness of providing mobile payment to determine which location should be used and whether mobile payment should be made available. Although this section uses only two factors, one can extend factorial designs to three or more factors (Gitlow, Melnyck, and Levine; Hicks and Turner; Levine; and Montgomery).

To analyze data from a two-factor factorial design, you use **two-way ANOVA**. The following definitions are needed to develop the two-way ANOVA procedure:

r = number of levels of factor A

c = number of levels of factor B

n' = number of values (replicates) for each cell (combination of a particular level of factor A and a particular level of factor B)

n = number of values in the entire experiment (where $n = rcn'$)

X_{ijk} = value of the kth observation for level i of factor A and level j of factor B

$$\overline{\overline{X}} = \frac{\sum_{i=1}^{r}\sum_{j=1}^{c}\sum_{k=1}^{n'}X_{ijk}}{rcn'} = \text{grand mean}$$

$$\overline{X}_{i..} = \frac{\sum_{j=1}^{c}\sum_{k=1}^{n'}X_{ijk}}{cn'} = \text{mean of the }i\text{th level of factor }A\text{ (where }i = 1, 2, \ldots, r)$$

$$\overline{X}_{.j.} = \frac{\sum_{i=1}^{r}\sum_{k=1}^{n'}X_{ijk}}{rn'} = \text{mean of the }j\text{th level of factor }B\text{ (where }i = 1, 2, \ldots, c)$$

$$\overline{X}_{ij.} = \frac{\sum_{k=1}^{n'}X_{ijk}}{n'} = \text{mean of the cell }ij\text{, the combination of the }i\text{th level of factor }A$$
$$\text{and the }j\text{th level of factor }B$$

Because of the complexity of these computations, one should only use computerized methods when performing this analysis. However, to help explain the two-way ANOVA, the decomposition of the total variation is illustrated. In this discussion, only cases in which there are an equal number of values (also called **replicates**) (sample sizes n') for each combination of the levels of factor A with those of factor B are considered. (See the Berenson, Kutner, and Montgomery references for a discussion of two-factor factorial designs with unequal sample sizes.)

Factor and Interaction Effects

There is an **interaction** between factors A and B if the effect of factor A is different for various levels of factor B. Thus, when dividing the total variation into different sources of variation, you need to account for a possible interaction effect, as well as for factor A, factor B, and random variation. To accomplish this, the total variation (SST) is subdivided into sum of squares due to factor A (or SSA), sum of squares due to factor B (or SSB), sum of squares due to the interaction effect of A and B (or $SSAB$), and sum of squares due to random variation (or SSE). Figure 11.9 on page 368 displays this decomposition of the total variation (SST).

FIGURE 11.9
Partitioning the total
variation in a two-factor
factorial design

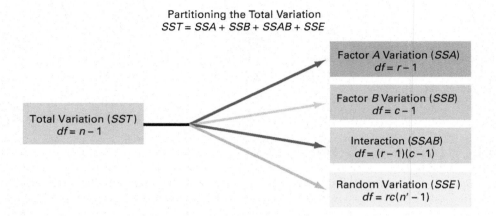

The sum of squares total (*SST*) represents the total variation among all the values around the grand mean. Equation (11.7) shows the computation for total variation.

TOTAL VARIATION IN TWO-WAY ANOVA

$$SST = \sum_{i=1}^{r}\sum_{j=1}^{c}\sum_{k=1}^{n'}(X_{ijk} - \bar{\bar{X}})^2 \tag{11.7}$$

The **sum of squares due to factor *A* (*SSA*)** represents the differences among the various levels of factor *A* and the grand mean. Equation (11.8) shows the computation for factor *A* variation.

FACTOR *A* VARIATION IN TWO-WAY ANOVA

$$SSA = cn'\sum_{i=1}^{r}(\bar{X}_{i..} - \bar{\bar{X}})^2 \tag{11.8}$$

The **sum of squares due to factor *B* (*SSB*)** represents the differences among the various levels of factor *B* and the grand mean. Equation (11.9) shows the computation for factor *B* variation.

FACTOR *B* VARIATION IN TWO-WAY ANOVA

$$SSB = rn'\sum_{j=1}^{c}(\bar{X}_{.j.} - \bar{\bar{X}})^2 \tag{11.9}$$

The **sum of squares due to interaction (*SSAB*)** represents the interacting effect of specific combinations of factor *A* and factor *B*. Equation (11.10) shows the computation for interaction variation.

INTERACTION VARIATION IN TWO-WAY ANOVA

$$SSAB = n'\sum_{i=1}^{r}\sum_{j=1}^{c}(\bar{X}_{ij.} - \bar{X}_{i..} - \bar{X}_{.j.} + \bar{\bar{X}})^2 \tag{11.10}$$

The **sum of squares error (SSE)** represents random variation—that is, the differences among the values within each cell and the corresponding cell mean. Equation (11.11) shows the computation for random variation.

RANDOM VARIATION IN TWO-WAY ANOVA

$$SSE = \sum_{i=1}^{r} \sum_{j=1}^{c} \sum_{k=1}^{n'} (X_{ijk} - \overline{X}_{ij.})^2 \qquad (11.11)$$

Because there are r levels of factor A, there are $r - 1$ degrees of freedom associated with SSA. Similarly, because there are c levels of factor B, there are $c - 1$ degrees of freedom associated with SSB. Because there are n' replicates in each of the rc cells, there are $rc(n' - 1)$ degrees of freedom associated with the SSE term. Because the method compares each value, X_{ijk}, to the grand mean, $\overline{\overline{X}}$, which is based on all n values, there are $n - 1$ degrees of freedom associated with the sum of squares total SST. Therefore, because the degrees of freedom for each of the sources of variation must add to the degrees of freedom for the total variation (SST), you can calculate the degrees of freedom for the interaction component ($SSAB$) by subtraction. The degrees of freedom for interaction are $(r - 1)(c - 1)$.

Dividing each sum of squares by its associated degrees of freedom calculates the four variances or **mean squares** (MSA, MSB, $MSAB$, and MSE). Equations (11.12a–d) present the mean square terms needed for a two-way ANOVA table.

student TIP

Mean square is another term for *variance*.

MEAN SQUARES IN TWO-WAY ANOVA

$$MSA = \frac{SSA}{r - 1} \qquad (11.12a)$$

$$MSB = \frac{SSB}{c - 1} \qquad (11.12b)$$

$$MSAB = \frac{SSAB}{(r - 1)(c - 1)} \qquad (11.12c)$$

$$MSE = \frac{SSE}{rc(n' - 1)} \qquad (11.12d)$$

Testing for Factor and Interaction Effects

There are three different tests to perform in a two-way ANOVA:

- A test of the hypothesis of no difference due to factor A
- A test of the hypothesis of no difference due to factor B
- A test of the hypothesis of no interaction of factors A and B

To test the hypothesis of no difference due to factor A, one defines the null and alternative hypotheses as

$$H_0: \mu_{1..} = \mu_{2..} = \cdots = \mu_{r..}$$

$$H_1: \text{Not all } \mu_{i..} \text{ are equal}$$

and uses the F_{STAT} test statistic in Equation (11.13) on page 370.

F TEST FOR FACTOR A EFFECT

$$F_{STAT} = \frac{MSA}{MSE} \qquad (11.13)$$

One rejects the null hypothesis at the α level of significance if

$$F_{STAT} = \frac{MSA}{MSE} > F_\alpha$$

where F_α is the upper-tail critical value from an F distribution with $r - 1$ and $rc(n' - 1)$ degrees of freedom.

To test the hypothesis of no difference due to factor B, one defines the null and alternative hypotheses as

$$H_0\colon \mu_{.1.} = \mu_{.2.} = \cdots = \mu_{.c.}$$
$$H_1\colon \text{Not all } \mu_{.j.} \text{ are equal}$$

and uses the F_{STAT} test statistic in Equation (11.14).

F TEST FOR FACTOR B EFFECT

$$F_{STAT} = \frac{MSB}{MSE} \qquad (11.14)$$

You reject the null hypothesis at the α level of significance if

$$F_{STAT} = \frac{MSB}{MSE} > F_\alpha$$

where F_α is the upper-tail critical value from an F distribution with $c - 1$ and $rc(n' - 1)$ degrees of freedom.

To test the hypothesis of no interaction of factors A and B, one defines the null and alternative hypotheses

$$H_0\colon \text{The interaction of } A \text{ and } B \text{ is equal to zero}$$
$$H_1\colon \text{The interaction of } A \text{ and } B \text{ is not equal to zero}$$

and uses the F_{STAT} test statistic in Equation (11.15).

F TEST FOR INTERACTION EFFECT

$$F_{STAT} = \frac{MSAB}{MSE} \qquad (11.15)$$

student TIP

In each of these F tests, the denominator of the F_{STAT} statistic is MSE.

One rejects the null hypothesis at the α level of significance if

$$F_{STAT} = \frac{MSAB}{MSE} > F_\alpha$$

where F_α is the upper-tail critical value from an F distribution with $(r - 1)(c - 1)$ and $rc(n' - 1)$ degrees of freedom.

Table 11.7 presents the entire two-way ANOVA table.

TABLE 11.7
Analysis of variance table for the two-factor factorial design

Source	Degrees of Freedom	Sum of Squares	Mean Square (Variance)	F
A	$r - 1$	SSA	$MSA = \dfrac{SSA}{r - 1}$	$F_{STAT} = \dfrac{MSA}{MSE}$
B	$c - 1$	SSB	$MSB = \dfrac{SSB}{c - 1}$	$F_{STAT} = \dfrac{MSB}{MSE}$
AB	$(r - 1)(c - 1)$	SSAB	$MSAB = \dfrac{SSAB}{(r - 1)(c - 1)}$	$F_{STAT} = \dfrac{MSAB}{MSE}$
Error	$rc(n' - 1)$	SSE	$MSE = \dfrac{SSE}{rc(n' - 1)}$	
Total	$n - 1$	SST		

To illustrate two-way ANOVA, return to the Arlingtons scenario on page 352. As a member of the sales team, you first explored how different in-store locations might affect the sales of mobile electronics items using one-way ANOVA. Now, to explore the effects of permitting mobile payment methods to buy mobile electronics items, you design an experiment that examines this second (B) factor as it studies the effects of in-store location (factor A) using two-way ANOVA. Two-way ANOVA will allow you to determine if there is a significant difference in mobile electronics sales among the four in-store locations *and* whether permitting mobile payment methods makes a difference.

To test the effects of the two factors, you conduct a 60-day experiment at 40 same-sized stores that have similar storewide net sales. You randomly assign ten stores to use the current in-aisle location, ten stores to use the special front location, ten stores to use the kiosk location, and ten stores to use the expert counter. In five stores in each of the four groups, you permit mobile payment methods (for the other five in each group, mobile payment methods are not permitted). At the end of the experiment, you organize the mobile electronics sales data by group and store the data in Mobile Electronics 2 . Table 11.8 presents the data of the experiment.

TABLE 11.8
Mobile electronics sales ($000) at four in-store locations with mobile payments permitted and not permitted

MOBILE PAYMENTS	IN-STORE LOCATION			
	In-Aisle	Front	Kiosk	Expert
No	30.06	32.22	30.78	30.33
No	29.96	31.47	30.91	30.29
No	30.19	32.13	30.79	30.25
No	29.96	31.86	30.95	30.25
No	27.74	32.29	31.13	30.55
Yes	30.66	32.81	31.34	31.03
Yes	29.99	32.65	31.80	31.77
Yes	30.73	32.81	32.00	30.97
Yes	30.72	32.42	31.07	31.43
Yes	30.73	33.12	31.69	30.72

Figure 11.10 presents the results for this example. In the Excel worksheet, the *A*, *B*, and Error sources of variation in Table 11.7 on page 371 are labeled Sample, Columns, and Within, respectively.

FIGURE 11.10

Excel two-way ANOVA results for the in-store location sales and mobile payment experiment

ANOVA: Two-Factor With Replication						
	A	B	C	D	E	F
SUMMARY		In-aisle	Front	Kiosk	Expert	Total
	No					
Count		5	5	5	5	20
Sum		147.91	159.97	154.56	151.67	614.11
Average		29.582	31.994	30.912	30.334	30.7055
Variance		1.0692	0.1124	0.0203	0.0157	1.0730
	Yes					
Count		5	5	5	5	20
Sum		152.83	163.81	157.9	155.92	630.46
Average		30.566	32.762	31.58	31.184	31.5230
Variance		0.1045	0.0656	0.1387	0.1722	0.7773
	Total					
Count		10	10	10	10	
Sum		300.74	323.78	312.46	307.59	
Average		30.074	32.378	31.246	30.759	
Variance		0.7906	0.2430	0.1946	0.2842	

ANOVA						
Source of Variation	**SS**	**df**	**MS**	**F**	**P-value**	**F crit**
Sample	6.6831	1	6.6831	31.4760	0.0000	4.1491
Columns	28.2274	3	9.4091	44.3154	0.0000	2.9011
Interaction	0.1339	3	0.0446	0.2103	0.8885	2.9011
Within	6.7943	32	0.2123			
Total	41.8388	39				
					Level of significance	0.05

To interpret the results, one starts by testing whether there is an interaction effect between factor *A* (mobile payments) and factor *B* (in-store locations). If the interaction effect is significant, further analysis will focus on this interaction. If the interaction effect is not significant, you can focus on the **main effects**—the potential effect of permitting mobile payment (factor *A*) and the potential differences in in-store locations (factor *B*).

Using the 0.05 level of significance to determine whether there is evidence of an interaction effect, one rejects the null hypothesis of no interaction between mobile payments and in-store locations if the computed F_{STAT} statistic is greater than 2.9011, the upper-tail critical value from the *F* distribution, with 3 and 32 degrees of freedom (see Figures 11.10 and 11.11).[2]

[2]Table E.5 does not provide the upper-tail critical values from the *F* distribution with 32 degrees of freedom in the denominator. When the desired degrees of freedom are not provided in the table, use the *p*-value computed by Excel.

FIGURE 11.11

Regions of rejection and nonrejection at the 0.05 level of significance, with 3 and 32 degrees of freedom

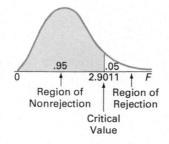

Because $F_{STAT} = 0.2103 < 2.9011$ or the *p*-value $= 0.8885 > 0.05$, one does not reject H_0. One concludes that there is insufficient evidence of an interaction effect between the two factors, mobile payment and in-store location. Finding insufficient evidence of an interaction effect between factors enables one to continue and focus on the main effects.

Using the 0.05 level of significance and testing whether there is an effect due to mobile payment options (yes or no) (factor *A*), one rejects the null hypothesis if the computed F_{STAT} test statistic is greater than 4.1491, the upper-tail critical value from the *F* distribution with 1 and 32 degrees of freedom (see Figures 11.10 and 11.12). Because $F_{STAT} = 31.4760 > 4.1491$ or the *p*-value $= 0.0000 < 0.05$, one rejects H_0. One concludes that there is evidence of a difference in the mean sales when mobile payment methods are permitted as compared to when they are not. Because the mean sales when mobile payment methods are permitted is 31.523 and is 30.7055 when they are not, one can conclude that permitting mobile payment methods has led to an increase in mean sales.

FIGURE 11.12
Regions of rejection and nonrejection at the 0.05 level of significance, with 1 and 32 degrees of freedom

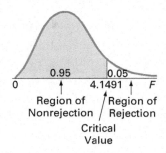

Using the 0.05 level of significance and testing for a difference among the in-store locations (factor B), one rejects the null hypothesis of no difference if the computed F_{STAT} test statistic is greater than 2.9011, the upper-tail critical value from the F distribution with 3 degrees of freedom in the numerator and 32 degrees of freedom in the denominator (see Figures 11.10 and 11.11). Because $F_{STAT} = 44.3154 > 2.9011$ or the p-value $= 0.0000 < 0.05$, one rejects H_0 and concludes that there is evidence of a difference in the mean sales among the four in-store locations.

Multiple Comparisons: The Tukey Procedure

If one or both of the factor effects are significant and there is no significant interaction effect, one can determine the particular levels that are significantly different by using the **Tukey multiple comparisons procedure for two-way ANOVA** when there are more than two levels of a factor (Kutner and Montgomery). Equation (11.16) gives the critical range for factor A.

CRITICAL RANGE FOR FACTOR A

$$\text{Critical range} = Q_\alpha \sqrt{\frac{MSE}{cn'}} \tag{11.16}$$

where Q_α is the upper-tail critical value from a Studentized range distribution having r and $rc(n' - 1)$ degrees of freedom. (Values for the Studentized range distribution are found in Table E.7.)

Equation (11.17) gives the critical range for factor B.

CRITICAL RANGE FOR FACTOR B

$$\text{Critical range} = Q_\alpha \sqrt{\frac{MSE}{rn'}} \tag{11.17}$$

where Q_α is the upper-tail critical value from a Studentized range distribution having c and $rc(n' - 1)$ degrees of freedom. (Values for the Studentized range distribution are found in Table E.7.)

To use the Tukey procedure, return to the Table 11.8 mobile electronics sales data on page 371. In the ANOVA summary table in Figure 11.10 on page 372, the interaction effect is not significant. Because there are only two categories for mobile payment (yes and no), there are no multiple comparisons to be constructed. Using $\alpha = 0.05$, there is evidence of a significant difference among the four in-store locations that comprise factor B. Thus, one can use the Tukey multiple comparisons procedure to determine which of the four in-store locations differ.

Because there are four in-store locations, there are $4(4 - 1)/2 = 6$ pairwise comparisons. Using the calculations presented in Figure 11.10, the absolute mean differences are as follows:

1. $|\overline{X}_{.1} - \overline{X}_{.2}| = |30.074 - 32.378| = 2.124$
2. $|\overline{X}_{.1} - \overline{X}_{.3}| = |30.074 - 31.246| = 1.172$
3. $|\overline{X}_{.1} - \overline{X}_{.4}| = |30.074 - 30.759| = 0.685$
4. $|\overline{X}_{.2} - \overline{X}_{.3}| = |32.378 - 31.246| = 1.132$
5. $|\overline{X}_{.2} - \overline{X}_{.4}| = |32.378 - 30.759| = 1.619$
6. $|\overline{X}_{.3} - \overline{X}_{.4}| = |31.246 - 30.759| = 0.487$

To determine the critical range, refer to Figure 11.10 to find $MSE = 0.2123$, $r = 2$, $c = 4$, and $n' = 5$. From Table E.7 [for $\alpha = 0.05$, $c = 4$, and $rc(n' - 1) = 32$], Q_α, the upper-tail critical value of the Studentized range distribution with 4 and 32 degrees of freedom is approximately 3.84. Using Equation (11.17),

$$\text{Critical range} = 3.84\sqrt{\frac{0.2123}{10}} = 0.5595$$

Because five of the six comparisons are greater than the critical range of 0.5595, one concludes that the population mean sales is different for the in-store locations except for the kiosk and expert locations. The front location is estimated to have higher mean sales than the other three in-store locations. The kiosk location is estimated to have higher mean sales than the in-aisle location. The expert location is estimated to have higher mean sales than the in-aisle location. Note that by using $\alpha = 0.05$, one can make all six comparisons with an overall error rate of only 5%. Consistent with the results of the one-factor experiment, one has additional evidence that selling mobile electronics items at the front location will increase sales the most. In addition, one also has evidence that enabling mobile payment will also lead to increased sales.

Visualizing Interaction Effects: The Cell Means Plot

One can gain a better understanding of the interaction effect by plotting the **cell means**, the means of all possible factor-level combinations. Figure 11.13 presents a cell means plot that uses the cell means for the mobile payments permitted/in-store location combinations shown in Figure 11.10 on page 372. From the plot of the mean sales for each combination of mobile payments permitted and in-store location, observe that the two lines (representing the two levels of mobile payments, yes and no) are roughly parallel. This indicates that the *difference* between the mean sales for stores that permit mobile payment methods and those that do not is virtually the same for the four in-store locations. In other words, there is no *interaction* between these two factors, as was indicated by the F test.

FIGURE 11.13

Excel cell means plot for mobile electronics sales based on mobile payments permitted and in-store location

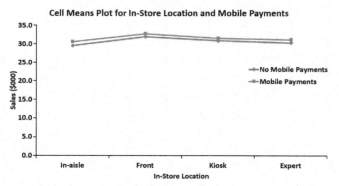

Interpreting Interaction Effects

How does one interpret an interaction? When there is an interaction, some levels of factor A respond better with certain levels of factor B. For example, with respect to mobile electronics sales, suppose that some in-store locations were better when mobile payment methods were permitted and other in-store locations were better when mobile payment methods were not permitted. If this were true, the lines of Figure 11.13 would not be nearly as parallel, and the interaction effect might be statistically significant. In such a situation, the difference between whether mobile payment methods were permitted is no longer the same for all in-store locations. Such an outcome would also complicate the interpretation of the *main effects* because differences in one factor (whether mobile payment methods were permitted) would not be consistent across the other factor (the in-store locations).

Example 11.2 illustrates a situation with a significant interaction effect.

EXAMPLE 11.2

Interpreting Significant Interaction Effects

A nationwide company specializing in preparing students for college and graduate school entrance exams, such as the SAT, ACT, GRE, and LSAT, had the business objective of improving its ACT preparatory course. Two factors of interest to the company are the length of the course (a condensed 10-day period or a regular 30-day period) and the type of course (traditional classroom or online distance learning). The company collected data by randomly assigning 10 clients to each of the four cells that represent a combination of length of the course and type of course. Table 11.9 presents the results, which are stored in ACT .

What are the effects of the type of course and the length of the course on ACT scores?

TABLE 11.9
ACT scores for different types and lengths of courses

TYPE OF COURSE	LENGTH OF COURSE			
	Condensed		Regular	
Traditional	26	18	34	28
Traditional	27	24	24	21
Traditional	25	19	35	23
Traditional	21	20	31	29
Traditional	21	18	28	26
Online	27	21	24	21
Online	29	32	16	19
Online	30	20	22	19
Online	24	28	20	24
Online	30	29	23	25

SOLUTION The Figure 11.14 cell means plot shows a strong interaction between the type of course and the length of the course. The nonparallel lines indicate that the effect of condensing the course depends on whether the course is taught in the traditional classroom or by online distance learning. The online mean score is higher when the course is condensed to a 10-day period, whereas the traditional mean score is higher when the course takes place over the regular 30-day period.

FIGURE 11.14
Excel cell means plot of mean ACT scores

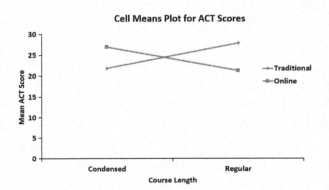

▶*(continued)*

To verify the visual analysis provided by interpreting the cell means plot, one tests whether there is a statistically significant interaction between factor A (length of course) and factor B (type of course). Using a 0.05 level of significance, one rejects the null hypothesis because $F_{STAT} = 24.2569 > 4.1132$ or the p-value equals $0.0000 < 0.05$ (see Figure 11.15 shown below). Thus, the hypothesis test confirms the interaction evident in the cell means plot.

FIGURE 11.15
Excel two-way ANOVA results for the ACT scores

	A	B	C	D	E	F	G
1	ANOVA: Two-Factor With Replication						
2							
3	SUMMARY	Condensed	Regular	Total			
4	*traditional*						
5	Count	10	10	20			
6	Sum	219	279	498			
7	Average	21.9	27.9	24.9			
8	Variance	11.2111	20.9889	24.7263			
9							
10	*online*						
11	Count	10	10	20			
12	Sum	270	213	483			
13	Average	27	21.3	24.15			
14	Variance	16.2222	8.0111	20.0289			
15							
16	*Total*						
17	Count	20	20				
18	Sum	489	492				
19	Average	24.45	24.6				
20	Variance	19.8395	25.2000				
21							
22							
23	ANOVA						
24	*Source of Variation*	SS	df	MS	F	P-value	F crit
25	Sample	5.6250	1	5.6250	0.3987	0.5318	4.1132
26	Columns	0.2250	1	0.2250	0.0159	0.9002	4.1132
27	Interaction	342.2250	1	342.2250	24.2569	0.0000	4.1132
28	Within	507.9000	36	14.1083			
29							
30	Total	855.9750	39				
31						*Level of significance*	0.05

The existence of this significant interaction effect complicates the interpretation of the hypothesis tests concerning the two main effects. One cannot directly conclude that there is no effect with respect to length of course and type of course, even though both have p-values > 0.05.

Given that the interaction is significant, one can reanalyze the data with the two factors collapsed into four groups of a single factor rather than a two-way ANOVA with two levels of each of the two factors. One reorganizes the data as follows: Group 1 is traditional condensed, Group 2 is traditional regular, Group 3 is online condensed, and Group 4 is online regular. Figure 11.16 shows the results for these data, stored in One-Way ACT.

FIGURE 11.16
Excel one-way ANOVA and Tukey-Kramer results for the ACT scores

	A	B	C	D	E	F	G
1	One-Way ANOVA (ANOVA: Single Factor)						
2							
3	SUMMARY						
4	*Groups*	Count	Sum	Average	Variance		
5	Group 1	10	219	21.9	11.2111		
6	Group 2	10	279	27.9	20.9889		
7	Group 3	10	270	27	16.2222		
8	Group 4	10	213	21.3	8.0111		
9							
10							
11	ANOVA						
12	*Source of Variation*	SS	df	MS	F	P-value	F crit
13	Between Groups	348.0750	3	116.0250	8.2239	0.0003	2.8663
14	Within Groups	507.9000	36	14.1083			
15							
16	Total	855.9750	39				
17						*Level of significance*	0.05

	A	B	C	D	E	F	G	H	I
1	Tukey Kramer Multiple Comparisons								
2									
3		Sample	Sample			Absolute	Std. Error	Critical	
4	Group	Mean	Size		Comparison	Difference	of Difference	Range	Results
5	1: Group 1	21.9	10		Group 1 to Group 2	6	1.1878	4.5373	Means are different
6	2: Group 2	27.9	10		Group 1 to Group 3	5.1	1.1878	4.5373	Means are different
7	3: Group 3	27	10		Group 1 to Group 4	0.6	1.1878	4.5373	Means are not different
8	4: Group 4	21.3	10		Group 2 to Group 3	0.9	1.1878	4.5373	Means are not different
9					Group 2 to Group 4	6.6	1.1878	4.5373	Means are different
10	Other Data				Group 3 to Group 4	5.7	1.1878	4.5373	Means are different
11	Level of significance	0.05							
12	Numerator d.f.	4							
13	Denominator d.f.	36							
14	MSW	14.10833							
15	Q Statistic	3.82							

▶*(continued)*

From Figure 11.16, because $F_{STAT} = 8.2239 > 2.8663$ or p-value $= 0.0003 < 0.05$, there is evidence of a significant difference in the four groups (traditional condensed, traditional regular, online condensed, and online regular). Using the Tukey-Kramer multiple comparisons procedure, traditional condensed is different from traditional regular and from online condensed. Traditional regular is also different from online regular, and online condensed is also different from online regular.

Therefore, whether condensing a course is a good idea depends on whether the course is offered in a traditional classroom or as an online distance learning course. To ensure the highest mean ACT scores, the company should use the traditional approach for courses that are given over a 30-day period but use the online approach for courses that are condensed into a 10-day period.

PROBLEMS FOR SECTION 11.2

LEARNING THE BASICS

11.15 Consider a two-factor factorial design with three levels for factor A, three levels for factor B, and four replicates in each of the nine cells.
a. How many degrees of freedom are there in determining the factor A variation and the factor B variation?
b. How many degrees of freedom are there in determining the interaction variation?
c. How many degrees of freedom are there in determining the random variation?
d. How many degrees of freedom are there in determining the total variation?

11.16 Assume that you are working with the results from Problem 11.15, and $SSA = 120$, $SSB = 110$, $SSE = 270$, and $SST = 540$.
a. What is $SSAB$?
b. What are MSA and MSB?
c. What is $MSAB$?
d. What is MSE?

11.17 Assume that you are working with the results from Problems 11.15 and 11.16.
a. What is the value of the F_{STAT} test statistic for the interaction effect?
b. What is the value of the F_{STAT} test statistic for the factor A effect?
c. What is the value of the F_{STAT} test statistic for the factor B effect?
d. Form the ANOVA summary table and fill in all values in the body of the table.

11.18 Given the results from Problems 11.15 through 11.17,
a. at the 0.05 level of significance, is there an effect due to factor A?
b. at the 0.05 level of significance, is there an effect due to factor B?
c. at the 0.05 level of significance, is there an interaction effect?

11.19 Given a two-way ANOVA with two levels for factor A, five levels for factor B, and four replicates in each of the 10 cells, with $SSA = 18$, $SSB = 64$, $SSE = 60$, and $SST = 150$,
a. form the ANOVA summary table and fill in all values in the body of the table.
b. at the 0.05 level of significance, is there an effect due to factor A?
c. at the 0.05 level of significance, is there an effect due to factor B?
d. at the 0.05 level of significance, is there an interaction effect?

11.20 Given a two-factor factorial experiment and the ANOVA summary table that follows, fill in all the missing results:

Source	Degrees of Freedom	Sum of Squares	Mean Square (Variance)	F
A	$r - 1 = 2$	$SSA = ?$	$MSA = 80$	$F_{STAT} = ?$
B	$c - 1 = ?$	$SSB = 220$	$MSB = ?$	$F_{STAT} = 11.0$
AB	$(r - 1)(c - 1) = 8$	$SSAB = ?$	$MSAB = 10$	$F_{STAT} = ?$
Error	$rc(n' - 1) = 30$	$SSE = ?$	$MSE = ?$	
Total	$n - 1 = ?$	$SST = ?$		

11.21 Given the results from Problem 11.20,
a. at the 0.05 level of significance, is there an effect due to factor A?
b. at the 0.05 level of significance, is there an effect due to factor B?
c. at the 0.05 level of significance, is there an interaction effect?

APPLYING THE CONCEPTS

11.22 An experiment was conducted to study the extrusion process of biodegradable packaging foam. Two of the factors considered for their effect on the unit density (mg/ml) were the die temperature (145°C vs. 155°C) and the die diameter (3 mm vs. 4 mm). The results are stored in Packaging Foam 1 .

Source: Data extracted from W. Y. Koh, K. M. Eskridge, and M. A. Hanna, "Supersaturated Split-Plot Designs," *Journal of Quality Technology*, 45, January 2013, pp. 61–72.

At the 0.05 level of significance,

a. is there an interaction between die temperature and die diameter?
b. is there an effect due to die temperature?
c. is there an effect due to die diameter?
d. Plot the mean unit density for each die temperature for each die diameter.
e. What can you conclude about the effect of die temperature and die diameter on mean unit density?

11.23 Referring to Problem 11.22, the effect of die temperature and die diameter on the foam diameter was also measured and the results stored in Packaging Foam 2 .

At the 0.05 level of significance,

a. is there an interaction between die temperature and die diameter?
b. is there an effect due to die temperature?
c. is there an effect due to die diameter?
d. Plot the mean foam diameter for each die temperature and die diameter.
e. What conclusions can you reach concerning the importance of each of these two factors on the foam diameter?

 11.24 A plastic injection molding process is often used in manufacturing because of its ability to mold complicated shapes. An experiment was conducted on the manufacture of a television remote part, and the warpage (mm) of the part was measured and stored in TV Remote .

Source: Data extracted from M. A. Barghash and F. A. Alkaabneh, "Shrinkage and Warpage Detailed Analysis and Optimization for the Injection Molding Process Using Multistage Experimental Design," *Quality Engineering*, 26, 2014, pp. 319–334.

Two factors were to be considered, the filling time (1, 2, or 3 sec) and the mold temperature (60, 72.5, or 85 °C).

At the 0.05 level of significance,

a. is there an interaction between filling time and mold temperature?
b. is there an effect due to filling time?
c. is there an effect due to mold temperature?
d. Plot the mean warpage for each filling time for each mold temperature.
e. If appropriate, use the Tukey multiple comparison procedure to determine which of the filling times and mold temperatures differ.
f. Discuss the results of (a) through (e).

11.25 A glass manufacturing company wanted to investigate the effect of breakoff pressure and stopper height on the percentage of breaking off chips. The results, stored in Glass 1 , were as follows:

| | STOPPER HEIGHT | |
BREAK OFF PRESSURE	Twenty	Twenty-Five
Two	1.75	0.75
Two	1.00	0.50

| | STOPPER HEIGHT | |
BREAK OFF PRESSURE	Twenty	Twenty-Five
Two	0.00	0.00
Two	1.00	0.25
Three	2.25	1.50
Three	1.50	1.25
Three	0.25	0.25
Three	0.75	0.75

Source: K. Kumar and S. Yadav, "Breakthrough Solution," *Six Sigma Forum Magazine*, November 2016, pp. 7–22.

At the 0.05 level of significance,

a. is there an interaction between the breakoff pressure and the stopper height?
b. is there an effect due to the breakoff pressure?
c. is there an effect due to the stopper height?
d. Plot the percentage breakoff for each breakoff pressure for each stopper height.
e. Discuss the results of (a) through (d).

11.26 A glass manufacturing company wanted to investigate the effect of zone 1 lower temperature (630 vs. 650) and zone 3 upper temperature (695 vs. 715) on the roller imprint of glass. The results stored in Glass 2 were as follows:

| | ZONE 3 UPPER | |
ZONE 1 LOWER	695	715
630	50	100
630	25	0
630	50	25
630	125	75
650	25	75
650	25	25
650	50	0
650	20	125

Source: K. Kumar and S. Yadav, "Breakthrough Solution," *Six Sigma Forum Magazine*, November 2016, pp. 7–22.

At the 0.05 level of significance,

a. is there an interaction between zone 1 lower and zone 3 upper?
b. is there an effect due to zone 1 lower?
c. is there an effect due to zone 3 upper?
d. Plot the roller imprint for each level of zone 1 lower for level of zone 3 upper.
e. Discuss the results of (a) through (d).

11.3 The Randomized Block Design

Section 10.2 discusses how to use the paired *t* test to evaluate the difference between the means of two groups when you have repeated measurements or matched samples. The randomized block design evaluates differences among more than two groups that contain matched samples or repeated measures that have been placed in blocks. The **Section 11.3 online topic** discusses this method and illustrates its use.

11.4 Fixed Effects, Random Effects, and Mixed Effects Models

Sections 11.1 through 11.3 do not consider the distinction between how the levels of a factor were selected. The equation for the *F* test depends on whether the levels of a factor were specifically selected or randomly selected from a population. The **Section 11.4 online topic** presents the appropriate *F* tests to use when the levels of a factor are either specifically selected or randomly selected from a population of levels.

▼ USING **STATISTICS**
The Means to Find Differences at Arlingtons, Revisited

In the Arlingtons scenario, you sought to determine whether there were differences in mobile electronics sales among four in-store locations as well as determine whether permitting mobile payments had an effect on those sales.

Using the one-way ANOVA, you determined that there was a difference in the mean sales for the four in-store locations. Using the two-way ANOVA, you determined that there was no interaction between in-store location and permitting mobile payment methods and that mean sales were higher when mobile payment methods were permitted than when such methods were not. In addition, you concluded that the population mean sales is different for the four in-store locations and reached these other conclusions:

- The front location is estimated to have higher mean sales than the other three locations.
- The kiosk location is estimated to have higher mean sales than the current in-aisle location.
- The expert location is estimated to have higher mean sales than the current in-aisle location.

Your next step as a member of the sales team might be to further investigate the differences among the sales locations as well as examine other factors that could influence mobile electronics sale.

▼ SUMMARY

This chapter explains the statistical procedures used to analyze the effect of one or two factors of interest. The chapter details the assumptions required for each procedure and emphasizes the need to investigate the validity of the assumptions underlying the hypothesis-testing procedures. Table 11.10 summarizes the designs and procedures that this chapter discusses.

TABLE 11.10
Chapter 11 designs
and procedures

Design	Procedure
Completely randomized design	One-way analysis of variance (Section 11.1)
Factorial design	Two-way analysis of variance (Section 11.2)
Randomized block design	Randomized block design (online Section 11.3)

▼ REFERENCES

Berenson, M. L., D. M. Levine, and M. Goldstein. *Intermediate Statistical Methods and Applications: A Computer Package Approach.* Upper Saddle River, NJ: Prentice Hall, 1983.

Corder, G. W., and D. I. Foreman. *Nonparametric Statistics: A Step-by-Step Approach.* New York: Wiley, 2014.

Daniel, W. W. *Applied Nonparametric Statistics*, 2nd ed. Boston: PWS Kent, 1990.

Gitlow, H. S., R. Melnyck, and D. Levine. *A Guide to Six Sigma and Process Improvement for Practitioners and Students*, 2nd ed. Old Tappan, NJ: Pearson Education, 2015.

Hicks, C. R., and K. Turner. *Fundamental Concepts in the Design of Experiments*, 5th ed. New York: Oxford University Press, 1999.

Kutner, M., J. Neter, C. Nachtsheim, and W. Li. *Applied Linear Statistical Models*, 5th ed. New York: McGraw-Hill-Irwin, 2005.

Levine, D. *Statistics for Six Sigma Green Belts.* Upper Saddle River, NJ: Financial Times/Prentice Hall, 2006.

Montgomery, D. M. *Design and Analysis of Experiments*, 8th ed. New York: Wiley, 2013.

▼ KEY EQUATIONS

Total Variation in One-Way ANOVA

$$SST = \sum_{j=1}^{c}\sum_{i=1}^{n_j} \left(X_{ij} - \overline{\overline{X}} \right)^2 \tag{11.1}$$

Among-Group Variation in One-Way ANOVA

$$SSA = \sum_{j=1}^{c} n_j \left(\overline{X}_j - \overline{\overline{X}} \right)^2 \tag{11.2}$$

Within-Group Variation in One-Way ANOVA

$$SSW = \sum_{j=1}^{c}\sum_{i=1}^{n_j} \left(X_{ij} - \overline{X}_j \right)^2 \tag{11.3}$$

Mean Squares in One-Way ANOVA

$$MSA = \frac{SSA}{c - 1} \tag{11.4a}$$

$$MSW = \frac{SSW}{n - c} \tag{11.4b}$$

$$MST = \frac{SST}{n - 1} \tag{11.4c}$$

One-Way ANOVA F_{STAT} Test Statistic

$$F_{STAT} = \frac{MSA}{MSW} \tag{11.5}$$

Critical Range for the Tukey-Kramer Procedure

$$\text{Critical range} = Q_\alpha \sqrt{\frac{MSW}{2}\left(\frac{1}{n_j} + \frac{1}{n_{j'}}\right)} \tag{11.6}$$

Total Variation in Two-Way ANOVA

$$SST = \sum_{i=1}^{r}\sum_{j=1}^{c}\sum_{k=1}^{n'} \left(X_{ijk} - \overline{\overline{X}} \right)^2 \tag{11.7}$$

Factor A Variation in Two-Way ANOVA

$$SSA = cn' \sum_{i=1}^{r} \left(\overline{X}_{i..} - \overline{\overline{X}} \right)^2 \tag{11.8}$$

Factor B Variation in Two-Way ANOVA

$$SSB = rn' \sum_{j=1}^{c} \left(\overline{X}_{.j.} - \overline{\overline{X}} \right)^2 \tag{11.9}$$

Interaction Variation in Two-Way ANOVA

$$SSAB = n' \sum_{i=1}^{r}\sum_{j=1}^{c} \left(\overline{X}_{ij.} - \overline{X}_{i..} - \overline{X}_{.j.} + \overline{\overline{X}} \right)^2 \tag{11.10}$$

Random Variation in Two-Way ANOVA

$$SSE = \sum_{i=1}^{r}\sum_{j=1}^{c}\sum_{k=1}^{n'} (X_{ijk} - \overline{X}_{ij.})^2 \tag{11.11}$$

Mean Squares in Two-Way ANOVA

$$MSA = \frac{SSA}{r - 1} \tag{11.12a}$$

$$MSB = \frac{SSB}{c - 1} \tag{11.12b}$$

$$MSAB = \frac{SSAB}{(r - 1)(c - 1)} \tag{11.12c}$$

$$MSE = \frac{SSE}{rc(n' - 1)} \tag{11.12d}$$

F Test for Factor A Effect

$$F_{STAT} = \frac{MSA}{MSE} \tag{11.13}$$

F Test for Factor B Effect

$$F_{STAT} = \frac{MSB}{MSE} \tag{11.14}$$

F Test for Interaction Effect

$$F_{STAT} = \frac{MSAB}{MSE} \tag{11.15}$$

Critical Range for Factor A

$$\text{Critical range} = Q_\alpha \sqrt{\frac{MSE}{cn'}} \tag{11.16}$$

Critical Range for Factor B

$$\text{Critical range} = Q_\alpha \sqrt{\frac{MSE}{rn'}} \tag{11.17}$$

▼ KEY TERMS

among-group variation   353
analysis of variance (ANOVA)   353
ANOVA summary table   357
cell means   374
completely randomized design   353
critical range   362
F distribution   356
factor   353
factorial design   353
grand mean, $\overline{\overline{X}}$ 354
groups   353
homogeneity of variance   360
interaction   367

levels   353
Levene test   361
main effect   372
mean squares   355, 369
multiple comparisons   362
normality   360
one-way ANOVA   353
randomized block design   353
randomness and independence   360
replicates   367
Studentized range distribution   362
sum of squares among groups (SSA)   355
sum of squares due to factor A (SSA)   368

sum of squares due to factor B (SSB)   368
sum of squares due to interaction
   (SSAB)   368
sum of squares error (SSE)   369
sum of squares total (SST)   353
sum of squares within groups (SSW)   355
total variation   354
Tukey multiple comparisons procedure for
   two-way ANOVA   373
Tukey-Kramer multiple comparisons
   procedure for one-way ANOVA   362
two-way ANOVA   367
within-group variation   355

▼ CHECKING YOUR UNDERSTANDING

11.27 In a one-way ANOVA, what is the difference between the among-groups variance MSA and the within-groups variance MSW?

11.28 What are the distinguishing features of the completely randomized design and two-factor factorial designs?

11.29 What are the assumptions of ANOVA?

11.30 Under what conditions should you use the one-way ANOVA F test to examine possible differences among the means of c independent populations?

11.31 When and how should you use multiple comparison procedures for evaluating pairwise combinations of the group means?

11.32 What is the difference between the one-way ANOVA F test and the Levene test?

11.33 Under what conditions should you use the two-way ANOVA F test to examine possible differences among the means of each factor in a factorial design?

11.34 What is meant by the concept of interaction in a two-factor factorial design?

11.35 How can you determine whether there is an interaction in the two-factor factorial design?

▼ CHAPTER REVIEW PROBLEMS

11.36 You are the production manager at a parachute manufacturing company. Parachutes are woven in your factory using a synthetic fiber purchased from one of four different suppliers. The strength of these fibers is an important characteristic that ensures quality parachutes. You need to decide whether the synthetic fibers from each of your four suppliers result in parachutes of equal strength. Furthermore, to produce parachutes your factory uses two types of looms, the Jetta and the Turk. You need to determine if the parachutes woven on each type of loom are equally strong. You also want to know if any differences in the strength of the parachute can be attributed to the four suppliers are dependent on the type of loom used. You conduct an experiment in which five different parachutes from each supplier are manufactured on each of the two different looms and collect and store the data in `Parachute Two-way`.

At the 0.05 level of significance,
a. is there an interaction between supplier and loom?
b. is there an effect due to loom?
c. is there an effect due to supplier?
d. Plot the mean strength for each supplier for each loom.
e. If appropriate, use the Tukey procedure to determine differences between suppliers.
f. Repeat the analysis, using the `Parachute One-way` file with suppliers as the only factor. Compare your results to those of (e).

11.37 Medical wires are used in the manufacture of cardiovascular devices. A study was conducted to determine the effect of several factors on the ratio of the load on a test specimen (YS) to the ultimate tensile strength (UTS). The file `Medical Wires 1` contains the study results, which examined factors including the machine (W95 vs. W96) and the reduction angle (narrow vs. wide).

Source: Data extracted from B. Nepal, S. Mohanty, and L. Kay, "Quality Improvement of Medical Wire Manufacturing Process," *Quality Engineering* 25, 2013, pp. 151–163.

At the 0.05 level of significance,
a. is there an interaction between machine type and reduction angle?
b. is there an effect due to machine type?
c. is there an effect due to reduction angle?
d. Plot the mean ratio of the load on a test specimen (YS) to the ultimate tensile strength (UTS) for each machine type for each reduction angle.
e. What can you conclude about the effects of machine type and reduction angle on the ratio of the load on a test specimen (YS) to the ultimate tensile strength (UTS)? Explain.
f. Repeat the analysis, using reduction angle as the only factor the and `Medical Wires 2` file. Compare your results to those of (c) and (e).

11.38 An operations manager wants to examine the effect of air-jet pressure (in pounds per square inch [psi]) on the breaking strength of yarn. Three different levels of air-jet pressure are to be considered: 30 psi, 40 psi, and 50 psi. A random sample of 18 yarns are selected from the same batch, and the yarns are randomly assigned, 6 each, to the 3 levels of air-jet pressure. The breaking strength scores are stored in Yarn .

a. Is there evidence of a significant difference in the variances of the breaking strengths for the three air-jet pressures? (Use $\alpha = 0.05$.)

b. At the 0.05 level of significance, is there evidence of a difference among mean breaking strengths for the three air-jet pressures?

c. If appropriate, use the Tukey-Kramer procedure to determine which air-jet pressures significantly differ with respect to mean breaking strength. (Use $\alpha = 0.05$.)

d. What should the operations manager conclude?

11.39 Suppose that, when setting up the experiment in Problem 11.38, the operations manager is able to study the effect of side-to-side aspect in addition to air-jet pressure. Thus, instead of the one-factor completely randomized design in Problem 11.38, a two-factor factorial design was used, with the first factor, side-to-side aspect, having two levels (nozzle and opposite) and the second factor, air-jet pressure, having three levels (30 psi, 40 psi, and 50 psi). A sample of 18 yarns is randomly assigned, three to each of the six side-to-side aspect and pressure level combinations. The breaking-strength scores, stored in Yarn , are as follows:

SIDE-TO-SIDE ASPECT	AIR-JET PRESSURE		
	30 psi	40 psi	50 psi
Nozzle	25.5	24.8	23.2
Nozzle	24.9	23.7	23.7
Nozzle	26.1	24.4	22.7
Opposite	24.7	23.6	22.6
Opposite	24.2	23.3	22.8
Opposite	23.6	21.4	24.9

At the 0.05 level of significance,

a. is there an interaction between side-to-side aspect and air-jet pressure?

b. is there an effect due to side-to-side aspect?

c. is there an effect due to air-jet pressure?

d. Plot the mean yarn breaking strength for each level of side-to-side aspect for each level of air-jet pressure.

e. If appropriate, use the Tukey procedure to study differences among the air-jet pressures.

f. On the basis of the results of (a) through (e), what conclusions can you reach concerning yarn breaking strength? Discuss.

g. Compare your results in (a) through (f) with those from the completely randomized design in Problem 11.38. Discuss fully.

11.40 A hotel wanted to develop a new system for delivering room service breakfasts. In the current system, an order form is left on the bed in each room. If the customer wishes to receive a room service breakfast, he or she places the order form on the doorknob before 11 P.M. The current system requires customers to select a 15-minute interval for desired delivery time (6:30–6:45 A.M., 6:45–7:00 A.M., etc.). The new system is designed to allow the customer to request a specific delivery time. The hotel wants to measure the difference (in minutes) between the actual delivery time and the requested delivery time of room service orders for breakfast. (A negative time means that the order was delivered before the requested time. A positive time means that the order was delivered after the requested time.) The factors included were the menu choice (American or Continental) and the desired time period in which the order was to be delivered (Early Time Period [6:30–8:00 A.M.] or Late Time Period [8:00–9:30 A.M.]). Ten orders for each combination of menu choice and desired time period were studied on a particular day. The data, stored in Breakfast , are as follows:

TYPE OF BREAKFAST	DESIRED TIME	
	Early Time Period	Late Time Period
Continental	1.2	-2.5
Continental	2.1	3.0
Continental	3.3	-0.2
Continental	4.4	1.2
Continental	3.4	1.2
Continental	5.3	0.7
Continental	2.2	-1.3
Continental	1.0	0.2
Continental	5.4	-0.5
Continental	1.4	3.8
American	4.4	6.0
American	1.1	2.3
American	4.8	4.2
American	7.1	3.8
American	6.7	5.5
American	5.6	1.8
American	9.5	5.1
American	4.1	4.2
American	7.9	4.9
American	9.4	4.0

At the 0.05 level of significance,

a. is there an interaction between type of breakfast and desired time?

b. is there an effect due to type of breakfast?

c. is there an effect due to desired time?

d. Plot the mean delivery time difference for each desired time for each type of breakfast.

e. On the basis of the results of (a) through (d), what conclusions can you reach concerning delivery time difference? Discuss.

11.41 Refer to the room service experiment in Problem 11.40. Now suppose that the results are as shown below and stored in Breakfast Results . Repeat (a) through (e), using these data, and compare the results to those of (a) through (e) of Problem 11.40.

TYPE OF BREAKFAST	DESIRED TIME	
	Early	Late
Continental	1.2	-0.5
Continental	2.1	5.0
Continental	3.3	1.8
Continental	4.4	3.2
Continental	3.4	3.2

	DESIRED TIME	
TYPE OF BREAKFAST	**Early**	**Late**
Continental	5.3	2.7
Continental	2.2	0.7
Continental	1.0	2.2
Continental	5.4	1.5
Continental	1.4	5.8
American	4.4	6.0
American	1.1	2.3
American	4.8	4.2
American	7.1	3.8
American	6.7	5.5
American	5.6	1.8
American	9.5	5.1
American	4.1	4.2
American	7.9	4.9
American	9.4	4.0

11.42 A pet food company has the business objective of having the weight of a can of cat food come as close to the specified weight as possible. Realizing that the size of the pieces of meat contained in a can and the can fill height could impact the weight of a can, a team studying the weight of canned cat food wondered whether the current larger chunk size produced higher can weight and more variability. The team decided to study the effect on weight of a cutting size that was finer than the current size. In addition, the team slightly lowered the target for the sensing mechanism that determines the fill height in order to determine the effect of the fill height on can weight.

Twenty cans were filled for each of the four combinations of piece size (fine and current) and fill height (low and current). The contents of each can were weighed, and the amount above or below the label weight of 3 ounces was recorded as the variable coded weight. For example, a can containing 2.90 ounces was given a coded weight of -0.10. Results were stored in Cat Food Cans .

Analyze these data and write a report for presentation to the team. Indicate the importance of the piece size and the fill height on the weight of the canned cat food. Be sure to include a recommendation for the level of each factor that will come closest to meeting the target weight and the limitations of this experiment, along with recommendations for future experiments that might be undertaken.

CHAPTER 11

▼ CASES

Managing Ashland MultiComm Services
PHASE 1

The computer operations department had a business objective of reducing the amount of time to fully update each subscriber's set of messages in a special secured email system. An experiment was conducted in which 24 subscribers were selected and three different messaging systems were used. Eight subscribers were assigned to each system, and the update times were measured. The results, stored in AMS11-1 , are presented in Table AMS11.1.

TABLE AMS11.1
Update times (in seconds) for three different systems

System 1	System 2	System 3
38.8	41.8	32.9
42.1	36.4	36.1
45.2	39.1	39.2
34.8	28.7	29.3
48.3	36.4	41.9
37.8	36.1	31.7
41.1	35.8	35.2
43.6	33.7	38.1

1. Analyze the data in Table AMS11.1 and write a report to the computer operations department that indicates your findings. Include an appendix in which you discuss the reason you selected a particular statistical test to compare the three email interfaces.

DO NOT CONTINUE UNTIL THE PHASE 1 EXERCISE HAS BEEN COMPLETED.

PHASE 2

After analyzing the data in Table AMS11.1, the computer operations department team decided to also study the effect of the connection media used (cable or fiber).

The team designed a study in which a total of 30 subscribers were chosen. The subscribers were randomly assigned to one of the three messaging systems so that there were five subscribers in each of the six combinations of the two factors—messaging system and media used. Measurements were taken on the updated time. Table AMS11.2 summarizes the results that are stored in AMS11-2 .

TABLE AMS11.2

Update times (in seconds), based on messaging system and media used

MEDIA	INTERFACE		
	System 1	System 2	System 3
Cable	45.6	41.7	35.3
	49.0	42.8	37.7
	41.8	40.0	41.0
	35.6	39.6	28.7
	43.4	36.0	31.8
Fiber	44.1	37.9	43.3
	40.8	41.1	40.0
	46.9	35.8	43.1
	51.8	45.3	39.6
	48.5	40.2	33.2

2. Completely analyze these data and write a report to the team that indicates the importance of each of the two factors and/ or the interaction between them on the update time. Include recommendations for future experiments to perform.

Digital Case

Apply your knowledge about ANOVA in this Digital Case, which continues the cereal-fill packaging dispute Digital Case from Chapters 7, 9, and 10.

After reviewing CCACC's latest document (see the Digital Case for Chapter 10 on page 345), Oxford Cereals has released **Second Analysis.pdf**, a press kit that Oxford Cereals has assembled to refute the claim that it is guilty of using selective data. Review the Oxford Cereals press kit and then answer the following questions.

1. Does Oxford Cereals have a legitimate argument? Why or why not?

2. Assuming that the samples Oxford Cereals has posted were randomly selected, perform the appropriate analysis to resolve the ongoing weight dispute.

3. What conclusions can you reach from your results? If you were called as an expert witness, would you support the claims of the CCACC or the claims of Oxford Cereals? Explain.

Sure Value Convenience Stores

You work in the corporate office for a nationwide convenience store franchise that operates nearly 10,000 stores. The per-store daily customer count (the mean number of customers in a store in one day) has been steady, at 900, for some time. To increase the customer count, the chain is considering cutting prices for coffee beverages. The question to be determined is how much to cut prices to increase the daily customer count without reducing the gross margin on coffee sales too much.

You decide to carry out an experiment in a sample of 24 stores where customer counts have been running almost exactly at the national average of 900. In 6 of the stores, the price of a small coffee will now be $0.59, in 6 stores the price of a small coffee will now be $0.69, in 6 stores, the price of a small coffee will now be $0.79, and in 6 stores, the price of a small coffee will now be $0.89. After four weeks of selling the coffee at the new price, the daily customer counts in the stores were recorded and stored in Coffee Sales.

1. Analyze the data and determine whether there is evidence of a difference in the daily customer count, based on the price of a small coffee.

2. If appropriate, determine which mean prices differ in daily customer counts.

3. What price do you recommend for a small coffee?

CardioGood Fitness

Return to the CardioGood Fitness case (stored in CardioGood Fitness) first presented on page 33.

1. Determine whether differences exist between customers based on the product purchased (TM195, TM498, TM798), in their age in years, education in years, annual household income ($), mean number of times the customer plans to use the treadmill each week, and mean number of miles the customer expects to walk or run each week.

2. Write a report to be presented to the management of Cardio-Good Fitness detailing your findings.

More Descriptive Choices Follow-Up

Follow up the Using Statistics scenario "More Descriptive Choices, Revisited" on page 147 by determining whether there is a difference between the small, mid-cap, and large market cap funds in the one-year return percentages, five-year return percentages, and ten-year return percentages (stored in Retirement Funds).

Clear Mountain State Student Survey

The Student News Service at Clear Mountain State University (CMSU) has decided to gather data about the undergraduate students who attend CMSU. They create and distribute a survey of 14 questions and receive responses from 111 undergraduates (stored in Student Survey).

1. At the 0.05 level of significance, is there evidence of a difference based on academic major in expected starting salary, number of social networking sites registered for, age, spending on textbooks and supplies, text messages sent in a week, and the wealth needed to feel rich?

2. At the 0.05 level of significance, is there evidence of a difference based on graduate school intention in grade point average, expected starting salary, number of social networking sites registered for, age, spending on textbooks and supplies, text messages sent in a week, and the wealth needed to feel rich?

▾EXCEL GUIDE

Analyzing Variation in One-Way ANOVA

Key Technique Use the Section EG2.5 instructions to construct scatter plots using stacked data. If necessary, change the levels of the factor to consecutive integers beginning with 1, as was done for the Figure 11.3 in-store location sales experiment data on page 357.

F Test for Differences Among More Than Two Means

Key Technique Use the **DEVSQ** (*cell range of data of all groups*) function to compute *SST*.
Use an expression in the form *SST –* **DEVSQ** (*group 1 data cell range*) *–* **DEVSQ** (*group 2 data cell range*)*… –* **DEVSQ** (*group n data cell range*) to compute *SSA*.

Example Perform the Figure 11.6 one-way ANOVA for the in-store location sales experiment on page 360.

PHStat Use One-Way ANOVA.

For the example, open to the **DATA worksheet** of the **Mobile Electronics workbook**. Select **PHStat → Multiple-Sample Tests → One-Way ANOVA**. In the procedure's dialog box (shown below):

1. Enter **0.05** as the **Level of Significance**.
2. Enter **A1:D6** as the **Group Data Cell Range**.
3. Check **First cells contain label**.
4. Enter a **Title**, clear the **Tukey-Kramer Procedure** check box, and click **OK**.

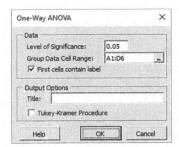

In addition to the worksheet shown in Figure 11.6, this procedure creates an **ASFData worksheet** to hold the data used for the test. See the following *Workbook* section for a complete description of this worksheet.

Workbook Use the **COMPUTE worksheet** of the **One-Way ANOVA workbook** as a template.

The COMPUTE worksheet uses the ASFDATA worksheet that already contains the data for the example. Modifying the COMPUTE worksheet for other problems involves multiple steps and is more complex than template modifications discussed in earlier chapters.

To modify the One-Way ANOVA workbook for other problems, first paste the data for the new problem into the ASF-Data worksheet, overwriting the in-store locations sales data. Then, in the COMPUTE worksheet (shown in Figure 11.6):

1. Edit the *SST* formula **= DEVSQ(ASFData!A1:D6)** in cell B16 to use the cell range of the new data just pasted into the ASFData worksheet.
2. Edit the cell B13 *SSA* formula so there are as many **DEVSQ**(*group column cell range*) terms as there are groups.
3. Change the level of significance in cell G17, if necessary.
4. If the problem contains three groups, select **row 8**, right-click, and select **Delete** from the shortcut menu.

   If the problem contains more than four groups, select **row 8**, right-click, and click **Insert** from the shortcut menu. Repeat this step as many times as necessary.
5. If you inserted new rows, enter (not copy) the formulas for those rows, using the formulas in row 7 as models.
6. Adjust table formatting as necessary.

To see the arithmetic formulas that the COMPUTE worksheet uses, not shown in Figure 11.6, open to the COMPUTE_FORMULAS worksheet.

Analysis ToolPak Use **Anova: Single Factor**.

For the example, open to the **DATA worksheet** of the **Mobile Electronics workbook** and:

1. Select **Data → Data Analysis**.
2. In the Data Analysis dialog box, select **Anova: Single Factor** from the **Analysis Tools** list and then click **OK**.

In the procedure's dialog box (shown on page 386):

3. Enter **A1:D6** as the **Input Range**.
4. Click **Columns**, check **Labels in First Row**, and enter **0.05** as **Alpha**.
5. Click **New Worksheet Ply**.
6. Click **OK**.

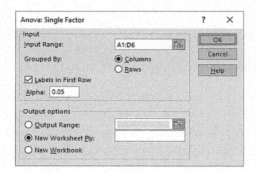

The Analysis ToolPak creates a worksheet that does not use formulas but is similar in layout to the Figure 11.6 worksheet on page 360.

Levene Test for Homogeneity of Variance

Key Technique Use the techniques for performing a one-way ANOVA.

Example Perform the Figure 11.7 Levene test for the in-store location sales experiment on page 361.

PHStat Use **Levene Test**.

For the example, open to the **DATA worksheet** of the **Mobile Electronics workbook**. Select **PHStat → Multiple-Sample Tests → Levene Test**. In the procedure's dialog box (shown below):

1. Enter **0.05** as the **Level of Significance**.
2. Enter **A1:D6** as the **Sample Data Cell Range**.
3. Check **First cells contain label**.
4. Enter a **Title** and click **OK**.

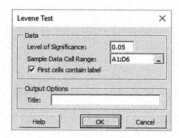

The procedure creates a worksheet that performs the Table 11.5 absolute differences computations (see page 363) as well as the Figure 11.7 worksheet. See the following *Workbook* section for a description of these worksheets.

Workbook Use the **COMPUTE worksheet** of the **Levene workbook** as a template.
The COMPUTE worksheet and the supporting AbsDiffs and DATA worksheets already contain the data for the example.

For other problems in which the absolute differences are already known, paste the absolute differences into the AbsDiffs worksheet. Otherwise, paste the problem data into the DATA worksheet, add formulas to compute the median for each group, and adjust the AbsDiffs worksheet as necessary. For example, for the in-store location sales experiment,

the following steps 1 through 7 were done with the workbook open to the DATA worksheet:

1. Enter the label **Medians** in **cellA7**, the first empty cell in column A.
2. Enter the formula **=MEDIAN(A2:A6)** in **cellA8**. (Cell range A2:A6 contains the data for the first group, in-aisle.)
3. Copy the cell A8 formula across through column D.
4. Open to the **AbsDiffs worksheet**.

In the AbsDiffs worksheet:

5. Enter row 1 column headings **AbsDiff1**, **AbsDiff2**, **AbsDiff3**, and **AbsDiff4** in columns A through D.
6. Enter the formula **=ABS(DATA!A2 – DATA!A8)** in cell A2. Copy this formula down through row 6.
7. Copy the formulas now in cell range A2:A6 across through column D. Absolute differences now appear in the cell range A2:D6.

Analysis ToolPak Use **Anova: Single Factor** with absolute difference data to perform the Levene test. If the absolute differences have not already been calculated, first use the preceding *Workbook* instructions to compute those values.

Multiple Comparisons: The Tukey-Kramer Procedure

Key Technique Use arithmetic formulas to compute the absolute mean differences and use the **IF** function to compare pairs of means.

Example Perform the Figure 11.8 Tukey-Kramer procedure for the in-store location sales experiment shown on page 363.

PHStat Use **One-Way ANOVA** with the **Tukey-Kramer procedure** option.

For the example, use the Section EG11.1 "*F* Test …" PHStat instructions, checking, not clearing, the **Tukey-Kramer Procedure** check box in step 4.

With this option, the procedure creates a second worksheet that is identical to the Figure 11.8 worksheet on page 363, other than missing a proper *Q* statistic value. Use Table E.7 to look up and enter the missing Studentized range *Q* statistic (4.05, for the example) for the level of significance and the numerator and denominator degrees of freedom that are given in the worksheet. (The second worksheet that the option creates will be identical to one of the "TK" worksheets discussed in the following *Workbook* instructions.)

Workbook Use the appropriate **"TK" worksheet** in the **One-Way ANOVA workbook** and manually look up and enter the appropriate Studentized range *Q* statistic value.

For the example, the **TK4 worksheet**, shown in Figure 11.8 on page 363, already has the appropriate *Q* statistic value (4.05) entered in cell B15. To see the arithmetic formulas that the TK4 worksheet uses, not shown in Figure 11.8, open to the TK4_FORMULAS worksheet.

For other problems, first modify the COMPUTE worksheet using the Section EG11.1 *Workbook* "*F* Test..." instructions. Then, open to the appropriate "TK" worksheet: TK3 (three groups), TK4 (four groups), TK5 (five groups), TK6 (six groups), or TK7 (seven groups). Use Table E.7 to look up the proper value of the Studentized range Q statistic for the level of significance and the numerator and denominator degrees of freedom for the problem.

When using the TK5, TK6, or TK7 worksheets, you must also enter the name, sample mean, and sample size for the fifth and subsequent, if applicable, groups.

Analysis ToolPak Modify the previous instructions to perform the Tukey-Kramer procedure in conjunction with using the **Anova: Single Factor** procedure. Transfer selected values from the Analysis ToolPak results worksheet to one of the TK worksheets in the **One-Way ANOVA workbook**.

For the example:

1. Use the Analysis ToolPak "*F* Test..." instructions on page 385 to create a worksheet that contains ANOVA results for the in-store locations experiment.
2. Record the name, **sample size** (in the **Count** column), and **sample mean** (in the **Average** column) of each group. Also record the *MSW* value, found in the cell that is the intersection of the **MS** column and **Within Groups** row, and the **denominator degrees of freedom**, found in the cell that is the intersection of the **df** column and **Within Groups** row.
3. Open to the **TK4 worksheet** of the **One-Way ANOVA workbook**.

In the TK4 worksheet:

4. Overwrite the formulas in cell range **A5:C8** by entering the name, sample mean, and sample size of each group into that range.
5. Enter **0.05** as the **Level of significance** in cell **B11**.
6. Enter **4** as the **Numerator d.f.** (equal to the number of groups) in cell **B12**.
7. Enter **16** as the **Denominator d.f.** in cell **B13**.
8. Enter **0.3044** as the **MSW** in cell **B14**.
9. Enter **4.05** as the **Q Statistic** in cell **B15**. (Use Table E.7 to look up the Studentized range Q statistic.)

EG11.2 The FACTORIAL DESIGN: TWO-WAY ANOVA

Key Technique Use the **DEVSQ** function to compute *SSA*, *SSB*, *SSAB*, *SSE*, and *SST*.

Example Perform the Figure 11.10 two-way ANOVA for the in-store location sales and mobile payment experiment on page 372.

PHStat Use **Two-Way ANOVA with replication**.

For the example, open to the **DATA worksheet** of the **Mobile Electronics 2 workbook**. Select **PHStat → Multiple-Sample Tests → Two-Way ANOVA with replication**. In the procedure's dialog box (shown below):

1. Enter **0.05** as the **Level of Significance**.
2. Enter **A1:E11** as the **Sample Data Cell Range**.
3. Check **First cells contain label**.
4. Enter a **Title** and click **OK**.

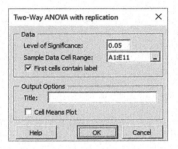

This procedure requires that the labels that identify factor *A* appear stacked in column A, followed by columns for factor *B*.

Workbook Use the **COMPUTE worksheet** of the **Two-Way ANOVA workbook** as a model.

For the example, the COMPUTE worksheet uses the ATFData worksheet that already contains the data to perform the test for the example.

For other problems in which $r = 2$ and $c = 4$, paste the data for the problem into the ATFData worksheet, overwriting the in-store location and mobile payments data and then adjust the factor level headings in the COMPUTE worksheet.

For problems with a different mix of factors and levels, consider using either the *PHStat* or *Analysis ToolPak* instructions. Modifying the COMPUTE worksheet for such problems requires inserting (or deleting) both rows and columns as well as editing several lengthy formulas found in the ANOVA table portion of the worksheet, operations that can be error-prone. The SHORT TAKES for Chapter 11 includes the instructions for these operations, should you choose to make such manual modifications.

To see the arithmetic formulas that the COMPUTE worksheet uses, not shown in Figure 11.10, open to the COMPUTE_FORMULAS worksheet.

Analysis ToolPak Use **Anova: Two-Factor With Replication**.

For the example, open to the **DATA worksheet** of the **Mobile Electronics 2 workbook** and:

1. Select **Data → Data Analysis**.
2. In the Data Analysis dialog box, select **Anova: Two-Factor With Replication** from the **Analysis Tools** list and then click **OK**.

In the procedure's dialog box (shown below):

3. Enter **A1:E11** as the **Input Range**.
4. Enter **5** as the **Rows per sample**.
5. Enter **0.05** as **Alpha**.
6. Click **New Worksheet Ply**.
7. Click **OK**.

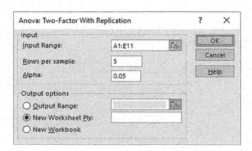

This procedure requires that the labels that identify factor *A* appear stacked in column A, followed by columns for factor *B*. The Analysis ToolPak creates a worksheet that does not use formulas but is similar in layout to the Figure 11.10 worksheet.

Visualizing Interaction Effects: The Cell Means Plot

Key Technique Create a worksheet that contains the means for each factor *B* level, by factor *A* level.

Example Construct the Figure 11.13 cell means plot for mobile electronics sales based on mobile payments permitted and in-store location on page 374.

PHStat Modify the *PHStat* instructions for the two-way ANOVA. In step 4, check **Cell Means Plot** before clicking **OK**.

Analysis ToolPak Use the *Workbook* instructions.

Workbook Create a cell means plot from a two-way ANOVA COMPUTE worksheet.

For the example, open to the **COMPUTE worksheet** of the **Two-Way ANOVA** workbook and:

1. Insert a new worksheet.
2. Copy cell range **B3:E3** of the COMPUTE worksheet (the factor *B* level names) to cell **B1** of the new worksheet, using the Paste Special **Values** option.
3. Copy the cell range **B7:E7** of the COMPUTE worksheet (the AVERAGE row for the factor *A* No level) and paste to cell **B2** of the new worksheet, using the Paste Special **Values** option.
4. Copy the cell range **B13:E13** of the COMPUTE worksheet (the AVERAGE row for the factor *A* Yes level) and paste to cell **B3** of a new worksheet, using the Paste Special **Values** option.
5. Enter **No** in cell **B3** and **Yes** in cell **A3** of the new worksheet as labels for the factor *A* levels.
6. Select the cell range **A1:E3**.
7. Select **Insert → Line** and select the **Line with Markers** gallery item.
8. Relocate the chart to a chart sheet, add axis titles, and modify the chart title by using the instructions in Appendix Section B.5.

For other problems, insert a new worksheet and first copy and paste the factor *B* level names to row 1 of the new worksheet and then copy and use Paste Special to transfer the values in the **Average** rows data for each factor *B* level to the new worksheet. (See Appendix B to learn more about the Paste Special command.)

Chi-Square and Nonparametric Tests

CONTENTS

OBJECTIVES

- Understand the chi-square
  test for contingency
  tables
- Understand application of
  the Marascuilo procedure
- Use nonparametric tests

▼ USING STATISTICS
Avoiding Guesswork About Resort Guests

You are the manager of T.C. Resort Properties, a collection of five upscale hotels located on two tropical islands. Guests who are satisfied with the quality of services during their stay are more likely to return on a future vacation and to recommend the hotel to friends and relatives. You have defined the business objective as improving the percentage of guests who choose to return to the hotels later. To assess the quality of services being provided by your hotels, your staff encourages guests to complete a satisfaction survey when they check out or via email after they check out.

You need to analyze the data from these surveys to determine the overall satisfaction with the services provided, the likelihood that the guests will return to the hotel, and the reasons some guests indicate that they will not return. For example, on one island, T.C. Resort Properties operates the Beachcomber and Windsurfer hotels. Is the perceived quality at the Beachcomber Hotel the same as at the Windsurfer Hotel? If there is a difference, how can you use this information to improve the overall quality of service at T.C. Resort Properties? Furthermore, if guests indicate that they are not planning to return, what are the most common reasons cited for this decision? Are the reasons cited unique to a certain hotel or common to all hotels operated by T.C. Resort Properties?

T he preceding three chapters discuss hypothesis-testing procedures to analyze both numerical and categorical data. This chapter extends hypothesis testing to analyze differences between population *proportions* based on two or more samples and to test the hypothesis of independence in the joint responses to two categorical variables. The chapter concludes with nonparametric tests as alternatives to several Chapter 10 and 11 hypothesis tests.

12.1 Chi-Square Test for the Difference Between Two Proportions

Section 10.3 describes the Z test for the difference between two proportions. Differences between two proportions can also be examined using a different hypothesis test. This second test uses a test statistic whose sampling distribution is approximated by a **chi-square (χ^2) distribution**, a right-skewed distribution whose shape depends solely on the number of degrees of freedom. The results of this χ^2 test are equivalent to those of the Z test that Section 10.3 describes.

The χ^2 **test for the difference between two proportions** requires that the counts of categorical responses between two independent groups be organized as a contingency table. As Section 2.1 first explains, contingency tables summarize the data of categorical variables. For this test, row variable categories display of the frequency of occurrence for the items of interest and items not of interest and column variable categories identify the two independent groups. Table 12.1 presents a generalized **2×2 contingency table** that defines all the table entries.

TABLE 12.1

Layout of a 2×2 contingency table

ROW VARIABLE	COLUMN VARIABLE		
	Group 1	Group 2	Totals
Items of interest	X_1	X_2	X
Items not of interest	$n_1 - X_1$	$n_2 - X_2$	$n - X$
Totals	n_1	n_2	n

where

$$X_1 = \text{number of items of interest in group 1}$$
$$X_2 = \text{number of items of interest in group 2}$$
$$n_1 - X_1 = \text{number of items that are not of interest in group 1}$$
$$n_2 - X_2 = \text{number of items that are not of interest in group 2}$$
$$X = X_1 + X_2, \text{the total number of items of interest}$$
$$n - X = (n_1 - X_1) + (n_2 - X_2), \text{the total number of items that are not of interest}$$
$$n_1 = \text{sample size in group 1}$$
$$n_2 = \text{sample size in group 2}$$
$$n = n_1 + n_2 = \text{total sample size}$$

To illustrate the chi-square test for the difference between two proportions, recall the T.C. Resort Properties scenario. On one island, T.C. Resort Properties operates the Beachcomber and the Windsurfer hotels. Using the DCOVA framework, managers have defined improving customer satisfaction at these hotels as an important business objective. In analyzing the data collected from the completed customer satisfaction surveys, managers decide to focus on the yes-or-no answers to the question "Are you likely to choose the hotel again?" Table 12.2 summarizes these responses as a 2×2 contingency table suitable for use with the chi-square test. This summary reveals that 163 of 227 Beachcomber guests and 154 of 262 Windsurfer guests responded yes to the question.

TABLE 12.2

2×2 contingency table for the hotel guest satisfaction survey

	HOTEL		
CHOOSE HOTEL AGAIN?	**Beachcomber**	**Windsurfer**	**Total**
Yes	163	154	317
No	64	108	172
Total	227	262	489

student TIP

Do not confuse this use of the Greek letter pi, π, to represent the population proportion with the mathematical constant that is approximately 3.14159.

Managers seek to determine if evidence exists of a significant difference in guest satisfaction between the two hotels (as measured by the responses). An analyst suggests using the χ^2 test for the difference between two proportions to test whether the population proportion of guests who would choose to return to the Beachcomber, π_1, is equal to the population proportion of guests who would choose to return to the Windsurfer, π_2. The null and alternative hypotheses for this test are

$H_0: \pi_1 = \pi_2$ (there is no difference between the two population proportions)

$H_1: \pi_1 \neq \pi_2$ (there is a difference between the two population proportions)

The test uses the χ^2_{STAT} test statistic that Equation (12.1) defines.

χ^2 TEST STATISTIC

$$\chi^2_{STAT} = \sum_{\text{all cells}} \frac{(f_o - f_e)^2}{f_e}$$ (12.1)

where

$f_o = $ **observed frequency** in a particular cell of a contingency table

$f_e = $ **expected frequency** in a particular cell if the null hypothesis is true

Table E.4 contains the cumulative probabilities for the chi-square distribution.

The χ^2_{STAT} test statistic approximately follows a chi-square distribution with 1 degree of freedom, a right-skewed distribution the lowest value of which is 0. (This is unlike the normal and t distributions that preceding chapters describe and use.) The test statistic follows the chi-square distribution with $(r - 1)$ times $(c - 1)$ degrees of freedom, where r is the number of rows and c is the number of columns. For the chi-square test for the difference between two proportions, there is 1 degree of freedom as $(2 - 1)$ times $(2 - 1)$ equals one.

If the null hypothesis is true, the proportion of items of interest from each of the two groups would differ only by chance. Either sample proportion would provide an estimate of the common population parameter, π. However, a statistic that combines these two separate estimates into one estimate of the population parameter provides a better estimate than either of the two estimates separately could provide. This statistic, the **estimated overall proportion for two groups**, $\bar{p}$, represents the total number of items of interest divided by the total sample size. The complement of $\bar{p}$, $1 - \bar{p}$, represents the estimated overall proportion of items that are not of interest in the two groups. Equation (12.2) defines $\bar{p}$.

THE ESTIMATED OVERALL PROPORTION FOR TWO GROUPS

$$\bar{p} = \frac{X_1 + X_2}{n_1 + n_2} = \frac{X}{n}$$ (12.2)

where

$X_1, X_2, n_1, n_2, X,$ and n are defined in Table 12.1 on page 390.

Table 12.3 shows expected frequencies calculations that use $\bar{p}$ and its complement $1 - \bar{p}$. Group sample sizes n_1 and n_2 are multiplied by $\bar{p}$, for the items of interest row, and $1 - \bar{p}$, for the items not of interest row.

TABLE 12.3
Expected frequencies calculations

| | COLUMN VARIABLE | |
ROW VARIABLE	Group 1	Group 2
Items of interest	$n_1\bar{p}$	$n_2\bar{p}$
Items not of interest	$n_1(1 - \bar{p})$	$n_2(1 - \bar{p})$

Using a level of significance α, one rejects the null hypothesis if the χ^2_{STAT} test statistic is greater than χ^2_{α}, the upper-tail critical value from the χ^2 distribution with 1 degree of freedom. The decision rule is

$$\text{Reject } H_0 \text{ if } \chi^2_{STAT} > \chi^2_{\alpha};$$

otherwise, do not reject H_0.

Figure 12.1 visualizes the decision rule.

FIGURE 12.1
Regions of rejection and nonrejection when using the chi-square test for the difference between two proportions, with level of significance α

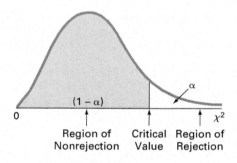

student TIP

The rejection region for this test is located only in the upper tail of the distribution because of the properties of the chi-square distribution that page 390 discusses.

If the null hypothesis is true, the χ^2_{STAT} test statistic should be close to zero because the squared difference between what is actually observed in each cell, f_o, and what is theoretically expected, f_e, should be very small. If H_0 is false, then there are differences in the population proportions, and the χ^2_{STAT} test statistic is expected to be large. However, what is a large difference in a cell is relative. Because calculating the test statistic includes division by the expected frequencies, the same actual difference between f_o and f_e from a cell with a small number of expected frequencies contributes more to the χ^2_{STAT} test statistic than a cell with a large number of expected frequencies.

To illustrate the use of the chi-square test for the difference between two proportions, recall the Table 12.2 contingency table for the T.C. Resort Properties example on page 391. The null hypothesis (H_0: $\pi_1 = \pi_2$) for this example states that there is no difference between the proportion of guests who are likely to choose either of these hotels again. For this example, $\bar{p}$, the estimate of the common parameter π, the population proportion of guests who are likely to choose either of these hotels again if the null hypothesis is true is

$$\bar{p} = \frac{X_1 + X_2}{n_1 + n_2} = \frac{163 + 154}{227 + 262} = \frac{317}{489} = 0.6483$$

The estimated proportion of guests who are *not* likely to choose these hotels again is the complement of $\bar{p}$, $1 - 0.6483 = 0.3517$. Multiplying these two proportions by the sample size for the Beachcomber Hotel group gives the number of guests expected to choose the Beachcomber Hotel again and the number not expected to choose this hotel again. In a similar manner, multiplying the two proportions by the sample size for the Windsurfer group yields the corresponding expected frequencies for that hotel.

EXAMPLE 12.1

Computing
the Expected
Frequencies

Calculate the expected frequencies for each of the four cells of Table 12.2 on page 391.

SOLUTION

Yes—Beachcomber: $\bar{p} = 0.6483$ and $n_1 = 227$, so $f_e = 147.16$
Yes—Windsurfer: $\bar{p} = 0.6483$ and $n_2 = 262$, so $f_e = 169.84$
No —Beachcomber: $1 - \bar{p} = 0.3517$ and $n_1 = 227$, so $f_e = 79.84$
No —Windsurfer: $1 - \bar{p} = 0.3517$ and $n_2 = 262$, so $f_e = 92.16$

Table 12.4 presents these expected frequencies next to the corresponding observed frequencies.

TABLE 12.4
Comparing the
observed (f_o)
and expected (f_e)
frequencies

| | HOTEL | | | | |
| | Beachcomber | | Windsurfer | | |
CHOOSE HOTEL AGAIN?	Observed	Expected	Observed	Expected	Total
Yes	163	147.16	154	169.84	317
No	64	79.84	108	92.16	172
Total	227	227.00	262	262.00	489

To test the null and alternative hypotheses of a test for difference in the population proportions

$$H_0: \pi_1 = \pi_2$$
$$H_1: \pi_1 \neq \pi_2$$

calculate the χ^2_{STAT} test statistic using Equation (12.1) on page 391 with the Table 12.5 observed and expected frequencies.

TABLE 12.5
Calculating the χ^2_{STAT}
test statistic for the
hotel guest satisfaction
survey

f_o	f_e	$(f_o - f_e)$	$(f_o - f_e)^2$	$(f_o - f_e)^2/f_e$
163	147.16	15.84	250.91	1.71
154	169.84	−15.84	250.91	1.48
64	79.84	−15.84	250.91	3.14
108	92.16	15.84	250.91	2.72
				9.05

From Table E.4, a portion of which Table 12.6 shows, using $\alpha = 0.05$, with 1 degree of freedom, the critical value of χ^2 is 3.841 (see Figure 12.2 on page 394). One degree of freedom is used for a 2×2 table because degrees of freedom are equal to (number of rows − 1) times (number of columns − 1).

TABLE 12.6
Finding the critical
value from the
chi-square distribution
with 1 degree of
freedom, using the 0.05
level of significance

	Cumulative Probabilities						
	.005	.01	. . .	.95	.975	.99	.995
	Upper-Tail Area						
Degrees of Freedom	.995	.99	. . .	.05	.025	.01	.005
1			. . .	3.841	5.024	6.635	7.879
2	0.010	0.020	. . .	5.991	7.378	9.210	10.597
3	0.072	0.115	. . .	7.815	9.348	11.345	12.838
4	0.207	0.297	. . .	9.488	11.143	13.277	14.860
5	0.412	0.554	. . .	11.071	12.833	15.086	16.750

FIGURE 12.2

Regions of rejection and nonrejection when finding the χ^2 critical value with 1 degree of freedom, at the 0.05 level of significance

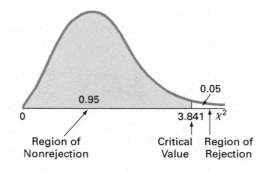

Table 12.7 summarizes the results of the chi-square test for the guest satisfaction survey for the Beachcomber and Windsurfer hotels using the calculations on page 393 and the Figure 12.3 software results. Based on the results, there is strong evidence to conclude that the two hotels are significantly different with respect to guest satisfaction, as measured by whether a guest is likely to return to the hotel again. Therefore, one concludes that a greater proportion of guests are likely to return to the Beachcomber than to the Windsurfer.

TABLE 12.7

Chi-square test summary for the guest satisfaction survey for two hotels

Results	Conclusions
$\chi^2_{STAT} = 9.0526$ is greater than 3.8415. The p-value $= 0.0026$ is less than the level of significance, $\alpha = 0.05$.	1. Reject the null hypothesis H_0. 2. Conclude that evidence exists that the two hotels are significantly different with respect to guest satisfaction. 3. The probability is 0.0026 that $\chi^2_{STAT} > 9.0526$.

FIGURE 12.3

Excel chi-square test worksheet results for the T.C. Resort Properties two-hotel guest satisfaction survey

	A	B	C	D	E	F	G
1	Chi-Square Test						
2							
3		Observed Frequencies				Calculations	
4			Hotel				
5	Choose Again?	Beachcomber	Windsurfer	Total		fo-fe	
6	Yes	163	154	317		15.8446	-15.8446
7	No	64	108	172		-15.8446	15.8446
8	Total	227	262	489			
9							
10		Expected Frequencies					
11			Hotel				
12	Choose Again?	Beachcomber	Windsurfer	Total		(fo-fe)^2/fe	
13	Yes	147.1554	169.8446	317		1.7060	1.4781
14	No	79.8446	92.1554	172		3.1442	2.7242
15	Total	227	262	489			
16							
17	Data						
18	Level of Significance	0.05					
19	Number of Rows	2					
20	Number of Columns	2					
21	Degrees of Freedom	1	=(B19 - 1) * (B20 - 1)				
22							
23	Results						
24	Critical Value	3.8415	=CHISQ.INV.RT(B18, B21)				
25	Chi-Square Test Statistic	9.0526	=SUM(F13:G14)				
26	p-Value	0.0026	=CHISQ.DIST.RT(B25, B21)				
27	Reject the null hypothesis		=IF(B26 < B18, "Reject the null hypothesis",				
28			"Do not reject the null hypothesis")				
29	Expected frequency assumption						
30	is met.		=IF(OR(B13 < 5, C13 < 5, B14 < 5, C14 < 5),				
			" is violated.", " is met.")				

In the Worksheet

The worksheet uses the CHISQ.INV.RT function to compute the cell B24 critical value and uses the CHISQ.DIST.RT function to compute the cell B26 p-value. The results of formulas that compute intermediate calculations in the first 15 rows provide the basis for computing the cell B25 test statistic.

The **COMPUTE worksheet** of the Excel Guide **Chi-Square workbook** is a copy of the Figure 12.3 worksheet.

Assumptions of the chi-square test For the χ^2 test to give accurate results for a 2×2 table, one must assume that each expected frequency is at least 5. If this assumption is not satisfied, one can use alternative procedures, such as Fisher's exact test (Corder, Daniel, and Hollander). In Section 12.2, the χ^2 test is extended to make comparisons and evaluate differences between the proportions among more than two groups. However, one cannot use the Z test if there are more than two groups.

Interrelationship of the standardized normal distribution and the chi-square distribution In the hotel guest satisfaction survey, both the Z test for the difference between two proportions (see Section 10.3) and the χ^2 test lead to the same conclusion. This result can be explained by the interrelationship between the standardized normal distribution and a chi-square distribution with 1 degree of freedom. For such situations, the χ^2_{STAT} test statistic is the square of the Z_{STAT} test statistic.

For example, in the guest satisfaction study, using Equation (10.5) on page 329, the calculated Z_{STAT} test statistic is $+3.0088$, and the calculated χ^2_{STAT} test statistic is 9.0526. Accounting for rounding differences, 9.0526 is the square of $+3.0088$. Also, at the 0.05 level of significance, the critical value of the χ^2 value with 1 degree of freedom is 3.841, the square of the Z value of ± 1.96. Furthermore, the p-values for both tests are equal. Therefore, when testing the null hypothesis of equality of proportions:

$$H_0: \pi_1 = \pi_2$$

against the alternative that the population proportions are not equal:

$$H_1: \pi_1 \neq \pi_2$$

the Z test and the χ^2 test are equivalent. If one seeks to determine whether there is evidence of a *directional* difference, such as $\pi_1 > \pi_2$, one *must* use the Z test, with the entire rejection region located in one tail of the standardized normal distribution.

PROBLEMS FOR SECTION 12.1

LEARNING THE BASICS

12.1 Determine the critical value of χ^2 with 1 degree of freedom in each of the following circumstances:
a. $\alpha = 0.01$
b. $\alpha = 0.005$
c. $\alpha = 0.10$

12.2 Determine the critical value of χ^2 with 1 degree of freedom in each of the following circumstances:
a. $\alpha = 0.05$
b. $\alpha = 0.025$
c. $\alpha = 0.01$

12.3 Use the following contingency table:

	A	B	Total
1	20	30	50
2	30	45	75
Total	50	75	125

a. Compute the expected frequency for each cell.
b. Compare the observed and expected frequencies for each cell.
c. Compute χ^2_{STAT}. Is it significant at $\alpha = 0.05$?

12.4 Use the following contingency table:

	A	B	Total
1	20	30	50
2	30	20	50
Total	50	50	100

a. Compute the expected frequency for each cell.
b. Compute χ^2_{STAT}. Is it significant at $\alpha = 0.05$?

APPLYING THE CONCEPTS

12.5 An Ipsos poll asked 1,004 adults, "If purchasing a used car made certain upgrades or features more affordable, what would be your preferred luxury upgrade?" The results indicated that 9% of the males and 14% of the females answered window tinting.

Source: Ipsos, "Safety Technology Tops the List of Most Desired Features Should They Be More Affordable When Purchasing a Used Car—Particularly Collision Avoidance," available at **bit.ly/2ufbS8Z**.

The poll description did not state the sample sizes of males and females. Suppose that both sample sizes were 502 and that 46 of 502 males and 71 of 502 females reported window tinting as their preferred luxury upgrade of choice.

a. Is there evidence of a difference between males and females in the proportion who said they prefer window tinting as a luxury upgrade at the 0.01 level of significance?
b. Find the *p*-value in (a) and interpret its meaning.
c. What are your answers to (a) and (b) if 60 males said they prefer window tinting as a luxury upgrade and 442 did not?
d. Compare the results of (a) through (c) to those of Problem 10.29 (a), (b), and (d) on page 334.

12.6 Does Cable Video on Demand (VOD D4+) increase ad effectiveness? A recent VOD study compared general TV and VOD D4+ audiences after viewing a brand ad. Whether the viewer indicated that the ad made them want to visit the brand website was collected and organized in the following table.

VIEWING AUDIENCE	MADE ME WANT TO VISIT THE BRAND WEBSITE	
	Yes	No
General TV	35	166
VOD D4 +	147	103

Source: Data extracted from *Understanding VOD Advertising Effectiveness*, **bit.ly/1JnmMup**.

a. Set up the null and alternative hypotheses to try to determine whether there is a difference in ad impact between general TV viewing and VOD D4+ viewing.
b. Conduct the hypothesis test defined in (a), using the 0.05 level of significance.
c. Compare the results of (a) and (b) to those of Problem 10.30 (a) and (b) on page 335.

12.7 Are you an impulse shopper? A survey of 500 grocery shoppers indicated that 29% of males and 40% of females make an impulse purchase every time they shop.

Source: Data extracted from "Women shoppers are impulsive while men snap up bargains," available at **bit.ly/2sLYmVx**.

Assume that the survey consisted of 250 males and 250 females.
a. At the 0.05 level of significance, is there evidence of a difference in the proportion of males and females who make an impulse purchase every time they shop?
b. Find the *p*-values and interpret its meaning.

✓SELF TEST **12.8** The Society for Human Resource Management (SHRM) collaborated with Globoforce on a series of organizational surveys with the goal of identifying challenges that HR leaders face and what strategies help them conquer those challenges. A recent survey indicates that employee retention/turnover (46%) and employee engagement (36%) were cited as the most important organizational challenges currently faced by HR professionals. One strategy that may have an impact on employee retention, turnover, and engagement is a successful employee recognition program. Surveying small organizations, those with 500 to 2,499 employees, and large organizations, those with 10,000 or more employees, SHRM and Globoforce showed

that 326 (77%) of the 423 small organizations have employee retention programs as compared to 167 (87%) of the 192 large organizations.

Source: Data extracted from *SHRM Survey Finding: Influencing Workplace Culture Through Employee Retention and Other Efforts*, available at **bit.ly/2rFvE9w**.

a. At the 0.01 level of significance, is there evidence of a significant difference between organizations with 500 to 2,499 employees and organizations with 10,000 or more employees with respect to the proportion that have employee recognition programs?
b. Find the *p*-value in (a) and interpret its meaning.
c. Compare the results of (a) and (b) to those of Problem 10.32 on page 335.

12.9 What social media tools do marketers commonly use? A survey by Social Media Examiner of B2B marketers (marketers that focus primarily on attracting businesses) and B2C marketers (marketers that primarily target consumers) reported that 267 (81%) of B2B marketers and 295 (44%) of B2C marketers commonly use LinkedIn as a social media tool. The study also revealed that 149 (45%) of B2B marketers and 308 (46%) of B2C marketers commonly use YouTube as a social media tool.

Source: Data extracted from *2017 Social Media Marketing Industry Report*, available at **bit.ly/2rFmLzh**.

Suppose the survey was based on 330 B2B marketers and 670 B2C marketers.
a. At the 0.05 level of significance, is there evidence of a difference between B2B marketers and B2C marketers in the proportion that commonly use LinkedIn as a social media tool?
b. Find the *p*-value in (a) and interpret its value.
c. At the 0.05 level of significance, is there evidence of a difference between B2B marketers and B2C marketers in the proportion that commonly use YouTube as a social media tool?
d. Find the *p*-value in (c) and interpret its meaning.

12.10 Does cobrowsing have positive effects on the customer experience? Cobrowsing refers to the ability to have a contact center agent and customer jointly navigate an application (e.g., web page, digital document, or mobile application) on a real time basis through the web. A study of businesses indicates that 81 of 129 cobrowsing organizations use skills-based routing to match the caller with the *right* agent, whereas 65 of 176 noncobrowsing organizations use skills-based routing to match the caller with the *right* agent.

Source: Data extracted from Cobrowsing Presents a 'Lucrative' Customer Service Opportunity, available at **bit.ly/1wwALWr**.

a. Construct a 2×2 contingency table.
b. At the 0.05 level of significance, is there evidence of a difference between cobrowsing organizations and noncobrowsing organizations in the proportion that use skills-based routing to match the caller with the *right* agent?
c. Find the *p*-value in (a) and interpret its meaning.
d. Compare the results of (a) and (b) to those of Problem 10.34 on page 335.

12.2 Chi-Square Test for Differences Among More Than Two Proportions

In the extended χ^2 test, the contingency table becomes a **2×c contingency table**, in which c is the number of independent populations being compared. In the extended test, the null and alternative hypotheses are

H_0: $\pi_1 = \pi_2 = \cdots = \pi_c$ (there are no differences among the c population proportions)

H_1: Not all π_j are equal (not all the c population proportions are equal)

where $j = 1, 2, \ldots, c$. The chi-square test for differences among more than two proportions uses Equation (12.1), repeated below, with $(c-1)$ degrees of freedom.[1]

$$\chi^2_{STAT} = \sum_{\text{all cells}} \frac{(f_o - f_e)^2}{f_e}$$

To calculate the expected frequency, f_e, for a cell in the first row of the contingency table, multiply the group sample size by $\bar{p}$. To calculate the expected frequency, f_e, for a cell in the second row in the contingency table, multiply the group sample size by $(1 - \bar{p})$.

If the null hypothesis is true and the proportions are equal across all c populations, the c sample proportions would differ only by chance. A statistic that combines the c separate estimates into one overall estimate of the population proportion, π, provides a better estimate than any one of the c estimates separately could provide. Equation (12.3) defines the statistic $\bar{p}$, the estimated overall proportion for all c groups combined, by extending Equation (12.2) on page 391.

THE ESTIMATED OVERALL PROPORTION FOR c GROUPS

$$\bar{p} = \frac{X_1 + X_2 + \cdots + X_c}{n_1 + n_2 + \cdots + n_c} = \frac{X}{n} \tag{12.3}$$

Using the level of significance α, one rejects the null hypothesis if the χ^2_{STAT} test statistic is greater than χ^2_α, the upper-tail critical value from a chi-square distribution with $c - 1$ degrees of freedom. The decision rule is

Reject H_0 if $\chi^2_{STAT} > \chi^2_\alpha$;
otherwise, do not reject H_0.

Figure 12.4 illustrates this decision rule.

FIGURE 12.4

Regions of rejection and nonrejection when testing for differences among c proportions using the χ^2 test

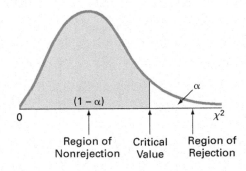

To illustrate the χ^2 test for differences among more than two groups, recall the T.C. Resort Properties scenario. On another island, T.C. Resort Properties operates the Golden Palm, Palm Royale, and Palm Princess hotels. Using the DCOVA framework, managers have the business objective of improving customer satisfaction at these hotels. In analyzing the data collected from the completed customer satisfaction surveys, managers decide to focus on the yes-or-no answers to the same question that was the focus of a two-hotel study on another island (see pages 390–395). Table 12.8 summarizes these responses as a 2×3 contingency table suitable for use with the chi-square test.

TABLE 12.8
2×3 contingency table for guest satisfaction survey

	HOTEL			
CHOOSE HOTEL AGAIN?	**Golden Palm**	**Palm Royale**	**Palm Princess**	**Total**
Yes	128	199	186	513
No	88	33	66	187
Total	216	232	252	700

The null hypothesis is that there are no differences among the three hotels in the proportion of guests who would likely choose the hotel again. Using Equation (12.3) to calculate an estimate of π, the population proportion of guests who would likely choose the hotel again is

$$\bar{p} = \frac{X_1 + X_2 + \cdots + X_c}{n_1 + n_2 + \cdots + n_c} = \frac{X}{n}$$
$$= \frac{128 + 199 + 186}{216 + 232 + 252} = \frac{513}{700}$$
$$= 0.733$$

The estimated overall proportion of guests who would *not* be likely to choose the hotel again is the complement, $(1 - \bar{p})$, or 0.267. Multiplying these two proportions by the sample size for each hotel yields the expected number of guests who would and would not likely return.

EXAMPLE 12.2

Calculating the Expected Frequencies

Calculate the expected frequencies for each of the six cells in Table 12.8.

SOLUTION

Yes—Golden Palm: $\bar{p} = 0.733$ and $n_1 = 216$, so $f_e = 158.30$

Yes—Palm Royale: $\bar{p} = 0.733$ and $n_2 = 232$, so $f_e = 170.02$

Yes—Palm Princess: $\bar{p} = 0.733$ and $n_3 = 252$, so $f_e = 184.68$

No —Golden Palm: $1 - \bar{p} = 0.267$ and $n_1 = 216$, so $f_e = 57.70$

No —Palm Royale: $1 - \bar{p} = 0.267$ and $n_2 = 232$, so $f_e = 61.98$

No —Palm Princess: $1 - \bar{p} = 0.267$ and $n_3 = 252$, so $f_e = 67.32$

Table 12.9 summarizes the Example 12.2 expected frequencies.

TABLE 12.9
Contingency table of expected frequencies from a guest satisfaction survey of three hotels

	HOTEL			
CHOOSE HOTEL AGAIN?	**Golden Palm**	**Palm Royale**	**Palm Princess**	**Total**
Yes	158.30	170.02	184.68	513
No	57.70	61.98	67.32	187
Total	216.00	232.00	252.00	700

To test the null hypothesis that the proportions are equal:

$$H_0: \pi_1 = \pi_2 = \pi_3$$

against the alternative that not all three proportions are equal:

$$H_1: \text{Not all } \pi_j \text{ are equal (where } j = 1, 2, 3)$$

use the Table 12.8 observed frequencies with software or use Table 12.8 and Table 12.9 expected frequencies to calculate the χ^2_{STAT} test statistic by using Equation (12.1) on page 391, with 2 degrees of freedom because $(2 - 1)(3 - 1) = 2$. Table 12.10 summarizes the calculations.

TABLE 12.10

Calculating the χ^2_{STAT} test statistic for the three-hotel guest satisfaction survey

f_o	f_e	$(f_o - f_e)$	$(f_o - f_e)^2$	$(f_o - f_e)^2/f_e$
128	158.30	−30.30	918.09	5.80
199	170.02	28.98	839.84	4.94
186	184.68	1.32	1.74	0.01
88	57.70	30.30	918.09	15.91
33	61.98	−28.98	839.84	13.55
66	67.32	−1.32	1.74	0.02
				40.23

Use Table E.4 to find the critical value of the χ^2 test statistic. Using $\alpha = 0.05$, the χ^2 critical value with 2 degrees of freedom is 5.991 as Figure 12.5 shows.

FIGURE 12.5

Regions of rejection and nonrejection when testing for differences in three proportions at the 0.05 level of significance, with 2 degrees of freedom

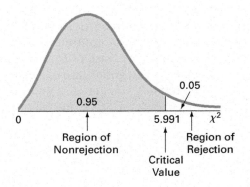

Table 12.11 summarizes the results of the chi-square test for the guest satisfaction survey for the Golden Palm, Palm Royale, and Palm Princess hotels using the Table 12.10 calculations and the Figure 12.6 software results. Based on the results, there is strong evidence to conclude that the three hotels are significantly different with respect to guest satisfaction, as measured by whether a guest is likely to return to the hotel again. Therefore, as part of the DCOVA framework, one can conclude that the hotels are different in terms of the proportion of guests who are likely to return.

TABLE 12.11

Chi-square test summary for the guest satisfaction survey for the three hotels

Results	Conclusions
$\chi^2_{STAT} = 40.23$ is greater than 5.9915.	1. Reject the null hypothesis H_0.
The p-value $= 0.0000$ is less than the level of significance, $\alpha = 0.05$.	2. Conclude that evidence exists that the three hotels are significantly different with respect to guest satisfaction.
	3. The probability is 0.0000 that $\chi^2_{STAT} > 40.23$.

FIGURE 12.6
Excel chi-square test worksheet results for the three-hotel guest satisfaction survey

	A	B	C	D	E	F	G	H	I
1	Chi-Square Test								
2									
3		Observed Frequencies							
4		Hotel						Calculations	
5	Choose Again?	Golden Palm	Palm Royale	Palm Princess	Total			fo - fe	
6	Yes	128	199	186	513		-30.2971	28.9771	1.32
7	No	88	33	66	187		30.2971	-28.9771	-1.32
8	Total	216	232	252	700				
9									
10		Expected Frequencies							
11		Hotel							
12	Choose Again?	Golden Palm	Palm Royale	Palm Princess	Total			(fo - fe)^2/fe	
13	Yes	158.2971	170.0229	184.68	513		5.7987	4.9386	0.0094
14	No	57.7029	61.9771	67.32	187		15.9077	13.5481	0.0259
15	Total	216	232	252	700				
16									
17	Data								
18	Level of Significance	0.05							
19	Number of Rows	2							
20	Number of Columns	3							
21	Degrees of Freedom	2	=(B19 - 1) * (B20 - 1)						
22									
23	Results								
24	Critical Value	5.9915	=CHISQ.INV.RT(B18, B21)						
25	Chi-Square Test Statistic	40.2284	=SUM(G13:I14)						
26	*p*-Value	0.0000	=CHISQ.DIST.RT(B25, B21)						
27	Reject the null hypothesis		=IF(B26 < B18, "Reject the null hypothesis",						
28			"Do not reject the null hypothesis")						
29	*Expected frequency assumption*								
30	*is met.*		=IF(OR(B13 < 1, C13 < 1, D13 < 1, B14 < 1, C14 < 1, D14 < 1), " is violated.", " is met.")						

In the Worksheet

The worksheet uses a design similar to the design of the COMPUTE worksheet that Figure 12.6 presents.

The **Chi-Square2×3 worksheet** of the Excel Guide **Chi-Square Worksheets workbook** is a copy of the Figure 12.6 worksheet.

Assumptions of the chi-square test for the 2×c contingency table

For the χ^2 test to give accurate results when dealing with $2 \times c$ contingency tables, all expected frequencies must be large. The definition of "large" has led to research among statisticians. Some statisticians (Hollander) have found that the test gives accurate results as long as all expected frequencies are at least 0.5. Other statisticians believe that no more than 20% of the cells should contain expected frequencies less than 5, and no cells should have expected frequencies less than 1 (Dixon). As a reasonable compromise between these points of view, to ensure the validity of the test, one should make sure that each expected frequency is at least 1. To do this, one may need to collapse two or more low-expected-frequency categories into one category in the contingency table before performing the test. If combining categories is undesirable, one can use one of the available alternative procedures (Corder, Daniel, and Marascuilo).

The Marascuilo Procedure

Rejecting the null hypothesis in a χ^2 test of equality of proportions in a $2 \times c$ table enables one to reach only the conclusion that not all c population proportions are equal. The multiple-comparisons **Marascuilo procedure** determines *which pairs* of sample proportions differ by comparing every pair of c sample proportions. This procedure first calculates the sample proportions and then calculates a critical range for each pair of sample proportions.[2] A specific pair is significantly different if the absolute difference in the sample proportions, $|p_j - p_{j'}|$, is greater than the critical range for the pair. Equation (12.4) defines the critical range for a pair of sample proportions.

[2]There will always be $c(c-1)$ pairs, by application of a Chapter 4 counting rule.

studentTIP
There is an α level of risk in the entire set of comparisons, not just a single comparison.

CRITICAL RANGE FOR THE MARASCUILO PROCEDURE

$$\text{Critical range} = \sqrt{\chi^2_{\alpha}}\sqrt{\frac{p_j(1-p_j)}{n_j} + \frac{p_{j'}(1-p_{j'})}{n_{j'}}} \quad \textbf{(12.4)}$$

where

p_j = proportion of items of interest in group j

$p_{j'}$ = proportion of items of interest in group j'

n_j = sample size in group j

$n_{j'}$ = sample size in group j'

To illustrate the Marascuilo procedure, recall the three-hotel guest satisfaction survey example. For the χ^2 test, there was evidence of a significant difference among the population proportions (see Table 12.11 on page 399). From Table 12.8 on page 398, the three sample proportions are

$$p_1 = \frac{X_1}{n_1} = \frac{128}{216} = 0.5926 \qquad p_2 = \frac{X_2}{n_2} = \frac{199}{232} = 0.8578 \qquad p_3 = \frac{X_3}{n_3} = \frac{186}{252} = 0.7381$$

For each of the three pairs, Table 12.12 shows the calculated absolute difference and the critical range. Using Table E.4 and an overall level of significance of 0.05, the upper-tail critical value for a chi-square distribution having $(c - 1) = 2$ degrees of freedom is 5.991. Therefore,

$$\sqrt{\chi^2_\alpha} = \sqrt{5.991} = 2.4477$$

TABLE 12.12
Marascuilo procedure calculations for the three-hotel guest satisfaction survey

Absolute Difference Between Pairs	Critical Range
$\lvert p_j - p_{j'} \rvert$	$2.4477\sqrt{\dfrac{p_j(1 - p_j)}{n_j} + \dfrac{p_{j'}(1 - p_{j'})}{n_{j'}}}$
$\lvert p_1 - p_2 \rvert = \lvert 0.5926 - 0.8578 \rvert = 0.2652$	$2.4477\sqrt{\dfrac{(0.5926)(0.4074)}{216} + \dfrac{(0.8578)(0.1422)}{232}} = 0.0992$
$\lvert p_1 - p_3 \rvert = \lvert 0.5926 - 0.7381 \rvert = 0.1455$	$2.4477\sqrt{\dfrac{(0.5926)(0.4074)}{216} + \dfrac{(0.7381)(0.2619)}{252}} = 0.1063$
$\lvert p_2 - p_3 \rvert = \lvert 0.8578 - 0.7381 \rvert = 0.1197$	$2.4477\sqrt{\dfrac{(0.8578)(0.1422)}{232} + \dfrac{(0.7381)(0.2619)}{252}} = 0.0880$

If an absolute difference for a pair is greater than the critical range for a pair, the pair of proportions are significantly different. At the 0.05 level of significance, one concludes that guest satisfaction is higher at the Palm Royale ($p_2 = 0.858$) than at either the Golden Palm ($p_1 = 0.593$) or the Palm Princess ($p_3 = 0.738$) and that guest satisfaction is also higher at the Palm Princess than at the Golden Palm. These results clearly suggest that T.C. Resort Properties management should investigate possible reasons for these differences. In particular, they should try to determine why satisfaction is significantly lower at the Golden Palm than at the other two hotels.

Figure 12.7 shows Excel results for this example.

FIGURE 12.7
Excel Marascuilo procedure results for the three-hotel guest satisfaction survey

	A	B	C	D
1	Marascuilo Procedure for Guest Satisfaction Analysis			
2				
3	Level of Significance	0.05	=ChiSquare2x3!B18	
4	Square Root of Critical Value	2.4477	=SQRT(ChiSquare2x3!B24)	
5				
6	Group Sample Proportions			
7	1: Golden Palm	0.5926	=ChiSquare2x3!B6/ChiSquare2x3!B8	
8	2: Palm Royale	0.8578	=ChiSquare2x3!C6/ChiSquare2x3!C8	
9	3: Palm Princess	0.7381	=ChiSquare2x3!D6/ChiSquare2x3!D8	
10				
11	MARASCUILO TABLE			
12	Proportions	Absolute Differences	Critical Range	
13	| Group 1 - Group 2 |	0.2652	0.0992	Significant
14	| Group 1 - Group 3 |	0.1455	0.1063	Significant
15				
16	| Group 2 - Group 3 |	0.1197	0.0880	Significant

In the Worksheet

The worksheet uses many of the results of the Figure 12.6 worksheet on page 400.

The **Marascuilo2×3 worksheet** of the Excel Guide **Chi-Square Worksheets workbook** is a copy of the Figure 12.7 worksheet.

The Analysis of Proportions (ANOP)

The analysis of proportions (ANOP) method provides a confidence interval approach that enables you to determine which, if any, of the c groups has a proportion significantly different from the overall mean of all the group proportions combined. The **ANOP online topic** discusses this method and illustrates its use.

PROBLEMS FOR SECTION 12.2

LEARNING THE BASICS

12.11 Consider a contingency table with two rows and five columns.
a. How many degrees of freedom are there in the contingency table?
b. Determine the critical value for $\alpha = 0.05$.
c. Determine the critical value for $\alpha = 0.01$.

12.12 Use the following contingency table:

	A	B	C	Total
1	10	30	50	90
2	40	45	50	135
Total	50	75	100	225

a. Compute the expected frequency for each cell.
b. Compute χ^2_{STAT}. Is it significant at $\alpha = 0.05$?

12.13 Use the following contingency table:

	A	B	C	Total
1	20	30	25	75
2	30	20	25	75
Total	50	50	50	150

a. Compute the expected frequency for each cell.
b. Compute χ^2_{STAT}. Is it significant at $\alpha = 0.05$?

APPLYING THE CONCEPTS

12.14 A growing share of Americans use smartphones as their primary means of online access at home. A survey of U.S. adults found that 58% of the 18- to 29-year-olds, 47% of the 30- to 49-year-olds, 27% of the 50- to 64-year-olds, and 15% of those age 65 or older mostly use a smartphone to go online.

Source: Data extracted from *Mobile Technology and Home Broadband 2019*, available at **bit.ly/2S5YFIw**.

Suppose the survey was based on 200 U.S. adults in each of the four age groups: 18 to 29, 30 to 49, 50 to 64, and 65 or older.
a. At the 0.05 level of significance, is there evidence of a difference among the age groups in the proportion of U.S. adults who mostly use a smartphone to go online?
b. Determine the p-value in (a) and interpret its meaning.
c. If appropriate, use the Marascuilo procedure and $\alpha = 0.05$ to determine which age groups differ.

12.15 Even though there has been an abundance of digital and analytics transformations occurring across the business landscape in recent years, few companies report achieving successful results. What has been the major objectives of company digital and analytics transformations? A McKinsey Global survey found that 42 of 52 (81%) retail companies sampled, 22 of 48 (45%) consumer packages goods companies sampled, 57 of 100 (57%) healthcare companies sampled, 41 of 75 (55%) automotive and assembly companies sampled, and 77 of 132 (58%) high tech companies sampled report *improve customer outcomes* as a major objective of their digital and analytics transformation.

Source: Data extracted from "Five moves to make during a digital transformation," available at **bit.ly/2XGkRPs**.

a. At the 0.05 level of significance, is there evidence of a difference among the industries with respect to the proportion of companies that report *improve customer outcomes* as a main objective of their digital and analytics transformation?
b. Compute the p-value and interpret its meaning.
c. If appropriate, use the Marascuilo procedure and $\alpha = 0.05$ to determine which industries differ.

✓SELF TEST **12.16** An Employee Value Proposition (EVP) is about defining the essence of a company. The EVP is the value an employee receives from the employer; it defines the commitment the company will make to develop the employee in exchange for the effort the employee puts in to benefit the company. But do all agree on what makes a unique and compelling EVP? A study showed that 14% of business executives, 38% of HR leaders, and 33% of employees say that compensation (pay and rewards) makes for a unique and compelling EVP.

Source: Data extracted from *Global Talent Trends 2019* available at **bit.ly/2sbrUzh**.

Assume that 200 individuals within each business group were surveyed.
a. Is there evidence of a difference among business groups with respect to the proportion that say compensation (pay and rewards) makes for a unique and compelling EVP?
b. Determine the p-value in (a) and interpret its meaning.
c. If appropriate, use the Marascuilo procedure and $\alpha = 0.05$ to determine which business groups differ in the proportion that say compensation (pay and rewards) makes for a unique and compelling EVP.

12.17 Repeat (a) and (b) of Problem 12.16, assuming that only 100 individuals from each business group were surveyed. Discuss the implications of sample size on the χ^2 test for differences among more than two populations.

12.18 What kinds of activities do you engage in when using a device while viewing video content on a TV screen? An IAB and MARU Matchbox study captured multitasking activities of adults who use different devices while watching TV. The study reported that 320 of 444 (72%) smartphone users sampled, 194 of 347 (56%) of computer users sampled, and 141 of 261 (54%) of tablet users sampled used their device to check social media unrelated to the video while watching TV.

Source: Data extracted from "The Changing TV Experience: 2017," available at **bit.ly/2sz4Mal.**

a. Is there evidence of a significant difference among the smartphone, computer, and tablet users with respect to the proportion who use their device to check social media unrelated to the video while watching TV? (Use $\alpha = 0.05$.)
b. Determine the p-value and interpret its meaning.
c. If appropriate, use the Marascuilo procedure and $\alpha = 0.05$ to determine which groups differ.

12.19 The MSCI 2018 Survey of Women on Boards showed that there continues to be a slow increase in the overall percentage of women on boards globally. The study reported that 72 of 72 (100%) French companies sampled, 49 of 62 (79%) German companies sampled, 14 of 24 (58.3%) Irish companies sampled, 15 of 21 (71.4%) Spanish companies sampled, and 18 of 42 (42.9%) Swiss companies sampled have at least three female directors on their boards.

Source: Data extracted from *Women on Boards Progress Report 2018*, available at **bit.ly/2YOor6D**.

a. Is there evidence of a significant difference among the countries with respect to the proportion of companies that have at least three female directors on their boards? (Use $\alpha = 0.05$.)
b. Determine the p-value and interpret its meaning.
c. If appropriate, use the Marascuilo procedure and $\alpha = 0.05$ to determine which groups differ.

12.3 Chi-Square Test of Independence

In Sections 12.1 and 12.2, you used the χ^2 test to evaluate potential differences among population proportions. For a contingency table that has r rows and c columns, you can generalize the χ^2 test as a *test of independence* for two categorical variables.

For the chi-square test of independence, the null and alternative hypotheses are

H_0: The two categorical variables are independent (there is no relationship between them)

H_1: The two categorical variables are dependent (there is a relationship between them)

The chi-square test of independence uses Equation (12.1):

$$\chi^2_{STAT} = \sum_{\text{all cells}} \frac{(f_o - f_e)^2}{f_e}$$

with $(r - 1)(c - 1)$ degrees of freedom.

Using the level of significance α, one rejects the null hypothesis if the χ^2_{STAT} test statistic is greater than χ^2_α, the upper-tail critical value from a chi-square distribution with $(r - 1)(c - 1)$ degrees of freedom. The decision rule is

$$\text{Reject } H_0 \text{ if } \chi^2_{STAT} > \chi^2_\alpha;$$
$$\text{otherwise, do not reject } H_0.$$

Figure 12.8 illustrates this decision rule.

FIGURE 12.8

Regions of rejection and nonrejection when testing for independence in an $r \times c$ contingency table, using the χ^2 test

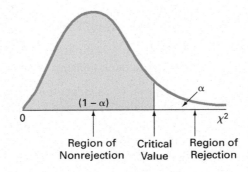

The **chi-square (χ^2) test of independence** is similar to the χ^2 test for equality of proportions. In this test, the null and alternative hypotheses are different, even as the test statistics and decision rules are the same. Because the test of independence uses different hypotheses, the conclusions one can reach differ as well. Also different between the two χ^2 tests is the method one uses to prepare samples.

In a test for equality of proportions, there is one factor of interest, with two or more levels. These levels represent samples selected from independent populations. The categorical responses in each group or level are classified into two categories, such as *an item of interest* and *not an item of interest*. The objective is to make comparisons and evaluate differences between the proportions of the *items of interest* among the various levels. However, in a test for independence, there are two factors of interest, each of which has two or more levels. One selects one sample and tallies the joint responses to the two categorical variables into the cells of a contingency table.

The test for equality of proportions for the guest satisfaction survey (see Sections 12.1 and 12.2) enables one to reach the conclusion that there is a significant relationship between the hotel and the likelihood that a guest would return. In contrast, the test of independence could be used to explore a possible relationship between the primary reason for not returning and the hotel the guest used. To illustrate the χ^2 test for independence, suppose that in the three-hotel guest satisfaction survey, respondents who stated that they were not likely to return also indicated one of four primary reasons for their unwillingness to return. Table 12.13 presents the contingency table. The table is a 4×3 table because there are the four levels of the primary reason factor and the three levels of the hotel factor.

TABLE 12.13
Contingency table of primary reason for not returning and hotel

PRIMARY REASON FOR NOT RETURNING	HOTEL			
	Golden Palm	Palm Royale	Palm Princess	Total
Amenities	23	7	37	67
Dining Options	13	5	13	31
Quality of Room	39	13	8	60
Staff/Service Issues	13	8	8	29
Total	88	33	66	187

In Table 12.13, observe that of the primary reasons for not planning to return to the hotel, 67 were due to amenities, 60 were due to quality of room, 31 were due to room dining options, and 29 were due to staff/service issues. In Table 12.8 on page 398, there were 88 guests at the Golden Palm, 33 guests at the Palm Royale, and 66 guests at the Palm Princess who were not planning to return. The observed frequencies in the cells of the 4×3 contingency table represent the joint tallies of the sampled guests with respect to primary reason for not returning and the hotel where they stayed. The null and alternative hypotheses are

H_0: There is no relationship between the primary reason for not returning and the hotel.
H_1: There is a relationship between the primary reason for not returning and the hotel.

For this test, to calculate the expected frequency, f_e, for a cell, first use the multiplication rule for independent events that Equation (4.7) defines on page 166 to calculate a cell probability. For any cell, the probability is the product of the probability of a level for the first factor times the probability of a level for the second factor. For example, the probability of responses expected in the (upper-left-corner) cell that represents the primary reason amenities for the Golden Palm is the product of the probabilities P(Amenities) and P(Golden Palm).

For this example, the proportion for the primary reason Amenities, P(Amenities), is $67/187 = 0.3583$, and the proportion of all Golden Palm responses, P(Golden Palm), is $88/187 = 0.4706$. If the null hypothesis is true, then the primary reason for not returning and the hotel are independent:

$$P(\text{Amenities } and \text{ Golden Palm}) = P(\text{Amenities}) \times P(\text{Golden Palm})$$

$$= (0.3583) \times (0.4706)$$

$$= 0.1686$$

The expected frequency is the product of the overall sample size, n, and this probability, $187 \times 0.1686 = 31.53$. Table 12.14 presents the f_e values for all table cells.

TABLE 12.14
Contingency table of expected frequencies of primary reason for not returning to hotel

PRIMARY REASON FOR NOT RETURNING	HOTEL			
	Golden Palm	**Palm Royale**	**Palm Princess**	**Total**
Amenities	31.53	11.82	23.65	67
Dining Options	14.59	5.47	10.94	31
Quality of Room	28.24	10.59	21.18	60
Staff/Service Issues	13.65	5.12	10.24	29
Total	88.00	33.00	66.00	187

Equation (12.5) defines an alternate method to calculate the expected frequency by taking the product of the row total and column total for a cell and dividing this product by the overall sample size.

CALCULATING THE EXPECTED FREQUENCY

$$f_e = \frac{\text{row total} \times \text{column total}}{n} \tag{12.5}$$

where

$$\text{row total} = \text{sum of the frequencies in the row}$$
$$\text{column total} = \text{sum of the frequencies in the column}$$
$$n = \text{overall sample size}$$

This alternate method results in simpler calculations. For example, using Equation (12.5) for the upper-left-corner cell (primary reason amenities for the Golden Palm),

$$f_e = \frac{\text{row total} \times \text{column total}}{n} = \frac{(67)(88)}{187} = 31.53$$

and for the lower-right-corner cell (staff/service issues for the Palm Princess),

$$f_e = \frac{\text{row total} \times \text{column total}}{n} = \frac{(29)(66)}{187} = 10.24$$

Table 12.15 on page 406 summarizes the calculations for the χ^2_{STAT} test statistic, using the Table 12.13 observed frequencies and the Table 12.14 expected frequencies.

Using the $\alpha = 0.05$ level of significance, the upper-tail critical value from the chi-square distribution with 6 degrees of freedom is 12.592 (see Table E.4). Because $\chi^2_{STAT} = 27.41 > 12.592$, one rejects the null hypothesis of independence (see Figure 12.9 on page 406).

TABLE 12.15
Calculating the χ^2_{STAT} test statistic for the test of independence

Cell	f_o	f_e	$(f_o - f_e)$	$(f_o - f_e)^2$	$(f_o - f_e)^2/f_e$
Amenities/Golden Palm	23	31.53	−8.53	72.76	2.31
Amenities/Palm Royale	7	11.82	−4.82	23.23	1.97
Amenities/Palm Princess	37	23.65	13.35	178.22	7.54
Dining Options/Golden Palm	13	14.59	−1.59	2.53	0.17
Dining Options/Palm Royale	5	5.47	−0.47	0.22	0.04
Dining Options/Palm Princess	13	10.94	2.06	4.24	0.39
Quality of Room/Golden Palm	39	28.24	10.76	115.78	4.10
Quality of Room/Palm Royale	13	10.59	2.41	5.81	0.55
Quality of Room/Palm Princess	8	21.18	−13.18	173.71	8.20
Staff/Service Issues/Golden Palm	13	13.65	−0.65	0.42	0.03
Staff/Service Issues/Palm Royale	8	5.12	2.88	8.29	1.62
Staff/Service Issues/Palm Princess	8	10.24	−2.24	5.02	0.49
					27.41

FIGURE 12.9
Regions of rejection and nonrejection when testing for independence in the three hotel guest satisfaction survey example at the 0.05 level of significance, with 6 degrees of freedom

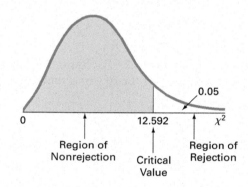

Table 12.16 summarizes the results of the chi-square test for the primary reason for not returning survey using the Table 12.15 calculations and the Figure 12.10 results. Based on the results, there is strong evidence to conclude that there is a relationship between the reason for not returning to the hotel again and the hotel that the guest stayed at. Therefore, T.C. Resort Properties managers can conclude that the hotels are different in terms of reasons guests state for not being likely to return. Amenities are underrepresented as a reason for not returning to the Golden Palm but are overrepresented at the Palm Princess. Guests are more satisfied with amenities at the Golden Palm than at the Palm Princess. Quality of room is overrepresented as a reason for not returning to the Golden Palm, but greatly underrepresented at the Palm Princess. Guests are much more satisfied with the quality of rooms of the Palm Princess than at the Golden Palm.

TABLE 12.16
Chi-square test summary for the primary reason for not returning survey

Results	Conclusions
$\chi^2_{STAT} = 27.41$ is greater than 12.592. The p-value $= 0.0001$ is less than the level of significance, $\alpha = 0.05$.	1. Reject the null hypothesis H_0. 2. Conclude that there is a relationship between the reason for not returning to the hotel again and the hotel that the guest stayed at. 3. The probability is 0.0001 that $\chi^2_{STAT} > 27.41$.

FIGURE 12.10
Excel chi-square test worksheet results for the Table 12.13 primary reason for not returning to hotel data

	A	B	C	D	E	F	G	H	I
1	Chi-Square Test of Independence								
2									
3			Observed Frequencies						
4			Hotel					Calculations	
5	Reason for Not Returning	Golden Palm	Palm Royale	Palm Princess	Total			fo - fe	
6	Amenities	23	7	37	67		-8.5294	-4.8235	13.3529
7	Dining Options	13	5	13	31		-1.5882	-0.4706	2.0588
8	Quality of Room	39	13	8	60		10.7647	2.4118	-13.1765
9	Staff/service Issues	13	8	8	29		-0.6471	2.8824	-2.2353
10	Total	88	33	66	187				
11									
12			Expected Frequencies						
13			Hotel						
14	Reason for Not Returning	Golden Palm	Palm Royale	Palm Princess	Total			(fo - fe)^2/fe	
15	Amenities	31.5294	11.8235	23.6471	67		2.3074	1.9678	7.5401
16	Dining Options	14.5882	5.4706	10.9412	31		0.1729	0.0405	0.3874
17	Quality of Room	28.2353	10.5882	21.1765	60		4.1040	0.5493	8.1987
18	Staff/service Issues	13.6471	5.1176	10.2353	29		0.0307	1.6234	0.4882
19	Total	88	33	66	187				
20									
21		Data							
22	Level of Significance	0.05							
23	Number of Rows	4							
24	Number of Columns	3							
25	Degrees of Freedom	6	=(B23 - 1) * (B24 - 1)						
26									
27		Results							
28	Critical Value	12.5916	=CHISQ.INV.RT(B22, B25)						
29	Chi-Square Test Statistic	27.4104	=SUM(G15:I18)						
30	p-Value	0.0001	=CHISQ.DIST.RT(B29, B25)						
31	Reject the null hypothesis		=IF(B30 < B22, "Reject the null hypothesis",						
32			"Do not reject the null hypothesis")						
33	Expected frequency assumption								
34	is met.		=IF(OR(B15 < 1, C15 < 1, D15 < 1, B16 < 1, C16 < 1, D16 < 1,						
			B17 < 1, C17 < 1, D17 < 1, B18 < 1, C18 < 1, D18 < 1),							
			" is violated.", " is met.")						

In the Worksheet

The worksheet uses the CHISQ.INV.RT function to compute the cell B24 critical value and uses the CHISQ.DIST.RT function to compute the cell B26 p-value. The results of formulas that compute intermediate calculations in the first 19 rows provide the basis for computing the cell B25 test statistic.

The **COMPUTE worksheet** of the Excel Guide **Chi-Square workbook** is a copy of the Figure 12.10 worksheet.

Assumptions of the chi-square test of independence To ensure accurate results, all expected frequencies need to be large in order to use the χ^2 test when dealing with $r \times c$ contingency tables. As in the case of $2 \times c$ contingency tables in Section 12.2, all expected frequencies should be at least 1. For contingency tables in which one or more expected frequencies are less than 1, one can use the chi-square test after collapsing two or more low-frequency rows into one row (or collapsing two or more low-frequency columns into one column). Merging rows or columns usually results in expected frequencies sufficiently large to ensure the accuracy of the χ^2 test.

PROBLEMS FOR SECTION 12.3

LEARNING THE BASICS

12.20 If a contingency table has three rows and four columns, how many degrees of freedom are there for the χ^2 test of independence?

12.21 When performing a χ^2 test of independence in a contingency table with r rows and c columns, determine the upper-tail critical value of the test statistic in each of the following circumstances:
a. $\alpha = 0.05$, $r = 4$ rows, $c = 5$ columns
b. $\alpha = 0.01$, $r = 4$ rows, $c = 5$ columns
c. $\alpha = 0.01$, $r = 4$ rows, $c = 6$ columns
d. $\alpha = 0.01$, $r = 3$ rows, $c = 6$ columns
e. $\alpha = 0.01$, $r = 6$ rows, $c = 3$ columns

APPLYING THE CONCEPTS

12.22 The owner of a restaurant serving Continental-style entrées has the business objective of learning more about the patterns of patron demand during the Friday-to-Sunday weekend time period.

Data were collected from 630 customers on the type of entrée and dessert ordered and organized into the following table:

TYPE OF DESSERT	TYPE OF ENTRÉE				
	Beef	Poultry	Fish	Pasta	Total
Ice cream	13	8	12	14	47
Cake	98	12	29	6	145
Fruit	8	10	6	2	26
None	124	98	149	41	412
Total	243	128	196	63	630

At the 0.05 level of significance, is there evidence of a relationship between type of dessert and type of entrée?

12.23 A Gallup survey across generations of workers gathered data on engagement at work. The results for a sample of 1,000 workers are as follows:

| | GENERATION | | | | |
LEVEL OF ENGAGEMENT	Millennials	Gen Xers	Baby Boomers	Traditionalists	Total
Engaged	102	109	93	14	318
Not Engaged	193	170	134	12	509
Actively Disengaged	55	61	53	4	173
Total	350	340	280	30	1,000

Source: Gallup, "What Millennials Want From Work and Live," available at **bit.ly/1T9dl7p**.

At the 0.05 level of significance, is there evidence of a significant relationship between generation and level of engagement in the workplace?

✓SELF TEST **12.24** How much time per week are employees willing to devote to improve their overall health and well-being by pursuing activities such as consistent exercise, researching healthy recipes, or engaging in wellness coaching? A study by UnitedHealthcare revealed the following results:

| | AGE GROUP | | | | | |
TIME PER WEEK	18–34	35–44	45–54	55–64	65+	Total
Less than 1 hour	35	25	33	10	5	108
1 to less than 3 hours	95	55	32	34	7	223
3 to less than 6 hours	60	44	42	33	5	184
6 to less than 8 hours	37	19	34	24	3	117
9 hours or more	34	22	20	20	1	97
Total	261	165	161	121	21	729

Source: Data extracted from *New survey: Employees say wellness programs improve health, productivity*," available at **bit.ly/2XGwzK7**.

At the 0.01 level of significance, is there evidence of a significant relationship between time willing to devote to health-related activities and age?

12.25 What makes sales leaders tick? Mercuri International conducted a study to explore sales strategies, processes, and support systems within businesses. Organizations were categorized by sales performance level (top performers vs. middle performers vs. bottom performers) and extent to which the organization invests in customer satisfaction. Results were organized into the following table.

| LEVEL OF INVESTMENT | SALES PERFORMANCE LEVEL | | | |
	Top	Middle	Bottom	Total
Annually	53	318	44	415
Every 2–4 years	40	245	23	308
Never	11	158	34	203
Total	104	721	101	926

Source: "Sales Excellence Survey 2017," available at **bit.ly/2kwjjW9**.

At the 0.05 level of significance, is there evidence of a significant relationship between sales performance level and level of investment in customer satisfaction?

12.26 PwC takes a closer look at what CEOs are looking for and are finding as new sources of value in their businesses and industries. Based on a recent Global CEO survey, CEOs are categorized by the main activity they identified that would strengthen their company in order to capitalize on new opportunities as well as the geographic region in which they are located. The results are as follows:

| | GEOGRAPHIC REGION | | | | |
IDENTIFIED MAIN ACTIVITY	Asia Pacific	Latin America	North America	Western Europe	Total
Innovation	117	41	27	66	251
Human capital	73	28	24	40	165
Competitive advantage	68	16	14	12	110
Digital and tech capabilities	54	19	21	60	154
Customer experience	39	14	14	42	109
M & A and partnerships	39	4	16	15	74
Trust and transparency	25	8	9	17	59
Funding growth	25	2	5	4	36
Big data and analytics	20	4	10	12	46
Cost containment	16	11	2	8	37
Navigating risk and regulation	6	7	5	4	22
Cybersecurity	6	2	2	2	12
Total	488	156	149	282	1,075

Source: "20th Annual Global CEO Survey," available at **pwc.to/2l3sSXh**.

At the 0.05 level of significance, is there evidence of a significant relationship between identified main activity and geographic region?

12.4 Wilcoxon Rank Sum Test for Two Independent Populations

Section 10.1 uses the t test for the difference between the means of two independent populations. If sample sizes are small and you cannot assume that the data in each sample are from normally distributed populations, one can use the pooled-variance t test, following a *normalizing transformation* on the data (Winer), or use a nonparametric method that does not depend on the assumption of normality for the two populations.

Nonparametric methods require few or no assumptions about the populations from which data are obtained (Hollander). The **Wilcoxon rank sum test** for whether there is a difference between two medians is one such method. One uses this method when the assumptions that the pooled-variance and separate-variance t tests that Section 10.1 discusses cannot be met. In such conditions, the Wilcoxon rank sum test is likely to have more statistical power (see Section 9.6) than those t tests. (When assumptions can be met, the test has almost as much power as the t tests.) One can also use the Wilcoxon rank sum test when only ordinal data exist, as often happens in consumer behavior and marketing research.

student TIP

You combine the two groups before you rank the values.

To perform the Wilcoxon rank sum test, one replaces the values in the two samples of sizes n_1 and n_2 with their combined ranks (unless the data contained the ranks initially). One begins by defining $n = n_1 + n_2$ as the total sample size. Next, one assigns the ranks so that rank 1 is given to the smallest of the n combined values, rank 2 is given to the second smallest, and so on, until rank n is given to the largest. If several values are tied, one assigns each value the average of the ranks that otherwise would have been assigned had there been no ties.

Whenever the two sample sizes are unequal, n_1 represents the smaller sample and n_2 the larger sample. The Wilcoxon rank sum test statistic, T_1, is defined as the sum of the ranks assigned to the n_1 values in the smaller sample. (For equal-sized samples, either sample may be used for determining T_1.) For any integer value n, the sum of the first n consecutive integers is $n(n + 1)/2$. Therefore, T_1 plus T_2, the sum of the ranks assigned to the n_2 items in the second sample, must equal $n(n + 1)/2$. You can use Equation (12.6) on page 410 to check the accuracy of your rankings.

CHECKING THE RANKINGS

$$T_1 + T_2 = \frac{n(n + 1)}{2} \qquad (12.6)$$

For the Figure 12.11 two-tail test, one rejects the null hypothesis if T_1 is greater than or equal to the upper critical value, or if T_1 is less than or equal to the lower critical value. Figure 12.11 also illustrates the two one-tail tests. For lower-tail tests which have the alternative hypothesis H_1: $M_1 < M_2$, that the median of population 1 (M_1) is less than the median of population 2 (M_2), one rejects the null hypothesis if the observed value of T_1 is less than or equal to the lower critical value. For upper-tail tests which have the alternative hypothesis H_1: $M_1 > M_2$, one rejects the null hypothesis if the observed value of T_1 equals or is greater than the upper critical value.

FIGURE 12.11

Regions of rejection and nonrejection using the Wilcoxon rank sum test

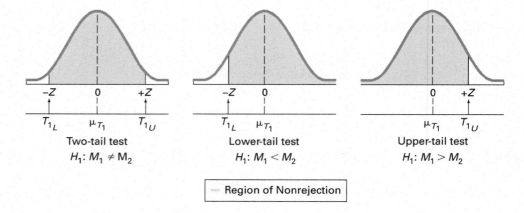

When n_1 and n_2 are each ≤ 10, use Table E.6 to find the critical values of the test statistic T_1. For large sample sizes, the test statistic T_1 is approximately normally distributed, with the mean, μ_{T_1}, equal to

$$\mu_{T_1} = \frac{n_1(n + 1)}{2}$$

and the standard deviation, σ_{T_1}, equal to

$$\sigma_{T_1} = \sqrt{\frac{n_1 n_2(n + 1)}{12}}$$

Therefore, Equation (12.7) defines the standardized Z test statistic for the Wilcoxon rank sum test.

LARGE-SAMPLE WILCOXON RANK SUM TEST

$$Z_{STAT} = \frac{T_1 - \dfrac{n_1(n + 1)}{2}}{\sqrt{\dfrac{n_1 n_2(n + 1)}{12}}} \qquad (12.7)$$

where the test statistic Z_{STAT} approximately follows a standardized normal distribution.

Use Equation (12.7) when the sample sizes are outside the range of Table E.6. Based on α, the level of significance selected, you reject the null hypothesis if the Z_{STAT} test statistic falls in the rejection region.

To study an application of the Wilcoxon rank sum test, recall the Chapter 10 Arlingtons scenario about VLABGo player monthly sales at the special front location and at the in-aisle location (stored in **VLABGo**). If one cannot assume that the populations are normally distributed,

[3] To test for differences in the median sales between the two locations, one must assume that the distributions of sales in both populations are identical except for differences in central tendency (i.e., the medians).

TABLE 12.17
Forming the combined rankings

the Wilcoxon rank sum test can be used to evaluate possible differences in the median sales for the two display locations.[3]

Table 12.17 presents VLABGo sales data and the combined ranks.

Special Front ($n_1 = 10$)	Combined Ranking	In-Aisle ($n_2 = 10$)	Combined Ranking
224	12	192	7
189	4.5	236	13
248	15	164	2
285	19	154	1
273	17	189	4.5
190	6	220	11
243	14	261	16
215	9	186	3
280	18	219	10
317	20	202	8

Source: Data are taken from Table 10.1 on page 313.

Because this study did not state in advance which display location is likely to have a higher median, one uses a two-tail test with the following null and alternative hypotheses:

$$H_0: M_1 = M_2 \text{ (the median sales are equal)}$$

$$H_1: M_1 \neq M_2 \text{ (the median sales are not equal)}$$

One begins by calculating T_1, the sum of the ranks assigned to the *smaller* sample. When the sample sizes are equal, as in this example, one can define either sample as the group from which to compute T_1. Choosing the special front location as the first group,

$$T_1 = 12 + 4.5 + 15 + 19 + 17 + 6 + 14 + 9 + 18 + 20 = 134.5$$

As a check on the ranking procedure, one calculates T_2,

$$T_2 = 7 + 13 + 2 + 1 + 4.5 + 11 + 16 + 3 + 10 + 8 = 75.5$$

and then uses Equation (12.6) on page 410 to show that the sum of the first $n = 20$ integers in the combined ranking is equal to $T_1 + T_2$:

$$T_1 + T_2 = \frac{n(n + 1)}{2}$$

$$134.5 + 75.5 = \frac{20(21)}{2} = 210$$

$$210 = 210$$

Next, one uses Table E.6 to determine the lower- and upper-tail critical values for the test statistic T_1. From Table 12.18, a portion of Table E.6, observe that for a level of significance of 0.05, the critical values are 78 and 132. The decision rule is

Reject H_0 if $T_1 \leq 78$ or if $T_1 \geq 132$;

otherwise, do not reject H_0.

TABLE 12.18

Finding the lower- and upper-tail critical values for the Wilcoxon rank sum test statistic, T, where $n_1 = 10$, $n_2 = 10$, and $\alpha = 0.05$

n_2	α One-tail	Two-tail	n_1 4	5	6	7	8	9	10
					(Lower, Upper)				
9	.05	.10	16,40	24,51	33,63	43,76	54,90	66,105	
	.025	.05	14,42	22,53	31,65	40,79	51,93	62,109	
	.01	.02	13,43	20,55	28,68	37,82	47,97	59,112	
	.005	.01	11,45	18,57	26,70	35,84	45,99	56,115	
10	.05	.10	17,43	26,54	35,67	45,81	56,96	69,111	82,128
	.025	.05	15,45	23,57	32,70	42,84	53,99	65,115	78,132
	.01	.02	13,47	21,59	29,73	39,87	49,103	61,119	74,136
	.005	.01	12,48	19,61	27,75	37,89	47,105	58,122	71,139

Source: Extracted from Table E.6.

Table 12.19 summarizes the results of the Wilcoxon rank sum test for VLABGo player monthly sales at the special front location and at the in-aisle location using the calculations on page 411 and the Figure 12.12 results. Based on the results, there is strong evidence to conclude that the two locations are significantly different in sales. Therefore, as part of the DCOVA framework, one can conclude that sales will be higher at the front location than the in-aisle location.

TABLE 12.19

Wilcoxon test summary for the monthly sales at two different locations

Results	Conclusions
$T_1 = 134.5$ is greater than 132. The p-value $= 0.0257$ is less than the level of significance, $\alpha = 0.05$.	1. Reject the null hypothesis H_0. 2. Conclude that evidence exists that the two locations are significantly different with respect to sales. 3. The probability is 0.0257 that $T_1 > 134.5$.

FIGURE 12.12

Excel Wilcoxon rank sum test worksheet results for VLABGo player monthly sales for two in-store locations

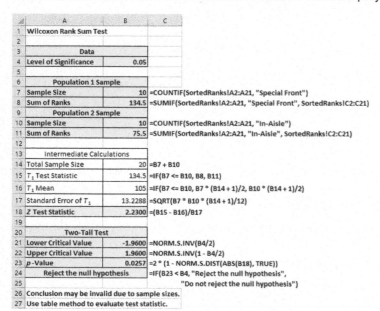

	A	B	C
1	Wilcoxon Rank Sum Test		
2			
3	Data		
4	Level of Significance	0.05	
5			
6	Population 1 Sample		
7	Sample Size	10	=COUNTIF(SortedRanks!A2:A21, "Special Front")
8	Sum of Ranks	134.5	=SUMIF(SortedRanks!A2:A21, "Special Front", SortedRanks!C2:C21)
9	Population 2 Sample		
10	Sample Size	10	=COUNTIF(SortedRanks!A2:A21, "In-Aisle")
11	Sum of Ranks	75.5	=SUMIF(SortedRanks!A2:A21, "In-Aisle", SortedRanks!C2:C21)
12			
13	Intermediate Calculations		
14	Total Sample Size	20	=B7 + B10
15	T_1 Test Statistic	134.5	=IF(B7 <= B10, B8, B11)
16	T_1 Mean	105	=IF(B7 <= B10, B7 * (B14 + 1)/2, B10 * (B14 + 1)/2)
17	Standard Error of T_1	13.2288	=SQRT(B7 * B10 * (B14 + 1)/12)
18	Z Test Statistic	2.2300	=(B15 - B16)/B17
19			
20	Two-Tail Test		
21	Lower Critical Value	-1.9600	=NORM.S.INV(B4/2)
22	Upper Critical Value	1.9600	=NORM.S.INV(1 - B4/2)
23	p-Value	0.0257	=2 * (1 - NORM.S.DIST(ABS(B18), TRUE))
24	Reject the null hypothesis		=IF(B23 < B4, "Reject the null hypothesis",
25			"Do not reject the null hypothesis")
26	Conclusion may be invalid due to sample sizes.		
27	Use table method to evaluate test statistic.		

In the Worksheet

The worksheet uses the Equation (12.7) large-sample approximation in computing the p-value. That value may be inaccurate when both sample sizes are less than or equal to 10 and the test statistic T_1 is less likely to be approximately normally distributed.

Formulas in cells A26 and A27 (not shown) use the IF function and the expression **AND(B7 <= 10, B10 <= 10)** to display a warning message if either or both sample sizes are 10 or less.

The **COMPUTE worksheet** of the Excel Guide **Wilcoxon workbook** is a copy of the Figure 12.12 worksheet.

Table E.6 shows the lower and upper critical values of the Wilcoxon rank sum test statistic, T_1, but only for situations in which both n_1 and n_2 are less than or equal to 10. If either one or both of the sample sizes are greater than 10, one *must* use the large-sample Z approximation formula [Equation (12.7) on page 410]. To demonstrate the large-sample Z approximation formula, consider the VLABGo player monthly sales data. Using Equation (12.7),

$$Z_{STAT} = \frac{T_1 - \dfrac{n_1(n+1)}{2}}{\sqrt{\dfrac{n_1 n_2(n+1)}{12}}}$$

$$= \frac{134.5 - \dfrac{(10)(21)}{2}}{\sqrt{\dfrac{(10)(10)(21)}{12}}}$$

$$= \frac{134.5 - 105}{13.2288} = 2.2300$$

student TIP

When the assumption of normality is met, use the pooled-variance or separate-variance test as these tests are more powerful. Use the Wilcoxon rank sum test only when you doubt the normality assumption.

Because $Z_{STAT} = 2.2300 > 1.96$, the critical value of Z at the 0.05 level of significance (or p-value $= 0.0257 < 0.05$), you reject H_0.

PROBLEMS FOR SECTION 12.4

LEARNING THE BASICS

12.27 Using Table E.6, determine the lower- and upper-tail critical values for the Wilcoxon rank sum test statistic, T_1, in each of the following two-tail tests:
a. $\alpha = 0.10$, $n_1 = 6$, $n_2 = 8$
b. $\alpha = 0.05$, $n_1 = 6$, $n_2 = 8$
c. $\alpha = 0.01$, $n_1 = 6$, $n_2 = 8$
d. Given the results in (a) through (c), what do you conclude regarding the width of the region of nonrejection as the selected level of significance, α, gets smaller?

12.28 Using Table E.6, determine the lower-tail critical value for the Wilcoxon rank sum test statistic, T_1, in each of the following one-tail tests:
a. $\alpha = 0.05$, $n_1 = 6$, $n_2 = 8$
b. $\alpha = 0.025$, $n_1 = 6$, $n_2 = 8$
c. $\alpha = 0.01$, $n_1 = 6$, $n_2 = 8$
d. $\alpha = 0.005$, $n_1 = 6$, $n_2 = 8$

12.29 The following information is available for two samples selected from independent populations:

   Sample 1: $n_1 = 7$ **Assigned ranks:** 4 1 8 2 5 10 11

   Sample 2: $n_2 = 9$ **Assigned ranks:** 7 16 12 9 3 14 13 6 15

What is the value of T_1 if you are testing the null hypothesis $H_0: M_1 = M_2$?

12.30 In Problem 12.29, what are the lower- and upper-tail critical values for the test statistic T_1 from Table E.6 if you use a 0.05 level of significance and the alternative hypothesis is $H_1: M_1 \neq M_2$?

12.31 In Problems 12.29 and 12.30, what is your statistical decision?

12.32 The following information is available for two samples selected from independent and similarly shaped right-skewed populations:

Sample 1: $n_1 = 5$ 1.1   2.3   2.9   3.6   14.7

Sample 2: $n_2 = 6$ 2.8   4.4   4.4   5.2   6   18.5

a. Replace the observed values with the corresponding ranks (where $1 =$ smallest value; $n = n_1 + n_2 = 13 =$ largest value) in the combined samples.
b. What is the value of the test statistic T_1?
c. Compute the value of T_2, the sum of the ranks in the larger sample.
d. To check the accuracy of your rankings, use Equation (12.6) on page 410 to demonstrate that $T_1 + T_2 = \dfrac{n(n+1)}{2}$.

12.33 From Problem 12.32, at the 0.05 level of significance, determine the lower-tail critical value for the Wilcoxon rank sum test statistic, T_1, if you want to test the null hypothesis, $H_0: M_1 \geq M_2$, against the one-tail alternative, $H_1: M_1 < M_2$.

12.34 In Problems 12.32 and 12.33, what is your statistical decision?

APPLYING THE CONCEPTS

12.35 A vice president for marketing recruits 20 college graduates for management training. Each of the 20 individuals is randomly assigned to one of two groups (10 in each group). A "traditional" method of training (T) is used in one group, and an "experimental" method (E) is used in the other. After the graduates spend six months on the job, the vice president ranks them on the basis of their performance, from 1 (worst) to 20 (best), with the following results, stored in Training Ranks .

   T: 1 2 3 5 9 10 12 13 14 15

   E: 4 6 7 8 11 16 17 18 19 20

Is there evidence of a difference in the median performance between the two methods? (Use $\alpha = 0.05$.)

12.36 Wine experts Gaiter and Brecher use a six-category scale when rating wines: Yech, OK, Good, Very Good, Delicious, and Delicious! Suppose Gaiter and Brecher tested wines from a random sample of eight inexpensive California Cabernets and a random sample of eight inexpensive Washington Cabernets, where *inexpensive* means wines with a U.S. suggested retail price of less than $20, and assigned the following ratings:

California—Good, Delicious, Yech, OK, OK, Very Good, Yech, OK

Washington—Very Good, OK, Delicious!, Very Good, Delicious, Good, Delicious, Delicious!

The ratings were then ranked and the ratings and the rankings stored in Cabernet.

Soruce: Data extracted from D. Gaiter and J. Brecher, "A Good U.S. Cabernet Is Hard to Find," *The Wall Street Journal*, May 19, 2006, p. W7.

a. Are the data collected by rating wines using this scale nominal, ordinal, interval, or ratio?

b. Why is the two-sample *t* test defined in Section 10.1 inappropriate to test the mean rating of California Cabernets versus Washington Cabernets?

c. Is there evidence of a significant difference in the median rating of California Cabernets and Washington Cabernets? (Use $\alpha = 0.05$.)

12.37 Is there a difference in the satisfaction rating of traditional cellphone providers who bill for service at the end of a month often under a contract and prepaid cellphone service providers who bill in advance without a contract? The file Cellphone Providers contains the satisfaction rating for 10 traditional cellphone providers and 13 prepaid cellphone service providers.

Source: Data extracted from "Carrier Ratings: Why It Pays to Think Small," *Consumer Reports*, February 2016, p. 51.

a. Is there evidence of a difference in the median satisfaction rating for traditional and prepaid providers? (Use $\alpha = 0.05$.)

b. What assumptions must you make in (a)?

c. Compare the results of (a) with those of Problem 10.9(a) on page 320.

✓SELF TEST **12.38** The management of a hotel has the business objective of increasing the return rate for hotel guests. One aspect of first impressions by guests relates to the time it takes to deliver a guest's luggage to the room after check-in to the hotel. A random sample of 20 deliveries on a particular day were selected in Wing A of the hotel, and a random sample of 20 deliveries were selected in Wing B. Delivery times were collected and stored in Luggage.

a. Is there evidence of a difference in the median delivery times in the two wings of the hotel? (Use $\alpha = 0.05$.)

b. Compare the results of (a) with those of Problem 10.65 on page 344.

12.39 The lengths of life (in hours) of a sample of 40 6-watt light emitting diode (LED) light bulbs produced by Manufacturer A and a sample of 40 6-watt LED light bulbs produced by Manufacturer B are stored in Bulbs.

a. Using a 0.05 level of significance, is there evidence of a difference in the median life of bulbs produced by the two manufacturers?

b. What assumptions must you make in (a)?

c. Compare the results of (a) with those of Problem 10.64 on page 344. Discuss.

12.40 Brand valuations are critical to CEOs, financial and marketing executives, security analysts, institutional investors, and others who depend on well-researched, reliable information for assessments and comparisons in decision making. Kantar Millward Brown annually publish the BrandZ Top Global Brands. The BrandZ rankings combine consumer measures of brand equity with financial measures to establish a *brand value* for each brand. The file BrandZ TechFin contains the 2019 brand values for the technology sector and the financial institutions sector.

Source: Data extracted from "BrandZ Top Global Brands," **bit.ly/BZUS2019DL**.

a. Using a 0.05 level of significance, is there evidence of a difference in the median brand value between the two sectors?

b. What assumptions must you make in (a)?

c. Compare the results of (a) with those of Problem 10.17 on page 321. Discuss.

12.41 A bank with a branch located in a commercial district of a city has developed an improved process for serving customers during the noon-to-1 P.M. lunch period. The bank has the business objective of reducing the waiting time (defined as the number of minutes that elapse from when the customer enters the line until he or she reaches the teller window) to increase customer satisfaction. A random sample of 15 customers is selected and waiting times are collected and stored in Bank Waiting. These waiting times (in minutes) are:

4.21 5.55 3.02 5.13 4.77 2.34 3.54 3.20

4.50 6.10 0.38 5.12 6.46 6.19 3.79

Another branch, located in a residential area, is also concerned with the noon-to-1 P.M. lunch period. A random sample of 15 customers is selected and waiting times are collected and stored in Bank Waiting 2. These waiting times (in minutes) are:

9.66 5.90 8.02 5.79 8.73 3.82 8.01 8.35

10.49 6.68 5.64 4.08 6.17 9.91 5.47

a. Is there evidence of a difference in the median waiting time between the two branches? (Use $\alpha = 0.05$.)

b. What assumptions must you make in (a)?

c. Compare the results (a) with those of Problem 10.12 (a) on page 321. Discuss.

12.42 The Super Bowl is a big viewing event watched by close to 200 million Americans that is also a big event for advertisers. The file Super Bowl Ad Ratings contains the rating of ads that ran between the opening kickoff and the final whistle.

Source: Data extracted from **www.admeter.usatoday.com/results/2019**.

a. Is there evidence of a difference in the median rating ads that ran before halftime and ads that ran at halftime or after (Use $\alpha = 0.05$.)

b. What assumptions must you make in (a)?

c. Compare the results of (a) with those of Problem 10.11 (a) on page 320. Discuss.

12.5 Kruskal-Wallis Rank Test for the One-Way ANOVA

If the normality assumption of the one-way ANOVA F test is violated, one can use the Kruskal-Wallis rank test. The **Kruskal-Wallis rank test** for differences among more than two medians is an extension of the Wilcoxon rank sum test for two independent populations that Section 12.4 discusses. The Kruskal-Wallis test has the same power relative to the one-way ANOVA F test that the Wilcoxon rank sum test has relative to the t test.

One uses the Kruskal-Wallis rank test to test whether c independent groups have equal medians. The null hypothesis is

$$H_0: M_1 = M_2 = \cdots = M_c$$

and the alternative hypothesis is

$$H_1: \text{Not all } M_j \text{ are equal (where } j = 1, 2, \ldots, c).$$

To use the Kruskal-Wallis rank test, one replaces the values in the c samples with their combined ranks. Rank 1 is given to the smallest of the combined values and rank n to the largest of the combined values (where $n = n_1 + n_2 + \cdots + n_c$). If any values are tied, one assigns each of them the mean of the ranks they would have otherwise been assigned if ties had not been present in the data.

The Kruskal-Wallis test is an alternative to the one-way ANOVA F test. Instead of comparing each of the c group means against the grand mean, the Kruskal-Wallis test compares the mean rank in each of the c groups against the overall mean rank, based on all n combined values. Equation (12.8) defines the Kruskal-Wallis test statistic, H.

KRUSKAL-WALLIS RANK TEST FOR DIFFERENCES AMONG c MEDIANS

$$H = \left[\frac{12}{n(n+1)} \sum_{j=1}^{c} \frac{T_j^2}{n_j} \right] - 3(n+1) \tag{12.8}$$

where

$\quad n = $ total number of values over the combined samples

$\quad n_j = $ number of values in the jth sample ($j = 1, 2, \ldots, c$)

$\quad T_j = $ sum of the ranks assigned to the jth sample

$\quad T_j^2 = $ square of the sum of the ranks assigned to the jth sample

$\quad c = $ number of groups

If there is a significant difference among the c groups, the mean rank differs considerably from group to group. In the process of squaring these differences, the test statistic H becomes large. If there are no differences present, the test statistic H is small because the mean of the ranks assigned in each group should be very similar from group to group.

As the sample sizes in each group get large (5 or larger), the sampling distribution of the test statistic, H, approximately follows the chi-square distribution with $c - 1$ degrees of freedom. Thus, one rejects the null hypothesis if the computed value of H is greater than the upper-tail critical value (see Figure 12.13). Therefore, the decision rule is

$$\text{Reject } H_0 \text{ if } H > \chi_\alpha^2;$$

$$\text{otherwise, do not reject } H_0.$$

FIGURE 12.13
Determining the rejection region for the Kruskal-Wallis test

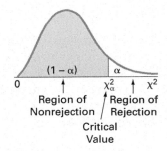

To illustrate the Kruskal-Wallis rank test for differences among c medians, return to the Arlingtons scenario on page 352 that concerns the in-store sales location experiment. If one cannot assume that the mobile electronics sales are normally distributed in all c groups, one can use the Kruskal-Wallis rank test.

The null hypothesis is that the median mobile electronics sales from each of the four in-store locations are equal. The alternative hypothesis is that at least one of these medians differs from the others:

$$H_0: M_1 = M_2 = M_3 = M_4$$

$$H_1: \text{Not all } M_j \text{ are equal (where } j = 1, 2, 3, 4).$$

Table 12.20 presents the data (stored in Mobile Electronics), along with the corresponding ranks of the in-store location sales experiment at Arlingtons.

TABLE 12.20

Mobile electronics sales and rank for four in-store locations

IN-AISLE		FRONT		KIOSK		EXPERT	
Sales	**Rank**	**Sales**	**Rank**	**Sales**	**Rank**	**Sales**	**Rank**
27.74	1	31.47	16	30.78	11	30.25	6.5
29.96	2.5	31.86	17	30.79	12	30.25	6.5
29.96	2.5	32.13	18	30.91	13	30.29	8
30.06	4	32.22	19	30.95	14	30.33	9
30.19	5	32.29	20	31.13	15	30.55	10

In assigning ranks to the sales, the lowest sales, the first in-aisle sales in Table 12.20, is assigned the rank of 1 and the highest sales, the fifth front sales, is assigned the rank of 20. Because the second and third in-aisle sales are tied for ranks 2 and 3, each is assigned the rank 2.5.

After all the ranks are assigned, you compute the sum of the ranks for each group:

$$\text{Rank sums: } T_1 = 15 \quad T_2 = 90 \quad T_3 = 65 \quad T_4 = 40$$

As a check on the rankings, recall from Equation (12.6) on page 410 that for any integer n, the sum of the first n consecutive integers is $n(n + 1)/2$. Therefore,

$$T_1 + T_2 + T_3 + T_4 = \frac{n(n + 1)}{2}$$

$$15 + 90 + 65 + 40 = \frac{(20)(21)}{2}$$

$$210 = 210$$

To test the null hypothesis of equal population medians, one calculates the test statistic H using Equation (12.8) on page 415:

$$H = \left[\frac{12}{n(n + 1)} \sum_{j=1}^{c} \frac{T_j^2}{n_j} \right] - 3(n + 1)$$

$$= \left\{ \frac{12}{(20)(21)} \left[\frac{(15)^2}{5} + \frac{(90)^2}{5} + \frac{(65)^2}{5} + \frac{(40)^2}{5} \right] \right\} - 3(21)$$

$$= \left(\frac{12}{420} \right)(2{,}830) - 63 = 17.8571$$

Table 12.22 on page 417 summarizes the results of the Kruskal-Wallis rank test test for the differences among the median mobile electronics sales for four in-store locations using the above calculations and the Figure 12.14 results. Based on the results, there is strong evidence that the four locations are significantly different in sales. Therefore, as part of the DCOVA framework, one can conclude that sales appear to be higher at the front location and the kiosk location.

TABLE 12.21

Finding χ_α^2, the upper-tail critical value for the Kruskal-Wallis rank test, at the 0.05 level of significance with 3 degrees of freedom

	Cumulative Area									
	.005	.01	.025	.05	.10	.25	.75	.90	.95	.975
	Upper-Tail Area									
Degrees of Freedom	.995	.99	.975	.95	.90	.75	.25	.10	.05	.025
1	—	—	0.001	0.004	0.016	0.102	1.323	2.706	3.841	5.024
2	0.010	0.020	0.051	0.103	0.211	0.575	2.773	4.605	5.991	7.378
3	0.072	0.115	0.216	0.352	0.584	1.213	4.108	6.251	7.815	9.348
4	0.207	0.297	0.484	0.711	1.064	1.923	5.385	7.779	9.488	11.143
5	0.412	0.554	0.831	1.145	1.610	2.675	6.626	9.236	11.071	12.833

Source: Extracted from Table E.4.

TABLE 12.22

Kruskal-Wallis rank test summary for median mobile electronics sales for four in-store locations

Results	Conclusions
$\chi_{STAT}^2 = 17.8571$ is greater than 7.815. The p-value $= 0.0005$ is less than the level of significance, $\alpha = 0.05$.	1. Reject the null hypothesis H_0. 2. Conclude that evidence exists that the four in-store locations are significantly different with respect to sales. (See the Daniel reference to simultaneously compare all four locations.) 3. The probability is 0.0005 that $\chi_{STAT}^2 > 17.8571$.

FIGURE 12.14

Excel Kruskal-Wallis rank test worksheet results for the differences among the median mobile electronics sales for four in-store locations

In the Worksheet

The worksheet uses the CHISQ.INV.RT function to compute the critical value in cell B13 and uses the CHISQ.DIST.RT function to compute the p-value in cell B14.

The **KruskalWallis4 worksheet** of the Excel Guide **Kruskal-Wallis Worksheets workbook** is a copy of the Figure 12.14 worksheet.

Assumptions of the Kruskal-Wallis Rank Test

To use the Kruskal-Wallis rank test, one makes these assumptions:

- The c samples are randomly and independently selected from their respective populations.
- The underlying variable is continuous.
- The data provide at least a set of ranks, both within and among the c samples.
- The c populations have the same variability.
- The c populations have the same shape.

The Kruskal-Wallis procedure makes less stringent assumptions than does the F test. If one ignores the last two assumptions (variability and shape), one can still use the Kruskal-Wallis rank test to determine whether at least one of the populations differs from the other populations in some characteristic—such as central tendency, variation, or shape.

To use the F test, one must assume that the c samples are from normal populations that have equal variances. When the more stringent assumptions of the F test hold, one should use the F test instead of the Kruskal-Wallis test because it has slightly more power to detect significant differences among groups. However, if the assumptions of the F test do not hold, one should use the Kruskal-Wallis test.

PROBLEMS FOR SECTION 12.5

LEARNING THE BASICS

12.43 What is the upper-tail critical value from the chi-square distribution if you use the Kruskal-Wallis rank test for comparing the medians in six populations at the 0.01 level of significance?

12.44 For this problem, use the results of Problem 12.43.
a. State the decision rule for testing the null hypothesis that all six groups have equal population medians.
b. What is your statistical decision if the computed value of the test statistic H is 13.77?

APPLYING THE CONCEPTS

12.45 A pet food company has the business objective of expanding its product line beyond its current kidney- and shrimp-based cat foods. The company developed two new products—one based on chicken livers and the other based on salmon. The company conducted an experiment to compare the two new products with its two existing ones, as well as a generic beef-based product sold at a supermarket chain.

For the experiment, a sample of 50 cats from the population at a local animal shelter was selected. Ten cats were randomly assigned to each of the five products being tested. Each of the cats was then presented with 3 ounces of the selected food in a dish at feeding time. The researchers defined the variable to be measured as the number of ounces of food that the cat consumed within a 10-minute time interval that began when the filled dish was presented. The results for this experiment are summarized in the table in the next column and stored in Cat Food .
a. At the 0.05 level of significance, is there evidence of a significant difference in the median amount of food eaten among the various products?
b. Compare the results of (a) with those of Problem 11.13 (a) on page 366.
c. Which test is more appropriate for these data: the Kruskal-Wallis rank test or the one-way ANOVA F test? Explain.

Kidney	Shrimp	Chicken Liver	Salmon	Beef
2.37	2.26	2.29	1.79	2.09
2.62	2.69	2.23	2.33	1.87
2.31	2.25	2.41	1.96	1.67
2.47	2.45	2.68	2.05	1.64
2.59	2.34	2.25	2.26	2.16
2.62	2.37	2.17	2.24	1.75
2.34	2.22	2.37	1.96	1.18
2.47	2.56	2.26	1.58	1.92
2.45	2.36	2.45	2.18	1.32
2.32	2.59	2.57	1.93	1.94

✓ SELF TEST 12.46 A hospital conducted a study of the waiting time in its emergency room. The hospital has a main campus, along with three affiliated locations. Management had a business objective of reducing waiting time for emergency room cases that did not require immediate attention. To study this, a random sample of 15 emergency room cases at each location were selected on a particular day, and the waiting time (recorded from check-in to when the patient was called into the clinic area) was measured. The results are stored in ER Waiting .
a. At the 0.05 level of significance, is there evidence of a difference in the median waiting times in the four locations?
b. Compare the results of (a) with those of Problem 11.9 (a) on page 365.

12.47 *QSR* magazine reports on the largest quick-serve and fast-casual brands in the United States. The file Quick-Serve Sales contains the food segment (burger, chicken, sandwich, or pizza/pasta) and U.S. mean sales per unit ($thousands) for each of 50 quick-service brands.

Source: Data extracted from "The QSR 50," **bit.ly/2xN4q4v**.

a. At the 0.05 level of significance, is there evidence of a difference in the median U.S. average sales per unit ($thousands) among the food segments?

b. Compare the results of (a) with those of Problem 11.11 (a) on page 366.

12.48 An advertising agency has been hired by a manufacturer of pens to develop an advertising campaign for the upcoming holiday season. To prepare for this project, the research director decides to initiate a study of the effect of advertising on product perception. An experiment is designed to compare five different advertisements. Advertisement A greatly undersells the pen's characteristics. Advertisement B slightly undersells the pen's characteristics. Advertisement C slightly oversells the pen's characteristics. Advertisement D greatly oversells the pen's characteristics. Advertisement E attempts to correctly state the pen's characteristics.

A sample of 30 adult respondents, taken from a larger focus group, is randomly assigned to the five advertisements (so that there are six respondents to each). After reading the advertisement and developing a sense of product expectation, all respondents unknowingly receive the same pen to evaluate. The respondents are permitted to test the pen and the plausibility of the advertising copy. The respondents are then asked to rate the pen from 1 to 7 on the product characteristic scales of appearance, durability, and writing performance. The *combined* scores of three ratings (appearance, durability, and writing performance) for the 30 respondents are stored in Pen . These data are:

A	B	C	D	E
15	16	8	5	12
18	17	7	6	19
17	21	10	13	18
19	16	15	11	12
19	19	14	9	17
20	17	14	10	14

a. At the 0.05 level of significance, is there evidence of a difference in the median ratings of the five advertisements?

b. Compare the results of (a) with those of Problem 11.10 (a) on page 365.

c. Which test is more appropriate for these data: the Kruskal-Wallis rank test or the one-way ANOVA *F* test? Explain.

12.49 A transportation strategist wanted to compare the traffic congestion levels across four continents: Asia, Europe, North America, and South America. The file Congestion Level contains congestion level, defined as the increase (%) in overall travel time when compared to a free flow situation (an uncongested situation) for 10 cities in each continent.

Source: Data extracted from "TomTom Traffic Index," **bit.ly/1RxyKAl**.

a. At the 0.05 level of significance, is there evidence of a difference in the median congestion levels across continents?

b. Compare the results of (a) with those of Problem 11.14 (a) on page 367.

12.50 The more costly and time consuming it is to export and import, the more difficult it is for local companies to be competitive and to reach international markets. As part of an initial investigation exploring foreign market entry, 10 countries were selected from each of four global regions. The cost associated with importing a standardized cargo of goods by sea transport in these countries (in $US per container) is stored in International Market 2 .

Source: Data extracted from **doingbusiness.org/data**.

a. At the 0.05 level of significance, is there evidence of a difference in the median cost across the four global regions associated with importing a standardized cargo of goods by sea transport?

b. Compare the results in (a) to those in Problem 11.8 (a) on page 365.

12.6 McNemar Test for the Difference Between Two Proportions (Related Samples)

Tests such as chi-square test for the difference between two proportions discussed in Section 12.1 require independent samples from each population. However, sometimes when you are testing differences between the proportion of items of interest, the data are collected from repeated measurements or matched samples.

To test whether there is evidence of a difference between the proportions when the data have been collected from two related samples, you can use the McNemar test. The **Section 12.6 online topic** discusses this test and illustrates its use.

12.7 Chi-Square Test for the Variance or Standard Deviation

When analyzing numerical data, sometimes you need to test a hypothesis about the population variance or standard deviation. Assuming that the data are normally distributed, you use the χ^2 test for the variance or standard deviation to test whether the population variance or standard deviation is equal to a specified value. The **Section 12.7 online topic** discusses this test and illustrates its use.

12.8 Wilcoxon Signed Ranks Test for Two Related Populations

Section 10.2 discusses using the paired *t* test to compare the means of two related populations. The paired *t* test assumes that the data are measured on an interval or a ratio scale and are normally distributed. When these assumptions cannot be made, the nonparametric Wilcoxon signed ranks test can be used to test for the median difference. The **Section 12.8 online topic** discusses this test and illustrates its use.

▼USING **STATISTICS**
Avoiding Guesswork ... , Revisited

In the Using Statistics scenario, you were the manager of T.C. Resort Properties, a collection of five upscale hotels located on two tropical islands. To assess the quality of services being provided by your hotels, guests are encouraged to complete a satisfaction survey at check-out time or later, via email. You analyzed the data from these surveys to determine the overall satisfaction with the services provided, the likelihood that the guests will return to the hotel, and the reasons given by some guests for not wanting to return.

On one island, T.C. Resort Properties operates the Beachcomber and Windsurfer hotels. You performed a chi-square test for the difference in two proportions and concluded that a greater proportion of guests are willing to return to the Beachcomber Hotel than to the Windsurfer. On the other island, T.C. Resort Properties operates the Golden Palm, Palm Royale, and Palm Princess hotels. To see if guest satisfaction was the same among the three hotels, you performed a chi-square test for the differences among more than two proportions. The test confirmed that the three

proportions are not equal, and guests seem to be most likely to return to the Palm Royale and least likely to return to the Golden Palm.

In addition, you investigated whether the reasons given for not returning to the Golden Palm, Palm Royale, and Palm Princess were unique to a certain hotel or common to all three hotels. By performing a chi-square test of independence, you determined that the reasons given for wanting to return or not depended on the hotel where the guests had been staying. By examining the observed and expected frequencies, you concluded that guests were more satisfied with the amenities at the Golden Palm and were much more satisfied with the quality of the Palm Princess rooms. Guest satisfaction with dining options was not significantly different among the three hotels.

▼REFERENCES

Corder, G. W., and D. I. Foreman, *Nonparametric Statistics: A Step-by-Step Approach*. New York: Wiley, 2014.

Daniel, W. W. *Applied Nonparametric Statistics*, 2nd ed. Boston: PWS Kent, 1990.

Dixon, W. J., and F. J. Massey, Jr. *Introduction to Statistical Analysis*, 4th ed. New York: McGraw-Hill, 1983.

Hollander, M., D. A. Wolfe, and E. Chicken, *Nonparametric Statistical Methods*, 3rd ed. New York: Wiley, 2014.

Lewontin, R. C., and J. Felsenstein. "Robustness of Homogeneity Tests in 2 × *n* Tables," *Biometrics*, 21(March 1965): 19–33.

Marascuilo, L. A. "Large-Sample Multiple Comparisons," *Psychological Bulletin*, 65(1966): 280–290.

Marascuilo, L. A., and M. McSweeney. *Nonparametric and Distribution-Free Methods for the Social Sciences*. Monterey, CA: Brooks/Cole, 1977.

Winer, B. J., D. R. Brown, and K. M. Michels. *Statistical Principles in Experimental Design*, 3rd ed. New York: McGraw-Hill, 1989.

▼SUMMARY

Figure 12.15 on page 421 presents a roadmap for this chapter. This chapter discusses hypothesis testing for analyzing categorical data from two independent samples and from more than two independent samples. In addition, the chapter applies the multiplication rule that Chapter 4 discusses to the hypothesis of independence in the joint responses to

two categorical variables. The chapter also discusses two nonparametric tests. One uses the Wilcoxon rank sum test when the assumptions of the *t* test for two independent samples were violated. One uses the Kruskal-Wallis test when the assumptions of the one-way ANOVA *F* test were violated.

FIGURE 12.15
Roadmap of Chapter 12

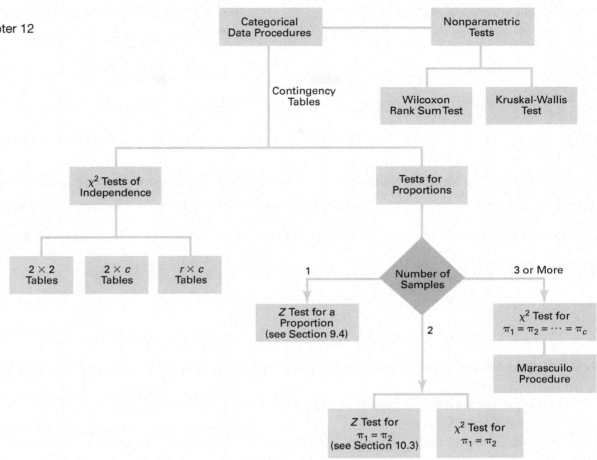

▼ KEY EQUATIONS

χ^2 Test Statistic

$$\chi^2_{STAT} = \sum_{\text{all cells}} \frac{(f_o - f_e)^2}{f_e}$$ **(12.1)**

The Estimated Overall Proportion for Two Groups

$$\bar{p} = \frac{X_1 + X_2}{n_1 + n_2} = \frac{X}{n}$$ **(12.2)**

The Estimated Overall Proportion for c Groups

$$\bar{p} = \frac{X_1 + X_2 + \cdots + X_c}{n_1 + n_2 + \cdots + n_c} = \frac{X}{n}$$ **(12.3)**

Critical Range for the Marascuilo Procedure

$$\text{Critical range} = \sqrt{\chi^2_\alpha}\sqrt{\frac{p_j(1 - p_j)}{n_j} + \frac{p_{j'}(1 - p_{j'})}{n_{j'}}}$$ **(12.4)**

Calculating the Expected Frequency

$$f_e = \frac{\text{row total} \times \text{column total}}{n}$$ **(12.5)**

Checking the Rankings

$$T_1 + T_2 = \frac{n(n + 1)}{2}$$ **(12.6)**

Large-Sample Wilcoxon Rank Sum Test

$$Z_{STAT} = \frac{T_1 - \dfrac{n_1(n + 1)}{2}}{\sqrt{\dfrac{n_1 n_2 (n + 1)}{12}}}$$ (12.7)

Kruskal-Wallis Rank Test for Differences Among
c Medians

$$H = \left[\frac{12}{n(n + 1)} \sum_{j=1}^{c} \frac{T_j^2}{n_j} \right] - 3(n + 1)$$ (12.8)

▼KEY TERMS

chi-square (χ^2) distribution   390
chi-square (χ^2) test for the difference
  between two proportions   390
chi-square (χ^2) test of independence   404
estimated overall proportion for two
  groups $\bar{p}$ 391

expected frequency (f_e)   391
Kruskal-Wallis rank test   415
Marascuilo procedure   400
nonparametric methods   409

observed frequency (f_o)   391
$2 \times c$ contingency table   397
2×2 contingency table   390
Wilcoxon rank sum test   409

▼CHECKING YOUR UNDERSTANDING

12.51 Under what conditions should you use the χ^2 test to determine whether there is a difference between the proportions of two independent populations?

12.52 Under what conditions should you use the χ^2 test to determine whether there is a difference among the proportions of more than two independent populations?

12.53 Under what conditions should you use the χ^2 test of independence?

12.54 Under what conditions should you use the Wilcoxon rank sum test instead of the t test for the difference between the means?

12.55 Under what conditions should you use the Kruskal-Wallis rank test instead of the one-way ANOVA?

▼CHAPTER REVIEW PROBLEMS

12.56 Undergraduate students at Miami University in Oxford, Ohio, were surveyed in order to evaluate the effect of gender and price on purchasing a pizza from Pizza Hut. Students were told to suppose that they were planning to have a large two-topping pizza delivered to their residence that evening. The students had to decide between ordering from Pizza Hut at a reduced price of $8.49 (the regular price for a large two-topping pizza from the Oxford Pizza Hut at the time was $11.49) and ordering a pizza from a different pizzeria. The following contingency table summarizes the student responses.

GENDER	Pizza Hut	Other	Total
Female	4	13	17
Male	6	12	18
Total	10	25	35

PIZZERIA

a. Using a 0.05 level of significance, is there evidence of a difference between males and females in their pizzeria selection?
b. What is your answer to (a) if nine of the male students selected Pizza Hut and nine selected another pizzeria?

The following contingency table summarizes a second survey that asked about purchase decisions at other prices.

PIZZERIA	PRICE $8.49	$11.49	$14.49	Total
Pizza Hut	10	5	2	17
Other	25	23	27	75
Total	35	28	29	92

c. Using a 0.05 level of significance and using the data in the second contingency table, is there evidence of a difference in pizzeria selection based on price?
d. Determine the p-value in (c) and interpret its meaning.

12.57 What social media tools do marketers commonly use? The Social Media Examiner surveyed marketers who commonly use an indicated social media tool. Surveyed were both B2B marketers, marketers that focus primarily on attracting businesses, and B2C marketers, marketers that primarily target consumers. Suppose the survey was based on 500 B2B marketers and 500 B2C marketers and yielded the results in the table at right.

Data extracted from *2017 Social Media Marketing Industry Report*, available at **bit.ly/2rFmLzh**.

SOCIAL MEDIA TOOL	BUSINESS FOCUS	
	B2B	**B2C**
Facebook	89%	97%
Twitter	75%	65%
LinkedIn	81%	44%
Pinterest	26%	32%

For *each social media tool*, at the 0.05 level of significance, determine whether there is a difference between B2B marketers and B2C marketers in the proportion who used each social media tool.

12.58 Business leaders around the world are becoming aware of the huge potential of digital transformation. Fujitsu conducted a global survey to find out more about how business leaders are responding to the digital transformation revolution. To assess the extent of business embarkment on digital transformation, a sample of 745 managers and key decision makers in mid- and large-sized companies was selected and asked whether their organization has embarked on digital transformation specific to their industry sector. following table summarizes results.

	INDUSTRY SECTOR					
EMBARKED?	**Finance, Insurance**	**Healthcare**	**Manufacturing**	**Transport (Logistic)**	**Wholesale, Retail Trade**	**Total**
Yes	70	51	149	26	50	346
No	68	49	165	30	87	399
Total	138	100	314	56	137	745

Source: "Global Digit Transformation Survey Report," available at **bit.ly/2qRXlLb**.

a. At the 0.05 level of significance, is there evidence of a difference in the proportion of organizations that have embarked on digital transformation on the basis of industry sector?

Respondents associated with organizations that have embarked on digital transformation were asked to describe the progress of the digital transformation. The following table summarizes results, cross-classified by industry sector.

	INDUSTRY SECTOR					
PROGRESS	**Finance, Insurance**	**Healthcare**	**Manufacturing**	**Transport (Logistic)**	**Wholesale, Retail Trade**	**Total**
Planning	7	15	15	5	10	52
Testing	31	12	24	5	11	83
Implementing	12	12	54	9	15	102
Outcomes delivered	20	12	56	7	14	109
Total	70	51	149	26	50	346

b. At the 0.05 level of significance, is there evidence of a relationship between digital transformation progress and industry sector?

12.59 Do Americans trust advertisements? A survey by YouGov asked Americans who view advertisements at least once a month how honest the advertisements that they see, read, and hear are. The following table presents the survey results.

	GEOGRAPHIC REGION				
HONEST?	**Northeast**	**Midwest**	**South**	**West**	**Total**
Yes	102	118	220	115	555
No	74	93	135	130	432
Total	176	211	355	245	987

Source: "Truth in advertising: 50% don't ...," **bit.ly/1ivIlLX**.

a. At the 0.05 level of significance, is there evidence of a difference in the proportion of Americans who say advertisements are honest on the basis of geographic region?

YouGov also asked Americans who view advertisements at least once a month how much they trust the advertisements that they see, read, and hear. The table at right presents the results of this second survey.

	GEOGRAPHIC REGION				
TRUST?	Northeast	Midwest	South	West	Total
Yes	88	108	202	93	491
No	88	103	153	152	496
Total	176	211	355	245	987

b. At the 0.05 level of significance, is there evidence of a difference in the proportion of Americans who say they trust advertisements on the basis of geographic region?

CHAPTER 12

▼ CASES

Managing Ashland MultiComm Services
PHASE 1

Reviewing the results of its research, the marketing department team concluded that a segment of Ashland households might be interested in a discounted trial subscription to the AMS *3-For-All* service. The team decided to test various discounts before determining the type of discount to offer during the trial period. It decided to conduct an experiment using three types of discounts plus a plan that offered no discount during the trial period:

1. No discount for the *3-For-All* service. Subscribers would pay $99.99 per month for the *3-For-All* service during the trial period.
2. Moderate discount for the *3-For-All* service. Subscribers would pay $79.99 per month for the *3-For-All* service during the trial period.
3. Substantial discount for the *3-For-All* service. Subscribers would pay $59.99 per month for the *3-For-All* service during the trial period.
4. Discount restaurant card. Subscribers would be given a special card providing a discount of 15% at selected restaurants in Ashland during the trial period.

Each participant in the experiment was randomly assigned to a discount plan. A random sample of 100 subscribers to each plan during the trial period was tracked to determine how many would continue to subscribe to the *3-For-All* service after the trial period. The following table summarizes the results.

CONTINUE SUBSCRIPTIONS AFTER TRIAL PERIOD	DISCOUNT PLANS				
	No Discount	Moderate Discount	Substantial Discount	Restaurant Card	Total
Yes	24	30	38	51	143
No	76	70	62	49	257
Total	100	100	100	100	400

1. Analyze the results of the experiment. Write a report to the team that includes your recommendation for which discount plan to use. Be prepared to discuss the limitations and assumptions of the experiment.

PHASE 2

The marketing department team discussed the results of the survey presented in Chapter 8, on pages 271–272. The team realized that the evaluation of individual questions was providing only limited information. In order to further understand the market for the *3-For-All* service, the data were organized in the following six contingency tables.

HAS AMS SMARTPHONE	HAS AMS INTERNET SERVICE		
	Yes	No	Total
Yes	55	28	83
No	207	128	335
Total	262	156	418

TYPE OF SERVICE	DISCOUNT TRIAL		
	Yes	No	Total
Basic or none	8	156	164
Enhanced	32	222	254
Total	40	378	418

SERVICE	WATCHES PREMIUM CONTENT				
	Almost Every Day	Several Times a Week	Almost Never	Never	Total
Basic or none	2	5	30	127	164
Enhanced	14	35	149	56	254
Total	16	40	179	183	418

DISCOUNT	WATCHES PREMIUM CONTENT				
	Almost Every Day	Several Times a Week	Almost Never	Never	Total
Yes	5	6	16	13	40
No	11	34	163	170	378
Total	16	40	179	183	418

	METHOD FOR CURRENT SUBSCRIPTION					
DISCOUNT	**Email/ Text**	**Toll-Free Number**	**AMS Website**	**In-Store Signup**	**MyTVLab Promo**	**Total**
Yes	5	14	12	4	5	40
No	65	50	224	32	7	378
Total	70	64	236	36	12	418

	METHOD FOR CURRENT SUBSCRIPTION					
GOLD CARD	**Email/ Text**	**Toll-Free Number**	**AMS Website**	**In-Store Signup**	**MyTVLab Promo**	**Total**
Yes	4	12	12	4	6	38
No	66	52	224	32	6	380
Total	70	64	236	36	12	418

2. Analyze the results of the contingency tables. Write a report for the marketing department team, discussing the marketing implications of the results for Ashland Multi-Comm Services.

Digital Case

Apply your knowledge of testing for the difference between two proportions in this Digital Case, which extends the T.C. Resort Properties Using Statistics scenario of this chapter.

As T.C. Resort Properties seeks to improve its customer service, the company faces new competition from SunLow Resorts. SunLow has recently opened resort hotels on the islands where T.C. Resort Properties has its five hotels. SunLow is currently advertising that a random survey of 300 customers revealed that about 60% of the customers preferred its "Concierge Class" travel reward program over the T.C. Resorts "TCRewards Plus" program.

Open and review **ConciergeClass.pdf**, an electronic brochure that describes the Concierge Class program and compares it to the T.C. Resorts program. Then answer the following questions:

1. Are the claims made by SunLow valid?
2. What analyses of the survey data would lead to a more favorable impression about T.C. Resort Properties?
3. Perform one of the analyses identified in your answer to step 2.
4. Review the data about the T.C. Resort Properties customers presented in this chapter. Are there any other questions that you might include in a future survey of travel reward programs? Explain.

Sure Value Convenience Stores

You work in the corporate office for a nationwide convenience store franchise that operates nearly 10,000 stores. The per-store daily customer count (i.e., the mean number of customers in a store in one day) has been steady, at 900, for some time. To increase the customer count, the chain is considering cutting prices for coffee beverages. Management needs to determine how much prices can be cut in order to increase the daily customer count without reducing the gross margin on coffee sales too much. You decide to carry out an experiment in a sample of 24 stores where customer counts have been running almost exactly at the national average of 900. In six of the stores, a small coffee will be $0.59, in another six stores the price will be $0.69, in a third group of six stores, the price will be $0.79, and in a fourth group of six stores, the price will now be $0.89. After four weeks, the daily customer count in the stores is stored in Coffee Sales.

At the 0.05 level of significance, is there evidence of a difference in the median daily customer count based on the price of a small coffee? What price should the stores sell the coffee for?

CardioGood Fitness

Return to the CardioGood Fitness case first presented on page 33. The data for this case are stored in CardioGood Fitness.

1. Determine whether differences exist in the median age in years, education in years, annual household income ($), number of times the customer plans to use the treadmill each week, and the number of miles the customer expects to walk or run each week based on the product purchased (TM195, TM498, TM798).
2. Determine whether differences exist in the relationship status (single or partnered), and the self-rated fitness based on the product purchased (TM195, TM498, TM798).
3. Write a report to be presented to the management of Cardio-Good Fitness, detailing your findings.

More Descriptive Choices Follow-Up

Follow up the "Using Statistics: More Descriptive Choices, Revisited" on page 147 by using the data that are stored in Retirement Funds to:

1. Determine whether there is a difference between the growth and value funds in the median one-year return percentages, five-year return percentages, and ten-year return percentages.
2. Determine whether there is a difference between the small, mid-cap, and large market cap funds in the median one-year return percentages, five-year return percentages, and ten-year return percentages.
3. Determine whether there is a difference in risk based on market cap, a difference in rating based on market cap, a difference in risk based on type of fund, and a difference in rating based on type of fund.
4. Write a report summarizing your findings.

Clear Mountain State Student Survey

The Student News Service at Clear Mountain State University (CMSU) has decided to gather data about the undergraduate students that attend CMSU. It creates and distributes a survey of 14 questions and receives responses from 111 undergraduates, which it stores in Student Survey .

1. Construct contingency tables using gender, major, plans to go to graduate school, and employment status. (You need to construct six tables, taking two variables at a time.) Analyze the data at the 0.05 level of significance to determine whether any significant relationships exist among these variables.

2. At the 0.05 level of significance, is there evidence of a difference between males and females in median grade point average, expected starting salary, number of social networking sites registered for, age, spending on textbooks and supplies, text messages sent in a week, and the wealth needed to feel rich?

3. At the 0.05 level of significance, is there evidence of a difference between students who plan to go to graduate school and those who do not plan to go to graduate school in median grade point average, expected starting salary, number of social networking sites registered for, age, and spending on textbooks and supplies.

▼EXCEL GUIDE

EG12.1 CHI-SQUARE TEST for the DIFFERENCE BETWEEN TWO PROPORTIONS

Key Technique Use the **CHISQ.INV.RT**(*level of significance, degrees of freedom*) function to compute the critical value.

Use the **CHISQ.DIST.RT**(*chi-square test statistic, degrees of freedom*) function to compute the *p*-value.

Example Perform the Figure 12.3 chi-square test for the two-hotel guest satisfaction data on page 394.

PHStat Use **Chi-Square Test for Differences in Two Proportions**.

For the example, select **PHStat → Two-Sample Tests (Summarized Data) → Chi-Square Test for Differences in Two Proportions**. In the procedure's dialog box, enter **0.05** as the **Level of Significance**, enter a **Title**, and click **OK**. In the new worksheet:

1. Read the yellow note about entering values and then press the **Delete** key to delete the note.
2. Enter **Hotel** in cell **B4** and **Choose Again?** in cell **A5**.
3. Enter **Beachcomber** in cell **B5** and **Windsurfer** in cell **C5**.
4. Enter **Yes** in cell **A6** and **No** in cell **A7**.
5. Enter **163**, **64**, **154**, and **108** in cells **B6**, **B7**, **C6**, and **C7**, respectively.

Workbook Use the **COMPUTE worksheet** of the **Chi-Square workbook** as a template.

The worksheet already contains the Table 12.2 two-hotel guest satisfaction data. For other problems, change the **Observed Frequencies** cell counts and row and column labels in rows 4 through 7.

EG12.2 CHI-SQUARE TEST for DIFFERENCES AMONG MORE THAN TWO PROPORTIONS

Key Technique Use the **CHISQ.INV.RT** and **CHISQ.DIST. RT** functions to compute the critical value and the *p*-value, respectively.

Example Perform the Figure 12.6 chi-square test for the three-hotel guest satisfaction data on page 400.

PHStat Use **Chi-Square Test**.

For the example, select **PHStat → Multiple-Sample Tests → Chi-Square Test**. In the procedure's dialog box (shown in right column):

1. Enter **0.05** as the **Level of Significance**.
2. Enter **2** as the **Number of Rows**.
3. Enter **3** as the **Number of Columns**.
4. Enter a **Title** and click **OK**.

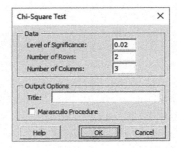

In the new worksheet:

5. Read the yellow note instructions about entering values and then press the **Delete key** to delete the note.
6. Enter the Table 12.8 data (on page 398), including row and column labels, in rows 4 through 7. The **#DIV/0!** error messages will disappear when you finish entering all the table data.

Workbook Use the **ChiSquare2×3 worksheet** of the **Chi-Square Worksheets workbook** as a model.

The worksheet already contains the page 398 Table 12.8 guest satisfaction data. For other 2 × 3 problems, change the **Observed Frequencies** cell counts and row and column labels in rows 4 through 7.

For 2 × 4 problems, use the **ChiSquare2×4 worksheet** and change the **Observed Frequencies** cell counts and row and column labels in that worksheet. For 2 × 5 problems, use the **ChiSquare2×5 worksheet** and change the **Observed Frequencies** cell counts and row and column labels in that worksheet.

The Marascuilo Procedure

Key Technique Use formulas to compute the absolute differences and the critical range.

Example Perform the Figure 12.7 Marascuilo procedure for the guest satisfaction survey on page 401.

PHStat Modify the *PHStat* instructions of the previous section. In step 4, check **Marascuilo Procedure** in addition to entering a **Title** and clicking **OK**.

Workbook Use the **Marascuilo2×3** of the **Chi-Square Worksheets workbook** as a template.

The worksheet requires no entries or changes to use. For 2 × 4 problems, use the **Marascuilo2×4 worksheet** and for 2 × 5 problems, use the **Marascuilo2×5 worksheet**.

Every Marascuilo worksheet uses values from the observed frequencies table in the companion ChiSquare worksheet to compute critical range values in the Marascuilo table area (rows 11 through 16 in Figure 12.7). In column D, the worksheet uses an IF function to compare the absolute difference to the critical range for each pair of groups and then displays either "Significant" or "Not Significant."

EG12.3 CHI-SQUARE TEST of INDEPENDENCE

Key Technique Use the **CHISQ.INV.RT** and **CHISQ. DIST.RT** functions to compute the critical value and the *p*-value, respectively.

Example Perform the Figure 12.10 chi-square test for the primary reason for not returning to hotel data on page 407.

PHStat Use **Chi-Square Test**.

For the example, select **PHStat→Multiple-Sample Tests→ Chi-Square Test**. In the procedure's dialog box (shown below):

1. Enter **0.05** as the **Level of Significance**.
2. Enter **4** as the **Number of Rows**.
3. Enter **3** as the **Number of Columns**.
4. Enter a **Title** and click **OK**.

In the new worksheet:

5. Read the yellow note about entering values and then press the **Delete key** to delete the note.
6. Enter the Table 12.13 data on page 404, including row and column labels, in rows 4 through 9. The **#DIV/0!** error messages will disappear when you finish entering all of the table data.

Workbook Use the **ChiSquare4×3 worksheet** of the **Chi-Square Worksheets workbook** as a model.

The worksheet already contains the page 404 Table 12.13 primary reason for not returning to hotel data. For other 4 × 3 problems, change the **Observed Frequencies** cell counts and row and column labels in rows 4 through 9.

For problems that use an *r* × *c* contingency table of a different size, use the appropriate ChiSquare worksheet. For example, for 3 × 4 problems, use the **ChiSquare3×4 worksheet** and for 4 × 3 problems, use the **ChiSquare4×3 worksheet**.

For each of these other worksheets, enter the contingency table data for the problem in the Observed Frequencies area.

EG12.4 WILCOXON RANK SUM TEST: A NONPARAMETRIC METHOD for TWO INDEPENDENT POPULATIONS

Key Technique Use the **NORM.S.INV**(*level of significance*) function to compute the upper and lower critical values and use **NORM.S.DIST**(*absolute value of the Z test statistic*) as part of a formula to compute the *p*-value.

For unsummarized data, use the **COUNTIF** and **SUMIF** functions (see Appendix Section F.2) to compute the sample size and the sum of ranks for each sample.

Example Perform the Figure 12.12 Wilcoxon rank sum test for the VLABGo player monthly sales for two in-store locations.

PHStat Use **Wilcoxon Rank Sum Test**.

For the example, open to the **DATA worksheet** of the **VLABGo workbook**. Select **PHStat→Two-Sample Tests (Unsummarized Data)→Wilcoxon Rank Sum Test**. In the procedure's dialog box (shown below):

1. Enter **0.05** as the **Level of Significance**.
2. Enter **A1:A11** as the **Population 1 Sample Cell Range**.
3. Enter **B1:B11** as the **Population 2 Sample Cell Range**.
4. Check **First cells in both ranges contain label**.
5. Click **Two-Tail Test**.
6. Enter a **Title** and click **OK**.

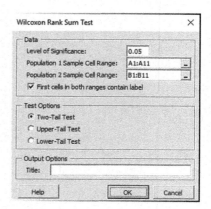

The procedure creates a SortedRanks worksheet that contains the sorted ranks in addition to the worksheet that Figure 12.12 presents. Both of these worksheets are discussed in the following *Workbook* instructions.

Workbook Use the **COMPUTE worksheet** of the **Wilcoxon workbook** as a template.

The worksheet already contains data and formulas to use the unsummarized data for the example. For other problems that use unsummarized data:

1. Open to the **SortedRanks worksheet**.
2. Enter the sorted values for both groups in stacked format, entering sample names in column A and sorted values in column B.
3. Assign a rank for each value and enter the ranks in column C.
4. If performing a two-tail test, open to the **COMPUTE worksheet**, otherwise open to the similar **COMPUTE_ALL worksheet** that includes the one-tail tests.
5. Edit the cell ranges in the formulas in cells **B7**, **B8**, **B10**, and **B11** to match the cell range of the new data.

For problems with summarized data, overwrite the formulas that compute the **Sample Size** and **Sum of Ranks** in the cell range **B7:B11**, with the values for these statistics.

EG12.5 KRUSKAL-WALLIS RANK TEST: A NONPARAMETRIC METHOD for the ONE-WAY ANOVA

Key Technique Use the **CHISQ.INV.RT**(*level of significance, number of groups:* **1**) function to compute the critical value and use the **CHISQ.DIST.RT**(*H test statistic, number of groups:* **1**) function to compute the *p*-value.

For unsummarized data, use the **COUNTIF** and **SUMIF** functions (see Appendix Section F.2) to compute the sample size and the sum of ranks for each sample.

Example Perform the Figure 12.14 Kruskal-Wallis rank test for differences among the median mobile electronics sales for four in-store locations on page 417.

PHStat Use **Kruskal-Wallis Rank Test**.

For the example, open to the **DATA worksheet** of the **Mobile Electronics workbook**. Select **PHStat → Multiple-Sample Tests → Kruskal-Wallis Rank Test**. In the procedure's dialog box (shown in right column):

1. Enter **0.05** as the **Level of Significance**.
2. Enter **A1:D6** as the **Sample Data Cell Range**.

3. Check **First cells contain label**.
4. Enter a **Title** and click **OK**.

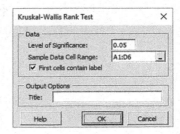

The procedure creates a SortedRanks worksheet that contains sorted ranks in addition to the worksheet shown in Figure 12.14 on page 417. Both of these worksheets are discussed in the following *Workbook* instructions.

Workbook Use the **KruskalWallis4 worksheet** of the **Kruskal-Wallis Worksheets workbook** as a template.

The worksheet already contains the data and formulas to use the unsummarized data for the example. For other problems with four groups and unsummarized data:

1. Open to the **SortedRanks worksheet**.
2. Enter the sorted values for both groups in stacked format, entering sample names in column A and sorted values in column B.
3. Assign a rank for each value and enter the ranks in column C.
4. Paste the unsummarized stacked data into this worksheet starting with Column E. (The first row of this pasted range should contain group names.)
5. Open to the **KruskalWallis4 worksheet** and edit the cell ranges in the formulas in columns **E** and **F**.

For other problems with four groups and summarized data, open to the **KruskalWallis4 worksheet** and overwrite the columns D, E, and F formulas with the name, sample size, and sum of ranks for each group.

For other problems with three groups, use the similar **KruskalWallis3 worksheet** and use the previous instructions for either unsummarized or summarized data, as appropriate.

13 Simple Linear Regression

OBJECTIVES

- Understand the meaning of the regression coefficients b_0 and b_1
- Understand the usefulness of regression analysis and how to properly perform this analysis.

▼USING STATISTICS
Knowing Customers at Sunflowers Apparel

Having survived recent economic slowdowns that have diminished their competitors, Sunflowers Apparel, a chain of upscale fashion stores for women, is in the midst of a companywide review that includes researching the factors that make their stores successful. Until recently, Sunflowers managers did not use data analysis to help select where to open stores, relying instead on subjective factors, such as the availability of an inexpensive lease or the perception that a particular location seemed ideal for one of their stores.

As the new director of planning, you have already consulted with marketing analytics firms that specialize in identifying and classifying groups of consumers. Based on such preliminary analyses, you have tentatively discovered that the profile of Sunflowers shoppers may not only be the upper middle class long suspected of being the chain's clientele but may also include younger, aspirational families with young children, and, surprisingly, urban hipsters who set trends and are mostly single.

You seek to develop a systematic approach that will lead to making better decisions during the site-selection process. As a starting point, you have asked one marketing analytics firm to collect and organize data for the number of people in the identified groups of interest who live within a fixed radius of each store. You believe that the greater numbers of profiled customers contribute to store sales, and you want to explore the possible use of this relationship in the decision-making process. How can you use statistics so that you can forecast the annual sales of a proposed store based on the number of profiled customers who reside within a fixed radius of a Sunflowers store?

430

The preceding four chapters focus on hypothesis testing methods. Chapter 9 discusses methods that allow you to make inferences about a population parameter. Chapters 10–12 present methods that look for differences among two or more populations. Beginning with this chapter, and continuing through Chapter 15, the focus shifts from examining differences among groups to predicting values of variables of interest.

Consider the data that a business generates as a by-product of ongoing operations, such as the Sunflowers Apparel sales data. To examine such data, one might look for possible relationships among variables. **Regression analysis** techniques help uncover such relationships.

Regression methods seek to discover how one or more X variables can predict the value of a Y variable. The Y variable is known as the **dependent variable** because its values depend on the X values in a regression model. X variables are also known as predictor variables or **independent variables**, in contrast to the dependent Y variable.[1]

[1]Independent variables are also known as **predictor** or **explanatory variables** and dependent variables are also known as **response variables**.

Regression methods first fit a **model** that describes the relationship between the X and Y variables and then evaluates the *goodness of fit*, how well the model describes the relationship. Decision makers then evaluate whether the mathematical assumptions that a model requires are valid for the data being analyzed. Should the assumptions hold, the regression model can then be used to make predictions about the Y variable for a given range of X values. Decision makers also use regression methods to help define or refine other models or to estimate values to be used in a model. Models also help one identify unusual values that may be outliers (see Hoaglin, Hocking, and Kutner).

Simple regression explores the relationship between one independent X variable and the dependent Y variable. *Multiple* regression, the subject of Chapters 14 and 15 explores the relationship between two or more independent X variables and the dependent Y variable. As the least complicated regression method to study, **simple linear regression** provides a good starting point for exploring and understanding regression, an important statistical technique that one often uses to understand data better and to help explore results of business analytics models.

Preliminary Analysis

Using a **scatter plot** (also known as **scatter diagram**) to visualize the X and Y variables, a technique that Section 2.5 discusses, can help suggest a starting point for regression analysis. The Figure 13.1 scatter plots illustrate six possible relationships between an X variable and a Y variable.

FIGURE 13.1
Six types of relationships found in scatter plots

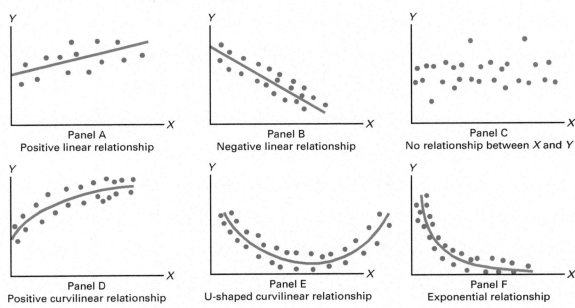

Panel A
Positive linear relationship

Panel B
Negative linear relationship

Panel C
No relationship between X and Y

Panel D
Positive curvilinear relationship

Panel E
U-shaped curvilinear relationship

Panel F
Exponential relationship

In Panel A, values of Y are generally increasing linearly as X increases. Figure 13.3 on page 433 shows another positive linear relationship, between the number of profiled customers of the store and the store's annual sales for the Sunflowers Apparel women's clothing store chain.

Panel B illustrates a negative linear relationship. As X increases, the values of Y are generally decreasing. An example of this type of relationship might be the price of a particular product and the amount of sales. As the price charged for the product increases, the amount of sales may tend to decrease.

Panel C shows a set of data in which there is very little or no relationship between X and Y. High and low values of Y appear at each value of X.

Panel D illustrates a positive curvilinear relationship between X and Y. The values of Y increase as X increases, but this increase tapers off beyond certain values of X. An example of a positive curvilinear relationship might be the age and maintenance cost of an automobile. As an automobile gets older, the maintenance cost may rise rapidly at first but then level off beyond a certain number of years.

Panel E illustrates a U-shaped relationship between X and Y. As X increases, at first Y generally decreases, but as X continues to increase, Y then increases above its minimum value. An example of this type of relationship might be entrepreneurial activity and levels of economic development as measured by GDP per capita. Entrepreneurial activity occurs more in the least and most developed countries.

Panel F illustrates an exponential relationship between X and Y. In this case, Y decreases very rapidly as X first increases, but then it decreases much less rapidly as X continues to increase. An example of an exponential relationship could be the value of an automobile and its age. The value drops drastically from its original price in the first year, but it decreases much less rapidly in subsequent years.

13.1 Simple Linear Regression Models

Simple linear regression models examine the straight line (*linear*) relationship between a dependent Y variable and a single independent X variable. Figure 13.2 presents a generalized **positive linear relationship** that has a positive slope.

FIGURE 13.2
Generalized positive linear relationship

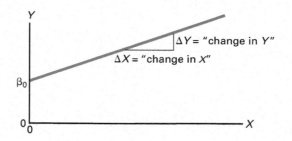

Equation (13.1) defines the simple linear regression model that expresses the relationship mathematically.

SIMPLE LINEAR REGRESSION MODEL

$$Y_i = \beta_0 + \beta_1 X_i + \varepsilon_i \qquad (13.1)$$

where

$\beta_0 = Y$ intercept for the population

$\beta_1 = $ slope for the population

$\varepsilon_i = $ random error in Y for observation i

$Y_i = $ dependent variable for observation i

$X_i = $ independent variable for observation i

The $Y_i = \beta_0 + \beta_1 X_i$ portion of the simple linear regression model expressed in Equation (13.1) is a straight line. The **slope** of the line, β_1, represents the expected change in Y per unit change in X. It represents the mean amount that Y changes (either positively or negatively) for a one-unit change in X. The **Y intercept**, β_0, represents the mean value of Y when X equals 0. The last component of the model, ε_i, represents the random error in Y for each observation, i. In other words, ε_i is the vertical distance of the actual value of Y_i above or below the expected value of Y_i on the line.

13.2 Determining the Simple Linear Regression Equation

As the new director of planning in the Sunflowers Apparel scenario, you suspect that the greater the number of profiled customers who reside within a fixed radius of a store, the greater the store sales will be. You wonder if a linear relationship between the number of profiled customers, as the numerical independent X variable, and annual store sales, as the dependent Y variable, exists. To examine this relationship, you collect data from a sample of 14 stores. Table 13.1, stored in Site Selection , presents these data.

TABLE 13.1

Number of profiled customers (in millions) and annual sales (in $millions) for a sample of 14 Sunflowers Apparel stores

Store	Profiled Customers (millions)	Annual Sales ($millions)	Store	Profiled Customers (millions)	Annual Sales ($millions)
1	3.7	5.7	8	3.1	4.7
2	3.6	5.9	9	3.2	6.1
3	2.8	6.7	10	3.5	4.9
4	5.6	9.5	11	5.2	10.7
5	3.3	5.4	12	4.6	7.6
6	2.2	3.5	13	5.8	11.8
7	3.3	6.2	14	3.0	4.1

Figure 13.3 displays the scatter plot for the data in Table 13.1. Observe the increasing relationship between profiled customers (X) and annual sales (Y). As the number of profiled customers increases, annual sales increase approximately as a straight line. Thus, you can assume that a straight line provides a useful mathematical model of this relationship. Now you need to determine the specific straight line that is the *best* fit to these data.

FIGURE 13.3

Scatter plot for the Sunflowers Apparel data

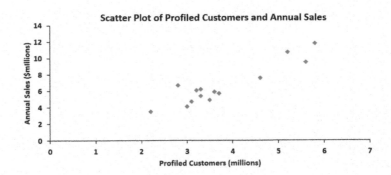

The Least-Squares Method

As the Sunflowers director of planning, you have hypothesized a statistical model to represent the relationship between two variables—number of profiled customers and sales—in the entire population of Sunflowers Apparel stores. However, as Table 13.1 shows, the data are collected from a *random sample* of stores. If certain assumptions are valid (see Section 13.4), you can use the sample Y intercept, b_0, and the sample slope, b_1, as estimates of the respective population parameters, β_0 and β_1. Equation (13.2) uses these estimates to form the **simple linear regression equation**. This straight line is often referred to as the **prediction line**.

student TIP

In mathematics, the symbol b is often used for the Y intercept instead of b_0, and the symbol m is often used for the slope instead of b_1.

SIMPLE LINEAR REGRESSION EQUATION: THE PREDICTION LINE

The predicted value of Y equals the Y intercept plus the slope multiplied by the value of X.

$$\hat{Y}_i = b_0 + b_1 X_i \tag{13.2}$$

where

$\hat{Y}_i$ = predicted value of Y for observation i

X_i = value of X for observation i

b_0 = sample Y intercept

b_1 = sample slope

Equation (13.2) requires establishing values for two **regression coefficients**—b_0 (the sample Y intercept) and b_1 (the sample slope). The most common approach to finding b_0 and b_1 is using the least-squares method. This method minimizes the sum of the squared differences between the actual values (Y_i) and the predicted values ($\hat{Y}_i$), using the simple linear regression equation [i.e., the prediction line; see Equation (13.2)]. This sum of squared differences is equal to

$$\sum_{i=1}^{n} (Y_i - \hat{Y}_i)^2$$

Because $\hat{Y}_i = b_0 + b_1 X_i$,

$$\sum_{i=1}^{n} (Y_i - \hat{Y}_i)^2 = \sum_{i=1}^{n} [Y_i - (b_0 + b_1 X_i)]^2$$

student TIP

A positive slope means that as X increases, Y is predicted to increase. A negative slope means that as X increases, Y is predicted to decrease.

Because this equation has two unknowns, b_0 and b_1, the sum of squared differences depends on the sample Y intercept, b_0, and the sample slope, b_1. The **least-squares method** determines the values of b_0 and b_1 that minimize the sum of squared differences around the prediction line. Any values for b_0 and b_1 other than those determined by the least-squares method result in a greater sum of squared differences between the actual values (Y_i) and the predicted values ($\hat{Y}_i$).

Figure 13.4 on page 435 presents results for the simple linear regression model for the Sunflowers Apparel data. Excel labels b_0 as Intercept and labels b_1 as Profiled Customers. In Figure 13.4, observe that $b_0 = -1.2088$ and $b_1 = 2.0742$. Using Equation (13.2), the prediction line for these data is

student TIP

Coefficients computed by handheld calculators may differ slightly from coefficients computed by software because of rounding errors.

$$\hat{Y}_i = -1.2088 + 2.0742 X_i$$

The slope, b_1, is $+2.0742$. This means that for each increase of 1 unit in X, the predicted mean value of Y is estimated to increase by 2.0742 units. In other words, for each increase of 1.0 million profiled customers within a fixed radius of the store, the predicted mean annual sales are estimated to increase by \$2.0742 million. Thus, the slope represents the portion of the annual sales that are estimated to vary according to the number of profiled customers.

FIGURE 13.4

Excel simple linear regression model worksheet results for the Sunflowers Apparel data

	A	B	C	D	E	F	G
1	Simple Linear Regression						
2							
3	*Regression Statistics*						
4	Multiple R	0.9208					
5	R Square	0.8479					
6	Adjusted R Square	0.8352					
7	Standard Error	0.9993					
8	Observations	14					
9							
10	ANOVA						
11		*df*	*SS*	*MS*	*F*	*Significance F*	
12	Regression	1	66.7854	66.7854	66.8792	0.0000	
13	Residual	12	11.9832	0.9986			
14	Total	13	78.7686				
15							
16		*Coefficients*	*Standard Error*	*t Stat*	*P-value*	*Lower 95%*	*Upper 95%*
17	Intercept	-1.2088	0.9949	-1.2151	0.2477	-3.3765	0.9588
18	Profiled Customers	2.0742	0.2536	8.1780	0.0000	1.5216	2.6268

In the Worksheet

The Section EG13.2 *Worksheet*, *PHStat*, and *Analysis ToolPak* instructions each create a regression results worksheet that looks like this worksheet. Using the *Worksheet* or *PHStat* instructions inserts formulas in columns L and M (shown below left) that use the LINEST function to compute regression coefficients and statistics.

The **COMPUTE worksheet** of the Excel Guide **Simple Linear Regression** is a copy of the Figure 13.4 worksheet.

	K	L	M
1	Intermediate Calculations		
2	b1, b0 Coefficients	2.0742	-1.2088
3	b1, b0 Standard Error	0.2536	0.9949
4	R Square, Standard Error	0.8479	0.9993
5	F, Residual df	66.8792	12.0000
6	Regression SS, Residual SS	66.7854	11.9832
7			
8	Confidence level	95%	
9	t Critical Value	2.1788	
10	Half Width b0	2.1676	
11	Half Width b1	0.5526	

Tableau displays similar information in the Describe Trend Model dialog box, as Section TG13.2 explains.

The Y intercept, b_0, is -1.2088. The Y intercept represents the predicted value of Y when X equals 0. Because the number of profiled customers of the store cannot be 0, this Y intercept has little or no practical interpretation. Also, the Y intercept for this example is outside the range of the observed values of the X variable, and therefore interpretations of the value of b_0 should be made cautiously. Figure 13.5 displays an Excel and Tableau scatter plot with the prediction line for Sunflowers Apparel data example. Note that Tableau shows the prediction line wrongly extending past the range of the dependent X variable, profiled customers.

FIGURE 13.5

Excel and Tableau scatter plot and prediction line for Sunflowers Apparel data

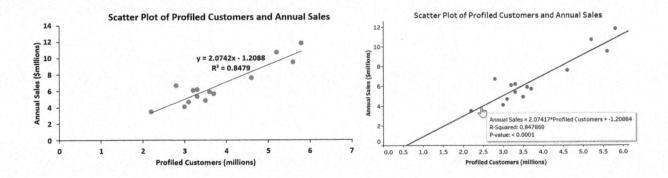

Example 13.1 on page 436 presents a problem in which a direct interpretation for the Y intercept exists. Example 13.2, that immediately follows Example 13.1, illustrates the use of a prediction line.

EXAMPLE 13.1

Interpreting the Y Intercept, b_0, and the Slope, b_1

A statistics professor wants to use the number of absences from class during the semester (X) to predict the final exam score (Y). A regression model is fit based on data collected from a class during a recent semester, with the following results:

$$\hat{Y}_i = 85.0 - 5X_i$$

What is the interpretation of the Y intercept, b_0, and the slope, b_1?

SOLUTION The Y intercept $b_0 = 85.0$ indicates that when the student does not have any absences from class during the semester, the predicted mean final exam score is 85.0. The slope $b_1 = -5$ indicates that for each increase of one absence from class during the semester, the predicted change in the mean final exam score is -5.0. In other words, the final exam score is predicted to decrease by a mean of 5 points for each increase of one absence from class during the semester.

EXAMPLE 13.2

Predicting Annual Sales Based on Number of Profiled Customers

Use the prediction line for the Sunflowers Apparel data to predict the annual sales for a store with 4 million profiled customers.

SOLUTION You can determine the predicted value of annual sales by substituting $X = 4$ (millions of profiled customers) into the simple linear regression equation:

$$\hat{Y}_i = -1.2088 + 2.0742X_i$$

$$\hat{Y}_i = -1.2088 + 2.0742(4) = 7.0879 \text{ or } \$7,087,900$$

Thus, a store with 4 million profiled customers has predicted mean annual sales of \$7,087,900.

Predictions in Regression Analysis: Interpolation Versus Extrapolation

One uses only the **relevant range** of the independent variable to make predictions. This relevant range represents all values from the smallest to the largest X used in developing the regression model. Hence, when predicting Y for a given value of X, one cannot extrapolate beyond this range of X values.

For example, in the Sunflowers Apparel scenario the number of profiled customers to predict annual sales varies from 2.2 to 5.8 million (see Table 13.1 on page 438). Therefore, you should predict annual sales *only* for stores that have between 2.2 and 5.8 million profiled customers. Any prediction of annual sales for stores outside this range wrongly assumes that the observed relationship between sales and the number of profiled customers for stores that have between 2.2 and 5.8 million profiled customers would be the same. For example, it would be improper to use the prediction line to forecast the sales for a new store that has 8 million profiled customers. The relationship between sales and the number of profiled customers might, for example, have a point of diminishing returns. If that was true, the effect that the number of profiled customers has on sales would be less, leading to an overestimation of the predicted sales.

Calculating the Slope, b_1, and the Y Intercept, b_0

For small data sets, using a handheld calculator can be a practical way of calculating the least-squares simple linear regression coefficients. Equations (13.3) and (13.4) present formulas for the b_1 and b_0 coefficients that minimize the complexity of operations.

$$\sum_{i=1}^{n} (Y_i - \hat{Y}_i)^2 = \sum_{i=1}^{n} [Y_i - (b_0 + b_1 X_i)]^2$$

COMPUTATIONAL FORMULA FOR THE SLOPE, b_1

$$b_1 = \frac{SSXY}{SSX} \tag{13.3}$$

where

$$SSXY = \sum_{i=1}^{n} (X_i - \overline{X})(Y_i - \overline{Y}) = \sum_{i=1}^{n} X_i Y_i - \frac{\left(\sum_{i=1}^{n} X_i\right)\left(\sum_{i=1}^{n} Y_i\right)}{n}$$

$$SSX = \sum_{i=1}^{n} (X_i - \overline{X})^2 = \sum_{i=1}^{n} X_i^2 - \frac{\left(\sum_{i=1}^{n} X_i\right)^2}{n}$$

COMPUTATIONAL FORMULA FOR THE Y INTERCEPT, b_0

$$b_0 = \overline{Y} - b_1 \overline{X} \tag{13.4}$$

where

$$\overline{Y} = \frac{\sum_{i=1}^{n} Y_i}{n}$$

$$\overline{X} = \frac{\sum_{i=1}^{n} X_i}{n}$$

EXAMPLE 13.3

Calculating the Slope, b_1, and the Y Intercept, b_0

▶(*continued*)

Calculate the slope, b_1, and the Y intercept, b_0, for the Sunflowers Apparel data.

SOLUTION Use Equations (13.3) and (13.4) with the sample size $n = 14$ and calculate: $\sum_{i=1}^{n} X_i$, the sum of the Profiled Customers X values; $\sum_{i=1}^{n} Y_i$, the sum of the Annual Sales Y values; $\sum_{i=1}^{n} X_i^2$, the sum of the squared X values; and $\sum_{i=1}^{n} X_i Y_i$, the sum of the product of X and Y. Table 13.2 on page 438 presents calculations necessary to determine these four quantities for the Sunflowers Apparel example. The table also includes $\sum_{i=1}^{n} Y_i^2$, the sum of the squared Y values that Section 13.3 uses to calculate SST.

TABLE 13.2
Calculations for the Sunflowers Apparel data

Store	X	Y	X^2	Y^2	XY
1	3.7	5.7	13.69	32.49	21.09
2	3.6	5.9	12.96	34.81	21.24
3	2.8	6.7	7.84	44.89	18.76
4	5.6	9.5	31.36	90.25	53.20
5	3.3	5.4	10.89	29.16	17.82
6	2.2	3.5	4.84	12.25	7.70
7	3.3	6.2	10.89	38.44	20.46
8	3.1	4.7	9.61	22.09	14.57
9	3.2	6.1	10.24	37.21	19.52
10	3.5	4.9	12.25	24.01	17.15
11	5.2	10.7	27.04	114.49	55.64
12	4.6	7.6	21.16	57.76	34.96
13	5.8	11.8	33.64	139.24	68.44
14	3.0	4.1	9.00	16.81	12.30
Totals	52.9	92.8	215.41	693.90	382.85

studentTIP

Although examples in this chapter show the manual evaluation of formulas to provide insight, best practice is to use software for all regression-related calculations.

Using Equations (13.3) and (13.4), calculate b_0 and b_1:

$$SSXY = \sum_{i=1}^{n}(X_i - \overline{X})(Y_i - \overline{Y}) = \sum_{i=1}^{n}X_iY_i - \frac{\left(\sum_{i=1}^{n}X_i\right)\left(\sum_{i=1}^{n}Y_i\right)}{n}$$

$$= 382.85 - \frac{(52.9)(92.8)}{14} = 382.85 - 350.65142$$

$$= 32.19858$$

$$SSX = \sum_{i=1}^{n}(X_i - \overline{X})^2 = \sum_{i=1}^{n}X_i^2 - \frac{\left(\sum_{i=1}^{n}X_i\right)^2}{n}$$

$$= 215.41 - \frac{(52.9)^2}{14} = 215.41 - 199.88642$$

$$= 15.52358$$

With these values, calculate b_1:

$$b_1 = \frac{SSXY}{SSX} = \frac{32.19858}{15.52358} = 2.07417$$

and:

$$\overline{Y} = \frac{\sum_{i=1}^{n}Y_i}{n} = \frac{92.8}{14} = 6.62857$$

$$\overline{X} = \frac{\sum_{i=1}^{n}X_i}{n} = \frac{52.9}{14} = 3.77857$$

calculate b_0:

$$b_0 = \overline{Y} - b_1\overline{X} = 6.62857 - 2.07417(3.77857) = -1.2088265$$

VISUAL EXPLORATIONS

Exploring Simple Linear Regression Coefficients

Open the **VE-Simple Linear Regression add-in workbook** to explore the coefficients. (For Excel technical requirements, see Appendix D.) When this workbook opens properly, it adds a **Simple Linear Regression** menu in either the Add-ins tab (Microsoft Windows) or the Apple menu bar (OS X).

To explore the effects of changing the regression coefficients, select **Simple Linear Regression→Explore Coefficients**. In the Explore Coefficients floating control panel (inset below), click the spinner buttons for b_1 **slope**

(the slope of the prediction line) and b_0 **intercept** (the Y intercept of the prediction line) to change the prediction line. Using the visual feedback of the chart, try to create a prediction line that is as close as possible to the prediction line defined by the least-squares estimates. In other words, try to make the **Difference from Target SSE** value as small as possible. (Section 13.3 defines *SSE*.)

At any time, click **Reset** to reset the b_1 and b_0 values or **Solution** to reveal the prediction line defined by the least-squares method. Click **Finish** when you are finished with this exercise.

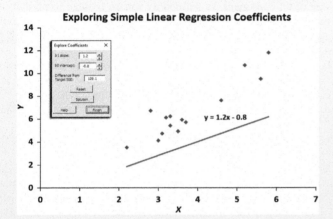

Using Your Own Regression Data

Open to the worksheet that contains your data and uses row 1 for variable names. Select **Simple Linear Regression using your worksheet data** from the **Simple Linear Regression** menu. In the procedure's dialog box, enter the cell range of your Y variable as the **Y Variable Cell Range** and the cell range of your X variable as the **X Variable Cell Range**. Click **First cells in both ranges contain a label**, enter a **Title**, and click **OK**. After the scatter plot appears, continue with the Explore Coefficients floating control panel.

PROBLEMS FOR SECTION 13.2

LEARNING THE BASICS

13.1 Fitting a straight line to a set of data yields the following prediction line:

$$\hat{Y}_i = 2 + 5X_i$$

a. Interpret the meaning of the Y intercept, b_0.
b. Interpret the meaning of the slope, b_1.
c. Predict the value of Y for $X = 3$.

13.2 If the values of X in Problem 13.1 range from 2 to 25, should you use this model to predict the mean value of Y when X equals
a. 3?   **b.** −3?   **c.** 0?   **d.** 24?

13.3 Fitting a straight line to a set of data yields the following prediction line:

$$\hat{Y}_i = 16 - 0.5X_i$$

a. Interpret the meaning of the Y intercept, b_0.
b. Interpret the meaning of the slope, b_1.
c. Predict the value of Y for $X = 6$.

APPLYING THE CONCEPTS

 13.4 The production of wine is a multibillion-dollar worldwide industry. In an attempt to develop a model of

wine quality as judged by wine experts, data were collected from red wine variants of Portuguese "Vinho Verde" wine.

Source: Data extracted from Cortez, P., Cerdeira, A., Almeida, F., Matos, T., and Reis, J., "Modeling Wine Preferences by Data Mining from Physiochemical Properties," *Decision Support Systems*, 47, 2009, pp. 547–553 and **bit.ly/9xKIEa**.

A sample of 50 wines is stored in Vinho Verde . Develop a simple linear regression model to predict wine quality, measured on a scale from 0 (very bad) to 10 (excellent), based on alcohol content (%).
a. Construct a scatter plot.
For these data, $b_0 = -0.3529$ and $b_1 = 0.5624$.
b. Interpret the meaning of the slope, b_1, in this problem.
c. Predict the mean wine quality for wines with a 10% alcohol content.
d. What conclusion can you reach based on the results of (a)–(c)?

13.5 Zagat's publishes restaurant ratings for various locations in the United States. The file Restaurants contains the Zagat rating for food, décor, service, and the cost per person for a sample of 100 restaurants located in the center of New York City and in an outlying area of New York City. Develop a regression model to predict the cost per person, based on a variable that represents the sum of the ratings for food, décor, and service.

Source: Extracted from *Zagat Survey 2016, New York City*

a. Construct a scatter plot.

b. Assuming a linear relationship, use the least-squares method to compute the regression coefficients b_0 and b_1.

c. Interpret the meaning of the Y intercept, b_0, and the slope, b_1, in this problem.

d. Predict the mean cost per person for a restaurant with a summated rating of 50.

e. What should you tell the owner of a group of restaurants in this geographical area about the relationship between the summated rating and the cost of a meal?

13.6 Is an MBA a golden ticket? Pursuing an MBA is a major personal investment. Tuition and expenses associated with business school programs are costly, but the high costs come with hopes of career advancement and high salaries. A prospective MBA student would like to examine the factors that impact starting salary upon graduation and decides to develop a model that uses program per-year tuition as a predictor of starting salary. Data were collected for 37 full-time MBA programs offered at private universities. The data are stored in **FTMBA** .

Source: Data extracted from "U.S. News Business School Compass," available at **premium.usnews.com/best-graduate-schools/top-business-schools/mba-rankings**.

a. Construct a scatter plot.

b. Assuming a linear relationship, use the least-squares method to determine the regression coefficients b_0 and b_1.

c. Interpret the meaning of the slope, b_1, in this problem.

d. Predict the mean starting salary upon graduation for a program that has a per-year tuition cost of $50,450.

e. What insights do you gain about the relationship between program per-year tuition and starting salary upon graduation?

13.7 Starbucks Coffee Co. uses a data-based approach to improve the quality and customer satisfaction of its products. When survey data indicated that Starbucks needed to improve its package-sealing process, an experiment was conducted to determine the factors in the bag-sealing equipment that might be affecting the ease of opening the bag without tearing the inner liner of the bag.

Source: Data extracted from L. Johnson and S. Burrows, "For Starbucks, It's in the Bag," *Quality Progress*, March 2011, pp. 17–23.

One factor that could affect the rating of the ability of the bag to resist tears was the plate gap on the bag-sealing equipment. Data were collected on 19 bags in which the plate gap was varied. The results are stored in **Starbucks** .

a. Construct a scatter plot.

b. Assuming a linear relationship, use the least-squares method to determine the regression coefficients b_0 and b_1.

c. Interpret the meaning of the slope, b_1, in this problem.

d. Predict the mean tear rating when the plate gap is equal to 0.

e. What should you tell management of Starbucks about the relationship between the plate gap and the tear rating?

13.8 The value of a sports franchise is directly related to the amount of revenue that a franchise can generate. The file **MLB Values** represents the current value in a recent year (in $millions) and the annual revenue (in $millions) for the 30 MLB baseball teams.

Source: Data extracted from **www.forbes.com/mlb-valuations/list**.

Suppose you want to develop a simple linear regression model to predict current value based on annual revenue generated.

a. Construct a scatter plot.

b. Use the least-squares method to determine the regression coefficients b_0 and b_1.

c. Interpret the meaning of b_0 and b_1 in this problem.

d. Predict the mean current value of a baseball team that generates $250 million of annual revenue.

e. What would you tell a group considering an investment in a MLB baseball team about the relationship between revenue and the current value of a team?

13.9 An agent for a residential real estate company in a suburb located outside of Washington, DC, has the business objective of developing more accurate estimates of the monthly rental cost for apartments. Toward that goal, the agent would like to use the size of an apartment, as defined by square footage to predict the monthly rental cost. The agent selects a sample of 57 one-bedroom apartments and collects and stores the data in **Silver Spring Rentals** .

a. Construct a scatter plot.

b. Use the least-squares method to determine the regression coefficients b_0 and b_1.

c. Interpret the meaning of b_0 and b_1 in this problem.

d. Predict the mean monthly rent for an apartment that has 800 square feet.

e. Why would it not be appropriate to use the model to predict the monthly rent for apartments that have 1,500 square feet?

f. Your friends Jim and Jennifer are considering signing a lease for a one-bedroom apartment in this residential neighborhood. They are trying to decide between two apartments, one with 800 square feet for a monthly rent of $1,130 and the other with 830 square feet for a monthly rent of $1,410. Based on (a) through (d), which apartment do you think is a better deal?

13.10 A box office analyst seeks to predict opening weekend box office gross for movies. Toward this goal, the analyst plans to use YouTube trailer views as a predictor. For each of 66 movies, the YouTube trailer view count, the number of YouTube trailer views from the release of the trailer through the Saturday before a movie opens, and the opening weekend box office gross (in $millions) are collected and stored in **Movie** .

Source: Data extracted from "Box Office Report," available at **bit.ly/2srM34F**.

For these data,

a. Construct a scatter plot.

b. Assuming a linear relationship, use the least-squares method to determine the regression coefficients b_0 and b_1.

c. Interpret the meaning of the slope, b_1, in this problem.

d. Predict the mean weekend box office gross for a movie that had 20 million YouTube trailer views.

e. What conclusions can you reach about predicting weekend box office gross from YouTube trailer views?

13.3 Measures of Variation

When using the least-squares method to determine the regression coefficients one needs to compute three measures of variation. The first measure, the **total sum of squares** (*SST*), is a measure of variation of the Y_i values around their mean, $\overline{Y}$. The **total variation**, or total sum of squares, is subdivided into **explained variation** and **unexplained variation**. The explained variation, or **regression sum of squares** (*SSR*), represents variation that is explained by the relationship between X and Y, and the unexplained variation, or **error sum of squares** (*SSE*), represents variation due to factors other than the relationship between X and Y. Figure 13.6 shows the different measures of variation for a single Y_i value.

FIGURE 13.6
Measures of variation

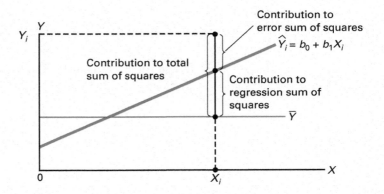

Computing the Sum of Squares

The regression sum of squares (*SSR*) is based on the difference between $\hat{Y}_i$ (the predicted value of Y from the prediction line) and $\overline{Y}$ (the mean value of Y). The error sum of squares (*SSE*) represents the part of the variation in Y that is not explained by the regression. It is based on the difference between Y_i and $\hat{Y}_i$. The total sum of squares (*SST*) is equal to the regression sum of squares (*SSR*) plus the error sum of squares (*SSE*). Equations (13.5), (13.6), (13.7), and (13.8) define these measures of variation and the total sum of squares (*SST*).

MEASURES OF VARIATION IN REGRESSION

The total sum of squares (*SST*) is equal to the regression sum of squares (*SSR*) plus the error sum of squares (*SSE*).

$$SST = SSR + SSE \tag{13.5}$$

TOTAL SUM OF SQUARES (*SST*)

The total sum of squares (*SST*) is equal to the sum of the squared differences between each observed value of Y and the mean value of Y.

$$SST = \text{Total sum of squares}$$

$$= \sum_{i=1}^{n} (Y_i - \overline{Y})^2 \tag{13.6}$$

REGRESSION SUM OF SQUARES (*SSR*)

The regression sum of squares (*SSR*) is equal to the sum of the squared differences between each predicted value of *Y* and the mean value of *Y*.

SSR = Explained variation or regression sum of squares

$$= \sum_{i=1}^{n} (\hat{Y}_i - \overline{Y})^2 \tag{13.7}$$

ERROR SUM OF SQUARES (*SSE*)

The error sum of squares (*SSE*) is equal to the sum of the squared differences between each observed value of *Y* and the predicted value of *Y*.

SSE = Unexplained variation or error sum of squares

$$= \sum_{i=1}^{n} (Y_i - \hat{Y}_i)^2 \tag{13.8}$$

Figure 13.7 shows the sum of squares portion of the Figure 13.4 results for the Sunflowers Apparel data. The total variation, *SST*, is equal to 78.7686. This amount is subdivided into the sum of squares explained by the regression (*SSR*), equal to 66.7854, and the sum of squares unexplained by the regression (*SSE*), equal to 11.9832. From Equation (13.5) on page 441:

$$SST = SSR + SSE$$

$$78.7686 = 66.7854 + 11.9832$$

FIGURE 13.7
Sum of squares portion of the Figure 13.4 Excel results

⊿	A	B	C	D	E	F	G
10	ANOVA						
11		df	SS	MS	F	Significance F	
12	Regression	1	66.7854	66.7854	66.8792	0.0000	
13	Residual	12	11.9832	0.9986			
14	Total	13	78.7686				
15							
16		Coefficients	Standard Error	t Stat	P-value	Lower 95%	Upper 95%
17	Intercept	-1.2088	0.9949	-1.2151	0.2477	-3.3765	0.9588
18	Profiled Customers	2.0742	0.2536	8.1780	0.0000	1.5216	2.6268

Equations (13.9) through (13.11) present computational formulas for calculating *SST*, *SSR* and *SSE*.

COMPUTATIONAL FORMULAS FOR *SST*, *SSR*, AND *SSE*

$$SST = \sum_{i=1}^{n} (Y_i - \overline{Y})^2 = \sum_{i=1}^{n} Y_i^2 - \frac{\left(\sum_{i=1}^{n} Y_i\right)^2}{n} \tag{13.9}$$

$$SSR = \sum_{i=1}^{n} (\hat{Y}_i - \overline{Y})^2 = b_0 \sum_{i=1}^{n} Y_i + b_1 \sum_{i=1}^{n} X_i Y_i - \frac{\left(\sum_{i=1}^{n} Y_i\right)^2}{n} \tag{13.10}$$

$$SSE = \sum_{i=1}^{n} (Y_i - \hat{Y}_i)^2 = \sum_{i=1}^{n} Y_i^2 - b_0 \sum_{i=1}^{n} Y_i - b_1 \sum_{i=1}^{n} X_i Y_i \tag{13.11}$$

The Coefficient of Determination

By themselves, *SSR*, *SSE*, and *SST* provide little information. However, the ratio of the regression sum of squares (*SSR*) to the total sum of squares (*SST*) measures the proportion of variation in *Y* that is explained by the linear relationship of the independent variable *X* with the dependent variable *Y* in the regression model. This ratio, called the coefficient of determination, r^2, is defined in Equation (13.12).

COEFFICIENT OF DETERMINATION

The coefficient of determination is equal to the regression sum of squares (i.e., explained variation) divided by the total sum of squares (i.e., total variation).

$$r^2 = \frac{\text{Regression sum of squares}}{\text{Total sum of squares}} = \frac{SSR}{SST} \quad \textbf{(13.12)}$$

studentTIP

r^2 must be a value between 0 and 1 inclusive. It cannot be negative.

The **coefficient of determination** measures the proportion of variation in *Y* that is explained by the variation in the independent variable *X* in the regression model. The range of r^2 is from 0 to 1, and the greater the value, the more the variation in *Y* in the regression model can be explained by the variation in *X*.

For the Sunflowers Apparel data, with $SSR = 66.7854$, $SSE = 11.9832$, and $SST = 78.7686$,

$$r^2 = \frac{66.7854}{78.7686} = 0.8479$$

Therefore, the variability in the number of profiled customers explains 84.79% of the variation in annual sales. This large r^2 indicates a strong linear relationship between these two variables because the regression model has explained 84.79% of the variability in predicting annual sales. (Only 15.21% of the sample variability in annual sales is due to factors not considered by the regression model.)

Figure 13.8 presents the regression statistics table portion of the Figure 13.4 results for the Sunflowers Apparel data. This table contains the coefficient of determination.

FIGURE 13.8
Regression statistics portion of the Figure 13.4 Excel results

	A	B
3	*Regression Statistics*	
4	Multiple R	0.9208
5	R Square	0.8479
6	Adjusted R Square	0.8352
7	Standard Error	0.9993
8	Observations	14

EXAMPLE 13.4

Calculating the Coefficient of Determination

Calculate the coefficient of determination, r^2, for the Sunflowers Apparel data.

SOLUTION The computational formulas that Equations 13.9 through 13.11 define can calculate the *SST*, *SSR*, and *SSE*.

These formulas can be used with the summary values in Table 13.2 on page 438 to calculate *SST*, *SSR*, and *SSE*.

$$SST = \sum_{i=1}^{n}(Y_i - \bar{Y})^2 = \sum_{i=1}^{n}Y_i^2 - \frac{\left(\sum_{i=1}^{n}Y_i\right)^2}{n}$$

$$= 693.9 - \frac{(92.8)^2}{14}$$

$$= 78.7686$$

▶(*continued*)

Using ways other than Excel to calculate these coefficients may result in slightly different results. Such results will be consistent with the conclusion that there is a strong linear relationship between the two variables.

$$SSR = \sum_{i=1}^{n}(\hat{Y}_i - \overline{Y})^2 = b_0\sum_{i=1}^{n}Y_i + b_1\sum_{i=1}^{n}X_iY_i - \frac{\left(\sum_{i=1}^{n}Y_i\right)^2}{n}$$

$$= (-1.2088265)(92.8) + (2.07417)(382.85) - \frac{(92.8)^2}{14}$$

$$= 66.7854$$

$$SSE = \sum_{i=1}^{n}(Y_i - \hat{Y}_i)^2 = \sum_{i=1}^{n}Y_i^2 - b_0\sum_{i=1}^{n}Y_i - b_1\sum_{i=1}^{n}X_iY_i$$

$$= 693.9 - (-1.2088265)(92.8) - (2.07417)(382.85)$$

$$= 11.9832$$

Therefore,

$$r^2 = \frac{66.7854}{78.7686} = 0.8479$$

Standard Error of the Estimate

Although the least-squares method produces the line that fits the data with the minimum amount of prediction error, unless all the observed data points fall on a straight line, the prediction line is not a perfect predictor. Just as all data values cannot be expected to be exactly equal to their mean, neither can all the values in a regression analysis be expected to be located exactly on the prediction line. Figure 13.5 on page 435 illustrates the variability around the prediction line for the Sunflowers Apparel data. In the Figure 13.5 scatter plot, many of the observed values of Y fall near the prediction line, but none of the values are exactly on the line.

The **standard error of the estimate** measures the variability of the observed Y values from the predicted Y values in the same way that the standard deviation in Chapter 3 measures the variability of each value around the sample mean. In other words, the standard error of the estimate is the standard deviation *around* the prediction line, whereas the standard deviation in Chapter 3 is the standard deviation *around* the sample mean. Equation (13.13) defines the standard error of the estimate, represented by the symbol S_{YX}.

STANDARD ERROR OF THE ESTIMATE

$$S_{YX} = \sqrt{\frac{SSE}{n-2}} = \sqrt{\frac{\sum_{i=1}^{n}(Y_i - \hat{Y}_i)^2}{n-2}} \tag{13.13}$$

where

$$Y_i = \text{observed value of } Y \text{ for a given } X_i$$

$$\hat{Y}_i = \text{predicted value of } Y \text{ for a given } X_i$$

$$SSE = \text{error sum of squares}$$

From Equation (13.8) and Figure 13.4 on page 435 or Figure 13.7 on page 442, $SSE = 11.9832$. Thus,

$$S_{YX} = \sqrt{\frac{11.9832}{14-2}} = 0.9993$$

This standard error of the estimate, equal to 0.9993 millions of dollars (i.e., $999,300), is labeled Standard Error in the Figure 13.8 Excel results. The standard error of the estimate represents a measure of the variation around the prediction line. It is measured in the same units as the dependent variable Y. The interpretation of the standard error of the estimate is similar to that of the standard deviation. Just as the standard deviation measures variability around the mean, the standard error of the estimate measures variability around the prediction line. For Sunflowers Apparel, the typical difference between actual annual sales at a store and the predicted annual sales using the regression equation is approximately $999,300.

PROBLEMS FOR SECTION 13.3

LEARNING THE BASICS

13.11 How do you interpret a coefficient of determination, r^2, equal to 0.80?

13.12 If $SSR = 36$ and $SSE = 4$, determine SST and then compute the coefficient of determination, r^2, and interpret its meaning.

13.13 If $SSR = 66$ and $SST = 88$, compute the coefficient of determination, r^2, and interpret its meaning.

13.14 If $SSE = 10$ and $SSR = 30$, compute the coefficient of determination, r^2, and interpret its meaning.

13.15 If $SSR = 120$, why is it impossible for SST to equal 110?

APPLYING THE CONCEPTS

✓ SELF TEST **13.16** In Problem 13.4 on page 439, the percentage of alcohol was used to predict wine quality (stored in Vinho Verde). For those data, $SSR = 21.8677$ and $SST = 64.0000$.
a. Determine the coefficient of determination, r^2, and interpret its meaning.
b. Determine the standard error of the estimate.
c. How useful do you think this regression model is for predicting sales?

13.17 In Problem 13.5 on page 439, you used the summated rating to predict the cost of a restaurant meal (stored in Restaurants).
a. Determine the coefficient of determination, r^2, and interpret its meaning.
b. Determine the standard error of the estimate.
c. How useful do you think this regression model is for predicting the cost of a restaurant meal?

13.18 In Problem 13.6 on page 440, a prospective MBA student wanted to predict starting salary upon graduation, based on program per-year tuition (stored in FTMBA). Using the results of that problem,
a. determine the coefficient of determination, r^2, and interpret its meaning.
b. determine the standard error of the estimate.
c. How useful do you think this regression model is for predicting starting salary?

13.19 In Problem 13.7 on page 440, you used the plate gap on the bag-sealing equipment to predict the tear rating of a bag of coffee (stored in Starbucks). Using the results of that problem,
a. determine the coefficient of determination, r^2, and interpret its meaning.
b. determine the standard error of the estimate.
c. How useful do you think this regression model is for predicting the tear rating based on the plate gap in the bag-sealing equipment?

13.20 In Problem 13.8 on page 440, you used annual revenues to predict the current value of a MLB baseball team, using the data stored in MLB Values. Using the results of that problem,
a. determine the coefficient of determination, r^2, and interpret its meaning.
b. determine the standard error of the estimate.
c. How useful do you think this regression model is for predicting the value of a MLB baseball team?

13.21 In Problem 13.9 on page 440, an agent for a real estate company wanted to predict the monthly rent for one-bedroom apartments, based on the size of the apartment (stored in Silver Spring Rentals). Using the results of that problem,
a. determine the coefficient of determination, r^2, and interpret its meaning.
b. determine the standard error of the estimate.
c. How useful do you think this regression model is for predicting the monthly rent?
d. Can you think of other variables that might explain the variation in monthly rent?

13.22 In Problem 13.10 on page 440, you used YouTube trailer views to predict movie weekend box office gross (stored in Movie). Using the results of that problem,
a. determine the coefficient of determination, r^2, and interpret its meaning.
b. determine the standard error of the estimate.
c. How useful do you think this regression model is for predicting movie weekend box office gross?
d. Can you think of other variables that might explain the variation in movie weekend box office gross?

13.4 Assumptions of Regression

Chapters 9 through 12 emphasize the importance of the assumptions to the validity of any conclusions based on hypothesis testing or analysis of variance results. The assumptions necessary for regression are similar to those of the analysis of variance because both are part of the general category of *linear models* (Kutner).

The four **assumptions of regression** (known by the acronym LINE) are:

- Linearity
- Independence of errors
- Normality of error
- Equal variance

The first assumption, **linearity**, states that the relationship between variables is linear. The Kutner reference discusses relationships between variables that are not linear.

The second assumption, **independence of errors**, requires that the errors (ε_i) be independent of one another. This assumption is particularly important when data are collected over a period of time. In such situations, the errors in a specific time period are sometimes correlated with those of the previous time period.

The third assumption, **normality**, requires that the errors (ε_i) be normally distributed at each value of X. Like the t test and the ANOVA F test, regression analysis is fairly robust against departures from the normality assumption. As long as the distribution of the errors at each level of X is not extremely different from a normal distribution, inferences about β_0 and β_1 are not seriously affected.

The fourth assumption, **equal variance**, or **homoscedasticity**, requires that the variance of the errors (ε_i) be constant for all values of X. In other words, the variability of Y values is the same when X is a low value as when X is a high value. The equal-variance assumption is important when making inferences about β_0 and β_1. If there are serious departures from this assumption, one can use either data transformations or weighted least-squares methods (Kutner).

13.5 Residual Analysis

Sections 13.2 and 13.3 develop a regression model using the least-squares method for the Sunflowers Apparel data. Is this the correct model for these data? Are the assumptions Section 13.4 discusses valid? **Residual analysis** visually evaluates the assumptions and helps one determine whether the regression model that has been selected is appropriate.

The **residual**, or estimated error value, e_i, is the difference between the observed (Y_i) and predicted $(\hat{Y}_i)$ values of the dependent variable for a given value of X_i. A residual appears on a scatter plot as the vertical distance between an observed value of Y and the prediction line. Equation (13.14) defines the residual.

RESIDUAL

The residual is equal to the difference between the observed value of Y and the predicted value of Y.

$$e_i = Y_i - \hat{Y}_i \qquad (13.14)$$

Evaluating the Assumptions

Recall from Section 13.4 that the four LINE assumptions of regression are linearity, independence, normality, and equal variance.

Linearity   To evaluate linearity, plot the residuals on the vertical axis against the corresponding X_i values of the independent variable on the horizontal axis. If the linear model is appropriate for the data, there will not be any apparent pattern in the residual plot. However, if the linear model is not appropriate, in the residual plot, there will be a relationship between the X_i values and the residuals, e_i.

Figure 13.9 shows such a pattern in the residuals. Panel A shows a situation in which, although there is an increasing trend in Y as X increases, the relationship seems curvilinear because the upward trend decreases for increasing values of X. This effect is even more apparent in Panel B, where there is a clear relationship between X_i and e_i. By removing the linear trend of X with Y, the residual plot has exposed the lack of fit in the simple linear model more clearly than the scatter plot in Panel A. For these data, a curvilinear model such as a quadratic model is a better fit and should be used instead of the simple linear model (Kutner et al.).

FIGURE 13.9
Studying the appropriateness of the simple linear regression model

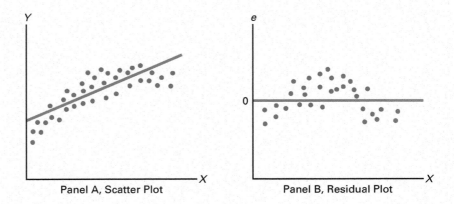

Panel A, Scatter Plot   Panel B, Residual Plot

To determine whether the simple linear regression model for the Sunflowers Apparel data is appropriate, one calculates the residuals. Figure 13.10 displays the predicted annual sales values and residuals for the Sunflowers Apparel data.

FIGURE 13.10
Table of residuals for the Sunflowers Apparel data

	A	B	C	D	E
1	Observation	Profiled Customers	Predicted Annual Sales	Annual Sales	Residuals
2	1	3.7	6.4656	5.7	-0.7656
3	2	3.6	6.2582	5.9	-0.3582
4	3	2.8	4.5988	6.7	2.1012
5	4	5.6	10.4065	9.5	-0.9065
6	5	3.3	5.6359	5.4	-0.2359
7	6	2.2	3.3543	3.5	0.1457
8	7	3.3	5.6359	6.2	0.5641
9	8	3.1	5.2211	4.7	-0.5211
10	9	3.2	5.4285	6.1	0.6715
11	10	3.5	6.0508	4.9	-1.1508
12	11	5.2	9.5769	10.7	1.1231
13	12	4.6	8.3324	7.6	-0.7324
14	13	5.8	10.8214	11.8	0.9786
15	14	3	5.0137	4.1	-0.9137

In the Worksheet

Column C formulas compute the predicted Y values by multiplying the column B X values by the b_1 coefficient (Figure 13.4 worksheet cell B18) and then adding the b_0 coefficient (Figure 13.4 worksheet cell B17). Column E formulas compute residuals by subtracting the predicted Y values from the column D Y values.

The **RESIDUALS worksheet** of the Excel Guide **Simple Linear Regression** is similar to the Figure 13.10 worksheet.

To assess linearity, one creates the residual plot of the residuals versus the independent variable (number of profiled customers) that Figure 13.11 presents. Although there is widespread scatter in the residual plot, there is no clear pattern or relationship between the residuals and X_i. The residuals appear to be evenly spread above and below 0 for different values of X. One can conclude that the linear model is appropriate for the Sunflowers Apparel data.

FIGURE 13.11
Excel plot of the residuals versus the profiled customers of a store for the Sunflowers Apparel data

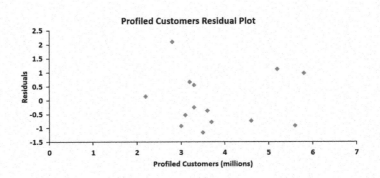

Independence   One can evaluate the assumption of independence of the errors by plotting the residuals in the order or sequence in which the data were collected. If the values of Y are part of a time series (see Section 2.5), a residual may sometimes be related to the residual that precedes it. If this relationship exists between consecutive residuals (which violates the assumption of independence), the plot of the residuals versus the time variable will often show a cyclical pattern. If it does, one would then use the alternative approaches that Kutner et al. discusses. (Because the Sunflowers Apparel data are not time-series data, one does not need to evaluate the independence assumption in the Sunflowers Apparel example.)

Normality   One can evaluate the assumption of normality in the errors by constructing a histogram (see Section 2.4), using a stem-and-leaf display (see Section 2.4), a boxplot (see Section 3.3), or a normal probability plot (see Section 6.3). To evaluate the normality assumption for the Sunflowers Apparel data, Table 13.3 organizes the residuals into a frequency distribution and Figure 13.12 is a normal probability plot.

TABLE 13.3

Frequency distribution of 14 residual values for the Sunflowers Apparel data

Residuals	Frequency
-1.25 but less than -0.75	4
-0.75 but less than -0.25	3
-0.25 but less than $+0.25$	2
$+0.25$ but less than $+0.75$	2
$+0.75$ but less than $+1.25$	2
$+1.25$ but less than $+1.75$	0
$+1.75$ but less than $+2.25$	1
	14

Although the small sample size makes it difficult to evaluate normality, from the normal probability plot of the residuals in Figure 13.12, the data do not appear to depart substantially from a normal distribution. The robustness of regression analysis with modest departures from normality enables one to conclude that one should not be overly concerned about departures from this normality assumption in the Sunflowers Apparel data.

FIGURE 13.12

Excel normal probability plot of the residuals for the Sunflowers Apparel data

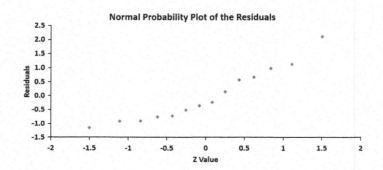

Equal Variance One can evaluate the assumption of equal variance from a plot of the residuals with X_i. One examines the plot to see if there is approximately the same amount of variation in the residuals at each value of X. For the Sunflowers Apparel data of Figure 13.11 on page 447, there do not appear to be major differences in the variability of the residuals for different X_i values. One can conclude that there is no apparent violation in the assumption of equal variance at each level of X.

To examine a case in which the equal-variance assumption is violated, observe Figure 13.13, which is a plot of the residuals with X_i for a hypothetical set of data. This plot is fan shaped because the variability of the residuals increases dramatically as X increases. Because this plot shows unequal variances of the residuals at different levels of X, the equal-variance assumption is invalid, and one would need to use the alternative approaches that the Kutner reference discusses.

FIGURE 13.13
Violation of equal variance

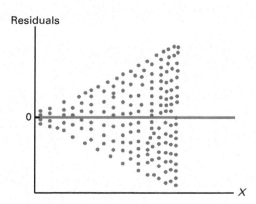

PROBLEMS FOR SECTION 13.5

LEARNING THE BASICS

13.23 The following results provide the X values, residuals, and a residual plot from a regression analysis:

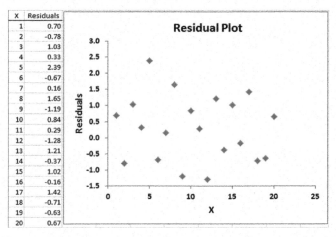

X	Residuals
1	0.70
2	-0.78
3	1.03
4	0.33
5	2.39
6	-0.67
7	0.16
8	1.65
9	-1.19
10	0.84
11	0.29
12	-1.28
13	1.21
14	-0.37
15	1.02
16	-0.16
17	1.42
18	-0.71
19	-0.63
20	0.67

Is there any evidence of a pattern in the residuals? Explain.

13.24 The following results show the X values, residuals, and a residual plot from a regression analysis:

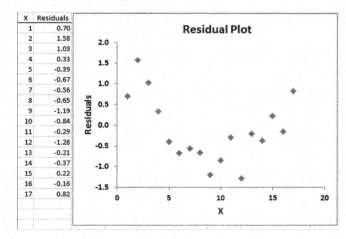

X	Residuals
1	0.70
2	1.58
3	1.03
4	0.33
5	-0.39
6	-0.67
7	-0.56
8	-0.65
9	-1.19
10	-0.84
11	-0.29
12	-1.28
13	-0.21
14	-0.37
15	0.22
16	-0.16
17	0.82

Is there any evidence of a pattern in the residuals? Explain.

APPLYING THE CONCEPTS

13.25 In Problem 13.5 on page 439, you used the summated rating to predict the cost of a restaurant meal. Perform a residual analysis for these data (stored in Restaurants). Evaluate whether the assumptions of regression have been seriously violated.

✓SELF TEST **13.26** In Problem 13.4 on page 439, you used the percentage of alcohol to predict wine quality. Perform a residual analysis for these data (stored in Vinho Verde). Evaluate whether the assumptions of regression have been seriously violated.

13.27 In Problem 13.7 on page 440, you used the plate gap on the bag-sealing equipment to predict the tear rating of a bag of coffee. Perform a residual analysis for these data (stored in Starbucks). Based on these results, evaluate whether the assumptions of regression have been seriously violated.

13.28 In Problem 13.6 on page 440, a prospective MBA student wanted to predict starting salary upon graduation, based on program per-year tuition. Perform a residual analysis for these data (stored in FTMBA). Based on these results, evaluate whether the assumptions of regression have been seriously violated.

13.29 In Problem 13.9 on page 440, an agent for a real estate company wanted to predict the monthly rent for one-bedroom apartments, based on the size of the apartments. Perform a residual analysis for these data (stored in Silver Spring Rentals). Based on these results, evaluate whether the assumptions of regression have been seriously violated.

13.30 In Problem 13.8 on page 440, you used annual revenues to predict the current value of a MLB baseball team. Perform a residual analysis for these data (stored in MLB Values). Based on these results, evaluate whether the assumptions of regression have been seriously violated.

13.31 In Problem 13.10 on page 440, you used YouTube trailer views to predict movie weekend box office gross. Perform a residual analysis for these data (stored in Movie). Based on these results, evaluate whether the assumptions of regression have been seriously violated.

13.6 Measuring Autocorrelation: The Durbin-Watson Statistic

One of the basic assumptions of the regression model is the independence of the errors. This assumption is sometimes violated when data are collected over sequential time periods because a residual at any one time period sometimes is similar to residuals at adjacent time periods. This pattern in the residuals is called **autocorrelation**. When a set of data has substantial autocorrelation, the validity of a regression model is in serious doubt.

Residual Plots to Detect Autocorrelation

As mentioned in Section 13.5, one way to detect autocorrelation is to plot the residuals in time order. If a positive autocorrelation effect exists, there will be clusters of residuals with the same sign, and one will readily detect an apparent pattern. If negative autocorrelation exists, residuals will tend to jump back and forth from positive to negative to positive, and so on. Because negative autocorrelation is very rarely seen in regression analysis, the example in this section illustrates positive autocorrelation.

To illustrate positive autocorrelation, consider the case of a package delivery store manager who wants to be able to predict weekly sales. In approaching this problem, the manager has decided to develop a regression model to use the number of customers making purchases as an independent variable. She collects data for a period of 15 weeks and then organizes and stores these data in Fifteen Weeks. Table 13.4 presents these data.

TABLE 13.4
Customers and sales for a period of 15 consecutive weeks

Week	Customers	Sales ($thousands)	Week	Customers	Sales ($thousands)
1	794	9.33	9	880	12.07
2	799	8.26	10	905	12.55
3	837	7.48	11	886	11.92
4	855	9.08	12	843	10.27
5	845	9.83	13	904	11.80
6	844	10.09	14	950	12.15
7	863	11.01	15	841	9.64
8	875	11.49			

Because the data are collected over a period of 15 consecutive weeks at the same store, the manager needs to determine whether there is autocorrelation. First, she can develop the simple linear regression model she can use to predict sales based on the number of customers assuming there is no autocorrelation in the residuals. Figure 13.14 presents results for these data.

FIGURE 13.14
Excel regression worksheet results for the Table 13.4 package delivery store data

⁞	A	B	C	D	E	F	G
1	Simple Linear Regression Analysis						
2							
3	*Regression Statistics*						
4	Multiple R	0.8108					
5	R Square	0.6574					
6	Adjusted R Square	0.6311					
7	Standard Error	0.9360					
8	Observations	15					
9							
10	ANOVA						
11		*df*	*SS*	*MS*	*F*	*Significance F*	
12	Regression	1	21.8604	21.8604	24.9501	0.0002	
13	Residual	13	11.3901	0.8762			
14	Total	14	33.2506				
15							
16		*Coefficients*	*Standard Error*	*t Stat*	*P-value*	*Lower 95%*	*Upper 95%*
17	Intercept	-16.0322	5.3102	-3.0192	0.0099	-27.5041	-4.5603
18	Customers	0.0308	0.0062	4.9950	0.0002	0.0175	0.0441

In the Worksheet

This worksheet shares its design with the Figure 13.4 worksheet on page 435.

From Figure 13.14, observe that r^2 is 0.6574, indicating that 65.74% of the variation in sales is explained by variation in the number of customers. In addition, the Y intercept, b_0, is -16.0322 and the slope, b_1, is 0.0308. However, before using this model for prediction, one must perform a residual analysis. Because the data have been collected over a consecutive period of 15 weeks, in addition to checking the linearity, normality, and equal-variance assumptions, one must investigate the independence-of-errors assumption. One plots the residuals versus time in Figure 13.15 in order to examine whether a pattern in the residuals exists. In Figure 13.15, one can see that the residuals tend to fluctuate up and down in a cyclical pattern. This cyclical pattern provides strong cause for concern about the existence of autocorrelation in the residuals and, therefore, a violation of the independence-of-errors assumption.

FIGURE 13.15
Excel residual plot for the Table 13.4 package delivery store data

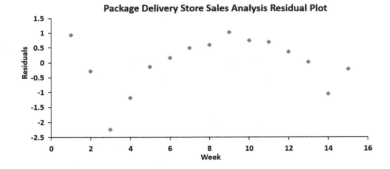

The Durbin-Watson Statistic

The **Durbin-Watson statistic** is used to measure autocorrelation. This statistic measures the correlation between each residual and the residual for the previous time period. Equation (13.15) defines the Durbin-Watson statistic.

DURBIN-WATSON STATISTIC

$$D = \frac{\sum_{i=2}^{n}(e_i - e_{i-1})^2}{\sum_{i=1}^{n}e_i^2}$$

(13.15)

where

e_i = residual at the time period i

In Equation (13.15), the numerator, $\sum_{i=2}^{n}(e_i - e_{i-1})^2$, represents the squared differ-ence between two successive residuals, summed from the second value to the nth value and the denominator, $\sum_{i=1}^{n}e_i^2$, represents the sum of the squared residuals. This means that the value of the Durbin-Watson statistic, D, will approach 0 if successive residuals are positively autocorrelated. If the residuals are not correlated, the value of D will be close to 2. (If the residuals are negatively autocorrelated, D will be greater than 2 and could even approach its maximum value of 4.)

From Figure 13.16, the Durbin-Watson statistic, D, is 0.8830 for the package delivery store data.

FIGURE 13.16
Excel Durbin-Watson statistic results for the package delivery store data

◢	A	B
1	Durbin-Watson Statistic	
2		
3	Sum of Squared Difference of Residuals	10.0575 =SUMXMY2(RESIDUALS!E3:E15, RESIDUALS!E2:E14)
4	Sum of Squared Residuals	11.3901 =SUMSQ(RESIDUALS!E2:E15)
5		
6	Durbin-Watson Statistic	0.8830 =B3/B4

In the Worksheet

The SUMXMY2 function in cell B3 computes the value of the Equation (13.15) numerator. The SUMSQ function in cell B4 computes the value of the Equation (13.15) denominator. These functions use the residuals data that are in column E of the RESIDUALS worksheet.

The **DURBIN_WATSON worksheet** of the Excel Guide **Simple Linear Regression** is similar to the Figure 13.16 worksheet.

One needs to determine when the autocorrelation is large enough to conclude that there is significant positive autocorrelation. To do so, one compares D to the critical values of the Durbin-Watson statistic found in Table E.8, a portion of which Table 13.5 shows. The critical values depend on α, the significance level chosen, n, the sample size, and k, the number of independent variables in the model (in simple linear regression, $k = 1$).

TABLE 13.5
Finding critical values of the Durbin-Watson statistic

	$\alpha = 0.05$									
	$k = 1$		$k = 2$		$k = 3$		$k = 4$		$k = 5$	
n	d_L	d_U	d_L	d_U	d_L	d_U	d_L	d_U	d_L	d_U
15	1.08	1.36	.95	1.54	.82	1.75	.69	1.97	.56	2.21
16	1.10	1.37	.98	1.54	.86	1.73	.74	1.93	.62	2.15
17	1.13	1.38	1.02	1.54	.90	1.71	.78	1.90	.67	2.10
18	1.16	1.39	1.05	1.53	.93	1.69	.82	1.87	.71	2.06

In Table 13.5, two values are shown for each combination of α (level of significance), n (sample size), and k (number of independent variables in the model). The first value, d_L, represents the lower critical value. If D is below d_L, one concludes that there is evidence of positive autocorrelation among the residuals. If this occurs, the least-squares method used in this chapter is inappropriate, and one should use alternative methods that the Kutner reference discusses. The second value, d_U, represents the upper critical value of D, above which one would conclude that there is no evidence of positive autocorrelation among the residuals. If D is between d_L and d_U, one is unable to arrive at a definite conclusion.

For the package delivery store data, with one independent variable ($k = 1$) and 15 values ($n = 15$), $d_L = 1.08$ and $d_U = 1.36$. Because $D = 0.8830 < 1.08$, one concludes that there is positive autocorrelation among the residuals. The least-squares regression analysis of the data that Figure 13.14 on page 451 presents is inappropriate because of the presence of significant positive autocorrelation among the residuals. In other words, the independence-of-errors assumption is invalid, and an alternative approach must be used.

PROBLEMS FOR SECTION 13.6

LEARNING THE BASICS

13.32 The residuals for 10 consecutive time periods are as follows:

Time Period	Residual	Time Period	Residual
1	−5	6	+1
2	−4	7	+2
3	−3	8	+3
4	−2	9	+4
5	−1	10	+5

a. Plot the residuals over time. What conclusion can you reach about the pattern of the residuals over time?
b. Based on (a), what conclusion can you reach about the autocorrelation of the residuals?

13.33 The residuals for 15 consecutive time periods are as follows:

Time Period	Residual	Time Period	Residual
1	+4	9	+6
2	−6	10	−3
3	−1	11	+1
4	−5	12	+3
5	+2	13	0
6	+5	14	−4
7	−2	15	−7
8	+7		

a. Plot the residuals over time. What conclusion can you reach about the pattern of the residuals over time?
b. Compute the Durbin-Watson statistic. At the 0.05 level of significance, is there evidence of positive autocorrelation among the residuals?
c. Based on (a) and (b), what conclusion can you reach about the autocorrelation of the residuals?

APPLYING THE CONCEPTS

13.34 In Problem 13.7 on page 440 concerning the bag-sealing equipment at Starbucks, you used the plate gap to predict the tear rating.
a. Is it necessary to compute the Durbin-Watson statistic in this case? Explain.
b. Under what circumstances is it necessary to compute the Durbin-Watson statistic before proceeding with the least-squares method of regression analysis?

13.35 What is the relationship between the price of crude oil and the price you pay at the pump for gasoline? The file Oil & Gasoline contains the price ($) for a barrel of crude oil (Cushing, Oklahoma, spot price) and a gallon of gasoline (U.S. average conventional spot price) for 236 weeks ending July 5, 2019.
Source: Data extracted from **www.eia.gov**.

a. Construct a scatter plot with the price of oil on the horizontal axis and the price of gasoline on the vertical axis.
b. Use the least-squares method to develop a simple linear regression equation to predict the price of a gallon of gasoline

using the price of a barrel of crude oil as the independent variable.
c. Interpret the meaning of the slope, b_1, in this problem.
d. Plot the residuals versus the time period.
e. Compute the Durbin-Watson statistic.
f. At the 0.05 level of significance, is there evidence of positive autocorrelation among the residuals?
g. Based on the results of (d) through (f), is there reason to question the validity of the model?
h. What conclusions can you reach concerning the relationship between the price of a barrel of crude oil and the price of a gallon of gasoline?

✓SELF TEST **13.36** A mail-order catalog business that sells personal computer supplies, software, and hardware maintains a centralized warehouse for the distribution of products ordered. Management is currently examining the process of distribution from the warehouse and has the business objective of determining the factors that affect warehouse distribution costs. Currently, a handling fee is added to the order, regardless of the amount of the order. Data that indicate the warehouse distribution costs and the number of orders received have been collected over the past 24 months and are stored in Warehouse Costs.
a. Assuming a linear relationship, use the least-squares method to find the regression coefficients b_0 and b_1.
b. Predict the monthly warehouse distribution costs when the number of orders is 4,500.
c. Plot the residuals versus the time period.
d. Compute the Durbin-Watson statistic. At the 0.05 level of significance, is there evidence of positive autocorrelation among the residuals?
e. Based on the results of (c) and (d), is there reason to question the validity of the model?
f. What conclusions can you reach concerning the factors that affect distribution costs?

13.37 A freshly brewed shot of espresso has three distinct components: the heart, body, and crema. The separation of these three components typically lasts only 10 to 20 seconds. To use the espresso shot in making a latte, a cappuccino, or another drink, the shot must be poured into the beverage during the separation of the heart, body, and crema. If the shot is used after the separation occurs, the drink becomes excessively bitter and acidic, ruining the final drink. Thus, a longer separation time allows the drink-maker more time to pour the shot and ensure that the beverage will meet expectations. An employee at a coffee shop hypothesized that the harder the espresso grounds were tamped down into the portafilter before brewing, the longer the separation time would be. An experiment using 24 observations was conducted to test this relationship. The independent variable Tamp measures the distance, in inches, between the espresso grounds and the top of the portafilter (i.e., the harder the tamp, the greater the distance). The dependent variable Time is the number of seconds the heart, body, and crema are separated (i.e., the amount of time after the shot is poured before it must be used for the customer's beverage). The data are stored in Espresso.
a. Use the least-squares method to develop a simple regression equation with Time as the dependent variable and Tamp as the independent variable.

b. Predict the separation time for a tamp distance of 0.50 inch.

c. Plot the residuals versus the time order of experimentation. Are there any noticeable patterns?

d. Compute the Durbin-Watson statistic. At the 0.05 level of significance, is there evidence of positive autocorrelation among the residuals?

e. Based on the results of (c) and (d), is there reason to question the validity of the model?

f. What conclusions can you reach concerning the effect of tamping on the time of separation?

13.38 The owners of a chain of ice cream stores have the business objective of improving the forecast of daily sales so that staffing shortages can be minimized during the summer season. As a starting point, the owners decide to develop a simple linear regression model to predict daily sales based on atmospheric temperature.

They select a sample of 21 consecutive days and store the results in ⬛ Ice Cream . (Hint: Determine which are the independent and dependent variables.)

a. Assuming a linear relationship, use the least-squares method to compute the regression coefficients b_0 and b_1.

b. Predict the sales for a day in which the temperature is 83°F.

c. Plot the residuals versus the time period.

d. Compute the Durbin-Watson statistic. At the 0.05 level of significance, is there evidence of positive autocorrelation among the residuals?

e. Based on the results of (c) and (d), is there reason to question the validity of the model?

f. What conclusions can you reach concerning the relationship between sales and atmospheric temperature?

13.7 Inferences About the Slope and Correlation Coefficient

Sections 13.1 through 13.3 use regression solely for descriptive purposes. These sections discuss how to determine the regression coefficients using the least-squares method and how to predict Y for a given value of X. In addition, these sections discuss how to calculate and interpret the standard error of the estimate and the coefficient of determination.

When the residual analysis that Section 13.5 discusses indicates that the assumptions of a least-squares regression model are not seriously violated and that the straight-line model is appropriate, one can make inferences about the linear relationship between the variables in the population.

t Test for the Slope

To determine the existence of a significant linear relationship between the X and Y variables, one tests whether β_1 (the population slope) is equal to 0. The null and alternative hypotheses are as follows:

$$H_0: \beta_1 = 0 \, [\text{There is no linear relationship (the slope is zero).}]$$
$$H_1: \beta_1 \neq 0 \, [\text{There is a linear relationship (the slope is not zero).}]$$

If one rejects the null hypothesis, one concludes that there is evidence of a linear relationship. Equation (13.16) defines the test statistic for the slope, which is based on the sampling distribution of the slope.

---

t TEST STATISTIC FOR
TESTING A HYPOTHESIS FOR A POPULATION SLOPE, β_1

The t_{STAT} test statistic equals the difference between the sample slope and hypothesized value of the population slope divided by S_{b_1}, the standard error of the slope.

$$t_{STAT} = \frac{b_1 - \beta_1}{S_{b_1}} \tag{13.16}$$

where

$$S_{b_1} = \frac{S_{YX}}{\sqrt{SSX}}$$

$$SSX = \sum_{i=1}^{n}(X_i - \overline{X})^2$$

The t_{STAT} test statistic follows a t distribution with $n - 2$ degrees of freedom.

Figure 13.17 presents the t test results for the Sunflowers Apparel scenario at the level of significance $\alpha = 0.05$.

FIGURE 13.17
Excel t test for the slope results for the Sunflowers Apparel data

▲	A	B	C	D	E	F	G
16		Coefficients	Standard Error	t Stat	P-value	Lower 95%	Upper 95%
17	Intercept	-1.2088	0.9949	-1.2151	0.2477	-3.3765	0.9588
18	Profiled Customers	2.0742	0.2536	8.1780	0.0000	1.5216	2.6268

From Figure 13.4 or Figure 13.17,

$$b_1 = +2.0742 \quad n = 14 \quad S_{b_1} = 0.2536$$

and

$$t_{STAT} = \frac{b_1 - \beta_1}{S_{b_1}} = \frac{2.0742 - 0}{0.2536} = 8.178$$

Using the 0.05 level of significance, the critical value of t with $n - 2 = 12$ degrees of freedom is 2.1788. Because $t_{STAT} = 8.178 > 2.1788$ or because the p-value is 0.0000, which is less than $\alpha = 0.05$, one rejects H_0 (see Figure 13.18). One concludes that there is a significant linear relationship between mean annual sales and the number of profiled customers.

FIGURE 13.18
Testing a hypothesis about the population slope at the 0.05 level of significance, with 12 degrees of freedom

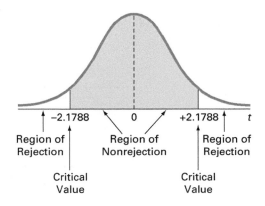

F Test for the Slope

As an alternative to the t test, in simple linear regression, one can use an F test to determine whether the slope is statistically significant. In previous chapters, Section 10.4 uses the F distribution to test the ratio of two variances and Section 11.1 uses the F distribution as part of the analysis of variance. Equation (13.17) defines the F test for the slope as the ratio of the variance that is due to the regression *(MSR)* divided by the error variance $(MSE = S_{YX}^2)$.

**F TEST STATISTIC FOR
TESTING A HYPOTHESIS FOR A POPULATION SLOPE, β_1**

The F_{STAT} test statistic is equal to the regression mean square *(MSR)* divided by the mean square error *(MSE)*.

$$F_{STAT} = \frac{MSR}{MSE} \tag{13.17}$$

where

$$MSR = \frac{SSR}{1} = SSR$$

$$MSE = \frac{SSE}{n - 2}$$

The F_{STAT} test statistic follows an F distribution with 1 and $n - 2$ degrees of freedom.

Using a level of significance α, the decision rule is

$$\text{Reject } H_0 \text{ if } F_{STAT} > F_{\alpha};$$

$$\text{otherwise, do not reject } H_0.$$

Table 13.6 organizes the complete set of results into an analysis of variance (ANOVA) table.

TABLE 13.6
ANOVA table for testing the significance of a regression coefficient

Source	df	Sum of Squares	Mean Square (variance)	F
Regression	1	SSR	$MSR = \dfrac{SSR}{1} = SSR$	$F_{STAT} = \dfrac{MSR}{MSE}$
Error	$n - 2$	SSE	$MSE = \dfrac{SSE}{n - 2}$	
Total	$n - 1$	SST		

Figure 13.19, the completed ANOVA table for the Sunflowers Apparel sales data (and part of Figure 13.4), shows that the computed F_{STAT} test statistic is 66.8792 and the p-value is 0.0000 (or less than 0.0001).

FIGURE 13.19
Excel, F test results for the Sunflowers Apparel data

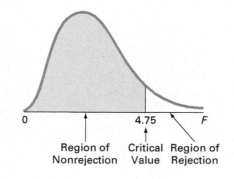

	A	B	C	D	E	F	G
10	ANOVA						
11		df	SS	MS	F	Significance F	
12	Regression	1	66.7854	66.7854	66.8792	0.0000	
13	Residual	12	11.9832	0.9986			
14	Total	13	78.7686				
15							
16		Coefficients	Standard Error	t Stat	P-value	Lower 95%	Upper 95%
17	Intercept	-1.2088	0.9949	-1.2151	0.2477	-3.3765	0.9588
18	Profiled Customers	2.0742	0.2536	8.1780	0.0000	1.5216	2.6268

Using a level of significance of 0.05, from Table E.5, the critical value of the F distribution, with 1 and 12 degrees of freedom, is 4.75 (see Figure 13.20). Because $F_{STAT} = 66.8792 > 4.75$ or because the p-value $= 0.0000 < 0.05$, one rejects H_0 and concludes that there is a significant linear relationship between the number of profiled customers and annual sales. Because the F test in Equation (13.17) on page 455 is equivalent to the t test in Equation (13.16) on page 454, one reaches the same conclusion using that other test.

FIGURE 13.20
Regions of rejection and nonrejection when testing for the significance of the slope at the 0.05 level of significance, with 1 and 12 degrees of freedom

Confidence Interval Estimate for the Slope

In addition to testing for the existence of a linear relationship between the variables, one can construct a confidence interval estimate of β_1 using Equation (13.18). Construct the confidence interval estimate for the population slope by taking the sample slope, b_1, and adding and subtracting the critical t value multiplied by the standard error of the slope.

CONFIDENCE INTERVAL ESTIMATE OF THE SLOPE, β_1

$$b_1 \pm t_{\alpha/2}S_{b_1}$$

$$b_1 - t_{\alpha/2}S_{b_1} \leq \beta_1 \leq b_1 + t_{\alpha/2}S_{b_1} \tag{13.18}$$

where

$t_{\alpha/2}$ = critical value corresponding to an upper-tail probability of $\alpha/2$ from the t distribution with $n - 2$ degrees of freedom (i.e., a cumulative area of $1 - \alpha/2$)

From the Figure 13.17 results on page 455,

$$b_1 = 2.0742 \quad n = 14 \quad S_{b_1} = 0.2536$$

To construct a 95% confidence interval estimate, $\alpha/2 = 0.025$, and from Table E.3, $t_{\alpha/2} = 2.1788$. Thus,

$$b_1 \pm t_{\alpha/2}S_{b_1} = 2.0742 \pm (2.1788)(0.2536)$$

$$= 2.0742 \pm 0.5526$$

$$1.5216 \leq \beta_1 \leq 2.6268$$

Therefore, one has 95% confidence that the population slope is between 1.5216 and 2.6268. The confidence interval indicates that for each increase of 1 million profiled customers, predicted annual sales are estimated to increase by at least $1,521,600 but no more than $2,626,800. Because both of these values are above 0, one has evidence of a significant linear relationship between annual sales and the number of profiled customers. Had the interval included 0, one would have concluded that there is no evidence of a significant linear relationship between the variables.

t Test for the Correlation Coefficient

Section 3.5 notes that the strength of the relationship between two numerical variables can be measured using the **correlation coefficient**, r. The values of the coefficient of correlation range from -1 for a perfect negative correlation to $+1$ for a perfect positive correlation. One uses the correlation coefficient to determine whether there is a statistically significant linear relationship between X and Y. To do so, one hypothesizes that the population correlation coefficient, ρ, is 0. Thus, the null and alternative hypotheses are

$$H_0: \rho = 0 \text{ (no correlation)}$$

$$H_1: \rho \neq 0 \text{ (correlation)}$$

Equation (13.19a) defines the test statistic for determining the existence of a significant correlation.

TESTING FOR THE EXISTENCE OF CORRELATION

$$t_{STAT} = \frac{r - \rho}{\sqrt{\dfrac{1 - r^2}{n - 2}}} \qquad (13.19a)$$

where

$$r = +\sqrt{r^2} \quad \text{if} \quad b_1 > 0$$

$$r = -\sqrt{r^2} \quad \text{if} \quad b_1 < 0$$

The t_{STAT} test statistic follows a t distribution with $n - 2$ degrees of freedom. Equation (3.17) on page 138 calculates r as

$$r = \frac{\text{cov}(X, Y)}{S_X S_Y} \qquad (13.19b)$$

where

$$\text{cov}(X, Y) = \frac{\displaystyle\sum_{i=1}^{n}(X_i - \overline{X})(Y_i - \overline{Y})}{n - 1}$$

$$S_X = \sqrt{\frac{\displaystyle\sum_{i=1}^{n}(X_i - \overline{X})^2}{n - 1}} \qquad S_Y = \sqrt{\frac{\displaystyle\sum_{i=1}^{n}(Y_i - \overline{Y})^2}{n - 1}}$$

In the Figure 13.4 Sunflowers Apparel results on page 435, $r^2 = 0.8479$ and $b_1 = +2.0742$. Because $b_1 > 0$, the correlation coefficient for annual sales and profiled customers is the positive square root of r^2—that is, $r = +\sqrt{0.8479} = +0.9208$. Using Equation (13.19a) to test the null hypothesis that there is no correlation between these two variables results in

$$t_{STAT} = \frac{r - 0}{\sqrt{\dfrac{1 - r^2}{n - 2}}} = \frac{0.9208 - 0}{\sqrt{\dfrac{1 - (0.9208)^2}{14 - 2}}} = 8.178$$

Using the 0.05 level of significance, because $t_{STAT} = 8.178 > 2.1788$, one rejects the null hypothesis. One concludes that there is a significant correlation between annual sales and the number of profiled customers. This t_{STAT} test statistic is the same value as the t_{STAT} test statistic calculated when testing whether the population slope, β_1, is equal to zero.

PROBLEMS FOR SECTION 13.7

LEARNING THE BASICS

13.39 You are testing the null hypothesis that there is no linear relationship between two variables, X and Y. From your sample of $n = 10$, you determine that $r = 0.80$.
a. What is the value of the t test statistic t_{STAT}?
b. At the $\alpha = 0.05$ level of significance, what are the critical values?
c. Based on your answers to (a) and (b), what statistical decision should you make?

13.40 You are testing the null hypothesis that there is no linear relationship between two variables, X and Y. From your sample of $n = 18$, you determine that $b_1 = +4.5$ and $S_{b_1} = 1.5$.
a. What is the value of t_{STAT}?
b. At the $\alpha = 0.05$ level of significance, what are the critical values?
c. Based on your answers to (a) and (b), what statistical decision should you make?
d. Construct a 95% confidence interval estimate of the population slope, β_1.

13.41 You are testing the null hypothesis that there is no linear relationship between two variables, X and Y. From your sample of $n = 20$, you determine that $SSR = 60$ and $SSE = 40$.

a. What is the value of F_{STAT}?

b. At the $\alpha = 0.05$ level of significance, what is the critical value?

c. Based on your answers to (a) and (b), what statistical decision should you make?

d. Compute the correlation coefficient by first computing r^2 and assuming that b_1 is negative.

e. At the 0.05 level of significance, is there a significant correlation between X and Y?

APPLYING THE CONCEPTS

✓SELF TEST **13.42** In Problem 13.4 on page 439, you used the percentage of alcohol to predict wine quality. The data are stored in Vinho Verde. From the results of that problem, $b_1 = 0.5624$ and $S_{b_1} = 0.1127$.

a. At the 0.05 level of significance, is there evidence of a linear relationship between the percentage of alcohol and wine quality?

b. Construct a 95% confidence interval estimate of the population slope, β_1.

13.43 In Problem 13.5 on page 439, you used the summated rating of a restaurant to predict the cost of a meal. The data are stored in Restaurants.

a. At the 0.05 level of significance, is there evidence of a linear relationship between the summated rating of a restaurant and the cost of a meal?

b. Construct a 95% confidence interval estimate of the population slope, β_1.

13.44 In Problem 13.6 on page 440, a prospective MBA student wanted to predict starting salary upon graduation, based on program per-year tuition. The data are stored in FTMBA. Use the results of that problem.

a. At the 0.05 level of significance, is there evidence of a linear relationship between the starting salary upon graduation and program per-year tuition?

b. Construct a 95% confidence interval estimate of the population slope, β_1.

13.45 In Problem 13.7 on page 440, you used the plate gap in the bag-sealing equipment to predict the tear rating of a bag of coffee. The data are stored in Starbucks. Use the results of that problem.

a. At the 0.05 level of significance, is there evidence of a linear relationship between the plate gap of the bag-sealing machine and the tear rating of a bag of coffee?

b. Construct a 95% confidence interval estimate of the population slope, β_1.

13.46 In Problem 13.8 on page 440, you used annual revenues to predict the current value of a MLB baseball team, using the data stored in MLB Values. Use the results of that problem.

a. At the 0.05 level of significance, is there evidence of a linear relationship between annual revenue and franchise value?

b. Construct a 95% confidence interval estimate of the population slope, β_1.

13.47 In Problem 13.9 on page 440, an agent for a real estate company wanted to predict the monthly rent for one-bedroom apartments, based on the size of the apartment. The data are stored in Silver Spring Rentals. Use the results of that problem.

a. At the 0.05 level of significance, is there evidence of a linear relationship between the size of the apartment and the monthly rent?

b. Construct a 95% confidence interval estimate of the population slope, β_1.

13.48 In Problem 13.10 on page 440, you used YouTube trailer views to predict movie weekend box office gross from data stored in Movie. Use the results of that problem.

a. At the 0.05 level of significance, is there evidence of a linear relationship between YouTube trailer views and movie weekend box office gross?

b. Construct a 95% confidence interval estimate of the population slope, β_1.

13.49 The volatility of a stock is often measured by its beta value. You can estimate the beta value of a stock by developing a simple linear regression model, using the percentage weekly change in the stock as the dependent variable and the percentage weekly change in a market index as the independent variable. The S&P 500 Index is a common index to use. For example, if you wanted to estimate the beta value for Disney, you could use the following model, which is sometimes referred to as a *market model*:

$$\% \text{ weekly change in Disney} = \beta_0$$
$$+ \beta_1(\text{percent weekly change in S\&P 500 index}) + \varepsilon$$

The least-squares regression estimate of the slope b_1 is the estimate of the beta value for Disney. A stock with a beta value of 1.0 tends to move the same as the overall market. A stock with a beta value of 1.5 tends to move 50% more than the overall market, and a stock with a beta value of 0.6 tends to move only 60% as much as the overall market. Stocks with negative beta values tend to move in the opposite direction of the overall market. The following table gives some beta values for some widely held stocks as of April 23, 2018.

Company	Ticker Symbol	Beta
Apple	AAPL	1.09
Disney	DIS	0.70
American Eagle Mines	AEM	−0.26
Marriott	MAR	1.25
Microsoft	MSFT	1.02
Procter & Gamble	PG	0.34

Source: Data extracted from finance.yahoo.com, July 16, 2019.

a. For each of the six companies, interpret the beta value.

b. How can investors use the beta value as a guide for investing?

13.50 Index funds are mutual funds that try to mimic the movement of leading indexes, such as the S&P 500 or the Russell 2000. The beta values (as described in Problem 13.49) for these funds are therefore approximately 1.0, and the estimated market models for these funds are approximately

$$\% \text{ weekly change in index fund} = 0.0$$
$$+ 1.0(\% \text{ weekly change in the index})$$

Leveraged index funds are designed to magnify the movement of major indexes. Direxion Funds is a leading provider of leveraged index and other alternative-class mutual fund products for investment advisors and sophisticated investors. Two of the company's funds are shown in the following table:

Name	Ticker Symbol	Description
Daily Small Cap Bull 3x Fund	TNA	300% of the Russell 2000 Index
Daily S&P 500 Bull 2x Fund	SPUU	200% of the S&P 500 Index

Source: Data extracted from **www.direxionfunds.com**.

The estimated market models for these funds are approximately

% daily change in TNA = 0.0
+ 3.0(% daily change in the Russell 2000)

% daily change in SPUU = 0.0
+ 2.0(% daily change in the S&P 500 Index)

Thus, if the Russell 2000 Index gains 10% over a period of time, the leveraged mutual fund TNA gains approximately 30%. On the downside, if the same index loses 20%, TNA loses approximately 60%.

a. The objective of the Direxion Funds Bull 2x Fund, SPUU, is 200% of the performance of the S&P 500 Index. What is its approximate market model?

b. If the S&P 500 Index gains 10% in a year, what return do you expect SPUU to have?

c. If the S&P 500 Index loses 20% in a year, what return do you expect SPUU to have?

d. What type of investors should be attracted to leveraged index funds? What type of investors should stay away from these funds?

13.51 The file Cereals contains the calories and sugar, in grams, in one serving of seven breakfast cereals:

Cereal	Calories	Sugar
Kellogg's All Bran	80	6
Kellogg's Corn Flakes	100	2
Wheaties	100	4
Nature's Path Organic Multigrain Flakes	110	4
Kellogg's Rice Krispies	130	4
Post Shredded Wheat Vanilla Almond	190	11
Kellogg's Mini Wheats	200	10

a. Compute and interpret the coefficient of correlation, r.

b. At the 0.05 level of significance, is there a significant linear relationship between calories and sugar?

13.52 Movie companies need to predict the gross receipts of an individual movie once the movie has debuted. The following results (stored in Potter Movies) are the first weekend gross, the U.S. gross, and the worldwide gross (in $millions) of the eight Harry Potter movies that debuted from 2001 to 2011:

Title	First Weekend	U.S. Gross	Worldwide Gross
Sorcerer's Stone	90.295	317.558	976.458
Chamber of Secrets	88.357	261.988	878.988
Prisoner of Azkaban	93.687	249.539	795.539
Goblet of Fire	102.335	290.013	896.013
Order of the Phoenix	77.108	292.005	938.469
Half-Blood Prince	77.836	301.460	934.601
Deathly Hallows Part I	125.017	295.001	955.417
Deathly Hallows Part II	169.189	381.001	1,328.11

Source: Data extracted from **www.the-numbers.com/interactive/comp-Harry-Potter.php**.

a. Compute the coefficient of correlation between first weekend gross and U.S. gross, first weekend gross and worldwide gross, and U.S. gross and worldwide gross.

b. At the 0.05 level of significance, is there a significant linear relationship between first weekend gross and U.S. gross, first weekend gross and worldwide gross, and U.S. gross and worldwide gross?

13.53 The file Mobile Speed contains the overall download and upload speeds in Mbps for eight carriers in the United States.
Source: Data extracted from "Tom's Guide, "Fastest Wireless Network 2019: It's Not Even Close," **bit.ly/2PcGiQE**.

a. Compute and interpret the coefficient of correlation, r.

b. At the 0.05 level of significance, is there a significant linear relationship between download and upload speed?

13.54 A survey by the Pew Research Center found that social networking is popular in many nations around the world. The file World Social Media contains the level of social media networking (measured as the percent of individuals polled who use social networking sites) and the GDP per capita based on purchasing power parity (PPP) for each of 28 emerging and developing countries.
Source: Data extracted from "2. Online Activities in Emerging and Developing Nations," **pewrsr.ch/1RX3Iqq**.

a. Compute and interpret the coefficient of correlation, r.

b. At the 0.05 level of significance, is there a significant linear relationship between GDP and social media usage?

c. What conclusions can you reach about the relationship between GDP and social media usage?

13.8 Estimation of Mean Values and Prediction of Individual Values

Section 13.2 discusses how a prediction line can be used to predict the mean value of Y for a given X. The Example 13.2 solution on page 436 uses this method to predict that the mean annual sales for stores that had 4 million profiled customers within a fixed radius is $7,087,900. This prediction is an example of a *point estimate* of the population mean. Chapter 8 introduces and explains confidence interval estimates, intervals around a point estimate. This section continues that discussion by presenting methods to develop a confidence interval estimate for the mean response for a given X and methods to develop a prediction interval for an individual response, Y, for a given value of X, a related concept.

The Confidence Interval Estimate for the Mean Response

Equation (13.20) defines the **confidence interval estimate for the mean response** for a given X.

CONFIDENCE INTERVAL ESTIMATE FOR THE MEAN OF Y

$$\hat{Y}_i \pm t_{\alpha/2} S_{YX} \sqrt{h_i}$$

$$\hat{Y}_i - t_{\alpha/2} S_{YX} \sqrt{h_i} \le \mu_{Y|X=X_i} \le \hat{Y}_i + t_{\alpha/2} S_{YX} \sqrt{h_i} \qquad (13.20)$$

where

$$h_i = \frac{1}{n} + \frac{(X_i - \bar{X})^2}{SSX}$$

$\hat{Y}_i$ = predicted value of Y; $\hat{Y}_i = b_0 + b_1 X_i$

S_{YX} = standard error of the estimate

n = sample size

X_i = given value of X

$\mu_{Y|X=X_i}$ = mean value of Y when $X = X_i$

$$SSX = \sum_{i=1}^{n} (X_i - \bar{X})^2$$

$t_{\alpha/2}$ = critical value corresponding to an upper-tail probability of $\alpha/2$ from the t distribution with $n - 2$ degrees of freedom (a cumulative area of $1 - \alpha/2$)

The width of the confidence interval in Equation (13.20) depends on several factors. Increased variation around the prediction line, as measured by the standard error of the estimate, results in a wider interval. As one would expect, increased sample size reduces the width of the interval. In addition, the width of the interval varies at different values of X. When one predicts Y for values of X close to $\bar{X}$, the interval is narrower than for predictions for X values farther away from $\bar{X}$.

In the Sunflowers Apparel example, suppose you want to construct a 95% confidence interval estimate of the mean annual sales (in $millions) for the entire population of stores that have 4 million profiled customers ($X = 4$). Using the simple linear regression equation

$$\hat{Y}_i = -1.2088 + 2.0742X_i$$

$$= -1.2088 + 2.0742(4) = 7.0879$$

$$\hat{Y}_i \pm t_{\alpha/2} S_{YX} \sqrt{h_i}$$

and given these

$$\bar{X} = 3.7786 \quad S_{YX} = 0.9993 \quad SSX = \sum_{i=1}^{n} (X_i - \bar{X})^2 = 15.5236$$

and, from Table E.3, $t_{\alpha/2} = 2.1788$, the confidence interval estimate is

$$\hat{Y}_i \pm t_{\alpha/2} S_{YX} \sqrt{h_i} = \hat{Y}_i \pm t_{\alpha/2} S_{YX} \sqrt{\frac{1}{n} + \frac{(X_i - \bar{X})^2}{SSX}}$$

$$= 7.0879 \pm (2.1788)(0.9993)\sqrt{\frac{1}{14} + \frac{(4 - 3.7786)^2}{15.5236}}$$

$$= 7.0879 \pm 0.5946$$

Therefore, the confidence interval estimate is

$$6.4932 \leq \mu_{Y|X=4} \leq 7.6825$$

Therefore, the 95% confidence interval estimate is that the population mean annual sales are between $6,493,200 and $7,682,500 for all stores with 4 million profiled customers.

The Prediction Interval for an Individual Response

In addition to constructing a confidence interval for the mean value of Y, one can also construct a prediction interval for an individual value of Y. Although the form of this interval is similar to that of the confidence interval estimate of Equation (13.20), the prediction interval is predicting an individual value, not estimating a mean. Equation (13.21) defines the **prediction interval for an individual response**, Y, at a given value, X_i, denoted by $Y_{X=X_i}$.

PREDICTION INTERVAL FOR AN INDIVIDUAL RESPONSE, Y

$$\hat{Y}_i \pm t_{\alpha/2} S_{YX} \sqrt{1 + h_i}$$

$$\hat{Y}_i - t_{\alpha/2} S_{YX} \sqrt{1 + h_i} \leq Y_{X=X_i} \leq \hat{Y}_i + t_{\alpha/2} S_{YX} \sqrt{1 + h_i} \qquad \textbf{(13.21)}$$

where

$Y_{X=X_i} =$ future value of Y when $X = X_i$

$t_{\alpha/2} =$ critical value corresponding to an upper-tail probability of $\alpha/2$ from the t distribution with $n - 2$ degrees of freedom (a cumulative area of $1 - \alpha/2$)

h_i, $\hat{Y}_i$, S_{YX}, n, and X_i as defined in Equation (13.20) on page 461.

To construct a 95% prediction interval of the annual sales for an individual store that has 4 million profiled customers ($X = 4$), first compute $\hat{Y}_i$. Using the prediction line

$$\hat{Y}_i = -1.2088 + 2.0742X_i$$

$$= -1.2088 + 2.0742(4)$$

$$= 7.0879$$

and given these

$$\overline{X} = 3.7786 \quad S_{YX} = 0.9993 \quad SSX = \sum_{i=1}^{n}(X_i - \overline{X})^2 = 15.5236$$

and, from Table E.3, $t_{\alpha/2} = 2.1788$, the prediction interval is

$$\hat{Y}_i \pm t_{\alpha/2} S_{YX} \sqrt{1 + \frac{1}{n} + \frac{(X_i - \overline{X})^2}{SSX}}$$

$$= 7.0879 \pm (2.1788)(0.9993)\sqrt{1 + \frac{1}{14} + \frac{(4 - 3.7786)^2}{15.5236}}$$

$$= 7.0879 \pm 2.2570$$

Therefore, the prediction interval is

$$4.8308 \leq Y_{X=4} \leq 9.3449$$

With 95% confidence, one predicts that the annual sales for an individual store with 4 million profiled customers is between \$4,830,800 and \$9,344,900.

The width of the prediction interval for annual sales at an individual store is much wider than the confidence interval estimate for the population mean annual sales. There is always much more variation in predicting an individual value than in estimating a mean value. Figure 13.21 presents the Excel results for the confidence interval estimate and the prediction interval for the Sunflowers Apparel data.

FIGURE 13.21

Excel worksheet results for the confidence interval estimate and prediction interval for the Sunflowers Apparel data

	A	B	
1	Confidence Interval Estimate and Prediction Interval		
2			
3	Data		
4	X Value	4	
5	Confidence Level	95%	
6			
7	Intermediate Calculations		
8	Sample Size	14	=COUNT(SLRData!A:A)
9	Degrees of Freedom	12	=B8 - 2
10	t Value	2.1788	=T.INV.2T(1 - B5, B9)
11	Sample Mean	3.7786	=AVERAGE(SLRData!A:A)
12	Sum of Squared Difference	15.5236	=DEVSQ(SLRData!A:A)
13	Standard Error of the Estimate	0.9993	=COMPUTE!B7
14	h Statistic	0.0746	=1/B8 + (B4 - B11)^2/B12
15	Predicted Y (YHat)	7.0879	=TREND(SLRData!B2:B15, SLRData!A2:A15, B4)
16			
17	For Average Y		
18	Interval Half Width	0.5946	=B10 * B13 * SQRT(B14)
19	Confidence Interval Lower Limit	6.4932	=B15 - B18
20	Confidence Interval Upper Limit	7.6825	=B15 + B18
21			
22	For Individual Response Y		
23	Interval Half Width	2.2570	=B10 * B13 * SQRT(1 + B14)
24	Prediction Interval Lower Limit	4.8308	=B15 - B23
25	Prediction Interval Upper Limit	9.3449	=B15 + B23

In the Worksheet

The T.INV.2T function in cell B10 computes the t critical value. The TREND function in cell B15 computes the predicted Y value from the regression data and the specified X value in cell B4.

AVERAGE, DEVSQ, which computes the SSX, and TREND all use data from a SLRData worksheet that contains the data for the independent X variable in column A and the data for the dependent Y variable in column B.

The **CIEandPI worksheet** of the Excel Guide **Simple Linear Regression** is a copy of the Figure 13.21 worksheet.

PROBLEMS FOR SECTION 13.8

LEARNING THE BASICS

13.55 Based on a sample of $n = 20$, the least-squares method was used to develop the following prediction line: $\hat{Y}_i = 5 + 3X_i$. In addition,

$$S_{YX} = 1.0 \quad \bar{X} = 2 \quad \sum_{i=1}^{n}(X_i - \bar{X})^2 = 20$$

a. Construct a 95% confidence interval estimate of the population mean response for $X = 2$.
b. Construct a 95% prediction interval of an individual response for $X = 2$.

13.56 Based on a sample of $n = 20$, the least-squares method was used to develop the following prediction line: $\hat{Y}_i = 5 + 3X_i$. In addition,

$$S_{YX} = 1.0 \quad \bar{X} = 2 \quad \sum_{i=1}^{n}(X_i - \bar{X})^2 = 20$$

a. Construct a 95% confidence interval estimate of the population mean response for $X = 4$.
b. Construct a 95% prediction interval of an individual response for $X = 4$.

c. Compare the results of (a) and (b) with those of Problem 13.55 (a) and (b). Which intervals are wider? Why?

APPLYING THE CONCEPTS

13.57 In Problem 13.5 on page 439, you used the summated rating of a restaurant to predict the cost of a meal. The data are stored in Restaurants .

a. Construct a 95% confidence interval estimate of the mean cost of a meal for restaurants that have a summated rating of 50.
b. Construct a 95% prediction interval of the cost of a meal for an individual restaurant that has a summated rating of 50.
c. Explain the difference in the results in (a) and (b).

✓ SELF TEST **13.58** In Problem 13.4 on page 439, you used the percentage of alcohol to predict wine quality. The data are stored in Vinho Verde . For these data, $S_{YX} = 0.9369$ and $h_i = 0.024934$ when $X = 10$.

a. Construct a 95% confidence interval estimate of the mean wine quality rating for all wines that have 10% alcohol.
b. Construct a 95% prediction interval of the wine quality rating of an individual wine that has 10% alcohol.
c. Explain the difference in the results in (a) and (b).

13.59 In Problem 13.7 on page 440, you used the plate gap on the bag-sealing equipment to predict the tear rating of a bag of coffee. The data are stored in Starbucks .
a. Construct a 95% confidence interval estimate of the mean tear rating for all bags of coffee when the plate gap is 0.
b. Construct a 95% prediction interval of the tear rating for an individual bag of coffee when the plate gap is 0.
c. Why is the interval in (a) narrower than the interval in (b)?

13.60 In Problem 13.6 on page 440, a prospective MBA student wanted to predict starting salary upon graduation, based on program per-year tuition. The data are stored in FTMBA .
a. Construct a 95% confidence interval estimate of the mean starting salary upon graduation of an individual program with per-year tuition cost of $50,450.
b. Construct a 95% prediction interval of the starting salary upon graduation of an individual program with per-year tuition cost of $50,450.
c. Why is the interval in (a) narrower than the interval in (b)?

13.61 In Problem 13.9 on page 440, an agent for a real estate company wanted to predict the monthly rent for one-bedroom apartments, based on the size of an apartment. The data are stored in Silver Spring Rentals .
a. Construct a 95% confidence interval estimate of the mean monthly rental for all one-bedroom apartments that are 800 square feet in size.

b. Construct a 95% prediction interval of the monthly rental for an individual one-bedroom apartment that is 800 square feet in size.
c. Explain the difference in the results in (a) and (b).

13.62 In Problem 13.8 on page 440, you used annual revenue to predict the value of a MLB baseball team, using the data stored in MLB Values .
a. Construct a 95% confidence interval estimate of the mean value of all baseball franchises that generate $250 million of annual revenue.
b. Construct a 95% prediction interval of the value of an individual baseball franchise that generates $250 million of annual revenue.
c. Explain the difference in the results in (a) and (b).

13.63 In Problem 13.10 on page 440, you used YouTube trailer views to predict movie weekend box office gross from data stored in Movie . A movie, about to be released, has 50 million YouTube trailer views.
a. What is the predicted weekend box office gross?
b. Which interval is more useful here, the confidence interval estimate of the mean or the prediction interval for an individual response? Explain.
c. Construct and interpret the interval you selected in (b).

13.9 Potential Pitfalls in Regression

Using regression analysis has several different types of potential pitfalls. Regression analysis requires knowledge of the subject matter, which, in turn, requires proper definition of the problem being solved or the goal being sought, the first task of the DCOVA framework (see the First Things First chapter). Without knowledge of the subject matter, important variables may be omitted from the regression model or nonsensical relationships among variables wrongly explored.

Many potential pitfalls arise from overlooking the issues that this chapter discusses. Being unaware of the assumptions of least-squares regression, not knowing how to evaluate the assumptions of least-squares regression, or extrapolating outside the relevant range are all common errors. Overlooking the alternatives that exist to least-squares regression if an assumption is violated or thinking that every relationship must be linear are errors seen in uninformed or incomplete analysis. In addition, not considering logical casuality, a basic principle of statistics (see page 6), can also occur when one gets too focused on the mechanics of performing a regression analysis.

Exhibit 13.1 lists the seven steps that helps one avoid the potential pitfalls of regression analysis.

EXHIBIT 13.1

Seven Steps for Avoiding the Potential Regression Pitfalls

1. Be clear about the problem or goal being investigated and the variables that need to be examined.
2. Construct a scatter plot to observe the possible relationship between X and Y.
3. Perform a residual analysis to check the assumptions of regression (linearity, independence, normality, equal variance):
   a. Plot the residuals versus the independent variable to determine whether the linear model is appropriate and to check for equal variance.

(*continued*)

    b. Construct a histogram, stem-and-leaf display, boxplot, or normal probability plot of the residuals to check for normality.

    c. Plot the residuals versus time to check for independence. (This step is necessary only if the data are collected over time.)

4. If there are violations of the assumptions, use alternative methods to least-squares regression or alternative least-squares models (Kutner).

5. If there are no violations of the assumptions, carry out tests for the significance of the regression coefficients and develop confidence and prediction intervals.

6. Refrain from making predictions and forecasts outside the relevant range of the independent variable.

7. Remember that the relationships identified in observational studies may or may not be due to cause-and-effect relationships. (While causation implies correlation, correlation does not imply causation.)

Someone not familiar with the assumptions of regression or how to evaluate those assumptions may reach wrong conclusions about the data being analyzed. For example, Table 13.7, stored in Anscombe, presents the Anscombe data set that illustrates the importance of using scatter plots and residual analysis to complement the calculation of the Y intercept, the slope, and r^2.

TABLE 13.7
Four sets of artificial data

Data Set A		Data Set B		Data Set C		Data Set D	
X_i	Y_i	X_i	Y_i	X_i	Y_i	X_i	Y_i
10	8.04	10	9.14	10	7.46	8	6.58
14	9.96	14	8.10	14	8.84	8	5.76
5	5.68	5	4.74	5	5.73	8	7.71
8	6.95	8	8.14	8	6.77	8	8.84
9	8.81	9	8.77	9	7.11	8	8.47
12	10.84	12	9.13	12	8.15	8	7.04
4	4.26	4	3.10	4	5.39	8	5.25
7	4.82	7	7.26	7	6.42	19	12.50
11	8.33	11	9.26	11	7.81	8	5.56
13	7.58	13	8.74	13	12.74	8	7.91
6	7.24	6	6.13	6	6.08	8	6.89

Source: Data extracted from F. J. Anscombe, "Graphs in Statistical Analysis," *The American Statistician*, 27 (1973), pp. 17–21.

Anscombe showed that all four data sets in Table 13.7 have the following identical results:

$$\hat{Y}_i = 3.0 + 0.5X_i \quad S_{YX} = 1.237 \quad S_{b_1} = 0.118 \quad r^2 = 0.667$$

$$SSR = 27.51 \quad SSE = 12.76 \quad SST = 41.27$$

Stopping the regression analysis of these data sets at this point would fail to observe the important differences among the sets. On page 466, the Figure 13.22 scatter plots and residual plots show how different the four data sets are!

FIGURE 13.22
Scatter plots and residual plots for the data sets A, B, C, and D

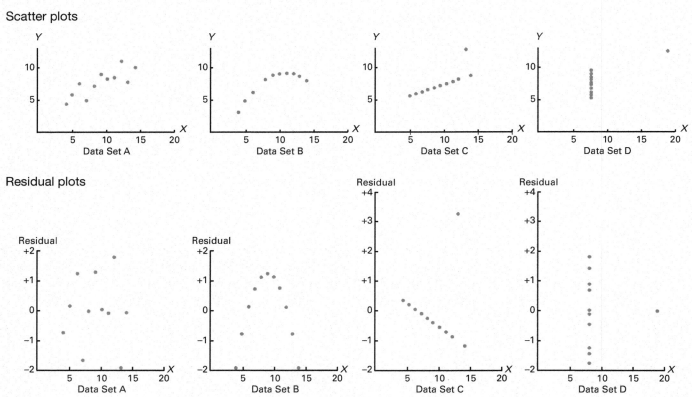

Scatter plots

Residual plots

Each data set has a different relationship between X and Y. The only data set that seems to approximately follow a straight line is data set A. The data set A residual plot does not show any obvious patterns or outlying residuals. This is not true for the residual plots for data sets B, C, and D. The data set B scatter plot shows that a curvilinear regression model is the more appropriate model and the data set B residual plot reinforces that conclusion.

The scatter plot and the residual plot for data set C clearly show an outlying observation. For this case, one can remove the outlier and reestimate the regression model (see Kutner). The scatter plot for data set D reveals a regression model that is heavily dependent on the outcome of a single data point ($X_8 = 19$ and $Y_8 = 12.50$). Any regression model with this characteristic should be used with caution.

▼USING **STATISTICS**
Knowing Customers ..., Revisited

In the Knowing Customers at Sunflowers Apparel scenario, you were the director of planning for a chain of upscale clothing stores for women. Until now, Sunflowers managers selected sites based on factors such as the availability of a good lease or a subjective opinion that a location seemed like a good place for a store. To make more objective decisions, you asked a marketing firm to identify and classify groups of consumers. After this first definitional step, the DCOVA framework was to develop a regression model to analyze the relationship between the number of profiled customers who live within a fixed radius of a Sunflowers store and the annual sales of the store. The model indicates that about 84.8% of the variation in sales was explained by the number of profiled customers who live within a fixed radius of a Sunflowers store. Furthermore, for each increase of 1 million profiled customers, mean annual sales were estimated to increase by $2.0742 million. With the LINE assumptions verified for this model, Sunflowers management can use the model to help make better decisions when selecting new sites for stores or to forecast sales for existing stores.

▼SUMMARY

This chapter develops the simple linear regression model and discusses the assumptions the model uses and how to evaluate them. Once assumptions are verified, the model can be used to predict values by using the prediction line and test for the significance of the slope. Figure 13.23 provides a roadmap for navigating through the process of applying a simple linear regresssion model to a set of data.

FIGURE 13.23
Roadmap for simple linear regression

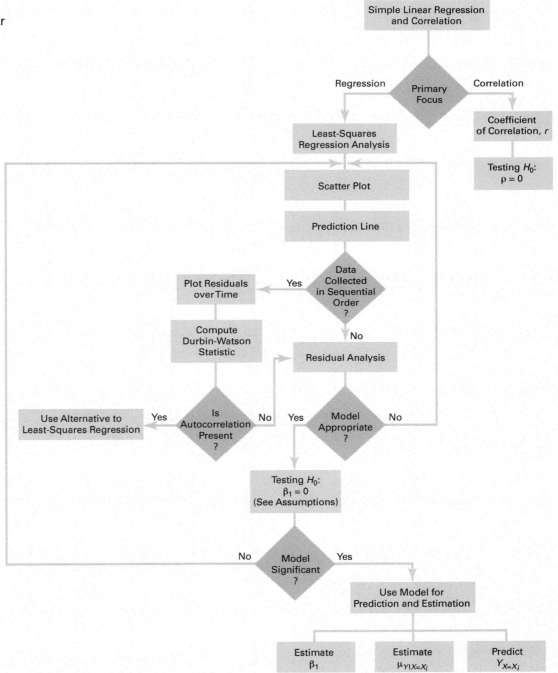

▼ REFERENCES

Anscombe, F. J. "Graphs in Statistical Analysis." *The American Statistician*, 27(1973): 17–21.

Hoaglin, D. C., and R. Welsch. "The Hat Matrix in Regression and ANOVA." *The American Statistician*, 32(1978): 17–22.

Hocking, R. R. "Developments in Linear Regression Methodology: 1959–1982." *Technometrics*, 25(1983): 219–250.

Kutner, M. H., C. J. Nachtsheim, J. Neter, and W. Li. *Applied Linear Statistical Models*, 5th ed. New York: McGraw-Hill/Irwin, 2005.

Montgomery, D. C., E. A. Peck, and G. G. Vining. *Introduction to Linear Regression Analysis*, 5th ed. New York, Wiley, 2012.

▼ KEY EQUATIONS

Simple Linear Regression Model

$$Y_i = \beta_0 + \beta_1 X_i + \varepsilon_i \tag{13.1}$$

Simple Linear Regression Equation: The Prediction Line

$$\hat{Y}_i = b_0 + b_1 X_i \tag{13.2}$$

Computational Formula for the Slope, b_1

$$b_1 = \frac{SSXY}{SSX} \tag{13.3}$$

Computational Formula for the Y Intercept, b_0

$$b_0 = \bar{Y} - b_1 \bar{X} \tag{13.4}$$

Measures of Variation in Regression

$$SST = SSR + SSE \tag{13.5}$$

Total Sum of Squares (SST)

$$SST = \text{Total sum of squares} = \sum_{i=1}^{n}(Y_i - \bar{Y})^2 \tag{13.6}$$

Regression Sum of Squares (SSR)

$SSR = $ Explained variation or regression sum of squares

$$= \sum_{i=1}^{n}(\hat{Y}_i - \bar{Y})^2 \tag{13.7}$$

Error Sum of Squares (SSE)

$SSE = $ Unexplained variation or error sum of squares

$$= \sum_{i=1}^{n}(Y_i - \hat{Y}_i)^2 \tag{13.8}$$

Computational Formula for SST

$$SST = \sum_{i=1}^{n}(Y_i - \bar{Y})^2 = \sum_{i=1}^{n}Y_i^2 - \frac{\left(\sum_{i=1}^{n}Y_i\right)^2}{n} \tag{13.9}$$

Computational Formula for SSR

$$SSR = \sum_{i=1}^{n}(\hat{Y}_i - \bar{Y})^2 = b_0\sum_{i=1}^{n}Y_i + b_1\sum_{i=1}^{n}X_iY_i - \frac{\left(\sum_{i=1}^{n}Y_i\right)^2}{n} \tag{13.10}$$

Computational Formula for SSE

$$SSE = \sum_{i=1}^{n}(Y_i - \hat{Y}_i)^2 = \sum_{i=1}^{n}Y_i^2 - b_0\sum_{i=1}^{n}Y_i - b_1\sum_{i=1}^{n}X_iY_i \tag{13.11}$$

Coefficient of Determination

$$r^2 = \frac{\text{Regression sum of squares}}{\text{Total sum of squares}} = \frac{SSR}{SST} \tag{13.12}$$

Standard Error of the Estimate

$$S_{YX} = \sqrt{\frac{SSE}{n-2}} = \sqrt{\frac{\sum_{i=1}^{n}(Y_i - \hat{Y}_i)^2}{n-2}} \tag{13.13}$$

Residual

$$e_i = Y_i - \hat{Y}_i \tag{13.14}$$

Durbin-Watson Statistic

$$D = \frac{\sum_{i=2}^{n}(e_i - e_{i-1})^2}{\sum_{i=1}^{n}e_i^2} \tag{13.15}$$

t Test Statistic for Testing a Hypothesis for a Population Slope, β_1

$$t_{STAT} = \frac{b_1 - \beta_1}{S_{b_1}} \tag{13.16}$$

F Test Statistic for Testing a Hypothesis for a Population Slope, β_1

$$F_{STAT} = \frac{MSR}{MSE} \tag{13.17}$$

Confidence Interval Estimate of the Slope, β_1

$$b_1 \pm t_{\alpha/2} S_{b_1}$$

$$b_1 - t_{\alpha/2} S_{b_1} \leq \beta_1 \leq b_1 + t_{\alpha/2} S_{b_1} \tag{13.18}$$

Testing for the Existence of Correlation

$$t_{STAT} = \frac{r - \rho}{\sqrt{\dfrac{1 - r^2}{n - 2}}} \qquad \text{(13.19a)}$$

$$r = \frac{\text{cov}(X, Y)}{S_X S_Y} \qquad \text{(13.19b)}$$

Confidence Interval Estimate for the Mean of Y

$$\hat{Y}_i \pm t_{\alpha/2} S_{YX} \sqrt{h_i}$$

$$\hat{Y}_i - t_{\alpha/2} S_{YX} \sqrt{h_i} \le \mu_{Y|X = X_i} \le \hat{Y}_i + t_{\alpha/2} S_{YX} \sqrt{h_i} \qquad \text{(13.20)}$$

Prediction Interval for an Individual Response, Y

$$\hat{Y}_i \pm t_{\alpha/2} S_{YX} \sqrt{1 + h_i}$$

$$\hat{Y}_i - t_{\alpha/2} S_{YX} \sqrt{1 + h_i} \le Y_{X = X_i} \le \hat{Y}_i + t_{\alpha/2} S_{YX} \sqrt{1 + h_i} \qquad \text{(13.21)}$$

▼ KEY TERMS

assumptions of regression   446
autocorrelation   450
coefficient of determination   443
confidence interval estimate for the mean response   461
correlation coefficient   457
dependent variable   431
Durbin-Watson statistic   451
equal variance   446
error sum of squares (SSE)   441
explained variation   441
explanatory variable   431
homoscedasticity   446
independence of errors   446

independent variable   431
least-squares method   434
linearity   446
model   431
normality   446
positive linear relationship   432
prediction interval for an individual response, Y   462
prediction line   433
regression analysis   431
regression coefficient   434
regression sum of squares (SSR)   441
relevant range   436

residual   446
residual analysis   446
response variable   431
scatter diagram   431
scatter plot   431
simple linear regression   431
simple linear regression equation   433
slope   433
standard error of the estimate   444
total sum of squares (SST)   441
total variation   441
unexplained variation   441
Y intercept   433

▼ CHECKING YOUR UNDERSTANDING

13.64 What is the interpretation of the Y intercept and the slope in the simple linear regression equation?

13.65 What is the interpretation of the coefficient of determination?

13.66 When is the unexplained variation (i.e., error sum of squares) equal to 0?

13.67 When is the explained variation (i.e., regression sum of squares) equal to 0?

13.68 Why should you always carry out a residual analysis as part of a regression model?

13.69 What are the assumptions of regression analysis?

13.70 How do you evaluate the assumptions of regression analysis?

13.71 When and how do you use the Durbin-Watson statistic?

13.72 What is the difference between a confidence interval estimate of the mean response, $\mu_{Y|X = X_i}$, and a prediction interval of $Y_{X = X_i}$?

▼CHAPTER REVIEW PROBLEMS

13.73 Can you use movie critics' opinions to forecast box office receipts on the opening weekend? The following data, stored in ` Tomatometer `, indicate the Tomatometer rating, the percentage of professional critic reviews that are positive, and the receipts per theater ($thousands) on the weekend a movie opened for 10 movies:

Movie	Tomatometer Rating	Receipts
The Mummy	16	7.8
Zookeeper's Wife	61	6.1
Beatriz at Dinner	80	28.4
The Hero	76	11.3
Wonder Woman	93	24.8
Baby Boss	52	13.3
The Circle	15	2.9
Dean	61	4.0
Baywatch	20	5.1
Churchill	38	1.9

Source: "Top Box Office Movies – Rotten Tomatoes," and "The Numbers – Weekend Box Office Chart for May 26th 2017," **bit.ly/2t0tqS6**.

a. Use the least-squares method to compute the regression coefficients b_0 and b_1.
b. Interpret the meaning of b_0 and b_1 in this problem.
c. Predict the mean receipts for a movie that has a Tomatometer rating of 55%.
d. Should you use the model to predict the receipts for a movie that has a Tomatometer rating of 5%? Why or why not?
e. Determine the coefficient of determination, r^2, and explain its meaning in this problem.
f. Perform a residual analysis. Is there any evidence of a pattern in the residuals? Explain.
g. At the 0.05 level of significance, is there evidence of a linear relationship between Tomatometer rating and receipts?
h. Construct a 95% confidence interval estimate of the mean receipts for a movie that has a Tomatometer rating of 55% and a 95% prediction interval of the receipts for a single movie that has a Tomatometer rating of 55%.
i. Based on the results of (a)–(h), do you think that Tomatometer rating is a useful predictor of receipts on the first weekend a movie opens? What issues about these data might make you hesitant to use Tomatometer rating to predict receipts?

13.74 Management of a soft-drink bottling company has the business objective of developing a method for allocating delivery costs to customers. Although one cost clearly relates to travel time within a particular route, another variable cost reflects the time required to unload the cases of soft drink at the delivery point. To begin, management decided to develop a regression model to predict delivery time based on the number of cases delivered. A sample of 20 deliveries within a territory was selected. The delivery times and the number of cases delivered were organized in the following table and stored in ` Delivery `.

Customer	Number of Cases	Delivery Time (minutes)	Customer	Number of Cases	Delivery Time (minutes)
1	52	32.1	11	161	43.0
2	64	34.8	12	184	49.4
3	73	36.2	13	202	57.2
4	85	37.8	14	218	56.8
5	95	37.8	15	243	60.6
6	103	39.7	16	254	61.2
7	116	38.5	17	267	58.2
8	121	41.9	18	275	63.1
9	143	44.2	19	287	65.6
10	157	47.1	20	298	67.3

a. Use the least-squares method to compute the regression coefficients b_0 and b_1.
b. Interpret the meaning of b_0 and b_1 in this problem.
c. Predict the mean delivery time for 150 cases of soft drink.
d. Should you use the model to predict the delivery time for a customer who is receiving 500 cases of soft drink? Why or why not?
e. Determine the coefficient of determination, r^2, and explain its meaning in this problem.
f. Perform a residual analysis. Is there any evidence of a pattern in the residuals? Explain.
g. At the 0.05 level of significance, is there evidence of a linear relationship between delivery time and the number of cases delivered?
h. Construct a 95% confidence interval estimate of the mean delivery time for 150 cases of soft drink and a 95% prediction interval of the delivery time for a single delivery of 150 cases of soft drink.
i. What conclusions can you reach from (a) through (h) about the relationship between the number of cases and delivery time?

13.75 Measuring the height of a California redwood tree is very difficult because these trees grow to heights of more than 300 feet. People familiar with these trees understand that the height of a California redwood tree is related to other characteristics of the tree, including the diameter of the tree at the breast height of a person. The data in ` Redwood ` represent the height (in feet) and diameter (in inches) at the breast height of a person for a sample of 21 California redwood trees.
a. Assuming a linear relationship, use the least-squares method to compute the regression coefficients b_0 and b_1. State the regression equation that predicts the height of a tree based on the tree's diameter at breast height of a person.
b. Interpret the meaning of the slope in this equation.
c. Predict the mean height for a tree that has a breast height diameter of 25 inches.
d. Interpret the meaning of the coefficient of determination in this problem.
e. Perform a residual analysis on the results and determine the adequacy of the model.

f. Determine whether there is a significant relationship between the height of redwood trees and the breast height diameter at the 0.05 level of significance.

g. Construct a 95% confidence interval estimate of the population slope between the height of the redwood trees and breast height diameter.

h. What conclusions can you reach about the relationship of the diameter of the tree and its height?

13.76 You want to develop a model to predict the asking price of homes based on their size. A sample of 61 single-family houses listed for sale in Silver Spring, Maryland, a suburb of Washington, DC, is selected to study the relationship between asking price (in $thousands) and living space (in square feet), and the data are collected and stored in Silver Spring Homes . (Hint: First determine which are the independent and dependent variables.)

a. Construct a scatter plot and, assuming a linear relationship, use the least-squares method to compute the regression coefficients b_0 and b_1.

b. Interpret the meaning of the Y intercept, b_0, and the slope, b_1, in this problem.

c. Use the prediction line developed in (a) to predict the mean asking price for a house whose living space is 2,000 square feet.

d. Determine the coefficient of determination, r^2, and interpret its meaning in this problem.

e. Perform a residual analysis on your results and evaluate the regression assumptions.

f. At the 0.05 level of significance, is there evidence of a linear relationship between asking price and living space?

g. Construct a 95% confidence interval estimate of the population slope.

h. What conclusions can you reach about the relationship between the living space and asking price?

13.77 You want to develop a model to predict the taxes of houses, based on asking price. A sample of 61 single-family houses listed for sale in Silver Spring, Maryland, a suburb of Washington, DC, is selected. The taxes (in $) and the asking price of the houses (in $thousands) are recorded and stored in Silver Spring Homes . (Hint: First determine which are the independent and dependent variables.)

a. Construct a scatter plot and, assuming a linear relationship, use the least-squares method to compute the regression coefficients b_0 and b_1.

b. Interpret the meaning of the Y intercept, b_0, and the slope, b_1, in this problem.

c. Use the prediction line developed in (a) to predict the mean taxes for a house whose asking price is $400,000.

d. Determine the coefficient of determination, r^2, and interpret its meaning in this problem.

e. Perform a residual analysis on your results and evaluate the regression assumptions.

f. At the 0.05 level of significance, is there evidence of a linear relationship between taxes and asking price?

g. What conclusions can you reach concerning the relationship between taxes and asking price?

13.78 An analyst has the objective of predicting the return on average tangible common equity (ROATCE) of banks. The analyst begins by using *efficiency ratio*, a measure of a bank's ability to turn resources into revenue. A sample of 100 American banks is selected and stored in American Banks .

Source: Data extracted from K. Badenhausen, "America's Best Banks 2017," available at **bit.ly/2tpw1Er**.

a. Construct a scatter plot and, assuming a linear relationship, use the least-squares method to compute the regression coefficients b_0 and b_1.

b. Interpret the meaning of the Y intercept, b_0, and the slope, b_1, in this problem.

c. Use the prediction line developed in (a) to predict the mean ROATCE for a bank with an efficiency ratio of 60%.

d. Determine the coefficient of determination, r^2, and interpret its meaning in this problem.

e. Perform a residual analysis on your results and evaluate the regression assumptions.

f. At the 0.05 level of significance, is there evidence of a linear relationship between efficiency ratio and ROATCE?

g. Construct a 95% confidence interval estimate of the mean ROATCE of banks with an efficiency ratio of 60% and a 95% prediction interval of the ROATCE for a particular bank with an efficiency ratio of 60%.

h. Construct a 95% confidence interval estimate of the population slope.

i. What conclusions can you reach concerning the relationship between efficiency ratio and ROATCE?

13.79 An accountant for a large department store has the business objective of developing a model to predict the amount of time it takes to process invoices. Data are collected from the past 32 working days, and the number of invoices processed and completion time (in hours) are stored in Invoice . (Hint: First determine which are the independent and dependent variables.)

a. Assuming a linear relationship, use the least-squares method to compute the regression coefficients b_0 and b_1.

b. Interpret the meaning of the Y intercept, b_0, and the slope, b_1, in this problem.

c. Use the prediction line developed in (a) to predict the mean amount of time it would take to process 150 invoices.

d. Determine the coefficient of determination, r^2, and interpret its meaning.

e. Plot the residuals against the number of invoices processed and also against time.

f. Based on the plots in (e), does the model seem appropriate?

g. Based on the results in (e) and (f), what conclusions can you reach about the validity of the prediction made in (c)?

h. What conclusions can you reach about the relationship between the number of invoices and the completion time?

13.80 On January 28, 1986, the space shuttle *Challenger* exploded, and seven astronauts lost their lives. Engineers for the manufacturer of the rocket motor prepared charts to make the case that the launch should not take place due to the cold weather that day. These arguments were rejected, and the launch tragically took place. Upon investigation after the tragedy, experts agreed that the disaster occurred because of leaky rubber O-rings that did not seal properly due to the cold temperature. Data indicating the atmospheric temperature at the time of 23 previous launches and the O-ring damage index are stored in O-Ring . (Data from flight 4 are omitted due to unknown O-ring condition.)

Sources: Data extracted from *Report of the Presidential Commission on the Space Shuttle Challenger Accident*, Washington, DC, 1986, Vol. II (H1–H3) and Vol. IV (664); and *Post-Challenger Evaluation of Space Shuttle Risk Assessment and Management*, Washington, DC, 1988, pp. 135–136.

a. Construct a scatter plot for the seven flights in which there was O-ring damage (O-ring damage index $\neq$ 0). What conclusions, if any, can you reach about the relationship between atmospheric temperature and O-ring damage?

b. Construct a scatter plot for all 23 flights.

c. Explain any differences in the interpretation of the relationship between atmospheric temperature and O-ring damage in (a) and (b).

d. Based on the scatter plot in (b), provide reasons why a prediction should not be made for an atmospheric temperature of 31°F, the temperature on the morning of the launch of the *Challenger*.

e. Although the assumption of a linear relationship may not be valid for the set of 23 flights, fit a simple linear regression model to predict O-ring damage, based on atmospheric temperature.

f. Include the prediction line found in (e) on the scatter plot developed in (b).

g. Based on the results in (f), do you think a linear model is appropriate for these data? Explain.

h. Perform a residual analysis. What conclusions do you reach?

13.81 A baseball analyst would like to study various team statistics for a recent season to determine which variables might be useful in predicting the number of wins achieved by teams during the season. He begins by using a team's earned run average (ERA), a measure of pitching performance, to predict the number of wins. He collects the team ERA and team wins for each of the 30 MLB baseball teams and stores these data in Baseball . (Hint: First determine which are the independent and dependent variables.)

a. Assuming a linear relationship, use the least-squares method to compute the regression coefficients b_0 and b_1.

b. Interpret the meaning of the Y intercept, b_0, and the slope, b_1, in this problem.

c. Use the prediction line developed in (a) to predict the mean number of wins for a team with an ERA of 4.50.

d. Compute the coefficient of determination, r^2, and interpret its meaning.

e. Perform a residual analysis on your results and determine the adequacy of the fit of the model.

f. At the 0.05 level of significance, is there evidence of a linear relationship between the number of wins and the ERA?

g. Construct a 95% confidence interval estimate of the mean number of wins expected for teams with an ERA of 4.50.

h. Construct a 95% prediction interval of the number of wins for an individual team that has an ERA of 4.50.

i. Construct a 95% confidence interval estimate of the population slope.

j. The 30 teams constitute a population. In order to use statistical inference, as in (f) through (i), the data must be assumed to represent a random sample. What "population" would this sample be drawing conclusions about?

k. What other independent variables might you consider for inclusion in the model?

l. What conclusions can you reach concerning the relationship between ERA and wins?

13.82 Can you use the annual revenues generated by NBA teams to predict current values? Figure 2.17 on page 66 shows a scatter plot of team revenue and current value, and Figure 3.12 on page 138, shows the correlation coefficient. Now, you want to develop a simple linear regression model to predict current values based on revenues. (Team current values and revenues are stored in NBA Financial .)

a. Assuming a linear relationship, use the least-squares method to compute the regression coefficients b_0 and b_1.

b. Interpret the meaning of the Y intercept, b_0, and the slope, b_1, in this problem.

c. Predict the mean current value of an NBA team that generates $150 million of annual revenue.

d. Compute the coefficient of determination, r^2, and interpret its meaning.

e. Perform a residual analysis on your results and evaluate the regression assumptions.

f. At the 0.05 level of significance, is there evidence of a linear relationship between the annual revenues generated and the current value of an NBA team?

g. Construct a 95% confidence interval estimate of the mean value of all NBA teams that generate $150 million of annual revenue.

h. Construct a 95% prediction interval of the value of an individual NBA team that generates $150 million of annual revenue.

i. Compare the results of (a) through (h) to those of baseball teams in Problems 13.8, 13.20, 13.30, 13.46, and 13.62 and European soccer teams in Problem 13.83.

13.83 In Problem 13.82 you used annual revenue to develop a model to predict the current value of NBA teams. Can you also use the annual revenues generated by European soccer teams to predict their current values? (European soccer team current values and revenues are stored in Soccer Values .)

a. Repeat Problem 13.82 (a) through (h) for the European soccer teams.

b. Compare the results of (a) to those of baseball teams in Problems 13.8, 13.20, 13.30, 13.46, and 13.62 and NBA teams in Problem 13.82.

13.84 During the fall harvest season in the United States, pumpkins are sold in large quantities at farm stands. Often, instead of weighing the pumpkins prior to sale, the farm stand operator will just place the pumpkin in the appropriate circular cutout on the counter. When asked why this was done, one farmer replied, "I can tell the weight of the pumpkin from its circumference." To determine whether this was really true, the circumference and weight of each pumpkin from a sample of 23 pumpkins were determined and the results stored in Pumpkin .

a. Assuming a linear relationship, use the least-squares method to compute the regression coefficients b_0 and b_1.

b. Interpret the meaning of the slope, b_1, in this problem.

c. Predict the mean weight for a pumpkin that is 60 centimeters in circumference.

d. Do you think it is a good idea for the farmer to sell pumpkins by circumference instead of weight? Explain.

e. Determine the coefficient of determination, r^2, and interpret its meaning.

f. Perform a residual analysis for these data and evaluate the regression assumptions.

g. At the 0.05 level of significance, is there evidence of a linear relationship between the circumference and weight of a pumpkin?

h. Construct a 95% confidence interval estimate of the population slope, β_1.

REPORT WRITING EXERCISE

13.85 In Problems 13.8, 13.20, 13.30, 13.46, 13.62, 13.82, and 13.83, you developed regression models to predict current value of MLB baseball, NBA basketball, and European soccer teams. Now, write a report based on the models you developed. Append to your report all appropriate charts and statistical information.

▾CASES

Managing Ashland MultiComm Services

To ensure that as many trial subscriptions to the *3-For-All* service as possible are converted to regular subscriptions, the marketing department works closely with the customer support department to accomplish a smooth initial process for the trial subscription customers. To assist in this effort, the marketing department needs to accurately forecast the monthly total of new regular subscriptions.

A team consisting of managers from the marketing and customer support departments was convened to develop a better method of forecasting new subscriptions. Previously, after examining new subscription data for the prior three months, a group of three managers would develop a subjective forecast of the number of new subscriptions. Livia Salvador, who was recently hired by the company to provide expertise in quantitative forecasting methods, suggested that the department look for factors that might help in predicting new subscriptions.

Members of the team found that the forecasts in the past year had been particularly inaccurate because in some months, much more time was spent on telemarketing than in other months. Livia collected data (stored in **AMS13**) for the number of new subscriptions and hours spent on telemarketing for each month for the past two years.

1. What criticism can you make concerning the method of forecasting that involved taking the new subscriptions data for the prior three months as the basis for future projections?

2. What factors other than number of telemarketing hours spent might be useful in predicting the number of new subscriptions? Explain.

3. **a.** Analyze the data and develop a regression model to predict the number of new subscriptions for a month, based on the number of hours spent on telemarketing for new subscriptions.

   b. If you expect to spend 1,200 hours on telemarketing per month, estimate the number of new subscriptions for the month. Indicate the assumptions on which this prediction is based. Do you think these assumptions are valid? Explain.

   c. What would be the danger of predicting the number of new subscriptions for a month in which 2,000 hours were spent on telemarketing?

Digital Case

Apply your knowledge of simple linear regression in this Digital Case, which extends the Sunflowers Apparel Using Statistics scenario from this chapter.

Leasing agents from the Triangle Mall Management Corporation have suggested that Sunflowers consider several locations in some of Triangle's newly renovated lifestyle malls that cater to shoppers with higher-than-mean disposable income. Although the locations are smaller than the typical Sunflowers location, the leasing agents argue that higher-than-mean disposable income in the surrounding community is a better predictor of higher sales than profiled customers. The leasing agents maintain that sample data from 14 Sunflowers stores prove that this is true.

Open **Triangle_Sunflower.pdf** and review the leasing agents' proposal and supporting documents. Then answer the following questions:

1. Should mean disposable income be used to predict sales based on the sample of 14 Sunflowers stores?

2. Should the management of Sunflowers accept the claims of Triangle's leasing agents? Why or why not?

3. Is it possible that the mean disposable income of the surrounding area is not an important factor in leasing new locations? Explain.

4. Are there any other factors not mentioned by the leasing agents that might be relevant to the store leasing decision?

Brynne Packaging

Brynne Packaging is a large packaging company, offering its customers the highest standards in innovative packaging solutions and reliable service. About 25% of the employees at Brynne Packaging are machine operators. The human resources department has suggested that the company consider using the Wesman Personnel Classification Test (WPCT), a measure of reasoning ability, to screen applicants for the machine operator job. In order to assess the WPCT as a predictor of future job performance, 25 recent applicants were tested using the WPCT; all were hired, regardless of their WPCT score. At a later time, supervisors were asked to rate the quality of the job performance of these 25 employees, using a 1-to-10 rating scale (where 1 = very low and 10 = very high). Factors considered in the ratings included the employee's output, defect rate, ability to implement continuous quality procedures, and contributions to team problem-solving efforts. The file **Brynne Packaging** contains the WPCT scores (WPCT) and job performance ratings (Ratings) for the 25 employees.

1. Assess the significance and importance of WPCT score as a predictor of job performance. Defend your answer.

2. Predict the mean job performance rating for all employees with a WPCT score of 6. Give a point prediction as well as a 95% confidence interval. Do you have any concerns using the regression model for predicting mean job performance rating given the WPCT score of 6?

3. Evaluate whether the assumptions of regression have been seriously violated.

EG13.2 DETERMINING the SIMPLE LINEAR REGRESSION EQUATION

Key Technique Use the **LINEST(*cell range of Y variable, cell range of X variable*, True, True)** array function to compute the b_1 and b_0 coefficients, the b_1 and b_0 standard errors, r^2 and the standard error of the estimate, the F test statistic and error *df*, and *SSR* and *SSE*.
Use the expression **T.INV.2T(1 – *confidence level, Error degrees of freedom*)** to compute the critical value for the *t* test.

Example Perform the Figure 13.4 analysis of the Sunflowers Apparel data on page 435.

PHStat Use **Simple Linear Regression**.

For the example, open to the **DATA worksheet** of the **Site Selection workbook**. Select **PHStat➔Regression➔Simple Linear Regression**. In the procedure's dialog box (shown below):

1. Enter **C1:C15** as the **Y Variable Cell Range**.
2. Enter **B1:B15** as the **X Variable Cell Range**.
3. Check **First cells in both ranges contain label**.
4. Enter **95** as the **Confidence level for regression coefficients**.
5. Check **Regression Statistics Table** and **ANOVA and Coefficients Table**.
6. Enter a **Title** and click **OK**.

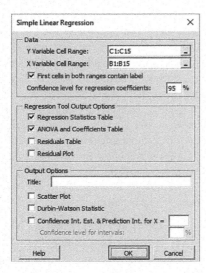

The procedure creates a worksheet that contains a copy of your data as well as the worksheet shown in Figure 13.4. For more information about these worksheets, read the following *Workbook* section.

To create a scatter plot that contains a prediction line and regression equation similar to Figure 13.5 on page 435, modify step 6 by checking **Scatter Plot** before clicking **OK**.

Workbook Use the **COMPUTE worksheet** of the **Simple Linear Regression workbook** as a template.

For the example, the worksheet uses the regression data already in the SLRData worksheet to perform the regression analysis. Worksheet columns A through I mimic the design of the Analysis ToolPak regression results even as the worksheet computes most values in columns L and M, unlike the ToolPak results, which do not include any cell formulas.

To perform simple linear regression for other data, paste the regression data into the SLRData worksheet, using column A for the *X* variable data and column B for the *Y* variable data. Then, open to the COMPUTE worksheet and:

1. Enter the confidence level in cell **L8**.
2. Select the gray-tinted cell range **L2:M6** (shown below).
3. In the formula bar, edit the **column A and B cell ranges** in the formula to reflect the range of the new regression data.
4. When finished editing, while holding down the **Control** and **Shift keys** (or **Command** on a Mac), press **Enter**.

Because the edited formula is an *array* formula (see Appendix Section B.2), the simple pressing of the Enter key without any other key being held down will not work as it would for entering simple formulas.

	K	L	M
1	Intermediate Calculations		
2	*b1, b0* Coefficients	2.0742	-1.2088
3	*b1, b0* Standard Error	0.2536	0.9949
4	R Square, Standard Error	0.8479	0.9993
5	*F*, Residual *df*	66.8792	12.0000
6	Regression SS, Residual SS	66.7854	11.9832
7			
8	Confidence level	95%	
9	*t* Critical Value	2.1788	
10	Half Width *b0*	2.1676	
11	Half Width *b1*	0.5526	

The gray-tinted cell range L2:M6 uses the LINEST function to compute the following statistics: the b_1 and b_0 coefficients in cells L2 and M2, the b_1 and b_0 standard errors in cells L3 and M3, r^2 and the standard error of the estimate in cells L4 and M4, the F test statistic and error *df* in cells L5 and M5, and SSR and SSE in cells L6 and M6.

Cell L9 uses the expression T.INV.2T(1 – confidence level, Error degrees of freedom) to compute the critical value for the t test. To see all of the formulas that the COMPUTE worksheet uses, open to the COMPUTE_FORMULAS worksheet.

Scatter Plot To create a scatter plot that contains both a prediction line and regression equation (similar to Figure 13.5 on page 435), first use the Section EG2.5 *Workbook* scatter plot instructions with the Table 13.1 Sunflowers Apparel data to create a scatter plot. Then select the chart and:

1. Select **Design** (or **Chart Design**)→**Add Chart Element**→**Trendline**→**More Trendline Options**.
2. Check the **Display Equation on chart** and **Display R-squared value on chart** check boxes near the bottom of the pane (shown below).

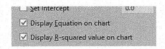

If the X axis of the scatter plot does not appear at the bottom of the plot, use the "Correcting the Display of the X Axis" instructions in Appendix Section B.5 to relocate the X axis to the bottom.

Analysis ToolPak Use **Regression**.

For the example, open to the **DATA worksheet** of the **SiteSelection workbook** and:

1. Select **Data**→**Data Analysis**.
2. In the Data Analysis dialog box, select **Regression** from the **Analysis** Tools list and then click **OK**.

In the Regression dialog box (shown below):

3. Enter **C1:C15** as the **Input Y Range** and enter **B1:B15** as the **Input X Range**.
4. Check **Labels** and check **Confidence Level** and enter **95** in its box.
5. Click **New Worksheet Ply** and then click **OK**.

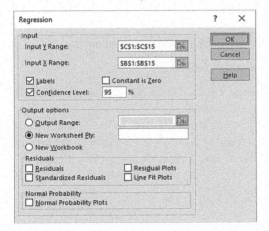

EG13.3 MEASURES of VARIATION

The measures of variation appear as part of the regression results worksheet that the Section EG13.2 instructions create.

If you use either Section EG13.2 *PHStat* or *Workbook* instructions, COMPUTE worksheet formulas compute these measures. The formulas in cells B5, B7, B13, C12, C13, D12, and E12 copy values computed by the array formula in cell range L2:M6.

EG13.5 RESIDUAL ANALYSIS

Key Technique Use arithmetic formulas to compute the residuals. To evaluate assumptions, use the Section EG2.5 scatter plot instructions for constructing residual plots and the Section EG6.3 instructions for constructing normal probability plots.

Example Compute the Figure 13.10 residuals for the Table 13.1 Sunflowers Apparel on page 447.

PHStat Use the Section EG13.2 *PHStat* instructions to compute the residuals. Use the Section EG6.3 *PHStat* instructions to construct a normal probability plot.

For the example, modify step 5 of the EG13.2 *PHStat* instructions by checking **Residuals Table** and **Residual Plot** in addition to checking the two other check boxes. PHStat creates a residual plot and a worksheet containing the residuals in addition to the COMPUTE worksheet described earlier.

To construct a normal probability plot, open to the residuals worksheet and modify the EG6.3 *PHStat* instructions by using the cell range of the residuals as the **Variable Cell Range** in step 1.

Workbook Use the **RESIDUALS worksheet** of the **Simple Linear Regression workbook** as a template.

The worksheet already computes the residuals for the example. For other problems, modify this worksheet by pasting the X values into column B and the Y values into column D. Then, for sample sizes smaller than 14, delete the extra rows. For sample sizes greater than 14, copy the column C and E formulas down through the row containing the last pair of X and Y values and add the new observation numbers in column A.

To construct a residual plot similar to Figure 13.11 on page 447, use the original X variable and the residuals (plotted as the Y variable) as the chart data and follow the Section EG2.5 scatter plot instructions. To construct a normal probability plot, follow the Section EG6.3 normal probability plot instructions, using the cell range of the residuals as the **Variable Cell Range**.

Analysis ToolPak Use the Section EG13.2 *Analysis ToolPak* instructions.

Modify step 5 by checking **Residuals** and **Residual Plots** before clicking **New Worksheet Ply** and then **OK**. To construct a residual plot or normal probability plot, use the *Workbook* instructions.

EG13.6   MEASURING AUTOCORRELATION: the DURBIN-WATSON STATISTIC

Key Technique Use the **SUMXMY2(*cell range of the second through last residual, cell range of the first through the second-to-last residual*)** function to compute the sum of squared difference of the residuals.

Use the **SUMSQ(*cell range of the residuals*)** function to compute the sum of squared residuals.

Example Compute the Durbin-Watson statistic for the package delivery data shown in the Figure 13.16 on page 452.

PHStat Use the *PHStat* instructions at the beginning of Section EG13.2. Modify step 6 by checking the **Durbin-Watson Statistic** output option before clicking **OK**.

Workbook Use the **DURBIN_WATSON worksheet** of the **Simple Linear Regression** workbook as a template.

The DURBIN_WATSON worksheet of the **Package Delivery workbook** already contains the proper cell formulas to compute the statistic for the example. (This workbook also uses the COMPUTE and RESIDUALS worksheet templates from the Simple Linear Regression workbook.)

To compute the Durbin-Watson statistic for other problems, first create the simple linear regression model and the residuals for the problem, using the Sections EG13.2 and EG13.5 *Workbook* instructions. Then open the DURBIN_WATSON worksheet and edit the formulas in cell B3 and B4 to point to the proper cell ranges of the new residuals.

EG13.7   INFERENCES ABOUT the SLOPE and CORRELATION COEFFICIENT

The *t* test for the slope, the *F* test for the slope, and the confidence interval estimate for the slope all appear in the worksheet created by using the Section EG13.2 instructions. The *t* test for the slope appears in cell D18, the *F* test for the slope appears in cell E18, and the confidence interval estimate for the slope appears in the cell range F18:G18 (and repeated in cell range H18:I18).

For the *PHStat* and *Workbook* worksheets (identical), cell D18 contains a formula that divides the cell B18 contents by cell C18 content. Cell E12 copies a value that the LINEST array function computes in cell L5. Cell F12 uses the F.DIST.RT function to compute the *p*-value for the *F* test for the slope. The cell range F18:G18 contains an arithmetic formula that uses the half-width of the b_1 that cell L11 computes.

EG13.8   ESTIMATION of MEAN VALUES and PREDICTION of INDIVIDUAL VALUES

Key Technique Use the **TREND(*Y variable cell range, X variable cell range, X value*)** function to compute the predicted *Y* value for the *X* value.

Use the **DEVSQ(*X variable cell range*)** function to compute the *SSX* value.

Example Compute the Figure 13.21 confidence interval estimate and prediction interval for the Sunflowers Apparel data that is shown on page 463.

PHStat Modify the Section EG13.2 *PHStat* instructions by replacing step 6 with these steps 6 and 7:

6. Check **Confidence Int. Est. & Prediction Int. for** *X*= and enter **4** in its box. Enter **95** as the percentage for **Confidence level for intervals**.
7. Enter a **Title** and click **OK**.

The additional worksheet created is discussed in the following *Workbook* instructions.

Workbook Use the **CIEandPI worksheet** of the **Simple Linear Regression workbook**, as a template.

The worksheet already contains the data and formulas for the example. To compute a confidence interval estimate and prediction interval for other problems:

1. Paste the regression data into the **SLRData worksheet**. Use column A for the *X* variable data and column B for the *Y* variable data.
2. Open to the **CIEandPI worksheet**.

In the CIEandPI worksheet:

3. Change values for the **X Value** and **Confidence Level**, as is necessary.
4. Edit the cell ranges used in the cell B15 formula that uses the TREND function to refer to the new cell ranges for the *Y* and *X* variables.

Use **scatter plots**.

For example, to perform a regression analysis equivalent to the Figure 13.4 analysis of the Sunflowers Apparel data on page 435, in a new Tableau workbook, click **Connect to Data** and open the **Site Selection Excel workbook**. In a new Tableau worksheet:

1. Drag **Profiled Customers** and drop it in the **Columns** shelf.

2. Drag **Annual Sales** and drop it in the **Rows** shelf.

3. Drag **Store** and drop it over the **Details icon** in the **Marks** card area.

4. Select **Analysis** and clear the **Aggregate Measures** checkmark.

5. Select **Analysis Trend Lines** and check **Show Trend Lines**.

6. Right-click the scatter plot and select **Trend Lines Describe Trend Model**.

7. In the Describe Trend Model dialog box click **Copy** to copy the summary results of the model.

8. Open to a new Microsoft Word document and press **Ctrl+V** (**Command+V** in macOS) to paste the summary results.

9. Optionally, delete the Row, Column, Line *p*-value, and DF columns in the table that appears at the end of the summary results.

10. Back in Tableau, click **Close** to close the Describe Trend Model dialog box.

Enter a worksheet title, turn off gridlines, and, optionally, adjust font and type characteristics using the Appendix Section B.5T instructions. To make the Figure 13.5 summary information box appear in the chart, move the mouse pointer over the prediction line. The summary information appears in a popup window that disappears when the mouse pointer is moved away from the prediction line.

Some of the measures of variation appear as part of the regression summary results that Section TG13.2 instructions create.

14

Introduction to Multiple Regression

CONTENTS

OBJECTIVES

- Develop multiple
  regression models and
  interpret the regression
  coefficients
- Determine which
  independent variables
  to include and identify
  the independent vari-
  ables most important to
  a model

▼USING **STATISTICS**
The Multiple Effects of OmniPower Bars

You are a marketing manager for OmniFoods, with oversight for nutrition bars and similar snack items. You seek to revive the sales of OmniPower, the company's primary product in this category. Originally marketed as a high-energy bar to runners, mountain climbers, and other athletes, OmniPower reached its greatest sales during an earlier time when high-energy bars were one of the most popular snack items with consumers. Now, you seek to reposition the product as a nutrition bar to benefit from the booming market for such bars.

Because the marketplace already contains several successful nutrition bars, you need to develop an effective marketing strategy. In particular, you need to determine the effect that price and in-store promotional expenses (special in-store coupons, signs, and displays as well as the cost of free samples) will have on sales of OmniPower. Before marketing the bar nationwide, you plan to conduct a test-market study of OmniPower sales, using a sample of 34 stores in a supermarket chain.

How can you extend the linear regression methods discussed in Chapter 13 to incorporate the effects of price *and* promotion into the same model? How can you use this model to improve the success of the nationwide introduction of OmniPower?

- Use categorical independent variables
- Understand logistic regression

Chapter 13 discusses simple linear regression models that use *one* numerical independent variable, *X*, to predict the value of a numerical dependent variable, *Y*. Often you can make better predictions by using *more than one* independent variable. This chapter introduces you to **multiple regression models** that use two or more independent variables to predict the value of a dependent variable.

14.1 Developing a Multiple Regression Model

In the OmniPower Bars scenario, your business objective, to determine the effect that price and in-store promotional expenses will have on sales, calls for examining a multiple regression model in which the price of an OmniPower bar in cents (X_1) and the monthly budget for in-store promotional expenses in dollars (X_2) are the independent variables and the number of OmniPower bars sold in a month (*Y*) is the dependent variable.

To develop this model, you collect data from a sample of 34 stores in a supermarket chain selected for the test-market study of OmniPower. You choose stores in a way to ensure that they all have approximately the same monthly sales volume. You organize and store the data collected in OmniPower. Table 14.1 presents these data.

TABLE 14.1
Monthly OmniPower sales, price, and promotional expenditures

Store	Sales	Price	Promotion	Store	Sales	Price	Promotion
1	4,141	59	200	18	2,730	79	400
2	3,842	59	200	19	2,618	79	400
3	3,056	59	200	20	4,421	79	400
4	3,519	59	200	21	4,113	79	600
5	4,226	59	400	22	3,746	79	600
6	4,630	59	400	23	3,532	79	600
7	3,507	59	400	24	3,825	79	600
8	3,754	59	400	25	1,096	99	200
9	5,000	59	600	26	761	99	200
10	5,120	59	600	27	2,088	99	200
11	4,011	59	600	28	820	99	200
12	5,015	59	600	29	2,114	99	400
13	1,916	79	200	30	1,882	99	400
14	675	79	200	31	2,159	99	400
15	3,636	79	200	32	1,602	99	400
16	3,224	79	200	33	3,354	99	600
17	2,295	79	400	34	2,927	99	600

Interpreting the Regression Coefficients

When there are several independent variables, one can extend the simple linear regression model of Equation (13.1) on page 432 by assuming a linear relationship between each independent variable and the dependent variable. Equation (14.1) defines the general case of a multiple regression model with *k* independent variables. Equation (14.2) defines the specific case of the multiple regression model with two independent variables, the simplest case.

MULTIPLE REGRESSION MODEL WITH k INDEPENDENT VARIABLES

$$Y_i = \beta_0 + \beta_1 X_{1i} + \beta_2 X_{2i} + \beta_3 X_{3i} + \ldots + \beta_k X_{ki} + \varepsilon_i \qquad \textbf{(14.1)}$$

where

Y_i = dependent variable for observation i

β_0 = Y intercept

β_1 = slope of Y with variable X_1, holding variables $X_2, X_3, \ldots, X_k$ constant

β_2 = slope of Y with variable X_2, holding variables $X_1, X_3, \ldots, X_k$ constant

$\vdots$

β_k = slope of Y with variable X_k holding variables $X_1, X_2, X_3, \ldots, X_{k-1}$ constant

ε_i = random error in Y for observation i

MULTIPLE REGRESSION MODEL WITH TWO INDEPENDENT VARIABLES

$$Y_i = \beta_0 + \beta_1 X_{1i} + \beta_2 X_{2i} + \varepsilon_i \qquad \textbf{(14.2)}$$

Equation (14.2) has three **net regression coefficients**: β_0, β_1, and β_2. As in simple linear regression, β_0 represents the Y intercept, the value of Y when $X = 0$. The other two terms are slopes defined as follows:

- β_1: the change in Y per unit change in X_1, taking into account the effect of X_2.
- β_2: the change in Y per unit change in X_2, taking into account the effect of X_1.

Each independent X variable always has its own β term. Therefore, a multiple regression model that has j X variables will always have $j+1$ β terms, β_1 through β_j plus the Y intercept, β_0.

The least-squares method that Section 13.1 introduces can also be used to calculate the sample regression coefficients b_0, b_1, and b_2 as estimates of the population parameters β_0, β_1, and β_2. Equation (14.3) defines the regression equation for a multiple regression model with two independent variables.

MULTIPLE REGRESSION EQUATION WITH TWO INDEPENDENT VARIABLES

$$\hat{Y}_i = b_0 + b_1 X_{1i} + b_2 X_{2i} \qquad \textbf{(14.3)}$$

Figure 14.1 on page 481 shows Excel results for the OmniPower sales data multiple regression model. In these results, the b_0 coefficient is labeled Intercept by Excel. From Figure 14.1, the computed values of the net regression coefficients are

$$b_0 = 5{,}837.5208 \quad b_1 = -53.2173 \quad b_2 = 3.6131$$

Therefore, the multiple regression equation is

$$\hat{Y}_i = 5{,}837.5208 - 53.2173 X_{1i} + 3.6131 X_{2i}$$

where

$\hat{Y}_i$ = predicted monthly sales of OmniPower bars for store i

X_{1i} = price of OmniPower bar (in cents) for store i

X_{2i} = monthly in-store promotional expenses (in $\$$) for store i

The sample Y intercept, b_0, estimates the number of OmniPower bars sold in a month if the price was zero cents and the total amount spent on promotional expenses was $\$0.00$. Because the price and promotion values are outside the range of price and promotion used in the test-market study, and because they make no logical sense for this problem, the value of b_0 has no useful interpretation.

Using the net regression coefficients b_1 and b_2, the effects of adding one cent to the price of OmniPower bars (X_1) or adding $\$1$ to monthly promotion expenditures (X_2) can be summarized

FIGURE 14.1

Excel worksheet results for the OmniPower sales multiple regression model

	A	B	C	D	E	F	G
1	Regression Analysis						
2							
3	*Regression Statistics*						
4	Multiple R	0.8705					
5	R Square	0.7577					
6	Adjusted R Square	0.7421					
7	Standard Error	638.0653					
8	Observations	34					
9							
10	ANOVA						
11		df	SS	MS	F	Significance F	
12	Regression	2	39472730.7730	19736365.3865	48.4771	0.0000	
13	Residual	31	12620946.6682	407127.3119			
14	Total	33	52093677.4412				
15							
16		Coefficients	Standard Error	t Stat	P-value	Lower 95%	Upper 95%
17	Intercept	5837.5208	628.1502	9.2932	0.0000	4556.3999	7118.6416
18	Price	-53.2173	6.8522	-7.7664	0.0000	-67.1925	-39.2421
19	Promotional Expenses	3.6131	0.6852	5.2728	0.0000	2.2155	5.0106

	I	J	K	L
1	Intermediate Calculations			
2	b2, b1, b0 intercepts	3.6131	-53.2173	5837.5208
3	b2, b1, b0 Standard Error	0.6852	6.8522	628.1502
4	R Square, Standard Error	0.7577	638.0653	#N/A
5	F, Residual df	48.4771	31	#N/A
6	Regression SS, Residual SS	39472730.77	12620946.67	#N/A
7				
8	Confidence level	95%		
9	t Critical Value	2.0395		
10	Half Width b0	1281.1208		
11	Half Width b1	13.9752		
12	Half Width b2	1.3975		

In the Worksheet

The Section EG14.1 *Worksheet*, *PHStat*, and *Analysis ToolPak* instructions each create a regression results worksheets that look like this worksheet. Using the *Worksheet* or *PHStat* instructions inserts formulas that start in column I (shown above) that use the LINEST function to compute regression coefficients and statistics.

The **COMPUTE worksheet** of the Excel Guide **Multiple Regression** is a copy of the Figure 14.1 worksheet.

for management in Table 14.2, which explains the effect of changing one independent variable while holding the value of the all other independent variables constant.

TABLE 14.2

Net effects table for the OmniPower sales multiple regression model

Independent Variable Change	Net Effect
A price increase of one cent	Predict mean OmniPower monthly sales to decrease by 53.2173 bars, holding constant the promotional expenditures.
An increase of $1 in monthly promotional expenditures	Predict mean OmniPower monthly sales to increase by 3.6131 bars, holding constant the price.

The Table 14.2 estimates will allow OmniFoods decision makers to better understand how pricing and promotional expenditures decisions are predicted to affect OmniPower sales. Using the tables, managers could predict that a 10-cent decrease in price would result in the mean monthly sales increasing by about 532 bars, holding promotional costs constant, or that a $100 increase in promotional expenditures would increase mean monthly sales by about 361 bars, holding price constant.

Because net regression coefficients always estimate the predicted mean change in Y per unit change in a specific X, holding constant the effect of the other X variables, net effects tables are always a good way to summarize multiple regression results for decision-making purposes.

Predicting the Dependent Variable *Y*

Use the multiple regression equation to predict values of the dependent variable. For example, what are the predicted mean sales for a store charging 79 cents during a month in which promotional expenses are $400? Using the multiple regression equation,

$$\hat{Y}_i = 5{,}837.5208 - 53.2173X_{1i} + 3.6131X_{2i}$$

with $X_{1i} = 79$ and $X_{2i} = 400$,

$$\hat{Y}_i = 5{,}837.5208 - 53.2173(79) + 3.6131(400)$$

$$= 3{,}078.57$$

studentTIP

You should only predict within the range of the values of all the independent variables.

Thus, you predict that stores charging 79 cents and spending $400 in promotional expenses will sell a mean of 3,078.57 OmniPower bars per month.

After developing the regression equation, doing a residual analysis (see Section 14.3), and determining the significance of the overall fitted model (see Section 14.2), one can construct a confidence interval estimate of the mean value and a prediction interval for an individual value. Figure 14.2 presents the Excel confidence interval estimates and prediction interval worksheet results for the OmniPower sales data.

FIGURE 14.2
Excel confidence interval estimate and prediction interval worksheet results for the OmniPower sales data

	A	B	C	D
1	Confidence Interval Estimate and Prediction Interval			
2				
3	Data			
4	Confidence Level	95%		
5		1		
6	Price given value	79		
7	Promotion given value	400		
8				
9	X'X	34	2646	13200
10		2646	214674	1018800
11		13200	1018800	6000000
12				
13	Inverse of X'X	0.9692	-0.0094	-0.0005
14		-0.0094	0.0001	0.0000
15		-0.0005	0.0000	0.0000
16				
17	X'G times Inverse of X'X	0.0121	0.0001	0.0000
18				
19	[X'G times Inverse of X'X] times XG	0.0298	=MMULT(B17:D17, B5:B7)	
20	t Statistic	2.0395	=T.INV.2T(1 - B4, COMPUTE!B13)	
21	Predicted Y (YHat)	3078.57	{=MMULT(TRANSPOSE(B5:B7), COMPUTE!B17:B19)}	
22				
23	For Average Predicted Y (YHat)			
24	Interval Half Width	224.50	=B20 * SQRT(B19) * COMPUTE!B7	
25	Confidence Interval Lower Limit	2854.07	=B21 - B24	
26	Confidence Interval Upper Limit	3303.08	=B21 + B24	
27				
28	For Individual Response Y			
29	Interval Half Width	1320.57	=B20 * SQRT(1 + B19) * COMPUTE!B7	
30	Prediction Interval Lower Limit	1758.01	=B21 - B29	
31	Prediction Interval Upper Limit	4399.14	=B21 + B29	

In the Worksheet

This worksheet uses the MMULT function in rows 9 through 11 and 17 to perform intermediate matrix calculations and uses the TRANSPOSE function to change a matrix column of data into a matrix row. Rows 13 through 15 use the MINVERSE function to compute the inverse of a matrix. Cell B21 contains an *array* formula that also uses the MMULT function to compute the predicted *Y* value. (Appendix F discusses the MMULT, TRANSPOSE, and MINVERSE functions.)

The **CIEandPI worksheet** of the Excel Guide **Multiple Regression** is a copy of the Figure 14.2 worksheet.

The 95% confidence interval estimate of the mean OmniPower sales for all stores charging 79 cents and spending $400 in promotional expenses is 2,854.07 to 3,303.08 bars. The prediction interval for an individual store is 1,758.01 to 4,399.14 bars.

PROBLEMS FOR SECTION 14.1

LEARNING THE BASICS

14.1 For this problem, use the following multiple regression equation:

$$\hat{Y}_i = 10 + 5X_{1i} + 3X_{2i}$$

a. Interpret the meaning of the slopes.
b. Interpret the meaning of the *Y* intercept.

14.2 For this problem, use the following multiple regression equation:

$$\hat{Y}_i = 50 - 2X_{1i} + 7X_{2i}$$

a. Interpret the meaning of the slopes.
b. Interpret the meaning of the *Y* intercept.

APPLYING THE CONCEPTS

14.3 A nonprofit analyst seeks to determine which variables should be used to predict nonprofit charitable commitment, a nonprofit organization commitment to its charitable purpose. Two independent variables under consideration are Revenue, a measurement of total revenue, in billions of dollars, as a measure of nonprofit size X_1 and Efficiency, a measurement of the percent of private donations remaining after fundraising expenses as a measure of nonprofit fundraising efficiency X_2. The dependent variable *Y* is Commitment, a measurement of the percent of total expenses that are allocated directly to charitable services. Data are collected from a random sample of 98 nonprofit organizations, with the following results:

Variable	Coefficients	Standard Error	T Statistic	p-Value
Intercept	11.002079	7.127101	1.54	0.1260
Revenue	0.6683647	0.320077	2.09	0.0395
Efficiency	0.8317339	0.077736	10.70	0.0001

a. State the multiple regression equation.
b. Interpret the meaning of the slopes, b_1 and b_2, in this problem.
c. What conclusions can you reach concerning nonprofit charitable commitment?

✓ SELF TEST **14.4** Profitability remains a challenge for banks and thrifts with less than $2 billion of assets. The business problem facing a bank analyst relates to the factors that affect return on average assets (ROAA), an indicator of how profitable a company is relative to its total assets. Data collected on a sample of 199 community banks and stored in `Community Banks` include the ROAA (%), the efficiency ratio (%) as a measure of bank productivity, and total risk-based capital (%) as a measure of capital adequacy.

Source: Data extracted from "All about scale: The Top 200 Publicly Traded Community Banks," **bit.ly/2tKX9in**.

a. State the multiple regression equation.
b. Interpret the meaning of the slopes, b_1 and b_2, in this problem.
c. Predict the mean ROAA when the efficiency ratio is 60% and the total risk-based capital is 15%.
d. Construct a 95% confidence interval estimate for the mean ROAA when the efficiency ratio is 60% and the total risk-based capital is 15%.
e. Construct a 95% prediction interval for the ROAA for a particular community bank when the efficiency ratio is 60% and the total risk-based capital is 15%.
f. Explain why the interval in (d) is narrower than the interval in (e).
g. What conclusions can you reach concerning ROAA?

14.5 The production of wine is a multibillion-dollar worldwide industry. In an attempt to develop a model of wine quality as judged by wine experts, data were collected from red wine variants of Portuguese "Vinho Verde" wine. A sample of 50 wines is stored in `Vinho Verde`.

Source: Data extracted from P. Cortez, Cerdeira, A., Almeida, F., Matos, T., and Reis, J., "Modeling Wine Preferences by Data Mining from Physiochemical Properties," *Decision Support Systems*, 47, 2009, pp. 547–553 and **bit.ly/9xKlEa**.

Develop a multiple linear regression model to predict wine quality, measured on a scale from 0 (very bad) to 10 (excellent) based on alcohol content (%) and the amount of chlorides.

a. State the multiple regression equation.
b. Interpret the meaning of the slopes, b_1 and b_2, in this problem.
c. Explain why the regression coefficient, b_0, has no practical meaning in the context of this problem.
d. Predict the mean wine quality rating for wines that have 10% alcohol and chlorides of 0.08.
e. Construct a 95% confidence interval estimate for the mean wine quality rating for wines that have 10% alcohol and chlorides of 0.08.
f. Construct a 95% prediction interval for the wine quality rating for an individual wine that has 10% alcohol and chlorides of 0.08.
g. What conclusions can you reach concerning this regression model?

14.6 Human resource managers face the business problem of assessing the impact of factors on full-time job growth. A human resource manager is interested in the impact of full-time voluntary turnover and total worldwide revenues on the number of full-time job openings at the beginning of a new year. Data are collected from a sample of 63 "best companies to work for." The total number of full-time job openings as of February 2017, the full-time voluntary turnover in the past year (in %), and the total worldwide revenue (in $billions) are recorded and stored in `Best Companies`.

Source: Data extracted from *100 Best Companies to Work For*, 2017, **fortune.com/best-companies**.

a. State the multiple regression equation.
b. Interpret the meaning of the slopes, b_1 and b_2, in this problem.
c. Interpret the meaning of the regression coefficient, b_0.
d. Which factor has the greatest effect on the number of full-time jobs added in the last year? Explain.

14.7 The business problem facing the director of broadcasting operations for a television station was the issue of standby hours (i.e., hours in which employees at the station are paid but are not actually involved in any activity) and what factors were related to standby hours. The study included the following variables:

Standby hours (Y)—Total number of standby hours in a week
Weekly staff count (X_1)—Weekly total of person-days
Remote engineering hours (X_2)—Total number of engineering hours worked by employees at locations away from the central plant

Data were collected for 26 weeks; these data are organized and stored in `Nickels 26Weeks`.

a. State the multiple regression equation.
b. Interpret the meaning of the slopes, b_1 and b_2, in this problem.
c. Explain why the regression coefficient, b_0, has no practical meaning in the context of this problem.
d. Predict the mean standby hours for a week in which the weekly staff count was 310 person-days and the remote engineering hours total was 400.
e. Construct a 95% confidence interval estimate for the mean standby hours for weeks in which the weekly staff count was 310 person-days and remote engineering hours total was 400.
f. Construct a 95% prediction interval for the standby hours for a single week in which the weekly staff count was 310 person-days and the engineering remote hours total was 400.
g. What conclusions can you reach concerning standby hours?

14.8 Nassau County is located approximately 25 miles east of New York City. The data organized and stored in `Glen Cove` include the fair market value (in $thousands), land area of the property in acres, and age, in years, for a sample of 30 single-family homes located in Glen Cove, a small city in Nassau County. Develop a multiple linear regression model to predict the fair market value based on land area of the property in acres and age in years.

a. State the multiple regression equation.
b. Interpret the meaning of the slopes, b_1 and b_2, in this problem.
c. Explain why the regression coefficient, b_0, has no practical meaning in the context of this problem.
d. Predict the mean fair market value for a house that has a land area of 0.25 acre and is 55 years old.
e. Construct a 95% confidence interval estimate for the mean fair market value for houses that have a land area of 0.25 acre and are 55 years old.
f. Construct a 95% prediction interval estimate for the fair market value for an individual house that has a land area of 0.25 acre and is 55 years old.

14.2 Evaluating Multiple Regression Models

Section 14.1 discusses developing a multiple regression model. Having developed a model, one proceeds to evaluating the entire model. Three ways of evaluating a multiple regression are using the coefficient of multiple determination, r^2; using the adjusted r^2; and performing the overall F test.

Coefficient of Multiple Determination, r^2

Section 13.3 explains that the coefficient of determination, r^2, measures the proportion of the variation in Y that is explained by the variability in the independent variable X in the simple linear regression model. In multiple regression, the **coefficient of multiple determination** represents the proportion of the variation in the dependent variable Y that is explained by all the variability in the independent X variables that the model includes. Equation (14.4) defines the coefficient of multiple determination for a multiple regression model with two or more independent variables.

> ### COEFFICIENT OF MULTIPLE DETERMINATION
>
> The coefficient of multiple determination is equal to the regression sum of squares (SSR) divided by the total sum of squares (SST).
>
> $$r^2 = \frac{\text{Regression sum of squares}}{\text{Total sum of squares}} = \frac{SSR}{SST} \qquad (14.4)$$

In the OmniPower example, from Figure 14.1 on page 481, $SSR = 39{,}472{,}730.77$ and $SST = 52{,}093{,}677.44$. Thus,

$$r^2 = \frac{SSR}{SST} = \frac{39{,}472{,}730.77}{52{,}093{,}677.44} = 0.7577$$

The coefficient of multiple determination, $r^2 = 0.7577$, indicates that 75.77% of the variation in sales is explained by the variation in the price and in the promotional expenses. In Figure 14.1, Excel labels the coefficient of multiple determination as R Square.

Adjusted r^2

When considering multiple regression models, some statisticians suggest using the **adjusted r^2** to take into account both the number of independent variables in the model and the sample size. Because a model that has additional independent variables will always have the same or higher regression sum of squares and r^2, using the adjusted r^2 provides a more appropriate interpretation when comparing models. Equation (14.5) defines the adjusted r^2.

> ### ADJUSTED r^2
>
> $$r_{\text{adj}}^2 = 1 - \left[(1 - r^2)\frac{n - 1}{n - k - 1} \right] \qquad (14.5)$$
>
> where
> k is the number of independent variables in the regression equation.

For the OmniPower sales data, because $r^2 = 0.7577$, $n = 34$, and $k = 2$,

$$r_{\text{adj}}^2 = 1 - \left[(1 - 0.7577)\frac{34 - 1}{34 - 2 - 1} \right] = 1 - \left[(0.2423)\frac{33}{31} \right]$$

$$= 1 - 0.2579$$

$$= 0.7421$$

Therefore, 74.21% of the variation in sales is explained by the multiple regression model—adjusted for the number of independent variables and sample size. In Figure 14.1 on page 481, Excel labels the adjusted r^2 as Adjusted R Square.

F Test for the Significance of the Overall Multiple Regression Model

One uses the **overall *F* test** to determine whether there is a significant relationship between the dependent variable and the entire set of independent variables, the *overall* multiple regression model. The null and alternative hypotheses:

$H_0: \beta_1 = \beta_2 = \cdots = \beta_k = 0$ (There is no linear relationship between the dependent variable and the independent variables.)

$H_1:$ At least one $\beta_j \neq 0$, where $j = 1, 2, \cdots, k$ (There is a linear relationship between the dependent variable and at least one independent variable.)

This test determines whether at least one independent variable has a linear relationship with the dependent variable. If one rejects H_0, one is *not* concluding that all the independent variables have a linear relationship with the dependent variable, only that *at least one* independent variable does. Equation (14.6) defines the overall *F* test statistic. Table 14.3 presents the ANOVA summary table.

OVERALL *F* TEST STATISTIC

The F_{STAT} test statistic is equal to the regression mean square (*MSR*) divided by the mean square error (*MSE*).

$$F_{STAT} = \frac{MSR}{MSE} \tag{14.6}$$

The F_{STAT} test statistic follows an *F* distribution with k and $n - k - 1$ degrees of freedom, where k is the number of independent variables in the regression model.

TABLE 14.3
ANOVA summary table for the overall *F* test

Source	Degrees of Freedom	Sum of Squares	Mean Squares (Variance)	F
Regression	k	SSR	$MSR = \dfrac{SSR}{k}$	$F_{STAT} = \dfrac{MSR}{MSE}$
Error	$n - k - 1$	SSE	$MSE = \dfrac{SSE}{n - k - 1}$	
Total	$n - 1$	SST		

The decision rule is

Reject H_0 at the α level of significance if $F_{STAT} > F_\alpha$; otherwise, do not reject H_0.

If one fails to reject the null hypothesis, one concludes that the model fit is *not* appropriate. If one rejects the null hypothesis, one proceeds with the model and uses methods that Sections 14.4 and 14.5 discuss to determine which independent variables should be included in the final regression model.

For the OmniPower sales study, using the 0.05 level of significance, α, and Table E.5, the critical value of the *F* distribution with 2 and 31 degrees of freedom is approximately 3.32. Figure 14.3 visualizes the regions of nonrejection and rejection using this critical value.

FIGURE 14.3
Testing for the significance of a set of regression coefficients at the 0.05 level of significance, with 2 and 31 degrees of freedom

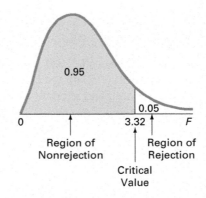

Figure 14.1 multiple regression results on page 481 includes the F_{STAT} test statistic in the ANOVA tables. Table 14.4 summarizes the results of the test for the set of regression coefficients. Based on the results, one concludes that either price or promotional expenses or both variables can be used to help predict mean monthly sales.

TABLE 14.4
Overall F test results and conclusions

student TIP

Using tables to summarize regression results and conclusions is a good way to communicate results to others.

Result	Conclusions
$F_{STAT} = 48.4771$ is greater than the F critical value, 3.32 p-value $= 0.0000$ is less than the level of significance, $\alpha = 0.05$	1. Reject the null hypothesis H_0. 2. Conclude that evidence exists for claiming that at least one of the independent X variables (price or promotional expenses) is related to the dependent Y variable, sales. 3. The probability is 0.0000 that $F_{STAT} > 48.4771$.

PROBLEMS FOR SECTION 14.2

LEARNING THE BASICS

14.9 The following ANOVA summary table is for a multiple regression model with two independent variables:

Source	Degrees of Freedom	Sum of Squares	Mean Squares	F
Regression	2	60		
Error	18	120		
Total	20	180		

a. Determine the regression mean square (*MSR*) and the mean square error (*MSE*).
b. Compute the overall F_{STAT} test statistic.
c. Determine whether there is a significant relationship between Y and the two independent variables at the 0.05 level of significance.
d. Compute the coefficient of multiple determination, r^2, and interpret its meaning.
e. Compute the adjusted r^2.

14.10 The following ANOVA summary table is for a multiple regression model with two independent variables:

Source	Degrees of Freedom	Sum of Squares	Mean Squares	F
Regression	2	30		
Error	10	120		
Total	12	150		

a. Determine the regression mean square (*MSR*) and the mean square error (*MSE*).
b. Compute the overall F_{STAT} test statistic.
c. Determine whether there is a significant relationship between Y and the two independent variables at the 0.05 level of significance.
d. Compute the coefficient of multiple determination, r^2, and interpret its meaning.
e. Compute the adjusted r^2.

APPLYING THE CONCEPTS

14.11 A financial analyst engaged in business valuation obtained financial data on 60 drug companies (Industry Group SIC 3 code: 283). The file Business Valuation contains the following variables:

Company—Drug Company name
PB fye—Price-to-book-value ratio (fiscal year ending)
ROE—Return on equity
SGrowth—Growth (GS5)

a. Develop a regression model to predict price-to-book-value ratio based on return on equity.
b. Develop a regression model to predict price-to-book-value ratio based on growth.
c. Develop a regression model to predict price-to-book-value ratio based on return on equity and growth.
d. Compute and interpret the adjusted r^2 for each of the three models.
e. Which of these three models do you think is the best predictor of price-to-book-value ratio?

✓SELF TEST **14.12** In Problem 14.3 on page 482, you predicted non-profit charitable commitment, based on nonprofit revenue and fundraising efficiency. The regression analysis resulted in this ANOVA table:

Source	Degrees of Freedom	Sum of Squares	Mean Squares	F	p-Value
Regression	2	3529.0718	1764.54	57.9410	<.0001
Error	95	2893.1323	30.45		
Total	97	6422.2041			

Determine whether there is a significant relationship between commitment and the two independent variables at the 0.05 level of significance.

14.13 In Problem 14.5 on page 483, you used the percentage of alcohol and chlorides to predict wine quality (stored in Vinho Verde). Use the results from that problem to do the following:

a. Determine whether there is a significant relationship between wine quality and the two independent variables (percentage of alcohol and chlorides) at the 0.05 level of significance.
b. Interpret the meaning of the p-value.
c. Compute the coefficient of multiple determination, r^2, and interpret its meaning.
d. Compute the adjusted r^2.

14.14 In Problem 14.4 on page 482, you used efficiency ratio and total risk-based capital to predict ROAA at a community bank (stored in Community Banks). Using the results from that problem,

a. determine whether there is a significant relationship between ROAA and the two independent variables (efficiency ratio and total risk-based capital) at the 0.05 level of significance.
b. interpret the meaning of the p-value.
c. compute the coefficient of multiple determination, r^2, and interpret its meaning.
d. compute the adjusted r^2.

14.15 In Problem 14.7 on page 483, you used the weekly staff count and remote engineering hours to predict standby hours (stored in Nickels 26Weeks). Using the results from that problem,

a. determine whether there is a significant relationship between standby hours and the two independent variables (total staff present and remote engineering hours) at the 0.05 level of significance.
b. interpret the meaning of the p-value.
c. compute the coefficient of multiple determination, r^2, and interpret its meaning.
d. compute the adjusted r^2.

14.16 In Problem 14.6 on page 483, you used full-time voluntary turnover (%) and total worldwide revenue ($billions) to predict number of full-time jobs added (stored in Best Companies). Using the results from that problem,

a. determine whether there is a significant relationship between number of full-time jobs added and the two independent variables (full-time voluntary turnover and total worldwide revenue) at the 0.05 level of significance.
b. interpret the meaning of the p-value.
c. compute the coefficient of multiple determination, r^2, and interpret its meaning.
d. compute the adjusted r^2.

14.17 In Problem 14.8 on page 483, you used the land area of a property and the age of a house to predict the fair market value (stored in Glen Cove). Using the results from that problem,

a. determine whether there is a significant relationship between fair market value and the two independent variables (land area of a property and age of a house) at the 0.05 level of significance.
b. interpret the meaning of the p-value.
c. compute the coefficient of multiple determination, r^2, and interpret its meaning.
d. compute the adjusted r^2.

14.3 Multiple Regression Residual Analysis

As with simple linear regression, an analysis of the residuals, differences between the actual and predicted Y values, determines whether a fitted model is the most appropriate model and can also assist in determining whether the assumptions of regression have been violated. Residual analyses for all multiple regression models require these residual plots:

studentTIP

A residual plot that does not contain any apparent patterns will look like a random scattering of points.

- residuals versus the predicted value of Y
- for each independent X variable, residuals versus the independent variable

Models that contain data that have been collected in time order also require a residual plot of the residuals versus time. Table 14.5 summarizes the significance of discovering a pattern in the three types of residual plots.

TABLE 14.5
Interpreting multiple regression residual plots

Residual Plot	Significance of Discovered Pattern
Residuals versus the predicted value of Y	Evidence of a possible curvilinear effect in at least one independent variable, a possible violation of the assumption of equal variance, or the need to transform the Y variable
Residuals versus an independent X variable	Suggests evidence of a curvilinear effect and, therefore, indicates the need to add a curvilinear independent variable to the multiple regression model (Cook)
Residuals versus data collected in time order	Evidence that the independence of errors assumption has been violated. Associated with this residual plot, as in Section 13.6, you can compute the Durbin-Watson statistic to determine the existence of positive autocorrelation among the residuals.

Figure 14.4 presents the residual plots for the OmniPower sales example. There is very little or no pattern in the relationship between the residuals and the predicted value of Y, the value of X_1, price, or the value of X_2, promotional expenses. One concludes that the multiple regression model is appropriate for predicting sales. There is no need to plot the residuals versus time because the data were not collected in time order.

FIGURE 14.4
Residual plots for the OmniPower sales data: residuals versus predicted Y, residuals versus price, and residuals versus promotional expenses

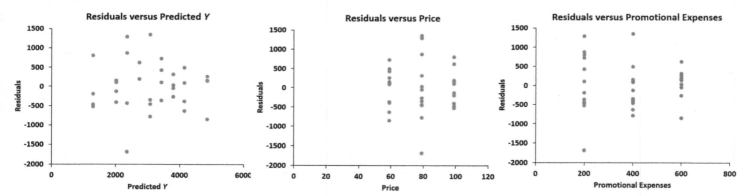

PROBLEMS FOR SECTION 14.3

APPLYING THE CONCEPTS

14.18 In Problem 14.4 on page 482, you used efficiency ratio and total risk-based capital to predict ROAA at a community bank (stored in Community Banks).

a. Plot the residuals versus $\hat{Y}_i$.
b. Plot the residuals versus X_{1i}.
c. Plot the residuals versus X_{2i}.
d. In the residual plots created in (a) through (c), is there any evidence of a violation of the regression assumptions? Explain.

14.19 In Problem 14.5 on page 485, you used the percentage of alcohol and chlorides to predict wine quality (stored in Vinho Verde).
a. Plot the residuals versus $\hat{Y}_i$
b. Plot the residuals versus X_{1i}.
c. Plot the residuals versus X_{2i}.
d. In the residual plots created in (a) through (c), is there any evidence of a violation of the regression assumptions? Explain.
e. Should you compute the Durbin-Watson statistic for these data? Explain.

14.20 In Problem 14.6 on page 483, you used full-time voluntary turnover (%), and total worldwide revenue ($billions) to predict number of full-time jobs added (stored in Best Companies).
a. Perform a residual analysis on your results.
b. If appropriate, perform the Durbin-Watson test, using $\alpha = 0.05$.
c. Are the regression assumptions valid for these data?

14.21 In Problem 14.7 on page 483, you used the weekly staff count and remote engineering hours to predict standby hours (stored in Nickels 26Weeks).

a. Perform a residual analysis on your results.
b. If appropriate, perform the Durbin-Watson test, using $\alpha = 0.05$.
c. Are the regression assumptions valid for these data?

14.22 In Problem 14.8 on page 483, you used the land area of a property and the age of a house to predict the fair market value (stored in Glen Cove).
a. Perform a residual analysis on your results.
b. If appropriate, perform the Durbin-Watson test, using $\alpha = 0.05$.
c. Are the regression assumptions valid for these data?

14.4 Inferences About the Population Regression Coefficients

Section 13.7 explains how the t test for the slope in a simple linear regression model can determine the significance of the relationship between the X and Y variables. That section also constructed a confidence interval estimate of the population slope. This section extends those procedures to multiple regression.

Tests of Hypothesis

In a simple linear regression model, to test a hypothesis concerning the population slope, β_1, you used Equation (13.16) on page 454:

$$t_{STAT} = \frac{b_1 - \beta_1}{S_{b_1}}$$

Equation (14.7) generalizes this equation for multiple regression.

TESTING FOR THE SLOPE IN MULTIPLE REGRESSION

$$t_{STAT} = \frac{b_j - \beta_j}{S_{b_j}} \qquad (14.7)$$

where

b_j = slope of variable j with Y, holding constant the effects of all other independent variables

S_{b_j} = standard error of the regression coefficient b_j

k = number of independent variables in the regression equation

β_j = hypothesized value of the population slope for variable j, holding constant the effects of all other independent variables

t_{STAT} = test statistic for a t distribution with $n - k - 1$ degrees of freedom

---

To determine whether variable X_2 (amount of promotional expenses) has a significant effect on sales, after taking into account the effect of the price of OmniPower bars, the null and alternative hypotheses are

$$H_0: \beta_2 = 0$$
$$H_1: \beta_2 \neq 0$$

From Equation (14.7) and Figure 14.1 on page 481,

$$t_{STAT} = \frac{b_2 - \beta_2}{S_{b_2}}$$

$$= \frac{3.6131 - 0}{0.6852} = 5.2728$$

If one selects the 0.05 level of significance, the critical values of t for 31 degrees of freedom from Table E.3 are -2.0395 and $+2.0395$ as Figure 14.5 illustrates.

FIGURE 14.5
Testing for significance of a regression coefficient at the 0.05 level of significance, with 31 degrees of freedom

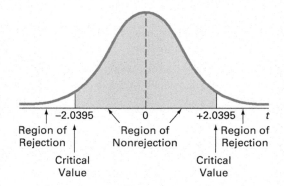

Table 14.6 summarizes the results of the test for the regression coefficient for promotional expenses (b_2) that appears as part of the Figure 14.1 OmniPower sales multiple regression results on page 481. Based on these conclusions, one concludes that promotional expenses has a significant effect on mean monthly sales.

TABLE 14.6
t test for the slope results and conclusions

Result	Conclusions
$t_{STAT} = 5.2728$ is greater than 2.0395 p-value $= 0.0000$ is less than the level of significance, $\alpha = 0.05$	1. Reject the null hypothesis H_0. 2. Conclude that strong evidence exists for claiming that promotional expenses is related to the dependent Y variable, sales, taking into account the price. 3. The probability is 0.0000 that $t_{STAT} < -5.2728$ or $t_{STAT} > 5.2728$.

Example 14.1 presents the test for the significance of β_1, the slope of sales with price.

EXAMPLE 14.1

Testing for the Significance of the Slope of Sales with Price

At the 0.05 level of significance, is there evidence that the slope of sales with price is different from zero?

SOLUTION Figure 14.1 on page 481, results show $t_{STAT} = -7.7664 < -2.0395$, the critical value for $\alpha = 0.05$ and that the p-value $= 0.0000 < 0.05$. One concludes that there is a significant relationship between price, X_1, and sales, taking into account the promotional expenses, X_2.

As shown with these two independent variables, the test of significance for a specific regression coefficient in multiple regression is a test for the significance of adding that variable into a regression model, given that the other variable is included. In other words, the t test for the regression coefficient is actually a test for the contribution of each independent variable.

Confidence Interval Estimation

Instead of testing the significance of a population slope, one might estimate the value of a population slope. Equation (14.8) defines the confidence interval estimate for a population slope in multiple regression.

CONFIDENCE INTERVAL ESTIMATE FOR THE SLOPE

$$b_j \pm t_{\alpha/2} S_{b_j} \tag{14.8}$$

where

$t_{\alpha/2}$ = the critical value corresponding to an upper-tail probability of $\alpha/2$ (a cumulative area of $1 - \alpha/2$) from the t distribution with $n - k - 1$ degrees of freedom

k = the number of independent variables

For the OmniPower bars example, one can construct a 95% confidence interval estimate of the population slope, β_1, the effect of price, X_1, on sales, Y, holding constant the effect of promotional expenses, X_2. From Table E.3, the critical value of t at the 95% confidence level with 31 degrees of freedom is 2.0395. Using Equation (14.8) and Figure 14.1 results on page 481

$$b_1 \pm t_{\alpha/2} S_{b_1}$$
$$-53.2173 \pm (2.0395)(6.8522)$$
$$-53.2173 \pm 13.9752$$
$$-67.1925 \le \beta_1 \le -39.2421$$

Taking into account the effect of promotional expenses, the estimated effect of a 1 cent increase in price is to reduce mean sales by approximately 39.2 to 67.2 bars. One has 95% confidence that this interval correctly estimates the relationship between these variables. From a hypothesis-testing viewpoint, because this confidence interval does not include 0, you conclude that the regression coefficient, for price, has a significant effect.

Example 14.2 constructs and interprets a confidence interval estimate for the slope of sales with promotional expenses.

EXAMPLE 14.2

Constructing a Confidence Interval Estimate for the Slope of Sales with Promotional Expenses

Construct a 95% confidence interval estimate of the population slope of sales with promotional expenses.

SOLUTION From Table E.3, the critical value of t at the 95% confidence level, with 31 degrees of freedom, is 2.0395. Using Equation (14.8) and Figure 14.1 on page 481,

$$b_2 \pm t_{\alpha/2} S_{b_2}$$
$$3.6131 \pm (2.0395)(0.6852)$$
$$3.6131 \pm 1.3975$$
$$2.2156 \le \beta_2 \le 5.0106$$

Thus, taking into account the effect of price, the estimated effect of each additional dollar of promotional expenses is to increase mean sales by approximately 2.22 to 5.01 bars. One has 95% confidence that this interval correctly estimates the relationship between these variables. From a hypothesis-testing viewpoint, because this confidence interval does not include 0, you can conclude that the regression coefficient, β_2, has a significant effect.

PROBLEMS FOR SECTION 14.4

LEARNING THE BASICS

14.23 Use the following information from a multiple regression analysis:

$$n = 25 \quad b_1 = 5 \quad b_2 = 10 \quad S_{b_1} = 2 \quad S_{b_2} = 8$$

a. Which variable has the largest slope, in units of a t statistic?
b. Construct a 95% confidence interval estimate of the population slope, β_1.
c. At the 0.05 level of significance, determine whether each independent variable makes a significant contribution to the regression model. On the basis of these results, indicate the independent variables to include in this model.

14.24 Use the following information from a multiple regression analysis:

$$n = 20 \quad b_1 = 4 \quad b_2 = 3 \quad S_{b_1} = 1.2 \quad S_{b_2} = 0.8$$

a. Which variable has the largest slope, in units of a t statistic?
b. Construct a 95% confidence interval estimate of the population slope, β_1.

c. At the 0.05 level of significance, determine whether each independent variable makes a significant contribution to the regression model. On the basis of these results, indicate the independent variables to include in this model.

APPLYING THE CONCEPTS

14.25 In Problem 14.3 on page 482, you predicted nonprofit charitable commitment, based on nonprofit revenue (Revenue) and fundraising efficiency (Efficiency) for a sample of 98 nonprofit organizations. Use the following results:

Variable	Coefficients	Standard Error	t Statistic	p-Value
Intercept	11.002079	7.127101	1.54	0.1260
Revenue	0.6683647	0.320077	2.09	0.0395
Efficiency	0.8317339	0.077736	10.70	0.0001

a. Construct 95% confidence interval estimates of the population slope between commitment and revenue and between commitment and efficiency.

b. At the 0.05 level of significance, determine whether each independent variable makes a significant contribution to the regression model. On the basis of these results, indicate the independent variables to include in this model.

✓ **SELF** **14.26** In Problem 14.4 on page 482, you used efficiency
TEST ratio and total risk-based capital to predict ROAA at a community bank (stored in Community Banks). Using the results from that problem,

a. construct a 95% confidence interval estimate of the population slope between ROAA and efficiency ratio.

b. at the 0.05 level of significance, determine whether each independent variable makes a significant contribution to the regression model. On the basis of these results, indicate the independent variables to include in this model.

14.27 In Problem 14.5 on page 483, you used the percentage of alcohol and chlorides to predict wine quality (stored in Vinho Verde). Using the results from that problem,

a. construct a 95% confidence interval estimate of the population slope between wine quality and the percentage of alcohol.

b. at the 0.05 level of significance, determine whether each independent variable makes a significant contribution to the regression model. On the basis of these results, indicate the independent variables to include in this model.

14.28 In Problem 14.6 on page 483, you used full-time voluntary turnover (%) and total worldwide revenue ($billions) to predict the

number of full-time job openings (stored in Best Companies). Using the results from that problem,

a. construct a 95% confidence interval estimate of the population slope between number of full-time job openings and total worldwide revenue.

b. at the 0.05 level of significance, determine whether each independent variable makes a significant contribution to the regression model. On the basis of these results, indicate the independent variables to include in this model.

14.29 In Problem 14.7 on page 483, you used the weekly staff present and remote engineering hours to predict standby hours (stored in Nickels 26Weeks). Using the results from that problem,

a. construct a 95% confidence interval estimate of the population slope between standby hours and weekly staff present.

b. at the 0.05 level of significance, determine whether each independent variable makes a significant contribution to the regression model. On the basis of these results, indicate the independent variables to include in this model.

14.30 In Problem 14.8 on page 483, you used land area of a property and age of a house to predict the fair market value (stored in Glen Cove). Using the results from that problem,

a. construct a 95% confidence interval estimate of the population slope between fair market value and land area of a property.

b. at the 0.05 level of significance, determine whether each independent variable makes a significant contribution to the regression model. On the basis of these results, indicate the independent variables to include in this model.

14.5 Testing Portions of the Multiple Regression Model

In developing a multiple regression model, one wants to use only those independent variables that significantly reduce the error in predicting the value of a dependent variable. If an independent variable does not improve the prediction, one can delete it from the multiple regression model and use a model with fewer independent variables.

The **partial F test** is an alternative to the t test that Section 14.4 discusses for determining the contribution of an independent variable. The partial F test determines the contribution to the regression sum of squares made by each independent variable after all the other independent variables have been included in the model. An independent variable is included only if it significantly improves the model.

To conduct partial F tests for the OmniPower sales example, one evaluates the contribution of promotional expenses (X_2) after price (X_1) has been included in the model and also evaluates the contribution of price (X_1) after promotional expenses (X_2) has been included in the model.

In general, if there are several independent variables, one determines the contribution of each independent variable by taking into account the regression sum of squares of a model that includes all independent variables except the one of interest, j. This regression sum of squares is denoted SSR (all Xs except j). Equation (14.9) determines the contribution of variable j, assuming that all other variables are already included.

> DETERMINING THE CONTRIBUTION OF AN INDEPENDENT VARIABLE TO THE REGRESSION MODEL
>
> $$SSR(X_j | \text{All } Xs \text{ except } j) = SSR(\text{All } Xs) - SSR(\text{All } Xs \text{ except } j) \qquad \textbf{(14.9)}$$

If there are two independent variables, Equations (14.10a) and (14.10b) are used to determine the contribution of each variable.

CONTRIBUTION OF VARIABLE X_1, GIVEN THAT X_2 HAS BEEN INCLUDED

$$SSR(X_1 \mid X_2) = SSR(X_1 \text{ and } X_2) - SSR(X_2) \qquad \textbf{(14.10a)}$$

CONTRIBUTION OF VARIABLE X_2, GIVEN THAT X_1 HAS BEEN INCLUDED

$$SSR(X_2 \mid X_1) = SSR(X_1 \text{ and } X_2) - SSR(X_1) \qquad \textbf{(14.10b)}$$

The term $SSR(X_2)$ represents the sum of squares due to regression for a model that includes only the independent variable X_2 (promotional expenses). Similarly, $SSR(X_1)$ represents the sum of squares due to regression for a model that includes only the independent variable X_1 (price). Figures 14.6 and 14.7 present results for these two models.

FIGURE 14.6
Excel worksheet results for the simple linear regression model of sales with promotional expenses, $SSR(X_2)$

	A	B	C	D	E	F	G
1	Sales and Promotional Expenses Analysis						
2							
3	*Regression Statistics*						
4	Multiple R	0.5351					
5	R Square	0.2863					
6	Adjusted R Square	0.2640					
7	Standard Error	1077.8721					
8	Observations	34					
9							
10	ANOVA						
11		*df*	*SS*	*MS*	*F*	*Significance F*	
12	Regression	1	14915814.1025	14915814.1025	12.8384	0.0011	
13	Residual	32	37177863.3387	1161808.2293			
14	Total	33	52093677.4412				
15							
16		*Coefficients*	*Standard Error*	*t Stat*	*P-value*	*Lower 95%*	*Upper 95%*
17	Intercept	1496.0161	483.9789	3.0911	0.0041	510.1843	2481.8480
18	Promotional Expenses	4.1281	1.1521	3.5831	0.0011	1.7813	6.4748

In the Worksheet

This worksheet uses the same template as the Figure 13.4 worksheet.

FIGURE 14.7
Excel worksheet results for the simple linear regression model of sales with price, $SSR(X_1)$

	A	B	C	D	E	F	G
1	Sales and Price Analysis						
2							
3	*Regression Statistics*						
4	Multiple R	0.7351					
5	R Square	0.5404					
6	Adjusted R Square	0.5261					
7	Standard Error	864.9457					
8	Observations	34					
9							
10	ANOVA						
11		*df*	*SS*	*MS*	*F*	*Significance F*	
12	Regression	1	28153486.1482	28153486.1482	37.6318	0.0000	
13	Residual	32	23940191.2930	748130.9779			
14	Total	33	52093677.4412				
15							
16		*Coefficients*	*Standard Error*	*t Stat*	*P-value*	*Lower 95%*	*Upper 95%*
17	Intercept	7512.3480	734.6189	10.2262	0.0000	6015.9796	9008.7164
18	Price	-56.7138	9.2451	-6.1345	0.0000	-75.5455	-37.8822

In the Worksheet

This worksheet uses the same template as the Figure 13.4 worksheet.

From Figure 14.6, $SSR(X_2) = 14,915,814.10$ and from Figure 14.1 on page 481 $SSR(X_1 \text{ and } X_2) = 39,472,730.77$. Then, using Equation (14.10a),

$$SSR(X_1 \mid X_2) = SSR(X_1 \text{ and } X_2) - SSR(X_2)$$

$$= 39,472,730.77 - 14,915,814.10$$

$$= 24,556,916.67$$

To determine whether X_1 significantly improves the model after X_2 has been included, divide the regression sum of squares into two component parts, as shown in Table 14.7.

TABLE 14.7
ANOVA table dividing the regression sum of squares into components to determine the contribution of variable X_1

Source	Degrees of Freedom	Sum of Squares	Mean Square (Variance)	F
Regression	2	39,472,730.77	19,736,365.39	
$\begin{cases} X_2 \\ X_1\|X_2 \end{cases}$	$\begin{cases} 1 \\ 1 \end{cases}$	$\begin{cases} 14,915,814.10 \\ 24,556,916.67 \end{cases}$	24,556,916.67	60.32
Error	31	12,620,946.67	407,127.31	
Total	33	52,093,677.44		

The null and alternative hypotheses to test for the contribution of X_1 to the model are:

H_0: Variable X_1 does not significantly improve the model after variable X_2 has been included.
H_1: Variable X_1 significantly improves the model after variable X_2 has been included.

Equation (14.11) defines the partial F test statistic for testing the contribution of an independent variable.

PARTIAL F TEST STATISTIC

$$F_{STAT} = \frac{SSR(X_j | \text{All } Xs \text{ except } j)}{MSE} \tag{14.11}$$

The partial F test statistic follows an F distribution with 1 and $n - k - 1$ degrees of freedom.

From Table 14.7,

$$F_{STAT} = \frac{24,556,916.67}{407,127.31} = 60.32$$

The partial F_{STAT} test statistic has 1 and $n - k - 1 = 34 - 2 - 1 = 31$ degrees of freedom. Using a level of significance of 0.05, the critical value from Table E.5 is approximately 4.17 as Figure 14.8 illustrates.

FIGURE 14.8
Testing for the contribution of a regression coefficient to a multiple regression model at the 0.05 level of significance, with 1 and 31 degrees of freedom

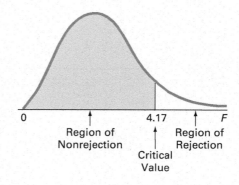

Because the computed partial F_{STAT} test statistic (60.32) is greater than the critical F value, 4.17, you reject H_0. You conclude that the addition of variable X_1, price, significantly improves a regression model that already contains variable X_2, promotional expenses.

To evaluate the contribution of variable X_2, promotional expenses, to a model in which variable X_1, price, has been included, use Equation (14.10b). First, from Figure 14.7 on page 493, observe that $SSR(X_1) = 28,153,486.15$. Second, from Table 14.7, observe that $SSR(X_1 \text{ and } X_2) = 39,472,730.77$. Then, using Equation (14.10b) on page 493,

$$SSR(X_2 | X_1) = 39,472,730.77 - 28,153,486.15$$

$$= 11,319,244.62$$

To determine whether X_2 significantly improves a model after X_1 has been included, you can divide the regression sum of squares into two component parts, as shown in Table 14.8.

TABLE 14.8
ANOVA table dividing the regression sum of squares into components to determine the contribution of variable X_2

Source	Degrees of Freedom	Sum of Squares	Mean Square (Variance)	F
Regression	2	39,472,730.77	19,736,365.39	
$\begin{Bmatrix} X_1 \\ X_2\mid X_1 \end{Bmatrix}$	$\begin{Bmatrix} 1 \\ 1 \end{Bmatrix}$	$\begin{Bmatrix} 28,153,486.15 \\ 11,319,244.62 \end{Bmatrix}$	11,319,244.62	27.80
Error	31	12,620,946.67	407,127.31	
Total	33	52,093,677.44		

The null and alternative hypotheses to test for the contribution of X_2 to the model are:

H_0: Variable X_2 does not significantly improve the model after variable X_1 has been included.
H_1: Variable X_2 significantly improves the model after variable X_1 has been included.

Using Equation (14.11) and Table 14.8,

$$F_{STAT} = \frac{11,319,244.62}{407,127.31} = 27.80$$

Again, using a 0.05 level of significance, the critical value of F, with 1 and 31 degrees of freedom, is approximately 4.17. Because the computed partial F_{STAT} test statistic (27.80) is greater than this critical value (4.17), you reject H_0. You conclude that the addition of variable X_2, promotional expenses, significantly improves the multiple regression model already containing X_1, price.

By testing for the contribution of each independent variable after the other independent variable has been included in the model, you determine that each of the two independent variables significantly improves the model. Therefore, the multiple regression model should include both X_1, price, and X_2, promotional expenses.

The partial F test statistic that this section discusses and the t test statistic of Equation (14.7) on page 489 are both used to determine the contribution of an independent variable to a multiple regression model. The hypothesis tests associated with these two statistics always result in the same decision (i.e., the p-values are identical). The t_{STAT} test statistics for the OmniPower regression model are -7.7664 and $+5.2728$, and the corresponding F_{STAT} test statistics are 60.32 and 27.80. Equation (14.12) states this relationship between t and F.[1]

[1] This relationship holds only when the F_{STAT} statistic has 1 degree of freedom in the numerator.

RELATIONSHIP BETWEEN A t STATISTIC AND AN F STATISTIC

$$t_{STAT}^2 = F_{STAT} \tag{14.12}$$

Coefficients of Partial Determination

Section 14.2 explains that the coefficient of multiple determination, r^2, measures the proportion of the variation in Y that is explained by variation in the independent variables. The **coefficients of partial determination** ($r_{Y1.2}^2$ and $r_{Y2.1}^2$) measure the proportion of the variation in the dependent Y variable that is explained by each independent X variable while controlling for, or holding constant, the other independent variable. These coefficients are different from the *coefficient of multiple determination* that measures the proportion of the variation in the dependent variable explained by the entire set of independent variables included in the model.

Equation (14.13) defines the coefficients of partial determination for a multiple regression model with two independent variables.

COEFFICIENTS OF PARTIAL DETERMINATION FOR A MULTIPLE REGRESSION MODEL CONTAINING TWO INDEPENDENT VARIABLES

$$r^2_{Y1.2} = \frac{SSR(X_1|X_2)}{SST - SSR(X_1 \text{ and } X_2) + SSR(X_1|X_2)} \tag{14.13a}$$

$$r^2_{Y2.1} = \frac{SSR(X_2|X_1)}{SST - SSR(X_1 \text{ and } X_2) + SSR(X_2|X_1)} \tag{14.13b}$$

where

$$
\begin{aligned}
SSR(X_1|X_2) &= \text{sum of squares of the contribution of variable } X_1 \text{ to the regression} \\
&\quad \text{model, given that variable } X_2 \text{ has been included in the model} \\
SST &= \text{total sum of squares for } Y \\
SSR(X_1 \text{ and } X_2) &= \text{regression sum of squares when variables } X_1 \text{ and } X_2 \text{ are both included} \\
&\quad \text{in the multiple regression model} \\
SSR(X_2|X_1) &= \text{sum of squares of the contribution of variable } X_2 \text{ to the regression} \\
&\quad \text{model, given that variable } X_1 \text{ has been included in the model}
\end{aligned}
$$

For the OmniPower sales example, the coefficient of partial determination, $r^2_{Y1.2}$, of variable Y with X_1 while holding X_2 constant is 0.6605. For a given (constant) amount of promotional expenses, 66.05% of the variation in OmniPower sales is explained by the variation in the price.

$$r^2_{Y1.2} = \frac{24{,}556{,}916.67}{52{,}093{,}677.44 - 39{,}472{,}730.77 + 24{,}556{,}916.67}$$

$$= 0.6605$$

For the OmniPower sales example, the coefficient of partial determination, $r^2_{Y2.1}$, of variable Y with X_2 while holding X_1 constant is 0.4728. For a given (constant) price, 47.28% of the variation in OmniPower sales is explained by variation in the amount of promotional expenses.

$$r^2_{Y2.1} = \frac{11{,}319{,}244.62}{52{,}093{,}677.44 - 39{,}472{,}730.77 + 11{,}319{,}244.62}$$

$$= 0.4728$$

Equation (14.14) defines the coefficient of partial determination for the jth variable in a multiple regression model containing several (k) independent variables.

COEFFICIENT OF PARTIAL DETERMINATION FOR A MULTIPLE REGRESSION MODEL CONTAINING k INDEPENDENT VARIABLES

$$r^2_{Yj.(\text{All variables } except\, j)} = \frac{SSR(X_j| \text{All } Xs\ except\ j)}{SST - SSR(\text{All } Xs) + SSR(X_j|\text{All } Xs\ except\ j)} \tag{14.14}$$

LEARNING THE BASICS

14.31 The following is the ANOVA summary table for a multiple regression model with two independent variables:

Source	Degrees of Freedom	Sum of Squares	Mean Squares	F
Regression	2	60		
Error	18	120		
Total	20	180		

If $SSR(X_1) = 45$ and $SSR(X_2) = 25$,
a. determine whether there is a significant relationship between Y and each independent variable at the 0.05 level of significance.
b. compute the coefficients of partial determination, $r^2_{Y1.2}$ and $r^2_{Y2.1}$, and interpret their meaning.

14.32 The following is the ANOVA summary table for a multiple regression model with two independent variables:

Source	Degrees of Freedom	Sum of Squares	Mean Squares	F
Regression	2	30		
Error	10	120		
Total	12	150		

If $SSR(X_1) = 20$ and $SSR(X_2) = 15$,
a. determine whether there is a significant relationship between Y and each independent variable at the 0.05 level of significance.
b. compute the coefficients of partial determination, $r^2_{Y1.2}$ and $r^2_{Y2.1}$, and interpret their meaning.

APPLYING THE CONCEPTS

14.33 In Problem 14.5 on page 483, you used alcohol percentage and chlorides to predict wine quality (stored in Vinho Verde). Using the results from that problem,
a. at the 0.05 level of significance, determine whether each independent variable makes a significant contribution to the regression model. On the basis of these results, indicate the most appropriate regression model for this set of data.
b. compute the coefficients of partial determination, $r^2_{Y1.2}$ and $r^2_{Y2.1}$, and interpret their meaning.

SELF TEST **14.34** In Problem 14.4 on page 482, you used efficiency ratio and total risk-based capital to predict ROAA at a community bank (stored in Community Banks). Using the results from that problem,
a. at the 0.05 level of significance, determine whether each independent variable makes a significant contribution to the regression model. On the basis of these results, indicate the most appropriate regression model for this set of data.
b. compute the coefficients of partial determination, $r^2_{Y1.2}$ and $r^2_{Y2.1}$, and interpret their meaning.

14.35 In Problem 14.7 on page 483, you used the weekly staff count and remote engineering hours to predict standby hours (stored in Nickels 26Weeks). Using the results from that problem,
a. at the 0.05 level of significance, determine whether each independent variable makes a significant contribution to the regression model. On the basis of these results, indicate the most appropriate regression model for this set of data.
b. compute the coefficients of partial determination, $r^2_{Y1.2}$ and $r^2_{Y2.1}$, and interpret their meaning.

14.36 In Problem 14.6 on page 483, you used full-time voluntary turnover (%), and total worldwide revenue ($billions) to predict the number of full-time job openings (stored in Best Companies). Using the results from that problem,
a. at the 0.05 level of significance, determine whether each independent variable makes a significant contribution to the regression model. On the basis of these results, indicate the most appropriate regression model for this set of data.
b. compute the coefficients of partial determination, $r^2_{Y1.2}$ and $r^2_{Y2.1}$, and interpret their meaning.

14.37 In Problem 14.8 on page 483, you used land area of a property and age of a house to predict the fair market value (stored in Glen Cove). Using the results from that problem,
a. at the 0.05 level of significance, determine whether each independent variable makes a significant contribution to the regression model. On the basis of these results, indicate the most appropriate regression model for this set of data.
b. compute the coefficients of partial determination, $r^2_{Y1.2}$ and $r^2_{Y2.1}$, and interpret their meaning.

14.6 Using Dummy Variables and Interaction Terms

The multiple regression models that Sections 14.1 through 14.5 discuss assumed that each independent variable is a numerical variable. For example, in Section 14.1, you used price and promotional expenses, two numerical independent variables, to predict the monthly sales of OmniPower nutrition bars. However, for some models, one needs to examine the effect of a categorical independent variable. In such cases, one uses a **dummy variable** to include a categorical independent variable in a regression model.

Dummy variables use the numeric values 0 and 1 to recode two categories of a categorical independent variable in a regression model. In general, the number of dummy variables one needs to define equals the number of categories −1. If a categorical independent variable has only two categories, one dummy variable, X_d, gets defined, and the values 0 and 1 represent the two categories. When the two categories represent the presence or absence of a characteristic, use 0 to represent the absence and 1 to represent the presence of the characteristic.

For example, to predict the monthly sales of the OmniPower bars, one might include the categorical variable location in the model to explore the possible effect on sales caused by displaying the OmniPower bars in the two different sales locations, a special front location and in the snack aisle, analogous to the locations used in the Chapter 10 Arlingtons scenario to sell streaming media players. In this case for the categorical variable location, the dummy variable, X_d, would have these values:

$$X_d = 0 \text{ if the value is the first category (special front location)}$$
$$X_d = 1 \text{ if the value is the second category (in-aisle location)}$$

To illustrate using dummy variables in regression, consider the case of a realtor who seeks to develop a model for predicting the asking price of houses listed for sale ($thousands) in Silver Spring, Maryland, using living space in the house (square feet) and whether the house has a fireplace as the independent variables. To include the categorical variable for the presence of a fireplace, the dummy variable X_2 is defined as

$$X_2 = 0 \text{ if the house does not have a fireplace}$$
$$X_2 = 1 \text{ if the house has a fireplace}$$

Assuming that the slope of asking price with living space is the same for houses that have and do not have a fireplace, the multiple regression model is

$$Y_i = \beta_0 + \beta_1 X_{1i} + \beta_2 X_{2i} + \varepsilon_i$$

where

Y_i = asking price, in thousands of dollars, for house i
β_0 = Y intercept
X_{1i} = living space, in thousands of square feet, for house i
β_1 = slope of asking price with living space, holding constant the presence or absence of a fireplace
X_{2i} = dummy variable that represents the absence or presence of a fireplace for house i
β_2 = net effect of the presence of a fireplace on asking price, holding constant the living space
ε_i = random error in Y for house i

Figure 14.9 presents the regression results for this model, using a sample of 61 Silver Spring houses listed for sale that was extracted from trulia.com and stored in Silver Spring Homes . In these results, the dummy variable X_2 is labeled as Fireplace.

FIGURE 14.9

Excel worksheet results for the regression model that includes Living Space and Fireplace

◢	A	B	C	D	E	F	G
1	Asking Price Analysis						
2							
3	*Regression Statistics*						
4	Multiple R	0.6842					
5	R Square	0.4681					
6	Adjusted R Square	0.4497					
7	Standard Error	66.8687					
8	Observations	61					
9							
10	ANOVA						
11		*df*	*SS*	*MS*	*F*	*Significance F*	
12	Regression	2	228210.1161	114105.0581	25.5187	0.0000	
13	Residual	58	259342.5606	4471.4235			
14	Total	60	487552.6767				
15							
16		*Coefficients*	*Standard Error*	*t Stat*	*P-value*	*Lower 95%*	*Upper 95%*
17	Intercept	302.2518	26.5548	11.3822	0.0000	249.0965	355.4071
18	Living Space	0.0765	0.0129	5.9179	0.0000	0.0507	0.1024
19	Fireplace	52.9674	19.1421	2.7671	0.0076	14.6504	91.2844

In the Worksheet

This worksheet uses the same template as the Figure 14.1 worksheet.

From Figure 14.9, the regression equation is

$$\hat{Y}_i = 302.2518 + 0.0765X_{1i} + 52.9674X_{2i}$$

For houses without a fireplace, one sets $X_2 = 0$:

$$\hat{Y}_i = 302.2518 + 0.0765X_{1i} + 52.9674X_{2i}$$
$$= 302.2518 + 0.0765X_{1i} + 52.9674(0)$$
$$= 302.2518 + 0.0765X_{1i}$$

For houses with a fireplace, one sets $X_2 = 1$:

$$\hat{Y}_i = 302.2518 + 0.0765X_{1i} + 52.9674X_{2i}$$
$$= 302.2518 + 0.0765X_{1i} + 52.9674(1)$$
$$= 355.2192 + 0.0765X_{1i}$$

Table 14.9 summarizes the results of the test for the regression coefficient for living space (b_1) and the regression coefficient for presence or absence of a fireplace (b_2) that appears as part of Figure 14.9, the Silver Spring houses multiple regression results on page 498. Based on these results, one can conclude that living space has a significant effect on mean asking price and the presence of a fireplace also has a significant effect.

TABLE 14.9
t test for the slope Results and conclusions for the Silver Spring houses multiple regression model

Result	Conclusions
$t_{STAT} = 5.9179$ is greater than 2.0017 p-value $= 0.0000$ is less than the level of significance, $\alpha = 0.05$	1. Reject the null hypothesis H_0. 2. Conclude that strong evidence exists for claiming that living space is related to the dependent Y variable, asking price, taking into account the presence or absence of a fireplace. 3. The probability is 0.0000 that $t_{STAT} < -5.9179$ or $t_{STAT} > 5.9179$
$t_{STAT} = 2.7671$ is greater than 2.0017 p-value $= 0.0076$ is less than the level of significance, $\alpha = 0.05$	1. Reject the null hypothesis H_0. 2. Conclude that strong evidence exists for claiming that presence of a fireplace is related to the dependent Y variable, asking price, taking into account the living space. 3. The probability is 0.0076 that $t_{STAT} < -2.7671$ or $t_{STAT} > 2.7671$.
$r^2 = 0.4681$	46.81% of the variation in the asking price can be explained by variation in living space and whether the house has a fireplace.

student TIP

Remember that an independent variable does not always make a significant contribution to a regression model.

Using the net regression coefficients b_1 and b_2, the Table 14.10 net effects table summarizes the effects of adding one square foot of living space (X_1) or the presence of a fireplace (X_2).

TABLE 14.10
Net Effects Table for the Silver Spring Houses Multiple Regression Model

Independent Variable Change	Net Effect
An increase of one square foot in living space	Predict mean asking price to increase by 0.0765 ($000) or $76.50 holding presence of a fireplace constant.
Presence of a fireplace	Predict mean asking price to increase by $52.9674 ($000) or $52,967.40 for a house with a fireplace holding living space constant.

In some situations, the categorical independent variable has more than two categories. When this occurs, two or more dummy variables are needed. Example 14.3 illustrates such situations.

EXAMPLE 14.3

Modeling a Three-Level Categorical Variable

Define a multiple regression model to predict the asking price of houses as the dependent variable, as was done in the previous example for the Silver Spring houses, and use living space and house type as independent variables. House type is a three-level categorical variable with the values colonial, ranch, and other.

SOLUTION To model the three-level categorical variable house type, two dummy variables, X_1 and X_2, are needed:

$$X_{1i} = 1 \text{ if the house type is colonial for house } i; 0 \text{ otherwise}$$

$$X_{2i} = 1 \text{ if the house type is ranch for house } i; 0 \text{ otherwise}$$

Thus, if house i is a colonial then $X_{1i} = 1$ and $X_{2i} = 0$; if house i is a ranch, then $X_{1i} = 0$ and $X_{2i} = 1$; and if house i is other (neither colonial nor ranch), then $X_{1i} = X_{2i} = 0$. Thus, house type other becomes the baseline category to which the effect of being a colonial or ranch house type is compared. A third independent variable is used for living space:

$$X_{3i} = \text{living space for observation } i$$

Thus, the regression model for this example is

$$Y_i = \beta_0 + \beta_1 X_{1i} + \beta_2 X_{2i} + \beta_3 X_{3i} + \varepsilon_i$$

where

$Y_i =$ asking price for house i

$\beta_0 = Y$ intercept

$\beta_1 =$ difference between the predicted asking price of house type colonial and the predicted asking price of house type other holding living space constant

$\beta_2 =$ difference between the predicted asking price of house type ranch and the predicted asking price of house type other holding living space constant

$\beta_3 =$ slope of asking price with living space, holding the house type constant

$\varepsilon_i =$ random error in y for observation i

Regression models can have several numerical independent variables along with a dummy variable. Example 14.4 illustrates a regression model in which there are two numerical independent variables and a categorical independent variable.

EXAMPLE 14.4

Studying a Regression Model That Contains a Dummy Variable and Two Numerical Independent Variables

A builder of suburban homes faces the business problem of seeking to predict heating oil consumption in single-family houses. The independent variables considered are atmospheric temperature (°F), X_1, insulation, the amount of attic insulation, inches, X_2, and ranch-style, whether the house is ranch-style, X_3. Data are collected from a sample of 15 single-family houses and stored in `Heating Oil`. Develop and analyze an appropriate regression model, using these three independent variables X_1, X_2, and X_3.

SOLUTION Define X_3, ranch-style, a dummy variable for ranch-style house, as follows:

$$X_3 = 0 \text{ if not a ranch-style house}$$

$$X_3 = 1 \text{ if a ranch-style house}$$

▶(continued)

Assuming that the slope between heating oil consumption and temperature, X_1, and between heating oil consumption and insulation, X_2, is the same for both styles of houses, the regression model is

$$Y_i = \beta_0 + \beta_1 X_{1i} + \beta_2 X_{2i} + \beta_3 X_{3i} + \varepsilon_i$$

where

Y_i = monthly heating oil consumption, in gallons, for house i

β_0 = Y intercept

β_1 = slope of heating oil consumption with temperature, holding constant the effect of insulation and ranch-style

β_2 = slope of heating oil consumption with insulation holding constant the effect of temperature and ranch-style

β_3 = incremental effect of the presence of a ranch-style house on heating oil consumption holding constant the effect of temperature and insulation

ε_i = random error in y for house i

Figure 14.10 presents results for this regression model.

FIGURE 14.10

Excel worksheet results for the regression model that includes temperature, insulation, and ranch-style for the heating oil data

	A	B	C	D	E	F	G
1	**Heating Oil Consumption Analysis**						
2							
3	*Regression Statistics*						
4	Multiple R	0.9942					
5	R Square	0.9884					
6	Adjusted R Square	0.9853					
7	Standard Error	15.7489					
8	Observations	15					
9							
10	ANOVA						
11		*df*	*SS*	*MS*	*F*	*Significance F*	
12	Regression	3	233406.9094	77802.3031	313.6822	0.0000	
13	Residual	11	2728.3200	248.0291			
14	Total	14	236135.2293				
15							
16		*Coefficients*	*Standard Error*	*t Stat*	*P-value*	*Lower 95%*	*Upper 95%*
17	Intercept	592.5401	14.3370	41.3295	0.0000	560.9846	624.0956
18	Temperature	-5.5251	0.2044	-27.0267	0.0000	-5.9751	-5.0752
19	Insulation	-21.3761	1.4480	-14.7623	0.0000	-24.5632	-18.1891
20	Ranch-style	-38.9727	8.3584	-4.6627	0.0007	-57.3695	-20.5759

From the results in Figure 14.10, the regression equation is

$$\hat{Y}_i = 592.5401 - 5.5251X_{1i} - 21.3761X_{2i} - 38.9727X_{3i}$$

For houses that are not ranch style, because $X_3 = 0$, the regression equation reduces to

$$\hat{Y}_i = 592.5401 - 5.5251X_{1i} - 21.3761X_{2i}$$

For houses that are ranch style, because $X_3 = 1$, the regression equation reduces to

$$\hat{Y}_i = 553.5674 - 5.5251X_{1i} - 21.3761X_{2i}$$

Table 14.11 on page 502 summarizes the results of the tests for the regression coefficient for temperature (b_1), insulation (b_2), and the regression coefficient for ranch-style, the presence or absence of a ranch-style house (b_3) that appears as part of Figure 14.10. Based on these results, the developer can conclude that temperature, insulation, and ranch-style each has a significant effect on mean monthly heating oil consumption.

▶(*continued*)

TABLE 14.11

t test for the slope results and conclusions for the Example 14.4 multiple regression model

Result	Conclusions
$t_{STAT} = -27.0267$ is less than -2.2010 p-value $= 0.0000$ is less than the level of significance, $\alpha = 0.05$	1. Reject the null hypothesis H_0. 2. Conclude that strong evidence exists for claiming that temperature is related to the dependent Y variable, heating oil consumption, holding insulation and ranch-style constant. 3. The probability is 0.0000 that $t_{STAT} < -27.0267$ or $t_{STAT} > 27.0267$
$t_{STAT} = -14.7623$ is less than -2.2010 p-value $= 0.0000$ is less than the level of significance, $\alpha = 0.05$	1. Reject the null hypothesis H_0. 2. Conclude that strong evidence exists for claiming that insulation is related to the dependent Y variable, heating oil consumption, holding temperature and ranch-style constant. 3. The probability is 0.0000 that $t_{STAT} < -14.7623$ or $t_{STAT} > 14.7623$
$t_{STAT} = -4.6627$ is less than -2.2010 p-value $= 0.0007$ is less than the level of significance, $\alpha = 0.05$	1. Reject the null hypothesis H_0. 2. Conclude that strong evidence exists for claiming that whether the house is a ranch-style is related to the dependent Y variable, heating oil consumption, holding temperature and insulation constant. 3. The probability is 0.0007 that $t_{STAT} < -4.6627$ or $t_{STAT} > 4.6627$
$r^2 = 0.9884$	98.84% of the variation in the heating oil consumption can be explained by variation in temperature, insulation, and whether the house is a ranch-style.

Using the net regression coefficients b_1, b_2, and b_3, the Table 14.12 net effect column summarizes the effects of an increase one degree of temperature (X_1), adding one inch to the insulation amount (X_2), and whether the house is a ranch-style (X_3). If the cost of adding one inch in attic insulation was equivalent to about 21 gallons of heating oil, the builder could predict that the new insulation would "pay for itself" by lowering heating oil consumption in about one month.

TABLE 14.12

Net effects table for the Example 14.4 multiple regression model

Independent Variable Change	Net Effect
An increase of one degree in temperature (°F)	Predict mean monthly heating oil consumption to decrease by 5.5251 gallons holding insulation and ranch-style constant.
An increase of one inch in attic insulation	Predict mean monthly heating oil consumption to decrease by 21.3761 gallons for each additional inch of attic insulation holding temperature and ranch-style constant.
Presence of a ranch-style house	Predict mean monthly heating oil consumption to decrease by 38.9727 gallons for a ranch-style house holding temperature and insulation constant.

Before anyone could use the model in Example 14.4 for decision making, one may want to determine whether the independent variables interact with each other.

Interactions

In the regression models discussed so far, the effect an independent variable has on the dependent variable has been assumed to be independent of the other independent variables in the model. An **interaction** occurs if the effect of an independent variable on the dependent variable changes according to the *value* of a second independent variable. For example, it is possible that advertising will have a large effect on the sales of a product when the price of a product is low. However, if the price of the product is too high, increases in advertising will not dramatically change sales. In this case, price and advertising are said to interact. In other words, one cannot make general statements about the effect of advertising on sales. The effect that advertising has on sales is *dependent* on the price. One uses an **interaction term**, also called a **cross-product term**, to model an interaction effect in a multiple regression model.

To illustrate the concept of interaction and use of an interaction term, recall the Section 14.6 asking price of homes example, the multiple regression results for which Figure 14.9 on page 498 presents. In the regression model, the realtor has assumed that the effect that living space has on the asking price is independent of whether the house has a fireplace. Stated in statistical language, the realtor has assumed that the slope of asking price with living space is the same for all houses, regardless of whether the house contains a fireplace. However, this assumption may not be true. If an interaction exists between living space and the presence or absence of a fireplace, the two slopes will be different.

To evaluate whether an interaction exists, one first defines an interaction term that is the product of the independent variable X_1 (living space) and the dummy variable X_2 (fireplace). Then, one tests whether this interaction variable makes a significant contribution to the regression model. If the interaction is significant, one cannot use the original model for prediction. For these data, one defines

$$X_3 = X_1 \times X_2$$

Figure 14.11 presents regression results for the model that includes the living space, X_1, the presence of a fireplace, X_2, and the interaction of X_1 and X_2, which has been defined as X_3 and labeled Living Space*Fireplace.

FIGURE 14.11

Excel worksheet results for the regression model that includes living space, fireplace, and interaction of living space and fireplace

	A	B	C	D	E	F	G
1	Asking Price Analysis						
2							
3	*Regression Statistics*						
4	Multiple R	0.6849					
5	R Square	0.4691					
6	Adjusted R Square	0.4411					
7	Standard Error	67.3907					
8	Observations	61					
9							
10	ANOVA						
11		*df*	*SS*	*MS*	*F*	*Significance F*	
12	Regression	3	228686.7174	76228.9058	16.7849	0.0000	
13	Residual	57	258865.9593	4541.5081			
14	Total	60	487552.6767				
15							
16		*Coefficients*	*Standard Error*	*t Stat*	*P-value*	*Lower 95%*	*Upper 95%*
17	Intercept	316.2350	50.7878	6.2266	0.0000	214.5341	417.9359
18	Living Space	0.0681	0.0292	2.3319	0.0233	0.0096	0.1265
19	Fireplace	34.8926	59.0359	0.5910	0.5568	-83.3248	153.1101
20	Living Space*Fireplace	0.0106	0.0326	0.3239	0.7472	-0.0548	0.0759

In the Worksheet

This worksheet uses the same template as the Figure 14.1 worksheet.

"LivingSpace*Fireplace" refers to the column that holds the cross-product term that models the interaction between the living space and fireplace variables.

student TIP

It is possible that the interaction between two independent variables will be significant even though one of the independent variables is not significant.

The null and alternate hypotheses to test for the existence of an interaction are

$$H_0: \beta_3 = 0$$
$$H_1: \beta_3 \neq 0.$$

Table 14.13 summarizes the results of the test for the interaction for living space (b_1) and presence of a fireplace (b_2) that appears as part of Figure 14.11. Based on these conclusions, one concludes that interaction of living space (b_1) and presence of a fireplace (b_2) is not significant. The interaction term should not be included in the regression model to predict asking price.

TABLE 14.13

t test for the Interaction for living space and presence of a fireplace results and conclusions

Result	Conclusions
$t_{STAT} = 0.3239$ is less than 2.0025 *p*-value $= 0.7472$ is greater than the level of significance, $\alpha = 0.05$	1. Do not reject the null hypothesis H_0. 2. Conclude that there is insufficient evidence of an interaction of living space (b_1) and presence of a fireplace (b_2). 3. The probability is 0.7472 that $t_{STAT} < -0.3239$ or $t_{STAT} > 0.3239$.

In Example 14.5, three interaction terms are added to the model.

EXAMPLE 14.5

Evaluating a Regression Model with Several Interactions

For the Example 14.4 data, determine whether adding interaction terms makes a significant contribution to the regression model.

SOLUTION To evaluate possible interactions between the independent variables, three interaction terms are constructed as follows: $X_4 = X_1 \times X_2$, $X_5 = X_1 \times X_3$, and $X_6 = X_2 \times X_3$. The regression model is now

$$Y_i = \beta_0 + \beta_1 X_{1i} + \beta_2 X_{2i} + \beta_3 X_{3i} + \beta_4 X_{4i} + \beta_5 X_{5i} + \beta_6 X_{6i} + \varepsilon_i$$

where X_1 is temperature, X_2 is insulation, X_3 is the dummy variable ranch-style, X_4 is the interaction between temperature and insulation, X_5 is the interaction between temperature and ranch-style, and X_6 is the interaction between insulation and ranch-style. Figure 14.12 presents the results for this regression model.

FIGURE 14.12

Excel worksheet results for the regression model that includes temperature, X_1; insulation, X_2; the dummy variable ranch-style, X_3; the interaction of temperature and insulation, X_4; the interaction of temperature and ranch-style, X_5; and the interaction of insulation and ranch-style, X_6

student TIP

For problems with more than a few independent variables, consider using either the PHStat or Analysis ToolPak add-ins.

	A	B	C	D	E	F	G
1	**Heating Oil Consumption Analysis**						
2							
3	*Regression Statistics*						
4	Multiple R	0.9966					
5	R Square	0.9931					
6	Adjusted R Square	0.9880					
7	Standard Error	14.2506					
8	Observations	15					
9							
10	ANOVA						
11		*df*	*SS*	*MS*	*F*	*Significance F*	
12	Regression	6	234510.5818	39085.0970	192.4607	0.0000	
13	Residual	8	1624.6475	203.0809			
14	Total	14	236135.2293				
15							
16		*Coefficients*	*Standard Error*	*t Stat*	*P-value*	*Lower 95%*	*Upper 95%*
17	Intercept	642.8867	26.7059	24.0728	0.0000	581.3027	704.4707
18	Temperature	-6.9263	0.7531	-9.1969	0.0000	-8.6629	-5.1896
19	Insulation	-27.8825	3.5801	-7.7882	0.0001	-36.1383	-19.6268
20	Ranch-style	-84.6088	29.9956	-2.8207	0.0225	-153.7788	-15.4389
21	Temperature*Insulation	0.1702	0.0886	1.9204	0.0911	-0.0342	0.3746
22	Temperature*Ranch-style	0.6596	0.4617	1.4286	0.1910	-0.4051	1.7242
23	Insulation*Ranch-style	4.9870	3.5137	1.4193	0.1936	-3.1156	13.0895

To test whether the three interactions significantly improve the regression model, one uses the partial *F* test. The null and alternative hypotheses are

$H_0: \beta_4 = \beta_5 = \beta_6 = 0$ (There are no interactions among X_1, X_2, and X_3.)
$H_1: \beta_4 \neq 0$ and/or $\beta_5 \neq 0$ and/or $\beta_6 \neq 0$ (X_1 interacts with X_2, and/or X_1 interacts with X_3, and/or X_2 interacts with X_3.)

From Figure 14.12,

$$SSR(X_1, X_2, X_3, X_4, X_5, X_6) = 234{,}510.5818 \text{ with 6 degrees of freedom}$$

▶(continued)

and from Figure 14.10 on page 501, $SSR(X_1, X_2, X_3) = 233{,}406.9094$ with 3 degrees of freedom. Thus,

$$SSR(X_1, X_2, X_3, X_4, X_5, X_6) - SSR(X_1, X_2, X_3) = 234{,}510.5818 - 233{,}406.9094 = 1{,}103.6724$$

The difference in degrees of freedom is $6 - 3 = 3$.

To use the partial F test for the simultaneous contribution of three variables to a model, one uses an extension of Equation (14.11) on page 494. The partial F_{STAT} test statistic is

$$F_{STAT} = \frac{[SSR(X_1, X_2, X_3, X_4, X_5, X_6) - SSR(X_1, X_2, X_3)]/3}{MSE(X_1, X_2, X_3, X_4, X_5, X_6)} = \frac{1{,}103.6724/3}{203.0809} = 1.8115$$

One compares the computed F_{STAT} test statistic to the critical F value for 3 and 8 degrees of freedom. Using a level of significance of 0.05, the critical F value from Table E.5 is 4.07. Because $F_{STAT} = 1.8115 < 4.07$, one concludes that the interactions do not make a significant contribution to the model, given that the model already includes temperature, X_1; insulation, X_2; and whether the house is ranch style, X_3. One can further conclude that the multiple regression model using X_1, X_2, and X_3 without any interaction terms is the better model. Had results indicated to reject the null hypothesis, one would then have needed to test the contribution of each interaction separately in order to determine which interaction terms to include in the model.

In general, if a model has several independent variables and one wants to test whether additional independent variables contribute to the model, the numerator of the F test is the SSR for all independent variables minus the SSR for the initial set of variables, divided by the number of independent variables whose contribution is being tested.

PROBLEMS FOR SECTION 14.6

LEARNING THE BASICS

14.38 Suppose X_1 is a numerical variable and X_2 is a dummy variable with two categories and the regression equation for a sample of $n = 20$ is

$$\hat{Y}_i = 6 + 4X_{1i} + 2X_{2i}$$

a. Interpret the regression coefficient associated with variable X_1.
b. Interpret the regression coefficient associated with variable X_2.
c. Suppose that the t_{STAT} test statistic for testing the contribution of variable X_2 is 3.27. At the 0.05 level of significance, is there evidence that variable X_2 makes a significant contribution to the model?

APPLYING THE CONCEPTS

14.39 The chair of the accounting department plans to develop a regression model to predict the grade point average in accounting for those students who are graduating and have completed the accounting major, based on a student's SAT score and whether the student received a grade of B or higher in the introductory statistics course (0 = no and 1 = yes).
a. Explain the steps involved in developing a regression model for these data. Be sure to indicate the particular models you need to evaluate and compare.
b. Suppose the regression coefficient for the variable whether the student received a grade of B or higher in the introductory statistics course is +0.30. How do you interpret this result?

14.40 A real estate association in a suburban community would like to study the relationship between the size of a single-family house (as measured by the number of rooms) and the selling price of the house (in $thousands). Two different neighborhoods are

included in the study, one on the east side of the community (=0) and the other on the west side (=1). A random sample of 20 houses was selected, with the results stored in Neighbor . For (a) through (k), do not include an interaction term.
a. State the multiple regression equation that predicts the selling price, based on the number of rooms and the neighborhood.
b. Interpret the regression coefficients in (a).
c. Predict the mean selling price for a house with nine rooms that is located in an east-side neighborhood. Construct a 95% confidence interval estimate and a 95% prediction interval.
d. Perform a residual analysis on the model and determine whether the regression assumptions are valid.
e. Is there a significant relationship between selling price and the two independent variables (rooms and neighborhood) at the 0.05 level of significance?
f. At the 0.05 level of significance, determine whether each independent variable makes a contribution to the regression model. Indicate the most appropriate regression model for this set of data.
g. Construct and interpret a 95% confidence interval estimate of the population slope for the relationship between selling price and number of rooms.
h. Construct and interpret a 95% confidence interval estimate of the population slope for the relationship between selling price and neighborhood.
i. Compute and interpret the adjusted r^2.
j. Compute the coefficients of partial determination and interpret their meaning.
k. What assumption do you need to make about the slope of selling price with number of rooms?

l. Add an interaction term to the model and, at the 0.05 level of significance, determine whether it makes a significant contribution to the model.

m. On the basis of the results of (f) and (l), which model is most appropriate? Explain.

n. What conclusions can the real estate association reach about the effect of the number of rooms and neighborhood on the selling price of homes?

14.41 In Problem 14.5 on page 483, you developed a multiple regression model to predict wine quality for red wines. Now, you wish to determine whether there is an effect on wine quality due to whether the wine is white (0) or red (1). These data are organized and stored in Red and White.

Develop a multiple regression model to predict wine quality based on the percentage of alcohol and the type of wine. For (a) through (m), do not include an interaction term.

a. State the multiple regression equation that predicts wine quality based on the percentage of alcohol and the type of wine.

b. Interpret the regression coefficients in (a).

c. Predict the mean quality for a red wine that has 10% alcohol. Construct a 95% confidence interval estimate and a 95% prediction interval.

d. Perform a residual analysis on the model and determine whether the regression assumptions are valid.

e. Is there a significant relationship between wine quality and the two independent variables (percentage of alcohol and the type of wine) at the 0.05 level of significance?

f. At the 0.05 level of significance, determine whether each independent variable makes a contribution to the regression model. Indicate the most appropriate regression model for this set of data.

g. Construct and interpret 95% confidence interval estimates of the population slope for the relationship between wine quality and the percentage of alcohol and between wine quality and the type of wine.

h. Compare the slope in (b) with the slope for the simple linear regression model of Problem 13.4 on page 439. Explain the difference in the results.

i. Compute and interpret the meaning of the coefficient of multiple determination, r^2.

j. Compute and interpret the adjusted r^2.

k. Compare r^2 with the r^2 value computed in Problem 13.16 (a) on page 445.

l. Compute the coefficients of partial determination and interpret their meaning.

m. What assumption about the slope of type of wine with wine quality do you need to make in this problem?

n. Add an interaction term to the model and, at the 0.05 level of significance, determine whether it makes a significant contribution to the model.

o. On the basis of the results of (f) and (n), which model is most appropriate? Explain.

p. What conclusions can you reach concerning the effect of alcohol percentage and type of wine on wine quality?

14.42 In mining engineering, holes are often drilled through rock, using drill bits. As a drill hole gets deeper, additional rods are added to the drill bit to enable additional drilling to take place. It is expected that drilling time increases with depth. This increased drilling time could be caused by several factors, including the mass of the drill rods that are strung together. The business problem relates to whether drilling is faster using dry drilling holes or wet drilling holes. Using dry drilling holes involves forcing compressed air down the drill rods to flush the cuttings and drive the hammer. Using wet drilling holes involves forcing water rather than air down the hole. Data have been collected from a sample of 50 drill holes that contains measurements of the time to drill each additional 5 feet (in minutes), the depth (in feet), and whether the hole was a dry drilling hole or a wet drilling hole. The data are organized and stored in Drill.

Source: Data extracted from R. Penner and D. G. Watts, "Mining Information," *The American Statistician*, 45, 1991, pp. 4–9.

Develop a model to predict additional drilling time, based on depth and type of drilling hole (dry or wet). For (a) through (k) do not include an interaction term.

a. State the multiple regression equation.

b. Interpret the regression coefficients in (a).

c. Predict the mean additional drilling time for a dry drilling hole at a depth of 100 feet. Construct a 95% confidence interval estimate and a 95% prediction interval.

d. Perform a residual analysis on the model and determine whether the regression assumptions are valid.

e. Is there a significant relationship between additional drilling time and the two independent variables (depth and type of drilling hole) at the 0.05 level of significance?

f. At the 0.05 level of significance, determine whether each independent variable makes a contribution to the regression model. Indicate the most appropriate regression model for this set of data.

g. Construct a 95% confidence interval estimate of the population slope for the relationship between additional drilling time and depth.

h. Construct a 95% confidence interval estimate of the population slope for the relationship between additional drilling time and the type of hole drilled.

i. Compute and interpret the adjusted r^2.

j. Compute the coefficients of partial determination and interpret their meaning.

k. What assumption do you need to make about the slope of additional drilling time with depth?

l. Add an interaction term to the model and, at the 0.05 level of significance, determine whether it makes a significant contribution to the model.

m. On the basis of the results of (f) and (l), which model is most appropriate? Explain.

n. What conclusions can you reach concerning the effect of depth and type of drilling hole on drilling time?

14.43 The owner of a moving company typically has his most experienced manager predict the total number of labor hours that will be required to complete an upcoming move. This approach has proved useful in the past, but the owner has the business objective of developing a more accurate method of predicting labor hours. In a preliminary effort to provide a more accurate method, the owner has decided to use the number of cubic feet moved and whether there is an elevator in the apartment building as the independent variables and has collected data for 36 moves in which the origin and destination were within the borough of Manhattan in New York City and the travel time was an insignificant portion of the hours worked. The data are organized and stored in Moving. For (a) through (k), do not include an interaction term.

a. State the multiple regression equation for predicting labor hours, using the number of cubic feet moved and whether there is an elevator.

b. Interpret the regression coefficients in (a).

c. Predict the mean labor hours for moving 500 cubic feet in an apartment building that has an elevator and construct a 95% confidence interval estimate and a 95% prediction interval.

d. Perform a residual analysis on the model and determine whether the regression assumptions are valid.

e. Is there a significant relationship between labor hours and the two independent variables (cubic feet moved and whether there is an elevator in the apartment building) at the 0.05 level of significance?

f. At the 0.05 level of significance, determine whether each independent variable makes a contribution to the regression model. Indicate the most appropriate regression model for this set of data.

g. Construct a 95% confidence interval estimate of the population slope for the relationship between labor hours and cubic feet moved.

h. Construct a 95% confidence interval estimate for the relationship between labor hours and the presence of an elevator.

i. Compute and interpret the adjusted r^2.

j. Compute the coefficients of partial determination and interpret their meaning.

k. What assumption do you need to make about the slope of labor hours with cubic feet moved?

l. Add an interaction term to the model, and at the 0.05 level of significance, determine whether it makes a significant contribution to the model.

m. On the basis of the results of (f) and (l), which model is most appropriate? Explain.

n. What conclusions can you reach concerning the effect of the number of cubic feet moved and whether there is an elevator on labor hours?

✓**SELF TEST** **14.44** In Problem 14.4 on page 482, you used efficiency ratio and total risk-based capital to predict ROAA at a community bank (stored in Community Banks). Develop a regression model to predict ROAA that includes efficiency ratio, total risk-based capital, and the interaction of efficiency ratio and total risk-based capital.

a. At the 0.05 level of significance, is there evidence that the interaction term makes a significant contribution to the model?

b. Which regression model is more appropriate, the one used in (a) or the one used in Problem 14.4? Explain.

14.45 Zagat's publishes restaurant ratings for various locations in the United States. The file Restaurants contains the Zagat rating for food, décor, service, and cost per person for a sample of 50 center city restaurants and 50 metro area restaurants.

Source: Data extracted from *Zagat Survey 2016, New York City Restaurants.*

Develop a regression model to predict the cost per person, based on a variable that represents the sum of the ratings for food, décor, and service and a dummy variable concerning location (center city versus metro area). For (a) through (m), do not include an interaction term.

a. State the multiple regression equation.

b. Interpret the regression coefficients in (a).

c. Predict the mean cost for a center city restaurant with a summated rating of 60 and construct a 95% confidence interval estimate and a 95% prediction interval.

d. Perform a residual analysis on the model and determine whether the regression assumptions are satisfied.

e. Is there a significant relationship between price and the two independent variables (summated rating and location) at the 0.05 level of significance?

f. At the 0.05 level of significance, determine whether each independent variable makes a contribution to the regression model. Indicate the most appropriate regression model for this set of data.

g. Construct a 95% confidence interval estimate of the population slope for the relationship between cost and summated rating.

h. Compare the slope in (b) with the slope for the simple linear regression model of Problem 13.5 on page 439. Explain the difference in the results.

i. Compute and interpret the meaning of the coefficient of multiple determination.

j. Compute and interpret the adjusted r^2.

k. Compare r^2 with the r^2 value computed in Problem 13.17 (b) on page 445.

l. Compute the coefficients of partial determination and interpret their meaning.

m. What assumption about the slope of cost with summated rating do you need to make in this problem?

n. Add an interaction term to the model and, at the 0.05 level of significance, determine whether it makes a significant contribution to the model.

o. On the basis of the results of (f) and (n), which model is most appropriate? Explain.

p. What conclusions can you reach about the effect of the summated rating and the location of the restaurant on the cost of a meal?

14.46 In Problem 14.6 on page 483, you used full-time voluntary turnover (%), and total worldwide revenue ($billions) to predict number of full-time job openings (stored in Best Companies). Develop a regression model to predict the number of full-time job openings that includes full-time voluntary turnover, total worldwide revenue, and the interaction of full-time voluntary turnover and total worldwide revenue.

a. At the 0.05 level of significance, is there evidence that the interaction term makes a significant contribution to the model?

b. Which regression model is more appropriate, the one used in this problem or the one used in Problem 14.6? Explain.

14.47 In Problem 14.5 on page 483, the percentage of alcohol and chlorides were used to predict the quality of red wines (stored in Vinho Verde). Develop a regression model that includes the percentage of alcohol, the chlorides, and the interaction of the percentage of alcohol and the chlorides to predict wine quality.

a. At the 0.05 level of significance, is there evidence that the interaction term makes a significant contribution to the model?

b. Which regression model is more appropriate, the one used in this problem or the one used in Problem 14.5? Explain.

14.48 In Problem 14.7 on page 483, you used weekly staff count and remote hours to predict standby hours (stored in Nickels 26Weeks). Develop a regression model to predict standby hours that includes total staff present, remote hours, and the interaction of total staff present and remote hours.

a. At the 0.05 level of significance, is there evidence that the interaction term makes a significant contribution to the model?

b. Which regression model is more appropriate, the one used in this problem or the one used in Problem 14.7? Explain.

14.49 The director of a training program for a large insurance company has the business objective of determining which training method is best for training underwriters. The three methods to be evaluated are classroom, online, and courseware app. The 30 trainees are divided into three randomly assigned groups of 10. Before the start of the training, each trainee is given a proficiency exam that measures mathematics and computer skills. At the end of the training, all students take the same end-of-training exam. The results are organized and stored in Underwriting .

Develop a multiple regression model to predict the score on the end-of-training exam, based on the score on the proficiency exam and the method of training used. For (a) through (k), do not include an interaction term.

a. State the multiple regression equation.
b. Interpret the regression coefficients in (a).
c. Predict the mean end-of-training exam score for a student with a proficiency exam score of 100 who had courseware app-based training.
d. Perform a residual analysis on the model and determine whether the regression assumptions are valid.
e. Is there a significant relationship between the end-of-training exam score and the independent variables (proficiency score and training method) at the 0.05 level of significance?
f. At the 0.05 level of significance, determine whether each independent variable makes a contribution to the regression model. Indicate the most appropriate regression model for this set of data.

g. Construct and interpret a 95% confidence interval estimate of the population slope for the relationship between the end-of-training exam score and the proficiency exam score.
h. Construct and interpret 95% confidence interval estimates of the population slope for the relationship between the end-of-training exam score and type of training method.
i. Compute and interpret the adjusted r^2.
j. Compute the coefficients of partial determination and interpret their meaning.
k. What assumption about the slope of proficiency score with end-of-training exam score do you need to make in this problem?
l. Add interaction terms to the model and, at the 0.05 level of significance, determine whether any interaction terms make a significant contribution to the model.
m. On the basis of the results of (f) and (l), which model is most appropriate? Explain.

CONSIDER THIS

What Is Not Normal? (Using a *categorical* dependent variable)

In Sections 14.1 through 14.6, every example of multiple regression uses a numerical dependent variable. For example, Section 14.1 considers estimating the value of the dependent numerical variable sales (of OmniPower bars), while Section 14.6 explores estimating the value of the dependent numerical variable asking price, based on several independent variables. Even the Sunflowers Apparel examples that Chapter 13 uses to discuss simple linear regression have a dependent numerical variable. Is using a dependent numerical variable the normal practice for regression? **Yes,** it is normal practice if one is using the least-squares method that Section 13.2 first explains.

Section 14.6 discusses a dummy variable to recode two categories of a categorical *independent* variable. Could a binary (two-category) categorical variable be similarly recoded as 0 and 1 to represent the two categories? Yes. Could that recoded variable be used as the dependent variable in a least-squares regression? **No!** Hmm, something is not *normal* here!

What is not normal is that using such a variable will cause the assumption of the normality of errors to be violated. Recall that the least-squares method comes includes assumptions about the data that Section 13.4 explains and that Section 14.3 extends to multiple regression. In order to have a usable regression model, these assumptions must not be violated. Therefore, the least-squares method cannot be used with a recoded categorical variable. Mistakenly using that method can also cause predicted values that are beyond the range of the values 0 and 1, and that, too, is not normal!

Using a (recoded) categorical variable requires a different regression approach that uses the concept of *likelihood*. For this context, **likelihood** asks the question, "What is the probability of obtaining the values of the regression data, given a specific set of regression coefficients?"

One can use the Figure 5.2 binomial worksheet on page 184 to understand analogously how likelihood works. In that worksheet, one enters the sample size in cell B4 and the probability of an event of interest in cell B5 (4 and 0.1, respectively, in Figure 5.2). Using those values, the worksheet computes the $P(X)$ for the five possible X events, 0 through 4, in the Binomial Probabilities Table worksheet area.

What if someone uses this worksheet and enters a new value for the probability of an event of interest in cell B5? The worksheet would update results, including the probabilities in the Binomial Probabilities Table. If that person somehow hid the value entered in cell B5, could others estimate that cell B5 value without unhiding the value? Yes! Others could make a copy of the worksheet and enter values in cell B5 until the probabilities in the second Binomial Probabilities Table matched the probabilities in the worksheet with the hidden cell B5. By doing that, others would have found the *likeliest* value for the probability of an event of interest and would have inverted the workings of the Figure 5.2 binomial worksheet. Likelihood inverts the calculation of probabilities that earlier chapters discuss and, not surprisingly, likelihood was once called *inverse probability*.

Logistic regression uses **maximum likelihood estimation (MLE)** to determine regression coefficients. MLE finds the regression coefficients that are associated with the highest probability of obtaining the regression data. (This is analogous to finding that likeliest value in the preceding example.) Note that MLE determines probabilities, not values. That means the dependent variable in logistic regression will always be a probability, a value between 0 and 1 inclusive, and that proves useful mathematically for reasons that interested readers can explore on their own.

14.7 Logistic Regression

Logistic regression, enables one to use regression models to predict the probability of a category of a dependent categorical variable for a given set of independent variables. The logistic regression model uses the **odds ratio**, which represents the probability of an event of interest compared with the probability of not having an event of interest. Equation (14.15) defines the odds ratio.

ODDS RATIO

$$\text{Odds ratio} = \frac{\text{probability of an event of interest}}{1 - \text{probability of an event of interest}} \tag{14.15}$$

Using Equation (14.15), if the probability of an event of interest is 0.50, the odds ratio is

$$\text{Odds ratio} = \frac{0.50}{1 - 0.50} = 1.0, \text{ or 1 to 1}$$

If the probability of an event of interest is 0.75, the odds ratio is

$$\text{Odds ratio} = \frac{0.75}{1 - 0.75} = 3.0, \text{ or 3 to 1}$$

The logistic regression model is based on the natural logarithm of the odds ratio, ln(odds ratio). Equation (14.16) defines the logistic regression model for k independent variables.

studentTIP

In is the symbol used for natural logarithms, also known as base *e* logarithms. In(*x*) is the logarithm of *x* having base *e*, where *e* ≅ 2.718282.

LOGISTIC REGRESSION MODEL

$$\ln(\text{Odds ratio}) = \beta_0 + \beta_1 X_{1i} + \beta_2 X_{2i} + \dots + \beta_k X_{ki} + \varepsilon_i \tag{14.16}$$

where

k = number of independent variables in the model

ε_i = random error in observation i

Sections 13.2 and 14.1 use the method of least squares to develop a regression equation. In logistic regression, a mathematical method called *maximum likelihood estimation* is typically used to develop a regression equation to predict the natural logarithm of this odds ratio. Equation (14.17) defines the logistic regression equation.

LOGISTIC REGRESSION EQUATION

$$\ln(\text{Estimated odds ratio}) = b_0 + b_1 X_{1i} + b_2 X_{2i} + \dots + b_k X_{ki} \tag{14.17}$$

Using Equation (14.18), one can compute the estimated odds ratio.

ESTIMATED ODDS RATIO

$$\text{Estimated odds ratio} = e^{\ln(\text{Estimated odds ratio})} \tag{14.18}$$

With an estimated odds ratio computed, one can compute the estimated probability of an event of interest using Equation (14.19).

ESTIMATED PROBABILITY OF AN EVENT OF INTEREST

$$\text{Estimated probability of an event of interest} = \frac{\text{estimated odds ratio}}{1 + \text{estimated odds ratio}} \tag{14.19}$$

To illustrate the use of logistic regression, consider the following case of the sales and marketing manager for the credit card division of a major financial company. The manager seeks to conduct a campaign to persuade existing holders of the bank's standard credit card to upgrade, for a nominal annual fee, to the bank's platinum card. The manager wonders, "Which of the existing standard credit cardholders should we target for this campaign?"

The manager has access to the results from a sample of 30 cardholders who were targeted during a pilot campaign last year. These results have been organized as three variables and stored in Cardholder Study . The three variables are upgraded, whether a cardholder upgraded to a premium card, Y (0 = no, 1 = yes); and two independent variables, purchases, the prior year's credit card purchases (in $thousands), X_1; and extra cards, whether the cardholder ordered additional credit cards for other authorized users, X_2 (0 = no, 1 = yes). Figure 14.13 presents the Excel results for the logistic regression model for this pilot study.

FIGURE 14.13

Excel logistic regression worksheet results for the cardholder pilot study data

	A	B	C	D	E
1	Cardholder Study Logistic Regression				
2					
3	Predictor	Coefficients	SE Coef	Z	p-Value
4	Intercept	-6.9394	2.9471	-2.3547	0.0185
5	Purchases	0.139458333	0.0681	2.049	0.0405
6	Extra Cards:1	2.7743271	1.1927	2.3261	0.0200
7					
8	Deviance	20.07690146			

In the Worksheet

This worksheet uses the NORM.S.DIST function in column E to compute p-values. Columns G through N (not shown) contain intermediate calculations and columns L and M use the IF function in ways much more complex than other worksheets that this book features.

The **COMPUTE worksheet** of the Excel Guide **Logistic Regression Example workbook** is a copy of the Figure 14.13 worksheet.

Using the net regression coefficients b_1 and b_2, the Table 14.14 net effects table summarizes the effects of the regression constant, purchases (X_1) and extra cards (X_2) for management.

TABLE 14.14

Net effects table for the cardholder pilot study multiple regression model

Net Effect	Interpretation
The regression constant -6.9394	The estimated natural logarithm of the odds ratio of purchasing the premium card is -6.9394 for a credit cardholder who did not charge any purchases last year and who does not have additional cards.
Each additional $1,000 in credit card purchases last year	The estimated natural logarithm of the odds ratio of purchasing the premium card increases by 0.1395 for each increase of $1,000 in annual credit card spending using the company's card, holding constant the effect of whether the credit cardholder has additional cards for other authorized users.
Whether additional credit cards are ordered for a member of the household	The estimated natural logarithm of the odds ratio of purchasing the premium card increases by 2.7743 for a credit cardholder who has additional cards for other authorized users compared with one who does not have additional cards, holding constant the annual credit card spending.

The Table 14.14 estimates will enable decision makers to better understand how spending and additional credit card ordering decisions are predicted to affect whether the cardholder will upgrade to a premium card. Managers can conclude that cardholders who charged more last year and possess additional cards for other authorized users are much more likely to upgrade to a premium credit card.

As is the case with least-squares regression models, a main purpose of performing logistic regression analysis is to provide predictions of a dependent variable. For example, consider a cardholder who charged $36,000 last year and possesses additional cards for members of the household. What is the probability the cardholder will upgrade to the premium card during the marketing campaign? Using $X_1 = 36$, $X_2 = 1$, Equation (14.17) on page 509, and the Figure 14.13 results on page 510,

$$\ln(\text{estimated odds of purchasing versus not purchasing}) = -6.9394 + (0.1395)(36) + (2.7743)(1)$$

$$= 0.8569$$

Then, using Equation (14.18) on page 509,

$$\text{estimated odds ratio} = e^{0.8569} = 2.3558$$

Therefore, the odds are 2.3558 to 1 that a credit cardholder who spent $36,000 last year and has additional cards will purchase the premium card during the campaign. Using Equation (14.19) on page 510, one can convert this odds ratio to a probability:

$$\text{estimated probability of purchasing premium card} = \frac{2.3558}{1 + 2.3558}$$

$$= 0.702$$

Thus, the estimated probability is 0.702 that a credit cardholder who spent $36,000 last year and has additional cards will purchase the premium card during the campaign. In other words, one predicts 70.2% of such individuals will purchase the premium card.

In order to make such predictions for decision making, one must determine whether or not the model is a good-fitting model. The **deviance statistic** is frequently used to determine whether the current model provides a good fit to the data. This statistic measures the fit of the current model compared with a model that has as many parameters as there are data points (what is called a *saturated* model). The deviance statistic follows a chi-square distribution with $n - k - 1$ degrees of freedom, where n is the sample size and k is the number of independent variables. The null and alternative hypotheses are

$$H_0: \text{The model is a good-fitting model.}$$

$$H_1: \text{The model is not a good-fitting model.}$$

studentTIP

Unlike other hypothesis tests, rejecting the null hypothesis for this test means that the model is *not* a good fit.

When using the deviance statistic for logistic regression, the null hypothesis represents a good-fitting model, which is the opposite of the null hypothesis when using the overall F test for the multiple regression model (see Section 14.2). Using the α level of significance, the decision rule is

$$\text{Reject } H_0 \text{ if deviance} > \chi_\alpha^2;$$

$$\text{otherwise, do not reject } H_o.$$

The critical value for a χ^2 statistic with $n - k - 1 = 30 - 2 - 1 = 27$ degrees of freedom is 40.113 (see Table E.4). From Figure 14.13 on page 510, the deviance $20.0769 < 40.113$. Thus, one does not reject H_0 and concludes that there is insufficient evidence that the model is not a good-fitting one.

With evidence that the model is a good-fitting one, you need to evaluate whether each of the independent variables makes a significant contribution to the model in the presence of others. Do that evaluation by examining either the Z test statistic (called the **Wald statistic** in this context).

Table 14.15 summarizes the results of the test for the regression coefficients for purchases (b_1) and extra cards (b_2) that appears as part of Figure 14.13. Based on these results, one can conclude that both the amount of purchases and whether the cardholder has additional cards for members of the household have a significant effect on whether the cardholder will upgrade to a premium card.

TABLE 14.15
Evaluating whether each of the independent variables makes a significant contribution

Result	Conclusions
$Z = 2.049$ is greater than 1.96 p-value $= 0.0405$ is less than the level of significance, $\alpha = 0.05$	1. Reject the null hypothesis H_0. 2. Conclude that evidence exists for claiming that the amount of purchases is related to whether the cardholder will upgrade to a premium card holding constant whether the cardholder has additional cards for members of the household. 3. The probability is 0.0405 that $Z < -2.049$ or $Z > 2.049$.
$Z = 2.3261$ is greater than 1.96 p-value $= 0.02$ is less than the level of significance, $\alpha = 0.05$	1. Reject the null hypothesis H_0. 2. Conclude that evidence exists for claiming that whether the cardholder will upgrade to a premium card is related to whether the cardholder has additional cards for members of the household holding constant the amount of purchases. 3. The probability is 0.02 that $Z < -2.3261$ or $Z > 2.3261$.

PROBLEMS FOR SECTION 14.7

LEARNING THE BASICS

14.50 Interpret the meaning of a slope coefficient equal to 2.2 in logistic regression.

14.51 Given an estimated odds ratio of 2.5, compute the estimated probability of an event of interest.

14.52 Given an estimated odds ratio of 0.75, compute the estimated probability of an event of interest.

14.53 Consider the following logistic regression equation:

$$\ln(\text{Estimated odds ratio}) = 0.1 + 0.5X_{1i} + 0.2X_{2i}$$

a. Interpret the meaning of the logistic regression coefficients.
b. If $X_1 = 2$ and $X_2 = 1.5$, compute the estimated odds ratio and interpret its meaning.
c. On the basis of the results of (b), compute the estimated probability of an event of interest.

APPLYING THE CONCEPTS

SELF TEST ✓ **14.54** Refer to Figure 14.13 on page 510.
a. Predict the probability that a cardholder who charged $36,000 last year and does not have any additional credit cards for other authorized users will purchase the platinum card during the marketing campaign.
b. Compare the results in (a) with those for a person with additional credit cards.

c. Predict the probability that a cardholder who charged $18,000 and does not have any additional credit cards for other authorized users will purchase the platinum card during the marketing campaign.
d. Compare the results of (a) and (c) and indicate what implications these results might have for the strategy for the marketing campaign.

14.55 A study was conducted to determine the factors involved in the rate of participation of discharged cardiac patients in a rehabilitation program. Data were collected from 516 treated patients.

Source: Data extracted from F. Van Der Meulen, T. Vermaat, and P. Williams, "Case Study: An Application of Logistic Regression in a Six Sigma Project in Health Care," *Quality Engineering*, 2011, pp. 113–124.

Among the variables used to predict participation ($0 = $ no, $1 = $ yes) were the distance traveled to rehabilitation in kilometers, whether the person had a car ($0 = $ no, $1 = $ yes), and the age of the person in years. The summarized data are:

	Estimate	Standard Error	Z Value	p-value
Intercept	5.7765	0.8619	6.702	0.0000
Distance	−0.0675	0.0111	−6.113	0.0000
Car	1.9369	0.2720	7.121	0.0000
Age	−0.0599	0.0119	−5.037	0.0000

a. State the logistic regression model.

b. Using the model in (a), predict the probability that a patient will participate in rehabilitation if he or she travels 20 km to rehabilitation, has a car, and is 65 years old.

c. Using the model in (a), predict the probability that a patient will participate in rehabilitation if he or she travels 20 km to rehabilitation, does not have a car, and is 65 years old.

d. Compare the results of (b) and (c).

e. At the 0.05 level of significance, is there evidence that the distance traveled, whether the patient has a car, and the age of the patient each make a significant contribution to the model?

f. What conclusions can you reach about the likelihood of a patient participating in the rehabilitation program?

14.56 Referring to Problem 14.41 on page 506, you have decided to analyze whether there are differences in fixed acidity, chlorides, and pH between white wines and red wines (0 = white 1 = red). Using the data stored in `Red and White`,

a. Develop a logistic regression model to predict whether the wine is red based on the fixed acidity, chlorides, and pH.

b. Explain the meaning of the regression coefficients in the model developed in (a).

c. Predict the probability that a wine is red if it has a fixed acidity of 7.0, chlorides of 0.04, and pH of 3.5.

d. At the 0.05 level of significance, is there evidence that the logistic regression model developed in (a) is a good-fitting model?

e. At the 0.05 level of significance, is there evidence that fixed acidity, chlorides, and pH each make a significant contribution to the model?

f. What conclusions concerning the probability of a wine selected being red can you reach?

14.57 Undergraduate students at Miami University in Oxford, Ohio, were surveyed in order to evaluate the effect of price on the purchase of a pizza from Pizza Hut. Students were first asked to imagine a situation in which they were planning to call and order for delivery a large two-topping pizza. Then they were asked to select from either Pizza Hut or another pizzeria of their choice. The price they would have to pay to get a Pizza Hut pizza differed from survey to survey. For example, some surveys used the price $11.49. Other prices investigated were $8.49, $9.49, $10.49, $12.49, $13.49, and $14.49. The dependent variable for this study is whether or not a student will select Pizza Hut. Possible independent variables are the price of a Pizza Hut pizza and the gender of the student. The file `Pizza Hut` contains responses from 220 students and includes these three variables:

Gender: 1 = male, 0 = female

Price: 8.49, 9.49, 10.49, 11.49, 12.49, 13.49, or 14.49

Purchase: 1 = the student selected Pizza Hut, 0 = the student selected another pizzeria

a. Develop a logistic regression model to predict the probability that a student selects Pizza Hut based on the price of the pizza. Is price an important indicator of purchase selection?

b. Develop a logistic regression model to predict the probability that a student selects Pizza Hut based on the price of the pizza and the gender of the student. Is price an important indicator of purchase selection? Is gender an important indicator of purchase selection?

c. Compare the results from (a) and (b). Which model would you choose? Discuss.

d. Using the model selected in (c), predict the probability that a student will select Pizza Hut if the price is $8.99.

e. Using the model selected in (c), predict the probability that a student will select Pizza Hut if the price is $11.49.

f. Using the model selected in (c), predict the probability that a student will select Pizza Hut if the price is $13.99.

14.58 An automotive insurance company wants to predict which filed stolen vehicle claims are fraudulent, based on the mean number of claims submitted per year by the policy holder and whether the policy is a new policy, that is, is one year old or less (coded as 1 = yes, 0 = no). Data from a random sample of 98 automotive insurance claims, organized and stored in `Insurance Fraud`, show that 49 are fraudulent (coded as 1) and 49 are not (coded as 0).

Source: Data extracted from A. Gepp *et al.*, "A Comparative Analysis of Decision Trees vis-à-vis Other Computational Data Mining Techniques in Automotive Insurance Fraud Detection," *Journal of Data Science*, 10 (2012), pp. 537–561.

a. Develop a logistic regression model to predict the probability of a fraudulent claim, based on the number of claims submitted per year by the policy holder and whether the policy is new.

b. Explain the meaning of the regression coefficients in the model in (a).

c. Predict the probability of a fraudulent claim given that the policy holder has submitted a mean of one claim per year and holds a new policy.

d. At the 0.05 level of significance, is there evidence that a logistic regression model that uses the mean number of claims submitted per year by the policy holder and whether the policy is new to predict the probability of a fraudulent claim is a good-fitting model?

e. At the 0.05 level of significance, is there evidence that the mean number of claims submitted per year by the policy holder and whether the policy is new each makes a significant contribution to the logistic model?

f. Develop a logistic regression model that includes only the number of claims submitted per year by the policy holder to predict the probability of a fraudulent claim.

g. Develop a logistic regression model that includes only whether the policy is new to predict a fraudulent claim.

h. Compare the models in (a), (f), and (g). Evaluate the differences among the models.

14.59 A marketing manager wants to predict customers with the risk of churning (switching their service contracts to another company) based on the number of calls the customer makes to the company call center and the number of visits the customer makes to the local service center. Data from a random sample of 30 customers, organized and stored in `Churn` show that 15 have churned (coded as 1) and 15 have not (coded as 0).

a. Develop a logistic regression model to predict the probability of churn, based on the number of calls the customer makes to the company call center and the number of visits the customer makes to the local service center.

b. Explain the meaning of the regression coefficients in the model in (a).

c. Predict the probability of churn for a customer who called the company call center 10 times and visited the local service center once.

d. At the 0.05 level of significance, is there evidence that a logistic regression model that uses the number of calls the customer makes to the company call center and the number of visits the customer makes to the local service center is a good-fitting model?

e. At the 0.05 level of significance, is there evidence that the number of calls the customer makes to the company call center and the number of visits the customer makes to the local service center each make a significant contribution to the logistic model?

f. Develop a logistic regression model that includes only the number of calls the customer makes to the company call center to predict the probability of churn.

g. Develop a logistic regression model that includes only the number of visits the customer makes to the local service center to predict churn.

h. Compare the models in (a), (f), and (g). Evaluate the differences among the models.

14.60 A local supermarket manager wants to use two independent variables, customer age (in years) and whether the customer subscribes to the supermarket chain's health/wellness e-newsletters (coded as $1 = $ yes and $0 = $ no) to predict which customers are likely to purchase a new line of organic products. Data from a random sample of 100 loyalty program customers, organized and stored in Organic Food , show that 65 have purchased the organic products (coded as 1) and 35 have not (coded as 0).

a. Develop a logistic regression model to predict the probability that a customer purchases the organic products, based on age and whether the customer subscribes to the supermarket chain's health/wellness e-newsletters.

b. Explain the meaning of the regression coefficients in the model in (a).

c. Predict the probability of purchasing the organic products for a 35-year-old customer who subscribes to the supermarket chain's health/wellness e-newsletters.

d. At the 0.05 level of significance, is there evidence that a logistic regression model that uses customer age and whether the customer subscribes to the supermarket chain's health/wellness e-newsletters to predict the probability of purchasing the organic products is a good-fitting model?

e. At the 0.05 level of significance, is there evidence that customer age and whether the customer subscribes to the supermarket chain's health/wellness e-newsletters each make a significant contribution to the logistic model?

f. What conclusions can you reach about which variables are affecting purchase of organic foods?

14.8 Cross-Validation

Overfitting describes any model that too-well describes a data set. Overfitting is not a problem if a model is being used for explanatory purposes. However, a model that too perfectly reflects all the variation, including quirks and random variation in the data, will be overfit and will most likely not be the best model when predicting **future data**, data that either does not exist as of today (but will exist "in the future") or data that *was not used* to construct the model.

A number of techniques guard against overfitting and help create a better prediction model. One commonly encountered technique, and one adaptable to any amount of data, is *cross-validation*. **Cross-validation** processes seek to create better-performing models for prediction. All these techniques split data in some way that is partially dependent on the needs of a decision maker as well as the sample size of the data being analyzed.

The simplest technique splits the data into two parts, the *training* and *test* sets. The **training set** is the data part used to create a model. The training set is always the larger part and typically 70% to 80% of the data being analyzed. The **test set** is the data part used to evaluate the predictive power of the model that the training sets helps create. Sometimes, the non-training set data is divided into a *validation* set and a test set. When so divided, the validation set is used to see if the error in the model increases when this second set is used, which suggests that the model is overfit.

Noting the qualities of the holdout method, some sources classify this method as something other than a proper type of cross- validation.

This simplest technique is sometimes called the holdout method because the test set is held out of the initial analysis. That name can be misleading, though, because all cross-validation techniques "hold out" data in some way. This simplest technique is computationally the simplest, and therefore the one often used with very large sets of data out of necessity. The technique can also be used for convenience for smaller data sets, subject to having two parts of sufficient sample size for statistical purposes. However, because only one test set is used, this technique will have higher variance in its results compared to techniques that use more than one test set, such as *k*-fold cross-validation.

The k-fold cross-validation technique splits the data into k equal parts, or *folds*, and runs k regression analyses in which a different part (fold) held out every time. Figure 14.14 illustrates k-fold cross-validation for an analysis in which $k = 4$.

FIGURE 14.14
k-fold cross-validation for $k = 4$

	Regression 1	**Regression 2**	**Regression 3**	**Regression 4**
Fold 1	left out	included	included	included
Fold 2	included	left out	included	included
Fold 3	included	included	left out	included
Fold 4	included	included	included	left out

No single "best" value of k can be determined, but experience suggests a value of 10 is a very good choice for larger data sets, assuming computational time is not a factor, although many use the value of 5 when considering the trade-off in computational time. Using the k-fold techniques enables a business analytics user to calculate the mean accuracy and standard deviation providing a better understanding of the variability in the data.

For small data sets, a special case of k-fold cross-validation called leave one out cross-validation (LOOCV) is used. LOOCV divides the data in the number of parts that equals the sample size, n. Each part consists of a single row of data left out, and each part is unique. LOOCV would be the cross-validation technique best used in the cardholder pilot study example that Section 14.7 discusses because the study only considers the data of 30 cardholders.

▼USING **STATISTICS**
The Multiple Effects..., Revisited

In the Using Statistics scenario, you were a marketing manager for OmniFoods, responsible for nutrition bars and similar snack items.

At the end of the one-month test-market study, you performed a multiple regression analysis on the data. Two independent variables were considered: the price of an OmniPower bar and the monthly budget for in-store promotional expenses. The dependent variable was the number of OmniPower bars sold in a month. The coefficient of determination indicated that 75.8% of the variation in sales was explained by knowing the price charged and the amount spent on in-store promotions. The model indicated that the predicted sales of OmniPower are estimated to decrease by 532 bars per month for each 10-cent increase in the price, and the predicted sales are estimated to increase by 361 bars for each additional $100 spent on promotions.

After studying the relative effects of price and promotion, OmniFoods needs to set price and promotion standards for a nationwide introduction (obviously, lower prices and higher promotion budgets lead to more sales, but they do so at a lower profit margin). You determined that if stores spend $400 a month for in-store promotions and charge 79 cents, the 95% confidence interval estimate of the mean monthly sales is 2,854 to 3,303 bars. OmniFoods can multiply the lower and upper bounds of this confidence interval by the number of stores included in the nationwide introduction to estimate total monthly sales. For example, if 1,000 stores are in the nationwide introduction, then total monthly sales should be between 2.854 million and 3.308 million bars.

▼SUMMARY

This chapter explains how to develop and fit multiple regression models that use two or more independent variables to predict the value of a dependent variable. The chapter also discusses how to include categorical independent variables and interaction terms in regression models and the logistic regression model that is used to predict a categorical dependent variable. Figure 14.15 summarizes how to apply a multiple regression model to a set of data.

FIGURE 14.15
Roadmap for multiple
regression

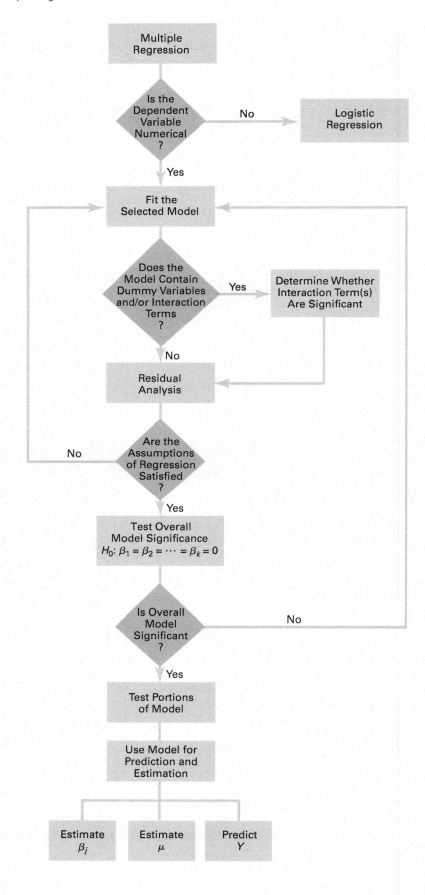

▼ REFERENCES

Andrews, D. F., and D. Pregibon. "Finding the Outliers that Matter." *Journal of the Royal Statistical Society* 40 (Ser. B., 1978): 85–93.

Atkinson, A. C. "Robust and Diagnostic Regression Analysis." *Communications in Statistics* 11 (1982): 2559–2572.

Belsley, D. A., E. Kuh, and R. Welsch. *Regression Diagnostics: Identifying Influential Data and Sources of Collinearity.* New York: Wiley, 1980.

Cook, R. D., and S. Weisberg. *Residuals and Influence in Regression.* New York: Chapman and Hall, 1982.

Gallistel, C. R. "Bayes for Beginners: Probability and Likelihood." *Observer* 28:7, September 2015. Available at **bit.ly/2xUrq2V.**

Hoaglin, D. C., and R. Welsch. "The Hat Matrix in Regression and ANOVA," *The American Statistician*, 32, (1978), 17–22.

Kutner, M., C. Nachtsheim, J. Neter, and W. Li. *Applied Linear Statistical Models*, 5th ed. New York: McGraw-Hill/Irwin, 2005.

Montgomery, D. C., E. A. Peck, and G. G. Vining. *Introduction to Linear Regression Analysis*, 5th ed. New York: Wiley, 2012.

▼ KEY EQUATIONS

Multiple Regression Model with k Independent Variables

$$Y_i = \beta_0 + \beta_1 X_{1i} + \beta_2 X_{2i} + \beta_3 X_{3i} + \ldots + \beta_k X_{ki} + \varepsilon_i \tag{14.1}$$

Multiple Regression Model with Two Independent Variables

$$Y_i = \beta_0 + \beta_1 X_{1i} + \beta_2 X_{2i} + \varepsilon_i \tag{14.2}$$

Multiple Regression Equation with Two Independent Variables

$$\hat{Y}_i = b_0 + b_1 X_{1i} + b_2 X_{2i} \tag{14.3}$$

Coefficient of Multiple Determination

$$r^2 = \frac{\text{Regression sum of squares}}{\text{Total sum of squares}} = \frac{SSR}{SST} \tag{14.4}$$

Adjusted r^2

$$r^2_{\text{adj}} = 1 - \left[(1 - r^2) \frac{n-1}{n-k-1} \right] \tag{14.5}$$

Overall F Test

$$F_{STAT} = \frac{MSR}{MSE} \tag{14.6}$$

Testing for the Slope in Multiple Regression

$$t_{STAT} = \frac{b_j - \beta_j}{S_{b_j}} \tag{14.7}$$

Confidence Interval Estimate for the Slope

$$b_j \pm t_{\alpha/2} S_{b_j} \tag{14.8}$$

Determining the Contribution of an Independent Variable to the Regression Model

$$SSR(X_j | \text{All } Xs \text{ except } j) = SSR(\text{All } Xs) - SSR(\text{All } Xs \text{ except } j) \tag{14.9}$$

Contribution of Variable X_1, Given That X_2 Has Been Included

$$SSR(X_1 | X_2) = SSR(X_1 \text{ and } X_2) - SSR(X_2) \tag{14.10a}$$

Contribution of Variable X_2, Given That X_1 Has Been Included

$$SSR(X_2 | X_1) = SSR(X_1 \text{ and } X_2) - SSR(X_1) \tag{14.10b}$$

Partial F Test Statistic

$$F_{STAT} = \frac{SSR(X_j | \text{All } Xs \text{ except } j)}{MSE} \tag{14.11}$$

Relationship Between a t Statistic and an F Statistic

$$t^2_{STAT} = F_{STAT} \tag{14.12}$$

Coefficients of Partial Determination for a Multiple Regression Model Containing Two Independent Variables

$$r^2_{Y1.2} = \frac{SSR(X_1 | X_2)}{SST - SSR(X_1 \text{ and } X_2) + SSR(X_1 | X_2)} \tag{14.13a}$$

and

$$r^2_{Y2.1} = \frac{SSR(X_2 | X_1)}{SST - SSR(X_1 \text{ and } X_2) + SSR(X_2 | X_1)} \tag{14.13b}$$

Coefficient of Partial Determination for a Multiple Regression Model Containing k Independent Variables

$$r^2_{Yj.(\text{All variables except } j)} = \frac{SSR(X_j | \text{All } Xs \text{ except } j)}{SST - SSR(\text{All } Xs) + SSR(X_j | \text{All } Xs \text{ except } j)} \tag{14.14}$$

Odds Ratio

$$\text{Odds ratio} = \frac{\text{probability of an event of interest}}{1 - \text{probability of an event of interest}} \tag{14.15}$$

Logistic Regression Model

$$\ln(\text{Odds ratio}) = \beta_0 + \beta_1 X_{1i} + \beta_2 X_{2i} + \cdots + \beta_k X_{ki} + \varepsilon_i$$
(14.16)

Logistic Regression Equation

$$\ln(\text{Estimated odds ratio}) = b_0 + b_1 X_{1i} + b_2 X_{2i} + \cdots + b_k X_{ki}$$
(14.17)

Estimated Odds Ratio

$$\text{Estimated odds ratio} = e^{\,\ln(\text{Estimated odds ratio})}$$ **(14.18)**

Estimated Probability of an Event of Interest

Estimated probability of an event of interest

$$= \frac{\text{estimated odds ratio}}{1 + \text{estimated odds ratio}}$$ **(14.19)**

▼ KEY TERMS

▼ CHECKING YOUR UNDERSTANDING

14.61 What is the difference between r^2 and adjusted r^2?

14.62 How does the interpretation of the regression coefficients differ in multiple regression and simple linear regression?

14.63 How does testing the significance of the entire multiple regression model differ from testing the contribution of each independent variable?

14.64 How do the coefficients of partial determination differ from the coefficient of multiple determination?

14.65 Why and how do you use dummy variables?

14.66 How can you evaluate whether the slope of the dependent variable with an independent variable is the same for each level of the dummy variable?

14.67 Under what circumstances do you include an interaction term in a regression model?

14.68 When a dummy variable is included in a regression model that has one numerical independent variable, what assumption do you need to make concerning the slope between the dependent variable, Y, and the numerical independent variable, X?

14.69 When do you use logistic regression?

14.70 What is the difference between least squares regression and logistic regression?

▼ CHAPTER REVIEW PROBLEMS

14.71 Increasing customer satisfaction typically results in increased purchase behavior. For many products, there is more than one measure of customer satisfaction. In many, purchase behavior can increase dramatically with an increase in just one of the customer satisfaction measures. Gunst and Barry ("One Way to Moderate Ceiling Effects," *Quality Progress*, October 2003, pp. 83–85) consider a product with two satisfaction measures, X_1 and X_2, that range from the lowest level of satisfaction, 1, to the highest level of satisfaction, 7. The dependent variable, Y, is a measure of purchase behavior, with the highest value generating the most sales. Consider the regression equation:

$$\hat{Y}_i = -3.888 + 1.449 X_{1i} + 1.462 X_{2i} - 0.190 X_{1i} X_{2i}$$

Suppose that X_1 is the perceived quality of the product and X_2 is the perceived value of the product. (Note: If the customer thinks the product is overpriced, he or she perceives it to be of low value and vice versa.)

a. What is the predicted purchase behavior when $X_1 = 2$ and $X_2 = 2$?

b. What is the predicted purchase behavior when $X_1 = 2$ and $X_2 = 7$?

c. What is the predicted purchase behavior when $X_1 = 7$ and $X_2 = 2$?

d. What is the predicted purchase behavior when $X_1 = 7$ and $X_2 = 7$?

e. What is the regression equation when $X_2 = 2$? What is the slope for X_1 now?

f. What is the regression equation when $X_2 = 7$? What is the slope for X_1 now?

g. What is the regression equation when $X_1 = 2$? What is the slope for X_2 now?

h. What is the regression equation when $X_1 = 7$? What is the slope for X_2 now?

i. Discuss the implications of (a) through (h) in the context of increasing sales for this product with two customer satisfaction measures.

14.72 The owner of a moving company typically has his most experienced manager predict the total number of labor hours that will be required to complete an upcoming move. This approach has proved useful in the past, but the owner has the business objective of developing a more accurate method of predicting labor hours. In a preliminary effort to provide a more accurate method, the owner has decided to use the number of cubic feet moved and the number of pieces of large furniture as the independent variables and has collected data for 36 moves in which the origin and destination were within the borough of Manhattan in New York City and the travel time was an insignificant portion of the hours worked. The data are organized and stored in `Moving`.

a. State the multiple regression equation.

b. Interpret the meaning of the slopes in this equation.

c. Predict the mean labor hours for moving 500 cubic feet with two large pieces of furniture.

d. Perform a residual analysis on your model and determine whether the regression assumptions are valid.

e. Determine whether there is a significant relationship between labor hours and the two independent variables (the number of cubic feet moved and the number of pieces of large furniture) at the 0.05 level of significance.

f. Determine the p-value in (e) and interpret its meaning.

g. Interpret the meaning of the coefficient of multiple determination in this problem.

h. Determine the adjusted r^2.

i. At the 0.05 level of significance, determine whether each independent variable makes a significant contribution to the regression model. Indicate the most appropriate regression model for this set of data.

j. Determine the p-values in (i) and interpret their meaning.

k. Construct a 95% confidence interval estimate of the population slope between labor hours and the number of cubic feet moved.

l. Compute and interpret the coefficients of partial determination.

m. What conclusions can you reach concerning labor hours?

14.73 Professional basketball has truly become a sport that generates interest among fans around the world. More and more players come from outside the United States to play in the National Basketball Association (NBA). You want to develop a regression model to predict the number of wins achieved by each NBA team, based on field goal (shots made) percentage and three-point field goal percentage for a recent season. The data are stored in `NBA`.

a. State the multiple regression equation.

b. Interpret the meaning of the slopes in this equation.

c. Predict the mean number of wins for a team that has a field goal percentage of 45% and a three-point field goal percentage of 35%.

d. Perform a residual analysis on your model and determine whether the regression assumptions are valid.

e. Is there a significant relationship between the number of wins and the two independent variables (field goal percentage and three-point field goal percentage) at the 0.05 level of significance?

f. Determine the p-value in (e) and interpret its meaning.

g. Interpret the meaning of the coefficient of multiple determination in this problem.

h. Determine the adjusted r^2.

i. At the 0.05 level of significance, determine whether each independent variable makes a significant contribution to the regression model. Indicate the most appropriate regression model for this set of data.

j. Determine the p-values in (i) and interpret their meaning.

k. Compute and interpret the coefficients of partial determination.

l. What conclusions can you reach concerning field goal percentage and three-point field goal percentage in predicting the number of wins?

14.74 A sample of 61 houses recently listed for sale in Silver Spring, Maryland, was selected with the objective of developing a model to predict the asking price (in $thousands), using the living space of the house (in square feet) and age (in years). The results are stored in `Silver Spring Homes`.

a. Fit a multiple regression model.

b. Interpret the meaning of the slopes in this model.

c. Predict the mean asking price for a house that has 2,000 square feet and is 55 years old.

d. Perform a residual analysis on your model and determine whether the regression assumptions are valid.

e. Determine whether there is a significant relationship between asking price and the two independent variables (house size and age) at the 0.05 level of significance.

f. Determine the p-value in (e) and interpret its meaning.

g. Interpret the meaning of the coefficient of multiple determination in this problem.

h. Determine the adjusted r^2.

i. At the 0.05 level of significance, determine whether each independent variable makes a significant contribution to the regression model. Indicate the most appropriate regression model for this set of data.

j. Determine the p-values in (i) and interpret their meaning.

k. Construct a 95% confidence interval estimate of the population slope between asking price and the living space of the house. How does the interpretation of the slope here differ from that in Problem 13.76 on page 471?

l. Compute and interpret the coefficients of partial determination.

m. What conclusions can you reach about the asking price?

14.75 Measuring the height of a California redwood tree is very difficult because these trees grow to heights over 300 feet. People familiar with these trees understand that the height of a California redwood tree is related to other characteristics of the tree, including the diameter of the tree at the breast height of a person (in inches) and the thickness of the bark of the tree (in inches). The file `Redwood` contains the height, diameter at breast height of a person, and bark thickness for a sample of 21 California redwood trees.

a. State the multiple regression equation that predicts the height of a tree, based on the tree's diameter at breast height and the thickness of the bark.

b. Interpret the meaning of the slopes in this equation.

c. Predict the mean height for a tree that has a breast height diameter of 25 inches and a bark thickness of 2 inches.

d. Interpret the meaning of the coefficient of multiple determination in this problem.

e. Perform a residual analysis on the model and determine whether the regression assumptions are valid.

f. Determine whether there is a significant relationship between the height of redwood trees and the two independent variables (breast-height diameter and bark thickness) at the 0.05 level of significance.

g. Construct a 95% confidence interval estimate of the population slope between the height of redwood trees and breast-height diameter and between the height of redwood trees and the bark thickness.

h. At the 0.05 level of significance, determine whether each independent variable makes a significant contribution to the regression model. Indicate the independent variables to include in this model.

i. Construct a 95% confidence interval estimate of the mean height for trees that have a breast-height diameter of 25 inches and a bark thickness of 2 inches, along with a prediction interval for an individual tree.

j. Compute and interpret the coefficients of partial determination.

k. What conclusions can you reach concerning the effect of the diameter of the tree and the thickness of the bark on the height of the tree?

14.76 A sample of 61 houses recently listed for sale in Silver Spring, Maryland, was selected with the objective of developing a model to predict the taxes (in $) based on the asking price of houses (in $thousands) and the age of the houses (in years) (stored in Silver Spring Homes):

a. State the multiple regression equation.

b. Interpret the meaning of the slopes in this equation.

c. Predict the mean taxes for a house that has an asking price of $400,000 and is 50 years old.

d. Perform a residual analysis on the model and determine whether the regression assumptions are valid.

e. Determine whether there is a significant relationship between taxes and the two independent variables (asking price and age) at the 0.05 level of significance.

f. Determine the p-value in (e) and interpret its meaning.

g. Interpret the meaning of the coefficient of multiple determination in this problem.

h. Determine the adjusted r^2.

i. At the 0.05 level of significance, determine whether each independent variable makes a significant contribution to the regression model. Indicate the most appropriate regression model for this set of data.

j. Determine the p-values in (i) and interpret their meaning.

k. Construct a 95% confidence interval estimate of the population slope between taxes and asking price. How does the interpretation of the slope here differ from that of Problem 13.77 on page 471?

l. Compute and interpret the coefficients of partial determination.

m. The real estate assessor's office has been publicly quoted as saying that the age of a house has no bearing on its taxes. Based on your answers to (a) through (l), do you agree with this statement? Explain.

14.77 A baseball analytics specialist wants to determine which variables are important in predicting a team's wins in a given season. He has collected data related to wins, earned run average (ERA), and runs scored per game for a recent season (stored in Baseball). Develop a model to predict the number of wins based on ERA and runs scored per game.

a. State the multiple regression equation.

b. Interpret the meaning of the slopes in this equation.

c. Predict the mean number of wins for a team that has an ERA of 4.50 and has scored 4.6 runs per game.

d. Perform a residual analysis on the model and determine whether the regression assumptions are valid.

e. Is there a significant relationship between the number of wins and the two independent variables (ERA and runs scored per game) at the 0.05 level of significance?

f. Determine the p-value in (e) and interpret its meaning.

g. Interpret the meaning of the coefficient of multiple determination in this problem.

h. Determine the adjusted r^2.

i. At the 0.05 level of significance, determine whether each independent variable makes a significant contribution to the regression model. Indicate the most appropriate regression model for this set of data.

j. Determine the p-values in (i) and interpret their meaning.

k. Construct a 95% confidence interval estimate of the population slope between wins and ERA.

l. Compute and interpret the coefficients of partial determination.

m. Which is more important in predicting wins—pitching, as measured by ERA, or offense, as measured by runs scored per game? Explain.

14.78 Referring to Problem 14.77, suppose that in addition to using ERA to predict the number of wins, the analytics specialist wants to include the league (0 = American, 1 = National) as an independent variable. Develop a model to predict wins based on ERA and league. For (a) through (k), do not include an interaction term.

a. State the multiple regression equation.

b. Interpret the slopes in (a).

c. Predict the mean number of wins for a team with an ERA of 4.50 in the American League.

d. Perform a residual analysis on the model and determine whether the regression assumptions are valid.

e. Is there a significant relationship between wins and the two independent variables (ERA and league) at the 0.05 level of significance?

f. At the 0.05 level of significance, determine whether each independent variable makes a contribution to the regression model. Indicate the most appropriate regression model for this set of data.

g. Construct a 95% confidence interval estimate of the population slope for the relationship between wins and ERA.

h. Construct a 95% confidence interval estimate of the population slope for the relationship between wins and league.

i. Compute and interpret the adjusted r^2.

j. Compute and interpret the coefficients of partial determination.

k. What assumption do you have to make about the slope of wins with ERA?

l. Add an interaction term to the model and, at the 0.05 level of significance, determine whether it makes a significant contribution to the model.

m. On the basis of the results of (f) and (l), which model is most appropriate? Explain.

14.79 You are a real estate broker who wants to compare property values in Glen Cove and Roslyn (which are located approximately 8 miles apart). In order to do so, you will analyze the data in GC and Roslyn, a file that includes samples of houses from Glen Cove and Roslyn. Making sure to include the dummy variable for location (Glen Cove or Roslyn), develop a regression model to predict fair market value, based on the land area of a property, the age of a house, and location. Be sure to determine whether any interaction terms need to be included in the model.

14.80 HR practitioners are increasing performing gender pay audits to understand whether a gender gap exists at their company. Practitioners examine payroll data for evidence of a gender pay gap. An HR practitioner collects data on base pay ($), gender (0 = female and 1 = male), and age (years) for 405 employees at his company and stores these data in HR .

Source: Data extracted from Chamberlain, A., *How to Analyze Your Gender Pay Gap: An Employer's Guide*, available at **bit.ly/2td7h33**.

Develop a multiple regression model that uses gender and age to predict employee base pay. Be sure to perform a thorough residual analysis. The HR practitioner suspected that there was a significant interaction between gender and age. Is there evidence to support the HR practitioner's suspicion?

14.81 Starbucks Coffee Co. uses a data-based approach to improving the quality and customer satisfaction of its products. When survey data indicated that Starbucks needed to improve its package sealing process, an experiment was conducted to determine the factors in the bag-sealing equipment that might be affecting the ease of opening the bag without tearing the inner liner of the bag.

Source: Data extracted from L. Johnson and S. Burrows, "For Starbucks, It's in the Bag," *Quality Progress*, March 2011, pp. 17–23.

Among the factors that could affect the rating of the ability of the bag to resist tears were the viscosity, pressure, and plate gap on the bag-sealing equipment.

Data were collected on 19 bags in which the plate gap was varied and stored in Starbucks . Develop a multiple regression model that uses the viscosity, pressure, and plate gap on the bag-sealing equipment to predict the tear rating of the bag. Be sure to perform a thorough residual analysis. Do you think that you need to use all three independent variables in the model? Explain.

14.82 An experiment was conducted to study the extrusion process of biodegradable packaging foam.

Source: Data extracted from W. Y. Koh, K. M. Eskridge, and M. A. Hanna, "Supersaturated Split-Plot Designs," *Journal of Quality Technology*, 45, January 2013, pp. 61–72.

Among the factors considered for their effect on the unit density (mg/ml) were the die temperature (145°C versus 155°C) and the die diameter (3 mm versus 4 mm). The results were stored in Packaging Foam 3 . Develop a multiple regression model that uses die temperature and die diameter to predict the foam density (mg/ml). Be sure to perform a thorough residual analysis. Do you think that you need to use both independent variables in the model? Explain.

14.83 Referring to Problem 14.82, instead of predicting the unit density, you now wish to predict the foam diameter from results stored in Packaging Foam 4 . Develop a multiple regression model that uses die temperature and die diameter to predict the foam diameter (mg/ml). Be sure to perform a thorough residual analysis. Do you think that you need to use both independent variables in the model? Explain.

CHAPTER

14

▾CASES

Managing Ashland MultiComm Services

In its continuing study of the *3-For-All* subscription solicitation process, a marketing department team wants to test the effects of two types of structured sales presentations (personal formal and personal informal) and the number of hours spent on telemarketing on the number of new subscriptions. The staff has recorded these data for the past 24 weeks in AMS14 .

Analyze these data and develop a multiple regression model to predict the number of new subscriptions for a week, based on the number of hours spent on telemarketing and the sales presentation type. Write a report, giving detailed findings concerning the regression model used.

Digital Case

Apply your knowledge of multiple regression models in this Digital Case, which extends the OmniFoods Using Statistics scenario from this chapter.

To ensure a successful test marketing of its OmniPower energy bars, the OmniFoods marketing department has contracted with In-Store Placements Group (ISPG), a merchandising consulting firm. ISPG will work with the grocery store chain that is conducting the test-market study. Using the same 34-store sample used in the test-market study, ISPG claims that the choice of shelf location and the presence of in-store OmniPower coupon dispensers both increase sales of the energy bars.

Open **Omni_ISPGMemo.pdf** to review the ISPG claims and supporting data. Then answer the following questions:

1. Are the supporting data consistent with ISPG's claims? Perform an appropriate statistical analysis to confirm (or discredit) the stated relationship between sales and the two independent variables of product shelf location and the presence of in-store OmniPower coupon dispensers.

2. If you were advising OmniFoods, would you recommend using a specific shelf location and in-store coupon dispensers to sell OmniPower bars?

3. What additional data would you advise collecting in order to determine the effectiveness of the sales promotion techniques used by ISPG?

▼EXCEL GUIDE

EG14.1 DEVELOPING a MULTIPLE REGRESSION MODEL

Interpreting the Regression Coefficients

Key Technique Use the **LINEST**(*cell range of Y variable, cell range of X variables*, **True, True**) function to compute the regression coefficients and related values.

Example Develop the Figure 14.1 multiple regression model for the OmniPower sales data on page 481.

PHStat Use **Multiple Regression**.

For the example, open to the **DATA worksheet** of the **OmniPower workbook**. Select **PHStat → Regression → Multiple Regression**, and in the procedure's dialog box (shown below):

1. Enter **A1:A35** as the **Y Variable Cell Range**.
2. Enter **B1:C35** as the **X Variables Cell Range**.
3. Check **First cells in both ranges contain label**.
4. Enter **95** as the **Confidence level for regression coefficients**.
5. Check **Regression Statistics Table** and **ANOVA and Coefficients Table**.
6. Enter a **Title** and click **OK**.

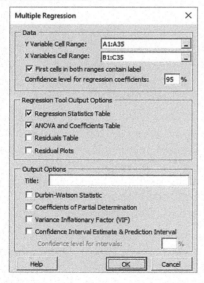

The procedure creates a worksheet that contains a copy of the data in addition to the Figure 14.1 worksheet.

Workbook Use the **COMPUTE worksheet** of the **Multiple Regression workbook** as a template.

For the example, the COMPUTE worksheet already uses the OmniPower sales data in the MRData worksheet to perform the regression analysis.

To perform multiple regression analyses for other data with two independent variables:

1. Paste the new regression data into the **MRData worksheet**, using column A for the *Y* variable data and subsequent columns, starting with B, for the *X* variable data.
2. Open to the **COMPUTE worksheet**.
3. Enter the **confidence level** in cell **L8**.
4. Edit the *array formula* in the cell range **L2:N6** to reflect the cell ranges of the data for the new *Y* and the new *X* variables.

These new cell ranges should start with row 2 so as to exclude the row 1 variable names, an exception to the usual practice in this book.

For problems with more than two independent variables, select, in step 4, a range wider than L2:N6, adding a column for each independent variable in excess of two. For example, with three independent variables, select the cell range **L2:O6**. Then continue with these steps 5 through 8:

5. Edit the labels in cells **K2** and **K3**.
6. Edit the ANOVA table formulas in columns **B** and **C**.
7. Select cell range **D18:I18**, right-click and select **Insert**. Repeat for as many times as necessary.
8. Select cell range **D17:I17** and copy down through all the rows of the ANOVA table (blank and nonblank).

The SHORT TAKES for Chapter 14 explain more about this Intermediate Calculations area. Steps 5 through 8 may be difficult for Excel novices to complete. If you are an Excel novice, consider using the *PHStat* or *Analysis ToolPak* instructions when your problem includes more than two independent *X* variables.

Analysis ToolPak Use **Regression**.

For the example, open to the **DATA worksheet** of the **OmniPower workbook** and:

1. Select **Data → Data Analysis**.
2. In the Data Analysis dialog box, select **Regression** from the **Analysis Tools** list and then click **OK**.

In the Regression dialog box (shown on page 523):

3. Enter **A1:A35** as the **Input Y Range** and enter **B1:C35** as the **Input X Range**.
4. Check **Labels** and check **Confidence Level** and enter **95** in its box.
5. Click **New Worksheet Ply**.
6. Click **OK**.

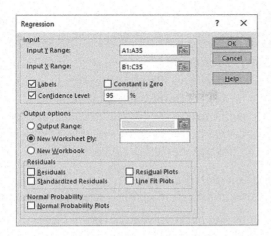

Predicting the Dependent Variable Y

Key Technique Use the **MMULT** array function and the **T.INV.2T** function to help compute intermediate values that determine the confidence interval estimate and prediction interval.

Example Compute the Figure 14.2 confidence interval estimate and prediction interval for the OmniPower sales data on page 482.

PHStat Use the *PHStat* "Interpreting the Regression Coefficients" instructions but replace step 6 with the following steps 6 through 8:

7. Check **Confidence Interval Estimate & Prediction Interval** and enter **95** as the percentage for **Confidence level for intervals**.
8. Enter a **Title** and click **OK**.
9. In the new worksheet, enter **79** in cell **B6** and enter **400** in cell **B7**.

These steps create a new worksheet that is similar to the CIEandPI worksheet that the following *Workbook* instructions discuss.

Workbook Use the **CIEandPI worksheet** of the **Multiple Regression workbook** as a template.

The worksheet already contains the data and formulas for the example. For other problems with two independent variables:

1. Paste the regression data for the independent variables into columns B and C of the **MRArray worksheet**.
2. Adjust the number of entries in column A, all of which are 1, to match the number of rows of the new data.
3. Use the "Interpreting the Regression Coefficients" *Worksheet* instructions to edit the COMPUTE worksheet to reflect the new data.
4. Open to the CIEandPI worksheet and edit the array formula in cell range **B9:D11** and the labels in cells **A6** and **A7** to reflect the new data.

Cell ranges in the array formula should start with row 2 so as to exclude the row 1 variable names, an exception to the usual practice in this book.

To learn more about the formulas that the CIEandPI worksheet uses, read the SHORT TAKES for Chapter 14.

EG14.2 EVALUATING MULTIPLE REGRESSION MODELS

The coefficient of multiple determination, r^2, the adjusted r^2, and the overall F test appear as part of the multiple regression results that the Section EG14.1 instructions create.

PHStat and the *Workbook* instructions use formulas to compute these results in the **COMPUTE worksheet**. Formulas in cells B5, B7, B13, C12, C13, D12, and E12 copy values computed by the array formula in cell range L2:N6. In cell F12, the expression **F.DIST.RT(F test statistic, 1, error degrees of freedom)** computes the p-value for the overall F test.

EG14.3 MULTIPLE REGRESSION RESIDUAL ANALYSIS

Key Technique Use arithmetic formulas and some results from the multiple regression COMPUTE worksheet to compute residuals.

Example Perform the residual analysis for the OmniPower sales data shown in Figure 14.4, starting on page 488.

PHStat Use the Section EG14.1 "Interpreting the Regression Coefficients" *PHStat* instructions. Modify step 5 by checking **Residuals Table** and **Residual Plots** in addition to checking **Regression Statistics Table** and **ANOVA and Coefficients Table**.

Workbook Use the **RESIDUALS worksheet** of the **Multiple Regression workbook** as a template. Then construct residual plots for the residuals and the predicted value of Y and for the residuals and each of the independent variables.

For the example, the RESIDUALS worksheet uses the Omni-Power sales data already in the **MRData worksheet** to compute the residuals. To compute residuals for other data, first use the EG14.1 "Interpreting the Regression Coefficients" *Workbook* instructions to modify the MRData and COMPUTE worksheets. Then, open to the **RESIDUALS worksheet** and:

1. If the number of independent variables is greater than 2, select column D, right-click, and click **Insert** from the shortcut menu. Repeat this step as many times as necessary to create the additional columns to hold all the X variables.
2. Paste the data for the X variables into columns, starting with column B and paste the Y values into the second-to-last column (column E if there are two X variables).
3. For sample sizes smaller than 34, delete the extra rows. For sample sizes greater than 34, copy the predicted Y and residuals formulas down through the row containing the last pair of X and Y values. Also, add the new observation numbers in column A.

To construct the residual plots, open to the RESIDUALS worksheet and select pairs of columns and then use the EG2.5 "The Scatter Plot" *Workbook* instructions. For example, to construct the residual plot for the residuals and the predicted value of *Y*, select columns D and F. (See Appendix B for help about selecting a noncontiguous cell range.)

To learn more about the formulas that the RESIDUAL worksheet uses, read the SHORT TAKES for Chapter 14.

Analysis ToolPak Use the Section EG14.1 *Analysis ToolPak* instructions. Modify step 5 by checking **Residuals** and **Residual Plots** before clicking **New Worksheet Ply** and then **OK**. The **Residuals Plots** option constructs residual plots only for each independent variable.

To construct a plot of the residuals and the predicted value of *Y*, select the predicted and residuals cells (in the RESIDUAL OUTPUT area of the regression results worksheet) and then apply the Section EG2.5 *Worksheet* "The Scatter Plot" instructions.

EG14.4 INFERENCES ABOUT the POPULATION REGRESSION COEFFICIENTS

The regression results worksheets that the Section EG14.1 instructions create include the information needed to make the inferences that Section 14.4 discusses.

EG14.5 TESTING PORTIONS of the MULTIPLE REGRESSION MODEL

Key Technique Adapt the Section EG14.1 "Interpreting the Regression Coefficients" instructions.

Example Test portions of the multiple regression model for the OmniPower sales data as discussed in Section 14.5, starting on page 492.

PHStat Use the Section EG14.1 *PHStat* "Interpreting the Regression Coefficients" instructions but modify step 6 by checking **Coefficients of Partial Determination** before you click **OK**.

Workbook Use one of the **CPD worksheets** of the **Multiple Regression workbook** as a template.

For the example, the **CPD_2 worksheet** already contains the data to compute the coefficients of partial determination. For other problems, first use the EG14.1 "Interpreting the Regression Coefficients" and EG13.2 *Worksheet* instructions to create all possible regression results worksheets.

For example, if you have two independent variables, you perform three regression analyses: *Y* with X_1 and X_2, *Y* with X_1, and *Y* with X_2, to create three regression results worksheets. Then, open to the **CPD worksheet** for the number of independent variables and follow the instructions in the worksheet to transfer values from the regression results worksheets you just created.

EG14.6 USING DUMMY VARIABLES and INTERACTION TERMS

Dummy Variables

Key Technique Use **Find and Replace** to create a dummy variable from a two-level categorical variable.

Example From the two-level categorical variable has fireplace, create the dummy variable fireplace that the Figure 14.9 regression model on page 498 uses.

Workbook For the example, open to the **OriginalData worksheet** of the **SilverSpringUncoded workbook** and:

1. Copy and paste the **Has Fireplace** values in column **M** to **column N** (the first empty column).
2. Enter **Fireplace** in cell **N1** and then select **column N**.
3. Press **Ctrl+H** (the keyboard shortcut for **Find and Replace**).

In the Find and Replace dialog box:

4. Enter **Yes** in the **Find what** box and enter **1** in the **Replace with** box.
5. Click **Replace All**. If a message box to confirm the replacement appears, click **OK** to continue.
6. Enter **No** in the **Find what** box and enter **0** in the **Replace with** box.
7. Click **Replace All**. If a message box to confirm the replacement appears, click **OK** to continue.
8. Click **Close**.

Categorical variables that have more than two levels require the use of formulas in multiple columns. For example, to create the Example 14.3 dummy variables for Example 14.3 on page 500, two columns are needed. Assume that the three-level house type variable in the example is in Column D. A first new column that contains formulas in the form =IF(**column D** *cell=first level*, **1, 0**) and a second new column that contains formulas in the form =IF(**column D** *cell=secondlevel*, **1, 0**) would properly create the two dummy variables that the example requires.

Interactions

To create an interaction term, add a column of formulas that multiply one independent variable by another. For example, if the first independent variable appeared in column B and the second independent variable appeared in column C, enter the formula =**B2*C2** in the row 2 cell of an empty new column and then copy the formula down through all rows of data to create the interaction.

EG14.7 LOGISTIC REGRESSION

Key Technique Use an automated process that incorporates the use of the Solver add-in to develop a logistic regression analysis model.

Example Develop the Figure 14.13 logistic regression model for the credit card pilot study data on page 510.

PHStat Use **Logistic Regression**.

For the example, open to the **DATA worksheet** of the **CardStudy workbook**. Select **PHStat→Regression→Logistic Regression**, and in the procedure's dialog box:

1. Enter **A1:A31** as the **Y Variable Cell Range**.
2. Enter **B1:C31** as the **X Variables Cell Range**.
3. Check **First cells in both ranges contain label**.
4. Enter a **Title** and click **OK**.

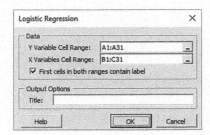

If the Solver add-in is not installed (see Appendix D), PHStat will display an error message instead of the Logistic Regression dialog box.

Workbook Use the **Logistic Regression add-in workbook**. *The Excel Solver add-in must be installed before using this add-in workbook* (see Appendix D).

For the example, the **COMPUTE worksheet** of the **Logistic Model workbook** already contains the logistic regression model. For other problems:

1. Open to the worksheet that contains the data for the problem. The worksheet *must* be part of a workbook saved in the current **.xlsx** format (not the older **.xls** format).
2. Open the Logistic Regression add-in workbook (as you would open any other Excel file).

If the add-in workbook opens properly, it adds a Logistic Add-in menu to the Add-ins tab in Microsoft Windows Excel or to the Apple menu bar in Excel for Mac.

3. Select **Logistic Add-in→Logistic Regression**.

In the Logistic Regression dialog box, (identical to the PHStat dialog box):

4. Use steps 1 through 4 of the *PHStat* instructions to complete the entries in the dialog box.

15

Multiple Regression Model Building

OBJECTIVES

- Use quadratic terms in a regression model
- Use transformed variables in a regression model
- Measure the correlation among independent variables
- Build a regression model using either the stepwise or best subsets approach
- Avoid the pitfalls involved in developing a multiple regression model

▼ USING STATISTICS
Valuing Parsimony at WSTA-TV

Nickels Broadcasting looks to minimize costs at its WSTA-TV News 37 broadcast center and has identified cutting *standby hours*, hours for which employees are scheduled but end up not being assigned any work, as a possible way to minimize costs. Nickels has already received an offer to move its broadcast center to Argleton, a locality that permits *on-call shifts*, a form of just-in-time scheduling. Using on-call shifts could eliminate many standby hours. However, relocating would create new costs and raise other concerns including the location of the proposed site, so management has deferred action on that offer.

Instead, Nickels management wonders if the numbers of staff hired each week and the number of weekly hours that the staff works in three job categories affects the weekly standby hours. Nickels hires you as an analyst and presents you with a 26-week sample of weekly standby hours as well as weekly staff count, remote engineering hours, graphics hours, and editorial production hours. You quickly establish that no single variable from the set of four independent variables can predict standby hours. How then do you build a multiple regression model that uses some or all of the four variables? How can you determine a "best" regression model without examining all possible models?

Threeither simple and multiple regression models that Chapters 13 and 14 discuss assume a linear relationship between the dependent Y variable and each independent X variable. This chapter extends the discussion of multiple regression to consider both nonlinear regression models as well as the methods that help efficiently develop the best model for any set of data, including those that have many independent X variables. Such methods can provide the means of identifying the "best" model for the WSTA-TV 26-week data.

15.1 The Quadratic Regression Model

One of the most common nonlinear relationships between variables is *curvilinear*. In the **curvilinear relationship**, the value of the dependent variable Y increases or decreases at a changing rate as the value of X changes. Figure 15.1 presents two scatter plots that illustrate two examples of curvilinear relationships.

FIGURE 15.1
Two curvilinear relationships

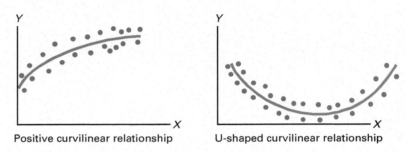

Positive curvilinear relationship   U-shaped curvilinear relationship

The simplest curvilinear relationship is a quadratic relationship in which there is a term in the regression model that is the square of the independent variable. Equation (15.1) defines this relationship as the **quadratic regression model**.

QUADRATIC REGRESSION MODEL

$$Y_i = \beta_0 + \beta_1 X_{1i} + \beta_2 X_{1i}^2 + \varepsilon_i \qquad (15.1)$$

where

$\beta_0 = Y$ intercept
$\beta_1 =$ coefficient of the linear effect on Y
$\beta_2 =$ coefficient of the quadratic effect on Y
$\varepsilon_i =$ random error in Y for observation i

The quadratic regression model is similar to the multiple regression model with two independent variables that Equation (14.2) on page 480 defines. However, in a quadratic model, the square of the first independent variable that serves as the second independent variable is called the **quadratic term**. As with a multiple regression model, you use the least-squares method to compute sample regression coefficients (b_0, b_1, and b_2) as estimates of the population parameters (β_0, β_1, and β_2). Equation (15.2) defines the regression equation for the quadratic model with an independent variable (X_1) and a dependent variable (Y).

student TIP

A quadratic regression model has an X term and an X squared term. Other curvilinear models can have additional X terms such as X cubed, X raised to the fourth power, and so on.

QUADRATIC REGRESSION EQUATION

$$\hat{Y}_i = b_0 + b_1 X_{1i} + b_2 X_{1i}^2 \qquad (15.2)$$

In Equation (15.2), the first regression coefficient, b_0, represents the Y intercept; the second regression coefficient, b_1, represents the linear effect; and the third regression coefficient, b_2, represents the quadratic effect.

Finding the Regression Coefficients and Predicting Y

To illustrate the quadratic regression model, consider a study that examined the business problem facing a concrete supplier of how adding fly ash affects the strength of concrete. (Fly ash is an inexpensive industrial waste by-product that can be used as a substitute for Portland cement, a more expensive ingredient of concrete.) Batches of concrete were prepared in which the percentage of fly ash ranged from 0% to 60%. Data were collected from a sample of 18 batches and organized and stored in Fly Ash. Table 15.1 summarizes the results.

TABLE 15.1

Fly ash percentage and strength of 18 batches of 28-day-old concrete

Fly Ash%	Strength (psi)	Fly Ash%	Strength (psi)
0	4,779	40	5,995
0	4,706	40	5,628
0	4,350	40	5,897
20	5,189	50	5,746
20	5,140	50	5,719
20	4,976	50	5,782
30	5,110	60	4,895
30	5,685	60	5,030
30	5,618	60	4,648

Creating the scatter plot in Figure 15.2 to visualize the relationship between the Fly Ash% and Strength variables enables one to better select the proper model for expressing the relationship between fly ash percentage and strength.

FIGURE 15.2

Scatter plot of fly ash percentage (X) and strength (Y)

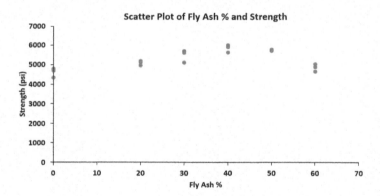

Figure 15.2 indicates an initial increase in the strength of the concrete as the percentage of fly ash increases. The strength appears to level off and then drop after achieving maximum strength at about 40% fly ash. Strength for 50% fly ash is slightly below strength at 40%, but strength at 60% fly ash is substantially below strength at 50%. Therefore, one should fit a quadratic model, not a linear model, to estimate strength based on fly ash percentage.

From Figure 15.3, which shows the Excel worksheet results for these data, the regression coefficients are

$$b_0 = 4,486.3611 \quad b_1 = 63.0052 \quad b_2 = -0.8765$$

Therefore, the quadratic regression equation is

$$\hat{Y}_i = 4,486.3611 + 63.0052X_{1i} - 0.8765X_{1i}^2$$

where

$$\hat{Y}_i = \text{predicted strength for sample } i$$
$$X_{1i} = \text{percentage of fly ash for sample } i$$

FIGURE 15.3
Excel multiple regression quadratic model worksheet results for the concrete strength data

	A	B	C	D	E	F	G
1	Concrete Strength Analysis						
2							
3	*Regression Statistics*						
4	Multiple R	0.8053					
5	R Square	0.6485					
6	Adjusted R Square	0.6016					
7	Standard Error	312.1129					
8	Observations	18					
9							
10	ANOVA						
11		*df*	*SS*	*MS*	*F*	*Significance F*	
12	Regression	2	2695473.4897	1347736.745	13.8351	0.0004	
13	Residual	15	1461217.0103	97414.4674			
14	Total	17	4156690.5000				
15							
16		*Coefficients*	*Standard Error*	*t Stat*	*P-value*	*Lower 95%*	*Upper 95%*
17	Intercept	4486.3611	174.7531	25.6726	0.0000	4113.8834	4858.8389
18	FlyAsh%	63.0052	12.3725	5.0923	0.0001	36.6338	89.3767
19	FlyAsh% ^2	-0.8765	0.1966	-4.4578	0.0005	-1.2955	-0.4574

In the Worksheet

This worksheet uses the same template as the Figure 14.1 worksheet. The regression coefficients for this model are in cells B17, B18, and B19.

Figure 15.4 is a scatter plot of this quadratic regression equation that shows the fit of the quadratic regression model to the original data.

FIGURE 15.4
Scatter plot showing the quadratic relationship between fly ash percentage and strength for the concrete data

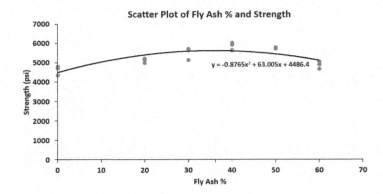

Scatter Plot of Fly Ash % and Strength

$y = -0.8765x^2 + 63.005x + 4486.4$

From the quadratic regression equation and Figure 15.4, the Y intercept (4,486.3611) is the predicted strength when the percentage of fly ash is 0. To interpret the coefficients b_1 and b_2, observe that after an initial increase, strength decreases as fly ash percentage increases. This nonlinear relationship is further demonstrated by predicting the strength for fly ash percentages of 20, 40, and 60. Using the quadratic regression equation,

$$\hat{Y}_i = 4{,}486.3611 + 63.0052X_{1i} - 0.8765X_{1i}^2$$

for $X_{1i} = 20$,

$$\hat{Y}_i = 4{,}486.3611 + 63.0052(20) - 0.8765(20)^2 = 5{,}395.865$$

for $X_{1i} = 40$,

$$\hat{Y}_i = 4{,}486.3611 + 63.0052(40) - 0.8765(40)^2 = 5{,}604.169$$

and for $X_{1i} = 60$,

$$\hat{Y}_i = 4{,}486.3611 + 63.0052(60) - 0.8765(60)^2 = 5{,}111.273$$

Thus, the predicted concrete strength for 40% fly ash is 208.304 psi above the predicted strength for 20% fly ash, but the predicted strength for 60% fly ash is 492.896 psi below the predicted strength for 40% fly ash. The concrete supplier should consider using a fly ash percentage of 40% and not using fly ash percentages of 20% or 60% because those percentages lead to reduced concrete strength.

Testing for the Significance of the Quadratic Model

After you calculate the quadratic regression equation, you can test whether there is a significant overall relationship between strength, Y, and fly ash percentage, X_1. The null and alternative hypotheses are as follows:

H_0: $\beta_1 = \beta_2 = 0$ (There is no overall relationship between X_1 and Y.)

H_1: β_1 and/or $\beta_2 \neq 0$ (There is an overall relationship between X_1 and Y.)

Equation (14.6) on page 485 defines the overall F_{STAT} test statistic used for this test:

$$F_{STAT} = \frac{MSR}{MSE}$$

From the Figure 15.3 results on page 529,

$$F_{STAT} = \frac{MSR}{MSE} = \frac{1{,}347{,}736.745}{97{,}414.4674} = 13.8351$$

Using a level of significance of 0.05 and Table E.5, the critical value of the F distribution, with 2 and 15 $(18 - 2 - 1)$ degrees of freedom, is 3.68 (see Figure 15.5).

FIGURE 15.5
Testing for the existence of the overall relationship at the 0.05 level of significance, with 2 and 15 degrees of freedom

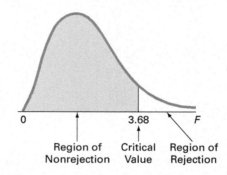

Table 15.2 summarizes the results of the test for the significance of the Figure 15.3 quadratic model on page 529. Based on the conclusions, there is strong evidence to conclude that strength is related to fly ash percentage. Therefore, one can state that fly ash percentage is useful in helping to determine the strength of the concrete.

TABLE 15.2
F test results for the significance of the quadratic model and conclusions

Result	Conclusions
$F_{STAT} = 13.8351$ is greater than 3.68 p-value $= 0.0004$ is less than the level of significance, $\alpha = 0.05$	1. Reject the null hypothesis H_0. 2. Conclude that evidence exists for claiming that fly ash percentage is related to the dependent Y variable, strength. 3. The probability is 0.0004 that $F_{STAT} > 13.8351$, given the null hypothesis is true.

Testing the Quadratic Effect

When using regression analysis to examine a relationship between two variables, the goal is to find the most accurate, as well as the *simplest*, model that expresses the relationship. Therefore, one must examine whether there is a significant difference between the quadratic model:

$$Y_i = \beta_0 + \beta_1 X_{1i} + \beta_2 X_{1i}^2 + \varepsilon_i$$

and the linear model:

$$Y_i = \beta_0 + \beta_1 X_{1i} + \varepsilon_i$$

Section 14.4 discusses the t test to determine whether each independent variable makes a significant contribution to the regression model. To test the significance of the contribution of the quadratic effect, one uses the following null and alternative hypotheses:

H_0: Including the quadratic effect does not significantly improve the model ($\beta_2 = 0$).

H_1: Including the quadratic effect significantly improves the model ($\beta_2 \neq 0$).

Equation (14.7) on page 489 defines the t_{STAT} test statistic for this test. The standard error of each regression coefficient and its corresponding t_{STAT} test statistic that this test needs appear in the regression results that Excel produces. For the fly ash example, using the values that appear in Figure 15.3:

$$t_{STAT} = \frac{b_2 - \beta_2}{S_{b_2}}$$

$$= \frac{-0.8765 - 0}{0.1966} = -4.4578$$

If one selects the 0.05 level of significance from Table E.3, the critical values for the t distribution with 15 degrees of freedom are -2.1315 and $+2.1315$ (see Figure 15.6).

FIGURE 15.6
Testing for the contribution of the quadratic effect to a regression model at the 0.05 level of significance, with 15 degrees of freedom

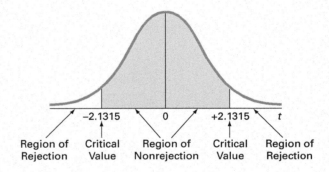

Table 15.3 summarizes the results of the test for the significance of the quadratic effect. Based on these conclusions, there is strong evidence to conclude that there is a quadratic effect of fly ash percentage and strength. Therefore, as part of the DCOVA framework, one can say that the quadratic effect of fly ash percentage can be used to help determine the strength of the concrete.

TABLE 15.3
t test results for the significance of the quadratic model and conclusions

Result	Conclusions
$t_{STAT} = -4.4578$ is less than -2.1315 p-value = 0.0005 is less than the level of significance, $\alpha = 0.05$	1. Reject the null hypothesis H_0. 2. Conclude that evidence exists for claiming that there is a quadratic effect of fly ash percentage with the dependent Y variable, strength. 3. The probability is 0.0005 that $t_{STAT} = \ < -4.4578$ or > 4.4578, given the null hypothesis is true.

Example 15.1 provides an additional illustration of a possible quadratic effect.

EXAMPLE 15.1

Studying the Quadratic Effect in a Multiple Regression Model

▶(*continued*)

A home builder studying the business problem of estimating the consumption of heating oil by single-family houses has decided to examine the effect of atmospheric temperature and the amount of attic insulation on heating oil consumption. Data are collected from a random sample of 15 single-family houses and stored in Heating Oil .

Figure 15.7 shows the regression results for a multiple regression model using the two independent variables: atmospheric temperature and attic insulation.

FIGURE 15.7
Excel multiple regression linear model worksheet results for predicting monthly consumption of heating oil

	A	B	C	D	E	F	G
1	**Heating Oil Consumption Analysis**						
2							
3	*Regression Statistics*						
4	Multiple R	0.9827					
5	R Square	0.9656					
6	Adjusted R Square	0.9599					
7	Standard Error	26.0138					
8	Observations	15					
9							
10	**ANOVA**						
11		*df*	*SS*	*MS*	*F*	*Significance F*	
12	Regression	2	228014.6263	114007.3132	168.4712	0.0000	
13	Residual	12	8120.6030	676.7169			
14	Total	14	236135.2293				
15							
16		*Coefficients*	*Standard Error*	*t Stat*	*P-value*	*Lower 95%*	*Upper 95%*
17	Intercept	562.1510	21.0931	26.6509	0.0000	516.1931	608.1089
18	Temperature	-5.4366	0.3362	-16.1699	0.0000	-6.1691	-4.7040
19	Insulation	-20.0123	2.3425	-8.5431	0.0000	-25.1162	-14.9084

The residual plot for attic insulation (not shown) contains some evidence of a quadratic effect. Therefore, the home builder reanalyzed the data by adding a quadratic term for attic insulation to the multiple regression model. At the 0.05 level of significance, is there evidence of a significant quadratic effect for attic insulation?

SOLUTION Figure 15.8 shows the results for this regression model.

FIGURE 15.8
Excel worksheet results for the multiple regression model with a quadratic term for attic insulation

	A	B	C	D	E	F	G
1	**Quadratic Effect for Insulation Variable?**						
2							
3	*Regression Statistics*						
4	Multiple R	0.9862					
5	R Square	0.9725					
6	Adjusted R Square	0.9650					
7	Standard Error	24.2938					
8	Observations	15					
9							
10	**ANOVA**						
11		*df*	*SS*	*MS*	*F*	*Significance F*	
12	Regression	3	229643.1645	76547.7215	129.7006	0.0000	
13	Residual	11	6492.0649	590.1877			
14	Total	14	236135.2293				
15							
16		*Coefficients*	*Standard Error*	*t Stat*	*P-value*	*Lower 95%*	*Upper 95%*
17	Intercept	624.5864	42.4352	14.7186	0.0000	531.1872	717.9856
18	Temperature	-5.3626	0.3171	-16.9099	0.0000	-6.0606	-4.6646
19	Insulation	-44.5868	14.9547	-2.9815	0.0125	-77.5019	-11.6717
20	Insulation ^2	1.8667	1.1238	1.6611	0.1249	-0.6067	4.3401

The multiple regression equation is

$$\hat{Y}_i = 624.5864 - 5.3626X_{1i} - 44.5868X_{2i} + 1.8667X_{2i}^2$$

To test for the significance of the quadratic effect:

H_0: Including the quadratic effect of insulation does not significantly improve the model ($\beta_3 = 0$).

H_1: Including the quadratic effect of insulation significantly improves the model ($\beta_3 \neq 0$).

From Figure 15.8 and Table E.3 with $11 \, (15 - 3 - 1)$ degrees of freedom, $-2.2010 < t_{STAT} = 1.6611 < 2.2010$ (or the p-value $= 0.1249 > 0.05$). Therefore, the home builder does not reject the null hypothesis. The home builder concludes that there is insufficient evidence that the quadratic effect for attic insulation is different from zero. In the interest of keeping the model as simple as possible, the home builder should use the Figure 15.7 multiple regression equation:

$$\hat{Y}_i = 562.1510 - 5.4366X_{1i} - 20.0123X_{2i}$$

The Coefficient of Multiple Determination

In the multiple regression model, the coefficient of multiple determination, r^2, which Section 14.2 explains, represents the proportion of variation in Y that is explained by variation in the independent variables. One computes r^2 by using Equation (14.4) on page 484:

$$r^2 = \frac{SSR}{SST}$$

Consider the quadratic regression model that predicts the strength of concrete using fly ash and fly ash squared. From Figure 15.3 on page 529,

$$SSR = 2{,}695{,}473.4897 \qquad SST = 4{,}156{,}690.5$$

Thus,

$$r^2 = \frac{SSR}{SST} = \frac{2{,}695{,}473.4897}{4{,}156{,}690.5} = 0.6485$$

This coefficient of multiple determination indicates that 64.85% of the variation in strength is explained by the quadratic relationship between strength and the percentage of fly ash. One should also compute r^2_{adj} to account for the number of independent variables and the sample size. In the quadratic regression model, $k = 2$ because there are two independent variables, X_1 and X_1^2. Thus, using Equation (14.5) on page 484,

$$r^2_{adj} = 1 - \left[(1 - r^2) \frac{(n-1)}{(n-k-1)} \right] = 1 - \left[(1 - 0.6485) \frac{17}{15} \right]$$

$$= 1 - 0.3984$$

$$= 0.6016$$

PROBLEMS FOR SECTION 15.1

LEARNING THE BASICS

15.1 The following is the quadratic regression equation for a sample of $n = 25$:

$$\hat{Y}_i = 5 + 3X_{1i} + 1.5X_{1i}^2$$

a. Predict Y for $X_1 = 2$.
b. Suppose that the computed t_{STAT} test statistic for the quadratic regression coefficient is 2.35. At the 0.05 level of significance, is there evidence that the quadratic model is better than the linear model?
c. Suppose that the computed t_{STAT} test statistic for the quadratic regression coefficient is 1.17. At the 0.05 level of significance, is there evidence that the quadratic model is better than the linear model?
d. Suppose the regression coefficient for the linear effect is -3.0. Predict Y for $X_1 = 2$.

APPLYING THE CONCEPTS

15.2 Businesses actively recruit business students with well-developed higher-order cognitive skills (HOCS) such as problem identification, analytical reasoning, and content integration skills. Researchers conducted a study to see if improvement in students' HOCS was related to the students' GPA.

Source: Data extracted from R. V. Bradley, C. S. Sankar, H. R. Clayton, V. W. Mbarika, and P. K. Raju, "A Study on the Impact of GPA on Perceived Improvement of Higher-Order Cognitive Skills," *Decision Sciences Journal of Innovative Education*, January 2007, 5(1), pp. 151–168.

The researchers conducted a study in which business students were taught using the case study method. Using data collected from 300 business students, the following quadratic regression equation was derived:

$$\text{HOCS} = -3.48 + 4.53(\text{GPA}) - 0.68(\text{GPA})^2$$

where the dependent variable HOCS measured the improvement in higher-order cognitive skills, with 1 being the lowest improvement in HOCS and 5 being the highest improvement in HOCS.

a. Construct a table of predicted HOCS, using GPA equal to 2.0, 2.1, 2.2, ..., 4.0.
b. Plot the values in the table constructed in (a), with GPA on the horizontal axis and predicted HOCS on the vertical axis.
c. Discuss the curvilinear relationship between students' GPA and their predicted improvement in HOCS.
d. The researchers reported that the model had an r^2 of 0.07 and an adjusted r^2 of 0.06. What does this tell you about the scatter of individual HOCS scores around the curvilinear relationship plotted in (b) and discussed in (c)?

15.3 The file **Engines** contains the data for a study that explored if automobile engine torque could be predicted from engine speed (in RPM, revolutions per minute).

Source: Data extracted from Y. Chen et al., "Cluster-Based Profile Analysis in Phase I," *Journal of Quality Technology*, 47, January 2015.

a. Construct a scatter plot for RPM and torque.
b. Fit a quadratic regression model and state the quadratic regression equation.
c. Predict the mean torque for an RPM of 3,000.
d. Perform a residual analysis on the results and determine whether the regression assumptions are valid.

e. At the 0.05 level of significance, is there a significant quadratic relationship between torque and RPM?

f. At the 0.05 level of significance, determine whether the quadratic model is a better fit than the linear model.

g. Interpret the meaning of the coefficient of multiple determination.

h. Compute the adjusted r^2.

i. What conclusions can you reach concerning the relationship between RPM and torque?

15.4 Is the number of calories in a beer related to the number of carbohydrates or the percentage of alcohol in the beer? Data concerning 157 of the best selling domestic beers in the United States are stored in [Domestic Beer]. The values for three variables are included: the number of calories per 12 ounces, the alcohol percentage, and the number of carbohydrates (in grams) per 12 ounces.

Source: Data extracted from **www.beer100.com/beercalories.htm**, April 3, 2019.

a. Perform a multiple linear regression analysis, using calories as the dependent variable and percentage alcohol and number of carbohydrates as the independent variables.

b. Add quadratic terms for alcohol percentage and the number of carbohydrates.

c. Which model is better, the one in (a) or (b)?

d. What conclusions can you reach concerning the relationship between the number of calories in a beer and the alcohol percentage and number of carbohydrates?

15.5 In the production of printed circuit boards, errors in the alignment of electrical connections are a source of scrap. The file [Registration Errors] contains the registration error and the temperature used in the production of circuit boards in an experiment in which higher cost material was used.

Source: Data extracted from C. Nachtsheim and B. Jones, "A Powerful Analytical Tool," *Six Sigma Forum Magazine*, August 2003, pp. 30–33.

a. Construct a scatter plot for temperature and registration error.

b. Fit a quadratic regression model to predict registration error and state the quadratic regression equation.

c. Perform a residual analysis on the results and determine whether the regression model is valid.

d. At the 0.05 level of significance, is there a significant quadratic relationship between temperature and registration error?

e. At the 0.05 level of significance, determine whether the quadratic model is a better fit than the linear model.

f. Interpret the meaning of the coefficient of multiple determination.

g. Compute the adjusted r^2.

h. What conclusions can you reach concerning the relationship between registration error and temperature?

✓**SELF TEST** **15.6** An automotive sales manager wishes to examine the relationship between age (years) and sales price ($) for used Honda automobiles. The file [Honda Prices] contains data

for a sample of Honda Civic LXs that were listed for sale at a car shopping website.

Source: Data extracted from **cargurus.com**.

a. Construct a scatter plot for age and price.

b. Fit a quadratic regression model to predict price and state the quadratic regression equation.

c. Predict the mean price of a Honda Civic LX that is five years old.

d. Perform a residual analysis on the results and determine whether the regression model is valid.

e. At the 0.05 level of significance, is there a significant quadratic relationship between age and price?

f. What is the p-value in (e)? Interpret its meaning.

g. At the 0.05 level of significance, determine whether the quadratic model is a better fit than the linear model.

h. What is the p-value in (g)? Interpret its meaning.

i. Interpret the meaning of the coefficient of multiple determination.

j. Compute the adjusted r^2.

k. What conclusions can you reach concerning the relationship between age and price?

15.7 Researchers wanted to investigate the relationship between employment and accommodation capacity in the European travel and tourism industry. The file [European Tourism] contains a sample of 28 European countries. Variables included are the number of jobs generated in the travel and tourism industry in 2018 and the number of establishments that provide overnight accommodation for tourists.

Source: Data extracted from "Database Eurostat," **bit.ly/2cyv0Vs**.

a. Construct a scatter plot of the number of jobs generated in the travel and tourism industry in 2018 (Y) and the number of establishments that provide overnight accommodation for tourists (X).

b. Fit a quadratic regression model to predict the number of jobs generated and state the quadratic regression equation.

c. Predict the mean number of jobs generated in the travel and tourism industry in 2018 for a country with 3,000 establishments that provide overnight accommodation for tourists.

d. Perform a residual analysis on the results and determine whether the regression model is valid.

e. At the 0.05 level of significance, is there a significant quadratic relationship between the number of jobs generated in the travel and tourism industry in 2018 and the number of establishments that provide overnight accommodation for tourists?

f. What is the p-value in (e)? Interpret its meaning.

g. At the 0.05 level of significance, determine whether the quadratic model is a better fit than the linear model.

h. Interpret the meaning of the coefficient of multiple determination.

i. Compute the adjusted r^2.

j. What conclusions can you reach concerning the relationship between the number of jobs generated in the travel and tourism industry in 2018 and the number of establishments that provide overnight accommodation for tourists?

15.2 Using Transformations in Regression Models

learnMORE

To learn more about logarithms, see Appendix Section A.3.

Transformations are mathematical alterations of data values made to either overcome violations of the assumptions of regression or to make a model whose form is not linear into a linear model. Transformations can be applied to the values of an independent X variable or the dependent Y variable or both. Among the many transformations available (see Kutner, Montgomery, and Peck), the square-root transformation and transformations involving the common logarithm (base 10) and the natural logarithm (base e) are the most commonly used.

The Square-Root Transformation

The **square-root transformation** often overcomes violations of the normality and equal-variance assumptions as well as transforms a model whose form is not linear into a linear model. When the error term is normally distributed and the errors are equal for all values of X, one uses a square-root transformation of X to make the linear model appropriate. When the errors are not equal for all values of X or when the errors are not normally distributed, one uses a square-root transformation of Y (see Kutner).

Equation (15.3) shows a regression model that uses a square-root transformation of the dependent variable.

REGRESSION MODEL WITH A SQUARE-ROOT TRANSFORMATION

$$\sqrt{Y_i} = \beta_0 + \beta_1 X_{1i} + \varepsilon_i \tag{15.3}$$

Example 15.2 illustrates the use of a square-root transformation of the Y variable.

EXAMPLE 15.2

Using the Square-Root Transformation of the Dependent Y Variable

Sales managers for Sure Value Convenience small-format stores have been mandated by senior management to increase revenues. One way to increase revenues is to conduct an experiment that will price the popular SVC CoffeePak of 10 coffee pods at various prices.

As a pilot experiment, the sales managers select 10 stores with similar CoffeePak sales. They select five different prices for the item, $3.99, $4.49, $4.99, $5.49, and $5.99, assigning unique groups of two stores to each price. The experiment runs for one month, and the sales data collected are stored in CoffeePak Study . The managers perform a regression analysis to estimate sales based on price and construct the scatter plot and residual plot that Figure 15.9 presents.

FIGURE 15.9

Scatter plot and residual plot for the CoffeePak study price and unit sales

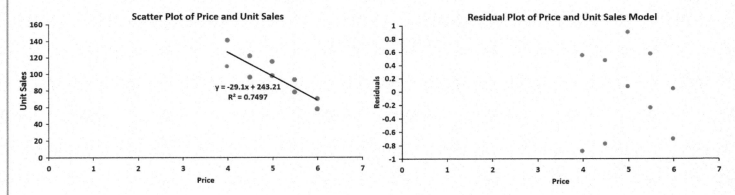

SOLUTION The regression model produced shows a linear relationship between price and sales of the CoffeePaks. For this model, $b_0 = 243.21$, $b_1 = -29.10$, and $r^2 = 0.7497$.

However, the residual plot shows much more variation in the residuals when the CoffeePaks are sold for $3.99 or $4.49. To overcome this lack of homogeneity of variance in the residuals, the sales managers transform the dependent Y variable unit sales by taking its square root. Figure 15.10 displays the scatter plot of price and the square root of unit sales and the residual plot for the simple linear regression model to estimate *the square root* of units sold.

▶(*continued*)

FIGURE 15.10
Scatter plot and residual plot for CoffeePak study price and the square root of unit sales

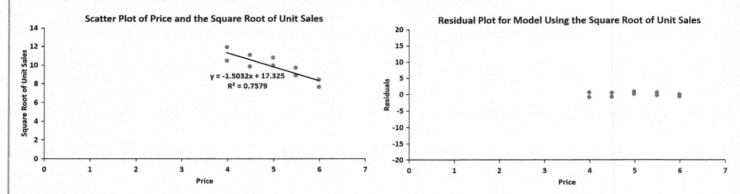

For this second model, $b_0 = 17.3251$ and $b_1 = -1.5032$, and $r^2 = 0.7579$. The managers note that Figure 15.10 residual plot shows much less variation in the residuals for different values of price, even though the r^2 values of the two models are not very different. Because the residuals differ much less than the linear model to predict units sold, the sales managers should use the model that predicts the square root of units sold as the basis for their decision making.

The Log Transformation

The **logarithmic transformation** often overcomes violations of the normality and equal-variance assumptions. One can also use the logarithmic transformation to change a nonlinear model into a linear model. Equation (15.4) shows a multiplicative model.

ORIGINAL MULTIPLICATIVE MODEL
$$Y_i = \beta_0 X_{1i}^{\beta_1} X_{2i}^{\beta_2} \varepsilon_i \tag{15.4}$$

By taking base 10 logarithms of both the dependent and independent variables, one transforms Equation (15.4) in to the model that Equation (15.5) defines.

TRANSFORMED MULTIPLICATIVE MODEL
$$\log Y_i = \log(\beta_0 X_{1i}^{\beta_1} X_{2i}^{\beta_2} \varepsilon_i)$$
$$= \log \beta_0 + \log(X_{1i}^{\beta_1}) + \log(X_{2i}^{\beta_2}) + \log \varepsilon_i$$
$$= \log \beta_0 + \beta_1 \log X_{1i} + \beta_2 \log X_{2i} + \log \varepsilon_i \tag{15.5}$$

The transformed model that Equation (15.5) defines is a *linear* model. Equations (15.6) and (15.7) illustrate that a similar transformation can be done for an exponential model using the natural logarithm of both sides of the equation. (The transformed exponential model that Equation 15.7 defines is a linear model.)

ORIGINAL EXPONENTIAL MODEL
$$Y_i = e^{\beta_0 + \beta_1 X_{1i} + \beta_2 X_{2i}} \varepsilon_i \tag{15.6}$$

(continued)

TRANSFORMED EXPONENTIAL MODEL

$$\ln Y_i = \ln(e^{\beta_0 + \beta_1 X_{1i} + \beta_2 X_{2i}} \varepsilon_i)$$
$$= \ln(e^{\beta_0 + \beta_1 X_{1i} + \beta_2 X_{2i}}) + \ln \varepsilon_i$$
$$= \beta_0 + \beta_1 X_{1i} + \beta_2 X_{2i} + \ln \varepsilon_i \qquad (15.7)$$

Example 15.3 illustrates the use of a natural log transformation.

EXAMPLE 15.3

Using the Natural Log Transformation

Social media comments about the popular Sure Value Convenience mini store in Oxford Glen complain about the long wait time to checkout with the store's single cashier. The Oxford Glen store manager wants to analyze this problem before the reputation of the store is affected by these comments. For each of 12 randomly selected customers, the manager records the waiting time (in seconds) of the customer and the number of people in line ahead of the customer and stores these data in Store Waiting Times . The manager performs a regression analysis to estimate waiting time based on persons in line ahead and constructs the Figure 15.11 plots.

FIGURE 15.11
Scatter plot and residual plot for the regression model for the store waiting time study

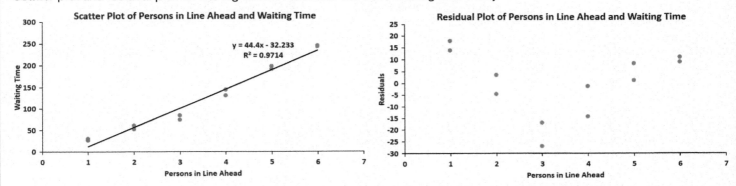

SOLUTION The regression model produced shows a linear relationship between the number of people in line ahead and waiting time. For this model, $b_0 = -32.233$, $b_1 = 44.4$, and $r^2 = 0.9714$. However, the residual plot shows a curvilinear relationship of the residuals with the number of people waiting. To eliminate this relationship in the residuals, the store manager transforms the dependent Y variable, waiting time, by taking its natural logarithm. Figure 15.12 displays the scatter plot of number of people in line ahead and the natural logarithm of waiting time and the residual plot for the simple linear regression model to estimate *the natural logarithm of* waiting time.

For this second model, b_0 value $= 2.9482$, $b_1 = 0.4781$, and $r^2 = 0.979$, similar to the r^2 value for the original model. Figure 15.12 residual plot does not show a pattern for different values of people in line ahead. Because of that fact, the store manager should use the second model for decision making.

FIGURE 15.12
Scatter plot and residual plot for the regression model of number of people in line ahead and the natural logarithm of waiting time

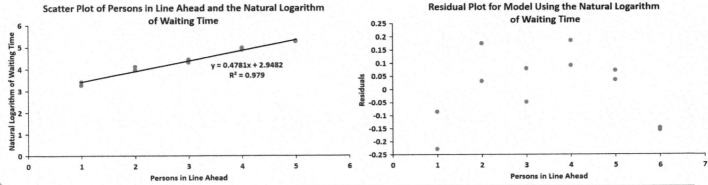

PROBLEMS FOR SECTION 15.2

LEARNING THE BASICS

15.8 Consider the following regression equation:

$$\log \hat{Y}_i = \log 3.07 + 0.9 \log X_{1i} + 1.41 \log X_{2i}$$

a. Predict the value of Y when $X_1 = 8.5$ and $X_2 = 5.2$.
b. Interpret the meaning of the regression coefficients b_0, b_1, and b_2.

15.9 Consider the following regression equation:

$$\ln \hat{Y}_i = 4.62 + 0.5X_{1i} + 0.7X_{2i}$$

a. Predict the value of Y when $X_1 = 8.5$ and $X_2 = 5.2$.
b. Interpret the meaning of the regression coefficients b_0, b_1, and b_2.

APPLYING THE CONCEPTS

✓ SELF TEST **15.10** Using the data of Problem 15.4 on page 534, stored in Domestic Beer , perform either a square-root transformation on the dependent variable (calories) or a square-root transformation on each of the independent variables (percentage alcohol and number of carbohydrates) depending on whether the residuals are normally distributed or vary across X values.
a. State the regression equation.
b. At the 0.05 level of significance, is there a significant relationship between calories and the percentage of alcohol and the number of carbohydrates?
c. Interpret the meaning of the coefficient of determination, r^2, in this problem.
d. Compute the adjusted r^2.
e. Compare your results with those in Problem 15.4. Which model is better? Why?

15.11 Using the data of Problem 15.4 on page 534, stored in Domestic Beer , perform a natural logarithmic transformation of the dependent variable (calories). Using the transformed dependent variable and the percentage of alcohol and the number of carbohydrates as the independent variables, perform a multiple regression analysis.
a. State the regression equation.
b. Perform a residual analysis of the results and determine whether regression assumptions are valid.

c. At the 0.05 level of significance, is there a significant relationship between the natural logarithm of calories and the percentage of alcohol and the number of carbohydrates?
d. Interpret the meaning of the coefficient of determination, r^2, in this problem.
e. Compute the adjusted r^2.
f. Compare your results with those in Problems 15.4 and 15.10. Which model is best? Why?

15.12 Using the data of Problem 15.6 on page 534, stored in Honda Prices , perform a natural logarithm transformation of the dependent variable (price). Using the transformed dependent variable and the age as the independent variable, perform a regression analysis.
a. State the regression equation.
b. Predict the mean price for a five-year-old Honda Civic LX.
c. Perform a residual analysis of the results and determine whether the regression assumptions are valid.
d. At the 0.05 level of significance, is there a significant relationship between the natural logarithm of price and age?
e. Interpret the meaning of the coefficient of determination, r^2, in this problem.
f. Compute the adjusted r^2.
g. Compare your results with those in Problem 15.6. Which model is better? Why?

15.13 Using the data of Problem 15.6 on page 534 stored in Honda Prices , perform a square-root transformation of the dependent variable (price). Using the square root of price as the dependent variable, perform a regression analysis.
a. State the regression equation.
b. Predict the mean price for a five-year-old Honda Civic LX.
c. Perform a residual analysis of the results and determine whether the regression model is valid.
d. At the 0.05 level of significance, is there a significant relationship between the square root of price and age?
e. Interpret the meaning of the coefficient of determination, r^2, in this problem.
f. Compute the adjusted r^2.
g. Compare your results with those of Problems 15.6 and 15.12. Which model is best? Why?

15.3 Collinearity

Collinearity of the independent variables exists when two or more of the independent variables are highly correlated with each other. When this occurs, collinear variables do not provide unique information, and it becomes difficult to separate the effects of such variables on the dependent variable. Collinearity may cause the values of the regression coefficients for the correlated variables to fluctuate drastically, depending on which independent variables are included in the model.

One method of measuring collinearity is to determine the **variance inflationary factor (VIF)** for each independent variable. Equation (15.8) defines VIF_j, the variance inflationary factor for variable j. The R_j^2 is the coefficient of multiple determination for a regression model, using variable X_j as the dependent variable and all other X variables as independent variables.

VARIANCE INFLATIONARY FACTOR

$$VIF_j = \frac{1}{1 - R_j^2} \qquad (15.8)$$

If there are only two independent variables, R_1^2 is the coefficient of determination between X_1 and X_2. It is identical to R_2^2, which is the coefficient of determination between X_2 and X_1. If there are three independent variables, then R_1^2 is the coefficient of multiple determination of X_1 with X_2 and X_3; R_2^2 is the coefficient of multiple determination of X_2 with X_1 and X_3; and R_3^2 is the coefficient of multiple determination of X_3 with X_1 and X_2.

If a set of independent variables is uncorrelated, each VIF_j is equal to 1. If the set is highly correlated, then a VIF_j might even exceed 10. Snee recommends using alternatives to least-squares regression if the maximum VIF_j exceeds 5.

Multiple regression models that have one or more large VIF values should be used with extreme caution. And because the independent variables contain overlapping information, you should always avoid interpreting the regression coefficient estimates separately because you cannot accurately estimate the individual effects of the independent variables. One approach in this situation is to delete the variable with the largest VIF value. The reduced model (the model with the independent variable with the largest VIF value deleted) is often free of collinearity problems. Keep eliminating and rerunning the regression analysis until no variables have a $VIF > 5$. If one determines that all the independent variables are needed in the model, one can use methods discussed in the Kutner reference.

In the OmniPower sales data (see Section 14.1), the correlation between the two independent variables, price and promotional expenditure, is -0.0968. Because this model has only two independent variables, from Equation (15.8):

$$VIF_1 = VIF_2 = \frac{1}{1 - (-0.0968)^2}$$

$$= 1.009$$

Thus, one can conclude that collinearity is not a concern when using the OmniPower sales data.

In models containing quadratic and interaction terms, collinearity is usually present. The linear and quadratic terms of an independent variable are usually highly correlated with each other, and an interaction term is often correlated with one or both of the independent variables making up the interaction. Thus, one cannot interpret individual regression coefficients separately. One needs to interpret the linear and quadratic regression coefficients together in order to understand the nonlinear relationship. Likewise, one needs to interpret an interaction regression coefficient in conjunction with the two regression coefficients associated with the variables comprising the interaction.

Finally, discovering large VIFs in quadratic or interaction models do not automatically disqualify those models for use. Large VIFs do mean that one must be careful when interpreting the regression coefficients.

PROBLEMS FOR SECTION 15.3

LEARNING THE BASICS

15.14 If the coefficient of determination between two independent variables is 0.20, what is the VIF?

15.15 If the coefficient of determination between two independent variables is 0.50, what is the VIF?

APPLYING THE CONCEPTS

✓ **SELF TEST** **15.16** Refer to Problem 14.4 on page 482. Perform a multiple regression analysis using the data in **Community Banks** and determine the VIF for each independent variable in the model. Is there reason to suspect the existence of collinearity?

15.17 Refer to Problem 14.5 on page 483. Perform a multiple regression analysis using the data in `Vinho Verde` and determine the *VIF* for each independent variable in the model. Is there reason to suspect the existence of collinearity?

15.18 Refer to Problem 14.6 on page 483. Perform a multiple regression analysis using the data in `Best Companies` and determine the *VIF* for each independent variable in the model. Is there reason to suspect the existence of collinearity?

15.19 Refer to Problem 14.7 on page 483. Perform a multiple regression analysis using the data in `Nickels 26Weeks` and determine the *VIF* for each independent variable in the model. Is there reason to suspect the existence of collinearity?

15.20 Refer to Problem 14.8 on page 483. Perform a multiple regression analysis using the data in `Glen Cove` and determine the *VIF* for each independent variable in the model. Is there reason to suspect the existence of collinearity?

15.4 Model Building

The techniques that Chapter 14 and this chapter discuss can be combined into a series of steps to identify the most appropriate regression model for a set of data.

Exhibit 15.1 lists these steps for successful model building.

EXHIBIT 15.1

Successful Model Building

1. Use the DCOVA framework to identify the business problem or goal to be examined, define variables, and collect data. Identify the variables that will serve as candidate independent X variables for the multiple regression model.
2. Develop a regression model that includes all candidate independent X variables.
3. Compute the *VIF* for each of the X variables. Apply the decision-making process that Section 15.3 discusses until no X variable has a $VIF > 5$.
4. Perform a best subsets analysis with the remaining independent variables and compute the C_p statistic or the adjusted r^2 for each subset regression model as this section discusses later.
5. Choose a best model from the models that have C_p close to or less than $k + 1$ and/or a high adjusted r^2.
6. Perform a complete analysis of that best model chosen, including a residual analysis.
7. Review the results of the residual analysis. If necessary, add quadratic or interaction terms or transform variables. Repeat steps 3 through 6.
8. Use the selected model for inference and prediction.

As this section later explains, performing a stepwise regression can be an alternate to steps 4 and 5, although at the cost of not examining all possible regression models.

These steps ensure that an appropriate model will be selected. That model may not be the optimal model, but the model will be one that decision-makers can use for prediction and inference. The **principle of parsimony** should cause you to choose the model with the fewest independent X variables that can predict the dependent Y variable adequately, should several different models in step 5 have a C_p statistic close to or less than $k + 1$ and/or a high adjusted r^2. Regression models with fewer independent variables are easier to interpret, particularly because they are less likely to be affected by the collinearity problems that Section 15.3 discusses.

To illustrate the model-building process, return to the Nickels Broadcasting scenario in which you were asked to build a regression model. Table 15.4 provides a 26-week sample that includes the weekly staff count (Staff), remote engineering hours (RemoteEng), graphics hours (Graphics), and editorial production hours (Production) as independent X variables to predict the dependent Y variable standby hours (Standby).

To begin analyzing the Table 15.4 26-week sample, stored in `Nickels 26Weeks`, calculate the variance inflationary factors (see Section 15.3) to measure the amount of collinearity among the independent variables. The four *VIF*s for the four independent variables appear in Figure 15.13 along with the results for the regression model that uses those variables. Observe that all the *VIF* values are relatively small, ranging from a high of 1.9993 for Production to a low of 1.2333 for RemoteEng. Using criteria developed by Snee that all *VIF* values should be less than 5.0, there is little evidence of collinearity among the set of independent variables.

TABLE 15.4

Predicting standby hours based on staff, remote engineering hours, graphics hours, and production hours

Week	Standby (Y)	Staff (X_1)	RemoteEng (X_2)	Graphics (X_3)	Production (X_4)
1	245	338	414	323	2001
2	177	333	598	340	2030
3	271	358	656	340	2226
4	211	372	631	352	2154
5	196	339	528	380	2078
6	135	289	409	339	2080
7	195	334	382	331	2073
8	118	293	399	311	1758
9	116	325	343	328	1624
10	147	311	338	353	1889
11	154	304	353	518	1988
12	146	312	289	440	2049
13	115	283	388	276	1796
14	161	307	402	207	1720
15	274	322	151	287	2056
16	245	335	228	290	1890
17	201	350	271	355	2187
18	183	339	440	300	2032
19	237	327	475	284	1856
20	175	328	347	337	2068
21	152	319	449	279	1813
22	188	325	336	244	1808
23	188	322	267	253	1834
24	197	317	235	272	1973
25	261	315	164	223	1839
26	232	331	270	272	1935

FIGURE 15.13

Excel multiple regression linear model worksheet results for predicting standby hours based on four independent variables (worksheets for Durbin-Watson statistic and *VIF*, inset)

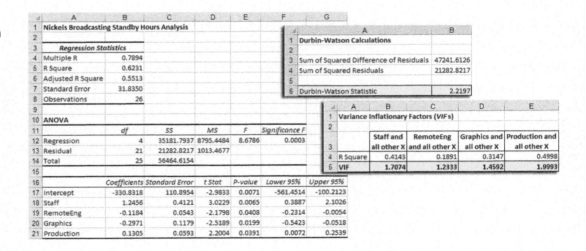

The Stepwise Regression Approach to Model Building

Whether a subset of all independent variables yields an adequate and appropriate model is the next step in model building. **Stepwise regression** is a model selection process that attempts to find the "best" regression model without examining all possible models.

The first step of stepwise regression is to find the best model that uses one independent variable. The next step is to find the best of the remaining independent variables to add to the model selected in the first step. An important feature of the stepwise approach is that an independent variable that has entered into the model at an early stage may subsequently be removed after other independent variables are considered. Therefore, in stepwise regression, variables are either added to or deleted from the regression model at each step of the model-building process. The *t* test for the slope (see Section 14.4) or the partial F_{STAT} test statistic (see Section 14.5) determines whether variables are added or deleted. The stepwise procedure terminates when no additional variables can be added to or deleted from the last model evaluated.

Figure 15.14 presents the stepwise regression results for the Nickels Broadcasting data.

FIGURE 15.14
Excel (PHStat) stepwise regression worksheet results for the Nickels Broadcasting data

	A	B	C	D	E	F	G	H
1	Stepwise Analysis for Nickels Broadcasting Standby Hours Analysis							
2	Table of Results for General Stepwise							
3								
4	Staff entered.							
5								
6			df	SS	MS	F	Significance F	
7		Regression	1	20667.3980	20667.3980	13.8563	0.0011	
8		Residual	24	35797.2174	1491.5507			
9		Total	25	56464.6154				
10								
11			Coefficients	Standard Error	t Stat	P-value	Lower 95%	Upper 95%
12		Intercept	-272.3816	124.2402	-2.1924	0.0383	-528.8008	-15.9625
13		Staff	1.4241	0.3826	3.7224	0.0011	0.6345	2.2136
14								
15								
16	RemoteEng entered.							
17								
18			df	SS	MS	F	Significance F	
19		Regression	2	27662.5429	13831.2714	11.0450	0.0004	
20		Residual	23	28802.0725	1252.2640			
21		Total	25	56464.6154				
22								
23			Coefficients	Standard Error	t Stat	P-value	Lower 95%	Upper 95%
24		Intercept	-330.6748	116.4802	-2.8389	0.0093	-571.6322	-89.7175
25		Staff	1.7649	0.3790	4.6562	0.0001	0.9808	2.5490
26		RemoteEng	-0.1390	0.0588	-2.3635	0.0269	-0.2606	-0.0173
27								
28								
29	No other variables could be entered into the model. Stepwise ends.							

In the Worksheet

This worksheet has no cell formulas, unlike most of the worksheets that this book features. This worksheet was created by the PHStat Stepwise Regression methods and contains pasted results transferred from other regression results worksheets that PHStat created and examined.

For this example, a significance level of 0.05 is used to enter a variable into the model or to delete a variable from the model. The first variable entered into the model is Staff, the variable that correlates most highly with the dependent variable Standby. Because the p-value $= 0.0011 < 0.05$, Staff is included in the regression model.

The next step involves selecting a second independent variable for the model. The second variable chosen is one that makes the largest contribution to the model, given that the first variable has been selected. For this model, the second variable is RemoteEng. Because the p-value for RemoteEng $= 0.0269 < 0.05$, RemoteEng is included in the regression model.

After RemoteEng has been entered into the model, the stepwise procedure determines whether Staff is still an important contributing variable or whether that variable can be eliminated from the model. Because the p-value of 0.0001 for Staff is less than 0.05, Staff remains in the regression model.

The next step involves selecting a third independent variable for the model. Because none of the other variables meets the 0.05 criterion for entry into the model, the stepwise procedure terminates with a model that includes the weekly staff count and the remote engineering hours.

This stepwise regression approach to model building was originally developed more than five decades ago, when regression computations were time-consuming and costly. Although stepwise regression limited the evaluation of alternative models, the method was deemed a good trade-off between evaluation and cost.

Given the ability of today's computers to perform regression computations at very low cost and high speed, stepwise regression has been superseded to some extent by the best subsets approach, which evaluates a larger set of alternative models. Stepwise regression is not obsolete, however. Today, stepwise regression can play an important role in helping to analyze big data when used with certain predictive analytics methods.

The Best Subsets Approach to Model Building

The **best subsets approach** evaluates all possible regression models for a given set of independent variables. Figure 15.15 presents best subsets regression results of all possible regression models for the Nickels Broadcasting data.

FIGURE 15.15

Excel best subsets regression worksheet results for the Nickels Broadcasting data

▲	A	B	C	D	E	F
1	Best Subsets Analysis for Standby Hours Analysis					
2						
3	Intermediate Calculations					
4	R2T	0.6231				
5	1 - R2T	0.3769				
6	n	26				
7	T	5				
8	n - T	21				
9						
10	**Model**	**Cp**	**k+1**	**R Square**	**Adj. R Square**	**Std. Error**
11	X1	13.3215	2	0.3660	0.3396	38.6206
12	X1X2	8.4193	3	0.4899	0.4456	35.3873
13	X1X2X3	7.8418	4	0.5362	0.4729	34.5029
14	X1X2X3X4	5.0000	5	0.6231	0.5513	31.8350
15	X1X2X4	9.3449	4	0.5092	0.4423	35.4921
16	X1X3	10.6486	3	0.4499	0.4021	36.7490
17	X1X3X4	7.7517	4	0.5378	0.4748	34.4426
18	X1X4	14.7982	3	0.3754	0.3211	39.1579
19	X2	33.2078	2	0.0091	-0.0322	48.2836
20	X2X3	32.3067	3	0.0612	-0.0205	48.0087
21	X2X3X4	12.1381	4	0.4591	0.3853	37.2608
22	X2X4	23.2481	3	0.2238	0.1563	43.6540
23	X3	30.3884	2	0.0597	0.0205	47.0345
24	X3X4	11.8231	3	0.4288	0.3791	37.4466
25	X4	24.1846	2	0.1710	0.1365	44.1619

In the Worksheet

This worksheet contains formulas that copy values from other regression results worksheets. Using the PHStat Best Subset procedure simplifies the preparation of this worksheet.

A criterion often used in model building is the adjusted r^2, which adjusts the r^2 of each model to account for the number of independent variables in the model as well as for the sample size (see Section 14.2). Because model building requires you to compare models with different numbers of independent variables, the adjusted r^2 is more appropriate than r^2. In Figure 15.15, the adjusted r^2 reaches a maximum value of 0.5513 when all four independent variables plus the intercept term (for a total of five estimated parameters) are included in the model.

A second criterion often used in the evaluation of competing models is the C_p **statistic** developed by Mallows (see Kutner). The C_p statistic, defined in Equation (15.9), measures the differences between a fitted regression model and a *true* model, along with random error.

C_P STATISTIC

$$C_p = \frac{(1 - R_k^2)(n - T)}{1 - R_T^2} - [n - 2(k + 1)] \tag{15.9}$$

where

k = number of independent variables included in a regression model

T = total number of parameters (including the intercept) to be estimated in the full regression model

R_k^2 = coefficient of multiple determination for a regression model that has k independent variables

R_T^2 = coefficient of multiple determination for a full regression model that contains all T estimated parameters

Using Equation (15.9) to compute C_p for the model containing Staff and RemoteEng,

$$n = 26 \quad k = 2 \quad T = 4 + 1 = 5 \quad R_k^2 = 0.4899 \quad R_T^2 = 0.6231$$

so that

$$C_p = \frac{(1 - 0.4899)(26 - 5)}{1 - 0.6231} - \left[26 - 2(2 + 1)\right]$$

$$= 8.42$$

When a regression model with k independent variables contains only random differences from a *true* model, the mean value of C_p is $k + 1$, the number of parameters. Thus, in evaluating many alternative regression models, the goal is to find models whose C_p is close to or less than $k + 1$. Figure 15.15 shows that only the model that includes all four independent variables has a C_p value close to or below $k + 1$. Therefore, using the C_p criterion, one would choose that model.

With many data sets, the C_p statistic often provides several alternative models for you to evaluate in greater depth. Moreover, the best model or models using the C_p criterion might differ from the model selected using the adjusted r^2 and/or the model selected using the stepwise procedure. For example, the Nickels Broadcasting model that stepwise regression selects has a C_p value of 8.4193, which is substantially above the suggested criterion of $k + 1 = 3$ for that model. Remember that there may be several equally appropriate models and no one uniquely best model. Final model selection often involves using subjective criteria, such as parsimony, interpretability, and departure from model assumptions, as evaluated by residual analysis.

When one has finished selecting the independent variables to include in the model, one needs to perform a residual analysis to evaluate the regression assumptions. For data collected in time order, one also needs to calculate the Durbin-Watson statistic to determine whether there is autocorrelation in the residuals (see Section 13.6). For the Nickels Broadcasting data (collected in time order), the Durbin-Watson statistic, D, is 2.2197 (see Figure 15.13 on page 541). Because D is greater than 2.0, there is no indication of positive correlation in the residuals.

The Figure 15.16 residual plots for the Nickels Broadcasting data reveal no apparent patterns. The Figure 15.17 plot of the residuals versus the predicted values of Y does not show evidence of unequal variance. Therefore, using the Figure 15.13 regression model results, one states the regression equation as

$$\hat{Y}_i = -330.8318 + 1.2456X_{1i} - 0.1184X_{2i} - 0.2971X_{3i} + 0.1305X_{4i}$$

FIGURE 15.16

Residual plots for the Nickels Broadcasting data

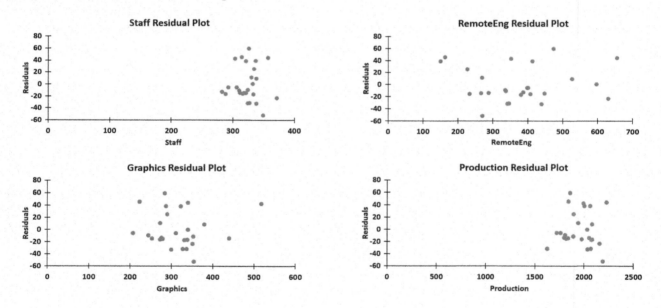

FIGURE 15.17

Scatter plot of the residuals versus the predicted values of Y

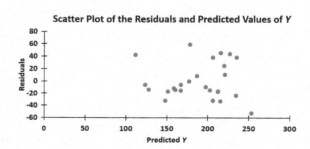

Example 15.4 presents a situation in which the C_p statistic is close to or less than $k + 1$ in more than one alternative model.

EXAMPLE 15.4

Choosing Among Alternative Regression Models

Table 15.5 shows results from a best subsets regression analysis of a regression model with seven independent variables. Determine which regression model you would choose as the *best* model.

TABLE 15.5
Partial results from best-subsets regression

Number of Variables	r^2	Adjusted r^2	C_p	Variables Included
1	0.121	0.119	113.9	X_4
1	0.093	0.090	130.4	X_1
1	0.083	0.080	136.2	X_3
2	0.214	0.210	62.1	X_3, X_4
2	0.191	0.186	75.6	X_1, X_3
2	0.181	0.177	81.0	X_1, X_4
3	0.285	0.280	22.6	X_1, X_3, X_4
3	0.268	0.263	32.4	X_3, X_4, X_5
3	0.240	0.234	49.0	X_2, X_3, X_4
4	0.308	0.301	11.3	X_1, X_2, X_3, X_4
4	0.304	0.297	14.0	X_1, X_3, X_4, X_6
4	0.296	0.289	18.3	X_1, X_3, X_4, X_5
5	0.317	0.308	8.2	X_1, X_2, X_3, X_4, X_5
5	0.315	0.306	9.6	X_1, X_2, X_3, X_4, X_6
5	0.313	0.304	10.7	X_1, X_3, X_4, X_5, X_6
6	0.323	0.313	6.8	$X_1, X_2, X_3, X_4, X_5, X_6$
6	0.319	0.309	9.0	$X_1, X_2, X_3, X_4, X_5, X_7$
6	0.317	0.306	10.4	$X_1, X_2, X_3, X_4, X_6, X_7$
7	0.324	0.312	8.0	$X_1, X_2, X_3, X_4, X_5, X_6, X_7$

SOLUTION From Table 15.5, you need to determine which models have C_p values that are less than or close to $k + 1$. Two models meet this criterion. The model with six independent variables ($X_1, X_2, X_3, X_4, X_5, X_6$) has a C_p value of 6.8, which is less than $k + 1 = 6 + 1 = 7$, and the full model with seven independent variables ($X_1, X_2, X_3, X_4, X_5, X_6, X_7$) has a C_p value of 8.0.

One way you can choose between the two models is to select the model with the largest adjusted r^2, which is the model with six independent variables. Another way to select a final model is to determine whether the models contain a subset of variables that are common. Then you test whether the contribution of the additional variables is significant. In this case, because the models differ only by the inclusion of variable X_7 in the full model, you test whether variable X_7 makes a significant contribution to the regression model, given that the variables $X_1, X_2, X_3, X_4, X_5,$ and X_6 are already included in the model. If the contribution is statistically significant, then you should include variable X_7 in the regression model. If variable X_7 does not make a statistically significant contribution, you should not include it in the model.

Figure 15.18 summarizes the model-building process.

FIGURE 15.18
The model-building process

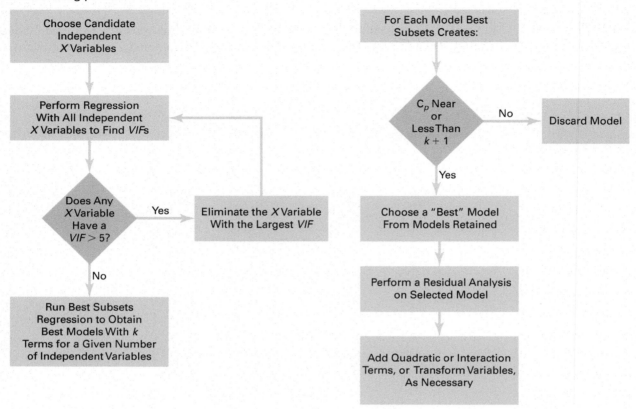

PROBLEMS FOR SECTION 15.4

LEARNING THE BASICS

15.21 You are considering four independent variables for inclusion in a regression model. You select a sample of $n = 30$, with the following results:

1. The model that includes independent variables A and B has a C_p value equal to 4.6.
2. The model that includes independent variables A and C has a C_p value equal to 2.4.
3. The model that includes independent variables A, B, and C has a C_p value equal to 2.7.
   a. Which models meet the criterion for further consideration? Explain.
   b. How would you compare the model that contains independent variables A, B, and C to the model that contains independent variables A and B? Explain.

15.22 You are considering six independent variables for inclusion in a regression model. You select a sample of $n = 40$, with the following results:

$$k = 2 \quad T = 6 + 1 = 7 \quad R_k^2 = 0.274 \quad R_T^2 = 0.653$$

a. Compute the C_p value for this two-independent-variable model.
b. Based on your answer to (a), does this model meet the criterion for further consideration as the best model? Explain.

APPLYING THE CONCEPTS

15.23 The file **FTMBA** contains data from a sample of full-time MBA programs offered by private universities. The variables collected for this sample are average starting salary upon graduation ($), the percentage of applicants to the full-time program who were accepted, the average GMAT test score of students entering the program, program per-year tuition ($), and percent of students with job offers at time of graduation.

Source: Data extracted from U.S. News & World Report Education, "Best Graduate Schools," **bit.ly/1E8MBcp**.

Develop the most appropriate multiple regression model to predict the mean starting salary upon graduation. Be sure to include a thorough residual analysis. In addition, provide a detailed explanation of the results, including a comparison of the most appropriate multiple regression model to the best simple linear regression model.

15.24 You need to develop a model to predict the asking price of houses listed for sale in Silver Spring, Maryland, based on the living space of the house, the lot size, and the age, whether it has a fireplace, the number of bedrooms, and the number of bathrooms. A sample of 61 houses is selected and the results are stored in **Silver Spring Homes**. Develop the most appropriate multiple regression model to predict asking price. Be sure to perform a thorough residual analysis. In addition, provide a detailed explanation of the results.

15.25 *Accounting Today* identified top public accounting firms in 10 geographic regions across the United States. The file **Regional Accounting Partners** contains data for public accounting firms in the Southeast, Gulf Coast, and Capital Regions. The variables are revenue ($millions), number of partners in the firm, number of offices in the firm, proportion of business dedicated to management advisory services (MAS%), whether the firm is located in the Southeast Region (0 = no, 1 = yes), and whether the firm is located in the Gulf Coast Region (0 = no, 1 = yes).

Source: Data extracted from "The 2019 Top 100 Firms and Regional Leaders," **bit.ly/2OjXtDo**.

Develop the most appropriate multiple regression model to predict firm revenue. Be sure to perform a thorough residual analysis. In addition, provide a detailed explanation of the results.

15.5 Pitfalls in Multiple Regression and Ethical Issues

Pitfalls in Multiple Regression

Model building is an art as well as a science. Different individuals may not always agree on the best multiple regression model. To develop a good regression model, use the process that Exhibit 15.1 and Figure 15.18 on pages 540 and 546 summarize. As you follow that process, you must avoid certain pitfalls that can interfere with the development of a useful model. Section 13.9 discussed pitfalls in simple linear regression and strategies for avoiding them. Multiple regression models require the following additional precautions to avoid common pitfalls:

- Interpret the regression coefficient for a particular independent variable from a perspective in which the values of all other independent variables are held constant.
- Evaluate residual plots for each independent variable.
- Evaluate interaction and quadratic terms.
- Compute the *VIF* for each independent variable before determining which independent variables to include in the model.
- Examine several alternative models, using best subsets regression.
- Use logistic regression instead of least squares regression when the dependent variable is categorical.
- Use cross-validation to improve accuracy of models for predicting future data.

Ethical Issues

Ethical issues arise when a user who wants to make predictions manipulates the development process of the multiple regression model. The key here is intent. In addition to the situations that Section 13.9 discusses, unethical behavior occurs when someone uses multiple regression analysis and *willfully fails* to remove from consideration independent variables that exhibit a high collinearity with other independent variables or *willfully fails* to use methods other than least-squares regression when the assumptions necessary for least-squares regression are seriously violated.

▼ USING **STATISTICS**
Valuing Parsimony..., Revisited

In the Using Statistics scenario, you were hired by Nickels Broadcasting to determine which variables have an effect on WSTA-TV standby hours. You were given a 26-week sample that contained the weekly standby hours, staff count, remote engineering hours, graphics hours, and production hours.

You performed a multiple regression analysis on the data. The coefficient of multiple determination indicated that 62.31% of the variation in standby hours can be explained by variation in the weekly staff count and the number of remote engineering, graphics, and editorial production hours. The model indicated that standby hours are estimated to increase by 1.2456 hours for each additional weekly staff member present, holding constant the other independent variables; to decrease by 0.1184 hour for each additional remote engineering hour, holding constant the other independent variables; to decrease by 0.2971 hour for each additional graphics hour, holding constant the other independent variables; and to increase by 0.1305 hour for each additional editorial production hour, holding constant the other independent variables.

Each of the four independent variables had a significant effect on Standby, holding constant the other independent variables. This regression model enables you to predict standby hours based on the weekly staff count, remote engineering hours, graphics hours, and editorial production hours. Any predictions developed by the model can then be carefully monitored, new data can be collected, and other variables may possibly be considered.

▼ SUMMARY

Figure 15.19 summarizes the several topics associated with multiple regression model building.

FIGURE 15.19
Roadmap for multiple regression

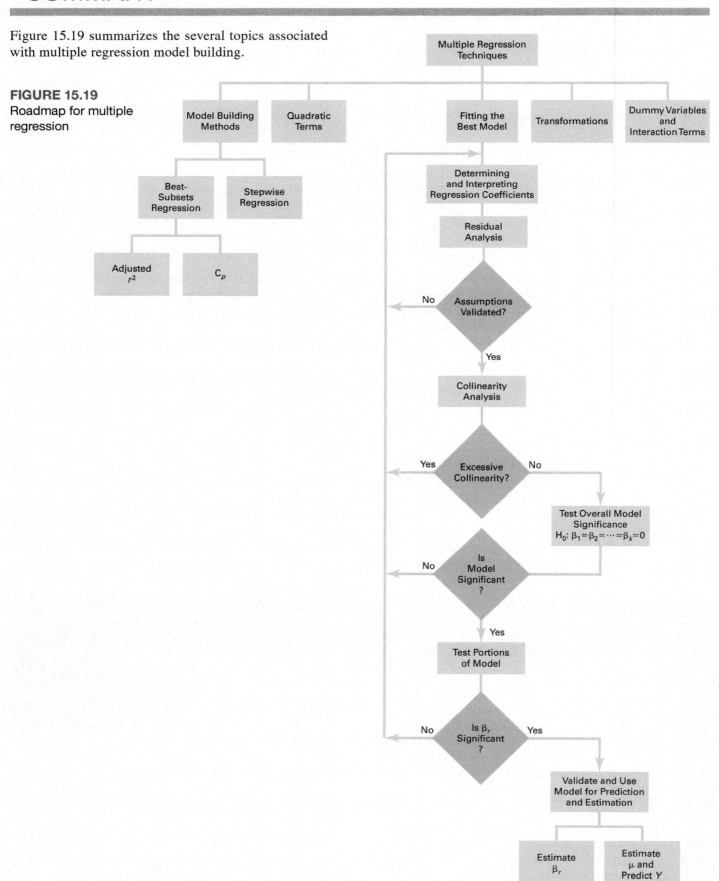

▼ REFERENCES

Kutner, M., C. Nachtsheim, J. Neter, and W. Li. *Applied Linear Statistical Models*, 5th ed. New York: McGraw-Hill/Irwin, 2005.

Montgomery, D. C., E. A. Peck, and G. G. Vining. *Introduction to Linear Regression Analysis*, 5th ed. New York: Wiley, 2012.

Snee, R. D. "Some Aspects of Nonorthogonal Data Analysis, Part I. Developing Prediction Equations." *Journal of Quality Technology* 5 (1973): 67–79.

▼ KEY EQUATIONS

Quadratic Regression Model

$$Y_i = \beta_0 + \beta_1 X_{1i} + \beta_2 X_{1i}^2 + \varepsilon_i \qquad (15.1)$$

Quadratic Regression Equation

$$\hat{Y}_i = b_0 + b_1 X_{1i} + b_2 X_{1i}^2 \qquad (15.2)$$

Regression Model with a Square-Root Transformation

$$\sqrt{Y_i} = \beta_0 + \beta_1 X_{1i} + \varepsilon_i \qquad (15.3)$$

Original Multiplicative Model

$$Y_i = \beta_0 X_{1i}^{\beta_1} X_{2i}^{\beta_2} \varepsilon_i \qquad (15.4)$$

Transformed Multiplicative Model

$$\log Y_i = \log\left(\beta_0 X_{1i}^{\beta_1} X_{2i}^{\beta_2} \varepsilon_i\right)$$
$$= \log \beta_0 + \log\left(X_{1i}^{\beta_1}\right) + \log(X_{2i}^{\beta_2}) + \log \varepsilon_i$$
$$= \log \beta_0 + \beta_1 \log X_{1i} + \beta_2 \log X_{2i} + \log \varepsilon_i \qquad (15.5)$$

Original Exponential Model

$$Y_i = e^{\beta_0 + \beta_1 X_{1i} + \beta_2 X_{2i}} \varepsilon_i \qquad (15.6)$$

Transformed Exponential Model

$$\ln Y_i = \ln(e^{\beta_0 + \beta_1 X_{1i} + \beta_2 X_{2i}} \varepsilon_i)$$
$$= \ln(e^{\beta_0 + \beta_1 X_{1i} + \beta_2 X_{2i}}) + \ln \varepsilon_i$$
$$= \beta_0 + \beta_1 X_{1i} + \beta_2 X_{2i} + \ln \varepsilon_i \qquad (15.7)$$

Variance Inflationary Factor

$$VIF_j = \frac{1}{1 - R_j^2} \qquad (15.8)$$

C_p Statistic

$$C_p = \frac{(1 - R_k^2)(n - T)}{1 - R_T^2} - \left[n - 2(k + 1)\right] \qquad (15.9)$$

▼ KEY TERMS

▼ CHECKING YOUR UNDERSTANDING

15.26 How can you evaluate whether collinearity exists in a multiple regression model?

15.27 What is the difference between stepwise regression and best subsets regression?

15.28 How do you choose among models according to the C_p statistic in best subsets regression?

▼ CHAPTER REVIEW PROBLEMS

15.29 A specialist in baseball analytics has expanded his analysis, presented in Problem 14.77 on page 520, of which variables are important in predicting a team's wins in a given baseball season. He has collected data in Baseball related to wins, ERA, saves, runs scored per game, batting average, home runs, and batting average against for a recent season.

Develop the most appropriate multiple regression model to predict a team's wins. Be sure to include a thorough residual analysis. In addition, provide a detailed explanation of the results.

15.30 In the production of printed circuit boards, errors in the alignment of electrical connections are a source of scrap. The file

Registration Errors Comparison contains the registration error, the temperature, the pressure, and the circuit board cost (low or high) used in the production of circuit boards.

Source: Data extracted from C. Nachtsheim and B. Jones, "A Powerful Analytical Tool," *Six Sigma Forum Magazine*, August 2003, pp. 30–33.

Develop the most appropriate multiple regression model to predict registration error.

15.31 Hemlock Farms is a community located in the Pocono Mountains area of eastern Pennsylvania. The file Hemlock Farms contains information on homes that were recently for sale. The variables included were

> List Price—Asking price of the house
> Hot Tub—Whether the house has a hot tub, with 0 = No and 1 = Yes
> Lake View—Whether the house has a lake view, with 0 = No and 1 = Yes
> Bathrooms—Number of bathrooms
> Bedrooms—Number of bedrooms
> Loft/Den—Whether the house has a loft or den, with 0 = No and 1 = Yes
> Finished basement—Whether the house has a finished basement, with 0 = No and 1 = Yes
> Acres—Number of acres for the property

Develop the most appropriate multiple regression model to predict the asking price. Be sure to perform a thorough residual analysis. In addition, provide a detailed explanation of your results.

15.32 Nassau County is located approximately 25 miles east of New York City. The file Glen Cove contains a sample of 30 single-family homes located in Glen Cove. Variables included are the fair market value, land area of the property (acres), interior size of the house (square feet), age (years), number of rooms, number of bathrooms, and number of cars that can be parked in the garage.
a. Develop the most appropriate multiple regression model to predict fair market value.
b. Compare the results in (a) with those of Problems 15.33 (a) and 15.34 (a).

15.33 Data similar to those in Problem 15.32 are available for homes located in Roslyn (approximately 8 miles from Glen Cove) and are stored in Roslyn .
a. Perform an analysis similar to that of Problem 15.32.
b. Compare the results in (a) with those of Problems 15.32 (a) and 15.34 (a).

15.34 Data similar to Problem 15.32 are available for homes located in Freeport (located approximately 20 miles from Roslyn) and are stored in Freeport .
a. Perform an analysis similar to that of Problem 15.32.
b. Compare the results in (a) with those of Problems 15.32 (a) and 15.33 (a).

15.35 You are a real estate broker who wants to compare property values in Glen Cove and Roslyn (which are located approximately 8 miles apart). Use the data in GC and Roslyn . Make sure to include the dummy variable for location (Glen Cove or Roslyn) in the regression model.
a. Develop the most appropriate multiple regression model to predict fair market value.
b. What conclusions can you reach concerning the differences in fair market value between Glen Cove and Roslyn?

15.36 You are a real estate broker who wants to compare property values in Glen Cove, Freeport, and Roslyn. Use the data in GC Freeport Roslyn .
a. Develop the most appropriate multiple regression model to predict fair market value.
b. What conclusions can you reach concerning the differences in fair market value between Glen Cove, Freeport, and Roslyn?

15.37 Financial analysts engage in business valuation to determine a company's value. A standard approach uses the multiple of earnings method: You multiply a company's profits by a certain value (average or median) to arrive at a final value. More recently, regression analysis has been demonstrated to consistently deliver more accurate predictions. A valuator has been given the assignment of valuing a drug company. She obtained financial data on 60 drug companies (Industry Group Standard Industrial Classification [SIC] 3 code 283), which included pharmaceutical preparation firms (SIC 4 code 2834), in vitro and in vivo diagnostic substances firms (SIC 4 code 2835), and biological products firms (SIC 4 2836). The file Business Valuation 2 contains these variables:

> COMPANY—Drug company name
> TS—Ticker symbol
> SIC 3—Standard Industrial Classification 3 code (industry group identifier)
> SIC 4—Standard Industrial Classification 4 code (industry identifier)
> PB fye—Price-to-book value ratio (fiscal year ending)
> PE fye—Price-to-earnings ratio (fiscal year ending)
> NL Assets—Natural log of assets (as a measure of size)
> ROE—Return on equity
> SGROWTH—Growth (GS5)
> DEBT/EBITDA—Ratio of debt to earnings before interest, taxes, depreciation, and amortization
> D2834—Dummy variable indicator of SIC 4 code 2834 (1 if 2834, 0 if not)
> D2835—Dummy variable indicator of SIC 4 code 2835 (1 if 2835, 0 if not)

Develop the most appropriate multiple regression model to predict the price-to-book value ratio. Perform a thorough residual analysis and provide a detailed explanation of your results.

15.38 The J. Conklin article, "It's a Marathon, Not a Sprint," *Quality Progress*, June 2009, pp. 46–49, discussed a metal deposition process in which a piece of metal is placed in an acid bath and an alloy is layered on top of it. The key quality characteristic is the thickness of the alloy layer. The file Thickness contains the following variables:

> Thickness—Thickness of the alloy layer
> Catalyst—Catalyst concentration in the acid bath
> pH—pH level of the acid bath
> Pressure—Pressure in the tank holding the acid bath
> Temp—Temperature in the tank holding the acid bath
> Voltage—Voltage applied to the tank holding the acid bath

Develop the most appropriate multiple regression model to predict the thickness of the alloy layer. Be sure to perform a thorough residual analysis. The article suggests that there is a significant interaction between the pressure and the temperature in the tank. Do you agree?

15.39 A molding machine that contains different cavities is used in producing plastic parts. The product characteristics of interest are the product length (in.) and weight (g). The mold cavities were filled with raw material powder and then vibrated during the experiment. The factors that were varied were the vibration time (seconds), the vibration pressure (psi), the vibration amplitude (%), the raw material density (g/mL), and the quantity of raw material (scoops). The experiment was conducted in two different cavities on the molding machine. The data are stored in Molding.

Source: Data extracted from M. Lopez and M. McShane-Vaughn, "Maximizing Product, Minimizing Costs," *Six Sigma Forum Magazine*, February 2008, pp. 18–23.

a. Develop the most appropriate multiple regression model to predict the product length in cavity 1. Be sure to perform a thorough residual analysis. In addition, provide a detailed explanation of your results.
b. Repeat (a) for cavity 2.
c. Compare the results for length in the two cavities.
d. Develop the most appropriate multiple regression model to predict the product weight in cavity 1. Be sure to perform a thorough residual analysis. In addition, provide a detailed explanation of your results.
e. Repeat (d) for cavity 2.
f. Compare the results for weight in the two cavities.

15.40 The file Cities contains a sample of 25 cities in the United States. Variables included are city average annual salary ($), unemployment rate (%), median home value ($thousands), number of violent crimes per 100,000 residents, average commuter travel time (minutes), and livability score, a rating on a scale of 0 to 100 that rates the overall livability of the city.

Source: Data extracted from "125 Best Places to Live in the USA," available at **bit.ly/2jYvtFz** and "AARP Livability Index," available at **bit.ly/1Qbd6oj**.

Develop the most appropriate multiple regression model to predict average annual salary ($). Be sure to perform a thorough residual analysis and provide a detailed explanation of the results as part of your answer.

15.41 A sports statistician seeks to evaluate NBA team statistics for a recent year. The statistician collects data for wins and points scored, field goal percentage, opponent field goal percentage, three point field goal percentage, free throw percentage, rebounds per game, assists per game, and turnovers per game, and stores these data in NBA Stats.

Develop the most appropriate multiple regression model to predict the number of wins for an NBA team. Be sure to perform a thorough residual analysis and provide a detailed explanation of the results as part of your answer.

REPORT WRITING EXERCISE

15.42 In Problems 15.32–15.36 you developed multiple regression models to predict the fair market value of houses in Glen Cove, Roslyn, and Freeport. Now write a report based on the models you developed. Append all appropriate charts and statistical information to your report.

CHAPTER 15

▾ CASES

The Mountain States Potato Company

Mountain States Potato Company sells a by-product of its potato-processing operation, called a filter cake, to area feedlots as cattle feed. The business problem faced by the feedlot owners is that the cattle are not gaining weight as quickly as they once were. The feedlot owners believe that the root cause of the problem is that the percentage of solids in the filter cake is too low.

Historically, the percentage of solids in the filter cakes ran slightly above 12%. Lately, however, the solids are running in the 11% range. What is actually affecting the solids is a mystery, but something has to be done quickly. Individuals involved in the process were asked to identify variables that might affect the percentage of solids. This review turned up the six variables (in addition to the percentage of solids) listed in the right column. Data collected by monitoring the process several times daily for 20 days are stored in Potato.

1. Thoroughly analyze the data and develop a regression model to predict the percentage of solids.

2. Write an executive summary concerning your findings to the president of the Mountain States Potato Company. Include specific recommendations on how to get the percentage of solids back above 12%.

Variable	Comments
SOLIDS	Percentage of solids in the filter cake.
PH	Acidity. This measure of acidity indicates bacterial action in the clarifier and is controlled by the amount of downtime in the system. As bacterial action progresses, organic acids are produced that can be measured using pH.
LOWER	Pressure of the vacuum line below the fluid line on the rotating drum.
UPPER	Pressure of the vacuum line above the fluid line on the rotating drum.
THICK	Filter cake thickness, measured on the drum.
VARIDRIV	Setting used to control the drum speed. May differ from DRUMSPD due to mechanical inefficiencies.
DRUMSPD	Speed at which the drum is rotating when collecting the filter cake. Measured with a stopwatch.

Sure Value Convenience Stores

You work in the corporate office for a nationwide convenience store franchise that operates nearly 10,000 stores. The per-store daily customer count (i.e., the mean number of customers in a store in one day) has been steady at 900 for some time. To increase the customer count, the chain is considering cutting prices for coffee beverages. The question to be answered is how much prices should be cut to increase the daily customer count without reducing the gross margin on coffee sales too much. You decide to carry out an experiment in a sample of 24 stores where customer counts have been running almost exactly at the national average of 900. In six of the stores, the price of a small coffee will now be $0.59, in six stores the price of a small coffee will now be $0.69, in six stores, the price of a small coffee will now be $0.79, and in six stores, the price of a small coffee will now be $0.89. After four weeks at the new prices, the daily customer count in the stores is determined and is stored in Coffee Sales 2 .

a. Construct a scatter plot for price and sales.

b. Fit a quadratic regression model and state the quadratic regression equation.

c. Predict the mean weekly sales for a small coffee priced at 79 cents.

d. Perform a residual analysis on the results and determine whether the regression model is valid.

e. At the 0.05 level of significance, is there a significant quadratic relationship between weekly sales and price?

f. At the 0.05 level of significance, determine whether the quadratic model is a better fit than the linear model.

g. Interpret the meaning of the coefficient of multiple determination.

h. Compute the adjusted r^2.

i. What price do you recommend the small coffee should be sold for?

Digital Case

Apply your knowledge of multiple regression model building in this Digital Case, which extends the Chapter 14 OmniPower Bars Using Statistics scenario.

Still concerned about ensuring a successful test marketing of its OmniPower bars, the marketing department of OmniFoods has contacted Connect2Coupons (C2C), another merchandising consultancy. C2C suggests that earlier analysis done by In-Store Placements Group (ISPG) was faulty because it did not use the correct type of data. C2C claims that its Internet-based viral marketing will have an even greater effect on OmniPower energy bar sales, as new data from the same 34-store sample will show. In response, ISPG says its earlier claims are valid and has reported to the OmniFoods marketing department that it can discern no simple relationship between C2C's viral marketing and increased OmniPower sales.

Open **OmniPowerForum15.pdf** to review all the claims made in a private online forum and chat hosted on the Omni-Foods corporate website. Then answer the following:

1. Which of the claims are true? False? True but misleading? Support your answer by performing an appropriate statistical analysis.

2. If the grocery store chain allowed OmniFoods to use an unlimited number of sales techniques, which techniques should it use? Explain.

3. If the grocery store chain allowed OmniFoods to use only one sales technique, which technique should it use? Explain.

The Craybill Instrumentation Company Case

The Craybill Instrumentation Company produces highly technical industrial instrumentation devices. The human resources (HR) director has the business objective of improving recruiting decisions concerning sales managers. The company has 45 sales regions, each headed by a sales manager. Many of the sales managers have degrees in electrical engineering, and due to the technical nature of the product line, several company officials believe that only applicants with degrees in electrical engineering should be considered.

At the time of their application, candidates are asked to take the Strong-Campbell Interest Inventory Test and the Wonderlic Personnel Test. Due to the time and money involved with the testing, some discussion has taken place about dropping one or both of the tests. To start, the HR director gathered information on each of the 45 current sales managers, including years of selling experience, electrical engineering background, and the scores from both the Wonderlic and Strong-Campbell tests. The HR director has decided to use regression modeling to predict a dependent variable of "sales index" score, which is the ratio of the regions' actual sales divided by the target sales. The target values are constructed each year by upper management, in consultation with the sales managers, and are based on past performance and market potential within each region. The file Managers contains information on the 45 current sales managers. The following variables are included:

Sales—Ratio of yearly sales divided by the target sales value for that region; the target values were mutually agreed-upon "realistic expectations"

Wonder—Score from the Wonderlic Personnel Test; the higher the score, the higher the applicant's perceived ability to manage

SC—Score on the Strong-Campbell Interest Inventory Test; the higher the score, the higher the applicant's perceived interest in sales

Experience—Number of years of selling experience prior to becoming a sales manager

Engineer—Dummy variable that equals 1 if the sales manager has a degree in electrical engineering and 0 otherwise

a. Develop the most appropriate regression model to predict sales.

b. Do you think that the company should continue administering both the Wonderlic and Strong-Campbell tests? Explain.

c. Do the data support the argument that electrical engineers outperform the other sales managers? Would you support the idea to hire only electrical engineers? Explain.

d. How important is prior selling experience in this case? Explain.

e. Discuss in detail how the HR director should incorporate the regression model you developed into the recruiting process.

More Descriptive Choices Follow-Up

Follow-up the Using Statistics scenario, "More Descriptive Choices, Revisited" on page 147, by developing regression models to predict the one-year return, the three-year return, the five-year return, and the ten-year return based on the assets, turnover ratio, expense ratio, beta, standard deviation, type of fund (growth versus value), and risk (stored in `Retirement Funds`). (For this analysis, combine low and average risk into the new category "not high.") Be sure to perform a thorough residual analysis. Provide a summary report that explains your results in detail.

▼EXCEL GUIDE

EG15.1 The QUADRATIC REGRESSION MODEL

Key Technique Use the exponential operator (^) in a column of formulas to create the quadratic term.

Example Create the quadratic term for the Section 15.1 concrete strength analysis.

PHStat, Workbook, *and* Analysis ToolPak For the example, open to the **DATA worksheet** of the **Fly Ash workbook**, which contains the independent *X* variable flyash% in column A and the dependent *Y* variable strength in column B and:

1. Select column B, right-click, and click **Insert** from the shortcut menu. This creates a new, blank column B, and changes strength to column C.
2. Enter the label **FlyAsh%^2** in cell **B1** and then enter the formula **=A2^2** in cell **B2**.
3. Copy this formula down column B through all the data rows (through row 19).

(Best practice places the quadratic term in a column that is contiguous to the columns of the other independent *X* variables.)

Adapt the Section EG14.1 instructions to perform a regression analysis using the quadratic term. For *PHStat*, use **C1:C19** as the **Y Variable Cell Range** and **A1:B19** as the **X Variables Cell Range**. For *Worksheet*, use **C2:C19** and **A2:B19** in step 4 as the new cell ranges. For *Analysis ToolPak*, use **C1:C19** as the **Input Y Range** and **A1:B19** as the **Input X Range**.

To create a scatter plot, adapt the EG2.5 "The Scatter Plot" instructions. For *PHStat*, use **C1:C19** as the **Y Variable Cell Range** and **A1:B19** as the **X Variable Cell Range**. For *Worksheet*, select the noncontiguous cell range **A1:A19, C1:C19** in step 1 and skip step 3. (Appendix B explains how to select a noncontiguous cell range.) Select the scatter chart and then:

1. Select **Design** (or **Chart Design**) → **Add Chart Element** → **Trendline** → **More Trendline Options**.

In the Format Trendline pane (parts shown below),

2. Click **Polynomial** (shown at top of right column).

3. Check **Display Equation on chart** (shown below).

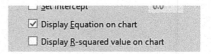

In older Excels, select **Layout** → **Trendline** → **More Trendline Options** in step 1 and in the Format Trendline dialog box, click **Trendline Options** in the left pane. In the Trendline Options right pane, click **Polynomial**, check **Display Equation on chart**, and click **OK**.

EG15.2 USING TRANSFORMATIONS in REGRESSION MODELS

The Square-Root Transformation

To the worksheet that contains your regression data, add a new column of formulas that computes the square root of the variable for which you want to create a square-root transformation. For example, to create a square-root transformation in a blank column D for a variable in a column C, enter the formula **=SQRT(C2)** in cell D2 of that worksheet and copy the formula down through all data rows.

If the column to the immediate right of the variable to be transformed is not empty, first select that column, right-click, and click **Insert** from the shortcut menu. Then place the transformation in the newly inserted blank column.

The Log Transformation

To the worksheet that contains your regression data, add a new column of formulas that computes the common (base 10) logarithm or natural logarithm (base *e*) of the dependent variable to create a log transformation. For example, to create a common logarithm transformation in a blank column D for a variable in column C, enter the formula **=LOG(C2)** in cell D2 of that worksheet and copy the formula down through all data rows. To create a natural logarithm transformation in a blank column D for a variable in column C, enter the formula **=LN(C2)** in cell D2 of that worksheet and copy the formula down through all data rows.

If the dependent variable appears in a column to the immediate right of the independent variable being transformed, first select the dependent variable column, right-click, and click **Insert** from the shortcut menu and then place the transformation of the independent variable in that new column.

EG15.3 COLLINEARITY

PHStat To compute the variance inflationary factor (*VIF*), use the EG14.1 "Interpreting the Regression Coefficients" *PHStat* instructions on page 522, but modify step 6 by checking **Variance Inflationary Factor** (*VIF*) before you click **OK**. The *VIF* will appear in cell B9 of the regression results worksheet, immediately following the Regression Statistics area.

Workbook To compute the variance inflationary factor, first use the EG14.1 "Interpreting the Regression Coefficients" *Workbook* instructions on page 522 to create regression results worksheets for every combination of independent variables in which one serves as the dependent variable. Then, in each of the regression results worksheets, enter the label *VIF* in cell **A9** and enter the formula $=1/(1 - B5)$ in cell **B9** to compute the *VIF*.

EG15.4 MODEL BUILDING

The Stepwise Regression Approach to Model Building

Key Technique Use PHStat to perform a stepwise analysis.
Example Perform the Figure 15.14 stepwise analysis for the Nickels Broadcasting data on page 542.

PHStat Use **Stepwise Regression**.

For the example, open to the **DATA worksheet** of the **Nickels 26Weeks workbook** and select **PHStat → Regression → Stepwise Regression**. In the procedure's dialog box (shown at the top of the right column):

1. Enter **A1:A27** as the **Y Variable Cell Range**.
2. Enter **B1:E27** as the **X Variables Cell Range**.
3. Check **First cells in both ranges contain label**.
4. Enter **95** as the **Confidence level for regression coefficients**.
5. Click **p values** as the **Stepwise Criteria**.
6. Click **General Stepwise** and keep the pair of **.05** values as the **p value to enter** and the **p value to remove**.
7. Enter a **Title** and click **OK**.

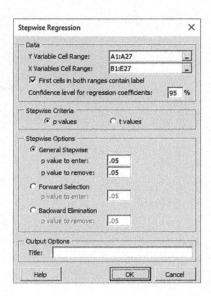

This procedure may take more than a few seconds to construct its results. The procedure finishes when the statement "Stepwise ends" is added to the stepwise regression results worksheet (in row 29 in Figure 15.14 on page 542).

The Best Subsets Approach to Model Building

Key Technique Use PHStat to perform a best subsets analysis.

Example Perform the Figure 15.15 best subsets analysis for the Nickels Broadcasting data on page 543.

PHStat Use **Best Subsets**.

For the example, open to the **DATA worksheet** of the **Nickels 26Weeks workbook**. Select **PHStat → Regression → Best Subsets**. In the procedure's dialog box (shown below):

1. Enter **A1:A27** as the **Y Variable Cell Range**.
2. Enter **B1:E27** as the **X Variables Cell Range**.
3. Check **First cells in each range contains label**.
4. Enter **95** as the **Confidence level for regression coefficients**.
5. Enter a **Title** and click **OK**.

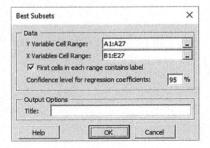

Because this procedure examines many different regression models, there may be a noticeable delay between when OK is clicked and results appear onscreen.

16

Time-Series Forecasting

OBJECTIVES

- Construct different time-series forecasting models for annual and seasonal data
- Choose the most appropriate time-series forecasting model

▼USING **STATISTICS**
Is the ByYourDoor Service Trending?

Senior managers at ByYourDoor, an online food delivery service, have asked you to analyze sales data. These managers would like to know if sales data can be used to estimate future sales. They already know that their business is thriving, but sales seem to be subject to periodic dips that make it hard to accurately estimate short-term physical and labor resources requirements for the company.

One manager wondered if a regression technique might be useful, but another manager recalls that simple and multiple regression models can only predict inside the range of the X values used to create the model. Looking forward would require going beyond the values in such a range. Is it even possible to make a useful estimation about a *future* value of a dependent Y variable?

Forecasting estimates future business conditions by monitoring changes that occur over time. Managers must be able to develop forecasts to anticipate likely changes their businesses will face. For example, retail marketing executives might forecast product demand, sales revenues, consumer preferences, and inventory, among other things, to make decisions regarding product promotions and strategic planning. **Time-series forecasting**, the focus of this chapter, uses a **time series**, a set of numerical data collected over time at regular intervals, as the basis for the estimation. Both government and business activities generate time series data. Some government examples include economic indicators such as a consumer price index or the quarterly gross domestic product (GDP) as well as measurements of real-world phenomena such as the mean monthly level of lakes, the levels of carbon dioxide in the air, or the daily high temperature for a locality. Businesses generate many types of time series and typically include annual measurements of sales revenues, net profits, and other accounting data in annual reports or similar documents.

Time-series forecasting is not the only type of forecasting that uses numerical data. **Causal forecasting methods** help determine the factors that relate to the variable being estimated. These methods include multiple regression analysis with lagged variables, econometric modeling, leading indicator analysis, and other economic barometers that are beyond the scope of this text, but which the Box, Hanke, and Montgomery references discuss.

Although time-series forecasting shares the goal of prediction with the regression methods that previous chapters discuss, time-series forecasting seeks to estimate a *future* value, a goal very different than from the goals of the regression methods that Chapters 13, 14, and 15 discuss. For ByTheDoor, an initial complication would be to establish the time interval that most makes sense for estimating future sales. The time interval can affect both the perception of the data as well as the statistical methods used to analyze the time series. Because the company buys supplies monthly and because customers use the service once a month, on average, collecting monthly data might make best sense, but other time intervals might also be appropriate depending on the goal of the senior managers at the firm.

16.1 Time-Series Component Factors

As Section 2.5 notes, a time-series plot, in which the X axis represents units of time and the Y axis represents the values of a numerical variable, can help visualize trends in data that occur over time. A **trend**, an overall long-term upward or downward movement, that exists in a time series, is one possible pattern, or component of a time series. Establishing whether a trend exists in a time series is an important early step in time-series analysis. Time-series plots can suggest whether a trend component exists in the time series. If a time series shows no trend, then the techniques of moving averages and exponential smoothing that Section 16.2 discusses can be used to analyze the time series. If a time series shows a trend, the various methods that Sections 16.3 through 16.5 discuss can be used if the time series represents annual data. Figure 16.1 Panel A shows a time series with a strong upward trend.

Time-series data may also show a combination of cyclical and irregular components. A **cyclical component** is up-and-down movement in the time series of medium duration, typically from two to ten years in length. Figure 16.1 Panel B shows a time series with four cycles of differing durations. These cycles often correlate with "business cycles" that are associated with certain types of economic activities.

Figure 16.1 Panel C visualizes a time series that has a strong irregular component. An **irregular component** reflects one-time changes to a time series that cannot be explained by the trend or cyclical components. For business decision makers, discovering an irregularity may signal an inflection point in which a significant business or economic change has occurred.

If a time series is collected at intervals of less than year, such as monthly or quarterly, the time series may also have a **seasonal component**. Figure 16.1 Panel D visualizes a seasonal component in which time series data values show the pattern of being very high during the summer months and very low during the winter months.

FIGURE 16.1

Trend, cyclical, irregular, and seasonal components of a time series

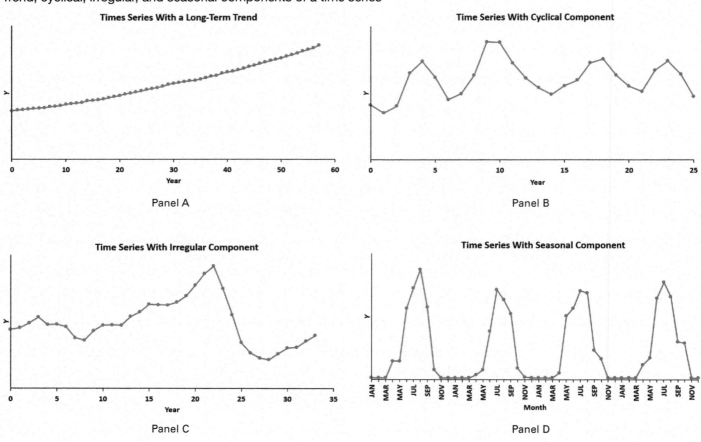

Panel A

Panel B

Panel C

Panel D

Figure 16.2 shows the time series of houses sold in the United States from June 1987 through June 2019. This time series contains two types of components. The irregular downward component centered on 2008 reflects the collapse of the U.S. housing market that led to the "Great Depression" of 2007–2009. Several monthly upward spikes can be seen, including one that occurs every March (red points). This seasonal effect persists through the irregular component centered on 2008, in which March 2007 and March 2008 sales represent temporary upswings. An analyst trying to predict future housing sales and not accounting for that seasonal effect might have been misled by these March changes during the 2007–2009 recessionary period. An analyst who understood that such spikes were seasonal would not expect these spikes to continue and therefore would have been less likely to overestimate the future housing sales.

FIGURE 16.2

Monthly sales of houses in the United States, June 1987 through June 2019 (red plots represent March sales)

16.2 Smoothing an Annual Time Series

Smoothing a time series, transforming the time series to show small-scale fluctuations, can help determine if a time series contains a trend because the smoothing minimizes the effects of the other time-series components. For example, Table 16.1 presents the annual U.S. and Canada movie attendance (in billions) from 2001 through 2018, as reflected by number of tickets sold. Figure 16.3 visualizes these data, stored in Movie Attendance .

TABLE 16.1
Annual movie attendance (in billions) from 2001 through 2018

Year	Attendance	Year	Attendance	Year	Attendance
2001	1.44	2007	1.40	2012	1.36
2002	1.58	2008	1.34	2013	1.34
2003	1.53	2009	1.41	2014	1.27
2004	1.51	2010	1.34	2015	1.32
2005	1.38	2011	1.28	2016	1.32
2006	1.41			2017	1.23
				2018	1.31

Source: Data extracted from **boxofficemojo.com/yearly**.

FIGURE 16.3
Time-series plot of movie attendance from 2001 through 2018

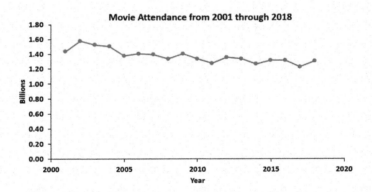

Figure 16.3 seems to show a slight downward trend in movie attendance for the time period plotted. However, the variation that exists from one time period to another can sometimes obscure a long-term trend, which can make an existing trend hard to identify. Using *moving averages* or *exponential smoothing* can smooth the data and better visualize a long-term trend that may be present.

Moving Averages

The **moving averages** method calculates means for sequences of consecutive time-series values for a time duration L. The sequences each differ by one time-series value, as the moving average method "moves" through the time-series. For example, for a three-year moving average for an annual time series of 11 years, the first calculated mean would be the mean of the time-series values for years 1 through 3, the second calculated mean would be the mean for years 2 through 4, and the ninth calculated mean would be the mean for years 9 through 11.

The moving averages method always reduces the number of values because moving averages cannot be calculated for the first $(L - 1)/2$ years and the last $(L - 1)/2$ years of the time series. For the example, in which $L = 3$, a moving average cannot be calculated for either the

student TIP

Remember that you cannot calculate moving averages at the beginning and at the end of the series.

first or last (eleventh) year. Although L could be any whole number, making L an odd number permits centering each moving average on a time value, which simplifies preparing tabular and visual summaries of a moving average. For example, if $L = 3$, the first moving average for an annual time series of 11 years would be centered on year 2. If $L = 5$, the first moving average would be centered on year 3. However, if $L = 4$, the moving average would be centered on year "2.5," a time value that is not part of the original time series.

For annual time-series data that do not contain an obvious cyclical component, using 3, 5, or 7 as the value of L are reasonable choices. If a cyclical component exists in a time series, the value of L should be a number that corresponds to or is a multiple of the estimated length of a cycle. Example 16.1 illustrates calculating moving averages for $L = 5$.

EXAMPLE 16.1

Calculating Five-Year Moving Averages

The following data represent revenue (in $millions) for a casual dining restaurant over the 11-year period 2008 to 2018.

$$4.0\ 5.0\ 7.0\ 6.0\ 8.0\ 9.0\ 5.0\ 7.0\ 7.5\ 5.5\ 6.5$$

Compute the five-year moving averages for this annual time series.

SOLUTION Five-year moving averages take the mean of five consecutive time-series values. The first of the five-year moving averages is

$$MA(5) = \frac{Y_1 + Y_2 + Y_3 + Y_4 + Y_5}{5} = \frac{4.0 + 5.0 + 7.0 + 6.0 + 8.0}{5} = \frac{30.0}{5} = 6.0$$

The second of the five-year moving averages is:

$$MA(5) = \frac{Y_2 + Y_3 + Y_4 + Y_5 + Y_6}{5} = \frac{5.0 + 7.0 + 6.0 + 8.0 + 9.0}{5} = \frac{35.0}{5} = 7.0$$

The third, fourth, fifth, sixth, and seventh moving averages are:

$$MA(5) = \frac{Y_3 + Y_4 + Y_5 + Y_6 + Y_7}{5} = \frac{7.0 + 6.0 + 8.0 + 9.0 + 5.0}{5} = \frac{35.0}{5} = 7.0$$

$$MA(5) = \frac{Y_4 + Y_5 + Y_6 + Y_7 + Y_8}{5} = \frac{6.0 + 8.0 + 9.0 + 5.0 + 7.0}{5} = \frac{35.0}{5} = 7.0$$

$$MA(5) = \frac{Y_5 + Y_6 + Y_7 + Y_8 + Y_9}{5} = \frac{8.0 + 9.0 + 5.0 + 7.0 + 7.5}{5} = \frac{36.5}{5} = 7.3$$

$$MA(5) = \frac{Y_6 + Y_7 + Y_8 + Y_9 + Y_{10}}{5} = \frac{9.0 + 5.0 + 7.0 + 7.5 + 5.5}{5} = \frac{34.0}{5} = 6.8$$

$$MA(5) = \frac{Y_7 + Y_8 + Y_9 + Y_{10} + Y_{11}}{5} = \frac{5.0 + 7.0 + 7.5 + 5.5 + 6.5}{5} = \frac{31.5}{5} = 6.3$$

---

student TIP

Using a time duration L that is an odd number facilitates the comparison of the moving averages with the original time series.

Figure 16.4 (left) visualizes the three-year and five-year moving averages for the movie attendance data stored in Movie Attendance . The moving average plots show a downward trend but, unlike the Figure 16.3 time-series plot, reveal that the trend has greatly slowed after 2004. Figure 16.4 (right), a redone plot that discards the time-series values for the early years 2001 through 2004, reveals a time series with a much weaker trend. That pattern suggests that using the shorter time series may lead to a more accurate short-term forecast of future movie attendance.

FIGURE 16.4

Time-series plots for the three- and five-year moving averages for the movie attendance for two time series, 2001 through 2018 and 2005 through 2018

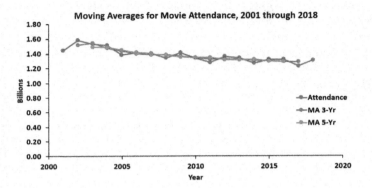
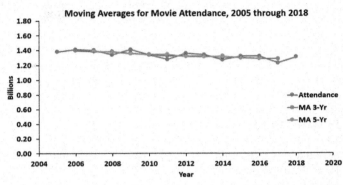

Later movie attendance examples in this chapter use the shorter 2005 through 2018 time series for the reasons this passage discusses.

While discarding data is almost never allowed in the inferential methods that earlier chapters discuss, discarding consecutive time series is an example of the partially subjective nature of time-series forecasting. Determining the proper length of a time series to be used for forecasting can be a mix of business experience and awareness of external or one-time, irregular factors. For the U.S. and Canadian movie attendance time series, further investigation reveals that the year 2002 was unusual in being the only year in which the popular *Star Wars*, *Harry Potter*, and the *Lord of the Rings* movie series all had releases. (And those three films were *only* the second, third, and fourth most popular movies that year, as 2002 also saw the release of the first modern-day *Spider-Man* movie.)

Because the shorter movie attendance time series shows no trend, the moving averages based on the shorter time series could be used for short-term forecasting. However, a second technique, *exponential smoothing*, typically offers better short-term forecasting.

Exponential Smoothing

Exponential smoothing consists of a series of *exponentially weighted* moving averages. The weights assigned to the values change so that the most recent (the last) value receives the highest weight, the previous value receives the second-highest weight, and so on, with the first value receiving the lowest weight. Therefore, the more recent a time-series value is, the more influence the value has on the smoothing function.

Exponential smoothing uses all previous values in contrast to the moving averages method, which uses only a subset of the time series to calculate individual moving averages. Exponential smoothing enables one to calculate short-term forecasts one period into the future, even when the presence and type of long-term trend in a time series is difficult to determine. Equation (16.1) defines an exponentially smoothed value for time period *i*. Note the special case that the smoothed value for time period 1 is the time period 1 value.

AN EXPONENTIALLY SMOOTHED VALUE FOR TIME PERIOD *i*

$$E_1 = Y_1$$

$$E_i = WY_i + (1 - W)E_{i-1} \tag{16.1}$$

where

$$i = 2, 3, 4, \ldots$$
E_i = exponentially smoothed series value for time period i
E_{i-1} = exponentially smoothed series calculated for time period $i - 1$
Y_i = time series value for period i
W = subjectively assigned weight or smoothing coefficient, where $0 < W < 1$

Choosing the weight or smoothing coefficient, W, that one assigns to the time series is both critical to the smoothing and somewhat subjective. One should select a small value for W, close to zero, if the goal is to smooth a series by eliminating unwanted cyclical and irregular variations in order to see the overall long-term tendency of the series. One should select a large value for W, close to 0.5, if the goal is forecasting future short-term directions.

Figure 16.5 presents the exponentially smoothed values (with smoothing coefficients $W = 0.50$ and $W = 0.25$), the movie attendance data from 2005 to 2018, and a plot of the original data and the two exponentially smoothed time series. Observe that exponential smoothing has smoothed some of the variation in the movie attendance.

FIGURE 16.5

Exponentially smoothed series ($W = 0.50$ and $W = 0.25$) worksheet and plot for the movie attendance data

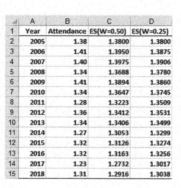

	A	B	C	D
1	Year	Attendance	ES(W=0.50)	ES(W=0.25)
2	2005	1.38	1.3800	1.3800
3	2006	1.41	1.3950	1.3875
4	2007	1.40	1.3975	1.3906
5	2008	1.34	1.3688	1.3780
6	2009	1.41	1.3894	1.3860
7	2010	1.34	1.3647	1.3745
8	2011	1.28	1.3223	1.3509
9	2012	1.36	1.3412	1.3531
10	2013	1.34	1.3406	1.3499
11	2014	1.27	1.3053	1.3299
12	2015	1.32	1.3126	1.3274
13	2016	1.32	1.3163	1.3256
14	2017	1.23	1.2732	1.3017
15	2018	1.31	1.2916	1.3038

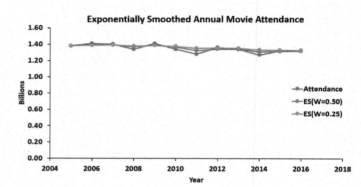

To illustrate these exponential smoothing calculations for a smoothing coefficient of $W = 0.25$, begin with the initial value $Y_{2005} = 1.38$ as the first smoothed value ($E_{2005} = 1.38$). Then, using the value of the time series for 2006 ($Y_{2006} = 1.41$), smooth the series for 2006 as follows:

$$E_{2006} = WY_{2006} + (1 - W)E_{2005}$$
$$= (0.25)(1.41) + (0.75)(1.38) = 1.3875$$

To smooth the series for 2007:

$$E_{2007} = WY_{2007} + (1 - W)E_{2006}$$
$$= (0.25)(1.40) + (0.75)(1.3875) = 1.3906$$

This smoothing would continue for each of the remaining years in the time series. (Figure 16.5 also contains results of this smoothing operation.)

Exponential smoothing is a weighted average of all previous time periods. Therefore, when using exponential smoothing for forecasting, one uses the smoothed value in the current time period as the forecast of the value in the following period $\left(\hat{Y}_{i+1} \right)$.

FORECASTING TIME PERIOD $i + 1$

$$\hat{Y}_{i+1} = E_i \tag{16.2}$$

To forecast the movie attendance in 2019, using a smoothing coefficient of $W = 0.25$, one uses the smoothed value for 2018 as its estimate.

$$\hat{Y}_{2018+1} = E_{2018}$$
$$\hat{Y}_{2019} = E_{2018}$$
$$\hat{Y}_{2019} = 1.3038$$

The exponentially smoothed forecast for 2019 is 1.3038 billion.

PROBLEMS FOR SECTION 16.2

LEARNING THE BASICS

16.1 If you are using exponential smoothing for forecasting an annual time series of revenues, what is your forecast for next year if the smoothed value for this year is $32.4 million?

16.2 Consider a nine-year moving average used to smooth a time series that was first recorded in 1984.
a. Which year serves as the first centered value in the smoothed series?
b. How many years of values in the series are lost when computing all the nine-year moving averages?

16.3 You are using exponential smoothing on an annual time series concerning total revenues (in $millions). You decide to use a smoothing coefficient of $W = 0.20$, and the exponentially smoothed value for 2019 is $E_{2019} = (0.20)(12.1) + (0.80)(9.4)$.
a. What is the smoothed value of this series in 2019?
b. What is the smoothed value of this series in 2020 if the value of the series in that year is $11.5 million?

APPLYING THE CONCEPTS

✓ SELF TEST **16.4** The data below, stored in **Computer Usage**, represent the hours per day spent by American desktop and laptop users from 2008 to 2017.

Year	Hours per Day	Year	Hours per Day
2008	2.2	2013	2.3
2009	2.3	2014	2.2
2010	2.4	2015	2.2
2011	2.6	2016	2.2
2012	2.5	2017	2.1

Source: Data extracted from R. Marvin, "Tech Addiction by the Numbers...," **bit.ly/2KECJAZ**.

a. Plot the time series.
b. Fit a three-year moving average to the data and plot the results.
c. Using a smoothing coefficient of $W = 0.50$, exponentially smooth the series and plot the results.
d. What is your exponentially smoothed forecast for 2018?
e. Repeat (c) and (d), using $W = 0.25$.
f. Compare the results of (d) and (e).
g. What conclusions can you reach about desktop/laptop use by American users?

16.5 The following data, stored in **Core Appliances**, list the number of shipments of core major household appliances in the United States from 2000 to 2017 (in millions).

Year	Shipments	Year	Shipments
2000	38.4	2009	36.5
2001	38.2	2010	38.2
2002	40.8	2011	36.0
2003	42.5	2012	35.8
2004	46.1	2013	39.2
2005	47.0	2014	41.5
2006	46.7	2015	42.9
2007	44.1	2016	44.7
2008	39.8	2017	46.4

Source: Data extracted from **www.statista.com**.

a. Plot the time series.
b. Fit a three-year moving average to the data and plot the results.
c. Using a smoothing coefficient of $W = 0.50$, exponentially smooth the series and plot the results.
d. What is your exponentially smoothed forecast for 2018?
e. Repeat (c) and (d), using $W = 0.25$.
f. Compare the results of (d) and (e).
g. What conclusions can you reach concerning the total number of shipments of core major household appliances in the United States from 2000 to 2017 (in millions)?

16.6 How have stocks performed in the past? The following table presents the data stored in **Stock Performance**, which show the performance of a broad measure of stock performance (by percentage) for each decade from the 1830s through the 2010s:

Decade	Performance (%)	Decade	Performance (%)
1830s	2.8	1920s	13.3
1840s	12.8	1930s	-2.2
1850s	6.6	1940s	9.6
1860s	12.5	1950s	18.2
1870s	7.5	1960s	8.3
1880s	6.0	1970s	6.6
1890s	5.5	1980s	16.6
1900s	10.9	1990s	17.6
1910s	2.2	2000s	1.2
		2010s*	11.0

* Through December 31, 2018.

Source: T. Lauricella, "Investors Hope the '10s Beat the '00s," *The Wall Street Journal*, December 21, 2009, pp. C1, C2 and Moneychimp. com, "CAGR of the Stock Market," **bit.ly/1aiqtiI**.

a. Plot the time series.
b. Fit a three-period moving average to the data and plot the results.
c. Using a smoothing coefficient of $W = 0.50$, exponentially smooth the series and plot the results.
d. What is your exponentially smoothed forecast for the 2020s?
e. Repeat (c) and (d), using $W = 0.25$.
f. Compare the results of (d) and (e).
g. What conclusions can you reach concerning how stocks have performed in the past?

16.7 The file **Coffee Exports** contains the coffee exports (in thousands of 60 kg bags) by Costa Rica from 2004 to 2018.
a. Plot the data.
b. Fit a three-year moving average to the data and plot the results.
c. Using a smoothing coefficient of $W = 0.50$, exponentially smooth the series and plot the results.
d. What is your exponentially smoothed forecast for 2019?
e. Repeat (c) and (d), using a smoothing coefficient of $W = 0.25$.
f. Compare the results of (d) and (e).
g. What conclusions can you reach about the exports of coffee in Costa Rica?

16.8 The file IPOs contains the number of initial public offerings (IPOs) issued from 2001 through 2018.

Source: Data extracted from Statista, "Number of IPOS in the United States from 1999 to 2018," **bit.ly/2uctnU4**.

a. Plot the data.

b. Fit a three-year moving average to the data and plot the results.

c. Using a smoothing coefficient of $W = 0.50$, exponentially smooth the series and plot the results.

d. What is your exponentially smoothed forecast for 2019?

e. Repeat (c) and (d), using a smoothing coefficient of $W = 0.25$.

f. Compare the results of (d) and (e).

16.3 Least-Squares Trend Fitting and Forecasting

To make intermediate and long-range forecasts requires identifying the trend component in a time series. Identifying the trend means being able to develop the most appropriate model that fits the trend. As with regression models that previous chapters discuss, time series data might fit a linear trend model (see Section 13.2), a quadratic trend model (see Section 15.1), or, if the time-series data increase at a rate such that the percentage difference from value to value is constant, an exponential trend model.

The Linear Trend Model

The **linear trend model**:

$$Y_i = \beta_0 + \beta_1 X_i + \varepsilon_i$$

is the simplest forecasting model. Equation (16.3) defines the linear trend forecasting equation.

> LINEAR TREND FORECASTING EQUATION
>
> $$\hat{Y}_i = b_0 + b_1 X_i \qquad\qquad (16.3)$$

Recall that in linear regression analysis, one uses the method of least squares to compute the sample slope, b_1, and the sample Y intercept, b_0. The values for X are then substituted into Equation (16.3) to predict Y.

When using the least-squares method for fitting trends in a time series, one can simplify the interpretation of the coefficients by assigning coded values to the X (time) variable. One assigns consecutively numbered integers, starting with 0, as the coded values for the time periods. For example, in time-series data that have been recorded annually for 8 years, one assigns the coded value 0 to the first year, the coded value 1 to the second year, the coded value 2 to the third year, and so on, concluding by assigning 7 to the eighth.

To illustrate model fitting, consider the Table 16.2 time series, stored in Alphabet, that lists Alphabet Inc.'s annual revenue (in $billions) from 2011 to 2018.

TABLE 16.2
Annual revenues for
Alphabet Inc. 2011–2018

Alphabet Inc. was formed in 2005 and is the holding company that owns Google.

Year	Revenue ($billions)
2011	37.905
2012	46.039
2013	55.519
2014	66.001
2015	74.989
2016	90.272
2017	110.855
2018	136.819

Source: Data extracted from J. Clement, "Annual revenue of Alphabet from 2011 to 2018," **bit.ly/2I7QJBq**.

Figure 16.6 presents the regression results for the simple linear regression model that uses the consecutive coded values 0 through 7 as the X (coded year) variable. These results produce the linear trend forecasting equation:

$$\hat{Y}_i = 30.2280 + 13.4491X_i$$

FIGURE 16.6

Excel regression worksheet results for the linear trend model to forecast revenue (in $billions) for Alphabet Inc.

▲	A	B	C	D	E	F	G
1	Linear Trend Model for Alphabet Inc. Annual Revenue						
2							
3	*Regression Statistics*						
4	Multiple R	0.9761					
5	R Square	0.9528					
6	Adjusted R Square	0.9449					
7	Standard Error	7.9211					
8	Observations	8					
9							
10	ANOVA						
11		*df*	*SS*	*MS*	*F*	*Significance F*	
12	Regression	1	7596.8963	7596.8963	121.0783	0.0000	
13	Residual	6	376.4621	62.7437			
14	Total	7	7973.3584				
15							
16		*Coefficients*	*Standard Error*	*t Stat*	*P-value*	*Lower 95%*	*Upper 95%*
17	Intercept	30.2280	5.1130	5.9119	0.0010	17.7168	42.7392
18	Coded Year	13.4491	1.2223	11.0036	0.0000	10.4584	16.4398

For this regression model, $X_1 = 0$ represents the year 2011, and the regression coefficients are interpreted as follows:

- The Y intercept, $b_0 = 30.2280$, is the predicted mean revenue (in $billions) at Alphabet Inc. during the origin, or base, year, 2011.
- The slope, $b_1 = 13.4491$, indicates that the mean revenue is predicted to increase by $13.4491 billion per year.

To project the trend in the revenue at Alphabet Inc. to 2019, one substitutes $X_9 = 8$, the code for 19 into the linear trend forecasting equation:

$$\hat{Y}_i = 30.2280 + 13.4491(8) = 137.8208 \text{ billions of dollars}$$

Figure 16.7 presents the linear trendline plotted with the time-series values. There is a strong upward linear trend, and r^2 is 0.9528, indicating that more than 95% of the variation in revenue is explained by the linear trend of the time series. However, observe that the early years are slightly above the trend line, but the middle years are below the trend line, and many of the later years are also above the trend line, but the last two years are above the trend line. To investigate whether a different trend model might provide a better fit, a *quadratic* trend model and an *exponential* trend model can be fitted.

FIGURE 16.7

Plot of the linear trend forecasting equation for Alphabet Inc. annual revenue

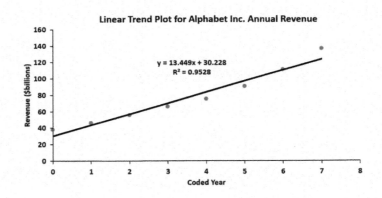

The Quadratic Trend Model

The **quadratic trend model**,

$$Y_i = \beta_0 + \beta_1 X_i + \beta_2 X_i^2 + \varepsilon_i$$

is a nonlinear model that contains a linear term and a curvilinear term in addition to a Y intercept. Using the least-squares method for a quadratic model that Section 15.1 describes, Equation (16.4) defines a quadratic trend forecasting equation.

> QUADRATIC TREND FORECASTING EQUATION
>
> $$\hat{Y}_i = b_0 + b_1 X_i + b_2 X_i^2 \tag{16.4}$$
>
> where
>
> $$b_0 = \text{estimated } Y \text{ intercept}$$
> $$b_1 = \text{estimated } \textit{linear} \text{ effect on } Y$$
> $$b_2 = \text{estimated } \textit{quadratic} \text{ effect on } Y$$

Figure 16.8 presents the regression results for the quadratic trend model to forecast annual revenue at Alphabet Inc. Using Equation (16.4) and the results from Figure 16.8,

$$\hat{Y}_i = 40.1296 + 3.5475 X_i + 1.4145 X_i^2$$

where the year coded 0 is 2011.

FIGURE 16.8

Excel regression worksheet results for the quadratic trend model to forecast annual revenue (in $billions) for Alphabet Inc.

	A	B	C	D	E	F	G
1	Quadratic Trend Model for Alphabet Inc. Annual Revenue						
2							
3	*Regression Statistics*						
4	Multiple R	0.9975					
5	R Square	0.9949					
6	Adjusted R Square	0.9929					
7	Standard Error	2.8396					
8	Observations	8					
9							
10	ANOVA						
11		*df*	*SS*	*MS*	*F*	*Significance F*	
12	Regression	2	7933.0409	3966.5204	491.9104	0.0000	
13	Residual	5	40.3175	8.0635			
14	Total	7	7973.3584				
15							
16		*Coefficients*	*Standard Error*	*t Stat*	*P-value*	*Lower 95%*	*Upper 95%*
17	Intercept	40.1296	2.3899	16.7913	0.0000	33.9862	46.2731
18	Coded Year	3.5475	1.5949	2.2242	0.0767	-0.5524	7.6474
19	Coded Year Squared	1.4145	0.2191	6.4566	0.0013	0.8513	1.9777

To compute a forecast using the quadratic trend equation, substitute the appropriate coded X value into this equation. For example, to forecast the trend in revenues for 2019 (i.e., $X = 8$),

$$\hat{Y}_i = 40.1296 + 3.5475(8) + 1.4145(8)^2 = 159.0376$$

Figure 16.9 plots the quadratic trend forecasting equation with the time-series data. From Figure 16.8, the t_{STAT} test statistic for the contribution of the quadratic term to the model is 6.4566 (p-value $= 0.0013$). Having an adjusted $r^2 = 0.9929$, this quadratic trend model provides a fit that is superior to the fit of the linear trend model.

FIGURE 16.9

Plot of the quadratic trend forecasting equation for Alphabet Inc. annual revenue

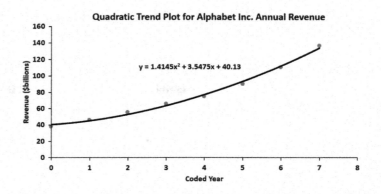

The Exponential Trend Model

When a time series increases at a rate such that the percentage difference from value to value is constant, an exponential trend is present. Equation (16.5) defines the **exponential trend model**.

EXPONENTIAL TREND MODEL

$$Y_i = \beta_0 \beta_1^{X_i} \varepsilon_i \tag{16.5}$$

where

$$\beta_0 = Y \text{ intercept}$$
$$(\beta_1 - 1) \times 100\% = \text{annual compound growth rate (\%)}$$

---

[1]Alternatively, you can use base e logarithms. For more information on logarithms, see Section A.3 in Appendix A.

　　The model in Equation (16.5) is not in the form of a linear regression model. To transform this nonlinear model to a linear model, one uses a base 10 logarithm transformation.[1] Taking the logarithm of each side of Equation (16.5) results in the transformed model that Equation (16.6) defines.

TRANSFORMED EXPONENTIAL TREND MODEL

$$\log(Y_i) = \log(\beta_0 \beta_1^{X_i} \varepsilon_i)$$
$$= \log(\beta_0) + \log(\beta_1^{X_i}) + \log(\varepsilon_i)$$
$$= \log(\beta_0) + X_i \log(\beta_1) + \log(\varepsilon_i) \tag{16.6}$$

---

student TIP

Log is the symbol used for base 10 logarithms. The log of a number is the power that 10 needs to be raised to equal that number.

　　Using the transformed model and the least-squares method, with $\log(Y_i)$ as the dependent variable and X_i as the independent variable produces the Equation (16.7a) forecasting equation.

EXPONENTIAL TREND FORECASTING EQUATION

$$\log(\hat{Y_i}) = b_0 + b_1 X_i \tag{16.7a}$$

where

$$b_0 = \text{estimate of } \log(\beta_0) \text{ and thus } 10^{b_0} = \hat{\beta}_0$$
$$b_1 = \text{estimate of } \log(\beta_1) \text{ and thus } 10^{b_1} = \hat{\beta}_1$$

therefore,

$$\hat{Y_i} = \hat{\beta}_0 \hat{\beta}_1^{X_i} \tag{16.7b}$$

where

$(\hat{\beta}_i - 1) \times 100\%$ is the estimated annual compound growth rate (%)

Figure 16.10 shows the regression results for an exponential trend model to forecast annual revenue at Alphabet Inc. Using Equation (16.7a) and the results from Figure 16.10,

$$\log(\hat{Y}_i) = 1.5814 + 0.0774X_i$$

where the year coded 0 is 2011.

FIGURE 16.10

Excel regression worksheet results for the exponential trend model to forecast annual revenue (in $billions) for Alphabet Inc.

	A	B	C	D	E	F	G
1	Exponential Trend Model for Alphabet Inc. Revenue						
2							
3	*Regression Statistics*						
4	Multiple R	0.9986					
5	R Square	0.9971					
6	Adjusted R Square	0.9967					
7	Standard Error	0.0109					
8	Observations	8					
9							
10	ANOVA						
11		*df*	*SS*	*MS*	*F*	*Significance F*	
12	Regression	1	0.2514	0.2514	2098.1125	0.0000	
13	Residual	6	0.0007	0.0001			
14	Total	7	0.2521				
15							
16		*Coefficients*	*Standard Error*	*t Stat*	*P-value*	*Lower 95%*	*Upper 95%*
17	Intercept	1.5814	0.0071	223.7972	0.0000	1.5641	1.5987
18	Coded Year	0.0774	0.0017	45.8052	0.0000	0.0732	0.0815

One calculates the values for $\hat{\beta}_0$ and $\hat{\beta}_1$ by taking the antilog of the regression coefficients b_0 and b_1:

$$\hat{\beta}_0 = \text{antilog}(b_0) = \text{antilog}(1.5814) = 10^{1.5814} = 38.1417$$

$$\hat{\beta}_1 = \text{antilog}(b_1) = \text{antilog}(0.0774) = 10^{0.0774} = 1.1951$$

Thus, using Equation (16.7b), the exponential trend forecasting equation is

$$\hat{Y}_i = (38.1417)(1.1951)^{X_i}$$

where the year coded 0 is 2011.

The Y intercept, $\hat{\beta}_0 = 38.1417$ billions of dollars, is the revenue forecast for the base year 2011. The value $(\hat{\beta}_1 - 1) \times 100\%, = 19.51\%$, is the annual compound growth rate in revenues at Alphabet Inc.

For forecasting purposes, one substitutes the appropriate coded X values into either Equation (16.7a) or Equation (16.7b). For example, to forecast revenues for 2019 ($X = 8$) using Equation (16.7a),

$$\log(\hat{Y}_i) = 1.5814 + 0.0774(8) = 2.2006$$

$$\hat{Y}_i = \text{antilog}(2.2006) = 10^{2.2006} = 158.708 \text{ billion of dollars}$$

Figure 16.11 plots the exponential trend forecasting equation, along with the time-series data. The adjusted r^2 for the exponential trend model (0.9967) is greater than the adjusted r^2 for the linear trend model (0.9449) and similar to the quadratic model (0.9929).

FIGURE 16.11

Plot of the exponential trend forecasting equation for Alphabet Inc. annual revenue

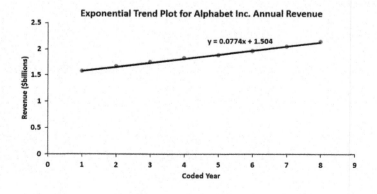

Model Selection Using First, Second, and Percentage Differences

Exhibit 16.1 summarizes how the first, second, and percentage differences in a time series helps determine which type of model is most appropriate for the time series. Although most time-series data will not perfectly fit any of the models, consider the first differences, second differences, and percentage differences as guides in choosing an appropriate model. Examples 16.2, 16.3, and 16.4 illustrate linear, quadratic, and exponential trend models that have perfect (or nearly perfect) fits to their respective data sets.

EXHIBIT 16.1

Model Selection Using First, Second, and Percentage Differences

- If a linear trend model provides a perfect fit to a time series, then the first differences will be constant:

$$(Y_2 - Y_1) = (Y_3 - Y_2) = \cdots = (Y_n - Y_{n-1})$$

- If a quadratic trend model provides a perfect fit to a time series, then the second differences will be constant:

$$[(Y_3 - Y_2) - (Y_2 - Y_1)] = [(Y_4 - Y_3) - (Y_3 - Y_2)] = \cdots = [(Y_n - Y_{n-1}) - (Y_{n-1} - Y_{n-2})]$$

- If an exponential trend model provides a perfect fit to a time series, then the percentage differences between consecutive values will be constant:

$$\frac{Y_2 - Y_1}{Y_1} \times 100\% = \frac{Y_3 - Y_2}{Y_2} \times 100\%, = \cdots = \frac{Y_n - Y_{n-1}}{Y_{n-1}} \times 100\%$$

EXAMPLE 16.2

A Linear Trend Model With a Perfect Fit

The following time series represents the number of customers per year (in thousands) at a branch of a fast-food chain:

	Year									
	2010	2011	2012	2013	2014	2015	2016	2017	2018	2019
Customers *Y*	200	205	210	215	220	225	230	235	240	245

Using first differences, show that the linear trend model provides a perfect fit to these data.

SOLUTION The following table shows the solution:

	Year									
	2010	2011	2012	2013	2014	2015	2016	2017	2018	2019
Customers *Y*	200	205	210	215	220	225	230	235	240	245
First differences		5.0	5.0	5.0	5.0	5.0	5.0	5.0	5.0	5.0

The differences between consecutive values in the series are the same throughout. Thus, the number of customers at the branch of the fast-food chain shows a linear growth pattern.

EXAMPLE 16.3

A Quadratic Trend Model With a Perfect Fit

The following time series represents the number of customers per year (in thousands) at another branch of a fast-food chain:

	Year									
	2010	2011	2012	2013	2014	2015	2016	2017	2018	2019
Customers Y	200	201	203.5	207.5	213	220	228.5	238.5	250	263

Using second differences, show that the quadratic trend model provides a perfect fit to these data.

SOLUTION The following table shows the solution:

	Year									
	2010	2011	2012	2013	2014	2015	2016	2017	2018	2019
Customers Y	200	201	203.5	207.5	213	220	228.5	238.5	250	263
First differences		1.0	2.5	4.0	5.5	7.0	8.5	10.0	11.5	13.0
Second differences			1.5	1.5	1.5	1.5	1.5	1.5	1.5	1.5

The second differences between consecutive pairs of values in the series are the same throughout. Thus, the number of customers at the branch of the fast-food chain shows a quadratic growth pattern. Its rate of growth is accelerating over time.

EXAMPLE 16.4

An Exponential Trend Model With an Almost Perfect Fit

The following time series represents the number of customers per year (in thousands) for another branch of the fast-food chain:

	Year									
	2010	2011	2012	2013	2014	2015	2016	2017	2018	2019
Customers Y	200	206	212.18	218.55	225.11	231.86	238.82	245.98	253.36	260.96

Using percentage differences, show that the exponential trend model provides almost a perfect fit to these data.

SOLUTION The following table shows the solution:

	Year									
	2010	2011	2012	2013	2014	2015	2016	2017	2018	2019
Customers Y	200	206	212.18	218.55	225.11	231.86	238.82	245.98	253.36	260.96
Percentage differences		3.0	3.0	3.0	3.0	3.0	3.0	3.0	3.0	3.0

The percentage differences between consecutive values in the series are approximately the same throughout. Thus, this branch of the fast-food chain shows an exponential growth pattern. Its rate of growth is approximately 3% per year.

Figure 16.12 shows a worksheet that compares the first, second, and percentage differences for Alphabet Inc revenue. Neither the first, second, or percentage differences are constant across the series, although the percentage differences are fairly consistent, save for one value. Therefore, other models, including the models that Section 16.5 discusses, may be more appropriate.

FIGURE 16.12

Excel worksheet template that computes first, second, and percentage differences in revenue (in $billions) for Alphabet Inc.

	A	B	C	D	E
1	Year	Revenue	First Difference	Second Difference	Percentage Difference
2	2011	37.905	#N/A	#N/A	#N/A
3	2012	46.039	8.134	#N/A	21.459
4	2013	55.519	9.480	1.346	20.591
5	2014	66.001	10.482	1.002	18.880
6	2015	74.989	8.988	-1.494	13.618
7	2016	90.272	15.283	6.295	20.380
8	2017	110.855	20.583	5.300	22.801
9	2018	136.819	25.964	5.381	23.422

PROBLEMS FOR SECTION 16.3

LEARNING THE BASICS

16.9 If you are using the method of least squares for fitting trends in an annual time series containing 25 consecutive yearly values,

a. what coded value do you assign to X for the first year in the series?

b. what coded value do you assign to X for the fifth year in the series?

c. what coded value do you assign to X for the most recent recorded year in the series?

d. what coded value do you assign to X if you want to project the trend and make a forecast five years beyond the last observed value?

16.10 The linear trend forecasting equation for an annual time series containing 22 values (from 1998 to 2019) on total revenues (in $millions) is

$$\hat{Y}_i = 4.0 + 1.5X_i$$

a. Interpret the Y intercept, b_0.

b. Interpret the slope, b_1.

c. What is the fitted trend value for the fifth year?

d. What is the fitted trend value for the most recent year?

e. What is the projected trend forecast three years after the last value?

16.11 The linear trend forecasting equation for an annual time series containing 42 values (from 1998 to 2019) on net sales (in $billions) is

$$\hat{Y}_i = 1.2 + 0.5X_i$$

a. Interpret the Y intercept, b_0.

b. Interpret the slope, b_1.

c. What is the fitted trend value for the tenth year?

d. What is the fitted trend value for the most recent year?

e. What is the projected trend forecast two years after the last value?

APPLYING THE CONCEPTS

✓SELF TEST **16.12** There has been much publicity about bonuses paid to workers on Wall Street. Just how large are these bonuses? The file **Bonuses** contains the bonuses paid (in $000) from 2000 to 2018.

Source: Data extracted from J. Spector, "Wall Street bonuses rise 1% to average $138,210," *USA Today*, March 15, 2017, and N. Chiwaya, "Wall Street bonuses declined 17 percent in 2018…," NBC News, March 26, 2019, available at **nbcnews .to/2Tch1Z9**.

a. Plot the data.

b. Compute a linear trend forecasting equation and plot the results.

c. Compute a quadratic trend forecasting equation and plot the results.

d. Compute an exponential trend forecasting equation and plot the results.

e. Using the forecasting equations in (b) through (d), what are your annual forecasts of the bonuses for 2019 and 2020?

f. How can you explain the differences in the three forecasts in (e)? What forecast do you think you should use? Why?

16.13 Gross domestic product (GDP) is a major indicator of a nation's overall economic activity. It consists of personal consumption expenditures, gross domestic investment, net exports of goods and services, and government consumption expenditures. The file **GDP** contains the GDP (in billions of current dollars) for the United States from 1980 to 2018.

Source: Data extracted from Bureau of Economic Analysis, U.S. Department of Commerce, **www.bea.gov**.

a. Plot the data.

b. Compute a linear trend forecasting equation and plot the trend line.

c. What are your forecasts for 2019 and 2020?

d. What conclusions can you reach concerning the trend in GDP?

16.14 The data in **Federal Receipts** represent federal receipts from 1978 through 2018 in billions of current dollars, from individual and corporate income tax, social insurance, excise tax, estate and gift tax, customs duties, and federal reserve deposits.

Source: Data extracted from "Historical Federal Receipt and Outlay Summary," Tax Policy Center, **tpc.io/1JMFKpo**.

a. Plot the series of data.

b. Compute a linear trend forecasting equation and plot the trend line.

c. What are your forecasts of the federal receipts for 2019 and 2020?

d. What conclusions can you reach concerning the trend in federal receipts?

16.15 The file **House Sales** contains the number of new, single-family houses sold in the United States from 1992 through 2018.

a. Plot the data.

b. Compute a linear trend forecasting equation and plot the trend line.

c. Compute a quadratic trend forecasting equation and plot the results.

d. Compute an exponential trend forecasting equation and plot the results.

e. Which model is the most appropriate?

f. Using the most appropriate model, forecast the number of new, single-family houses sold in the United States in 2019.

16.16 The data shown in the following table and stored in `Solar Power` represent the yearly amount of solar power generated by utilities (in millions of kWh) in the United States from 2002 through 2018:

Year	Solar Power Generated (millions of kWh)	Year	Solar Power Generated (millions of kWh)
2002	555	2010	1,212
2003	534	2011	1,818
2004	575	2012	4,327
2005	551	2013	9,253
2006	508	2014	18,321
2007	612	2015	26,473
2008	864	2016	36,754
2009	892	2017	52,958
		2018	66,604

Source: Data extracted from **en.wikipedia.org/wiki/Solar_power_in__the_ United_States**.

a. Plot the data.
b. Compute a linear trend forecasting equation and plot the trend line.
c. Compute a quadratic trend forecasting equation and plot the results.
d. Compute an exponential trend forecasting equation and plot the results.
e. Using the models in (b) through (d), what are your annual trend forecasts of the yearly amount of solar power generated by utilities (in millions of kWh) in the United States in 2019 and 2020?

16.17 The file `Auto Production` contains the number of automobiles assembled in the United States (in thousands) from 1999 to 2018.

Source: Data extracted from FRED, "Domestic Auto Production," **bit.ly/2KHKCpz**.

a. Plot the data.
b. Compute a linear trend forecasting equation and plot the trend line.
c. Compute a quadratic trend forecasting equation and plot the results.
d. Compute an exponential trend forecasting equation and plot the results.
e. Which model is the most appropriate?
f. Using the most appropriate model, forecast the U.S. automobiles assembled for 2019.

16.18 The average salary of Major League Baseball players on opening day from 2000 to 2019 is stored in `MLB Salaries` and in the following table.

Year	Salary ($millions)	Year	Salary ($millions)
2000	1.99	2010	3.27
2001	2.29	2011	3.32
2002	2.38	2012	3.38
2003	2.58	2013	3.62
2004	2.49	2014	3.81
2005	2.63	2015	4.25
2006	2.83	2016	4.40
2007	2.92	2017	4.70
2008	3.13	2018	4.63
2009	3.26	2019	4.70

Source: Data extracted from "Baseball Salaries," *USA Today*, April 6, 2009, p. 6C; **mlb.com**, *USA Today*, April 2, 2017, and "MLB Salaries - MLB Baseball - USA TODAY," **bit.ly/1kj4E4N**.

a. Plot the data.
b. Compute a linear trend forecasting equation and plot the trend line.
c. Compute a quadratic trend forecasting equation and plot the results.
d. Compute an exponential trend forecasting equation and plot the results.
e. Which model is the most appropriate?
f. Using the most appropriate model, forecast the average salary for 2020.

16.19 The file `Silver` contains the following prices in London for an ounce of silver (in US$) on the last day of the year from 1999 to 2018:

Year	Price (US$/ounce)	Year	Price (US$/ounce)
1999	5.330	2009	16.990
2000	4.570	2010	30.630
2001	4.520	2011	28.180
2002	4.670	2012	29.950
2003	5.965	2013	19.500
2004	6.815	2014	15.970
2005	8.830	2015	13.820
2006	12.900	2016	15.990
2007	14.760	2017	16.865
2008	10.790	2018	15.490

Source: Data extracted from JM Bullion, "Silver Spot Price & Charts," **bit.ly/2w4YPYI**.

a. Plot the data.
b. Compute a linear trend forecasting equation and plot the trend line.
c. Compute a quadratic trend forecasting equation and plot the results.
d. Compute an exponential trend forecasting equation and plot the results.
e. Which model is the most appropriate?
f. Using the most appropriate model, forecast the price of silver at the end of 2019.

16.20 The data in `CPI-U` reflect the annual values of the consumer price index for all urban consumers (CPI-U) in the United States over the 54-year period 1965 through 2018, using 1982 through 1986 as the base period. This index measures the average change in prices over time in a fixed "market basket" of goods and services purchased by all urban consumers, including urban clerical, professional, managerial, and technical workers; self-employed individuals; short-term workers; unemployed individuals; and retirees.

Source: Data extracted from Bureau of Labor Statistics, U.S. Department of Labor, **www.bls.gov**.

a. Plot the data.
b. Describe the movement in this time series over the 54-year period.
c. Compute a linear trend forecasting equation and plot the trend line.
d. Compute a quadratic trend forecasting equation and plot the results.
e. Compute an exponential trend forecasting equation and plot the results.
f. Which model is the most appropriate?
g. Using the most appropriate model, forecast the CPI for 2019 and 2020.

16.21 Although you should not expect a perfectly fitting model for any time-series data, you can consider the first differences, second differences, and percentage differences for a given series as guides in choosing an appropriate model. For this problem, use each of the time series presented in the following table and stored in Time Series A.

Year	Series I	Series II	Series III
2007	10.0	30.0	60.0
2008	15.1	33.1	67.9
2009	24.0	36.4	76.1
2010	36.7	39.9	84.0
2011	53.8	43.9	92.2
2012	74.8	48.2	100.0
2013	100.0	53.2	108.0
2014	129.2	58.2	115.8
2015	162.4	64.5	124.1
2016	199.0	70.7	132.0
2017	239.3	77.1	140.0
2018	283.5	83.9	147.8

a. Determine the most appropriate model.
b. Compute the forecasting equation.
c. Forecast the value for 2019.

16.22 A time-series plot often helps you determine the appropriate model to use. For this problem, use each of the time series presented in the following table and stored in Time Series B.

Year	Series I	Series II
2007	100.0	100.0
2008	115.2	115.2
2009	130.1	131.7
2010	144.9	150.8
2011	160.0	174.1
2012	175.0	200.0
2013	189.8	230.8
2014	204.9	266.1
2015	219.8	305.5
2016	235.0	351.8
2017	249.8	403.0
2018	264.9	469.2

a. Plot the observed data Y over time X and plot the logarithm of the observed data (log Y) over time X to determine whether a linear trend model or an exponential trend model is more appropriate. (Hint: If the plot of log Y versus X appears to be linear, an exponential trend model provides an appropriate fit.)
b. Compute the appropriate forecasting equations.
c. Forecast the values for 2019.

16.4 Autoregressive Modeling for Trend Fitting and Forecasting

Frequently, the values of a time series at particular points in time are highly correlated with the values that precede and succeed them. This type of correlation is called *autocorrelation*. When the autocorrelation exists between values that are in consecutive periods in a time series, the time series displays **first-order autocorrelation**. When the autocorrelation exists between values that are two periods apart, the time series displays **second-order autocorrelation**. For the general case in which the autocorrelation exists between values that are p periods apart, the time series displays *pth-order autocorrelation*.

Autoregressive modeling uses a set of *lagged predictor variables* to overcome the problems that autocorrelation causes with other models. A **lagged predictor variable** takes its value from the value of a predictor variable for a previous time period. To analyze pth-order autocorrelation, you create a set of p lagged predictor variables. The first lagged predictor variable takes its value from the value of a predictor variable that is one time period away, the *lag*; the second lagged predictor variable takes its value from the value of a predictor variable that is two time periods away; and so on until the pth lagged predictor variable that takes its value from the value of a predictor variable that is p time periods away. Note that each subsequent lagged predictor variable contains one less time-series value. In the general case, a p lagged variable will contain p less values.

Equation (16.8) defines the **pth-order autoregressive model**. In the equation, $A_0, A_1, \ldots, A_p$ represent the parameters and $a_0, a_1, \ldots, a_p$ represent the corresponding regression coefficients. This is similar to the multiple regression model, Equation (14.1) on page 480, in which $\beta_0, \beta_1, \ldots, \beta_k$, represent the regression parameters and $b_0, b_1, \ldots, b_k$ represent the corresponding regression coefficients.

learnMORE

The exponential smoothing model that Section 16.3 describes and the autoregressive models that Section 16.4 describes are special cases of autoregressive integrated moving average (ARIMA) models developed by Box and Jenkins. To learn more about such models, see Bisgaard and Kulahci; Box, Jenkins, Reinsel and Leung; and Pecar.

pTH-ORDER AUTOREGRESSIVE MODELS

$$Y_i = A_0 + A_1 Y_{i-1} + A_2 Y_{i-2} + \cdots + A_p Y_{i-p} + \delta_i \tag{16.8}$$

where

$$
\begin{aligned}
Y_i &= \text{observed value of the series at time i} \\
Y_{i-1} &= \text{observed value of the series at time } i - 1 \\
Y_{i-2} &= \text{observed value of the series at time } i - 2 \\
Y_{i-p} &= \text{observed value of the series at time } i - p \\
p &= \text{number of autoregression parameters (not including a } Y \text{ intercept)} \\
&\quad \text{to be estimated from least-squares regression analysis} \\
A_0, A_1, A_2, \ldots, A_p &= \text{autoregression parameters to be estimated from least-squares} \\
&\quad \text{regression analysis} \\
\delta_i &= \text{a nonautocorrelated random error component (with mean } = 0 \\
&\quad \text{and constant variance)}
\end{aligned}
$$

student TIP

δ is the Greek letter delta.

Equations (16.9) and (16.10) define two specific autoregressive models. Equation (16.9) defines the **first-order autoregressive model** and is similar in form to the simple linear regression model, Equation (13.1) on page 432. Equation (16.10) defines the **second-order autoregressive model** and is similar to the multiple regression model with two independent variables, Equation (14.2) on page 480.

FIRST-ORDER AUTOREGRESSIVE MODEL

$$Y_i = A_0 + A_1 Y_{i-1} + \delta_i \tag{16.9}$$

SECOND-ORDER AUTOREGRESSIVE MODEL

$$Y_i = A_0 + A_1 Y_{i-1} + A_2 Y_{i-2} + \delta_i \tag{16.10}$$

Selecting an Appropriate Autoregressive Model

Selecting an appropriate autoregressive model can be complicated. One must weigh the advantages of using a simpler model against the concern of using a model that does not take into account important autocorrelation in the data. On the other hand, selecting a higher-order model that requires estimates of numerous parameters may contain some unnecessary parameters, especially if the time series is short (n is small). Recall that when computing an estimate of A_p, p out of the n time series values are lost due to the lagging of values. Examples 16.5 and 16.6 illustrate this loss.

EXAMPLE 16.5

Comparison Schema for a First-Order Autoregressive Model

Consider the following series of $n = 7$ consecutive annual values:

				Year			
	1	**2**	**3**	**4**	**5**	**6**	**7**
Series	31	34	37	35	36	43	40

Show the comparisons needed for a first-order autoregressive model.

▶(*continued*)

SOLUTION

Year i	First-Order Autoregressive Model (Lag1: Y_i versus Y_{i-1})
1	$31 \leftrightarrow \ldots$
2	$34 \leftrightarrow 31$
3	$37 \leftrightarrow 34$
4	$35 \leftrightarrow 37$
5	$36 \leftrightarrow 35$
6	$43 \leftrightarrow 36$
7	$40 \leftrightarrow 43$

Because Y_1 is the first value and there is no value prior to it, Y_1 is not used in the regression analysis. Therefore, the first-order autoregressive model would be based on six pairs of values.

EXAMPLE 16.6

Comparison Schema for a Second-Order Autoregressive Model

Consider the following series of $n = 7$ consecutive annual values:

	Year						
	1	**2**	**3**	**4**	**5**	**6**	**7**
Series	31	34	37	35	36	43	40

Show the comparisons needed for a second-order autoregressive model.

SOLUTION

Year i	Second-Order Autoregressive Model Lag2: Y_i vs. Y_{i-1} and Y_i vs. Y_{i-2}
1	$31 \leftrightarrow \ldots$ and $31 \leftrightarrow \ldots$
2	$34 \leftrightarrow 31$ and $34 \leftrightarrow \ldots$
3	$37 \leftrightarrow 34$ and $37 \leftrightarrow 31$
4	$35 \leftrightarrow 37$ and $35 \leftrightarrow 34$
5	$36 \leftrightarrow 35$ and $36 \leftrightarrow 37$
6	$43 \leftrightarrow 36$ and $43 \leftrightarrow 35$
7	$40 \leftrightarrow 43$ and $40 \leftrightarrow 36$

Because no value is recorded prior to Y_1, the first two comparisons, each of which requires a value prior to Y_1, cannot be used when performing regression analysis. Therefore, the second-order autoregressive model would be based on five pairs of values.

Determining the Appropriateness of a Selected Model

After selecting a model and using the least-squares method to compute the regression coefficients, one must determine the appropriateness of the model. One either selects a particular pth-order autoregressive model based on previous experiences with similar data or starts with a model that contains several autoregressive parameters and then eliminates the higher-order parameters that do not significantly contribute to the model. In this latter approach, one uses a t test for the significance of A_p, the highest-order autoregressive parameter in the current model under consideration. The null and alternative hypotheses are

$$H_0: A_p = 0$$

$$H_1: A_p \neq 0$$

Equation (16.11) defines the test statistic.

---

t TEST FOR SIGNIFICANCE OF THE HIGHEST-ORDER AUTOREGRESSIVE PARAMETER, A_p

$$t_{STAT} = \frac{a_p - A_p}{S_{a_p}}$$

(16.11)

where

A_p = hypothesized value of the highest-order parameter, A_p, in the autoregressive model

a_p = regression coefficient that estimates the highest-order parameter, A_p, in the autoregressive model

S_{a_p} = standard deviation of a_p

The t_{STAT} test statistic follows a t distribution with $n - 2p - 1$ degrees of freedom. In addition to the degrees of freedom lost for each of the p population parameters being estimated, an additional p degrees of freedom are lost because there are p fewer comparisons to be made from the original n values in the time series.

For a given level of significance, α, you reject the null hypothesis if the t_{STAT} test statistic is greater than the upper-tail critical value from the t distribution or if the t_{STAT} test statistic is less than the lower-tail critical value from the t distribution. Thus, the decision rule is

Reject H_0 if $t_{STAT} < -t_\alpha/2$ or if $t_{STAT} > t_{a/2}$;

otherwise, do not reject H_0.

Figure 16.13 illustrates the decision rule and regions of rejection and nonrejection.

FIGURE 16.13
Rejection regions for a two-tail test for the significance of the highest-order autoregressive parameter A_p

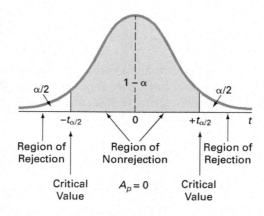

If one does not reject the null hypothesis that $A_p = 0$, one concludes that the selected model contains too many estimated autoregressive parameters. One discards the highest-order term and develops an autoregressive model of order $p - 1$, using the least-squares method. Then, one repeats the test of the hypothesis that the new highest-order parameter is 0. This cycle of testing and modeling continues until one rejects H_0. When this occurs, one can conclude that the remaining highest-order parameter is significant and that the model is acceptable for forecasting purposes.

Equation (16.12) defines the fitted *p*th-order autoregressive equation.

FITTED *p*TH-ORDER AUTOREGRESSIVE EQUATION

$$\hat{Y}_i = a_0 + a_1 Y_{i-1} + a_2 Y_{i-2} + \cdots + a_p Y_{i-p} \tag{16.12}$$

where

$\hat{Y}_i$ = fitted values of the series at time i

Y_{i-1} = observed value of the series at time $i - 1$

Y_{i-2} = observed value of the series at time $i - 2$

Y_{i-p} = observed value of the series at time $i - p$

p = number of autoregression parameters (not including a Y intercept) to be estimated from least-squares regression analysis

$a_0, a_1, a_2, \ldots, a_p$ = regression coefficients

One uses Equation (16.13) to forecast j years into the future from the current nth time period.

*p*TH-ORDER AUTOREGRESSIVE FORECASTING EQUATION

$$\hat{Y}_{n+j} = a_0 + a_1 \hat{Y}_{n+j-1} + a_2 \hat{Y}_{n+j-2} + \cdots + a_p \hat{Y}_{n+j-p} \tag{16.13}$$

where

$a_0, a_1, a_2, \ldots, a_p$ = regression coefficients that estimate the parameters

p = number of autoregression parameters (not including a Y intercept) to be estimated from least-squares regression analysis

j = number of years into the future

$\hat{Y}_{n+j-p}$ = forecast of Y_{n+j-p} from the current year for $j - p > 0$

$\hat{Y}_{n+j-p}$ = observed value for Y_{n+j-p} for $j - p \leq 0$

Thus, to make forecasts j years into the future, using a third-order autoregressive model, one needs only the most recent $p = 3$ values (Y_n, Y_{n-1}, and Y_{n-2}) and the regression estimates a_0, a_1, a_2, and a_3.

To forecast one year ahead, Equation (16.13) becomes

$$\hat{Y}_{n+1} = a_0 + a_1 Y_n + a_2 Y_{n-1} + a_3 Y_{n-2}$$

To forecast two years ahead, Equation (16.13) becomes

$$\hat{Y}_{n+2} = a_0 + a_1 \hat{Y}_{n+1} + a_2 Y_n + a_3 Y_{n-1}$$

To forecast three years ahead, Equation (16.13) becomes

$$\hat{Y}_{n+3} = a_0 + a_1 \hat{Y}_{n+2} + a_2 \hat{Y}_{n+1} + a_3 Y_n$$

and so on.

Autoregressive modeling is a powerful forecasting technique for time series that have auto-correlation. Exhibit 16.2 summarizes the steps to construct an autoregressive model.

EXHIBIT 16.2

Autoregressive Modeling Steps

1. Choose a value for p, the highest-order parameter in the autoregressive model to be evaluated, remembering that the t test for significance is based on $n - 2p - 1$ degrees of freedom.
2. Create a set of p lagged predictor variables. (See Figure 16.14 for an example.)
3. Perform a least-squares analysis of the multiple regression model containing all p lagged predictor variables.
4. Test for the significance of A_p, the highest-order autoregressive parameter in the model.
5. If one does not reject the null hypothesis, discard the pth variable and repeat steps 3 and 4 with a revised degrees of freedom that correspond to the revised number of predictors.

   If one rejects the null hypothesis, select the autoregressive model with all p predictors for fitting [see Equation (16.12)] and forecasting [see Equation (16.13)].

student TIP

Remember that in an autoregressive model, the independent variable(s) are equal to the dependent variable lagged by a certain number of time periods.

To demonstrate the autoregressive modeling approach, consider the Coca-Cola Company annual revenue for the years 2000 through 2018 (see Figure 16.14). Figure 16.14 presents a worksheet template that computes three lagged predictor variables, Lag1, Lag2, and Lag3, that can be used for the first-order, second-order, and third-order autoregressive models, using the Coca-Cola Company annual revenue data.

FIGURE 16.14

Excel worksheet template for computing lagged predictor variables for the first-order, second-order, and third-order autoregressive models for the Coca-Cola Company annual revenue, 2000–2018

	A	B	C	D	E
1	Year	Revenues	Lag1	Lag2	Lag3
2	2000	20.5	#N/A	#N/A	#N/A
3	2001	20.1	20.5	#N/A	#N/A
4	2002	19.6	20.1	20.5	#N/A
5	2003	21.0	19.6	20.1	20.5
6	2004	21.9	21.0	19.6	20.1
7	2005	23.1	21.9	21.0	19.6
8	2006	24.1	23.1	21.9	21.0
9	2007	28.9	24.1	23.1	21.9
10	2008	31.9	28.9	24.1	23.1
11	2009	31.0	31.9	28.9	24.1
12	2010	35.1	31.0	31.9	28.9
13	2011	46.5	35.1	31.0	31.9
14	2012	48.0	46.5	35.1	31.0
15	2013	46.7	48.0	46.5	35.1
16	2014	45.9	46.7	48.0	46.5
17	2015	44.3	45.9	46.7	48.0
18	2016	41.9	44.3	45.9	46.7
19	2017	35.4	41.9	44.3	45.9
20	2018	31.9	35.4	41.9	44.3

To fit the third-order autoregressive model, all three lagged predictor variables are used. To fit the second-order autoregressive model, only the Lag1 and Lag2 variables are used. To fit the first-order autoregressive model, only the Lag1 variable is used.

Selecting the autoregressive model that best fits an annual time series begins with the highest order autoregressive model to be evaluated. Choosing which model will be the highest order model to be evaluated can be based on prior experience using a time series. When no past experience or insights exists, one often chooses to use a third-order model as the starting point, thereby making $p = 3$. Because the Coca-Cola Company revenue time series is being evaluated without prior experience, the highest-order model to evaluate is the third-order autoregressive model.

From Figure 16.15, the fitted third-order autoregressive equation is

$$\hat{Y}_i = 5.1923 + 1.4023Y_{i-1} - 0.6557Y_{i-2} + 0.1128Y_{i-3}$$

where the first year in the series is 2003.

FIGURE 16.15

Excel regression worksheet results for a third-order autoregressive model for the Coca-Cola Company annual revenue

	A	B	C	D	E	F	G
1	Third-Order Autoregressive Model for The Coca Cola Company Annual Revenue						
2							
3	*Regression Statistics*						
4	Multiple R	0.9441					
5	R Square	0.8913					
6	Adjusted R Square	0.8641					
7	Standard Error	3.5505					
8	Observations	16					
9							
10	ANOVA						
11		*df*	*SS*	*MS*	*F*	*Significance F*	
12	Regression	3	1240.0478	413.3493	32.7899	0.0000	
13	Residual	12	151.2722	12.6060			
14	Total	15	1391.3200				
15							
16		*Coefficients*	*Standard Error*	*t Stat*	*P-value*	*Lower 95%*	*Upper 95%*
17	Intercept	5.1923	3.1945	1.6254	0.1300	-1.7679	12.1524
18	Lag1	1.4023	0.2816	4.9795	0.0003	0.7887	2.0159
19	Lag2	-0.6557	0.4782	-1.3714	0.1954	-1.6976	0.3861
20	Lag3	0.1128	0.3096	0.3645	0.7219	-0.5618	0.7875

One evaluates this model by testing for the significance of A_3, the highest-order parameter. The null and alternative hypotheses are

$$H_0: A_3 = 0$$
$$H_1: A_3 \neq 0$$

The highest-order regression coefficient, a_3, for the fitted third-order autoregressive model is 0.1128, with a standard error of 0.3096. Using Equation (16.11) on page 576 and the worksheet results given in Figure 16.15,

$$t_{STAT} = \frac{a_3 - A_3}{S_{a_3}} = \frac{0.1128 - 0}{0.3096} = 0.3645$$

Using a 0.05 level of significance, the two-tail t test with 12 degrees of freedom has critical values of ± 2.1788. Because $-2.1788 < t_{STAT} = 0.3645 < 2.1788$ or because the p-value $= 0.7219 > 0.05$, one does not reject H_0. One concludes that the third-order parameter of the autoregressive model is not significant and should not remain in the model. Therefore, one continues by fitting the Figure 16.16 second-order autoregressive model.

The fitted second-order autoregressive equation is

$$\hat{Y}_i = 4.5737 + 1.3856Y_{i-1} - 0.5156, Y_{i-2}$$

where the first year of the series is 2002.

FIGURE 16.16

Excel regression worksheet results for the second-order autoregressive model for the Coca-Cola Company annual revenue

	A	B	C	D	E	F	G
1	Second-Order Autoregressive Model for The Coca Cola Company Annual Revenue						
2							
3	*Regression Statistics*						
4	Multiple R	0.9494					
5	R Square	0.9013					
6	Adjusted R Square	0.8872					
7	Standard Error	3.3697					
8	Observations	17					
9							
10	ANOVA						
11		*df*	*SS*	*MS*	*F*	*Significance F*	
12	Regression	2	1451.2332	725.6166	63.9032	0.0000	
13	Residual	14	158.9691	11.3549			
14	Total	16	1610.2024				
15							
16		*Coefficients*	*Standard Error*	*t Stat*	*P-value*	*Lower 95%*	*Upper 95%*
17	Intercept	4.5737	2.7725	1.6497	0.1213	-1.3727	10.5200
18	Lag1	1.3856	0.2329	5.9502	0.0000	0.8861	1.8850
19	Lag2	-0.5158	0.2240	-2.3032	0.0371	-0.9962	-0.0355

One evaluates this model by testing for the significance of the A_2 parameter. The null and alternative hypotheses are:

$$H_0: A_2 = 0$$
$$H_1: A_2 \neq 0$$

From Figure 16.15, the highest-order parameter estimate is $a_2 = -0.5158$, with a standard error of 0.2240. Using Equation (16.11) on page 576,

$$t_{STAT} = \frac{a_2 - A_2}{S_{a_2}} = \frac{-0.5158 - 0}{0.2240} = -2.3032$$

Using the 0.05 level of significance, the two-tail t test with 14 degrees of freedom has critical values of ± 2.1448. Because $t_{STAT} = -2.3032 < -2.1488$ or because the p-value $= 0.0371 < 0.05$, one rejects H_0. One concludes that the second-order parameter of the autoregressive model is significant and should be included in the model.

The model-building approach has led to the selection of the second-order autoregressive model as the most appropriate for these data. Using the estimates $a_0 = 4.5737$, and $a_1 = 1.3856$, as well as the most recent data value $Y_{19} = 31.9$, the forecasts of revenue at the Coca-Cola Company for 2019 from Equation (16.13) on page 577 are

$$\hat{Y}_{n+j} = 4.5737 + 1.3856\hat{Y}_{n+j-1} + -0.5158\hat{Y}_{n+j-2}$$

Therefore,

2019: 1 year ahead, $\hat{Y}_{20} = 4.5737 + 1.3856(31.9) - 0.5158(35.4) = 30.525$ billions of dollars

Figure 16.17 displays the actual and predicted Y values from the second-order autoregressive model.

FIGURE 16.17

Plot of actual and predicted revenue from a second-order autoregressive model at the Coca-Cola Company

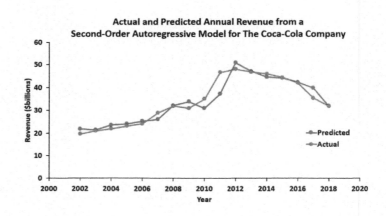

Actual and Predicted Annual Revenue from a
Second-Order Autoregressive Model for The Coca-Cola Company

EXAMPLE 16.7

Fitting a First-Order Autoregressive Model for Alphabet Inc. Annual Revenue

▶(*continued*)

In Section 16.3, linear, quadratic, and exponential models were fitted to Alphabet Inc. annual revenue from 2011 to 2018. Fit an autoregressive model to these data.

SOLUTION　Because the sample size consists of only eight years, only a first-order autoregressive model should be considered. Figure 16.18 presents that model.

FIGURE 16.18
Excel regression work-sheet results for the first-order autoregressive model for Alphabet Inc. annual revenue

▲	A	B	C	D	E	F	G
1	First-Order Autoregressive Model for Alphabet Inc. Annual Revenue						
2							
3	*Regression Statistics*						
4	Multiple R	0.9978					
5	R Square	0.9957					
6	Adjusted R Square	0.9948					
7	Standard Error	2.3203					
8	Observations	7					
9							
10	ANOVA						
11		*df*	*SS*	*MS*	*F*	*Significance F*	
12	Regression	1	6172.7764	6172.7764	1146.5968	0.0000	
13	Residual	5	26.9178	5.3836			
14	Total	6	6199.6942				
15							
16		*Coefficients*	*Standard Error*	*t Stat*	*P-value*	*Lower 95%*	*Upper 95%*
17	Intercept	-3.3511	2.6947	-1.2436	0.2688	-10.2781	3.5758
18	Lag1	1.2541	0.0370	33.8614	0.0000	1.1589	1.3493

The fitted regression model is

$$\text{Predicted } Y_i = -3.3511 + 1.2541 Y_{i-1}$$

One evaluates this model by testing for the significance of the A_1 parameter. The null and alternative hypotheses are

$$H_0: A_1 = 0$$
$$H_1: A_1 \neq 0$$

Using Equation (16.11) on page 576 and the 0.05 level of significance, the two-tail t test with 5 degrees of freedom has critical values of -2.5706 and $+2.5706$. Because $t_{STAT} = 33.8614 > 2.5706$ or because the p-value $= 0.0000 < 0.05$, one concludes that the first-order parameter of the autoregressive model is significant and should remain in the model. One also notes that $r^2 = 0.9957$, which indicates the 99.57% of the variation in revenue is explained by the model.

PROBLEMS FOR SECTION 16.4

LEARNING THE BASICS

16.23 You are given an annual time series with 40 consecutive values and asked to fit a fifth-order autoregressive model.
a. How many comparisons are lost in developing the autoregressive model?
b. How many parameters do you need to estimate?
c. Which of the original 40 values do you need for forecasting?
d. State the fifth-order autoregressive model.
e. Write an equation to indicate how you would forecast j years into the future.

16.24 A third-order autoregressive model is fitted to an annual time series with 17 values and has the following estimated parameters and standard errors:

$$a_0 = 4.50 \quad a_1 = 1.80 \quad a_2 = 0.80 \quad a_3 = 0.24$$
$$S_{a_1} = 0.50 \quad S_{a_2} = 0.30 \quad S_{a_3} = 0.10$$

At the 0.05 level of significance, test the appropriateness of the fitted model.

16.25 Refer to Problem 16.24. The three most recent values are

$$Y_{15} = 23 \quad Y_{16} = 28 \quad Y_{17} = 34$$

Forecast the values for the next year and the following year.

16.26 Refer to Problem 16.24. Suppose, when testing for the appropriateness of the fitted model, the standard errors are

$$S_{a_1} = 0.45 \quad S_{a_2} = 0.35 \quad S_{a_3} = 0.15$$

a. What conclusions can you reach?
b. Discuss how to proceed if forecasting is still your main objective.

APPLYING THE CONCEPTS

16.27 Using the data for Problem 16.15 on page 571 that represent the number of new, single-family houses sold in the United States from 1992 through 2018 (stored in House Sales),

a. fit a third-order autoregressive model to the new single-family homes sold and test for the significance of the third-order autoregressive parameter. (Use $\alpha = 0.05$.)

b. if necessary, fit a second-order autoregressive model to the new single-family homes sold and test for the significance of the second-order autoregressive parameter. (Use $\alpha = 0.05$.)

c. if necessary, fit a first-order autoregressive model to the new single-family homes sold and test for the significance of the first-order autoregressive parameter. (Use $\alpha = 0.05$.)

d. if appropriate, forecast the new single-family homes sold in 2019.

✓**SELF TEST** **16.28** Using the data for Problem 16.12 on page 571 concerning the bonuses paid to workers on Wall Street from 2000 to 2018 (stored in Bonuses),

a. fit a third-order autoregressive model to the bonuses paid and test for the significance of the third-order autoregressive parameter. (Use $\alpha = 0.05$.)

b. if necessary, fit a second-order autoregressive model to the bonuses paid and test for the significance of the second-order autoregressive parameter. (Use $\alpha = 0.05$.)

c. if necessary, fit a first-order autoregressive model to the bonuses paid and test for the significance of the first-order autoregressive parameter. (Use $\alpha = 0.05$.)

d. if appropriate, forecast the bonuses paid in 2019.

16.29 Using the data for Problem 16.17 on page 572 concerning the number of automobiles assembled in the United States from 1999 to 2018 (stored in Auto Production),

a. fit a third-order autoregressive model to the number of automobiles assembled in the United States and test for the significance of the third-order autoregressive parameter. (Use $\alpha = 0.05$.)

b. if necessary, fit a second-order autoregressive model to the number of automobiles assembled in the United States and test for the significance of the second-order autoregressive parameter. (Use $\alpha = 0.05$.)

c. if necessary, fit a first-order autoregressive model to the number of automobiles assembled in the United States and test for the significance of the first-order autoregressive parameter. (Use $\alpha = 0.05$.)

d. forecast the number of automobiles assembled in the United States for 2019.

16.30 Using the average baseball salary from 2000 through 2019 data for Problem 16.18 on page 572 (stored in MLB Salaries),

a. fit a third-order autoregressive model to the average baseball salary and test for the significance of the third-order autoregressive parameter. (Use $\alpha = 0.05$.)

b. if necessary, fit a second-order autoregressive model to the average baseball salary and test for the significance of the second-order autoregressive parameter. (Use $\alpha = 0.05$.)

c. if necessary, fit a first-order autoregressive model to the average baseball salary and test for the significance of the first-order autoregressive parameter. (Use $\alpha = 0.05$.)

d. forecast the average baseball salary for 2020.

16.31 Using the yearly amount of solar power generated by utilities (in millions of kWh) in the United States from 2002 through 2018 data for Problem 16.16 on page 572 (stored in Solar Power),

a. fit a third-order autoregressive model to the amount of solar power installed and test for the significance of the third-order autoregressive parameter. (Use $\alpha = 0.05$.)

b. if necessary, fit a second-order autoregressive model to the amount of solar power installed and test for the significance of the second-order autoregressive parameter. (Use $\alpha = 0.05$.)

c. if necessary, fit a first-order autoregressive model to the amount of solar power installed and test for the significance of the first-order autoregressive parameter. (Use $\alpha = 0.05$.)

d. forecast the yearly amount of solar power generated by utilities (in millions of kWh) in the United States in 2019.

16.5 Choosing an Appropriate Forecasting Model

The previous two sections discuss six time-series methods for forecasting: the linear trend model, the quadratic trend model, and the exponential trend model (Section 16.3) and the first-order, second-order, and pth-order autoregressive models (Section 16.4). To choose which one of the six models should be used for forecasting, one should consider these four criteria:

- The results from a residual analysis.
- The magnitude of the residuals through squared differences.
- The magnitude of the residuals through absolute differences.
- The principle of parsimony.

Residual Analysis

Sections 13.5 and 14.3 define residuals as the differences between observed and predicted values. After fitting a particular model to a time series, one plots the residuals over the n time periods. As shown in Figure 16.19 Panel A, if the particular model fits adequately, the residuals represent the irregular component of the time series. Therefore, they should be randomly distributed throughout the series. However, as illustrated in the three remaining

FIGURE 16.19
Residual analysis for studying patterns of errors in regression models

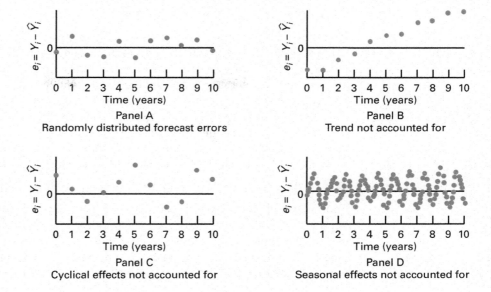

Panel A
Randomly distributed forecast errors

Panel B
Trend not accounted for

Panel C
Cyclical effects not accounted for

Panel D
Seasonal effects not accounted for

panels of Figure 16.19, if the particular model does not fit adequately, the residuals may show a systematic pattern, such as a failure to account for trend (Panel B), a failure to account for cyclical variation (Panel C), or, with monthly or quarterly data, a failure to account for seasonal variation (Panel D).

The Magnitude of the Residuals Through Squared or Absolute Differences

If, after performing a residual analysis, you still believe that two or more models appear to fit the data adequately, one can use additional methods for model selection. Numerous measures based on the residuals are available (see Bowerman, O'Connell, and Koehler; Box et al.).

In regression analysis (see Section 13.3), you have already used the standard error of the estimate S_{YX} as a measure of variation around the predicted values. For a particular model, this measure is based on the sum of squared differences between the actual and predicted values in a time series. If a model fits the time-series data perfectly, then the standard error of the estimate is zero. If a model fits the time-series data poorly, then S_{YX} is large. Thus, when comparing the adequacy of two or more forecasting models, you can select the model with the smallest S_{YX} as most appropriate.

However, a major drawback to using S_{YX} when comparing forecasting models is that whenever there is a large difference between even a single Y_i and $\hat{Y}_i$, the value of S_{YX} becomes overly inflated because the differences between Y_i and $\hat{Y}_i$ are squared. For this reason, many statisticians prefer the **mean absolute deviation (MAD)**. Equation (16.14) defines the *MAD* as the mean of the absolute differences between the actual and predicted values in a time series.

MEAN ABSOLUTE DEVIATION

$$MAD = \frac{\sum_{i=1}^{n} |Y_i - \hat{Y}_i|}{n}$$

(16.14)

If a model fits the time-series data perfectly, the *MAD* is zero. If a model fits the time-series data poorly, the *MAD* is large. When comparing two or more forecasting models, one can select the model with the smallest *MAD* as the most appropriate.

The Principle of Parsimony

If, after performing a residual analysis and comparing the S_{YX} and *MAD* measures, one still believes that two or more models appear to adequately fit the data, one can use the principle of parsimony for model selection. As Section 15.4 first explains, **parsimony** guides one to select the regression model with the fewest independent variables that can predict the dependent variable adequately. In general, the principle of parsimony guides one to select the least complex regression model. Among the six forecasting models studied in this chapter, most statisticians consider the least-squares linear and quadratic models and the first-order autoregressive model as simpler than the second and *p*th-order autoregressive models and the least-squares exponential model.

A Comparison of Four Forecasting Methods

To illustrate the model selection process, one can compare four of the forecasting models that Sections 16.3 and 16.4 discuss: the linear model, the quadratic model, the exponential model, and the first-order autoregressive model. Figure 16.20 presents the residual plots for the four models for a time series being analyzed. In reaching conclusions from these residual plots, you must use caution because there are only 19 values for the linear model, the quadratic model, and the exponential model and only 18 values for the first-order autoregressive model.

In Figure 16.20, observe that the residuals in the linear model, quadratic model, and exponential model are positive for the early years, negative for the intermediate years, and positive again for the later years, and negative again for the latest years. For the first-order autoregressive models the residuals do not exhibit any clear systematic pattern although one residual is highly positive.

FIGURE 16.20
Residual plots for four forecasting models

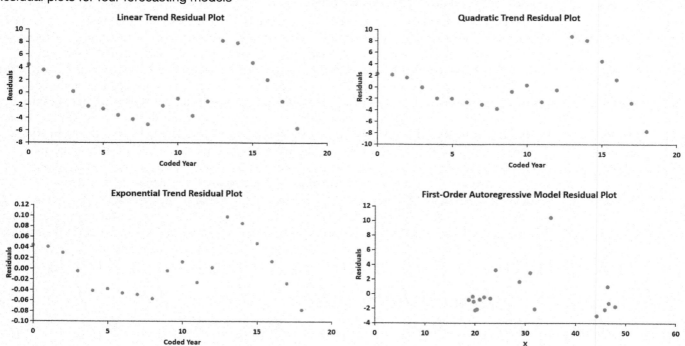

To summarize, on the basis of the residual analysis of all four forecasting models, it appears that the first-order autoregressive model is the most appropriate, and the linear, quadratic, and exponential models are not appropriate.

For further verification, one can compare the magnitude of the residuals in the four models. Figure 16.21 shows the actual values (Y_i) along with the predicted values $\hat{Y}_i$, the residuals (e_i), the error sum of squares (SSE), the standard error of the estimate (S_{YX}), and the mean absolute deviation (MAD) for each of the four models.

For the Alphabet Inc. annual revenue time series, the first-order autoregressive model provides the smallest SSE, followed by the exponential, quadratic, and the linear model. The first-order autoregressive model also contains the smallest S_{YX}. The MAD for the first-order autoregressive model is also less than the MAD for the other three models. Therefore, based on these results and the residual plots (not shown), one would select the first-order autoregressive model.

FIGURE 16.21

Comparison of four forecasting models using SSE, S_{YX}, and MAD

		Linear		Quadratic		Exponential		First-Order AR	
Year	Revenue	Predicted	Residual	Predicted	Residual	Predicted	Residual	Predicted	Residual
2011	37.9050	30.2280	7.6770	40.1296	-2.2246	38.1385	-0.2335	#N/A	#N/A
2012	46.0390	43.6771	2.3619	45.0916	0.9474	45.5757	0.4633	44.1857	1.8533
2013	55.5190	57.1262	-1.6072	52.8827	2.6363	54.4632	1.0558	54.3866	1.1324
2014	66.0010	70.5753	-4.5743	63.5027	2.4983	65.0839	0.9171	66.2755	-0.2745
2015	74.9890	84.0244	-9.0354	76.9518	-1.9628	77.7756	-2.7866	79.4211	-4.4321
2016	90.2720	97.4735	-7.2015	93.2300	-2.9580	92.9423	-2.6703	90.6930	-0.4210
2017	110.8550	110.9226	-0.0676	112.3372	-1.4822	111.0666	-0.2116	109.8594	0.9956
2018	136.8190	124.3718	12.4473	134.2734	2.5456	132.7252	4.0938	135.6727	1.1463
		SSE	376.4621	SSE	40.3175	SSE	33.9247	SSE	26.9178
		S_{YX}	7.9211	S_{YX}	2.8396	S_{YX}	2.3778	S_{YX}	2.3203
		MAD	5.6215	MAD	2.1569	MAD	1.5540	MAD	1.4650

After one selects a particular forecasting model, one continually monitors and updates the model. Once new data become available, one can use these data points to refine and improve the model.

Forecasting methods that this chapter discuss assume that patterns in the data seen in historical data will continue for future data. If large errors between forecasted and actual values occur, the underlying structure of the time series may have changed and the assumption may no longer be valid.

PROBLEMS FOR SECTION 16.5

LEARNING THE BASICS

16.32 The following residuals are from a linear trend model used to forecast sales:

2.0 −0.5 1.5 1.0 0.0 1.0 −3.0 1.5 −4.5 2.0 0.0 −1.0

a. Compute S_{YX} and interpret your findings.
b. Compute the MAD and interpret your findings.

16.33 Refer to Problem 16.32. Suppose the first residual is 12.0 (instead of 2.0) and the last residual is −11.0 (instead of −1.0).

Compute S_{YX} and interpret your findings Compute the MAD and interpret your findings.

APPLYING THE CONCEPTS

16.34 Using the yearly amount of solar power generated by utilities (in millions of kWh) in the United States data for Problem 16.16 on page 572 and Problem 16.31 on page 582 (stored in `Solar Power`),
a. perform a residual analysis.
b. compute the standard error of the estimate (S_{YX}).

c. compute the MAD.
d. On the basis of (a) through (c), and the principle of parsimony, which forecasting model would you select? Discuss.

16.35 Using the new, single-family house sales data for Problem 16.15 on page 571 and Problem 16.27 on page 582 (stored in `House Sales`),
a. perform a residual analysis for each model.
b. compute the standard error of the estimate (S_{YX}) for each model.
c. compute the MAD for each model.
d. On the basis of (a) through (c) and the principle of parsimony, which forecasting model would you select? Discuss.

✓**SELF TEST** **16.36** Using the bonuses paid to workers on Wall Street data for Problem 16.12 on page 571 and Problem 16.28 on page 582 (stored in `Bonuses`),
a. perform a residual analysis for each model.
b. compute the standard error of the estimate (S_{YX}) for each model.
c. compute the MAD for each model.
d. On the basis of (a) through (c) and the principle of parsimony, which forecasting model would you select? Discuss.

16.37 Using the data for the number of automobiles assembled in the United States for Problem 16.17 on page 572 and Problem 16.29 on page 582 (stored in Auto Production),
a. perform a residual analysis for each model.
b. compute the standard error of the estimate (S_{YX}) for each model.
c. compute the *MAD* for each model.
d. On the basis of (a) through (c) and the principle of parsimony, which forecasting model would you select? Discuss.

16.38 Using the average baseball salary data for Problem 16.18 on page 572 and Problem 16.30 on page 582 (stored in MLB Salaries),
a. perform a residual analysis for each model.
b. compute the standard error of the estimate (S_{YX}) for each model.

c. compute the *MAD* for each model.
d. On the basis of (a) through (c) and the principle of parsimony, which forecasting model would you select? Discuss.

16.39 Referring to the results for Problem 16.13 on page 571 that used the file GDP ,
a. perform a residual analysis.
b. compute the standard error of the estimate (S_{YX}).
c. compute the *MAD*.
d. On the basis of (a) through (c), are you satisfied with your linear trend forecasts in Problem 16.13? Discuss.

16.6 Time-Series Forecasting of Seasonal Data

As Section 16.1 first mentions, time-series data that are collected in intervals more frequently than annually, such as quarterly, monthly, weekly, or daily time series, may contain a seasonal component. To illustrate forecasting with seasonal data, consider the Table 16.3 time series that presents six years' worth of quarterly revenues for Amazon.com Inc., the global technology company based in Seattle. Figure 16.22 visualizes this time series that is stored in Amazon .

TABLE 16.3
Quarterly revenue (in $billions) for Amazon.com, Inc., 2013–2018

	Year					
Quarter	**2013**	**2014**	**2015**	**2016**	**2017**	**2018**
1	16.070	19.741	22.717	25.128	35.714	51.042
2	15.704	19.340	23.185	30.404	37.955	52.888
3	17.092	20.579	25.358	32.714	43.744	55.576
4	25.586	29.328	35.746	43.741	60.453	72.383

Source: Data extracted from **ycharts.com/companies/AMZN/revenues**.

FIGURE 16.22
Time-series plot of quarterly revenue for Amazon.com, Inc., 2013–2018

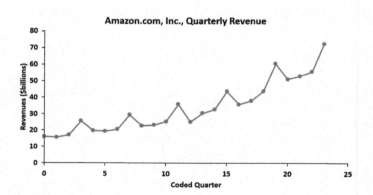

Amazon.com, Inc., Quarterly Revenue

Least-Squares Forecasting with Monthly or Quarterly Data

To develop a least-squares regression model that includes a seasonal component, the least-squares exponential trend fitting method used in Section 16.3 is combined with dummy variables to represent the quarters (see Section 14.6) to model the seasonal component.

Equation (16.15) defines the exponential trend model for quarterly data.

EXPONENTIAL MODEL WITH QUARTERLY DATA

$$Y_i = \beta_0 \beta_1^{X_i} \beta_2^{Q_1} \beta_3^{Q_2} \beta_4^{Q_3} \varepsilon_i \tag{16.15}$$

where

$$X_i = \text{coded quarterly value, } i = 0, 1, 2, \ldots$$
$$Q_1 = 1 \text{ if first quarter, 0 if not first quarter}$$
$$Q_2 = 1 \text{ if second quarter, 0 if not second quarter}$$
$$Q_3 = 1 \text{ if third quarter, 0 if not third quarter}$$
$$\beta_0 = Y \text{ intercept}$$
$$(\beta_1 - 1) \times 100\% = \text{quarterly compound growth rate (in \%)}$$
$$\beta_2 = \text{multiplier for first quarter relative to fourth quarter}$$
$$\beta_3 = \text{multiplier for second quarter relative to fourth quarter}$$
$$\beta_4 = \text{multiplier for third quarter relative to fourth quarter}$$
$$\varepsilon_i = \text{value of the irregular component for time period } i$$

[2]One could also choose to use base e logarithms. For more information on logarithms, see Appendix Section A.3.

The model in Equation (16.15) is not in the form of a linear regression model. To transform this nonlinear model to a linear model, one uses a base 10 logarithmic transformation.[2] Taking the logarithm of each side of Equation (16.15) results in Equation (16.16).

TRANSFORMED EXPONENTIAL MODEL WITH QUARTERLY DATA

$$\log(Y_i) = \log\left(\beta_0 \beta_1^{X_i} \beta_2^{Q_1} \beta_3^{Q_2} \beta_4^{Q_3} \varepsilon_i\right) \tag{16.16}$$

$$= \log(\beta_0) + \log(\beta_1^{X_i}) + \log(\beta_2^{Q_1}) + \log(\beta_3^{Q_2}) + \log(\beta_4^{Q_3}) + \log(\varepsilon_i)$$

$$= \log(\beta_0) + X_i \log(\beta_1) + Q_1 \log(\beta_2) + Q_2 \log(\beta_3) + Q_3 \log(\beta_4) + \log(\varepsilon_i)$$

One uses the Equation (16.16) linear model for least-squares regression estimation. Performing the regression analysis using $\log(Y_i)$ as the dependent variable and X_i, Q_1, Q_2, and Q_3 as the independent variables results in Equation (16.17).

EXPONENTIAL GROWTH WITH QUARTERLY DATA FORECASTING EQUATION

$$\log(\hat{Y}_i) = b_0 + b_1 X_i + b_2 Q_1 + b_3 Q_2 + b_4 Q_3 \tag{16.17}$$

where

$$b_0 = \text{estimate of } \log(\beta_0) \text{ and thus } 10^{b_0} = \hat{\beta}_0$$
$$b_1 = \text{estimate of } \log(\beta_1) \text{ and thus } 10^{b_1} = \hat{\beta}_1$$
$$b_2 = \text{estimate of } \log(\beta_2) \text{ and thus } 10^{b_2} = \hat{\beta}_2$$
$$b_3 = \text{estimate of } \log(\beta_3) \text{ and thus } 10^{b_3} = \hat{\beta}_3$$
$$b_4 = \text{estimate of } \log(\beta_4) \text{ and thus } 10^{b_4} = \hat{\beta}_4$$

One uses Equation (16.18) for monthly data.

EXPONENTIAL MODEL WITH MONTHLY DATA

$$Y_i = \beta_0 \beta_1^{X_i} \beta_2^{M_1} \beta_3^{M_2} \beta_4^{M_3} \beta_5^{M_4} \beta_6^{M_5} \beta_7^{M_6} \beta_8^{M_7} \beta_9^{M_8} \beta_{10}^{M_9} \beta_{11}^{M_{10}} \beta_{12}^{M_{11}} \varepsilon_i \tag{16.18}$$

where

$$X_i = \text{coded monthly value, } i = 0, 1, 2, \ldots$$
$$M_1 = 1 \text{ if January, 0 if not January}$$
$$M_2 = 1 \text{ if February, 0 if not February}$$
$$M_3 = 1 \text{ if March, 0 if not March}$$
$$\vdots$$
$$M_{11} = 1 \text{ if November, 0 if not November}$$
$$\beta_0 = Y \text{ intercept}$$
$$(\beta_1 - 1) \times 100\% = \text{monthly compound growth rate (in \%)}$$
$$\beta_2 = \text{multiplier for January relative to December}$$
$$\beta_3 = \text{multiplier for February relative to December}$$
$$\beta_4 = \text{multiplier for March relative to December}$$
$$\vdots$$
$$\beta_{12} = \text{multiplier for November relative to December}$$
$$\varepsilon_i = \text{value of the irregular component for time period } i$$

The model in Equation (16.18) is not in the form of a linear regression model. To transform this nonlinear model into a linear model, one can use a base 10 logarithm transformation. Taking the logarithm of each side of Equation (16.18) results in Equation (16.19).

TRANSFORMED EXPONENTIAL MODEL WITH MONTHLY DATA

$$\log(Y_i) = \log(\beta_0 \beta_1^{X_i} \beta_2^{M_1} \beta_3^{M_2} \beta_4^{M_3} \beta_5^{M_4} \beta_6^{M_5} \beta_7^{M_6} \beta_8^{M_7} \beta_9^{M_8} \beta_{10}^{M_9} \beta_{11}^{M_{10}} \beta_{12}^{M_{11}} \varepsilon_i) \tag{16.19}$$
$$= \log(\beta_0) + X_i \log(\beta_1) + M_1 \log(\beta_2) + M_2 \log(\beta_3)$$
$$+ M_3 \log(\beta_4) + M_4 \log(\beta_5) + M_5 \log(\beta_6) + M_6 \log(\beta_7)$$
$$+ M_7 \log(\beta_8) + M_8 \log(\beta_9) + M_9 \log(\beta_{10}) + M_{10} \log(\beta_{11})$$
$$+ M_{11} \log(\beta_{12}) + \log(\varepsilon_i)$$

One can use the Equation (16.19) linear model for least-squares estimation. Performing the regression analysis using $\log(Y_i)$ as the dependent variable and $X_i, M_1, M_2, \ldots,$ and M_{11} as the independent variables results in Equation (16.20).

EXPONENTIAL GROWTH WITH MONTHLY DATA FORECASTING EQUATION

$$\log(\hat{Y}_i) = b_0 + b_1 X_i + b_2 M_1 + b_3 M_2 + b_4 M_3 + b_5 M_4 + b_6 M_5 + b_7 M_6$$
$$+ b_8 M_7 + b_9 M_8 + b_{10} M_9 + b_{11} M_{10} + b_{12} M_{11} \tag{16.20}$$

where

$$b_0 = \text{estimate of } \log(\beta_0) \text{ and thus } 10^{b_0} = \hat{\beta}_0$$
$$b_1 = \text{estimate of } \log(\beta_1) \text{ and thus } 10^{b_1} = \hat{\beta}_1$$
$$b_2 = \text{estimate of } \log(\beta_2) \text{ and thus } 10^{b_2} = \hat{\beta}_2$$
$$b_3 = \text{estimate of } \log(\beta_3) \text{ and thus } 10^{b_3} = \hat{\beta}_3$$
$$\vdots$$
$$b_{12} = \text{estimate of } \log(\beta_{12}) \text{ and thus } 10^{b_{12}} = \hat{\beta}_{12}$$

Q_1, Q_2, and Q_3 are the three dummy variables needed to represent the four quarter periods in a quarterly time series. M_1, M_2, M_3, ..., M_{11} are the 11 dummy variables needed to represent the 12 months in a monthly time series. In building the model, one uses $\log(Y_i)$ instead of Y_i values and then finds the regression coefficients by taking the antilog of the regression coefficients developed from Equations (16.17) and (16.20).

Although at first glance these regression models look imposing, when fitting or forecasting for any one time period, the values of all or all but one of the dummy variables in the model are equal to zero, and the equations simplify dramatically. In establishing the dummy variables for quarterly time-series data, the fourth quarter is the base period and has a coded value of zero for each dummy variable. With a quarterly time series, Equation (16.17) reduces as follows:

For any first quarter: $\log(\hat{Y}_i) = b_0 + b_1 X_i + b_2$

For any second quarter: $\log(\hat{Y}_i) = b_0 + b_1 X_i + b_3$

For any third quarter: $\log(\hat{Y}_i) = b_0 + b_1 X_i + b_4$

For any fourth quarter: $\log(\hat{Y}_i) = b_0 + b_1 X_i$

When establishing the dummy variables for each month, December serves as the base period and has a coded value of 0 for each dummy variable. For example, with a monthly time series, Equation (16.20) reduces as follows:

For any January: $\log(\hat{Y}_i) = b_0 + b_1 X_i + b_2$

For any February: $\log(\hat{Y}_i) = b_0 + b_1 X_i + b_3$

For any November: $\log(\hat{Y}_i) = b_0 + b_1 X_i + b_{12}$

For any December: $\log(\hat{Y}_i) = b_0 + b_1 X_i$

To demonstrate the process of model building and least-squares forecasting with a quarterly time series, recall the Amazon.com, Inc., revenue data (in billions of dollars) which Table 16.3 on page 586 presents. The data are from the first quarter of 2013 through the last quarter of 2018. Figure 16.23 presents the regression results for the quarterly exponential trend model.

FIGURE 16.23
Excel regression worksheet results for the quarterly revenue data for Amazon.com, Inc.

	A	B	C	D	E	F	G
1	Quarterly Exponential Trend Model for Amazon.com, Inc., Revenue						
2							
3	Regression Statistics						
4	Multiple R	0.9912					
5	R Square	0.9826					
6	Adjusted R Square	0.9789					
7	Standard Error	0.0279					
8	Observations	24					
9							
10	ANOVA						
11		df	SS	MS	F	Significance F	
12	Regression	4	0.8355	0.2089	267.7254	0.0000	
13	Residual	19	0.0148	0.0008			
14	Total	23	0.8503				
15							
16		Coefficients	Standard Error	t Stat	P-value	Lower 95%	Upper 95%
17	Intercept	1.2957	0.0157	82.3179	0.0000	1.2628	1.3287
18	Coded Quarter	0.0248	0.0008	29.7404	0.0000	0.0231	0.0266
19	Q1	-0.1242	0.0163	-7.6083	0.0000	-0.1583	-0.0900
20	Q2	-0.1299	0.0162	-8.0118	0.0000	-0.1638	-0.0960
21	Q3	-0.1184	0.0161	-7.3347	0.0000	-0.1522	-0.0846

From Figure 16.23, the model fits the data very well. The coefficient of determination $r^2 = 0.9826$, the adjusted $r^2 = 0.9789$, and the overall F test results in an F_{STAT} test statistic of 267.7254 (p-value $= 0.000$). At the 0.05 level of significance, each regression coefficient is highly statistically significant and contributes to the model. Table 16.4 lists the antilogs of all the regression coefficients.

Regression Coefficient	$b_i = \log \hat{\beta}_i$	$\hat{\beta}_i = \text{antilog } (b_i) = 10^{b_i}$
b_0: Y intercept	1.2957	19.7560
b_1: coded quarter	0.0248	1.0588
b_2: first quarter	−0.1242	0.7513
b_3: second quarter	−0.1299	0.7415
b_4: third quarter	−0.1184	0.6943

The interpretations for $\hat{\beta}_0$, $\hat{\beta}_1$, $\hat{\beta}_2$, $\hat{\beta}_3$, and $\hat{\beta}_4$ are as follows:

- The Y intercept, $\hat{\beta}_0 = 19.7560$ (in \$billions), is the *unadjusted* forecast for quarterly revenues in the first quarter of 2013, the initial quarter in the time series. *Unadjusted* means that the seasonal component is not incorporated in the forecast.
- The value $(\hat{\beta}_1 - 1) \times 100\%$, $= 0.0588$, or 5.88%, is the estimated *quarterly compound growth rate* in revenues, after adjusting for the seasonal component.
- $\hat{\beta}_2 = 0.7513$ is the seasonal multiplier for the first quarter relative to the fourth quarter; it indicates that there is $1 - 0.7513 = 24.87\%$ less revenue for the first quarter than for the fourth quarter.
- $\hat{\beta}_3 = 0.7415$ is the seasonal multiplier for the second quarter relative to the fourth quarter; it indicates that there is $1 - 0.7415 = 25.85\%$ less revenue for the second quarter than for the fourth quarter.
- $\hat{\beta}_4 = 0.6943$ is the seasonal multiplier for the third quarter relative to the fourth quarter; it indicates that there is $1 - 0.6943 = 30.57\%$ less revenue for the third quarter than for the fourth quarter. Thus, the fourth quarter, which includes the holiday shopping season, has the strongest sales.

Using the regression coefficients b_0, b_1, b_2, b_3, and b_4, and Equation (16.17) on page 587, you can make forecasts for selected quarters. As an example, to predict revenues for the fourth quarter of 2018 ($X_i = 23$),

$$\log(\hat{Y}_i) = b_0 + b_1 X_i$$
$$= 1.2957 + (0.0248)(23)$$
$$= 1.8661$$

Thus,

$$\log(\hat{Y}_i) = 10^{1.8658} = 73.4176$$

The predicted revenue for the fourth quarter of fiscal 2018 is \$73.4176 billion. To make a forecast for a future time period, such as the first quarter of fiscal 2019 ($X_i = 24$, $Q_1 = 1$),

$$\log(\hat{Y}_i) = b_0 + b_1 X_i + b_2 Q_1$$
$$= 1.2957 + (0.0248)(24) + (-0.1242)(1)$$
$$= 1.7667$$

Thus,

$$\hat{Y}_i = 10^{1.7667} = 58.4386$$

The predicted revenue for the first quarter of fiscal 2019 is 58.4386 billion.

PROBLEMS FOR SECTION 16.6

LEARNING THE BASICS

16.40 In forecasting a monthly time series over a five-year period from January 2014 to December 2018, the exponential trend forecasting equation for January is

$$\log \hat{Y}_i = 2.0 + 0.01X_i + 0.10 \text{ (January)}$$

Take the antilog of the appropriate coefficient from this equation and interpret the
a. Y intercept, $\hat{b}_0$.
b. monthly compound growth rate.
c. January multiplier.

16.41 In forecasting daily time-series data, how many dummy variables are needed to account for the seasonal component day of the week?

16.42 In forecasting a quarterly time series over the five-year period from the first quarter of 2014 through the fourth quarter of 2018, the exponential trend forecasting equation is given by

$$\log \hat{Y}_i = 3.0 + 0.10X_i - 0.25Q_1 + 0.20Q_2 + 0.15Q_3$$

where quarter zero is the first quarter of 2014. Take the antilog of the appropriate coefficient from this equation and interpret the
a. Y intercept, $\hat{b}_0$.
b. quarterly compound growth rate.
c. second-quarter multiplier.

16.43 Refer to the exponential model given in Problem 16.42.
a. What is the fitted value of the series in the fourth quarter of 2018?
b. What is the fitted value of the series in the first quarter of 2018?
c. What is the forecast in the fourth quarter of 2019?
d. What is the forecast in the first quarter of 2019?

APPLYING THE CONCEPTS

 16.44 The data in `Target` are quarterly revenues (in $millions) for Target from 2000-Q1 through 2019-Q1.

Source: Data extracted from **ycharts.com/companies/TGT/revenues**.

a. Do you think that the revenues for Target are subject to seasonal variation? Explain.
b. Plot the data. Does this chart support your answer in (a)?
c. Develop an exponential trend forecasting equation with quarterly components.
d. Interpret the quarterly compound growth rate.
e. Interpret the quarterly multipliers.
f. What are the forecasts for 2019-Q2, 2019-Q3, 2019-Q4, and all four quarters of 2020?

16.45 Are gasoline prices higher during the height of the summer vacation season than at other times? The file `Gas Prices` contains the mean monthly prices (in $/gallon) for unleaded gasoline in the United States from January 2006 to July 2019.

Source: Data extracted from U.S. Energy Information Administration, "Monthly Energy Review," **bit.ly/2wYUEtV**.

a. Construct a time-series plot.
b. Develop an exponential trend forecasting equation with monthly components.

c. Interpret the monthly compound growth rate.
d. Interpret the monthly multipliers.
e. Write a short summary of your findings.

16.46 The file `Crude Oil` contains the volume (in barrels) of Canadian crude oil exports by rail from January 2012 to May 2019.

Source: Data extracted from National Energy Board, "Canadian Crude Oil Exports by Rail - Monthly Data," **bit.ly/1RJTx6c**.

a. Plot the time-series data.
b. Develop an exponential trend forecasting equation with monthly components.
c. What is the fitted value in May 2019?
d. What are the forecasts for the last four months of 2019?
e. Interpret the monthly compound growth rate.
f. Interpret the July multiplier.

16.47 The file `Call Center` contains the monthly call volume for an existing product.

Source: Data extracted from S. Madadevan and J. Overstreet, "Use of Warranty and Reliability Data to Inform Call Center Staffing," *Quality Engineering* 24 (2012): 386–399.

a. Construct the time-series plot.
b. Describe the monthly pattern in the data.
c. In general, would you say that the overall call volume is increasing or decreasing? Explain.
d. Develop an exponential trend forecasting equation with monthly components.
e. Interpret the monthly compound growth rate.
f. Interpret the January multiplier.
g. What is the predicted call volume for month 60?
h. What is the predicted call volume for month 61?
i. How can this type of time-series forecasting benefit the call center?

16.48 The file `Silver-Q` contains the price in London for an ounce of silver (in US$) at the end of each quarter from 2004 through 2018.

Source: Data extracted from USAGold, "Daily Silver Price History," **bit.ly/2OBoVuB**.

a. Plot the data.
b. Develop an exponential trend forecasting equation with quarterly components.
c. Interpret the quarterly compound growth rate.
d. Interpret the first quarter multiplier.
e. What is the fitted value for the last quarter of 2018?
f. What are the forecasts for all four quarters of 2019?
g. Are the forecasts in (f) accurate? Explain.

16.49 The file `Gold` contains the price in London for an ounce of gold (in US$) at the end of each quarter from 2004 through 2018.

Source: Data extracted from USAGold, "Daily Gold Price History," **bit.ly/2w8iBSl**.

a. Plot the data.
b. Develop an exponential trend forecasting equation with quarterly components.
c. Interpret the quarterly compound growth rate.
d. Interpret the first quarter multiplier.
e. What is the fitted value for the last quarter of 2018?
f. What are the forecasts for all four quarters of 2019?
g. Are the forecasts in (f) accurate? Explain.

16.7 Index Numbers

An index number measures the value of an item (or group of items) at a particular point in time as a percentage of the value of an item (or group of items) at another point in time. The **Section 16.7 online topic** discusses this concept and illustrates its application.

CONSIDER THIS

Let the Model User Beware

When using a model, one must always review the assumptions built into the model and think about how novel or changing circumstances may render the model less useful.

Implicit in the time-series models developed in this chapter is that past data can be used to help predict the future. While using past data in this way is a legitimate application of time-series models, every so often, a crisis in financial markets illustrates that using models that rely on the past to predict the future is not without risk.

For example, during August 2007, many hedge funds suffered unprecedented losses. Apparently, many hedge fund managers used models that based their investment strategy on trading patterns over long time periods. These models did not—and could not—reflect trading patterns contrary to historical patterns (G. Morgenson, "A Week When Risk Came Home to Roost," *The New York Times*, August 12,

2007, pp. B1, B7). When fund managers in early August 2007 needed to sell stocks due to losses in their fixed income portfolios, stocks that were previously stronger became weaker, and weaker ones became stronger—the reverse of what the models expected. Making matters worse, many fund managers were using similar models and rigidly made investment decisions solely based on what those models said. These similar actions multiplied the effect of the selling pressure, an effect that the models had not considered and that therefore could not be seen in the models' results.

This example illustrates that using models does not absolve you of the responsibility of being a thoughtful decision maker. Go ahead and use models—when appropriately used, they will enhance your decision making. But always remember that no model can completely remove the risk involved in making a business decision.

▼USING **STATISTICS**
Is the ByYourDoor Service Trending? Revisited

In the ByYourDoor scenario, you were asked to analyze time-series sales data for the online food delivery service. You researched time-series forecasting methods and learned how to make short-term estimates of future time-series values.

You learned when to use moving averages and exponential smoothing methods to develop forecasts. You predicted that the movie attendance in 2019 would be 1.3038 billion.

For Alphabet Inc., you used least-squares linear, quadratic, and exponential models and a first-order autoregressive model to develop revenue forecasts. You evaluated these alternative models and determined that the first-order autoregressive model gave the best forecast, according to several

criteria. You predicted that the revenue of Alphabet Inc. would be $137.8208 billion in 2019.

You realized that the ByYourDoor time series has a seasonal component and helped managers decide that you should be analyzing monthly data. You practiced for your task by using a least-squares regression model with a seasonal component to forecast revenues for Amazon.com, Inc. You predicted that Amazon would have revenue of $58.4386 billion in the first quarter of fiscal 2019.

▼SUMMARY

This chapter discusses smoothing techniques, least-squares trend fitting, autoregressive models, and forecasting of seasonal data. Figure 16.24 summarizes the time-series methods discussed in this chapter.

When using time-series forecasting, plot the time series and answer the following question: Is there a trend in the data? If there is a trend, then you can use the autoregressive model or the linear, quadratic, or exponential trend models. If there is no obvious trend in the time-series plot, then you should use moving averages or exponential smoothing to smooth out the effect of random effects and possible cyclical effects. After smoothing the data, if a trend is still not present, then you can use exponential smoothing to forecast short-term future values. If smoothing the data reveals a trend, then you can use the autoregressive model, or the linear, quadratic, or exponential trend models.

FIGURE 16.24
Summary chart of time-series forecasting methods

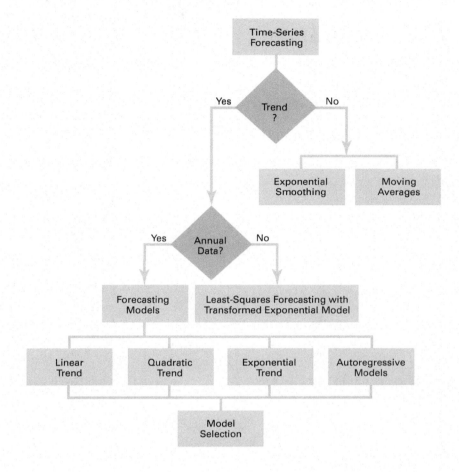

▼ REFERENCES

Bisgaard, S., and M. Kulahci. *Time Series Analysis and Forecasting by Example*. Hoboken, NJ: John Wiley and Sons, 2011.

Bowerman, B. L., R. T. O'Connell, and A. Koehler. *Forecasting, Time Series, and Regression*, 4th ed. Belmont, CA: Duxbury Press, 2005.

Box, G. E. P., G. M. Jenkins, G. C. Reinsel, and G. M. Leung. *Time Series Analysis: Forecasting and Control*, 4th ed. Hoboken, NJ: John Wiley and Sons, 2015.

Hanke, J. E., D. W. Wichern, and A. G. Reitsch. *Business Forecasting*, 7th ed. Upper Saddle River, NJ: Prentice Hall, 2001.

Montgomery, D. C., C. L. Jenning, and M. Kulahci. *Introduction to Time Series Analysis and Forecasting*, 2nd ed. Hoboken, NJ: John Wiley and Sons, 2016.

Pecar, B. *Box-Jenkins ARIMA Modelling in Excel*. Seattle, WA: Amazon Digital Services, 2017.

▼ KEY EQUATIONS

An Exponentially Smoothed Value for Time Period i

$$E_1 = Y_1$$

$$E_i = WY_i + (1 - W)E_{i-1} \text{ for } i = 2, 3, 4, \ldots \qquad \textbf{(16.1)}$$

Forecasting Time Period $i + 1$

$$\hat{Y}_{i+1} = E_i \qquad \textbf{(16.2)}$$

Linear Trend Forecasting Equation

$$\hat{Y}_i = b_0 + b_1 X_i \qquad \textbf{(16.3)}$$

Quadratic Trend Forecasting Equation

$$\hat{Y}_i = b_0 + b_1 X_i + b_2 X_i^2 \qquad (16.4)$$

Exponential Trend Model

$$Y_i = \beta_0 \beta_1^{X_i} \varepsilon_i \qquad (16.5)$$

Transformed Exponential Trend Model

$$
\begin{aligned}
\log(Y_i) &= \log(\beta_0 \beta_1^{X_i} \varepsilon_i) \\
&= \log(\beta_0) + \log(\beta_1^{X_i}) + \log(\varepsilon_i) \\
&= \log(\beta_0) + X_i \log(\beta_1) + \log(\varepsilon_i) \qquad (16.6)
\end{aligned}
$$

Exponential Trend Forecasting Equation

$$\log(\hat{Y}_i) = b_0 + b_1 X_i \qquad (16.7a)$$

$$\hat{Y}_i = \hat{\beta}_0 \hat{\beta}_1^{X_i} \qquad (16.7b)$$

pth-Order Autoregressive Models

$$Y_i = A_0 + A_1 Y_{i-1} + A_2 Y_{i-2} + \cdots + A_p Y_{i-p} + \delta_i \qquad (16.8)$$

First-Order Autoregressive Model

$$Y_i = A_0 + A_1 Y_{i-1} + \delta_i \qquad (16.9)$$

Second-Order Autoregressive Model

$$Y_i = A_0 + A_1 Y_{i-1} + A_2 Y_{i-2} + \delta_i \qquad (16.10)$$

t Test for Significance of the Highest-Order Autoregressive Parameter, A_p

$$t_{STAT} = \frac{a_p - A_p}{S_{a_p}} \qquad (16.11)$$

Fitted pth-Order Autoregressive Equation

$$\hat{Y}_i = a_0 + a_1 Y_{i-1} + a_2 Y_{i-2} + \cdots + a_p Y_{i-p} \qquad (16.12)$$

pth-Order Autoregressive Forecasting Equation

$$\hat{Y}_{n+j} = a_0 + a_1 \hat{Y}_{n+j-1} + a_2 \hat{Y}_{n+j-2} + \cdots + a_p \hat{Y}_{n+j-p} \qquad (16.13)$$

Mean Absolute Deviation

$$MAD = \frac{\sum_{i=1}^{n} |Y_i - \hat{Y}_i|}{n} \qquad (16.14)$$

Exponential Model With Quarterly Data

$$Y_i = \beta_0 \beta_1^{X_i} \beta_2^{Q_1} \beta_3^{Q_2} \beta_4^{Q_3} \varepsilon_i \qquad (16.15)$$

Transformed Exponential Model With Quarterly Data

$$
\begin{aligned}
\log(Y_i) &= \log(\beta_0 \beta_1^{X_i} \beta_2^{Q_1} \beta_3^{Q_2} \beta_4^{Q_3} \varepsilon_i) \\
&= \log(\beta_0) + \log(\beta_1^{X_i}) + \log(\beta_2^{Q_1}) + \log(\beta_3^{Q_2}) \\
&\quad + \log(\beta_4^{Q_3}) + \log(\varepsilon_i) \\
&= \log(\beta_0) + X_i \log(\beta_1) + Q_1 \log(\beta_2) \\
&\quad + Q_2 \log(\beta_3) + Q_3 \log(\beta_4) + \log(\varepsilon_i) \qquad (16.16)
\end{aligned}
$$

Exponential Growth With Quarterly Data Forecasting Equation

$$\log(\hat{Y}_i) = b_0 + b_1 X_i + b_2 Q_1 + b_3 Q_2 + b_4 Q_3 \qquad (16.17)$$

Exponential Model With Monthly Data

$$Y_i = \beta_0 \beta_1^{X_i} \beta_2^{M_1} \beta_3^{M_2} \beta_4^{M_3} \beta_5^{M_4} \beta_6^{M_5} \beta_7^{M_6} \beta_8^{M_7} \beta_9^{M_8} \beta_{10}^{M_9} \beta_{11}^{M_{10}} \beta_{12}^{M_{11}} \varepsilon_i \qquad (16.18)$$

Transformed Exponential Model With Monthly Data

$$
\begin{aligned}
\log(Y_i) &= \log(\beta_0 \beta_1^{X_i} \beta_2^{M_1} \beta_3^{M_2} \beta_4^{M_3} \beta_5^{M_4} \beta_6^{M_5} \beta_7^{M_6} \beta_8^{M_7} \beta_9^{M_8} \beta_{10}^{M_9} \beta_{11}^{M_{10}} \beta_{12}^{M_{11}} \varepsilon_i) \\
&= \log(\beta_0) + X_i \log(\beta_1) + M_1 \log(\beta_2) + M_2 \log(\beta_3) \\
&\quad + M_3 \log(\beta_4) + M_4 \log(\beta_5) + M_5 \log(\beta_6) + M_6 \log(\beta_7) \\
&\quad + M_7 \log(\beta_8) + M_8 \log(\beta_9) + M_9 \log(\beta_{10}) + M_{10} \log(\beta_{11}) \\
&\quad + M_{11} \log(\beta_{12}) + \log(\varepsilon_i) \qquad (16.19)
\end{aligned}
$$

Exponential Growth With Monthly Data Forecasting Equation

$$
\begin{aligned}
\log(\hat{Y}_i) &= b_0 + b_1 X_i + b_2 M_1 + b_3 M_2 + b_4 M_3 + b_5 M_4 + b_6 M_5 \\
&\quad + b_7 M_6 + b_8 M_7 + b_9 M_8 + b_{10} M_9 + b_{11} M_{10} + b_{12} M_{11} \qquad (16.20)
\end{aligned}
$$

▼ KEY TERMS

▼CHECKING YOUR UNDERSTANDING

16.50 What is a time series?

16.51 What are the different components of a time-series model?

16.52 What is the difference between moving averages and exponential smoothing?

16.53 Under what circumstances is the exponential trend model most appropriate?

16.54 How does the least-squares linear trend forecasting model developed in this chapter differ from the least-squares linear regression model considered in Chapter 13?

16.55 How does autoregressive modeling differ from the other approaches to forecasting?

16.56 What are the different approaches to choosing an appropriate forecasting model?

16.57 What is the major difference between using S_{YX} and MAD for evaluating how well a particular model fits the data?

16.58 How does forecasting for monthly or quarterly data differ from forecasting for annual data?

▼CHAPTER REVIEW PROBLEMS

16.59 The data in the following table, stored in Polio , represent the annual incidence rates (per 100,000 persons) of reported acute poliomyelitis recorded over five-year periods from 1915 to 1955:

Year	1915	1920	1925	1930	1935	1940	1945	1950	1955
Rate	3.1	2.2	5.3	7.5	8.5	7.4	10.3	22.1	17.6

Source: Data extracted from B. Wattenberg, Ed., *The Statistical History of the United States: From Colonial Times to the Present*, ser. B303.

a. Plot the data.
b. Compute the linear trend forecasting equation and plot the trend line.
c. What are your forecasts for 1960, 1965, and 1970?
d. Using a library or the Internet, find the actually reported incidence rates of acute poliomyelitis for 1960, 1965, and 1970. Record your results.
e. Why are the forecasts you made in (c) not useful? Discuss.

16.60 The U.S. Department of Labor gathers and publishes statistics concerning the labor market. The file Workforce contains the size of the U.S. civilian noninstitutional population of people 16 years and over (in thousands) and the U.S. civilian noninstitutional workforce of people 16 years and over (in thousands) for 1984–2018. The workforce variable reports the number of people in the population who have a job or are actively looking for a job.

Source: Data extracted from Bureau of Labor Statistics, U.S. Department of Labor, **www.bls.gov**.

a. Plot the time series for the U.S. civilian noninstitutional population of people 16 years and older.
b. Compute the linear trend forecasting equation.
c. Forecast the U.S. civilian noninstitutional population of people 16 years and older for 2019 and 2020.
d. Repeat (a) through (c) for the U.S. civilian noninstitutional workforce of people 16 years and older.

16.61 The monthly commercial and residential prices for natural gas (dollars per thousand cubic feet) in the United States from January 2008 through May 2019 are stored in Natural Gas .

Source: Data extracted from Energy Information Administration, U.S. Department of Energy, **www.eia.gov**, *Natural Gas Monthly*, July 31, 2019.

For the commercial price and the residential price,

a. do you think the price for natural gas has a seasonal component?
b. plot the time series. Does this chart support your answer in (a)?
c. compute an exponential trend forecasting equation for monthly data.
d. interpret the monthly compound growth rate.
e. interpret the monthly multipliers. Do the multipliers support your answers in (a) and (b)?
f. compare the results for the commercial prices and the residential prices.

16.62 The file McDonalds contains the gross revenues (in billions of current dollars) of McDonald's Corporation from 1975 through 2018:

a. Plot the data.
b. Compute the linear trend forecasting equation.
c. Compute the quadratic trend forecasting equation.
d. Compute the exponential trend forecasting equation.
e. Determine the best-fitting autoregressive model, using $\alpha = 0.05$.
f. Perform a residual analysis for each of the models in (b) through (e).
g. Compute the standard error of the estimate (S_{YX}) and the MAD for each corresponding model in (f).
h. On the basis of your results in (f) and (g), along with a consideration of the principle of parsimony, which model would you select for purposes of forecasting? Discuss.
i. Using the selected model in (h), forecast gross revenues for 2019.

16.63 Teachers' Retirement System of the City of New York offers several types of investments for its members. Among the choices are investments with fixed and variable rates of return. There are several categories of variable-return investments. The Diversified Equity Fund consists of investments that are primarily made in stocks, and the Balanced Fund consists of investments in corporate bonds and other types of lower risk instruments. The data in `TRS NYC` represent the value of a unit of each type of variable-return investment at the beginning of each year from 1984 to 2019.

Source: Data extracted from "Total investments Value," **bit.ly/2yVE6Wt**.

For each of the two time series,
a. plot the data.
b. compute the linear trend forecasting equation.
c. compute the quadratic trend forecasting equation.
d. compute the exponential trend forecasting equation.
e. determine the best-fitting autoregressive model, using $\alpha = 0.05$.
f. Perform a residual analysis for each of the models in (b) through (e).
g. Compute the standard error of the estimate (S_{YX}) and the *MAD* for each corresponding model in (f).

h. On the basis of your results in (f) and (g), along with a consideration of the principle of parsimony, which model would you select for purposes of forecasting? Discuss.
i. Using the selected model in (h), forecast the unit values for 2020.
j. Based on the results of (a) through (i), what investment strategy would you recommend for a member of the Teachers' Retirement System of the City of New York? Explain.

REPORT WRITING EXERCISE

16.64 As a consultant to an investment company trading in various currencies, you have been assigned the task of studying long-term trends in the exchange rates of the Canadian dollar, the Japanese yen, and the English pound. Data from 1980 to 2018 are stored in `Currency`, where the Canadian dollar, the Japanese yen, and the English pound are expressed in units per U.S. dollar.

Develop a forecasting model for the exchange rate of each of these three currencies and provide forecasts for 2019 and 2020 for each currency. Write an executive summary for a presentation to be given to the investment company. Append to this executive summary a discussion regarding possible limitations that may exist in these models.

CHAPTER
16

▾CASES

Managing Ashland MultiComm Services

As part of the continuing strategic initiative to increase subscribers to the *3-For-All* cable/phone/Internet services, the marketing department is closely monitoring the number of subscribers. To help do so, forecasts are to be developed for the number of subscribers in the future. To accomplish this task, the number of subscribers for the most recent 24-month period has been determined and is stored in `AMS16`.

1. Analyze these data and develop a model to forecast the number of subscribers. Present your findings in a report that includes the assumptions of the model and its limitations. Forecast the number of subscribers for the next four months.

2. Would you be willing to use the model developed to forecast the number of subscribers one year into the future? Explain.

3. Compare the trend in the number of subscribers to the number of new subscribers per month stored in `AMS13`. What explanation can you provide for any differences?

Digital Case

Apply your knowledge about time-series forecasting in this Digital Case.

The *Ashland Herald* competes for readers in the Tri-Cities area with the newer *Oxford Glen Journal* (*OGJ*). Recently, the circulation staff at the *OGJ* claimed that their newspaper's circulation and subscription base is growing faster than that of the *Herald* and that local advertisers would do better if they transferred their advertisements from the *Herald* to the *OGJ*. The circulation department of the *Herald* has complained to the Ashland Chamber of Commerce about *OGJ*'s claims and has asked the chamber to investigate, a request that was welcomed by *OGJ*'s circulation staff.

Open **ACC_Mediation216.pdf** to review the circulation dispute information collected by the Ashland Chamber of Commerce. Then answer the following:

1. Which newspaper would you say has the right to claim the fastest growing circulation and subscription base? Support your answer by performing and summarizing an appropriate statistical analysis.

2. What is the single most positive fact about the *Herald*'s circulation and subscription base? What is the single most positive fact about the *OGJ*'s circulation and subscription base? Explain your answers.

3. What additional data would be helpful in investigating the circulation claims made by the staffs of each newspaper?

▼EXCEL GUIDE

EG16.2 SMOOTHING an ANNUAL TIME SERIES

Moving Averages

Key Technique Use the **AVERAGE(*cell range of L consecutive values*)** function to compute a moving average. Use the special value **#N/A** (not available) for time periods in which no moving average can be computed.

Example Calculate the Figure 16.3 three- and five-year moving averages for the movie attendance data on page 559.

Workbook Use the **COMPUTE worksheet** of the **Moving Averages workbook** as a template.

The worksheet already contains the data and formulas for the example. For other problems, paste the time-series data into columns A and B and:

1. For data that contain more than 12 time periods, copy the formulas in cell range **C13:D13** down through the new table rows; otherwise, delete rows as necessary.

2. Enter the special value **#N/A** in columns C and D for the first and last time periods.

3. Enter **#N/A** in the second and second-to-last time periods in column D.

To construct a moving average plot for other problems, open to the adjusted COMPUTE worksheet and:

1. Select the cell range of the time-series and the moving averages (**A1:D13** for the example).

2. Select **Insert → Scatter (X, Y) or Bubble Chart** and select the **Scatter** gallery item.
   Select **Insert → X Y (Scatter)** and select the **Scatter** gallery item.

3. Relocate the chart to a chart sheet, turn off the gridlines, add axis titles, and modify the chart title by using the instructions in Appendix Section B.5.

Exponential Smoothing

Key Technique Use arithmetic formulas to compute exponentially smoothed values.

Example Calculate the Figure 16.5 exponentially smoothed series ($W = 0.50$ and $W = 0.25$) for the movie attendance data on page 562.

Workbook Use the **COMPUTE worksheet** of the **Exponential Smoothing workbook**, as a template.

The worksheet already contains the data and formulas for the example. In this worksheet, cells C2 and D2 contain the formula **=B2** that copies the initial value of the time

series. The exponential smoothing begins in row 3, with cell C3 formula **= 0.5 * B3 + 0.75 * C2**, and cell **D3** formula **= 0.25 * B3 + 0.75 * D2**. Note that these formulas simplify the Equation (16.1) expression $1 - W$ as the values 0.5 and 0.75.

For other problems, paste the time-series data into columns A and B and adjust the entries in columns C and D. For problems with fewer than 12 time periods, delete the excess rows. For problems with more than 12 time periods, select cell range **C13:D13** and copy down through the new table rows.

To construct a plot of exponentially smoothed values for other problems, open to the adjusted COMPUTE worksheet and:

1. Select the cell range of the time-series data and the exponentially smoothed values (**A1:D13** for the example).

2. Select **Insert → Scatter (X, Y) or Bubble Chart** (or **Scatter**) and select the Scatter gallery item.
   Select **Insert → X Y (Scatter)** and select the **Scatter** gallery item.

3. Relocate the chart to a chart sheet, turn off the gridlines, add axis titles, and modify the chart title by using the instructions in Appendix Section B.5.

Analysis ToolPak Use **Exponential Smoothing**.

For the example, open to the **DATA worksheet** of the **Movie Attendance workbook** and:

1. Select **Data → Data Analysis**.

2. In the Data Analysis dialog box, select **Exponential Smoothing** from the **Analysis Tools** list and then click **OK**.

In the Exponential Smoothing dialog box (shown below):

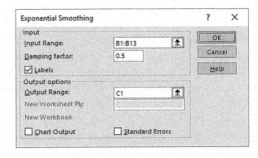

3. Enter **B1:B13** as the **Input Range**.

4. Enter **0.5** as the **Damping factor**. (The damping factor is equal to $1 - W$.)

5. Check **Labels**, enter **C1** as the **Output Range**, and click **OK**.

In the new column C:

6. Copy the last formula in cell **C12** to cell **C13**.
7. Enter the column heading **ES(W=.50)** in cell **C1**, replacing the **#N/A** value.

To create the exponentially smoothed values that use a smoothing coefficient of $W = 0.25$, repeat steps 1 through 7 but enter **0.75** as the **Damping factor** in step 4, enter **D1** as the **Output Range** in step 5, and enter **ES(W=.25)** as the column heading in step 7.

EG16.3 LEAST-SQUARES TREND FITTING and FORECASTING

The Linear Trend Model

Key Technique Modify the Section EG13.2 instructions on page 474.

Use the cell range of the coded variable as the X variable cell range (called the **X Variable Cell Range** in the *PHStat* instructions, called the *cell range of X variable* in the *Workbook* instructions, and called the **Input X Range** in the *Analysis ToolPak* instructions).

To enter many coded values, use **Home → Fill** (in the Editing group) **→ Series** and in the Series dialog box, click **Columns** and **Linear**, and select appropriate values for **Step value** and **Stop value**.

The Quadratic Trend Model

Key Technique Modify the Section EG15.1 instructions on page 554.

Use the cell range of the coded variable and the squared coded variable as the X variables cell range, called the **X Variables Cell Range** in the *PHStat* instructions and the **Input X Range** in the *Analysis ToolPak* instructions.

The Exponential Trend Model

Key Technique Modify the Section EG15.2 instructions on page 554 and the EG13.5 instructions on page 475.

Use the **POWER(10,** *predicted log(Y)*$)$ function to compute the predicted Y values from the predicted $\log(Y)$ results.

To create an exponential trend model, first convert the values of the dependent variable Y to $\log(Y)$ values using the Section EG15.2 instructions. Then perform a simple linear regression analysis with residual analysis using the $\log(Y)$ values. Modify the Section EG13.5 instructions using the cell range of the $\log(Y)$ values as the Y variable cell range and the cell range of the coded variable as the X variable cell range.

If you use the *PHStat* or *Workbook* instructions, residuals will appear in a residuals worksheet. If you use the Analysis ToolPak instructions, residuals will appear in the RESIDUAL OUTPUT area of the regression results worksheet. Because you use $\log(Y)$ values for the regression, the predicted Y and residuals listed are *log values* that need to be converted. [The Analysis ToolPak incorrectly labels the new column for the logs of the residuals *Residuals*, and not *LOG(Residuals)*.]

In an empty column in the residuals worksheet (*PHStat* or *Workbook*) or an empty column range to the right of RESIDUALS OUTPUT area (*Analysis ToolPak*):

1. Add a column of formulas that use the POWER function to compute the predicted Y values.
2. Copy the original Y values to the next empty column.
3. In the next empty (third new) column, enter formulas in the form **=Y value cell-predicted Y cell** to compute the residuals.

Use columns G through I of the **RESIDUALS worksheet** of the **Exponential Trend workbook** as a model for these three columns. The worksheet already contains the values and formulas needed to create the Figure 16.11 plot that fits an exponential trend forecasting equation for Alphabet Inc. annual revenue.

To construct an exponential trend plot, first select the cell range of the time-series data and then use the Section EG2.5 instructions to construct a scatter plot. (For Alphabet Inc. revenue example, use the cell range is **B1:C9** in the **Data worksheet** of the **Alphabet workbook**.) Select the chart and:

1. Select **Design** (or **Chart Design**) **→ Add Chart Element → Trendline → More Trendline Options**.
2. In the Format Trendline pane, click **Exponential**.

Model Selection Using First, Second, and Percentage Differences

Key Technique Use the **COMPUTE worksheet** of the **Differences workbook** (see Figure 16.12 page 571), as a model for developing a differences worksheet.

Use arithmetic formulas to compute the first, second, and percentage differences. Use division formulas to compute the percentage differences and use subtraction formulas to compute the first and second differences. Open to the **COMPUTE_FORMULAS worksheet** to review the formulas the COMPUTE worksheet uses.

EG16.4 AUTOREGRESSIVE MODELING for TREND FITTING and FORECASTING

Creating Lagged Predictor Variables

Key Technique Use the **COMPUTE worksheet** of the **Lagged Predictors workbook** as a model for developing lagged predictor variables for the first-order, second-order, and third-order autoregressive models.

Create lagged predictor variables by creating a column of formulas that refer to a previous row's (previous time period's) Y value. Enter the special worksheet value **#N/A** (not available) for the cells in the column to which lagged values do not apply.

When specifying cell ranges for a lagged predictor variable, you include only rows that contain lagged values. Contrary to the usual practice in this book, you do not include rows that contain **#N/A**, nor do you include the row 1 column heading.

Open to the **COMPUTE_FORMULAS** worksheet to review the formulas that the worksheet template in Figure 16.14 on page 578 uses.

Autoregressive Modeling

Key Technique To create a third-order or second-order autoregressive model, modify the Section EG14.1 instructions on page 552. Use the cell range of the first-order, second-order, and third-order lagged predictor variables as the *X* variables cell range for the third-order model. Use the cell range of the first-order and second-order lagged predictor variables as the *X* variables cell range for the second-order model.

If you use the *PHStat* instructions, modify step 3 to *clear*, not *check*, **First cells in both ranges contain label**. If using the *Workbook* instructions, use the **COMPUTE3 worksheet** in lieu of the COMPUTE worksheet for the third-order model. If using the *Analysis ToolPak* instructions, do not check **Labels** in step 4.

To create a first-order autoregressive model, modify the Section EG13.2 instructions on page 474. Use the cell range of the first-order lagged predictor variable as the *X* variable cell range (called the **X Variable Cell Range** in the *PHStat* instructions, the *cell range of X variable* in the *Workbook* instructions, and the **Input X Range** in the *Analysis ToolPak* instructions). If using the *PHStat* instructions, modify step 3 to *clear*, not *check*, **First cells in both ranges contain label**. If using the *Analysis ToolPak* instructions, do not check **Labels** in step 4.

EG16.5 CHOOSING an APPROPRIATE FORECASTING MODEL

Performing a Residual Analysis

To create residual plots for the linear trend model or the first-order autoregressive model, use the Section EG13.5 instructions on page 475.

To create residual plots for the quadratic trend model or second-order autoregressive model, use the Section EG14.3 instructions on page 523.

To create residual plots for the exponential trend model, use the instructions Section EG16.4 on page 523.

To create residual plots for the third-order autoregressive model, modify the Section EG14.3 instructions to use the **RESIDUALS3** worksheet instead of the RESIDUALS worksheet if you use the *Workbook* instructions.

Measuring the Magnitude of the Residuals Through Squared or Absolute Differences

Key Technique Use the functions **SUMPRODUCT** and **COUNT** to compute the mean absolute deviation (*MAD*).

To compute the mean absolute deviation (*MAD*), first perform a residual analysis. Then, in an empty cell, add the formula **=SUMPRODUCT(ABS(*residuals cell range*)) / COUNT(*residuals cell range*)**. When entering the *residuals*

cell range, do not include the column heading in the cell range. (See Appendix Section F.2 to learn more about the application of **SUMPRODUCT** function in this formula.)

Cell I10 of the **RESIDUALS_FORMULAS worksheet** of the **Exponential Trend workbook** contains the *MAD* formula for the Alphabet Inc. annual revenue example.

A Comparison of Four Forecasting Methods

Key Technique Use the **COMPARE** worksheet of the **Forecasting Comparison workbook** as a model.

Construct a model comparison worksheet similar to the Figure 16.21 worksheet on page 585 by using **Paste Special values** (see Appendix Section B.5) to transfer results from the regression results worksheets.

For the *SSE* values (row 22 in Figure 16.21), copy the regression results worksheet cell C13, the *SS* value for Residual in the ANOVA table. For the S_{YX} values (row), copy the regression results worksheet cell B7, labeled Standard Error, for all but the exponential trend model. For the *MAD* values, add formulas as discussed in the previous section.

For the S_{YX} value for the exponential trend model, enter a formula in the form **=SQRT(*exponential SSE cell* / (COUNT(*cell range of exponential residuals*) - 2))**. In the COMPARE worksheet, this formula is **=SQRT(H22/ (COUNT(H3: H21) - 2))**.

Open to the **COMPARE_FORMULAS** worksheet to discover how the COMPARE worksheet uses the SUMSQ function as an alternate way of displaying the *SSE* values.

EG16.6 TIME-SERIES FORECASTING of SEASONAL DATA

Least-Squares Forecasting With Monthly or Quarterly Data

To develop a least-squares regression model for monthly or quarterly data, add columns of formulas that use the **IF** function (see Appendix Section F.2) to create dummy variables for the quarterly or monthly data. Enter all formulas in the form **=IF(*comparison*,1, 0)**.

Shown at right are the first five rows of columns F through K of a data worksheet that contains dummy variables. In the first illustration, columns F, G, and H contain the quarterly dummy variables Q1, Q2, and Q3 that are based on column B coded quarter values (not shown). In the second illustration, columns J and K contain the two monthly variables M1 and M6 that are based on column C month values (also not shown).

	F	G	H
1	Q1	Q2	Q3
2	=IF(B2 = 1, 1, 0)	=IF(B2 = 2, 1, 0)	=IF(B2 = 3, 1, 0)
3	=IF(B3 = 1, 1, 0)	=IF(B3 = 2, 1, 0)	=IF(B3 = 3, 1, 0)
4	=IF(B4 = 1, 1, 0)	=IF(B4 = 2, 1, 0)	=IF(B4 = 3, 1, 0)
5	=IF(B5 = 1, 1, 0)	=IF(B5 = 2, 1, 0)	=IF(B5 = 3, 1, 0)

	J	K
1	M1	M6
2	=IF(C2 ="January", 1, 0)	=IF(C2 = "June", 1, 0)
3	=IF(C3 ="January", 1, 0)	=IF(C3 = "June", 1, 0)
4	=IF(C4 ="January", 1, 0)	=IF(C4 = "June", 1, 0)
5	=IF(C5 ="January", 1, 0)	=IF(C5 = "June", 1, 0)

17

Business Analytics

OBJECTIVES

- Understand fundamental business analytics concepts
- Gain experience with selected analytics methods
- Understand the variety of predictive analytics methods

▼ USING STATISTICS
Back to Arlingtons for the Future

Through sales experiments that the Using Statistics scenarios in Chapter 10 describe, Arlingtons discovered how the location of items in a store can affect the in-store sales. While making store placement decisions and charging varying store placement fees based on those experiments did increase revenues, long-term retailing trends toward online commerce continued to hurt the overall financial health of Arlingtons. When a private equity firm made an unsolicited bid for Arlingtons, senior management and the board of directors at Arlingtons reluctantly agreed to a buyout.

The new owners believe that with advanced data analysis, they can grow the business, especially in the online marketplace where Arlingtons has been a weak competitor. Just as multiple regression allows consideration of several independent variables, they believe that other methods associated with *business analytics* will allow them to analyze many more relevant variables. For example, the new owners look to track customer buying habits and to be able to answer questions such as "Who were those customers that were most likely to buy the VLABGo players from the special front of store sales location?" and "What else could one expect those customers to buy at Arlingtons?" The new owners also believe that they will be able to start getting answers to more fundamental questions such as "Should we even be selling mobile electronics?" and "Should we invest more in online sales and less in brick-and-mortar (physical) stores?"

To introduce business analytics to existing store managers, the new owners have hired you to prepare notes for a management seminar that would introduce business analytics to these managers, each of whom already have a knowledge of introductory business statistics. What do you say to such a group?

17.1 Business Analytics Overview

Business statistics first gained widespread usage in an age of manual filing systems and limited computerization. The first wave of business computers made practical the calculations of advanced inferential methods that previous chapters discuss, but data handling and storage was often limited or clumsy or both. As information technology and management matured, the application of business statistics grew within organizations and was applied to larger and larger sets of data. In today's world, where even mobile devices surpass the functionality of supercomputers that existed 30 years ago, much more can be done to support fact-based decision making.

This "much more" is the practical realization of techniques long imagined but that could not be implemented due to the limitations of information technology in the past. This much more combines statistics, information systems, and management science. This much more often uses well-known methods but extends those methods into more functional areas or provides the means to analyze large volumes of data. This much more is business analytics that Section FTF.2 on page 4 first defines.

Section FTF.2 describes business analytics as "the changing face of statistics," but these sets of techniques could also be called "the changing face of business." Just as business students today typically take at least one course in business statistics, business students of tomorrow (and some even today) will be taking at least one course in business analytics. This chapter serves as an introduction and bridge to that future.

Business Analytics Categories

Business analytics methods help management decision makers answer what has happened or has been happening in the business, what could happen in the business, or what should happen based on a recommended course of action. These three kinds of management questions define the three main categories of business analytics (see Table 17.1).

TABLE 17.1
Management questions that define business analytics categories

Question	Business analytics category
What has happened or has been happening?	Descriptive analytics
What could happen?	Predictive analytics
What should happen?	Prescriptive analytics

Descriptive analytics answer "What has happened or has been happening?" questions. Descriptive analytics methods summarize historical data to identify patterns to the data that might be worthy of investigation or provide decision makers with new insights about business operations. Many methods contain the ability for decision makers to *drill down*, or reveal, the details of data that were summarized and most are related to or extensions of methods that Chapter 2 discusses.

Predictive analytics answer "What could happen?" questions. Several subtypes of this category exist. **Prediction methods** use historical data to explain relationships among variables or to estimate ("predict") values of a dependent Y variable. Multiple regression, which Chapter 14 discusses, is an example of such a method. **Classification methods** assign items in a collection to target categories or classes. Logistic regression, which Chapter 14 discusses, assigns items to one of two target categories. **Clustering methods** find groupings in data being analyzed. **Association methods** find items that tend to occur together or specify the rules that explain such co-occurrences.

Prescriptive analytics answer "What should happen?" questions. Prescriptive methods seek to optimize the performance of a business and offer decision-making recommendations for how to respond to and manage business circumstances in the future. These methods often evaluate models that predictive analytics methods build to determine new ways to operate a business while balancing constraints and considering business objectives. Prescriptive methods "can take processes that were once expensive, arduous, and difficult, and complete them in a cost-effective and effortless manner" (Morgan).

Business Analytics Vocabulary

Using business analytics, one discovers that a vocabulary has been partially invented from suppliers of computer services or derived from fields other than statistics, such as artificial intelligence. Although this new vocabulary can make business analytics sound difficult to learn, many terms relate directly to the key terms of earlier chapters or have direct parallels to concepts that earlier chapters discuss. Therefore, a foundation based on statistical learning serves one well when facing business analytics.

Perhaps the most commonly encountered term is **data mining**. Thinking about natural resources mining, where valuable minerals are extracted from the earth, a marketer saw an analogy to "mining data" to extract valuable information from corporate data stores and coined the term. Because the term was invented by thinking about corporate data, data mining usually implies processing large volumes of data or even big data that might combine corporate data (a primary source) with external, secondary sources of data. However, smaller sets of data can be data mined, including certain sets that other chapters use to demonstrate concepts.

While "data mining" varies depending on the software and information system used, data mining typically combines what Section 14.5 calls model building with the application of one or more descriptive or predictive analytics methods. This means one element of data mining can be understood as similar to stepwise or best subsets regression that Section 14.5 discusses.

A conversation about business analytics today often includes the terms *artificial intelligence* (*AI*) and *machine learning*. While "AI" suggests to some a future of talking cyborgs, dysfunctional computers, or the like, **artificial intelligence** is a branch of computer science more than 60 years old and seeks to use software to mimic human expertise, reasoning, or knowledge. **Machine learning** are the AI methods that help automate model building. Using Excel to perform regression analysis illustrates a simple case of machine learning: Excel computes the coefficients that could be otherwise be manually calculated by a human with knowledge of the defining equations found in Chapters 13 and 14.

Classifying a business analytics method as either *supervised* or *unsupervised* is also borrowed from artificial intelligence. **Supervised methods** use explicit facts, called **labeled data**, to better "learn" relationships among variables and build models. All regression data are labeled because the data includes not only values of the independent X variables, but values for the dependent Y variable. Therefore, all regression methods are examples of supervised methods.

Note that anyone who has acquired language using an illustrated vocabulary book, which labels pictures with vocabulary words, has learned language using labeled data, too.

In contrast, **unsupervised methods** uncover relationships among variables with unlabeled data. Models produced by unsupervised methods can sometimes be hard for business analyst users or decision makers to understand because the model can be based on abstract mathematical properties of the data.

CONSIDER THIS

What's My Major If I Want to Be a Data Miner?

To be most effective as a data miner, you need a broad base of business skills, as one would get majoring in any business subject. Most critically, you need to know how to define problems and requirements using a problem-solving framework such as the DCOVA model and have an awareness for basic concepts of statistics, goals of this book. You might supplement your knowledge with a course that fully introduces business analytics and gives you additional practice in the data preparation tasks that Chapter 1 summarizes. But you do not need to major in data mining to be a data miner, just as you do not need to major in statistics to apply statistical methods to fact-based decision making.

If you are, or plan to be, a graduate student, consider a concentration in business analytics that more closely examines the application of data mining to a functional area. Whatever choices you make, the points made in Section FTF.1 about using a framework and understanding that analytical skills are more important than arithmetic (and other mathematical) skills will always hold. Ironically, as data mining/business analytics software gets more capable and gains the ability to analyze more and more data in ever increasing sophisticated ways, the points that Section FTF.1 emphasizes will become increasingly important.

Inferential Statistics and Predictive Analytics

Inferential statistics form an important foundation for predictive analytics, and one can consider inferential methods such as multiple regression as examples of predictive analytics, too. Therefore, anyone who has studied multiple regression has already begun studying business analytics!

Chapter 13 explains that "regression methods seek to discover how one or more independent X variables can predict the value of a dependent Y variable." Most regression examples that Chapters 13 through 15 consider uncover and explain the relationship among the variables being analyzed. Regression methods for those cases produce **explanatory models** that help better explain the relationship among variables. For example, in the Chapter 14 OmniPower scenario, the resulting explanatory model helps one better estimate the value of the dependent sales variable, given specific values of the price and promotional expenses variables. The value of the dependent sales variable is the predicted *mean* sales, that is, an average of what will be expected over time given the explanatory model.

One could say that the model *predicts* the predicted mean sales, instead of *estimates*. In fact, people—and this book—sometimes use the verbs *estimate* and *predict* interchangeably when discussing how one uses a regression model. That usage is not wrong, but it does obscure the important distinction: One uses predictive analytics methods to produce **prediction models**, models that *predict* individual cases, not estimate the general case. A method can produce a model that, in one context, is an explanatory model and, in another context, is a prediction model. Logistic regression, which Section 14.7 discusses, best illustrates this.

Section 14.7 uses a cardholder study and logistic regression to develop a regression equation that enables one to estimate the probability that a cardholder will upgrade to a premium card based on specific values for purchases made last year and whether the cardholder possesses additional cards. In the specific worked-out example, the result was a probability of 0.702, which is equivalent to saying that, in the general case, one expects 70.2% of all individuals with those specific values of purchases and additional cards to upgrade to a premium card.

Section 14.7 *uses* the results of logistic regression as an explanatory model. While a perfectly appropriate use, logistic regression becomes a predictive analytics method when the logistic regression model is used as (part of) a prediction model. After all, logistic regression uses a binary categorical variable, so each cardholder upgraded or did not upgrade to a premium card. No cardholder upgraded "70.2% of a premium card" nor did "70.2% of a cardholder" upgrade to a premium card.

Recall that the data in Cardholder Study , analyzed to create the logistic model, uses the values of 1 and 0 for the upgraded variable to represent the two categories, or **classes**, "upgraded to a premium card" and "did not upgrade to a premium card." Could the logistic regression model be used as a prediction model to *predict* individual values for the upgraded variable? Yes, if one uses a **decision criterion**, a value set using business experience or insight or by experimentation.

One establishes a **decision rule** that compares the criterion to the estimated probability to decide whether to predict the value 1 or 0 for the upgraded variable. Figure 17.1 diagrams how applying a decision rule turns logistic regression into a method of classification.

FIGURE 17.1
Using estimated probability for classification

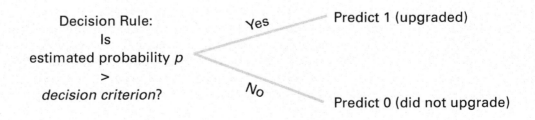

Note that stepwise regression and best subset regression require using decision rules, too, even as the phrase never appears in Chapter 15. In Exhibit 15.1 "Successful Model Building" on page 540, step 5 specifies to choose the best model based the values of adjusted r^2 or the C_p statistic, thereby stating a decision rule. When a regression method includes the automated selection of a model based on a decision rule, the method becomes a true example of predictive analytics.

Microsoft Excel and Business Analytics

Microsoft Excel plays mostly a supporting role when using business analytics. Many use Excel as a convenient means to select and organize data or to perform various data preparation tasks to ready data for later analysis by business analytics applications. Reflecting that choice, many analytics programs, including Tableau and Microsoft Power BI Desktop, include the ability to import Excel workbook files for analysis.

Excel can be used to perform statistical analyses that help better explain, support, or explore business analytics results. One can also modify Excel, adding Microsoft-supplied or third-party add-ins that give Excel new analytics-oriented functionality. Add-ins that perform basic descriptive analytics or implement predictive analytics methods such as tree induction (see Section 17.3) are available on either a free-to-use, free-to-try, or for-sale basis.

Remainder of This Chapter

Because Excel is most likely to be used in conjunction with descriptive analytics, Section 17.2 discusses descriptive analytics and presents an example that uses the Microsoft Power BI Desktop, an application that complements and works with Microsoft Excel. Section 17.3 features regression and classification trees, a predictive analytics method implemented in a variety of free-to-try and for-sale add-ins. (As Section 17.3 discusses, trees are also related to the regression methods that Chapters 13 through 15 discuss.)

Section 17.4 provides examples of clustering, and Section 17.5 presents an example of association analysis, two other types of predictive analytics. Section 17.6 introduces readers to text analytics, a much-discussed application of business analytics, and Section 17.7 provides readers with an introduction to prescriptive analytics.

17.2 Descriptive Analytics

Chapters 2 and 3 discuss descriptive methods that organize and visualize previously collected data. What if current data could be organized and visualized as it gets collected? That would change descriptive methods from being summaries of the status of a business at some point in the past into a tool that could be used for day-to-day, if not minute-by-minute, business monitoring. Giving decision makers this ability is one of the goals of descriptive analytics.

Descriptive analytics provide the means to monitor business activities in *near real time*, very quickly after a transaction or other business event has occurred. Being able to do this monitoring can be useful for a business that handles perishable inventory. As the First Things First Chapter Using Statistics scenario notes, empty seats on an airplane or in a concert hall or theater cannot be sold after a certain time. Descriptive analytics allows for a continuously updated display of the inventory, informing late-to-buy customers of the current availability of seats as well as visualizing patterns of sold and unsold seats for managers.

Descriptive analytics can help manage sets of interrelated flows of people or objects as those flows occur. For example, managers of large sports complexes use descriptive analytics to monitor the flow of cars in parking facilities, the flow of arriving patrons into the stadium, as well as the flow of patrons inside the stadium. Summaries generated by descriptive methods can highlight trends as they occur, such as points of growing congestion. By being provided with such information in a timely manner, stadium managers can redirect personnel to trouble spots in the complex and redirect patrons to entrances or facilities that are underused.

Dashboards

Dashboards are comprehensive summary displays that enable decision makers to monitor a business or business activity. Dashboards present the most important pieces of information, typically, in a visual format that allows decision makers to quickly perceive the overall status of an activity. Dashboards present these key indicators in a way that provides drill-down abilities that can reveal progressive levels of detail interactively.

Dashboards can be of any size, from a single desktop computer display, to wall-mounted displays or even larger, such as the nearly 800-square-foot NASDAQ MarketSite Video Wall at Times Square, which can be configured as a NASDAQ stock market dashboard that provides current stock market trends for passersby and viewers of financial programming (see Paczkowski).

Figure 17.2 presents a Microsoft Power BI dashboard that the new managers at Arlingtons might use to monitor national sales. The dashboard uses word tiles and clickable tabular summaries to present sales summaries at different levels of detail: by store category and then by the subcategories of the hardlines department that include mobile electronics sales, the subject of Chapter 10 sales experiments. In Figure 17.2, managers have decided to focus on mobile electronics sales and are currently viewing mobile electronics sales from one of the four national sales regions, while monitoring total mobile electronics sales nationwide (6544). By viewing a dashboard, the new owners of Arlingtons have a clearer and more immediate picture of current sales throughout the Arlingtons chain. That may help them better react to changes as they seek to manage the retailer to better success.

Figure 17.2 illustrates that dashboards can visually present drilled down data and act as complements to the data exploration techniques that organize and visualize a mix of variables that Sections 2.6 and 2.7 summarize. While Figure 17.2 contains simple data visualizations, the more complex visualizations that Section 2.7 describe, such as treemaps and colored scatter plots, can also appear in dashboards. For dashboards designed for individual users, multidimensional contingency tables that permit drill-down (see Sections 2.6) are also found.

FIGURE 17.2

National sales dashboard for the Arlingtons retail chain

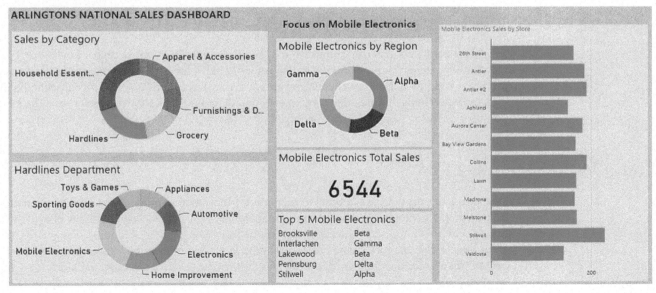

Data Dimensionality and Descriptive Analytics

Descriptive analytics also help visualize data that have a higher **data dimensionality**, the number of variables that comprise a data item, by using color, size, or motion to represent additional dimensions. This higher dimensionality overcomes the limits of standard business display technologies, such as screens and paper, that are two-dimensional surfaces.

A **bubble chart** uses filled-in circles, called bubbles, the color and size (diameter) of which add additional data dimensions. Typically, color represents a categorical variable and size represents a numerical variable, but either of these attributes can be used in the other way. **Dynamic bubble charts**, also known as motion charts, extend bubble charts by using motion to represent one additional data dimension, typically time. These charts take the form of animations in which bubbles change over time. The changing position of bubbles over time often reveal complex trends and interactions better than an equivalent time-series plot of the data.

Figure 17.3 shows a time-lapse image from a Tableau dynamic bubble chart animation that visualizes domestic movie revenues, by the MPAA ratings G, PG, PG-13, and R, for the years 2002 through 2018. The time-lapse image shows only the animation for the even years in this time series. The animation reveals that as revenues of G-rated movies increase in a year, those revenues tend to depress the revenues of PG-rated movies, suggesting some relationship. The animation also shows how revenues for G-rated movies shrink over time and revenues for PG-13 movies increase over time.

FIGURE 17.3
Time-lapse of Tableau dynamic bubble chart for domestic movie revenues by MPPA rating, for the years 2002 through 2018, showing even years only

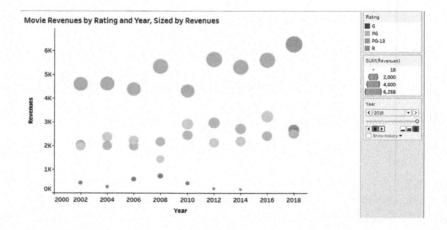

For the new owners of Arlingtons, a dynamic bubble chart might reveal how store, region, merchandise department, or merchandise category sales have changed over time. Such a chart might be used as part of an executive summary that introduces changes in merchandising or geographical focus that the new owners may decide to undertake as well as serve as the starting point for deciding to make such changes.

17.3 Decision Trees

Decision trees provide an alternate to regression for prediction models that use labeled data, values for both a dependent Y variable and independent X variables. Decision trees provide insight into the relative importance of variables, can be used as a basis for recoding categorical X variables into fewer categories, and can handle data that contain missing values. Like stepwise and best subset regression methods, decision trees can also be used for selecting variables for a model.

Tree induction methods treat all of the data as the **root node** and iteratively use a decision criterion to split the data, most commonly into two parts, forming a **binary tree**. In a binary tree, each split made corresponds to a child node that links to tree *branches* that represent the two parts of the split. Typically, methods apply the *greedy algorithm* to split nodes and seek to make the best split of the data. In seeking the best split for the node, that process does not consider the effect of the split on other parts of the tree. After each split, the split must be evaluated to decide whether to continue the tree building by permitting the method to make another split.

When using a decision tree method, one typically permits the method to make as many splits as possible. Because decision tree methods use labeled data to form prediction models, an overfitted tree model is always possible. Using a decision criterion or subjective judgment, one **prunes** the tree, eliminating branches and nodes to avoid overfitting. One can also uses cross-validation methods (see Section 14.8) in the same way as one uses cross-validation to improve a logistic regression prediction model.

The JMP data analysis software produced the decision tree examples in this section.

While Excel does not contain a decision tree chart type, many free-to-try Excel add-ins do. Power BI Premium, a superset of the Power BI desktop that works with Excel and that Section SG17.2 discusses, also includes the ability to create decision trees.

Regression Trees

Regression trees are decision trees that model a numerical dependent Y variable. Regression trees analyze the data that least-squares multiple regression uses. One evaluates a split using a decision criterion such as the **logworth statistic**. The logworth statistic is equal to -1 times base 10 log of a special p-value that is used to evaluate a split. Recall that examples in this book use a p-value of 0.05 as a decision criterion, making the analogous logworth statistic 1.3013, -1 times log (0.05). For more stringent p-value criterion of 0.01, the analogous logworth statistic is 2. One compares the value of the logworth statistic computed for a split with a predetermined decision value, perhaps 1.3013 or 2, to decide if a split is a good split.

Figure 17.4 presents the regression tree for the Chapter 14 OmniPower regression example that uses price and promotion expenses to estimate sales of OmniPower bars. The split of the root node has a logworth statistic of 4.2667, making the split a good split. The second of the node "price less than 99 cents" has a logworth statistic of 2.1843, making for another good split.

FIGURE 17.4

Regression tree for the Chapter 14 OmniPower sales data

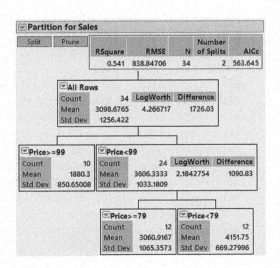

After this second split, the tree method could not find any other split to make and therefore stopped. Using either 1.3013 or 2 as the logworth decision value, this tree does not need to be pruned because both splits are good splits and the tree is ready for use. The tree reveals that price is more important than promotional expenses in determining sales and that promotional expenses is never used as the basis of a split, suggesting its relative unimportance to sales.

Recalling that the price in the data is always 99, 79, or 59 cents, the tree can be interpreted by price. Selling bars at 99 cents generates mean sales about one-half that of the sales of bars sold at 79 and 59 cents ($1,880 versus $3,606). Likewise, selling bars at 59 cents generates about one-third more mean sales than selling bars at 79 cents ($4,152 versus $3,061). An OmniPower marketer looking at this tree might conclude to conduct sales experiments with additional pricing options to further explore this observed effect.

Figure 17.5 presents the regression tree for the Chapter 15 Nickels Broadcasting standby hours analysis. Before stopping, the tree method created two splits, with the second split having a logworth value of 1.1603, a low value that indicates a poor split. One would prune this tree of the second pair of branches (circled in red in Figure 17.5) before using the tree for insights and decision making. The revised tree suggests that when staffing hours are 315 or more, the mean standby hours needed will be about 50% more than when staffing hours are less than 315 (207.6 hours versus 139.4 hours).

FIGURE 17.5

Regression tree induction results for the Chapter 15 Nickels Broadcasting analysis (before the results of the second split, circled in red, are pruned)

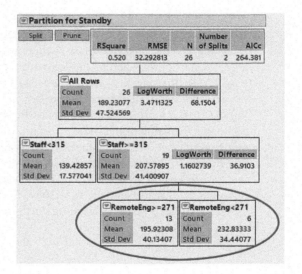

Classification Trees

Classification trees are decision trees that model a categorical Y variable. Classification trees analyze the data that logistic regression uses. Classification tree methods split nodes using various techniques that compute probabilities. Some methods seek to minimize the **Gini impurity**, G^2, which estimates the probability of a misclassification of a randomly selected occurrence selected from the data that a node represents. By definition, a Gini impurity cannot be less than zero, with zero representing the case in which each data occurrence has the same classification (such data are sometimes called *homogenous*). Although measures such as the Gini impurity determine splits, one evaluates splits using other decision criteria, such as the logworth statistic.

Figure 17.6 presents the classification for the Section 14.7 cardholder study analysis. That analysis examines the effect that a cardholder's credit card purchases and decision to request extra cards has on whether the cardholder will upgrade to a premium version of the credit card. In the classification tree, the red bars represent the count of cardholders who did not upgrade, and the blue bars represent the count of cardholders who did upgrade.

The classification tree method stops after making three splits and finding two groups with a Gini impurity of 0, one all "blue" and the other all "red." However, while all splits lowered the Gini impurity, not all splits are good splits. The splits that divide the ordered-extra-cards (value 1) and did-not-order-extra-cards (value 0) are poor splits with very low values. The four bottom branches (circled in red in Figure 17.6) should be pruned before the tree can be used as a prediction model.

FIGURE 17.6
Classification tree induction results for the Section 14.7 cardholder study data (before any pruning occurs)

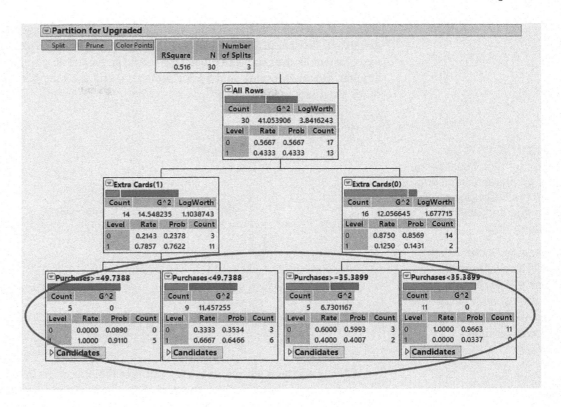

However, knowing that cardholders who ordered extra cards and had credit card purchases of greater than $49,738.80 has a valid explanatory purpose and suggests that those who charge the most are likeliest to upgrade. Using this tree as an explanatory model might led one not to prune the four branches.

Subjectivity and Interpretation

Note that using decision trees requires a degree of subjectivity and interpretation greater than that required to use the results of inferential statistics method this book discusses. For example, there is no one "correct" criterion value for the logworth statistic, and one may decide to prune, or not to prune, branches based on one's use of a model.

Supplying subjectivity and interpretation and understanding the difference between using a model for explanatory or predictive (in the sense that Section 14.8 uses) purposes, is a prerequisite for using business analytics. For brevity's sake, examples that follow do not explicitly consider this issue, which is always present when one uses predictive analytics.

17.4 Clustering

Clustering groups items into sets that have similar items. Values for each variable in an item are compared to other items' values, and a calculated *distance* determines similarity of items. For two-dimensional data, one distance calculation, the **Euclidean distance**, calculates the square root of the sum of the squared differences among corresponding pairs of values in each item.

Clustering methods vary based on how they compute and apply distance when forming **clusters**, sets of similar items. Some methods, such as *k*-**means clustering**, work best with purely numerical variables, while others, such as **hierarchical clustering**, work equally with a mix of numerical and categorical variables. When analyzing items with many data dimensions, the basis for clustering may not be obvious to a business analytics user because the basis may be based on abstract mathematical relationships that do not have direct analogues to business processes.

Although typically used for data with many dimensions, clustering can be demonstrated using data that have two data dimensions. Figure 17.7 visualizes a *k*-means clustering as a colored scatter plot that represents the upload and download speeds of eight cellphone providers, using the data in Mobile Speed. The *k*-means clustering uses a decision rule that identifies two clusters as the preferred clustering for these data: The orange color cluster consists of the Cricket, Boost (Mobile), and Sprint providers, and the blue color cluster consists of the five other providers in the sample.

FIGURE 17.7
Scatter plot visualization of a *k*-means clustering for the cellphone provider upload and download speed data

Because Cricket clusters with Boost Mobile and Sprint, one can infer that upload speed has a greater influence on clustering than download speed.

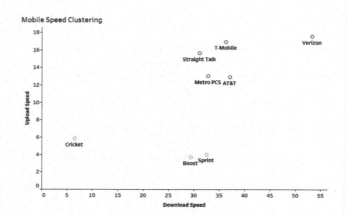

Clustering such simple data may not provide any more insight into the data than a simple scatter plot, but consider the more complex sample of 306 growth funds, stored in Growth Funds that Chapter 2 introduces and contains the 10 numerical variables (10 data dimensions): 1YrReturn, 3YrReturn, 5YrReturn, 10YrReturn, Assets, Beta, Expense Ratio, SD, Sharpe Ratio, and Turnover Ratio. Figure 17.8 left displays a (partial) visualization of the *k*-means clustering for *k* = 3 for these data. Figure 17.8 right displays a scatter plot of 10YrReturn and Expense Ratio variables, colored by cluster. The scatter plot suggests that lower expense ratios and higher ten-year rates of return are similarities found in cluster 3 (red color).

FIGURE 17.8
Left: Partial bar chart visualization of the *k*-means clustering for *k* = 3 for the growth funds sample
Right: Scatter plot of 10YrReturn and Expense Ratio variables, colored by cluster

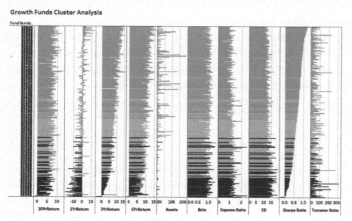 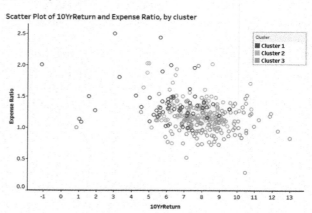

17.5 Association Analysis

In contrast to clustering that looks for similarity across items in a sample being analyzed, association analysis looks for similarity within the set of values that comprise one item. Association methods support many of the most-discussed applications of business analytics, such as market-basket analysis, a technique that can identify buying patterns and habits of different categories of consumers, and recommendation services that recommend additional products or services based on a consumer's past behavior. (Online retailers or streaming services such as Amazon and Netflix use recommendation services to offer customers additional choices.)

As with other types of predictive analytics methods, some association methods analyze numerical variables, while others, such as **multiple correspondence analysis** (**MCA**) analyze categorical variables. MCA examines similarity of items, as reflected in underlying contingency tables of items, to discover associations among categories of multiple categorical variables. Managers in the T.C. Resort Properties scenario (Chapter 12) might use MCA to provide additional insights into guest satisfaction.

Table 12.13 on page 404 summarizes guest satisfaction survey responses, by hotel, for reasons for not returning to a hotel. Not shown in Table 12.13 are responses to additional questions that asked guests their booking source (the T.C. Resorts website, an agent or third-party site, or walk-in) and their relationship status (single, couple, or family).

Figure 17.9 presents a biplot that summarizes these variables as well as the primary reason for not returning variable and reveals a number of associations. Those who gave the quality-of-room reason for not returning were associated with those who booked a room on the company's website. Using a third-party travel site or agent to book a room is associated with guests who stayed at the Palm Princess. Couples are more closely associated than either families or single people with the dining options reason for not returning.

FIGURE 17.9

Biplot of the multiple correspondence analysis for the T.C. Resorts guest satisfaction survey for the primary reason for not returning to a hotel

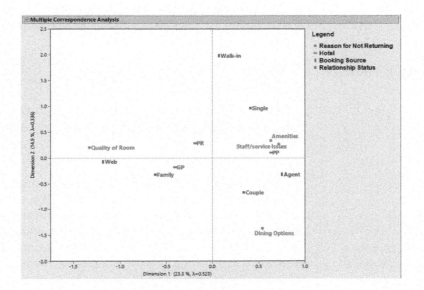

Biplots are graphs that plot the relationships among the rows and columns of a contingency table. Distances between points in biplots are not linear. In Figure 17.9, saying that because the Palm Princess Resort (PP) is at twice the distance to the quality-of-room reason than the Palm Royale Resort (PR), the Palm Princess is only "half as associated" with this reason as the Palm Royale would be incorrect. Likewise, assigning meanings to the two dimensions is pointless because they are mathematical abstractions.

These associations could raise new questions for the managers to explore, such as what attributes of the Palm Princess cause this hotel to be associated with the use of third-party agents. The associations might result in business changes, such as revising room descriptions on the company's website to see if complaints about the room quality decrease over time.

17.6 Text Analytics

The First Things First Chapter defines *unstructured data* as data that are not comprehensible without additional interpretation. In the past, the interpretation of such data was done manually, limiting the effectiveness and timeliness of such data. Today, business analytics (and related techniques) can automate as well as analyze such data. While fictional portrayals of applications such as facial or voice processing systems sometimes exaggerate the capabilities of current technology, every day many customer response centers use voice-related statistical and analytics techniques to recognize and respond to language spoken by customers calling a help line—without the need for human intervention.

Techniques that use *unstructured text* are among the methods most evolved for business use today. The first business information systems used *structured* text, which can be translated in a row and column entries. Entries in standard business forms are structured text because each form can be represented easily as a row of data with column values corresponding to a filled-in response. **Unstructured text** are words, phrases, passages, or any type of writing that cannot be made to fit a template easily. Sending a text message, posting a comment to a social media website, or writing answers to an essay test are examples of unstructured text. Collections of big data typically contain unstructured text that results from trying to combine data in different forms from different sources.

While unstructured text has always existed, the growing use of big data and the growing influence of social media has made being able to interpret and analyze this form of unstructured data increasingly important. **Text analytics** is the blend of descriptive and prescriptive analytics that automates that interpretation and makes analysis possible. Text analytics takes many forms, including some techniques related to the application of Bayes' theorem that the Consider This feature in Chapter 4 explores.

As an example, consider an online retailer that allows customers to post reviews of products bought. Table 17.2 contains three such reviews about an unspecified product.

TABLE 17.2
Three reviews of a product

Customer	Comment
Jill from Wynnewood	Great—I love this product and highly recommend it.
Bill from Woodwynn	Great—if you love a product that breaks after its first use!
Bryn from Billwood	Seller shipped fast. Five Stars!

With manual interpretation, most would recognize that Jill is positive about the product and Bill is not, while most would suspect that Bryn reviewed the seller of the product and not the product itself. Using methods known as **sentiment analysis**, frequencies of words that an analyst has classified as being positive words could be tallied and comments that have many positive terms separated and *classified*. Newer techniques called **semantic analysis** use *clustering* methods that operate on word *associations*. Semantic analysis combines predictive analytics with computer science natural language processing methods and does not rely on an analyst's classification of words as being positive, negative, or neutral.

Because of the complexity of text analytics, a practical demonstration of the concepts in this section is beyond the scope of this book.

One current focus of managers and researchers alike is *latent semantic analysis*. **Latent semantic analysis** creates clusters based on the "latent," dimensions of similarity in the unstructured text that exist implicitly. Note that latent semantic analysis (LSA) discovers clusters—not meanings of words. LSA calculations are complex and only fairly recently practical in business computing. As with clustering in general, using LSA does not guarantee creating clusters that have practical use for a decision maker.

17.7 Prescriptive Analytics

Prescriptive analytics seek to optimize the performance of a business and offer decision-making recommendations for how to respond to and manage business circumstances in the future. Prescriptive analytics often builds on the results of predictive analytics methods, which themselves are built on inferential statistics and combines those results with management science techniques while using data handling and processing capabilities of current information systems to access large data sets or run multiple analyses. To fully understand prescriptive analytics requires exposure to management science, which is beyond the scope of this book. Generally, prescriptive methods are based on one of two approaches: optimization or simulation.

Optimization and Simulation

In **optimization**, a decision maker sets constraints, which reflect resource limitations that a business process faces, or numerical goals, to learn how the process can work most effectively. Prescriptive optimization methods result in a single solution, known as the *decision model*, that represents the best way to manage the business process. Managers in the T.C. Resort Properties scenario, using the results of multiple correspondence analysis (see Section 17.5) and other predictive analytics methods, might assign spending allocations for items such as room improvements, staff training, dining facilities, website expenses, travel agent payments, or reception services, or any combination of these items first, and then determine how best to allocate the rest of their budget.

In contrast, in **simulation**, a decision maker repeatedly runs a predictive analytics model while varying the assumptions or data of the model to create a set of results that offer choices about the business process being modeled. A decision maker then uses decision criteria to choose a specific run of the model, which is not guaranteed to be optimal, to guide decision making. Simulation offers an alternative to optimization when the business process under study is not well understood or is subject to the unforeseen. Prescriptive simulation methods, sometimes called *simulation optimization*, automates this process of choosing and sometimes enhances the choice by varying an analytics method itself, such as varying the value of k or the measure of distance used in a clustering analysis (see Section 17.4).

▼USING **STATISTICS**
Back to Arlingtons ... , Revisited

I n the Using Statistics scenario, you were asked to prepare notes for a management seminar that would introduce business analytics to store managers at Arlingtons. You decide to explain how descriptive analytics can help managers know the status of current business activities and how dashboards, specifically, can be the mode of presentation for such information. You decide to explain that business analytics also includes the categories of predictive and prescriptive analytics and that methods of predictive analytics can be further classified as methods involving prediction, classification, clustering, or association. You decide that the managers should understand how predictive analytics extends and builds on inferential methods such as regression. You also realize that managers should appreciate how clustering and association methods can help identify groups of customers of interest and gain insights into customer buying habits. You decide that the concluding section of your notes should explain how prescriptive analytics can build on the results of predictive analytics to start to answer questions about what *should* happen that can guide future managerial decision making.

▼REFERENCES

Breiman, L., J. Friedman, C. J. Stone, and R. A. Olshen. *Classification and Regression Trees*. London: Chapman and Hall, 1984.

Cox, T. F., and M. A. Cox. *Multidimensional Scaling*, 2nd ed. Boca Raton, FL: CRC Press, 2010.

Doron Cohen, as quoted in Morgan, L. "8 Smart Ways to Use Prescriptive Analytics." *InformationWeek* 6/28/2016, available at **ubm.io/293ZMoy**.

Everitt, B. S., S. Landau, and M. Leese. *Cluster Analysis*, 5th ed. New York: John Wiley, 2011.

Few, S. *Information Dashboard Design: Displaying Data for At-a-Glance Monitoring*, 2nd ed. Burlingame, CA: Analytics Press, 2013.

Koren, Y. "The BellKor Solution to the Netflix Grand Prize," available at **bit.ly/2vZVAkZ**.

Levine, D., D. Stephan, and K. Szabat. *Understanding Business Analytics*. Boston: Pearson, forthcoming 2020.

Loh, W. Y. "Fifty Years of Classification and Regression Trees." *International Statistical Review*, 2013.

"NASDAQ Wall Capabilities," **bit.ly/1ubnLGQ**.

Paczkowski, W. *Market Data Analysis Using JMP*. Cary, NC: SAS institute, 2016.

Provost, F., and T. Fawcett. *Data Science for Business*. Sebastopol, CA: O'Reilly Media, 2013.

▼KEY TERMS

artificial intelligence   602
association methods   601
binary tree   607
biplot   611
bubble chart   606
classes   603
classification methods   601
classification tree   608
clustering   609
clustering methods   601
clusters   609
dashboard   605
data dimensionality   606
data mining   602
decision criterion   603
decision rule   603

decision trees   607
descriptive analytics   601
dynamic bubble charts   606
Euclidean distance   609
explanatory models   602
hierarchical clustering   609
Gini impurity, G^2 608
k-means clustering   609
labeled data   602
latent semantic analysis   612
logworth statistic   607
machine learning   602
multiple correspondence analysis
   (MCA)   611
optimization   613
prediction methods   601

prediction models   603
predictive analytics   601
prescriptive analytics   601
prune   607
regression tree   607
root node   607
semantic analysis   612
sentiment analysis   612
simulation   613
supervised methods   602
text analytics   612
tree induction methods   607
unstructured text   612
unsupervised methods   602

▼CHECKING YOUR UNDERSTANDING

17.1 What are the major categories of business analytics?

17.2 What prevented the widespread adoption of business analytics in the past?

17.3 What is data mining?

17.4 What is artificial intelligence, and what is its relationship to machine learning?

17.5 Explain the difference between explanatory and prediction models.

17.6 How are decision criteria used in decision rules?

17.7 How do dashboards enable business analytics users to explore data?

17.8 What is meant by "data dimensionality?"

17.9 How do regression trees differ from classification trees?

17.10 How does clustering differ from association analysis?

17.11 What is text analytics relationship to clustering and association methods?

17.12 What are the two main approaches in prescriptive analytics?

▾SOFTWARE GUIDE

INTRODUCTION

This software guide combines Excel and Tableau instructions into one guide.

SG17.2 DESCRIPTIVE ANALYTICS

Dashboards

Excel Use copy-and-paste commands or the PowerBI Desktop program.

Selecting Excel visualizations, copying them, and then pasting them into Word documents, PowerPoint slides, or other compatible formats can create results that mimic a dashboard. For a more professional result that includes the ability to interact with results use Power BI Desktop, a Microsoft business analytics program that can be downloaded at **powerbi.microsoft.com/desktop**.

With the Power BI Desktop installed and opened, open the **Arlingtons National Sales Power BI file (.pbix)** that contains the Figure 17.2 dashboard. The Power BI Desktop display includes a panel in which visualizations can be selected, formatted, and assigned data, as well as a Fields panel, which lists the tables that provide the source data for the visualizations in the dashboard (shown below).

For the Arlingtons dashboard, the tables correspond to worksheets in the Arlingtons Dashboard Data workbook that were imported to Power BI Desktop. Some tables are linked. For example, clicking the Alpha sector in the "Mobile Electronics by Region" doughnut chart, selects only those stores that are in the Alpha region, which changes the "Top 5 Mobile Electronics" list. Such interactions can also drill down data, although the dashboard does not illustrate that feature.

To create a new dashboard, select **File→New**. To import Excel data, select **Home→Get Data** and in the Get Data dialog box, select **All** from the left list, **Excel** from the right list, and click **Connect**. In the standard Open dialog box, select the Excel workbook to import and click **Open**. Power BI Desktop displays a Navigator dialog box that lists the names of the worksheets in the selected workbook. Check the worksheets to be imported and then click **Load**. Power BI Desktop imports the worksheets as additional tables and lists them in the Fields panel.

Click the icon for a visualization to insert that visualization. Visualizations can be resized and repositioned in the dashboard report. Visualizations can be associated with data by dragging fields from the Fields pane into various field boxes that appear in the Visualization pane. Clicking the paint roller icon allows custom formatting of a selected visualization.

Power BI Desktop is one part of the Power BI family that includes online service and mobile components. The Power BI website (**powerbi.microsoft.com/desktop**) provides a complete summary as well as complete documentation for using Power BI and several worked-out business analytics examples. Note that Power BI considers Power BI files as reports and reserves the term dashboard for descriptive reports that can be *shared* by many users. In spite of that distinction, a Power BI file fully demonstrates the concept of a descriptive analytics dashboard.

Tableau Use **New Dashboard**.

To create a dashboard, first create the tabular and visual summaries that comprise the dashboard as separate worksheets in the same Tableau workbook. With the workbook open, select **Dashboard New Dashboard**. Tableau inserts a new dashboard into the workbook and displays the Dashboard tab in place of the Data tab. In the Dashboard tab (shown below in two parts):

1. Click the **Size** pull-down to reveal more settings. Click the **Range** pull-down menu and select **Automatic** (shown below left).

2. Click anywhere in the white space of the window to hide the revealed size settings and to be able to see the Sheets and Objects lists unobscured, an example of which is shown below right.

3. At the bottom of the Dashboard tab, check **Show dashboard title**.

Tableau displays the current dashboard name in the dashboard area.

4. Double-click the dashboard title and in the Edit Title dialog box (not shown), change the title to a more descriptive name and then click **OK**.
5. For each worksheet to appear in the dashboard, drag the worksheet name from the Sheets list and drop it in the dashboard area.

To adjust the appearance of individual worksheet elements, click the drop-down icon on the right edge of the worksheet frame and select the appropriate choice from the drop-down menu. For example, to suppress the display of an unnecessary chart legend, select **Legends** and clear **Color** (or **Size**) **Legend**.

If dashboard worksheet elements are poorly fitted into the dashboard, each worksheet frame can be adjusted for a better fit. To adjust a worksheet frame, click the drop-down icon on the edge of the worksheet frame and select **Fit → Fit Width** from the pull-down menu to fit the width of the worksheet in the dashboard or select **Fit → Entire View** to display the entire worksheet in the frame.

Dynamic Bubble Charts

Example Create a dynamic bubble chart for domestic movie revenues by MPPA rating, for the years 2002 through 2018, that is the basis of the Figure 17.3 time-lapse illustration on page 606.

Tableau Use **circle views** with the **Pages** feature.

Open a new Tableau workbook. Click **Connect to Data** and open the **Movie Revenues by MPAA Rating Excel workbook** to establish the data source. In a new Tableau worksheet:

1. Drag **Rating** and drop it in the **Rows** shelf.
2. Drag **Revenues** and drop it in the **Columns** shelf.

Tableau creates a bar chart that displays the total revenues for the period for the years 2002 through 2018, by four MPAA ratings. To create a dynamic bubble chart, first change visualization to a circle views chart.

3. If necessary, click the **Show Me** tab to make the contents of the tab visible.
4. Select the **circle views icon** in the Show Me gallery (shown below).

The visualization changes from the default bar chart to a circle views chart. Note that in the Marks card area Rating has been associated with the Color attribute of the circle views chart.

5. Drag **Revenues** and drop it over the **Size** icon in the Marks card area. Tableau associates the SUM(Revenues) with the Size attribute of the circle views chart.
6. Drag **Year** and drop it in the **Pages** shelf.
7. Drag **Year** a second time and drop it in the **Columns** shelf.
8. Click the **Size** icon and slide the **Size Slider** to its midpoint.
9. Click the **Shape** icon. From the Shape gallery, select the **filled-in circle icon**.
10. Edit the worksheet title.
11. If necessary, click the **Show Me** tab to hide the tab and to reveal the legends for color and size (shown below left) and the playback controls (shown below right).

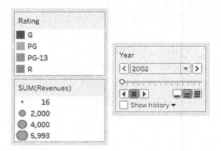

Press the **play button icon** to play the time-series animation. The playback controls also include a set of three playback speed icons for slow, normal, or fast playback. (Normal is selected in the illustration above.)

SG17.5 PREDICTIVE ANALYTICS for CLUSTERING

Example Perform the Figure 17.8 k-means clustering for $k = 3$ for the sample of growth funds on page 610.

Tableau Use the **Cluster** model in the Analytics tab.

For the example, to perform a clustering similar to the Figure 17.8 k-means clustering $k = 3$ for the growth funds on page 610 open a new Tableau workbook. Click **Connect to Data** and open the **Growth Funds Excel workbook** to establish the data source. In a new Tableau worksheet:

1. Drag **Fund Number** and drop it in the **Rows** shelf.
2. Select the **ten field measures** (10YrReturn, 1YrReturn, 3YRReturn, 5YrReturn, Assets, Beta, Expense Ratio, SD, Sharpe Ratio, and Turnover Ratio) and drop them in the **Columns** shelf.

Tableau creates bar charts for each measure by fund number. To cluster the data, open to the worksheet that contains the bar charts and:

1. Select the **Analytics** tab (shown below).
2. Drag the **Cluster** model from the tab and drop it over the bar chats.

3. In the Cluster dialog box (shown below), enter **3** as the **Number of Clusters** in the columns list and then close the dialog box.

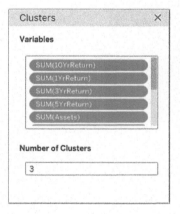

Tableau associates Clusters with the Color attribute in the Marks card area (shown below left) and adds a Clusters legend to the worksheet and colors of the bars of each fund number by its cluster membership (shown below right).

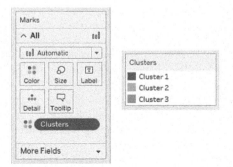

To see summary measures of the clusters, click the Clusters pill in the Marks Card area and select **Describe clusters** from the menu. Tableau opens a new dialog box (not shown) that contains summary measures about the clusters.

To view the complete cluster analysis, use the Section SG17.2 Tableau instructions to insert a new dashboard. Drag the cluster analysis worksheet into the dashboard area and adjust the frame to display the entire worksheet.

When examining the complete cluster analysis, the sparkline-like bars begin to reveal some patterns in the three clusters. One cluster, primarily comprising funds and presented near the bottom of the display seems to be characterized by high turnover ratios and very poor one-year returns.

To discover information about specific retirement funds, mouse over the bars to pop up a window containing the data for the fund and its cluster membership. Shown below is the pop-up window for retirement fund RF008.

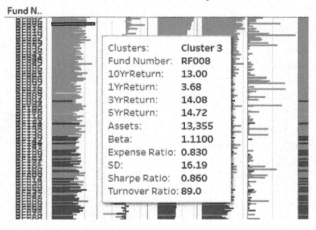

18

Getting Ready to Analyze Data in the Future

OBJECTIVES

- Identify the questions to ask when choosing which statistical methods to use to conduct data analysis
- Generate rules for applying statistics in future studies and analyses

▼USING **STATISTICS**
Mounting Future Analyses

Learning and applying business statistics methodology has some similarities with planning and executing a mountain climbing expedition. Initially, what might seem intimidating, or even overwhelming, can be conquered by applying methods and techniques using a framework that identifies and guides you through a series of tasks. In Section FTF.1, you first learned how the **DCOVA framework** can help apply statistical methods to business problems. After learning methods in early chapters to Define, Collect, and Organize data, you have spent most of your time studying ways to Visualize and Analyze data.

Determining which methods to use to organize, visualize, and analyze your data may have seemed straightforward when you worked out examples or problems from a particular chapter in which the data had already been defined and collected (and perhaps even organized) for you. The defined and possibly organized data gave clues about which methods to choose, as did the chapter itself. For example, while studying the descriptive statistics chapter, you could deduce properly that applying the inferential methods of other chapters would not be part of any example or problem.

But what should you do when you find yourself in new situations such as needing to analyze data for another course or to help solve a problem in a real business setting? You will not be studying a chapter of this book, so the methods to choose and apply will not necessarily be as obvious to you as they may have been when working out a specific problem from a specific chapter. How then can you guide yourself to choosing appropriate statistical methods as you seek to mount future analyses?

S electing the appropriate methods to use with data turns out to be the single most difficult thing you do when you apply business statistics to real situations. This is also the single most important task you face. Recall that when using the DCOVA approach, you first define the variables that you want to study in order to solve a business problem or meet a business objective. To do this, you identify the type of business problem, such as trying to describe a group or trying to make inferences about a group, and then determine the type of variable—*numerical* or *categorical*—you will be analyzing.

That act of defining a variable provides the starting point for selecting appropriate statistical methods to use. Once you know the type of variable, you can ask yourself a series of questions about what you seek to do with that variable. The questions can guide you to the appropriate methods to select as surely as a mountain guide can help you to the summit of a mountain. Therefore, the *answer* to the question "How can you guide yourself to choosing appropriate statistical methods as you seek to mount future analyses?" is "Ask more questions."

In the following two sections, this chapter presents two sets of questions, one for numerical variables and the other for categorical variables, that you can ask yourself once you have defined your variable. Unlike other chapters, this chapter introduces a Using Statistics scenario to raise a completely different type of question.

studentTIP

Recall that *numerical variables* have values that represent quantities, while *categorical variables* have values that represent categories.

18.1 Analyzing Numerical Variables

To analyze a numerical variable, choose the appropriate Exhibit 18.1 question and then read the answer to the question in this section.

EXHIBIT 18.1

Questions to Ask When Analyzing Numerical Variables

Do you want to

- describe the characteristics of the variable (possibly broken down into several groups)?
- reach conclusions about the mean or the standard deviation of the variable in a population?
- determine whether the mean and/or standard deviation of the variable differs depending on the group?
- determine which factors affect the value of a variable?
- predict the value of the variable based on the values of other variables?
- classify items into groups or look for patterns of association among items?
- determine whether the values of the variable are stable over time?

Describe the Characteristics of a Numerical Variable?

You develop tables and charts and compute descriptive statistics to describe characteristics such as central tendency, variation, and shape. Specifically, you can create a stem-and-leaf display, percentage distribution, histogram, polygon, boxplot, normal probability plot, and treemap (see Sections 2.2, 2.4, 2.7, 3.3, and 6.3), and you can compute statistics such as the mean, median, mode, quartiles, range, interquartile range, standard deviation, variance, coefficient of variation, skewness, and kurtosis (see Sections 3.1, 3.2, and 3.3).

Reach Conclusions About the Population Mean or the Standard Deviation?

You have several different choices, and you can use any combination of these choices. To estimate the mean value of the variable in a population, you construct a confidence interval estimate of the mean (see Section 8.2). To determine whether the population mean is equal to a specific value, you conduct a t test of hypothesis for the mean (see Section 9.2). To determine whether the population standard deviation or variance is equal to a specific value, you conduct a χ^2 test of hypothesis for the standard deviation or variance (see online Section 12.7).

Determine Whether the Mean and/or Standard Deviation Differs Depending on the Group?

When examining differences between groups, you first need to establish which categorical variable to use to divide your data into groups. You then need to know whether this grouping variable divides your data into two groups (such as male and female groups for a gender variable) or whether the variable divides your data into more than two groups (such as the four in-store locations for mobile electronics discussed in Section 11.1). Finally, you must ask whether your data set contains independent groups or matched or repeated measurements.

If the Grouping Variable Defines Two Independent Groups and You Are Interested in Central Tendency   Which hypothesis tests you use depends on the assumptions you make about your data.

If you assume that your numerical variable is normally distributed and that the variances are equal, you conduct a pooled t test for the difference between the means (see Section 10.1). If you cannot assume that the variances are equal, you conduct a separate-variance t test for the difference between the means (see Section 10.1). In either case, if you believe that your numerical variables are not normally distributed, you can perform a Wilcoxon rank sum test (see Section 12.4) and compare the results of this test to those of the t test.

To evaluate the assumption of normality that the pooled t test and separate-variance t test include, you can construct boxplots and normal probability plots for each group (see Sections 3.3 and 6.3).

If the Grouping Variable Defines Two Groups of Matched Samples or Repeated Measurements and You Are Interested in Central Tendency   If you can assume that the paired differences are normally distributed, you conduct a paired t test (see Section 10.2).

If the Grouping Variable Defines Two Independent Groups and You Are Interested in Variability   If you can assume that your numerical variable is normally distributed, you conduct an F test for the difference between two variances (see Section 10.4).

If the Grouping Variable Defines More Than Two Independent Groups and You Are Interested in Central Tendency   If you can assume that the values of the numerical variable are normally distributed, you conduct a one-way analysis of variance (see Section 11.1); otherwise, you conduct a Kruskal-Wallis rank test (see Section 12.5). You can use the Levene test (see Section 11.1) to test for the homogeneity of variance between the groups.

If the Grouping Variable Defines More Than Two Groups of Matched Samples or Repeated Measurements and You Are Interested in Central Tendency   Suppose that you have a design where the rows represent the blocks and the columns represent the levels of a factor. If you can assume that the values of the numerical variable are normally distributed, you conduct a randomized block design F test (see online Section 11.3).

Determine Which Factors Affect the Value of a Variable?

If two factors are to be examined to determine their effect on the values of a variable, you develop a two-factor factorial design (see Section 11.2).

Predict the Value of a Variable Based on the Values of Other Variables?

When predicting the values of a numerical dependent variable, you conduct least-squares regression analysis. The least-squares regression model you develop depends on the number of independent variables in your model. If only one independent variable is being used to predict the numerical dependent variable of interest, you develop a simple linear regression model (see Chapter 13); otherwise, you develop a multiple regression model (see Chapters 14 and 15) or a regression tree (see Chapter 17).

If you have values over a period of time and you want to forecast the variable for future time periods, you can use moving averages, exponential smoothing, least-squares forecasting, and autoregressive modeling (see Chapter 16). If you want to visualize many variables simultaneously, you can use sparklines (see Section 2.7).

Classify or Associate Items?

You use an appropriate predictive analytics classification or association method (see Chapter 17).

Determine Whether the Values of a Variable Are Stable Over Time?

If you are studying a process and have collected data on the values of a numerical variable over a time period, you construct R and $\overline{X}$ charts (see online Section 19.5). If you have collected data in which the values are counts of the number of nonconformities, you construct a c chart (see online Section 19.4).

18.2 Analyzing Categorical Variables

To analyze a categorical variable, choose the appropriate Exhibit 18.2 question and then read the answer to the question in this section.

EXHIBIT 18.2

Questions to Ask When Analyzing Categorical Variables

Do you want to

- describe the proportion of items of interest in each category (possibly broken down into several groups)?
- reach conclusions about the proportion of items of interest in a population?
- determine whether the proportion of items of interest differs depending on the group?
- predict the proportion of items of interest based on the values of other variables?
- cluster items into groups or look for patterns of association among items?
- determine whether the proportion of items of interest is stable over time?

Describe the Proportion of Items of Interest in Each Category?

You create summary tables and use these charts: bar chart, pie chart, doughnut chart, Pareto chart, or side-by-side bar chart (see Sections 2.1 and 2.3).

Reach Conclusions About the Proportion of Items of Interest?

You have two different choices. You can estimate the proportion of items of interest in a population by constructing a confidence interval estimate of the proportion (see Section 8.3). Or you can determine whether the population proportion is equal to a specific value by conducting a Z test of hypothesis for the proportion (see Section 9.4).

Determine Whether the Proportion of Items of Interest Differs Depending on the Group?

When examining this difference, you first need to establish the number of categories associated with your categorical variable and the number of groups in your analysis. If your data contain two groups, you must also ask if your data contain independent groups or if your data contain matched samples or repeated measurements.

For Two Categories and Two Independent Groups   You conduct either the Z test for the difference between two proportions (see Section 10.3) or the χ^2 test for the difference between two proportions (see Section 12.1).

For Two Categories and Two Groups of Matched or Repeated Measurements   You conduct the McNemar test (see online Section 12.6).

For Two Categories and More Than Two Independent Groups   You conduct a χ^2 test for the difference among several proportions (see Section 12.2).

For More Than Two Categories and More Than Two Groups   You develop contingency tables, use multidimensional contingency tables to drill down to examine relationships among two or more categorical variables, and map the categories of several categorical variables (Sections 2.1, 2.6, and Chapter 17). When you have two categorical variables, you conduct a χ^2 test of independence (see Section 12.3).

Predict the Proportion of Items of Interest Based on the Values of Other Variables?

You develop a logistic regression model (see Section 14.7) or a classification tree (see Chapter 17).

Cluster or Associate Items?

You use an appropriate predictive analytics clustering or association method (see Chapter 17).

Determine Whether the Proportion of Items of Interest Is Stable Over Time?

If you are studying a process and have collected data over a time period, you can create the appropriate control chart. If you have collected the proportion of items of interest over a time period, you develop a p chart (see online Section 19.2).

▼ USING **STATISTICS**
The Future to Be Visited

This chapter summarizes the statistical methods that the book discusses in previous chapters as exhibits that list questions that help you determine the appropriate methods with which to analyze your data. As the First Things First Chapter notes, business statistics is an important part of your business education. Consider the topics and concepts that this book discusses as only the beginning of your business statistics and data analysis education. In the future, pay attention to new developments such as business analytics becoming the changing face of statistics. Consider taking follow-up courses in business analytics or advanced business statistics as one way of better anticipating that future.

▼ CHAPTER REVIEW PROBLEMS

18.1 In many manufacturing processes, the term *work-in-process* (often abbreviated WIP) is used. At the LSS Publishing book manufacturing plants, WIP represents the time it takes for sheets from a press to be folded, gathered, sewn, tipped on end sheets, bound together to form a book, and the book placed in a packing carton. The operational definition of the variable of interest, processing time, is the number of days (measured in hundredths) from when the sheets come off the press to when the book is placed in a packing carton. The company has the business objective of determining whether there are differences in the WIP between plants. Data have been collected from samples of 20 books at each of two production plants. The data, stored in WIP, are as follows:

Plant A

5.62	5.29	16.25	10.92	11.46	21.62	8.45	8.58	5.41	11.42
11.62	7.29	7.50	7.96	4.42	10.50	7.58	9.29	7.54	8.92

Plant B

9.54	11.46	16.62	12.62	25.75	15.41	14.29	13.13	13.71	10.04
5.75	12.46	9.17	13.21	6.00	2.33	14.25	5.37	6.25	9.71

Completely analyze the data.

18.2 The file Travel and Tourism contains 2018 travel and tourism industry data for a sample of 28 European countries. Variables included are travel and tourism jobs (in thousands), nights spent by European residents at tourism establishments (in millions), euros spent by European residents at tourism establishments (in thousands), and the number of tourism establishments that provide overnight accommodations.

Source: Data extracted from Eurostat, **bit.ly/2cyv0Vs.**

Using this data, you seek to predict the number of jobs the travel and tourism industry generates. Completely analyze the data.

18.3 The file Philly contains a sample of 25 neighborhoods in Philadelphia. Variables included are neighborhood population, median sales price of homes in the second quarter of 2017, mean number of days homes were on the market in the second quarter of a recent year, number of homes sold in the second quarter of a recent year, median neighborhood household income, percentage of residents in the neighborhood with a bachelor's degree or higher, and whether the neighborhood is considered "hot" (coded as 1 = yes, 0 = no).

Data extracted from **bit.ly/33wKukl, bit.ly/2smOyVu, bit.ly/2v4mqZd,** and **bit.ly/2n0RNPW.**

Using this data, you seek to predict median sales price of homes. Completely analyze the data.

18.4 Many factors affect attendance at Major League Baseball (MLB) games. Among the factors are those relating to the capacity of the stadium, how the team performed during the season, and characteristics about the team itself. For each MLB team in a recent year, the file MLB Attendance Study includes the number of wins, season attendance, mean attendance per game, stadium capacity, mean percentage of capacity, mean batter age, mean pitcher age, number of all-star players on team, and team payroll ($millions).

You want to predict attendance and determine the factors that influence attendance. Completely analyze the data.

18.5 The file Used Cars contains attributes of cars that are currently part of an inventory of a used car dealership. The variables included are car, year, age, price ($), mileage, and fuel mileage (mpg).

Source: Data extracted from **www.truecar.com/used-cars-for-sale/.**

You want to describe each of these variables, and you would like to predict the price of the used cars. Analyze the data.

18.6 A study was conducted to determine whether any gender bias existed in an academic science environment. Faculty from several universities were asked to rate candidates for the position of undergraduate laboratory manager based on the candidate's application. The gender of the applicant was given in the applicant's materials. The raters were from either biology, chemistry, or physics departments. Each rater was to give a competence rating to the applicant's materials on a 7-point scale, with 1 being the lowest and 7 being the highest. In addition, the rater supplied a starting salary that should be offered to the applicant. These data (which have been altered from an actual study to preserve the anonymity of the respondents) are stored in Candidate Assessment.

Analyze the data. Do you think that there is any gender bias in the evaluations? Support your point of view with specific references to your data analysis.

18.7 The file Restaurant Ratings contains restaurant ratings for food, décor, service, cost per person, and popularity index (popularity points the restaurant received divided by the number of people who voted for that restaurant) for various types of restaurants in a large city.

You want to study differences in the cost of a meal for the different types of cuisines and also want to be able to predict the cost of a meal. Completely analyze the data.

18.8 Churn occurs when customers stop doing business with a company or stop subscribing to a service. The file Bank Churn Study contains data from a churn study conducted by an international bank.

Source: Data extracted from S. Dixit, "Building your own Artificial Neural Network…," **bit.ly/2HhnwFh**.

For each bank customer in the sample, the study includes ID number, credit score, domicile, gender, age, tenure (number of years that the customer has been a client of the bank), bank balance, number of bank products the customer has purchased through the bank, whether the customer has a credit card (0 = no, 1 = yes), whether the customer is an active member (0 = no, 1 = yes), salary, and whether the customer left the bank (0 = no, 1 = yes).

Analyze the data and assess the likelihood that the customer will leave the bank.

18.9 A mining company operates a large heap-leach gold mine in the western United States. The gold mined at this location consists of ore that is very low grade, having about 0.0032 ounce of gold in 1 ton of ore. The process of heap-leaching involves the mining, crushing, stacking, and leaching of millions of tons of gold ore per year. In the process, ore is placed in a large heap on an impermeable pad. A weak chemical solution is sprinkled over the heap and is collected at the bottom after percolating through the ore. As the solution percolates through the ore, the gold is dissolved and is later recovered from the solution. This technology, which has been used for more than 30 years, has made the operation profitable. Due to the large amount of ore that is handled, the company is continually exploring ways to improve the process.

As part of an expansion several years ago, the stacking process was automated with the construction of a computer-controlled stacker. This stacker was designed to load 35,000 tons of ore per day at a cost that was less than the previous process that used manually operated trucks and bulldozers. However, since its installation, the stacker has not been able to achieve these results consistently. Data for a recent 35-day period that indicate the amount stacked (tons) and the downtime (minutes) are stored in the file Mining. Other data that indicate the causes for the downtime are stored in Mining Downtime Causes.

Analyze the data, making sure to present conclusions about the daily amount stacked and the causes of the downtime. In addition, be sure to develop a model to predict the amount stacked based on downtime.

18.10 Wally's Discount Stores sell basic essentials for food, home, health, apparel, and home. Wally's produces and sells private-label goods under its ShowGo label in addition to select nationally advertised brands. The file Wallys contains a sample of 410 Wally's shoppers who purchased ShowGo products during a recent month. For each customer in the sample, the file contains an ID number, the amount of ShowGo products purchased ($), and whether the customer has a Wally's credit card (No or Yes), as well as the following demographic data: weekly household income ($), gender, age (years), customer self-rated healthy eating habits (1-to-10 scale, in which 1 is poor and 10 is excellent), and customer self-rated active lifestyle (1-to-10 scale, in which 1 is sedentary and 10 is very active).

Analyze these data and prepare a report describing your conclusions.

18.11 The file Hybrid Sales contains the number of domestic and imported hybrid vehicles sold in the United States from 1999 to 2018.

Source: Data extracted from "Gasoline Hybrid and Electric Vehicle Sales," **bit.ly/2nLkvXf**, and "Advanced Technology Vehicle Sales Dashboard," **bit.ly/32gnNAK**.

You want to be able to predict the number of domestic and imported hybrid vehicles sold in the United States in 2019 and 2020. Completely analyze the data.

APPENDICES

Basic Math Concepts and Symbols

A.1 Operators

Operators express a calculation or a logical comparison. Operators are building blocks for the equations that define statistical concepts and for formulas, statements that process data in Excel and Minitab worksheets and JMP data tables.

$+$	add	$=$	equal to	$>$	greater than
$-$	subtract	$\neq$	not equal to	$\geq$	greater than or equal to
$\times$	multiply	$\cong$	approximately equal to	$<$	less than
$\div$	divide			$\leq$	less than or equal to

A.2 Rules for Arithmetic Operations

Rule	Example
1. $a + b = c$ and $b + a = c$	$2 + 1 = 3$ and $1 + 2 = 3$
2. $a + (b + c) = (a + b) + c$	$5 + (7 + 4) = (5 + 7) + 4 = 16$
3. $a - b = c$ but $b - a \neq c$	$9 - 7 = 2$ but $7 - 9 \neq 2$
4. $(a)(b) = (b)(a)$	$(7)(6) = (6)(7) = 42$
5. $(a)(b + c) = ab + ac$	$(2)(3 + 5) = (2)(3) + (2)(5) = 16$
6. $a \div b \neq b \div a$	$12 \div 3 \neq 3 \div 12$
7. $\dfrac{a + b}{c} = \dfrac{a}{c} + \dfrac{b}{c}$	$\dfrac{7 + 3}{2} = \dfrac{7}{2} + \dfrac{3}{2} = 5$
8. $\dfrac{a}{b + c} \neq \dfrac{a}{b} + \dfrac{a}{c}$	$\dfrac{3}{4 + 5} \neq \dfrac{3}{4} + \dfrac{3}{5}$
9. $\dfrac{1}{a} + \dfrac{1}{b} = \dfrac{b + a}{ab}$	$\dfrac{1}{3} + \dfrac{1}{5} = \dfrac{5 + 3}{(3)(5)} = \dfrac{8}{15}$
10. $\left(\dfrac{a}{b}\right)\left(\dfrac{c}{d}\right) = \left(\dfrac{ac}{bd}\right)$	$\left(\dfrac{2}{3}\right)\left(\dfrac{6}{7}\right) = \left(\dfrac{(2)(6)}{(3)(7)}\right) = \dfrac{12}{21}$
11. $\dfrac{a}{b} \div \dfrac{c}{d} = \dfrac{ad}{bc}$	$\dfrac{5}{8} \div \dfrac{3}{7} = \left(\dfrac{(5)(7)}{(8)(3)}\right) = \dfrac{35}{24}$

A.3 Rules for Algebra: Exponents and Square Roots

Rule	Example
1. $(X^a)(X^b) = X^{a+b}$	$(4^2)(4^3) = 4^5$
2. $(X^a)^b = X^{ab}$	$(2^2)^3 = 2^6$
3. $(X^a/X^b) = X^{a-b}$	$\dfrac{3^5}{3^3} = 3^2$
4. $\dfrac{X^a}{X^a} = X^0 = 1$	$\dfrac{3^4}{3^4} = 3^0 = 1$
5. $\sqrt{XY} = \sqrt{X}\sqrt{Y}$	$\sqrt{(25)(4)} = \sqrt{25}\sqrt{4} = 10$
6. $\sqrt{\dfrac{X}{Y}} = \dfrac{\sqrt{X}}{\sqrt{Y}}$	$\sqrt{\dfrac{16}{100}} = \dfrac{\sqrt{16}}{\sqrt{100}} = 0.40$

A.4 Rules for Logarithms

Base 10

Log is the symbol used for base-10 logarithms:

Rule	Example
1. $\log(10^a) = a$	$\log(100) = \log(10^2) = 2$
2. If $\log(a) = b$, then $a = 10^b$	If $\log(a) = 2$, then $a = 10^2 = 100$
3. $\log(ab) = \log(a) + \log(b)$	$\log(100) = \log[(10)(10)] = \log(10) + \log(10)$
	$= 1 + 1 = 2$
4. $\log(a^b) = (b)\log(a)$	$\log(1{,}000) = \log(10^3) = (3)\log(10) = (3)(1) = 3$
5. $\log(a/b) = \log(a) - \log(b)$	$\log(100) = \log(1{,}000/10) = \log(1{,}000) - \log(10)$
	$= 3 - 1 = 2$

EXAMPLE

Take the base-10 logarithm of each side for the equation: $Y = \beta_0 \beta_1^X \varepsilon$

SOLUTION Apply rules 3 and 4:

$$\log(Y) = \log(\beta_0 \beta_1^X \varepsilon)$$
$$= \log(\beta_0) + \log(\beta_1^X) + \log(\varepsilon)$$
$$= \log(\beta_0) + X\log(\beta_1) + \log(\varepsilon)$$

Base e

In is the symbol used for base e logarithms, commonly referred to as natural logarithms. e is Euler's number, and $e \cong 2.718282$:

Rule	Example
1. $\ln(e^a) = a$	$\ln(7.389056) = \ln(e^2) = 2$
2. If $\ln(a) = b$, then $a = e^b$	If $\ln(a) = 2$, then $a = e^2 = 7.389056$
3. $\ln(ab) = \ln(a) + \ln(b)$	$\ln(100) = \ln[(10)(10)]$
	$= \ln(10) + \ln(10)$
	$= 2.302585 + 2.302585 = 4.605170$
4. $\ln(a^b) = (b)\ln(a)$	$\ln(1{,}000) = \ln(10^3) = 3\ln(10)$
	$= 3(2.302585) = 6.907755$
5. $\ln(a/b) = \ln(a) - \ln(b)$	$\ln(100) = \ln(1{,}000/10) = \ln(1{,}000) - \ln(10)$
	$= 6.907755 - 2.302585 = 4.605170$

EXAMPLE

Take the base e logarithm of each side for the equation: $Y = \beta_0 \beta_1^X \varepsilon$

SOLUTION Apply rules 3 and 4:

$$\ln(Y) = \ln(\beta_0 \beta_1^X \varepsilon)$$
$$= \ln(\beta_0) + \ln(\beta_1^X) + \ln(\varepsilon)$$
$$= \ln(\beta_0) + X\ln(\beta_1) + \ln(\varepsilon)$$

A.5 Summation Notation

The symbol Σ, the Greek capital letter sigma, represents "taking the sum of." Consider a set of n values for variable X. The expression $\sum\limits_{i=1}^{n} X_i$ means to take the sum of the X_i values from X_1 through X_n:

$$\sum_{i=1}^{n} X_i = X_1 + X_2 + X_3 + \cdots + X_n$$

To illustrate the use of the symbol Σ, consider five values of a variable X: $X_1 = 2$, $X_2 = 0$, $X_3 = -1$, $X_4 = 5$, and $X_5 = 7$. Thus:

$$\sum_{i=1}^{5} X_i = X_1 + X_2 + X_3 + X_4 + X_5 = 2 + 0 + (-1) + 5 + 7 = 13$$

In statistics, the squared values of a variable are often summed. Thus:

$$\sum_{i=1}^{n} X_i^2 = X_1^2 + X_2^2 + X_3^2 + \cdots + X_n^2$$

and, in the example above:

$$\sum_{i=1}^{5} X_i^2 = X_1^2 + X_2^2 + X_3^2 + X_4^2 + X_5^2$$

$$= 2^2 + 0^2 + (-1)^2 + 5^2 + 7^2 = 4 + 0 + 1 + 25 + 49 = 79$$

$\sum\limits_{i=1}^{n} X_i^2$, the summation of the squares, is *not* the same as $\left(\sum\limits_{i=1}^{n} X_i \right)^2$, the square of the sum:

$$\sum_{i=1}^{n} X_i^2 \neq \left(\sum_{i=1}^{n} X_i \right)^2$$

In the example given above, the summation of squares is equal to 79. This is not equal to the square of the sum, which is $13^2 = 169$.

Another frequently used operation involves the summation of the product. Consider two variables, X and Y, each having n values. Then:

$$\sum_{i=1}^{n} X_i Y_i = X_1 Y_1 + X_2 Y_2 + X_3 Y_3 + \cdots + X_n Y_n$$

Continuing with the previous example, suppose there is a second variable, Y, whose five values are $Y_1 = 1$, $Y_2 = 3$, $Y_3 = -2$, $Y_4 = 4$, and $Y_5 = 3$. Then,

$$\sum_{i=1}^{n} X_i Y_i = X_1 Y_1 + X_2 Y_2 + X_3 Y_3 + X_4 Y_4 + X_5 Y_5$$

$$= (2)(1) + (0)(3) + (-1)(-2) + (5)(4) + (7)(3)$$

$$= 2 + 0 + 2 + 20 + 21$$

$$= 45$$

In calculating $\sum\limits_{i=1}^{n} X_i Y_i$, the first value of X is multiplied by the first value of Y, the second value of X is multiplied by the second value of Y, and so on. These products are then summed in order to compute the desired result. However, the summation of products is *not* equal to the product of the individual sums:

$$\sum_{i=1}^{n} X_i Y_i \neq \left(\sum_{i=1}^{n} X_i \right)\left(\sum_{i=1}^{n} Y_i \right)$$

In this example,

$$\sum_{i=1}^{5} X_i = 13$$

and

$$\sum_{i=1}^{5} Y_i = 1 + 3 + (-2) + 4 + 3 = 9$$

so that

$$\left(\sum_{i=1}^{5} X_i\right)\left(\sum_{i=1}^{5} Y_i\right) = (13)(9) = 117$$

However,

$$\sum_{i=1}^{5} X_i Y_i = 45$$

The following table summarizes these results.

Value	X_i	Y_i	$X_i Y_i$
1	2	1	2
2	0	3	0
3	−1	−2	2
4	5	4	20
5	7	3	21
	$\sum_{i=1}^{5} X_i = 13$	$\sum_{i=1}^{5} Y_i = 9$	$\sum_{i=1}^{5} X_i Y_i = 45$

Rule 1 The summation of the values of two variables is equal to the sum of the values of each summed variable:

$$\sum_{i=1}^{n}(X_i + Y_i) = \sum_{i=1}^{n} X_i + \sum_{i=1}^{n} Y_i$$

Thus,

$$\sum_{i=1}^{5}(X_i + Y_i) = (2 + 1) + (0 + 3) + (-1 + (-2)) + (5 + 4) + (7 + 3)$$
$$= 3 + 3 + (-3) + 9 + 10$$
$$= 22$$

$$\sum_{i=1}^{5} X_i + \sum_{i=1}^{5} Y_i = 13 + 9 = 22$$

Rule 2 The summation of a difference between the values of two variables is equal to the difference between the summed values of the variables:

$$\sum_{i=1}^{n}(X_i - Y_i) = \sum_{i=1}^{n} X_i - \sum_{i=1}^{n} Y_i$$

Thus,

$$\sum_{i=1}^{5}(X_i - Y_i) = (2 - 1) + (0 - 3) + (-1 - (-2)) + (5 - 4) + (7 - 3)$$
$$= 1 + (-3) + 1 + 1 + 4$$
$$= 4$$

$$\sum_{i=1}^{5} X_i - \sum_{i=1}^{5} Y_i = 13 - 9 = 4$$

Rule 3 The sum of a constant times a variable is equal to that constant times the sum of the values of the variable:

$$\sum_{i=1}^{n} cX_i = c\sum_{i=1}^{n} X_i$$

where c is a constant. Thus, if $c = 2$,

$$\sum_{i=1}^{5} cX_i = \sum_{i=1}^{5} 2X_i = (2)(2) + (2)(0) + (2)(-1) + (2)(5) + (2)(7)$$

$$= 4 + 0 + (-2) + 10 + 14$$

$$= 26$$

$$c\sum_{i=1}^{5} X_i = 2\sum_{i=1}^{5} X_i = (2)(13) = 26$$

Rule 4 A constant summed n times will be equal to n times the value of the constant.

$$\sum_{i=1}^{n} c = nc$$

where c is a constant. Thus, if the constant $c = 2$ is summed 5 times,

$$\sum_{i=1}^{5} c = 2 + 2 + 2 + 2 + 2 = 10$$

$$nc = (5)(2) = 10$$

EXAMPLE

Suppose there are six values for the variables X and Y, such that $X_1 = 2, X_2 = 1, X_3 = 5, X_4 = -3, X_5 = 1, X_6 = -2$ and $Y_1 = 4, Y_2 = 0, Y_3 = -1, Y_4 = 2, Y_5 = 7$, and $Y_6 = -3$. Compute each of the following:

a. $\sum_{i=1}^{6} X_i$

b. $\sum_{i=1}^{6} Y_i$

c. $\sum_{i=1}^{6} X_i^2$

d. $\sum_{i=1}^{6} Y_i^2$

e. $\sum_{i=1}^{6} X_i Y_i$

f. $\sum_{i=1}^{6} (X_i + Y_i)$

g. $\sum_{i=1}^{6} (X_i - Y_i)$

h. $\sum_{i=1}^{6} \left(X_i - 3Y_i + 2X_i^2 \right)$

i. $\sum_{i=1}^{6} (cX_i)$, where $c = -1$

j. $\sum_{i=1}^{6} (X_i - 3Y_i + c)$, where $c = +3$

Answers

a. 4   b. 9   c. 44   d. 79   e. 10   f. (13)   g. −5   h. 65   i. −4   j. −5

▼ REFERENCES

1. Bashaw, W. L., *Mathematics for Statistics* (New York: Wiley, 1969).

2. Lanzer, P., *Basic Math: Fractions, Decimals, Percents* (Hicksville, NY: Video Aided Instruction, 2006).

3. Levine, D. and A. Brandwein, *The MBA Primer: Business Statistics*, 3rd ed. (Cincinnati, OH: Cengage Publishing, 2011).

4. Levine, D., *Statistics* (Hicksville, NY: Video Aided Instruction, 2006).

5. Shane, H., *Algebra 1* (Hicksville, NY: Video Aided Instruction, 2006).

A.6 Greek Alphabet

Greek Letter		Name	Greek Letter		Name
A	α	alpha	N	ν	nu
B	β	beta	Ξ	ξ	xi
Γ	γ	gamma	O	o	omicron
Δ	δ	delta	Π	π	pi
E	ε	epsilon	P	ρ	rho
Z	ζ	zeta	Σ	σ	sigma
H	η	eta	T	τ	tau
Θ	θ	theta	Y	υ	upsilon
I	ι	iota	Φ	ϕ	phi
K	κ	kappa	X	χ	chi
Λ	λ	lambda	Ψ	ψ	psi
M	μ	mu	Ω	ω	omega

Important Software Skills and Concepts

B.1 Identifying the Software Version

Using an outdated version of Microsoft Excel with this book can make learning about business statistics harder and confound a reader following Guide instructions. Programs change over time in both their functionality and user interfaces, so using an out-of-date version of one of the programs that the book discusses could result in frustration or failure to complete tasks. When using Microsoft Excel, differences arise based on how Excel was distributed (disk or subscription download) or whether periodic updates have been applied to Excel. Use this section to determine the version number of the software being used. Having a properly updated and current version of Excel is the best way to proceed with this book.

Excel

Excel Guide instructions in this book work best with an up-to-date Excel 365 for Microsoft Windows or Excel 365 for Mac. Instructions are also compatible with Microsoft Windows Excel 2019 and 2016, and Excel for Mac 2019 and 2016, the versions of Excel that Microsoft fully supported at the time of publication. Users of disk-based Excels, Excel 2016, or the deprecated Excel 2013 will be able to do all but some of the newer charts that the Chapter 2 Excel Guide describes.

When slight variations among versions occur, the variations appear in parentheses or explanatory sentences that identify those variations. For example, a number of charting instructions begin **Design** (or **Chart Design**)→**Add Chart Element** because the Design tab is called Chart Design in Excel for Mac. When Excel for Mac differs greatly from its Windows counterpart, Excel for Mac instructions appear in this color.

Identify the build number Excel has both a version number and a build number which identifies the extent to which the Excel copy has been updated. Knowing both can identify if an Excel copy is up-to-date and can also be helpful if technical support is needed. To identify the build number, open Excel and follow the appropriate instructions.

In Microsoft Windows Excel, select **File**→**Account** and, in the Account pane that appears, click the **About Excel icon**. In the dialog box that appears, note the build number that appears in parentheses, after the words Microsoft Excel. The

Account pane may contain an **Update Options** pull-down list from which **Update Now** can be selected to have Excel check for updates.

In Excel for Mac, select **Excel**→**About Excel** and in the dialog box that appears, note the build number. Check for updates by selecting **Help**→**Check for Updates**.

Tableau (Public version)

To identify the current version, open Tableau and select **Help**→**About Tableau Public**. Tableau automatically checks for updates and will prompt users to download updates as they become available.

B.2 Formulas

Formulas are programming-like instructions that process data found in worksheets and data tables. Formulas can compute intermediate calculations, generate new data or statistics, retrieve data from other cells, or use a logical comparison to make a decision, among other things. In Excel, each worksheet cell can have its own formula. Cells that contain formulas show the result of their formulas and not the formulas themselves.

In Excel, the keyboard shortcut **Ctrl+`** (grave accent) acts as a toggle to turn on and off the display of formulas.

Formulas make possible reusable templates such as the Figure 6.16 normal probabilities template on page 209. Users of the Excel Guide workbooks will discover that most workbooks contain one or more worksheets that present the formulas that the workbook uses to calculate results.

Entering a Formula

Guide instructions discuss the specifics of entering a formula as the need arises. For the general case in Excel, typing an equals sign ($=$) followed by the combination of arithmetic operators and cell references and pressing the **Enter key**, enters a formula for a specific cell.

Functions simplify arithmetic operations or provide access to advanced processing or statistical calculations. Functions can simplify formulas. In Excel, formulas often contain cell ranges, a shorthand way to refer to a group of cells. For example, in Excel, the formula **=A1+A2+A3+A4+A5+A6+A7** that sums the first seven cells in column A can be simplified using the SUM function **=SUM(A1:A7)** that uses the cell range A1:A7. (Section B.3 further explains cell ranges.)

Entering an Array Formula (Excel)

In Excel, an array formula defines a formula for a rectangular group of cells (the "array" of cells). To enter an array formula, first select the cell range and then type the formula, and then, while holding down the **Ctrl** and **Shift** keys, press **Enter** to enter the array formula into all of the cells of the cell range. (In Excel for Mac, pressing **Command+Enter** also enters an array formula.)

To edit an array formula, first select the cells that contain the array formula, then edit the formula and then press **Enter** while holding down **Ctrl+Shift** (or press **Command+Enter**). When selecting a cell that contains an array formula, Excel adds a pair of curly braces { } to the display of the formula in the formula bar to indicate that the formula is an array formula. These curly braces disappear when the formula is being edited. (Never type the curly braces when entering an array formula.)

Pasting with Paste Special (Excel)

While the keyboard shortcuts **Ctrl+C** and **Ctrl+V** to copy and paste cell contents will often suffice, pasting data from one worksheet to another can sometimes cause unexpected side effects when the source worksheet contains formulas. When the two worksheets are in different workbooks, a simple paste creates an external link to the original workbook that can lead to possible errors at a later time. Even pasting between worksheets in the same workbook can lead to problems if what is being pasted is a cell range of formulas. Use **Paste Special** to avoid these complications.

To use this command, copy the source cell range using **Ctrl+C** and then right-click the cell (or cell range) that is the target of the paste and click **Paste Special** from the shortcut menu.

In the Paste Special dialog box (shown below), click **Values** and then click **OK**. Paste Special Values pastes the current values of the cells in the first workbook and not formulas that use cell references to the first workbook.

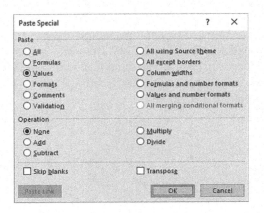

Paste Special can paste other types of information, including cell formatting information. In some copying contexts, placing the mouse pointer over Paste Special in the shortcut menu will reveal a gallery of shortcuts to the choices presented in the Paste Special dialog box.

To use PHStat with data in the form of formulas, first copy your data and then use Paste Special to paste columns of equivalent *values*. (Click **Values** in the Paste Special dialog box to create the values.) PHStat will not function properly if the data for a procedure are in the form of formulas.

Verifying Formulas

After entering all formulas or before using a worksheet or data table template that contains formulas, best practice suggests verifying the formulas for accuracy. In JMP and Minitab, the verification can be as simple as temporarily adding new data rows that contain simple numbers to verify that all formulas compute the correct results.

In Excel, additionally, relationships among cells can be examined visually. Selecting **Formulas → Trace Precedents** reveals relationships between a formula and its *precedents*, the cells that a formula references. Selecting **Formulas → Trace Dependents** reveals the relationship between a cell and its *dependents* cells that contain formulas that reference that cell.

B.3 Excel Cell References

Every Excel worksheet cell has its own **cell reference**, an address that identifies the cell based on the lettered column and numbered row of the cell. For example, the cell A1 is the cell in the first column and first row, A3 is the cell in the first column and third row, and C1 is the cell in the third column and first row.

Cell references can be a **cell range** that refers to a rectangular group of cells. A cell range names the upper-left cell and the lower-right cell of the group, using the form *UpperLeftCell:LowerRightCell*. For example, the cell range C1:C12 refers to the first 12 cells in column C while the cell range A1:D3 refers to all the cells in columns A through D in rows 1 through 3. Cell ranges can also name one or more columns or rows such as A:A, all the cells in column A, and 4:6, all the cells in rows 4 through 6.

In workbooks that contain more than one worksheet, appending a worksheet name in the form *WorksheetName!* as a prefix to a cell reference uniquely identifies a cell or cell range. For example, the cell reference COMPUTE!B8 uniquely identifies cell B8 of the COMPUTE worksheet, and the cell reference DATA!A:A uniquely identifies all the cells in column A of the DATA worksheet. If the name of a worksheet contains spaces or special characters, such as CITY DATA_1, you must enclose the worksheet name in a pair of single quotes as part of the prefix, such as 'CITY DATA_1'!A2.

When Excel encounters a cell reference without a worksheet prefix, Excel assumes that the reference identifies cells that are in the same worksheet as the formula being entered, a data entry shortcut that Excel Guide instructions use. Occasionally, an Excel feature requires that one use a worksheet prefix and instructions note such exceptions as necessary.

Although this book does not use them, cell references can include a workbook prefix in the form *[WorkbookName] WorksheetName!* If you discover workbook prefixes in the formulas you create using the instructions in this book, you may have committed an inadvertent error when transferring data from one workbook to another. Review your work and make sure you intended to include a workbook name prefix in your formula.

Absolute and Relative Cell References

To avoid the drudgery of typing many similar formulas, a formula can be entered once and then copied to other cells. For example, to copy a formula that has been entered in cell C2 down the column through row 12:

1. Right-click cell C2 and press **Ctrl+C** to copy the formula. A movie marquee–like highlight appears around cell C2.
2. Select the cell range **C3:C12**.
3. With the cell range highlighted, press **Ctrl+V** to paste the formula into the cells of the cell range.

During this copy-and-paste operation, Excel adjusts these **relative cell references** in formulas so that copying the formula **=A2+B2** from cell C2 to cell C3 results in the formula **=A3+B3** being pasted into cell C3, the formula **=A4+B4** being pasted into cell C4, and so on.

Sometimes, this automatic adjustment is unwanted. For example, when copying the cell C2 formula **=(A2+B2)/B15**, if cell B15 contained the divisor to be used in all formulas, that reference should not be adjusted to B16, B17, and so on. To prevent Excel from adjusting a cell reference, use **absolute cell references** by inserting dollar signs ($) before the column and row references of a relative cell reference. For example, the absolute cell reference **$B$15** in the copied cell C2 formula **=(A2+B2)/$B$15** will cause Excel to paste the formula **=(A3+B3)/$B$15** into cell C3.

Do not confuse the use of the dollar sign symbol with the worksheet formatting operation that displays numbers as dollar currency amounts.

Selecting Cell Ranges for Charts

Cell ranges can be entered in Excel dialog boxes in one of several ways. Cell ranges can be typed (most Excel Guide instructions use this method) or selected using the mouse pointer. Likewise, most of the time cell ranges can be entered using either relative or absolute references. Two important exceptions to these general rules are the Axis Labels and Edit Series dialog boxes, associated with chart labels and data series.

To enter a cell range into these two dialog boxes, enter the cell range as a *formula* that uses absolute cell references in the form *WorksheetName!UpperLeftCell:LowerRightCell*, as the examples below illustrate. Entering these cell ranges is best done using the mouse-pointer method. Typing the cell range in these dialog boxes will often be frustrating as keys such as the cursor keys do not function as they do in other dialog boxes.

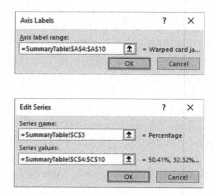

Selecting Non-contiguous Cell Ranges

In the general case, enter a non-contiguous cell range such as the cells A1:A11 and C1:C11 by typing each cell range, separated by commas. For the example, type **A1:A11, C1:C11**. To enter a non-contiguous cell range for the Axis Labels and Edit Series dialog boxes that the previous section discusses, use the mouse pointer method. To use the mouse-pointer method with such ranges, first, select the cell range of the first group of cells and then, while holding down **Ctrl**, select the cell range of the other groups of cells that form the non-contiguous cell range.

B.4 Excel Worksheet Formatting

Format the contents of worksheet cells by either making entries in the Format Cells dialog box or clicking shortcut icons in the Home tab.

Format Cells Method

To use the Format Cells dialog box method, right-click a cell or cell range and click Format Cells in the shortcut menu. In the Format Cells dialog box, select the **Number** tab. Clicking a **Category** changes the panel to the right of the list. For example, clicking **Number** displays a panel (shown below) in which the number of decimal places to display can be specified.

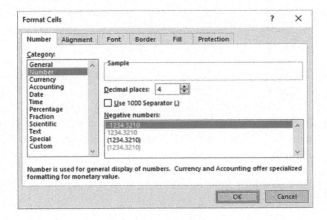

Click the **Alignment** tab of the Format Cells dialog box (partially shown below), to display a panel in which the horizontal and vertical positioning of cell contents can be specified as well as whether the cell contents can be wrapped to a second line if the contents are longer than the cell width.

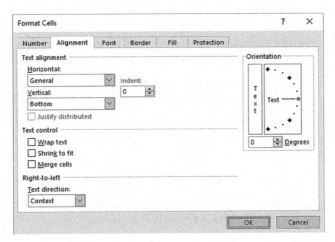

Home Tab Shortcuts Method

You can also format the contents of worksheets by using shortcuts on the Home tab. In Microsoft Windows Excel, these shortcuts are divided into the groups that the following instructions name. In Excel for Mac, the groups are implicit and group names are not shown on the Home tab.

Use the Font **group** shortcuts (shown at top below) to change the typeface, point size, color, and styling such as roman, bold, or italic of the text a cell displays or the background color of a cell. Use the **fill icon** in the same group to change the background color for a cell (shown as yellow in the illustration below). Click the drop-down button to the right of the fill icon to display a gallery of colors from which you can select a color or click **More Colors** for more choices.

Click the **A icon** drop-down button (not in Excel for Mac) to display a palette of color choices for changing the color of the text being displayed (shown at bottom below).

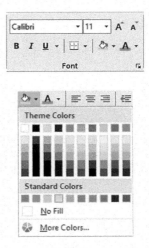

Use the shortcuts in the **Number** group (shown below) to change the formatting of numeric values, including the formatting changes the discussion of the Format Cells dialog box mentions.

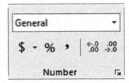

To adjust the width of a column to an optimal size, select the column and then select **Format → Autofit Column Width** in the Cells group (shown below). Excel will adjust the width of the column to accommodate the width of the widest value in the column.

Many Home tab shortcuts contain a drop-down arrow that, when clicked, displays a gallery of choices. For **Merge & Center**, the gallery (shown below) displays all cell merging operations.

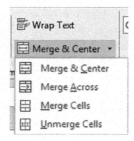

B.5E Excel Chart Formatting

Excel often produces charts that contain elements that need reformatting or changes to enhance chart presentation. To enhance a chart in ways that the Excel Guide instructions suggest, first select the chart to be corrected. (If Chart Tools or PivotChart Tools appear above the Ribbon tabs, a chart has been selected.) Then, apply the instructions to make the necessary changes.

If a chart on a chart sheet is either too large to be fully seen or too small and surrounded by a frame mat that is too large, click the **Zoom Out** or **Zoom In icons**, located in the lower-right of the Excel window frame, to adjust the chart display.

Most Commonly Made Changes

To relocate a chart to its own chart sheet:

1. Click the chart background and click **Move Chart** from the shortcut menu.
2. In the Move Chart dialog box, click **New Sheet**, enter a name for the new chart sheet, and click **OK**.

To turn off improper horizontal gridlines:

Design (or **Chart Design**)→**Add Chart Element**→ **Gridlines**→**Primary Major Horizontal**

Layout (or **Chart Layout**)→**Gridlines**→**Primary Horizontal Gridlines**→**None**

To turn off improper vertical gridlines:

Design (or **Chart Design**)→**Add Chart Element**→ **Gridlines**→**Primary Major Vertical**

Layout (or **Chart Layout**)→**Gridlines**→ **Primary Vertical Gridlines**→**None**

To turn off a chart legend:

Design (or **Chart Design**)→**Add Chart Element**→ **Legend**→**None**

Layout (or **Chart Layout**)→ **Legend**→**None** (or **No Legend**)

Chart and Axis Titles

To add a chart title to a chart missing a title:

1. Select **Design** (or **Chart Design**)→**Add Chart Element**→**Chart Title**→**Above Chart**. Otherwise, click on the chart and then select **Layout** (or **Chart Layout**)→**Chart Title**→**Above Chart**.
2. In the box that is added to the chart, select the words "Chart Title" and enter an appropriate title.

To add a title to a horizontal axis missing a title:

1. **Design** (or **Chart Design**)→**Add Chart Element**→**Axis Titles**→**Primary Horizontal**. In the new text box in the chart, replace the words Axis Title with an appropriate title.
2. **Layout** (or **Chart Layout**)→**Axis Titles**→**Primary Horizontal Axis Title**→**Title Below Axis**. In the new text box in the chart, replace the words Axis Title with an appropriate title.

To add a title to a vertical axis missing a title:

1. **Design** (or **Chart Design**)→**Add Chart Element**→**Axis Titles**→**Primary Vertical**. In the new text box in the chart, replace the words Axis Title with an appropriate title.
2. **Layout** (or **Chart Layout**)→**Axis Titles**→**Primary Vertical Axis Title**→**Rotated Title**. In the new text box in the chart, replace the words Axis Title with an appropriate title.

Chart Axes

To turn on the display of the X axis, if not already shown:

Design (or **Chart Design**)→**Add Chart Element**→ **Axes**→**Primary Horizontal**

Layout (or **Chart Layout**)→**Axes**→**Primary Horizontal Axis**→**Show Left to Right Axis** (or **Show Default Axis** or **Primary Default Axis**)

To turn on the display of the Y axis, if not already shown:

Design (or **Chart Design**)→**Add Chart Element**→ **Axes**→**Primary Vertical**

Layout (or **Chart Layout**)→**Axes**→**Primary Vertical Axis**→**Show Default Axis**

For a chart that contains secondary axes, to turn off the secondary horizontal axis title:

Design (or **Chart Design**)→**Add Chart Element**→ **Axis Titles**→**Secondary Horizontal**

Layout (or **Chart Layout**)→**Axis Titles**→**Secondary Horizontal Axis Title**→**None (or No Axis Title)**

For a chart that contains secondary axes, to turn on the secondary vertical axis title:

Design (or **Chart Design**)→**Add Chart Element**→ **Axis Titles**→**Secondary Vertical**

Layout (or **Chart Layout**)→**Axis Titles**→**Secondary Vertical Axis Title**→**Rotated Title**

Correcting the Display of the *X* Axis

In scatter plots and related line charts, Microsoft Excel displays the *X* axis at the *Y* axis origin ($Y = 0$). When plots have negative values, this causes the *X* axis not to appear at the bottom of the chart.

To relocate the *X* axis to the bottom of a scatter plot or line chart, open to the chart sheet that contains the chart, right-click the **Y axis**, and click **Format Axis** from the shortcut menu. In the Format Axis pane click **Axis value** and, in its box, enter the value shown in the **Minimum box** in the same pane.

Emphasizing Histogram Bars

To better emphasize each bar in a histogram, open to the chart sheet containing the histogram, right-click over one of the histogram bars, and click **Format Data Series** in the shortcut menu. In the Format Data Series pane, click the bucket icon. In the Border group, click **Solid line** (Click **Border** to reveal settings, if necessary.). From the **Color drop-down list**, select the darkest color in the same column as the currently selected (highlighted) color. Then, enter **2** (for 2 pt) as the **Width**.

B.5T Tableau Chart Formatting

Tableau produces charts with elements that need to be changed to enhance presentation or that may not reproduce well when printed or otherwise transferred from the Tableau worksheet. To change chart elements, first open to the worksheet that contains the elements that need to be changed. Then, select the element and make the necessary changes. Although default preferences cannot be set in Tableau Desktop Public Edition, a shortcut does exist when making the same changes to two or more charts of the same type. To use the shortcut, make changes to one chart and select **Format→Copy Formatting**. Then open to another chart and select **Format→Paste Formatting**.

By default, Tableau assigns the worksheet tab name as the worksheet title. To enter a different title or change the text attributes of a title, double-click the current title. In the Edit Title dialog box (shown at top right):

1. Select **<Sheet Name>** (the default worksheet title).
2. Type the new title, replacing <Sheet Name>.
3. Select the new title.
4. As necessary, use the **font face**, **font size**, and **color pull-down lists** to change the font, size, or color.
5. As necessary, click the **bold icon, italic icon**, or **underline icon** to turn on (or off) these attributes.
6. As necessary, click one of the tree justification icons.
7. Click **OK**.

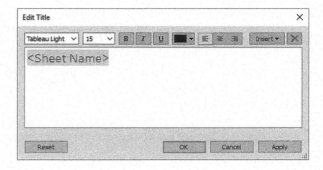

Changes only apply to selected text, not the contents of the Edit Title dialog box. Skipping step 3 means none of the actions in step 4 or 5 will affect the text.

The most common other changes to chart formatting use either the Font gallery or the Lines gallery (shown below).

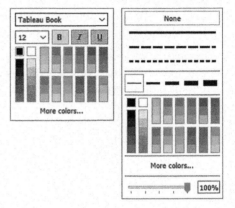

To turn off unnecessary grid lines:

1. Select **Format→Lines**.
2. In the Sheet tab of the Format Lines panel (shown below), select **None** from the **Grid Lines** pull-down **Lines gallery**.
3. Click the **panel close icon** (the "X" in the upper right of the Format Lines panel) to close the Format Lines panel to redisplay the Data and Analytics tabs.

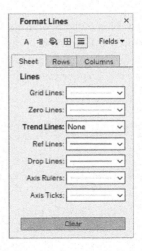

To adjust an axis line:

1. Select **Format → Lines.**
2. In the Sheet tab of the Format Lines panel (shown above), select the line type, line width, and color from the **Axis Rulers** pull-down **Lines gallery.**
3. Click the **panel close icon** (the "X" icon in the upper right of the Format Lines panel) to close the Format Lines panel to redisplay the Data and Analytics tabs.

To adjust the text attributes of a chart label:

1. Right-click the text to be adjusted and select **Format** from the shortcut menu.
2. In the Axis tab of the Format panel, select text attributes from the Default group **Font** pull-down **Font gallery** to adjust an axis label.
3. Click the **panel close icon** (the "X" icon in the upper right of the Format panel) to close the Format Lines panel to redisplay the Data and Analytics tabs.

To change the formatting of axis tick labels, select attributes from the Ticks, Numbers, or Alignment pull-down menus (not shown below) as part of step 2.

Tableau also enables one to change chart borders or shading. Use **Format→Alignment** or **Format→Borders**, commands that lead to additional tabs and galleries that share similarities to the panels and galleries that this section discusses.

B.6 Creating Histograms for Discrete Probability Distributions (Excel)

Create a histogram for a discrete probability distribution based on a discrete probabilities table. For example, to create the Figure 5.3 histogram of the binomial probability distribution on page 184, open to the **COMPUTE worksheet** of the **Binomial workbook**. Select the cell range

B14:B18, the probabilities in the Binomial Probabilities Table, and:

1. Select **Insert** (or **Charts**)→**Column** and select the **Clustered Column** gallery item.
2. Right-click the chart and click **Select Data** in the shortcut menu.

In the Select Data Source dialog box:

3. Click **Edit** under the **Horizontal (Categories) Axis Labels** heading. In the Axis Labels display, drag the mouse to select and enter the cell range A14:A18 with a worksheet prefix (see Section B.3), *as a formula* in the **Axis label range** box.
   In Excel for Mac, in the Select Data Source display, click the icon inside the **Horizontal (Category) axis labels** [or **Category (X) axis labels**] box and drag the mouse to select and enter the same cell range, A14:A18.
4. Click **OK**.

In the chart:

5. Right-click inside a bar and click **Format Data Series** in the shortcut menu.
6. In the Format Data Series display, click **Series Options**. In the Series Options, click **Series Options**, enter **0** as the **Gap Width** and then close the display. (To see the second Series Options, you may have to first click the chart [third] icon near the top of the task pane.) In Excel for Mac, there is only one Series Options label, and the Gap Width setting is displayed without having to click Series Options.

Relocate the chart to a chart sheet and adjust the chart formatting by using the instructions in Section B.5.

B.7 Deleting the "Extra" Histogram Bar (Excel)

As "Classes and Excel Bins" on page 46 explains, in Excel bins approximate classes. This approximation creates an "extra" bin that will have a frequency of zero. To delete the histogram bar associated with this extra bin, edit the cell range that Excel uses to construct the histogram.

Right-click the histogram background and click **Select Data**. In the Select Data Source dialog box in Microsoft Windows Excel:

1. Click **Edit** under the **Legend Entries (Series)** heading.
2. In the Edit Series dialog box, edit the **Series values** cell range formula to begin with the second cell of the original cell range and click OK.
3. Click **Edit** under the **Horizontal (Categories) Axis Labels** heading.
4. In the Axis Labels dialog box, edit the **Axis label range** formula to begin with the second cell of the original cell range and click **OK**.

In the Select Data Source dialog box in Excel for Mac:

1. Edit the **Y values** cell range formula to begin with the second cell of the original cell range and click OK.
2. Edit the **Horizontal (Category) axis labels** [or **Category (X) axis labels**] formula.
3. In the Axis Labels dialog box, edit the **Axis label range** formula to begin with the second cell of the original cell range and click **OK**.

Using Non-numeric Labels in a Time-Series Plot

As the Excel Guide for Chapter 2 explains in detail, one uses the Excel scatter plot chart type to create a time-series plot. Excel scatter plots require numerical data for the X-axis labels. For the case of calendar data in a series such as January 2020, February 2020, March 2020, etc., Excel stores the data internally as a number which enables one to use the series to label the X-axis. Figure 16.2 illustrates a use of such *date data*.

However, sometimes one wants to use non-numerical labels to describe time periods or needs to use calendar data that will note be interpreted properly by Excel. For example,

one might want to use a repeating set of month labels, January through December, to emphasize the seasonal trend in a time series plot. While the months are date data, Excel will wrongly interpret every occurrence of a month as the same X value or create a proper time series plot but invent a coded month value that counts up from 1 for the repeating series of January through December X values.

The Figure 16.1 Panel D time series plot uses a repeating set of month labels January through December to emphasize the seasonal nature of the trend seen. Panel D is not a scatter plot, but an Excel line chart that uses a formatting workaround to allow those month labels to appear as non-numeric labels. To use this workaround:

1. Right-click the axis and select **Axis Settings** in the shortcut menu.
2. Select **Insert→Insert Line or Area Chart** (#2 in the labeled Charts group on page 88) and select the **Line with Markers** gallery item in the 2-D Line group.
3. Right-click the X-axis and select **Format Axis** from the shortcut menu.
4. In the Format Axis display, click **Axis Options**, if necessary, to reveal axis options and then click the **On tick marks** option button under the Axis position heading.

Online Resources

C.1 About the Online Resources for This Book

Online resources complement and extend the study of business statistics with this book. Some resources, such as the data files, are integral to learning with this book, while other resources, such as online sections and chapters, are optional and can be skipped without one losing comprehension about the concepts and methods that this book discusses.

For using Microsoft Excel, this book fully integrates both a set of Excel Guide workbooks that contain templates or model solutions for applying Excel to specific statistical methods and PHStat, the Pearson statistics Excel add-in, that the authors developed to support this book. (See Section C.3 for more about these choices.)

Online resources also include the optional online pamphlets, chapters, sections, and "short takes" as PDF files; documents that support the end-of-chapter cases, also as PDF files; and the Visual Explorations Excel macro workbooks that interactively demonstrate selected statistical concepts.

Access the Online Resources

To access the online resources for this book, visit the public download page for this book:

1. Open a web browser and go to **www.pearsonhighered .com/levine**.
2. In that web page, locate this book, *Statistics for Managers Using Microsoft Excel, 9/e*, and click **Student Download Page** that appears below the title.
3. In the student download page, click the links of interest.

Registered users of a MyLab Statistics course for this book can also use the Tools for Success page:

1. Open the MyLab Statistics course for this book.
2. Click **Tools for Success** in the left pane.
3. In the Tools for Success page, click the links of interest.

Note that the Tools for Success page contains a number of items that were not prepared by the authors of this book but which Pearson offers MyLab Statistics users as additional supplements. Those additional supplements do not appear on the student download page.

In either method, clicking most item links will trigger a prompt to save a file. Some files are zip archives, collections of files that need to be "unzipped" or expanded before use. Clicking the PHStat link will redirect a browser to a separate PHStat home page, from which one can obtain the PHStat add-in. Appendix H discusses obtaining PHStat in detail.

C.2 Data Files

This section presents the alphabetized list of the data files identified in examples and problems. As Section FTF.4 first explains, the names of such data files appear in a special inverted color typeface such as `Retirement Funds`.

Entries for each data file define the variables found in the file, category definitions for categorical variables in the file, and list the chapters that reference the file.

Unless otherwise noted, data files are provided as Excel workbooks, most of which contain a single worksheet named DATA. Entries for data files intended only for use with Tableau include the code (T).

311 Call Center Day and abandonment rate percentage (Chapter 3)

Accounting Partners Firm and number of partners (Chapter 3)

Accounting Partners 2 Region and number of partners (Chapter 10)

Regional Accounting Partners Region, revenue ($millions), number of partners, number of professionals, MAS (%), southeast (0 = no, 1 = yes), and Gulf Coast southeast (0 = no, 1 = yes) (Chapter 15)

ACT Method (online or traditional), ACT scores for condensed course, and ACT scores for regular course (Chapter 11)

One-Way ACT Group 1 ACT scores, group 2 ACT scores, group 3 ACT scores, and group 4 ACT scores (Chapter 11)

Adindex Respondent, cola A Adindex, and cola B Adindex (Chapter 10)

Alphabet Year, coded year, and revenues (Chapter 16)

Amazon Quarter and revenue (Chapter 16)

American Banks ROATCE (%) and efficiency rating percentage (Chapter 13)

AMS2-1 Types of errors and frequency, types of errors and cost, and types of wrong billing errors and cost (as three separate worksheets) (Chapter 2)

AMS2-2 Days and number of calls (Chapter 2)

AMS8 Rate willing to pay ($) (Chapter 8)

AMS9 Upload speed (Chapter 9)

AMS10 Update times for email interface 1 and email interface 2 (Chapter 10)

AMS11-1 Update time for system 1, system 2, and system 3 (Chapter 11)

AMS11-2 Technology (cable or fiber) and interface (system 1, system 2, or system 3) (Chapter 11)

AMS13 Number of hours spent telemarketing and number of new subscriptions (Chapter 13)

AMS14 Week, number of new subscriptions, hours spent tele- marketing, and type of presentation (formal or informal) (Chapter 14)

AMS16 Month and number of home delivery subscriptions (Chapter 16)

Anscombe Data sets A, B, C, and D, each with 11 pairs of X and Y values (Chapter 13)

Arlingtons Dashboard Data nine worksheets containing table data about Arlingtons stores and sales (Chapter 18)

ATM Transactions Cause, frequency, and percentage (Chapter 2)

Auto Production Year, coded year, and number of units produced (Chapter 16)

Bank Waiting Waiting time (in minutes) of 15 customers at a bank located in a commercial district (Chapters 3, 9, 10, and 12)

Bank Waiting 2 Waiting time (in minutes) of 15 customers at a bank located in a residential area (Chapters 3, 10, and 12)

Baseball Game Cost Team and average total cost for two people ($) (Chapter 2)

Best Companies Company, full-time jobs job openings, total worldwide revenues ($billions), and total voluntary turnover percentage (Chapters 14 and 15)

Bonuses Year and bonuses ($000) (Chapter 16)

BrandZ TechFin Brand, brand value in 2019 ($millions), percent change in brand value from 2018, and product sector (Chapters 10 and 12)

Breakfast Type (Continental or American), delivery time difference for early time period, and delivery time difference for late time period (Chapter 11)

Breakfast Results Type (Continental or American), delivery time difference for early time period, and delivery time difference for late time period (Chapter 11)

Brynne Packaging WPCT score and rating (Chapter 13)

Bulbs Manufacturer (1 = A, 2 = B) and length of life (hours) (Chapters 2, 10, and 12)

Bundle Restaurant, bundle score, and typical cost ($) (Chapter 2)

Business Analytics Usage and percentage (Chapter 2)

Business Valuation Drug company name, PB fye (price-to-book-value ratio), ROE, and sgrowth percentage (Chapter 14)

Business Valuation 2 Company, ticker symbol, SIC3 code, SIC4 code, PB fye (price-to-book-value ratio), PE fye (price-to-earnings ratio), natural log of assets (as a measure of size), ROE, sgrowth percentage, debt-to-EBITDA ratio, dummy variable for SIC4 code 2834 (1 = 2834, 0 = not 2824), and dummy variable for SIC4 code 2835 (1 = 2835, 0 = not 2835) (Chapter 15)

Call Center Month and call volume (Chapter 16)

Call Duration Time (seconds) spent by agent talking with a customer (Chapters 2 and 3)

Candidate Assessment Salary, competence rating, gender of candidate (F or M), gender of rater (F or M), rater/candidate gender (F to F, F to M, M to M, M to M), school (Private, Public), department (Biology, Chemistry, Physics), and age of rater (Chapter 18)

CardioGood Fitness Product purchased (TM195, TM498, TM798), age in years, gender (Male or Female), education in years, relationship status (Single or Partnered), average number of times the customer plans to use the treadmill each week, self-rated fitness on a 1-to-5 ordinal scale (1 = poor to 5 = excellent), annual household income ($), and average number of miles the customer expects to walk/run each week (Chapters 2, 3, 6, 8, 10, 11, and 12)

Cardholder Study Upgraded (0 = no, 1 = yes), purchases ($000), and extra cards (0 = no, 1 = yes) (Chapters 14 and 17)

Cat Food Ounces eaten of kidney, shrimp, chicken liver, salmon, and beef cat food (Chapters 11 and 12)

Cat Food Cans Piece size (F = fine, C = chunky), coded weight for low fill height, and coded weight for current fill height (Chapter 11)

CD Rates Bank, 1-year CD rate, and 5-year CD rate (Chapters 2, 3, 6, and 8)

Cellphone Providers Name of provider, type (traditional or prepaid), and rating (Chapters 10 and 12)

Cereals Cereal, calories, carbohydrates, and sugar (Chapters 3 and 13)

CFPB Categories Category and number of complaints (Chapter 2)

CFPB Companies Company and number of complaints (Chapter 2)

Churn Customer ID, churn coded (0 = no, 1 = yes), churn, calls, and visits (Chapter 14)

Cigarette Tax State and cigarette tax rate (Chapter 3)

Cities City, average annual salary ($), unemployment rate (%), median home value ($000), violent crime rate per 100,000 residents, average commuting time (minutes), livability score (Chapter 15)

City Internet Speed City, download speed (Mbps), and upload speed (Mbps) (Chapters 2 and 3)

Coaches Pay School, total pay ($000), and major conference member (No or Yes) (Chapters 2 and 3)

Coca-Cola Year, coded year, and revenues ($billions) (Chapter 16)

Coffee Exports Year and exports in thousands of 60-kg bags (Chapter 16)

Coffee Sales Coffee sales at $0.59, $0.69, $0.79, and $0.89 (Chapters 11 and 12)

Coffee Sales 2 Coffee sales and price (Chapter 15)

CoffeePak Study Persons in line ahead and waiting time (Chapter 15)

College Debt Option and average debt at graduation ($) (Chapter 10)

Community Banks Institution, location, ROA (%), efficiency ratio (%), and total risk-based capital percentage (Chapters 14 and 15)

Commuting Time City and average weekly commuting time (minutes) (Chapters 2, 3, and 8)

Computer Usage Year and hours (Chapter 16)

Concrete Strength Sample number and compressive strength after two days and seven days (Chapter 10)

Congestion Level Asian, European, and North American percentage increases in overall travel time compared to a free flow situation (Chapters 11 and 12)

Connection Speed Country, average connection speed in Mbps, average peak connection speed, percent above 4 Mbps, and percent above 10 Mbps (Chapter 3)

Core Appliances Year and shipments in millions (Chapter 16)

Corn Flakes thickness (mm) for four different toasting times (seconds) (Chapter 11)

CPI-U Year, coded year, and value of the consumer price index (CPI-U) (Chapter 16)

Credit Scores City, state, and average credit score (Chapters 2 and 3)

Crude Oil Month and volume (barrels) (Chapter 16)

Currency Year, coded year, and exchange rates (against the U.S. dollar) for the Canadian dollar, Japanese yen, and English pound sterling (Chapters 2 and 16)

Delivery Customer, number of cases, and delivery time (Chapter 13)

Dirty Data ID, gender (male or female), age (years), class (Sophomore, Junior, or Senior), Major (Accounting, CIS, Economics/Finance, International Business, Management, Retailing/Marketing, and Other) (Chapter 1)

Domestic Beer Brand, alcohol percentage, calories, and carbohydrates (Chapters 2, 3, 6, and 15)

Dow Dogs Stock and 1-year return (Chapter 3)

Dow Market Cap Company and market capitalization ($billions) (Chapters 3 and 6)

Drill Depth, time to drill additional 5 feet, and type of hole (dry or wet) (Chapter 14)

Drink Amount of soft drink filled in 2-liter bottles (Chapters 2 and 9)

Employee Actions Action taken and percentage (Chapter 2)

Energy State and per capita kilowatt hour use (Chapter 3)

Energy Sources Source and billion kWh produced (Chapter 2)

Engines RPM and torque (Chapter 15)

Entree Type and number served (Chapter 2)

ER Waiting Emergency room waiting time (in minutes) at the main facility and at satellite 1, satellite 2, and satellite 3 (Chapters 11 and 12)

Espresso Tamp (inches) and time (seconds) (Chapter 13)

European Tourism Country, tourism employment in 2018, and tourism establishments (Chapter 15)

Fast Casual Amount spent on fast food ($) (Chapters 2, 8, and 9)

Federal Receipts Year, coded year, and federal receipts ($billions current) (Chapter 16)

Fifteen Weeks Week number, number of customers, and sales ($000) over a period of 15 consecutive weeks (Chapter 13)

FlyAsh Fly ash percentage and strength (PSI) (Chapter 15)

Food Consumption Food, types, per capital consumption for United States, Japan, and Russia (Chapter 2)

Freeport Address, fair market value ($000), property size (acres), house size, age, number of rooms, number of bathrooms, and number of cars that can be parked in the garage (Chapter 15)

FTMBA School number, tuition per year ($), GMAT score, acceptance rate (%), graduates employed at graduation (%), and mean starting salary and bonus ($) (Chapters 13 and 15)

Furniture Days between receipt and resolution of complaints regarding purchased furniture (Chapters 2, 3, 8, and 9)

Gas Prices Month and price per gallon ($) (Chapter 16)

GC Freeport Roslyn Address, location (Glen Cove, Freeport, or Roslyn), fair market value ($000), property size (acres), age, house size (sq. ft.), number of rooms, number of bathrooms, and number of cars that can be parked in the garage (Chapter 15)

GC and Roslyn Address, location (Glen Cove or Roslyn), fair market value ($000), property size (acres), age, house size (sq. ft.), number of rooms, number of bathrooms, and number of cars that can be parked in the garage (Chapters 14 and 15)

GDP Year and gross domestic product (Chapter 16)

Glass 1 Breakoff pressure, percentage of chips breaking off at stopper height of 20, percentage of chips breaking off at stopper height of 25 (Chapter 11)

Glass 2 Zone 1 lower, effect on imprint at temperature of 695, effect on imprint at temperature of 715 (Chapter 11)

Glen Cove Address, fair market value ($000), property size (acres), age, house size (sq. ft.), number of rooms, number of bathrooms, and number of cars that can be parked in the garage (Chapters 14 and 15)

Gold Quarter, coded quarter, price ($), Q1, Q2, and Q3 (Chapter 16)

Granule Granule loss in Boston and Vermont shingles (Chapters 3, 8, 9, and 10)

Growth Funds Subset of Retirement Funds where fund type is growth (Chapter 17) (T)

Heating Oil Monthly consumption of heating oil (gallons), temperature (degrees Fahrenheit), attic insulation (inches), and ranch-style (0 = not ranch-style, 1 = ranch-style) (Chapters 14 and 15)

Hemlock Farms Asking price, hot tub (0 = no, 1 = yes), rooms, lake view (0 = no, 1 = yes), bathrooms, bedrooms, loft/ den (0 = no, 1 = yes), finished basement (0 = no, 1 = yes), and number of acres (Chapter 15)

Honda Prices Age in years and price ($) (Chapter 15)

House Sales Year, coded year, and total sales (Chapter 16)

Hybrid Cars Year and number sold (Chapter 18)

Ice Cream Daily temperature (in degrees Fahrenheit) and sales ($000) for 21 days (Chapter 13)

Indices Year, change in DJIA, S&P 500, and NASDAQ indices (Chapter 3)

Insurance Processing time in days for insurance policies (Chapters 3, 8, and 9)

Insurance Claims Claims, buildup (0 = buildup not indicated, 1 = buildup indicated), and excess payment ($) (Chapter 8)

Insurance Fraud ID, fraud coded (0 = No, 1 = Yes), fraud (No or Yes), new business coded (0 = No, 1 = Yes), new business (No or Yes), and claims/year (Chapter 14)

International Market Country, level of development (Emerging or Developed), and time required to start a business (days) (Chapter 10)

International Market 2 Country, region, cost to export container (US$), and cost to import container (US$) (Chapters 11 and 12)

Internet Provider Speeds City and upload and download speeds (Mbps) for AT&T, T-Mobile, Sprint, and Verizon (Chapter 10)

Invoice Number of invoices processed and amount of time (hours) for 30 days (Chapter 13)

Invoices Amount recorded (in dollars) from sales invoices (Chapters 8 and 9)

IPOs Year and number of IPOs (Chapter 16)

Luggage Delivery time (in minutes) for luggage in Wing A and Wing B of a hotel (Chapters 10 and 12)

Machine Learning Usage and percentage (Chapter 2)

Managers Sales (ratio of yearly sales divided by the target sales value for that region), Wonderlic Personnel Test score, Strong-Campbell Interest Inventory Test score, number of years of selling experience prior to becoming a sales manager, and whether the sales manager has a degree in electrical engineering (No or Yes) (Chapter 15)

Market Basket Product, Costco cost, and Walmart cost (Chapter 10)

McDonalds Year, coded year, and McDonald's Corporation annual gross revenue ($billions) (Chapter 16)

McDonalds Stores State and number of stores (Chapter 3)

Medical Wires 1 Machine type, narrow, and wide (Chapter 11)

Medical Wires 2 Narrow and wide (Chapter 11)

Metals Year and the total rate of return percentage for platinum, gold, and silver (Chapter 3)

Mining Day, amount stacked, and downtime (Chapter 18)

Mining Downtime Causes Day; hours of downtime due to mechanical, electrical, tonnage restriction, operator, and no feed; and total downtime hours (Chapter 18)

MLB Attendance Study Team, wins season attendance, mean attendance per game, stadium capacity, mean capacity percentage, mean batter age, mean pitcher age, all-star players, and team payroll (Chapter 18)

MLB Salaries Year and average major league baseball salary ($millions) (Chapter 16)

MLB Values Team, revenue ($millions), and value ($millions) (Chapter 13)

Mobile Commerce Country and mobile commerce penetration percentage (Chapters 3 and 8)

Mobile Device Study Time in minutes spent per day using the Internet from a mobile device (Chapter 9)

Mobile Device Study 2 Time in minutes spent per day using the Internet from a mobile device (Chapter 10)

Mobile Electronics In-aisle sales, front sales, kiosk sales, and expert area sales (Chapters 11 and 12)

Mobile Electronics 2 Mobile payments (No or Yes), in-aisle sales, front sales, kiosk sales, and expert area sales (Chapter 11)

Mobile Speed Carrier, download speed in Mbps, and upload speed in Mbps (Chapters 2, 3, and 13)

Moisture Moisture content of Boston shingles and Vermont shingles (Chapter 9)

Molding Vibration time (seconds), vibration pressure (psi), vibration amplitude (%), raw material density (g/ml), quantity of raw material (scoops), product length in cavity 1 (in.), product length in cavity 2 (in.), product weight in cavity 1 (gr.), and product weight in cavity 2 (gr.) (Chapter 15)

Monthly Movie Revenues Month and monthly revenues from 2005 through 2018 (Chapter 2)

Movie Title, box office gross ($millions), and DVD revenue ($millions) (Chapter 13)

Movie Attendance Year and movie attendance (billions) (Chapters 2 and 16)

Movie Revenues Year and revenue ($billions) (Chapter 2)

Movie Revenues by MPAA Rating Year, rating (G, PG, PG-13, or R), and revenues (Chapter 17) (T)

Moving Labor hours, cubic feet, number of large pieces of furniture, and availability of an elevator (Chapters 14 and 17)

Natural Gas Month, wellhead price ($/thousands cu. ft.), and residential price ($/thousands cu. ft.) (Chapters 2 and 16)

NBA Financial Team code, annual revenue ($millions), and current value ($millions) (Chapters 2, 3, and 13)

NBA Stats Team, wins, points per game, field goal percentage, opponent field goal percentage, three-point percentage, free throw percentage, rebounds per game, assists per game, and turnovers per game (Chapters 14 and 15)

NBAValues Team name, team code, annual revenue ($millions), and current value ($millions) (Chapter 2)

Nickels 26Weeks Standby hours, staff present, remote engineering hours, graphics hours, and production labor hours (Chapters 14 and 15)

New Home Sales Month, sales in thousands, and mean price ($000) (Chapter 2)

OAEP Complaints by Industry Industry group and number of complaints (Chapter 2)

OAEP Complaints by Category Complaint category and number of complaints (Chapter 2)

Oil&Gasoline Week, price of a gallon of gasoline ($), and price of oil per barrel ($) (Chapter 13)

OmniPower Bars sold, price (cents), and promotion expenses ($) (Chapters 14 and 17)

Online Shopping How purchases were made and percentage (Chapter 2)

Organic Food Customer, organic food purchaser (0 = no, 1 = yes), age, and online health wellness e-newsletters subscriber (0 = no, 1 = yes), (Chapter 14)

O-Ring Flight number, temperature, and O-ring damage index (Chapter 13)

Packaging Foam 1 Die temperature, 3 mm. diameter, and 4 mm. diameter (Chapter 11)

Packaging Foam 2 Die temperature, 3 mm. diameter, and 4 mm. diameter (Chapter 11)

Packaging Foam 3 Die temperature, die diameter, and foam density (Chapter 14)

Packaging Foam 4 Die temperature, die diameter, and foam diameter (Chapter 14)

Pallet Weight of Boston shingles and weight of Vermont shingles (Chapters 2, 8, 9, and 10)

Parachute One-way Tensile strength of parachutes from suppliers 1, 2, 3, and 4 (Chapter 11)

Parachute Two-way Loom and tensile strength of parachutes from suppliers 1, 2, 3, and 4 (Chapter 11)

Pen Ad and product rating (Chapters 11, 12)

Philly Zip code, population, median sales price 2012 ($000), average days on market 2012, units sold 2012, median household income ($), percentage of residents with a BA or higher, and hotness (0 = not hot, 1 = hot) (Chapter 17)

Pizza Hut Gender coded (0 = Female, 1 = Male), gender (Female or Male), price ($), and purchase (0 = student selected another pizzeria, 1 = student selected Pizza Hut) (Chapter 14)

Pizza Delivery Time period, delivery time for local restaurant, and delivery time for national chain (Chapter 10)

Polio Year and incidence rates per 100,000 persons of reported poliomyelitis (Chapter 16)

Potato Percentage of solids content in filter cake, acidity (pH), lower pressure, upper pressure, cake thickness, varidrive speed, and drum speed setting (Chapter 15)

Potter Movies Title, first weekend gross ($millions), U.S. gross ($millions), and worldwide gross ($millions) (Chapters 2, 3, and 13)

Property Taxes State and property taxes per capita ($) (Chapters 2, 3, and 6)

Protein Type of food, calories (in grams), protein, percentage of calories from fat, percentage of calories from saturated fat, and cholesterol (mg) (Chapters 2 and 3)

Pumpkin Circumference and weight of pumpkins (Chapter 13)

Python Time to complete program (Chapter 10)

Quick-Serve Sales Mean sales per unit ($000) at burger, chicken, sandwich, and snack chains (Chapters 11 and 12)

Red and White Fixed acidity, volatile acidity, citric acid, residual sugar, chlorides, free sulfur dioxide, total sulfur dioxide, density, pH, sulphates, alcohol, wine type coded (0 = White, 1 = Red), wine type (Red or White), and quality (Chapter 14)

Redwood Height (ft.), breast height diameter (in.), and bark thickness (in.) (Chapters 13 and 14)

Registration Errors Comparison Registration error, temperature, pressure, and supplier (Chapter 15)

Registration Errors Registration error and temperature (Chapter 15)

Restaurant Ratings Location, food rating, decor rating, service rating, cost of a meal, popularity index, and cuisine [American (New), Chinese, French, Indian, Italian, Japanese, or Mexican] (Chapter 17)

Restaurants Location (Center City or Metro Area), meal cost, food rating, decor rating, service rating, summated rating, and coded location (0 = Center City, 1 = Metro Area); also contains Unstacked worksheet (Chapters 2, 3, 10, 13 and 14)

Restaurants for Tableau Tableau version of the Restaurants DATA worksheet (Chapters 2, 3, and 13) (T)

Retirement Funds Fund number, market cap (Small, MidCap, or Large), fund type (Growth or Value), risk level (Low, Average, or High), assets ($millions), turnover ratio, beta (volatility measure), standard deviation (measure of returns relative to 36-month average), one-year return, three-year return, five-year return, ten-year return, expense ratio, and star rating; also contains the unstacked, StackedSummary, and StackedHierarchy worksheets (Chapters 2, 3, 6, 8, 10, 11, 12, and 15)

Retirement Funds for Tableau Tableau version of the Retirement Funds DATA worksheet (Chapters 2, 3, and 13) (T)

Roslyn Address, fair market value ($000), property size (acres), house size (sq. ft.), age, number of rooms, number of bathrooms, and number of cars that can be parked in the garage (Chapter 15)

Self-Learning Robots Usage and percentage (Chapter 2)

Service Level Time to answer (Chapters 2, 8, and 9)

Silver Year and price of silver ($) (Chapter 16)

Silver-Q Quarter; coded quarter; price of silver ($); and Q1, Q2, and Q3 (all 0 = no, 1 = yes) (Chapter 16)

Silver Spring Homes Address, asking price ($000), lot size (acres), yearly taxes ($), central a/c (0 = no, 1 = yes), number of bedrooms, number of bathrooms, age (years), number of parking spaces, finished basement (0 = no, 1 = yes), brick (0 = no, 1 = yes), and fireplace (0 = no, 1 = yes) (Chapters 13, 14, and 15)

Silver Spring Homes Uncoded Silver Spring Homes with these uncoded variables: has central a/c, has finished basement, has brick, and has fireplace (all N or Y) (Chapter 14)

Silver Spring Rentals Apartment size (sq. ft.) and monthly rental cost ($) (Chapter 13)

Site Selection Store number, profiled customers, and sales ($millions) (Chapter 13)

Smartphone Sales Type, and market share percentage for five quarters in 2017 and 2018 (Chapter 2)

Smartphones Price ($) (Chapter 3)

Soccer Values Team, country, revenue ($millions), and value ($millions) (Chapter 13)

Solar Power Year and amount of solar power generated (megawatts) (Chapter 16)

Starbucks Tear, viscosity, pressure, and plate gap (Chapters 13 and 14)

Steel Error in actual length and specified length (Chapters 2, 6, 8, and 9)

Stock Performance Decade and stock performance percentage (Chapters 2 and 16)

Store Waiting Time Number waiting and waiting time (seconds) (Chapter 15)

Student Survey ID, gender (Female or Male), age (as of last birthday), class designation (Sophomore, Junior, or Senior), major (Accounting, CIS, Economics/Finance, International Business, Management, Retail/Marketing,

Other, or Undecided), graduate school intention (No, Yes, or Undecided), cumulative GPA, current employment status (Full-Time, Part-Time, or Unemployed), expected starting salary ($000), number of social networking sites joined, satisfaction with student advisement services on campus, amount spent on books and supplies this semester, type of computer preferred (Desktop, Laptop, or Tablet), text messages per week, and wealth accumulated to feel rich (Chapters 2, 3, 6, 8, 10, 11, and 12)

Super Bowl Ad Costs Year and cost of 30-second advertisement ($millions) (Chapter 2)

Super Bowl Ad Ratings Company, rating, time of ad (First Quarter, Second Quarter, Halftime, Third Quarter, or Fourth Quarter), and first half (No or Yes) (Chapters 2, 3, 6, 8, 10, and 12)

Super Bowl Ad Scores Brand advertised, ad time (seconds), and score (Chapters 3, 6, 8, 10, and 12)

Target Quarter; coded quarter; revenue; and Q1, Q2, and Q3 (all 0 = no, 1 = yes) (Chapter 16)

Taxi Delays Delay time in minutes for driver A and delay time in minutes for driver B (Chapter 10)

Teabags Teabag weight (ounces) (Chapters 3, 8, and 9)

Technologies Technologies and frequency (Chapter 2)

Telecomm Provider, TV rating, and phone rating (Chapter 10)

Textbook Costs Revenue category, detail, and percentage (Chapter 2)

Thickness Thickness, catalyst, pH, pressure, temperature, and voltage (Chapters 14 and 15)

Three Hotel Responses Reason for not returning, hotel (GP, PR, or PP), booking source (agent, walk-in, or web), relationship status (couple, family, or single (Chapter 17)

Time Series A Year, coded year, and three time series (I, II, and III) (Chapter 16)

Time Series B Year, coded year, and two time series (I and II) (Chapter 16)

Times Get-ready times (Chapter 3)

Tomatometer Movie, tomato meter rating, and receipts ($000) (Chapter 13)

Training Ranks Rank and method (Chapter 12)

Travel and Tourism Country, 2018 travel and tourism jobs, 2018 nights spent by European residents at tourism establishments (millions), 2018 amount spent by European residents at tourism establishments (€000), and 2018 tourism establishments that provide overnight accommodations (Chapter 18)

Trough Width of trough (Chapter 9)

TRS NYC Year, unit value of diversified equity funds, and unit value of balanced funds (Chapter 16)

TV Remote Filling time, 60, 72.5, and 85 degrees (Chapter 11)

Underwriting End-of-training exam score, proficiency exam score, and training method (classroom, courseware app, or online) (Chapter 14)

Used Cars Car, year, age, price ($), mileage, and fuel mileage (mpg) (Chapter 18)

Utility Utilities charges ($) for 50 one-bedroom apartments (Chapters 2 and 6)

Vinho Verde Fixed acidity, volatile acidity, citric acid, residual sugar, chlorides, free sulfur dioxide, total sulfur dioxide, density, pH, sulphates, alcohol, and quality (Chapters 13, 14, and 15)

VLABGo Storefront and in-aisle sales (Chapters 10 and 12)

Wait Waiting time and seating time (Chapter 6)

Wallys ID number, ShowGo products purchased ($), Wallys credit card (No or Yes), weekly household income ($), gender (Female or Male), age, healthy eating rating, and active lifestyle rating (Chapter 18)

Warehouse Costs Distribution cost ($000), sales ($000), and number of orders (Chapter 13)

WIP Processing times at each of two plants (1 = A, 2 = B) (Chapter 18)

Work Hours Needed Team name, work hours needed (Chapters 2, 3, and 6)

Workforce Year, population, and size of the workforce (Chapter 16)

World Smartphone Country, GDP, and social media usage percentage (Chapters 2 and 3)

Yarn Side-by-side aspect and breaking strength scores for 30 psi, 40 psi, and 50 psi (Chapter 11)

C.3 Microsoft Excel Files Integrated With This Book

This book integrates both a set of Excel Guide workbooks that contain templates or model solutions for applying Excel to statistical methods and PHStat, the Pearson statistics Excel add-in, that the authors developed to support this book. "In the Worksheet" notes for many Excel illustrations in this book reference individual worksheets that are found in specific Excel Guide Workbooks or created by PHStat.

PHStat can duplicate the all of the results of the Excel Guide workbooks, but the reverse is not true. For a very few methods, such as stepwise or best subsets regression, PHStat uses background processing to create results that are partially calculated off-worksheet. For such methods, PHStat is the best way to create results.

The book also integrates the Visual Explorations Excel macro workbooks. These workbooks interactively visualize probability distribution, sampling distribution, and simple linear regression concepts. These workbooks work best with Microsoft Windows versions of Excel.

Excel Guide Workbooks

Excel Guide workbooks contain templates or model solutions for applying Excel to a particular statistical method. Chapter examples and the Excel Guide *Workbook* instructions feature worksheets from these workbooks.

Most workbooks include a **COMPUTE worksheet** (often shown in this book) and a **COMPUTE_FORMULAS worksheet** that allows you to examine all of the formulas that the worksheet uses. The Excel Guide workbooks (with chapter references) are:

Dirty Data (1)	Discrete Variable (5)
Recoded (1)	Binomial (5)
Challenging (2)	Poisson (5)
Summary Table (2)	Portfolio (5)
Contingency Table (2)	Hypergeometric (5)
Distributions (2)	Normal (6)
Pareto (2)	NPP (6)
Histogram (2)	Exponential (6)
Polygons (2)	SDS (7)
Scatter Plot (2)	CIE sigma known (8)
Time Series (2)	CIE sigma unknown (8)
MCT (2)	CIE Proportion (8)
Slicers (2)	Sample Size Mean (8)
Sparklines (2)	Sample Size Proportion (8)
Central Tendency (3)	Z Mean (9)
Descriptive (3)	T Mean (9)
Quartiles (3)	Z Proportion (9)
Boxplot (3)	Pooled-Variance T (10)
Parameters (3)	Separate-Variance T (10)
Covariance (3)	Paired T (10)
Correlation (3)	Z Two Proportions (10)
Probabilities (4)	F Two Variances (10)
Bayes (4)	One-Way ANOVA (11)

Levene (11)	Logistic Regression Example (14)
Two-Way ANOVA (11)	
Randomized Block (11)	Logistic Regression add-in (14)
Chi-Square (12)	
Chi-Square Worksheets (12)	Moving Averages (16)
Wilcoxon (12)	Exponential Smoothing (16)
Kruskal-Wallis Worksheets (12)	Exponential Trend (16)
	Differences (16)
Simple Linear Regression (13)	Lagged Predictors (16)
	Forecasting Comparison (16)
Package Delivery (13)	Arlingtons National Sales .pbix (17)
Multiple Regression (14)	

PHStat

PHStat is the Pearson Education statistics add-in for Microsoft Excel that simplifies the task of using Excel as you learn business statistics. PHStat comes packaged as a zip file archive that you download and unzip to the folder of your choice. The archive contains:

PHStat.xlam, the main add-in workbook.

PHStat readme.pdf   Explains the technical requirements, and setup and troubleshooting procedures for PHStat (PDF format).

PHStatHelp.chm   The integrated help system for users of Microsoft Windows Excel.

PHStatHelp.pdf   The help system as a PDF format file.

PHStatHelp.epub   The help system in Open Publication Structure eBook format.

For more information about PHStat, see Appendix H.

Visual Explorations

Visual Explorations are add-in workbooks that interactively demonstrate various key statistical concepts. To use these workbooks with Microsoft Windows Excel, first verify the Excel security settings (see step 4 in Appendix Section D.1). The Visual Explorations workbooks are:

VE-Normal Distribution
VE-Sampling Distribution
VE-Simple Linear Regression

C.4 Supplemental Files

Over three dozen online supplemental files provide opportunities for additional learning with this book. This set of files include two additional chapters and numerous additional sections that customize learning and which are optional to the main content of this book. Supplemental files also include the SHORT TAKES that expand on in-chapter explanations and the files that support the end-of-chapter cases.

All supplemental files use the Portable Document Format (PDF) that are best viewed using the latest version of Adobe Acrobat Reader (**get.adobe.com/reader/**) or Acrobat Pro. Files that support the Digital Cases use advanced PDF features and *require* the use of Acrobat Reader or Pro.

Configuring Software

D.1 Microsoft Excel Configuration

Step 1: Update Excel

Proper configuration begins by ensuring the copy of Excel to be used with this book has been properly updated. For Microsoft Windows Excel, with any workbook open (even a blank one), select **File→Account** and in the Account panel select **Update Now** from the **Update Options** pull-down list. For Excel for Mac, select **Help→Check for Updates** to load the separate Microsoft AutoUpdate program that handles the downloading and installation of Office updates.

Step 2: Verify Microsoft Add-Ins

To use the *Analysis ToolPak* Excel Guide instructions, requires the Analysis ToolPak add-in. To use the Excel Guide instructions for logistic regression (Section EG14.7), requires the Solver add-in. Microsoft supplies these add-ins as part of any Excel installation, but the add-ins may not have been previously activated. (Readers who will not be using the Analysis ToolPak instructions and the Section EG14.7 instructions should skip to step 3.)

To check for the presence of the Analysis ToolPak or Solver add-ins in Microsoft Windows Excel:

1. Select **File→Options**.

In the Excel Options dialog box:

2. Click **Add-Ins** in the left pane and look for the entry **Analysis ToolPak** (or **Solver Add-in**) in the right pane, under **Active Application Add-ins**.
3. If the entry appears, click **OK**.
4. If the entry does not appear in the **Active Application Add-ins** list, select **Excel Add-ins** from the **Manage** drop-down list and then click **Go**.
5. In the Add-Ins dialog box, check **Analysis ToolPak** (or **Solver Add-in**) in the **Add-Ins available** list and click **OK**.

If the Add-Ins available list does not include a Microsoft-supplied add-in that you need, rerun the Microsoft Office setup program to install the missing add-in.

To check for the presence of the Analysis ToolPak or Solver add-ins in Excel for Mac:

1. Select **Tools→Options**.
2. In the Adds-Ins dialog box, check **Analysis ToolPak** (or **Solver Add-In**) in the **Add-Ins available** list and click **OK**.

If the Add-Ins available list does not include a Microsoft-supplied add-in that you need, click **Browse** to locate the add-in. If a message appears that states that the add-in is not currently installed on your Mac, click **Yes** to install the add-in. Then exit Excel and restart Excel.

Step 3: Verify Excel Security Settings

Using Microsoft Windows Excel requires verifying Excel security settings to use either PHStat or one of the Visual Explorations add-in workbooks (see Section C.3). (Excel for Mac has no security settings and readers using Excel for Mac should skip to step 4.)

To properly configure the Microsoft Windows Excel security settings:

1. Select **File→Options**.

In the Excel Options dialog box (shown below):

2. Click **Trust Center** in the left pane and then click **Trust Center Settings** in the right pane.

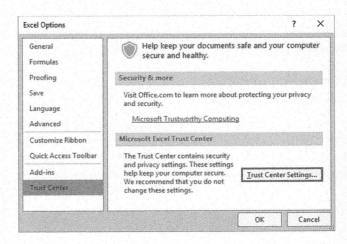

In the Trust Center dialog box:

3. Click **Add-ins** in the next left pane, and in the Add-ins right pane, clear all of the checkboxes (shown below).

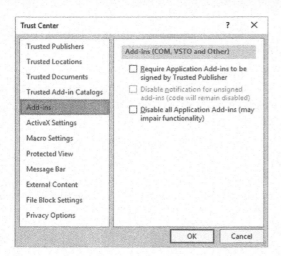

4. Click **Macro Settings** in the left pane, and in the Macro Settings right pane (shown below), click **Disable all macros with notification** and check **Trust access to the VBA object model**.

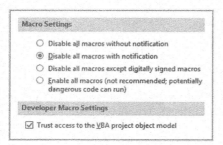

5. Click **OK** to close the Trust Center dialog box.

Back in the Excel Options dialog box:

6. Click **OK** to finish.

On some systems that have stringent security settings, you might need to modify step 4. For such systems, in step 4, also click **Trusted Locations** in the left pane and then, in the Trusted Locations right pane, click **Add new location** to add

the folder path that you chose to store the PHStat or Visual Explorations add-in files.

Step 4: Opening Add-ins

Opening PHStat or one of the Visual Explorations add-in workbooks (see Section C.3) will cause Excel to display a security notice that will be similar to the security notices for Microsoft Windows Excel and Excel for Mac notices shown below.

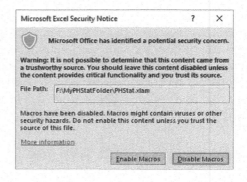

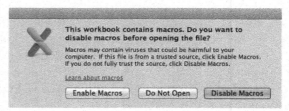

In these dialog boxes, click **Enable Macros**, which is *not* the default choice, to enable and use the add-in.

(Because Microsoft supplies the Analysis ToolPak and Solver add-ins, using either of those add-ins will *not* cause Excel to display a security notice.)

D.2 Supplemental Files

Tableau Public, also known as Tableau Desktop Public Edition, does not require a special initial set up. To review settings, select **Help → Settings and Performance** and review the entries in the submenu.

Table

TABLE E.1
Table of Random Numbers

Row	00000 12345	00001 67890	11111 12345	11112 67890	22222 12345	22223 67890	33333 12345	33334 67890
01	49280	88924	35779	00283	81163	07275	89863	02348
02	61870	41657	07468	08612	98083	97349	20775	45091
03	43898	65923	25078	86129	78496	97653	91550	08078
04	62993	93912	30454	84598	56095	20664	12872	64647
05	33850	58555	51438	85507	71865	79488	76783	31708
06	97340	03364	88472	04334	63919	36394	11095	92470
07	70543	29776	10087	10072	55980	64688	68239	20461
08	89382	93809	00796	95945	34101	81277	66090	88872
09	37818	72142	67140	50785	22380	16703	53362	44940
10	60430	22834	14130	96593	23298	56203	92671	15925
11	82975	66158	84731	19436	55790	69229	28661	13675
12	30987	71938	40355	54324	08401	26299	49420	59208
13	55700	24586	93247	32596	11865	63397	44251	43189
14	14756	23997	78643	75912	83832	32768	18928	57070
15	32166	53251	70654	92827	63491	04233	33825	69662
16	23236	73751	31888	81718	06546	83246	47651	04877
17	45794	26926	15130	82455	78305	55058	52551	47182
18	09893	20505	14225	68514	47427	56788	96297	78822
19	54382	74598	91499	14523	68479	27686	46162	83554
20	94750	89923	37089	20048	80336	94598	26940	36858
21	70297	34135	53140	33340	42050	82341	44104	82949
22	85157	47954	32979	26575	57600	40881	12250	73742
23	11100	02340	12860	74697	96644	89439	28707	25815
24	36871	50775	30592	57143	17381	68856	25853	35041
25	23913	48357	63308	16090	51690	54607	72407	55538
26	79348	36085	27973	65157	07456	22255	25626	57054
27	92074	54641	53673	54421	18130	60103	69593	49464
28	06873	21440	75593	41373	49502	17972	82578	16364
29	12478	37622	99659	31065	83613	69889	58869	29571
30	57175	55564	65411	42547	70457	03426	72937	83792
31	91616	11075	80103	07831	59309	13276	26710	73000
32	78025	73539	14621	39044	47450	03197	12787	47709
33	27587	67228	80145	10175	12822	86687	65530	49325
34	16690	20427	04251	64477	73709	73945	92396	68263
35	70183	58065	65489	31833	82093	16747	10386	59293
36	90730	35385	15679	99742	50866	78028	75573	67257
37	10934	93242	13431	24590	02770	48582	00906	58595
38	82462	30166	79613	47416	13389	80268	05085	96666
39	27463	10433	07606	16285	93699	60912	94532	95632
40	02979	52997	09079	92709	90110	47506	53693	49892
41	46888	69929	75233	52507	32097	37594	10067	67327
42	53638	83161	08289	12639	08141	12640	28437	09268
43	82433	61427	17239	89160	19666	08814	37841	12847
44	35766	31672	50082	22795	66948	65581	84393	15890
45	10853	42581	08792	13257	61973	24450	52351	16602
46	20341	27398	72906	63955	17276	10646	74692	48438
47	54458	90542	77563	51839	52901	53355	83281	19177
48	26337	66530	16687	35179	46560	00123	44546	79896
49	34314	23729	85264	05575	96855	23820	11091	79821
50	28603	10708	68933	34189	92166	15181	66628	58599

TABLE E.1
Table of Random
Numbers (*continued*)

Row	00000 12345	00001 67890	11111 12345	11112 67890	22222 12345	22223 67890	33333 12345	33334 67890
51	66194	28926	99547	16625	45515	67953	12108	57846
52	78240	43195	24837	32511	70880	22070	52622	61881
53	00833	88000	67299	68215	11274	55624	32991	17436
54	12111	86683	61270	58036	64192	90611	15145	01748
55	47189	99951	05755	03834	43782	90599	40282	51417
56	76396	72486	62423	27618	84184	78922	73561	52818
57	46409	17469	32483	09083	76175	19985	26309	91536
58	74626	22111	87286	46772	42243	68046	44250	42439
59	34450	81974	93723	49023	58432	67083	36876	93391
60	36327	72135	33005	28701	34710	49359	50693	89311
61	74185	77536	84825	09934	99103	09325	67389	45869
62	12296	41623	62873	37943	25584	09609	63360	47270
63	90822	60280	88925	99610	42772	60561	76873	04117
64	72121	79152	96591	90305	10189	79778	68016	13747
65	95268	41377	25684	08151	61816	58555	54305	86189
66	92603	09091	75884	93424	72586	88903	30061	14457
67	18813	90291	05275	01223	79607	95426	34900	09778
68	38840	26903	28624	67157	51986	42865	14508	49315
69	05959	33836	53758	16562	41081	38012	41230	20528
70	85141	21155	99212	32685	51403	31926	69813	58781
71	75047	59643	31074	38172	03718	32119	69506	67143
72	30752	95260	68032	62871	58781	34143	68790	69766
73	22986	82575	42187	62295	84295	30634	66562	31442
74	99439	86692	90348	66036	48399	73451	26698	39437
75	20389	93029	11881	71685	65452	89047	63669	02656
76	39249	05173	68256	36359	20250	68686	05947	09335
77	96777	33605	29481	20063	09398	01843	35139	61344
78	04860	32918	10798	50492	52655	33359	94713	28393
79	41613	42375	00403	03656	77580	87772	86877	57085
80	17930	00794	53836	53692	67135	98102	61912	11246
81	24649	31845	25736	75231	83808	98917	93829	99430
82	79899	34061	54308	59358	56462	58166	97302	86828
83	76801	49594	81002	30397	52728	15101	72070	33706
84	36239	63636	38140	65731	39788	06872	38971	53363
85	07392	64449	17886	63632	53995	17574	22247	62607
86	67133	04181	33874	98835	67453	59734	76381	63455
87	77759	31504	32832	70861	15152	29733	75371	39174
88	85992	72268	42920	20810	29361	51423	90306	73574
89	79553	75952	54116	65553	47139	60579	09165	85490
90	41101	17336	48951	53674	17880	45260	08575	49321
91	36191	17095	32123	91576	84221	78902	82010	30847
92	62329	63898	23268	74283	26091	68409	69704	82267
93	14751	13151	93115	01437	56945	89661	67680	79790
94	48462	59278	44185	29616	76537	19589	83139	28454
95	29435	88105	59651	44391	74588	55114	80834	85686
96	28340	29285	12965	14821	80425	16602	44653	70467
97	02167	58940	27149	80242	10587	79786	34959	75339
98	17864	00991	39557	54981	23588	81914	37609	13128
99	79675	80605	60059	35862	00254	36546	21545	78179
100	72335	82037	92003	34100	29879	46613	89720	13274

Source: Partially extracted from the Rand Corporation, *A Million Random Digits with 100,000 Normal Deviates* (Glencoe, IL, The Free Press, 1955).

TABLE E.2
The Cumulative Standardized Normal Distribution

Entry represents area under the cumulative standardized
normal distribution from $-\infty$ to Z

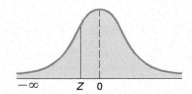

Z	Cumulative Probabilities									
	0.00	0.01	0.02	0.03	0.04	0.05	0.06	0.07	0.08	0.09
−6.0	0.000000001									
−5.5	0.000000019									
−5.0	0.000000287									
−4.5	0.000003398									
−4.0	0.000031671									
−3.9	0.00005	0.00005	0.00004	0.00004	0.00004	0.00004	0.00004	0.00004	0.00003	0.00003
−3.8	0.00007	0.00007	0.00007	0.00006	0.00006	0.00006	0.00006	0.00005	0.00005	0.00005
−3.7	0.00011	0.00010	0.00010	0.00010	0.00009	0.00009	0.00008	0.00008	0.00008	0.00008
−3.6	0.00016	0.00015	0.00015	0.00014	0.00014	0.00013	0.00013	0.00012	0.00012	0.00011
−3.5	0.00023	0.00022	0.00022	0.00021	0.00020	0.00019	0.00019	0.00018	0.00017	0.00017
−3.4	0.00034	0.00032	0.00031	0.00030	0.00029	0.00028	0.00027	0.00026	0.00025	0.00024
−3.3	0.00048	0.00047	0.00045	0.00043	0.00042	0.00040	0.00039	0.00038	0.00036	0.00035
−3.2	0.00069	0.00066	0.00064	0.00062	0.00060	0.00058	0.00056	0.00054	0.00052	0.00050
−3.1	0.00097	0.00094	0.00090	0.00087	0.00084	0.00082	0.00079	0.00076	0.00074	0.00071
−3.0	0.00135	0.00131	0.00126	0.00122	0.00118	0.00114	0.00111	0.00107	0.00103	0.00100
−2.9	0.0019	0.0018	0.0018	0.0017	0.0016	0.0016	0.0015	0.0015	0.0014	0.0014
−2.8	0.0026	0.0025	0.0024	0.0023	0.0023	0.0022	0.0021	0.0021	0.0020	0.0019
−2.7	0.0035	0.0034	0.0033	0.0032	0.0031	0.0030	0.0029	0.0028	0.0027	0.0026
−2.6	0.0047	0.0045	0.0044	0.0043	0.0041	0.0040	0.0039	0.0038	0.0037	0.0036
−2.5	0.0062	0.0060	0.0059	0.0057	0.0055	0.0054	0.0052	0.0051	0.0049	0.0048
−2.4	0.0082	0.0080	0.0078	0.0075	0.0073	0.0071	0.0069	0.0068	0.0066	0.0064
−2.3	0.0107	0.0104	0.0102	0.0099	0.0096	0.0094	0.0091	0.0089	0.0087	0.0084
−2.2	0.0139	0.0136	0.0132	0.0129	0.0125	0.0122	0.0119	0.0116	0.0113	0.0110
−2.1	0.0179	0.0174	0.0170	0.0166	0.0162	0.0158	0.0154	0.0150	0.0146	0.0143
−2.0	0.0228	0.0222	0.0217	0.0212	0.0207	0.0202	0.0197	0.0192	0.0188	0.0183
−1.9	0.0287	0.0281	0.0274	0.0268	0.0262	0.0256	0.0250	0.0244	0.0239	0.0233
−1.8	0.0359	0.0351	0.0344	0.0336	0.0329	0.0322	0.0314	0.0307	0.0301	0.0294
−1.7	0.0446	0.0436	0.0427	0.0418	0.0409	0.0401	0.0392	0.0384	0.0375	0.0367
−1.6	0.0548	0.0537	0.0526	0.0516	0.0505	0.0495	0.0485	0.0475	0.0465	0.0455
−1.5	0.0668	0.0655	0.0643	0.0630	0.0618	0.0606	0.0594	0.0582	0.0571	0.0559
−1.4	0.0808	0.0793	0.0778	0.0764	0.0749	0.0735	0.0721	0.0708	0.0694	0.0681
−1.3	0.0968	0.0951	0.0934	0.0918	0.0901	0.0885	0.0869	0.0853	0.0838	0.0823
−1.2	0.1151	0.1131	0.1112	0.1093	0.1075	0.1056	0.1038	0.1020	0.1003	0.0985
−1.1	0.1357	0.1335	0.1314	0.1292	0.1271	0.1251	0.1230	0.1210	0.1190	0.1170
−1.0	0.1587	0.1562	0.1539	0.1515	0.1492	0.1469	0.1446	0.1423	0.1401	0.1379
−0.9	0.1841	0.1814	0.1788	0.1762	0.1736	0.1711	0.1685	0.1660	0.1635	0.1611
−0.8	0.2119	0.2090	0.2061	0.2033	0.2005	0.1977	0.1949	0.1922	0.1894	0.1867
−0.7	0.2420	0.2388	0.2358	0.2327	0.2296	0.2266	0.2236	0.2206	0.2177	0.2148
−0.6	0.2743	0.2709	0.2676	0.2643	0.2611	0.2578	0.2546	0.2514	0.2482	0.2451
−0.5	0.3085	0.3050	0.3015	0.2981	0.2946	0.2912	0.2877	0.2843	0.2810	0.2776
−0.4	0.3446	0.3409	0.3372	0.3336	0.3300	0.3264	0.3228	0.3192	0.3156	0.3121
−0.3	0.3821	0.3783	0.3745	0.3707	0.3669	0.3632	0.3594	0.3557	0.3520	0.3483
−0.2	0.4207	0.4168	0.4129	0.4090	0.4052	0.4013	0.3974	0.3936	0.3897	0.3859
−0.1	0.4602	0.4562	0.4522	0.4483	0.4443	0.4404	0.4364	0.4325	0.4286	0.4247
−0.0	0.5000	0.4960	0.4920	0.4880	0.4840	0.4801	0.4761	0.4721	0.4681	0.4641

TABLE E.2

The Cumulative Standardized Normal Distribution (*continued*)

Entry represents area under the cumulative standardized
normal distribution from $-\infty$ to Z

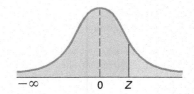

	Cumulative Probabilities									
Z	0.00	0.01	0.02	0.03	0.04	0.05	0.06	0.07	0.08	0.09
0.0	0.5000	0.5040	0.5080	0.5120	0.5160	0.5199	0.5239	0.5279	0.5319	0.5359
0.1	0.5398	0.5438	0.5478	0.5517	0.5557	0.5596	0.5636	0.5675	0.5714	0.5753
0.2	0.5793	0.5832	0.5871	0.5910	0.5948	0.5987	0.6026	0.6064	0.6103	0.6141
0.3	0.6179	0.6217	0.6255	0.6293	0.6331	0.6368	0.6406	0.6443	0.6480	0.6517
0.4	0.6554	0.6591	0.6628	0.6664	0.6700	0.6736	0.6772	0.6808	0.6844	0.6879
0.5	0.6915	0.6950	0.6985	0.7019	0.7054	0.7088	0.7123	0.7157	0.7190	0.7224
0.6	0.7257	0.7291	0.7324	0.7357	0.7389	0.7422	0.7454	0.7486	0.7518	0.7549
0.7	0.7580	0.7612	0.7642	0.7673	0.7704	0.7734	0.7764	0.7794	0.7823	0.7852
0.8	0.7881	0.7910	0.7939	0.7967	0.7995	0.8023	0.8051	0.8078	0.8106	0.8133
0.9	0.8159	0.8186	0.8212	0.8238	0.8264	0.8289	0.8315	0.8340	0.8365	0.8389
1.0	0.8413	0.8438	0.8461	0.8485	0.8508	0.8531	0.8554	0.8577	0.8599	0.8621
1.1	0.8643	0.8665	0.8686	0.8708	0.8729	0.8749	0.8770	0.8790	0.8810	0.8830
1.2	0.8849	0.8869	0.8888	0.8907	0.8925	0.8944	0.8962	0.8980	0.8997	0.9015
1.3	0.9032	0.9049	0.9066	0.9082	0.9099	0.9115	0.9131	0.9147	0.9162	0.9177
1.4	0.9192	0.9207	0.9222	0.9236	0.9251	0.9265	0.9279	0.9292	0.9306	0.9319
1.5	0.9332	0.9345	0.9357	0.9370	0.9382	0.9394	0.9406	0.9418	0.9429	0.9441
1.6	0.9452	0.9463	0.9474	0.9484	0.9495	0.9505	0.9515	0.9525	0.9535	0.9545
1.7	0.9554	0.9564	0.9573	0.9582	0.9591	0.9599	0.9608	0.9616	0.9625	0.9633
1.8	0.9641	0.9649	0.9656	0.9664	0.9671	0.9678	0.9686	0.9693	0.9699	0.9706
1.9	0.9713	0.9719	0.9726	0.9732	0.9738	0.9744	0.9750	0.9756	0.9761	0.9767
2.0	0.9772	0.9778	0.9783	0.9788	0.9793	0.9798	0.9803	0.9808	0.9812	0.9817
2.1	0.9821	0.9826	0.9830	0.9834	0.9838	0.9842	0.9846	0.9850	0.9854	0.9857
2.2	0.9861	0.9864	0.9868	0.9871	0.9875	0.9878	0.9881	0.9884	0.9887	0.9890
2.3	0.9893	0.9896	0.9898	0.9901	0.9904	0.9906	0.9909	0.9911	0.9913	0.9916
2.4	0.9918	0.9920	0.9922	0.9925	0.9927	0.9929	0.9931	0.9932	0.9934	0.9936
2.5	0.9938	0.9940	0.9941	0.9943	0.9945	0.9946	0.9948	0.9949	0.9951	0.9952
2.6	0.9953	0.9955	0.9956	0.9957	0.9959	0.9960	0.9961	0.9962	0.9963	0.9964
2.7	0.9965	0.9966	0.9967	0.9968	0.9969	0.9970	0.9971	0.9972	0.9973	0.9974
2.8	0.9974	0.9975	0.9976	0.9977	0.9977	0.9978	0.9979	0.9979	0.9980	0.9981
2.9	0.9981	0.9982	0.9982	0.9983	0.9984	0.9984	0.9985	0.9985	0.9986	0.9986
3.0	0.99865	0.99869	0.99874	0.99878	0.99882	0.99886	0.99889	0.99893	0.99897	0.99900
3.1	0.99903	0.99906	0.99910	0.99913	0.99916	0.99918	0.99921	0.99924	0.99926	0.99929
3.2	0.99931	0.99934	0.99936	0.99938	0.99940	0.99942	0.99944	0.99946	0.99948	0.99950
3.3	0.99952	0.99953	0.99955	0.99957	0.99958	0.99960	0.99961	0.99962	0.99964	0.99965
3.4	0.99966	0.99968	0.99969	0.99970	0.99971	0.99972	0.99973	0.99974	0.99975	0.99976
3.5	0.99977	0.99978	0.99978	0.99979	0.99980	0.99981	0.99981	0.99982	0.99983	0.99983
3.6	0.99984	0.99985	0.99985	0.99986	0.99986	0.99987	0.99987	0.99988	0.99988	0.99989
3.7	0.99989	0.99990	0.99990	0.99990	0.99991	0.99991	0.99992	0.99992	0.99992	0.99992
3.8	0.99993	0.99993	0.99993	0.99994	0.99994	0.99994	0.99994	0.99995	0.99995	0.99995
3.9	0.99995	0.99995	0.99996	0.99996	0.99996	0.99996	0.99996	0.99996	0.99997	0.99997
4.0	0.999968329									
4.5	0.999996602									
5.0	0.999999713									
5.5	0.999999981									
6.0	0.999999999									

TABLE E.3
Critical Values of *t*

For a particular number of degrees of freedom, entry represents the critical value of *t* corresponding to the cumulative probability $(1 - \alpha)$ and a specified upper-tail area (α).

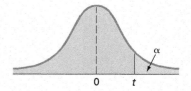

Degrees of Freedom	Cumulative Probabilities					
	0.75	0.90	0.95	0.975	0.99	0.995
	Upper-Tail Areas					
	0.25	0.10	0.05	0.025	0.01	0.005
1	1.0000	3.0777	6.3138	12.7062	31.8207	63.6574
2	0.8165	1.8856	2.9200	4.3027	6.9646	9.9248
3	0.7649	1.6377	2.3534	3.1824	4.5407	5.8409
4	0.7407	1.5332	2.1318	2.7764	3.7469	4.6041
5	0.7267	1.4759	2.0150	2.5706	3.3649	4.0322
6	0.7176	1.4398	1.9432	2.4469	3.1427	3.7074
7	0.7111	1.4149	1.8946	2.3646	2.9980	3.4995
8	0.7064	1.3968	1.8595	2.3060	2.8965	3.3554
9	0.7027	1.3830	1.8331	2.2622	2.8214	3.2498
10	0.6998	1.3722	1.8125	2.2281	2.7638	3.1693
11	0.6974	1.3634	1.7959	2.2010	2.7181	3.1058
12	0.6955	1.3562	1.7823	2.1788	2.6810	3.0545
13	0.6938	1.3502	1.7709	2.1604	2.6503	3.0123
14	0.6924	1.3450	1.7613	2.1448	2.6245	2.9768
15	0.6912	1.3406	1.7531	2.1315	2.6025	2.9467
16	0.6901	1.3368	1.7459	2.1199	2.5835	2.9208
17	0.6892	1.3334	1.7396	2.1098	2.5669	2.8982
18	0.6884	1.3304	1.7341	2.1009	2.5524	2.8784
19	0.6876	1.3277	1.7291	2.0930	2.5395	2.8609
20	0.6870	1.3253	1.7247	2.0860	2.5280	2.8453
21	0.6864	1.3232	1.7207	2.0796	2.5177	2.8314
22	0.6858	1.3212	1.7171	2.0739	2.5083	2.8188
23	0.6853	1.3195	1.7139	2.0687	2.4999	2.8073
24	0.6848	1.3178	1.7109	2.0639	2.4922	2.7969
25	0.6844	1.3163	1.7081	2.0595	2.4851	2.7874
26	0.6840	1.3150	1.7056	2.0555	2.4786	2.7787
27	0.6837	1.3137	1.7033	2.0518	2.4727	2.7707
28	0.6834	1.3125	1.7011	2.0484	2.4671	2.7633
29	0.6830	1.3114	1.6991	2.0452	2.4620	2.7564
30	0.6828	1.3104	1.6973	2.0423	2.4573	2.7500
31	0.6825	1.3095	1.6955	2.0395	2.4528	2.7440
32	0.6822	1.3086	1.6939	2.0369	2.4487	2.7385
33	0.6820	1.3077	1.6924	2.0345	2.4448	2.7333
34	0.6818	1.3070	1.6909	2.0322	2.4411	2.7284
35	0.6816	1.3062	1.6896	2.0301	2.4377	2.7238
36	0.6814	1.3055	1.6883	2.0281	2.4345	2.7195
37	0.6812	1.3049	1.6871	2.0262	2.4314	2.7154
38	0.6810	1.3042	1.6860	2.0244	2.4286	2.7116
39	0.6808	1.3036	1.6849	2.0227	2.4258	2.7079
40	0.6807	1.3031	1.6839	2.0211	2.4233	2.7045
41	0.6805	1.3025	1.6829	2.0195	2.4208	2.7012
42	0.6804	1.3020	1.6820	2.0181	2.4185	2.6981
43	0.6802	1.3016	1.6811	2.0167	2.4163	2.6951
44	0.6801	1.3011	1.6802	2.0154	2.4141	2.6923
45	0.6800	1.3006	1.6794	2.0141	2.4121	2.6896
46	0.6799	1.3002	1.6787	2.0129	2.4102	2.6870
47	0.6797	1.2998	1.6779	2.0117	2.4083	2.6846
48	0.6796	1.2994	1.6772	2.0106	2.4066	2.6822
49	0.6795	1.2991	1.6766	2.0096	2.4049	2.6800
50	0.6794	1.2987	1.6759	2.0086	2.4033	2.6778

TABLE E.3

Critical Values of *t* (*continued*)

For a particular number of degrees of freedom, entry represents the critical value of *t* corresponding to the cumulative probability $(1 - \alpha)$ and a specified upper-tail area (α).

Degrees of Freedom	Cumulative Probabilities					
	0.75	0.90	0.95	0.975	0.99	0.995
	Upper-Tail Areas					
	0.25	0.10	0.05	0.025	0.01	0.005
51	0.6793	1.2984	1.6753	2.0076	2.4017	2.6757
52	0.6792	1.2980	1.6747	2.0066	2.4002	2.6737
53	0.6791	1.2977	1.6741	2.0057	2.3988	2.6718
54	0.6791	1.2974	1.6736	2.0049	2.3974	2.6700
55	0.6790	1.2971	1.6730	2.0040	2.3961	2.6682
56	0.6789	1.2969	1.6725	2.0032	2.3948	2.6665
57	0.6788	1.2966	1.6720	2.0025	2.3936	2.6649
58	0.6787	1.2963	1.6716	2.0017	2.3924	2.6633
59	0.6787	1.2961	1.6711	2.0010	2.3912	2.6618
60	0.6786	1.2958	1.6706	2.0003	2.3901	2.6603
61	0.6785	1.2956	1.6702	1.9996	2.3890	2.6589
62	0.6785	1.2954	1.6698	1.9990	2.3880	2.6575
63	0.6784	1.2951	1.6694	1.9983	2.3870	2.6561
64	0.6783	1.2949	1.6690	1.9977	2.3860	2.6549
65	0.6783	1.2947	1.6686	1.9971	2.3851	2.6536
66	0.6782	1.2945	1.6683	1.9966	2.3842	2.6524
67	0.6782	1.2943	1.6679	1.9960	2.3833	2.6512
68	0.6781	1.2941	1.6676	1.9955	2.3824	2.6501
69	0.6781	1.2939	1.6672	1.9949	2.3816	2.6490
70	0.6780	1.2938	1.6669	1.9944	2.3808	2.6479
71	0.6780	1.2936	1.6666	1.9939	2.3800	2.6469
72	0.6779	1.2934	1.6663	1.9935	2.3793	2.6459
73	0.6779	1.2933	1.6660	1.9930	2.3785	2.6449
74	0.6778	1.2931	1.6657	1.9925	2.3778	2.6439
75	0.6778	1.2929	1.6654	1.9921	2.3771	2.6430
76	0.6777	1.2928	1.6652	1.9917	2.3764	2.6421
77	0.6777	1.2926	1.6649	1.9913	2.3758	2.6412
78	0.6776	1.2925	1.6646	1.9908	2.3751	2.6403
79	0.6776	1.2924	1.6644	1.9905	2.3745	2.6395
80	0.6776	1.2922	1.6641	1.9901	2.3739	2.6387
81	0.6775	1.2921	1.6639	1.9897	2.3733	2.6379
82	0.6775	1.2920	1.6636	1.9893	2.3727	2.6371
83	0.6775	1.2918	1.6634	1.9890	2.3721	2.6364
84	0.6774	1.2917	1.6632	1.9886	2.3716	2.6356
85	0.6774	1.2916	1.6630	1.9883	2.3710	2.6349
86	0.6774	1.2915	1.6628	1.9879	2.3705	2.6342
87	0.6773	1.2914	1.6626	1.9876	2.3700	2.6335
88	0.6773	1.2912	1.6624	1.9873	2.3695	2.6329
89	0.6773	1.2911	1.6622	1.9870	2.3690	2.6322
90	0.6772	1.2910	1.6620	1.9867	2.3685	2.6316
91	0.6772	1.2909	1.6618	1.9864	2.3680	2.6309
92	0.6772	1.2908	1.6616	1.9861	2.3676	2.6303
93	0.6771	1.2907	1.6614	1.9858	2.3671	2.6297
94	0.6771	1.2906	1.6612	1.9855	2.3667	2.6291
95	0.6771	1.2905	1.6611	1.9853	2.3662	2.6286
96	0.6771	1.2904	1.6609	1.9850	2.3658	2.6280
97	0.6770	1.2903	1.6607	1.9847	2.3654	2.6275
98	0.6770	1.2902	1.6606	1.9845	2.3650	2.6269
99	0.6770	1.2902	1.6604	1.9842	2.3646	2.6264
100	0.6770	1.2901	1.6602	1.9840	2.3642	2.6259
110	0.6767	1.2893	1.6588	1.9818	2.3607	2.6213
120	0.6765	1.2886	1.6577	1.9799	2.3578	2.6174
∞	0.6745	1.2816	1.6449	1.9600	2.3263	2.5758

TABLE E.4
Critical Values of χ^2

For a particular number of degrees of freedom, entry represents the critical value of χ^2 corresponding to the cumulative probability $(1 - \alpha)$ and a specified upper-tail area (α).

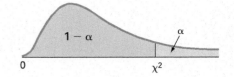

	Cumulative Probabilities											
	0.005	0.01	0.025	0.05	0.10	0.25	0.75	0.90	0.95	0.975	0.99	0.995
Degrees of Freedom	**Upper-Tail Areas (α)**											
	0.995	0.99	0.975	0.95	0.90	0.75	0.25	0.10	0.05	0.025	0.01	0.005
1			0.001	0.004	0.016	0.102	1.323	2.706	3.841	5.024	6.635	7.879
2	0.010	0.020	0.051	0.103	0.211	0.575	2.773	4.605	5.991	7.378	9.210	10.597
3	0.072	0.115	0.216	0.352	0.584	1.213	4.108	6.251	7.815	9.348	11.345	12.838
4	0.207	0.297	0.484	0.711	1.064	1.923	5.385	7.779	9.488	11.143	13.277	14.860
5	0.412	0.554	0.831	1.145	1.610	2.675	6.626	9.236	11.071	12.833	15.086	16.750
6	0.676	0.872	1.237	1.635	2.204	3.455	7.841	10.645	12.592	14.449	16.812	18.548
7	0.989	1.239	1.690	2.167	2.833	4.255	9.037	12.017	14.067	16.013	18.475	20.278
8	1.344	1.646	2.180	2.733	3.490	5.071	10.219	13.362	15.507	17.535	20.090	21.955
9	1.735	2.088	2.700	3.325	4.168	5.899	11.389	14.684	16.919	19.023	21.666	23.589
10	2.156	2.558	3.247	3.940	4.865	6.737	12.549	15.987	18.307	20.483	23.209	25.188
11	2.603	3.053	3.816	4.575	5.578	7.584	13.701	17.275	19.675	21.920	24.725	26.757
12	3.074	3.571	4.404	5.226	6.304	8.438	14.845	18.549	21.026	23.337	26.217	28.299
13	3.565	4.107	5.009	5.892	7.042	9.299	15.984	19.812	22.362	24.736	27.688	29.819
14	4.075	4.660	5.629	6.571	7.790	10.165	17.117	21.064	23.685	26.119	29.141	31.319
15	4.601	5.229	6.262	7.261	8.547	11.037	18.245	22.307	24.996	27.488	30.578	32.801
16	5.142	5.812	6.908	7.962	9.312	11.912	19.369	23.542	26.296	28.845	32.000	34.267
17	5.697	6.408	7.564	8.672	10.085	12.792	20.489	24.769	27.587	30.191	33.409	35.718
18	6.265	7.015	8.231	9.390	10.865	13.675	21.605	25.989	28.869	31.526	34.805	37.156
19	6.844	7.633	8.907	10.117	11.651	14.562	22.718	27.204	30.144	32.852	36.191	38.582
20	7.434	8.260	9.591	10.851	12.443	15.452	23.828	28.412	31.410	34.170	37.566	39.997
21	8.034	8.897	10.283	11.591	13.240	16.344	24.935	29.615	32.671	35.479	38.932	41.401
22	8.643	9.542	10.982	12.338	14.042	17.240	26.039	30.813	33.924	36.781	40.289	42.796
23	9.260	10.196	11.689	13.091	14.848	18.137	27.141	32.007	35.172	38.076	41.638	44.181
24	9.886	10.856	12.401	13.848	15.659	19.037	28.241	33.196	36.415	39.364	42.980	45.559
25	10.520	11.524	13.120	14.611	16.473	19.939	29.339	34.382	37.652	40.646	44.314	46.928
26	11.160	12.198	13.844	15.379	17.292	20.843	30.435	35.563	38.885	41.923	45.642	48.290
27	11.808	12.879	14.573	16.151	18.114	21.749	31.528	36.741	40.113	43.194	46.963	49.645
28	12.461	13.565	15.308	16.928	18.939	22.657	32.620	37.916	41.337	44.461	48.278	50.993
29	13.121	14.257	16.047	17.708	19.768	23.567	33.711	39.087	42.557	45.722	49.588	52.336
30	13.787	14.954	16.791	18.493	20.599	24.478	34.800	40.256	43.773	46.979	50.892	53.672

For larger values of degrees of freedom (df) the expression $Z = \sqrt{2\chi^2} - \sqrt{2(df) - 1}$ may be used and the resulting upper-tail area can be found from the cumulative standardized normal distribution (Table E.2).

TABLE E.5
Critical Values of F

For a particular combination of numerator and denominator degrees of freedom, entry represents the critical values of F corresponding to the cumulative probability $(1 - \alpha)$ and a specified upper-tail area (α).

$\alpha = 0.05$

Cumulative Probabilities = 0.95

Upper-Tail Areas = 0.05

Denominator, df_2	Numerator, df_1																		
	1	2	3	4	5	6	7	8	9	10	12	15	20	24	30	40	60	120	∞
1	161.40	199.50	215.70	224.60	230.20	234.00	236.80	238.90	240.50	241.90	243.90	245.90	248.00	249.10	250.10	251.10	252.20	253.30	254.30
2	18.51	19.00	19.16	19.25	19.30	19.33	19.35	19.37	19.38	19.40	19.41	19.43	19.45	19.45	19.46	19.47	19.48	19.49	19.50
3	10.13	9.55	9.28	9.12	9.01	8.94	8.89	8.85	8.81	8.79	8.74	8.70	8.66	8.64	8.62	8.59	8.57	8.55	8.53
4	7.71	6.94	6.59	6.39	6.26	6.16	6.09	6.04	6.00	5.96	5.91	5.86	5.80	5.77	5.75	5.72	5.69	5.66	5.63
5	6.61	5.79	5.41	5.19	5.05	4.95	4.88	4.82	4.77	4.74	4.68	4.62	4.56	4.53	4.50	4.46	4.43	4.40	4.36
6	5.99	5.14	4.76	4.53	4.39	4.28	4.21	4.15	4.10	4.06	4.00	3.94	3.87	3.84	3.81	3.77	3.74	3.70	3.67
7	5.59	4.74	4.35	4.12	3.97	3.87	3.79	3.73	3.68	3.64	3.57	3.51	3.44	3.41	3.38	3.34	3.30	3.27	3.23
8	5.32	4.46	4.07	3.84	3.69	3.58	3.50	3.44	3.39	3.35	3.28	3.22	3.15	3.12	3.08	3.04	3.01	2.97	2.93
9	5.12	4.26	3.86	3.63	3.48	3.37	3.29	3.23	3.18	3.14	3.07	3.01	2.94	2.90	2.86	2.83	2.79	2.75	2.71
10	4.96	4.10	3.71	3.48	3.33	3.22	3.14	3.07	3.02	2.98	2.91	2.85	2.77	2.74	2.70	2.66	2.62	2.58	2.54
11	4.84	3.98	3.59	3.36	3.20	3.09	3.01	2.95	2.90	2.85	2.79	2.72	2.65	2.61	2.57	2.53	2.49	2.45	2.40
12	4.75	3.89	3.49	3.26	3.11	3.00	2.91	2.85	2.80	2.75	2.69	2.62	2.54	2.51	2.47	2.43	2.38	2.34	2.30
13	4.67	3.81	3.41	3.18	3.03	2.92	2.83	2.77	2.71	2.67	2.60	2.53	2.46	2.42	2.38	2.34	2.30	2.25	2.21
14	4.60	3.74	3.34	3.11	2.96	2.85	2.76	2.70	2.65	2.60	2.53	2.46	2.39	2.35	2.31	2.27	2.22	2.18	2.13
15	4.54	3.68	3.29	3.06	2.90	2.79	2.71	2.64	2.59	2.54	2.48	2.40	2.33	2.29	2.25	2.20	2.16	2.11	2.07
16	4.49	3.63	3.24	3.01	2.85	2.74	2.66	2.59	2.54	2.49	2.42	2.35	2.28	2.24	2.19	2.15	2.11	2.06	2.01
17	4.45	3.59	3.20	2.96	2.81	2.70	2.61	2.55	2.49	2.45	2.38	2.31	2.23	2.19	2.15	2.10	2.06	2.01	1.96
18	4.41	3.55	3.16	2.93	2.77	2.66	2.58	2.51	2.46	2.41	2.34	2.27	2.19	2.15	2.11	2.06	2.02	1.97	1.92
19	4.38	3.52	3.13	2.90	2.74	2.63	2.54	2.48	2.42	2.38	2.31	2.23	2.16	2.11	2.07	2.03	1.98	1.93	1.88
20	4.35	3.49	3.10	2.87	2.71	2.60	2.51	2.45	2.39	2.35	2.28	2.20	2.12	2.08	2.04	1.99	1.95	1.90	1.84
21	4.32	3.47	3.07	2.84	2.68	2.57	2.49	2.42	2.37	2.32	2.25	2.18	2.10	2.05	2.01	1.96	1.92	1.87	1.81
22	4.30	3.44	3.05	2.82	2.66	2.55	2.46	2.40	2.34	2.30	2.23	2.15	2.07	2.03	1.98	1.94	1.89	1.84	1.78
23	4.28	3.42	3.03	2.80	2.64	2.53	2.44	2.37	2.32	2.27	2.20	2.13	2.05	2.01	1.96	1.91	1.86	1.81	1.76
24	4.26	3.40	3.01	2.78	2.62	2.51	2.42	2.36	2.30	2.25	2.18	2.11	2.03	1.98	1.94	1.89	1.84	1.79	1.73
25	4.24	3.39	2.99	2.76	2.60	2.49	2.40	2.34	2.28	2.24	2.16	2.09	2.01	1.96	1.92	1.87	1.82	1.77	1.71
26	4.23	3.37	2.98	2.74	2.59	2.47	2.39	2.32	2.27	2.22	2.15	2.07	1.99	1.95	1.90	1.85	1.80	1.75	1.69
27	4.21	3.35	2.96	2.73	2.57	2.46	2.37	2.31	2.25	2.20	2.13	2.06	1.97	1.93	1.88	1.84	1.79	1.73	1.67
28	4.20	3.34	2.95	2.71	2.56	2.45	2.36	2.29	2.24	2.19	2.12	2.04	1.96	1.91	1.87	1.82	1.77	1.71	1.65
29	4.18	3.33	2.93	2.70	2.55	2.43	2.35	2.28	2.22	2.18	2.10	2.03	1.94	1.90	1.85	1.81	1.75	1.70	1.64
30	4.17	3.32	2.92	2.69	2.53	2.42	2.33	2.27	2.21	2.16	2.09	2.01	1.93	1.89	1.84	1.79	1.74	1.68	1.62
40	4.08	3.23	2.84	2.61	2.45	2.34	2.25	2.18	2.12	2.08	2.00	1.92	1.84	1.79	1.74	1.69	1.64	1.58	1.51
60	4.00	3.15	2.76	2.53	2.37	2.25	2.17	2.10	2.04	1.99	1.92	1.84	1.75	1.70	1.65	1.59	1.53	1.47	1.39
120	3.92	3.07	2.68	2.45	2.29	2.17	2.09	2.02	1.96	1.91	1.83	1.75	1.66	1.61	1.55	1.50	1.43	1.35	1.25
∞	3.84	3.00	2.60	2.37	2.21	2.10	2.01	1.94	1.88	1.83	1.75	1.67	1.57	1.52	1.46	1.39	1.32	1.22	1.00

(continued)

TABLE E.5

Critical Values of F (continued)

For a particular combination of numerator and denominator degrees of freedom, entry represents the critical values of F corresponding to the cumulative probability $(1 - \alpha)$ and a specified upper-tail area (α).

$\alpha = 0.025$

Cumulative Probabilities = 0.975

Upper-Tail Areas = 0.025

Numerator, df_1

Denominator, df_2	1	2	3	4	5	6	7	8	9	10	12	15	20	24	30	40	60	120	∞
1	647.80	799.50	864.20	899.60	921.80	937.10	948.20	956.70	963.30	968.60	976.70	984.90	993.10	997.20	1,001.00	1,006.00	1,010.00	1,014.00	1,018.00
2	38.51	39.00	39.17	39.25	39.30	39.33	39.36	39.39	39.39	39.40	39.41	39.43	39.45	39.46	39.46	39.47	39.48	39.49	39.50
3	17.44	16.04	15.44	15.10	14.88	14.73	14.62	14.54	14.47	14.42	14.34	14.25	14.17	14.12	14.08	14.04	13.99	13.95	13.90
4	12.22	10.65	9.98	9.60	9.36	9.20	9.07	8.98	8.90	8.84	8.75	8.66	8.56	8.51	8.46	8.41	8.36	8.31	8.26
5	10.01	8.43	7.76	7.39	7.15	6.98	6.85	6.76	6.68	6.62	6.52	6.43	6.33	6.28	6.23	6.18	6.12	6.07	6.02
6	8.81	7.26	6.60	6.23	5.99	5.82	5.70	5.60	5.52	5.46	5.37	5.27	5.17	5.12	5.07	5.01	4.96	4.90	4.85
7	8.07	6.54	5.89	5.52	5.29	5.12	4.99	4.90	4.82	4.76	4.67	4.57	4.47	4.42	4.36	4.31	4.25	4.20	4.14
8	7.57	6.06	5.42	5.05	4.82	4.65	4.53	4.43	4.36	4.30	4.20	4.10	4.00	3.95	3.89	3.84	3.78	3.73	3.67
9	7.21	5.71	5.08	4.72	4.48	4.32	4.20	4.10	4.03	3.96	3.87	3.77	3.67	3.61	3.56	3.51	3.45	3.39	3.33
10	6.94	5.46	4.83	4.47	4.24	4.07	3.95	3.85	3.78	3.72	3.62	3.52	3.42	3.37	3.31	3.26	3.20	3.14	3.08
11	6.72	5.26	4.63	4.28	4.04	3.88	3.76	3.66	3.59	3.53	3.43	3.33	3.23	3.17	3.12	3.06	3.00	2.94	2.88
12	6.55	5.10	4.47	4.12	3.89	3.73	3.61	3.51	3.44	3.37	3.28	3.18	3.07	3.02	2.96	2.91	2.85	2.79	2.72
13	6.41	4.97	4.35	4.00	3.77	3.60	3.48	3.39	3.31	3.25	3.15	3.05	2.95	2.89	2.84	2.78	2.72	2.66	2.60
14	6.30	4.86	4.24	3.89	3.66	3.50	3.38	3.29	3.21	3.15	3.05	2.95	2.84	2.79	2.73	2.67	2.61	2.55	2.49
15	6.20	4.77	4.15	3.80	3.58	3.41	3.29	3.20	3.12	3.06	2.96	2.86	2.76	2.70	2.64	2.59	2.52	2.46	2.40
16	6.12	4.69	4.08	3.73	3.50	3.34	3.22	3.12	3.05	2.99	2.89	2.79	2.68	2.63	2.57	2.51	2.45	2.38	2.32
17	6.04	4.62	4.01	3.66	3.44	3.28	3.16	3.06	2.98	2.92	2.82	2.72	2.62	2.56	2.50	2.44	2.38	2.32	2.25
18	5.98	4.56	3.95	3.61	3.38	3.22	3.10	3.01	2.93	2.87	2.77	2.67	2.56	2.50	2.44	2.38	2.32	2.26	2.19
19	5.92	4.51	3.90	3.56	3.33	3.17	3.05	2.96	2.88	2.82	2.72	2.62	2.51	2.45	2.39	2.33	2.27	2.20	2.13
20	5.87	4.46	3.86	3.51	3.29	3.13	3.01	2.91	2.84	2.77	2.68	2.57	2.46	2.41	2.35	2.29	2.22	2.16	2.09
21	5.83	4.42	3.82	3.48	3.25	3.09	2.97	2.87	2.80	2.73	2.64	2.53	2.42	2.37	2.31	2.25	2.18	2.11	2.04
22	5.79	4.38	3.78	3.44	3.22	3.05	2.93	2.84	2.76	2.70	2.60	2.50	2.39	2.33	2.27	2.21	2.14	2.08	2.00
23	5.75	4.35	3.75	3.41	3.18	3.02	2.90	2.81	2.73	2.67	2.57	2.47	2.36	2.30	2.24	2.18	2.11	2.04	1.97
24	5.72	4.32	3.72	3.38	3.15	2.99	2.87	2.78	2.70	2.64	2.54	2.44	2.33	2.27	2.21	2.15	2.08	2.01	1.94
25	5.69	4.29	3.69	3.35	3.13	2.97	2.85	2.75	2.68	2.61	2.51	2.41	2.30	2.24	2.18	2.12	2.05	1.98	1.91
26	5.66	4.27	3.67	3.33	3.10	2.94	2.82	2.73	2.65	2.59	2.49	2.39	2.28	2.22	2.16	2.09	2.03	1.95	1.88
27	5.63	4.24	3.65	3.31	3.08	2.92	2.80	2.71	2.63	2.57	2.47	2.36	2.25	2.19	2.13	2.07	2.00	1.93	1.85
28	5.61	4.22	3.63	3.29	3.06	2.90	2.78	2.69	2.61	2.55	2.45	2.34	2.23	2.17	2.11	2.05	1.98	1.91	1.83
29	5.59	4.20	3.61	3.27	3.04	2.88	2.76	2.67	2.59	2.53	2.43	2.32	2.21	2.15	2.09	2.03	1.96	1.89	1.81
30	5.57	4.18	3.59	3.25	3.03	2.87	2.75	2.65	2.57	2.51	2.41	2.31	2.20	2.14	2.07	2.01	1.94	1.87	1.79
40	5.42	4.05	3.46	3.13	2.90	2.74	2.62	2.53	2.45	2.39	2.29	2.18	2.07	2.01	1.94	1.88	1.80	1.72	1.64
60	5.29	3.93	3.34	3.01	2.79	2.63	2.51	2.41	2.33	2.27	2.17	2.06	1.94	1.88	1.82	1.74	1.67	1.58	1.48
120	5.15	3.80	3.23	2.89	2.67	2.52	2.39	2.30	2.22	2.16	2.05	1.94	1.82	1.76	1.69	1.61	1.53	1.43	1.31
∞	5.02	3.69	3.12	2.79	2.57	2.41	2.29	2.19	2.11	2.05	1.94	1.83	1.71	1.64	1.57	1.48	1.39	1.27	1.00

TABLE E.5
Critical Values of F (continued)

For a particular combination of numerator and denominator degrees of freedom, entry represents the critical values of F corresponding to the cumulative probability $(1 - \alpha)$ and a specified upper-tail area (α).

Cumulative Probabilities = 0.99

Upper-Tail Areas = 0.01

Denominator, df_2	Numerator, df_1																		
	1	2	3	4	5	6	7	8	9	10	12	15	20	24	30	40	60	120	∞
1	4,052.00	4,999.50	5,403.00	5,625.00	5,764.00	5,859.00	5,928.00	5,982.00	6,022.00	6,056.00	6,106.00	6,157.00	6,209.00	6,235.00	6,261.00	6,287.00	6,313.00	6,339.00	6,366.00
2	98.50	99.00	99.17	99.25	99.30	99.33	99.36	99.37	99.39	99.40	99.42	99.43	44.45	99.46	99.47	99.47	99.48	99.49	99.50
3	34.12	30.82	29.46	28.71	28.24	27.91	27.67	27.49	27.35	27.23	27.05	26.87	26.69	26.60	26.50	26.41	26.32	26.22	26.13
4	21.20	18.00	16.69	15.98	15.52	15.21	14.98	14.80	14.66	14.55	14.37	14.20	14.02	13.93	13.84	13.75	13.65	13.56	13.46
5	16.26	13.27	12.06	11.39	10.97	10.67	10.46	10.29	10.16	10.05	9.89	9.72	9.55	9.47	9.38	9.29	9.20	9.11	9.02
6	13.75	10.92	9.78	9.15	8.75	8.47	8.26	8.10	7.98	7.87	7.72	7.56	7.40	7.31	7.23	7.14	7.06	6.97	6.88
7	12.25	9.55	8.45	7.85	7.46	7.19	6.99	6.84	6.72	6.62	6.47	6.31	6.16	6.07	5.99	5.91	5.82	5.74	5.65
8	11.26	8.65	7.59	7.01	6.63	6.37	6.18	6.03	5.91	5.81	5.67	5.52	5.36	5.28	5.20	5.12	5.03	4.95	4.86
9	10.56	8.02	6.99	6.42	6.06	5.80	5.61	5.47	5.35	5.26	5.11	4.96	4.81	4.73	4.65	4.57	4.48	4.40	4.31
10	10.04	7.56	6.55	5.99	5.64	5.39	5.20	5.06	4.94	4.85	4.71	4.56	4.41	4.33	4.25	4.17	4.08	4.00	3.91
11	9.65	7.21	6.22	5.67	5.32	5.07	4.89	4.74	4.63	4.54	4.40	4.25	4.10	4.02	3.94	3.86	3.78	3.69	3.60
12	9.33	6.93	5.95	5.41	5.06	4.82	4.64	4.50	4.39	4.30	4.16	4.01	3.86	3.78	3.70	3.62	3.54	3.45	3.36
13	9.07	6.70	5.74	5.21	4.86	4.62	4.44	4.30	4.19	4.10	3.96	3.82	3.66	3.59	3.51	3.43	3.34	3.25	3.17
14	8.86	6.51	5.56	5.04	4.69	4.46	4.28	4.14	4.03	3.94	3.80	3.66	3.51	3.43	3.35	3.27	3.18	3.09	3.00
15	8.68	6.36	5.42	4.89	4.56	4.32	4.14	4.00	3.89	3.80	3.67	3.52	3.37	3.29	3.21	3.13	3.05	2.96	2.87
16	8.53	6.23	5.29	4.77	4.44	4.20	4.03	3.89	3.78	3.69	3.55	3.41	3.26	3.18	3.10	3.02	2.93	2.81	2.75
17	8.40	6.11	5.18	4.67	4.34	4.10	3.93	3.79	3.68	3.59	3.46	3.31	3.16	3.08	3.00	2.92	2.83	2.75	2.65
18	8.29	6.01	5.09	4.58	4.25	4.01	3.84	3.71	3.60	3.51	3.37	3.23	3.08	3.00	2.92	2.84	2.75	2.66	2.57
19	8.18	5.93	5.01	4.50	4.17	3.94	3.77	3.63	3.52	3.43	3.30	3.15	3.00	2.92	2.84	2.76	2.67	2.58	2.49
20	8.10	5.85	4.94	4.43	4.10	3.87	3.70	3.56	3.46	3.37	3.23	3.09	2.94	2.86	2.78	2.69	2.61	2.52	2.42
21	8.02	5.78	4.87	4.37	4.04	3.81	3.64	3.51	3.40	3.31	3.17	3.03	2.88	2.80	2.72	2.64	2.55	2.46	2.36
22	7.95	5.72	4.82	4.31	3.99	3.76	3.59	3.45	3.35	3.26	3.12	2.98	2.83	2.75	2.67	2.58	2.50	2.40	2.31
23	7.88	5.66	4.76	4.26	3.94	3.71	3.54	3.41	3.30	3.21	3.07	2.93	2.78	2.70	2.62	2.54	2.45	2.35	2.26
24	7.82	5.61	4.72	4.22	3.90	3.67	3.50	3.36	3.26	3.17	3.03	2.89	2.74	2.66	2.58	2.49	2.40	2.31	2.21
25	7.77	5.57	4.68	4.18	3.85	3.63	3.46	3.32	3.22	3.13	2.99	2.85	2.70	2.62	2.54	2.45	2.36	2.27	2.17
26	7.72	5.53	4.64	4.14	3.82	3.59	3.42	3.29	3.18	3.09	2.96	2.81	2.66	2.58	2.50	2.42	2.33	2.23	2.13
27	7.68	5.49	4.60	4.11	3.78	3.56	3.39	3.26	3.15	3.06	2.93	2.78	2.63	2.55	2.47	2.38	2.29	2.20	2.10
28	7.64	5.45	4.57	4.07	3.75	3.53	3.36	3.23	3.12	3.03	2.90	2.75	2.60	2.52	2.44	2.35	2.26	2.17	2.06
29	7.60	5.42	4.54	4.04	3.73	3.50	3.33	3.20	3.09	3.00	2.87	2.73	2.57	2.49	2.41	2.33	2.23	2.14	2.03
30	7.56	5.39	4.51	4.02	3.70	3.47	3.30	3.17	3.07	2.98	2.84	2.70	2.55	2.47	2.39	2.30	2.21	2.11	2.01
40	7.31	5.18	4.31	3.83	3.51	3.29	3.12	2.99	2.89	2.80	2.66	2.52	2.37	2.29	2.20	2.11	2.02	1.92	1.80
60	7.08	4.98	4.13	3.65	3.34	3.12	2.95	2.82	2.72	2.63	2.50	2.35	2.20	2.12	2.03	1.94	1.84	1.73	1.60
120	6.85	4.79	3.95	3.48	3.17	2.96	2.79	2.66	2.56	2.47	2.34	2.19	2.03	1.95	1.86	1.76	1.66	1.53	1.38
∞	6.63	4.61	3.78	3.32	3.02	2.80	2.64	2.51	2.41	2.32	2.18	2.04	1.88	1.79	1.70	1.59	1.47	1.32	1.00

(continued)

TABLE E.5
Critical Values of F (continued)

$\alpha = 0.005$

For a particular combination of numerator and denominator degrees of freedom, entry represents the critical values of F corresponding to the cumulative probability $(1 - \alpha)$ and a specified upper-tail area (α).

Cumulative Probabilities = 0.995

Upper-Tail Areas = 0.005

Denominator, df_2	Numerator, df_1																		
	1	2	3	4	5	6	7	8	9	10	12	15	20	24	30	40	60	120	∞
1	16,211.00	20,000.00	21,615.00	22,500.00	23,056.00	23,437.00	23,715.00	23,925.00	24,091.00	24,224.00	24,426.00	24,630.00	24,836.00	24,910.00	25,044.00	25,148.00	25,253.00	25,359.00	25,465.00
2	198.50	199.00	199.20	199.20	199.30	199.30	199.40	199.40	199.40	199.40	199.40	199.40	199.40	199.50	199.50	199.50	199.50	199.50	199.50
3	55.55	49.80	47.47	46.19	45.39	44.84	44.43	44.13	43.88	43.69	43.39	43.08	42.78	42.62	42.47	42.31	42.15	41.99	41.83
4	31.33	26.28	24.26	23.15	22.46	21.97	21.62	21.35	21.14	20.97	20.70	20.44	20.17	20.03	19.89	19.75	19.61	19.47	19.32
5	22.78	18.31	16.53	15.56	14.94	14.51	14.20	13.96	13.77	13.62	13.38	13.15	12.90	12.78	12.66	12.53	12.40	12.27	12.11
6	18.63	14.54	12.92	12.03	11.46	11.07	10.79	10.57	10.39	10.25	10.03	9.81	9.59	9.47	9.36	9.24	9.12	9.00	8.88
7	16.24	12.40	10.88	10.05	9.52	9.16	8.89	8.68	8.51	8.38	8.18	7.97	7.75	7.65	7.53	7.42	7.31	7.19	7.08
8	14.69	11.04	9.60	8.81	8.30	7.95	7.69	7.50	7.34	7.21	7.01	6.81	6.61	6.50	6.40	6.29	6.18	6.06	5.95
9	13.61	10.11	8.72	7.96	7.47	7.13	6.88	6.69	6.54	6.42	6.23	6.03	5.83	5.73	5.62	5.52	5.41	5.30	5.19
10	12.83	9.43	8.08	7.34	6.87	6.54	6.30	6.12	5.97	5.85	5.66	5.47	5.27	5.17	5.07	4.97	4.86	4.75	4.61
11	12.23	8.91	7.60	6.88	6.42	6.10	5.86	5.68	5.54	5.42	5.24	5.05	4.86	4.75	4.65	4.55	4.44	4.34	4.23
12	11.75	8.51	7.23	6.52	6.07	5.76	5.52	5.35	5.20	5.09	4.91	4.72	4.53	4.43	4.33	4.23	4.12	4.01	3.90
13	11.37	8.19	6.93	6.23	5.79	5.48	5.25	5.08	4.94	4.82	4.64	4.46	4.27	4.17	4.07	3.97	3.87	3.76	3.65
14	11.06	7.92	6.68	6.00	5.56	5.26	5.03	4.86	4.72	4.60	4.43	4.25	4.06	3.96	3.86	3.76	3.66	3.55	3.41
15	10.80	7.70	6.48	5.80	5.37	5.07	4.85	4.67	4.54	4.42	4.25	4.07	3.88	3.79	3.69	3.58	3.48	3.37	3.26
16	10.58	7.51	6.30	5.64	5.21	4.91	4.69	4.52	4.38	4.27	4.10	3.92	3.73	3.64	3.54	3.44	3.33	3.22	3.11
17	10.38	7.35	6.16	5.50	5.07	4.78	4.56	4.39	4.25	4.14	3.97	3.79	3.61	3.51	3.41	3.31	3.21	3.10	2.98
18	10.22	7.21	6.03	5.37	4.96	4.66	4.44	4.28	4.14	4.03	3.86	3.68	3.50	3.40	3.30	3.20	3.10	2.99	2.87
19	10.07	7.09	5.92	5.27	4.85	4.56	4.34	4.18	4.04	3.93	3.76	3.59	3.40	3.31	3.21	3.11	3.00	2.89	2.78
20	9.94	6.99	5.82	5.17	4.76	4.47	4.26	4.09	3.96	3.85	3.68	3.50	3.32	3.22	3.12	3.02	2.92	2.81	2.69
21	9.83	6.89	5.73	5.09	4.68	4.39	4.18	4.02	3.88	3.77	3.60	3.43	3.24	3.15	3.05	2.95	2.84	2.73	2.61
22	9.73	6.81	5.65	5.02	4.61	4.32	4.11	3.94	3.81	3.70	3.54	3.36	3.18	3.08	2.98	2.88	2.77	2.66	2.55
23	9.63	6.73	5.58	4.95	4.54	4.26	4.05	3.88	3.75	3.64	3.47	3.30	3.12	3.02	2.92	2.82	2.71	2.60	2.48
24	9.55	6.66	5.52	4.89	4.49	4.20	3.99	3.83	3.69	3.59	3.42	3.25	3.06	2.97	2.87	2.77	2.66	2.55	2.43
25	9.48	6.60	5.46	4.84	4.43	4.15	3.94	3.78	3.64	3.54	3.37	3.20	3.01	2.92	2.82	2.72	2.61	2.50	2.38
26	9.41	6.54	5.41	4.79	4.38	4.10	3.89	3.73	3.60	3.49	3.33	3.15	2.97	2.87	2.77	2.67	2.56	2.45	2.33
27	9.34	6.49	5.36	4.74	4.34	4.06	3.85	3.69	3.56	3.45	3.28	3.11	2.93	2.83	2.73	2.63	2.52	2.41	2.29
28	9.28	6.44	5.32	4.70	4.30	4.02	3.81	3.65	3.52	3.41	3.25	3.07	2.89	2.79	2.69	2.59	2.48	2.37	2.25
29	9.23	6.40	5.28	4.66	4.26	3.98	3.77	3.61	3.48	3.38	3.21	3.04	2.86	2.76	2.66	2.56	2.45	2.33	2.21
30	9.18	6.35	5.24	4.62	4.23	3.95	3.74	3.58	3.45	3.34	3.18	3.01	2.82	2.73	2.63	2.52	2.42	2.30	2.18
40	8.83	6.07	4.98	4.37	3.99	3.71	3.51	3.35	3.22	3.12	2.95	2.78	2.60	2.50	2.40	2.30	2.18	2.06	1.93
60	8.49	5.79	4.73	4.14	3.76	3.49	3.29	3.13	3.01	2.90	2.74	2.57	2.39	2.29	2.19	2.08	1.96	1.83	1.69
120	8.18	5.54	4.50	3.92	3.55	3.28	3.09	2.93	2.81	2.71	2.54	2.37	2.19	2.09	1.98	1.87	1.75	1.61	1.43
∞	7.88	5.30	4.28	3.72	3.35	3.09	2.90	2.74	2.62	2.52	2.36	2.19	2.00	1.90	1.79	1.67	1.53	1.36	1.00

TABLE E.6
Lower and Upper Critical Values, T_1, of the Wilcoxon Rank Sum Test

n_2	α One-tail	Two-tail	n_1 4	5	6	7	8	9	10
4	0.05	0.10	11,25						
	0.025	0.05	10,26						
	0.01	0.02	—,—						
	0.005	0.01	—,—						
5	0.05	0.10	12,28	19,36					
	0.025	0.05	11,29	17,38					
	0.01	0.02	10,30	16,39					
	0.005	0.01	—,—	15,40					
6	0.05	0.10	13,31	20,40	28,50				
	0.025	0.05	12,32	18,42	26,52				
	0.01	0.02	11,33	17,43	24,54				
	0.005	0.01	10,34	16,44	23,55				
7	0.05	0.10	14,34	21,44	29,55	39,66			
	0.025	0.05	13,35	20,45	27,57	36,69			
	0.01	0.02	11,37	18,47	25,59	34,71			
	0.005	0.01	10,38	16,49	24,60	32,73			
8	0.05	0.10	15,37	23,47	31,59	41,71	51,85		
	0.025	0.05	14,38	21,49	29,61	38,74	49,87		
	0.01	0.02	12,40	19,51	27,63	35,77	45,91		
	0.005	0.01	11,41	17,53	25,65	34,78	43,93		
9	0.05	0.10	16,40	24,51	33,63	43,76	54,90	66,105	
	0.025	0.05	14,42	22,53	31,65	40,79	51,93	62,109	
	0.01	0.02	13,43	20,55	28,68	37,82	47,97	59,112	
	0.005	0.01	11,45	18,57	26,70	35,84	45,99	56,115	
10	0.05	0.10	17,43	26,54	35,67	45,81	56,96	69,111	82,128
	0.025	0.05	15,45	23,57	32,70	42,84	53,99	65,115	78,132
	0.01	0.02	13,47	21,59	29,73	39,87	49,103	61,119	74,136
	0.005	0.01	12,48	19,61	27,75	37,89	47,105	58,122	71,139

Source: Adapted from TABLE 1 of F. Wilcoxon and R. A. Wilcox, *Some Rapid Approximate Statistical Procedures* (Pearl River, NY: Lederle Laboratories, 1964), with permission of the American Cyanamid Company.

TABLE E.7
Critical Values of the Studentized Range, Q

Upper 5% Points ($\alpha = 0.05$)

Denominator, df	Numerator, df																		
	2	3	4	5	6	7	8	9	10	11	12	13	14	15	16	17	18	19	20
1	18.00	27.00	32.80	37.10	40.40	43.10	45.40	47.40	49.10	50.60	52.00	53.20	54.30	55.40	56.30	57.20	58.00	58.80	59.60
2	6.09	8.30	9.80	10.90	11.70	12.40	13.00	13.50	14.00	14.40	14.70	15.10	15.40	15.70	15.90	16.10	16.40	16.60	16.80
3	4.50	5.91	6.82	7.50	8.04	8.48	8.85	9.18	9.46	9.72	9.95	10.15	10.35	10.52	10.69	10.84	10.98	11.11	11.24
4	3.93	5.04	5.76	6.29	6.71	7.05	7.35	7.60	7.83	8.03	8.21	8.37	8.52	8.66	8.79	8.91	9.03	9.13	9.23
5	3.64	4.60	5.22	5.67	6.03	6.33	6.58	6.80	6.99	7.17	7.32	7.47	7.60	7.72	7.83	7.93	8.03	8.12	8.21
6	3.46	4.34	4.90	5.31	5.63	5.89	6.12	6.32	6.49	6.65	6.79	6.92	7.03	7.14	7.24	7.34	7.43	7.51	7.59
7	3.34	4.16	4.68	5.06	5.36	5.61	5.82	6.00	6.16	6.30	6.43	6.55	6.66	6.76	6.85	6.94	7.02	7.09	7.17
8	3.26	4.04	4.53	4.89	5.17	5.40	5.60	5.77	5.92	6.05	6.18	6.29	6.39	6.48	6.57	6.65	6.73	6.80	6.87
9	3.20	3.95	4.42	4.76	5.02	5.24	5.43	5.60	5.74	5.87	5.98	6.09	6.19	6.28	6.36	6.44	6.51	6.58	6.64
10	3.15	3.88	4.33	4.65	4.91	5.12	5.30	5.46	5.60	5.72	5.83	5.93	6.03	6.11	6.20	6.27	6.34	6.40	6.47
11	3.11	3.82	4.26	4.57	4.82	5.03	5.20	5.35	5.49	5.61	5.71	5.81	5.90	5.99	6.06	6.14	6.20	6.26	6.33
12	3.08	3.77	4.20	4.51	4.75	4.95	5.12	5.27	5.40	5.51	5.62	5.71	5.80	5.88	5.95	6.03	6.09	6.15	6.21
13	3.06	3.73	4.15	4.45	4.69	4.88	5.05	5.19	5.32	5.43	5.53	5.63	5.71	5.79	5.86	5.93	6.00	6.05	6.11
14	3.03	3.70	4.11	4.41	4.64	4.83	4.99	5.13	5.25	5.36	5.46	5.55	5.64	5.72	5.79	5.85	5.92	5.97	6.03
15	3.01	3.67	4.08	4.37	4.60	4.78	4.94	5.08	5.20	5.31	5.40	5.49	5.58	5.65	5.72	5.79	5.85	5.90	5.96
16	3.00	3.65	4.05	4.33	4.56	4.74	4.90	5.03	5.15	5.26	5.35	5.44	5.52	5.59	5.66	5.72	5.79	5.84	5.90
17	2.98	3.63	4.02	4.30	4.52	4.71	4.86	4.99	5.11	5.21	5.31	5.39	5.47	5.55	5.61	5.68	5.74	5.79	5.84
18	2.97	3.61	4.00	4.28	4.49	4.67	4.82	4.96	5.07	5.17	5.27	5.35	5.43	5.50	5.57	5.63	5.69	5.74	5.79
19	2.96	3.59	3.98	4.25	4.47	4.65	4.79	4.92	5.04	5.14	5.23	5.32	5.39	5.46	5.53	5.59	5.65	5.70	5.75
20	2.95	3.58	3.96	4.23	4.45	4.62	4.77	4.90	5.01	5.11	5.20	5.28	5.36	5.43	5.49	5.55	5.61	5.66	5.71
24	2.92	3.53	3.90	4.17	4.37	4.54	4.68	4.81	4.92	5.01	5.10	5.18	5.25	5.32	5.38	5.44	5.50	5.54	5.59
30	2.89	3.49	3.84	4.10	4.30	4.46	4.60	4.72	4.83	4.92	5.00	5.08	5.15	5.21	5.27	5.33	5.38	5.43	5.48
40	2.86	3.44	3.79	4.04	4.23	4.39	4.52	4.63	4.74	4.82	4.91	4.98	5.05	5.11	5.16	5.22	5.27	5.31	5.36
60	2.83	3.40	3.74	3.98	4.16	4.31	4.44	4.55	4.65	4.73	4.81	4.88	4.94	5.00	5.06	5.11	5.16	5.20	5.24
120	2.80	3.36	3.69	3.92	4.10	4.24	4.36	4.48	4.56	4.64	4.72	4.78	4.84	4.90	4.95	5.00	5.05	5.09	5.13
∞	2.77	3.31	3.63	3.86	4.03	4.17	4.29	4.39	4.47	4.55	4.62	4.68	4.74	4.80	4.85	4.89	4.93	4.97	5.01

TABLE E.7

Critical Values of the Studentized Range, Q (continued)

Upper 1% Points ($\alpha = 0.01$)

Denominator, df	Numerator, df																		
	2	3	4	5	6	7	8	9	10	11	12	13	14	15	16	17	18	19	20
1	90.03	135.00	164.30	185.60	202.20	215.80	227.20	237.00	245.60	253.20	260.00	266.20	271.80	277.00	281.80	286.30	290.40	294.30	298.00
2	14.04	19.02	22.29	24.72	26.63	28.20	29.53	30.68	31.69	32.59	33.40	34.13	34.81	35.43	36.00	36.53	37.03	37.50	37.95
3	8.26	10.62	12.17	13.33	14.24	15.00	15.64	16.20	16.69	17.13	17.53	17.89	18.22	18.52	18.81	19.07	19.32	19.55	19.77
4	6.51	8.12	9.17	9.96	10.58	11.10	11.55	11.93	12.27	12.57	12.84	13.09	13.32	13.53	13.73	13.91	14.08	14.24	14.40
5	5.70	6.98	7.80	8.42	8.91	9.32	9.67	9.97	10.24	10.48	10.70	10.89	11.08	11.24	11.40	11.55	11.68	11.81	11.93
6	5.24	6.33	7.03	7.56	7.97	8.32	8.61	8.87	9.10	9.30	9.49	9.65	9.81	9.95	10.08	10.21	10.32	10.43	10.54
7	4.95	5.92	6.54	7.01	7.37	7.68	7.94	8.17	8.37	8.55	8.71	8.86	9.00	9.12	9.24	9.35	9.46	9.55	9.65
8	4.75	5.64	6.20	6.63	6.96	7.24	7.47	7.68	7.86	8.03	8.18	8.31	8.44	8.55	8.66	8.76	8.85	8.94	9.03
9	4.60	5.43	5.96	6.35	6.66	6.92	7.13	7.32	7.50	7.65	7.78	7.91	8.03	8.13	8.23	8.33	8.41	8.50	8.57
10	4.48	5.27	5.77	6.14	6.43	6.67	6.87	7.06	7.21	7.36	7.49	7.60	7.71	7.81	7.91	7.99	8.08	8.15	8.23
11	4.39	5.15	5.62	5.97	6.25	6.48	6.67	6.84	6.99	7.13	7.25	7.36	7.47	7.56	7.65	7.73	7.81	7.88	7.95
12	4.32	5.04	5.50	5.84	6.10	6.32	6.51	6.67	6.81	6.94	7.06	7.17	7.26	7.36	7.44	7.52	7.59	7.66	7.73
13	4.26	4.96	5.40	5.73	5.98	6.19	6.37	6.53	6.67	6.79	6.90	7.01	7.10	7.19	7.27	7.35	7.42	7.49	7.55
14	4.21	4.90	5.32	5.63	5.88	6.09	6.26	6.41	6.54	6.66	6.77	6.87	6.96	7.05	7.13	7.20	7.27	7.33	7.40
15	4.17	4.84	5.25	5.56	5.80	5.99	6.16	6.31	6.44	6.56	6.66	6.76	6.85	6.93	7.00	7.07	7.14	7.20	7.26
16	4.13	4.79	5.19	5.49	5.72	5.92	6.08	6.22	6.35	6.46	6.56	6.66	6.74	6.82	6.90	6.97	7.03	7.09	7.15
17	4.10	4.74	5.14	5.43	5.66	5.85	6.01	6.15	6.27	6.38	6.48	6.57	6.66	6.73	6.81	6.87	6.94	7.00	7.05
18	4.07	4.70	5.09	5.38	5.60	5.79	5.94	6.08	6.20	6.31	6.41	6.50	6.58	6.66	6.73	6.79	6.85	6.91	6.97
19	4.05	4.67	5.05	5.33	5.55	5.74	5.89	6.02	6.14	6.25	6.34	6.43	6.51	6.59	6.65	6.72	6.78	6.84	6.89
20	4.02	4.64	5.02	5.29	5.51	5.69	5.84	5.97	6.09	6.19	6.29	6.37	6.45	6.52	6.59	6.65	6.71	6.77	6.82
24	3.96	4.55	4.91	5.17	5.37	5.54	5.69	5.81	5.92	6.02	6.11	6.19	6.26	6.33	6.39	6.45	6.51	6.56	6.61
30	3.89	4.46	4.80	5.05	5.24	5.40	5.54	5.65	5.76	5.85	5.93	6.01	6.08	6.14	6.20	6.26	6.31	6.36	6.41
40	3.83	4.37	4.70	4.93	5.11	5.27	5.39	5.50	5.60	5.69	5.76	5.84	5.90	5.96	6.02	6.07	6.12	6.17	6.21
60	3.76	4.28	4.60	4.82	4.99	5.13	5.25	5.36	5.45	5.53	5.60	5.67	5.73	5.79	5.84	5.89	5.93	5.97	6.02
120	3.70	4.20	4.50	4.71	4.87	5.01	5.12	5.21	5.30	5.38	5.44	5.51	5.56	5.61	5.66	5.71	5.75	5.79	5.83
∞	3.64	4.12	4.40	4.60	4.76	4.88	4.99	5.08	5.16	5.23	5.29	5.35	5.40	5.45	5.49	5.54	5.57	5.61	5.65

Source: Extracted from H. L. Harter and D. S. Clemm, "The Probability Integrals of the Range and of the Studentized Range—Probability Integral, Percentage Points, and Moments of the Range," *Wright Air Development Technical Report 58–484*, Vol. 1, 1959.

TABLE E.8
Critical Values, d_L and d_U, of the Durbin-Watson Statistic, D (Critical Values Are One-Sided)[a]

	α = 0.05										α = 0.01									
	k = 1		k = 2		k = 3		k = 4		k = 5		k = 1		k = 2		k = 3		k = 4		k = 5	
n	d_L	d_U	d_L	d_U	d_L	d_U	d_L	d_U	d_L	d_U	d_L	d_U	d_L	d_U	d_L	d_U	d_L	d_U	d_L	d_U
15	1.08	1.36	.95	1.54	.82	1.75	.69	1.97	.56	2.21	.81	1.07	.70	1.25	.59	1.46	.49	1.70	.39	1.96
16	1.10	1.37	.98	1.54	.86	1.73	.74	1.93	.62	2.15	.84	1.09	.74	1.25	.63	1.44	.53	1.66	.44	1.90
17	1.13	1.38	1.02	1.54	.90	1.71	.78	1.90	.67	2.10	.87	1.10	.77	1.25	.67	1.43	.57	1.63	.48	1.85
18	1.16	1.39	1.05	1.53	.93	1.69	.82	1.87	.71	2.06	.90	1.12	.80	1.26	.71	1.42	.61	1.60	.52	1.80
19	1.18	1.40	1.08	1.53	.97	1.68	.86	1.85	.75	2.02	.93	1.13	.83	1.26	.74	1.41	.65	1.58	.56	1.77
20	1.20	1.41	1.10	1.54	1.00	1.68	.90	1.83	.79	1.99	.95	1.15	.86	1.27	.77	1.41	.68	1.57	.60	1.74
21	1.22	1.42	1.13	1.54	1.03	1.67	.93	1.81	.83	1.96	.97	1.16	.89	1.27	.80	1.41	.72	1.55	.63	1.71
22	1.24	1.43	1.15	1.54	1.05	1.66	.96	1.80	.86	1.94	1.00	1.17	.91	1.28	.83	1.40	.75	1.54	.66	1.69
23	1.26	1.44	1.17	1.54	1.08	1.66	.99	1.79	.90	1.92	1.02	1.19	.94	1.29	.86	1.40	.77	1.53	.70	1.67
24	1.27	1.45	1.19	1.55	1.10	1.66	1.01	1.78	.93	1.90	1.04	1.20	.96	1.30	.88	1.41	.80	1.53	.72	1.66
25	1.29	1.45	1.21	1.55	1.12	1.66	1.04	1.77	.95	1.89	1.05	1.21	.98	1.30	.90	1.41	.83	1.52	.75	1.65
26	1.30	1.46	1.22	1.55	1.14	1.65	1.06	1.76	.98	1.88	1.07	1.22	1.00	1.31	.93	1.41	.85	1.52	.78	1.64
27	1.32	1.47	1.24	1.56	1.16	1.65	1.08	1.76	1.01	1.86	1.09	1.23	1.02	1.32	.95	1.41	.88	1.51	.81	1.63
28	1.33	1.48	1.26	1.56	1.18	1.65	1.10	1.75	1.03	1.85	1.10	1.24	1.04	1.32	.97	1.41	.90	1.51	.83	1.62
29	1.34	1.48	1.27	1.56	1.20	1.65	1.12	1.74	1.05	1.84	1.12	1.25	1.05	1.33	.99	1.42	.92	1.51	.85	1.61
30	1.35	1.49	1.28	1.57	1.21	1.65	1.14	1.74	1.07	1.83	1.13	1.26	1.07	1.34	1.01	1.42	.94	1.51	.88	1.61
31	1.36	1.50	1.30	1.57	1.23	1.65	1.16	1.74	1.09	1.83	1.15	1.27	1.08	1.34	1.02	1.42	.96	1.51	.90	1.60
32	1.37	1.50	1.31	1.57	1.24	1.65	1.18	1.73	1.11	1.82	1.16	1.28	1.10	1.35	1.04	1.43	.98	1.51	.92	1.60
33	1.38	1.51	1.32	1.58	1.26	1.65	1.19	1.73	1.13	1.81	1.17	1.29	1.11	1.36	1.05	1.43	1.00	1.51	.94	1.59
34	1.39	1.51	1.33	1.58	1.27	1.65	1.21	1.73	1.15	1.81	1.18	1.30	1.13	1.36	1.07	1.43	1.01	1.51	.95	1.59
35	1.40	1.52	1.34	1.58	1.28	1.65	1.22	1.73	1.16	1.80	1.19	1.31	1.14	1.37	1.08	1.44	1.03	1.51	.97	1.59
36	1.41	1.52	1.35	1.59	1.29	1.65	1.24	1.73	1.18	1.80	1.21	1.32	1.15	1.38	1.10	1.44	1.04	1.51	.99	1.59
37	1.42	1.53	1.36	1.59	1.31	1.66	1.25	1.72	1.19	1.80	1.22	1.32	1.16	1.38	1.11	1.45	1.06	1.51	1.00	1.59
38	1.43	1.54	1.37	1.59	1.32	1.66	1.26	1.72	1.21	1.79	1.23	1.33	1.18	1.39	1.12	1.45	1.07	1.52	1.02	1.58
39	1.43	1.54	1.38	1.60	1.33	1.66	1.27	1.72	1.22	1.79	1.24	1.34	1.19	1.39	1.14	1.45	1.09	1.52	1.03	1.58
40	1.44	1.54	1.39	1.60	1.34	1.66	1.29	1.72	1.23	1.79	1.25	1.34	1.20	1.40	1.15	1.46	1.10	1.52	1.05	1.58
45	1.48	1.57	1.43	1.62	1.38	1.67	1.34	1.72	1.29	1.78	1.29	1.38	1.24	1.42	1.20	1.48	1.16	1.53	1.11	1.58
50	1.50	1.59	1.46	1.63	1.42	1.67	1.38	1.72	1.34	1.77	1.32	1.40	1.28	1.45	1.24	1.49	1.20	1.54	1.16	1.59
55	1.53	1.60	1.49	1.64	1.45	1.68	1.41	1.72	1.38	1.77	1.36	1.43	1.32	1.47	1.28	1.51	1.25	1.55	1.21	1.59
60	1.55	1.62	1.51	1.65	1.48	1.69	1.44	1.73	1.41	1.77	1.38	1.45	1.35	1.48	1.32	1.52	1.28	1.56	1.25	1.60
65	1.57	1.63	1.54	1.66	1.50	1.70	1.47	1.73	1.44	1.77	1.41	1.47	1.38	1.50	1.35	1.53	1.31	1.57	1.28	1.61
70	1.58	1.64	1.55	1.67	1.52	1.70	1.49	1.74	1.46	1.77	1.43	1.49	1.40	1.52	1.37	1.55	1.34	1.58	1.31	1.61
75	1.60	1.65	1.57	1.68	1.54	1.71	1.51	1.74	1.49	1.77	1.45	1.50	1.42	1.53	1.39	1.56	1.37	1.59	1.34	1.62
80	1.61	1.66	1.59	1.69	1.56	1.72	1.53	1.74	1.51	1.77	1.47	1.52	1.44	1.54	1.42	1.57	1.39	1.60	1.36	1.62
85	1.62	1.67	1.60	1.70	1.57	1.72	1.55	1.75	1.52	1.77	1.48	1.53	1.46	1.55	1.43	1.58	1.41	1.60	1.39	1.63
90	1.63	1.68	1.61	1.70	1.59	1.73	1.57	1.75	1.54	1.78	1.50	1.54	1.47	1.56	1.45	1.59	1.43	1.61	1.41	1.64
95	1.64	1.69	1.62	1.71	1.60	1.73	1.58	1.75	1.56	1.78	1.51	1.55	1.49	1.57	1.47	1.60	1.45	1.62	1.42	1.64
100	1.65	1.69	1.63	1.72	1.61	1.74	1.59	1.76	1.57	1.78	1.52	1.56	1.50	1.58	1.48	1.60	1.46	1.63	1.44	1.65

[a] n = number of observations; k = number of independent variables.

Source: Computed from TSP 4.5 based on R. W. Farebrother, "A Remark on Algorithms AS106, AS153, and AS155: The Distribution of a Linear Combination of Chi-Square Random Variables," *Journal of the Royal Statistical Society*, Series C (Applied Statistics), (1984), 29, pp. 323–333.

TABLE E.9
Control Chart Factors

Number of Observations in Sample/Subgroup (n)	d_2	d_3	D_3	D_4	A_2
2	1.128	0.853	0	3.267	1.880
3	1.693	0.888	0	2.575	1.023
4	2.059	0.880	0	2.282	0.729
5	2.326	0.864	0	2.114	0.577
6	2.534	0.848	0	2.004	0.483
7	2.704	0.833	0.076	1.924	0.419
8	2.847	0.820	0.136	1.864	0.373
9	2.970	0.808	0.184	1.816	0.337
10	3.078	0.797	0.223	1.777	0.308
11	3.173	0.787	0.256	1.744	0.285
12	3.258	0.778	0.283	1.717	0.266
13	3.336	0.770	0.307	1.693	0.249
14	3.407	0.763	0.328	1.672	0.235
15	3.472	0.756	0.347	1.653	0.223
16	3.532	0.750	0.363	1.637	0.212
17	3.588	0.744	0.378	1.622	0.203
18	3.640	0.739	0.391	1.609	0.194
19	3.689	0.733	0.404	1.596	0.187
20	3.735	0.729	0.415	1.585	0.180
21	3.778	0.724	0.425	1.575	0.173
22	3.819	0.720	0.435	1.565	0.167
23	3.858	0.716	0.443	1.557	0.162
24	3.895	0.712	0.452	1.548	0.157
25	3.931	0.708	0.459	1.541	0.153

Source: Reprinted from *ASTM-STP 15D* by kind permission of the American Society for Testing and Materials. Copyright ASTM International, 100 Barr Harbor Drive, Conshohocken, PA 19428.

TABLE E.10
The Standardized Normal Distribution

Entry represents area under the standardized normal
distribution from the mean to Z

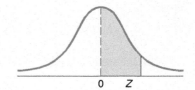

Z	.00	.01	.02	.03	.04	.05	.06	.07	.08	.09
0.0	.0000	.0040	.0080	.0120	.0160	.0199	.0239	.0279	.0319	.0359
0.1	.0398	.0438	.0478	.0517	.0557	.0596	.0636	.0675	.0714	.0753
0.2	.0793	.0832	.0871	.0910	.0948	.0987	.1026	.1064	.1103	.1141
0.3	.1179	.1217	.1255	.1293	.1331	.1368	.1406	.1443	.1480	.1517
0.4	.1554	.1591	.1628	.1664	.1700	.1736	.1772	.1808	.1844	.1879
0.5	.1915	.1950	.1985	.2019	.2054	.2088	.2123	.2157	.2190	.2224
0.6	.2257	.2291	.2324	.2357	.2389	.2422	.2454	.2486	.2518	.2549
0.7	.2580	.2612	.2642	.2673	.2704	.2734	.2764	.2794	.2823	.2852
0.8	.2881	.2910	.2939	.2967	.2995	.3023	.3051	.3078	.3106	.3133
0.9	.3159	.3186	.3212	.3238	.3264	.3289	.3315	.3340	.3365	.3389
1.0	.3413	.3438	.3461	.3485	.3508	.3531	.3554	.3577	.3599	.3621
1.1	.3643	.3665	.3686	.3708	.3729	.3749	.3770	.3790	.3810	.3830
1.2	.3849	.3869	.3888	.3907	.3925	.3944	.3962	.3980	.3997	.4015
1.3	.4032	.4049	.4066	.4082	.4099	.4115	.4131	.4147	.4162	.4177
1.4	.4192	.4207	.4222	.4236	.4251	.4265	.4279	.4292	.4306	.4319
1.5	.4332	.4345	.4357	.4370	.4382	.4394	.4406	.4418	.4429	.4441
1.6	.4452	.4463	.4474	.4484	.4495	.4505	.4515	.4525	.4535	.4545
1.7	.4554	.4564	.4573	.4582	.4591	.4599	.4608	.4616	.4625	.4633
1.8	.4641	.4649	.4656	.4664	.4671	.4678	.4686	.4693	.4699	.4706
1.9	.4713	.4719	.4726	.4732	.4738	.4744	.4750	.4756	.4761	.4767
2.0	.4772	.4778	.4783	.4788	.4793	.4798	.4803	.4808	.4812	.4817
2.1	.4821	.4826	.4830	.4834	.4838	.4842	.4846	.4850	.4854	.4857
2.2	.4861	.4864	.4868	.4871	.4875	.4878	.4881	.4884	.4887	.4890
2.3	.4893	.4896	.4898	.4901	.4904	.4906	.4909	.4911	.4913	.4916
2.4	.4918	.4920	.4922	.4925	.4927	.4929	.4931	.4932	.4934	.4936
2.5	.4938	.4940	.4941	.4943	.4945	.4946	.4948	.4949	.4951	.4952
2.6	.4953	.4955	.4956	.4957	.4959	.4960	.4961	.4962	.4963	.4964
2.7	.4965	.4966	.4967	.4968	.4969	.4970	.4971	.4972	.4973	.4974
2.8	.4974	.4975	.4976	.4977	.4977	.4978	.4979	.4979	.4980	.4981
2.9	.4981	.4982	.4982	.4983	.4984	.4984	.4985	.4985	.4986	.4986
3.0	.49865	.49869	.49874	.49878	.49882	.49886	.49889	.49893	.49897	.49900
3.1	.49903	.49906	.49910	.49913	.49916	.49918	.49921	.49924	.49926	.49929
3.2	.49931	.49934	.49936	.49938	.49940	.49942	.49944	.49946	.49948	.49950
3.3	.49952	.49953	.49955	.49957	.49958	.49960	.49961	.49962	.49964	.49965
3.4	.49966	.49968	.49969	.49970	.49971	.49972	.49973	.49974	.49975	.49976
3.5	.49977	.49978	.49978	.49979	.49980	.49981	.49981	.49982	.49983	.49983
3.6	.49984	.49985	.49985	.49986	.49986	.49987	.49987	.49988	.49988	.49989
3.7	.49989	.49990	.49990	.49990	.49991	.49991	.49992	.49992	.49992	.49992
3.8	.49993	.49993	.49993	.49994	.49994	.49994	.49994	.49995	.49995	.49995
3.9	.49995	.49995	.49996	.49996	.49996	.49996	.49996	.49996	.49997	.49997

Useful Knowledge

The useful knowledge in this Appendix simplifies using Excel or Tableau and further explains select features not otherwise explained by the software guides. Using Excel of Tableau with this book does not require mastery of the content of this appendix.

F.1 Keyboard Shortcuts

Editing Shortcuts

Ctrl+C copies a worksheet entry and **Ctrl+V** pastes that entry into the place that the editing cursor or worksheet cell highlight indicates. Pressing **Ctrl+X** cuts the currently selected entry for pasting somewhere else. In Excel, **Ctrl+C** and **Ctrl+V** (or **Ctrl+X** and **Ctrl+V**) can also be used to copy (or cut) and paste workbook objects such as charts.

Pressing **Ctrl+Z** undoes the last operation, and **Ctrl+Y** redoes the last operation. Pressing **Enter** or **Tab** finalizes an entry typed into a worksheet cell.

Excel Formatting & Utility Shortcuts

Pressing **Ctrl+B** toggles on (or off) boldface text style for the currently selected object. Pressing **Ctrl+I** toggles on (or off) italic text style for the currently selected object. Pressing **Ctrl+Shift+%** formats numeric values as a percentage with no decimal places.

Pressing **Ctrl+F** finds a **Find what** value, and pressing **Ctrl+H** replaces a **Find what** value with the **Replace with** value. Pressing **Ctrl+A** selects the entire current worksheet (useful as part of a worksheet copy or format operation). Pressing **Esc** cancels an action or a dialog box. Pressing **F1** displays the Microsoft Excel help system.

Tableau Utility Commands

Tableau contains a number of keyboard shortcut alternatives to menu selections and other actions. **Ctrl+D** (or **Command+D** on a Mac) displays the Connect (to data source) panel. **Ctrl+Tab** (or **Shift+Command+]**) cycles through the worksheets in a workbook. **Ctrl+1** (or **Command+1**) opens the Show Me gallery. **Ctrl+M** (or **Command+T**) opens a new worksheet. **F12** (or **Option+Command+E**) reverts a workbook to the last saved state.

F.2 Understanding the Nonstatistical Excel Functions

Various Excel Guide and PHStat worksheets use nonstatistical functions that either compute an intermediate result or perform a mathematical or programming operation.

CEILING(*cell, round-to value*) takes the numeric value in *cell* and rounds it to the next multiple of the *round-to value*. For example, if the *round-to value* is **0.5**, as it is in several column B formulas in the COMPUTE worksheet of the Quartiles workbook, then the numeric value will be rounded either to an integer or a number that contains a half such as 1.5.

COUNT(*cell range*) counts the number of cells in a cell range that contain a numeric value. This function is often used to compute the sample size, n, for example, in cell B9 of the COMPUTE worksheet of the Correlation workbook. When seen in the worksheets presented in this book, the *cell range* will typically be the cell range of variable column, such as **DATA!A:A**.

COUNTIF(*cell range for all values, value to be matched*) counts the number of occurrences of a value in a cell range. For example, the COMPUTE worksheet of the Wilcoxon workbook uses **COUNTIF(SortedRanks!A2:A21, "Special Front")** in cell B7 to compute the sample size of the Population 1 Sample by counting the number of occurrences of the sample name Special Front in column A of the SortedRanks worksheet.

DEVSQ(*variable cell range*) computes the sum of the squares of the differences between a variable value and the mean of that variable.

FLOOR(*cell, 1*) takes the numeric value in *cell* and rounds it down to the nearest integer.

IF(*logical comparison, what to display if comparison holds, what to display if comparison is false*) uses the *logical comparison* to make a choice between two alternatives. In the worksheets shown in this book, the IF function typically chooses from two text values, such as **Reject the null hypothesis** and **Do not reject the null hypothesis**, to display.

MMULT(*cell range 1, cell range 2*) treats both *cell range 1* and *cell range 2* as matrices and computes the matrix product of the two matrices. When each of the two cell ranges is either a single row or a single column, MMULT can be used as part of a regular formula. If the cell ranges each represent rows and columns, then MMULT must be used as part of an array formula (see Appendix Section B.2).

ROUND(*cell*, **0**) takes the numeric value in *cell* and rounds to the nearest whole number.

SMALL(*cell range, k*) selects the *k*th smallest value in *cell range*.

SQRT(*value*) computes the square root of *value*, where *value* is either a cell reference or an arithmetic expression.

SUMIF(*cell range for all values, value to be matched, cell range in which to select cells for summing*) sums only those rows in *cell range in which to select cells for summing* in which the value in *cell range for all values* matches the *value to be matched*. SUMIF provides a convenient way to compute the sum of ranks for a sample in a worksheet that contains stacked data.

SUMPRODUCT(*cell range 1, cell range 2*) multiplies each cell in *cell range 1* by the corresponding cell in *cell range 2* and then sums those products. If *cell range 1* contains a column of differences between an X value and the mean of the variable X, and *cell range 2* contains a column of differences between a Y value and the mean of the variable Y, then this function would compute the value of the numerator in Equation (3.16) that defines the sample covariance.

TRANSPOSE(*horizontal or vertical cell range*) takes the *cell range*, which must be either a horizontal cell range (cells all in the same row) or a vertical cell range (cells all in the same column) and transposes, or rearranges, the cell in the other orientation such that a horizontal cell range becomes a vertical cell range and vice versa. When used inside another function, Excel considers the results of this function to be an *array*, not a cell range.

VLOOKUP(*lookup value cell, table of lookup values, table column to use*) function displays a value that has been looked up in a *table of lookup values*, a rectangular cell range. In the ADVANCED worksheet of the Recoded workbook, the function uses the values in the second column of *table of lookup values* (an example of which is shown below) to look up the Honors values based on the GPA of a student (the *lookup value cell*). Numbers in the first column of *table of lookup values* are implied ranges such that No Honors is the value displayed if the GPA is at least 0, but less than 3; Honor Roll is the value displayed if the GPA is at least 3, but less than 3.3; and so on:

0	No Honors
3	Honor Roll
3.3	Dean's List
3.7	President's List

Software FAQs

G.1 Microsoft Excel FAQs

Which Microsoft Excel version should be used with this book?

Use the most current Excel version. Using the most current version will provide you with the best learning experience and give you the most up-to-date practical knowledge about Excel.

What is Office 365? Excel 365?

Office 365 is the subscription-based service that offers the latest version of Microsoft Office programs for download and installation. Office 365 provides access to Excel 365, which is always the most current version of Microsoft Excel available.

What does "Compatibility Mode" in the title bar mean?

Excel displays "Compatibility Mode" when you open and use a workbook that was stored using the older **.xls** Excel workbook file format or using the **.xlsx** format in a copy of Excel that is not fully up-to-date. Compatibility Mode does not affect Excel functionality but will cause Excel to review your workbook for exclusive-to-xlsx formatting properties.

To convert a **.xls** workbook to the **.xlsx** format, use **Save As** to save (re-save) the workbook in **.xlsx** format. One quirk in Microsoft Excel is that when you convert a workbook by using **Save As**, the newly converted .xlsx workbook stays temporarily in Compatibility Mode. To avoid possible complications and errors, close the newly converted workbook and then reopen it.

Using Compatibility Mode can cause minor differences in the objects such as charts and PivotTables that Excel creates and can cause problems when you seek to transfer data from other workbooks.

What is the Microsoft Office Store?

The Office Store is an Internet-based service that distributes enhancements to Microsoft Office programs such as Excel. In *some* Excel versions, the store can be used to add functionality to Excel.

In compatible Excel versions, the Insert tab contains links to the Office Store as well as the added functionality that was installed previously. Using the Office Store requires a Microsoft account and not every item in the Store is complimentary.

In the Insert tab, what are Recommended PivotTables and Recommended Charts? Should I use these features?

These features display "recommended" PivotTables or charts as shortcuts. Unfortunately, the recommended PivotTables can include statistical errors such as treating the categories of a categorical variable as zero values of a numerical variable and the recommended charts often do not conform to best practices.

Can I use a mobile version of Microsoft Excel such as the Microsoft Excel app for Android with this book?

You can use mobile versions of Excel to open and review any of the data workbooks and Excel Guide workbooks that this book uses.

G.2 PHStat FAQs

Where can I learn more about PHStat?

Appendix H presents a full description of PHStat, the Pearson statistics add-in that provides a software assist for creating Excel solutions to statistical problems. Visit the PHStat home page, www.pearsonhighered.com/phstat, for news updates about PHStat that may have occurred after the publication of this book.

Which versions of Excel are compatible with PHStat?

PHStat is compatible with all Excel versions that Microsoft supported at the time of publication of this book. Those versions include Excel 365; Microsoft Windows Excels 2013, 2016, and 2019; and Excel for Mac 2016 and 2019.

How do I download a copy of PHStat?

You use an access code to download PHStat through the PHStat home page, as fully explained in Appendix H. Before downloading PHStat, download the PHStat readme file that is available (without an access code) from the PHStat home page as well as from the student download page and the MyLab Statistics "Tools for Success" page for this book.

How do I get PHStat ready for use?

Section D.1 explains how to get PHStat ready for use. You should also review the PHStat readme file (available for download as discussed in Appendix C) for any late-breaking news or changes that might affect this process.

When I use a particular PHStat procedure, I get an error message that includes the words "unexpected error." What should I do?

"Unexpected error" messages are typically caused by improperly prepared data. Review your data to ensure that you have organized your data according to the conventions PHStat expects, as explained in the PHStat help system.

Where can I get further news and information about PHStat? Where can I get further assistance about using PHStat?

Several online sources can provide you with news and information or provide you with assistance that supplements the readme file and help system included with PHStat.

www.pearsonhighered.com/phstat is the official PHStat home page. The page will contain late-breaking news about PHStat as well as provide a link to Pearson Support website, **support.pearson.com/getsupport/**.

softwareforbusstat.org is a new website that discusses using software, including PHStat and Excel, in business statistics courses. On the home page of that website, you will find links to the latest news and developments about PHStat.

How can I get updates to PHStat when they become available?

PHStat is subject to continuous improvement. When enhancements are made, a new PHStat zip archive is posted for download through the official web page for PHStat. If you hold a valid access code, you can download that archive and overwrite your older version. To discover which version of PHStat you have, select **About PHStat** from the PHStat menu.

G.3 Tableau FAQs

Which version of Tableau should be used with this book?

Use Tableau Public, also known as Tableau Desktop Public Edition. Tableau Public version 2019 was the source for all instructions and screen illustrations presented. Occasionally in the past, Tableau has altered the contents or the appearances of some dialog boxes, so future Tableau Public versions may show slight variations in low-level dialog boxes.

Can I use other Tableau versions with this book?

Tableau Server or the paid-subscription version (not Public Edition) of Tableau Desktop can also be used with the instructions in this book. When using those editions, some menu selections to connect to data or save and retrieve Tableau workbooks will be slightly different even as the worksheet visualization creation will be identical or very similar.

As of mid-2018, Tableau changed its licensing of products and reorganized its product offerings for corporate users.

In a corporate environment, a Tableau Server or Desktop user must hold a "Creator" license in order to fully use the Tableau Guide instructions in this book.

Tableau also offers Tableau Reader as a complimentary download. This version *cannot* be used with the Tableau Guide instructions.

Why do Tableau Guide instructions ask readers to open *Excel* workbooks?

Tableau Public contains a number of restrictions including that *Tableau* workbooks being opened must have been created by the Tableau Public user or shared by others in a public space provided by Tableau on its website. By writing instructions that begin with Excel workbooks, readers are spared the complication of linking the shared spaces, and the instructions can serve readers of both the Tableau Desktop Public Edition and the paid-subscription version of Tableau Desktop.

Why are Tableau Guide instructions provided for only selected chapters?

As an advanced data visualization tool, Tableau has its own strengths and weaknesses. Although programming capabilities built into Tableau Desktop (all versions) enable one to produce all types of statistical results, information on Tableau's own website note that some results are more simply produced using other programs such as Microsoft Excel.

Readers interested in learning more about advanced Tableau functionality can review the Section TG2.7 "Sparklines" instructions that contain ten steps to create a set of sparklines and include nine additional steps that illustrate the use of the Calculation Editor to create a calculated field. However, note that the Excel instructions for creating sparklines (see Section EG2.7) contain only five steps, suggesting that using Excel might be the better choice. (And Excel instructions to color high and low values in each sparkline would be as simple as checking two check boxes and selecting two colors, if a set of sparklines was being created from scratch, without the benefit of a template.)

Generally, readers of this book should see Tableau as a compliment to the other software, especially Microsoft Excel. Understanding complimentary uses of a second program reflects a growing business practice in which multiple programs are used together to visualize and communicate the results of data analyses.

All About PHStat

H.1 What is PHStat?

PHStat helps create Excel worksheet solutions to statistical problems. Users supply the necessary data in dialog boxes and PHStat executes the low-level menu selection and data entry tasks needed to create a solution. By freeing learners from distractions such as typographical errors, PHStat allows learners to focus on statistical results and avoid getting frustrated or lost in the operational details of a program with which they may not be totally familiar.

PHStat uses Excel building blocks to create worksheet solutions. These worksheet solutions illustrate Excel techniques such as proper worksheet design and proper organization and application of formulas and functions. Users can examine solutions and gain new Excel skills and insights into creating worksheet solutions. Additionally, many solutions serve as what-if templates in which the effects of changing data on the results can be explored. Such PHStat templates are fully reusable and *transportable*, meaning that they can be reused on any academic, home, or business computer on which Excel has been installed.

With its focus on assisting the learning of statistics, PHStat is not intended as a replacement for commercial statistical applications. To support learning, PHStat typically implements manual methods of calculation that follow textbook mathematical definitions of statistical procedures. Such methods allow PHStat users to match intermediate results to textbook concepts, reinforcing learning. However, those methods of calculation may be ill-suited for real-world data sets that have unusual numerical properties or that contain a large number (many thousands) of rows of data. Exercise caution when using PHStat with data sets not supplied with a textbook or by a statistics instructor.

How PHStat Works

PHStat executes the low-level menu selection and data entry tasks needed to use Excel for statistical analyses. For most analyses, PHStat:

- retrieves a model template that is similar or identical to an Excel Guide workbook model template and solution.
- fills in the template with user-supplied data or cell ranges.
- makes minor adjustments to worksheet formulas, as necessary.
- adds the template and any supporting worksheets or chart sheets to the currently open workbook.

In the worksheets that PHStat creates, user-changeable worksheet cells are tinted light turquoise and the cells that contain the results are tinted in light yellow. For most analyses, the yellow-tinted results cells as well as the cells that display intermediate results are minimally formatted to reveal the true value that Microsoft Excel has computed. For some procedures, these values will have a large number of (seemingly) significant digits that can be reformatted for presentation purposes.

Preparing Data for PHStat Analysis

Prepare data for analysis by PHStat by placing the data in columns in a new worksheet, beginning with column A and row 1. Use row 1 to enter column labels. Due to the technical limitations of Excel, avoid using numbers as a row 1 labels. If you must enter a number, enter the number preceded by an apostrophe, for example, enter '2020. Make sure that all cells that display numbers contain *numeric values* and not formulas that display a number. If necessary, use Paste Special (see Appendix Section B.2) to convert any cell contents to numeric values.

For procedures that require two or more cell ranges, such as the regression procedures, make sure that all cell ranges are from the worksheet that PHStat will use. If Excel displays "Compatibility Mode" in the title bar or if the workbook that contains the data to be used has been saved in the older **.xls** format, save the opened workbook using the newer **.xlsx** format, close the workbook, and then reopen the workbook before using PHStat.

H.2 Obtaining and Setting Up PHStat

For computer systems in which PHStat is not already available, obtaining and setting up PHStat requires steps that Exhibit H.1 summarizes. Obtaining PHStat requires Internet access.

EXHIBIT   **H.1**

Steps to Obtain and Set Up PHStat

Step 1 Verify that the computer system on which PHStat is to be set up contains a current version of Microsoft Excel. Visit the PHStat home page (**www.pearsonhighered .com/phstat**) and download and review the PHStat readme file to learn which Excel versions are considered current at the time of the visit to that web page.

Step 2 Obtain a valid PHStat access code. This access code may have been bundled with this book, previously purchased separately, or obtained online through the PHStat home page.

Step 3 Create or log into a Pearson Education account. Users who have previously used a Pearson MyLab product such as MyLab Statistics already have a Pearson Education account. Users who need to create an account can do so through the PHStat home page.

Step 4 Associate the PHStat access code with the Pearson Education account. This step can also be done through the PHStat home page.

Step 5 Download PHStat. As Appendix Section C.3 describes, PHStat comes packaged as a zip archive file.

Step 6 Unzip the PHStat zip archive and place the files in the archive together in any existing or new folder that is not on the Microsoft Windows or OS X Desktop.

Because the exact details of Steps 1 through 4 may change during the lifetime of this edition, visit the PHStat home page (**www.pearsonhighered.com/phstat**) for information about any such changes that occurred since this book was published. Also, download and review the PHStat readme file, mentioned in Step 1, for any late-breaking changes to PHStat, including new functionality. (The PHStat readme file is also available on the Student Download Page and the Tools for Success page that Appendix Section C.1 discusses.)

During the time for which a PHStat access code is valid, periodically visit the PHStat home page to see if a newer version of PHStat is available for download. Newer versions are posted as necessary to respond to changes that Microsoft makes to Excel or to add new functionality to PHStat.

Pearson Education accounts are complimentary and a person needs only one account, which holds information about all of Pearson learning products that the person may have licensed for use. For example, readers that use MyLab Statistics and PHStat with this book should use the same Pearson Education account to register access to both.

H.3 Using PHStat

PHStat takes the form of an Excel add-in workbook. To use PHStat, first open Excel. Then use the Excel (File) Open dialog box to open the PHStat workbook (**PHStat.xlam**). As PHStat begins to load, Excel displays a security notice dialog box (see Appendix Section D.1). Click **Enable Macros** in the dialog box to permit PHStat to be opened.

When properly loaded, PHStat adds its set of menu choices to the Excel user interface. How these choices appear, depends on the Excel version being used:

- In Microsoft Windows Excels, PHStat adds a PHStat tab to the Office Ribbon and also adds a PHStat pull-down menu to the Add-Ins tab (for compatibility to earlier versions of PHStat that did use an Office Ribbon tab).
- In Excel for Mac 2016, PHStat adds a PHStat tab to the Office Ribbon. (This tab is similar to the Windows Ribbon tab).

Microsoft Windows Excel users can use either the PHStat or the Add-in tab, which have identical functionality. (When following an instruction to select **PHStat**, Microsoft Windows Excel users can click either the PHStat tab or the PHStat pull-down menu in the Add-Ins tab.)

To perform an analysis, first open to the worksheet that contains the data for analysis. Then select **PHStat** and make a selection. The top-level selections include ten categories that lead to a submenu of specific statistical methods. Once a specific method has been chosen, PHStat either inserts a worksheet template for the user to fill in or, more commonly, displays a dialog box in which a user makes entries and selections. Click **OK** in a dialog box to instruct PHStat to complete the analysis. Worksheets and chart sheets that PHStat generates are inserted into the currently opened workbook, the workbook that contains the data for the analysis.

H.4 PHStat Procedures, by Category

PHStat includes over 60 statistical and utility procedures, grouped into 13 categories. By category, these procedures are:

Data preparation: stack and unstack data
Descriptive Statistics: boxplot, descriptive summary, dot scale diagram, frequency distribution, histogram and polygons, Pareto diagram, scatter plot, sparklines, stem-and-leaf display, treemap, one-way tables and charts, and two-way tables and charts
Probability and probability distributions: simple and joint probabilities, normal probability plot, and binomial, exponential, hypergeometric, and Poisson probability distributions
Sampling: sampling distributions simulation
Confidence interval estimation: for the mean, sigma unknown; for the mean, sigma known, for the population variance, for the proportion, and for the total difference
Sample size determination: for the mean and the proportion
One-sample tests: Z test for the mean, sigma known; t test for the mean, sigma unknown; chi-square test for the variance; and Z test for the proportion
Two-sample tests (unsummarized data): pooled-variance t test, separate-variance t test, paired t test, F test for differences in two variances, and Wilcoxon rank sum test
Two-sample tests (summarized data): pooled-variance t test, separate-variance t test, paired t test, Z test for the differences in two means, F test for differences in two variances, chi-square test for differences in two proportions, Z test for the difference in two proportions, and McNemar test
Multiple-sample tests: chi-square test, Marascuilo procedure, Kruskal-Wallis rank test, Levene test, one-way ANOVA, Tukey-Kramer procedure, randomized block design, and two-way ANOVA with replication
Regression: simple linear regression, multiple regression, best subsets, stepwise regression, and logistic regression
Control charts: p chart, c chart, and R and Xbar charts
Decision-making: covariance and portfolio management, expected monetary value, expected opportunity loss, and opportunity loss

Self-Test Solutions and Answers to Selected Even-Numbered Problems

The following sections present worked-out solutions to Self-Test Problems and brief answers to most of the even-numbered problems in the text. For more detailed solutions, including explanations, interpretations, and Excel results, see the *Student Solutions Manual*.

CHAPTER 1

1.2 Small, medium, and large sizes imply order but do not specify how much the size of the business increases at each level.

1.4 (a) The number of cellphones is a numerical variable that is discrete because the outcome is a count. It is ratio scaled because it has a true zero point. **(b)** Monthly data usage is a numerical variable that is continuous because any value within a range of values can occur. It is ratio scaled because it has a true zero point. **(c)** Number of text messages exchanged per month is a numerical variable that is discrete because the outcome is a count. It is ratio scaled because it has a true zero point. **(d)** Voice usage per month is a numerical variable that is continuous because any value within a range of values can occur. It is ratio scaled because it has a true zero point. **(e)** Whether a cellphone is used for streaming video is a categorical variable because the answer can be only yes or no. This also makes it a nominal-scaled variable.

1.6 (a) Categorical, nominal scale **(b)** Numerical, continuous, ratio scale **(c)** Categorical, nominal scale **(d)** Numerical, discrete, ratio scale **(e)** Categorical, nominal scale.

1.8 Type of data: **(a)** Numerical, continuous **(b)** Numerical, discrete **(c)** Numerical, continuous **(d)** Categorical scale. Measurement scale: **(a)** ratio scale **(b)** ratio scale **(c)** ratio scale **(d)** nominal scale.

1.10 The underlying variable, ability of the students, may be continuous, but the measuring device, the test, does not have enough precision to distinguish between the two students.

1.12 (a) Data distributed by an organization or individual **(b)** sample.

1.18 Sample without replacement: Read from left to right in three-digit sequences and continue unfinished sequences from the end of the row to the beginning of the next row:

Row 05: 338 505 855 551 438 855 077 186 579 488 767 833 170
Rows 05–06: 897
Row 06: 340 033 648 847 204 334 639 193 639 411 095 924
Rows 06–07: 707
Row 07: 054 329 776 100 871 007 255 980 646 886 823 920 461
Row 08: 893 829 380 900 796 959 453 410 181 277 660 908 887
Rows 08–09: 237
Row 09: 818 721 426 714 050 785 223 801 670 353 362 449
Rows 09–10: 406
Note: All sequences above 902 and duplicates are discarded.

1.20 A simple random sample would be less practical for personal interviews because of travel costs (unless interviewees are paid to go to a central interviewing location).

1.22 Here all members of the population are equally likely to be selected, and the sample selection mechanism is based on chance. But selection of two elements is not independent; for example, if *A* is in the sample, we know that *B* is also and that *C* and *D* are not.

1.24 (a)
Row 16: 2323 6737 5131 8888 1718 0654 6832 4647 6510 4877
Row 17: 4579 4269 2615 1308 2455 7830 5550 5852 5514 7182
Row 18: 0989 3205 0514 2256 8514 4642 7567 8896 2977 8822
Row 19: 5438 2745 9891 4991 4523 6847 9276 8646 1628 3554
Row 20: 9475 0899 2337 0892 0048 8033 6945 9826 9403 6858
Row 21: 7029 7341 3553 1403 3340 4205 0823 4144 1048 2949
Row 22: 8515 7479 5432 9792 6575 5760 0408 8112 2507 3742
Row 23: 1110 0023 4012 8607 4697 9664 4894 3928 7072 5815
Row 24: 3687 1507 7530 5925 7143 1738 1688 5625 8533 5041
Row 25: 2391 3483 5763 3081 6090 5169 0546
Note: All sequences above 5,000 are discarded. There were no repeating sequences.

(b)

0089	0189	0289	0389	0489	0589	0689	0789	0889	0989
1089	1189	1289	1389	1489	1589	1689	1789	1889	1989
2089	2189	2289	2389	2489	2589	2689	2789	2889	2989
3089	3189	3289	3389	3489	3589	3689	3789	3889	3989
4089	4189	4289	4389	4489	4589	4689	4789	4889	4989

(c) With the single exception of invoice 0989, the invoices selected in the simple random sample are not the same as those selected in the systematic sample. It would be highly unlikely that a simple random sample would select the same units as a systematic sample.

1.26 (a) For the third value, Apple is spelled incorrectly. The twelfth value should be Blackberry not Blueberry. The fifteenth value, APPLE, may lead to an irregularity. The eighteenth value should be Samsung not Samsun. **(b)** The eighth value is a missing value.

1.28 (a) The times for each of the hotels would be arranged in separate columns. **(b)** The hotel names would be in one column and the times would be in a second column.

1.30 Before accepting the results of a survey of college students, you might want to know, for example: Who funded the survey? Why was it conducted? What was the population from which the sample was selected? What sampling design was used? What mode of response was used: a personal interview, a telephone interview, or a mail survey? Were interviewers trained? Were survey questions field-tested? What questions were asked? Were the questions clear, accurate, unbiased, and valid? What operational definition of immediately and effortlessly was used? What was the response rate?

1.32 The results are based on a survey of bank executives. If the frame is supposed to be banking institutions, how is the population defined? There is no information about the response rate, so there is an undefined nonresponse error.

1.34 Before accepting the results of the survey, you might want to know, for example: Who funded the study? Why was it conducted? What was the population from which the sample was selected? What sampling design was used? What mode of response was used: a personal interview, a telephone interview, or a mail survey? Were interviewers trained? Were survey questions field-tested? What other questions were asked? Were the questions clear, accurate, unbiased, and valid? What was the response rate? What was the margin of error? What was the sample size? What frame was used?

1.52 (a) All benefitted employees at the university. **(b)** The 3,095 employees who responded to the survey. **(c)** Gender, marital status, and employment are categorical. Age (years), education level (years completed), and household income ($) are numerical.

CHAPTER 2

2.2 (a) Table of frequencies for all student responses:

| | Student Major Categories | | | |
Gender	A	C	M	Totals
Male	14	9	2	25
Female	6	6	3	15
Totals	20	15	5	40

(b) Table based on total percentages:

| | Student Major Categories | | | |
Gender	A	C	M	Totals
Male	35.0%	22.5%	5.0%	62.5%
Female	15.0%	15.0%	7.5%	37.5%
Totals	50.0%	37.5%	12.5%	100.0%

Table based on row percentages:

| | Student Major Categories | | | |
Gender	A	C	M	Totals
Male	56.0%	36.0%	8.0%	100.0%
Female	40.0%	40.0%	20.0%	100.0%
Totals	50.0%	37.5%	12.5%	100.0%

Table based on column percentages:

| | Student Major Categories | | | |
Gender	A	C	M	Totals
Male	70.0%	60.0%	40.0%	62.5%
Female	30.0%	40.0%	60.0%	37.5%
Totals	100.0%	100.0%	100.0%	100.0%

2.4 (a) The percentage of complaints for each category:

Category	Total	Percentage
Bank Account or Service	202	9.330%
Consumer Loan	132	6.097%
Credit Card	175	8.083%
Credit Reporting	581	26.836%
Debt Collection	486	22.448%
Mortgage	442	20.416%
Other	72	3.326%
Student Loan	75	3.464%
Grand Total	2,165	

(b) There are more complaints for credit reporting, debt collection, and mortgage than the other categories. These categories account for about 70% of all the complaints.

(c) The percentage of complaints for each company:

Company	Total	Percentage
Bank of America	42	3.64%
Capital One	93	8.07%
Citibank	59	5.12%
Ditech Financial	31	2.69%
Equifax	217	18.82%
Experian	177	15.35%
JPMorgan	128	11.10%
Nationstar Mortgage	39	3.38%
Navient	38	3.30%
Ocwen	41	3.56%
Synchrony	43	3.73%
Trans-Union	168	14.57%
Wells Fargo	77	6.68%
Grand Total	1,153	

(d) Equifax, Trans-Union, and Experion, all of which are credit score companies, have the most complaints.

2.6 The largest sources of power generation in the United States in July, 2018, were coal, natural gas, and nuclear, followed by much smaller contributions from hydroelectric and renewable sources other than hydroelectric and solar. Solar, petroleum coke and liquids, other gas, and all other sources ("other") contributed very little.

2.8 (a) Table of row percentages:

| | Gender | | |
Overloaded	Male	Female	Total
Yes	44.08%	55.92%	100%
No	53.54%	46.46%	100%
Total	51.64%	48.36%	100%

Table of column percentages:

| | Gender | | |
Overloaded	Male	Female	Total
Yes	17.07%	23.13%	20.00%
No	82.93%	76.87%	80.00%
Total	100.00%	100.00%	100.00%

Table of total percentages:

| | Gender | | |
Overloaded	Male	Female	Total
Yes	8.82%	11.18%	20.00%
No	42.82%	37.18%	80.00%
Total	51.64%	48.36%	100.00%

(b) A higher percentage of females feel information overload.

2.10 There is a small difference in the percentage between males and females who would choose window tinting as their most preferred luxury upgrade.

2.12 73 78 78 78 85 88 91.

2.14 (a) $60, 000 – under $100,000, $100, 000 – under $140,000, $140, 000 – under $180,000, $180, 000 – under $220,000, $220, 000 – under $260,000, $260, 000 – under $300,000 **(b)** $40,000 **(c)** $80,000, $120,000, $160,000, $200,000, $240,000, $280,000

2.16 (a)

Electricity Costs	Frequency	Percentage
$80 but less than $100	4	8%
$100 but less than $120	7	14%
$120 but less than $140	9	18%
$140 but less than $160	13	26%
$160 but less than $180	9	18%
$180 but less than $200	5	10%
$200 but less than $220	3	6%

(b)

Electricity Costs	Frequency	Percentage	Cumulative %
$99	4	8.00%	8.00%
$119	7	14.00%	22.00%
$139	9	18.00%	40.00%
$159	13	26.00%	66.00%
$179	9	18.00%	84.00%
$199	5	10.00%	94.00%
$219	3	6.00%	100.00%

(c) The majority of utility charges are clustered between $120 and $180.

2.18 (a), (b)

Credit Score	Frequency	Percent (%)	Cumulative Percent (%)
560 – under 580	4	0.16	0.16
580 – under 600	24	0.93	1.09
600 – under 620	68	2.65	3.74
620 – under 640	290	11.28	15.02
640 – under 660	548	21.32	36.34
660 – under 680	560	21.79	58.13
680 – under 700	507	19.73	77.86
700 – under 720	378	14.71	92.57
720 – under 740	168	6.54	99.11
740 – under 760	22	0.86	99.96
760 – under 780	1	0.04	100.00

(c) The average credit scores are concentrated between 620 and 720.

2.20 (a)

Time in Seconds	Frequency	Percentage
5 – under 10	8	16%
10 – under 15	8	30%
15 – under 20	8	36%
20 – under 25	8	12%
25 – under 30	8	6%

(b)

Time in Seconds	Percentage Less Than
5	0
10	16
15	46
20	82
25	94
30	100

(c) The target is being met since 82% of the calls are being answered in less than 20 seconds.

2.22 (a)

Bulb Life (hours)	Percentage, Mftr A	Percentage, Mftr B
46,500 but less than 47,500	7.5%	0.0%
47,500 but less than 48,500	12.5%	5.0%
48,500 but less than 49,500	50.0%	20.0%
49,500 but less than 50,500	22.5%	40.0%
50,500 but less than 51,500	7.5%	22.5%
51,500 but less than 52,500	0.0%	12.5%

(b)

% Less Than	Percentage Less Than, Mftr A	Percentage Less Than, Mftr B
47,500	7.5%	0.0%
48,500	20.0%	5.0%
49,500	70.0%	25.0%
50,500	92.5%	65.0%
511,500	100.0%	87.5%
52,500	100.0%	100.0%

(c) Manufacturer B produces bulbs with longer lives than Manufacturer A. The cumulative percentage for Manufacturer B shows that 65% of its bulbs lasted less than 50,500 hours, contrasted with 92.5% of Manufacturer A's bulbs. None of Manufacturer A's bulbs lasted at least 51,500 hours, but 12.5% of Manufacturer B's bulbs lasted at least 51,500 hours. At the same time, 7.5% of Manufacturer A's bulbs lasted less than 47,500 hours, whereas none of Manufacturer B's bulbs lasted less than 47,500 hours.

2.24 (b) The Pareto chart is best for portraying these data because it not only sorts the frequencies in descending order but also provides the cumulative line on the same chart. **(c)** You can conclude that searching and buying online was the highest category and the other three were equally likely.

2.26 (b) 81.98%. **(d)** The Pareto chart allows you to see which sources account for most of the electricity.

2.28 (b) Because energy use is spread over many types of appliances, a bar chart may be best in showing which types of appliances used the most energy. **(c)** Space cooling, space heating, and water heating accounted for 40% of the residential energy use in the United States.

2.30 (b) Females are more likely to be overloaded with information

2.32 (b) There is a small difference in the percentage between males and females who would choose window tinting as their most preferred luxury upgrade.

2.34 50 74 74 76 81 89 92.

2.36 (a)

Stem Unit	Household Hours
4	0
5	0 0 5 6 8 8 9
6	0 1 3 3 4 5 7
7	3 3 8 8 8 9 9
8	0 4 8
9	
10	2 5
11	
12	8
13	3
14	3

(b) The results are concentrated between 5 and 8 hours.

2.38 (c) The majority of utility charges are clustered between $120 and $180.

2.40 Property taxes on a $176K home seem concentrated between $700 and $2,200 and also between $3,200 and $3,700.

2.42 The average credit scores are concentrated between 620 and 720.

2.44 The target is being met since 82% of the calls are being answered in less than 20 seconds.

2.46 (c) Manufacturer B produces bulbs with longer lives than Manufacturer A.

2.48 (b) Yes, there is a strong positive relationship between X and Y. As X increases, so does Y.

2.50 (c) There appears to be a linear relationship between the first weekend gross and either the U.S. gross or the worldwide gross of Harry Potter movies. However, this relationship is greatly affected by the results of the last movie, *Deathly Hallows, Part II*.

2.52 (a), (c) There appears to be a positive relationship between the download speed and the upload speed. Yes, this is borne out by the data especially at high speeds.

2.54 (b) There is a great deal of variation in the returns from decade to decade. Most of the returns are between 5% and 15%. The 1950s, 1980s, and 1990s had exceptionally high returns, and only the 1930s had a negative return.

2.56 (b) There was a decline in movie attendance between 2001 and 2018. During that time, movie attendance increased from 2002 to 2004 but then decreased to a level below that in 2001.

2.58 Multidimensional contingency table showing PivotTable percentages.

Count of Type	Star Rating					
Type	One	Two	Three	Four	Five	Grand Total
Growth	**5.43%**	**17.12%**	**27.35%**	**11.27%**	**2.71%**	**63.88%**
Large	3.76%	7.72%	13.57%	5.43%	1.67%	32.15%
Mid-Cap	1.25%	5.43%	7.52%	3.13%	0.63%	17.96%
Small	0.42%	3.97%	6.26%	2.71%	0.42%	13.78%
Value	**2.92%**	**10.65%**	**13.99%**	**7.31%**	**1.25%**	**36.12%**
Large	2.09%	6.68%	9.19%	3.97%	1.25%	23.18%
Mid-Cap	0.63%	2.09%	2.71%	1.04%	0.00%	6.47%
Small	0.21%	1.88%	2.09%	2.30%	0.00%	6.48%
Grand Total	**8.35%**	**27.77%**	**41.34%**	**18.58%**	**3.97%**	**100.00%**

(b) The growth and value funds have similar patterns in terms of star rating and type. Both growth and value funds have more funds with a rating of three. Very few funds have ratings of five.
(c) Multidimensional contingency table showing PivotTable average three-year rates of return.

Count of Type	Star Rating					
Type	One	Two	Three	Four	Five	Grand Total
Growth	**5.41**	**7.04**	**8.94**	**10.14**	**12.83**	**8.51**
Large	6.97	9.43	10.62	11.83	14.25	10.30
Mid-Cap	2.27	5.07	7.93	8.77	11.22	6.93
Small	0.78	5.09	6.52	8.35	9.53	6.39
Value	**4.43**	**5.49**	**7.29**	**8.34**	**10.23**	**6.84**
Large	5.23	6.05	7.58	8.85	10.23	7.29
Mid-Cap	2.79	5.77	7.32	9.26	-	6.69
Small	1.33	3.20	5.93	7.04	-	5.39
Grand Total	**5.07**	**6.45**	**8.38**	**9.43**	**12.01**	**7.91**

(d) There are 65 large cap growth funds with a rating of three. Their average three-year return percentage is 10.62.

2.60 Multidimensional contingency table showing PivotTable tallies as percentages.

Count of Type	Star Rating					
Type	One	Two	Three	Four	Five	Grand Total
Growth	**5.43%**	**17.12%**	**27.35%**	**11.27%**	**2.71%**	**63.88%**
Low	1.25%	2.09%	4.80%	3.55%	1.46%	13.15%
Average	1.67%	7.72%	15.87%	6.05%	0.42%	31.73%
High	2.51%	7.31%	6.68%	1.67%	0.84%	19.00%
Value	**2.92%**	**10.65%**	**13.99%**	**7.31%**	**1.25%**	**36.12%**
Low	0.84%	4.38%	7.10%	4.38%	0.84%	17.54%
Average	1.25%	4.80%	5.85%	2.71%	0.42%	15.03%
High	0.84%	1.46%	1.04%	0.21%	0.00%	3.55%
Grand Total	**8.35%**	**27.77%**	**41.34%**	**18.58%**	**3.96%**	**100.00%**

(b) Patterns of star rating conditioned on risk:
For the growth funds as a group, most are rated as three-star, followed by two-star, four-star, one-star, and five-star. The pattern of star rating is different among the various risk growth funds.
For the value funds as a group, most are rated as three-star, followed by two-star, four-star, one-star and five-star. Among the high-risk value funds, more are two-star than three-star.
Most of the growth funds are rated as average-risk, followed by high-risk and then low-risk. The pattern is not the same among all the rating categories.
Most of the value funds are rated as low-risk, followed by average-risk and then high-risk. The pattern is the same among the three-star, four-star, and five-star value funds. Among the one-star and two-star funds, there are more average risk funds than low risk funds.
(c)

Average of 3YrReturn%	Star Rating					
Type	One	Two	Three	Four	Five	Grand Total
Growth	**5.41**	**7.04**	**8.94**	**10.14**	**12.83**	**8.51**
Low	7.53	8.60	9.89	10.29	12.64	9.87
Average	6.17	7.99	9.28	10.43	11.96	9.06
High	3.83	5.59	7.45	8.76	13.59	6.64
Value	**4.43**	**5.49**	**7.29**	**8.34**	**10.23**	**6.84**
Low	5.29	7.00	7.66	8.57	10.74	7.76
Average	5.01	4.98	6.97	7.96	9.23	6.41
High	2.71	2.63	6.53	8.39		4.13
Grand Total	**5.07**	**6.45**	**8.38**	**9.43**	**12.01**	**7.91**

The average three-year return percentages for growth funds are higher than for value funds. The return is higher for funds with higher ratings than lower ratings. This pattern holds for the growth funds for each risk level. For the low risk and average risk value funds, the return is lowest for the funds with a two-star rating.

(d) There are 32 growth funds with high risk with a rating of three. These funds have an average three-year return percentage of 7.45.

2.62 The fund with the highest five-year return of 15.72% is a large cap growth fund that has a four-star rating and low risk.

2.64 Funds 479, 471, 347, 443, and 477 have the lowest five-year return percentages.

2.66 The five funds with the lowest five-year return percentages have (1) midcap growth, average risk, one-star rating, (2) midcap growth, high risk, two-star rating, (3) large value, average risk, two-star rating, (4) midcap growth, high risk, one-star rating, and (5) small value, average risk, two-star rating.

2.68 There has been a decline in the price of natural gas over time. However, there is no pattern within the years. For some years, the price is higher in the beginning of the year. For other years, the price is higher in the latter part of the year. Sometimes, there is little variation within the year.

2.88 (c) The publisher gets the largest portion (66.06%) of the revenue. 24.93% is editorial production manufacturing costs. The publisher's marketing accounts for the next largest share of the revenue, at 11.6%. Author and bookstore personnel each account for around 11 to 12% of the revenue, whereas the publisher and bookstore profit and income account for more than 26% of the revenue. Yes, the bookstore gets almost twice the revenue of the authors.

2.90 (b) The pie chart or the Pareto chart would be best. The pie chart would allow you to see each category as part of the whole, while the Pareto chart would enable you to see that Small marketing/content marketing team is the dominant category. **(d)** The pie chart or the Pareto chart would be best. The pie chart would allow you to see each category as part of the whole while the Pareto chart would enable you to see that very committed to content marketing is the dominant category. **(e)** Most organizations have a small marketing/content marketing team and are very committed to content marketing.

2.92 (a)

Dessert Ordered	Gender		Total
	Male	Female	
Yes	66%	34%	100%
No	48%	52%	100%
Total	52%	48%	100%

Dessert Ordered	Gender		Total
	Male	Female	
Yes	29%	17%	23%
No	71%	83%	77%
Total	100%	100%	100%

Dessert Ordered	Gender		Total
	Male	Female	
Yes	15%	8%	23%
No	37%	40%	77%
Total	52%	48%	100%

Dessert Ordered	Beef Entrée		Total
	Yes	No	
Yes	52%	48%	100%
No	25%	75%	100%
Total	31%	69%	100%

Dessert Ordered	Beef Entrée		Total
	Yes	No	
Yes	38%	16%	23%
No	62%	84%	77%
Total	100%	100%	100%

Dessert Ordered	Beef Entrée		Total
	Yes	No	
Yes	11.75%	10.79%	22.54%
No	19.52%	57.94%	77.46%
Total	31.27%	68.73%	100%

(b) If the owner is interested in finding out the percentage of males and females who order dessert or the percentage of those who order a beef entrée and a dessert among all patrons, the table of total percentages is most informative. If the owner is interested in the effect of gender on ordering of dessert or the effect of ordering a beef entrée on the ordering of dessert, the table of column percentages will be most informative. Because dessert is usually ordered after the main entrée, and the owner has no direct control over the gender of patrons, the table of row percentages is not very useful here. **(c)** 29% of the men ordered desserts, compared to 17 of the women; men are almost twice as likely to order dessert as women. Almost 38% of the patrons ordering a beef entrée ordered dessert, compared to 16% of patrons ordering all other entrées. Patrons ordering beef are more than 2.3 times as likely to order dessert as patrons ordering any other entrée.

2.94 (a) Most of the complaints were against U.S. airlines.
(b) More of the complaints were due to flight and baggage problems.

2.96 (c) The alcohol percentage is concentrated between 4% and 6%, with more between 4% and 5%. The calories are concentrated between 140 and 160. The carbohydrates are concentrated between 12 and 15. There are outliers in the percentage of alcohol in both tails. There are a few beers with alcohol content as high as around 11.5%. There are a few beers with calorie content as high as around 313 and carbohydrates as high as 32.1. There is a strong positive relationship between percentage of alcohol and calories and between calories and carbohydrates, and there is a moderately positive relationship between percentage alcohol and carbohydrates.

2.98 (b) The yield of one-year CDs shows that most values are at least 1.0. The yield of five-year CDs shows that most values are at least 1.6. **(d)** There appears to be a strong positive relationship between the yield of the one-year CD and the five-year CD.

2.100 (a)

Frequency (Boston)		
Weight (Boston)	Frequency	Percentage
3,015 but less than 3,050	2	0.54%
3,050 but less than 3,085	44	11.96%
3,085 but less than 3,120	122	33.15%
3,120 but less than 3,155	131	35.60%
3,155 but less than 3,190	58	15.76%
3,190 but less than 3,225	7	1.90%
3,225 but less than 3,260	3	0.82%
3,260 but less than 3,295	1	0.27%

(b)

Frequency (Vermont)		
Weight (Vermont)	Frequency	Percentage
3,550 but less than 3,600	4	1.21%
3,600 but less than 3,650	31	9.39%
3,650 but less than 3,700	115	34.85%
3,700 but less than 3,750	131	39.70%
3,750 but less than 3,800	36	10.91%
3,800 but less than 3,850	12	3.64%
3,850 but less than 3,900	1	0.30%

(d) 0.54% of the Boston shingles pallets are underweight, and 0.27% are overweight. 1.21% of the Vermont shingles pallets are underweight, and 3.94% are overweight.

2.102 (a)

Calories	Frequency	Percentage	Limit	Percentage Less Than
50 but less than 100	3	12%	100	12%
100 but less than 150	3	12%	150	24%
150 but less than 200	9	36%	200	60%
200 but less than 250	6	24%	250	84%
250 but less than 300	3	12%	300	96%
300 but less than 350	0	0%	350	96%
350 but less than 400	1	4%	400	100%

(b)

Cholesterol	Frequency	Percentage	Limit	Percentage Less Than
0 but less than 50	2	8%	50	8%
50 but less than 100	17	68%	100	76%
100 but less than 150	4	16%	150	92%
150 but less than 200	1	4%	200	96%
200 but less than 250	0	0%	250	96%
250 but less than 300	0	0%	300	96%
300 but less than 350	0	0%	350	96%
350 but less than 400	0	0%	400	96%
400 but less than 450	0	0%	450	96%
450 but less than 500	1	4%	500	100%

(e) There is very little relationship between calories and cholesterol. **(f)** The sampled fresh red meats, poultry, and fish vary from 98 to 397 calories per serving, with the highest concentration between 150 and 200 calories. One protein source, spareribs, with 397 calories, is more than 100 calories above the next-highest-caloric food. Spareribs and fried liver are both very different from other foods sampled—the former on calories and the latter on cholesterol content.

2.104 (b) There is a downward trend in the amount filled. **(c)** The amount filled in the next bottle will most likely be below 1.894 liters. **(d)** The scatter plot of the amount of soft drink filled against time reveals the trend of the data, whereas a histogram only provides information on the distribution of the data.

2.106 (a) The percentage of downloads is 9.64% for the original call-to-action button and 13.64% for the new call-to-action button. **(c)** The new call-to-action button has a higher percentage of downloads at 13.64% when compared to the original call-to-action button with a 9.64% of downloads. **(d)** The percentage of downloads is 8.90% for the original web design and 9.41% for the new web design. **(f)** The new web design has only a slightly higher percentage of downloads at 9.41% when compared to the original web design with an 8.90% of downloads. **(g)** The new web design is only slightly more successful than the original web design while the new call-to-action button is much more successful than the original call-to-action button with about 41% higher percentage of downloads.

(h)

Call-to-Action Button	Web Design	Percentage of Downloads
Old	Old	8.30%
New	Old	13.70%
Old	New	9.50%
New	New	17.00%

(i) The new call-to-action button and the new web design together had a higher percentage of downloads. **(j)** The new web design is only slightly more successful than the original web design while the new call-to-action button is much more successful than the original call-to-action button with about 41% higher percentage of downloads. However, the combination of the new call-to-action button and the new web design results in more than twice as high a percentage of downloads than the combination of the original call-to-action button and the original web design.

CHAPTER 3

3.2 (a) Mean = 7, median = 7, mode = 7. **(b)** Range = 9, S^2 = 10.8, S = 3.286, CV = 46.948%. **(c)** Z scores: 0, -0.913, 0.609, 0, -1.217, 1.522. None of the Z scores are larger than 3.0 or smaller than -3.0. There is no outlier. **(d)** Symmetric because mean = median.

3.4 (a) Mean = 2, median = 7, mode = 7. **(b)** Range = 17, S^2 = 62 S = 7.874, CV = 393.7%. **(c)** 0.635, -0.889, -1.270, 0.635, 0.889. There are no outliers. **(d)** Left-skewed because mean < median.

3.6 -0.0835

3.8 (a)

	Grade X	Grade Y
Mean	575	575.4
Median	575	575
Standard deviation	6.40	2.07

(b) If quality is measured by central tendency, Grade X tires provide slightly better quality because X's mean and median are both equal to the expected value, 575 mm. If, however, quality is measured by consistency, Grade Y provides better quality because, even though Y's mean is only slightly larger than the mean for Grade X, Y's standard deviation is much smaller. The range in values for Grade Y is 5 mm compared to the range in values for Grade X, which is 16 mm.

(c)

	Grade X	Grade Y, Altered
Mean	575	577.4
Median	575	575
Standard deviation	6.40	6.11

When the fifth Y tire measures 588 mm rather than 578 mm, Y's mean inner diameter becomes 577.4 mm, which is larger than X's mean inner diameter, and Y's standard deviation increases from 2.07 mm to 6.11 mm. In this case, X's tires are providing better quality in terms of the mean inner diameter, with only slightly more variation among the tires than Y's.

3.10 (a), (b)

	Download Speed (Mbps)	Upload Speed (Mbps)
Mean	32.375	11.175
Median	32.65	12.95
Minimum	6.5	3.7
Maximum	53.3	17.5
Range	46.8	13.8
Variance	165.1107	33.5593
Standard deviation	12.8495	5.7930
Coefficient of variation	39.69%	51.84%
Skewness	-0.7204	-0.3956
Kurtosis	3.1589	-1.9516
Sample size	8	8

(c) The mean is about the same as the median for the download speed. The upload speed is indicating a left or negative skewed distribution (the skewness statistic is also negative). The kurtosis statistic is positive for the upload speed, indicating a distribution that is more peaked than a normal (bell-shaped) distribution. The kurtosis statistic is negative for download speed, indicating a distribution that is less peaked than a normal distribution.

(d) The mean download speed is much higher than the mean upload speed. The median download speed indicates that half the carriers have a download speed of at least 32.65 Mbps as compared to a median upload speed of 12.95 Mbps that indicates that half the carriers have an upload speed of at least 12.95 Mbps. There is much more variation in the download speed than the upload speed because the standard deviation is 12.8495 as compared to 5.7930.

3.12 (a), (b)

	Household Hours
Mean	7.5667
Median	7
Minimum	4
Maximum	14.3
Range	10.3
Variance	6.1444
Standard deviation	2.4788
Coefficient of variation	32.76%
Skewness	1.3521
Kurtosis	1.5387
Sample size	30

(c) (d) The mean household hours needed is greater than the median, indicating a right or positive skewed distribution (the skewness statistic is also positive). The kurtosis statistic is positive, indicating a distribution that is more peaked than a normal (bell-shaped) distribution.

3.14 (a), (b)

Mobile Commerce Penetration (%)	
Mean	45.0833
Median	45
Mode	46
Minimum	30
Maximum	60
Range	30
Variance	58.5145
Standard Deviation	7.6495
Coefficient of Variation	16.97%
Count	24

Country	Mobile Commerce Penetration (%)	Z Score
Australia	46	0.1198
Austria	44	−0.1416
Belgium	38	−0.9260
Brazil	43	−0.2723
Canada	33	−1.5796
Denmark	51	−0.7735
Finland	49	−0.5120
France	39	−0.7953
Germany	50	−0.6427
Italy	41	−0.5338
Japan	55	−1.2964
Netherlands	49	−0.5120

Country	Mobile Commerce Penetration (%)	Z Score
New Zealand	44	−0.1416
Norway	57	−1.5578
Poland	33	−1.5796
Russia	30	−1.9718
South Korea	47	−0.2506
Spain	48	−0.3813
Sweden	60	1.9500
Switzerland	43	−0.2723
Taiwan	42	−0.4031
Turkey	46	0.1198
United Kingdom	55	1.2964
United States	39	−0.7953

Because there are no Z values below −3.0 or above 3.0, there are no outliers. **(c)** The mean is approximately the same the median, so Mobile Commerce Penetration is symmetrical. **(d)** The mean Mobile Commerce Penetration is 45.0833% and half the countries have values greater than or equal to 45%. The average scatter around the mean is 7.6495%. The lowest value is 30% (Russia), and the highest value is 60% (Sweden).

3.16 (a), (b)

Time (seconds)	
Mean	232.76
Median	228
Mode	243
Range	1,076
Variance	25,181.44
Standard Deviation	158.6866

(c) The mean time is 232.78 seconds, and half the calls last greater than or equal to 228 seconds, so call duration is slightly right-skewed. The average scatter around the mean is 158.6866 seconds. The shortest call lasted 65 seconds, and the longest call lasted 1,141 seconds.

3.18 (a) Mean = 7.11, median = 6.68. **(b)** Variance = 4.336, standard deviation = 2.082, range = 6.67, CV = 29.27%.

Waiting Time	Z Score	Waiting Time	Z Score
9.66	1.222431	10.49	1.62105
5.90	−0.58336	6.68	−0.20875
8.02	0.434799	5.64	−0.70823
5.79	−0.63619	4.08	−1.45744
8.73	0.775786	6.17	−0.45369
3.82	−1.58231	9.91	1.342497
8.01	0.429996	5.47	−0.78987
8.35	0.593286		

Because there are no Z values below −3.0 or above 3.0, there are no outliers.

(c) Because the mean is greater than the median, the distribution is right-skewed. **(d)** The mean and median are both greater than five minutes. The distribution is right-skewed, meaning that there are some unusually high values. Further, 13 of the 15 bank customers sampled (or 86.7%) had waiting times greater than five minutes. So the customer is likely to experience a waiting time in excess of five minutes. The manager overstated the bank's service record in responding that the customer would "almost certainly" not wait longer than five minutes for service.

3.20 (a) Geometric rate of return is 14.29% **(b)** $1,142.90 **(c)** The rate of return was much better than that of GE, which declined in price.

3.22 (a) Platinum = −0.0274% Gold = −5.36% Silver = −2.32%. **(b)** Gold had the best rate of return followed by silver. Only platinum had a negative rate of return. **(c)** The return on the metals was much worse than the rate of the stock indices.

3.24 (a)

Mean of 3YrReturn%	Rating					
Type	**One**	**Two**	**Three**	**Four**	**Five**	**Grand Total**
Growth	**5.41**	**7.04**	**8.94**	**10.14**	**12.83**	**8.51**
Large	6.97	9.43	10.62	11.83	14.25	10.30
Mid-Cap	2.27	5.07	7.93	8.77	11.22	6.93
Small	0.78	5.09	6.52	8.35	9.53	6.39
Value	**4.43**	**5.49**	**7.29**	**8.34**	**10.23**	**6.84**
Large	5.23	6.05	7.58	8.85	10.23	7.29
Mid-Cap	2.79	5.77	7.32	9.26	–	6.69
Small	1.33	3.20	5.93	7.04	–	5.39

(b)

StdDev of 3Yr Return%	Rating					
Type	**One**	**Two**	**Three**	**Four**	**Five**	**Grand Total**
Growth	**3.72**	**2.85**	**2.71**	**2.23**	**2.12**	**3.19**
Large	2.86	1.34	2.23	1.43	0.89	2.56
Mid-Cap	3.49	2.04	2.08	1.03	1.02	2.86
Small	0.84	2.40	2.08	2.11	0.62	2.52
Value	**2.07**	**2.40**	**1.20**	**2.09**	**1.32**	**2.33**
Large	1.81	1.68	0.98	1.63	1.32	1.93
Mid-Cap	1.00	2.90	1.13	0.99	–	2.51
Small	–	2.88	1.36	2.62	–	2.35
Grand Total	**3.24**	**2.78**	**2.44**	**2.34**	**2.24**	**3.02**

(c) The mean three-year return of small-cap funds is much lower than mid-cap and large funds. Five-star funds for all market cap categories show the highest mean three-year returns. The mean three-year returns for all combinations of type and market cap rises as the star rating rises, consistent to the mean three-year returns for all growth and value funds.

The standard deviations of the three-year return for large-cap and mid-cap value funds vary greatly among star rating categories.

3.26 (a)

Mean of 3Yr Return%	Rating					
Type	**One**	**Two**	**Three**	**Four**	**Five**	**Grand Total**
Growth	**5.41**	**7.04**	**8.94**	**10.14**	**12.83**	**8.51**
Low	7.53	8.60	9.89	10.29	12.64	9.87
Average	6.17	7.99	9.28	10.43	11.96	9.06
High	3.83	5.59	7.45	8.76	13.59	6.64
Value	**4.43**	**5.49**	**7.29**	**8.34**	**10.23**	**6.84**
Low	5.29	7.00	7.66	8.57	10.74	7.76
Average	5.01	4.98	6.97	7.96	9.23	6.41
High	2.71	2.63	6.53	8.39	–	4.13
Grand Total	**5.07**	**6.45**	**8.38**	**9.43**	**12.01**	**7.91**

(b)

StdDev of 3Yr Return%	Rating					
	One	**Two**	**Three**	**Four**	**Five**	**Grand Total**
Growth	**3.72**	**2.85**	**2.71**	**2.23**	**2.12**	**3.19**
Low	3.27	1.57	2.02	2.05	2.04	2.42
Average	4.37	2.43	2.67	2.42	2.51	2.86
High	2.98	2.92	2.73	1.43	2.47	3.39

StdDev of 3Yr Return%	Rating					
	One	**Two**	**Three**	**Four**	**Five**	**Grand Total**
Value	**2.07**	**2.40**	**1.20**	**2.09**	**1.32**	**2.33**
Low	1.46	1.12	1.00	2.15	0.85	1.72
Average	2.11	2.43	1.25	2.09	1.87	2.27
High	–	2.88	1.36	2.62	–	2.35
Grand Total	**3.24**	**2.78**	**2.44**	**2.34**	**2.24**	**3.02**

(c) The mean three-year return of high-risk funds is much lower than the other risk categories except for five-star funds. In all risk categories, five-star funds have the highest mean three-year return. The mean three-year returns for high-risk growth and value funds for one-, two-, and three-star rating funds are lower than the means for the other risk categories.

The standard deviations of the three-year return for low-risk funds show the most consistency across star rating categories and the standard deviations of the three-year return for low-risk funds are the lowest across categories. They also vary greatly among star rating categories.

3.28 (a) 4, 9, 5. **(b)** 3, 4, 7, 9, 12. **(c)** The distances between the median and the extremes are close, 4 and 5, but the differences in the tails are different (1 on the left and 3 on the right), so this distribution is slightly right-skewed. **(d)** In Problem 3.2 (d), because mean = median, the distribution is symmetric. The box part of the graph is slightly left skewed, but the tails show right-skewness.

3.30 (a) −6.5, 8, 14.5. **(b)** −8, −6.5, 7, 8, 9. **(c)** The shape is left-skewed. **(d)** This is consistent with the answer in Problem 3.4 (d).

3.32 (a), (b) Minimum = 30 Q_1 = 39, Median = 45 Q_3 = 50 Maximum = 60, Interquartile range = 11 **(c)** the boxplot is approximately symmetric.

3.34 (a), (b) Halftime or after Q_1 = 4.96, Q_3 = 6.41, Minimum = 3.63, Median = 5.59, Maximum = 7.69 Interquartile range = 1.45 First or Second Quarter: Q_1 = 4.81, Q_3 = 6.14, Interquartile range = 1.33 Minimum = 4.22, Median = 5.34, Maximum = 6.51 **(c)** The boxplot plot for halftime or after is approximately symmetrical and the boxplot for the first or second quarter is also approximately symmetric.

3.36 (a) Commercial district five-number summary: 0.38 3.2 4.5 5.55 6.46. Residential area five-number summary: 3.82 5.64 6.68 8.73 10.49. **(b)** Commercial district: The distribution is left-skewed. Residential area: The distribution is slightly right-skewed. **(c)** The central tendency of the waiting times for the bank branch located in the commercial district of a city is lower than that of the branch located in the residential area. There are a few long waiting times for the branch located in the residential area, whereas there are a few exceptionally short waiting times for the branch located in the commercial area.

3.38 (a) Population mean, μ = 6. **(b)** Population standard deviation, σ = 1.673, population variance, σ^2 = 2.8.

3.40 (a) 68%. **(b)** 95%. **(c)** At least 0%, 75%, 88.89%. **(d)** $\mu - 4\sigma$ to $\mu + 4\sigma$ or −2.8 to 19.2.

3.42 (a) Mean = $\dfrac{67.33}{51}$ = 13.4771 variance = 11.6792, standard deviation = $\sqrt{11.6792}$ = 3.4175 **(b)** 74.51%, 96.08%, and 98.04% of these locations have mean per capita energy consumption within 1, 2, and 3 standard deviations of the mean, respectively. **(c)** This is slightly different from 68%, 95%, and 99.7%, according to the empirical rule.

3.44 (a) Covariance = 65.2909, **(b)** r = +1.0. **(c)** there is a perfect positive relationship.

3.46 (a) $\text{cov}(X, Y) = \dfrac{\sum_{i=1}^{n}(X_i - \bar{X})(Y_i - \bar{Y})}{n - 1} = \dfrac{800}{6} = 133.3333.$

(b) $r = \dfrac{\text{cov}(X, Y)}{S_X S_Y} = \dfrac{133.3333}{(46.9042)(3.3877)} = 0.8391.$

(c) The correlation coefficient is more valuable for expressing the relationship between calories and sugar because it does not depend on the units used to measure calories and sugar. **(d)** There is a strong positive linear relationship between calories and sugar.

3.48 (a) $\text{cov}(X, Y)$ = 45.5036 **(b)** r = 0.61113 **(c)** There is a positive linear relationship between download and upload speed.

3.66 (a) Mean = 45.22, median = 45, 1st quartile = 25, 3rd quartile = 63. **(b)** Range = 83, interquartile range = 38, variance = 535.7949, standard deviation = 23.1472, CV = 51.19%. **(c)** The distribution is approximately symmetric. **(d)** The mean approval process takes 45.22, days, with 50% of the policies being approved in less than 45 days. 50% of the applications are approved between 25 and 63 days. About 25% of the applications are approved in no more than 25 days.

3.68 (a) Mean = 14.98, median = 15 range = 23, S = 5.5567. The mean and median width virtually equal. The range of the answer time is 23 seconds, and the average scatter around the mean is 5.5567 seconds. **(b)** 5 12 15 18 28. **(c)** Even though the mean = median, the right tail is longer, so the distribution is right-skewed. **(d)** The service level is being met because 75% of the calls are answered in less than 18 seconds.

3.70 (a), (b)

	Bundle Score	Typical Cost ($)
Mean	54.775	24.175
Median	62	20
Mode	75	8
Standard Deviation	27.6215	18.1276
Sample Variance	762.9481	328.6096
Range	98	83
Minimum	2	5
Maximum	100	88
First Quartile	34	9
Third Quartile	75	31
Interquartile Range	41	22
CV	50.43%	74.98%

(c) The typical cost is right-skewed, while the bundle score is left-skewed. **(d)** r = 0.3465. **(e)** The mean typical cost is $24.18, with an average spread around the mean equaling $18.13. The spread between the lowest and highest costs is $83. The middle 50% of the typical cost fall over a range of $22 from $9 to $31, while half of the typical cost is below $20. The mean bundle score is 54.775, with an average spread around the mean equaling 27.6215. The spread between the lowest and highest scores is 98. The middle 50% of the scores fall over a range of 41 from 34 to 75, while half of the scores are below 62. The typical cost is right-skewed, while the bundle score is left-skewed. There is a weak positive linear relationship between typical cost and bundle score.

3.72 (a) Boston: 0.04, 0.17, 0.23, 0.32, 0.98; Vermont: 0.02, 0.13, 0.20, 0.28, 0.83. **(b)** Both distributions are right-skewed. **(c)** Both sets of shingles did well in achieving a granule loss of 0.8 gram or less. Only two Boston shingles had a granule loss greater than 0.8 gram. The next highest to these was 0.6 gram. These two values can be considered outliers. Only 1.176% of the shingles failed the specification. Only one of the Vermont shingles had a granule loss greater than 0.8 gram. The next highest was 0.58 gram. Thus, only 0.714% of the shingles failed to meet the specification.

3.74 (a) The correlation between calories and protein is 0.4644. **(b)** The correlation between calories and cholesterol is 0.1777. **(c)** The correlation between protein and cholesterol is 0.1417. **(d)** There is a weak positive linear relationship between calories and protein, with a correlation coefficient of 0.46. The positive linear relationships between calories and cholesterol and between protein and cholesterol are very weak.

3.76 (a), (b)

	Annual Taxes on $176K Home	Median Home Value ($000)
Mean	1,979.490196	195.6509804
Median	1,763	165.9
Mode	#N/A	#N/A
Minimum	489	100.2
Maximum	4,029	504.5
Range	3,540	404.3
Variance	11,065.8549	7,418.7265
Standard Deviation	900.5919	86.1320
Coeff. of Variation	45.50%	44.02%
Skewness	0.6423	1.6988
Kurtosis	−0.5014	3.3069
Count	51	51
Standard Error	126.1081	12.0609

(c) The box plot shows that taxes are right skewed and the median value of homes is highly right skewed.**(d)** The coefficient of correlation is −0.041. **(e)** There is a large variation in taxes and the median value of homes from state to state..

3.78 (a), (b)

Abandonment Rate in % (7:00 AM–3:00 PM)	
Mean	13.8636
Median	10
Mode	9
Standard Deviation	7.6239
Sample Variance	58.1233
Range	29
Minimum	5
Maximum	34
First Quartile	9
Third Quartile	20
Interquartile Range	11
CV	54.99%

(c) The data are right-skewed. **(d)** $r = 0.7575$ **(e)** The mean abandonment rate is 13.86%. Half of the abandonment rates are less than 10%. One-quarter of the abandonment rates are less than 9%, while another

one-quarter are more than 20%. The overall spread of the abandonment rates is 29%. The middle 50% of the abandonment rates are spread over 11%. The average spread of abandonment rates around the mean is 7.62%. The abandonment rates are right-skewed.

3.80 (a), (b)

Average Credit Score	
Mean	673.24
Median	672.02
Mode	684.52
Standard Deviation	31.7156
Sample Variance	1,005.8784
Range	214.51
Minimum	565.00
Maximum	779.51
Count	2,570
First Quartile	649.82
Third Quartile	697.21
Interquartile Range	47.39
Skewness	−0.0071
Kurtosis	−0.3710
CV	4.71%

(c) The data are symmetrical. **(d)** The mean of the average credit scores is 673.24. Half of the average credit scores are less than 672.02. One-quarter of the average credit scores are less than 649.82, while another one-quarter is more than 697.21. The overall spread of average credit scores is 214.51. The middle 50% of the average credit scores spread over 47.39. The average spread of average credit scores around the mean is 31.7156.

CHAPTER 4

4.2 (a) Simple events include selecting a red ball. **(b)** Selecting a white ball. **(c)** The sample space consists of the 12 red balls and the 8 white balls.

4.4 (a) 0.6. **(b)** 0.10. **(c)** 0.35. **(d)** 0.90.

4.6 (a) Mutually exclusive, not collectively exhaustive. **(b)** Not mutually exclusive, not collectively exhaustive. **(c)** Mutually exclusive, not collectively exhaustive. **(d)** Mutually exclusive, collectively exhaustive.

4.8 (a) Is a millennial. **(b)** Is a millennial and has a retirement account. **(c)** Does not have a retirement account. **(d)** Is a millennial and has a retirement account is a joint event because it consists of two characteristics.

4.10 (a) A marketer who plans to increase use of LinkedIn. **(b)** A B2B marketer who plans to increase use of LinkedIn. **(c)** A marketer who does not plan to increase use of LinkedIn. **(d)** A marketer who plans to increase use of LinkedIn and is a B2C marketer is a joint event because it consists of two characteristics, plans to increase use of LinkedIn and is a B2C marketer.

4.12 (a) 686/1,803. (b) 206/1,803. (c) 717/1,803. (d) the probability in (c) includes those who fully support increased use of educational technologies in higher education and those who are digital learning leaders.

4.14 (a) 1,106/1,261. **(b)** 911/1,261. **(c)** 1,196/1,206. **(d)** 1,261/1,261.

4.16 (a) 0.33. **(b)** 0.33. **(c)** 0.67. **(d)** Because $P(A|B) = P(A) = 1/3$, events A and B are independent.

4.18 0.50.

4.20 Because $P(A \text{ and } B) = 0.20$ and $P(A)P(B) = 0.12$, events A and B are not independent.

4.22 (a) 0.6502. **(b)** 0.4300. **(c)** probability (increased use of LinkedIn) = 0.5049, which is not equal to P(Increased use of LinkedIn | $B2B$) = 0.6502. Therefore, increased use of LinkedIn and business focus are not independent.

4.24 (a) $511/1,597 = 0.3200$. **(b)** $1,086/1,597 = 0.6800$. **(c)** $175/206 = 0.8495$. **(d)** $31/206 = 0.1515$.

4.26 (a) 0.0417. **(b)** 0.0375. **(c)** Because P(Needs warranty repair | Manufacturer based in the United States) = 0.0417 and P(Needs warranty repair) = 0.04, the two events are not independent.

4.28 (a) 0.0045. **(b)** 0.012. **(c)** 0.0059. **(d)** 0.0483.

4.36 (a)

	Generation		
Prefer Hybrid Advice	**Baby Boomers**	**Millennials**	**Total**
Yes	140	320	460
No	360	180	540
Total	500	500	1,000

(b) Preferring hybrid investment advice; being a baby boomer and preferring hybrid investment advice. **(c)** 0.46. **(d)** 0.14. **(e)** They are not independent because baby boomers and millennials have different probabilities of preferring hybrid investment advice.

4.38 (a) $82/276 = 0.2971$. **(b)** $115/276 = 0.4167$. **(c)** $142/276 = 0.5145$. **(d)** $32/276 = 0.1159$. **(e)** $4/147 = 0.0272$.

CHAPTER 5

5.2 (a)
$$\mu = 0(0.10) + 1(0.20) + 2(0.45) + 3(0.15) + 4(0.05) + 5(0.05) = 2.0.$$
(b) $\sigma = \sqrt{\begin{array}{l}(0-2)^2(0.10) + (1-2)^2(0.20) + (2-2)^2(0.45) + \\ (3-2)^2(0.15) + (4-2)^2(0.05) + (5-2)^2(0.05)\end{array}} = 1.183.$
(c) $0.45 + 0.15 + 0.05 + 0.05 = 0.70$.

5.4 (a)

X	$P(X)$
$\$ -1$	$21/36$
$\$ +1$	$15/36$

(b)

X	$P(X)$
$\$ -1$	$21/36$
$\$ +1$	$15/36$

(c)

X	$P(X)$
$\$ -1$	$30/36$
$\$ +4$	$6/36$

(d) $-\$0.167$ for each method of play.

5.6 (a) 2.1058. **(b)** 1.4671. **(c)** $66/104 = 0.6346$.

5.8 (a) E(Bond Fund) = $58.20; E(Common Stock Fund) = $63.01. **(b)** $\sigma_X = \$61.55$; $\sigma_Y = \$195.22$. **(c)** Based on the expected value criteria, you would choose the common stock fund. However, the common stock fund also has a standard deviation more than three times higher than that for the corporate bond fund. An investor should carefully weigh the increased risk. **(d)** If you chose the common stock fund, you would need to assess your reaction to the small possibility that you could lose virtually all of your entire investment.

5.10 (a) 0.40, 0.60. **(b)** 1.60, 0.98. **(c)** 4.0, 0.894. **(d)** 1.50, 0.866.

5.12 (a) 0.2056. **(b)** 0.0108. **(c)** 0.2893. **(d)** $\mu = 2.82$, $\sigma = 1.2225$. **(e)** That each American adult owns an iPhone or does not own an iPhone and that each person is independent of all other persons.

5.14 (a) 0.7374. **(b)** 0.2281. **(c)** 0.9972. **(d)** 0.0028.

5.16 (a) 0.7511. **(b)** 0.0008. **(c)** 0.9767. **(d)** $\mu = 2.727$, $\sigma = 0.4982$. **(e)** McDonald's has a slightly higher probability of filling orders correctly, and Wendy's has a slightly lower probability.

5.18 (a) 0.2565. **(b)** 0.1396. **(c)** 0.3033. **(d)** 0.0247.

5.20 (a) 0.0337. **(b)** 0.0067. **(c)** 0.9596. **(d)** 0.0404.

5.22 (a)
$$\begin{aligned} P(X < 5) &= P(X = 0) + P(X = 1) + P(x = 2) + P(X = 3) \\ &\quad + P(X = 4) \\ &= \frac{e^{-6}(6)^0}{0!} + \frac{e^{-6}(6)^1}{1!} + \frac{e^{-6}(6)^2}{2!} + \frac{e^{-6}(6)^3}{3!} + \frac{e^{-6}(6)^4}{4!} \\ &= 0.002479 + 0.014873 + 0.044618 + 0.089235 + 0.133853 \\ &= 0.2851. \end{aligned}$$
(b) $P(X = 5) = \dfrac{e^{-6}(6)^5}{5!} = 0.1606$.
(c) $P(X \geq 5) = 1 - P(X < 5) = 1 - 0.2851 = 0.7149$.
(d) $P(X = 4 \text{ or } X = 5) = P(X = 4) + P(X = 5) = \dfrac{e^{-6}(6)^4}{4!} + \dfrac{e^{-6}(6)^5}{5!}$
$$= 0.2945.$$

5.24 (a) 0.7261. **(b)** 0.2739. **(c)** 0.0415.

5.26 (a) 0.0302. **(b)** 0.1057. **(c)** 0.8641. **(d)** 0.1359.

5.28 (a) 0.3396. **(b)** 0.9044. **(c)** Because Ford had a higher mean rate of problems per car than Toyota, the probability of a randomly selected Ford having zero problems and the probability of no more than two problems are both lower than for Toyota.

5.34 (a) 0.664. **(b)** 0.664. **(c)** 0.3266. **(d)** 0.0043. **(e)** The assumption of independence may not be true.

5.36 (a) 0.0287. **(b)** 0.5213.

5.38 (a) 0.0060. **(b)** 0.2007. **(c)** 0.1662. **(d)** Mean = 4.0, standard deviation = 1.5492. **(e)** Because the percentage of bills containing an error is lower in this problem, the probability is higher in (a) and (b) of this problem and lower in (c).

5.40 (a) 9.2. **(b)** 2.2289. **(c)** 0.1652. **(d)** 0.0461. **(e)** 0.9848.

5.42 (a) 0.0000. **(b)** 0.0564. **(c)** 0.9720. **(d)** Based on the results in (a)–(c), the probability that the Standard & Poor's 500 Index will increase if there is an early gain in the first five trading days of the year is very likely to be close to 0.90 because that yields a probability of 97.20% that at least 36 of the 44 years the Standard & Poor's 500 Index will increase the entire year.

5.44 (a) The assumptions needed are (i) the probability that a questionable claim is referred by an investigator is constant, (ii) the probability that a questionable claim is referred by an investigator approaches 0 as the interval gets smaller, and (iii) the probability that a questionable claim is referred by an investigator is independent from interval to interval. **(b)** 0.1277. **(c)** 0.9015. **(d)** 0.0985.

CHAPTER 6

6.2 (a) 0.9089. **(b)** 0.0911. **(c)** +1.96. **(d)** −1.00 and +1.00.

6.4 (a) 0.1401. **(b)** 0.4168. **(c)** 0.3918. **(d)** +1.00.

6.6 (a) 0.9599. **(b)** 0.0228. **(c)** 43.42. **(d)** 46.64 and 53.36.

6.8 (a) $P(34 < X < 50) = P(-1.33 < Z < 0) = 0.4082$.
(b) $P(X < 30) + P(X > 60) = P(Z < -1.67) + P(Z > 0.83)$
$= 0.0475 + (1.0 - 0.7967) = 0.2508$. **(c)** $P(Z < -0.84) \cong 0.20$,
$Z = -0.84 = \dfrac{X - 50}{12}$, $X = 50 - 0.84(12) = 39.92$ thousand miles, or
39,920 miles. **(d)** The smaller standard deviation makes the absolute
Z values larger. **(a)** $P(34 < X < 50) = P(-1.60 < Z < 0) = 0.4452$.
(b) $P(X < 30) + P(X > 60) = P(Z < -2.00) + P(Z > 1.00)$
$= 0.0228 + (1.0 - 0.8413) = 0.1815$. **(c)** $X = 50 - 0.84(10) = 41.6$
thousand miles, or 41,600 miles.

6.10 (a) 0.9878. **(b)** 0.8185. **(c)** 86.16%. **(d)** Option 1: Because your score
of 81% on this exam represents a Z score of 1.00, which is below the minimum Z score of 1.28, you will not earn an A grade on the exam under
this grading option. Option 2: Because your score of 68% on this exam
represents a Z score of 2.00, which is well above the minimum Z score of
1.28, you will earn an A grade on the exam under this grading option. You
should prefer Option 2.

6.12 (a) 0.0591. **(b)** 0.5636. **(c)** 0.0003. **(d)** 56.1108 gallons.

6.14 With 39 values, the smallest of the standard normal quantile values
covers an area under the normal curve of 0.025. The corresponding Z
value is -1.96. The middle (20th) value has a cumulative area of 0.50
and a corresponding Z value of 0.0. The largest of the standard normal
quantile values covers an area under the normal curve of 0.975, and its
corresponding Z value is $+1.96$.

6.16 Super Bowl Ad Ratings First and Second Period: **(a)** Mean $= 5.38$,
median $= 5.34$, $S = 0.7153$, range $= 2.29$, $6S = 4.2918$, interquartile
range $= 1.33$, $1.33(0.7153) = 0.9513$. The mean is approximately the
same as the median. The range is much less than $6S$, and the interquartile
range is more than $1.33S$. The skewness statistic is 0.0358, indicating a
symmetric distribution, and the kurtosis statistic is -1.2567, indicating a
platykurtic distribution.
Super Bowl Ad Ratings Halftime and Afterward: **(a)** Mean $= 5.65$,
median $= 5.59$, $S = 1.0411$, range $= 4.06$, $6S = 6.2466$, interquartile
range $= 1.45$, $1.33(1.0411) = 1.3817$. The mean is approximately the
same as the median. The range is much less than $6S$, and the interquartile
range is slightly more than $1.33S$. The skewness statistic is 0.1406, indicating an approximately symmetric distribution, and the kurtosis statistic
is -0.5910, indicating a platykurtic distribution.

6.18 (a) Mean $= \$1,979.49$, median $= \$1,763$, $S = \$900.8957$,
range $= \$3,540$, $6S = 6(900.5919) = \$5,403.5514$, interquartile
range $= \$1,333$, $1.33(900.5919) = 1,197.7872$. The mean is greater
than the median. The range is much less than $6S$, and the interquartile
range is less than $1.33S$. **(b)** The normal probability plot appears to be
right skewed. The skewness statistic is 0.6408. The kurtosis is -0.5058,
indicating some departure from a normal distribution.

6.20 (a) Interquartile range $= 0.0025$, $S = 0.0017$, range $= 0.008$,
$1.33(S) = 0.0023$, $6(S) = 0.0102$. Because the interquartile range is
close to $1.33S$ and the range is also close to $6S$, the data appear to be
approximately normally distributed. **(b)** The normal probability plot
suggests that the data appear to be approximately normally distributed.

6.22 (a) Five-number summary: 82 127 148.5 168 213; mean $= 147.06$,
mode $= 130$, range $= 131$, interquartile range $= 41$, standard deviation
$= 31.69$. The mean is very close to the median. The five-number
summary suggests that the distribution is approximately symmetric around
the median. The interquartile range is very close to $1.33S$. The range is
about \$50 below $6S$. In general, the distribution of the data appears to
closely resemble a normal distribution. **(b)** The normal probability plot
confirms that the data appear to be approximately normally distributed.

6.24 (a) 0.1667. **(b)** 0.1667. **(c)** 0.7083. **(d)** Mean $= 60$,
standard deviation $= 34.641$.

6.26 (a) 0.5000. **(b)** 0.2500. **(c)** 0.6000. **(d)** Mean $= 3$,
standard deviation $= 0.4082$.

6.34 (a) 0.4772. **(b)** 0.9544. **(c)** 0.0456. **(d)** 1.8835. **(e)** 1.8710 and 2.1290.

6.36 (a) 0.0228. **(b)** 0.1524. **(c)** \$275.63. **(d)** \$224.37 to \$275.63.

6.38 (a) Waiting time will more closely resemble an exponential
distribution. **(b)** Seating time will more closely resemble a normal
distribution. **(c)** Both the histogram and normal probability plot suggest
that waiting time more closely resembles an exponential distribution.
(d) Both the histogram and normal probability plot suggest that seating
time more closely resembles a normal distribution.

6.40 (a) 0.4602. **(b)** 0.3812. **(c)** 0.0808. **(d)** \$5,009.46. **(e)** \$5,156.01 and
6,723.99.

CHAPTER 7

7.2 (a) Virtually 0. **(b)** 0.1587. **(c)** 0.0139. **(d)** 50.195.

7.4 (a) Both means are equal to 6. This property is called unbiasedness.
(c) The distribution for $n = 3$ has less variability. The larger sample size
has resulted in sample means being closer to μ. **(d)** Same answer as in (c).

7.6 (a) The probability that an *individual* energy bar has a weight below
42.05 grams is 0.2743. **(b)** The probability that the *mean* of a sample of
four energy bars has a weight below 42.05 grams is 0.1151. **(c)** The probability that the *mean* of a sample of 25 energy bars has a weight below
42.05 grams is 0.00135. **(d)** (a) refers to an individual energy bar while
(c) refers to the mean of a sample of 25 energy bars. There is a 27.43%
chance that an individual energy bar will have a weight below 42.05
grams but only a chance of 0.135% that a mean of 25 energy bars will
have a weight below 42.05 grams. **(e)** Increasing the sample size from
four to 25 reduced the probability the mean will have a weight below
42.05 grams from 11.51% to 0.135%.

7.8 (a) When $n = 4$, because the mean is larger than the median, the
distribution of the sales price of new houses is skewed to the right, and
so is the sampling distribution of $\overline{X}$ although it will be less skewed than
the population. **(b)** If you select samples of $n = 100$, the shape of the
sampling distribution of the sample mean will be very close to a normal
distribution, with a mean of \$382,700 and a standard error of the mean of
\$9,000. **(c)** 0.0791. **(d)** 0.0057.

7.10 (a) 0.8413. **(b)** 16.0364. **(c)** To be able to use the standardized
normal distribution as an approximation for the area under the curve, you
must assume that the population is approximately symmetrical.
(d) 15.5182.

7.12 (a) 0.40. **(b)** 0.0704.

7.14

(a) $\pi = 0.501$, $\sigma_p = \sqrt{\dfrac{\pi(1 - \pi)}{n}} = \sqrt{\dfrac{0.501(1 - 0.501)}{100}} = 0.05$
$P(p > 0.55) = P(Z > 0.98) = 1.0 - 0.8365 = 0.1635$.

(b) $\pi = 0.60$, $\sigma_p = \sqrt{\dfrac{\pi(1 - \pi)}{n}} = \sqrt{\dfrac{0.6(1 - 0.6)}{100}} = 0.04899$
$P(p > 0.55) = P(Z > -1.021) = 1.0 - 0.1539 = 0.8461$.

(c) $\pi = 0.49$, $\sigma_p = \sqrt{\dfrac{\pi(1 - \pi)}{n}} = \sqrt{\dfrac{0.49(1 - 0.49)}{100}} = 0.05$
$P(p > 0.55) = P(Z > 1.20) = 1.0 - 0.8849 = 0.1151$.

(d) Increasing the sample size by a factor of 4 decreases the standard error by a factor of 2.
(a) $P(p > 0.55) = P(Z > 1.96) = 1.0 - 0.9750 = 0.0250$.
(b) $P(p > 0.55) = P(Z > -2.04) = 1.0 - 0.0207 = 0.9793$.
(c) $P(p > 0.55) = P(Z > 2.40) = 1.0 - 0.9918 = 0.0082$.

7.16 **(a)** 0.8522. **(b)** 0.7045. **(c)** 0.1478. **(d)** **(a)** 0.9820. **(b)** 0.9640. **(c)** 0.0180.

7.18 **(a)** 0.3594. **(b)** The probability is 90% that the sample percentage will be contained between 0.3244 to 0.4496. **(c)** The probability is 95% that the sample percentage will be contained between 0.3344 to 0.4596.

7.20 **(a)** 0.2367. **(b)** The probability is 90% that the sample percentage will be contained between 0.7534 and 0.8466. **(c)** The probability is 90% that the sample percentage will be contained between 0.7445 and 0.8555.

7.26 **(a)** 0.4999. **(b)** 0.00009. **(c)** 0.0000. **(d)** 0.0000. **(e)** 0.7518.

7.28 **(a)** 0.8944. **(b)** 4.617; 4.783. **(c)** 4.641.

7.30 **(a)** 0.0000 **(b)** 0.0126 **(c)** 0.9874.

CHAPTER 8

8.2 $114.68 \le \mu \le 135.32$.

8.4 Yes, it is true because 5% of intervals will not include the population mean.

8.6 **(a)** You would compute the mean first because you need the mean to compute the standard deviation. If you had a sample, you would compute the sample mean. If you had the population mean, you would compute the population standard deviation. **(b)** If you have a sample, you are computing the sample standard deviation, not the population standard deviation needed in Equation (8.1). If you have a population and have computed the population mean and population standard deviation, you don't need a confidence interval estimate of the population mean because you already know the mean.

8.8 Equation (8.1) assumes that you know the population standard deviation. Because you are selecting a sample of 100 from the population, you are computing a sample standard deviation, not the population standard deviation.

8.10 **(a)** $\overline{X} \pm Z \cdot \dfrac{\sigma}{\sqrt{n}} = 49{,}875 \pm 1.96 \cdot \dfrac{1{,}500}{\sqrt{64}}$;
$49{,}507.51 \le \mu \le 50{,}242.49$.
(b) Yes, because the confidence interval includes 50,000 hours the manufacturer can support a claim that the bulbs have a mean of 50,000 hours. **(c)** No. Because σ is known and $n = 64$, from the Central Limit Theorem, you know that the sampling distribution of $\overline{X}$ is approximately normal. **(d)** The confidence interval is narrower, based on a population standard deviation of 500 hours rather than the original standard $49{,}752.50 \le \mu \le 49{,}997.50$. No, because the confidence interval does not include 50,000 hours.

8.12 **(a)** 2.2622. **(b)** 3.2498. **(c)** 2.0395. **(d)** 1.9977. **(e)** 1.7531.

8.14 $-0.12 \le \mu \le 11.84$, $2.00 \le \mu \le 6.00$. The presence of the outlier increases the sample mean and greatly inflates the sample standard deviation.

8.16 **(a)** $87 \pm (1.9781)(9)/\sqrt{87}$; $85.46 \le \mu \le 88.54$. **(b)** You can be 95% confident that the population mean amount of one-time gift is between $85.46 and $88.54.

8.18 **(a)** $6.31 \le \mu \le 7.87$. **(b)** You can be 95% confident that the population mean amount spent for lunch at a fast-food restaurant is between $6.31 and $7.87. **(c)** That the population distribution is normally distributed. **(d)** The assumption of normality is not seriously violated and, with a sample of 15, the validity of the confidence interval is not seriously impacted.

8.20 **(a)** For first and second quarter ads: $5.12 \le \mu \le 5.65$. For halftime and second half ads: $5.24 \le \mu \le 6.06$. **(b)** You are 95% confident that the mean rating for first and second quarter ads is between 5.12 and 5.65. You are 95% confident that the mean rating for halftime and second half ads is between 5.24 and 6.06. **(c)** The confidence intervals for the two groups of ads are similar. **(d)** You need to assume that the distributions of the rating for the two groups of ads are normally distributed. **(e)** The distribution of the each group of ads appears approximately normally distributed.

8.22 **(a)** $31.12 \le \mu \le 54.96$. **(b)** The number of days is approximately normally distributed. **(c)** No, the outliers skew the data. **(d)** Because the sample size is fairly large, at $n = 50$, the use of the t distribution is appropriate.

8.24 **(a)** $41.85 \le \mu \le 48.31$. **(b)** That the population distribution is normally distributed. **(c)** The normal probability plot appears to be approximately normally distributed.

8.26 $0.19 \le \pi \le 0.31$.

8.28 **(a)**
$$p = \frac{X}{n} = \frac{135}{500} = 0.27, p \pm Z\sqrt{\frac{p(1-p)}{n}} = 0.27 \pm 2.58\sqrt{\frac{0.27(0.73)}{500}};$$
$0.2189 \le \pi \le 0.3211$. **(b)** The manager in charge of promotional programs can infer that the proportion of households that would upgrade to an improved cellphone if it were made available at a substantially reduced cost is somewhere between 0.22 and 0.32, with 99% confidence.

8.30 **(a)** $0.2328 \le \pi \le 0.2872$. **(b)** No, you cannot because the interval estimate includes 0.25 (25%). **(c)** $0.2514 \le \pi \le 0.2686$. Yes, you can, because the interval is above 0.25 (25%). **(d)** The larger the sample size, the narrower the confidence interval, holding everything else constant.

8.32 **(a)** $0.8632 \le \pi \le 0.8822$. **(b)** $0.1770 \le \pi \le 0.2007$. **(c)** Because almost 90% of adults have purchased something online, but only about 20% are weekly online shoppers, the director of e-commerce sales may want to focus on those adults who are weekly online shoppers.

8.34 $n = 35$.

8.36 $n = 1{,}041$.

8.38 **(a)** $n = \dfrac{Z^2 \sigma^2}{e^2} = \dfrac{(1.96)^2 (400)^2}{50^2} = 245.86$. Use $n = 246$.

(b) $n = \dfrac{Z^2 \sigma^2}{e^2} = \dfrac{(1.96)^2 (400)^2}{25^2} = 983.41$. Use $n = 984$.

8.40 $n = 55$.

8.42 **(a)** $n = 107$. **(b)** $n = 62$.

8.44 **(a)** $n = 246$. **(b)** $n = 385$. **(c)** $n = 554$. **(d)** When there is more variability in the population, a larger sample is needed to accurately estimate the mean.

8.46 **(a)** $6209 \le \pi \le 0.7878$. **(b)** $0.5015 \le \pi \le 0.6812$. **(c)** $0.0759 \le \pi \le 0.2024$. **(d)** **(a)** $n = 2{,}017$, **(b)** $n = 2{,}324$, **(c)** $n = 1{,}157$.

8.48 **(a)** If you conducted a follow-up study, you would use $\pi = 0.38$ in the sample size formula because it is based on past information on the proportion. **(b)** $n = 1{,}006$.

8.54 (a) PC/laptop: $0.8173 \leq \pi \leq 0.8628$.

Smartphone: $0.8923 \leq \pi \leq 0.9277$.

Tablet: $0.4690 \leq \pi \leq 0.5310$.

Smart watch: $0.0814 \leq \pi \leq 0.1186$.

(b) Most adults have a PC/laptop and a smartphone. Some adults have a tablet computer and very few have a smart watch.

8.56 (a) $49.88 \leq \mu \leq 52.12$. **(b)** $0.6760 \leq \pi \leq 0.9240$. **(c)** $n = 25$. **(d)** $n = 267$. **(e)** If a single sample were to be selected for both purposes, the larger of the two sample sizes ($n = 267$) should be used.

8.58 (a) $3.19 \leq \mu \leq 9.21$. **(b)** $0.3242 \leq \pi \leq 0.7158$. **(c)** $n = 110$. **(d)** $n = 121$. **(e)** If a single sample were to be selected for both purposes, the larger of the two sample sizes ($n = 121$) should be used.

8.60 (a) $0.2562 \leq \pi \leq 0.3638$. **(b)** $3.22 \leq \mu \leq \$3.78$. **(c)** $\$17,581.68 \leq \mu \leq \$18,418.32$.

8.62 (a) $\$36.66 \leq \mu \leq \$40.42$. **(b)** $0.2027 \leq \pi \leq 0.3973$. **(c)** $n = 110$. **(d)** $n = 423$. **(e)** If a single sample were to be selected for both purposes, the larger of the two sample sizes ($n = 423$) should be used.

8.64 (a) $0.4643 \leq \pi \leq 0.6690$. **(b)** $\$136.28 \leq \mu \leq \$502.21$.

8.66 (a) $13.40 \leq \mu \leq 16.56$. **(b)** With 95% confidence, the population mean answer time is somewhere between 13.40 and 16.56 seconds. **(c)** The assumption is valid as the answer time is approximately normally distributed.

8.68 (a) $0.2425 \leq \mu \leq 0.2856$. **(b)** $0.1975 \leq \mu \leq 0.2385$. **(c)** The amounts of granule loss for both brands are skewed to the right, but the sample sizes are large enough. **(d)** Because the two confidence intervals do not overlap, it appears that the mean granule loss of Boston shingles is higher than that of Vermont shingles.

CHAPTER 9

9.2 Because $Z_{STAT} = +2.21 > 1.96$, reject H_0.

9.4 Reject H_0 if $Z_{STAT} < -2.58$ or if $Z_{STAT} > 2.58$.

9.6 p-value $= 0.0456$.

9.8 p-value $= 0.1676$.

9.10 H_0: Defendant is guilty; H_1: Defendant is innocent. A Type I error would be not convicting a guilty person. A Type II error would be convicting an innocent person.

9.12 H_0: $\mu = 20$ minutes. 20 minutes is adequate travel time between classes. H_1: $\mu \neq 20$ minutes. 20 minutes is not adequate travel time between classes.

9.14 (a) $Z_{STAT} = \dfrac{49,875 - 50,000}{\dfrac{1,000}{\sqrt{64}}} = -0.6667$. Because

$-1.96 < Z_{STAT} = -0.6667 < 1.96$, do not reject H_0. **(b)** p-value $= 0.5050$. **(c)** $49,507.51 \leq \mu \leq 50,242.49$. **(d)** The conclusions are the same.

9.16 (a) Because $-2.58 < Z_{STAT} = -1.7678 < 2.58$, do not reject H_0. **(b)** p-value $= 0.0771$. **(c)** $0.9877 \leq \mu \leq 1.0023$. **(d)** The conclusions are the same.

9.18 $t_{STAT} = 2.00$.

9.20 ± 2.1315.

9.22 No, you should not use a t test because the original population is left-skewed, and the sample size is not large enough for the t test to be valid.

9.24 (a) $t_{STAT} = (3.57 - 3.70)/(0.8/\sqrt{64}) = -1.30$. Because $-1.9983 < t_{STAT} = -1.30 < 1.9983$ and p-value $= 0.1984 > 0.05$, do not reject H_0. There is insufficient evidence that the population mean waiting time is different from 3.7 minutes. **(b)** Because $n = 64$, the sampling distribution of the t test statistic is approximately normal. In general, the t test is appropriate for this sample size except for the case where the population is extremely skewed or bimodal.

9.26 (a) $-1.9842 < t_{STAT} = 1.25 < 1.9842$, do not reject H_0. There is insufficient evidence that the population mean spent by Amazon Prime customers is different from $1,475. **(b)** p-value $= 0.2142 > 0.05$. The probability of getting a t_{STAT} statistic greater than $+1.25$ or less than -1.25, given that the null hypothesis is true, is 0.2142.

9.28 (a) Because $-2.1448 < t_{STAT} = 1.6344 < 2.1448$, do not reject H_0. There is not enough evidence to conclude that the mean amount spent for lunch at a fast-food restaurant, is different from $6.50. **(b)** The p-value is 0.1245. If the population mean is $6.50, the probability of observing a sample of fifteen customers that will result in a sample mean farther away from the hypothesized value than this sample is 0.1245. **(c)** The distribution of the amount spent is normally distributed. **(d)** With a sample size of 15, it is difficult to evaluate the assumption of normality. However, the distribution may be fairly symmetric because the mean and the median are close in value. Also, the boxplot appears only slightly skewed so the normality assumption does not appear to be seriously violated.

9.30 (a) Because $-2.0096 < t_{STAT} = 0.114 < 2.0096$, do not reject H_0. There is no evidence that the mean amount is different from 2 liters. **(b)** p-value $= 0.9095$. **(c)** Yes, the data appear to have met the normality assumption. **(e)** The amount of fill is decreasing over time so the values are not independent. Therefore, the t test is invalid.

9.32 (a) Because $t_{STAT} = -5.9355 < -2.0106$, reject H_0. There is enough evidence to conclude that mean widths of the troughs is different from 8.46 inches. **(b)** The population distribution is normal. **(c)** Although the distribution of the widths is left-skewed, the large sample size means that the validity of the t test is not seriously affected. The large sample size allows you to use the t distribution.

9.34 (a) Because $-2.68 < t_{STAT} = 0.094 < 2.68$, do not reject H_0. There is no evidence that the mean amount is different from 5.5 grams. **(b)** $5.462 \leq \mu \leq 5.542$. **(c)** The conclusions are the same.

9.36 p-value $= 0.0228$.

9.38 p-value $= 0.0838$.

9.40 p-value $= 0.9162$.

9.42 2.7638.

9.44 -2.5280.

9.46 (a) $t_{STAT} = 2.6880 > 1.6694$, reject H_0. There is evidence that the population mean bus miles is greater than 8,000 miles. **(b)** p-value $= 0.0046 < 0.05$. The probability of getting a t_{STAT} statistic greater than 2.6880, given that the null hypothesis is true, is 0.0046.

9.48 (a) $t_{STAT} = (24.05 - 30)/(16.5/\sqrt{860}) = -10.5750$. Because $t_{STAT} = -10.5750 < -2.3307$, reject H_0. p-value $= 0.0000 < 0.01$, reject H_0. **(b)** The probability of getting a sample mean of 24 minutes or less if the population mean is 30 minutes is 0.000.

9.50 (a) $t_{STAT} = 1.9221 < 2.3549$, do not reject H_0. There is insufficient evidence that the population mean one-time gift donation is greater than $85.50. **(b)** The probability of getting a sample mean of $87 or more if the population mean is $85.50 is 0.0284.

9.52 $p = 0.22$.

9.54 Do not reject H_0.

9.56 (a) $Z_{STAT} = -2.0681$, p-value $= 0.0193$. Because $Z_{STAT} = -2.0681 < 1.645$ or p-value $= 0.0193 < 0.05$, reject H_0. There is evidence to show that less than 69.52% of students at your university use the Chrome web browser. **(b)** $Z_{STAT} = -5.0658$, p-value $= 0.0000$. Because $Z_{STAT} = -5.0658 < 1.645$, or p-value $= 0.0000 < 0.05$, reject H_0. There is evidence to show that less than 69.52% of students at your university use the Chrome web browser. **(c)** The sample size had an effect on being able to reject the null hypothesis. **(d)** You would be very unlikely to reject the null hypothesis with a sample of 20.

9.58 H_0: $\pi = 0.60$; H_1: $\pi \neq 0.60$. Decision rule: If $Z_{STAT} > 1.96$ or $Z_{STAT} < -1.96$, reject H_0.

$$p = \frac{464}{703} = 0.6600$$

Test statistic:

$$Z_{STAT} = \frac{p - \pi}{\sqrt{\dfrac{\pi(1 - \pi)}{n}}} = \frac{0.6600 - 0.60}{\sqrt{\dfrac{0.60(1 - 0.60)}{703}}} = 3.2488.$$

Because $Z_{STAT} = 3.2488 > 1.96$ or p-value $= 0.0012 < 0.05$, reject H_0 and conclude that there is evidence that the proportion of all talent acquisition professionals who report competition is the biggest obstacle to attracting the best talent at their company is different from 60%.

9.60 (a) H_0: $\pi \geq 0.294$. H_1: $\pi < 0.294$.
(b) $Z_{STAT} = -0.5268 > -1.645$; p-value $= 0.2992$. Because $Z_{STAT} = -0.5268 > -1.645$ or p-value $= 0.2992 > 0.05$, do not reject H_0. There is insufficient evidence that the percentage is less than 29.4%.

9.70 (a) Concluding that a firm will go bankrupt when it will not.
(b) Concluding that a firm will not go bankrupt when it will go bankrupt.
(c) Type I. **(d)** If the revised model results in more moderate or large Z scores, the probability of committing a Type I error will increase. Many more of the firms will be predicted to go bankrupt than will go bankrupt. On the other hand, the revised model that results in more moderate or large Z scores will lower the probability of committing a Type II error because few firms will be predicted to go bankrupt than will actually go bankrupt.

9.72 (a) Because $t_{STAT} = 3.3197 > 2.0010$, reject H_0. **(b)** p-value $= 0.0015$. **(c)** Because $Z_{STAT} = 0.2582 < 1.645$, do not reject H_0. **(d)** Because $-2.0010 < t_{STAT} = -1.1066 < 2.0010$, do not reject H_0. **(e)** Because $Z_{STAT} = 2.3238 > 1.645$, reject H_0.

9.74 (a) Because $t_{STAT} = -1.69 > -1.7613$, do not reject H_0. **(b)** The data are from a population that is normally distributed. **(d)** With the exception of one extreme value, the data are approximately normally distributed. **(e)** There is insufficient evidence to state that the waiting time is less than five minutes.

9.76 (a) Because $t_{STAT} = -1.47 > -1.6896$, do not reject H_0. **(b)** p-value $= 0.0748$. If the null hypothesis is true, the probability of obtaining a t_{STAT} of -1.47 or more extreme is 0.0748. **(c)** Because $t_{STAT} = -3.10 < -1.6973$, reject H_0. **(d)** p-value $= 0.0021$. If the null hypothesis is true, the probability of obtaining a t_{STAT} of -3.10 or more extreme is 0.0021. **(e)** The data in the population are assumed to be normally distributed. **(g)** Both boxplots suggest that the data are skewed slightly to the right, more so for the Boston shingles. However, the very large sample sizes mean that the results of the t test are relatively insensitive to the departure from normality.

9.78 (a) $t_{STAT} = -3.2912$, reject H_0. **(b)** p-value $= 0.0012$. The probability of getting a t_{STAT} value below -3.2912 or above $+3.2912$ is 0.0012.

(c) $t_{STAT} = -7.9075$, reject H_0. **(d)** p-value $= 0.0000$. The probability of getting a t_{STAT} value below -7.9075 or above $+7.9075$ is 0.0000.
(e) Because of the large sample sizes, you do not need to be concerned with the normality assumption.

CHAPTER 10

10.2 (a) $t = 3.8959$. **(b)** $df = 21$. **(c)** 2.5177. **(d)** Because $t_{STAT} = 3.8959 > 2.5177$, reject H_0.

10.4 $3.73 \leq \mu_1 - \mu_2 \leq 12.27$.

10.6 Because $t_{STAT} = 2.6762 < 2.9979$ or p-value $= 0.0158 > 0.01$, do not reject H_0. There is no evidence that the mean of population one is greater than the mean of population 2.

10.8 (a) Because $t_{STAT} = 2.8990 > 1.6620$ or p-value $= 0.0024 < 0.05$, reject H_0. There is evidence that the mean amount of Walker Crisps eaten by children who watched a commercial featuring a long-standing sports celebrity endorser is higher than for those who watched a commercial for an alternative food snack.
(b) $3.4616 \leq \mu_1 - \mu_2 \leq 18.5384$. **(c)** The results cannot be compared because (a) is a one-tail test and (b) is a confidence interval that is comparable only to the results of a two-tail test. **(d)** You would choose the commercial featuring a long-standing celebrity endorser.

10.10 (a) H_0: $\mu_1 = \mu_2$, where Populations: 1 = Southeast, 2 = Gulf Coast. H_1: $\mu_1 \neq \mu_2$. Decision rule: $df = 42$. If $t_{STAT} < -2.0181$ or $t_{STAT} > 2.0181$, reject H_0.
Test statistic:

$$S_p^2 = \frac{(n_1 - 1)(S_1^2) + (n_2 - 1)(S_2^2)}{(n_1 - 1) + (n_2 - 1)}$$

$$= \frac{(21)(51.8684^2) + (21)(62.2588^2)}{21 + 21} = 3{,}283.2435$$

$$t_{STAT} = \frac{(\overline{X}_1 - \overline{X}_2) - (\mu_1 - \mu_2)}{\sqrt{S_p^2\left(\dfrac{1}{n_1} + \dfrac{1}{n_2}\right)}}$$

$$= \frac{(45.0455 - 40.8182) - 0}{\sqrt{3{,}283.2435\left(\dfrac{1}{22} + \dfrac{1}{22}\right)}} = 0.2447.$$

Decision: Because $-2.0181 < t_{STAT} = 0.2447 < 2.0181$, do not reject H_0. There is not enough evidence to conclude that the mean number of partners between the Southeast and Gulf Coast is different.
(b) p-value $= 0.8079$. **(c)** In order to use the pooled-variance t test, you need to assume that the populations are normally distributed with equal variances.

10.12 (a) Because $t_{STAT} = -4.1343 < -2.0484$, reject H_0.
(b) p-value $= 0.0003 < 0.05$, reject H_0. **(c)** The populations of waiting times are approximately normally distributed.
(d) $-4.2292 \leq \mu_1 - \mu_2 \leq -1.4268$.

10.14 (a) Because $t_{STAT} = 2.7349 > 2.0484$, reject H_0. There is evidence of a difference in the mean time to start a business between developed and emerging countries. **(b)** p-value $= 0.0107$. The probability that two samples have a mean difference of 14.62 or more is 0.0107 if there is no difference in the mean time to start a business between developed and emerging countries. **(c)** You need to assume that the population distribution of the time to start a business of both developed and emerging countries is normally distributed. **(d)** $3.6700 \leq \mu_1 - \mu_2 \leq 25.5700$.

10.16 (a) Because $t_{STAT} = -2.1554 < -2.0017$ or p-value $= 0.03535 < 0.05$, reject H_0. There is evidence of a difference in the mean time per day accessing the Internet via a mobile device between males and females. **(b)** You must assume that each of the two independent populations is normally distributed.

10.18 $df = 19$.

10.20 (a) $t_{STAT} = (-1.5566)/(1.424/\sqrt{9}) = -3.2772$. Because $t_{STAT} = -3.2772 < -2.306$ or p-value $= 0.0112 < 0.05$, reject H_0. There is enough evidence of a difference in the mean summated ratings between the two brands. **(b)** You must assume that the distribution of the differences between the two ratings is approximately normal. **(c)** p-value $= 0.0112$. The probability of obtaining a mean difference in ratings that results in a test statistic that deviates from 0 by 3.2772 or more in either direction is 0.0112 if there is no difference in the mean summated ratings between the two brands. **(d)** $-2.6501 \le \mu_D \le -0.4610$. You are 95% confident that the mean difference in summated ratings between brand A and brand B is somewhere between -2.6501 and -0.4610.

10.22 (a) Because $t_{STAT} = -4.1372 < -2.0244$, reject H_0. There is evidence to conclude that the mean download speed at AT&T is lower than at Verizon Wireless. **(b)** You must assume that the distribution of the differences between the ratings is approximately normal. **(d)** The confidence interval is from -2.9359 to -1.0067.

10.24 (a) Because $t_{STAT} = 1.8425 < 1.943$, do not reject H_0. There is not enough evidence to conclude that the mean bone marrow microvessel density is higher before the stem cell transplant than after the stem cell transplant. **(b)** p-value $= 0.0575$. The probability that the t statistic for the mean difference in microvessel density is 1.8425 or more is 5.75% if the mean density is not higher before the stem cell transplant than after the stem cell transplant. **(c)** $-28.26 \le \mu_D \le 200.55$. You are 95% confident that the mean difference in bone marrow microvessel density before and after the stem cell transplant is somewhere between -28.26 and 200.55. **(d)** That the distribution of the difference before and after the stem cell transplant is normally distributed.

10.26 (a) Because $t_{STAT} = -9.3721 < -2.4258$, reject H_0. There is evidence that the mean strength is lower at two days than at seven days. **(b)** The population of differences in strength is approximately normally distributed. **(c)** $p = 0.000 < 0.05$, reject H_0.

10.28 (a) Because $-2.58 \le Z_{STAT} = -0.58 \le 2.58$, do not reject H_0. **(b)** $-0.273 \le \pi_1 - \pi_2 \le 0.173$.

10.30 (a) $H_0: \pi_1 \le \pi_2$. $H_1: \pi_1 > \pi_2$. Populations: $1 =$ VOD D4+ $2 =$ general TV. **(b)** Because $Z_{STAT} = 8.9045 > 1.6449$ or p-value $= 0.0000 < 0.05$, reject H_0. There is evidence to conclude that the population proportion of those who viewed the brand on VOD D4+ were more likely to visit the brand website. **(c)** Yes, the result in (b) makes it appropriate to claim that the population proportion of those who viewed the brand on VOD D4+ were more likely to visit the brand website than those who viewed the brand on general TV.

10.32 (a) $H_0: \pi_1 = \pi_2$. $H_1: \pi_1 \ne \pi_2$. Decision rule: If $|Z_{STAT}| > 2.58$, reject H_0.

Test statistic: $\bar{p} = \dfrac{X_1 + X_2}{n_1 + n_2} = \dfrac{326 + 167}{423 + 192} = 0.8016$

$Z_{STAT} = \dfrac{(p_1 - p_2) - (\pi_1 - \pi_2)}{\sqrt{\bar{p}(1-\bar{p})\left(\dfrac{1}{n_1} + \dfrac{1}{n_2}\right)}} = \dfrac{(0.7707 - 0.8698) - 0}{\sqrt{0.8016(1-0.8016)\left(\dfrac{1}{423} + \dfrac{1}{192}\right)}}$.

$Z_{STAT} = -2.8560 < -2.58$, reject H_0. There is evidence of a difference in the proportion of organizations with recognition programs between organizations that have between 500 and 2,499 employees and organizations that have 10,000+ employees **(b)** p-value $= 0.0043$.

The probability of obtaining a difference in proportions that gives rise to a test statistic below -2.8516 or above $+2.8516$ is 0.0043 if there is no difference in the proportion based on the size of the organization. **(c)** $-0.1809 \le \pi_1 - \pi_2 \le -0.0173$. You are 99% confident that the difference in the proportion based on the size of the organization is between 1.73% and 18.09%.

10.34 (a) Because $Z_{STAT} = 4.4662 > 1.96$, reject H_0. There is evidence of a difference in the proportion of cobrowsing organizations and non-cobrowsing organizations that use skills-based routing to match the caller with the *right* agent. **(b)** p-value $= 0.0000$. The probability of obtaining a difference in proportions that is 0.2586 or more in either direction is 0.0000 if there is no difference between the proportion of cobrowsing organizations and non-cobrowsing organizations that use skills-based routing to match the caller with the *right* agent.

10.36 (a) 2.20. **(b)** 2.57. **(c)** 3.50.

10.38 (a) Population B: $S^2 = 25$. **(b)** 1.5625.

10.40 $df_{\text{numerator}} = 24$, $df_{\text{denominator}} = 24$.

10.42 Because $F_{STAT} = 1.2109 < 2.27$, do not reject H_0.

10.44 (a) Because $F_{STAT} = 1.2995 < 3.18$, do not reject H_0. **(b)** Because $F_{STAT} = 1.2995 < 2.62$, do not reject H_0.

10.46 (a) $H_0: \sigma_1^2 = \sigma_2^2$. $H_1: \sigma_1^2 \ne \sigma_2^2$.

Decision rule: If $F_{STAT} > 2.4086$, reject H_0.

Test statistic: $F_{STAT} = \dfrac{S_1^2}{S_2^2} = \dfrac{3,876.1558}{2,690.3312} = 1.4408$.

Decision: Because $F_{STAT} = 1.4408 < 2.4086$, do not reject H_0. There is insufficient evidence to conclude that the two population variances are different. **(b)** p-value $= 0.4096 > 0.05$, do not reject H_0. **(c)** The test assumes that each of the two populations is normally distributed. **(d)** Based on (a) and (b), a pooled-variance t test should be used.

10.48 (a) Because $F_{STAT} = 2.1187 > 2.1124$ or p-value $= 0.0491 = 0.05$, reject H_0. There is sufficient evidence of a difference in the variability of the ratings between the two groups. **(b)** p-value $= 0.05$. The probability of obtaining a sample that yields a test statistic more extreme than 2.1187 is 0.049 if there is no difference in the two population variances. **(c)** The test assumes that each of the two populations are normally distributed. **(d)** Based on (a) and (b), a separate-variance t test should be used.

10.50 (a) Because $F_{STAT} = 69.50001 > 1.9811$ or p-value $= 0.0000 < 0.05$, reject H_0. There is evidence of a difference in the variance of the delay times between the two drivers. **(b)** You assume that the delay times are normally distributed. **(c)** From the boxplot and the normal probability plots, the delay times appear to be approximately normally distributed. **(d)** Because there is a difference in the variance of the delay times between the two drivers, you should use the separate variance t-test to determine whether there is evidence of a difference in the mean delay time between the two drivers.

10.58 (a) Because $F_{STAT} = 2.9736 > 1.9288$, or p-value $= 0.0016 < 0.05$, reject H_0. There is a difference in the variance of the salary of Black Belts and Green Belts. **(b)** The separate-variance t test. **(c)** Because $t_{STAT} = 5.9488 > 1.6639$ or p-value $= 0.0000 < 0.05$, reject H_0. There is evidence that the mean salary of Black Belts is greater than the mean salary of Green Belts.

10.60 (a) Because $F_{STAT} = 1.3611 < 1.6854$, do not reject H_0. There is insufficient evidence to conclude that there is a difference between the variances in the online time per week between women and men. **(b)** It is more appropriate to use a pooled-variance t test. Using the pooled-variance t test, because $t_{STAT} = -9.7619 < -2.6009$, reject H_0. There is evidence of a difference in the mean online time per week between women and men. **(c)** Because $F_{STAT} = 1.7778 > 1.6854$, reject H_0. There is evidence to conclude that there is a difference between the variances in the time spent playing games between women and men. **(d)** Using the separate-variance t test, because $t_{STAT} = -26.4 < -2.603$, reject H_0. There is evidence of a difference in the mean time spent playing games between women and men.

10.62 (a) Because $t_{STAT} = 3.3282 > 1.8595$, or the p-value $= 0.0052 < 0.05$ reject H_0. There is enough evidence to conclude that the introductory computer students required more than a mean of 10 minutes to write and run a program in Python. **(b)** Because $t_{STAT} = 1.3636 < 1.8595$, do not reject H_0. There is not enough evidence to conclude that the introductory computer students required more than a mean of 10 minutes to write and run a Python program. **(c)** Although the mean time necessary to complete the assignment increased from 12 to 16 minutes as a result of the increase in one data value, the standard deviation went from 1.8 to 13.2, which reduced the value of t statistic. **(d)** Because $F_{STAT} = 1.2308 < 4.2951$, do not reject H_0. There is not enough evidence to conclude that the population variances are different for the Introduction to Computers students and computer majors. Hence, the pooled-variance t test is a valid test to determine whether computer majors can write a Python program in less time than introductory students, assuming that the distributions of the time needed to write a Python program for both the Introduction to Computers students and the computer majors are approximately normally distributed. Because $t_{STAT} = 4.0666 > 1.7341$, reject H_0. There is enough evidence that the mean time is higher for Introduction to Computers students than for computer majors. **(e)** p-value $= 0.0052$. If the true population mean amount of time needed for Introduction to Computer students to write a Python program is no more than 10 minutes, the probability of observing a sample mean greater than the 12 minutes in the current sample is 0.0362%. Hence, at a 5% level of significance, you can conclude that the population mean amount of time needed for Introduction to Computer students to write a Python program is more than 10 minutes. As illustrated in (d), in which there is not enough evidence to conclude that the population variances are different for the Introduction to Computers students and computer majors, the pooled-variance t test performed is a valid test to determine whether computer majors can write a Python program in less time than introductory students, assuming that the distribution of the time needed to write a Python program for both the Introduction to Computers students and the computer majors are approximately normally distributed.

10.64 From the boxplot and the summary statistics, both distributions are approximately normally distributed. $F_{STAT} = 1.056 < 1.89$. There is insufficient evidence to conclude that the two population variances are significantly different at the 5% level of significance. $t_{STAT} = -5.084 < -1.99$. At the 5% level of significance, there is sufficient evidence to reject the null hypothesis of no difference in the mean life of the bulbs between the two manufacturers. You can conclude that there is a significant difference in the mean life of the bulbs between the two manufacturers.

10.66 (a) Because $Z_{STAT} = 3.6911 > 1.96$, reject H_0. There is enough evidence to conclude that there is a difference in the proportion of men and women who order dessert. **(b)** Because $Z_{STAT} = 6.0873 > 1.96$, reject H_0. There is enough evidence to conclude that there is a difference in the proportion of people who order dessert based on whether they ordered a beef entree.

10.68 The normal probability plots suggest that the two populations are not normally distributed. An F test is inappropriate for testing the difference in the two variances. The sample variances for Boston and Vermont shingles are 0.0203 and 0.015, respectively. Because $t_{STAT} = 3.015 > 1.967$ or p-value $= 0.0028 < \alpha = 0.05$, reject H_0. There is sufficient evidence to conclude that there is a difference in the mean granule loss of Boston and Vermont shingles.

CHAPTER 11

11.2 (a) $SSW = 150$. **(b)** $MSA = 15$. **(c)** $MSW = 5$. **(d)** $F_{STAT} = 3$.

11.4 (a) 2. **(b)** 18. **(c)** 20.

11.6 (a) Reject H_0 if $F_{STAT} > 2.95$; otherwise, do not reject H_0. **(b)** Because $F_{STAT} = 4 > 2.95$, reject H_0. **(c)** The table does not have 28 degrees of freedom in the denominator, so use the next larger critical value, $Q_\alpha = 3.90$. **(d)** Critical range $= 6.166$.

11.8 (a) $H_0:\mu_A = \mu_B = \mu_C = \mu_D$ and H_1: At least one mean is different.

$$MSA = \frac{SSA}{c-1} = \frac{1,151,016.4750}{3} = 383,672.1583.$$

$$MSW = \frac{SSW}{n-c} = \frac{2,961,835.3000}{36} = 82,273.2028.$$

$$F_{STAT} = \frac{MSA}{MSW} = \frac{383,672.1583}{82,273.2028} = 4.6634.$$

Because the p-value is 0.0075 and $F_{STAT} = 5.7121 > 4.6634$, reject H_0. There is sufficient evidence of a difference in the mean import cost across

the four global regions. **(b)** Critical range $= Q_\alpha \sqrt{\dfrac{MSW}{2}\left(\dfrac{1}{n_j} + \dfrac{1}{n_{j'}}\right)}$

$$= 3.81\sqrt{\frac{82,273.2028}{2}\left(\frac{1}{10} + \frac{1}{10}\right)} = 90.7046.$$

From the Tukey-Kramer procedure, there is a difference in the mean import cost among the East Asia and Pacific region, Latin America and the Caribbean, Eastern Europe and Central Asia, and Latin American and Caribbean. None of the other regions are different. **(c)** ANOVA output for Levene's test for homogeneity of variance:

$$MSA = \frac{SSA}{c-1} = \frac{191890.4750}{3} = 63,630.1583$$

$$MSW = \frac{SSW}{n-c} = \frac{1,469,223.4}{36} = 40,811.7611$$

$$F_{STAT} = \frac{MSA}{MSW} = \frac{63,630.1583}{40,811.7611} = 1.5591$$

Because p-value $= 0.2161 > 0.05$ and $F_{STAT} = 1.5591 < 2.8663$, do not reject H_0. There is insufficient evidence to conclude that the variances in the import cost are different. **(d)** From the results in (a) and (b), the mean import cost for the East Asia and Pacific region and eastern Europe and Central Asia is lower than for Latin America and the Caribbean.

11.10 (a) Because $F_{STAT} = 12.56 > 2.76$, reject H_0. **(b)** Critical range $= 4.67$. Advertisements A and B are different from Advertisements C and D. Advertisement E is only different from Advertisement D. **(c)** Because $F_{STAT} = 1.927 < 2.76$, do not reject H_0. There is no evidence of a significant difference in the variation in the ratings among the five advertisements. **(d)** The advertisements underselling the pen's characteristics had the highest mean ratings, and the advertisements overselling the pen's characteristics had the lowest mean ratings. Therefore, use an advertisement that undersells the pen's characteristics and avoid advertisements that oversell the pen's characteristics.

11.12 (a)

Source	Degrees of Freedom	Sum of Squares	Mean Squares	F
Among groups	2	15,671,226,037.72	3,134,245,208	0.6808
Within groups	50	230,181,430,044.96	4603,628,601	
Total	52	245,852,656,082.6800		

(b) Because $F_{STAT} = 0.6808 < 2.41$, do not reject H_0. There is insufficient evidence of a difference in the mean brand value of the different groups.
(c) Because there was no significant difference among the groups, none of the critical ranges were significant.

11.14 (a) Because $F_{STAT} = 5.3495 > 2.8663$; p-value $= 0.0038 < 0.05$, reject H_0. **(b)** Critical range $= 9.3401$ (using 36 degrees of freedom and interpolating). Asia is different from North America.
(c) The assumptions are that the samples are randomly and independently selected (or randomly assigned), the original populations of congestion are approximately normally distributed, and the variances are equal.
(d) Because $F_{STAT} = 0.4321 < 2.8663$; p-value $= 0.7333 > 0.05$, do not reject H_0. There is insufficient evidence of a difference in the variation in the mean congestion level among the continents.

11.16 (a) 40. **(b)** 60 and 55. **(c)** 10. **(d)** 10.

11.18 (a) Because $F_{STAT} = 6.00 > 3.35$, reject H_0. **(b)** Because $F_{STAT} = 5.50 > 3.35$, reject H_0. **(c)** Because $F_{STAT} = 1.00 < 2.73$, do not reject H_0.

11.20 $df_B = 4$, $df_{TOTAL} = 44$, $SSA = 160$, $SSAB = 80$, $SSE = 150$, $SST = 610$, $MSB = 55$, $MSE = 5$. For A: $F_{STAT} = 16$.
For B: $F_{STAT} = 11$. For AB: $F_{STAT} = 2$. **(a)** Because $F_{STAT} = 16 > 3.32$, reject H_0. Factor A is significant. (b) Because $F_{STAT} = 11 > 2.69$, reject H_0. Factor B is significant. (c) Because $F_{STAT} = 2.0 < 2.27$, do not reject H_0. The AB interaction is not significant.

11.22 (a) Because $F_{STAT} = 3.4032 < 4.3512$, do not reject H_0.
(b) Because $F_{STAT} = 1.8496 < 4.3512$, do not reject H_0. **(c)** Because $F_{STAT} = 9.4549 > 4.3512$, reject H_0. **(e)** Die diameter has a significant effect on density, but die temperature does not. However, the cell means plot shows that the density seems higher with a 3 mm die diameter at 155°C but that there is little difference in density with a 4 mm die diameter. This interaction is not significant at the 0.05 level of significance.

11.24 (a) H_0: There is no interaction between filling time and mold temperature. H_1: There is an interaction between filling time and mold temperature.

Because $F_{STAT} = \dfrac{0.1136}{0.05} = 2.27 < 2.9277$ or the p-value $=$

$0.1018 > 0.05$, do not reject H_0. There is insufficient evidence of interaction between filling time and mold temperature. **(b)** $F_{Stat} = 9.0222 > 3.5546$, reject H_0. There is evidence of a difference in the warpage due to the filling time. **(c)** $F_{Stat} = 4.2305 > 3.5546$, reject H_0. There is evidence of a difference in the warpage due to the mold temperature. **(e)** The warpage for a three-second filling time seems to be much higher at 60°C and 72.5°C but not at 85°C.

11.26 (a) $F_{STAT} = 0.8325$, p-value $= 0.3725 > 0.05$, do not reject H_0. There is not enough evidence to conclude that there is an interaction between zone lower and zone 3 upper. **(b)** $F_{STAT} = 0.3820$, p-value is $0.5481 > 0.05$, do not reject H_0. There is insufficient evidence to conclude that there is an effect due to zone 1 lower. **(c)** $F_{STAT} = 0.1048$, p-value $= 0.7517 > 0.05$, do not reject H_0. There is inadequate evidence to conclude that there is an effect due to zone 3 upper.
(d) A large difference at a zone 3 upper of 695°C but only a small difference at zone 3 upper of 715°C. **(e)** Because this difference appeared on the cell means plot but the interaction was not statistically significant because of the large MSE, further testing should be done with larger sample sizes.

11.36 (a) Because $F_{STAT} = 0.0111 < 2.9011$, do not reject H_0.
(b) Because $F_{STAT} = 0.8096 < 4.1491$, do not reject H_0. **(c)** Because $F_{STAT} = 5.1999 > 2.9011$, reject H_0. **(e)** Critical range $= 3.56$. Only the means of Suppliers 1 and 2 are different. You can conclude that the

mean strength is lower for Supplier 1 than for Supplier 2, but there are no statistically significant differences between Suppliers 1 and 3, Suppliers 1 and 4, Suppliers 2 and 3, Suppliers 2 and 4, and Suppliers 3 and 4.
(f) $F_{STAT} = 5.6998 > 2.8663$ (p-value $= 0.0027 < 0.05$).
There is evidence that the mean strength of suppliers is different.
Critical range $= 3.359$. Supplier 1 has a mean strength that is less than suppliers 2 and 3.

11.38 (a) Because $F_{STAT} = 0.075 < 3.68$, do not reject H_0. **(b)** Because $F_{STAT} = 4.09 > 3.68$, reject H_0. **(c)** Critical range $= 1.489$. Breaking strength is significantly different between 30 and 50 psi.

11.40 (a) Because $F_{STAT} = 0.1899 < 4.1132$, do not reject H_0.
There is insufficient evidence to conclude that there is any interaction between type of breakfast and desired time. **(b)** Because $F_{STAT} = 30.4434 > 4.1132$, reject H_0. There is sufficient evidence to conclude that there is an effect due to type of breakfast. **(c)** Because $F_{STAT} = 12.4441 > 4.1132$, reject H_0. There is sufficient evidence to conclude that there is an effect due to desired time. **(e)** At the 5% level of significance, both the type of breakfast ordered and the desired time have an effect on delivery time difference. There is no interaction between the type of breakfast ordered and the desired time.

11.42 Interaction: $F_{STAT} = 0.2169 < 3.9668$ or p-value $= 0.6428 > 0.05$. There is insufficient evidence of an interaction between piece size and fill height. Piece size: $F_{STAT} = 842.2242 > 3.9668$ or p-value $= 0.0000 < 0.05$. There is evidence of an effect due to piece size. The fine piece size has a lower difference in coded weight. Fill height: $F_{STAT} = 217.0816 > 3.9668$ or p-value $= 0.0000 < 0.05$. There is evidence of an effect due to fill height. The low fill height has a lower difference in coded weight.

CHAPTER 12

12.2 (a) For $df = 1$ and $\alpha = 0.05$, $\chi_\alpha^2 = 3.841$. **(b)** For $df = 1$ and $\alpha = 0.025$, $\chi^2 = 5.024$. **(c)** For $df = 1$ and $\alpha = 0.01$, $\chi_\alpha^2 = 6.635$.

12.4 (a) All $f_e = 25$. **(b)** Because $\chi_{STAT}^2 = 4.00 > 3.841$, reject H_0.

12.6 (a) $H_0: \pi_1 = \pi_2$. $H_1: \pi_1 \neq \pi_2$. **(b)** Because $\chi_{STAT}^2 = 79.29 > 3.841$, reject H_0. There is evidence to conclude that the population proportion of those who viewed the brand on general TV was different from those who viewed the brand on VOD D4+. p-value $= 0.0000$. The probability of obtaining a test statistic of 79.29 or larger when the null hypothesis is true is 0.0000. **(c)** You should not compare the results in (a) to those of Problem 10.30 (b) because that was a one-tail test.

12.8 (a) $H_0: \pi_1 = \pi_2$. $H_1: \pi_1 \neq \pi_2$. Because $\chi_{STAT}^2 = (326 - 339.0878)^2/339.0878 + (97 - 83.9122)^2/83.9122 + (167 - 153.9122)^2/153.9122 + (25 - 38.0878)^2/38.0878 = 8.1566 > 6.635$, reject H_0.
There is evidence of a difference in the proportion of organizations with 500 to 2,499 employees and organizations with 2,500+ employees with respect to the proportion that have employee recognition programs.
(b) p-value $= 0.0043$. The probability of obtaining a difference in proportions that gives rise to a test statistic above 8.1566 is 0.0043 if there is no difference in the proportion in the two groups. **(c)** The results of (a) and (b) are exactly the same as those of Problem 10.32. The χ^2 in (a) and the Z in Problem 10.32 (a) satisfy the relationship that $\chi^2 = 8.1566 = Z^2 = (-2.856)^2$, and the p-value in (b) is exactly the same as the p-value computed in Problem 10.32 (b).

12.10 (b) Because $\chi_{STAT}^2 = 19.9467 > 3.841$, reject H_0. There is evidence that there is a significant difference between the proportion of cobrowsing organizations and non-cobrowsing organizations that use skills-based routing to match the caller with the *right* agent. **(c)** p-value is virtually zero. The probability of obtaining a test statistic of 19.9467 or larger when the null hypothesis is true is 0.0000. **(d)** The results are identical because $(4.4662)^2 = 19.9467$.

12.12 (a) The expected frequencies for the first row are 20, 30, and 40. The expected frequencies for the second row are 30, 45, and 60. **(b)** Because $\chi^2_{STAT} = 12.5 > 5.991$, reject H_0.

12.14 (a) Because the calculated test statistic 96.7761 is greater than the critical value of 7.8147, you reject H_0 and conclude that there is evidence of a difference among the age groups in the proportion of smartphone owners who mostly use smartphones to go online. **(b)** p-value = 0.0000. The probability of obtaining a data set that gives rise to a test statistic of 96.7761 or more is 0.0000 if there is no difference in the proportion who mostly use smartphones to go online. **(c)** There is a significant difference between 18- to 29-year-olds and 50- to 64-years-olds and those 65 and older. There is a significant difference between 30- to 49-year-olds and 50- to 64-year-olds and those 65 and older. There is a significant difference between those who are between 50 and 64 years old and those 65 years old or older.

12.16 (a) $H_0: \pi_1 = \pi_2 = \pi_3$. H_1: At least one proportion differs.

Observed Frequencies

Compensation value	Group BE	HR	Employees	Total
Yes	28	76	66	170
No	172	124	134	430
Total	200	200	200	600

Expected Frequencies

Investing?	Global Region NA	E	A	Total
Yes	56.6667	56.6667	56.6667	170
No	143.3333	143.3333	143.3333	430
Total	200	200	200	600

Data

Level of Significance	0.05
Number of Rows	2
Number of Columns	3
Degrees of Freedom	2

Results

Critical Value	5.9915
Chi-Square Test Statistic	31.5841
p-Value	0.0000
Reject the null hypothesis	

Because $31.5841 > 5.9915$, reject H_0.
There is a significant difference among business groups with respect to the proportion that say compensation (pay and rewards) makes for a unique and compelling EVP. **(b)** p-value = 0.0000. The probability of a test statistic greater than 31.5841 is 0.0000. **(c)**

Level of Significance	0.05
Square Root of Critical Value	2.4477

Sample Proportions

Group 1	0.14
Group 2	0.38
Group 3	0.33

Marascuilo Table

Proportions	Absolute Differences	Critical Range	
|Group 1 − Group 2|	0.124	0.1033	Significant
|Group 1 − Group 3|	0.19	0.1011	Significant
|Group 2 − Group 3|	0.05	0.1170	Not significant

Business executives are different from HR leaders and from employees.

12.18 (a) Because $\chi^2_{STAT} = 31.6888 > 5.9915$, reject H_0. There is evidence of a difference in the percentage who use their device to check social media while watching TV between the groups. **(b)** p-value = 0.0000. **(c)** Smartphone versus computer: $0.1616 > 0.0835$. Significant. Smartphone versus tablet: $0.1805 > 0.0917$. Significant. Computer versus tablet: $0.0188 < 0.0998$. Not significant. The smartphone group is different from the computer and tablet groups.

12.20 $df = (r − 1)(c − 1) = (3 − 1)(4 − 1) = 6$.

12.22 Because $\chi^2_{STAT} = 92.1028 > 16.919$, reject H_0 and conclude that there is evidence of a relationship between the type of dessert ordered and the type of entrée ordered.

12.24 Because $\chi^2_{STAT} = 27.06509 < 31.9993$, do not reject H_0. There is no evidence of a relationship between time willing to devote to health and age.

12.26 Because $\chi^2_{STAT} = 81.6061 > 47.3999$ reject H_0. There is evidence of a relationship between identified main opportunity and geographic region.

12.28 (a) 31. **(b)** 29. **(c)** 27. **(d)** 25.

12.30 40 and 79.

12.32 (a) The ranks for Sample 1 are 1, 2, 4, 5, and 10. The ranks for Sample 2 are 3, 6.5, 6.5, 8, 9, and 11. **(b)** 22. **(c)** 44.

12.34 Because $T_1 = 22 > 20$, do not reject H_0.

12.36 (a) The data are ordinal. **(b)** The two-sample t test is inappropriate because the data can only be placed in ranked order. **(c)** Because $Z_{STAT} = −2.2054 < −1.96$, reject H_0. There is evidence of a significance difference in the median rating of California Cabernets and Washington Cabernets.

12.38 (a) $H_0: M_1 = M_2$, where Populations: $1 =$ Wing A, $2 =$ Wing B. $H_1: M_1 \neq M_2$.

Population 1 sample: Sample size 20, sum of ranks 561

Population 2 sample: Sample size 20, sum of ranks 259

$$\mu_{T_1} = \frac{n_1(n+1)}{2} = \frac{20(40+1)}{2} = 410$$

$$\sigma_{T_1} = \sqrt{\frac{n_1 n_2(n+1)}{12}} = \sqrt{\frac{20(20)(40+1)}{12}} = 36.9685$$

$$Z_{STAT} = \frac{T_1 - \mu_{T_1}}{S_{T_1}} = \frac{561 - 410}{36.9685} = 4.0846$$

Decision: Because $Z_{STAT} = 4.0846 > 1.96$ (or p-value = $0.0000 < 0.05$), reject H_0. There is sufficient evidence of a difference in the median delivery time in the two wings of the hotel. **(b)** The results of (a) are consistent with the results of Problem 10.51.

12.40 (a) Because $Z_{STAT} = -0.5100 < -1.96$, do not reject H_0. There is not enough evidence to conclude that there is a difference in the median brand value between the two sectors. **(b)** You must assume approximately equal variability in the two populations. **(c)** Using the pooled-variance t test and the separate-variance t test, you also did not reject the null hypothesis. From all three tests, you conclude that there is no difference in median brand value for the two sectors.

12.42 (a) Because $-1.96 < Z_{STAT} = 0.9743 < 1.96$ (or the p-value $= 0.3299 > 0.05$), do not reject H_0. There is not enough evidence to conclude that there is a difference in the median rating of ads that play before and after halftime. **(b)** You must assume approximately equal variability in the two populations. **(c)** Using the pooled-variance t-test, you do not reject the null hypothesis ($t = -2.0032 < t_{STAT} = 1.1541 < 2.0032$; p-value $= 0.2534 > 0.05$) and conclude that there is insufficient evidence of a difference in the mean rating of ads before and after halftime. in Problem 10.11 (a).

12.44 (a) Decision rule: If $H > \chi_U^2 = 15.086$, reject H_0. **(b)** Because $H = 13.77 < 15.806$, do not reject H_0.

12.46 (a) $H = 13.517 > 7.815$, p-value $= 0.0036 < 0.05$, reject H_0. There is sufficient evidence of a difference in the median waiting time in the four locations. **(b)** The results are consistent with those of Problem 11.9.

12.48 (a) $H = 19.3269 > 9.488$, reject H_0. There is evidence of a difference in the median ratings of the ads. **(b)** The results are consistent with those of Problem 11.10. **(c)** Because the combined scores are not true continuous variables, the nonparametric Kruskal-Wallis rank test is more appropriate because it does not require that the scores are normally distributed.

12.50 (a) Because $H = 13.0522 > 7.815$ or the p-value is 0.0045, reject H_0. There is sufficient evidence of a difference in the median cost associated with importing a standardized cargo of goods by sea transport across the global regions. **(b)** The results are the same.

12.56 (a) Because $\chi_{STAT}^2 = 0.412 < 3.841$, do not reject H_0. There is insufficient evidence to conclude that there is a relationship between a student's gender and pizzeria selection. **(b)** Because $\chi_{STAT}^2 = 2.624 < 3.841$, do not reject H_0. There is insufficient evidence to conclude that there is a relationship between a student's gender and pizzeria selection. **(c)** Because $\chi_{STAT}^2 = 4.956 < 5.991$, do not reject H_0. There is insufficient evidence to conclude that there is a relationship between price and pizzeria selection. **(d)** p-value $= 0.0839$. The probability of a sample that gives a test statistic equal to or greater than 4.956 is 8.39% if the null hypothesis of no relationship between price and pizzeria selection is true.

12.58 (a) Because $\chi_{STAT}^2 = 7.4298 < 9.4877$; p-value $= 0.1148 > 0.05$, do not reject H_0. There is not enough evidence to conclude that there is evidence of a difference in the proportion of organizations that have embarked on digital transformation on the basis of industry sector. **(b)** Because $\chi_{STAT}^2 = 38.09 > 21.0261$; p-value $= 0.0001 < 0.05$, reject H_0. There is evidence of a relationship between digital transformation progress and industry sector.

CHAPTER 13

13.2 (a) Yes. **(b)** No. **(c)** No. **(d)** Yes.

13.4 (a) The scatter plot shows a positive linear relationship. **(b)** For each increase in alcohol percentage of 1.0, mean predicted mean wine quality is estimated to increase by 0.5624. **(c)** $\hat{Y} = 5.2715$. **(d)** Wine quality appears to be affected by the alcohol percentage. Each increase of 1% in alcohol leads to a mean increase in wine quality of a little more than half a unit.

13.6 (b) $b_0 = -13,130.6592$, $b_1 = 2.4218$. **(c)** For each increase of $1,000 in tuition, the mean starting salary is predicted to increase by $2,421.80. **(d)** $109,047.01 **(e)** Starting salary seems higher for those schools that have a higher tuition.

13.8 (b) $b_0 = -1,039.5317$, $b_1 = 8.5816$. **(c)** For each additional million-dollar increase in revenue, the mean value is predicted to increase by an estimated $8.5816 million. Literal interpretation of b_0 is not meaningful because a team cannot have zero revenue. **(d)** $1,105.864 million. **(e)** That the current value of the team can be expected to increase as revenue increases.

13.10 (b) $b_0 = -0.7744$, $b_1 = 1.4030$. **(c)** For each increase of million YouTube trailer views, the predicted weekend box office gross is estimated to increase by $1.4030 million. **(d)** $27.2847 million. **(e)** You can conclude that the mean predicted increase in weekend box office gross is $1.4030 million for each million increase in YouTube trailer views.

13.12 $SST = 40$, $r^2 = 0.90$. 90% of the variation in the dependent variable can be explained by the variation in the independent variable.

13.14 $r^2 = 0.75$. 75% of the variation in the dependent variable can be explained by the variation in the independent variable.

13.16 (a) $r^2 = \dfrac{SSR}{SST} = \dfrac{21.8677}{64.0000} = 0.3417$, 34.17% of the variation in wine quality can be explained by the variation in the percentage of alcohol.

(b) $S_{YX} = \sqrt{\dfrac{SSE}{n-2}} = \sqrt{\dfrac{\sum_{i=1}^{n}(Y_i - \hat{Y}_i)^2}{n-2}} = \sqrt{\dfrac{42.1323}{48}} = 0.9369.$

(c) Based on (a) and (b), the model should be somewhat useful for predicting wine quality.

13.18 (a) $r^2 = 0.7665$. 76.65% of the variation in starting salary can be explained by the variation in tuition. **(b)** $S_{YX} = 15,944.3807$. **(c)** Based on (a) and (b), the model should be very useful for predicting the starting salary.

13.20 (a) $r^2 = 0.9612$, 96.12% of the variation in the current value of a MLB baseball team can be explained by the variation in its annual revenue. **(b)** $S_{YX} = 140.8188$. **(c)** Based on (a) and (b), the model should be very useful for predicting the current value of a baseball team.

13.22 (a) $r^2 = 0.6676$, 66.76% of the variation in weekend box office gross can be explained by the variation in YouTube trailer views. **(b)** $S_{YX} = 19.4447$. **(c)** Based on (a) and (b), the model should be useful for predicting weekend box office gross. **(d)** Other variables that might explain the variation in weekend box office gross could be the amount spent on advertising, the timing of the release of the movie, and the type of movie.

13.24 A residual analysis of the data indicates a pattern, with sizable clusters of consecutive residuals that are either all positive or all negative. This pattern indicates a violation of the assumption of linearity. A curvilinear model should be investigated.

13.26 There does not appear to be a pattern in the residual plot. The assumptions of regression do not appear to be seriously violated.

13.28 Based on the residual plot, the assumption of equal variance may be violated.

13.30 Based on the residual plot, there is no evidence of a pattern.

13.32 (a) An increasing linear relationship exists. **(b)** There is evidence of a strong positive autocorrelation among the residuals.

13.34 (a) No, because the data were not collected over time. **(b)** If data were collected at a single store and studied over a period of time, you would compute the Durbin-Watson statistic.

13.36 (a)

$$b_1 = \frac{SSXY}{SSX} = \frac{201,399.05}{12,495,626} = 0.0161$$

$$b_0 = \overline{Y} - b_1\overline{X} = 71.2621 - 0.0161\,(4,393) = 0.4576.$$

(b) $\hat{Y} = 0.458 + 0.0161X = 0.4576 + 0.0161(4,500) = 72.9867$, or $72,987. **(c)** There is no evidence of a pattern in the residuals over time.

(d) $D = \dfrac{\sum\limits_{i=2}^{n}(e_i - e_{i-1})^2}{\sum\limits_{i=1}^{n} e_i^2} = \dfrac{1,243.2244}{599.0683} = 2.08 > 1.45.$ There is no

evidence of positive autocorrelation among the residuals. **(e)** Based on a residual analysis, the model appears to be adequate.

13.38 (a) $b_0 = -2.535$, $b_1 = 0.06073$. **(b)** $2,505.40. **(d)** $D = 1.64 > d_U = 1.42$, so there is no evidence of positive autocorrelation among the residuals. **(e)** The plot shows some nonlinear pattern, suggesting that a non-linear model might be better. Otherwise, the model appears to be adequate.

13.40 (a) 3.00. **(b)** ± 2.1199. **(c)** Reject H_0. There is evidence that the fitted linear regression model is useful. **(d)** $1.32 \le \beta_1 \le 7.68$.

13.42 (a) $t_{STAT} = \dfrac{b_1 - \beta_1}{S_{b_1}} = \dfrac{0.5624}{0.1127} = 4.9913 > 2.0106.$ Reject H_0.

There is evidence of a linear relationship between the percentage of alcohol and wine quality.
(b) $b \pm t_{\alpha/2}S_{b_1} = 0.5624 \pm 2.0106\,(0.1127)\,0.3359 \le \beta_1 \le 0.7890.$

13.44 (a) $t_{STAT} = 10.7174 > 2.0301$; $p\text{-value} = 0.0000 < 0.05$. Reject H_0. There is evidence of a linear relationship between tuition and starting salary. **(b)** $1.963 \le \beta_1 \le 2.8805$.

13.46 (a) $t_{STAT} = 26.3347 > 2.0484$ or because the p-value is 0.0000, reject H_0 at the 5% level of significance. There is evidence of a linear relationship between annual revenue and current value.
(b) $7.9141 \le \beta_1 \le 9.2491$.

13.48 (a) $t_{STAT} = 11.3381 > 1.9977$ or because the p-value = $0.0000 < 0.05$; reject H_0. There is evidence of a linear relationship between YouTube trailer views and weekend box office gross.
(b) $1.1558 \le \beta_1 \le 1.6501$.

13.50 (a) (% daily change in SPUU) = $b_0 + 2.0$ (% daily change in S&P 500 index). **(b)** If the S&P 500 gains 10% in a year, SPUU is expected to gain an estimated 20%. **(c)** If the S&P 500 loses 20% in a year, SPUU is expected to lose an estimated 40%. **(d)** Risk takers will be attracted to leveraged funds, and risk-averse investors will stay away.

13.52 (a), (b) First weekend and U.S. gross: $r = 0.7284$, $t_{STAT} = 2.6042 > 2.4469$, p-value = $0.0404 < 0.05$. Reject H_0. At the 0.05 level of significance, there is evidence of a linear relationship between first weekend sales and U.S. gross. First weekend and worldwide gross: $r = 0.8233$, $t_{STAT} = 3.5532 > 2.4469$, p-value = $0.0120 < 0.05$. Reject H_0. At the 0.05 level of significance, there is evidence of a linear relationship between first weekend sales and worldwide gross. U.S. gross and worldwide gross: $r = 0.9642$, $t_{STAT} = 8.9061 > 2.4469$, p-value = $0.0001 < 0.05$. Reject H_0. At the 0.05 level of significance, there is evidence of a linear relationship between U.S gross and worldwide gross.

13.54 (a) $r = 0.5674$. There is a moderate linear relationship between social media networking and the GDP per capita.
(b) $t_{STAT} = 3.4454$, p-value = $0.0020 < 0.05$. Reject H_0. At the 0.05 level of significance, there is significant evidence of a linear relationship between social media networking and the GDP per capita.
(c) There appears to be a linear relationship.

13.56 (a) $15.95 \le \mu_{Y|X=4} \le 18.05.$ **(b)** $14.651 \le Y_{X=4} \le 19.349.$
(c) The intervals in this problem are wider than in Problem 13.55 because they involve X values that are different from the mean.

13.58 (a) $\hat{Y} = -0.3529 + (0.5624)(10) = 5.2715$ $\hat{Y} \pm t_{\alpha/2}S_{YX}\sqrt{h_i}$
$$= 5.2715 \pm 2.0106(0.9369)\sqrt{0.0249}$$
$$4.9741 \le \mu_{Y|X=10} \le 5.5690.$$
(b) $\hat{Y} \pm t_{\alpha/2}S_{YX}\sqrt{1 + h_i}$
$$= 5.2715 \pm 2.0106(0.9369)\sqrt{1 + 0.0249}$$
$$3.3645 \le Y_{X=10} \le 7.1786.$$
(c) Part (b) provides a prediction interval for the individual response given a specific value of the independent variable, and part (a) provides a confidence interval estimate for the mean value, given a specific value of the independent variable. Because there is much more variation in predicting an individual value than in estimating a mean value, a prediction interval is wider than a confidence interval estimate.

13.60 (a) $103,638.95 \le \mu_{Y|X=50,450} \le $114,455.06.
(b) $76,229.52 \le Y_{X=50,450} \le $141,864.49. **(c)** You can estimate a mean more precisely than you can predict a single observation.

13.62 (a) $1,043.1911 \le \mu_{Y|X=250} \le 1,168.5370.$ **(b)** $810.6799 \le Y_{X=250} \le 1,401.0480$ **(c)** Because there is much more variation in predicting an individual value than in estimating a mean, the prediction interval is wider than the confidence interval.

13.74 (a) $b_0 = 24.84$, $b_1 = 0.14$. **(b)** For each additional case, the predicted delivery time is estimated to increase by 0.14 minute. The interpretation of the Y intercept is not meaningful because the number of cases delivered cannot be 0. **(c)** 45.84. **(d)** No, 500 is outside the relevant range of the data used to fit the regression equation. **(e)** $r^2 = 0.972$. **(f)** There is no obvious pattern in the residuals, so the assumptions of regression are met. The model appears to be adequate. **(g)** $t_{STAT} = 24.88 > 2.1009$; reject H_0.
(h) $44.88 \le \mu_{Y|X=150} \le 46.80.$ $41.56 \le Y_{X=150} \le 50.12.$
(i) The number of cases explains almost all of the variation in delivery time.

13.76 (a) $b_0 = 326.5935$, $b_1 = 0.0835$. **(b)** For each additional square foot of living space in the house, the mean asking price is predicted to increase by $83.50. The estimated asking price of a house with 0 living space is 326.5935 thousand dollars. However, this interpretation is not meaningful because the living space of the house cannot be 0. **(c)** $\hat{Y} = 493.6769$ thousand dollars. **(d)** $r^2 = 0.3979$. So 39.79% of the variation in asking price is explained by the variation in living space. **(e)** Neither the residual plot nor the normal probability plot reveals any potential violation of the linearity, equal variance, and normality assumptions. **(f)** $t_{STAT} = 6.2436 > 2.0010$, p-value is 0.0000. Because p-value < 0.05, reject H_0. There is evidence of a linear relationship between asking price and living space.
(g) $0.0568 \le \beta_1 \le 0.1103.$ **(h)** The living space in the house is somewhat useful in predicting the asking price, but because only 39.79% of the variation in asking price is explained by variation in living space, other variables should be considered.

13.78 (a) $b_0 = 21.2034$, $b_1 = -0.1517$. **(b)** For each additional point on the efficiency ratio, the predicted mean tangible common equity (ROATCE) is estimated to decrease by 0.1517. For an efficiency of 0, the predicted mean tangible common equity (ROATCE) is 21.2034. **(c)** 12.0989.
(d) $r^2 = 0.1882$. **(e)** There is no obvious pattern in the residuals, so the assumptions of regression are met. The model appears to be adequate.
(f) $t_{STAT} = -4.7662 < -1.9845$; reject H_0. There is evidence of a linear relationship between efficiency ratio and tangible common equity (ROATCE).
(g) $11.4060 \le \mu_{Y|X=60} \le 12.7918,$ $5.1534 \le Y_{X=60} \le 19.0444.$
(h) $-0.2149 \le \beta_1 \le -0.0886.$ **(i)** There is a small relationship between efficiency ratio and tangible common equity (ROATCE).

13.80 (a) There is no clear relationship shown on the scatter plot. **(c)** Looking at all 23 flights, when the temperature is lower, there is likely to be some O-ring damage, particularly if the temperature is below 60 degrees. **(d)** 31 degrees is outside the relevant range, so a prediction should not be made. **(e)** Predicted $Y = 18.036 - 0.240X$, where X = temperature and Y = O-ring damage. **(g)** A nonlinear model would be more appropriate. **(h)** The appearance on the residual plot of a nonlinear pattern indicates that a nonlinear model would be better. It also appears that the normality assumption is invalid.

13.82 (a) $b_0 = -1.3477$, $b_1 = 0.0121$. **(b)** For each additional million-dollar increase in revenue, the current value will increase by an estimated 0.0121 billion. Literal interpretation of b_0 is not meaningful because an operating franchise cannot have zero revenue. **(c)** 469.117 million. **(d)** $r^2 = 0.8288$. 82.88% of the variation in the current value of an NBA basketball team can be explained by the variation in its annual revenue. **(e)** There does not appear to be a pattern in the residual plot. The assumptions of regression do not appear to be seriously violated. **(f)** $t_{STAT} = 11.6473 > 2.0484$ or because the p-value is 0.0000, reject H_0 at the 5% level of significance. There is evidence of a linear relationship between annual revenue and current value. **(g)** $0.1913 \leq \mu_{Y|X=150} \leq 0.74697$ billions. **(h)** $-0.2569 \leq Y_{X=150} \leq 1.1951$ billions. **(i)** The strength of the relationship between revenue and current value is approximately the same for NBA basketball teams and for European soccer teams but lower than for MLB baseball teams.

13.84 (a) $b_0 = -2,629.222$, $b_1 = 82.472$. **(b)** For each additional centimeter in circumference, the weight is estimated to increase by 82.472 grams. **(c)** 2,319.08 grams. **(d)** Yes, because circumference is a very strong predictor of weight. **(e)** $r^2 = 0.937$. **(f)** There appears to be a nonlinear relationship between circumference and weight. **(g)** p-value is virtually $0 < 0.05$; reject H_0. **(h)** $72.7875 \leq \beta_1 \leq 92.156$.

CHAPTER 14

14.2 (a) For each one-unit increase in X_1, you estimate that the mean of Y will decrease 2 units, holding X_2 constant. For each one-unit increase in X_2, you estimate that the mean of Y will increase 7 units, holding X_1 constant. **(b)** The Y intercept, equal to 50, estimates the value of Y when both X_1 and X_2 are 0.

14.4 (a) $\hat{Y} = 1.3960 - 0.0117X_1 + 0.0286X_2$. **(b)** For a given capital adequacy, for each increase of 1% in efficiency ratio, ROAA decreases by 0.0117%. For a given efficiency ratio, for each increase of 1% in capital adequacy, ROAA increases by 0.0286% **(c)** $\hat{Y} = 1.1214$. **(d)** $1.0798 \leq \mu_{Y|X} \leq 1.1629$. **(e)** $0.5679 \leq Y_X \leq 1.6749$. **(f)** The interval in (e) is narrower because it is estimating the mean value, not an individual value. **(g)** The model uses both the efficiency ratio and capital adequacy to predict ROA. This may produce a better model than if only one of these independent variables is included.

14.6 (a) $\hat{Y} = 301.78 + 3.4771X_1 + 41.041X_2$. **(b)** For a given amount of voluntary turnover, for each increase of \$1 billion in worldwide revenue, the mean number of full-time jobs added is predicted to increase by 3.4771. For a given \$1 billion in worldwide revenue, for each increase of 1% in voluntary turnover, the mean number of full-time jobs added is predicted to increase by 41.041. **(c)** The Y intercept has no meaning in this problem. **(d)** Holding the other independent variable constant, voluntary turnover has a higher slope than worldwide revenue.

14.8 (a) $\hat{Y} = 532.2883 + 407.1346X_1 - 2.8257X_2$, where $X_1 =$ land area, $X_2 =$ age. **(b)** For a given age, each increase by one acre in land area is estimated to result in an increase in the mean fair market value by \$407.1346 thousands. For a given land area, each increase of one year in age is estimated to result in a decrease in the mean fair market value by \$2.8257 thousands. **(c)** The interpretation of b_0 has no practical

meaning here because it would represent the estimated fair market value of a new house that has no land area. **(d)** $\hat{Y} = \$478.6577$ thousands. **(e)** $446.8367 \leq \mu_{Y|X} \leq 510.4788$. **(f)** $307.2577 \leq Y_X \leq 650.0577$.

14.10 (a) $MSR = 15$, $MSE = 12$. **(b)** 1.25. **(c)** $F_{STAT} = 1.25 < 4.10$; do not reject H_0. **(d)** 0.20. 20% of the variation in Y is explained by variation in X. **(e)** 0.04.

14.12 p-value for revenue is $0.0395 < 0.05$ and the p-value for efficiency is less than $0.0001 < 0.05$. Reject H_0 for each of the independent variables. There is evidence of a significant linear relationship with each of the independent variables.

14.14 (a) $F_{STAT} = 37.8384 > 3.00$; reject H_0. **(b)** p-value = 0.0000. The probability of obtaining an F_{STAT} value > 37.8384 if the null hypothesis is true is 0.0000. **(c)** $r^2 = 0.2785$. 27.85% of the variation in ROAA can be explained by variation in efficiency ratio and variation in risk-based capital. **(d)** $r^2_{adj} = 0.2712$.

14.16 (a) $F_{STAT} = 1.95 < 3.15$; Do not reject H_0. There is insufficient evidence of a significant linear relationship. **(b)** p-value = 0.1512. The probability of obtaining an F_{STAT} value > 1.95 if the null hypothesis is true is 0.1512. **(c)** $r^2 = 0.0610$. 6.10% of the variation in full-time jobs added can be explained by variation in worldwide revenue and variation in full-time voluntary turnover. **(d)** $r^2_{adj} = 0.0297$.

14.18 (a) – (e) Based on a residual analysis, there is no evidence of a violation of the assumptions of regression.

14.20 (a) There is no evidence of a violation of the assumptions. **(b)** Because the data are not collected over time, the Durbin-Watson test is not appropriate. **(c)** They are valid.

14.22 (a) The residual analysis reveals no patterns. **(b)** Because the data are not collected over time, the Durbin-Watson test is not appropriate. **(c)** There are no apparent violations in the assumptions.

14.24 (a) Variable X_2 has a larger slope in terms of the t statistic of 3.75 than variable X_1, which has a smaller slope in terms of the t statistic of 3.33. **(b)** $1.46824 \leq \beta_1 \leq 6.53176$. **(c)** For $X_1 : t_{STAT} = 3.33 > 2.1098$. Reject H_0. There is evidence that X_1 contributes to a model already containing X_2. For $X_2 : t_{STAT} = 3.75 > 2.1098$. Reject H_0. There is evidence that X_2 contributes to a model already containing X_1. Both X_1 and X_2 should be included in the model.

14.26 (a) 95% confidence interval on $\beta_1 : b_1 \pm tS_{b_1}$, -0.0117 ± 1.98 (0.0022), $-0.0161 \leq \beta_1 \leq -0.0074$. **(b)** For $X_1 : t_{STAT} = b_1/S_{b_1} = -0.0177/0.0022 = -5.3415 < -1.98$. Reject H_0. There is evidence that X_1 contributes to a model already containing X_2. For $X_2 : t_{STAT} = b_2/S_{b_2} = 0.0286/0.0054 = 5.2992 > 1.98$. Reject H_0. There is evidence that X_2 contributes to a model already containing X_1. Both X_1 (efficiency ratio) and X_2 (total risk-based capital) should be included in the model.

14.28 (a) $-5.8682 \leq \beta_1 \leq 12.8225$. **(b)** For $X_1 : t_{STAT} = 0.7443 < 2.0003$. Don't reject H_0. There is insufficient evidence that X_1 contributes to a model already containing X_2. For $X_2 : t_{STAT} = 1.8835 < 2.0003$. Do not reject H_0. There is insufficient evidence that X_2 contributes to a model already containing X_1. Neither variable contributes to a model that includes the other variable. You should consider using only a simple linear regression model.

14.30 (a) $274.1702 \leq \beta_1 \leq 540.0990$. **(b)** For $X_1 : t_{STAT} = 6.2827$ and p-value = 0.0000. Because p-value < 0.05, reject H_0. There is evidence that X_1 contributes to a model already containing X_2. For $X_2 : t_{STAT} = -4.1475$ and p-value = 0.0003. Because p-value < 0.05 reject H_0. There is evidence that X_2 contributes to a model already containing X_1. Both X_1 (land area) and X_2 (age) should be included in the model.

14.32 (a) For X_1: $F_{STAT} = 1.25 < 4.96$; do not reject H_0. For X_2: $F_{STAT} = 0.833 < 4.96$; do not reject H_0. **(b)** 0.1111, 0.0769.

14.34 (a) For X_1: $SSR(X_1|X_2) = SSR(X_1 \text{ and } X_2) - SSR(X_2) =$

$5.9271 - 3.6923 = 2.2348$ $F_{STAT} = \dfrac{SSR(X_1|X_2)}{MSE} =$

$\dfrac{2.2348}{15.3521/196} = 28.5227 > 3.897$. Reject H_0. There is evidence that X_1 contributes to a model already containing X_2. For X_2: $SSR(X_2|X_1) = SSR(X_1 \text{ and } X_2) - SSR(X_1) = 5.9271 - 3.7275 = 2.1996$,

$F_{STAT} = \dfrac{SSR(X_2|X_1)}{MSE} = \dfrac{2.1996}{15.3521/196} = 28.0823 > 3.897$.

Reject H_0. There is evidence that X_2 contributes to a model already containing X_1. Because both X_1 and X_2 make a significant contribution to the model in the presence of the other variable, both variables should be included in the model.

(b) $r_{Y1.2}^2 = \dfrac{SSR(X_1|X_2)}{SST - SSR(X_1 \text{ and } X_2) + SSR(X_1|X_2)}$

$= \dfrac{2.2348}{21.2791 - 5.9271 + 2.2348} = 0.1271$.

Holding constant the effect of the total risk based capital, 12.71% of the variation in ROAA can be explained by the variation in efficiency ratio.

$r_{Y2.1}^2 = \dfrac{SSR(X_2|X_1)}{SST - SSR(X_1 \text{ and } X_2) + SSR(X_2|X_1)}$

$= \dfrac{2.1996}{21.2791 - 5.9271 + 2.1996} = 0.1253$

Holding constant the effect of efficiency ratio 12.53% of the variation in ROAA can be explained by the variation in the total risk-based capital.

14.36 (a) For X_1: $F_{STAT} = 0.554 < 4.00$; Don't reject H_0. There is insufficient evidence that X_1 contributes to a model containing X_2. For X_2: $F_{STAT} = 3.5476 < 4.00$. Do not reject H_0. There is insufficient evidence that X_2 contributes to a model already containing X_1. Because only X_1 makes a significant contribution to the model in the presence of the other variable, only X_i should be included in the model. **(b)** $r_{Y1.2}^2 = 0.0091$. Holding constant the effect of full-time voluntary turnover, 0.91% of the variation in full-time jobs added be explained by the variation in total worldwide revenue . $r_{Y2.1}^2 = 0.0558$. Holding constant the effect of total worldwide revenue, 5.58% of the variation in full-time jobs created can be explained by the variation in full-time voluntary turnover.

14.38 (a) Holding constant the effect of X_2, for each increase of one unit of X_1, Y increases by 4 units. **(b)** Holding constant the effect of X_1, for each increase of one unit of X_2, Y increases by 2 units. **(c)** Because $t_{STAT} = 3.27 > 2.1098$, reject H_0. Variable X_2 makes a significant contribution to the model.

14.40 (a) $\hat{Y} = 243.7371 + 9.2189X_1 + 12.6967X_2$, where $X_1 =$ number of rooms and $X_2 =$ neighborhood (east $= 0$). **(b)** Holding constant the effect of neighborhood, for each additional room, the mean selling price is estimated to increase by 9.2189 thousands of dollars, or \$9,218.9. For a given number of rooms, a west neighborhood is estimated to increase the mean selling price over an east neighborhood by 12.6967 thousands of dollars, or \$12,696.7. **(c)** $\hat{Y} = 326.7076$, or \$326,707.6. \$309,560.04 $\leq Y_X \leq$ 343,855.1. \$321,471.44 $\leq \mu_{Y|X} \leq$ \$331,943.71. **(d)** Based on a residual analysis, the model appears to be adequate. **(e)** $F_{STAT} = 55.39$, the p-value is virtually 0. Because p-value < 0.05, reject H_0. There is evidence of a significant relationship between selling price and the two independent variables (rooms and neighborhood). **(f)** For X_1: $t_{STAT} = 8.9537$, the p-value is virtually 0. Reject H_0. Number of rooms makes a significant contribution and should be included in the model. For X_2: $t_{STAT} = 3.5913$, p-value $= 0.0023 < 0.05$. Reject H_0. Neighborhood makes a significant contribution and should be included in the model. Based

on these results, the regression model with the two independent variables should be used. **(g)** $7.0466 \leq \beta_1 \leq 11.3913$. **(h)** $5.2378 \leq \beta_2 \leq 20.1557$. **(i)** $r_{adj}^2 = 0.851$. **(j)** $r_{Y1.2}^2 = 0.825$. Holding constant the effect of neighborhood, 82.5% of the variation in selling price can be explained by variation in number of rooms. $r_{Y2.1}^2 = 0.431$. Holding constant the effect of number of rooms, 43.1% of the variation in selling price can be explained by variation in neighborhood. **(k)** The slope of selling price with number of rooms is the same, regardless of whether the house is located in an east or west neighborhood. **(l)** $\hat{Y} = 253.95 + 8.032X_1 - 5.90X_2 + 2.089X_1X_2$. For X_1X_2, p-value $= 0.330$. Do not reject H_0. There is no evidence that the interaction term makes a contribution to the model. **(m)** The model in (b) should be used. **(n)** The number of rooms and the neighborhood both significantly affect the selling price, but the number of rooms has a greater effect.

14.42 (a) Predicted time $= 8.01 + 0.00523$ Depth $- 2.105$ Dry. **(b)** Holding constant the effect of type of drilling, for each foot increase in depth of the hole, the mean drilling time is estimated to increase by 0.00523 minutes. For a given depth, a dry drilling hole is estimated to reduce the drilling time over wet drilling by a mean of 2.1052 minutes. **(c)** 6.428 minutes, $6.210 \leq \mu_{Y|X} \leq 6.646$, $4.923 \leq Y_X \leq 7.932$. **(d)** The model appears to be adequate. **(e)** $F_{STAT} = 111.11 > 3.09$; reject H_0. **(f)** $t_{STAT} = 5.03 > 1.9847$; reject H_0. $t_{STAT} = -14.03 < -1.9847$; reject H_0. Include both variables. **(g)** $0.0032 \leq \beta_1 \leq 0.0073$. **(h)** $-2.403 \leq \beta_2 \leq -1.808$. **(i)** 69.0%. **(j)** 0.207, 0.670. **(k)** The slope of the additional drilling time with the depth of the hole is the same, regardless of the type of drilling method used. **(l)** The p-value of the interaction term $= 0.462 > 0.05$, so the term is not significant and should not be included in the model. **(m)** The model in part (b) should be used. Both variables affect the drilling time. Dry drilling holes should be used to reduce the drilling time.

14.44 (a) $\hat{Y} = 1.1079 - 0.0070X_1 + 0.0448X_2 - 0.0003X_1X_2$, where $X_1 =$ efficiency ratio, $X_2 =$ total risk-based capital, p-value $= 0.4593 > 0.05$. Do not reject H_0. There is not enough evidence that the interaction term makes a contribution to the model. **(b)** Because there is insufficient evidence of any interaction effect between efficiency ratio and total risk-based capital, the model in Problem 14.4 should be used.

14.46 (a) The p-value of the interaction term $= 0.1650 < 0.05$, so the term is not significant and should be not included in the model. **(b)** Use the model developed Problem 14.6.

14.48 (a) For X_1X_2, p-value $= 0.2353 > 0.05$. Do not reject H_0. There is insufficient evidence that the interaction term makes a contribution to the model. **(b)** Because there is not enough evidence of an interaction effect between total staff present and remote hours, the model in Problem 14.7 should be used.

14.50 Holding constant the effect of other variables, the natural logarithm of the estimated odds ratio for the dependent categorical response will increase by 2.2 for each unit increase in the particular independent variable.

14.52 0.4286.

14.54 (a) ln(estimated odds ratio) $= -6.9394 + 0.1395X_1 + 2.7743X_2 = -6.9394 + 0.1395(36) + 2.7743(0) = -1.91908$. Estimated odds ratio $= 0.1470$. Estimated Probability of Success $=$ Odds Ratio/(1 + Odds Ratio) $= 0.1470/(1 + 0.1470) = 0.1260$. **(b)** From the text discussion of the example, 70.2% of the individuals who charge \$36,000 per annum and possess additional cards can be expected to purchase the premium card. Only 12.60% of the individuals who charge \$36,000 per annum and do not possess additional cards can be expected to purchase the premium card. For a given amount of money charged per annum, the likelihood of purchasing a premium card is substantially higher among individuals who already possess additional cards than for those who do not possess additional cards. **(c)** ln (estimated odds ratio)

$= -6.9394 + 0.13957X_1 + 2.7743X_2 = -6.9394 + 0.1395(18) +$
$2.7743(0) = -4.4298.$ Estimated odds ratio $= e^{-4.4298} = 0.0119.$
Estimated Probability of Success $=$ Odds Ratio/(1 + Odds Ratio) $=$
$0.0119/(1 + 0.0119) = 0.01178.$ **(d)** Among individuals who do not
purchase additional cards, the likelihood of purchasing a premium card
diminishes dramatically with a substantial decrease in the amount
charged per annum.

14.56 (a) ln(estimated odds) $= -47.4723 + 1.3099$ fixed acidity $+$
90.5722 chlorides $+ 9.777$ pH. **(b)** Holding constant the effect of
chlorides and pH, for each increase of one point in fixed acidity, ln
(estimated odds) increases by an estimate of 1.3099. Holding constant the
effect of fixed acidity and pH, for each increase of one point in chlorides,
ln(estimated odds) increases by an estimate of 90.5722. Holding constant
the effect of fixed acidity and chlorides, for each increase of one point in
pH, ln(estimated odds) increases by an estimate of 9.777. **(c)** 0.3686.
(d) Deviance $= 54.456,$ p-value $= 1.0000,$ do not reject $H_0,$ so model
is adequate. **(e)** For fixed acidity: $Z_{STAT} = 3.17 > 1.96,$ reject $H_0.$ For
chlorides: $Z_{STAT} = 4.00 > 1.96,$ reject $H_0.$ For pH: $Z_{STAT} = 3.29 > 1.96,$
reject $H_0.$ Each variable makes a significant contribution to the model.
(f) Fixed acidity, chlorides, and pH are all important factors in distinguishing between white and red wines.

14.58 (a) ln(estimated odds) $= -0.6048 + 0.0938$ claims/year $+$
1.8108 new business. **(b)** Holding constant the effects of whether the
policy is new, for each increase of the number of claims submitted
per year by the policy holder, ln(odds) increases by an estimate of
0.0938. Holding constant the number of claims submitted per year
by the policy holder, ln(odds) is estimated to be 1.8108 higher when
the policy is new as compared to when the policy is not new.
(c) ln(estimated odds ratio) $= 1.2998.$ Estimated odds ratio $= 3.6684$
Estimated probability of a fraudulent claim $= 0.7858.$ **(d)** The
deviance statistic is 119.4353 with a p-value $= 0.0457 < 0.05.$
Reject $H_0.$ The model is not a good fitting model. **(e)** For claims/year:
$Z_{STAT} = 0.1865,$ p-value $= 0.8521 > 0.05.$ Do not reject $H_0.$ There is
insufficient evidence that the number of claims submitted per year by the
policy holder makes a significant contribution to the logistic regression
model. For new business: $Z_{STAT} = 2.2261,$ p-value $= 0.0260 < 0.05.$
Reject $H_0.$ There is sufficient evidence that whether the policy is new
makes a significant contribution to the logistic model regression.
(f) ln(estimated odds) $= -1.0125 + 0.9927$ claims/year.
(g) ln(estimated odds) $= -0.5423 + 1.9286$ new business.
(h) The deviance statistic for (f) is 125.0102 with a
p-value $= 0.0250 < 0.05.$ Reject $H_0.$ The model is not a good
fitting model. The deviance statistic for (g) is 119.4702 with a
p-value $= 0.0526 > 0.05.$ Do not reject $H_0.$ The model is a good fitting
model. The model in (g) should be used to predict a fraudulent claim.

14.60 (a) ln(estimated odds) $= 1.252 - 0.0323$ Age $+ 2.2165$
subscribes to the wellness newsletters. **(b)** Holding constant the
effect of subscribes to the wellness newsletters, for each increase of
one year in age, ln(estimated odds) decreases by an estimate of 0.0323.
Holding constant the effect of age, for a customer who subscribes to
the wellness newsletters, ln(estimated odds) increases by an estimate of
2.2165. **(c)** 0.912. **(d)** Deviance $= 102.8762,$ p-value $= 0.3264.$ Do not
reject H_0 so model is adequate. **(e)** For Age: $Z = -1.8053 > -1.96,$
Do not reject $H_0.$ For subscribes to the wellness newsletters:
$Z = 4.3286 > 1.96,$ Reject $H_0.$ **(f)** Only subscribes to wellness
newsletters is useful in predicting whether a customer will purchase
organic food.

14.72 (a) $\hat{Y} = -3.9152 + 0.0319X_1 + 4.2228X_2,$ where $X_1 =$ number
cubic feet moved and $X_2 =$ number of pieces of large furniture.
(b) Holding constant the number of pieces of large furniture, for each
additional cubic foot moved, the mean labor hours are estimated to

increase by 0.0319. Holding constant the amount of cubic feet moved,
for each additional piece of large furniture, the mean labor hours are estimated to increase by 4.2228. **(c)** $\hat{Y} = 20.4926.$ **(d)** Based on a residual
analysis, the errors appear to be normally distributed. The equal-variance
assumption might be violated because the variances appear to be larger
around the center region of both independent variables. There might
also be violation of the linearity assumption. A model with quadratic
terms for both independent variables might be fitted. **(e)** $F_{STAT} = 228.80,$
p-value is virtually $0 < 0.05,$ reject $H_0.$ There is evidence of a significant
relationship between labor hours and the two independent variables (the
amount of cubic feet moved and the number of pieces of large furniture).
(f) The p-value is virtually 0. The probability of obtaining a test statistic
of 228.80 or greater is virtually 0 if there is no significant relationship
between labor hours and the two independent variables (the amount of
cubic feet moved and the number of pieces of large furniture).
(g) $r^2 = 0.9327.$ 93.27% of the variation in labor hours can be explained
by variation in the number of cubic feet moved and the number of pieces
of large furniture. **(h)** $r^2_{adj} = 0.9287.$ **(i)** For X_1: $t_{STAT} = 6.9339,$ the
p-value is virtually 0. Reject $H_0.$ The number of cubic feet moved makes
a significant contribution and should be included in the model. For
X_2: $t_{STAT} = 4.6192,$ the p-value is virtually 0. Reject $H_0.$ The number
of pieces of large furniture makes a significant contribution and should be
included in the model. Based on these results, the regression model with
the two independent variables should be used. **(j)** For X_1: $t_{STAT} = 6.9339,$
the p-value is virtually 0. The probability of obtaining a sample that
will yield a test statistic greater than 6.9339 is virtually 0 if the number of cubic feet moved does not make a significant contribution,
holding the effect of the number of pieces of large furniture constant.
For X_2: $t_{STAT} = 4.6192,$ the p-value is virtually 0. The probability of
obtaining a sample that will yield a test statistic greater than 4.6192 is
virtually 0 if the number of pieces of large furniture does not make a
significant contribution, holding the effect of the amount of cubic feet
moved constant. **(k)** $0.0226 \leq \beta_1 \leq 0.0413.$ **(l)** $r^2_{Y1.2} = 0.5930.$ Holding
constant the effect of the number of pieces of large furniture, 59.3% of
the variation in labor hours can be explained by variation in the amount
of cubic feet moved. $r^2_{Y2.1} = 0.3927.$ Holding constant the effect of the
number of cubic feet moved, 39.27% of the variation in labor hours can
be explained by variation in the number of pieces of large furniture.
(m) Both the number of cubic feet moved and the number of large pieces
of furniture are useful in predicting the labor hours, but the cubic feet
moved is more important.

14.74 (a) $\hat{Y} = 360.2158 + 0.0775X_1 - 0.4122X_2,$ where $X_1 =$ house
size and $X_2 =$ age. **(b)** Holding constant the age, for each additional
square foot in the size of the house, the mean asking price is estimated to
increase by 77.50 thousand dollars. Holding constant the living space of
the house, for each additional year in age, the asking price is estimated to
decrease by 0.4122 thousand dollars. **(c)** $\hat{Y} = 492.5316$ thousand dollars.
(d) Based on a residual analysis, the model appears to be adequate.
(e) $F_{STAT} = 19.4909,$ the p-value $= 0.0000 < 0.05,$ reject $H_0.$ There
is evidence of a significant relationship between asking price and the
two independent variables (size of the house and age). **(f)** The p-value is
0.0000. The probability of obtaining a test statistic of 19.4909 or greater is
virtually 0 if there is no significant relationship between asking price and
the two independent variables (living space of the house and age).
(g) $r^2 = 0.4019.$ 40.19% of the variation in asking price can be explained
by variation in the size of the house and age. **(h)** $r^2_{adj} = 0.3813.$ **(i)** For
X_1: $t_{STAT} = 4.6904,$ the p-value is 0.0000. Reject $H_0.$ The living space
of the house makes a significant contribution and should be included in
the model. For X_2: $t_{STAT} = -0.6304,$ p-value $= 0.5309 > 0.05.$ Do not
reject $H_0.$ Age does not make a significant contribution and should not be
included in the model. Based on these results, the regression model with
only the size of the house should be used. **(j)** For X_1: $t_{STAT} = 4.6904.$ The
probability of obtaining a sample that will yield a test statistic farther away

than 4.6904 is 0.0000 if the living space does not make a significant contribution, holding age constant. For X_2: $t_{STAT} = -0.6304$. The probability of obtaining a sample that will yield a test statistic further away than 0.6304 is 0.5309 if the age does not make a significant contribution holding the effect of the living space constant. **(k)** $0.0444 \leq \beta_1 \leq 0.1106$. You are 95% confident that the asking price will increase by an amount somewhere between \$44.40 thousand and \$110.60 thousand for each additional thousand square foot increase in living space, holding constant the age of the house. In Problem 13.76, you are 95% confident that the assessed value will increase by an amount somewhere between \$56.8 thousand and \$110.30 thousand for each additional 1,000 square foot increase in living space, regardless of the age of the house. **(l)** $r^2_{Y1.2} = 0.2750$. Holding constant the effect of the age of the house, 27.50% of the variation in asking price can be explained by variation in the living space of the house. $r^2_{Y2.1} = 0.0068$. Holding constant the effect of the size of the house, 0.68% of the variation in asking price can be explained by variation in the age of the house. **(m)** Only the living space of the house should be used to predict asking price.

14.76 (a) $\hat{Y} = -90.2166 + 9.2169X_1 + 2.5069X_2$, where $X_1 = $ asking price and $X_2 = $ age. **(b)** Holding age constant, for each additional \$1,000 in asking price, the taxes are estimated to increase by a mean of \$9.2169 thousand. Holding asking price constant, for each additional year, the taxes are estimated to increase by \$2.5069. **(c)** $\hat{Y} = \$3,721.90$. **(d)** Based on a residual analysis, the errors appear to be normally distributed. The equal-variance assumption appears to be valid. However, there is one very large residual that is from the house that is 107 years old. Removing this point, still leaves a residual for the house that has an asking price of \$550,000 and is 52 years old. However, because this model is an almost perfect fit, you may want to use this model. In this model, age is no longer significant. **(e)** $F_{STAT} = 1,677.8619$, p-value $= 0.0000 < 0.05$, reject H_0. There is evidence of a significant relationship between taxes and the two independent variables (asking price and age). **(f)** p-value $= 0.0000$. The probability of obtaining an F_{STAT} test statistic of 1,677.8619 or greater is virtually 0 if there is no significant relationship between taxes and the two independent variables (asking price and age). **(g)** $r^2 = 0.9830$, 98.30% of the variation in taxes can be explained by variation in asking price and age. **(h)** $r^2_{adj} = 0.9824$. **(i)** For X_1: $t_{STAT} = 53.7184$, p-value $= 0.0000 < 0.05$. Reject H_0. The asking price makes a significant contribution and should be included in the model. For X_2: $t_{STAT} = 2.7873$, p-value $= 0.0072 < 0.05$. Reject H_0. The age of a house makes a significant contribution and should be included in the model. Based on these results, the regression model with asking price and age should be used. **(j)** For X_1: p-value $= 0.0000$. The probability of obtaining a sample that will yield a test statistic greater than 53.7184 is 0.0000 if the asking price does not make a significant contribution, holding age constant. For X_2: p-value $= 0.0072$. The probability of obtaining a sample that will yield a test statistic greater than 2.7873 is 0.0072 if the age of a house doezs not make a significant contribution, holding the effect of the asking price constant. **(k)** $8.8735 \leq \beta_1 \leq 9.5604$. You are 95% confident that the mean taxes will increase by an amount somewhere between \$8.87 and \$9.56 for each additional \$1,000 increase in the asking price, holding constant the age. In Problem 13.77, you are 95% confident that the mean taxes will increase by an amount somewhere between \$5.968 and \$11.03 for each additional \$1,000 increase in asking price, regardless of the age. **(l)** $r^2_{Y1.2} = 0.9803$. Holding constant the effect of age, 98.03% of the variation in taxes can be explained by variation in the asking price. $r^2_{Y2.1} = 0.1181$. Holding constant the effect of the asking price, 11.81% of the variation in taxes can be explained by variation in the age. **(m)** Based on your answers to (b) through (k), the age of a house has an effect on its taxes. However, given the results when the 107-year-old house is not included, the assessor can state that for houses that are not that old, that age does not have an effect on taxes.

14.78 (a) $\hat{Y} = 160.6120 - 18.7181X_1 - 2.8903X_2$, where $X_1 = $ ERA and $X_2 = $ league (American $= 0$ National $= 1$). **(b)** Holding constant the effect of the league, for each additional earned run, the number of wins is estimated to decrease by 18.7181. For a given ERA, a team in the National League is estimated to have 2.8903 fewer wins than a team in the American League. **(c)** 76.3803 wins. **(d)** Based on a residual analysis, there is no pattern in the errors. There is no apparent violation of other assumptions. **(e)** $F_{STAT} = 24.306 > 3.35$, p-value $= 0.0000 < 0.05$, reject H_0. There is evidence of a significant relationship between wins and the two independent variables (ERA and league). **(f)** For X_1: $t_{STAT} = -6.9184 < -2.0518$, the p-value $= 0.0000$. Reject H_0. ERA makes a significant contribution and should be included in the model. For X_2: $t_{STAT} = -1.1966 > -2.0518$, p-value $= 0.2419 > 0.05$. Do not reject H_0. The league does not make a significant contribution and should not be included in the model. Based on these results, the regression model with only the ERA as the independent variable should be used. **(g)** $-24.2687 \leq \beta_1 \leq -13.1676$. **(h)** $-7.8464 \leq \beta_2 \leq 2.0639$. **(i)** $r^2_{adj} = 0.6165$. 61.65% of the variation in wins can be explained by the variation in ERA and league after adjusting for number of independent variables and sample size. **(j)** $r^2_{Y1.2} = 0.6394$. Holding constant the effect of league, 63.94% of the variation in number of wins can be explained by the variation in ERA. $r^2_{Y2.1} = 0.0504$. Holding constant the effect of ERA, 5.04% of the variation in number of wins can be explained by the variation in league. **(k)** The slope of the number of wins with ERA is the same, regardless of whether the team belongs to the American League or the National League. **(l)** For X_1X_2: $t_{STAT} = 1.175 < 2.0555$ the p-value is $0.2506 > 0.05$. Do not reject H_0. There is no evidence that the interaction term makes a contribution to the model. **(m)** The model with one independent variable (ERA) should be used.

14.80 The multiple regression model is Predicted base salary $= 48,091.7853 + 8,249.2156$ (gender) $+ 1,061.4521$ (age). Holding constant the age of the person, the mean base salary is predicted to be \$8,249.22 higher for males than for females. Holding constant the gender of the person, for each addition year of age, the mean base salary is predicted to be \$1,061.45 higher. The regression model with the two independent variables has $F = 118.0925$ and a p-value $= 0.0000$. So, you can conclude that at least one of the independent variable makes a significant contribution to the model to predict base pay. Each independent variable makes a significant contribution to the regression model given that the other variable is included. ($t_{STAT} = 3.9937$, p-value $= 0.0001$ for gender and $t_{STAT} = 14.8592$, p-value $= 0.0000$ for age). Both independent variables should be included in the model. 37.01% of the variation in base salary can be explained by gender and age. There is no pattern in the residuals and no other violations of the assumptions, so the model appears to be appropriate. Including an interaction term of gender and age does not significantly improve the model ($t_{stat} = -0.2371$, p-value $= 0.8127 > 0.05$). You can conclude that females are paid less than males holding constant the age of the person. Perhaps other variables such as department, seniority, and score on a performance evaluation can be included in the model to see if the model is improved.

14.82 $b_0 = 18.2892$ (die temperature), $b_1 = 0.5976$, (die diameter), $b_2 = -13.5108$. The r^2 of the multiple regression model is 0.3257 so 32.57% of the variation in foam density can be explained by the variation of die temperature and die diameter. The F test statistic for the combined significance of die temperature and die diameter is 5.0718 with a p-value of 0.0160. Hence, at a 5% level of significance, there is enough evidence to conclude that die temperature and die diameter affect foam density. The p-value of the t test for the significance of die temperature is 0.2117, which is greater than 5%. Hence, there is insufficient evidence to conclude that die temperature affects foam density holding constant the effect of die diameter. The p-value of the t test for the significance of die diameter is 0.0083, which is less than 5%.There is enough evidence to conclude that die diameter affects foam density at the 5% level of significance holding constant the effect of die temperature. After removing die temperature from the model, $b_0 = 107.9267$ (die diameter), $b_1 = -13.5108$. The r^2 of the multiple regression is 0.2724. So 27.24% of the variation in foam density can be explained by the variation of die diameter. The p-value of the t test for the significance of die diameter is 0.0087, which is less than 5%. There is enough evidence to conclude that die diameter affects foam density at the 5% level of significance. There is some lack of equality in the residuals and some departure from normality.

CHAPTER 15

15.2 (a) Predicted HOCS is 2.8600, 3.0342, 3.1948, 3.3418, 3.4752, 3.5950, 3.7012, 3.7938, 3.8728, 3.9382, 3.99, 4.0282, 4.0528, 4.0638, 4.0612, 4.045, 4.0152, 3.9718, 3.9148, 3.8442, and 3.76. **(c)** The curvilinear relationship suggests that HOCS increases at a decreasing rate. It reaches its maximum value of 4.0638 at GPA = 3.3 and declines after that as GPA continues to increase. **(d)** An r^2 of 0.07 and an adjusted r^2 of 0.06 tell you that GPA has very low explanatory power in identifying the variation in HOCS. You can tell that the individual HOCS scores are scattered widely around the curvilinear relationship.

15.4 (a) $\hat{Y} = -5.5476 + 21.6158X_1 + 3.9112X_2$ where $X_1 =$ alcohol % and $X_2 =$ carbohydrates. $F_{STAT} = 2{,}272.9982$ p-value $= 0.0000 < 0.05$, so reject H_0. At the 5% level of significance, the linear terms are significant together. **(b)** $\hat{Y} = 10.3937 + 14.9055X_1 + 4.6043X_2 + 0.5054X_1^2 - 0.0029X_2^2$, where $X_1 =$ alcohol % and $X_2 =$ carbohydrates. **(c)** $F_{STAT} = 1{,}165.9130$ p-value $= 0.0000 < 0.05$, so reject H_0. At the 5% level of significance, the model with quadratic terms are significant. $t_{STAT} = 2.2394$, and the p-value $= 0.0212$. Reject H_0. There is enough evidence that the quadratic term for alcohol % is significant at the 5% level of significance. $t_{STAT} = -1.3930$, p-value $= 0.1657$. Do not reject H_0. There is insufficient evidence that the quadratic term for carbohydrates is significant at the 5% level of significance. Hence, because the quadratic term for alcohol % is significant, the model in **(b)** that includes this term is better. **(d)** The number of calories in a beer depends quadratically on the alcohol percentage but linearly on the number of carbohydrates. The alcohol percentage and number of carbohydrates explain about 96.84% of the variation in the number of calories in a beer.

15.6 (b) price $= 18{,}029.9837 - 1{,}812.9389$ age $+ 63.2116$ age^2. **(c)** $18{,}029.9837 - 1{,}812.9389(5) + 63.2116(5)^2 = \$10{,}545.58$. **(d)** There are no patterns in any of the residual plots. **(e)** $F_{STAT} = 243.5061 > 3.27$. Reject H_0. There is a significant quadratic relationship between age and price. **(f)** p-value $= 0.0000$. The probability of $F_{STAT} = 243.5061$ or higher is 0.0000, given the null hypothesis is true. **(g)** $t_{STAT} = 4.8631 > 2.0281$. Reject H_0. **(h)** The probability of $t_{STAT} < -4.8631$ or > 4.8631 is 0.0000, given the null hypothesis is true. **(i)** $r^2 = 0.9312$. 93.12% of the variation in price can be explained by the quadratic relationship between age and price. **(j)** adjusted $r^2 = 0.9273$. **(k)** There is a strong quadratic relationship between age and price.

15.8 (a) 215.37. **(b)** For each additional unit of the logarithm of X_1, the logarithm of Y is estimated to increase by 0.9 unit, holding all other variables constant. For each additional unit of the logarithm of X_2, the logarithm of Y is estimated to increase by 1.41 units, holding all other variables constant.

15.10 (a) $\sqrt{\hat{Y}} = 6.2448 + 0.7796X_1 + 0.1661X_2$, where $X_1 =$ alcohol % and $X_2 =$ carbohydrates. **(b)** The normal probability plot of the linear model showed departure from a normal distribution, so a square-root transformation of calories was done. $F_{STAT} = 1{,}732.8312$. Because the p-value is 0.0000, reject H_0 at the 5% level of significance. There is evidence of a significant linear relationship between the square root of calories and the percentage of alcohol and the number of carbohydrates. **(c)** $r^2 = 0.9575$. So 95.75% of the variation in the square root of calories

can be explained by the variation in the percentage of alcohol and the number of carbohydrates. **(d)** Adjusted $r^2 = 0.9569$. **(e)** The model in Problem 15.10 is better because the residual plot is not right skewed.

15.12 (a) Predicted ln(Price) $= 9.7771 - 0.10622$ Age. **(b)** \$10,573.4350. **(c)** The model is adequate. **(d)** $t_{STAT} = -19.4814 < -2.0262$; reject H_0. **(e)** 91.12%. 91.12% of the variation in the natural log of price can be explained by the age of the auto. **(f)** 90.88%. **(g)** Choose the model from Problem 15.6. That model has a higher adjusted r^2 of 92.73%.

15.14 1.25.

15.16 $R_1^2 = 0.0634$, $VIF_1 = \dfrac{1}{1 - 0.0634} = 1.0677$, $R_2^2 = 0.0634$, $VIF_2 = \dfrac{1}{1 - 0.0634} = 1.0677$. There is no evidence of collinearity because both VIFs are < 5.

15.18 $VIF = 1.0066 < 5$. There is no evidence of collinearity.

15.20 $VIF = 1.0105$. There is no evidence of collinearity.

15.22 (a) 35.04. **(b)** $C_p > 3$. This does not meet the criterion for consideration of a good model.

15.24 Let $Y =$ asking price, $X_1 =$ lot size, $X_2 =$ living space, and $X_3 =$ number of bedrooms. $X_4 =$ number of bathrooms, $X_5 =$ age, and $X_6 =$ fireplace (0 = No, 1 = Yes). Based on a full regression model involving all of the variables, all the VIF values (1.3953, 2.1175, 2.0878, 2.3537, 1.7807, and 1.0939, respectively) are less than 5. There is no reason to suspect the existence of collinearity. Based on a best subsets regression and examination of the resulting C_p values, the best model appear to be a model with variables X_2 and X_6, which has $C_p = 0.8701$. Models that add other variables do not change the results very much. Based on a stepwise regression analysis with all the original variables, only variables X_2 and X_6 make a significant contribution to the model at the 0.05 level. Thus, the best model is the model using the living area of the house (X_2) and fireplace X_6 should be included in the model. This was the model developed in Section 14.6.

15.30 (a) An analysis of the linear regression model with all of the three possible independent variables reveals that the highest VIF is only 1.06. A stepwise regression model selects only the supplier dummy variable for inclusion in the model. A best subsets regression produces only one model that has a C_p value less than or equal to $k + 1$, the model that includes pressure and the supplier dummy variable. This model is $\hat{Y} = -31.5929 + 0.7879X_2 + 13.1029X_3$. This model has $F = 5.1088$ with a p-value $= 0.027$. $r^2 = 0.4816$, $r_{adj}^2 = 0.3873$. A residual analysis does not reveal any strong patterns. The errors appear to be normally distributed.

15.32 (a) Best model: $C_p = 2.1558$, predicted fair market value $= 260.6791 + 362.8318$ land $+ 0.1109$ house size (sq ft) $- 1.7543$ age. **(b)** The adjusted r^2 for the best model in 15.32(a), 15.33(a), and 15.34(a) are, respectively, 0.8242, 0.9047, and 0.8481. The model in 15.33(a) has the highest explanatory power after adjusting for the number of independent variables and sample size.

15.34 (a) Predicted fair maket value $= 145.1217 + 149.9337$ land $+ 0.0913$ house size (sq. ft.). **(b)** The adjusted r^2 for the best model in 15.32(a), 15.33(a), and 15.34(a) are, respectively, 0.8242, 0.9047, and 0.8481. The model in 15.33(a) has the highest explanatory power after adjusting for the number of independent variables and sample size.

15.36 Let $Y =$ fair market value, $X_1 =$ land area, $X_2 =$ interior size, $X_3 =$ age, $X_4 =$ number of rooms, $X_5 =$ number of bathrooms, $X_6 =$ garage size, $X_7 = 1$ if Glen Cove and 0 otherwise, and $X_8 = 1$ if Roslyn and 0 otherwise. **(a)** The VIFs of X_2, X_3, and X_7 are greater than 5.

Dropping X_2 with the largest *VIF*, X_3 still has a *VIF* greater than 5. After dropping X_2 and X_3, all remaining *VIF*s are less than 5 so there is no reason to suspect collinearity between any pair of variables. The following is the multiple regression model that has the smallest $C_p(4.3211)$ and the highest adjusted $r^2(0.6815)$:

Fair Market Value = 49.2379 + 579.0105 Land + 109.5767 Baths
+ 48.2282 Garage + 213.2326 Roslyn

The individual *t* test for the significance of each independent variable at the 5% level of significance concludes that only property size, baths, and the dummy variable Roslyn are significant given that the others are in the model. The following is the multiple regression result for the model chosen by stepwise regression:

Fair Market Value = 30.3016 + 611.6910 Land + 130.7788 Baths
+ 214.2567 Roslyn

All the variables are significant individually at the 5% level of significance. Combining the stepwise regression and the best subsets regression results along with the individual *t* test results, the most appropriate multiple regression model for predicting the fair market value is the stepwise regression model.
(b) The estimated fair market value in Roslyn is \$214.2567 thousands above Glen Cove or Freeport for two otherwise identical properties.

15.38 In the multiple regression model with catalyst, pH, pressure, temperature, and voltage as independent variables, none of the variables has a *VIF* value of 5 or larger. The best subsets approach showed that only the model containing X_1, X_2, X_3, X_4, and X_5 should be considered, where X_1 = catalyst, X_2 = pH, X_3 = pressure, X_4 = temp, and X_5 = voltage. Looking at the *p*-values of the *t* statistics for each slope coefficient of the model that includes X_1 through X_5 reveals that pH level is not significant at the 5% level of significance (*p*-value = 0.2862). The multiple regression model with pH level deleted shows that all coefficients are significant individually at the 5% level of significance. The best linear model is determined to be $\hat{Y} = 3.6833 + 0.1548X_1 - 0.04197X_3 - 0.4036X_4 + 0.4288X_5$. The overall model has $F = 77.0793$, with a *p*-value that is virtually 0. $r^2 = 0.8726$, $r^2_{adj} = 0.8613$. The normal probability plot does not suggest possible violation of the normality assumption. A residual analysis reveals a potential nonlinear relationship in temperature. The *p*-value of the squared term for temperature (0.1273) in the following quadratic transformation of temperature does not support the need for a quadratic transformation at the 5% level of significance. The *p*-value of the interaction term between pressure and temperature (0.0780) indicates that there is not enough evidence of an interaction at the 5% level of significance. The best model is the one that includes catalyst, pressure, temperature, and voltage, which explains 87.26% of the variation in thickness.

15.40 Best subset regression produced several models that had $C_p \leq k + 1$. They were $X_2X_3 = 3.9$, $X_2X_3X_4 = 3.3$, and $X_1X_2X_3X_4 = 4.7$. Stepwise regression produced a model that included only X_2 (median home value) and X_4 (average commuting time). Because X_2 (median home value), X_3 (violent crime rate), and average commuting time (X_4) had a low C_p, this model was chosen for further analysis. The residual plot for all the independent variables showed only random patterns and no violations in the assumptions. The model is

Median average annual salary = 16,830 + 38.256 median

home value (\$000) − 9.534

violent crime/100,000 residents

+ 1,053 average commuting

time in minutes

The r^2 of this model is 0.847, meaning that 84.7% of the variation in average annual salary can be explained by variation in median home value, variation in violent crime, and variation in average commuting time.

CHAPTER 16

16.2

(a) Because you need data from four prior years to obtain the centered nine-year moving average for any given year and the first recorded value is for 1984, the first centered moving average value you can calculate is for 1988.
(b) You would lose four years for the period 1984–87 because you do not have enough past values to compute a centered moving average. You will also lose the final four years of recorded time series because you do not have enough later values to compute a centered moving average. Therefore, you will lose a total of eight years in computing a series of nine-year moving averages.

16.4
(b)

↓	C1	C2	C3
	Year	Hours	AVER1
1	2008	2.2	*
2	2009	2.3	2.30000
3	2010	2.4	2.43333
4	2011	2.6	2.50000
5	2012	2.5	2.46667
6	2013	2.3	2.33333
7	2014	2.2	2.23333
8	2015	2.2	2.20000
9	2016	2.2	2.16667
10	2017	2.1	*

(c)

↓	C1	C2	C3
	Year	Hours	ES(W=0.50)
1	2008	2.2	2.20000
2	2009	2.3	2.25000
3	2010	2.4	2.32500
4	2011	2.6	2.46250
5	2012	2.5	2.48125
6	2013	2.3	2.39063
7	2014	2.2	2.29531
8	2015	2.2	2.24766
9	2016	2.2	2.22383
10	2017	2.1	2.16191

(d) $\hat{Y}_{2018} = E_{2017} = 2.16191$
(e)

↓	C1	C2	C4
	Year	Hours	ES (W=0.25)
1	2008	2.2	2.20000
2	2009	2.3	2.22500
3	2010	2.4	2.26875
4	2011	2.6	2.35156
5	2012	2.5	2.38867
6	2013	2.3	2.36650
7	2014	2.2	2.32488
8	2015	2.2	2.29366
9	2016	2.2	2.27024
10	2017	2.1	2.22768

$\hat{Y}_{2018} = E_{2017} = 2.22768$.

(f) The exponential smoothing with $W = 0.5$ assigns more weight to the more recent values and is better for short-term forecasting, while the exponential smoothing with $W = 0.25$, which assigns more weight to more distant values, is better suited for identifying long-term tendencies. **(g)** There is no perceptible trend in the number of hours spent per day by American desktop/laptop users from 2008–17. American desktop/laptop users have consistently spent between 2 and 2.5 hours per day on their computers.

16.6
(b)

C1-T	C2	C3
Decade	Performance(%)	MA 3-VR
1830s	2.8	*
1840s	12.8	7.4000
1850s	6.6	10.6333
1860s	12.5	8.8667
1870s	7.5	8.6667
1880s	6.0	6.3333
1890s	5.5	7.4667
1900s	10.9	6.2000
1910s	2.2	8.8000
1920s	13.3	4.4333
1930s	−2.2	6.9000
1940s	9.6	8.5333
1950s	18.2	12.0333
1960s	8.3	11.0333
1970s	6.6	10.5000
1980s	16.6	13.6000
1990s	17.6	11.8000
2000s	1.2	9.9333
2010s	11.0	*

(c)

C1-T	C2	C4
Decade	Performance(%)	ES(W = 0.50)
1830s	2.8	2.8000
1840s	12.8	7.8000
1850s	6.6	7.2000
1860s	12.5	9.8500
1870s	7.5	8.6750
1880s	6.0	7.3375
1890s	5.5	6.4188
1900s	10.9	8.6594
1910s	2.2	5.4297
1920s	13.3	9.3648
1930s	−2.2	3.5824
1940s	9.6	6.5912
1950s	18.2	12.3956
1960s	8.3	10.3478
1970s	6.6	8.4739
1980s	16.6	12.5370
1990s	17.6	15.0685
2000s	1.2	8.1342
2010s	11.0	9.5671

(d) $\hat{Y}_{2020s} = E_{2010s} = 9.5671$.

(e)

C1-T	C2	C5
Decade	Performance(%)	ES (W = 0.25)
1830s	2.8	2.8000
1840s	12.8	5.3000
1850s	6.6	5.6250
1860s	12.5	7.3438
1870s	7.5	7.3828
1880s	6.0	7.0371
1890s	5.5	6.6528
1900s	10.9	7.7146
1910s	2.2	6.3360
1920s	13.3	8.0770
1930s	−2.2	5.5077
1940s	9.6	6.5308
1950s	18.2	9.4481
1960s	8.3	9.1611
1970s	6.6	8.5208
1980s	16.6	10.5406
1990s	17.6	12.3055
2000s	1.2	9.5291
2010s	11.0	9.8968

$\hat{Y}_{2020s} = E_{2010s} = 9.8968$.

(f) The exponentially smoothed forecast for 2020s is lower with a W of 0.50 compared to a W of 0.25. The exponential smoothing with $W = 0.5$ assigns more weight to the more recent values and is better for short-term forecasting, while the exponential smoothing with $W = 0.25$, which assigns more weight to more distant values, is better suited for identifying long-term tendencies.
(g) Exponential smoothing with a W of 0.25 reveals a general upward trend in the performance of stocks over the last several decades.

16.8
(b)

Year	IPOs	MA 3-YR
2001	84	*
2002	70	75.000
2003	71	122.333
2004	226	167.667
2005	206	210.333
2006	199	206.000
2007	213	141.000
2008	11	95.667
2009	63	76.000
2010	154	114.000
2011	125	135.667
2012	128	158.333
2013	222	208.333
2014	275	222.333
2015	170	183.333
2016	105	145.000
2017	160	151.667
2018	190	*

(c)

Year	IPOs	ES($W = 0.50$)
2001	84	84.000
2002	70	77.000
2003	71	74.000
2004	226	150.000
2005	206	178.000
2006	199	188.500
2007	213	200.750
2008	11	105.875
2009	63	84.438
2010	154	119.219
2011	125	122.109
2012	128	125.055
2013	222	173.527
2014	275	224.264
2015	170	197.132
2016	105	151.066
2017	160	155.533
2018	190	172.766

(d) $\hat{Y}_{2019} = E_{2018} = 172.766$ IPOs.

(e)

Year	IPOs	ES ($W = 0.25$)
2001	84	84.000
2002	70	80.500
2003	71	78.125
2004	226	115.094
2005	206	137.820
2006	199	153.115
2007	213	168.086
2008	11	128.815
2009	63	112.361
2010	154	122.771
2011	125	123.328
2012	128	124.496
2013	222	148.872
2014	275	180.404
2015	170	177.803
2016	105	159.602
2017	160	159.702
2018	190	167.276

$\hat{Y}_{2019} = E_{2018} = 167.276$ IPOs.

(f) The exponentially smoothed 2019 forecast for IPOs is lower with a W of 0.25 compared to a W of 0.50. The exponential smoothing with $W = 0.5$ assigns more weight to the more recent values and is better for short-term forecasting, while the exponential smoothing with $W = 0.25$, which assigns more weight to more distant values, is better suited for identifying long-term tendencies. There appears to be a cyclical component every several years with up and down cycles of the number of IPOs.

16.10 (a) The Y-intercept $b_0 = 4.0$ is the fitted trend value reflecting the real total revenues (in millions of dollars) during the origin or base year 1998.
(b) The slope $b_1 = 1.5$ indicates that the real total revenues are increasing at an estimated rate of 1.5 million per year.

(c) Year is 2002, $X = 2002 - 1998 = 4$.

$$\hat{Y}_{2002} = 4.0 + 1.5(4) = 10.0 \text{ million dollars.}$$

(d) Year is 2019, $X = 2019 - 1998 = 21$.

$$\hat{Y}_{2019} = 4.0 + 1.5(21) = 35.5 \text{ million dollars.}$$

(e) Year is 2022, $X = 2022 - 1998 = 24$.

$$\hat{Y}_{2022} = 4.0 + 1.5(24) = 40 \text{ million dollars.}$$

16.12
(b) Regression Analysis: Bonus($000) versus Coded Year

Analysis of Variance

Source	DF	Adj SS	Adj MS	F-Value	P-Value
Regression	1	9506	9505.6	10.95	0.004
Coded Year	1	9506	9505.6	10.95	0.004
Error	17	14760	868.2		
Total	18	24265			

Model Summary

S	R-sq	R-sq(adj)	R-sq(pred)
29.4654	39.17%	35.60%	26.37%

Coefficients

Term	Coef	SE Coef	T-Value	P-Value	VIF
Constant	98.2	13.0	7.55	0.000	
Coded Year	4.08	12.3	3.31	0.004	1.00

Regression Equation

$$\text{Bonus}(\$000) = 98.2 - 4.08 \text{ Coded Year}$$
$$\hat{Y} = 98.2 + 4.08X, \text{ where } X = \text{ years relative to 2000.}$$

(c) Regression Analysis: Bonus($000) versus Coded Year, Coded Year Sq

Analysis of Variance

Source	DF	Adj SS	Adj MS	F-Value	P-Value
Regression	2	10836	5418.0	6.46	0.009
Coded Year	1	3686	3685.5	4.39	0.052
Coded Year Sq	1	1330	1330.4	1.59	0.226
Error	16	13429	339.3		
Total	18	24265			

Model Summary

S	R-sq	R-sq(adj)	R-sq(pred)
28.9711	44.66%	37.74%	25.25%

Coefficients

Term	Coef	SE Coef	T-Value	P-Value	VIF
Constant	82.2	18.0	4.57	0.000	
Coded Year	9.72	4.64	2.10	0.052	14.61
Coded Year Sq	-0.313	0.249	-1.26	0.226	14.61

Regression Equation

$$\text{Bonus}(\$000) = 82.2 - 9.72 \text{ Coded Year} - 0.313 \text{ Coded Year Sq}$$
$$\hat{Y} = 82.2 + 9.72X - 0.313X^2, \text{ where } X = \text{ years relative to 2000.}$$

(d) Regression Analysis: log Bonus versus Coded Year

Analysis of Variance

Source	DF	Adj SS	Adj MS	F-Value	P-Value
Regression	1	0.1372	0.13716	12.41	0.003
Coded Year	1	0.1372	0.13716	12.41	0.003
Error	17	0.1879	0.01105		
Total	18	0.3251			

Model Summary

S	R-sq	R-sq(adj)	R-sq(pred)
0.105130	42.20%	38.80%	28.52%

Coefficients

Term	Coef	SE Coef	T-Value	P-Value	VIF
Constant	1.9727	0.0464	42.52	0.000	
Coded Year	0.01551	0.00440	3.52	0.003	1.00

Regression Equation

$$\text{log Bonus} = 1.9727 + 0.01551 \text{ Coded Year}$$

$\log_{10}\hat{Y} = 1.9727 + 0.01551(X)$, where X = years relative to 2000.
(e) Linear: $\hat{Y}_{2019} = 98.2 + 4.08(19) = 175.72$.

$$\hat{Y}_{2020} = 98.2 + 4.08(20) = 179.80.$$

Quadratic: $\hat{Y}_{2019} = 82.2 + 9.72(19) - 0.313(19)^2 = 153.8870$.
$$\hat{Y}_{2020} = 82.2 + 9.72(20) - 0.313(20)^2 = 151.40.$$
Exponential: $\log_{10}\hat{Y}_{2019} = 1.9727 + 0.01551(19) = 2.26739$.

$$\hat{Y}_{2019} = 10^{2.26739} = 185.093.$$

$\log_{10}\hat{Y}_{2020} = 1.9727 + 0.01551(20) = 2.2829$.
$$\hat{Y}_{2020} = 10^{2.2829} = 191.8227.$$

(f)

Year	Revenues	First Difference	Second Difference	Percentage Difference
2000	100.5	#N/A	#N/A	#N/A
2001	74.1	−26.4	#N/A	−26.27%
2002	60.9	−13.2	13.2	−17.81%
2003	99.9	39.0	52.2	64.04%
2004	113.5	13.6	−25.4	13.61%
2005	149.8	36.3	22.7	31.98%
2006	191.4	41.6	5.3	27.77%
2007	177.8	−13.6	−55.2	−7.11%
2008	100.9	−76.9	−63.3	−43.25%
2009	140.6	39.7	116.6	39.35%
2010	139.0	−1.6	−41.3	−1.14%
2011	111.4	−27.6	−26.0	−19.86%
2012	142.9	31.5	59.1	28.28%
2013	169.8	26.9	−4.6	18.82%
2014	160.3	−9.5	−36.4	−5.59%
2015	136.8	−23.5	−14.0	−14.66%
2016	156.8	20.0	43.5	14.62%
2017	184.4	27.6	7.6	17.60%
2018	153.7	−30.7	−58.3	−16.65%

A review of first, second, and percentage differences reveals no particular model is more appropriate than the other. Based on the principle of parsimony, one might choose the linear model.

16.14

(b) Regression Analysis: Receipts versus Coded Year

Analysis of Variance

Source	DF	Adj SS	Adj MS	F-Value	P-Value
Regression	1	30230638	30280638	916.97	0.000
Coded Year	1	30280638	30280638	916.97	0.000
Error	39	1287872	33022		
Total	40	31568511			

Model Summary

S	R-sq	R-sq(adj)	R-sq(pred)
181.721	95.92%	95.82%	95.42%

Coefficients

Term	Coef	SE Coef	T-Value	P-Value	VIF
Constant	225.8	55.7	4.05	0.000	
Coded Year	72.63	2.40	30.28	0.000	1.00

Regression Equation

$$\text{Receipts} = 225.8 + 72.63 \text{ Coded Year}$$

$\hat{Y} = 225.8 + 72.63(X)$, where X = years relative to 1978.

(c) $\hat{Y}_{2019} = 225.8 + 72.63(41) = 3{,}203.63$ billion.
$\hat{Y}_{2020} = 225.8 + 72.63(42) = 3{,}276.26$ billion.
(d) There is an upward trend in federal receipts from 1978 through 2018, which appears to be linear.

16.16

(b) Linear:

Regression Analysis: Solar Power Generated versus Coded Year

Analysis of Variance

Source	DF	Adj SS	Adj MS	F-Value	P-Value
Regression	1	4447549704	4447549704	28.12	0.000
Coded Year	1	4447549704	4447549704	28.12	0.000
Error	15	2372666419	158177761		
Total	16	6820216122			

Model Summary

S	R-sq	R-sq(adj)	R-sq(pred)
12576.9	65.21%	62.89%	51.28%

Coefficients

Term	Coef	SE Coef	T-Value	P-Value	VIF
Constant	−13307	5841	−2.28	0.038	
Coded Year	3302	623	530	0.000	1.00

Regression Equation

$$\text{Solar Power Generated} = -13307 + 3302 \text{ Coded Year}$$

$\hat{Y} = -13{,}307 + 3{,}302(X)$, where X = years relative to 2002

(c) Quadratic:

Regression Analysis: Solar Power Generated versus . . . r, Coded Year Sq

Analysis of Variance

Source	DF	Adj SS	Adj MS	F-Value	P-Value
Regression	2	6516476292	3258238146	150.18	0.000
Coded Year	1	694663208	694663208	32.02	0.000
Coded Year Sq	1	2068926589	2068926589	95.36	0.000
Error	14	303739830	21695702		
Total	16	6820216122			

Model Summary

S	R-sq	R-sq(adj)	R-sq(pred)
4657.86	95.55%	94.91%	91.93%

Coefficients

Term	Coef	SE Coef	T-Value	P-Value	VIF
Constant	7358	3026	2.43	0.029	
Coded Year	−4964	877	−5.66	0.000	14.47
Coded Year Sq	516.6	52.9	9.77	0.000	1447

Regression Equation

Solar Power Generated $= 7358 - 4964$ Coded Year $+ 516.6$ Coded Year Sq

$\hat{Y} = 7{,}358 - 4{,}964X + 516.6X^2$, where $X =$ years relative to 2002.

(d) Exponential:

Regression Analysis: Solar Power Log 10 versus Coded Year

Analysis of Variance

Source	DF	Adj SS	Adj MS	F-Value	P-Value
Regression	1	9.218	9.21783	124.71	0.000
Coded Year	1	9.218	9.21783	124.71	0.000
Error	15	1.109	0.07391		
Total	16	10.327			

Model Summary

S	R-sq	R-sq(adj)	R-sq(pred)
0.271872	89.26%	88.55%	85.92%

Coefficients

Term	Coef	SE Coef	T-Value	P-Value	VIF
Constant	2.274	0.126	18.01	0.000	
Coded Year	0.1503	0.0135	11.17	0.000	1.00

Regression Equation

Solar Power Log 10 $= 2.274 + 0.1503$ Coded Year

$\log_{10}\hat{Y} = 2.274 + 0.1503(X)$, where $X =$ years relative to 2002.

(d) Linear:

$\hat{Y}_{2019} = -13{,}307 + 3{,}302(17) = 42{,}827$ million kWh.

$\hat{Y}_{2020} = -13{,}307 + 3{,}302(18) = 46{,}129$ million kWh.

Quadratic:

$\hat{Y} = 7{,}358 - 4{,}964(17) + 516.6(17)^2 = 72{,}267.4$ million kWh.

$\hat{Y} = 7{,}358 - 4{,}964(18) + 516.6(18)^2 = 85{,}384.4$ million kWh.

Exponential:

$\log_{10}\hat{Y}_{2019} = 2.274 + 0.1503(17) = 4.8291$.

$\hat{Y}_{2019} = 10^{4.8291} = 67{,}468.34$ million kWh.

$\log_{10}\hat{Y}_{2020} = 2.274 + 0.1503(18) = 4.9794$.

$\hat{Y}_{2020} = 10^{4.9794} = 95{,}367.41$ million kWh.

16.18

(b) Linear:

Regression Analysis: Salary ($mil) versus Coded Year

Analysis of Variance

Source	DF	Adj SS	Adj MS	F-Value	P-Value
Regression	1	13.3042	13.3042	529.35	0.000
Coded Year	1	13.3042	13.3042	529.35	0.000
Error	18	0.4524	0.0251		
Total	19	13.7566			

Model Summary

S	R-sq	R-sq(adj)	R-sq(pred)
0.158534	96.71%	96.53%	95.96%

Coefficients

Term	Coef	SE Coef	T-Value	P-Value	VIF
Constant	1.9853	0.0683	29.06	0.000	
Coded Year	0.14144	0.00615	23.01	0.000	1.00

Regression Equation

Salary ($mil) $= 1.9353 + 0.14144$ Coded Year

$\hat{Y} = 1.9853 + 0.14144(X)$, where $X =$ years relative to 2000.

(c) Quadratic:

Regression Analysis: Salary ($mil) versus Coded Year, Coded Year Sq

Analysis of Variance

Source	DF	Adj SS	Adj MS	F-Value	P-Value
Regression	2	13.4579	6.72893	382.93	0.000
Coded Year	1	0.3292	0.32920	18.73	0.000
Coded Year Sq	1	0.1537	0.15367	8.74	0.009
Error	17	0.2987	0.01757		
Total	19	13.7566			

Model Summary

S	R-sq	R-sq(adj)	R-sq(pred)
0.132560	97.83%	97.57%	96.77%

Coefficients

Term	Coef	SE Coef	T-Value	P-Value	VIF
Constant	2.1539	0.0807	26.68	0.000	
Coded Year	0.0852	0.0197	4.33	0.000	14.67
Coded Year Sq	0.00296	0.00100	2.96	0.009	14.67

Regression Equation

Salary ($mil) $= 2.1539 - 0.0352$ Coded Year $+ 0.00296$ Coded Year Sq

$\hat{Y} = 2.1539 + 0.0852X + 0.00296X^2$, where $X =$ years relative to 2000.

(d) Exponential:
Regression Analysis: Salary log 10 versus Coded Year

Analysis of Variance

Source	DF	Adj SS	Adj MS	F-Value	P-Value
Regression	1	0.232769	0.232769	845.09	0.000
Coded Year	1	0.232769	0.232769	845.09	0.000
Error	18	0.004958	0.000275		
Total	19	0.237726			

Model Summary

S	R-sq	R-sq(adj)	R-sq(pred)
0.0165963	97.91%	97.80%	97.32%

Coefficients

Term	Coef	SE Coef	T-Value	P-Value	VIF
Constant	0.33101	0.00715	46.28	0.000	
Coded Year	0.018709	0.000644	29.07	0.000	1.00

Regression Equation

Salary log 10 = 0.33101 + 0.018705 Coded Year

$\log_{10}\hat{Y} = 0.33101 + 0.018709(X)$, where X = years relative to 2000.

(e)

Year	Salary ($mil)	First Difference	Second Difference	Percentage Difference
2000	1.99	#N/A	#N/A	#N/A
2001	2.29	0.3	#N/A	15.08%
2002	2.38	0.1	−0.2	3.93%
2003	2.58	0.2	0.1	3.40%
2004	2.49	−0.1	−0.3	−3.49%
2005	2.63	0.1	0.2	5.62%
2006	2.83	0.2	0.1	7.60%
2007	2.92	0.1	−0.1	3.18%
2008	3.13	0.2	0.1	7.19%
2009	3.26	0.1	−0.1	4.15%
2010	3.27	0.0	−0.1	0.31%
2011	3.32	0.0	0.0	1.53%
2012	3.38	0.1	0.0	1.81%
2013	3.62	0.2	0.2	7.10%
2014	3.81	0.2	−0.1	5.25%
2015	4.25	0.4	0.3	11.55%
2016	4.40	0.2	−0.3	3.53%
2017	4.70	0.3	0.1	6.82%
2018	4.63	−0.1	−0.4	−1.49%
2019	4.70	0.1	0.1	1.51%

The first and second differences are relatively consistent across the series. This is not the case for percentage differences. Based on the principle of parsimony, one might choose the linear model.
(e) Linear forecast: $\hat{Y}_{2020} = 1.9853 + 0.14144(20) = 4.8141$ million.

16.20 (b) There has been an upward trend in the CPI-U in the United States from 1965 through 2018.

(c) Linear:
Regression Analysis: CPI versus Coded Year

Analysis of Variance

Source	DF	Adj SS	Adj MS	F-Value	P-Value
Regression	1	261479	261479	11402.09	0.000
Coded Year	1	261479	261479	11402.09	0.000
Error	52	1192	23		
Total	53	262671			

Model Summary

S	R-sq	R-sq(adj)	R-sq(pred)
4.78879	99.55%	99.54%	99.50%

Coefficients

Term	Coef	SE Coef	T-Value	P-Value	VIF
Constant	16.71	1.29	13.00	0.000	
Coded Year	4.4647	0.0418	106.78	0.000	1.00

Regression Equation

CPI = 16.71 − 4.4647 Coded Year

$\hat{Y} = 16.71 + 4.4647(X)$, where X = years relative to 1965.

(d) Quadratic:
Regression Analysis: CPI versus Coded Year, Coded Year Sq

Analysis of Variance

Source	DF	Adj SS	Adj MS	F-Value	P-Value
Regression	2	261504	130752	5714.90	0.000
Coded Year	1	15653	15653	684.17	0.000
Coded Year Sq	1	26	25	1.12	0.295
Error	51	1167	23		
Total	53	262671			

Model Summary

S	R-sq	R-sq(adj)	R-sq(pred)
4.78322	99.56%	99.54%	99.48%

Coefficients

Term	Coef	SE Coef	T-Value	P-Value	VIF
Constant	18.17	1.88	9.65	0.000	
Coded Year	4.296	0.164	26.16	0.000	15.47
Coded Year Sq	0.00317	0.00300	1.06	0.295	15.47

Regression Equation

CPI = 18.17 + 4.296 Coded Year + 0.00317 Coded Year Sq

$\hat{Y} = 18.17 + 4.296X + 0.00317X^2$, where X = years relative to 1965.

(e) Exponential:
Regression Analysis: CPI log 10 versus Coded Year

Analysis of Variance

Source	DF	Adj SS	Adj MS	F-Value	P-Value
Regression	1	4.0304	4.03041	775.31	0.000
Coded Year	1	4.0304	4.03041	775.31	0.000
Error	52	0.2703	0.00520		
Total	53	4.3007			

Model Summary

S	R-sq	R-sq(adj)	R-sq(pred)
0.0721000	93.71%	93.59%	93.15%

Coefficients

Term	Coef	SE Coef	T-Value	P-Value	VIF
Constant	1.5883	0.0194	82.07	0.000	
Coded Year	0.017529	0.000630	27.84	0.000	1.00

Regression Equation

$$\text{CPI log 10} = 1.5883 + 0.017529 \text{ Coded Year}$$

$\log_{10}\hat{Y} = 1.5883 + 0.017529(X)$, where X = years relative to 1965.

(f) A review of the first, second, and percentage differences revealed similar levels of variation across the time series. Based on the principle of parsimony, one might choose the linear model.

(g) Linear forecast:

$$\hat{Y}_{2019} = 16.71 + 4.4647(54) = 257.8038.$$
$$\hat{Y}_{2020} = 16.71 + 4.4647(55) = 262.2685.$$

16.22 (a) For Time Series I, the graph of Y versus X appears to be more linear than the graph of log Y versus X, so a linear model appears to be more appropriate. For Time Series II, the graph of log Y versus X appears to be more linear than the graph of Y versus X, so an exponential model appears to be more appropriate.

(b) Time Series I: $\hat{Y} = 100.0731 + 14.9776(X)$, where X = years relative to 2007.

Time Series II: $\hat{Y} = 10^{1.9982 + 0.0609(X)}$, where X = years relative to 2007.

(c) $X = 12$ for year 2019 in all models. Forecasts for the year 2019:

Time Series I: $\hat{Y} = 100.0731 + 14.9776(12) = 279.8043.$

Time Series II $\hat{Y} = 10^{1.9982 + 0.0609(12)} = 535.7967.$

16.24 $t_{STAT} = \dfrac{a_3}{S_{a_3}} = \dfrac{0.24}{0.10} = 2.4$ is greater than the critical bound of 2.2281. Reject H_0.

There is sufficient evidence that the third-order regression parameter is significantly different from zero. A third-order autoregressive model is appropriate.

16.26 (a) $t_{STAT} = \dfrac{a_3}{S_{a_3}} = \dfrac{0.24}{0.15} = 1.6$ is less than the critical bound of 2.2281. Do not reject H_0. There is not sufficient evidence that the third-order regression parameter is significantly different than zero. A third-order autoregressive model is not appropriate.

(b) Fit a second-order autoregressive model, and test to see if it is appropriate.

16.28 (a)

Regression Analysis: Bonus($000) versus lag1, lag2, lag3

Method

Rows unused 3

Analysis of Variance

Source	DF	Adj SS	Adj MS	F-Value	P-Value
Regression	3	3461.9	1154.0	1.60	0.240
lag1	1	3269.4	3269.4	4.55	0.054
lag2	1	959.3	959.3	1.33	0.271
lag3	1	467.8	467.8	0.65	0.436
Error	12	8628.3	719.0		
Total	15	12090.2			

Model Summary

S	R-sq	R-sq(adj)	R-sq(pred)
26.8147	28.63%	10.79%	0.00%

Coefficients

Term	Coef	SE Coef	T-Value	P-Value	VIF
Constant	91.1	34.2	2.66	0.021	
lag1	0.531	0.249	2.13	0.054	1.61
lag2	−0.336	0.291	−1.16	0.271	2.38
lag3	0.193	0.239	0.81	0.436	1.63

Regression Equation

$$\text{Bonus}(\$000) = 91.1 + 0.531 \text{ lag1} - 0.336 \text{ lag2} + 0.193 \text{ lag3}$$

For the third-order term, $t_{STAT} = 0.81$ with a p-value of 0.436. The third term can be dropped because it is not significant at the 0.05 significance level.

(b)

Regression Analysis: Bonus($000) versus lag1, lag2

Method

Rows unused 2

Analysis of Variance

Source	DF	Adj SS	Adj MS	F-Value	P-Value
Regression	2	7041.4	3520.7	4.18	0.038
lag1	1	6232.8	6232.8	7.40	0.017
lag2	1	653.1	653.1	0.78	0.393
Error	14	11794.9	342.5		
Total	16	18836.3			

Model Summary

S	R-sq	R-sq(adj)	R-sq(pred)
29.0257	37.38%	28.44%	0.00%

Coefficients

Term	Coef	SE Coef	T-Value	P-Value	VIF
Constant	79.8	30.0	2.66	0.019	
lag1	0.661	0.243	2.72	0.017	1.59
lag2	−0.221	0.251	−0.88	0.393	1.59

Regression Equation

$$\text{Bonus}(\$000) = 79.3 + 0.661 \text{ lag1} - 0.221 \text{ lag2}$$

Fits and Diagnostics for Unusual Observations

Obs	Bonus($000)	Fit	Resid	Std Resid
9	100.9	154.9	−54.0	−2.12 R

For the second-order term, $t_{STAT} = -0.88$ with a p-value of 0.393. The second-order term can be dropped because it is not significant at the 0.05 significance level.

(c) Regression Analysis: Bonus($000) versus lag1

Method

Rows unused 1

Analysis of Variance

Source	DF	Adj SS	Adj MS	F-Value	P-Value
Regression	1	8520	8519.6	9.41	0.007
lag1	1	8520	8519.6	9.41	0.007
Error	16	14491	905.7		
Total	17	23011			

Model Summary

S	R-sq	R-sq(adj)	R-sq(pred)
30.0948	37.02%	33.09%	19.94%

Coefficients

Term	Coef	SE Coef	T-Value	P-Value	VIF
Constant	56.9	27.0	2.11	0.051	
lag1	0.597	0.195	3.07	0.007	1.00

Regression Equation

$$\text{Bonus}(\$000) = 56.9 + 0.597\,\text{lag1}$$

For the first-order term, $t_{STAT} = 3.07$ with a p-value of 0.007. The first-order term should be retained because it is significant at the 0.05 significance level.

(d) $\hat{Y}_{2019} = 56.9 + 0.597(153.7) = 148.6589$.

16.30 (a) Regression Analysis: Salary ($mil) versus lag1, lag2, lag3

Method

Rows unused 3

Analysis of Variance

Source	DF	Adj SS	Adj MS	F-Value	P-Value
Regression	3	9.07633	3.02544	153.58	0.000
lag1	1	0.27113	0.27113	13.76	0.003
lag2	1	0.01087	0.01087	0.55	0.471
lag3	1	0.01921	0.01921	0.98	0.341
Error	13	0.25609	0.01970		
Total	16	9.33242			

Model Summary

S	R-sq	R-sq(adj)	R-sq(pred)
0.140355	97.26%	96.62%	95.56%

Coefficients

Term	Coef	SE Coef	T-Value	P-Value	VIF
Constant	0.134	0.164	0.82	0.429	
lag1	1.020	0.275	3.71	0.003	34.30
lag2	0.275	0.371	0.74	0.471	57.79
lag3	-0.310	0.314	-0.99	0.341	36.81

Regression Equation

$$\text{Salary}(\$mil) = 0.134 + 1.020\,\text{lag1} + 0.275\,\text{lag2} - 0.310\,\text{lag3}$$

For the third-order term, $t_{STAT} = -0.99$ with a p-value of 0.341. The third term can be dropped because it is not significant at the 0.05 significance level.

(b)

Regression Analysis: Salary ($mil) versus lag1, lag2

Method

Rows unused 2

Analysis of Variance

Source	DF	Adj SS	Adj MS	F-Value	P-Value
Regression	2	10.2940	5.14698	279.70	0.000
lag1	1	0.2725	0.27249	14.81	0.002
lag2	1	0.0009	0.00093	0.05	0.825
Error	15	0.2760	0.01840		
Total	17	10.5700			

Model Summary

S	R-sq	R-sq(adj)	R-sq(pred)
0.135652	97.39%	97.04%	96.26%

Coefficients

Term	Coef	SE Coef	T-Value	P-Value	VIF
Constant	0.102	0.146	0.70	0.495	
lag1	0.955	0.248	3.85	0.002	33.77
lag2	0.057	0.252	0.23	0.825	33.77

Regression Equation

$$\text{Salary}(\$mil) = 0.102 + 0.955\,\text{lag1} + 0.057\,\text{lag2}$$

For the second-order term, $t_{STAT} = 0.23$ with a p-value of 0.825. The second term can be dropped because it is not significant at the 0.05 significance level.

(c) Regression Analysis: Salary ($mil) versus lag1

Method

Rows unused 1

Analysis of Variance

Source	DF	Adj SS	Adj MS	F-Value	P-Value
Regression	1	11.5661	11.5661	648.48	0.000
lag1	1	11.5661	11.5661	648.48	0.000
Error	17	0.3032	0.0178		
Total	18	11.8693			

Model Summary

S	R-sq	R-sq(adj)	R-sq(pred)
0.133550	97.45%	97.30%	96.69%

Coefficients

Term	Coef	SE Coef	T-Value	P-Value	VIF
Constant	0.172	0.130	1.32	0.204	
lag1	0.9910	0.0389	25.47	0.000	1.00

Regression Equation

$$\text{Salary}(\$mil) = 0.172 + 0.9910\,\text{lag1}$$

For the first-order term, $t_{STAT} = 25.47$ with a p-value of 0.000. The first-order term should be retained because it is significant at the 0.05 significance level.

(d) $\hat{Y}_{2020} = 0.172 + 0.9910(4.7) = 4.8297$ million.

16.32

(a) $S_{YX} = \sqrt{\dfrac{\sum_{i=1}^{n}(Y_i - \hat{Y}_i)^2}{n - p - 1}} = \sqrt{\dfrac{45}{12 - 1 - 1}} = 2.121$. The standard error of the estimate is 2.121.

(b) $MAD = \dfrac{\sum_{i=1}^{n}|Y_i - \hat{Y}_i|}{n} = \dfrac{18}{12} = 1.5$. The mean absolute deviation is 1.5.

16.34

(b)(c)

Solar Power

	S_{yx}	MAD
Linear	12576.9	9870.772
Quadratic	4657.86	3861.937
Exponential	8317.774	4526.413
AR-First Order	2342.68	1766.126

(d) The residual plots for linear, quadratic, and exponential reveal cyclical patterns across coded year. The first-order autoregressive model has no clear pattern but does have one outlier. The first-order autoregressive model has the smallest S_{xy} and MAD values. On the basis of (a) through (c) and the principle of parsimony, the first-order autoregressive model would be the best model for forecasting.

16.36
(b) (c)
Bonuses

	S_{yx}	MAD
Linear	29.465	20.936
Quadratic	28.971	20.509
Exponential	30.581	21.803
AR-First Order	30.095	23.097

(d) The residual plots for the linear, quadratic, and exponential reveal a slight cyclical pattern for the first part of the time series followed by no clear pattern for the remainder of the series. The first-order autoregressive revealed no clear pattern throughout the time series. Each of the models had similar S_{xy} and MAD values. On the basis of the residual plots and the principle of parsimony, the first-order autoregressive model might be the best choice for forecasting.

16.38
(b) (c)
Salary

	S_{yx}	MAD
Linear	0.159	0.120
Quadratic	0.133	0.104
Exponential	0.128	0.104
AR-First Order	0.134	0.100

(d) The residual plots for linear, quadratic, and exponential reveal cyclical patterns across coded year. The first-order autoregressive model has no clear pattern. The S_{xy} and MAD values are similar across each of the models. On the basis of (a) through (c) and the principle of parsimony, the first-order autoregressive model might be the best model for forecasting.

16.40 (a) $\log \hat{B}_0 = 2$, $\hat{B}_0 = 100$.
This is the unadjusted forecast.
(b) $\log \hat{B}_1 = 0.01$, $\hat{B}_1 = 1.0233$.
The estimated monthly compound growth rate is 2.33%.
(c) $\log \hat{B}_2 = 0.10$, $\hat{B}_2 = 1.2589$.
The January values in the time series are estimated to have a mean 25.89% higher than the December values.

16.42 (a) $\log \hat{B}_0 = 3$, $\hat{B}_0 = 1,000$.
This is the unadjusted forecast.
(b) $\log \hat{B}_1 = 0.10$, $\hat{B}_1 = 1.2589$.
The estimated quarterly compound growth rate is $(\hat{B}_1 - 1)100\% = 25.89\%$.
(c) $\log \hat{B}_3 = 0.20$, $\hat{B}_3 = 1.5849$.

16.44 (a) The revenues for Target appear to be subject to seasonal variation given that revenues are consistently higher in the fourth quarter, which includes several substantial holidays.

The plot confirms the answer for (a) by clearly revealing a seasonal component to revenues.

(c)

Coefficients

Term	Coef	SE Coef	T-Value	P-Value	VIF
Constant	1.1091	0.0130	85.63	0.000	
Coded Q	0.004177	0.000218	19.13	0.000	1.00
Q1	−0.1346	0.0136	−9.88	0.000	1.52
Q2	−0.1233	0.0138	−8.93	0.000	1.51
Q3	−0.1233	0.0138	−8.93	0.000	1.51

Regression Equation

LogRevenue $= 1.1091 + 0.004177$ Coded Q $- 0.1346$ Q1
$- 0.1233$ Q2 $- 0.1233$ Q3

(d) $\log_{10} \hat{B}_1 = 0.0042$; $\hat{B}_1 = 10^{0.0042} = 1.0097$.
The estimated quarterly compound growth rate is
$(\hat{B}_1 - 1)100\% = 0.97\%$.
(e)

Quarter	$b_i = \log \hat{B}_i$	$\hat{B}_i = 10^{b_i}$	$(\hat{B}_i - 1)100\%$
First	−0.1346	0.7335	−26.65%
Second	−0.1233	0.7528	−24.72%
Third	−0.1233	0.7528	−24.72%

The first, second, and third multipliers are −26.65%, −24.72%, and −24.72% relative to fourth quarter values, respectively.

(f) 2019 Revenues ($millions)

Quarter	$\hat{Y}$
Second	20.2936
Third	20.4932
Fourth	27.4827

2020 Revenues ($millions)

Quarter	$\hat{Y}$
First	20.3512
Second	21.0892
Third	21.2971
Fourth	28.5601

16.46
(b)

Coefficients

Term	Coei	SE Coei	T-Value	P-Value	VIF
Constant	4.9069	0.0944	52.00	0.000	
Coded Month	0.006291	0.000918	6.85	0.000	1.00
Month_Jan	−0.167	0.115	−1.45	0.151	1.95
Month_Feb	−0.203	0.115	−1.77	0.081	1.95
Month_Mar	−0.134	0.115	−1.17	0.246	1.95
Month_Apr	−0.125	0.115	−1.09	0.279	1.95
Month_May	−0.137	0.115	−1.19	0.238	1.95
Month_Jun	−0.204	0.119	−1.72	0.089	1.85
Month_Jul	−0.175	0.119	−1.48	0.144	1.85
Month_Aug	−0.139	0.119	−1.17	0.245	1.84
Month_Sep	−0.075	0.119	−0.63	0.529	1.84
Month_Oct	−0.054	0.119	−0.45	0.652	1.84
Month_Nov	−0.073	0.119	−0.62	0.540	1.84

Regression Equation

Log Volume = 4.9069 + 0.006291 Coded Month

$\qquad$ − 0.167 Month_Jan − 0.203 Month_Feb

$\qquad$ − 0.134 Month_Mar − 0.125 Month_Apr

$\qquad$ − 0.137 Month_May − 0.204 Month_Jun

$\qquad$ − 0.175 Month_Jul − 0.139 Month_Aug − 0.075 Month_Sep

$\qquad$ − 0.054 Month_Oct − 0.073 Month_Nov

(c) December 2019 Forecast: $\log \hat{Y}_{95} = 5.5045$, $\hat{Y}_{95} = 319{,}521.4$ barrels.

(d) Forecast for last four months of 2019:

Month	Coded Month	$\hat{Y}$
September	92	257,335.7
October	93	274,283.7
November	94	266,164.4
December	95	319,521.4

(e) $\log_{10} \hat{\beta}_1 = 0.006291$; $\hat{\beta}_1 = 10^{0.006291} = 1.0146$.
The estimated monthly compound growth rate is
$(\hat{\beta}_1 - 1)100\% = 1.46\%$ after adjusting for the seasonal component.
(f) The July values are estimated to have a mean
of 33.17% below the December values.

16.48 (b)

Coefficients

Term	Coef	SE Coef	T-Value	P-Value	VIF
Constant	1.0212	0.0575	17.75	0.000	
Coded Quarter	0.00574	0.00124	4.63	0.000	1.00
Q1	0.0375	0.0607	0.62	0.539	1.51
Q2	0.0043	0.0606	0.07	0.944	1.50
Q3	0.0074	0.0606	0.12	0.903	1.50

Regression Equation

Log Price = 1.0212 + 0.00574 Coded Quarter
+ 0.0375 Q1 + 0.0043 Q2 + 0.0074 Q3

(c) $\log_{10} \hat{\beta}_1 = 0.00574$; $\hat{\beta}_1 = 10^{0.00574} = 1.0133$.
The estimated quarterly compound growth rate is
$(\hat{\beta}_1 - 1)100\% = 1.33\%$.
(d) $\log_{10} \hat{\beta}_2 = 0.0375$; $\hat{\beta}_2 = 10^{0.0375} = 1.0902$.
The first quarter values are estimated to have a mean of 9.02% above
the fourth quarter values. A review of the p-values associated with the
t test on the slope of the coefficients reveals that the slope coefficients
for Quarter 1, Quarter 2, and Quarter 3 are not significant at the 0.05
significance level.
(e) The fitted value is 22.91 (US$).
(f) 2019 Forecasts:

Quarter	Coded Quarter	$\hat{Y}$
First	60	25.3070
Second	61	23.7531
Third	62	24.2443
Fourth	63	24.1513

(g) The forecasts are not likely to be accurate given that that the
quarterly exponential trend model did not fit the data particularly well. The adjusted $r^2 = 0.2308$. In addition, the time series
contained an irregular component from 2010 through 2013.

16.60

(b)

Regression Analysis: Population versus Coded Year

Analysis of Variance

Source	DF	Adj SS	Adj MS	F-Value	P-Value
Regression	1	21635022177	21635022177	10461.18	0.000
Coded Year	1	21635022177	21635022177	10461.18	0.000
Error	33	68248128	2068125		
Total	34	21703270305			

Model Summary

S	R-sq	R-sq(adj)	R-sq(pred)
1438.10	99.69%	99.68%	99.65%

Coefficients

Term	Coef	SE Coef	T-Value	P-Value	VIF
Constant	173763	476	365.10	0.000	
Coded Year	2461.8	24.1	102.28	0.000	1.00

Regression Equation

$\qquad$ Population = 173763 + 2461.8 Coded Year

$\hat{Y} = 173{,}763 + 2{,}461.8(X)$ where $X =$ years relative to 1984.
(c) $\hat{Y}_{2019} = 173{,}763 + 2{,}461.8(35) = 259{,}926$ (thousands).
$\hat{Y}_{2020} = 173{,}763 + 2{,}461.8(36) = 262{,}387.8$ (thousands).
(d)

Regression Analysis: Workforce versus Coded Year

Analysis of Variance

Source	DF	Adj SS	Adj MS	F-Value	P-Value
Regression	1	7019320655	7019320655	1312.93	0.000
Coded Year	1	7019320655	7019320655	1312.93	0.000
Error	33	176427811	5346297		
Total	34	7195748466			

Model Summary

S	R-sq	R-sq(adj)	R-sq(pred)
2312.21	97.55%	97.47%	97.21%

Coefficients

Term	Coef	SE Coef	T-Value	P-Value	VIF
Constant	117084	765	153.01	0.000	
Coded Year	1402.2	38.7	36.23	0.000	1.00

Regression Equation

$\qquad$ Workforce = 117084 + 1402.2 Coded Year

$\hat{Y} = 117{,}084 + 1{,}402.2(X)$ where $X =$ years relative to 1984.
(c) $\hat{Y}_{2019} = 117{,}084 + 1402.2(35) = 166{,}161$ (thousands).
$\hat{Y}_{2020} = 117{,}084 + 1402.2(35) = 167{,}563.2$ (thousands).

16.62

(b) Linear:

Regression Analysis: Revenues versus Coded Year

Analysis of Variance

Source	DF	Adj SS	Adj MS	F-Value	P-Value
Regression	1	3352.7	3352.65	619.03	0.000
Coded Year	1	3352.7	3352.65	619.03	0.000
Error	42	227.5	542		
Total	43	3580.1			

Model Summary

S	R-sq	R-sq(adj)	R-sq(pred)
2.32723	93.65%	93.49%	92.82%

Coefficients

Term	Coef	SE Coef	T-Value	P-Value	VIF
Constant	−2.120	0.690	−3.07	0.004	
Coded Year	0.6874	0.0276	24.88	0.000	1.00

Regression Equation

$$\text{Revenues} = -2.120 + 0.6874 \text{ Coded Year}$$

$\hat{Y} = -2.120 + 0.6874(X)$, where $X = $ years relative to 1975.

(c) Quadratic:

Regression Analysis: Revenues versus Coded Year, Coded Year Sq

Analysis of Variance

Source	DF	Adj SS	Adj MS	F-Value	P-Value
Regression	2	3370.63	1685.32	329.84	0.000
Coded Year	1	113.98	113.98	22.31	0.000
Coded Year Sq	1	17.98	17.98	3.52	0.068
Error	41	209.49	5.11		
Total	43	3580.13			

Model Summary

S	R-sq	R-sq(adj)	R-sq(pred)
2.26043	94.15%	93.86%	92.48%

Coefficients

Term	Coef	SE Coef	T-Value	P-Value	VIF
Constant	−0.785	0.978	−0.80	0.427	
Coded Year	0.497	0.105	4.72	0.000	15.36
Coded Year Sq	0.00444	0.00236	1.88	0.068	15.36

Regression Equation

$$\text{Revenues} = -0.785 + 0.497 \text{ Coded Year} + 0.00444 \text{ Coded Year Sq}$$

$\hat{Y} = -0.785 + 0.497X + 0.00444X^2$, where $X = $ years relative to 1975.

(d) Exponential:

Regression Analysis: Log Revenues versus Coded Year

Analysis of Variance

Source	DF	Adj SS	Adj MS	F-Value	P-Value
Regression	1	7.4983	7.49834	607.96	0.000
Coded Year	1	7.4983	7.49834	607.96	0.000
Error	42	0.5180	0.01233		
Total	43	8.0164			

Model Summary

S	R-Sq	R-sq(adj)	R-sq(pred)
0.111057	93.54%	93.38%	92.59%

Coefficients

Term	Coef	SE Coef	T-Value	P-Value	VIF
Constant	0.2393	0.0329	7.27	0.000	
Coded Year	0.03251	0.00132	24.66	0.000	1.00

Regression Equation

$$\text{Log Revenues} = 0.2393 + 0.03251 \text{ Coded Year}$$

$\log_{10}\hat{Y} = 0.2393 + 0.03251(X)$, where $X = $ years relative to 1975.

(e) Autoregressive:

For the third-order term, $t_{STAT} = 0.08$ with a p-value of 0.935. The third-order term can be dropped because it is not significant at the 0.05 significance level.

Regression Analysis: Revenues versus lag1, lag2

Method

Rows unused 2

Analysis of Variance

Source	DF	Adj SS	Adj MS	F-Value	P-Value
Regression	2	3278.23	1639.11	2914.83	0.000
lag1	1	85.42	85.42	151.90	0.000
lag2	1	14.02	14.02	24.93	0.000
Error	39	21.93	0.56		
Total	41	3300.16			

Model Summary

S	R-sq	R-sq(adj)	R-sq(pred)
0.749890	99.34%	99.30%	99.16%

Coefficients

Term	Coef	SE Coef	T-Value	P-Value	VIF
Constant	0.344	0.212	1.62	0.113	
lag1	1.648	0.134	12.32	0.000	107.30
lag2	−0.665	0.133	−4.99	0.000	107.30

Regression Equation

$$\text{Revenues} = 0.344 + 1.648 \text{ lag1} - 0.665 \text{ lag2}$$

For the second-order term, $t_{STAT} = -4.99$ with a p-value of 0.000. The second-order term cannot be dropped because it is significant at the 0.05 significance level. The second-order model is appropriate.

$$\hat{Y}_i = 0.344 + 1.648(Y_{i-1}) - 0.665(Y_{i-2}).$$

(g)

Bonuses

	S_{yx}	MAD
Linear	2.3272	1.8713
Quadratic	2.2604	1.6022
Exponential	5.3568	2.7731
AR - Second Order	0.7499	0.5072

(h) The residuals plots reveal clear cyclical patterns for the linear, quadratic, and exponential models. The residual plot for second-order autoregressive model revealed no cyclical pattern. However, the residual plot did reveal a potential equal variance assumption violation discussed in an earlier chapter. The second-order autoregressive model also had the smallest values for the standard error of the estimate and MAD. Based on the results from (f), (g), and the principle of parsimony, the second-order autoregressive model would be best suited for forecasting.

(i) $\hat{Y}_{2019} = 0.344 + 1.648(21) - 0.665(22.8) = 19.79$ billion.

16.64 A time series analysis of the Canadian dollar reveals a moderate component with up and down cycles of varying durations. Although the currency rate varies in cycles, the 1980 exchange rate is fairly similar to the 2019 exchange rate. A review of residual plots reveals that the linear, quadratic, and exponential models may be problematic due to cyclical variation in the residuals. In contrast, a first-order autoregressive model revealed a random pattern of residuals. In addition, the first-order autoregressive model had the smallest standard of the estimate and *MAD* values. The first-order autoregressive was the most appropriate model to use for forecasting. Using this model, the forecasted exchange rate is 1.3504 (units per U.S. dollar) and 1.3388 (units per U.S. dollar) for 2020 and 2021, respectively.

A time series analysis of the Japanese yen exchange rate revealed a steep drop in the rate beginning in 1986. The rate dropped from 238.47 (units per U.S. dollar) in 1985 to 128.17 (units per U.S. dollar) in 1988. The exchange rate had a cyclical component from that point forward with an overall declining trend. A review of the residual plots, standard of the estimate and *MAD* values, revealed that the first-order autoregressive model was the most appropriate for forecasting. Using this model, the forecasted exchange rate is 109.278 (units per U.S. dollar) and 109.000 (units per U.S. dollar) for 2020 and 2021, respectively.

A time series analysis of the English pound reveals a moderate component with up and down cycles of varying durations. Although the currency rate varies in cycles, the 1980 exchange rate has increased from 0.4302 (units per U.S. dollar) in 1980 to 0.7847 (units per U.S. dollar) in 2019. A review of the residual plots and the standard of the estimate and *MAD* values revealed that the first-order autoregressive model was the most appropriate for forecasting. Using this model, the forecasted exchange rate is 0.7430 (units per U.S. dollar) and 0.7146 (units per U.S. dollar) for 2020 and 2021, respectively.

An unexpected irregular component in the future could not be anticipated by the autoregressive models used for each of the currencies.

Regression Analysis: Canadian$ versus Coded Year

Analysis of Variance

Source	DF	Adj SS	Adj MS	F-Value	P-Value
Regression	1	0.02300	0.02300	0.99	0.327
Coded Year	1	0.02300	0.02300	0.99	0.32.7
Error	38	0.88559	0.02331		
Total	39	0.90860			

Model Summary

S	R-sq	R-sq(adj)	R-sq(pred)
0.152660	2.53%	0.00%	0.00%

Coefficients

Term	Coef	SE Coef	T-Value	P-Value	VIF
Constant	1.3010	0.0474	27.46	0.000	
Coded Year	−0.00208	0.00209	−0.99	0.327	1.00

Regression Equation

Canadians $= 1.3010 - 0.00208$ Coded Year

Quadratic:

Regression Analysis: Canadian$ versus Coded Year, Coded Year Sq

Analysis of Variance

Source	DF	Adj SS	Adj MS	F-Value	P-Value
Regression	2	0.05118	0.02559	1.10	0.342
Coded Year	1	0.01525	0.01525	0.66	0.422
Coded Year Sq	1	0.02818	0.02818	1.22	0.277
Error	37	0.85742	0.02317		
Total	39	0.90860			

Model Summary

S	R-sq	R-sq(adj)	R-sq(pred)
0.152228	5.63%	0.53%	0.00%

Coefficients

Term	Coef	SE Coei	T-Value	P-Value	VIF
Constant	1.2459	0.0687	18.12	0.000	
Coded Year	0.00662	0.00815	0.81	0.422	15.30
Coded Year Sq	−0.000223	0.000202	−1.10	0.277	15.30

Regression Equation

Canadian$ $= 1.2459 + 0.00662$ Coded Year $- 0.000223$ Coded Year Sq

Exponential:

Regression Analysis: LogCanadian versus Coded Year

Analysis of Variance

Source	DF	Adj SS	Adj MS	F-Value	P-Value
Regression	1	0.003711	0.003711	1.31	0.259
Coded Year	1	0.003711	0.003711	1.31	0.259
Error	38	0.107412	0.002827		
Total	39	0.111124			

Model Summary

S	R-sq	R-sq(adj)	R-sq(pred)
0.0531662	3.34%	0.80%	0.00%

Coefficients

Term	Coef	SE Coef	T-Value	P-Value	VIF
Constant	0.1136	0.0165	6.89	0.000	
Coded Year	−0.000834	0.000728	−1.15	0.259	1.00

Regression Equation

LogCanadian $= 0.1136 - 0.000834$ Coded Year

Regression Analysis: Canadian$ versus CanLag1

Analysis of Variance

Source	DF	Adj SS	Adj MS	F-Value	P-Value
Regression	1	0.6037	0.603656	75.35	0.000
CanLag1	1	0.6037	0.603656	75.35	0.000
Error	37	0.2964	0.008011		
Total	38	0.9001			

Model Summary

S	R-sq	R-sq(adj)	R-sq(pred)
0.0895059	67.07%	66.18%	62.12%

Coefficients

Term	Coef	SE Coef	T-Value	P-Value	VIF
Constant	0.231	0.120	1.93	0.061	
CanLag1	0.8201	0.0945	8.68	0.000	1.00

Regression Equation

$$\text{Canadian\$} = 0.231 + 0.8201 \text{ CanLag1}$$

Canadian

	s_{yx}	MAD
Linear	0.1527	0.1245
Quadratic	0.1522	0.1224
Exponential	0.153	0.1247
AR-First Order	0.0895	0.07132

Regression Analysis: Yen versus Coded Year

Analysis of Variance

Source	DF	Adj SS	Adj MS	F-Value	P-Value
Regression	1	46637	46637	46.54	0.000
Coded Year	1	46637	46637	46.54	0.000
Error	38	38083	1002		
Total	39	84720			

Model Summary

S	R-sq	R-sq(adj)	R-sq(pred)
31.6572	55.05%	53.87%	48.91%

Coefficients

Term	Coef	SE Coef	T-Value	P-Value	VIF
Constant	191.37	9.83	19.48	0.000	
Coded Year	−2.958	0.434	−6.82	0.000	1.00

Regression Equation

$$\text{Yen} = 191.37 - 2.958 \text{ Coded Year}$$

Regression Analysis: Yen versus Coded Year, Coded Year Sq

Analysis of Variance

Source	DF	Adj SS	Adj MS	F-Value	P-Value
Regression	2	69339	34669.6	83.40	0.000
Coded Year	1	40354	40353.8	97.08	0.000
Coded Year Sq	1	22702	22702.2	54.61	0.000
Error	37	15381	415.7		
Total	39	84720			

Model Summary

S	R-sq	R-sq(adj)	R-sq(pred)
20.3885	81.85%	80.86%	78.83%

Coefficients

Term	Coef	SE Coef	T-Value	P-Value	VIF
Constant	240.79	9.21	26.15	0.000	
Coded Year	−10.76	1.09	−9.85	0.000	15.30
Coded Year Sq	0.2001	0.0271	7.39	0.000	15.30

Regression Equation

$$\text{Yen} = 240{,}79 - 10.76 \text{ Coded Year} + 0.2001 \text{ Coded Year Sq}$$

Exponential:

Regression Analysis: LogYen versus Coded Year

Analysis of Variance

Source	DF	Adj SS	Adj SS	F-Value	P-Value
Regression	1	0.3822	0.382229	50.24	0.000
Coded Year	1	0.3822	0.382229	50.24	0.000
Error	38	0.2891	0.007608		
Total	39	0.6713			

Model Summary

S	R-sq	R-sq(adj)	R-sq(pred)
0.0872215	56.94%	55.80%	51.23%

Coefficients

Term	Coef	SE Coef	T-Value	P-Value	VIF
Constant	2.2701	0.0271	83.85	0.000	
Coded Year	−0.00847	0.00119	−7.09	0.000	1.00

Regression Equation

$$\text{LogYen} = 2.2701 - 0.00847 \text{ Coded Year}$$

Autoregressive First Order:

Regression Analysis: Yen versus YenLag1

Method

Rows unused 1

Analysis of Variance

Source	DF	Adj SS	Adj MS	F-Value	P-Value
Regression	1	66819	66818.8	273.42	0.000
YenLag1	1	66819	66818.8	273.42	0.000
Error	37	9042	244.4		
Total	38	75861			

Model Summary

S	R-sq	R-sq(adj)	R-sq(pred)
15.6326	88.08%	87.76%	85.23%

Coefficients

Term	Coef	SE Coef	T-Value	P-Value	VIF
Constant	11.61	7.66	1.52	0.138	
VenLag1	0.8912	0.0539	16.54	0.000	1.00

Regression Equation

$$\text{Yen} = 11.61 + 0.3912 \text{ YenLag1}$$

Yen

	s_{yx}	MAD
Linear	31.6572	25.9902
Quadratic	20.3885	16.3877
Exponential	29.9896	22.8734
AR-First Order	15.6326	11.3533

English Pound, Linear:

Regression Analysis: English Pound versus Coded Year

Analysis of Variance

Source	DF	Adj SS	Adj MS	F-Value	P-Value
Regression	1	0.03801	0.038013	6.09	0.018
Coded Year	1	0.03801	0.038013	6.09	0.018
Error	38	0.23708	0.006239		
Total	39	0.27510			

Model Summary

S	R-sq	R-sq(adj)	R-sq(pred)
0.0789874	13.82%	11.55%	2.13%

Coefficients

Term	Coef	SE Coef	T-Value	P-Value	VIF
Constant	0.5767	0.0245	23.52	0.000	
Coded Year	0.00267	0.00108	2.47	0.018	1.00

Regression Equation

$$\text{English Pound} = 0.5767 + 0.00267 \text{ Coded Year}$$

Quadratic:

Regression Analysis: English Pound versus Coded Year, Coded Year Sq

Analysis of Variance

Source	DF	Adj SS	Adj MS	F-Value	P-Value
Regression	2	0.056616	0.028308	4.79	0.014
Coded Vear	1	0.006725	0.006725	1.14	0.293
Coded Year Sq	1	0.018603	0.018603	3.15	0.084
Error	37	0.218479	0.005905		
Total	39	0.275096			

Model Summary

S	R-sq	R-sq(adj)	R-sq(pred)
0.0768430	20.58%	16.29%	0.29%

Coefficients

Term	Coef	SE Coef	T-Value	P-Value	VIF
Constant	0.6215	0.0347	17.91	0.000	
Coded Year	−0.00439	0.00412	−1.07	0293	15.30
Coded Year Sq	0.000181	0.000102	1.77	0.084	15.30

Regression Equation

$$\text{English Pound} = 0.6215 - 0.00439 \text{ Coded Year} - 0.000181 \text{ Coded Year Sq}$$

Exponential:

Regression Analysis: LogEnglish Pound versus Coded Year

Analysis of Variance

Source	DF	Adj SS	Adj MS	F-Value	P-Value
Regression	1	0.01880	0.018797	6.14	0.018
Coded Year	1	0.01880	0.018797	6.14	0.018
Error	38	0.11626	0.003060		
Total	39	0.13506			

Model Summary

S	R-sq	R-sq(adj)	R-sq(pred)
0.0553133	13.92%	11.65%	2.20%

Coefficients

Term	Coef	SE Coef	T-Value	P-Value	VIF
Constant	−0.2419	0.0172	−14.09	0.000	
Coded Year	0.001878	0.000758	2.48	0.018	1.00

Regression Equation

$$\text{LogEnglish Pound} = -0.2419 + 0.001878 \text{ Coded Year}$$

Autoregressive First Order:

Regression Analysis: English Pound versus EngLag1

Method

Rows unused 1

Analysis of Variance

Source	DF	Adj SS	Adj MS	F-Value	P-Value
Regression	1	0.1165	0.116495	36.49	0.000
EngLagl	1	0.1165	0.116495	36.49	0.000
Error	37	0.1181	0.003193		
Total	38	0.2346			

Model Summary

S	R-sq	R-sq(adj)	R-sq(pred)
0.0565043	49.65%	48.29%	43.74%

Coefficients

Term	Coef	SE Coef	T-Value	P-Value	VIF
Constant	0.2075	0.0712	2.92	0.006	
EngLagl	0.682	0.113	6.04	0.000	1.00

Regression Equation

$$\text{English Pound} = 0.2075 + 0.682 \text{ EngLagl}$$

English Pound

	S_{yx}	MAD
Linear	0.079	0.0613
Quadratic	0.0768	0.0585
Exponential	0.0789	0.0607
AR-First Order	0.0565	0.0443

Index

Credits

Photos

Cover
Andrey Tolkachev/Shutterstock

Chapter 00
Page 1, wallix/Getty Images

Chapter 1
Pages 16 and 30, Haveseen/YAY Micro/AGE Fotostock

Chapter 2
Page 38, scanrail/123RF

Chapter 3
Pages 108 and 141, gitanna/Fotolia

Chapter 4
Pages 152 and 171, vectorfusionart/Shutterstock

Chapter 5
Pages 176 and 191, Hongqi Zhang/123RF

Chapter 6
Pages 198 and 218, cloki/Shutterstock

Chapter 7
Pages 224 and 239, bluecinema/Getty Images

Chapter 8
Pages 244 and 267, Shutterstock

Chapter 9
Pages 275 and 304, ahmettozar/Getty Images

Chapter 10
Pages 311 and 340, GUNDAM_Ai/Shutterstock

Chapter 11
Pages 352 and 379, fotoinfot/Shutterstock

Chapter 12
Pages 389 and 420, Vibrant Image Studio/Shutterstock

Chapter 13
Pages 430 and 466, pixfly/Shutterstock

Chapter 14
Pages 478 and 515, maridav/123RF

Chapter 15
Pages 526 and 547, Anthony Brown/Fotolia

Chapter 16
Pages 556 and 592, stylephotographs/123rf

Chapter 17
Pages 600 and 613, Rawpixel.com/Shutterstock

Chapter 18
Pages 618 and 623, Sharyn Rosenberg

Chapter 19
Pages 19-1 and 19-30, zest_marina/Fotolia

Chapter 20
Pages 20-1 and 20-22, Ken Mellot/Shutterstock